International Handbook of Research in History,
Philosophy and Science Teaching

Michael R. Matthews
Editor

International Handbook of Research in History, Philosophy and Science Teaching

Volume I

Editor
Michael R. Matthews
School of Education
University of New South Wales
Sydney, NSW, Australia

ISBN 978-94-007-7653-1   ISBN 978-94-007-7654-8 (eBook)
DOI 10.1007/978-94-007-7654-8
Springer Dordrecht Heidelberg New York London

Library of Congress Control Number: 2013958394

© Springer Science+Business Media Dordrecht 2014
This work is subject to copyright. All rights are reserved by the Publisher, whether the whole or part of the material is concerned, specifically the rights of translation, reprinting, reuse of illustrations, recitation, broadcasting, reproduction on microfilms or in any other physical way, and transmission or information storage and retrieval, electronic adaptation, computer software, or by similar or dissimilar methodology now known or hereafter developed. Exempted from this legal reservation are brief excerpts in connection with reviews or scholarly analysis or material supplied specifically for the purpose of being entered and executed on a computer system, for exclusive use by the purchaser of the work. Duplication of this publication or parts thereof is permitted only under the provisions of the Copyright Law of the Publisher's location, in its current version, and permission for use must always be obtained from Springer. Permissions for use may be obtained through RightsLink at the Copyright Clearance Center. Violations are liable to prosecution under the respective Copyright Law.
The use of general descriptive names, registered names, trademarks, service marks, etc. in this publication does not imply, even in the absence of a specific statement, that such names are exempt from the relevant protective laws and regulations and therefore free for general use.
While the advice and information in this book are believed to be true and accurate at the date of publication, neither the authors nor the editors nor the publisher can accept any legal responsibility for any errors or omissions that may be made. The publisher makes no warranty, express or implied, with respect to the material contained herein.

Printed on acid-free paper

Springer is part of Springer Science+Business Media (www.springer.com)

Contents

Volume 1

**1 Introduction: The History, Purpose and Content
of the Springer** *International Handbook of Research
in History, Philosophy and Science Teaching* 1
Michael R. Matthews

Part I Pedagogical Studies: Physics

**2 Pendulum Motion: A Case Study in How History
and Philosophy Can Contribute to Science Education** 19
Michael R. Matthews

3 Using History to Teach Mechanics 57
Colin Gauld

4 Teaching Optics: A Historico-Philosophical Perspective 97
Igal Galili

**5 Teaching and Learning Electricity:
The Relations Between Macroscopic
Level Observations and Microscopic Level Theories** 129
Jenaro Guisasola

**6 The Role of History and Philosophy in Research
on Teaching and Learning of Relativity** 157
Olivia Levrini

**7 Meeting the Challenge: Quantum Physics
in Introductory Physics Courses** 183
Ileana M. Greca and Olival Freire

v

8	Teaching Energy Informed by the History and Epistemology of the Concept with Implications for Teacher Education	211

Manuel Bächtold and Muriel Guedj

9 Teaching About Thermal Phenomena and Thermodynamics: The Contribution of the History and Philosophy of Science 245

Ugo Besson

Part II Pedagogical Studies: Chemistry

10 Philosophy of Chemistry in Chemical Education: Recent Trends and Future Directions 287

Sibel Erduran and Ebru Z. Mugaloglu

11 The Place of the History of Chemistry in the Teaching and Learning of Chemistry 317

Kevin C. de Berg

12 Historical Teaching of Atomic and Molecular Structure 343

José Antonio Chamizo and Andoni Garritz

Part III Pedagogical Studies: Biology

13 History and Philosophy of Science and the Teaching of Evolution: Students' Conceptions and Explanations 377

Kostas Kampourakis and Ross H. Nehm

14 History and Philosophy of Science and the Teaching of Macroevolution 401

Ross H. Nehm and Kostas Kampourakis

15 Twenty-First-Century Genetics and Genomics: Contributions of HPS-Informed Research and Pedagogy 423

Niklas M. Gericke and Mike U. Smith

16 The Contribution of History and Philosophy to the Problem of Hybrid Views About Genes in Genetics Teaching 469

Charbel N. El-Hani, Ana Maria R. de Almeida,
Gilberto C. Bomfim, Leyla M. Joaquim,
João Carlos M. Magalhães, Lia M.N. Meyer,
Maiana A. Pitombo, and Vanessa C. dos Santos

Part IV Pedagogical Studies: Ecology

17 Contextualising the Teaching and Learning of Ecology: Historical and Philosophical Considerations 523

Ageliki Lefkaditou, Konstantinos Korfiatis,
and Tasos Hovardas

Contents vii

Part V Pedagogical Studies: Earth Sciences

18 **Teaching Controversies in Earth Science:
The Role of History and Philosophy of Science**................................ 553
Glenn Dolphin and Jeff Dodick

Part VI Pedagogical Studies: Astronomy

19 **Perspectives of History and Philosophy
on Teaching Astronomy** 603
Horacio Tignanelli and Yann Benétreau-Dupin

Part VII Pedagogical Studies: Cosmology

20 **The Science of the Universe: Cosmology
and Science Education**.. 643
Helge Kragh

Part VIII Pedagogical Studies: Mathematics

21 **History of Mathematics in Mathematics Education** 669
Michael N. Fried

22 **Philosophy and the Secondary School
Mathematics Classroom** 705
Stuart Rowlands

23 **A Role for Quasi-Empiricism in Mathematics Education**................. 731
Eduard Glas

24 **History of Mathematics in Mathematics Teacher Education** 755
Kathleen M. Clark

25 **The Role of Mathematics in Liberal Arts Education**........................ 793
Judith V. Grabiner

Volume 2

26 **The Role of History and Philosophy in University
Mathematics Education**.. 837
Tinne Hoff Kjeldsen and Jessica Carter

27 **On the Use of Primary Sources in the Teaching
and Learning of Mathematics**... 873
Uffe Thomas Jankvist

Part IX Theoretical Studies: Features of Science and Education

28 Nature of Science in the Science Curriculum: Origin, Development, Implications and Shifting Emphases 911
Derek Hodson

29 The Development, Use, and Interpretation of Nature of Science Assessments 971
Norman G. Lederman, Stephen A. Bartos, and Judith S. Lederman

30 New Directions for Nature of Science Research 999
Gürol Irzik and Robert Nola

31 Appraising Constructivism in Science Education 1023
Peter Slezak

32 Postmodernism and Science Education: An Appraisal 1057
Jim Mackenzie, Ron Good, and James Robert Brown

33 Philosophical Dimensions of Social and Ethical Issues in School Science Education: Values in Science and in Science Classrooms 1087
Ana C. Couló

34 Social Studies of Science and Science Teaching 1119
Gábor Kutrovátz and Gábor Áron Zemplén

35 Generative Modelling in Physics and in Physics Education: From Aspects of Research Practices to Suggestions for Education 1143
Ismo T. Koponen and Suvi Tala

36 Models in Science and in Learning Science: Focusing Scientific Practice on Sense-making 1171
Cynthia Passmore, Julia Svoboda Gouvea, and Ronald Giere

37 Laws and Explanations in Biology and Chemistry: Philosophical Perspectives and Educational Implications 1203
Zoubeida R. Dagher and Sibel Erduran

38 Thought Experiments in Science and in Science Education 1235
Mervi A. Asikainen and Pekka E. Hirvonen

Part X Theoretical Studies: Teaching, Learning and Understanding Science

39 Philosophy of Education and Science Education: A Vital but Underdeveloped Relationship 1259
Roland M. Schulz

Contents

40 Conceptions of Scientific Literacy: Identifying and Evaluating Their Programmatic Elements 1317
Stephen P. Norris, Linda M. Phillips, and David P. Burns

41 Conceptual Change: Analogies Great and Small and the Quest for Coherence 1345
Brian Dunst and Alex Levine

42 Inquiry Teaching and Learning: Philosophical Considerations 1363
Gregory J. Kelly

43 Research on Student Learning in Science: A Wittgensteinian Perspective 1381
Wendy Sherman Heckler

44 Science Textbooks: The Role of History and Philosophy of Science 1411
Mansoor Niaz

45 Revisiting School Scientific Argumentation from the Perspective of the History and Philosophy of Science 1443
Agustín Adúriz-Bravo

46 Historical-Investigative Approaches in Science Teaching 1473
Peter Heering and Dietmar Höttecke

47 Science Teaching with Historically Based Stories: Theoretical and Practical Perspectives 1503
Stephen Klassen and Cathrine Froese Klassen

48 Philosophical Inquiry and Critical Thinking in Primary and Secondary Science Education 1531
Tim Sprod

49 Informal and Non-formal Education: History of Science in Museums 1565
Anastasia Filippoupoliti and Dimitris Koliopoulos

Part XI Theoretical Studies: Science, Culture and Society

50 Science, Worldviews and Education 1585
Michael R. Matthews

51 What Significance Does Christianity Have for Science Education? 1637
Michael J. Reiss

x   Contents

Volume 3

**52 Rejecting Materialism: Responses to Modern Science
in the Muslim Middle East** 1663
Taner Edis and Saouma BouJaoude

**53 Indian Experiences with Science: Considerations
for History, Philosophy, and Science Education** 1691
Sundar Sarukkai

**54 Historical Interactions Between Judaism and Science
and Their Influence on Science Teaching and Learning** 1721
Jeff Dodick and Raphael B. Shuchat

**55 Challenges of Multiculturalism in Science Education:
Indigenisation, Internationalisation, and *Transkulturalität*** 1759
Kai Horsthemke and Larry D. Yore

**56 Science, Religion, and Naturalism: Metaphysical
and Methodological Incompatibilities** 1793
Martin Mahner

Part XII   Theoretical Studies: Science Education Research

**57 Methodological Issues in Science Education Research:
A Perspective from the Philosophy of Science** 1839
Keith S. Taber

**58 History, Philosophy, and Sociology of Science
and Science-Technology-Society Traditions
in Science Education: Continuities and Discontinuities** 1895
Veli-Matti Vesterinen, María-Antonia Manassero-Mas,
and Ángel Vázquez-Alonso

**59 Cultural Studies in Science Education:
Philosophical Considerations** 1927
Christine L. McCarthy

60 Science Education in the Historical Study of the Sciences 1965
Kathryn M. Olesko

Part XIII   Regional Studies

**61 Nature of Science in the Science Curriculum
and in Teacher Education Programs in the United States** 1993
William F. McComas

Contents   xi

62 The History and Philosophy of Science in Science Curricula and Teacher Education in Canada 2025
Don Metz

63 History and Philosophy of Science and the Teaching of Science in England 2045
John L. Taylor and Andrew Hunt

64 Incorporation of HPS/NOS Content in School and Teacher Education Programmes in Europe 2083
Liborio Dibattista and Francesca Morgese

65 History in Bosnia and Herzegovina Physics Textbooks for Primary School: Historical Accuracy and Cognitive Adequacy 2119
Josip Slisko and Zalkida Hadzibegovic

66 One Country, Two Systems: Nature of Science Education in Mainland China and Hong Kong 2149
Siu Ling Wong, Zhi Hong Wan, and Ka Lok Cheng

67 Trends in HPS/NOS Research in Korean Science Education 2177
Jinwoong Song and Yong Jae Joung

68 History and Philosophy of Science in Japanese Education: A Historical Overview 2217
Yuko Murakami and Manabu Sumida

69 The History and Philosophy of Science and Their Relationship to the Teaching of Sciences in Mexico 2247
Ana Barahona, José Antonio Chamizo, Andoni Garritz, and Josip Slisko

70 History and Philosophy of Science in Science Education, in Brazil 2271
Roberto de Andrade Martins, Cibelle Celestino Silva, and Maria Elice Brzezinski Prestes

71 Science Teaching and Research in Argentina: The Contribution of History and Philosophy of Science 2301
Irene Arriassecq and Alcira Rivarosa

Part XIV   Biographical Studies

72 Ernst Mach: A Genetic Introduction to His Educational Theory and Pedagogy 2329
Hayo Siemsen

73 Frederick W. Westaway and Science Education: An Endless Quest 2359
William H. Brock and Edgar W. Jenkins

74 E. J. Holmyard and the Historical Approach to Science Teaching 2383
Edgar W. Jenkins

75 John Dewey and Science Education 2409
James Scott Johnston

76 Joseph J. Schwab: His Work and His Legacy 2433
George E. DeBoer

Name Index 2459

Subject Index 2505

Chapter 1
Introduction: The History, Purpose and Content of the Springer *International Handbook of Research in History, Philosophy and Science Teaching*

Michael R. Matthews

This is the first handbook to be published that is devoted to the field of historical and philosophical research in science and mathematics education (HPS&ST). Given that science and mathematics through their long history have always been engaged with philosophy and that for over a century it has been recognised that science and mathematics curriculum development, teaching, assessment and learning give rise to many historical and philosophical questions, it is unfortunate that such a handbook has been so long coming.

This work is an international endeavour with its 76 chapters being written by 125 authors from 30 countries. Each chapter has benefited from reviews by up to six scholars and has undergone multiple revisions. More than 300 reviewers, from the disciplines of history, philosophy, education, psychology, mathematics and natural science were willing to contribute their time and expertise to the project. Volunteer copyeditors, with command of both the subject area and English expression, also contributed to the final form of the chapters. A great debt is owed by authors, the research community and readers to these reviewers and copyeditors for their anonymous and unrewarded work. The handbook has grown directly from the Springer journal *Science & Education: Contributions from History and Philosophy of Science and Mathematics.*[1]

---

[1] The journal was the first such research journal devoted exclusively to HPS-informed research in science and mathematics education. Nearly all of the 125 authors have published in the journal and the 300+ reviewers have been drawn from the journal's pool of 900+ reviewers (these can be seen at http://ihpst.net/journal/reviewers/list-of-reviewers/). But the century of research covered by the contributors extends far beyond the pages of the journal, as can be seen by looking at the Reference lists of the chapters.

M.R. Matthews (✉)
School of Education, University of New South Wales, Sydney, NSW, Australia
e-mail: m.matthews@unsw.edu.au

M.R. Matthews (ed.), *International Handbook of Research in History, Philosophy and Science Teaching*, DOI 10.1007/978-94-007-7654-8_1,
© Springer Science+Business Media Dordrecht 2014

1.1 The International History, Philosophy and Science Teaching Group

The journal in turn is associated with the International History, Philosophy and Science Teaching Group that held its first conference in 1989 at Florida State University, with subsequent conferences held biennially.[2] The conferences are attended by historians, philosophers, cognitive psychologists, scientists, mathematicians, education researchers and teachers all of whom have contributed greatly to the formation of a vibrant, congenial, multidisciplinary, international research community. This community forms the core of the authors and reviewers for the handbook; the handbook is a concrete expression of the interests and scholarly work of this IHPST community.

The structure, contents and rationale of the handbook have a lineage that goes back to the very beginnings of the IHPST group, and thus there is benefit in giving an account of its early history. In 1987, I took sabbatical leave at the Philosophy Department of Florida State University in order to pursue with David Gruender some research on Galileo's pendulum discoveries. While in Tallahassee, I attended a large Newton celebration sponsored by the AAAS to honour the tri-centenary of the publication of Newton's *Principia*. Returning from that Washington meeting, I casually mentioned to Jaakko Hintikka that 'it is a pity that science teachers do not attend such meetings, there was so much there that would have been of interest and use to them'. In response, Hintikka, the editor of *Synthese*, the major Kluwer philosophy of science journal, suggested that I edit a special issue of journal on the subject of 'History, Philosophy and Science Teaching' (HPS&ST). This casual exchange was to be the seed of the IHPST group, the journal *Science & Education* and, 25 years later, this handbook.

I began writing, at a time before email and the web, to scholars I knew who had HPS&ST interests and asking them to send me names of others they knew; this was a sort of academic 'pyramid' scheme. The result was a very large and impressive collection of manuscripts written by historians, philosophers, scientists, cognitive scientists and educators.[3] With far too many manuscripts for a single issue of *Synthese*, I reached agreement with other journal editors to publish eight special

---

[2] These have been Queen's University Kingston (1992), University of Minnesota (1995), University of Calgary (1997), University of Pavia (1999), Denver (2001), University of Manitoba (2003), University of Leeds (2005), University of Calgary (2007), University of Notre Dame (2009), Aristotle University Thessaloniki (2011) and University of Pittsburgh (2013). Since 2010, these international conferences have been augmented by regional conferences in Latin America: Maresias Beach, Brazil, in 2010, Mendoza Argentina (2012) and Asia: Seoul National University (2012) and National Taiwan Normal University (2014).

[3] Among those who contributed manuscripts were Joan Solomon, Rodger Bybee, Manuel Sequeia, Laurinda Leite, Harvey Siegel, Martin Eger, Nancy Nersessian, Ernst von Glasersfeld, Joseph Pitt, Jim Garrison, Ian Winchester, Michael Ruse, Arthur Stinner, James Cushing, Stephen Brush, Arnold Arons, Michael Otte, Dimiter Ginev, Derek Hodson, Fritz Rohrlich, Mansoor Niaz, George Kauffman, Pinchas Tamir and Wim van der Steen.

1 Introduction: The History, Purpose and Content of the Springer...

issues of different journals devoted to the subject. These together constituted the first ever journal issues with the title 'History, Philosophy and Science Teaching'.[4]

David Gruender, and Ken Tobin who was newly appointed to Science Education at Florida State University, suggested bringing authors and readers together for an HPS&ST conference. The resulting meeting, with the generous support of the National Science Foundation and of Florida State University, was held in November 1989. There were 180 participants including nearly all of the above-listed journal contributors. Two large volumes of Proceedings – *The History and Philosophy of Science in Science Teaching*, edited by Don Herget and containing 75 papers – were produced.[5] Others gave papers or contributed to the conference.[6] With the special issue articles, the *Proceedings* and other papers, there was an abundance of material with which participants could engage.

Fortunately, in the process of 'networking' for the conference, contact was made with Fabio Bevilacqua from the University of Pavia and who was chairperson of the Interdivisional Group on History of Physics of the European Physical Society.[7] Although from a Physics Department, Bevilacqua had completed his PhD in the History and Philosophy of Science Department at Cambridge University, with a thesis supervised by Mary Hesse and Gerd Buchdahl.

The European Group's Pavia conference was held under the auspices of the International Commission on Physics Education (ICPE), and it explicitly tried to build on an earlier ICPE conference (1970) on 'History in the Teaching of Physics' whose published Proceedings were edited by Stephen G. Brush and Allen L. King. The 1983 Pavia conference organisers, Fabio Bevilacqua and Peter Kennedy, wrote in the *Pavia Conference Proceedings* that 'we began to feel that to confine the discussion only to the history of physics was unduly restrictive and that philosophy and sociology had much to contribute in seeking to show a more complete picture of physics'. From the beginning, the IHPST group had the same conviction but applied to all the sciences.

---

[4] The journals were *Educational Philosophy and Theory* 20(2), (1988); *Synthese* 80(1), (1989); *Interchange* 20(2), (1989); *Studies in Philosophy and Education* 10(1), (1990); *Science Education* 75(1), (1991); *Journal of Research in Science Teaching* 29(4), (1992); *International Journal of Science Education* 12(3), (1990); and *Interchange* 24(1–2), (1993).

[5] The *Proceedings* included papers written by, among others, Sandra Abell, Angelo Collins, Jere Confrey, George Cossman, Zoubeida Dagher, Peter Davson-Galle, Arthur Lucas, Michael Akeroyd, James Gallagher, Teresa Levy, Richard Duschl, Thomas Settle, Hugh Petrie, Robert Hatch, Jane Martin, Joseph Nussbaum, Stellan Ohlsson, Luise Prior McCarty, Edgar Jenkins, Jacques Désautels, Marie Larochelle, Thomas Wallenmaier, Alberto Cordero, Sharon Bailin, Jim Stewart and Carolyn Carter.

[6] Among these were Peter Slezak, Robert Carson, Douglas Allchin, Judith Kinnear, Michael Clough, Hans O. Anderson, Penny Gilmer, Richard Grandy, Jack Lochhead, Zofia Golab-Meyer, James Wandersee, Matilde Vicentini, Peter Taylor, Brian Woolnough and Joseph Novak.

[7] The European group had already held education conferences in Pavia (1983), Munich (1986) and Paris (1988). Subsequently, it would hold conferences in Cambridge (1990), Madrid (1992), Szombathely (1994) and Bratislava (1996) with printed Proceedings being produced for each of these meetings. In 1999, the Group's conference was held jointly with the IHPST conference in Pavia and Lake Como.

Bevilacqua attended the Tallahassee meeting (and is remembered for his commanding role as the scarlet-cloaked Cardinal Bellarmine in Joan Solomon's conference production of 'The Trial of Galileo' in which Michael Ruse is remembered for his Galileo performance). Connection with the European group contributed greatly to making IHPST less a US-Anglo grouping and more robustly an international group. On account of the uncommon spread of disciplines represented and its conviviality, Tallahassee was an overwhelmingly successful and much-remembered meeting. The participants constituted an informal IHPST group for which I became the newsletter editor.

There are many things that can be said about the background and deliberations of the Tallahassee meeting. The first is that although the bulk of the conference was concerned with the traditional liberal education agenda of how HPS can enhance and improve the teaching of science, it did occur at the same time as the 'Science Wars' were erupting in the HPS and Science Studies communities; it was an intellectually exciting and polarising time. The wars erupted on many fronts - in sociology of science the Edinburgh 'Strong Programme' was gaining academic traction fuelled in part by relativist and constructivist interpretations of Thomas Kuhn; many feminist and multicultural critiques of science and of orthodox philosophy of science had been published; postmodernist outlooks were being manifested in many departments.[8]

To some degree, the Science Wars, Postmodernism and Realist versus Constructivist debates were played out at the conference. A plenary session was devoted to the Constructivist debate; it was chaired by Ken Tobin and contributed to by Jaques Désautels, Ernst von Glasersfeld and David Gruender. Gruender's paper was titled: 'Some Philosophical Reflections on Constructivism', and he wrote: 'It is impossible to look at current literature dealing with the education of teachers, especially in science and mathematics, without noticing the galvanizing effects of the newly introduced theory of "constructivism".' He went on to caution that: 'this whole approach of defining knowledge in terms of environmental feedback leading to constructs which better enable the knower to survive in the environment raises serious theoretical issues of its own. And this is so whether one prefers the version offered by Piaget or by Dewey'.

There were divisions at the conference about the epistemological, ontological and pedagogical merits of constructivism, a division between two intellectual tendencies, loosely labelled Realism and Constructivism, yet pleasingly the conference was marked by convivial and congenial exchanges on the subject. There was wide agreement about the benefit of constructivist pedagogy, but disagreement

---

[8] By the time of the conference, the work of Jean-François Lyotard, Michel Foucault, Michael Mulkay, Bruno Latour, Harry Collins, Sandra Harding, Evelyn Fox Keller, Andrew Pickering, David Bloor, Michael Lynch, Steve Woolgar, Donna Haraway, Sal Restivo, Mary Belenky and Jacques Derrida had been published, much read and having some influence on theorists in education circles. Ernst von Glasersfeld, the 'radical constructivist', was an energetic participant at the conference and a contributor to the *Synthese* special issue.

about its commonly related epistemological and ontological claims. This tension has carried through the subsequent history of the group and the journal. For the journal debate began with Wallis Suchting's severe paper 'Constructivism Deconstructed' and Ernst von Glasersfeld's 'Reply' both in the first volume (1992), and continued through a special double issue on the subject in the sixth volume (1997), and into subsequent volumes right through to the present handbook chapter.

A second noteworthy thing about the Tallahassee conference and in the collection of journal special issues is the part played by cross-disciplinary training of individuals involved. In particular, the conference and journal special issues came about because an Italian Physics lecturer had completed an HPS degree at Cambridge, and an Australian Education lecturer had completed a philosophy degree at the University of Sydney and had taken sabbatical leaves in the Boston University and FSU Philosophy departments. Other participants had comparable cross-disciplinary backgrounds. For everyone, the value of scientists and science and mathematics educators working with philosophers, historians, cognitive psychologists and others was immediately apparent.

The value of cross-disciplinary training, or at least cooperation, was a lasting lesson that has informed the subsequent history of IHPST, the journal *Science & Education* and, 25 years later, the organisation of this handbook. It is a lesson that perhaps should inform the training and preparation of science educators where too often the standard trajectory is Science followed by Education and then educational research without mastering any other foundation discipline such as Philosophy, Psychology, History or Sociology.

After 20 years of productive but informal existence without office bearers, the IHPST group was formalised in 2007 at its Calgary conference. A constitution was adopted, elections for a governing council were held and the following aims adopted:

(a) The utilisation of historical, philosophical and sociological scholarship to clarify and deal with the many curricular, pedagogical and theoretical issues facing contemporary science education. Among the latter are serious educational questions raised by Religion, Multiculturalism, Worldviews, Feminism and teaching the Nature of Science.
(b) Collaboration between the communities of scientists, historians, philosophers, cognitive psychologists, sociologists, and science educators, and school and college teachers.
(c) The inclusion of appropriate history, philosophy and sociology of science courses in science teacher-education programmes.
(d) The dissemination of accounts of lessons, units of work and programmes in science, at all levels, that have successfully utilised history, philosophy and sociology.
(e) Discussion of the philosophy and purposes of science education, and its contribution to the intellectual and ethical development of individuals and cultures.

This handbook contributes to realising these aims.

1.2 *Science & Education* Journal

The journal began during a conversation at a US Philosophy of Education conference in 1990 with Peter de Liefde, then Kluwer Education Editor. Kluwer did not then have a presence in science education, and he saw the possibility of building on the IHPST newsletter and community in creating a new scholarly journal. With a great deal of assistance from many people who agreed to be on the editorial committee, the journal commenced publication in 1992. In its beginnings, the journal tried to meet the highest standards; pleasingly, it was able to publish research by deservedly well-known scholars from the fields of science education, mathematics education and history and philosophy of science.[9] It is no exaggeration to say that the disciplinary spread and quality of authors had not before been seen in education journals. The multidisciplinary pattern and high standards were maintained in the following 20+ years where well-known scholars have been published who may not otherwise have addressed issues in science and mathematics education.[10]

Since its beginning in 1992 with four numbers per year, the journal has grown both in size and in scholarly recognition. In 1997, it moved to six numbers, in 2003 to eight numbers and in 2007 to ten numbers per volume; in 2011, there were 108,650 article downloads from its Springer site.

1.3 The Handbook Project

The handbook project began in 2010 during discussion with Bernadette Ohmer, the Springer Education Editor (Springer having taken over Kluwer in 2005) about how best to celebrate the 20th anniversary of the founding of *Science & Education*. It was soon obvious to both of us that a HPS and Science Teaching Handbook was the best and most useful way to mark the journal's publication milestone. This began

---

[9] In the first year, papers by, among others, Wallis Suchting, Paul Kirschner, Mark Silverman, Derek Hodson, Martin Eger, Helge Kragh, Maryvonne Hallez, Israel Scheffler, Alberto Cordero, Creso Franco and Dominique Colinvaux-de-Dominguez were published. In the second year, papers by, among others, Richard Kitchener, Gerd Buchdahl, Jack Rowell, Walter Jung, Henry Nielsen, Harvey Siegel, Lewis Pyenson, Victor Katz, Bernard Cohen, Nancy Brickhouse and Enrico Giannetto. The third year saw papers by, among others, John Heilbron, Peter Machamer, Michael Martin, Robert S. Cohen, Peter Slezak, Andrea Woody, James Garrison and Jane Martin. A number of these papers had their origins in conferences of the Interdivisional Group on History of Physics of the European Physical Society.

[10] Philosophers who have published in the journal include John Worrall, Alan Musgrave, Hasok Chang, Peter Machamer, Michael Martin, Noretta Koertge, Robert Crease, Patrick Heelan, Robert Nola, Alan Chalmers, Mario Bunge, Robert Pennock, Steve Fuller, Jane Roland Martin, Howard Sankey, Demetris Portides, Hugh Lacey, Gürol Irzik, Cassandra Pinnick, Joseph Agassi, Michael Ruse, David Depew, Massimo Pigliucci and many more. Historians whose work has been published have included John Heilbron, Lewis Pyenson, Roger Stuewer, William Carroll, Stephen Brush, Roberto de Andrade Martins, Bernadette Bensaude-Vincent, Ronald Numbers, John Hedley Brooke, Diane Paul and many more.

the three-year process of contacting, inviting, structuring, writing, reviewing, revising, more reviewing and writing that has led to the 2014 publication of the handbook.

For the historic record and for understanding the contents of the Handbook, it is worth repeating the initial invitation to authors:

> The guiding principle for the *Handbook* chapters is to review and document HPS-influenced scholarship in the specific field, to indicate any strengths and weaknesses in the tradition of research, to draw some lessons from the history of this research tradition, and to suggest fruitful ways forward. … The expectation is that the handbook will demonstrate that HPS contributes significantly to the understanding and resolution of the numerous theoretical, curricular and pedagogical questions and problems that arise in science and mathematics education.

Authors accepting the invitation to contribute received a reply saying:

> The expectation is that [the Handbook] will make the history and philosophy of science (and mathematics) a more routine and expected part of science and mathematics teaching, teacher education and graduate research programmes.
>
> My own view is that much the same arguments developed in the handbook will apply to teaching and research in any discipline – economics, history, geography, psychology, theology, music, art, cognitive science, literature and so on. That is, to educate someone in any discipline requires a grasp of the history and philosophy of the discipline; and to conduct serious research in the teaching and learning of any discipline will likewise require historical knowledge and philosophical competence. Hopefully this handbook might inspire others to repeat the exercise for other disciplines.

It will be for readers to judge how significant the handbook's contribution is to science and mathematics education. Readers will have their own view on whether teaching a subject requires some knowledge of the history and philosophy of the subject, and they will also have their own view on the degree to which research in the teaching and learning of science and mathematics requires historical and philosophical competence. Handbook authors affirm both positions. If their arguments are convincing, then they have clear implications for teacher education and for doctoral programmes that prepare education researchers.

1.4 Handbook Structure

Focussed discussion of HPS&ST questions was given a significant boost in the nineteenth century when Ernst Mach, the great German physicist, philosopher, historian and educator, founded in 1887 the world's second science education journal - *Zeitschrift für den physikalischen und chemischen Unterricht.*[11] In the USA, John Dewey in the 1920s explicitly addressed HPS&ST issues, later taken up in the 1950s and 1960s by, among others, James Conant, Gerald Holton, Stephen G. Brush, Leo Klopfer, Robert S. Cohen, Joseph Schwab and Arnold Arons.

---

[11] The first such journal was *Zeitschrift für mathematischen und naturwissenschaftlichen Unterricht* which began publication in 1870. It was edited by J. C. V. Hoffmann, a secondary school teacher in the Saxony mining town of Freiberg (thanks to Kathryn Olesko for this information).

In the UK, HPS&ST issues were addressed from the 1920s in books and articles by Frederick Westaway, Eric Holmyard and James Partington and subsequently by John Bradley, Joan Solomon and others. The same questions have been investigated in Spanish, Portuguese, French, German, Italian, Finnish and other traditions. So there is an abundance of material to be covered and appraised in an HPS&ST handbook.

The first question in putting the handbook together was how to structure its contents. My choice was to group extant research into four sections:

Pedagogical Studies
Theoretical Studies
Regional Studies
Biographical Studies

1.4.1 Pedagogical Studies

The Pedagogical section was straightforward. Since Mach's time, educators have looked to history and philosophy in order to improve and make more interesting and engaging the classroom teaching of science and mathematics. Curriculum writers have likewise turned to the history and philosophy of both disciplines for guidance about the philosophical structure and epistemology of the subjects, and suggestions about the best order, from a psychological or maturation perspective, in which to present the subjects. For over a century, these endeavours have been pursued in Physics, Chemistry, Biology, Mathematics and more recently in the Earth Sciences, Astronomy, Cosmology and Ecology. Since, for instance, the 1920s HPS-informed articles have appeared in *The Journal of Chemical Education, The School Science Review* and *Science Education*; they might also be found at this early time in *The American Journal of Physics* and *Physics Education*.

The research literature on HPS and physics teaching is voluminous. This is perhaps to be expected given that Ernst Mach is the founder of formal, organised, published HPS&ST research and that all of the prominent physicists of the nineteenth and twentieth centuries were, like Mach, engaged by philosophy and wrote books on the subject. Handbook chapters cover each of the areas of Mechanics, Optics, Electricity, Relativity, Quantum theory, Energy and Thermodynamics. One need only mention these science fields to be reminded that major historical figures contributed to their development, and in each there were, and still are, serious philosophical issues and controversies. The specific case of pendulum motion is included as an example of how the understanding and teaching of even mundane areas of science can be illuminated and energised by knowledge of the history and philosophy of topic.

For over a century, there has been insightful writing on the history of chemistry and of course on some of the major advances and controversies in the discipline such as the phlogiston versus oxygen theory of combustion, formulation of the periodic table, uncovering of atomic structure and resultant theory, and organic compounds and their creation. Much has been written on the work of Priestley, Lavoisier, Dalton, Mendeleev, Davy, Kehulé, Pauling and other major contributors. There has also been a long history, since Edward Frankland and Henry Armstrong in the nineteenth century and Eric Holmyard between the wars, of serious efforts to utilise the history of chemistry in creating chemistry curriculum and improving chemistry teaching. Two chapters here deal with this research. In contrast, philosophers have not paid the same attention to chemistry, but over the past three decades this has changed, and there is now at least one journal dedicated to the subject, *Foundations of Chemistry*, and there have been important books published in the field. Philosophy was mostly implicit in the long decades of utilising history in chemistry education; it was made explicit in the 1960s by John Bradley, the Machian chemist, in his debates with Nuffield Scheme 'atomic modelists'. In this debate he lamented that: 'The young people of this country come hopefully to school asking for the bread of experience; we give them the stones of atomic models'.[12] Pleasingly, a handbook chapter deals with the now more conscious efforts to explicate philosophy of chemistry and to connect this with issues in chemistry education.

History and philosophy have a far more public face in the teaching of biology, this is especially so for the teaching of evolution and of genetics and four handbook chapters are devoted to these topics. Macroevolution, or the evolution of new species, has been seen since Darwin as a difficult biological problem, and one that has philosophical overtones. The philosopher Karl Popper famously asserted that the core Darwinian thesis – natural selection operates to separate the best adaptations in an environment – far from being a scientific insight is simply non-scientific as it is a hollow tautology (the best adapted species means that it is the species that survives). And the whole question of creation of new species demands a definition of species, something that is harder to do than it sounds. Can such definitions be given without recourse to Aristotelian essentialism? Leaving aside the powerful religious and cultural constraints in learning evolution, there are well-documented psychological constraints to mastery of the theory. The foremost of these is deep-seated, inborn, teleological mental outlooks that we all have; the animal and even vegetable world are understood as intentional and goal-driven. This is a basic Aristotelianism that is close to the surface in Lamarckian accounts of evolution and on the surface of many cultures' understanding of the natural world. This is something against which Darwin struggled, and it is inside the heads of all students. The two Evolution chapters deal with, among other things, this range of questions.

One of the genetics chapters establishes that it is a very difficult subject to teach and discusses how the history of genetics is related to important philosophical issues such as reductionism, genetic determinism and the relationship between biological function and structure. The chapter documents empirical studies where HPS considerations can improve the teaching and learning of the subject. The second genetics chapter reports results on how ideas about genes and gene function are treated in

---

[12] *The School Science Review*, 1964 vol. 45, p. 366. Obviously, teachers require some understanding of debates about instrumentalism, realism and positivism to appreciate Bradley's charge.

textbooks and appear in students' views; it also reports on a teaching strategy for improving students' understanding of scientific models in genetics.

HPS has contributed to the sciences of ecology, astronomy and geology. The handbook chapters on these fields of study appraise the large bodies of research that have appealed to HPS for their better teaching and better student learning. In the cosmology chapter, we are reminded that the subject differs in some respects significantly from other sciences, primarily because of its intimate association with issues of a conceptual and philosophical nature. Because cosmology in the broader sense relates to the students' world views, it provides a means for bridging the gap between the teaching of science and the teaching of humanistic subjects, and clearly philosophical matters of time, causation and creation are germane for any informed teaching and learning of the subject.

It is worth drawing attention to the inclusion of mathematics in this first section. Unfortunately, science education handbooks too often ignore research in mathematics education. In the editorial of the first number (1992) of *Science & Education*, I wrote that: 'One major division that *Science & Education* seeks to overcome is that between researchers in mathematics education and researchers in science education. Seldom, particularly in the Anglo world, do these two groups meet or read each-others' work ... The history and philosophy of science and of mathematics are interwoven disciplines, they are a natural vehicle for bringing the two communities together. Many problems in science education have their origins in the quantitative side of science, and many problems in mathematics education have their origins in the supposed irrelevance of mathematical formalism' (p. 2). Science cannot be done without mathematics, and science even from the earliest ages cannot be learnt without learning relevant mathematics; so the divorce between the two research communities is unfortunate and ultimately to the detriment of teachers and learners. The seven mathematical papers in this handbook flesh out this claim and appraise aspects of the long tradition of HPM&MT scholarship.

1.4.2 Theoretical Studies

Many topics included in the Theoretical section were straightforward; they were obvious choices. Science teachers, curriculum writers, examiners and textbook authors clearly have to address larger philosophical matters about, for example, religion, multiculturalism, indigenous knowledge systems, nature of science, scientific method and inquiry, argumentation, constructivism, evolution education, postmodernism, scientific literacy and the relation of science to personal and cultural world views. And where such questions are not addressed educators frequently need to justify their failure to do so.

Issues, for instance, about teaching and assessing the nature of science have been put on national curricular and assessment tables across the world. These NOS matters are so extensive and the research so voluminous that they are addressed in three papers. The same applies to religion where religious traditions have had centuries of

1 Introduction: The History, Purpose and Content of the Springer...

engagement with science and science education and so of course does atheism. Seven papers in the handbook deal with these bodies of research and debate. There are also chapters on how the HPS&ST tradition connects to the science-technology-society (STS) tradition and more recently the cultural studies tradition in education. Examination of these connections and divergences benefits from historical and philosophical elaboration.

Other theoretical topics might not be so apparent, but nevertheless they are important; they have historical and philosophical dimensions and are covered in handbook chapters. All involved in science and mathematics education need to understand then explain core features of the subject they are teaching: what scientific explanation is, what laws are, what scientific method is or is not, what proof is, what models are, how values enter or do not enter scientific investigation and decision making, how thought experiments have functioned in science and can function in classrooms and so on. Handbook contributions discuss these topics and research on how they are best taught.

Also discussed is the topic of student learning and how research on it can be illuminated by philosophy. Many, following Dewey and Piaget, have pointed out that the psychology of learning and the epistemology of what is learnt need to be better connected. One of the biggest fields in science education research over the past four decades has been conceptual change research, yet in the famous foundational 1982 article by Posner and associates, they point out that they are proposing a theory of *rational* or *reasonable* conceptual change and assuredly the promotion of rationality and reasonable thinking is at least one aim of science education. Once this is appreciated, then it is clear that historians and philosophers can fruitfully be involved with educators; investigating rationality, its shades and alternatives, is central to their disciplines.

Likewise, when cognitive scientists say that knowledge is 'what can be retrieved from long-term memory', philosophers can draw on the long history of epistemology to point out serious problems with this formulation: not everything remembered is knowledge, and claims are not knowledge because they are remembered; other things are involved. Since Plato established that merely true belief is not knowledge, philosophers have discussed the 'other things' involved. Cross-disciplinary engagement between educators, psychologists and philosophers is the way forward here. The conceptual change and Wittgenstein chapters appraise research in this field.

Narrative teaching, informal learning and the long tradition of 'historical investigative teaching' which is based on student 'reproduction' of classical experiments and engagement in the debates occasioned by these experiments – all give rise to philosophical questions and can be illuminated by historical studies. Everyone recognises that without science teaching, there would be no science, but this core reality is oft left unexamined. The chapters here on the role of textbooks in instruction, and on the attention given, and not given, to science education by historians of science examine the literature and arguments on this nexus between science and science teaching.

One of the most important elements that guided the development of the handbook, that energised *Science & Education* journal and that fostered a good deal of the century-plus of HPS&ST writing and research is an underlying conviction about what science and mathematics education should be; that is, what personal and social goals they should pursue, what kind of teaching and assessment is appropriate and what curriculum is justified. When spelt out, this amounts to an underlying philosophy of science and mathematics education. What has animated this work is a conception of *liberal* education, but such an idea needs to be elaborated and defended against alternatives. Philosophy of education is the discipline where, since Mach and Dewey, these debates have occurred; it is a discipline with which teachers need to engage. Without doubt, the most formative influence on my own teaching and educational engagements was the work of the philosophers of education Richard Peters and Israel Scheffler, with some of Peters' arguments being the 'most practical' thing I learnt in my teacher education programme at Sydney University.

Fortunately, the handbook includes a chapter detailing and appraising the fruits of this long connection of philosophy of education with practical and theoretical issues in science and mathematics education. The specific chapter, and more broadly the 34 papers in the Theoretical section, of the handbook provides evidence for the usefulness of having Philosophy or other Foundation studies included in teacher education programmes, and for researchers having them included in doctoral programmes. As has been pointed out, without such exposure or training, educators too often adopt 'slogan-like' positions in philosophy, psychology and sociology.

1.4.3 Regional Studies

Having a regional studies section in the handbook was also straightforward. HPS&ST issues and associated research have occupied teachers and educators in many countries. By detailing for selected countries and regions, these debates and research something can be gleaned about the international extent of concern about the place of history and philosophy, or nature of science, in science teaching; and the particular ways in which teachers, academics and educational administrators in different countries have responded to this concern. The USA, England and Brazil have had the longest and most public engagement with these issues and have generated the most public and scholarly argument. Other countries have had similar debates and their history is discussed here. Of particular note is the inclusion of chapters dealing with how HPS&ST questions have been addressed in three Asian countries – Japan, China and Korea – for whom modern science was, initially, an imported body of beliefs and practices. On this matter, it is worth relating that Asia is now the 'gold medalist' for *Science & Education* article downloads, edging out both North America and Europe.

The Regional chapters can minimise the extent to which the educational wheel has to be reinvented; provincial and national decision making can be informed by the successes and failures of what has occurred elsewhere. For each country, one can see debates about curriculum construction and authority, about appropriate teacher education and about appropriate assessment. These chapters are a contribution to Comparative Education, as well as to science and mathematics education. But for space and time constraints, other countries and regions could have been included; they have their own HPS&ST histories that could be told. Certainly, more individual European countries could have been included – at least France, Spain, Greece, the Nordic countries and Turkey.

1.4.4 Biographical Studies

The fourth, Biographical studies, section is of special importance to the handbook and to HPS&ST research. Current scholarship is part of a tradition that stretches back over a century, something not often enough appreciated. Too often the arguments, analyses and conceptual distinctions of important scholars of the past, which can be a source of enlightenment in the present, are neglected. Also lost is the good example of scholarship and engagement with educational issues, processes and institutions that such writers and researchers provide and that can inspire and be emulated.

In an effort to mitigate this tendency, *Science & Education* in its early volumes reproduced each year a 'Golden Oldie', a good paper that had been published 40, 50 or 60 years earlier. These included classic papers by Israel Scheffler, Robert S. Cohen, I. Bernard Cohen, John Dewey and Walter Jung. The idea was to show that a good argument or a useful conceptual distinction stands the test of time and can be fruitfully engaged with by current researchers. Newton famously remarked that he could see further because he stood upon the shoulders of giants; this is also possible in education provided we know who and what has gone before. Unfortunately, neither teacher education nor doctoral programmes do much to spread such knowledge and consequent sense of engagement in a tradition.

Consider the opening pages of a 1929 text for UK science teachers where a successful science teacher is described as one who:

> knows his own subject ... is widely read in other branches of science ... knows how to teach ... is able to express himself lucidly ... is skilful in manipulation ... is resourceful both at the demonstration table and in the laboratory ... is a logician to his finger-tips ... is something of a philosopher ... is so far an historian that he can sit down with a crowd of [students] and talk to them about the personal equations, the lives, and the work of such geniuses as Galileo, Newton, Faraday and Darwin. More than this he is an enthusiast, full of faith in his own particular work. (F. W. Westaway, *Science Teaching*, 1929, p. 3)

After 80 years of research and debate, it is a challenge to think of what else needs adding to this account. The author, Frederick W. Westaway, was a remarkable man who himself was something of a historian and philosopher with major books published in both fields; he was also a science teacher; and perhaps above all he was an HMI, a Her Majesty's Inspector for School Science. He did not live and work in an ivory tower, but was an administrator and held for decades a crucial bureaucratic position in UK education. He is all but unknown by current science education researchers. By good fortune in 1993, I stumbled over his 440-page 1929 book on the shelf of an Auckland second-hand book shop. The handbook chapter on Westaway will do something to correct his undeserved neglect.

The five chapters in this section – on Mach, Dewey, Schwab, Westaway and Holmyard – deal with the foundation figures of HPS&ST scholarship. Chapter authors were asked to explicate the view of HPS held by their subjects and how their views connected to then-extant HPS positions, indicate how this HPS understanding had connection with educational practice, describe what impact the subject's writings had at the time and provide some hindsight evaluation of the person's place in the history of science education. A demanding task, but marvellously well done here by the chapter authors.

Others who appealed to history and philosophy of science to illuminate theoretical, curricular and pedagogical issues in science and mathematics education could have been added to the section, but space constraints intervened. Among these would be at least James Conant, Arnold Arons, Martin Wagenschein, Walter Jung, Eino Kaila and Fabio Bevilacqua. Gerald Holton whose many HPS books and articles, HPS-informed physics texts and above all his long engagement in development and promotion of the Harvard Project Physics course, has a special place in the field of HPS&ST scholarship and would be added to the Biographical section if practicalities allowed. Many others had well-developed HPS&ST ideas, but less sustained educational engagements so were not considered for inclusion. These would include J. D. Bernal, Philipp Frank, Herbert Feigl and Martin Eger. In mathematics education, comparable 'classics' lists can be provided of scholars who have consciously appealed to the history and philosophy of mathematics to address theoretical, curricular and pedagogical questions. Teachers, graduate students and professors can benefit from engaging with the writing of any of the researchers named here.

1.5   Writing and Communication

The editorial for the first issue of *Science & Education* (1992) stated that the journal will: 'encourage clear and intelligible writing that is well argued and contains a minimum of jargon' (p. 8). Frederick Westaway in his 1926 book *The Writing of Clear English* and George Orwell in his 1945 essay 'Politics and the English Language' both stressed the connection between clear writing and clear thinking. Too often in education, jargon and lazy 'eduspeak' occurs; where it does, clear and useful communication, thinking and analysis are imperilled. Different chapters pick out different examples of this malady. Effort has been made to have the handbook conform to ideals of good writing and clear communication.

Acknowledgement   Springer editorial staff should be thanked: Bernadette Ohmer for suggesting, encouraging and preparing the initial path for the project and Marianna Pascale and Sathiamoorthy Rajeswari for guiding it through its complex production stage. Inevitably with such a big project, one could expect tensions and disappointments, but pleasingly there have been few. Although time consuming, my editorial duties have been personally and professionally rewarding. I have learnt much by working with the large group of contributors from many countries and many disciplines. Much is owed to these scholars, and to the large group of reviewers who diligently commented on and corrected drafts of the chapters, and to the unsung copyeditors. Hopefully, the writing and editorial labours have reinforced the importance of 'laying out the past and current state of historical and philosophical research in science and mathematics education' and have contributed usefully to graduate students and researchers who will advance the HPS&ST programme.

Michael R. Matthews is an honorary associate professor in the School of Education at the University of New South Wales. He has bachelor's and master's degrees from the University of Sydney in science, philosophy, psychology, history and philosophy of science, and education. He was awarded the Ph.D. degree in philosophy of education from the University of New South Wales. He has taught high school science, and lectured at Sydney Teachers' College and the University of New South Wales. He was the Foundation Professor of Science Education at The University of Auckland (1992–1993). He has published in philosophy of education, history and philosophy of science, and science education journals, handbooks, encyclopaedias and anthologies. His books include *The Marxist Theory of Schooling: A Study of Epistemology and Education* (Humanities Press, 1980), *Pyrmont & Ultimo: A History* (1982), *Science Teaching: The Role of History and Philosophy of Science* (Routledge, 1994), *Challenging New Zealand Science Education* (Dunmore Press, 1995), and *Time for Science Education* (Plenum Publishers, 2000). He has edited *The Scientific Background to Modern Philosophy* (Hackett Publishing Company, 1989), *History, Philosophy and Science Teaching: Selected Readings* (Teachers College Press, 1991), *Constructivism in Science Education: A Philosophical Examination* (Kluwer Academic Publishers, 1998), *Science Education and Culture: The Role of History and Philosophy of Science* (with F. Bevilacqua & E. Giannetto, Kluwer Academic Publishers, 2001), *The Pendulum: Scientific, Historical, Philosophical and Educational Perspectives* (with C.F. Gauld & A. Stinner, Springer, 2005) and *Science, Worldviews and Education* (2009). He is Foundation Editor of the journal *Science & Education*. Outside of the academy, he served two terms as an alderman on Sydney City Council (1980–1986).

Part I
Pedagogical Studies: Physics

Chapter 2
Pendulum Motion: A Case Study in How History and Philosophy Can Contribute to Science Education

Michael R. Matthews

2.1 Introduction

The pendulum has played a major role in the development of Western society, science and culture. The pendulum was central to the studies of Galileo, Huygens, Newton, Hooke and all the leading figures of the Scientific Revolution. The study and manipulation of the pendulum established many things: an accurate method of timekeeping, leading to solving the longitude problem; discovery of the conservation and collision laws; ascertainment of the value of the acceleration due to gravity g, showing the variation of g from equatorial to polar regions and hence determining the oblate shape of the earth; provided the crucial evidence for Newton's synthesis of terrestrial and celestial mechanics, showing that fundamental laws are universal in the solar system; a dynamical proof for the rotation of the earth on its axis; the equivalence of inertial and gravitational mass; an accurate measurement of the density and hence mass of the earth; and much more. The historian, Domenico Bertoloni Meli, wrote:

> Starting with Galileo, the pendulum was taking a prominent place in the study of motion and mechanics, both as a time-measuring device and as a tool for studying motion, force, gravity and collision. (Meli 2006, p. 206)

Another historian, Bertrand Hall, attested that:

> In the history of physics the pendulum plays a role of singular importance. From the early years of the seventeenth century, when Galileo announced his formulation of the laws governing pendular motion, to the early years of this century, when it was displaced by devices of superior accuracy, the pendulum was either an object of study or a means to study questions in astronomy, gravitation and mechanics. (Hall 1978, p. 441)

No surprise that James Gleick nominated that 'the pendulum is the emblem of classical mechanics' (Gleick 1987, p. 39).

M.R. Matthews (✉)
School of Education, University of New South Wales, Sydney, NSW, Australia
e-mail: m.matthews@unsw.edu.au

M.R. Matthews (ed.), *International Handbook of Research in History, Philosophy and Science Teaching*, DOI 10.1007/978-94-007-7654-8_2,
© Springer Science+Business Media Dordrecht 2014

Unfortunately, the centrality and importance of the pendulum for the development of modern science is not reflected in textbooks and school curricula where it appears as an 'exceedingly arid' subject and is mostly, even in the best classes, dismissed with well-remembered formulae [$T = 2\pi \sqrt{l/g}$] and some routine mathematical exercises and maybe some practical classes. This represents a missed opportunity for enriched physics teaching and for cultivating wider appreciation of the nature of science and its contribution to society and culture. Also missed is the opportunity to give students the sense of participation in the scientific tradition of procedures, experiments, theoretical debate and understanding that has been forged by creative and diligent thinkers, and that with good reason many have considered as the very model for intelligent investigation of the natural and social worlds.

2.2 Galileo's Pendulum Analysis

Galileo in his final work, *The Two New Sciences*, written during the period of house arrest after the trial that, for many, marked the beginning of the Modern Age, wrote:

> We come now to the other questions, relating to pendulums, a subject which may appear to many exceedingly arid, especially to those philosophers who are continually occupied with the more profound questions of nature. Nevertheless, the problem is one which I do not scorn. I am encouraged by the example of Aristotle whom I admire especially because he did not fail to discuss every subject which he thought in any degree worthy of consideration. (Galileo 1638/1954, pp. 94–95)

Galileo's comment that pendulum investigations appear 'exceedingly arid' has been echoed by science students over the following 400 years. The pendulum is regularly voted the 'most boring subject' in physics. But this need not be so: pendulum studies and investigations that are informed by the history and philosophy of the subject matter can be deeply engaging and can introduce to students the wide vista of interplay between science, technology, society, philosophy, mathematics and culture.

While the youthful Galileo was briefly a medical student at Pisa, he utilised the pendulum to make a simple diagnostic instrument for measuring pulse beats. This was the *pulsilogium*. Medical practitioners in Galileo's day realised that pulse rate was of great significance, but there was no objective, let alone accurate, measurement of pulse beat. Galileo's answer to the problem was ingenious and simple: he suspended a lead weight on a short length of string, mounted the string on a scaled board, set the pendulum in motion, and then moved his finger down the board from the point of suspension (thus effectively shortening the pendulum) until the pendulum oscillated in time with the patient's pulse. As the period of oscillation depended only on the length of the string, and not on the amplitude of swing, or the weight of the bob, the length of string provided an objective and repeatable measure of pulse speed that could be communicated between doctors and patients, and kept as a record.

The *pulsilogium* provides a useful epistemological lesson. Initially, something subjective, the pulse, was used to measure the passage of time – occurrences, especially in music, were spoken of as taking so many pulse-beats. With Galileo's *pulsilogium*, this subjective measure itself becomes subject to an external, objective, public measure – the length of the *pulsilogium*'s string. This was a small step in the direction of objective and precise measurement upon which scientific advance in the seventeenth and subsequent centuries would depend.[1]

After three decades of work, Galileo's well-known pendulum claims were the following:

LAW OF WEIGHT INDEPENDENCE: period is independent of weight
LAW OF AMPLITUDE INDEPENDENCE: period is independent of amplitude
LAW OF LENGTH: period varies directly as length; specifically the square root of length
LAW OF ISOCHRONY: for any pendulum, all swings take the same time; pendulum motion is isochronous;

After his appointment to a lectureship in mathematics at the University of Pisa in 1588, Galileo quickly became immersed in the mathematics and mechanics of the 'Superhuman Archimedes', whom he never mentions 'without a feeling of awe' (Galileo 1590/1960, p. 67). Galileo's major Pisan work is his *On Motion* (1590/1960). In it he deals with the full range of problems being discussed among natural philosophers – free-fall, motion on balances, motion on inclined planes and circular motions. In these discussions, the physical circumstances are depicted geometrically, and mathematical reasoning is used to establish various conclusions in physics: Galileo here begins the mathematising of physics. Galileo's genius was to see that all of the above motions could be dealt with in one geometrical construction. That is, motions which appeared so different in the world could all be depicted and dealt with mathematically in a common manner.[2]

Galileo develops this line of analysis in an unpublished work, *On Mechanics* (Galileo 1600/1960), a work that followed in the decade after his *On Motion*.[3] It is in *On Mechanics* that the pendulum situation which is implicit in the 1590 construction is made explicit. He deals with all the standard machines – the screw, plane, lever, pulley – and makes the very un-Aristotelian theoretical claim that:

> … heavy bodies, all external and adventitious impediments being removed, can be moved in the plane of the horizon by any minimum force. (Galileo 1600/1960, p. 171)

This is un-Aristotelian because the core of Aristotle's philosophy is that physics is to deal with the world as it is, not with an idealised or mathematical world where,

---

[1] Making a *pulsilogium* is a simple and rewarding class exercise. The basic lesson of science, the move from subjective experience to objective measurement, can be well illustrated.

[2] This will be a recurrent theme in the history of pendulum-related science where it is seen that many different mechanical, biological and chemical processes will manifest the mathematical formulae for Simple Harmonic Motion.

[3] Maurice Clavelin provided the foundational analysis of Galileo's early mechanics, including *De Motu* and *Le Mecaniche* in his *The Natural Philosophy of Galileo* (Clavelin 1974, Chap. 3).

contra-reality, all external and adventitious impediments are removed. For Aristotelian scientists, this is fantasy land. Nevertheless this claim of Galileo's puts him on the track towards a doctrine of circular inertia. He says that Pappus of Alexandria 'missed the mark' in his discussion of forces on bodies, because he made the assumption that 'a weight would have to be moved in a horizontal plane by a given force' (Galileo 1600/1960, p. 172). Galileo says that this assumption is false 'because no sensible force is required (neglecting accidental impediments which are not considered by the theoretician)' (Galileo 1600/1960, p. 172). He uses the following construction, and argument, to make his point, and in so doing sets up the situation that enables him to analyse pendulum motion in terms of circular motion and motion along chords of a circle.

This is a most fruitful construction. It will allow Galileo to analyse pendulum motion as motion in a circular rim and as motion on a suspended string. By considering initial, infintisimal, motions, he is able to consider pendulum motion as a series of tangential motions down inclined planes. Two years later, he will write an important letter to his patron Guidobaldo del Monte about these propositions.

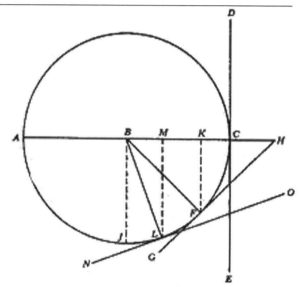

Galileo's 1600 composite diagram of lever, inclined plane, vertical fall and pendulum

In his later work, Galileo expresses these pendulum claims as follows. In the First Day of his 1638 *Discorsi* Galileo expresses his law of weight independence as:

> Accordingly I took two balls, one of lead and one of cork, the former more than a hundred times heavier than the latter, and suspended them by means of two equal fine threads, each four or five cubits long. Pulling each ball aside from the perpendicular, I let them go at the same instant, and they, falling along the circumferences of circles having these equal strings for semi-diameters, passed beyond the perpendicular and returned along the same path. This free vibration repeated a hundred times showed clearly that the heavy body maintains so nearly the period of the light body that neither in a hundred swings nor even in a thousand will the former anticipate the latter by as much as a single moment, so perfectly do they keep step. (Galileo 1638/1954, p. 84)

In the Fourth Day of the 1633 *Dialogue*, Galileo states his law of amplitude independence, saying:

> ... truly remarkable ... that the same pendulum makes its oscillations with the same frequency, or very little different – almost imperceptibly – whether these are made through large arcs or very small ones along a given circumference. I mean that if we remove the pendulum from the perpendicular just one, two, or three degrees, or on the other hand seventy degrees or eighty degrees, or even up to a whole quadrant, it will make its vibrations when it is set free with the same frequency in either case. (Galileo 1633/1953, p. 450)

In the First Day of his 1638 *Discourse*, Galileo states his law of length when, in discussing the tuning of musical instruments, saying:

> As to the times of vibration of bodies suspended by threads of different lengths, they bear to each other the same proportion as the square roots of the lengths of the thread; or one might say the lengths are to each other as the squares of the times; so that if one wishes to make the vibration-time of one pendulum twice that of another, he must make its suspension four times as long. In like manner, if one pendulum has a suspension nine times as long as another, this second pendulum will execute three vibrations during each one of the first; from which it follows that the lengths of the suspending cords bear to each other the [inverse] ratio of the squares of the number of vibrations performed in the same time. (Galileo 1638/1954, p. 96)

Isochrony is of the greatest importance for the subsequent scientific and social utilisation of the pendulum. In the late fifteenth century, the great observer Leonardo da Vinci extensively examined, manipulated and drew pendula, but as one commentator remarks, 'He failed, however, to recognize the fundamental properties of the pendulum, the isochronism of its oscillation, and the rules governing its period' (Bedini 1991, p. 5).[4]

In the Fourth Day of the 1633 *Dialogue*, Galileo approaches his law of isochrony by saying

Take an arc made of a very smooth and polished kjconcave hoop bending along the curvature of the circumference *ADB* [Fig.6], so that a well-rounded and smooth ball can run freely in it (the rim of a sieve is well suited for this experiment). Now I say that wherever you place the ball, whether near to or far from the ultimate limit *B* ... and let it go, it will arrive at the point *B* in equal times ... a truly remarkable phenomenon. (Galileo 1633/1953, p. 451)	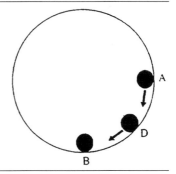

In the First Day of the 1638 *Discorsi*, Galileo writes of his law of isochrony that:

> But observe this: having pulled aside the pendulum of lead, say through an arc of fifty degrees, and set it free, it swings beyond the perpendicular almost fifty degrees, thus describing an arc of nearly one hundred degrees; on the return swing it describes a little

---

[4] So not surprising that children do not just 'see' these properties. For medieval and scholastic treatments of the pendulum see Hall (1978).

smaller arc; and after a large number of such vibrations it finally comes to rest. Each vibration, whether of ninety, fifty, twenty, ten or four degrees occupies the same time: accordingly the speed of the moving body keeps diminishing since in equal intervals of time, it traverses arcs which grow smaller and smaller. ... Precisely the same thing happens with the pendulum of cork. (Galileo 1638/1954, p. 84)

In the Third Day of the *Discorsi*, he uses his law of chords to move towards a demonstration of isochrony:

If from the highest or lowest point in a vertical circle there be drawn any inclined planes meeting the circumference the times of descent along these chords are equal to each other. (Galileo 1638/1954, pp. 188–189)

The circle on which these chords are drawn represents the path of a pendulum whose pivot is the circle's centre. Equality of time for movement on arcs will give him his sought-for isochrony proof.

Although now routine and repeated in textbooks and 'replicated' in school practical classes, these claims when made were, with good reason, very contentious and disputed. Much about science and the nature of science can be learned from the disputes.

2.3 Galileo's Methodological Innovation

The seventeenth century's analysis of pendulum motion is a particularly apt window through which to view the methodological heart of the scientific revolution. Thomas Kuhn in his *Structure of Scientific Revolutions* used Galileo's account of the pendulum to mark the epistemological transformation from the old to the new science. Kuhn wrote:

Since remote antiquity most people have seen one or another heavy body swinging back and forth on a string or chain until it finally comes to rest. To the Aristotelians, who believed that a heavy body is moved by its own nature from a higher position to a state of natural rest at a lower one, the swinging body was simply falling with difficulty. ...Galileo, on the other hand, looking at the swinging body, saw a pendulum, a body that almost succeeded in repeating the same motion over and over again ad infinitum. And having seen that much, Galileo observed other properties of the pendulum as well and constructed many of the most significant and original parts of his new dynamics around them. (Kuhn 1970, pp. 118–119)

The debate between the Aristotelian Guidobaldo del Monte (Galileo's own patron) and Galileo over the latter's pendular claims represents, in microcosm, the larger methodological struggle between Aristotelianism and the new science. This struggle is about the legitimacy of idealisation in science, and the utilisation of mathematics in the construction and interpretation of experiments (Matthews 2004; Nola 2004). All students through to the present day are in the position of

2 Pendulum Motion: A Case Study in How History and Philosophy Can Contribute...

da Vinci and del Monte. Without methodological assistance they do not see what Galileo 'saw'.[5]

Del Monte was a prominent mathematician, engineer and patron of Galileo (Matthews 2000, pp. 100–108; Meli 1992; Renn et al. 2000). In a 1580 letter to Giacomo Contarini, del Monte writes:

> Briefly speaking about these things you have to know that before I have written anything about mechanics I have never (in order to avoid errors) wanted to determine anything, be it as little as it may, if I have not first seen by an effect that the experience confronts itself precisely with the demonstration, and of any little thing I have made its experiment. (Renn et al. 2000, p. 339)

This then is the methodological basis for del Monte's criticism of Galileo's mathematical treatment of pendulum motion. It echoes Aristotle's empiricism: his view that 'if we cannot believe our eyes, what can we believe?' The crucial surviving document in the exchange between Galileo and his patron is a 29th November 1602 letter where Galileo writes of his discovery of the isochrony of the pendulum and conveys his mathematical proofs of the proposition.[6]

The long letter is a milestone in the history of scientific methodology. Galileo writes:

> The experiment you [del Monte] tell me you made in the [rim of a vertical] sieve may be very inconclusive, perhaps by reason of the surface not being perfectly circular, and again because in a single passage one cannot well observe the precise beginning of motion. But if you will take the same concave surface and let ball B go freely from a great distance, as at point B, it will go through a large distance at the beginning of its oscillations and a small one at the end of these, yet it will not on that account make the latter more frequently than the former. Then as to its appearing unreasonable that given a quadrant 100 miles long, one of two equal moveables might traverse the whole and [in the same time] another but a single span, I say that it is true that this contains something of the wonderful (Drake 1978, p. 69)

Thus in 1602, Galileo claimed two things about motion on chords within a circle:

1. That in a circle, the time of descent of a body free-falling along all chords terminating at the nadir, is the same regardless of the length of the chord.
2. In the same circle, the time of descent along a chord is longer than along its composite chords, even though the direct route is shorter than the composite route.

---

[5] Importantly, Galileo literally saw what da Vinci, del Monte and everyone else saw; what was in front of his eyes was the same as was in front of everyone else's; what was behind his eyes was the difference. He constructed a different model of the pendulum phenomenon. On this see Giere (1988, pp. 68–80, 1994)

[6] The letter was written in October 1602 (*Opere*, Edizione Nazionale, Florence 1934, vol. 10, pp. 97–100), and a translation has been provided by Stillman Drake (Drake 1978, pp. 69–71) and it is also translated in Renn et al. (1998, pp. 104–106). Ronald Naylor (1980, pp. 367–371) and W.C. Humphreys (Humphreys 1967, pp. 232–234) discuss the letter in the context of Galileo's work on the law of fall.

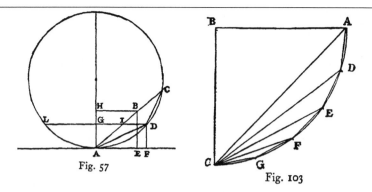

Galileo's geometrical constructions for his law of chords

This gets him tantalisingly close to a claim about motion along the *arcs* of the circle, the pendulum case, but not quite there. He is not prepared to make the leap, saying 'But I cannot manage to demonstrate that arcs SIA and IA are passed in equal times, which is what I am seeking'.[7] Galileo 'sees' that they are passed in equal times; he has empirical proof – if one can take hypothetical behaviour in ideal situations as empirical proof, but he lacks a 'demonstration'. This is something that he believes only mathematics can provide. Galileo needed and sought certainty in his physics as this was required for his endorsement of the Copernican view of the solar system; biblical heliocentrism could not be overturned for merely 'probable' or 'possible' science, only for demonstrable science.

The empirical problems were examples where the world did not 'correspond punctually' to the events demonstrated mathematically by Galileo. In his more candid moments, Galileo acknowledged that events do not always correspond to his theory; that the material world and his so-called 'world on paper', the theoretical world, did not correspond. Immediately after mathematically establishing his famous law of parabolic motion of projectiles, he remarks that:

> I grant that these conclusions proved in the abstract will be different when applied in the concrete and will be fallacious to this extent, that neither will the horizontal motion be uniform nor the natural acceleration be in the ratio assumed, nor the path of the projectile a parabola. (Galileo 1638/1954, p. 251)

One can imagine the reaction of del Monte and other hardworking Aristotelian natural philosophers and mechanicians when presented with such a qualification. When baldly stated, it confounded the basic Aristotelian and empiricist objective of science, namely to tell us about the world in which we live. Consider, for instance, the surprise of Giovanni Renieri, a gunner who attempted to apply Galileo's theory to his craft, who, when he complained in 1647 to Torricelli that his guns did not behave according to Galileo's predictions, was told by Torricelli that 'his teacher

---

[7] On Galileo's geometrical constructions of pendula movement and his physical interpretations of them, see especially Machamer and Hepburn (2004).

spoke the language of geometry and was not bound by any empirical result' (Segre 1991, p. 43). Not surprisingly, del Monte said that Galileo was a great mathematician, but a hopeless physicist. His complaint was the methodological kernel of the scientific revolution in which abstraction, idealisation and mathematical analysis typified the New Science and separated it from the old science.[8]

2.4 Galileo, Experimentation and Measurement

It should not be inferred from the foregoing that Galileo was indifferent to experimental evidence. He was a most careful experimenter. His insight was to have a mathematical model of motion and then to compare real motions with this model; his 'world on paper' preceded his 'real world'. This of course was the method employed by astronomers since before Plato's time. The philosophically interesting point is what adjustments, if any and for what reasons, one makes to the mathematical model in the light of experimental results. How much does reason and how much does measurement contribute to theory development?

Galileo, as with other natural philosophers since Aristotle, was interested in understanding free fall, the most obvious terrestrial 'natural' motion. Thus, in Day Three of his *New Sciences*, Salvati (Galileo's alter ego) says: 'thus we may picture to our mind a motion as uniformly and continuously accelerated when, during any equal intervals of time whatever, equal increments of speed are given to it.' (Galileo 1638/1954, p. 161). Sagredo, the supposedly impartial bystander to the discussion, then says: 'I may nevertheless without offense be allowed to doubt whether such a definition as the above, established in an abstract manner, corresponds to and describes that kind of accelerated motion which we meet in nature in the case of freely falling bodies.' (Galileo 1638/1954, p. 162). As has often been pointed out, Galileo was interested in the 'how' of such natural motions, as well as the 'why' questions which preoccupied the Aristotelian tradition. But free fall was too fast for direct investigation and measurement, so Galileo turned to two motions that embody free fall, but in which the motion can be more easily manipulated and investigated: motion on an inclined plane and pendulum motion.

In Galileo's justly famous inclined plane experiment, he noticed initially by chanting that the distances travelled along the plane in equal time (chant) intervals were (taking the first distance to be one unit) 1, 3, 5, 7, 9, 11, etc. He then saw that the sums of these distances (total distance travelled – 1, 4, 9, 16, 25, 36) were in the proportion to the squares of the times elapsed. He proceeded to replace these chant-based, hence fixed, time-interval measurements with his ingenuous water-weighing measure of time.[9]

---

[8] Some especially insightful discussions of Galileo's methodological revolution are McMullin (1978, 1990), Machamer (1998), and Mittelstrass (1972).

[9] Discussions of Galileo's inclined plane investigations are in Costabel (1975), Humphreys (1967), and Palmieri (2011).

In 1604, he tried to time accurately pendulum swings of different amplitudes using his weight of flowing water method. He took a 1740 *punto* (1.635 m) pendulum and let it swing to a vertical board from 90° and from 10°, while taking his finger off the outflow (when releasing the pendulum) and then putting it on the outflow tube of the water bowl (when the pendulum struck the board). For the full quadrant of fall he got 1,043 grains (2.17 oz) of water, for the small amplitude he got 945 grains (1.96 oz). The difference in time (9.4 %) for large and small amplitude swings is very close to the 10 % that can now be calculated for this particular length (Drake 1990, p. 22). Galileo was meticulous about his experimental results: in this case it was his interpretation that was flawed. He attributed the difference in time to *impediments*, not realising that it was the *circular* arc that was the fundamental disturbing cause. Huygens would discover this geometrically, and substitute the *cycloidal* curve for his isochronic pendulum.

One most important measurement that Galileo performed in these early years (1600–1604) was the ratio of the distance fallen from rest to the length of a pendulum whose period was the time of fall of the body. A pendulum was held out from a vertical board and released simultaneously with the dropping of a weight. He adjusted the pendulum length until the 'thud' of the pendulum hitting the wall coincided with the 'thud' of the weight hitting the floor. If t is the time from release to 'thud', then $4t$ will be the period of the pendulum (from release to vertical is one quarter of its period). This ratio is a constant for all heights of fall, and we now know it is the same at all places on earth and even on the moon; it is equal to $\pi^2/8$.

Galileo measured the time of free fall through a given length to the time of the swing of a pendulum of the same length from release to a vertical board (one quarter of its period, T). He let a ball drop 2000 *punto* (1.88 m) and timed its fall using his weight of water method (850 grains). He then took a 2000 *punto* pendulum and timed its quarter-oscillation (942 grains). The ratio of the two times was 1.108. And it is a constant for all lengths of fall. That is, for all heights, the time of fall compared to the time for a quarter oscillation of a pendulum whose length is equal to the particular height, is constant. Using modern methods, we can calculate this ratio to be equal to $\pi/2\sqrt{2}$, or 1.1107. Galileo's result of 1.108, by weighing water released during the fall on a beam balance, indicates how careful Galileo was in his experiments and measurements.[10]

2.5 Contemporary Reproductions of Galileo's Experiments

Beginning with his first biographer Vicenzio Viviani (1622–1703), through the work of Mach in the late nineteenth century, and up to Stillman Drake's compendious studies this century, Galileo has been depicted as a patient experimentalist who examined nature rather than books about nature, as Aristotelians were supposedly

---

[10] Stillman Drake discusses these measurements in his *Galileo: Pioneer Scientist* (Drake 1990, pp. 23–25). So also does James MacLachlan in his *Galileo Galilei* (MacLachlan 1997, pp. 114–117).

2 Pendulum Motion: A Case Study in How History and Philosophy Can Contribute... 29

doing. This empiricist interpretation has been the dominant tradition in Galilean historiography. Alexandre Koyré's rationalist, intellectualist, neo-Platonic interpretation of Galileo burst like a thunder-clap over this tradition (Koyré 1943, 1953, 1960). Koyré wrote of his work that:

> I have tried to describe and justify Galileo's use of the method of imaginary experiment concurrently with, and even in preference to, real experiment. In fact, it is an extremely fruitful method which incarnates, as it were, the demands of theory in imaginary objects, thereby allowing the former to be put in concrete form, and enables us to understand tangible reality as a deviation from the perfect model which it provides. (Koyré 1960, p. 82)

Against this contested historiographical background, it is also not surprising that scholars have scrutinised Galileo's pendulum experiments and have endeavoured to replicate them[11] (Ariotti 1968, 1972; MacLachlan 1976; Naylor 1974, 1976; Settle 1961, 1967). Concerning Galileo's claims for weight independence, Ronald Naylor, for instance, found that:

> Using two 76 inch pendulums, one having a brass bob, the other cork, both swinging initially through a total arc of 30°, the brass bob was seen to lead the cork by one quarter of an oscillation after only twenty-five completed swings. (Naylor 1974, p. 33)

Of the claims about amplitude independence, James MacLachlan wrote:

> Now, if anyone swings two equal pendulums through such unequal arcs [Galileo's 80° and 5°] it is easy for him to observe that the more widely swinging one takes a longer time to complete the first oscillation, and after a few more it will have fallen considerably behind the other. (MacLachlan 1976, p. 178)

Indeed the difference in period between a 90° and a 3° swing is 18 %. On the mass independence claims, MacLachlan writes:

> As for Galileo's remark that they [cork and lead balls] would not differ even in a thousand oscillations, I have found that a cork ball 10 cm. in diameter is needed just to continue oscillating 500 times. However, for the lead to be 100 times heavier than that, it would have to be more than 10 cm. in diameter, and it would make even fewer oscillations (perhaps only 93) in the time that the cork bob made a hundred. (MacLachlan 1976, p. 181)

The discrepant results of Naylor, MacLachlan and others, do not mean that Galileo's work was just 'imaginary' as Koyré suggests. Undoubtedly these experiments were conducted, but equally undoubtedly the results were 'embellished'. Ronald Naylor provides a reasonable summary of the historical evidence:

> This paper suggests that while Galileo did undoubtedly devise and use experiments similar to those described in the *Discorsi*, it seems evident that in publication he idealised and simplified the results of these researches. Thus some experimental accounts in the *Discorsi* appear to contain the essential distillate of many experiments rather than the description of any actual experiment. Ultimately, the idealised versions of the experiments seem to fuse with Galileo's theoretical model. Even so, it seems likely that at times Galileo was just as capable of providing a totally imaginary experiment in order to support his case. (Naylor 1974, p. 25)

---

[11] See Ariotti (1968, 1972), MacLachlan (1976), Naylor (1974, 1976), Palmieri (2009), and Settle (1961, 1967).

Ignoring discrepant data and embellishing results is a methodological two-edged sword: it allows the experimenter to keep their eyes on the main game, but sometimes the discrepant, or outlying, results are not the product of 'accidents' and 'impediments' (Galileo's terminology), but of basic mechanisms in the world. This was the case with Galileo's continued commitment to the circle being the isochronous curve. There is an epistemological lesson to be learnt here about how experimental results relate to theoretical commitments. That is, we need, in the beginning, to distinguish the *theorised* objects of science, and their properties, from the *material* objects of the world, and their behaviour (Matthews 2000, Chap. 10), or our models of physical processes and the processes themselves (Giere 1988, Chap. 3). This distinction has important implications for pedagogical programmes of Discovery Learning, Experiential Learning and Radical Constructivism. The worst of these programmes *confine* students to their experiential world.

2.6 The Pendulum and Timekeeping

The pendulum played more than a scientific role in the formation of the modern world. The pendulum was central to the horological revolution that was intimately tied to the scientific revolution. Huygens in 1673, following Galileo's epochal analysis of pendulum motion, utilised the pendulum in clockwork and so provided the world's first accurate measure of time (Yoder 1988). The accuracy of mechanical clocks went, in the space of a couple of decades, from plus or minus half-an-hour per day to a few seconds per day.[12] This quantum increase in accuracy of timing enabled hitherto unimagined degrees of precision measurement in mechanics, navigation and astronomy. It ushered in the world of precision characteristic of the scientific revolution (Wise 1995). Time could then confidently be expressed as an independent variable in the investigation of nature.

Christiaan Huygens (1629–1695) refined Galileo's pendulum laws and was the first to use these refined laws in creating a pendulum clock. Huygens stands out among the great scientific minds of the seventeenth century who addressed themselves to the improvement of time measurement and the solution of the longitude problem. Huygens possessed both manual and intellectual skills of the highest order. He was the son of a well-connected and wealthy Dutch diplomat, who, with good reason, called his son *mon Archimède*. Upon Huygens death, the great Leibniz wrote: 'The loss of the illustrious Monsieur Huygens is inestimable; few people knew him as well as I; in my opinion he equalled the reputation of Galileo and Descartes and aided theirs because he surpassed the discoveries that they made; in a word, he was one of the premier ornaments of our time' (Yoder 1991, p. 1). Given

---

[12] Among many excellent books on the history of timekeeping, see Barnett (1998), Landes (1983), and van Rossum (1996).

that the 'times' contained Galileo, Descartes, Pascal, Boyle, Newton as well as Leibniz himself, this was no small praise.

There was an element of metaphysics in Galileo's adherence to the circle as the tautochrone or path of isochronous motion. The same conviction perhaps that lead him to discuss and defend Copernicus's theory of *circular* planetary orbits, despite Kepler's *elliptical* refinement of Copernicus's views being published in 1619, 14 years before Galileo's great *Dialogue*, and Galileo having a copy of the work in his library. The same conviction perhaps led Galileo to the doctrine of *circular* inertia.[13]

Huygens modified Galileo's analysis by showing, mathematically, that it was movement on the cycloid, not the circle that was isochronous. He provides the following account of this discovery:

> We have discovered a line whose curvature is marvellously and quite rationally suited to give the required equality to the pendulum. . . . This line is the path traced out in air by a nail which is fixed to the circumference of a rotating wheel which revolves continuously. The geometers of the present age have called this line a cycloid and have carefully investigated its many other properties. Of interest to us is what we have called the power of this line to measure time, which we found not by expecting this but only by following in the footsteps of geometry. (Huygens 1673/1986, p. 11)

The cycloid is the curve described by a point P rigidly attached to a circle C that rolls, without sliding, on a fixed line AB. The full arc ABD has a length equal to 8r (r = the radius of the generating circle). A heavy point which travels along an arc of cycloid placed in a vertical position with the concavity pointing upwards will always take the same amount of time to reach the lowest point, independent of the point from which it was released.

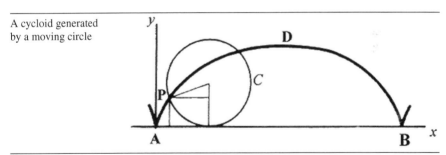

A cycloid generated by a moving circle

Having shown mathematically that the cycloid was isochronous, Huygens then devised a simple way of making a suspended pendulum swing in a cycloidal path – he made two metal cycloidal cheeks and caused the pendulum to swing between them. Huygens first pendulum clock was accurate to one minute per day; working with the best clockmakers, he soon made clocks accurate to one second per day.

---

[13] Alexandre Koyré (Koyré 1943) and Edwin Burtt (Burtt 1932, pp. 61–95) regarded this metaphysical conviction as evidence of Galileo's Platonism.

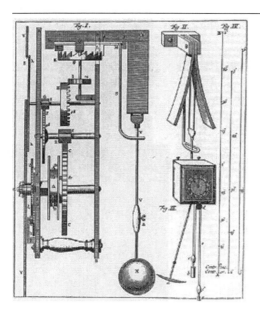

| Huygens' own drawings for a clock, showing the cycloidal cheeks constraining the pendulum | Cut-away pendulum clock, the slowly falling weight provided small impulses to the pendulum to maintain its motion |

After showing that the period of a simple pendulum varied as the square root of its length, Huygens then derived the familiar equation:

$$T = 2\pi\sqrt{l/g}$$

The time of one small oscillation is related to the time of perpendicular fall from half the height of the pendulum as the circumference of a circle is related to its diameter [π] (Huygens 1673/1986, p.171).

2.7 The Pendulum in Newton's Mechanics

The pendulum played a comparable role in Newton's work to what it had for Galileo and Huygens. Newton used the pendulum to determine the gravitational constant **g**, to improve timekeeping, to disprove the existence of the mechanical philosophers' aether presumption, to show the proportionality of mass to weight, to determine the coefficient of elasticity of bodies, to investigate the laws of impact and to determine the speed of sound. Richard Westfall, a Newtonian scholar of great distinction, wrote: "the pendulum became the most important instrument of seventeenth-century science ... Without it the seventeenth century could not have begot the world of precision" (Westfall 1990, p. 67). Concerning the pendulum's role in Newton's science, Westfall has said that "It is not too much to assert that without the pendulum there would have been no *Principia*" (Westfall 1990, p. 82).

2.7.1 The Demonstration of Newton's Laws

Three properties made the pendulum an ideal vehicle for the demonstration of Newton's laws and the investigation of collisions: pendulums of the same length reach their nadir at the same time irrespective of where they are released; they reach their nadir simultaneously regardless of their mass; and their velocity at the nadir is proportional to the length of the chord joining the nadir to the point of release. Additionally, the paths of the colliding bodies could be constrained. Newton sets up pendulum collision experiments to demonstrate his laws of motion and to explicate is nascent conservation law. He concludes one demonstration by writing:

> By the meeting and collision of bodies, the quantity of motion, obtained from the sum of the motions directed towards the same way, or from the difference of those that were directed towards contrary ways, was never changed. ... (Newton 1729/1934, pp. 22–24)

Newton does not label this the conservation of momentum. He speaks ill-definedly of conservation of 'motion', but it is modern momentum, mv, a vector quality, that he is describing. In current terminology, his conclusion is given by the formula:

$$m_1 u_1 + m_2 u_2 = m_1 v_2 + m_2 v_2$$

The pendulum has here made a significant contribution to the foundation of classical mechanics. The law of conservation of momentum was true for both elastic (where there is no energy absorbed in the collision itself) and inelastic collisions (where energy is absorbed). For instance, if the above pendulums were made of putty, then when they collided, they would deform and simply come to a halt, there would be no motion after the collision. In this situation one could hardly talk of conservation of 'motion' – although, Descartes, for instance, resolutely maintained that the quantity mv was the basic measure of 'motion', and mv was the basic entity that was conserved in the world.

Newton's third law, 'action is equal to reaction', was demonstrated by Newton using two long (3–4 m) pendula and having them collide. He used a result of Galileo (that the speed of a pendulum at its lowest point is proportional to the chord of its arc) and applied it to the collision by comparing the quantities mass multiplied by chord length, before and after collision (Gauld 1998, 2004a). For centuries, Newton's Cradle apparatus, which wonderfully manifests this conservation law, has intrigued students and citizens (Gauld 2006).

2.7.2 Unifying Terrestrial and Celestial Mechanics

The major question for Newton and natural philosophers was whether Newton's postulated *attractive force* between bodies was truly *universal*; that is, did it apply not only to bodies on earth, but also between bodies in the solar system? Aristotle, as with all ancient philosophers, made a clear distinction between the heavenly and terrestrial (sub-lunar) realms; the former being eternal, unchanging and perfect, the realm of the Gods; the latter being changeable, imperfect, and corruptible, the realm of man. It was thus 'natural' that the science of both realms would be different; and, to speak anachronistically, laws applying to the terrestrial realm would not apply to the celestial realm. This cosmic divide lasted for 2000 years.

It was the analysis of pendulum motion that rendered untenable the celestial/terrestrial distinction, and enabled the move from 'the closed world to the infinite universe' (Koyré 1957). The same laws governing the pendulum were extended to the moon, and then to the planets. The long-standing celestial/terrestrial distinction in physics was dissolved. The same laws were seen to apply in the heavens as on earth: there was just one world, a unitary cosmos.

At 22 years of age, while ensconced in Lincolnshire to avoid London's Great Plague, Newton began to speculate that the moon's orbit and an apple's fall might have a common cause (Herivel 1965, pp. 65–69). He was able to calculate that in 1 s, while travelling about 1 km in its orbit, the moon deviates from a straight-line path by about a twentieth of an inch. In the same period of time an object projected horizontally on the earth would fall about 16 feet. The ratio of the moon's 'fall' to the apple's fall is then about 1:3,700. This was very close to the ratio of the square of the apple's distance from the earth's centre (the earth's radius), to the square of the moon's distance from the earth's centre, 1:3,600. Was this a cosmic coincidence? Or did the earth's gravitational attraction apply equally to the apple and the moon?

Following the dictates of his own method, Newton then experimentally investigates whether the derived consequences are seen in reality. He defers to Huygens' experimental measurement, saying:

> And with this very force we actually find that bodies here upon earth do really descend; for a pendulum oscillating seconds in the latitude of *Paris* will be 3 *Paris* feet, and 8 lines ½ in length, as Mr *Huygens* has observed. And the space which a heavy body describes by falling in one second of time is to half the length of this pendulum as the square of the ratio of the circumference of a circle to its diameter (as Mr *Huygens* has also shown), and is therefore 15 *Paris* feet, 1 inch, 1 line $^7/_9$ And therefore the force by which the moon is retained in its orbit becomes, at the very surface of the earth, equal to the force of gravity which we observe in heavy bodies there. (Newton 1729/1934, p. 408)

Newton then draws his conclusion:

> And therefore the force by which the moon is retained in its orbit is that very same force which we commonly call gravity.

The pendulum had brought heaven down to earth.[14]

---

[14] On the pendulum's role in this unification, see especially Boulos (2006).

2.8 Huygens' Proposal of an International Standard of Length

Huygens saw that in his pendulum equation $T = 2\pi \sqrt{l/g}$ he only variable was l, as π was a constant and, provided one stayed near to sea level, g was also constant, and mass did not figure in the equation at all. So all pendula of a given length will have the same period, whether they be in France, England, Russia, Latin America, China or Australia. Huygens was clever enough to see that the pendulum would solve not only the timekeeping and longitude problems, but an additional vexing problem, namely establishing an international length standard, and in 1673 he proposed the length of a seconds pendulum (a pendulum that beats in seconds; that is, whose period is 2 s) to be the international unit. The length of the seconds pendulum was experimentally determined by adjusting a pendulum so that it oscillated $24 \times 60 \times 60$ times in a sideral day; that is, between successive transits of a fixed star across the centre of a graduated telescope lens (the sideral day being slightly longer than a solar day).

This seems like a daunting task, but it was not so overwhelming. Huygens and others knew that the length l of a pendulum varied as the square of period or T^2. So the length of a seconds pendulum is to the length of any arbitrary pendulum as $1/T^2$. But $1/T^2$ is as $n^2/60^2$, or the square of the number of swings or beats in 3,600 seconds or one hour. As Meli observes: 'therefore, by counting the number of oscillations and the length of his pendulum, Huygens could determine the length of the seconds pendulum' (Meli 2006, p. 205).

Having an international unit of length, or even a national unit, would have been a major contribution to simplifying the chaotic state of measurement existing in science and everyday life. Within France, as in other countries, the unit of length varied from city to city, and even within cities. This was a significant problem for commerce, trade, construction, military hardware and technology; to say nothing of science. Many attempts had been made to simplify and unify the chaotic French system. One estimate is that in France alone there were 250,000 different, local, measures of length, weight and volume (Alder 1995, p. 43). And each European state had comparable confusion of abundance, as did, of course, all other nations and cultures.

In the standard formulae, it is easy to show that the length of a seconds pendulum will be one metre.

$$T = 2\pi \sqrt{l/g}$$
$$\text{So } T^2 = 4\pi^2 l/g$$
$$\text{So } l = T^2 g / 4\pi^2$$
$$\text{Substitute } T = 2s \left(\text{beat is a second}\right), \ g = 9.8 \text{ms}^{-2}, \pi^2 = 9.8$$
$$\text{Then } l = 1m$$

And this result can reliably be demonstrated with even the crudest one-metre pendulum, a heavy nut on a piece of string suffices: 10 complete swings will take 20 s, 20 complete swings will take 40 s. A great virtue of the seconds pendulum as the international length standard was that it was a fully 'natural' standard; it was something fixed by nature unlike standards based on the length of a king's arm or foot. And of course an international length standard would provide a related volume standard and hence a mass standard when the unit volume was filled with rain water. A kilogram is the weight of one litre (1,000 cc) of water. All of this can be engagingly reproduced with classes.

It is not accidental that 200 years after Huygens, the General Conference on Weights and Measures meeting in Paris defined the standard universal metre as 'the length of path travelled by light in vacuum during a time interval of 1/299,792,458 of a second'.[15] This seemingly bizarre and arbitrary figure is within a millimetre of Huygen's original and entirely natural length standard, and it was so chosen precisely to replicate the length of the seconds pendulum. Unfortunately it is the former, not the latter, that students meet in the opening pages of their science texts, and so confirms their worst fears about the 'strangeness' of science. In the definition of standards, science gets off to a bad pedagogical start.

2.9 The Pendulum and Determining the Shape of the Earth

Huygen's proposal depended on g being constant around the world (at least at sea level); it depended on the earth being spherical. This seemed a most reasonable assumption. Indeed to say that the earth was not regular and spherical was tantamount to casting aspersions on the Creator: surely God the Almighty would not make a misshapen earth. But in 1673, contrary to all expectation, this assumption was brought into question by the behaviour of the pendulum.

When Jean-Dominique Cassini (1625–1712) became director of the French *Académie Royale des Sciences* in 1669 he began sending expeditions into the different parts of the world to observe the longitudes of localities for the perfection of geography and navigation. The second such voyage was Jean Richer's to Cayenne in 1672–1673 (Olmsted 1942). Cayenne was in French Guiana, at latitude approximately 5°N. It was chosen as a site for astronomical observations because equatorial observations were minimally affected by refraction of light passing through the earth's atmosphere – the observer, the sun and the planets were all in the same plane.

The primary purpose of Richer's voyage was to ascertain the value of solar parallax and to correct the tables of refraction used by navigators and astronomers. A secondary consideration was checking the reliability of marine pendulum clocks which were being carried for the purpose of establishing Cayenne's exact longitude.

The voyage was spectacularly successful in its primary purposes: the obliquity of the ecliptic was determined, the timing of solstices and equinoxes was refined and, most importantly, a new and far more accurate value for the parallax of the sun was ascertained – 9.5″ of arc. But it was an unexpected consequence of Richer's voyage

---

[15] Accounts of the development of the standard metre can be found in Alder (1995, 2002), Kula (1986, Chaps. 21–23).

which destroyed Huygens' vision of a universal standard of length 'for all nations and all ages'.

Richer found that a pendulum set to swing in seconds at Paris, had to be shortened in order to swing in seconds at Cayenne, not much – 2.8 mm, about the thickness of a matchstick – but nevertheless shortened. Richer found that a Paris seconds-clock apparently lost 2½ min daily at Cayenne. And the only apparent explanation for slowing of the pendulum at the equator was that g is less at the equator than at Paris and the poles – in other words, that the earth is a 'flattened' sphere, an oblate, with the equator being further from the centre than the poles.[16]

2.10   The Testing of Scientific Theories

Richer's claim that the pendulum clock slows in equatorial regions nicely illustrates some key methodological matters about science, and about theory testing. The entrenched belief since Erastosthenes in the second century BC was that the earth was spherical (theory T), and on the assumption that gravity alone affects the period of a constant length pendulum, the observational implication was that period at Paris and the period at Cayenne of Huygens' seconds-pendulum would be the same (O). Thus T implies O:

$$T \rightarrow O$$

But Richer seemingly found that the period at Cayenne was longer (~ O). Thus, on simple, falsificationist views of theory testing such as were enunciated first by Huygens himself, and famously developed by the philosopher Karl Popper early in the twentieth century (Popper 1934/1959), we have:

$$T \rightarrow O$$
$$\sim O$$
$$\therefore \sim T$$

But theory testing is never so simple – a matter that was recognised by Popper, and articulated by Thomas Kuhn (1970) and Imre Lakatos (1970). In the seventeenth century, many upholders of T just denied the second premise, ~O. The astronomer Jean Picard, for instance, did not accept Richer's findings. Rather than accept the message of varying gravitation, he doubted the messenger. Similarly, Huygens did not think highly of Richer as an experimentalist.

Others saw that theories did not confront evidence on their own, there was always an 'other things being equal' assumption made in theory test; there were *ceteris paribus* clauses (C) that accompanied the theory into the experiment. These clauses

---

[16] On the history of debate about the shape of the earth, see Chapin (1994), Greenberg (1995) and Heiskanen and Vening Meinesz (1958).

characteristically included statements about the reliability of the instruments, the competence of the observer, the assumed empirical state of affairs, theoretical and mathematical devices used in deriving O, and so on. Thus:

$$T + C \rightarrow O$$
$$\sim O$$
$$\therefore \sim T \text{ or } \sim C$$

People who maintained belief in T, reasonably said that the assumption that other things were equal was mistaken – perhaps humidity had interfered with the swings, heat had lengthened the pendulum, friction at the pivot increased in the tropics and so on. These, in principle, were legitimate concerns. But more and more evidence came in, and from other experimenters including Sir Edmund Halley, confirming Richer's observations. Thus $\sim O$ became established as a scientific fact, to use Fleck's terminology (Fleck 1935/1979), and upholders of T, the spherical earth hypothesis, had to adjust to it. This was not easy; giving up established theories in science is never easy, especially as the alternative was to accept that the earth was oblate in shape, an ungainly shape for the Creator to have fashioned.

There were a number of obvious items in **C** that could be pointed to as the cause of the pendulum slowing:

C^1 The experimenter was incompetent.
C^2 Humidity in the tropics caused the pendulum to slow because the air was denser.
C^3 Heat in the tropics caused the pendulum to expand, hence it beat slower.
C^4 The tropical environment caused increased friction in the moving parts of the clock.

Each of these could account for the slowing, and hence preserve the truth of the spherical earth theory. But each of them was in turn ruled out by progressively better controlled and conducted experiments. Many of course would say that adjustment of the thickness of a match (3 mm) as a proportion of a metre (1,000 mm) was so minimal that it could just be attributed to experimental error, or simply ignored. And if the theory is important, then that is an understandable tendency. But for more tough-minded scientists it seemed that the long held, and religiously endorsed, theory of the spherical earth had to be rejected.

But Huygens could see a more sophisticated explanation for the lessening of g at the equator, whilst still maintaining T, the theory of a spherical earth. He argued that:

C^5 Objects at the equator rotated faster than at Paris and hence the centrifugal force at the equator was greater, this countered the centripetal force of gravity, hence diminishing the nett downwards force (gravity) at the equator, hence decreasing the speed of oscillation of the pendulum; that is, increasing its period.

This final explanation for the slowing of equatorial pendula was quite legitimate and appeared to save the theory. Many would be happy to just pick up this 'get out of jail free' card and continue to believe that the earth was spherical. Huygens did not do so. He calculated the actual centrifugal force at the equator and determined

2 Pendulum Motion: A Case Study in How History and Philosophy Can Contribute...

that a shortening of 1.5 mm was required to make up for the spinning earth effect.[17] But this left 1.5 mm not accounted for. This is less than the thickness of a match, yet for such a minute discrepancy Huygens and Newton were prepared to abandon the spherical earth theory and claim that the true shape of the earth was an oblate. For the new quantitative science, 'near enough' was not 'good enough', something that students can be taught to appreciate.

This episode did not escape the attention of Voltaire, a populariser of Newtonian science and a key figure in the European Enlightenment who, in 1738, wrote:

> At last in 1672, Mr Richer, in a Voyage to Cayenna, near the Line, undertaken by Order of Lewis XIV under the protection of Colbert, the Father of all Arts; Richer, I say, among many Observations, found that the Pendulum of his Clock no longer made its Vibrations so frequently as in the Latitude of Paris, and that it was absolutely necessary to shorten it by a Line, that is, eleventh Part of our Inch, and about a Quarter more.
>
> Natural Philosophy and Geometry were not then, by far, so much cultivated as at present. Who could have believed that from this Remark, so trifling in Appearance, that from the Difference of the eleventh of our Inch, or thereabouts, could have sprung the greatest of physical Truths? It was found, at first, that Gravity must needs be less under the Equator, than in the Latitude of France, since Gravity alone occasions the Vibration of a Pendulum.
>
> In Consequence of this it was discovered, that, whereas the Gravity of Bodies is by so much the less powerful, as these Bodies are farther removed from the Centre of the Earth, the Region of the Equator must absolutely be much more elevated than that of France; and so must be farther removed from the Centre; and therefore, that the Earth could not be a Sphere. (Fauvel and Gray 1987, p. 420)

He dryly commented that:

> Many Philosophers, on occasion of these Discoveries, did what Men usually do, in Points concerning which it is requisite to change their Opinion; they opposed the new-discovered Truth. (Fauvel and Gray 1987, p. 420)

Voltaire and proponents of the Enlightenment thought that the way that the Shape of the Earth debate was resolved could be emulated in other fields of hotly contested debate and disagreement – especially in politics, religion, ethics and law – and instead of doing what 'men usually do' in these fields, they would do what the natural philosophers did, namely change their opinions when contrary evidence accrued and was verified.[18]

2.11 Some Social and Cultural Impacts of Timekeeping

The advent of accurate timekeeping in the eighteenth century had enormous impact on European social and cultural life, and by extension on the rest of the globe.

---

[17] For the physics and mathematics of these calculations, see Holton and Brush (2001, pp. 128–129).

[18] This is a wonderful episode in the history of science. A great story can be made, even a drama. All the elements are there: powerful and prestigious figures, 'no name' outsiders, struggles over a big issue, mathematics and serious calculations, religion, final decisions and ample opportunity to preserve the status quo. But sadly the episode is little known and hardly ever taught.

2.11.1 Solving the Longitude Problem

Accurate time measurement was long seen as the solution to the problem of longitude determination which had vexed European maritime nations in their efforts to sail beyond Europe's shores. If an accurate and reliable clock was carried on voyages from London, Lisbon, Genoa, or any other port, then by comparing its time with local noon (as determined by noting the moment of an object's shortest shadow or, more precisely, by using optical instruments to determine when the sun passes the location's north–south meridian), the longitude of any place in the journey could be ascertained. The physics was simple. The earth rotates 360° in 24 h, or 15° in one hour, or one degree each 4 min. So if at destination the clock at origin is set to 12 at noon then if at destination at noon it reads 10 am at local noon, then the destination is 30° east of origin. As latitude could already be determined, this enabled the world to be mapped. In turn, this provided a firm base on which European trade and colonisation could proceed. The chances of being lost at sea were greatly decreased. John Harrison's marine chronometer, which followed on his extensive pendulum clock constructions, solved the longitude problem.[19]

2.11.2 A Clockwork Society

The clock transformed social life and customs: patterns of daily life could be 'liberated' from natural chronology (the seasonally varying rising and setting of the sun) and subjected to artificial chronology; labour could be regulated by clockwork and, because time duration could be measured, there could be debate and struggle about the length of the working day and the wages that were due to agricultural and urban workers; timetables for stage and later train and ship transport could be enacted; the starting time for religious and cultural events could be specified; punctuality could become a virtue; and so on. The transition from 'natural' to 'artificial' hours was of great social and psychological consequence: technology, a human creation, begins to govern its creator.[20] Lewis Mumford, the social historian, has commented that

> The clock, not the steam-engine, is the key-machine of the modern industrial age. … by its essential nature it dissociated time from human events and helped create the belief in an independent world of mathematically measurable sequences: the special world of science. (Mumford 1934, pp. 14–15)

---

[19] Dava Sobel has given the Longitude Problem enormous exposure (Sobel 1995). Other more detailed and wide-ranging treatments are in Andrewes (1998), Gould (1923) and Howse (1980).

[20] Many books deal with the social and cultural history of timekeeping, among them are: Cipolla (1967), Landes (1983), Macey (1980), and Rossum (1996).

2.11.3 A Clockwork Universe and Its Maker

The clock did duty in philosophy. It was a metaphor for the new mechanical world-view that was challenging the entrenched Aristotelian, organic and teleological, view of the world that has sustained so much of European intellectual and religious life. In theology, the clock was appealed to in the influential argument from design for God's existence – if the world functions regularly like a clock, as Newton and the Newtonians maintained, then there must be a cosmic clockmaker.[21]

Leibniz closes his famous 'world as clock' correspondence with the Newtonian Samuel Clark by writing:

> I maintained that the dependence of the machine of the world upon its divine author, is rather a reason why there can be no such imperfection in it; and that the work of God does not want to be set right again; that it is not liable to be disordered; and lastly, that it cannot lessen in perfection. (Alexander 1956, p. 89)

2.11.4 Foucault's Pendulum Makes Visible the Earth's Rotation

The pendulum provided the first ever visible and dynamic 'proof' of the rotation of the earth. On Newton's theory, a pendulum set swinging in a particular plane, should continue to swing indefinitely in that same plane. The only forces on the bob being the tension in the cord, and its weight directed vertically downwards. Léon Foucault – described as 'a mediocre pupil at school, [but] a natural physicist and an incomparable experimenter' (Dugas 1988, p. 380) – 'saw' that if a pendulum were placed exactly at the north pole, and suspended in such a way that the point of suspension was free to rotate (i.e. it did not constrain the pendulum's movement by applying torque), then:

> if the oscillations can continue for twenty-four hours, in this time the plane will execute a whole revolution about the vertical through the point of suspension. ... at the pole, the experiment must succeed in all its purity. (Dugas 1988, p. 380)

As the pendulum is moved from the pole to the equator, Foucault easily showed that if T^1 is the time in which the plane of the pendulum rotates $360°$, and T is the period of rotation of the earth, and β is the latitude where the experiment is being conducted, then:

$$T^1 = T / \sin\beta$$

From the formula, it can be seen that at the poles, $T^1 = T$ (as $\sin\beta = \sin 90° = 1$); whereas at the equator $T^1 = \infty$ (or infinity, as $\sin 0° = 0$), thus there is no rotation of the plane of oscillation at the equator.

---

[21] Macey (1980), Pt. II is a nice introduction to the utilisation of the clock in eighteenth-century philosophy and theology.

On February 2, 1851, Foucault invited the French scientific community to 'come see the Earth turn, tomorrow, from three to five, at Meridian Hall of the Paris Observatory'. His eponymously named long massive pendulum provided an experimental 'proof' of the Copernican theory; something that eluded Galileo, Newton and all the other mathematical and scientific luminaries who sought it (Tobin 2003; Aczel 2003, 2004).

Until Foucault's demonstration, all astronomical observations could be fitted, with suitable adjustments such as those made by Tycho Brahe, to the stationary earth theory of the Christian tradition. The 'legitimacy' of such ad hoc adjustments in order to preserve the geocentric model of the solar system was exploited by the Catholic Church that kept the works of Copernicus and Galileo on the *Index of Prohibited Books* up until 1835 (Fantoli 1994, p. 473). To most nineteenth-century physicists, the manifest rotation of Foucault's pendulum shown in the successive knocking down of markers placed in a circle, was a dramatic proof of the earth's rotation. Around the world tens of thousands read accounts of the pendulum, and thousands attended demonstrations; scores of newspapers editorialised on the subject; cartoonists had a picnic with it (Conlin 1999).

2.12   The Pendulum in the Classroom

It has been long recognised that much of Newtonian physics could be demonstrated, and properties of the world determined, by experimental manipulation of the pendulum. The conical pendulum, for instance, representing idealised planetary circular motion with constant velocity and yet a constant force and acceleration towards the centre. In the century after Newton, the pendulum was used widely in illustrating or 'proving' the fundamentals of Newtonian science, or classical physics, as we know it. Mathematics gave a description of the pendulum's movement (the phenomenon) without causal explanation, while Newton's laws identify the dynamic factors responsible for the observed motion. This is how physics is done: first represent or model the idealised phenomenon in mathematical terms, then explain it using the best causal theory (Newburgh 2004, pp. 297–299).[22] Then progressively try to mathematically represent less and less idealised and more and more realistic versions of the phenomenon and seek to identify the secondary causes or interfering factors with this progression going hand-in-hand with refinement of experimental situations. This is the progression from the idealised simple pendulum to the damped friction-affected realistic pendulum.

---

[22] The pendulum, and all physical phenomena, can be represented by different mathematical devices: geometry, Hamiltonian equations and so on. Geometry has the advantage of connecting more immediately and intuitively to the physics of the phenomena; a not inconsiderable advantage and so a step that students should pass through on their way to algebraic representation of the pendulum.

The simple pendulum became more sophisticated – conical, compound, cradle, torsion, reversible, ballistic, coupled pendula were all crafted – and thus extending the range and accuracy of classical mechanics. The pendulum was recognised as a case of Simple Harmonic Motion where the displacing force and motion are directly and inversely related, and this motion was recognised as ubiquitous in nature.

Following Galileo, Huygens and Newton, numerous famous physicists have been associated with this history: Robert Hooke, Henry Kater, Count Rumford, George Atwood, George Stokes, Roland von Eötvös, Henry Cavendish, and others. In the past century, further developments occurred with the chaotic and quantum pendula. And at every stage the intimate dependency of physics on mathematics is apparent.[23] Two physicists have commented that:

> There is a quite unexpected connection between the classical pendulum – chaotic or otherwise – and quantum mechanics when it functions on a macroscopic scale, as happens in superconductors. More specifically, the connection arises through something known as the Josephson effect. ... there is an exact correspondence between the dynamics of the Josephson devices and the dynamics of the classical pendulum. (Baker and Blackburn 2005, p. 211)

The pendulum has long been part of the physics curriculum, a fact well documented in Colin Gauld's structured bibliography of nearly 300 pendulum articles that have appeared over the past 50 years in four major physics teaching journals (Gauld 2004b). Teachers have used the simple pendulum, swinging through small angles, to teach the skills of measurement and graphical techniques for deriving the relationship between dependent (in this case, period) and independent variables (length of the string). More complex types of pendulums (such as the physical, spring-mass, torsional and Wilberforce pendulums) have been used to demonstrate dramatically a wide range of physical phenomena and provide a context in which students can become acquainted with the process of mathematical modelling in science. In the classroom, pendulum motion provides a model for many everyday oscillatory phenomena such as walking and the movement of a child's swing. At the tertiary level there has been renewed interest in the pendulum to demonstrate chaotic behaviour. For these investigations, the pendulum amplitude is unrestricted and the point of suspension is vibrated at varying amplitudes and frequencies. By removing the requirement that the amplitude be small, the behaviour of the pendulum as a non-linear oscillator can clearly be seen (Weltner et al. 2004). And the pendulum has been used to facilitate students moving from classical understanding to quantum physics (Barnes et al. 2004).

The history of pendulum investigations contains almost everything required to teach the fundamentals of kinematics, dynamics and classical physics, along with scientific methodology, epistemology and process skills. Nevertheless this history is sadly under-utilised in schools; much more could be made, beginning at the

---

[23] Gregory Baker and James Blackburn provide an excellent account of the role played by the pendulum in the development of physics from Galileo to superconductivity (Baker and Blackburn 2005). Randall Peters discusses largely unexplored uses of the pendulum in investigating the science of material deformation and creep (Peters 2004).

kindergarten level, of the pendulum as a device for teaching physics content, scientific methodology, relationships of science, technology, society and culture, and more broadly for teaching the nature of science. This is especially so when repeatedly curriculum documents speak of the need to teach 'science in context', to teach 'the connection of science and technology', to teach 'the relationship of science to everyday life', to teach 'the big picture of science'. The pendulum allows all of these liberal educational goals to be advanced if not achieved. This contrast between the pendulum's scientific and social importance and its educational neglect highlights an increasingly recognised deficiency in science education: There is little sense of students being introduced to and appreciating a tradition of thought. Music, Art, Literature, Philosophy and Theology students are given this sense of tradition and appreciation of the major contributors to it; science students only barely, if at all. Education does little to make more general Newton's sense of 'standing on the shoulders of giants'.

2.13 The Pendulum and Textbooks

Science textbooks pay very little attention to the historical, methodological and cultural dimension of pendulum motion. It is sometimes given a cameo appearance in the story of Galileo who supposedly during a church sermon observed a swaying chandelier and timed its swings with his pulse, and 'hey presto' there was the law of isochronic motion. This account is found in Fredrick Wolf's physics text:

> When he [Galileo] was barely seventeen years old, he made a passive observation of a chandelier swinging like a pendulum in the church at Pisa where he grew up. He noticed that it swung in the gentle breeze coming through the half-opened church door. Bored with the sermon, he watched the chandelier carefully, then placed his fingertips on his wrist, and felt his pulse. He noticed an amazing thing. . . . Sometimes the chandelier swings widely and sometimes it hardly swings at all . . . [yet] it made the same number of swings every sixty pulse beats. (Wolf 1981, p. 33)

Wolf's story, *sans* boredom, appears in the opening pages of the most widely used high-school physics text in the world – the Physical Science Study Committee's *Physics* (PSSC Physical Science Study Committee 1960).

Whatever the problems with Wolf and the PSSC text might be, the Galileo story is at least presented. However, more often the pendulum appears in physics texts without any historical context; as a standard it is introduced merely as an instance of simple harmonic motion. The extent to which the pendulum has been plucked from its historical and cultural roots can be seen in the Harvard Project Physics text, an excellent and most contextual of texts where, nevertheless, the equation:

$$T = 2\pi\sqrt{l/g}$$

is abruptly introduced for the period of the pendulum, and students are told 'you may learn in a later physics course how to derive the formula' (Holton et al. 1974, p. 98).

2.14 The Pendulum and Recent US Science Education Reform Proposals

It is instructive, if sobering, to look at the utilisation of the pendulum in the past three decades of intense efforts to improve US school science programmes. These efforts have involved thousands of individuals in bodies such as the American Association for the Advancement of Science, the National Research Council, the National Academy of Science, the National Academy of Engineering, the National Science Foundation; peak disciplinary bodies in physics, chemistry, biology, earth science; and all major national and state science education organisations including the National Science Teachers Association and the National Association for Research in Science Teaching.

The reform efforts have their origin in the 1983 Reagan-era Report *A Nation at Risk: An Imperative for Educational Reform* (NCEE 1983).[24] Concerning science education, the Commission recommended that:

> The teaching of science in high school should provide graduates with an introduction to: (a) the concepts, laws, and processes of the physical and biological sciences; (b) the methods of scientific inquiry and reasoning; (c) the application of scientific knowledge to everyday life; and (d) the social and environmental implications of scientific and technological development. Science courses must be revised and updated for both the college-bound and those not intending to go to college. (NCEE 1983, p. 25)

This 'liberal' or contextual approach to science education has been followed-through in all subsequent major US curricular reform proposals. Clearly the pendulum is tailor made to contribute to the realisation of each of the four stated goals of reformed science education. Needless to say, this opportunity has been under-utilised, to put not too fine a point on it. Everyone recognises that there is a gulf between the content of curriculum documents and the content of classroom practice; but here the gulf begins in the documents themselves, between the stated 'liberal' objectives and the curriculum content.

2.14.1 Scope, Sequence and Coordination

The large-scale and influential curriculum proposal of the US National Science Teachers Association (NSTA) – *Scope, Sequence and Coordination* (Aldridge 1992) – highlights the pendulum to illustrate its claims for sequencing and coordination in science instruction. Yet nowhere in its discussion of the pendulum is history, philosophy or technology mentioned. That such a huge and well-funded body as NSTA could, in the early 1990s, write a national science curriculum

---

[24] Gerald Holton, a member of National Commission for Excellence in Education (NCEE) that prepared the report, has provided an account of its disturbing contents that chart the 'tide of mediocrity' in US education, and its recommendations for turning the tide (Holton 1986).

proposal without the participation of historians or philosophers of science is a sad commentary on the gulf between the science education and the HPS communities.[25]

2.14.2 Project 2061

In 1989, the American Association for the Advancement of Science (AAAS) published its wonderfully comprehensive *Science for All Americans* report (AAAS 1989). It acknowledged that 'schools do not need to be asked to teach more and more content, but rather to focus on what is essential for scientific literacy and to teach it more effectively' (AAAS 1989, p. 4). The report saw that students need to learn about 'The Nature of Science', and hence that was the title of the report's first chapter. The report recognised the importance of learning about the interrelationship of science and mathematics, saying: 'The alliance between science and mathematics has a long history dating back many centuries. … Mathematics is the chief language of science' (AAAS 1989, p. 34). And it acknowledged that some episodes in the history of science should be appreciated because 'they are of surpassing significance to our cultural heritage' (AAAS 1989, p. 111). Among the ten such episodes it picks out is Newton's demonstration that the same laws apply to motion in the heavens and on earth' (AAAS 1989, p. 113). It provides a very rich elaboration of this episode and its scientific, philosophical and cultural impacts. Unfortunately, there is no mention of what enabled Newton to achieve this unification, namely the pendulum; had such mention been made, this 'big idea' could have been connected to something tangible in all students' experience, the place of mathematics in science could have again been underlined, and a wonderful case study in the nature of science could have been built upon.

2.14.3 The US National Standards

The underutilisation of the pendulum can be gauged from looking at the recently adopted US National Science Education Standards (NRC 1996). The *Standards* adopt the same liberal or expansive view of scientific literacy as the NCEE did in 1983 saying that it 'includes understanding the nature of science, the scientific enterprise, and the role of science in society and personal life' (NRC 1996, p. 21). The *Standards* devote two pages to the pendulum (pp. 146–147). However there is no mention of the history, philosophy, or cultural impact of pendulum motion studies; no mention of the pendulum's connection with timekeeping; no

---

[25] This observation was made in 1992 by a senior NSTA official in private correspondence with the author.

mention of the longitude problem; and no mention of Foucault's pendulum. Astonishingly in the suggested assessment exercise, the obvious opportunity to connect standards of length (the metre) with standards of time (the second) is not taken. Rather, students are asked to construct a pendulum that makes six swings in 15 s. This is a largely pointless exercise, especially when they could have been asked to make one that beats in seconds and then measure its length and inquire about the coincidence between their seconds pendulum and the metre (Matthews 1998).

Depressingly the *Standards* document was reviewed in draft form by tens of thousands of teachers and educators. It is clear that if even a few of the readers had a little historical and philosophical knowledge about the pendulum, this could have transformed the treatment of the subject in the *Standards* and would have encouraged teachers to realise the liberal goals of the document through their treatment of the pendulum. This would have resulted in a much richer and more meaningful science education for US students. That this historical and philosophical knowledge is not manifest in the *Standards* indicates the amount of work that needs to be done in having science educators become more familiar with the history and philosophy of the subject they teach, and of having the US science education community more engaged with the communities of historians and philosophers of science.

2.14.4 *America's Lab Report*

The US National Research Council commissioned a large study on practical work in US schools which was published as *America's Lab Report: Investigations in High School Science* (NRC 2006). The book has 236 pages, seven chapters, and hundreds of references. The pendulum has three entries in the Index. On its first appearance, it is said to be regrettable that teachers simplify pendulum experiments and ignore the 'host of variables that may affect its operation' (NRC 2006, p. 117). Teachers are advised to recognise these 'impediments' such as friction and air resistance but the writers go on to say that this 'can quickly become overwhelming to the student and the instructor' (NRC 2006, p. 118). This is not very helpful. It could have been an occasion to say something about the fundamental importance of idealisation and abstraction to the very enterprise of science, of not letting the trees get in the way of seeing the forest. This was the problem identified by Thomas Kuhn in his discussion of the pendulum and faced by da Vinci; it is the heart of the debate between Galileo and his patron Guidabaldo del Monte. But the *Lab Report* says nothing about this fundamental scientific procedure much less provide some historical background to its resolution. The pendulum allows in a tangible way for students to begin seeing the effect of 'impediments' and 'accidents' (Koertge 1977), or 'errors' in contemporary language, on the manifestation or 'visibility' of core natural processes.

On the pendulum's second appearance, the 'typical pendulum experiment' is criticised because it is 'cleaned up' and used just to teach science content – that the 'period of a pendulum depends on the length of the string and the force of gravity' – and not scientific process skills (NRC 2006, p. 126). In contrast to these 'bad' pendulum practical classes, on the pendulum's third appearance a 'good' class is described over two pages in a highlighted box. In this class, teachers are first advised to demonstrate swinging pendulums, then in a very guided fashion to have students graph the relationships between period and mass, period and amplitude, and period and length, and finally it is suggested that the teacher discuss the importance of obtaining adequate amount of data over a range of the independent variables (NRC 2006, pp. 128–129).

There does not seem to be much especially good or noteworthy about this. Everything about the rich history of the pendulum has been stripped out: no mention of Galileo, Huygens, Newton, Hooke's universal gravitation, timekeeping, clocks, length standards, longitude, shape of the earth, or conservation laws. No connection intimated between science, technology and society; no sense of participation in a scientific tradition. Nothing. Teachers are not even told to talk about these great scientists and their pendulum-based discoveries.

And in this set-piece, nationally distributed, 'model pendulum lesson' teachers and students are told to graph period against length. This is a task with only minimal useful outcomes; such a graph provides a scatter of points that merely establishes a trend. US physics students in the final year of high school, 17–18 years old, could have been so easily asked to additionally plot period against the square root of length. When this is done, nothing is inconclusive: a straight line is obtained from the data, not a scatter of points. Period is seen, as it was by Galileo and Huygens, to not just vary as length, but to vary directly as the square root of length; the conclusion from the data moves from inconclusive $T \alpha L$ to conclusive $T = k\sqrt{L}$. The model lesson tells teachers to 'avoid introducing the formal pendulum equation, because the laboratory activity is not designed to verify this known relationship' (NRC 2006, p. 129). Final-year students in Japan, Korea, Singapore and a good deal of the rest of the world have no such problem, neither should US students.

The graph of period against square root of length shows in a manageable way the dramatic impact of mathematics on physics; without the mathematical notion of square root we see qualitative trends, utilising the square root we see a precise quantitative relationship. Further, this precise relationship will allow the pendulum to be connected with free fall where distance of fall varies as the square of time. All of this is missed in the *Lab Report*, and also missed is the opportunity for richer pendulum-informed teaching of physics. What appears to have happened is what the NRC recognises in another publication:

> As educators, we are underestimating what young children are capable of as students of science – the bar is almost always set too low. (NRC 2007, p. vii)

2.14.5 The Next Generation Science Standards

For the past three years in the US, a new national science education standards document, called the *Next Generation Science Standards* (NRC 2012) has been progressively developed.[26] As the NGSS says:

> The impetus for this project grew from the recognition that, although the existing national documents on science content for grades K-12 (developed in the early to mid-1990s) were an important step in strengthening science education, there is much room for improvement. Not only has science progressed, but the education community has learned important lessons from 10 years of implementing standards-based education, and there is a new and growing body of research on learning and teaching in science that can inform a revision of the standards and revitalize science education. (NRC 2012, p. ix)

The NGSS incorporate and build on the 'existing national documents' but a novel feature is the conscious effort to connect science learning to engineering, to scientific practices, and to make it progressive and cumulative from the beginning of elementary school. These are seen as its *differentia* from 'the existing national documents'. An NGSS press release (10 April 2013) says that instead of students learning by rote, their focus would be on:

> learning how science is done: how ideas are developed and tested, what counts as strong or weak evidence and how insights from many scientific disciplines fit together into a coherent picture of the world.

The pendulum 'ticks all of the NGSS boxes' so as to speak. Very young children, as shown in Japan, Korea and numerous other countries, can profitably and enjoyably engage with pendulum activities (Sumida 2004; Kwon et al. 2006).[27] It is not accidental that Jean Piaget used the pendulum for his investigation of the progressive development of children's scientific reasoning ability, especially their identification and control of variables (Bond 2004). The sophistication of pendulum activities and their relation with other areas and topics in science can be enhanced with progression through school; obvious connections with mathematics, technology and engineering can be made, and even connections with chemistry (De Berg 2006). The full range of process skills (data collection and representation, hypothesis generation), methodological skills (generating hypotheses, evaluating these against evidence, theory testing and so on) and model construction can all be cultivated using pendulum classes (Kwon et al. 2006; Stafford 2004; Zachos 2004).

---

[26] The 320pp draft is available free from the National Academies Press website; it is titled *A Framework for K-12 Science Education*. Background studies for the NGSS are in NRC (2007).

[27] An excellent pendulum booklet is produced for Japanese elementary students. Galileo's image occupies the entire front cover while Huygens' image occupies the entire rear cover – a nice comment on the universality of science and its ability to be embraced by cultures beyond its original European home. Japanese students, at least, can gain some sense of participation in the scientific tradition and their indebtedness to those that have gone before.

It remains to be seen how the pendulum will feature, in the final NGSS document but the signs are not good. In the current (2012) draft the pendulum is mentioned four times and each time it is in connection with the transformation of energy from potential to kinetic forms. This is a level of abstraction way beyond what is needed or called for; it is beyond the life experience of the students; and it reifies the role played by the pendulum in the history of physics and in its social utilisation. The draft document mentions Newton's laws, his theory of gravitation, the conservation of momentum, but no mention of the pendulum that could so easily be used to manifest and make experiential each of these learning goals.

2.15 The International Pendulum Project

The *International Pendulum Project* (IPP) had its origins with the publication of the book *Time for Science Education: How Teaching the History and Philosophy of Pendulum Motion Can Contribute to Science Literacy* (Matthews 2000). The book was about a decade in gestation, has 13 chapters, 1,200 references, and ranges widely over the history, methodology, cultural impact and pedagogy of pendulum studies. Interest in the subject matter of the book was sufficient to bring a large international group of scholars together for conferences at the University of New South Wales in 2002 and again in 2005. Participants recognised the need for teachers and students to be more aware of the important role played by the pendulum in the history of science, and to investigate and promote better and more enriched pendulum teaching in schools.

Scholars from 20 countries contributed to the IPP, and their research appeared in three special issues of the journal *Science & Education* (vol.13 nos. 4–5, 7–8, 2004, vol.15 no.6, 2006). Thirty-three papers from these issues were published in the anthology *The Pendulum: Scientific, Historical, Philosophical and Educational Perspectives* (Matthews et al. 2005). Importantly the contributors came from education, physics, cognitive science, philosophy and history. It is the cross-disciplinary input that gave the IPP its distinctive strength.

2.16 Conclusion

The NGSS gives three reasons for producing updated standards in the USA, one of which is that there is a 'growing body of research on learning and teaching in science' that can be utilised. Due caution should be exercised about such claims. In the 1950s and 1960s, the 'growing body of research on learning and teaching' gave us Behaviourism, which has now disappeared without educational trace; in the 1980s and 1990s, the 'growing body of research on learning and teaching' gave us Constructivism with all its well-known philosophical and pedagogical problems (Matthews 2000a, 2012). Good understanding of teaching and learning is certainly needed, but the improvement of curricula does not flow just from knowledge about *how* to better teach and learn material, but rather it flows from knowledge of *what* material to teach and learn, and *where* to place the topics and concepts in state and national standards. This is where a richer understanding of the history and philosophy of pendulum studies and utilisation (and of course of all other topics) can well contribute to science education. It can make for better curricula and for better connections between disciplinary strands in curricula.

The following diagram, where the columns represent curriculum subjects and the circles topics within subjects, displays the integrative curricular function of history and philosophy.[28]

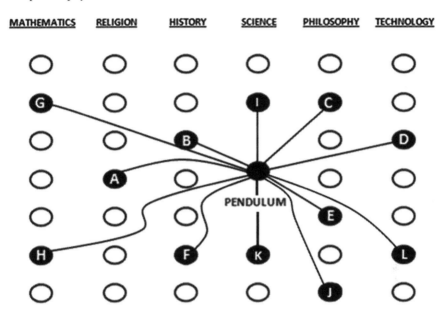

A. The design argument
B. European voyages of discovery
C. Aristotelian physics and methodology
D. Pendulum clock
E. Idealisation and theory testing
F. Timekeeping and social regulation
G. Geometry of the circle
H. Applied mathematics
I. Measurement and standards
J. Time
K. Energy
L. Geodesy

The content of the school day, or at least year, can be more of a tapestry, rather than a curtain of unconnected curricular beads. The latter is a well-documented problem with US science education, with its fabled 'mile wide and one inch thick' curricula (Kesidou and Roseman 2002). But pendulum motion, if taught from a

---

[28] The idea for this visual representation of the argument comes from a AAAS lecture of Gerald Holton, subsequently published as Holton (1995). For elementary schools, READING could be added as a column.

historical and philosophical perspective, allows connections to be made with topics in religion, history, mathematics, philosophy, music and literature, as well as other topics in the science programme. And such teaching promotes greater understanding of science, its methodology, and its contribution to society and culture. But such connections first need to be recognised by curriculum writers and by teachers charged with implementing curricula or achieving standards; this raises the whole question of HPS in pre-service or in-service teacher education, but that is a different subject for a different chapter.

References

(AAAS) American Association for the Advancement of Science: 1989, *Project 2061: Science for All Americans*, AAAS, Washington, DC. Also published as Rutherford and Ahlgren (1990).

(NCEE) National Commission on Excellence in Education: 1983, *A Nation At Risk: The Imperative for Education Reform*, US Department of Education, Washington DC.

(NRC) National Research Council: 1996, *National Science Education Standards*, National Academies Press, Washington DC.

(NRC) National Research Council: 2006, *America's Lab Report: Investigations in High School Science*, National Academies Press, Washington DC.

(NRC) National Research Council: 2007, *Taking Science to School. Learning and Teaching Science in Grades K-8*, National Academies Press, Washington DC.

(NRC) National Research Council: 2012, *A Framework for K-12 Science Education*, National Academies Press, Washington DC.

(PSSC) Physical Science Study Committee: 1960, *Physics*, D.C. Heath & Co., Boston.

Aczel, A.D.: 2003, *Pendulum: Léon Foucault and the Triumph of Science*, Atria Books, New York.

Aczel, A.D.: 2004, 'Leon Foucault: His Life, Times and Achievements', *Science & Education* 13(7–8), 675–687.

Alder, K.: 1995, 'A Revolution to Measure: The Political Economy of the Metric System in France'. In M.N. Wise (ed.), *The Values of Precision*, Princeton University Press, Princeton, NJ, pp. 39–71.

Alder, K.: 2002, *The Measure of All Things: The Seven-Year Odyssey that Transformed the World*, Little Brown, London.

Aldridge, B.G.: 1992, 'Project on *Scope, Sequence, and Coordination*: A New Synthesis for Improving Science Education', *Journal of Science Education and Technology* 1(1), 13–21.

Alexander, H.G. (ed.): 1956, *The Leibniz-Clarke Correspondence*, Manchester University Press, Manchester.

Andrewes, W.J.H. (ed.): 1998, *The Quest for Longitude: The Proceedings of the Longitude Symposium, Harvard University, Cambridge, Massachusetts, November 4–6, 1993*, 2nd Edition, Collection of Historical Scientific Instruments, Harvard University, Cambridge, MA.

Ariotti, P.E.: 1968, 'Galileo on the Isochrony of the Pendulum', *Isis* 59, 414–426.

Ariotti, P.E.: 1972, 'Aspects of the Conception and Development of the Pendulum in the 17th Century', *Archive for History of the Exact Sciences* 8, 329–410.

Baker, G.L. & Blackburn, J.A.: 2005, *The Pendulum: A Case Study in Physics*, Oxford University Press, Oxford.

Barnes, M.B., Garner, J. & Reid, D.: 2004, 'The Pendulum as a Vehicle for Transitioning from Classical to Quantum Physics: History, Quantum Concepts and Educational Challenges', *Science & Education* 13(4–5), 417–436.

Barnett, J.E.: 1998, *Time's Pendulum: From Sundials to Atomic Clocks, the Fascinating History of Timekeeping and How Our Discoveries Changed the World*, Harcourt Brace & Co., New York.

2 Pendulum Motion: A Case Study in How History and Philosophy Can Contribute... 53

Bedini, S.A.: 1991, *The Pulse of Time: Galileo Galilei, the Determination of Longitude, and the Pendulum Clock*, Olschki, Florence.

Bond, T.G.: 2004, 'Piaget and the Pendulum', *Science & Education* 13(4–5), 389–399.

Boulos, P.J.: 2006, 'Newton's Path to Universal Gravitation: The Role of the Pendulum', *Science & Education* 15(6), 577–595.

Burtt, E.A.: 1932, *The Metaphysical Foundations of Modern Physical Science* (second edition), Routledge & Kegan Paul, London.

Chapin, S.L.: 1994, 'Geodesy'. In I. Grattan-Guinness (ed.), *Companion Encyclopedia of the History and Philosophy of the Mathematical Sciences*, Routledge, London, pp. 1089–1100.

Cipolla, C.: 1967, *Clocks and Culture: 1300–1700*, Collins, London.

Clavelin, M.: 1974, *The Natural Philosophy of Galileo. Essay on the Origin and Formation of Classical Mechanics*, MIT Press, Cambridge.

Conlin, M.F.: 1999, 'The Popular and Scientific Reception of the Foucault Pendulum in the United States', *Isis* 90(2), 181–204.

Costabel, P.: 1975, 'Mathematics and Galileo's Inclined Plane Experiments'. In M.L.R. Bonelli & W.R. Shea (eds.), *Reason, Experiment, and Mysticism*, Macmillan, London, pp. 177–187.

De Berg, K.C.: 2006, 'Chemistry and the Pendulum: What Have They to do With Each Other?', *Science & Education* 15(6), 619–641.

Drake, S.: 1978, *Galileo at Work*, University of Chicago Press, Chicago. Reprinted Dover Publications, New York, 1996.

Drake, S.: 1990, *Galileo: Pioneer Scientist*, University of Toronto Press, Toronto.

Dugas, R.: 1988, *A History of Mechanics*, Dover, New York. (orig. 1955).

Fantoli, A.: 1994, *Galileo: For Copernicanism and for the Church*, (G.V. Coyne trans.), Vatican Observatory Publications, Vatican City. (Distributed by University of Notre Dame Press.)

Fauvel, J. & Gray, J. (eds.): 1987, *The History of Mathematics: A Reader*, Macmillan, London.

Fleck, L.: 1935/1979, *Genesis and Development of a Scientific Fact*, T.J. Trenn and R.K. Merton (eds.), University of Chicago Press, Chicago.

Galileo, G.: 1590/1960, *De Motu*. In I.E. Drabkin & S. Drake (eds), *Galileo Galilei On Motion and On Mechanics*, University of Wisconsin Press, Madison, pp. 13–114.

Galileo, G.: 1600/1960, *On Mechanics*. In I.E. Drabkin & S. Drake (eds), *Galileo Galilei On Motion and On Mechanics*, University of Wisconsin Press, Madison, pp. 147–182.

Galileo, G.: 1633/1953, *Dialogue Concerning the Two Chief World Systems*, S. Drake (trans.), University of California Press, Berkeley. (second revised edition, 1967)

Galileo, G.: 1638/1954, *Dialogues Concerning Two New Sciences*, trans. H. Crew & A. de Salvio, Dover Publications, New York (orig. 1914).

Gauld, C.F.: 1998, 'Solutions to the Problem of Impact in the 17th and 18th Centuries and Teaching Newton's Third Law Today', *Science & Education* 7(1), 49–67.

Gauld, C.F.: 2004a, 'The Treatment of Cycloidal Pendulum Motion in Newton's *Principia*', *Science & Education* 13(7–8), 663–673.

Gauld, C.F.: 2004b, 'Pendulums in Physics Education Literature: A Bibliography', *Science & Education* 13(7–8), 811–832.

Gauld, C.F.: 2006, 'Newton's Cradle in Physics Education', *Science & Education* 15(6), 597–617.

Giere, R.N.: 1988, *Explaining Science: A Cognitive Approach*, University of Chicago Press, Chicago.

Giere, R.N.: 1994, 'The Cognitive Structure of Scientific Theories', *Philosophy of Science* 64, 276–296.

Gleick, J.: 1987, *Chaos: Making a New Science*, Penguin, London.

Gould, R. T.: 1923, *The Marine Chronometer, Its History and Development*, J.D. Potter, London. Reprinted by The Holland Press, London, 1978.

Greenberg, J.L.: 1995, *The Problem of the Earth's Shape from Newton to Clairaut: The Rise of Mathematical Science in Eighteenth-Century Paris and the Fall of 'Normal' Science*, Cambridge University Press, Cambridge.

Hall, B.S.: 1978, 'The Scholastic Pendulum', *Annals of Science* 35, 441–462.

Heiskanen, W.A. & Vening Meinesz, F.A.: 1958, *The Earth and its Gravity Field*, McGraw, NY.

Herivel, J: 1965, *The Background to Newton's 'Principia'*, Clarendon Press, Oxford.

Holton, G., Rutherford, F.J. & Watson, F.G.: 1974, *The Project Physics Course: Motion*, Horwitz Group, Sydney.

Holton, G.: 1986, '"A Nation At Risk" Revisited'. In his *The Advancement of Science and Its Burdens*, Cambridge University Press, Cambridge, pp. 253–278.

Holton, G.: 1995, 'How Can Science Courses Use the History of Science?' In his *Einstein, History and Other Passions*, American Institute of Physics, Woodbury, NY, pp. 257–264.

Howse, D.: 1980, *Greenwich Time and the Discovery of Longitude*, Oxford University Press, Oxford.

Humphreys, W.C.: 1967, 'Galileo, Falling Bodies and Inclined Planes: An Attempt at Reconstructing Galileo's Discovery of the Law of Squares', *British Journal for the History of Science* 3(11), 225–244.

Huygens, C.: 1673/1986, *Horologium Oscillatorium. The Pendulum Clock or Geometrical Demonstrations Concerning the Motion of Pendula as Applied to Clocks*, R.J. Blackwell, trans., Iowa State University Press, Ames.

Kesidou, S. & Roseman, J.E.: 2002, 'How well do Middle School Science Programs Measure Up? Findings from Project 2061's Curriculum Review', *Journal of Research in Science Teaching* 39(6), 522–549.

Koertge, N.: 1977, 'Galileo and the Problem of Accidents', *Journal of the History of Ideas* 38, 389–408.

Koyré, A.: 1943/1968, 'Galileo and Plato', *Journal of the History of Ideas* 4, 400–428. Reprinted in his *Metaphysics and Measurement,* 1968, pp. 16–43.

Koyré, A.: 1953/1968, 'An Experiment in Measurement', *Proceedings of the American Philosophical Society* 7, 222–237. Reproduced in his *Metaphysics and Measurement,* 1968, pp. 89–117.

Koyré, A.: 1957, *From the Closed World to the Infinite Universe*, The Johns Hopkins University Press, Baltimore.

Koyré, A.: 1960, 'Galileo's Treatise "De Motu Gravium": The Use and Abuse of Imaginary Experiment', *Revue d'Histoire des Sciences* 13, 197–245. Reprinted in his *Metaphysics and Measurement*, 1968, pp. 44–88.

Kuhn, T.S.: 1970, *The Structure of Scientific Revolutions* (2nd edition), Chicago University Press, Chicago. (First edition, 1962).

Kula, W.: 1986, *Measures and Man*, Princeton University Press, Princeton NJ.

Kwon, Y.-J., Jeong, J.-S. & Park, Y.-B.: 2006, 'Roles of Abductive Reasoning and Prior Belief in Children's Generation of Hypotheses about Pendulum Motion', *Science & Education* 15(6), 643–656.

Lakatos, I.: 1970, 'Falsification and the Methodology of Scientific Research Programmes'. In I. Lakatos & A. Musgrave (eds.) *Criticism and the Growth of Knowledge*, Cambridge University Press, Cambridge, pp. 91–196.

Landes, D.S.: 1983, *Revolution in Time. Clocks and the Making of the Modern World*, Harvard University Press, Cambridge, MA.

Macey, S.L.: 1980, *Clocks and Cosmos: Time in Western Life and Thought*, Archon Books, Hamden, CT.

Machamer, P. & Hepburn, B.: 2004, 'Galileo and the Pendulum: Latching on to Time', *Science & Education* 13(4–5), 333–347.

Machamer, P.: 1998, 'Galileo's Machines, His Mathematics, and His Experiments'. In P. Machamer (ed.) *The Cambridge Companion to Galileo*, Cambridge University Press, pp. 53–79.

MacLachlan, J.: 1976, 'Galileo's Experiments with Pendulums: Real and Imaginary', *Annals of Science* **33**, 173–185.

MacLachlan, J.: 1997, *Galileo Galilei: First Physicist*, Oxford University Press, New York.

Matthews, M.R., Gauld, C.F. & Stinner, A. (eds.): 2005, *The Pendulum: Scientific, Historical, Philosophical and Educational Perspectives*, Springer, Dordrecht.

Matthews, M.R.: 1998, 'Opportunities Lost: The Pendulum in the USA National Science Education Standards', *Journal of Science Education and Technology* 7(3), 203–214.

Matthews, M.R.: 2000, *Time for Science Education: How Teaching the History and Philosophy of Pendulum Motion Can Contribute to Science Literacy*, Plenum Press, New York.

Matthews, M.R.: 2000a, 'Constructivism in Science and Mathematics Education'. In D.C. Phillips (ed.) *National Society for the Study of Education 99th Yearbook*, National Society for the Study of Education, Chicago, pp. 161–192.

Matthews, M.R.: 2001, 'Methodology and Politics in Science: The Case of Huygens' 1673 Proposal of the Seconds Pendulum as an International Standard of Length and Some Educational Suggestions', *Science & Education* 10(1–2), 119–135

Matthews, M.R.: 2004, 'Idealisation in Galileo's Pendulum Discoveries: Historical, Philosophical and Pedagogical Considerations', Science & Education 13(7–8), 689–715.

Matthews, M.R.: 2012, 'Philosophical and Pedagogical Problems with Constructivism in Science Education', *Tréma* 38, 41–56.

McMullin, E.: 1978, 'The Conception of Science in Galileo's Work'. In R.E. Butts & J.C. Pitt (eds.) *New Perspectives on Galileo*, Reidel Publishing Company, Dordrecht, pp. 209–258.

McMullin, E.: 1990, 'Conceptions of Science in the Scientific Revolution'. In D.C. Lindberg & R.S. Westman (eds.) *Reappraisals of the Scientific Revolution*, Cambridge University Press, Cambridge.

Meli, D.B.: 1992, 'Guidobaldo del Monte and the Archimedean Revival', *Nuncius* 7, 3–34.

Meli, D.B.: 2006, *Thinking with Objects*, The Johns Hopkins University Press, Baltimore.

Mittelstrass, J.: 1972, 'The Galilean Revolution: The Historical Fate of a Methodological Insight', *Studies in the History and Philosophy of Science* 2, 297–328.

Mumford, L.: 1934, *Technics and Civilization*, Harcourt Brace Jovanovich, New York.

Naylor, R.H.: 1974, 'Galileo's Simple Pendulum', *Physics* 16, 23–46.

Naylor, R.H.: 1976, 'Galileo: Real Experiment and Didactic Experiment', *Isis* 67(238), 398–419.

Naylor, R.H.: 1980, 'The Role of Experiment in Galileo's Early Work on the Law of Fall', *Annals of Science* 37, 363–378.

Newburgh, R.: 2004, 'The Pendulum: A Paradigm for the Linear Oscillator', *Science & Education* 13(4–5), 297–307.

Newton, I.: 1729/1934, *Mathematical Principles of Mathematical Philosophy*, (translated A. Motte, revised F. Cajori), University of California Press, Berkeley.

Nola, R.: 2004, 'Pendula, Models, Constructivism and Reality', *Science & Education* 13(4–5), 349–377.

Olmsted, J.W.: 1942, 'The Scientific Expedition of Jean Richer to Cayenne (1672–1673)', *Isis* 34, 117–128.

Palmieri, P.: 2009, 'A Phenomenology of Galileo's Experiments with Pendulums', *British Journal for History of Science* 42(4), 479–513.

Palmieri, P.: 2011, *A History of Galileo's Inclined Plane Experiment and Its Philosophical Implications*, The Edwin Mellen Press, Lewiston, NY.

Peters, R.D.: 2004, 'The Pendulum in the 21st Century: Relic or Trendsetter?', *Science & Education* 13(4–5), 279–295.

Popper, K.R.: 1934/1959, *The Logic of Scientific Discovery*, Hutchinson, London.

Renn, J., Damerow, P., Rieger, S. & Camerota, M.: 1998, *Hunting the White Elephant: When and How did Galileo Discover the Law of Fall*, Max Planck Institute for the History of Science, Preprint 97, Berlin.

Renn, J., Damerow, P. & Rieger, S.: 2000, 'Hunting the White Elephant: When and How did Galileo Discover the Law of Fall?', *Science in Context* 13(3–4), 299–422.

Segre, M.: 1991, *In the Wake of Galileo*, Rutgers University Press, New Brunswick, NJ.

Settle, T.B.: 1961, 'An Experiment in the History of Science', *Science* 133, 19–23.

Settle, T.B.: 1967, 'Galileo's Use of Experiment as a Tool of Investigation'. In E. McMullin (ed.) *Galileo: Man of Science*, Basic Books, New York, pp. 315–337.

Sobel, D.: 1995, *Longitude: The True Story of a Lone Genius Who Solved the Greatest Scientific Problem of His Time*, Walker Publishing Company, New York.

Stafford, E.: 2004, 'What the Pendulum can Tell Educators about Children's Scientific Reasoning', *Science & Education* 13(7–8), 757–790.

Sumida, M.: 2004, 'The Reproduction of Scientific Understanding about Pendulum Motion in the Public', *Science & Education* 13(4–5), 473–492.

Tobin, W.: 2003, *The Life and Science of Léon Foucault: The Man Who Proved the Earth Rotates*, Cambridge University Press, Cambridge.

van Rossum, G.: 1996, *History of the Hour: Clocks and Modern Temporal Orders*, Chicago University Press, Chicago.

Weltner, K., Esperidião, A.S.C., Andrade, R.F.S. & Miranda, P.: 2004, 'Introduction to the Treatment of Non-Linear Effects Using a Gravitational Pendulum', *Science & Education* 13(7–8), 613–630.

Westfall, R.S.: 1990, 'Making a World of Precision: Newton and the Construction of a Quantitative Physics'. In F. Durham & R.D Purrington (eds.), *Some Truer Method. Reflections on the Heritage of Newton*, Columbia University Press, New York, pp. 59–87.

Wise, M.N. (ed.): 1995, *The Values of Precision*, Princeton University Press, Princeton.

Wolf, F.A.: 1981, *Taking the Quantum Leap*, Harper & Row, New York.

Yoder, J.G.: 1988, *Unrolling Time: Christiaan Huygens and the Mathematization of Nature*, Cambridge University Press, Cambridge.

Yoder, J.G.: 1991, 'Christian Huygens' Great Treasure', *Tractrix* 3, 1–13.

Zachos, P.: 2004, 'Pendulum Phenomena and the Assessment of Scientific Inquiry Capabilities', *Science & Education* 13(7–8), 743–756.

Michael R. Matthews is an honorary associate professor in the School of Education at the University of New South Wales. He has bachelor's and master's degrees from the University of Sydney in science, philosophy, psychology, history and philosophy of science, and education. He was awarded the Ph.D. degree in philosophy of education from the University of New South Wales. He has taught high school science, and lectured at Sydney Teachers' College and the University of New South Wales. He was the Foundation Professor of Science Education at The University of Auckland (1992–1993). He has published in philosophy of education, history and philosophy of science, and science education journals, handbooks, encyclopaedias and anthologies. His books include *The Marxist Theory of Schooling: A Study of Epistemology and Education* (Humanities Press, 1980), *Pyrmont & Ultimo: A History* (1982), *Science Teaching: The Role of History and Philosophy of Science* (Routledge, 1994), *Challenging New Zealand Science Education* (Dunmore Press, 1995), and *Time for Science Education* (Plenum Publishers, 2000). He has edited *The Scientific Background to Modern Philosophy* (Hackett Publishing Company, 1989), *History, Philosophy and Science Teaching: Selected Readings* (Teachers College Press, 1991), *Constructivism in Science Education: A Philosophical Examination* (Kluwer Academic Publishers, 1998), *Science Education and Culture: The Role of History and Philosophy of Science* (with F. Bevilacqua & E. Giannetto, Kluwer Academic Publishers, 2001), *The Pendulum: Scientific, Historical, Philosophical and Educational Perspectives* (with C.F. Gauld & A. Stinner, Springer, 2005) and *Science, Worldviews and Education* (2009). He is Foundation Editor of the journal *Science & Education*. In 2010 he was awarded the Joseph H. Hazen Education Prize of the History of Science Society (USA) in recognition of his contributions to the teaching of history of science. Outside of the academy, he served two terms as an alderman on Sydney City Council (1980–1986).

Chapter 3
Using History to Teach Mechanics

Colin Gauld

3.1 Introduction

The history of mechanics serves a number of functions in science education.[1] The first is its *cultural* function in which appeal to history is used to teach about the changing role science has played in society in the past and the nature of science as it is portrayed in the activities of scientists of old. There is also cultural value in simply knowing about the past and allowing it to inform our attitude to the present progress of science. The second is its *disciplinary* function in which the history of science is used to teach the concepts of science more effectively (Gauld 1977). The similarity between the concepts in the history of mechanics and the ideas which students appear to adopt now has been frequently commented upon, and many have expected history to provide clues about how better to teach those concepts which are difficult for students to learn.

In this chapter a survey of the history of mechanics from the time of Aristotle is presented followed by some of the contributions which this history can make to the teaching of mechanics at various levels of education.

---

[1] I thank the anonymous reviewers whose insightful comments led to significant improvements in the content and the structure of this chapter.

C. Gauld (✉)
School of Education, University of New South Wales, Sydney 2052, Australia
e-mail: colin@daydesign.com.au

M.R. Matthews (ed.), *International Handbook of Research in History, Philosophy and Science Teaching*, DOI 10.1007/978-94-007-7654-8_3,
© Springer Science+Business Media Dordrecht 2014

3.2 A Brief History of Mechanics from Aristotle to Newton and Beyond

3.2.1 Aristotle

The history of mechanics right up to the time of Isaac Newton partly involved the untangling of three components of motion in the real world: the nature of the body moved, the cause of the motion and the resistance against which the motion of the body took place. The dominant view over the whole of this period was that of Aristotle who lived from 384 to 322 B.C.

Aristotle distinguished between two types of motion[2] – natural and violent (Aristotle no date a, hereafter referred to as *Physics*, 4.8). The downward motion of heavy bodies or the upward motion of light bodies is natural as they move towards their natural place at the centre of the universe (the centre of the earth) or the upper spheres respectively. Any departure from natural motion was called "violent" or "compulsory" motion and was the result of a cause outside of the body (*Physics*, 8.4). This meant that a body could move off in other directions than up or down at a speed other than that which was "natural" if an external mover impelled it.

Rest occurred when a body reached its natural place or when its tendency to do so was impeded by a cause which prevented this natural motion. The first Aristotle called "natural rest" and the second "unnatural rest" (*Physics*, 5.6).

In Aristotle's world all motion took place against resistance of different sizes (*Physics*, 4.8) since motion occurred as bodies moved through corporeal substances of different density (or viscosity) or over surfaces of varying degrees of roughness.

A fundamental axiom in Aristotle's views about motion was that all that moves is moved by something else (*Physics*, 7.1). This meant that even the natural motion of bodies was caused by something else and, as far as it is clear in his writing, he saw the tendency for heavy bodies to move downwards as the essential meaning of the term "heavy" (*Physics*, 8.4). Weight was not a property of the body as such but only an expression of this tendency (Aristotle no date b, hereafter referred to as *Heavens*, 3.2). For a heavy body at rest, one of the causes of its later downward movement was the *removal* of the external impediment preventing its natural motion (*Physics*, 8.4). For violent motion the cause was always outside of the body and it was necessary that there be contact between the mover and the body moved.

Aristotle's notion of universal resistance to motion was closely related to his belief in the non-existence of the void (Gregory 1999). One of Aristotle's objections was that since the void was nothing, it possessed no properties – not even properties with zero value (*Physics*, 4.8); a body entering one "side" of a "void" would instantly appear at the opposite "side" and so the body would be in two places at the same time.

---

[2] For Aristotle "motion" was a term that covered all types of change and what we call "motion" was a change of place – movement from one position to another – called by him "local motion" or "locomotion".

3.2.2 Projectile Motion

The motion of projectiles posed a particular problem for Aristotle. It was easy to identify the cause of its beginning to move since the mover was in contact with the projectile. However, it was not so easy to understand why it continued to move once it had lost contact with the mover. One proposal posited by Aristotle was that the original force moved the projectile which then moved the air which then moved the projectile (*Physics,* 4.8; 8.10; *Heavens,* 3.2). This occurred in successive stages with different portions of air first being pushed and then themselves pushing. Over time the force from the successive portions decreased (*Physics,* 8.10). Another of Aristotle's objections to motion through the void was based on the view that since a medium was necessary for the motion of projectiles and the void contained no such medium, then motion in the void was impossible (*Physics,* 4.8).

Philoponus in the sixth century A.D. saw no sense at all in Aristotle's explanation for the motion of a projectile after contact with the projector was broken.[3] For example, he could see no reason why, when air was pushed forward by the projectile, it then reversed its direction and moved back towards the end of the projectile and then again reversed its direction to push forward on the back of the projectile. He also argued that as the projectile moved forward, air from behind the projectile would immediately fill the space previously occupied by the projectile, so that the air pushed from the front of the projectile would have no space to occupy. Philoponus had a great deal of difficulty believing that its continued motion could be due to this activity of air since one could not initiate the motion of a projectile simply by setting in motion the air behind it.

The solution Philoponus offered was that a motive force was imparted to the projectile by the projector, and he claimed that a projectile would move more quickly in the void than through some medium. Over time the strength of the motive force imparted by the projector decreased and eventually became zero (Franco 2004; Moody 1951b, p. 390).

In dealing with the motion of projectiles, Buridan in the fourteenth century argued against a number of alternative explanations for the continuation of the motion after the projectile had left the projector's hand (Buridan in Clagett 1959, pp. 532–40). He was not happy about Aristotle's notion of the role of the air since a mill wheel continued spinning after the turning force was removed. He reported that when a boat which was being hauled along by ropes was released. It also continued moving although those on the boat felt no wind pushing the sails. Buridan adopted a solution in which a non-decaying impetus was transferred from the projector to the moving body.

Early in his career Galileo (1564–1642) believed that when a projectile left the hand which moved it, it received an impressed force which began to decay (like the heat in a body or the ringing of a bell) as the projectile moved (Galileo 1590/1960, hereafter referred to as *Motion,* pp. 76–80). Thus, a projectile thrown upwards would gradually decrease in speed until the remaining size of the impressed force

---

[3] See Cohen and Drabkin (1958, pp. 221–3) and Wolff (1987).

became equal to the weight of the body, and then it would move downwards with increasing speed as the difference between the weight and the impressed force increased. Eventually, when the size of the impressed force was zero, the body would travel at its natural speed until it reached its final destination (*Motion*, pp. 76–89). Something of this notion occurred when a body was allowed to fall from rest since the force which initially restrained the body transferred an impressed force to the body which again allowed the body to accelerate before reaching its natural speed (*Motion*, pp. 90–2).

By the time Galileo had published in 1638 his last work, *Discourses on the Two New Sciences* (Galileo 1638/1974, hereafter referred to as *Discourses*), he had turned from investigating the causes of motion to describing it (although there is some evidence that he now accepted the existence of a non-decaying impetus). On day 4 of his *Discourses,* he presented his theory of projectile motion developed on the basis of the assumption that such motion consisted of two independent perpendicular motions – a uniform horizontal speed and a uniform vertical acceleration. Galileo carried out experiments to determine the trajectory of a projectile which was projected horizontally from a curved section at the bottom of an inclined plane.

3.2.3 Free Fall

One relationship found in Aristotle's writing is that the speed of a moving body is proportional to the impelling force and inversely proportional to the total resistance to its motion (*Physics,* 4.8; 7.2; *Heavens*, 1.6).[4] When falling, the weight functions as the force responsible for the motion. When he referred to speed during a fall, he usually meant something like our concept of average speed. He apparently believed that as a body fell downwards because of the tendency to move towards its natural place, this tendency increased as it neared the centre of the earth and so its weight and its speed increased (*Physics,* 1.8). In other words he was well aware that falling bodies accelerated.[5]

Philoponus accepted the possible existence of the void, and he justified this notion through use of a different relationship between motive force and consequent speed to that used by Aristotle. He argued that a body would move through the void with its maximum speed proportional to its weight. If the void were then to be filled with a medium, this would resist the motion and so decrease the speed by an amount proportional to the density of the medium. Thus his relationship can be expressed in our terms as $S \propto W - R$.[6]

In the early twelfth century, the position of the Arabic philosopher Avempace was very much that of Philoponus. Avempace believed that bodies were moved by

---

[4] If the void were thought of as space with zero density, Aristotle's relationship would imply that bodies would move through it with infinite speed which Aristotle considered to be impossible.

[5] See Aristotle, *Physics*, 5.6; 6.7; 8.9 and *Heavens*, 1.8; 2.6.

[6] See Cohen and Drabkin (1958, pp. 217–21), Moody (1951b, p 360), and Wolff (1987).

their own nature and that weight was an intrinsic property of a heavy body. Bodies moved through the void with natural speeds that were proportional to their densities and, for him, the only resistance in this case was the distance to be traversed by the body through the void. Motive force was measured by the time taken to traverse a given distance. A medium acted to reduce this natural speed and was therefore an accidental aspect of motion rather than an essential one as Aristotle believed. Avempace had some difficulty accounting for the acceleration exhibited by falling bodies (Moody 1951a, b; Grant 1964, 1965).

In the later twelfth century, Averroes' position was a refinement of Aristotle's. He rejected the notion that bodies moved by their own nature. Motion for him was the overcoming of resistance and motive force was measured by the product SR. He rejected the possibility of motion in the void on the grounds that, because S was proportional to F/R, the speed in the void would be infinite.

The Aristotelian Bradwardine (1290–1349) accepted the arguments that were levelled by Avempace and others against Aristotle's notion that the speed of a body through a resisting medium was proportional to the ratio of the motive force to the resistance. For example, Archimedes had shown that for a body falling through a medium, if the downward motive force was equal to the upthrust (i.e. the resistance to downward motion), the body should neither float nor sink (Moody 1951b, p. 399). On the other hand, Bradwardine was also not happy with the proposal that the speed was determined by the difference between the motive force and the resistance which provided some support for the possibility of a void against Aristotle's strong objections. He argued that when Aristotle referred to proportionality, he did not mean simple proportionality but geometric proportionality.[7]

Buridan considered gravity to be a force which added impetus to the falling body. This impetus was an internal motive force which was directly proportional to the speed. Gravity added impetus so that the speed increased; this, in turn increased the motive force which impelled the falling body so increasing the impetus and thus the body accelerated (Drake 1975c).

One of the problems discussed by medieval thinkers was how to describe the motion of bodies as clearly as possible. While much energy was expended on trying to explain the causes of motion, it became clear, in the fourteenth century, that this question could be placed to one side and that one could concentrate on the kinematical rather than on the dynamical aspect of motion. An obstacle to this was the lack of a precise way of characterising speed. Aristotle (and many of his commentators) only referred to the time taken to travel a certain distance (so that a faster object covered the distance in a shorter time) or the distance travelled in a certain time (so that a faster object covered a greater distance in the time). In our terminology they were using the notion of average speed. In the fourteenth century the notion of instantaneous speed was introduced for a body of which the speed was changing.

---

[7] See Dijksterhuis (1961, pp. 190–1). This relationship means, in our terms, the following: when the speed, S, doubles, the ratio, motive force (F)/resistance (R), is squared or, more generally, when

$$K = S_1/S_2, \quad \frac{F_2}{R_2} = \left(\frac{F_1}{R_1}\right)^k \quad . \text{ In modern terms this means } \quad S \propto \log\left(\frac{F}{R}\right).$$

This was first understood as the distance travelled if the body continued, from that time, for a given duration, with that (now constant) speed.

With this concept of instantaneous speed came the notion of uniformly accelerated motion in which the speed increased in equal increments in equal intervals of time, and, in Merton College, Oxford, during the first half of the fourteenth century, the mean value theorem was discovered. This stated that if a body moved so that its speed changed uniformly from zero to S in a time, T, then the distance travelled would be the same as for a body which travelled for a time, T, at half that speed ($S/2$).

Nicole Oresme introduced a method of presenting information about qualities (such as hotness and whiteness) using two perpendicular axes with information about extension in space or time along the horizontal axis and vertical lines distributed along this axis to represent the intensity of the quality (such as speed) of interest (Durand 1941). Uniform speed was thus represented by a rectangle where the constant height represented a constant speed during the time, T, while uniformly accelerated motion was represented by a right-angled triangle. It became clear that in this representation, the area of the figure so produced (at least in these two cases) was equal to the distance travelled by the body. However, at this stage, there was no indication that such a motion might be related to what occurred in free fall.

In thinking about the variation in the speed of a falling object, Isaac Beeckman (1588–1637), 200 years after Oresme and a younger contemporary of Galileo, took the work of Oresme as his starting point and envisaged that the line which represented time was divided into n segments. He imagined that instead of the speed increasing steadily throughout the time interval, it proceeded in a series of "jerks" (Dijksterhuis 1961; Beeckman in Clagett 1959, pp. 417–8).

Beeckman illustrated his theory with the diagram shown in Fig. 3.1 in which AE represents time and the horizontal lines represent speed.

The distance travelled by the body is proportional to the sum of the areas of the four rectangles in the diagram. If the number of intervals into which AE has been divided is increased, the total area of the small triangles equivalent to a, b, c and d decreases so that in the limit the distance travelled by the body during the time A to E is proportional to the area of the triangle AEF. Thus Beeckman was able to reason in this way without having to introduce a definition of instantaneous speed as the ratio between two vanishing quantities.

Early in his life Galileo was very much influenced by the works of "superhuman Archimedes" (*Motion*, p. 67), and he developed the notion that for falling bodies, it was not the weight that was the determining factor but the density so that the speed in this case was proportional to the difference between the density of the body and the density of the medium. As a result he argued against Aristotle's concept of absolutely heavy bodies (i.e. bodies which always, regardless of the circumstances, fell towards the centre of the universe) and absolutely light bodies (i.e. bodies which always moved away from that centre). Instead, if the density of a body was less than that of the medium (e.g. wood in water), the body would move upwards, while if its density was more than that of the medium (e.g. wood in air), the body would move downwards (*Motion*, pp. 23–6). He claimed that bodies in a medium did not weigh their natural weight but only the difference between that weight and the

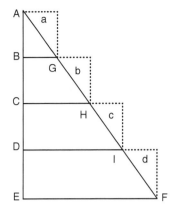

Fig. 3.1 Beeckman's treatment of uniformly accelerated motion. Line *AE* represents the time which has been divided into four equal intervals *AB*, *BC*, *CD* and *DE*. The speed between *A* and *B* is represented by a line of constant length equal to *BG*; between time *B* and *C*, the speed is represented by a line of constant length equal to *CH*; between *C* and *D*, the speed is represented by lines equal in length to *DI*; and between *D* and *E*, the speed is represented by lines equal in length to *EF*

weight of a volume of the medium equal to that of the body. In a vacuum the weight of a body would be its natural weight (*Motion*, p. 46). One of Galileo's conclusions at this stage in his life was that bodies with the same density would move downwards with the same speed in a vacuum, while bodies with different densities would move with speeds proportional to their densities.

At one time Galileo believed that the motion of a falling body was a uniform increase in speed with equal intervals of distance, but in his *Discourses* (pp. 159–60), published towards the end of his life, he argued against this notion and opted for a uniform increase in speed with equal intervals of time. This concept was verified through his experiments with balls rolling down inclined planes (Hahn 2002). By this time Galileo was convinced that all bodies fell in a vacuum with the same acceleration.[8]

3.2.4 Forced Motion

Aristotle developed a relationship between the external force, the weight, the distance traversed over a surface and the time taken to move this distance. For Aristotle (*Physics*, 7.5) but in our modern terms

---

[8] In days 3 and 4 of his *Discourses* (pp. 158–9), Galileo indicated a lack of interest in extrinsic, efficient causes such as forces (Machamer 1978) and sought firstly to describe the motion of a falling body. For Descartes and Newton the search for such causes was much more central to their investigations.

$$\frac{D}{T} \propto \frac{F}{W} \ (\text{as long as } F > R)$$

In this relationship, W is a measure of the resistance to the motion which is directly related to the weight. This relates to the difference in the motions of heavier bodies (such as ships) and lighter bodies (such as boxes), the former requiring many people to pull with a certain speed while the latter requires fewer people to pull with the same speed across sand. Aristotle was also aware of the fact that a certain force was required to commence the motion and that, below this force, the body would remain at rest. Bradwardine's modification, mentioned previously, was also applied in the Middle Ages to the situation of forced or violent motion.

3.2.5 Circular Motion

Aristotle believed that circular motion was the only motion which could be eternal since there is no starting or finishing point with a circle. Motion along a straight line must cease when the end is reached because, for Aristotle, an infinite straight line was an impossibility (*Physics,* 8.8; 8.9). Circular motion was perfect and eternal and was therefore appropriate for the planets and the planetary spheres, the motive force for which was "the unmoved mover" (Koestler 1968, p. 61; Aristotle, *Heavens*, 3.2). A body undergoing circular motion, in one sense at least, did not change its place.

The difference between motion on the earth and in the heavens which Aristotle emphasised was downplayed by Buridan who believed that the planets and stars could also be impelled by his impetus rather than by the "intelligences" which Aristotle presumed.

Galileo argued that if a ball ran down one inclined plane and then up another, it would reach the same height from which it began. If the second inclined plane was lowered, the ball would have to travel further before this height was reached. In the limit, Galileo reasoned, the ball would continue forever along a horizontal plane at the speed it had reached after its initial fall (Galileo 1632/1967, hereafter referred to as *Dialogue*, pp. 145–8). However, for Galileo, this "plane" was not flat but followed a circle around the earth. In this way he reached his concept of circular inertial motion.

René Descartes (1596–1650) realised that in the light of his second natural law, which referred to a body's tendency to move in a straight line, motion in a circle would involve a tendency (*conatus*) for the body to pull away from the centre (Descartes 1966, pp. 217–8).

The conical pendulum played a central role in the thought of both Descartes and Christiaan Huygens (1629–1695) as they considered the nature of circular motion. Since a force is necessary to draw a pendulum aside from the vertical (see Fig. 3.2b), there must be a similar force in the case of a conical pendulum rotating in a circle with the string at an angle to the vertical (see Fig. 3.2c). This Huygens called a centrifugal force. Huygens viewed uniform circular motion (along with uniform

Fig. 3.2 The role of centrifugal force in the conical pendulum

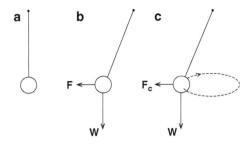

rectilinear motion) as a type of inertial motion. He developed Descartes' view and showed mathematically that the tendency to flee from the centre was given by mv^2/r (Dijksterhuis 1961, pp. 368–70).

3.2.6 Impact

Towards the end of his life, Galileo showed interest in the fact that a moving hammer could drive a nail into wood further than if the hammer simply rested on the nail, and he sought to find the relationship between the effect of the moving hammer and its mass and speed (*Discourses*, pp. 285–8).[9]

Descartes proposed seven laws of impact by which to predict the outcome of a collision between two bodies when the initial conditions were known. Because of some of the premises of Descartes' philosophical system (Descartes 1966, pp. 216–8; Blackwell 1966; Hall 1960–1962), a number of these laws gave faulty outcomes. For example, Descartes believed that no matter how fast it was travelling, a smaller body could not cause a larger stationary body to move.

Huygens was very critical of Descartes' laws of impact and used a methodological device to show that they were inconsistent among themselves. In the case of a smaller body striking a stationary larger body, Huygens imagined an observer on a boat moving alongside the smaller body so that it appeared to be at rest (Dijksterhuis 1961, pp. 373–6). From the observer's point of view, the collision was now transformed into one in which a larger body collided with a stationary smaller body in which Descartes claimed motion was transferred. Huygens worked within a purely kinematic framework and he had no use for the concept of force. He assumed that, in a collision, if the speed of one body was reversed unchanged in magnitude, then the speed of the other body would also be reversed with no change in magnitude (Erlichson 1997b).

Gottfried Leibniz (1646–1716) was another follower of Descartes who was critical of Descartes' laws of impact. He used his principle of continuity (Leibniz 1692/1969, pp. 397–404) to show that Descartes' laws were inconsistent among themselves. This principle says that if two trends are associated then, if the changes in one are continuous, the changes in the other are also continuous. For example, in collisions

---

[9] A younger contemporary of Galileo, Marcus Marci, also developed a theory of impact (Aiton 1970).

the outcome speed should be associated with the incoming speed in this way. Leibniz argued that, for the collision between a smaller body with a stationary larger body, if one gradually increased the weight of the smaller body, a point would be reached when the weights of the two bodies were equal in which case Descartes claimed that the moving body would stop and the stationary body would move off with the same speed as the originally moving body. This would mean that a gradual increase in weight of the first body would mean a sudden change, for the second body, from no speed to a speed equal to that of the first. This behaviour was contradicted by Leibniz's principle of continuity.

Leibniz went on to develop his own laws of impact based on three axioms (Garber 1995):

1. The relative velocity of the two bodies is the same before and after the collision.
2. The sum of vector mv is conserved in the collision.
3. The sum of mv^2 is conserved in the collision.

Unlike Huygens, Leibniz approached impact from a dynamical point of view and held a positive concept of force (*vis*) which he argued was found in two forms – dead and living. Dead force (*vis mortua*) was that which acted when bodies were at rest in equilibrium; living force (*vis viva*) was the force in moving bodies and which was transferred during collisions. Leibniz claimed that *vis viva* was measured by mv^2 rather than by scalar $|mv|$ as Descartes had believed.

3.2.7 Pendulum Motion

Another type of motion of interest to Galileo was that of the pendulum (*Motion*, p. 108). Early on, by watching the swing of a censer in the cathedral at Pisa, he concluded that the time of oscillation was constant regardless of the amplitude and was also independent of the weight of the suspended body (*Dialogue*, p. 450; Matthews 2000, pp. 95–107). The difference the weight made was that the pendulum would swing for a much longer time if the suspended body was heavy.

Galileo's conclusion that the period of a pendulum was independent of its amplitude was probably an inference from his observations of pendulum motion with small amplitude, and his theoretical demonstration that the time for a body to move down a chord to the lowest point on a circle was the same regardless of the starting point.[10]

Huygens showed that a pendulum moving along a cycloidal path would be isochronous regardless of the amplitude of the motion.[11]

---

[10] See Ariotti (1968), Erlichson (1994, 2001), Gauld (1999), MacLachlan (1976), and Naylor (2003).

[11] Further details about the pendulum in history and teaching can be found in Matthews et al. (2005).

3.2.8 Isaac Newton

In Newton's day laws of motion were designed to help explain the collision process between two bodies,[12] and this enables us to understand the meaning of his three laws (which function as axioms in his system). In his *Principia* (1729/1960) *Law 1* tells us the state of a body when no interaction occurs, that is, Newton described the natural state of the body, so that, when a force was exerted, its action could be recognised. *Law 2* tells us that, when a collision takes place, the change in the product, mv, for one body is proportional to the force of impact, while *Law 3* indicates that the forces which the two bodies exert on each other during the collision are equal and opposite. It is interesting to note that there is no mention of time in Newton's statement of his second law. However, in other parts of the *Principia*, Newton makes clear that the change in vector mv is also proportional to the duration of the action of the force.[13] For example, in his discussion of the motion of a body which experiences a resistance proportional to the square of the speed (*Principia*, Book 2, Proposition 9), the accelerating force is proportional to the change in velocity divided by the time interval.[14]

Early in his career Newton (1642–1727) used a body bouncing along a polygonal path inside a circular container beginning with four segments and increasing the number until the body moved along a circular path to show that the centrifugal force was the same size as Huygens had found, namely, mv^2/r.[15] In the *Principia* Newton tackled the problem of motion under the action of a central force. This led him to the conclusion (*Principia*, Book 1, Proposition 4, Corollary 1) that it was not a centrifugal force that acted on a body moving in a circle to pull it from the centre but a centripetal force which continually caused a departure from the straight line path defining the body's inertial motion (see *Principia*, Book 1, Proposition 1; Erlichson 1991).

Projectile motion is treated in the *Principia* not only when the projectile meets no resistance from the medium through which it is travelling but also (in Book 2) when it moves through a resisting medium. To test his assumptions about the effect of resisting mediums, Newton used pendulums and falling bodies moving through water and air.[16]

---

[12] See Arons and Bork (1964), Dijksterhuis (1961, pp. 464–77), and Erlichson (1995).

[13] Time does not appear in Newton's second law because in a collision the duration of the forces on the two bodies involved is the same.

[14] See Gauld (2010, equation (3)), Pourciau (2011), and Westfall (1971, pp. 481–91). In Book 1 of the *Principia*, Newton developed the implications of his three laws for the action of central forces on bodies which experience no resistance (other than that of the "force of inertia"). It is in this book that he derived Kepler's laws of planetary motion from an inverse square of force. In Book 3 Newton applied the insights of Book 1 to observation made on the motion of planets, the moon and comets. In Book 2 his Laws are applied to a variety of other situations including the motion of bodies through resisting mediums, fluid flow and waves. Densmore (1995) is an excellent guide to the *Principia*. Pourciau (2011) presents a different view of the nature of Newton's second law than that presented here.

[15] See Herival (1965), Newton and Henry (2000), Stinner (2001), and Westfall (1971, pp. 353–55).

[16] See *Principia*, Book 2, Sect. 6, Gauld (2009, 2010).

Pendulum motion was dealt with by Newton in Book 1 of the *Principia* (Propositions 50 and 52) in which he showed much more elegantly than Huygens did that a pendulum moving along a cycloidal path took the same time for each oscillation regardless of the amplitude. He also implied that, for small oscillations, the periodic times of a cycloidal and a circular pendulum of the same length were equal (Gauld 2004).

Thus Newton's system successfully dealt with all of those problems which occupied his predecessors and provided one theory by which free fall, forced motion, motion down an incline, impact, projectile motion, circular motion and the motion of the pendulum could be understood.

3.2.9 Beyond Newton

After the publication of the *Principia*, a debate arose between those who followed Newton and those who followed Leibniz as to what was the appropriate measure of the force of a moving body, $|mv|$, mv or mv^2. The Leibnizians referred to experiments in which a falling or colliding ball left an impression in soft clay to support their case that it was mv^2. People, such as Jean d'Alembert, John Desaguliers and Thomas Reid claimed that the dispute was simply a terminological one and depended on what aspect of motion was the focus of interest – time or distance (Boudri 2002, p. 110).

However, for Leibniz himself the issue of importance was his own view of the nature of matter and force (Gale 1973). He, like Descartes, believed that the amount of force in the world was constant but, unlike Descartes, he believed that inelastic collisions showed that the scalar $|mv|$ of Descartes was not conserved. On the other hand he argued that the vector mv of Newton could either be positive or negative rather than always positive as was the case with Leibniz's mv^2. Of course, mv^2 was not conserved in inelastic collisions, but Leibniz claimed that it was conserved in the elastic particles which he believed made up the colliding bodies (Gauld 1998a).

The dispute lingered on until the late 1800s and eventually petered out without any definite contribution resolving the issue.[17] Boudri (2002) claimed that the change which contributed most to its resolution was from a dependence on a metaphysics of substance to that of a metaphysics of relations. The various aspects of the past concept of force – $|mv|$, mv and mv^2 – were seen to be constructs of the quantities involved in the study of motion (such as distance and time) and the notion of force as a cause became less dominant (see also Coelho 2010). In addition, the move from sole consideration of forces of impact in which the "force of motion" was of prime importance to an acceptance of force-at-a-distance as an external impressed force which was central to Newton's system (though not so much for his followers) meant that arguments about the force of motion became of decreasing significance (Papineau 1977). Lagrange's formulation of the laws of mechanics was

---

[17] See Hankins (1965), Iltis (1970, 1971, 1973), Laudan (1968), and Papineau (1977).

in terms of kinetic energy (T) and potential energy (V) which depend only on mass, position and time. The concept of force does not explicitly appear but is implicitly present as $\partial V/\partial x$ (e.g. see Hanc et al. 2003). Ernst Mach attempted to eliminate the notion of force altogether from Newton's system of mechanics.[18]

3.3 History of Mechanics and the Nature of Science

3.3.1 Some Issues in the History of Mechanics

Throughout the history of mechanics, the concepts of distance and time have been basic and generally presumed to be directly apprehended in experience. On the other hand those of force and mass have not and could only be experienced through their effects. Both concepts have undergone significant change in the minds of early scientists and in their role in mechanics. For some, the notion of force as a cause is itself a source of concern as it was seen as an unnecessary metaphysical intrusion into a science which should be free of metaphysics.

A major contribution to the development of mechanics was the use of mathematics to represent motion, and changes to this mathematical representation helped to promote mechanical research. The logical framework of Euclid's geometry was a significant factor in determining the way in which mechanical ideas were to be presented. Both Galileo and Newton used the definition, axiom, proposition, theorem and problem structure in their major works. These issues will be dealt with in more detail in what follows.[19]

3.3.1.1 Force

One of the major difficulties in the development of mechanics was the lack of a clear definition of force. The notion arose from the ideas of cause and effect with force as the cause and motion as the effect. However, it became clear that the concept of force was difficult to usefully define apart from its effects and, for some people, force itself became an effect of motion since motion was used to define it. Some adopted the view that, since force was such a difficult concept to define, they would limit their study to that of motion itself and so, for example, Galileo referred rarely in his *Discourses* to the concept of force. On the other hand both Leibniz and Newton adopted a view that force was a real (although difficult to define) entity which was to be measured by the effects which it produced.

---

[18] See Mach (1893/1960, pp. 303–07; 319–24), but see also Bunge (1966).

[19] Coelho (2012) provides a detailed analysis of conceptual issues in mechanics.

The history of mechanics to the time of Newton (and beyond) has largely been concerned with drawing distinctions between concepts related to the notion of force. Some of these are:

- Force as a cause of velocity or as a cause of acceleration
- External force and internal force
- Force as a cause of acceleration and inertia as a resistance to acceleration
- Inertia as a force of persistence (momentum) or as a force of resistance (mass)
- Inertia as a force within matter or as a property of matter
- Contact force and force-at-a-distance
- The effect of a force over time (momentum) and the effect of a force over distance (kinetic energy)

Early in the development of mechanics, the concept of force covered a number of ideas that we now distinguish. For example, it included concepts of power, work, kinetic energy, momentum and action. One difficulty in the history of mechanics was that of seeing the importance of these distinctions and encouraging the community of scholars to accept the need for establishing and maintaining them. For example, Huygens worked with the expression mv^2 without seeing its importance, while Leibniz took it as the fundamental measure of the force of motion. On the other hand, Newton made no use of mv^2 but related force to the change in mv.

This issue came to a head when the dispute over the true nature of force – Newton's mv or Leibniz's mv^2 – was in full swing. Newton's followers referred to the collisions between bodies in which mv was conserved, while Leibniz's followers referred to experiments in which springs were compressed or soft clay was depressed by the action of moving bodies in which equal deformations occurred for bodies with equal mv^2.

For a long time the free fall of a body was attributed (e.g. by Aristotle) to its tendency to move to its natural place and was not considered to be, like violent motion, the result of an external force. Neither Galileo nor Descartes saw free fall as the paradigmatic example of forced motion but simply as an example of uniformly accelerated motion.

A barrier to the progress of mechanics was the implicit assumption that there were two types of natural motion – rectilinear and circular. Galileo believed that inertial motion when no force acted was in a circle about the centre of the earth. Descartes, while believing that circular motion was unnatural and involved the action of a centrifugal force, still saw it as a state of equilibrium (Westfall 1971, pp. 81–2). Huygens saw a close analogy between uniform rectilinear motion and uniform circular motion (Westfall 1971, pp. 170–1). Newton on the other hand was clear about the proposition that inertial motion was rectilinear while circular motion required a centripetal force. In the eighteenth century d'Alembert treated circular motion as inertial and reintroduced the concept of centrifugal force which was required so that the resultant force on a body undergoing uniform circular motion was zero.[20] The generalised mechanics of Lagrange and modern rotational

---

[20] d'Alembert's approach to the solution of problems of motion is still alive today in some engineering contexts (see Newburgh et al. 2004). See also the discussion of inertial forces by Coelho (2012), and Galili (2012).

3 Using History to Teach Mechanics

mechanics also treats the circular motion of solid bodies about a fixed axis as inertial and replaces the concepts of force, velocity and acceleration with those of torque, angular velocity and angular acceleration.

3.3.1.2 Inertial Mass

Another serious obstacle to progress in the development of mechanics was, initially, lack of an adequate concept of (inertial) mass and the lack of a distinction between mass and weight (Franklin 1976). Weight is an easily experienced characteristic of bodies but was attributed to different things by different people. For Aristotle and most of those influenced by him right up to the seventeenth century, weight was simply an expression of the tendency of bodies away from their natural place to move towards that place. This was associated with volume which was the property of the body most directly related to weight. There was an awareness that, for bodies made of different material, different weights could be associated with the same volume thus giving rise to the notion of density (see Biener and Smeenk 2004). Certainly the difference between the densities of water and air was known to Aristotle (*Physics,* 4.8). Without a notion of mass, the resistance of bodies to change of motion (and especially to change from rest to motion) was attributed to other factors than inertial mass.

Huygens possessed an embryonic concept of the resistance which mass presented to attempt to change the motion of a body while Newton, in the *Principia* at least, was clear about this idea. He defined in his list of definitions what he called the *vis insita* or the *vis inertiae* (*Principia,* Definition 3) by which a body maintained its state of rest or uniform velocity in a straight line and resisted the actions of external forces.[21] Not until the time of Newton did mass take on the character of inertia which we attribute to it today.

3.3.1.3 Mathematics

In the Middle Ages the concept of speed was distinguished from that of distance or time which could be measured. Speed was a quality more like charity or wisdom which could be more or less intense but which could not be measured.

The development of an adequate framework for the study of kinematics was hindered by the lack of a clear definition of instantaneous speed. Up to the fourteenth century speed was defined in terms of the distance travelled in a certain time or the time taken to travel a certain distance. If one of these variables was not the same for the bodies, it was not generally possible to compare speeds. The most this notion could lead to was a qualitative version of our concept of average speed.

---

[21]Gabbey (1980) argues that, in Newton's *Principia,* there are two concepts of *vis inertiae* that associated with the persistence of motion and measured by mv and that associated with the change of velocity and measured by $m\Delta v$.

Another problem was that a dependence on the Euclidean theory of proportions restricted ratios to those between-like quantities, and so the notion of speed as a ratio between distance and time was prohibited. In the fourteenth century interest was turned to uniformly accelerated motion in which the velocity changed from moment to moment. The closest to a definition of instantaneous speed at that time was the constant speed which an accelerating body had at a particular point in time if it were to cease accelerating and continue with that speed. This was the definition which Galileo used in his experiments on the parabolic trajectory of horizontally projected falling bodies. The development of the calculus by Leibniz and Newton was the necessary tool for defining instantaneous speed in a more useful way.

A broader issue was the role of mathematics in understanding the nature of mechanics. Galileo was the first to develop a comprehensive system of mechanics which was thoroughly mathematical in nature and which was presented as a series of logical deductions resting on axioms which were not always self-evident.[22] In both Galileo's *Discourses* and Newton's *Principia*, geometrical representations dominate and the structure of each parallels that of Euclid's mathematics.

3.3.2 Some Philosophical Issues

There are many philosophical issues arising out of the history of mechanics. Some of these relate to the nature of the concepts employed in the subject, while others are related to deeper concerns with the nature of the reality (if any) underlying the phenomena under investigation. There are also questions relating to the way scientific activity is carried out and scientific knowledge validated. Some of these issues are discussed below.

3.3.2.1 Meaning Matters

Even within the history of mechanics, a similar form of words may represent quite different meanings because of the particular philosophical or cultural framework in which the statement is embedded. Descartes (1966, pp. 216–7) and Newton (1729/1960) presented laws which state that, in the absence of external forces, a body will move with constant speed along a straight line. Gabbey (1980) pointed out that although the law as a description of what happens might be the same for these two writers, its meaning for them was significantly different. For Descartes the law provided the explanatory basis for understanding, for example, what

---

[22] Following his definition of naturally accelerated motion, Galileo's postulate in his *Discourses* related to motion down inclined planes: *I assume that the degrees of speed acquired by the same moveable over different inclinations of planes are equal whenever the heights of those planes are equal* (*Discourses*, p. 162). This postulate would not have been self-evident to Galileo's contemporaries and later in the *Discourses* he deduced it as a theorem!

happens in collisions between bodies. He believed that rest was a different state from motion and that a force was necessary to maintain rest as well as to maintain motion. Descartes denied the existence of inertia and claimed that the apparent inertial properties of bodies were simply due to the redistribution of "motion" among colliding bodies. The force exerted by a moving body was *size x speed*. For Newton, rather than being explanatory, the law simply provided a norm, departure from which indicated the existence of a mechanical process which was then explained in terms of his second and third laws. He believed that rest and uniform rectilinear motion were equivalent states. The force exerted on a moving body was equal to *mass x change of velocity*. The meaning of the laws depends on the whole framework within which each worked.

It is also important to note that the *Physics* of Aristotle also contains a similar statement of Newton's first law: "a thing will either be at rest [in a void] or must be moved ad infinitum, unless something more powerful gets in the way" (4.8). He based this conclusion on an argument that, in the void, one point is no different from any other point so there was no reason for the body to stop here rather than there. However, in Aristotle's scheme of things, his conclusion only provided a reason for rejecting the existence of the void because he believed that the statement above was absurd since bodies always stopped.[23]

3.3.2.2 Idealisation in Mechanics

Science is essentially an attempt to understand the natural world, and progress is generally made when what is experienced is explained by what is not experienced. The matter in Aristotle's world was understood in terms of the four basic (but unseen) ideal elements: earth, fire, water and air. In later science mechanical phenomena were understood as consisting of ideal, law-like behaviour along with impediments or intrusions which caused departures from this law-like behaviour (see Matthews 2004). For example, while a vacuum did not actually exist in the experience of the Greek thinkers (although it could be conceptualised), Philoponus believed that if bodies moved through a vacuum their speeds would be proportional to their weights. The mediums through which they travelled imposed a resistance which caused departure from this behaviour. Galileo's experiments were designed to show this law-like behaviour in spite of the existence of impediments which caused the results to be other than ideal.

What is considered to be the ideal behaviour, of course, depends on one's view of the world and can sometimes be shown to be mistaken. Philoponus' view was eventually replaced by that of the mature Galileo who argued that, in a vacuum, all bodies would fall with the same speed. Newton claimed that an isolated body, free from the influence of all other bodies, would travel in a straight line at a constant

---

[23] Galileo's notion of inertial motion was expressed in similar terms (see *Discourses*, p. 197), but for him the path was not a straight line but a circle around the earth (*Dialogue*, pp. 147–8). On the status of Newton's first law among physicists over the last two centuries, see Whitrow (1950).

velocity, while Mach suggested that if the inertial properties of the body were determined by the overall matter in the universe, removing the influence of this matter would also destroy the inertial properties of the remaining body.[24,25]

3.3.2.3 Empiricism Versus Realism in Mechanics

Throughout the history of science, there have been two main trends in the way in which the purpose of scientific activity has been conceived, namely, empiricism, that is, understanding the phenomena perceived by our senses (and nothing more), or realism, that is, understanding the nature of the reality which lies behind what we perceive. For empiricists, many of the constructs of physics (including the idealisation mentioned above) which cannot be directly experienced are simply devices for relating in an economical way aspects of phenomena we can experience. For the realist, such constructs possess a real existence even though they may not be experienced directly.[26] Moody (1951a, p. 190) pointed out that during the Middle Ages, the dispute between Averroes and Avempace was essentially one between empiricism and realism. Avempace considered it more reasonable to consider natural motion as that which took place without impediments while "to define the natural as that which never happens, seems to Averroes absurd" (Moody 1951a, p. 189). Avempace was asserting the reality of things which could never be experienced, while Averroes believed that such things were simply figments of the imagination.

Bunge (1966) criticised Mach's empiricist attempt to eliminate metaphysics from mechanics by eliminating concepts of force and mass and argued that metaphysics was an inevitable component of mechanics.

3.3.2.4 The Role of Observation and Experiment

Aristotle was a keen observer of nature and sought to explain natural phenomena in terms of self-evident truths and conclusions derived from them by a series of widely accepted forms of argument. The empirical base on which his understanding rested was those actual experiences we become aware of through our senses, and this understanding greatly influenced the medieval thinkers who followed Aristotle.

Galileo on the other hand modified what naturally occurred by designing experiments in which the natural impediments (as far as he was aware of them)

---

[24] Hanson (1963) pointed out that it is impossible to consider the motion of an isolated body without a fixed reference frame and, for this, one needs the existence of at least one other body. However, as soon as this second body is introduced, the first is no longer isolated so that it appears that Newton's first law refers to an impossible state of affairs.

[25] The device known as Newton's cradle provides another example of illegitimate idealisation (Gauld 2006; Hutzler et al. 2004).

[26] Matthews (1994, pp. 163–74) has given a number of examples of contemporaries who were on opposite sides of this divide.

were reduced as far as possible or else dealt with in some other way (Koertge 1977; Segré 1980). Thus, through his inclined plane experiments, he was able to conclude that if all the impediments were removed, free fall motion would be uniformly accelerated.[27]

Newton operated in much the same way as Galileo, and from his three axioms and eight definitions was developed a deductive system describing the way things would move if those axioms were true. In Book 1 of his *Principia*, there are no experiments in the Galilean sense but he did provide some empirical evidence to support his third axiom or law. In Book 3 Newton showed that his conclusions deduced in Book 1 explained phenomena in our solar system. In Book 2 Newton carried out experiments into the resistance which various mediums presented to moving bodies and dealt with discrepancies between his results and his expectations in a rather cavalier manner having more confidence in his theory than in his results (Gauld 2009, 2010).

3.3.3 Frontier Science

Science is often taught as a completed self-consistent body of knowledge supported by evidence from demonstrations or experiments, and science is seen as this final, fully justified body of knowledge found by the use of methods appropriate to activities labelled as "scientific". The focus in teaching is on the final product of the process of scientific thought such as the expositions found in Galileo's *Discourses* or Newton's *Principia* or in most modern-day mechanics textbooks. In fact, publications such as these are generally all that are available for the teacher.

However, if one focuses not on this body of knowledge but on the processes which have led to its formation, the activity labelled "scientific" becomes much messier. The history of mechanics demonstrates clearly the unruliness of this process, and the study of laboratory worksheets, notebooks and correspondence such as those of Galileo[28] and Newton[29] shows very clearly something of the processes of thought which led up to that body of knowledge. This "frontier science" is part of the scientific process along with the dead ends and side tracks which accompany it. In Book 2 on his *Principia*, Newton explored new areas in

---

[27] It is interesting to note that in the discussion which took place on day 4 of *Discourses*, Galileo dealt with two-dimensional trajectory motion but presented no experimental data although it was evident from his working papers that he had carried out a series of experiments to show the parabolic nature of these trajectories. While he tried as far as possible to reduce impediments, he was not aware of the effect of rotation on the acceleration of a rolling ball and no doubt noticed the rather significant discrepancies between his results in his unpublished working papers and what he expected to find (see Sect. 3.5.3.2).

[28] Galileo's working papers can be viewed at the website: http://mpiwg-berlin.mpg.de/Galileo_Prototype/index.htm.

[29] See Hall and Tilling (1975–1977), Herival (1965), Scott (1967), Turnbull (1959–1961), and Whiteside (1967–1981).

which to apply his theory and often (especially in his fluid mechanics) moved along on the basis of assumptions with little foundation.[30] Such procedures show how scientific knowledge often emerges from ignorance, error, adherence to particular world views, the absence of appropriate equipment and the lack of adequate analytical tools.

The works of Drake,[31] Hill (1979, 1988), Lindberg (1965), MacLachlan (1976), Naylor,[32] Segré (1980), and Settle (1961) on attempting to make sense of the worksheets of Galileo and of Herival (1965) and Westfall (1971) on interpreting Newton's notebooks and correspondence show something of the difficulties in these processes but also open up the rich world of the human endeavour of scientific discovery. It also reveals more clearly than most textbooks do the tentative nature of scientific knowledge.

3.3.4 Mechanics and Technology

There has been a close link between science (concerned with explanations) and technology (concerned with know-how) throughout history although it has never been a one-way relationship (Price and Cross 1997). While it is true that the development of science has led to new technological devices and processes, it has also been the case that technology has often developed independently of science and has led to new areas of scientific investigation. For example, simple machines like the lever and the inclined plane have been known since antiquity but were central to the investigations of Galileo into motion. The needs of navigation stimulated the search for more and more accurate clocks (Matthews 2000, Chap. 2), and military technology led to increased interest in the motion of projectiles. This shows clearly the way in which science influences and, in turn, is influenced by society.

3.4 History of Mechanics and Student Conceptions

A great deal of research has been carried out into the conceptions which people possess and which relate to mechanics.[33] One of the main findings of this research is that many people possess the concept found in Aristotle that motion requires a mover. As a result a body not experiencing the action of a force will be at rest.

---

[30] See Gauld (2010), Herival (1965), Smith (2001), and Westfall (1971).

[31] See Drake (1973, 1974, 1975a, b, 1978, 1990).

[32] See Naylor (1974a, b, 1976, 1977, 1980, 1983).

[33] See, for example, Brown (1989), Clement (1982), Doménech et al. (1993), Galili and Bar (1992), Gunstone (1984), Halloun and Hestenes (1985), Ioannides and Vosniadou (2002), Lythcott (1985), McCloskey (1983), Montanero et al. (1995), Steinberg et al. (1990), Twigger et al. (1994), Viennot (1979), and Whitaker (1983).

In many cases people believe that, when a projectile is thrown, a force is imparted to it by the thrower which enables it to move after it has left the hand of the thrower. In the case of collisions, this notion transfers to the view that the body which has the greatest value of *mv* exerts the greatest force so that the forces exerted by the two bodies on each other are not equal (as they are in Newton's third law) but depend on the mass and speed of the bodies.

It has been pointed out that this concept of force is very like that held by Buridan and Piaget and Garcia (1989, pp. 30–87) argued that the development of the conception of force in young people follows very much the same path as that in the history of mechanics between the times of Aristotle and Newton.[34] This has suggested to many science educators that reference to the way in which the history of mechanics progressed may assist teachers in encouraging students to make the transition to more developed concepts such as those advocated by Newton.

However, it must be acknowledged that the contexts in which students and early physicists work are vastly different (Gauld 1991). Physicists possess a wider range of skills and interests relevant to their work than modern students. They consciously pursue understanding as they solve problems which arise for them in the process of increasingly articulating and discussing these problems and solutions with one another. On the other hand, until they are asked by an educational researcher, the modern student may probably never have consciously considered the question of the nature of force except in so far as they are taught about it in school.

In spite of the possible differences in meaning between statements of students and statements of early scientists about motion and force (cf. Sect. 3.3.2.1), there is still sufficient similarity to expect that history might supply resources which are able to make more plausible those concepts which are to be taught but which students find difficult to accept.

3.5 Some Historical Resources for Teaching Mechanics

Four potential resources from history to help students learn about the nature of science or to understand those concepts important in mechanics are discussed in what follows. These are (a) explanations and illustrations, (b) thought experiments, (c) experiments, instruments and technological devices and (d) anecdotes, vignettes and stories.

3.5.1 *Explanations and Illustrations*

History provides a source of explanations and illustrations which may assist present-day students to understand those things which they find difficult. Posner and his colleagues (1982) argued that, for a student to accept what the teacher is communicating,

---

[34] See also Eckstein (1997), Nersessian and Resnick (1989), and Ioannides and Vosniadou (2002).

the student must at least find what is being said to be plausible. It must make some sense whether or not the student at first believes it to be the case. The intellectual environment of the past in which problems closer to the everyday experiences of the student were being investigated could be a source of alternative explanations to those in the textbook or provided by the teacher.

One simple example is with common misunderstandings of Newton's third law which is often stated as: "To every action there is an equal and opposite reaction". This gives no indication about the meaning of action and reaction and many students, for example, regard the weight of a book resting on a table and the force of the table on the book to be the action and reaction of Newton's third law. Thus the students link Newton's law with equilibrium. Newton himself stated the law in another, more helpful and less misleading, way: "The mutual actions of two bodies upon each other are always equal, and directed to contrary parts". This directs attention to the role of the two bodies and their actions on each other.

Because of their belief that the force of a moving body is determined by its speed and its mass, students have difficulty in believing that the forces between two colliding bodies are equal as Newton's third law of motion states. This belief was prevalent up to and beyond the time of Newton, and there were many attempts by Newton and those who promoted his system to render this law more plausible with detailed explanations and illustrations. Gauld (1998a, b) has discussed many of these contributions and shown that they often demonstrated that the third law of motion followed from ideas which were presumed to be self-evident (or at least accepted by the audience).

For example, Mach (1893/1960, pp. 247–8) developed an illustration the purpose of which was to make Newton's third law more plausible. His starting point was the self-evident truth that if one of two identical bodies, A and B, exerted a force, F, on the other then, because of symmetry, the second would exert an equal and opposite force, $-F$, on the first. Imagine now that there were three identical bodies, A, B and C and that B and C were combined to make a body, D, with twice the size and mass of A. The force experienced by A would now be $2F$ because B and C in body D would both exert forces equal to F on it. The force on the body D would now be $-2F$ because A would exert a force equal to F on each of B and C. This argument can be extended to explain why Newton's third law applies to bodies of unequal mass but each made up of a number of different identical bodies, but it is not so easy to extend it to bodies which are not identical since the necessary symmetry no longer exists.

3.5.2   Thought Experiments

Another way in which new ideas have been made more plausible in the early history of mechanics has been to use thought experiments. Procedures are carried out in thought which would normally be difficult or impossible to carry out in actual practice, and the audience is led from some generally accepted premise to a

3   Using History to Teach Mechanics

previously unknown or unexpected conclusion.[35] It is probable that such thought experiments could serve a similar role in today's classrooms (Matthews 1994, pp. 99–105). A number of thought experiments relating to mechanics are discussed.[36]

3.5.2.1   Galileo and the Speed of Falling Bodies

One of Galileo's older contemporaries, Giovanni Benedetti (1530–1590), had proposed a thought experiment to show that two bodies of different weight but the same density would fall in a vacuum with the same speed (Dijksterhuis 1961, p. 269). Everyone agreed that two bodies of equal size and density would fall together to the earth. If they were loosely tied together with a string they would therefore exert no force on the string. Thus they could equally be joined so that they form one body, now of twice the weight, which should therefore fall with the same speed as the two tied loosely.

Galileo modified this thought experiment to involve two bodies of different densities and different weights (*Discourses*, pp. 66–7). According to Aristotle the heavier would fall with a greater speed than the lighter. If the two were tied together with a string, the heavier one would pull the lighter one down so that its speed increased, while the lighter one would pull the heavier one up so that its speed decreased. Thus the speed of the two would be somewhere between the speeds of the two separately. However, the combined body, according to Aristotle, should fall with a speed proportional to its weight and so travel faster than the heavier of the two. Thus, Galileo claimed to demonstrate that Aristotle's conclusion was invalid.

3.5.2.2   Stevin and the Inclined Plane

Another of Galileo's older contemporaries, Simon Stevin (1548–1620), presented a thought experiment in which a body resting on an inclined plane was connected by a string passing over a pulley to a body hanging vertically. Of interest to Stevin and his contemporaries was the relationship between the weights of these two bodies when they were at rest (Dijksterhuis 1961, pp. 326–7). Stevin imagined a long chain loop hung motionless over a frictionless inclined plane so that part of it draped the vertical section, part of it draped the inclined section and the rest hung in curve below the plane (see Fig. 3.3). He argued that the part of the chain below the plane hung symmetrically and so exerted equal forces on the other two parts. In fact it

---

[35] See Galili (2009), Gendler (1998), and Helm and Gilbert (1985). Of course the truth of the outcome of a thought experiment depends on the truth of the premise.

[36] Other thought experiments not discussed here include Galileo's use of two inclined planes to show that an unimpeded moving body would continue to move with undiminished speed on a "horizontal plane" around the earth (see *Dialogue*, pp. 145–8) and Archimedes', Galileo's and Mach's thought experiment to establish the principle of the lever (Galileo, *Discourses*, pp. 109–12; Goe 1972; Mach 1893/1960, pp. 13–8).

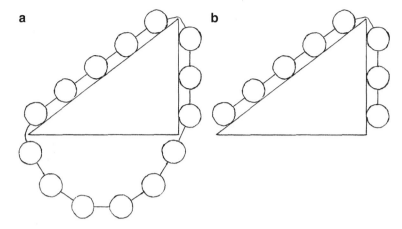

Fig. 3.3 Stevin's thought experiment to compare effective weights of bodies on an inclined plane

could be removed without disturbing the balance of the chain. The other two parts therefore balanced each other. Since the weight per unit length of the chain was constant, the weights of these two parts were therefore proportional to their lengths, that is, to the lengths of the incline and the vertical section, respectively. Thus a vertically hanging weight would balance a larger weight on the incline in the same proportion that the length of the incline is larger than the vertical height.[37]

3.5.3 *Experiments, Instruments and Technological Devices*

In the early history of science, equipment and experiments were much simpler and so provide a source of activities for use in classrooms which can help students deal with some of the problems they experience with the concepts they are learning. Many of the historical experiments have been used both to understand in more detail what the early scientists achieved and also to assist present-day students to understand the concepts which those scientists developed. A few well-known experiments are considered below, but many others, suitable for classroom use, can be found in the historical and educational literature.[38]

---

[37] Another possibility for Stevin's arrangement is that the chain moves from one state to another identical state and so keeps moving forever. Stevin ruled this out because he denied that perpetual motion was possible. Of course, perpetual motion is impossible because energy is dissipated through friction, but in Stevin's arrangement friction is necessarily absent.

[38] These include the investigation of the motion of the simple pendulum (MacLachlan 1976; Matthews 2000, pp. 245–8), the use of Escriche's inclined pendulum to vary the effective value of the acceleration due to gravity (Vaquero and Gallego 2000; Mach 1893/1960, pp. 207–9) and experiments based on Newton's investigation of the resistance to motion of air and water (Gauld 2009, 2010). Blair (2001) and O'Connell (2001) encourage the use of ancient technological devices as aids in teaching about motion in the classroom.

3 Using History to Teach Mechanics

3.5.3.1 The Inclined Plane Experiment

A number of people have carried out replications of Galileo's experiment with inclined planes.[39] Drake and Settle provided detailed accounts of Galileo's possible procedure (Drake 1990, pp. 9–15; Settle 1961) and, while some speculation about the finer details was required, the basic outline was Galileo's own (*Discourses*, pp. 169–70). The purposes of these investigations have differed. For some (e.g. Settle 1961; Drake 1973), it was to determine whether Galileo could actually have carried out the experiment with the equipment which he had and with the accuracy he claimed. For others it was as an exercise for students who were learning about motion with an eye to its history (Straulino 2008; Erlichson 1997a; Sherman 1974).

Drake (1975b, 1990) claimed that in the first instance, Galileo may have used the regular metre of a song to arrange markers across the plane so that, as the ball rolled over each marker, it gave an audible click which coincided with the beats of the song. The distance between markers would then be in the ratio of 1, 3, 5, 7, 9 and so on.

The more normal procedure is to use a large can of water which is allowed to flow through a small bore pipe (*Discourses*, p. 170). The end of the pipe is uncovered at the beginning of the required time interval and covered up at the end. The mass of water is the measure of the size of the time interval (Settle 1961; Drake 1973).[40]

3.5.3.2 The Parabolic Path of Trajectories and the Law of Free Fall

Galileo accepted the notion that the horizontal motion of a trajectory continued at a constant speed, while the vertical motion was accelerated uniformly. These two motions were independent of each other.

His experiments to determine the shape of the trajectory were described and replicated by Drake (1990, pp. 109–15), Naylor[41] and Teichmann (1999). A ball was allowed to run down an inclined plane towards the top of a table. At the end of the plane, a small curve allowed the ball to be projected horizontally so that it then travelled along a curved path to the floor. If the ball always began from the same vertical height, H, above the table, it would leave the top of the table with the same speed, S. The horizontal range, D, from the point of projection was measured by observing the point of impact on the floor. Galileo believed (correctly) that the speed at the bottom of the incline was proportional to the square root of the vertical height from which the ball began. Because this horizontal speed was maintained as the ball moved from the end of the incline to the floor, the range was proportional to S. By using various values of H, Galileo was able to show that D was proportional to $\sqrt{(H)}$.[42]

---

[39] See Drake (1973, 1990, pp. 9–15), MacLachlan (1976), Naylor (1974b), Settle (1961), Sherman (1974), and Straulino (2008).

[40] Use of this method in a classroom is shown on the website http://www.youtube.com/watch?v=ZUgYcbBi46w.

[41] See Naylor (1974b, 1976, 1980, 1983).

[42] The problem Galileo apparently encountered with this experiment was that when H was equal to the height of the table on which the inclined plane was situated, he expected that D would equal $2H$.

3.5.3.3 Newton's Colliding Pendulums

To illustrate his third law, Newton described a series of experiments in which pendulum bobs collided. He used a property of the pendulum well known to Galileo and to Newton's contemporaries, namely, that the speed at the bottom of the swing was proportional to the length of the chord from its starting point to the bottom point (Erlichson 2001; Gauld 1998a, 1999). For a long pendulum with a small amplitude, this means that the speed at the bottom was almost proportional to the distance it was pulled back before the bob was released.

Bifilar pendulums allow the pendulum to swing in only one plane and with two such pendulums the speed of the bobs for two colliding bodies can easily be controlled by the distance they are initially pulled back. Using bob masses and speeds which are in simple ratios, the application of Newton's third law in a dynamic context can easily be explored in the classroom.

3.5.4 Anecdotes, Vignettes and Stories

Many textbooks contain small inclusions in which the history of science is acknowledged (Shrigley and Koballa 1989). These are often there simply to lighten up what is the more serious development of the concepts being taught or to provide some relief from what is often considered to be the boring stuff of science teaching. However, when used in this way and by teachers who are not historians of science, the history often becomes distorted and the only purpose is to provide motivation to the students in their learning of scientific concepts (Gauld 1977; Whitaker 1979a, b). The historical integrity of the accounts is often lost. Wandersee (1990) has explored the more systematic use of small-scale historical vignettes in science lessons and has demonstrated their effectiveness (especially as far as fostering interest is concerned) when used with some topics. Klassen (2009) outlined the nature of stories in teaching science and has described how they should be developed from knowledge of the historical context in which they are set.[43]

Stinner and his colleagues[44] emphasised the importance of the role of large-scale contexts in teaching the story of mechanics. This provides students with the opportunity of becoming familiar with the overall structure of thought of a period and with the way in which investigation of motion was embedded in this context. In this way, for example, the difference between the Aristotelian, Galilean and Newtonian frameworks can be pointed out to students and the nature of progress in science – both in methodology and concepts – be better appreciated by them (Rosenblatt 2011).

Instead, his value for D was only 80% of what he expected. Today we can account for this discrepancy by appealing to the rotational kinetic energy of the ball which reduces D to 5/7 (or 85%) of Galileo's expected value.

[43] Nersessian (1992, p. 71) suggested that one of the reasons for the success of thought experiments is that they are set in an attractive narrative context.

[44] See, for example, Stinner (1989, 1990, 1994, 1995, 1996, 2001) and Stinner and Williams (1993).

3 Using History to Teach Mechanics

Two stories related to the history of mechanics and often found in physics textbooks are those concerning the claim that Galileo defeated his Aristotelian opponents with his experiment from the Leaning Tower of Pisa and the claim that Newton's conception of universal gravitation arose from his observation of a falling apple.

3.5.4.1 Galileo and the Leaning Tower of Pisa

In 1935 Lane Cooper concluded that the story of Galileo confounding the Aristotelians of his day by dropping bodies of different weights from the top of the Leaning Tower of Pisa was untrue. This conclusion was based on the fact that the various versions of the story contained discrepancies. The original account was by Viviani, one of Galileo's students, who was known to be unreliable in some aspects of his account of Galileo's life. Cooper also based it on what was known of Galileo's views of falling bodies at that time and of the views of the Aristotelians. Aristotle believed that the speed of falling bodies was proportional to their weight, while Galileo, when in Pisa, believed that their speed was proportional to their densities. Thus, for Aristotle heavier bodies always fell faster than lighter bodies, while, for Galileo, bodies with the same density fell at the same speed although bodies of different densities fell with different speeds.[45]

Cooper's third argument was that nowhere in Galileo's writing does he refer to this experiment even though it was deemed to be so devastating a rebuttal of Aristotle's position. In addition, Galileo's opponents are silent about the apparent damage Galileo did to their case.

However, Drake (1978, pp. 18–21; 413–6) expressed a different opinion of the veracity of the experiment and believed that it might have occurred, not "in the presence of all the other teachers and philosophers, and the whole assembly of students" as Viviani claimed (Cooper 1935, p. 26) but in the presence of Galileo's own students at Pisa.

This lack of consensus about the story provides some incentive for students to investigate the likelihood of both views based on a consideration of the views of Galileo and his opponents at the time.[46]

3.5.4.2 Newton and the Falling Apple

This well-known story which claims that Newton's law of universal gravitation was discovered when an apple fell on his head as he sat under an apple tree is usually written off as untrue. However, McKie and de Beer (1951–2a, b) traced the ancestry of the story to two sources during the last year of Newton's life and to Newton

---

[45] See Adler and Coulter (1978) and Moody (1951a, b); for an alternative reconstruction, see Franklin (1979).

[46] A re-enactment of this story can be seen at the website http://youtube.com/watch?v=_Kv-U5tjNCY. The discussion by Erlichson (1993) is also helpful in considering the truth of this story.

himself. However, one might ask, if true, what does the story tell us of importance about Newton's thought? Westfall (1993, pp. 51–2) claimed that

> the story vulgarizes universal gravitation by treating it as a bright idea. A bright idea cannot shape a scientific tradition … Universal gravitation did not yield to Newton at his first effort. He hesitated and floundered, baffled for the moment by overwhelming complexities, which were enough in mechanics alone and were multiplied sevenfold by the total context.

Westfall's point about the implications of the usual telling of the story was that, from the evidence provided by his notebooks, the development of Newton's idea of universal gravitation was much more protracted than the story suggests (see also Smith 1997).

3.6 Some Curriculum Examples

3.6.1 *Mach and* The Science of Mechanics

A pioneering work with the intention of teaching mechanics through its history was *The Science of Mechanics* by Ernst Mach (1893/1960). In this book Mach presented his view of how and why mechanics developed in the way it did, and he argued his claim that mechanics should be rightly conceived as the outcome of an empiricist methodology. In spite of its philosophical bias, this book is an accessible and a rich source of information about the historical context of mechanics, some of which has been cited in previous sections of this chapter.

3.6.2 *Taylor and* Physics: The Pioneer Science

In the twentieth century the textbook, *Physics: The Pioneer Science*, by Lloyd Taylor (1941) was a notable effort to bring the study of physics to a wider audience. Its table of contents reads like that of a conventional physics textbook "but the conventional subject matter, along with its mathematical treatment, has been embedded in a historical matrix" (Taylor 1941, p. vi). Chapters 1–18 are devoted to mechanics, and the concepts are developed with an eye to the issues of importance to those involved in its history. Along with the treatment of the concepts of mechanics, Taylor interpolates comments on the nature of science and of the position on these matters taken by the historical figures he appeals to.

3.6.3 *Holton and His Legacy*

From the early 1950s to the present, Gerald Holton promoted the use of history to teach physics in widely used textbooks[47] and especially in the *Project Physics* course (Holton et al. 1970).

---

[47] See Holton (1952), Holton and Brush (2001), and Holton and Roller (1958).

3 Using History to Teach Mechanics

In Holton and Roller's book, *Foundations of Modern Physical Science*, Chaps. 1–18 deal with mechanics (including planetary motion). The table of contents in this book is somewhat different to that of Taylor's in that some of the chapters have a specifically historical (e.g. Chap. 2: "Galileo and the kinematics of free fall") or philosophical (e.g. Chap. 8: "On the nature of scientific theory") focus. Section IV (Chaps. 13–15) is concerned with "structure and method in physical theory".

The *Project Physics* course was based on similar principles to those guiding textbooks of Holton and his colleagues but, in addition, contained a great deal of additional material to assist the teacher. The course was presented in six separate units of which two, *Concepts of Motion* and *The Triumph of Mechanics*, were related to the focus of this chapter. Historical information and practical activities were included, and each unit was accompanied by a reader which contained other writings relevant to the topic of the unit. The course was widely adapted for use in other countries and was recently updated.[48]

3.6.4 Contextual Teaching and Curriculum Structure in Mechanics

In the texts discussed above, what goes on in the classroom is determined largely by the written material provided by the teacher. However, there is considerable evidence that students often end up learning a series of "facts" which make little sense to them beyond the classroom because of the alternative conceptions which they more strongly adhere to and because they are not being taught in a credible way how scientific knowledge has been gained and justified. As a result many argue that a new approach is long overdue.[49] The case made by such educators is that students will learn science better by subjecting their own ideas and those they are being taught to greater scrutiny in order to understand better the origins of scientific knowledge, the way this knowledge has changed and the reasons for these changes. In this enterprise the history of science has a major role to play.

The story of the development of mechanics begins, both for Aristotle and for present-day students, with the idea that all motion requires a mover. Students will therefore be expected to have problems, as Aristotle did, with understanding why, for example, projectiles continue to move after contact with their mover is broken. Without consciously arguing the case, students today often adopt a version of the medieval impetus solution to this problem. Respect for the ideas of students links what they are learning with where the students are at present, while respect for the history of science relates these ideas to the wider context of science itself. In this way, at least in the area of mechanics, students are able to see that ideas somewhat like their own have been held by well-known scientists, and therefore efforts to understand why they changed (or need to be changed) are encouraged.

---

[48] See Cassidy et al. (2002); see also Holton (2003).

[49] For example, Arons (1988); Galili (2012); Monk and Osborne (1997); Rosenblatt (2011); Stinner (1989, 1994, 1995).

3.7 Teaching Mechanics in the Science Education Research Literature

Contributions from the science education research literature are many and varied and include (a) presentations of periods of history or analyses of historical development of particular areas of mechanics without any reference to their use in the classroom, (b) such analyses accompanied by suggestions about their potential use for teaching mechanics, (c) more explicit advice about how a particular event or sequence of events in history could overcome problems in teaching mechanics, (d) descriptions of programmes used for teaching a particular topic in mechanics using history and (e) evaluations of courses which have employed a historical approach to teaching mechanics.

Over a 20-year period, 105 papers dealing with the history of mechanics were published in nine science education journals,[50] and of these 26 made no explicit reference to classroom matters (e.g. see Hecht 2003). Of the 79 papers which do discuss the relevance of the history for the classroom, most of these discussions are brief.[51]

On the other hand, the whole of Stinner's (1994) outline of the history of force from Aristotle to Einstein was presented in a form specifically designed to be used by teachers. Espinoza's (2005) survey of ideas about motion and force both throughout history and in the thought of modern-day students was published with the express purpose of helping students overcome conceptual problems which they apparently shared with early scientists. Possible aids in overcoming widespread difficulties in understanding and accepting Newton's third law were presented by Gauld (1998b) who assembled the approaches used by Newton and those who followed him to render this law more plausible to those attempting to learn it. In order to demonstrate the effect of varying the acceleration due to gravity on the period of the pendulum, Vaquero and Gallego (2000) recommended the use of Escriche's pendulum (invented in 1876) in which the plane of oscillation of the pendulum can be varied to change the size of the force responsible for the oscillation without changing the mass of the bob (see also Mach 1893/1960, pp. 207–8).

In only 16 papers is reference made of actual classroom implementation. There have been complete courses in which the study of motion is taught both to demonstrate the way in which science progresses and to help students understand the concepts involved. Erlichson (1997a, 1999a, b, 2001) described a course for nonscience majors called "Galileo to Newton and Beyond" in which the students repeated historical experiments (including Galileo's inclined plane and pendulum experiments and elastic collision experiments when dealing with Huygens) and discussed the development of concepts of motion in the seventeenth century.

---

[50] Journals included in the analysis were *American Journal of Physics* (17), *International Journal of Science Education* (4), *Journal of Research in Science Teaching* (1), *Physics Education* (16), *Research in Science Education* (3), *Science & Education* (41), *Science Education* (0), *The Physics Teacher* (20) and *The Science Teacher* (2). The period covered was 1992–2011 and the number of articles in each journal is shown in parentheses.

[51] See, for example, Galili and Tzeitlin (2003) and Wörmer (2007).

3 Using History to Teach Mechanics

The students also discussed the implications of some of the propositions in Galileo's *Discourses* and Book I of Newton's *Principia*. Many of Erlichson's papers in physics teaching journals (for example Erlichson 1997b, 1999b, 2001) serve to provide background to such courses as this. Fowler (2003) briefly described a similar course called "Galileo to Einstein". Stinner and his colleagues[52] proposed an approach to teaching mechanics which takes seriously the context within which questions are asked. One of their contexts is history, and awareness of this enables choices to be made about how information from the history of mechanics can be employed in the classroom. Teichmann (1999) described how he used one of Galileo's working manuscripts (*f.*116*v* apparently containing diagrams and the results of theoretical calculations and experimental results for Galileo's trajectory experiments) with physics teachers attending in-service courses at the Deutsches Museum in Munich. This enabled them to investigate the meaning of the symbols on the page and to attempt to understand the possible process of Galileo's thinking.

Within the group of publications in science education journals over the last 20 years, there have been only six in which evaluations of courses designed to use history to teach mechanics are reported (although some of these evaluations are only informal). In some cases[53] these reports have been very cursory, while in other cases[54] the reports are much more comprehensive. Seker and Welsh (2006) investigated the teaching of two units on motion and on force with the specific purpose of addressing three different aims for using physics history, namely, learning of concepts, learning about the nature of science and development of interest. They devised three different instructional procedures each of which was designed to address a different aim along with a traditional procedure in which history played no part. They showed that in all classes students improved in their understanding of the concepts involved, that use of stories did improve some aspects of interest, that certain features of the nature of science improved and that there were differences in the outcomes in these areas for the two units. This last result suggests that using history might produce different outcomes depending on the topic which is being taught.

3.8 The Next Step

A survey of the literature shows that there is a great deal of information about the history of mechanics in historical, philosophical and science educational publications. However, before it substantially influences the teaching of mechanics, three problems must be overcome: teachers have to be convinced that changing their approach to teaching mechanics by introducing more history has advantages for them and their students; the material must be translated into a form which is more

---

[52] See Stinner (1989, 1994, 1995, 2001).

[53] See, for example, Kokkotas et al. (2009), Kubli (1999), and Teichmann (1999).

[54] See Kalman and Aulls (2003) and Seker and Welsh (2006).

easily accessible to teachers; and the effectiveness of using these resources needs to be clearly demonstrated through well-designed evaluations.

3.8.1   Teacher Use of Historical Curriculum Resources

The focus in secondary school physics is generally on the concepts of physics, and the main roles of the history of mechanics are in assisting in the teaching of these concepts and, often to a lesser extent, helping the student to understand something of the nature of science (see Wandersee 1990).[55]

Because of this focus on the teaching of science concepts, the task of encouraging teachers of mechanics to introduce historical material is not an easy one (Monk and Osborne 1997). Typical of the attitude of many physics teachers is that of Paul Hewitt, a widely respected US secondary school physics teacher, who complained:

> that time spent on the 17th-century physics of Galileo is time not spent on the physics of Newton, Kepler, Pascal, Joule, Kelvin, Faraday, Ampere, Coulomb, Ohm, Maxwell, Huygens, Bohr, Planck, Einstein, de Broglie, Heisenberg, Rutherford, Curie, Fermi, Feynman and others. (2003)

Monk and Osborne (1997) argued that the constraint of the focus on teaching concepts should be a major factor in the design of programmes which take into account the history of science and they suggested that neglect of this factor has been the reason why many suggestions about using history in teaching physics have not been implemented. They caution against moving too rapidly but advocate a historical approach to selected units. They propose the presentation of the history as a story with all those features which make storytelling motivational.[56]

3.8.2   Historical Resources for Teaching Mechanics

Much of the information about the history of mechanics is spread rather thinly through the literature, and a great deal of it is in technical, historical or philosophical journals. Because these publications are not primarily concerned with educational implications, the material is often unsuitable for use by teachers in the classrooms. Even where historical information appears in science education journals and is accompanied by advice about classroom use, teachers would still have to work hard to adapt it in appropriate ways.

---

[55] There are, of course, more sophisticated contexts in which the history of mechanics can be used such as in tertiary courses on the history of science, but the greatest exposure to mechanics occurs in the secondary school whether for future specialists of physics or for science for non-specialists.

[56] See also Arons (1988), Rosenblatt (2011, Chap. 6), Stinner (1994), and Stinner and Williams (1993),

3 Using History to Teach Mechanics

Teacher specialties tend to be in their science area and not in history or philosophy, and many of the incidental historical or philosophical contributions of physics textbook are often unsatisfactory. Apart from resources which accompany a fully developed course such as *Project Physics* or those designed by Erlichson (1997a) or Fowler (2003), there is little that is available in a form which teachers could use with only a reasonable effort to incorporate it into their teaching programmes. A collection similar to that produced by Sutton (1938) or Meiners (1970) for physics demonstrations would make a useful beginning.

3.8.3 Evaluation of Programmes Which Use Historical Material

A number of claims have been made about the value of using historical material in the mechanics classroom. Most frequently these refer to the use of history not for its intrinsic value in informing mechanics students about the development of their subject but for the role of history in motivating students, helping them to understand better the concepts of mechanics or developing students' appreciation of the nature of science.

While there are many (mainly informal) suggestions about the value of a particular historical presentation in the teaching of mechanics, there have been very few serious attempts at evaluation to see whether these suggestions possess any substance. Comprehensive evaluation of the *Project Physics* course was carried out (Welch 1973), but since the 1970s the only published report in the area of mechanics is that by Seker and Welsh (2006). Research is needed to establish clearly the characteristics of effective programmes in which history is used to teach mechanics.

References

Adler, C.G & Coulter, B.L. (1978). Galileo and the Tower of Pisa experiment. *American Journal of Physics*, 46(3), 199–201

Aiton, E.J. (1970). Ionnes Marcus Marci. *Annals of Science*, 26, 153–164.

Ariotti, P. (1968). Galileo on the isochrony of the pendulum. *Isis*, 59(4), 414–426.

Aristotle (no date a). *Physics* (translated by R. P. Hardie and R. K. Gaye). Available at http://classics.mit.edu//Aristotle/physics.html. Accessed 27 March 2011.

Aristotle, (no date b). *On the Heavens* (translated by J. L. Stocks). Available at: http://classics.mit.edu//Aristotle/heavens.html. Accessed 15 April 2011.

Arons, A.B. (1988). Historical and philosophical perspectives attainable in introductory physics. *Educational Philosophy and Theory*, 20(2), 13–23.

Arons, A.B. & Bork, A.M. (1964). Newton's laws of motion and the 17th century laws of impact. *American Journal of Physics*, 32, 313–317.

Biener, Z. & Smeenk, C. (2004). Pendulums. pedagogy, and matter: Lessons from the editing of Newton's *Principia. Science & Education*, 13(4–5), 309–320.

Blackwell, R.J. (1966). Descartes' laws of motion. *Isis*, 57(2), 220–234.

Blair, M. (2001). Applying age-old physics. *The Science Teacher;* 68(9); 32–37.

Boudri, J.C. (2002). *What was Mechanical about Mechanics?* (translated by S. McGlinn). Kluwer, Dordrecht.

Brown, D.E. (1989). Students' concept of force: The importance of understanding Newton's third law. *Physics Education*, 24, 353–358.

Bunge, M. (1966). Mach's critique of Newton's mechanics, *American Journal of Physics*, 34, 585–596.

Cassidy, D., Holton, G. & Rutherford, J. (2002). *Understanding Physics*. Springer, New York.

Clagett, M. (1959). *The Science of Mechanics in the Middle Ages*. University of Wisconsin Press, Madison.

Clement, J. (1982). Students' preconceptions in introductory mechanics. *American Journal of Physics*, 50(1), 66–71.

Coelho, L. (2010). On the concept of force: How the understanding of history can improve physics teaching. *Science & Education*, 19(1), 91–113.

Coelho, R.L. (2012). Conceptual problems in the foundations of mechanics. *Science & Education*, 21, 1337–1356.

Cohen, M.R. & Drabkin, I.E. (1958) *A Source Book in Greek Science*. Harvard University Press, Cambridge, MA.

Cooper, L. (1935). *Aristotle, Galileo, and the Tower of Pisa*. Cornell University Press, Ithaca.

Densmore, D. (1995). *Newton's* Principia: *The Central Argument*. Green Lion Press, Santa Fe.

Descartes, R. (1966). *Philosophical Writings*. Translated and edited by E. Anscombe & P. Geach, Nelson, Melbourne.

Dijksterhuis, E.J. (1961). *The Mechanization of the World Picture*. Oxford University Press, Oxford.

Doménech, A., Casasús, E. Doménech, M.T. & Buñol, I.B. (1993). The classical concept of mass: Theoretical difficulties and students' definitions. *International Journal of Science Education*, 15(2), 163–173.

Drake, S. (1973). Galileo's discovery of the law of free fall. *Scientific American*, 228:85–92

Drake, S. (1974). Mathematics and discovery in Galileo's physics. *Historia Mathematica*, 1(2), 129–150.

Drake, S. (1975a). Galileo's discovery of the parabolic trajectory. *Scientific American*, 232(3), 102–110

Drake, S. (1975b). The role of music in Galileo's experiments. *Scientific American*, 232(6), 98–104.

Drake, S. (1975c). Impetus theory reappraised. *Journal of the History of Ideas*, 36(1), 27–46.

Drake, S. (1978). *Galileo at Work: His Scientific Biography*. University of Chicago Press, Chicago.

Drake, S. (1990). *Galileo: Pioneer Scientist*. University of Toronto Press, Toronto.

Durand, D.B. (1941). Nicole Oresme and the mediaeval origins of modern science. *Speculum*, 16(2) 167–185.

Eckstein, S.G. (1997). Parallelism in the development of children's ideas and the historical development of projectile motion theories. *International Journal of Science Education*, 19(9), 1057–1073.

Erlichson, H. (1991). Motive force and centripetal force in Newton's mechanics. *American Journal of Physics*, 59(9), 842–849.

Erlichson, H. (1993). Galileo and high tower experiments. *Centaurus*, 36, 33–45.

Erlichson, H. (1994). Galileo's pendulums and planes. *Annals of Science*, 51(3), 263–272.

Erlichson, H. (1995). Newton's strange collisions. *Physics Teacher*, 33(3), 169–171.

Erlichson, H. (1997a). Galileo to Newton – A liberal-arts physics course. *The Physics Teacher*, 35, 532–535.

Erlichson, H. (1997b). The young Huygens solves the problem of elastic collisions. *American Journal of Physics*, 65(2), 149–154.

Erlichson, H. (1999a). Science for generalists. *American Journal of Physics*, 67(2), 103.

Erlichson, H. (1999b). Galileo's pendulum. *The Physics Teacher*, 378: 478–479.

Erlichson, H. (2001). A proposition well known to geometers. *The Physics Teacher*, 39, 152–153.

Espinoza, F. (2005). An analysis of the historical development of ideas about motion and its implications for teaching. *Physics Education*, 40, 139–146.

Fowler, M. (2003). Galileo and Einstein: Using history to teach basic physics to nonscientists. *Science & Education*, 12(2), 229–231.

3 Using History to Teach Mechanics

Franco, A.B. (2004). Avempace, projectile motion, and impetus theory. *Journal of the History of Ideas*. 64, 521–546.

Franklin, A. (1976). Principle of inertia in the Middle Ages. *American Journal of Physics*, 44(6), 529–545.

Franklin, A. (1979). Galileo and the leaning tower: An Aristotelian interpretation. *Physics Education*, 14(1), 60–63.

Gabbey, A. (1980). Force and inertia in the seventeenth century: Descartes and Newton. In S. Gaukroger (ed.) *Descartes: Philosophy, Mathematics and Physics*, Harvester: Brighton, 230–320.

Gale, G. (1973). Leibniz's dynamical metaphysics and the origins of the vis viva controversy. *Systematics*, 11, 184–207.

Galilei, Galileo (1590/1960). On Motion (translated by I.E. Drabkin). In S. Drake & I.E. Drabkin (eds), *Galileo Galilei On Motion and On Mechanics*, pp. 13–131, University of Wisconsin Press, Madison.

Galilei, Galileo (1632/1967). [*Dialogue*] *Dialogue Concerning the Two Chief World Systems*. (Translated by S. Drake) University of California Press, Berkeley.

Galilei, Galileo (1638/1974). [*Discourses*] *Discourses on the Two New Sciences*. (Translated by S. Drake) University of Wisconsin Press, Madison.

Galili, I. (2009). Thought experiments: Determining their meaning. *Science & Education*, 18(1), 1–23.

Galili (2012). Promotion of cultural content knowledge through the use of the history and philosophy of science. *Science & Education*, 21, 1233–1316.

Galili, I. & Bar, V. (1992). Motion implies force: Where to expect vestiges of the misconception? *International Journal of Science Education*, 14(1), 63–81.

Galili, I. & Tzeitlin, M. (2003). Newton's first law: Text, translations, interpretations and physics education. *Science & Education*, 12(1), 45–73.

Garber, D. (1995). Leibniz: Physics and philosophy. In N. Jolley (ed.) *The Cambridge Companion to Leibniz*, Cambridge University Press, Cambridge, pp. 271–352.

Gauld, C.F. (1977). The role of history in the teaching of science, *Australian Science Teachers Journal*, 23(3), 47–52.

Gauld, C.F. (1991). History of science, individual development and science teaching, *Research in Science Education*, 21, 133–140.

Gauld, C.F. (1998a). Solutions to the problem of impact in the 17th and 18th centuries and teaching Newton's third law today, *Science & Education*, 7(1), 49–67.

Gauld, C.F. (1998b). Making more plausible what is hard to believe. Historical justifications and illustrations of Newton's third law. *Science & Education*, 7(2), 159–172.

Gauld, C.F. (1999). Using colliding pendulums to teach Newton's third law. *The Physics Teacher*, 37(2), 116–119.

Gauld, C.F. (2004). The treatment of cycloidal pendulum motion in Newton's *Principia*. *Science & Education*, 13(7–8), 663–673.

Gauld, C. F. (2006). Newton's cradle in physics education, *Science & Education*, 15(6), 597–617.

Gauld, C.F. (2009). Newton's use of the pendulum to investigate fluid resistance: A case study and some implications for teaching about the nature of science. *Science & Education*, 18(3–4), 383–400.

Gauld, C.F. (2010). Newton's investigation of the resistance to moving bodies in continuous fluids and the nature of 'frontier science'. *Science & Education*, 19(10), 939–961.

Gendler, T.S. (1998). Galileo and the indispensability of scientific thought experiments. *British Journal for the Philosophy of Science*, 49, 397–424.

Goe, G. (1972). Archimedes' theory of the lever and Mach's critique. *Studies in History and Philosophy of Science*, 2, 329–345.

Grant, E. (1964). Motion in the void and the principle of inertia in the Middle Ages. *Isis*, 55(3), 265–292.

Grant, E. (1965). Aristotle, Philoponus, Avempace, and Galileo's Pisan dynamics. *Centaurus*, 11(2), 79–95.

Gregory, A. (1999). Ancient science and the vacuum. *Physics Education*, 34(4), 209–213.

Gunstone, R. (1984). Circular motion: Some pre-instructional alternative frameworks. *Research in Science Education*, 14, 125–135.

Hahn, A.J. (2002). The pendulum swings again: A mathematical reassessment of Galileo's experiments with inclined planes. *Archive for History of the Exact Sciences*. 56, 339–361.

Hall, A.R. (1960–1962). Cartesian dynamics, *Archive for the History of the Exact Sciences*, 1, 172–178.

Hall, A.R. & Tilling, L. (1975–1977) *The Correspondence of Isaac Newton*. Volumes 5–7. Cambridge University Press, Cambridge.

Halloun, I.A. & Hestenes, D. (1985). Common sense concepts about motion. *American Journal of Physics*, 53(11), 1056–1065.

Hanc, J., Slavomir, T. & Hancova, M. (2003). Simple derivation of Newtonian mechanics from the principle of least action. *American Journal of Physics*, 71(4), 386–391.

Hankins, T.L. (1965). Eighteenth-century attempts to resolve the *vis viva* controversy. *Isis*, 56(3), 281–297.

Hanson, N. (1963). The Law of Inertia: A philosopher's touchstone. *Philosophy of Science*, 30(2) 107–121.

Hecht, E. (2003). An historico-critical account of potential energy: Is *PE* really real? *The Physics Teacher*, 41, 486–493.

Helm, P. & Gilbert, J.K. (1985). Thought experiments in physics education – Part 1. *Physics Education*, 20(3), 124–131.

Herival, J. (1965). *The Background to Newton's Principia: A Study of Newton's Dynamical Researches in the Years 1664–84*. Oxford University Press, Oxford.

Hewitt, P.G. (2003). Overtime on Galilean physics. *The Physics Teacher*, 41, 444.

Hill, D.K. (1979). A note on a Galilean worksheet. *Isis*, 70(2) 269–271.

Hill, D.K. (1988). Dissecting trajectories: Galileo's early experiments on projectile motion and the law of fall. *Isis*, 79(4) 646–668.

Holton, G. (1952). *Introduction to Concepts and Theories in Physical Science*. Addison-Wesley, Cambridge, MA.

Holton, G. (2003). The Project Physics Course, Then and now. *Science & Education*, 12(8), 779–786.

Holton, G. & Brush, S.G (2001). *Physics: The Human Adventure*. Rutgers University Press, New Brunswick, NJ.

Holton, G. & Roller, D. (1958). *Foundations of Modern Physical Science*. Addison-Wesley, Reading, MA.

Holton, G., Rutherford, F.J. & Watson, F.G. (1970). *Project Physics*. Holt, Rinehart & Winston, New York.

Hutzler, S., Delaney, G., Weaire, D. & MacLeod, F. (2004). Rocking Newton's cradle, *American Journal of Physics*, 72(12), 1508–1516.

Iltis, C. (1970). D'Alembert and the vis viva controversy. *Studies in History and Philosophy of Science*, 1(2), 135–144.

Iltis, C. (1971). Leibniz and the *vis viva* controversy. *Isis*, 62, 21–35.

Iltis, C. (1973). The decline of Cartesianism in mechanics: The Leibnizian-Cartesian debates. *Isis*, 64, 356–373.

Ioannides, C. & Vosniadou, S. (2002). The changing meanings of force. *Cognitive Science Quarterly*, 2, 5–61.

Kalman, C.S. & Aulls, M.W. (2003). Can an analysis of the contrast between pre-Galilean and Newtonian theoretical frameworks help students develop a scientific mindset? *Science & Education*, 12(8), 761–772.

Klassen, S. (2009). The construction and analysis of a science story: A proposed methodology. *Science & Education*, 18(3–4), 401–423,

Koertge, N. (1977). Galileo and the problem of accidents. *Journal of the History of Ideas*, 38, 389–408.

Koestler, A. (1968). *The Sleepwalkers*. Penguin, Harmondsworth.

Kokkotas, P., Piliouras, P., Malamitsa, K. & Stamoulis, E. (2009). Teaching physics to in-service primary school teachers in the context of history of science: The case of falling bodies. *Science & Education*, 18(5), 609–629.

Kubli, F. (1999). Historical aspects in physics teaching: Using Galileo's work in a new Swiss project. *Science & Education*, 8(2), 137–150.

3 Using History to Teach Mechanics 93

Laudan, L.L. (1968). The *vis viva* controversy, a post-mortem. *Isis*, 59, 131–143.

Leibniz, G.W. (1692/1969). Critical thoughts on the general part of the principles of Descartes. In L. Loemker (ed.) *Gottfried Wilhelm Leibniz: Philosophical Papers and Letters*. Reidel, Dordrecht, pp. 397–404.

Lindberg, D.C. (1965). Galileo's experiments on falling bodies. *Isis*, 56(3) 352–354

Lythcott, J. (1985). "Aristotelian" was the answer, but what was the question? *American Journal of Physics*, 53(5), 428–432.

Mach, E. (1893/1960). *The Science of Mechanics*. Open Court, New York.

Machamer, P. (1978). Galileo and the causes. In R.E. Butts & J.C. Pitts (eds) *New Perspectives on Galileo*. Reidel, Dordrecht, pp. 161–180.

MacLachlan, J. (1976). Galileo's experiments with pendulums: Real and imaginary. *Annals of Science*, 33, 173–185.

Matthews, M.R. (1994). *Science Teaching: The Role of History and Philosophy of Science*. Routledge, New York.

Matthews, M.R. (2000). *Time for Science Education*. Kluwer, New York.

Matthews, M.R. (2004). Idealisation and Galileo's pendulum discoveries: Historical, philosophical and pedagogical considerations. *Science & Education*, 13(7–8), 689–715.

Matthews, M.R., Gauld C.F. & Stinner, A. (eds) (2005). *The Pendulum: Scientific, Historical, Philosophical & Educational Perspectives*. Springer, Dordrecht.

McCloskey, M. (1983). Intuitive physics. *Scientific American*, 248(4), 114–122.

McKie, D. & de Beer, G.R. (1951–2a). Newton's apple. *Notes and Records of the Royal Society*, 9, 46–54.

McKie, D. & de Beer, G.R. (1951–2b). Newton's apple: An addendum, *Notes and Records of the Royal Society*, 9, 333–335.

Meiners, H.F. (ed.) (1970). *Physics Demonstration Experiments*. Ronald Press, New York.

Monk, M. & Osborne, J. (1997). Placing the history and philosophy of science on the curriculum: A model for the development of pedagogy. *Science Education*, 81, 405–424.

Montanero, M., Perez, A.L. & Suero, M.I. (1995). A survey of students' understanding of colliding bodies. *Physics Education*, 30, 277–283

Moody, E.A. (1951a). Galileo and Avempace: The dynamics of the Leaning Tower experiment 1. *Journal for the History of Ideas*, 12, 163–193.

Moody, E.A. (1951b). Galileo and Avempace: The dynamics of the Leaning Tower experiment 2. *Journal for the History of Ideas*, 12, 375–422.

Naylor, R.H. (1974a). Galileo and the problem of free fall. *British Journal for the History of Science*, 7, 105–134.

Naylor, R.H. (1974b). Galileo: Real experiment and didactic demonstration. *Isis*, 67, 398–419.

Naylor, R.H. (1976). Galileo: The search for the parabolic trajectory. *Annals of Science*, 33, 153–172.

Naylor, R.H. (1977). Galileo's theory of motion: Processes of conceptual change in the period 1604–1610. *Annals of Science*, 34, 365–392.

Naylor, R.H. (1980). Galileo's theory of projectile motion. *Isis*, 71, 550–570.

Naylor, R.H. (1983). Galileo's early experiments on projectile trajectories. *Annals of Science*, 40, 391–396.

Naylor, R.H. (2003). Galileo, Copernicanism and the origins of the new science of motion. *British Journal for the History of Science*, 36(2), 151–181.

Nersessian, N. (1992). The procedural turn: or, why do thought experiments work? In R.N. Giere (ed.) *Cognitive Models of Science*. University of Minnesota Press, Minneapolis, pp. 45–76.

Nersessian, N.J. & Resnick, L.B. (1989). Comparing historical and intuitive explanations of motion: Does "naïve physics" have a structure? *Proceedings of the Cognitive Science Society*, 11, 412–420.

Newburgh, R., Peidle, J. & Rueckner, W. (2004). When equal masses don't balance. *Physics Education*, 39(3), 289–292.

Newton, I. (1729/1960). *Mathematical Principles of Natural Philosophy*. (translated from the third edition by Andrew Motte, revised by Florian Cajori), University of California Press, Berkeley, CA.

Newton, I. & Henry, R.C. (2000). Circular motion. *American Journal of Physics*, 68(7), 637–639.

O'Connell, J. (2001). Dynamics of a medieval missile launcher: The trebuchet. *The Physics Teacher*, 39, 471–473.

Papineau, D. (1977). The *vis viva* controversy: Do meanings matter? *Studies in the History and Philosophy of Science*, 8(2), 111–142.

Piaget, J. & Garcia, R. (1989). *Psychogenesis and the History of Science* (trans. H. Feider). Columbia University Press, New York.

Posner, G.J., Strike, K.A., Hewson, P.W. & Gertzog, W.A. (1982). Accommodation of a scientific conception: Towards a theory of conceptual change. *Science Education*, 66(2), 211–227.

Pourciau, B. (2011). Is Newton's second law really Newton's. *American Journal of Physics*, 79(10), 1015–1022.

Price, R. & Cross, R. (1997). Conceptions of science and technology clarified: Improving the teaching of science. *International Journal of Science Education*, 17(3), 285–293.

Rosenblatt, L. (2011). *Rethinking the Way We Teach Science*. Routledge, New York.

Scott, J.F. (1967). *The Correspondence of Isaac Newton*. Volume 4. Cambridge University Press, Cambridge.

Segré, M. (1980). The role of experiment in Galileo's physics. *Archive for History of Exact Sciences*, 23, 227–252.

Seker, H. & Welsh, L.C. (2006). The use of history of mechanics in teaching motion and force units. *Science & Education*, 15(1), 55–89.

Settle, T.B. (1961). An experiment in the history of science. *Science*, 133, 19–23.

Sherman, P.D. (1974). Galileo and the inclined plane controversy. *The Physics Teacher*, 12, 343–348.

Shrigley, R.L. & Koballa, T.R. (1989). Anecdotes: What research suggests about their use in the science classroom. *School Science & Mathematics*, 89(4), 293–298.

Smith, D.S. (1997). Newton's apple. *Physics Education*, 32(2), 129–131.

Smith, G.E. (2001). The Newtonian style in Book II of the *Principia*. In I.B. Cohen & J. Buchwald (eds), *Newton's Natural Philosophy*. MIT Press, Cambridge, MA, pp. 240–313.

Steinberg, M.S., Brown, D.E. & Clement, J. (1990). Genius is not immune to persistent misconceptions: Conceptual difficulties impeding Isaac Newton and contemporary physics students. *International Journal of Science Education*, 12(3), 265–273.

Stinner, A. (1989). The teaching of physics and the contexts of inquiry: From Aristotle to Einstein. *Science Education*, 73(5), 591–605.

Stinner, A. (1990). Philosophy, thought experiments and large context problems in secondary school physics courses. *International Journal of Science Education*, 12(3), 244–257.

Stinner, A. (1994). The story of force: From Aristotle to Einstein. *Physics Education*, 29(2), 77–86.

Stinner, A. (1995). Contextual settings, science stories, and large scale context problems: Toward a more humanistic science education. *Science Education*, 79(5), 555–581.

Stinner, A. (1996). Providing a contextual base and a theoretical structure to guide the teaching of science from early to senior years. *Science & Education*, 5(3), 247–266.

Stinner, A. (2001). Linking 'The Book of Nature' and 'The Book of Science': Using circular motion as an exemplar beyond the textbook. *Science & Education*, 10(4), 323–344.

Stinner, A. & Williams, H. (1993). Conceptual change, history, and science stories. *Interchange*, 24(1–2), 87–103.

Straulino, S. (2008). Reconstruction of Galileo Galilei's experiment: The inclined plane. *Physics Education*, 43(3), 316–321.

Sutton, R.M. (1938). *Demonstration Experiments in Physics*. McGraw-Hill, New York. Now available on the internet from The Physical Instructional Resource Association at http://physicslearning.colorado.edu/PiraHome/Sutton/Sutton.htm.

Taylor, L.W. (1941). *Physics: the Pioneer Science*. Houghton Mifflin, Boston.

Teichmann, J. (1999). Studying Galileo at secondary school: A reconstruction of his 'jumping hill' experiment and the process of discovery. *Science & Education*, 8(2), 121–136.

Turnbull, H.W. (1959-61). *The Correspondence of Isaac Newton*. Volumes 1–3. Cambridge University Press, Cambridge.

Twigger, D. et al. (1994). The conception of force and motion of students between 10 and 15 years: An interview study designed to guide instruction. *International Journal of Science Education*, 16(2), 215–229.

Vaquero, J.M. & Gallego, M.C. (2000). An old apparatus for physics teaching: Escriche's pendulum. *The Physics Teacher*, 38, 424–425.

Viennot, L. (1979). Spontaneous learning in elementary dynamics. *European Journal of Science Education*, 1(2), 205–221.

Wandersee, J.H. (1990). On the value and use of the history of science in teaching today's science: Constructing historical vignettes. In D.E. Herget (ed.) *More History and Philosophy of Science in Science Teaching*. Florida State University, Talahassee, 278–283.

Welch, W.W. (1973). Review of the research and evaluation program of Harvard Project Physics. *Journal of Research in Science Teaching*, 10(4), 365–378.

Westfall, R.S. (1971). *Force in Newton's Physics: The Science of Dynamics in the Seventeenth Century*. McDonald, London.

Westfall, R.S. (1993). *The Life of Isaac Newton*. Cambridge University Press, Cambridge.

Whitaker, M.A.B. (1979a). History and quasi-history in physics education – Part 1. *Physics Education*, 14(2), 108–112.

Whitaker, M.A.B. (1979b). History and quasi-history in physics education – Part 2. *Physics Education*, 14(3), 239–242.

Whitaker, R.J. (1983). Aristotle is not dead: Student understanding of trajectory motion. *American Journal of Physics*, 51(4), 352–357.

Whiteside, D.T. (1967–1981). *The Mathematical Papers of Isaac Newton*, Cambridge University Press, Cambridge.

Whitrow, G.J. (1950). On the foundations of dynamics. *British Journal for the Philosophy of Science*, 1(2), 92–107.

Wolff, M. (1987). Philoponus and the rise of preclassical dynamics. In R. Sorabji (ed.) *Philoponus and the Rejection of Aristotelianism*. Cornell University Press, Ithaca, NY. (pp. 84–120).

Wörmer, C.H. (2007). Galileo's method proves useful in today's classroom. *Physics Education*, 47(5), 437–438.

Colin Gauld is a visiting member in the School of Education at the University of New South Wales, Sydney. He has a Ph.D. in physics from the University of Sydney and, after completing a Diploma in Education, taught physics and mathematics in secondary schools before taking up a position in teacher education at UNSW. His areas of specialisation in research include physics education, the role of history and philosophy of science in science teaching, and the relationship between religion and science. He has recently published papers in *Science & Education* on the historical context of Newton's third law and on possible implications of Newton's experiments on fluid resistance in Book 2 of the *Principia* for teaching physics. He was a participant in the International Pendulum Project and co-editor of *The Pendulum: Scientific, Historical, Philosophical and Educational Perspectives*.

Chapter 4
Teaching Optics: A Historico-Philosophical Perspective

Igal Galili

4.1 Introduction

This chapter reviews the attempts to include the history and philosophy of science (HPS) in the teaching of light and vision and the lessons learned from these attempts. This kind of curricular innovation requires special effort and draws on extensive research in learning theory and cognitive psychology and culturology, all applied to a science curriculum on light.

Light is traditionally seen in science as one of the two entities that comprise physical reality: light and matter. The dichotomy stems from the difference between photons and other elementary particles, which possess mass. Our scientific knowledge of light is organized in the form of the *Theory of Light* – Optics.[1] The history of science provides an astonishing story of transformations of this knowledge through different periods and levels of complexity, before the appearance of the modern theory of light and matter. This history can be represented as a discourse of theories in which a certain theory dominated during each period (Fig. 4.1).

Within the liberal tradition of science education, the major question is how to represent this knowledge in order to give the students an inclusive and essentially representative big picture of human knowledge about light. As part of this effort, it is important to identify significant periods of knowledge transformation regarding light and in this way create a structure to be addressed in the course of learning.

---

[1] *Theory* is used here in the inclusive sense of a collection of knowledge elements about reality in a particular domain. A fundamental theory in physics includes principles, laws, concepts, models, experiments, problems, practical applications, apparatus, and other elements, all conforming to the same set of basic principles. The broad structure of this knowledge will be specified below.

I. Galili (✉)
The Amos de-Shalit Science Teaching Center, The Hebrew University of Jerusalem, Jerusalem 91904, Israel
e-mail: igal.galili@mail.huji.ac.il

M.R. Matthews (ed.), *International Handbook of Research in History, Philosophy and Science Teaching*, DOI 10.1007/978-94-007-7654-8_4,
© Springer Science+Business Media Dordrecht 2014

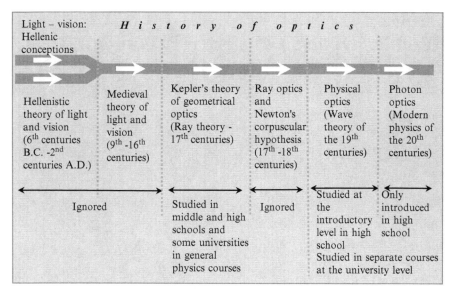

Fig. 4.1 The historical development of optical theories and their appearance in high school and university curricula

As seen in Fig. 4.1, some theories are not usually presented to students in regular physics courses. Moreover, the three parts of optics which are presented usually appear in isolated educational settings: (i) Geometrical Optics (the theory of rays), (ii) Physical Optics (the theory of waves), and (iii) Modern Optics (the theory of photons). The latter is often incorporated into modern physics courses. Optics contributes to physics education all the way to the university level, even if it is chopped into pieces along the way.

The history of science presents a big picture that unifies the various theories of light. This perspective includes the pre-theory conceptions of light and vision in the Hellenic period as well as the theory of rays developed during the Hellenistic, medieval European, and Muslim periods prior to the scientific revolution of the seventeenth century. Usually, teaching of optics does not include any developments prior to the ray theory of the 17th century and even then, Newton's color rays, Huygens' pressure waves, the particle-wave debate which led to the domination of Newton's conception of particles in the 18th century are scarcely mentioned.

I will now show how elements of HPS related to optics have been incorporated into teaching materials. First comes students' knowledge of light, followed by a discussion of the available resources of the history of optics. Next come the views regarding teaching the knowledge of optics and the nature of this scientific knowledge. The recently developed framework of the *cultural knowledge* of optics is then reviewed, framing the contribution of HPS to the teaching of science.

4.2 Studies of Students' Knowledge of Optics

Educational constructivism (e.g., Duit et al. 2005) sees students' knowledge of light and vision, before and after instruction, as being of great importance. Numerous students' conceptions in optical knowledge have been investigated and reported in an organized manner (e.g., Galili and Hazan 2000a). The abundance of misconceptions results from the counterintuitiveness of light theory and the process of seeing objects which requires a nonobvious explanation: a diverging light flux emanated from each point of the observed object converges inside the eye to a correspondent point on the retina. The image created in this way is then interpreted by the mind. This conception is usually replaced by "commonsense" ideas, while the process described in theoretical terms is not actually what one can "see." Scientific knowledge in the area of optics as developed by scholars can be seen as a puzzle resolved over hundreds of years.

Here are some aspects of the complexity. The "passive" nature of vision – its intermissive character – is not obvious. The observer *receives* light, but the impact of single photons on the retina is never perceived. Only the impact of many photons can start a faint visual perception, making vision continuous. Visual perception is analyzed unconsciously and "informs" us that light is static and fills space, that it can be observed "from aside" as an object rather than an event or process.

The speed of light is enormous and never perceived as finite; it seems that light expands instantly the moment we press the switch. The wavelength of light is much shorter than that of any water wave, and thus the wave nature of light was revealed only through delicate experiments showing tiny deviations of light rays from straight paths. The obscure nature of light led to speculations, some of which were extremely inventive (Aristotle), which scientists used to describe and explain the phenomena of light prior to the presently adopted accounts. It is thus natural to realize that people spontaneously produce alternatives to the scientific account.

Researches who have revealed and documented students' conceptions and views on light and vision usually share the epistemologies of educational constructivism (e.g., Driver and Bell 1986) and cognitive psychology (Ausubel 1968). Numerous studies have shown that students' conceptions of light and vision show a certain consistency and similarity across educational and cultural backgrounds.[2] This universality indicates the "objective" origin of naïve knowledge, which is stronger than differences of psychological, social, ethnic, educational, and curricular factors.

---

[2] See, for example, Andersson and Karrqvist (1983), Beaty (1987), Bendall et al. (1993), Bouwens (1987), Boyes and Stanisstreet (1991), Colin and Viennot (2001), Colin (2001), Feher and Rice (1988, 1992), Fetherstonhaugh et al. (1987), Fetherstonhaugh and Treagust (1992), Fleer (1996), Galili (1996), Galili et al. (1993), Goldberg and McDermott (1986, 1987), Guesne (1985), Jung (1981, 1982, 1987), La Rosa et al. (1984), Langley et al. (1997), Olivieri et al. (1988), Osborne et al. (1993), Perales et al. (1989), Ramadas and Driver (1989), Reiner et al. (1995), Reiner (1992), Rice and Feher (1987), Ronen and Eylon (1993), Saxena (1991), Schnepps and Sadler (1989), Segel and Cosgrove (1993), Selley (1996a, b), Singh and Butler (1990), Stead and Osborne (1980), and Watts (1985).

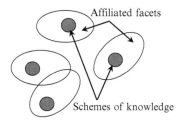

Fig. 4.2 Schematic representation of the Scheme-Facets-of-Knowledge structure of students' knowledge. The small circles designate schemes and the oval areas, clusters of affiliated facets. Since the same facet may match more than one scheme, the clusters may overlap

The alternative conceptions of light phenomena were found at all levels, from kindergarten (e.g., Guesne 1985; Shapiro 1994) to Ph.D. graduates (e.g., Schnepps and Sadler 1989), teachers (e.g., Atwood and Atwood 1996) and even in textbooks (e.g., Beaty 1987).

The abundance of persistent misconceptions suggests consideration of the structure of students' knowledge and the changes this knowledge undergoes during the course of learning. Some researchers have suggested considering students' knowledge as alternative *theories* (McCloskey 1983) parallel to science, even if inferior in inclusiveness and coherency. diSessa (1993) introduced the inclusive mental constructs of *phenomenological primitives* (p-prims) and considered students' knowledge to be fragmented and governed by p-prims, "naïve sense of mechanism" (ibid.). For example, the p-prim "force as a continuous mover" suggests that each motion is affiliated with an agent (force) and is reminiscent of the account of "violent" motion in Aristotelian theory. Another cognitive construct, *facets-of-knowledge* (Minstrell 1992), describes students' reasoning and strategies in concrete physical situations. Facets represent stable conceptual, operational, and representative ideas and beliefs. For example, students may consider a mirror image to be an entity that leaves an object, travels to a mirror, and remains there, to be observed later by a viewer (Bendall et al. 1993).

Based on these perspectives, students' knowledge in optics has been organized in a two-level structure of Scheme-Facets-of-Knowledge (e.g., Galili and Lavrik 1998). A *scheme of knowledge* represents general concepts, certain mechanisms, and cause-effect relationships between physical factors, such as "Light is comprised of light rays" or "An image is transferred as a whole entity" (Galili and Hazan 2000a, b). *Schemes* manifest themselves in context-specific *facets*: concrete realizations of the correspondent schemes. For example, the Image Holistic Scheme is related to a cluster of facets each of which applies the same idea to the various contexts of vision, mirrors, lenses, pinholes, and prisms. In all these cases, an image moves and stays as a whole. This two-level theoretical approach reduces the multitude of conceptions to less numerous schemes with affiliated clusters of facets (Fig. 4.2).

4.3 Resources in the History of Optics

Together with mechanics and astronomy, optics is one of the oldest areas of scientific exploration. Its history goes back to the dawn of science and illustrates how human knowledge of nature evolved. Mechanics, astronomy, and optics address the reality experienced directly through sense perception. Other domains of physics address more hidden reality and draw on the established concepts of mechanics and optics.

For the purposes of education, text resources in the history of optics may be classified in the following way:

1. Texts on *the history of science*, usually addressing great variety of specific cases from the history of optics.[3]
2. In-depth monographs, studies and detailed investigations of the history of optics.[4]
3. Textbooks in optics sometimes provide a historical presentation of disciplinary knowledge.[5] However, unlike the previous groups, these texts often ignore alternative and currently abandoned concepts from the past of science.
4. Books providing a narrative history of optics, written for the general public (e.g., Park 1997). These authors face a special challenge: to remain conceptually valid without the formal mathematical account and present the ideas, laws, and concepts of optics qualitatively. As a result, these books can be used in a very limited way in schools. However, they remain a highly stimulating source to satisfy their readers' curiosity, interest, historical awareness, imagination, and intellectual desire – all of crucial importance in overcoming the barrier of formalism on the way to genuine comprehension.
5. Original treatises remain indispensible, although novices may have troubles to understand them unaided. Pioneers in optics often explained their claims and related them to previous knowledge, displaying the continuity of the disciplinary discourse. Despite their archaic notions, style, and worldview, the originals preserve their validity.[6] Some originals are available in sourcebooks such as Cohen and Drabkin (eds.) (1966) and Magie (ed.) (1969).

---

[3] See, for example, Crombie (1959, 1990), Dijksterhuis (1986), Forbes and Dijksterhuis (1963), Gliozzi (1965), Lindberg (1992), Mason (1962), Pedersen and Phil (1974), Sambursky (1959), Steneck (1976), Whittaker (1960), and Wolf (1968).

[4] See, for example, Boyer (1987), Dijksterhuis (2004), Emmott (1961), Endry (1980), Gaukroger (1995), Hakfoort (1995), Herzberger (1966), Kipnis (1991), Lauginie (2012), Lindberg (1976, 1978, 1985, 2002), Middleton (1961, 1963), Rashed (2002), Ronchi (1970, 1991), Russell (2002), Sabra (1981, 1989, 2003), Sambursky (1958), Shapiro (1973, 1993), Smith (1996, 1999), and Westfall (1962, 1989).

[5] See, for example, Arons (1965), Galili and Hazan (2004, 2009), Kipnis (1992), Mach (1913/1926), Taylor (1941), and Hecht (1998).

[6] Important original texts in optics include Aristotle (1952), Bragg (1959), Descartes (1637/1965), 1998, Fresnel (1866), Goethe (1810), Huygens (1690/1912), Kepler (1610/2000), Newton (1671/1974, 1704/1952), Ross (2008), and Young (1804, 1807).

4.4 Perspectives on the Involvement of HPS in the Teaching of Optics

The major subject of this review is the teaching of optics through the use of HPS-based materials. The idea of the use of history in developing an understanding of science has been analyzed by different scholars. For instance, Collingwood categorically argued in his *The Idea of Nature*:

> I conclude that natural science as a form of thought exists and always has existed in a context of history, and depends on historical thought for its existence. From this I venture to infer that no one can understand natural science unless he understands history and that no one can answer the question what nature is unless he knows what history is. (Collingwood 1949, p. 177)

Matthews (1989, 1994, 2000) has refined this claim in the framework of modern perspectives on science teaching and has listed and discussed the advantages of using HPS to achieve that goal. Seroglou and Koumaras (2001) have provided a review of the research on this subject available at the time. The following presents the approaches to the use of HPS in teaching optics:

1. The first proponents of using the history of science in science education based their argument on the tradition of liberal education, that is, the value of broad scientific literacy, and provided historical reviews at the beginning of their monographs (e.g., Lagrange 1788; Mach 1913). Pedrotti and Pedrotti (1998) and Hecht (1998) did the same in modern optical textbooks. There is, however, a norm which distinguishes the history found in such reviews from that found in historical studies. The former only address the elements of "correct" knowledge (Type-A). For instance, there are the specular reflection of light and its refraction, investigated by the heroes of Hellenistic optics, Heron and Archimedes (Cohen and Drabkin 1966). Heron argued for mirror reflection using the principle of "minimal path" (ibid, p. 263); Archimedes treated the same phenomenon using the idea of light path reversibility (Russo 2004, p. 63; Kipnis 1992, p. 27). The restriction of the discussion to Type-A knowledge, however, makes Heron's consideration irrelevant. Heron argued that the infinite speed of rays did not allow any deviation from the minimum path. Similarly, when this treatment quotes Fermat's corrections of "minimal" to "extreme" (maximal or minimal) and replaced "distance" to "time" as more "correct" and in accordance with scientific knowledge, Fermat's motivation, the *intention of nature* seeking the "simplest way" (Ross 2008, p. vi), is ignored.[7]

   Type-A reconstructions of the history of science as a method of teaching optics have led to development of special curriculum units (e.g., Mihas 2008; Mihas and Andrealis 2005; Andreou and Raftopoulos 2011). Mihas (2008) and Mihas and Andrealis (2005) have reconstructed the experiments from the Hellenistic (Ptolemy) and medieval Islamic (Al-Haytham) periods. The authors supported their teaching by using computer simulations.

---

[7] The restriction of discussions to Type-A knowledge may be connected to the positivist philosophy seemingly prevailing in science classes (e.g., Benson 1989). This approach, however, does not adequately present controversies in scientific discourse or the educational complexity in facing specific misconceptions.

Kipnis (1992) developed a special course of optics for science teachers employing selected historical experiments where he applied the *historical-investigative* method. In his course, Kipnis suggested reproducing historical experiments using apparatus similar to the historical ones. Students discussed the results of the experiments and were guided to the theoretical implications leading to the conceptions and laws of optics. His discursive pedagogy, which is reminiscent of Galilean discourses, is close to the modern idea of "guided discovery" as a method of knowledge construction. This approach has been described as enhancing students' and teachers' interest in science, developing their initiative and inventiveness, and providing them with insights into the process of doing science (Kipnis 1996, 1998).

However, the history of optics contains more than the elements described above; it possesses other elements, Type-B knowledge, knowledge which emerged and was later refuted, being replaced by more advanced accounts. This knowledge is often seen as irrelevant and undesirable in science classes (Galili and Hazan 2001), as "incorrect" ideas may be seen as confusing the students, who, being immature, are unable to resolve discrepancies in the subject matter. In this view, novices require definite, correct, and unequivocal information, so, even when educators do state:

> If I were endowed with dictatorial powers, I would require everyone receiving a degree in a scientific subject to know its history and to have read the classical papers relating to it. Historical knowledge is important because it stimulates creative thinking. (Herzberger 1966, p. 1383)

they may only address knowledge of a certain type. Ironically, however, the restriction to Type-A knowledge frequently found in popular books about science practically excludes a need for history in education. Indeed, the laws of reflection and refraction of light do not require reference to Heron, Archimedes, Fermat, or Descartes to grasp their meaning and application. Indeed, numerous textbooks in optics do not mention these heroes of optical history.

2. Another idea regarding the use of HPS in educational materials has appeared in connection to research in science education. Its roots stem from the idea of *recapitulation* (ontogeny recapitulates phylogeny) as applied to education. Piaget was among the early proponents of this view (Jardine 2006), which has been discussed in psychology in the past (Kofka 1925). Leaving aside the presently rejected extreme – "*every* individual passes through *all* the stages of collective development" – certain fundamental ideas in the history of science (the *phylogeny* of the scientific knowledge) are similar to certain ideas and conceptions which students demonstrate during the course of learning (the *ontogeny* of the individual knowledge). Ideas such as "motion implies force," "motion implies impetus," "light fills space," and "the image moves from object to observer" are repeatedly shared by the history of science and by numerous students across nations, countries, and ages. Similar clear parallels may be found between students' understanding of vision and optical imagery (Galili and Hazan 2000b; Dedes 2005; De Hosson and Kaminski 2007). Therefore, even if this is not a case of recapitulation in the literal sense, there is a definite similarity between historical and individual progression from simpler erroneous to more complex ideas regarding light-vision conceptions. Despite the obvious differences from the past

resulting from metaphysical, sociological, and technological factors, the cognitive sameness of conceptual restructuring in science and education is suggestive (Nersessian 1989). Modern pedagogy may benefit from the history of science by using it to anticipate students' ideas and misconceptions during the course of learning (Wandersee 1986) and by learning from failures and errors in science (Kipnis 2010).[8] This interpretation legitimizes addressing Type-B knowledge in teaching using the history of optics.

The constructivist educational dictum of addressing students' conceptions in order to allow meaningful learning enhanced the remedial influence of addressing, analyzing, and discussing Type-B knowledge from the past. Consider, for instance, the conception developed by pre-Socratic scholars that an exact replica of an object comprised of atoms (*eidolon*) was continuously shed by the object in all directions (e.g., Russell 2002); this replica moved toward the observers and entered their eyes. Addressing and criticizing this kind of "holistic scheme" could assist those students who develop such ideas, causing a *cognitive resonance* leading to conceptual change and scientific understanding. Monk and Osborn (1997) and Duit et al. (2005) have recommended this pedagogy, and Galili and Hazan (2000b, 2004) have employed it in teaching optics.

3. Another view of the use of HPS materials in the teaching of optics makes use of *conceptual variation* in successful pedagogy. This approach was developed by Marton et al. (2004), who argued for a *space of learning* created by the variation of the target subject in teaching mathematics. It revived the scholastic method of analysis: to know a certain concept means to appreciate the different and possible alternatives. This method clearly calls for using Type-B historical knowledge to create the variations of the concept learned. For example, the history informs about the conceptions of optical image, such as holistic transfer (the Atomists), active vision (the Pythagoreans, Euclid, Ptolemy, Al-Kindi), point-to-point projection by light rays (Al-Haytham). All these conceptions are spontaneously produced by students (Galili and Hazan 2000b). All these conceptions can establish a space of learning from which students can be encouraged to discern the correct depiction of image creation as made by Kepler.

4. Finally, the fourth perspective, suggested by Tseitlin and Galili (2005), considers scientific knowledge to be a *culture of rules* (Lotman 2001). This perspective must be distinguished from the culture of science as social functioning (Latour 1987), from "science and culture" that placed scientific contents (physics, biology, astronomy) in the historical development of mathematics and natural philosophy (e.g., Fehl 1965), or from addressing ethnical aspects of knowledge in education (e.g., Aikenhead 1997; Aikenhead and Jegede 1999), and from considering the relationship of science and society (Bevilacqua et al. 2001) or art (e.g., Galili and Zinn 2007). Unlike all of these perspectives, Tseitlin and Galili consider the knowledge of physics *itself* as a culture.

---

[8] The proponents of this approach quote "Those who forget the past are doomed to repeat it," attributed to George Santayana, and "Those who fail to *learn* from history are doomed to repeat it," Winston Churchill.

4 Teaching Optics: A Historico-Philosophical Perspective

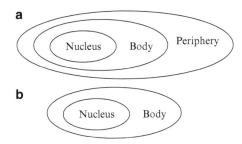

Fig. 4.3 (**a**) Schematic representation of the discipline-culture structure of a scientific theory. The elements of knowledge are located in three different areas. (**b**) Schematic representation of the discipline structure of a scientific theory

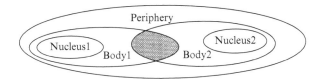

Fig. 4.4 Two fundamental theories in physics structured in DC form. Each nucleus is placed in the periphery of the other theory, thus representing the conceptual *incommensurability* of the nuclei, while the overlay of the bodies (*the shaded area*) represents their possible practical *commensurability*

Within this approach, fundamental physical theories possess a triadic structure termed as *discipline-culture* (DC): nucleus-body-periphery (Fig. 4.3a). This is instead of the regular disciplinary structure: nucleus-body (Fig. 4.3b).[9] The nucleus includes the fundamentals, the paradigmatic model, principles, and concepts, while the body is made up of various applications of the nucleus: solved problems, working models, explained phenomena and experiments, and developed apparatus. The periphery is the area incorporating the elements which contradict and challenge the nucleus, such as problems/phenomena which cannot be resolved/explained by a particular nucleus. Within this perspective, Type-A knowledge elements contribute to the nucleus and body, while the periphery accumulates Type-B knowledge elements contradicting the nucleus.

With respect to optical knowledge, there are four basic theories which emerged, developed, and were dominant in different historical periods: rays, classical particles, waves, and quantum particles (photons). The bodies of the theories define their areas of validity and partly overlap when different nuclei successfully explain the same phenomenon or experiment. Thus, if Fig. 4.4 represents the theories of waves (Fresnel) and classical particles (Newton), then the phenomena of reflection and refraction belong to the shaded area, strict geometrical shadow solely to the body of

---

[9] Lakatos (1978) considered a similar structure when he described *scientific research programs*. However, the contents of all areas become different when one represents the knowledge of a fundamental theory as a culture.

particles, and diffraction solely to the body of waves. Photoelectric phenomenon would belong to the common periphery in this case. Teaching optics in a cultural way seeks to transform the naïve knowledge structured in scheme-facets (Fig. 4.2) into knowledge structured in DC form (Fig. 4.3a). This knowledge is defined as cultural content knowledge (CCK) and can be applied to teaching through the use of historical excurses (Galili 2012).

This suggested tripartite structure of knowledge cannot be linked in a simple way to the views of a certain scientist. Just as it is often impossible to identify an individual scientist with a single philosophical position (Galileo is a good example of such "inconsistency": empiricist in some cases and rationalist in others). The conceptual knowledge of a scientist may not allow full identification with one of these four theories of light. Descartes presents an illustrative example, as his view of light as a "successive propagation in space of a tendency to motion requiring no transport of matter" (Descartes 1637) is suggestive of a pressure wave in a medium of fine matter particles excluding the possibility of void in space. At the same time, he also actively used light rays to account for other phenomena (rainbow, vision, etc.) in a precise way. Thus, the identification of fundamental theories is valuable despite the specific, sometimes contradictory, positions of different individuals, as a way of determining the basis in which any system of views can be resolved as a vector to its components.

4.4.1 Example: Teaching the Concept of Optical Image Using a Historical Approach

Teaching the concept of optical image using history may take the form of an examination of several accounts of image created by means of light (Galili 2012). These accounts include the conceptions of "active vision" by the Pythagoreans, the Atomists' *eidola*, Plato's hybrid model, and Aristotle's transmission of tension through medium, all from the pre-theory period. Within the framework of the first optical theory – the theory of rays – established by Euclid, Al-Haytham in the eleventh century correctly explained optical image in a camera obscura and incorrectly explained the visual image created in a human eye. Kepler, in the seventeenth century, used flux of light rays and provided the explanation of vision which is currently taught in schools (Fig. 4.5).

The mentioned conceptions of visual image constitute a diachronic dialogue of scientific ideas. By teaching them to students, literacy in the history of science is improved (teaching approach 1 above), and the teacher is made alert to alternative conceptions spontaneously produced by students: holistic image (intromission), "active" vision (extramission), image projection point-by-point by single light rays, etc. (approach 2). A discussion of these conceptions of image formation may cause cognitive resonance, helping students to overcome their misconceptions (approach 2). The cluster of conceptions regarding optical image creates a specific space of learning in which the students may discern through comparison and contrast the scientific account of optical image (approach 3). Finally, contrasting and comparing selected historical accounts (Fig. 4.5) provides meaning for the light image conception to be learned – its cultural knowledge (approach 4). The inclusion of several

4 Teaching Optics: A Historico-Philosophical Perspective

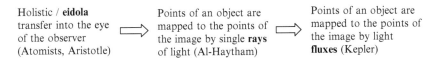

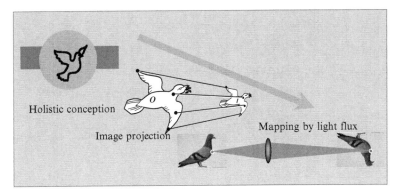

Fig. 4.5 Historical sequence of conceptions within the intromission understanding of optical images

alternatives of image understanding is meant to encourage students to appreciate the progress of theoretical fundamentals of any scientific issue and its continuous upgrading. This method of teaching surpasses instruction for "puzzle solvers" within the given paradigm (in the nucleus of a certain theory) (Kuhn 1977, p. 192) and suggests an awareness and appreciation of scientific progress.

4.5 Learning from Optics About the Nature of Science

Coping with the factors impeding understanding of optics creates an opportunity to learn about the nature of science and scientific knowledge. The latter includes *syntactic* knowledge, the ways in which the validity of knowledge is established, and knowledge *organization,* the structure of the subject matter (Schwab 1964, 1978). Both types of knowledge determine the *nature of science* (e.g., Kipnis 1998; McComas 2005, 2008). I will now review the possible impact of teaching optics in a culturally rich perspective involving the history of science with respect to several aspects of *scientific knowledge.*

4.5.1 The Role of Theory

Physics curricula often stress modeling as the major feature of physics, leaving the fundamental theories of physics in the shade. Theory is often mentioned as part of the opposition between theory and experiment, thus missing the centrality of the concept of theory and the fact that no experiment can be conceived without a theory.

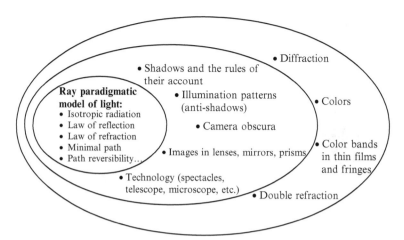

Fig. 4.6 Discipline-culture structure of the light theory of rays at the time of the scientific revolution of the seventeenth century. The contents of all three areas are illustrated

Experiments in science are normally *theory driven* and *theory laden* (e.g., Hanson 1958). Knowledge of optics demonstrates the organization of physics in terms of fundamental theories on a historical route from less to more advanced, thus creating a big picture of physics knowledge (Fig. 4.1).[10] Several accounts of light and vision reveal the polyphonic features of the scientific discourse, making scientific knowledge *cultural*.

Thus, at the beginning of science, scholars argued about intro- versus extramission conceptions of vision, the nature of light, and its speed, action, and behavior. Gradually, the first optical theory, the theory of rays, was developed through the efforts of many scholars from Euclid (third century B.C.) to Kepler (seventeenth century). They established Geometrical Optics.

The issues of color, light diffraction, and double refraction then emerged as challenging the theory of rays and those scholars who tried to save it, a situation which can be presented using a certain discipline-culture structure (Fig. 4.6). Newton's efforts to tackle the problem of light by "deducing" theory "from the phenomena" and avoiding the use of "hypothesis" (Newton 1729/1999, p. 943) were made within ray theory. He defined the light ray operationally in the first lines of his *Opticks* and proceeded from this concept:

> The least Light or part of Light, which may be stopp'd alone without the rest of the Light, or propagated alone, or do or suffer any thing alone, which the rest of the Light doth not or suffers not, I call a Ray of Light. (Newton 1704/1952, p. 1)

The strategy of staying with rays as the directly observed entity may lead people to consider Newton's *Opticks* to be a "neutral" investigation, especially the modern

---

[10] One may locate the laws of reflection and refraction in the nucleus of the theory (as Newton did, see in the following) or in the body of knowledge, that is, being proved basing on the principles of light path being minimal/extremal and its being reversible (as Heron and Archimedes did with reflection in the Hellenistic physics and Fermat – in the seventeenth century, with refraction). Both ways are educationally valid given that they are supported in the course of teaching-learning.

4 Teaching Optics: A Historico-Philosophical Perspective

reader who looks for Newton to take a side in the particle-wave controversy (Raftopoulos et al. 2005). It is indicative in this regard to compare Newton's accounts of light in the *Opticks* and the *Principia*. In the latter, he used the mechanistic theory of particles to *demonstrate* the law of (specular) reflection and Snell's law of refraction by considering particles interacting with matter (Newton 1727/1999, pp. 623–625), while in the former he *postulated* the same laws, thus placing them in the nucleus of the theory of rays (Newton 1704/1952, p. 5).

In order to explain color dispersion, Newton introduced the idea of light rays varying in refrangibility – color rays. Applying the classical method of *resolution* and *composition*, used by Aristotle, medieval scholars, and Galileo (Losee 2001, p. 28), Newton decomposed sunlight into the color spectrum and then, to remove speculations about the "creative" role of the prism, resynthesized white light by combining colored lights (Boyer 1987; Gaukroger 1995, p. 265).

Newton proceeded to use rays to explain the pattern of color rings as due to a thin layer of air of varying thickness between lens and plate (Newton's rings). Today, this is clearly an interference phenomenon, but it was not so for Newton, who explained it within the ray theory and without interference by ascribing to each ray a periodicity of "fits," predispositions to the reflection of the light ray from or penetration into the transparent medium (Tyndall 1877; Westfall 1989; Kipnis 1991; Shapiro 1993).[11]

Newton then turned to the diffraction of light and meticulously reproduced and refined the experiments of Grimaldi (1665). He rejected the suggested by Grimaldi splitting light into regular and extraordinary components in order to explain the light fringes next to the edge of geometrical shadows (Gliozzi 1965, pp. 121–122; Taylor 1941, p. 516) and replaced Grimaldi's *diffraction* with a *inflection* of rays. Yet, he failed to produce a theoretical account of light inflection. After a detailed description of the phenomena in numerous settings, he abruptly stopped because of "being interrupted" – a dramatic turn in a scientific treatise:

> When I made the foregoing Observations, I design'd to repeat most of them with more care and exactness, and to make some new ones for determining the manner how the Rays of Light are bent in their passage by Bodies, for making the Fringes of Colours with the dark lines between them. But I was then interrupted, and cannot now think of taking these things into farther Consideration. (Newton 1704/1952, pp. 338–339)

Some years before, Huygens, in his *Treatise on Light*, had tried to explain double refraction within his wave theory. He succeeded in using his inventive geometrical account of anisotropic expansion of light in a crystal to describe light beam splitting in a single crystal of calcite (Iceland crystal), but he failed to explain the behavior of the light beam passing through two crystals placed one after another. The beams amazingly change their refraction in the second crystal. Somewhat similar to Newton he quit:

> Before finishing the treatise on this Crystal, I will add one more marvelous phenomenon which I discovered after having written all the foregoing. For though I have not been able till now to find its cause, I do not for that reason wish to desist from describing it,

---

[11] Newton's numerical results on ray periodicity were of unprecedented accuracy for his time: for yellow-orange ray it was 1/89,000 in. (Newton 1704/1952, p. 285), well conforming to the half wavelength known today.

in order to give opportunity to others to investigate it. It seems that it will be necessary to make still further suppositions besides those which I have made .... (Huygens 1690/1912, p. 92)

Newton, after the main text of the *Opticks*, added *Queries*, where he described his considerations and hypotheses regarding the nature of light. Exactly as Huygens before him, Newton addressed the future researchers:

And since I have not finish'd this part of my Design, I shall conclude with proposing only some Queries, in order to a farther search to be made by others. (Newton 1704/1952, pp. 339–406)

Only there, in the *Queries*, did Newton allow himself to speculate: "Are not rays of light small particles emitted by shining substances ...?" and argued for the advantages of the *corpuscular* nature of light over the *wave* theory suggested by Huygens. There, addressing double refraction (birefringence), Newton stretched the ray theory even further and introduced *sides* to the light rays – a primitive version of the polarization of light (Mach 1913/1926, p. 189). In this way, he suggested a qualitative explanation of light passing through two consecutive crystals, the phenomenon which had puzzled Huygens (Newton 1704/1952, Query 26, pp. 358–361).[12] Newton finally quit, but not before he expressed his preference for the *particle* nature of light.

Thus, in the contest between two seventeenth-century theories, particles and waves each presented its successes in accounting for light (the body of the theory) and admitted its failures (the periphery). Newton's conjecture of light particles was definitely not treated as a theory Newton would prefer (as he did in mechanics[13]), but even so, it was preferred to its rival – Huygens' wave theory. In the end, though, neither scholar managed to produce an overall theory of light. Hakfoort states that:

From about 1700 the *Traité* [*Treatise on Light*] was almost completely ignored even in research reports from within the medium tradition. (Hakfoort 1995, p. 53)

Newton's conception of light particles remained dominant throughout the eighteenth century (Britannica Encyclopaedia 1770/1979), until Thomas Young and Augustin Jean Fresnel accounted for several new experiments and succeeded in demonstrating the clear superiority of the modified wave theory by introducing the principle of interference (Lipson 1968; Kipnis 1991). The nineteenth century witnessed the triumph of Fresnel's wave theory, which seemed to be unlimitedly true beyond any doubt, but not for a long time. In the twentieth century, new problems emerged to challenge the wave theory. To account for them Plank in 1900 and Einstein in 1905 produced heuristic models of light quanta. Placed in the periphery of the wave theory (Fig. 4.7), these constructs led to the new theory of light – the quantum theory, or the theory of photons.

---

[12] The quantitative account of the polarization of light was provided much later by Malus in the nineteenth century (Malus' law), who introduced and described the polarization of light particles instead of Newton's *sides* of light rays.

[13] The list of Newton's successes should also include the dynamic account of light behavior in the *Principia* and Newton's polemics there with Descartes' paradigm of plenum.

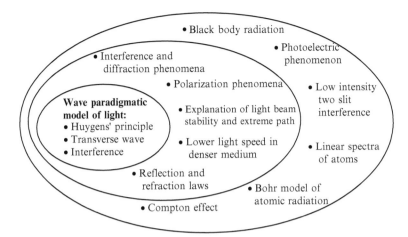

Fig. 4.7 Discipline-culture structure of the light theory of waves prior to the scientific revolution of the twentieth century

At each moment of history, the dominant theory of light can be represented as possessing the discipline-culture structure exemplified in Figs. 4.6 and 4.7. This presentation visualizes teaching about the transition between the successive theories of light and the competition between them. The dramatic contest of theories of light reveals the big picture of optical knowledge organized in terms of theories. Such a picture could be presented to the students in the form of summary ("vista point") lectures following a regular course (Levrini et al. 2014) or as a part of class instruction in a specially designed curriculum (Galili and Hazan 2000b).

4.5.2 The Role of Experiment

Teaching optics historically can be used to illustrate the role of experiment in physics knowledge. This topic is often addressed in the context of the opposition between theory and experiment.[14] A class discussion may begin from an examination of the famous image by Raphael Sanzio (Fig. 4.8) manifesting the symbiosis of rationalist and empiricist approaches as the essence of science.

For instance, the two optical theories of the seventeenth century differed with regard to the speed of light in transparent media. Newton implied that the speed of light in a dense transparent medium (water) would be higher than in a rare one (air); the wave theory of Huygens stated the opposite (e.g., Sabra 1981, pp. 217, 302). Both theories produced Snell's law of refraction and equated the refraction index to v_1/v_2 or, alternatively, to v_2/v_1.

---

[14] In the cultural approach, experiment may be affiliated to either body or periphery area within the triadic structure of theory knowledge.

Fig. 4.8 Plato and Aristotle in Raphael Sanzio's fresco *The School of Athens* (c. 1511). The gesturing of the two philosophers is commonly interpreted as emblematic of the epistemological dispute between rationalism (theory first) and empiricism (experience first) (See, for example, Galili (2013))

The opposing conclusions regarding the speed of light in different media could not be resolved using theoretical tools. New experiments were required. Augustin Fresnel and François Arago, as early as 1816, demonstrated experimentally that light traveled slower in glass (mica) than in air (Kipnis 1991, p. 178). Born (1962, p. 97) has emphasized the experimental confirmation of this result by Leon Foucault in 1850 (Foucault 1854). In both cases, however, the voice of experiment was decisive; theories could not manage without experiment. Given the central role of theory,[15] the students may learn to appreciate the fundamental complementarity between theoretical and experimental considerations as the essential feature of scientific account of any subject. Specific cases where experiment surpasses theory (Brush 1974) may further illustrate the reciprocity of the theory-experiment relationship. All in all, the idea that the validity of a theory in physics demands the firm basis of pertinent experiments is comprehensively represented.

The victory of the theory of relativity may be used to further refine the relationship between theory and experiment, which is often presented as drawing on a single experiment by Michelson and Morley: light interference in the moving interferometer. A cultural historical presentation reveals that rival theories (aether drag, emission theory, length contraction) provided alternative explanations for the same experiment. The victory of the special theory of relativity emerged from a series of different experiments where only Einstein's theory was able to account for all the results (Panofsky and Phillips 1962, p. 240). This way one physical theory replaces another.

The same context is appropriate for a discussion of the historically popular idea of the "critical experiment" (*experimentum crucis*). The history of light contains a series of such experiments, but a closer look often displays a more complex picture.

---

[15] See Sect. 4.5.1.

Texts often mention Newton's prism experiment as proof that light is composed of colored rays (e.g., Wolf 1968, pp. 264–271; Boyer 1987, pp. 200–268), Poisson's spot experiment is seen as proof of the wave theory of light (e.g., Kipnis 1991, pp. 220–222), and Foucault's experiment on the speed of light in water is also taken as proof of the wave theory. Although the need for concise teaching is understandable, the teacher may mention a mature understanding of "critical experiments" as indicating and suggesting, but never addressing all the possibilities in clarifying the truth about nature. Duhem (1982, pp. 188–190) made just this point when he addressed the nature of light and wrote: "The truth of a physical theory is not decided by heads or tails." Popper (1965, pp. 54–55) followed him and stated that the power of experimental evidence may serve as a strong argument in theory refutation, rather than proof of its being "true." The physics teacher should moderate the strong convincing appeal of "critical experiments" within the cultural teaching of optics. Yes, sunlight includes all colors, but is it composed of *rays*? Yes, light possesses a wave nature, but can one qualitatively explain the Poisson spot by a sort of Newtonian inflection of light rays? Similarly, can Foucault's results on the speed of light in water be explained by Fresnel's drag coefficient of the aether? An experiment may serve as a great step forward, but science always proceeds by seeking a variety of evidence for the same theory.

Finally, one of the most impressive illustrations of the decisive role of a real experiment may be the famous EPR experiment on a pair of entangled photons. Designed as a thought experiment by Einstein in 1935 with particles (later translated by Bohm to the experiment with photons), it was meant to demonstrate the "incomplete" nature of quantum mechanics. The experiment drew on a theoretical assumption which was considered to be unquestionably true – the principle of the locality of physical events (e.g., Cushing 1998, p. 325). For years this *thought* experiment remained an open problem kept in the periphery of the quantum theory in order to question its major paradigm. The *real* experiment carried out by Aspect in 1981 with pairs of entangled photons radiating in opposite directions provided the verdict: quantum physics is correct, Einstein was wrong. Microscopic objects are subject to the principle of nonlocality (e.g., Cushing 1994, pp. 14–16; Penrose 1997, pp. 64–66). From the periphery of quantum theory, the EPR thought experiment migrated to its body as a real experiment, but not before the principle of nonlocality was added to the nucleus of quantum theory. The historical teaching of light, thus, clarifies the complex and reciprocal relationship between theory and experiment. This complementarity implies an essential entanglement which can be expressed in the terms of the Bohr principle (e.g., Migdal 1990).

4.5.3 Cumulative Nature of Scientific Knowledge

Among the advantages of the historical perspective is its ability to illustrate the cumulative character of scientific knowledge. Kuhn's (1962/1970) thesis about periods with conceptually incommensurable paradigms (nuclei of the dominating theories) is often interpreted excessively as a renouncement of the idea of cumulative knowledge in science, despite Kuhn's own clarification in his postscript of 1970 (ibid, pp. 205–207). A historical view of optics allows the refinement of this subject while displaying the continuous progress of knowledge accumulation and use. At each stage, scientific research is inherently related to previously obtained results. Even though there are different theoretical frameworks in each historical period, similar or even identical questions were tackled, drawing on information from previous studies either positively or negatively.

Thus, a teacher may introduce the old problem from Aristotle's *Problems* (Aristotle 1952, pp. 334–335): in the camera obscura a small opening leads to a circular image of the sun, while a large opening produces an illuminated area in the shape of the opening (Lindberg 1968). More than 1,300 years later, Al-Haytham resolved this mystery when he applied a concept introduced by Al-Kindi about 200 years earlier: light expands from *each* point of a light source in *all* directions. Al-Haytham's account for the image in the camera obscura was adopted by della Porta a few hundred years later; in the sixteenth century he suggested that the human eye is similar to the camera obscura but with a lens in the hole. Kepler, in the seventeenth century, resolved the enigma of vision in dialogue with the scholars of the past. He used his knowledge of Euclid, Al-Haytham, and others to resolve the problem they failed to solve. The power of knowledge accumulation is that which Bernard of Chartres (twelfth century) epitomized in his famous pronouncement so relevant to scientific knowledge and science education:

> We are like dwarfs standing on the shoulders of the giants, so that we can see more things than them, and see farther, not because our vision is sharper or our stature higher, but because we can raise ourselves up thanks to their giant stature. (Crombie 1959, p. 27)

In another example, an optics instructor may follow up the principle of the propagation of light in space starting from the statement of *minimal* path length in specular reflection by Heron the Hellenist. Fermat later refined the same principle for the *extreme* path in terms of the time required by light. In 1871 Lord Rayleigh used the Fresnel zone plate to experimentally demonstrate the Huygens-Fresnel principle of the propagation of light (Mach 1913/1926, p. 287; Hecht 1998, p. 487). He demonstrated that light moves not only in straight lines between any two points but in all other ways between them. The old argument of Newton (1729/1999, pp. 762–765) against the wave nature of light was thus removed. The idea of multiple paths of light led Feynman (1948, 1985) to the most fundamental principle of quantum electrodynamics – multiple intermediate events or paths, in general sense, between any two states. Over the centuries, scholars working within different paradigms constructed physics knowledge about the expansion of light.

In a sense, scientists have maintained conceptual discourse, diachronic and synchronic, on research questions and drawn on previously attained understandings. As Collingwood stated in his *The Idea of History*:

> The two phases [of science] are related not merely by way of succession, but by way of continuity, and continuity of a peculiar kind. If Einstein makes an advance on Newton, he does it by knowing Newton's thought and retaining it within his own, ... He might have done this, no doubt, without having read Newton in the original for himself; but not without

4 Teaching Optics: A Historico-Philosophical Perspective

having received Newton's doctrine from someone. ... It is only in so far as Einstein knows that theory, as a fact in the history of science, that he can make an advance upon it. Newton thus lives in Einstein in the way in which any past experience lives in the mind of the historian, ... re-enacted here and now together with a development of itself that is partly constructive or positive and partly critical or negative. (Collingwood 1956, p. 127)

An awareness of this aspect of the scientific knowledge promotes cultural content knowledge.

4.5.4 Objective Nature of Scientific Knowledge

Traditionally, the objectiveness of an account of nature is understood by scientists to mean the independence of this knowledge from will, mood, desires, etc: "the way whether and how the dropped object falls is independent of our attitude to that"; scientific accounts of phenomena should be objective in this sense, and thus scientists often state that physics is *objective* (e.g., Weinberg 2001). *Only* such objective knowledge can be the subject of critical discourse maintained by science, indicating its "health" (Popper 1981; Holton 1985). However, as mentioned by Einstein (1987), the particular system of concepts scientists use in their account of nature is the result of their free decision further tested by experiment. It is not surprising, therefore, that some researchers have highlighted the subjective features of scientific knowledge, its dependence on imagination, beliefs, worldviews, social constraints, etc. (e.g., McComas 1998, 2005). Their emphasis on the interaction between the "instrument" (the scientists) and the object (the nature) has been illustrated by Kierkegaard:

> ... The speculative [philosophers] in our time are stupidly objective. They completely forget that the thinker himself is simultaneously the musical instrument, the flute, on which he plays. (Kierkegaard 1952, quoted in Migdal 1985)

In contrast, many philosophers and historians of science believe that "the metaphysical tenets of individual scientists, though often quite strong, are generally so varied, so vague, and so technically inept that in a sense they cancel out, made ineffectual by the lack of a basis for general acceptance and agreement of such tenets" (Holton 1985). That is, variation among individuals is essential in the search for objectiveness. The history of science enhances this claim.

Within the historical teaching of optics, the issue of objectivity may be illustrated, discussed, and provided with operational meaning. Consider the principle regulating light path. Heron of Alexandria in his *Catoptrics* demonstrated the rule of the specular reflection of light: the light path presents the shortest trajectory between any two points, including mirror reflection (Cohen and Drabkin 1966, p. 263). This rule is an example of objective knowledge. However, the interpretation of this result as nature seeks the most "economical" way to go, or that nature does nothing in vain *(natura frustra nihil agit)*, is a subjective, metaphysical one. Furthermore, using the method of *Maxima and Minima*, Fermat in the seventeenth century advocated the *extreme* temporal rather than spatial path of light (Ross 2008), an objective truth. However, he also claimed that his finding expressed a "natural intention," a subjective view. Measurement confirmed the law of refraction as the sine ratio of the angles of incidence and refraction, objective knowledge. Descartes believed that this empirical law of refraction was not sufficient because it did not *explain* the phenomenon. He suggested an ad hoc mechanism of light refraction (Descartes 1637/1965, p. 79), claiming an analogy between light and the motion of a ball being hit downward by a tennis racket at a water surface (Ross 2008, p. v). This analogy was used to explain the increased vigor of the ball in water. Given that Descartes did not ascribe velocity to light, the artificial and subjective nature of this analogy is obvious. Mach called it "unintelligible and unscientific" (Sabra 1981, p. 104), but the approach of Fermat, Descartes' opponent, was unsatisfactory as well: how and why could light possibly "decide" on the extreme path?

Only in the nineteenth century were the subjective speculations regarding light propagation removed, following the introduction of wave interference by Fresnel as a tool to apply Huygens' principle. The experiment by Raleigh who covered odd (or even) Fresnel zones on the screen placed on the light way demonstrated that light did not "decide" which way to go from one point to another but went in all ways between these two points. The interference of all the beams produces the familiar phenomena of light reflection and refraction. Thus, the *subjective* part associated with the Fermat principle was dismissed, and the *objective* one remained. Feynman (1948, 1985) further developed this new understanding to include particles with mass. An introductory optics course can, and in a way should, display a qualitative account of the full story.

The objectivity of physics knowledge was framed within the idea of the "third world"[16]: the virtual intellectual space incorporating physical theories (Popper 1978, 1981). Disconnected from individuals, it contains objective knowledge of the world. Holton (1985) introduced science-1 and science-2 for the same purpose – to distinguish between the objective core and subjective elements of physical theories. The four theories of light illustrate an area of optics knowledge in the third world. Though very different in validity, they share the property of objectivity, remaining human, that is, a subject for refinement and falsification.

4.5.5 The Role of Mathematics

Mathematics provides one of the central features of physics knowledge. Its role in physics is important, complex, and many faceted. The history of optics is eloquent in this regard. Euclid, Archimedes, and Ptolemy were the first to introduce mathematics into the optics of Hellenistic science, addressing the features of light and vision (e.g., Smith 1982; Russo 2004). In this way they rebelled against the previously dominant perspective of the Hellenic philosophers who argued for a qualitative conceptual account of nature as the major agenda of physics. Euclid introduced

---

[16] To be distinguished from the real world (the first one) and the personal world (the second one).

the central mathematical tool of optics – rays of light and vision – and developed the method of *perspective* (two-dimensional representations of three-dimensional reality). In his hands, optical theory became a brunch of geometry. Great mathematical skill, however, was not enough to allow him to explain the nature of light and vision and prevent fundamental confusion. Indeed, Euclid, Ptolemy, Al-Kindi, and many others mastered perspective but held and skillfully argued for the faulty ideas of active vision ("eyes radiate vision rays …," scan the reality) (Lindberg 1976). In physics class, this combination of advanced and erroneous views by the same scholars may contribute to understanding features of the relationship between mathematics and physics.

Another example of this kind is the history of the sine law of refraction. Ptolemy was the first to tackle the problem (Ptolemy 1940; Smith 1982; Mihas 2008). His data did not fit the constant proportionality between the angles of incidence and refraction of visual rays. Ptolemy tried to "adjust" their behavior to a quadratic dependence (Russo 2004, p. 64). Light rays behaved in the same way, as they were postulated to follow the same path as vision but in the opposite direction. However, the true ratio known as the sine law was not obtained by Ptolemy. Smith (1982) has explained this failure by the fact that Ptolemy, like Heron, only used spatial (geometrical) considerations of the vision-light path, while the key to the true account of refraction, the explanation, was to treat the problem using temporal (kinetic, physical) considerations, as Descartes and others did much later (e.g., Sabra 1981, pp. 105–116). Thus, obtaining the correct mathematical account in Hellenistic physics was impeded by an inappropriate physical approach: geometry and numbers were not enough.

For centuries scientists kept trying to find the mathematical form of the refraction law. Even in the seventeenth century, a skillful mathematician like Kepler, who had famously proved himself by his demonstration of the elliptical orbits of planets and who wrote prolifically on optics and vision, failed to reveal the refraction law, perhaps because he trusted the not sufficiently accurate tables of Vitello (thirteenth century) (Herzberger 1966). Kepler continued to use the linear dependence of angles ratio as a good approximation of refraction at small angles.[17] Eventually, several scholars worked out the correct law: Ibn Sahl (c. 984), Harriot (c. 1602), Snell (c. 1621), and Descartes (c. 1637) (Sabra 1981; Rashed 2002, p. 313; Kwan et al. 2002). As mentioned, the first explicit explanations of this law involved physical considerations regarding the speed of light. Ironically, it was Descartes who first elaborated the demonstration of the law, despite his own denial of the finite speed of light.

History tells us that the mathematical form itself and even its deduction from Fermat's principle of extreme time of light trajectory (the "easiest course" and the principle of economy) did not suffice for physicists. Scientists wanted to know the *mechanism* which caused this particular form of the refraction, the mechanism that underpins the rather unusual mathematical form of the sine law. Descartes' artificial

---

[17] This simplified law of refraction can be used in teaching optical phenomena presented qualitatively (Galili and Goldberg 1996).

analogy of the ball entering the water was not at all persuasive.[18] In contrast, Huygens and Newton were more convincing. Both scholars reproduced Snell's law theoretically, even though they based themselves on the contrasting models of waves and particles. Drawing on the particle model, Newton supported Descartes' conjecture; the velocity of light in a denser medium increases, perhaps due to the gravitational attraction of the medium. Huygens, however, deduced the same law from his principle of secondary waves and inferred the opposite – the lower velocity of light in a denser medium (Sabra 1981, pp. 300–302). Mathematics falsified all the other options but could not help to choose between the two remaining theories. This choice was made by physicists in the nineteenth century.[19]

In the contest of theories of light, Huygens surpassed Newton in the mathematical accuracy of his account of double refraction in a single crystal of Island spar, but, as already mentioned, he failed to explain the behavior of light in two successive crystals. Newton, addressing the same phenomenon, correctly suggested the transverse polarization of light by his assumption of light ray *sides* but did not provide a mathematical account of double refraction. The framework of the wave optics of Fresnel in the nineteenth century was necessary for this account. In retrospect, in the seventeenth century researchers lacked the essential mathematical tools required to account for the transverse running wave: two-variable functions to depict a running wave, equations in partial derivatives for a wave equation, and calculus of the kind Fresnel applied to the principle of interference. Newton, who performed a revolution in mathematics by inventing calculus to account for gravitation, did not instigate another mathematical upheaval required to account for optics.

The examples of the history of optics mentioned here clearly demonstrate that mathematics and physics are fundamentally entwined. Like theory and experiment, the relationship between mathematics and physics can be expressed in terms of complementarity in the sense introduced by Bohr (e.g., Migdal 1990, p. 16). Einstein expressed a very close idea, saying "As far as the laws of mathematics refer to reality, they are not certain; and as far as they are certain, they do not refer to reality." In summary, Einstein's addressing the relationship between physics and philosophy may be paraphrased to conclude that mathematics without physics is blind (i.e., unable to provide qualitative understanding and causal meaning), while physics without mathematics is empty (i.e., destined to produce unresolved speculations which might be conceptually valid but untestable).

4.5.6 Commonsense Complexity

Another important feature of scientific knowledge to reveal by teaching optics in a cultural historical perspective is its relationship with common sense. The history of optics is eloquent in this respect as well. Thus, for years, many scholars could not

---

[18] See Sect. 4.5.4.

[19] See Sect. 4.5.2.

accept the idea of the "upside-down" image created on the retina of the human eye, despite the fact that since antiquity it had been known that the retina, and no other part of the eye, is connected to the brain. The inverted image observed in the camera obscura and on the screen placed behind a convex lens did not seem to be relevant, apparently contradicting common sense. This "obviousness" misled great minds. Al-Haytham erroneously placed the "correct," right-side-up image on the surface of the eye lens. Later, Leonardo da Vinci painstakingly searched for two successive inversions which would provide a right-side-up image. Only Kepler, in the seventeenth century, removed the enigma: common sense is lying – the image in the eye is inverted and is dealt with as such by the mind – another stage in the process of vision (Lindberg 1976).

Another example deals with the nature of light. Since people do not usually feel light entering the eye, common sense conceives of light not as a moving agent but rather as a state or medium; light "fills" and "stays" in space. Light as a static entity fits the biblical description of the creation of the world and the commonsense conceptions of students. The historical teaching of optics may address this topic in relation to the difference between the visual perception of light and its objective existence – an epistemological issue (Gregory 1979; Linn et al. 2003). This interesting topic can be related to the historical split in the concept of light: lux seems to be close to our concept of illumination, while lumen seems to represent light as physical entity (Steneck 1976; Galili and Hazan 2004). Lux and lumen are Latin terms introduced in the translation of the Bible to Latin (the original text in Hebrew has no such split, though the split in the meaning of light is discussed in religious sources).

Discussing the existing versus perceived dichotomy in physics may lead students to a critique of naïve common sense and an understanding of its difference from scientific knowledge (Cromer 1993; Wolpert 1994) and yet its necessity in doing science (Conant 1961; Bronowski 1967). Koyré (1943) has stated that the role of the founders of modern science was "to replace a pretty natural approach, that of common sense, by another, which is not natural at all." Einstein wisely moderated this extreme claim by saying that science continuously refines and corrects common sense. Science "upgrades" common sense and transmits the benefits of the changes to all through public education.

4.5.7 Teaching Modern Physics

There is an interesting pedagogical phenomenon: unlike the teaching of classical physics, which often ignores the history of science, the teaching of introductory modern physics is normally historical and reproduces step by step the transition from classical physical theories to modern ones. A historical narrative mitigates the overwhelming conceptual novelty of relativistic or quantum physics and the significantly more complex mathematics which they employ. The teaching of modern physics usually starts by addressing the problems of the light theory of waves from the periphery of that theory (Fig. 4.7). The events of the beginning of the twentieth century gradually introduce the student to modern theoretical and epistemological perspectives on the objective reality with regard to light. In this context, the discoveries become interwoven with the construction of the new theory.[20]

Fizeau's experiment on the propagation of light in moving water (Fizeau 1851) can illustrate the particular teaching potential of the culturally-framed teaching of modern physics already-mentioned[21] with respect to the Michelson-Morley and EPR experiments. This example well illustrates the principle of the correspondence of physical theories. Initially, Fresnel's model of the partial drag of the aether, the "drag coefficient" (Fresnel 1818), saved the appearance of Fizeau's experiment by providing an explanation of the experimental results. However, the theory of special relativity, based on fundamentally different ideas, reproduced the drag coefficient as the first approximation of the more accurate relativistic account (e.g., French 1968, pp. 46–49). It was thus shown that the new theory was able to explain the previous accounts and in a more accurate way.

Another item of optical history illustrating the quantum nature of light is the two-slit experiment in which a very low intensity light beam creates an illumination pattern of interference as emerging from the accumulation of numerous separate spots – the experiment by Geoffrey Ingram Taylor (1909). When there is a much larger number of photons, and after a considerable amount of time, the regular interference pattern of two-slit experiments emerges, reproducing the classical result of Young (1804) and Fresnel (1866). Taylor's experiment demonstrates the intimate relationship between two theories of light – quantum and classical electromagnetic. When students are presented with the discipline-culture structure, the two theories possess distinguished nuclei and partially overlaid bodies of knowledge (Fig. 4.4).

4.6 Conclusion

Teaching optics using HPS is possible and beneficial. This pedagogy clarifies the disciplinary knowledge of light and vision as well as the nature of scientific knowledge. The validity of the *cultural content knowledge* established in this way in the student goes beyond the mere correction of misconceptions. A curriculum enriched by HPS addresses the creation of knowledge which took place in both diachronic and synchronic scientific discourse. Both essentially involve elements of incorrect in disciplinary sense knowledge which is inherently connected to and elucidates the meaning of correct knowledge. This approach takes advantage of the similarity,

---

[20] Educators may use an artistic metaphor to represent the transition from the epistemological credo of classical physics to that of the modern theories. The relief on the Nobel Prize medal for physics can be seen as representing the epistemology of classical physics, while a sketch depicting the myth of *Pygmalion and Galatea* may do the same for modern physics (Levrini et al. 2014; Galili 2013).

[21] See Sect. 4.5.2.

though not identity, between students' ideas regarding light and vision and those developed in the course of the history of science, causing cognitive resonance in the students.

The beneficial impact of dealing with the historical knowledge of optics expands on understanding of the nature of science and the fundamental features of scientific knowledge. This impact deserves more investigation in the perspective of considering physics knowledge as a culture. Introducing a discipline-culture structured curriculum allows understanding of the role of HPS in optics knowledge representation, and the dynamic relationship between the four basic optical theories otherwise often remains disconnected for being taught in different courses. The introduced integrated picture emphasizes the image of science as theory-based knowledge (e.g., Bunge 1973) and removes the oversimplified perception of incommensurable theories of light, thus upgrading the meaning of their relationship in a big picture.

References

Aristotle (1952). On the Soul. In *The Works of Aristotle*. Chicago: Encyclopedia Britannica, vol. 1, pp. 627–668.

Aikenhead, G.S. (1997). Towards a first nations cross-cultural science and technology curriculum. *Science Education*, 81(2), 217–238.

Aikenhead, G.S. & Jegede, O.J. (1999). Cross-cultural science education: A cognitive explanation of a cultural phenomenon. *Journal of Research in Science Education*, 36(3), 267–287.

Andersson, B. & Karrqvist, C. (1983). How Swedish pupils, aged 12–15 years, understand light & its properties. *European Journal of Science Education*, 5(4), 387–402.

Andreou, C. & Raftopoulos, A. (2011). Lessons from the History of the Concept of the Ray for Teaching Geometrical Optics. *Science & Education*, 20(10), 1007–1037.

Arons, A. (1965). *Development of concepts of physics*. Reading, Mass: Addison-Wesley.

Atwood, R. K. & Atwood, V. A. (1996). Preservice elementary teachers' conceptions of the causes of seasons. *Journal of Research in Science Teaching*, 33, 553–563.

Ausubel, D.P. (1968). *Educational Psychology: A Cognitive View*. New York: Holt, Rinehart & Winston.

Beaty, W. (1987). The origin of misconceptions in optics? *American Journal of Physics*, 55(10), 872–873.

Bendall, S., Goldberg, F., & Galili, I. (1993). Prospective elementary teachers' prior knowledge about light. *Journal of Research in Science Teaching*, 30(9), 1169–1187.

Benson, G. (1989). Epistemology and Science Curriculum. *Journal of Curriculum Studies* 21(4), 329–344.

Bevilacqua, F., Giannetto, E., & Matthews, M. (2001). *Science education and culture. The contribution of history and philosophy of science*. Dordrecht: Kluwer.

Born, M. (1962). *Einstein's Theory of Relativity*. New York: Dover.

Bouwens, R. (1987). Misconceptions among pupils regarding geometrical optics. In J.D. Novak (ed.), *Proceedings of the Second International Seminar on Misconceptions and Educational Strategies in Science & Mathematics*. Ithaca: Cornell University.

Boyer, C.B. (1987). *The Rainbow: From Myth to Mathematics*, Princeton: Princeton University Press.

Boyes, E. & Stanisstreet, M. (1991). Development of Pupils' Ideas of Hearing and Seeing-the Path of Light and Sound. *Research in Science and Technology Education*, 9(2), 223–244.

Bragg, W. (1959). *The Universe of Light*. New York: Dover.

Britannica Encyclopaedia (1770/1979). Edinburgh, Society of Gentlemen in Scotland, The First Edition.

Bronowski, J. (1967). *The Common Sense of Science*. Cambridge, Mass: Harvard University Press.

Brush, S.G. (1974). Should the History of Science Be Rated X? *Science*, 183, 1164–1172.

Bunge, M. (1973). *Philosophy of Science*. Dordrecht: Reidel.

Cohen, R.M. & Drabkin, E.I. (1966). *A Source Book in Greek Science*. New York: McGraw-Hill Book Company, Inc.

Colin, P. & Viennot, L. (2001). Using two models in optics: students' difficulties & suggestions for teaching. *American Journal of Physics, Physics Education Research Supplement,* 69(7), S36–44.

Colin, P. (2001). Two models in a physical situation: the case of optics. Students' difficulties, teachers' viewpoints and guidelines for a didactical structure. In H. Behrendt, H. Dahncke, R. Duit, W. Graeber, M. Komorek, A. Kross, P. Reiska (eds.), *Research in Science Education – Past, Present & Future* (pp. 241–246). Dordrecht: Kluwer Academic Publishers.

Collingwood, R.G. (1949). *The Idea of Nature*. Oxford: Clarendon Press.

Collingwood, R.G. (1956). *The Idea of History*. New York: Oxford University Press.

Conant, J.B. (1961). *Science and Common Sense*. New Haven: Yale University Press.

Crombie, A.C. (1959). *Medieval and Early Modern Science*. New York: Doubleday Anchor Books.

Crombie, A.C. (1990). *Science, Optics and Music in the Medieval and Early Modern Thought*. London: The Hambledon Press.

Cromer, A. (1993). *Uncommon Sense*. New York: Oxford University Press.

Cushing, J. (1994). *Quantum Mechanics: Historical Contingency and the Copenhagen Hegemony*. Chicago: University of Chicago Press.

Cushing, J. (1998). *Philosophical concepts in physics*. Cambridge, UK: Cambridge University Press.

De Hosson, C. & Kaminski, W. (2007). Historical controversy as an educational tool: Evaluating elements of a teaching-learning sequence conducted with the text "Dialogue on the ways that vision operates". *International Journal of Science Education*, 29(2), 617–642.

Dedes, C. (2005). The mechanism of vision: Conceptual similarities between historical models and children's representations. *Science & Education*, 14, 699–712.

Descartes, R. (1637/1965). *Discourse on Method, Optics, Geometry and Meteorology. Second Discourse – of Refraction*. New York: Bobbs-Merrill.

Descartes, R. (1998). *The World and Other Writings*. Cambridge, UK: Cambridge University Press.

Dijksterhuis, E.J. (1986). *The Mechanization of the World Picture, Pythagoras to Newton*. Princeton: Princeton University Press.

Dijksterhuis, F.E. (2004). *Lenses and Waves: Christian Huygens and the Mathematical Science of Optics in the Seventeenth Century*. Dordrecht: Kluwer Academic Publishers.

diSessa, A. (1993). Toward an epistemology of physics. *Cognition & Instruction,* 10, 105–225.

Driver, R. & Bell, B. (1986). Students' Thinking and the Learning of Science: A Constructivist View. *School Science Review*, 67, 443–456.

Duhem, P. (1906/1982). *The Aim and Structure of Physical Theory*, Princeton University Press, Princeton, NJ.

Duit, R., Gropengießer, H., & Kattmann, U. (2005). Towards science education research that is relevant for improving practice: The model of educational reconstruction. In H.E. Fischer (ed.), *Developing standards in research on science education* (pp. 1–9). London: Taylor & Francis.

Einstein, A. (1987). *Letters to Solovine: 1906–1955* (May 7, 1952). New York: Open Road, Integrated Media.

Emmott, W. (1961). Some Early Experiments in Physical Optics, *Optician*, 142 (Aug), 138–140; (Sept.) 189–193; 211–215; 296–298; (Nov) 449–456.

Endry, J. (1980). Newton's Theory of Colour. *Centaurus,* 23(3), 230–251.

Feher, E. & Rice, K. (1988). Shadows and anti-images: children's conception of light and vision II. *Science Education*, 72(5), 637–649.

Feher, E. & Rice, K. (1992). Children' s conceptions of color. *Journal of Research in Science Teaching*, 29(5), 505–520.

4 Teaching Optics: A Historico-Philosophical Perspective

Fehl, N. E. (1965). *Science and Culture.* Hong Kong: Chung Chi, The Chinese University of Hong Kong.

Fetherstonhaugh, A., Happs, J., & Treagust, D. (1987). Student misconceptions about light: A comparative study of prevalent views found in Western Australia, France, New Zealand, Sweden and the United States. *Research in Science Education,* 17(1), 156–164.

Fetherstonhaugh, T. & Treagust, D. (1992). Students' Understanding of Light and its Properties: Teaching to Engender Conceptual Change. *Science Education,* 76(6), 653–672.

Feynman, R. (1948). The Space-Time Formulation of Nonrelativistic Quantum Mechanics. *Reviews of Modern Physics,* 20(2), 367–387.

Feynman, R. (1985). *QED – The Strange Theory of Light and Matter* (Lecture 2). Princeton: Princeton University Press.

Fizeau, H. (1851). Sur les hypotheses relatives à l'éther lumineux, et sur une expérience qui paraît démontrer que le mouvement des corps change la vitesse avec laquelle la lumière se propage dans leur intérieur. *Comptes-rendus hebdomadaires de l'Académie des sciences,* 33, 349–355.

Fleer, M. (1996). Early learning about light: mapping preschool children's thinking about light before, during & after involvement in a two week teaching program. *International Journal of Science Education,* 18(7), 819–836.

Forbes, R.J. & Dijksterhuis, E.J. (1963). *A History of Science and Technology. 'Nature Obeyed and Conquered'.* Baltimore: Penguin Books.

Foucault, L. (1854). Sur les vitesses relatives de la lumière dans l'air et dans l'eau (On the relative Velocities of Light in Air and in Water). *Annales de Chimie et de Physique,* 41(3), 129–164.

French, A. (1968). *Special Relativity.* The MIT Introductory Physics Series. New York: Norton.

Fresnel, A. (1818). Lettre de M. Fresnel à M. Arago, sur l'influence du mouvement terrestre dans quelques phénomènes d'optique. *Annales de Chimie et de Physique,* 9(1), 57–66.

Fresnel, A. (1866–70). *Oeuvres complètes d'Augustin Fresnel,* 3 vols. Paris: Imprimerie Impériale.

Galili, I. (1996). Student's Conceptual Change in Geometrical Optics. *International Journal in Science Education,* 18(7), 847–868.

Galili, I. (2012). Promotion of Content Cultural Knowledge through the Use of History and Philosophy of Science, *Science & Education,* 21(9), 1283–1316, doi: 10.1007/s111910119376.

Galili, I. (2013). On the Power of Fine Arts Pictorial Imagery in Science Education in Science Education. *Science & Education,* doi: 10.1007/s11191-013-9593-6.

Galili, I. & Goldberg, F. (1996). Using a linear approximation for single-surface refraction to explain some virtual image phenomena. *American Journal of Physics,* 64(3), 256–264.

Galili, I. & Hazan, A. (2000a). Learners' knowledge in optics: Interpretation, structure, and analysis. *International Journal in Science Education,* 22(1), 57–88.

Galili, I. & Hazan, A. (2000b). The influence of historically oriented course on students' content knowledge in optics evaluated by means of facets-schemes analysis. *Physics Education Research, American Journal of Physics,* 68(7), S3-S15.

Galili, I. & Hazan, A. (2001). Experts' views on using history and philosophy of science in the practice of physics instruction. *Science & Education,* 10(4), 345–367.

Galili, I. & Hazan, A. (2004). Optics – The theory of light and vision in the broad cultural approach. Jerusalem, Israel: Science Teaching Center, The Hebrew University of Jerusalem.

Galili, I. & Hazan, A. (2009). *Physical Optics - the theory of light in the broad cultural approach,* Parts II and III (Physical Optics of Waves and the Modern Theory of Light). Jerusalem, Israel: Science Teaching Center, The Hebrew University of Jerusalem.

Galili, I. & Lavrik, V. (1998). Flux concept in learning about light: A critique of the present situation. *Science Education,* 82, 591–613.

Galili, I. & Zinn, B. (2007). Physics and art – A cultural symbiosis in physics education. *Science & Education,* 16(3–5), 441–460.

Galili, I., Bendall, S., & Godberg, S. (1996). The effects of prior knowledge and instruction on understanding image formation. *Journal of Research in Science Teaching,* 30(3), 271–301.

Gaukroger, S. (1995). *Descartes. An Intellectual Biography.* Oxford: Clarendon Press.

Gliozzi, M. (1965). *Storia della Fisica,* Vol. II. Storia della Scienze, Torino, Italy.

Goethe, J. W. (1810). Zur Farbenlehre. See J. Pawlik (1974), *Goethe's Farbenlehre*. Verlag Dumont Schauberg, Cologne. English translation by C.L. Eastlake (1970). *Theory of Colors*. Cambridge, Mass: The MIT Press.

Goldberg, F. & McDermott, L.C. (1986). Student difficulties in understanding image formation by a plane mirror. *Physics Teacher,* 24(8), 472–480.

Goldberg, F. & McDermott, L.C. (1987). An investigation of students' understanding of the real image formed by a converging lens or concave mirror. *American Journal of Physics,* 55(2), 108–119.

Gregory, R. L. (1979). *Eye and Brain*. Princeton: Princeton University Press.

Grimaldi, F. M. (1665). *Physico-Mathesis de lumine, coloribus, et iride*. Bologna: Vittorio Bonati.

Guesne, E. (1985). Light. In R. Driver, E. Guesne, & A. Tiberghien (eds.), *Children's ideas in science* (pp. 11–32). Milton Keynes: Open University Press.

Hakfoort, C. (1995). *Optics in the Age of Euler. Conceptions of the Nature of Light, 1700–1795*. Cambridge, UK: Cambridge University Press.

Hecht, E. (1998). *Optics*. Reading, Mass: Addison-Wesley.

Herzberger, M. (1966). Optics from Euclid to Huygens. *Applied Optics,* 5(9), 1383–1393.

Holton, G. (1985). *Introduction to concepts and theories in physical science.* Second edition revised by Brush, S.G. Princeton University Press, Princeton, NJ.

Huygens, Ch. (1690/1912). *Treatise on light: In which are explained the causes of that which occurs in reflection & in refraction, and particularly in the strange refraction of Iceland crystal.* London: McMillan. In French: Huygens, Ch. (1992). *Traité de la lumière*. Paris: Dunod.

Jardine, D.W. (2006). *Piaget and Education*. New York: Peter Lang.

Jung, W. (1981). Erhebungen zu Schülervorstellungen in Optik. *Physica Didactica,* 8, 137.

Jung, W. (1982). Ergebnisse einer Optik-Erhebung. *Physica Didactica,* 9, 19.

Jung, W. (1987). Understanding students' understanding: the case of elementary optics. *Proceedings of the Second International Seminar: Misconceptions and Educational Strategies in Science and Mathematics,* vol. 3, pp. 268–277. Ithaca: Cornell University Press.

Kepler, J. (1610/2000). *Optics: Paralipomena to Witelo and the Optical Part of Astronomy*. Santa Fe, New Mexico: Green Lion Press.

Kierkegaard, S. (1952). *Gesammelte Werke*. Dusseldorf, Köln, p. 1951.

Kipnis, N. (1991). *History of the Principle of Interference of Light*. Basel: Birkhauser Verlag.

Kipnis, N. (1992). *Rediscovering Optics*. Minneapolis: BENA Press.

Kipnis, N. (1996). The historical-investigative approach to teaching science. *Science & Education,* 5, 277–292.

Kipnis, N. (1998). A history of science approach to the nature of science: Learning science by rediscovering it. In W.F. McComas (ed.), *The nature of science in science education* (pp. 177–196). Dordrecht: Kluwer Academic Publisher.

Kipnis, N. (2010). Scientific errors. *Science & Education,* 10, 33–49.

Kofka, K. (1925). *The Growth of Mind* (p. 44). New York: Harcourt, Brace & Co.

Koyré, A. (1943). In G. Holton (1952). *Introduction to Concepts and Theories in Physical Science* (pp. 21–22). Reading, Mass.: Addison-Wesley.

Kuhn, T. (1962/1970). *The structure of the scientific revolution.* Chicago: The University of Chicago Press.

Kuhn, T. (1977). The Function of Measurement in Modern Physical Science. In *The Essential Tension*. Chicago: The University of Chicago Press.

Kwan, A., Dudley, J., & Lantz, E. (2002). Who really discovered Snell's law? *Physics World*, April 2002, 64.

Lagrange, J.L. (1788/1938). *Mechanique Analitique*. Paris, Desaint.

La Rosa, C., Mayer, M., Patrizi, P., & Vicentini-Missoni, M. (1984). Commonsense knowledge in optics: Preliminary results of an investigation into the properties of light. *European Journal of Science Education*, 6(4), 387–397.

Lakatos, I. (1978). *The Methodology of Scientific Research Programs*. Cambridge, UK: Cambridge University Press.

Langley, D., Ronen, M., & Eylon, B. (1997). Light propagation and visual patterns: Pre-instruction learners' conceptions. *Journal of Research in Science Teaching*, 34, 399–424.

Latour, B. (1987). *Science in Action*. Cambridge, Mass: Harvard University Press.

Lauginie, P. (2012). How did Light Acquire a Velocity? *Science & Education* (in press).

Levrini, O., Bertozzi, E., Gagliardi, M., Grimellini-Tomasini, N., Pecori, B., Tasquier, G., & Galili, I. (2014). Meeting the Discipline-Culture Framework of Physics Knowledge: An Experiment in Italian Secondary Schools. *Science & Education* (in press).

Lindberg, D.C. (1968). The theory of pinhole images from antiquity to the thirteenth century. *Archive for History of Exact Sciences*, 5(2), 154–176.

Lindberg, D.C. (1976). *Theories of vision form Al-Kindi to Kepler*. Chicago: The University of Chicago Press.

Lindberg, D.C. (1978). *The Science of Optics*. In D.C. Lindberg (ed.), *Science in the Middle Ages* (pp. 338–368). Chicago: The University of Chicago Press.

Lindberg, D.C. (1985). Laying the Foundations of Geometrical Optics: Maurolico, Kepler, and the Medieval Tradition. In *Discourse of Light from the Middle Ages to the Enlightenment* (pp. 1–65). Los Angeles: The University of California Los Angeles.

Lindberg, D.C. (1992). *The Beginnings of Western Science: the European Scientific Tradition in Philosophical, Religious, and Institutional Context, 600 B.C. to A.D. 1450*. Chicago: The University of Chicago Press.

Lindberg, D.C. (2002). The Western reception of Arabic Optics. In R. Rashed (ed.). *Encyclopedia of the History of Arabic Science* (vol. 2, pp. 363–371). Florence, KY: Routledge.

Linn, M.C., Clark, D., & Slotta, J.D. (2003). WISE Design for Knowledge Integration. *Science Education* 87, 517–538.

Lipson, H. (1968). *The Great Experiments in Physics*. Edinburgh: Oliver & Boyd.

Losee, J. (2001). *A Historical Introduction to the Philosophy of Science*. New York: Oxford University Press.

Lotman, Yu. (2001). *Universe of Mind. A Semiotic Theory of Culture*. Great Britain: I.B. Tauris & Co.

Mach, E. (1913/1926). *The Principles of Physical Optics. An Historical and Philosophical Treatment*. New York: Dover.

Magie, W.F. (1969). Light. In *A Source Book in Physics* (pp. 265–386). Cambridge, Mass: Harvard University Press.

Marton, F., Runesson, U., & Tsui, A.B.M. (2004). The Space of Learning. In F. Marton & A.B.M. Tsui (eds.), *Classroom Discourse and the Space of Learning* (pp. 3–40). Mahwah, NJ: Lawrence Erlbaum.

Mason, S.F. (1962). *A History of the Sciences*. New York: Collier Books, Macmillan.

Matthews, M.R. (1989). A role for history and philosophy of science in science teaching. *Interchange*, 20, 3–15.

Matthews, M.R. (1994). *Science Teaching. The Role of History, Philosophy and Science,* New York: Routledge.

Matthews, M.R. (2000). *Time for Science Education: How Teaching the History and Philosophy of Pendulum Motion Can Contribute to Science Literacy*. New York: Plenum Press.

McCloskey, M. (1983). Intuitive physics. *Scientific American*, 248(4), 114–122 and McCloskey, M. (1983), Naive Theories of Motion. In D. Genter & A.L. Stevens (eds.), *Mental models* (pp. 299–324). Hillsdale, NJ: Erlbaum.

McComas, W.F. (1998). *The nature of science in science education*. Dordrecht: Kluwer.

McComas, W. (2005). Teaching the Nature of Science: What Illustrations and Examples Exist in Popular Books on the Subject? Paper presented at the IHPST Conference, Leeds (UK) July 15–18, 2005. http://www.ihpst2005.leeds.ac.uk/papers/McComas.pdf. Accessed 1 August, 2012.

McComas, W.F. (2008). Seeking historical examples to illustrate key aspects of the nature of science. *Science & Education,* 17(2–3), 249–263.

Middleton, W.E.K. (1961). Archimedes, Kircher, Buffon, and the Burning-Mirrors. *Isis*, 52(4), 533–543.

Middleton, W.E.K. (1963). Note on the invention of photometry. *American Journal of Physics,* 31(2), 177–181.

Migdal, A.B. (1985). Niels Bohr and Quantum Physics. Soviet Physics Uspekhi, 28(10), 910–934, doi:10.1070/PU1985v028n10ABEH003951

Migdal, A.B. (1990). Physics and Philosophy. *Voprosi Philosophii,* 1, 5–32 (In Russian).

Mihas, P. (2008). Developing Ideas of Refraction, Lenses and Rainbow Through the Use of Historical Resources, *Science & Education,* 17(7), 751–777.

Mihas, P. & Andreadis, P. (2005). A historical approach to the teaching of the linear propagation of light, shadows and pinhole cameras. *Science & Education,* 14, 675–697.

Minstrell, J. (1992). Facets of students' knowledge and relevant instruction. In R. Duit, F. Goldberg & H. Niedderer (eds.), *Research in physics learning: Theoretical issues and empirical studies* (pp. 110–128). Kiel: IPN.

Monk, M. & Osborne, J. (1997). Placing the history and philosophy of science on the curriculum: A Model for the development of pedagogy. *Science Education,* 81(4), 405–424.

Nersessian, N. (1989). Conceptual Change in Science and in Science Education. *Synthese,* 8(1), 163–183.

Newton, I. (1671/1974). The new theory about light & colours. In H.S. Thyer (ed.), *Newton's philosophy of nature.* New York: Hafner Press.

Newton, I. (1704/1952). *Opticks.* New York: Dover.

Newton, I. (1727/1999). *Mathematical principles of natural philosophy.* University of California Press, Berkeley.

Newton, I. (1729/1999). *Philosophiae Naturalis Principia Mathematica.* Third Edition. Berkeley: University of California Press.

Olivieri, G., Torosantucci, G., & Vicentini, M. (1988). Colored shadows. *International Journal of Science of Education,* 10(5), 561–569.

Osborne, J.F., Black, P., Meadows, J., & Smith, M. (1993). Young children's (7–11) ideas about light and their development. *International Journal of Science Education,* 15, 89–93.

Panofsky, W.K.H. & Phillips, M. (1962). *Classical Electricity and Magnetism* (p. 282). Reading, Mass: Addison-Wesley.

Park, D. (1997). *The fire within the eye. A historical essay on the nature and meaning of light.* Princeton: Princeton University Press.

Pedersen, O. & Phil, M. (1974). *Early physics and astronomy.* London: Macdonald & Janes.

Pedrotti, L.S. & Pedrotti, F.L. (1998). *Optics and Vision* (pp. 1–11). Upper Saddle River, New Jersey: Prentice-Hall.

Penrose, R. (1997). *The Large the Small and the Human Mind.* Cambridge, UK: Cambridge University Press.

Perales, F.J., Nievas, F., & Cervantes, A. (1989). Misconceptions on Geometric Optics and their association with relevant educational variables. *International Journal of Science Education,* 11, 273–286.

Popper, K.R. (1965). *Conjectures and Refutations: The Growth of Scientific Knowledge.* New York: Harper Torchbooks.

Popper, K.R. (1978). Three worlds. The Tanner Lecture on Human Values. The University of Michigan. http://www.tannerlectures.utah.edu/lectures/documents/popper80.pdf. Accessed on August 1, 2012.

Popper, K.R. (1981). *Objective knowledge.* Oxford: Clarendon Press.

Ptolemy, C. (1940/1966). Refraction. In M.R. Cohen & I.E. Drabkin (eds.), *A Source Book in Greek Science.* New York: McGraw-Hill Book, pp. 271–281.

Raftopoulos, A., Kalyfommatou, N., & Constantinou, C.P. (2005). The Properties and the Nature of Light: The Study of Newton's Work and the Teaching of Optics. *Science & Education,* 14(6), 649–673.

Ramadas, J. & Driver, R. (1989). *Aspects of Secondary Students' Ideas about Light.* Leeds: University of Leeds, Center for Studies in Science and Mathematics Education.

Rashed, R. (2002). Geometrical optics. In R. Rashed (ed.), *Encyclopedia of the History of Arabic Science* (vol. 2, pp. 299–324). Florence, KY: Routledge.

Reiner, M. (1992). Patterns of thought on light and underlying commitments. In R. Duit, F. Goldberg & H. Niedderer (eds.), *Research in Physics Learning: Theoretical Issues and Empirical Studies* (pp. 99–109). Kiel: IPN.

Reiner, M., Pea, R.D., & Shulman, D.J. (1995). Impact of Simulator-Based Instruction on Diagramming in Geometrical Optics by Introductory Physics Students. *Journal of Science Education and Technology*, 4(3), 199–226.

Rice, K. & Feher, E. (1987). Pinholes and images: children's conceptions of light and vision. I. *Science Education*, 71(6), 629–639.

Ronchi, V. (1970). *The Nature of Light: An Historical Survey*. London: Heinemann.

Ronchi, V. (1991). *Optics. The science of vision*. New York: Dover.

Ronen, M. & Eylon, B. (1993). To see or not to see: the eye in geometrical optics: when and how. *Physics Education,* 28, 52–59.

Ross, J. (2008). *Fermat's Complete Correspondence on Light*. Dynamis. http://science.larouchepac.com/fermat/

Russell, G.A. (2002). The Emergence of the Physiological Optics. In R. Rashed (ed.). *Encyclopedia of the History of Arabic Science* (vol. 2, pp. 325–350). Florence, KY: Routledge.

Russo, L. (2004). *The Forgotten Revolution: How Science Was Born in 300 BC and Why It Had to Be Reborn*. Berlin: Springer Verlag.

Sabra, I.A. (1981). *Theories of Light. From Descartes to Newton*. Cambridge, UK: Cambridge University Press.

Sabra, I.A. (1989). The optics of ibn Al-Haytham. Books I–III, on direct vision. Translation with commentary (vol. I). London: The Warburg Institute, University of London.

Sabra, I.A. (2003). Ibn Al- Haytham's revolutionary project in optics: The achievement and the obstacle. In J.P. Hodgedijk & I. Sabra (eds.), *The enterprise of science in Islam, new perspectives* (pp. 85–118). Cambridge, Mass.: The MIT Press.

Sambursky, S. (1958). Philoponus' interpretation of Aristotle's theory of light. *Osiris*, 13, 114–126.

Sambursky, S. (1959). *Physics of the stoics*. London: Routledge & Kegan.

Saxena, A.B. (1991). The understanding of the properties of light by students in India. *International Journal of Science Education*, 13, 283–290.

Schnepps, M.H. & Sadler, P.M. (1989). *A private universe – preconceptions that block learning* [Videotape]. Cambridge, Mass.: Harvard University/Smithsonian Institution.

Schwab, J.J. (1964). Problems, Topics, and Issues. In S. Elam (ed.), *Education and the Structure of Knowledge* (pp. 4–47). Chicago: Rand McNally.

Schwab, J.J. (1978). Education and the Structure of the Discipline. In I. Westbury & N.J. Wilkof (eds.), *Science, Curriculum, and Liberal Education* (pp. 229–272). Chicago: The University of Chicago Press.

Segel, G. & Cosgrove, M. (1993). The sun is sleeping now: early learning about light and shadows. *Research in Science Education,* 23(2), 276–285.

Selley, N. J. (1996a). Children's ideas on light and vision. *International Journal of Science Education*, 18(6), 713–723.

Selley, N. J. (1996b). Towards a phenomenography of light and vision. *International Journal of Science Education*, 18(8), 836–845.

Seroglou, F. & Koumaras, P. (2001). The contribution of the history of physics in physics education: A review. *Science & Education*, 10(1–2), 153–172.

Shapiro, A.E. (1973). Kinematic Optics: A Study of the Wave Theory of Light in the Seventeenth Century, *Archive for History of Exact Sciences,* 11, 134–266.

Shapiro, A.E. (1993). *Fits, Passions, and Paroxysms*. Cambridge, UK: Cambridge University Press.

Shapiro, B.L. (1994). *What Children Bring to Light*. New York: Teachers' College Press.

Singh, A. & Butler, P. (1990). Refraction: Conceptions and Knowledge Structure. *International Journal of Science Education,* 12(4), 429–442.

Smith, A.M. (1982). Ptolemy's Search for a Law of Refraction: A Case-Study in the Classical Methodology of "Saving the Appearances" and its Limitations. *Archive for History of Exact Sciences,* 26 (3), 221–240.

Smith, A.M. (1996). Ptolemy's Theory of Visual Perception: An English Translation of the Optics with Introduction and Commentary. *Transactions of the American Philosophical Society,* 86(2), 1–300.

Smith, A.M. (1999). Ptolemy and the Foundations of Ancient Mathematical Optics: A Source Based Guided Study, *Transactions of the American Philosophical Society,* 89(3), 1–172.

Stead, B.F. & Osborne, R.J. (1980). Exploring science students' conception of light. *Australian Science Teaching Journal,* 26(1), 84–90.

Steneck, N.H. (1976). *Science and Creation in the Middle Ages.* South Bend: University of Notre Dame Press.

Taylor, G.I. (1909). Interference fringes with feeble light. *Proceedings of the Cambridge Philosophical Society,* 15, 114–115.

Taylor, L.W. (1941). *Physics. The Pioneer Science.* New York: Dover.

Tseitlin, M. & Galili I. (2005). Teaching physics in looking for its self: from a physics-discipline to a physics-culture. *Science & Education,* 14 (3–5), 235–261.

Tyndal, J. (1877). *Six Lectures on Light.* New York: Appleton & Co.

Wandersee, J.H. (1986). Can the History of Science Help Science Educators Anticipate Students' Misconceptions? *Journal of Research in Science Teaching,* 23(7), 581–597.

Watts, D. M. (1985). Students' conceptions of light: A case study. *Physics Education,* 20(2), 183–187.

Weinberg, S. (2001). *Facing Up – Science and Its Cultural Adversaries.* Cambridge, Mass.: Harvard University Press.

Westfall, R.S. (1962). The Development of Newton's Theory of Color. *Isis,* 53(3), 339–358.

Westfall, R.S. (1989). *Mechanical Science in the Construction of Modern Science* (pp. 50–64). Cambridge, UK: Cambridge University Press.

Whittaker, E. (1960). *A History of the Theories of Aether and Electricity.* New York: Harper.

Wolf, A. (1968). *A History of Science, Technology and Philosophy in the 16th & 17th Centuries.* Gloucester, Mass.: Smith.

Wolpert, L. (1994). *The Unnatural Nature of Science.* Cambridge, Mass.: Harvard University Press.

Young, T. (1804). Experiments and calculations relative to physical optics (The 1803 Bakerian Lecture) *Philosophical Transactions of the Royal Society of London,* 94, 1–16.

Young, T. (1807). *A Course of Lectures on Natural Philosophy and the Mechanical Arts.* London: J. Johnson.

Igal Galili is professor of science education at the Amos de-Shalit Science Teaching Center in the Faculty of Mathematics and Natural Sciences of the Hebrew University of Jerusalem, Israel. His Ph.D. is in theoretical physics from Racah Institute of Physics at the Hebrew University. His research interests include the structure of students' knowledge of physics, the structure and nature of physics knowledge, and its representation for learning. He argues for the framework of discipline-culture for the fundamental theories in physics as creating cultural content knowledge. That structure makes explicit the essential role of the history and philosophy of science in providing a necessary foundation for meaningful learning and understating of physics. Among his products are an introductory course on optics using the cultural approach, *Fundamentals of Physics* and *Modern Physics* for school students in Israel. Several historical excurses on physical concepts produced in the European project HIPST were published in the collection *The Pleasure of Understanding*.

Chapter 5
Teaching and Learning Electricity: The Relations Between Macroscopic Level Observations and Microscopic Level Theories

Jenaro Guisasola

One of the areas of research in physics education boasting the highest number of works over the last three decades is electricity. Numerous studies on the teaching and learning of electricity have been conducted (Duit 2009). Two reasons for this high number of studies in the area of electricity will be discussed next.

First, electrical phenomena and its properties are an important part of physics instruction at many different levels. Students learn about the idea of charge and electrical circuits in elementary school and gradually integrate more complex ideas to interpret electrical phenomena. Studying the models needed to interpret electromagnetic phenomena is a productive area: it provides a solid background to understand issues that range from the electromagnetic nature of matter to the foundation of contemporary technology. The structure of the electromagnetic nature of matter is both beautiful and useful.

Furthermore, electromagnetic theories provide a good context for teaching scientific reasoning skills such as model-building and model-drawing relations between macroscopic level description phenomena and microscopic level theories. As research shows, many times learners need to have the ability to reason holistically. Psillos (1998) shows the necessity of global reasoning for analysing the components of the electric circuit. Viennot (2001) explains that overcoming the "causal reasoning" and/or the "reasoning based on the formula" is a necessary condition for understanding electric circuits and other areas of electricity.

Second, electricity is an area of physics that students find significantly more difficult to understand than mechanics. Comprehension levels for electricity concepts are highly idiosyncratic. Moreover, literature shows confusion between electricity concepts and the terminology used in everyday life (e.g. electricity energy, voltage, electric power). This comes as no surprise due to the complexity of the concepts involved, but it is more disconcerting that this lack of understanding remains almost unchanged by teaching (McDermott and Shaffer 1992; Wandersee et al. 1994).

J. Guisasola (✉)
Department of Applied Physics, University of the Basque Country, San Sebastián, Spain
e-mail: jenaro.guisasola@ehu.es

M.R. Matthews (ed.), *International Handbook of Research in History,*
Philosophy and Science Teaching, DOI 10.1007/978-94-007-7654-8_5,
© Springer Science+Business Media Dordrecht 2014

Research carried out on new proposals to improve the situation offers unequal results (Mulhall et al. 2001). Some studies present specific progress whilst others do not. As a consequence two main problems can be identified: (1) students' prior knowledge interacts strongly with the teaching strategies used, producing a wide variety of learning achievements (Saglam and Millar 2005), and (2) teaching strategies have to combine the macroscopic level electric phenomena and the microscopic level theory (Chabay and Sherwood 2006; Young and Freedman 2008).

5.1 Issues Emerging from Physics Education Research

The main line of research on teaching and learning electricity over the last few decades has focused on studying students' alternative conceptions (Driver et al. 1994, Wandersee et al. 1994). In the case of DC circuits and electrostatics, research suggests a consensus about the main learning difficulties.

Current thinking suggests that prior knowledge and students' conceptions interfere and affect their learning in new contexts (Ausubel 1978; Duit and Treagust 1998). These assumptions set up the students' scientific skills, and as Etkina and colleagues (2006) state, "… these (scientific skills) are not automatic skills, but are instead processes that students need to use reflectively and critically" (Vosniadou 2002, p. 1). Conclusions without evidence or employing a single strategy, which generally involves specific and direct application of a "recipe", are common occurrences (Guisasola et al. 2008; Viennot 2001).

Prior knowledge and students' conceptions interfere and affect their learning in new contexts (Ausubel 1978; Duit and Treagust 1998). These assumptions set up the students' scientific skills and, as Etkina and colleagues (2006) states "… these (scientific skills) are not automatic skills, but are instead processes that students need to use reflectively and critically" (p. 1). Certain forms of reasoning characteristic of everyday life such as making conclusions without evidence or employing a single strategy, which generally involves specific and direct application of a "recipe", appear frequently (Guisasola et al. 2008; Viennot 2001).

Alternative ideas about electromagnetic phenomena can arise from the academic context, as it requires elaborate knowledge far removed from daily life. As the physics contents within a teaching programme and the textbooks are part of the academic context, it is particularly necessary to carry out research on them and their effect on learning.

In order to present the main learning difficulties and alternative ideas detected by the research, they are grouped into different conceptual aspects from electricity. It does not follow a historical development of the research findings but criteria relating to the conceptual physics framework and its relation to the electricity curriculum. However, different research projects have shown that alternative conceptions are not random ideas, but they have some internal cohesion, structured in "alternative conceptual framework" (Oliva 1999; Watts and Taber 1996). The findings presented here in each subsection most often refer to related aspects, so their view of the whole concept should not be lost.

5.1.1 Students' Difficulties in Learning Electrostatic Phenomena and Electric Fields

University physics textbooks present electrostatics before DC circuits, different from secondary courses (ages 12–16) that start with DC circuits without explicitly analysing the electrostatic electrification phenomena or explaining a model for the electrical nature of matter (Stocklmayer and Treagust 1994).

Until the present, few studies have addressed learning difficulties in electrostatics. Problems exist in the failure to learn about the scientific models that are used to interpret basic electrostatic phenomena such as interactions between point charges, phenomena involving friction charging or electrical induction charging.

Galili (1995) shows that in Israel students aged 16–18 and future secondary teachers apply Newton's third law superficially (a difficulty observed in work on alternative conceptions in the field of mechanics) and have serious difficulties when analysing the polarisation of a metal under electrical interactions, difficulties that derive from alternative conceptions of mechanics.

According to Furio and colleagues (2004), Spanish students aged 17–18 demonstrate different alternative conceptions when interpreting electrification phenomena due to friction or induction. A majority of students consider electricity as a fluid composed of particles that can be transmitted through conductors. For electrical interaction to take place, the fluid must go from one body to another. Students in this category are not able to scientifically explain phenomena that involve remote actions, such as electrical induction and polarisation of matter. Only a minority of students use the Newtonian model of action at a distance to explain the phenomenon of remote electrification and, consequently, to give a scientific explanation of induction and polarisation phenomena. In addition, in a study with Korean middle school and college students on their ideas regarding electrostatic induction, Park and colleagues (2001) found that many students show a lack of understanding about dielectric polarisation, even though this concept is basic to everyday experiences, such as the attraction of a piece of paper through the rubbing of a comb. Moreover, some middle school students misunderstood the role of an electroscope and were not sure about which material was a conductor or a non-conductor.

In the study of electrostatics, electric field and electric potential are two important concepts. Research into students' difficulties shows that the vast majority do not have a scientific grasp and make incorrect applications of these concepts, stemming from a badly understood recall of the information received during instruction.

In a psychogenetic study of the ideas that influence the concept of field, Nardi and Carvalho (1990) interviewed 45 Brazilian students on four cases of electrostatic interaction phenomena, one of which consists of an electrostatic pendulum attracting a positively charged rod. They show that the students' answers can be classified into three levels: (a) students that do not understand action at a distance and do not manage to relate the results of the experiment to a single cause, (b) students that attribute the action to the existence of forces at all points around a generating force that depends on the distance and that consider the field as represented by a vector magnitude with direction and meaning and (c) students who recognise that the field is a vector and discuss it correctly as such in different contexts and also use a scientific language that coincides with the theory taught at secondary level (ages 16–18). The authors suggest similarities between the classification obtained and the historical development of the field theory. They propose using historical development of field ideas both to identify issues in learning as well as a guide to helping students move from one interpretation to another.

Törnkvist and colleagues (1993) administered a questionnaire related to the electric field and its graphic and mathematical representations to over 500 university students. They found that 85 % of the students believed it possible that two field lines could intersect, 49 % believe that field lines can form an angle and 29 % consider that electrical field lines can be circular. The authors suggest that these poor results are a consequence of students' "naive" conceptions of the electric field which are based on intuition rather than on what has been explained in class. According to the authors, students tend to treat the field lines like isolated entities in a Euclidian field rather than as a set of curves that represent a physical property of space that has a mathematical representation. This "naive" way of reasoning and representing the electric field is also supported by the results found in the study by Galili (1995), previously mentioned. Pocovi and Finley (2002) studied the conceptions of 39 college students regarding field lines that support the aforementioned conclusions. They found that many students considered field lines as material entities that are capable of transporting charges and impose the path that the charges must follow.

Studies by Viennot and Rainson (1992) and Rainson and colleagues (1994), involving over 100 university students from France and Algeria, showed that the vast majority had difficulties in applying the superposition principle and in interpreting the electric field in a material medium. The authors believe that these difficulties are due mainly to a limited comprehension of the mathematical equations and mistaken reasoning relating to the causality of the phenomena. A high percentage of students related the fact that the charges did not move or the insulating nature of the matter with the fact that an electric field did not exist. The results suggest that students need to imagine an effect (movement of charges) to accept a cause. Another detected difficulty is linked to a causal interpretation of the formula. The study was carried out in relation to the expression of electric field around a conductor given by Coulomb's theory. The authors partly attribute the students' learning difficulties to a deficient and confusing pedagogic treatment of electric fields. In particular, they insist that the superposition principle for electrical interaction is a long way from being clear for the students and that it is useful to work on it in static situations before analysing electric circuits (Viennot and Rainson 1999). In addition, teaching must highlight the explanation of the causal aspects of both electric field and electric force. The authors propose that this focus should be unified both in electrostatics and electrical circuits.

Furio and Guisasola (1998) investigated Spanish students' difficulties on understanding the concept of electric field at high school and university level. They based their work on the hypothesis that historical problems in the development of the electric field theory relate to students' difficulties in understanding this concept. They found that most students do not use correctly the field concept and instead reason based on the Newtonian model of "action at a distance". These difficulties can be due to a linear, accumulative presentation of electrostatics in traditional teaching, teaching that does not consider qualitative leaps within the development of the theory. In a later study, Saarelainen et al. (2007) show that some students' difficulties are also related to the mathematical methods required and to the meaning of the field concept in electricity and in magnetism.

Kenosen and colleagues (2011) found that students do not include the vector nature of field quantities in their reasoning. In addition, students described the direction of the force interaction instead of the electric field. Some of the students were unwilling to apply the field concept in their reasoning. Authors state that this problem may result from the well-known difficulty that involves shifting students' understanding from the particle-based Coulombian conceptual profile to the field-based Maxwellian one.[1]

5.1.2 Students' Difficulties in Learning Electrical Potential and Electrical Capacitance

One of the most researched concepts in teaching electricity is electrical potential. The majority of these studies have focused on learning this concept in the context of electrical circuits, but from the 1990s onwards, more works emerged analysing this concept in the context of electrostatics and how it relates to electrical circuits. Eylon and Ganiel (1990) found that electrical potential is one of the concepts giving students the greatest learning difficulties when interpreting electrical circuits. They attribute this difficulty to the fact that the electrical circuits are described in terms of "macroscopic" variables (current, resistance, voltage measured by the voltmeter) whilst the explanations use models (charges, fields, potential). Such models relate concepts studied in electrostatics with those coming into play when analysing electrical circuits. The study shows that it would be a good idea to explicitly relate concepts such as electric field and electric potential studied in electrostatics with those in DC circuits.

It is well documented that secondary and first year university students are not capable of establishing relations between the concept of potential in electrostatics and their application in electrical circuits (Benseghir and Closset 1996; Cohen

---

[1] Furio and Guisasola (1998), Rainson et al. (1994), and Viennot and Rainson (1992).

et al. 1983). This lack of relation means that the concept of electrical potential remains vague and is only used as a calculational convenience. Many students use the concepts of electrical potential and potential difference without a consistent meaning in an explanatory model (Shaffer and McDermott 1992; Shipstone et al. 1988), and, based on an incorrect use of Ohm's law, consider that if there is no current between two points in a circuit, there is no potential difference. This suggests that students think that the potential difference is a consequence of the flow of charges rather than its cause (Periago and Bohigas 2005; Steinberg 1992). In addition, Cohen et al. (1983) found that the battery of a DC circuit is conceived by the students as a device that supplies "constant current" rather than one that constantly maintains the potential difference between its poles. Many students confuse the concept of potential difference with the quantity of electrical charge (Thacker et al. 1999).

Guruswamy and colleagues (1997) show that students analyse the passing of charges between two conductors joined by a conducting wire looking at the quantity of charge in each conductor and not taking into account the potential difference between them. Guisasola and colleagues (2002) carried out a study on how freshmen students learn the concept of electrical capacity. The majority of students do not grasp the concept of potential of a charged body and identify its capacity with the quantity of charges that it accumulates. This prevents them from giving the right explanation of the phenomena such as bodies being charged by induction. In a study with secondary and university students, Benseghir and Closset (1996) show that for some students the potential difference between the terminals on a battery and the current circulation are not related: the potential difference is considered strictly numerically as a characteristic of the battery, and the electrical current is analysed from an electrostatic point of view (attraction between charges, different sign of the charges on either end, etc.). A significant number of students only consider that there exists a potential difference between the points of the circuit whenever the difference in signs is perceived (positive pole and negative pole) or when there is a variation of the quantity of charges between the points (within the resistor there is no variation in the quantity of charge). Students consider the potential difference as "abstract" and prefer a much more accessible concept such as electrical charge. It seems that when the students do not attribute meaning to these concepts, they take refuge in their operative definitions and base their reasoning on formulas with no meaning (Viennot 2001). In summary, the concept of potential is frequently presented in a purely operative way, and students are asked to make a leap through formal maths. This is where many of them fail.

High school and university students have difficulties when learning about electrical potential, due to the absence of analysis of electric circuits and its energetic balance. Most 3rd year physics students still do not clearly understand the usefulness of the concepts of potential difference and emf (Guisasola and Montero 2010). This shows the need to present potential difference and electromotive force to show that these measure different kinds of actions produced by radically different causes (Jimenez and Fernández 1998, Roche 1987; Varney and Fisher 1980).

5.1.3 Students' Difficulties in Learning DC Circuits

Research into teaching and learning about DC circuits points to students presented with a typical model for electric current as one of the charges moving between two points at a different electrical potential (charge flow model). Textbooks do not agree on the type of charges (positives or negatives) that are involved in the current. Charge flow presents serious difficulties for students. Closset (1983) showed that many secondary and university students analyse the circuit using "sequential" reasoning. Those students think that there are different entities ("current", "electrons" or "electricity") associated with intensity and tension that come out of the battery and are more or less affected as they pass through each element of the circuit, for example, "the current is used in the resistor" or "the current is spent in the bulb" without reference to what might have happened to the "current" before the element under analysis. They also do not consider how the "current" returns to the other pole of the battery. In addition, other studies found that secondary students believe that the current is spent as it passes through a bulb or that the current provided by the battery is independent of the topology of the circuit. Secondary students have difficulties interpreting the behaviour of resistors connected in series and in parallel in a complete DC circuit. The students find it difficult to accept that when the number of resistors in parallel increases, the total resistance decreases. They also fail to understand the relations between current and resistors and resistors and potential difference (Liegeois and Mullet 2002).

Concepts of electrical potential and potential difference are frequently confused with current intensity or energy. These concepts are taken to represent the "strength" of a battery. In addition, students frequently do not understand that the potential difference between two points in a circuit depends on its topology. Smith and van Kampen (2011) investigated pre-service science teachers' qualitative understanding of circuits consisting of multiple batteries in single and multiple loops. They found that most students were unable to explain the effects of adding batteries in single and multiple loops, as they tended to use reasoning based on current and resistance instead as on voltage, that thinking of the battery as a source of constant current resurfaced in this new context and that answers given were inconsistent with current conservation.

Borges and Gilbert (1999) study of explanatory models of electrical circuits showed that secondary students present alternative models such as "electricity as flow" and "electricity as opposing currents" where the electrical charges that make up the current are not taken individually. These models are barely concerned with the nature of electricity and essentially descriptive. Both models are very limited in terms of predicting the behaviour of the electrical current in the circuit. Borges and Gilbert (1999) show that we find more complex explanatory models in students in their last years of secondary school and at university level. For example, they think of electricity as "moving charges" or as "field". These models are capable of explaining some phenomena related to electrical currents, such as a relationship between the intensity of the current and the battery's potential difference. Greca and

Moreira (1997) show that the students' models for explaining electricity become more complex over the years of instruction but for most continue to be far from the scientific model.[2,3]

5.1.4 Summary of Research Findings on Students' Difficulties

Most of the mentioned difficulties seem resistant to traditional teaching of electrical circuits. Therefore, over the past decades, a great effort has been devoted to understanding students' conceptions before and after instruction. As a result, we have today some conceptual understanding on key electrical concepts:

– Students find it difficult to interpret electrical induction and polarisation phenomena using Coulomb's explanatory model for action at a distance.
– Most students from the last years of secondary school and university do not understand the ontological difference between "action at a distance" and the "field model". This leads to confuse the concept of field with that of the force that is exerted on the electrical charges and therefore not taking into account the medium where the interaction takes place.
– Students use inappropriate causal analysis to interpret equations such as the superposition principle.
– Many students confuse the electric field with the imaginary field lines that are used to represent it. Students state that the electric field exists only along the field lines and do not think of it as existing in every point in space.
– Students have a confused meaning of the concept of electrical potential and potential difference, leading them to avoid to use these concepts to analyse the movement of charges in a conducting wire.
– Students attribute the passing of electrical current to the difference in quantity of charge between the ends of a conductor.
– Students "take refuge" in operative definitions ("formulaic solutions") to analyse electrical phenomena. They usually base their reasoning on a literal description of the "formula" or an incorrect causal analysis of it.
– Most students think that electric potential as defined in electrostatics is different from the electric potential as defined in electric circuits.
– Most students do not relate macroscopic phenomena (electrical attractions and repulsions, electrical current, voltage of battery, etc.) with the microscopic concepts which build the explanatory theory (field, potential difference, polarisation, etc.).
– Most students in their analysis of simple electric circuits think that current is used up in a resistance; drifting electrons push each other through a wire just as water molecules push each other through a pipe; they confuse the Kirchhoff

---

[2] Barbas and Psillos (1997), Cohen et al. (1983), Dupin and Joshua (1987), McDermott and Shafer (1992), and Shipstone et al. (1988).

[3] Duit and von Rhöneck (1998), Psillos et al. (1988), and Testa et al. (2006).

loop rule and Ohm's law (although Kirchhoff's law is a much more general principle), assuming that electromotive force and potential difference are synonymous.

As a result of these efforts to identify students' learning difficulties, greater attention should be given to conceptual understanding in physics programmes and textbooks (see, e.g. Engelhard and Beichner 2004; Halloun and Hestanes 1985; Maloney et al. 2001; www.ncsu.edu/per/testinfo.html).

More research is needed on students' conceptions in other areas such as capacitance and its relations with electrical potential, the movement of charges and electrical potential in a more complex electrical system.

5.1.5 Possible Reasons Underlying Students' Alternative Conceptions

Traditional teaching does not appear to improve students' lack of understanding of the electric field. There is a wide gap between the student's thoughts and this abstract concept (Furio and Guisasola 1998; Viennot and Rainson 1999). Some other reasons frequently suggested are:

– Poor knowledge of the mathematical tools demanded by the operative definition and its application (vectors, derivates and integrals)
– Poor knowledge of the basic concepts in the area of mechanics (force, work and energy)

In teachers' "spontaneous thinking" (Hewson and Hewson 1988), students are often blamed for these problems. Most teachers refer only to students' deficiencies to account for the general failure in learning, but the sort of teaching responsible should also be considered. Research into students' ideas and ways of reasoning identifies subject matter that must be better taught so as to improve understanding (Furio et al. 2003; Viennot and Rainson 1999).

Teachers' conceptions show a wide range of viewpoints concerning the teaching of DC circuits. Some viewpoints are consistent with alternative views of the students themselves like "straightforward" ideas on how circuits operate (Gunstone et al. 2009). In a study on the metaphors that experts and amateurs use when explaining electricity, Stocklmayer and Treagust (1996) found that "the teachers had a mechanical model which gave rise to images of electrons as small balls moving along tunnel-like wires" (ibid., p. 171). This conception contrasts with the experts' mental image that "was more global and holistic than the mechanical electron view. Essentially, these practitioners were concerned with the circuit as a whole" (ibid. p. 174). On the other hand, teachers frequently use scientific vocabulary to refer to how electrical phenomena work, but concepts such as electrical potential, potential difference or electrical field are avoided or misunderstood (Mulhall et al. 2001).

Many textbook presentations of these concepts are dominated by mathematical instrumentalism and simplification in justification. For decades (Moreau and Ryan 1985), studies on presentation of the concepts of electricity in textbooks indicate that many books do not pay attention or emphasise the connection between electrostatics and electrical circuits. For example, the fact that electrical potential in circuits is exactly the same as in electrostatics is not highlighted, assuming that the students make the connection themselves. Heald (1984) claimed that there is a discontinuity in the presentation of the topics of electrostatics and DC circuits in introductory physics courses. In electrostatics, the analysis focuses explicitly on the electrical charges in the bodies and on the electrical field and potential in space and matter. In the following chapter on DC circuits, attention focuses on batteries, resistors, conductors and condensers. In the 1990s, Stocklmayer and Treagust (1994) carried out a study on the ways that textbooks presented the concepts of electricity in the period 1891–1991 and found that despite the fact that the historical development of the electromagnetic theory made important qualitative jumps towards a modern understanding of electrical current in a circuit, there are few changes in this regard in the analysed texts. Most represent the electrical current as the movement of a fluid – a pre-Faraday image. In addition, Bagno and Eylon (1997) found that many textbooks present an electrical field as a force to be applied on the electrical charges; this idea might lead students to misunderstand this difficult and nonintuitive concept.

Researchers agree on students' scant learning concerning electricity; however, there is a lack of consensus about specific learning targets for electricity. For example, Shaffer and McDermott (1992) focused on electric current, Licht (1991), Psillos (1998) and Psillos and colleagues (1988), emphasised potential difference whilst Eylon and Ganiel (1990) and Sherwood and Chabay (1999) focused on a microscopic-oriented approach. The lack of consensus results from the vast number of aspects such as the nature of the models and analogies that are considered appropriate when teaching electricity, the nature of the very concepts that can be used at each level and the relationship between the world of phenomena at a macroscopic level and the explanatory theories at a microscopic level (Dupin and Joshua 1989; Härtel 1982). The nature of the models and analogies that each teacher chooses are intrinsically linked to the teacher's understanding of the concept (Duit and von Röneck 1998; Pintó 2005). Therefore, it is necessary to define the conceptual and methodological aims of the teaching sequences in electricity. A careful reflection is required in order to justify, from a theoretical framework of physics teaching, "what" to teach and "how" to go about it. Both aspects are interrelated and require conceptual and epistemological analysis even in elementary aspects. Contributions to these problems from history and philosophy of science and the current theoretical framework of physics are reviewed in the next section.

5.2 The Contribution of History and Epistemology of Science to Teaching Electricity

Scientific concepts and theories do not emerge miraculously but are the result of an arduous process of problem solving and a rigorous testing of initial hypotheses (Nersseian 1995). In science, dynamic change and alteration are the rule rather than the exception (Thackray 1980). Quoting Kuhn (1984): "I was drawn … to history of science by a totally unanticipated fascination with the reconstruction of old scientific ideas and of the processes by which they were transformed to more recent ones" (ibid. p. 31). Knowing how explanatory ideas lead towards the current scientific model can provide important information when setting learning targets and selecting knowledge that helps to design teaching sequences (Duschl 1994; Wandersee 1992). The history of science is a useful instrument when teaching sciences, specifically electricity, to identify problems encountered in building concepts and theories, which epistemological barriers had to be overcome and which ideas led to progress. Furthermore, history of science can show the social context where theories were developed and the technological repercussions that resulted from the acquired knowledge.

Current consensus states that understanding concepts and theories requires knowledge not only of the current state of understanding of a particular topic, but also of the way that knowledge has been developed and refined over time. Moreover, educational standards developed in the last decades (National Research Council 1996; Rocard et al. 2007) call for a presentation of concepts and theories involving not only a historical perspective but also a meaningful introduction of terms and an appropriate representation of the social and scientific context of the origin of the key ideas and solutions.

The structure of science, the nature of the scientific method and the validation of scientists' judgements are some of the areas in which history and philosophy of science can enrich the teaching of science. There are many arguments defending the inclusion of history of science in the curriculum, particularly its integration in teaching strategies. This section considers the history of science as a useful instrument for identifying problems in the construction of concepts and theories and for indicating the epistemological barriers that had to be overcome and the ideas that permitted progress to be made (Furio et al. 2003). By building on this information, teaching objectives can be drawn up that might help in designing teaching sequences that will significantly improve the teaching and learning of concepts and theories (Mäntylä 2011; Niaz 2008). Nevertheless, in order for this information to be useful in the design of a didactic sequence, it requires a historical and epistemological study to be carried out with "pedagogical intentionality" and knowledge of students' learning difficulties. A critical study of the history and epistemology of science (where history is seen as a source of solved problems leading to advances in scientific knowledge) is likely to show teachers and researchers qualitative leaps in the evolution of a concept. To consider these "discontinuities" between meanings of concepts and models may help to clarify, explain and explore physics concepts and understand students' learning difficulties.

The history of electricity shows the most important epistemological and ontological difficulties in the development of the theory of electricity that researchers had to overcome to arrive at today's conceptual framework of electricity. Because the concepts of electricity are abstract and quite remote from students' spontaneous ideas, the historical perspective can be important in terms of making decisions regarding teaching sequences and objectives. Conceptual changes in science could provide some insight for contemporary instruction of science. However, there are obvious differences in the reasoning processes of current students and past physicists to be taken into account in teaching.

5.3 The History of the Evolution of Theories About Electricity During the Eighteenth and Nineteenth Centuries

William Gilbert, partly compiling J. Cardan's ideas published in "De subtilitate" (1550), established a clear division between the effect of amber and magnetism in his book "De Magnete, Magneticisque corporibus, et de magno magnete tellure" (1600). With the use of the "versorium", Gilbert carried out the first classification of "electric" and "nonelectric" materials. Gilbert explained that these phenomena were due to "material nature" freed on rubbing "electric" bodies such as glass or amber. At that time, Gilbert's "effluvia" model was used to explain electric attraction between bodies charged by rubbing. It was also used in the classification of bodies into "electric" and "nonelectric" depending on whether they became charged when rubbed and electrical discharges in rarefied gases or induced "glows" (Whittaker 1987). This explanatory model of "electric effluvia" failed to give plausible explanations for new electric phenomena such as electric repulsion or electric transmission. After Gray's discovery of electrical movement, it was not possible to accept the effluvia were inseparably joined to the bodies from which they had flowed through rubbing. It had to be admitted that outflows had an independent existence, as it was possible that they were transferred from one body to another. Therefore, these effluvia were acknowledged under the name of "electric fluid" as one of the substances that made up the world. Du Fay's and Franklin's contributions, among others, were to confirm a model that described electricity as an electric fluid made up of extremely subtle particles. The "electric fluid" model did not explain why two bodies which lack fluid (negatively charged) repel each other and it also had some difficulty in explaining induction. Towards the final third of the eighteenth century, it was thought that quantitative foundations were needed to advance the study of electricity. Thus, researchers such as Cavendish, Priestley and Coulomb looked for a theory similar to gravitation, under the clear influence of Newtonian mechanics (Conant et al. 1962; Harman 1982).[4,5]

---

[4] Duschl (2000), Matthews (1994), McComas et al. (2000), Rudge and Home (2004), and Wandersee (1992).

[5] Clough and Olson (2004), Izquierdo and Aduriz-Bravo (2003), Seroglou et al. (1998), and Solomon (2002).

The new model that emerged at the beginning of nineteenth century is coherent with Newton's physics, in the sense that it introduces notion of "action at a distance" forces, which operate instantaneously between charged bodies. The interactions are central forces, calculable by means of Coulomb's law. With the law on the conservation of charge and Coulomb's law on the attraction of charged bodies, electricity was raised to the level of "modern science". The result was that the "action at a distance" theories became almost the only focus of attention, until much later, when Faraday led electrical theory towards more complex and fruitful explanations using the concept of field lines (Whittaker 1987).

During the nineteenth century, different discoveries showed that the Coulomb model of interpreting electromagnetic phenomena had to be rethought. Oersted showed experimentally that "transverse actions" existed between an electric current and a compass, as opposed to the concept of central forces for all actions at a distance Wise (1990). The role of the surroundings in which the interaction took place also began to be emphasised by new experimental facts, e.g. it was found that containers with air pressure maintained the charges better on a conductor (Berkson 1974; Cantor et al. 1991). Volta's discovery produced a continuous electric current by setting up different materials in a certain order. These findings, among others, were to provide evidence to support the unity of natural forces (Sutton 1981). The law of the conservation of energy, formulated in 1840, placed the phenomena of light, heat, electricity and magnetism into a framework of general principles.

After Volta, different explanations appeared as to how an electrical circuit might operate. For decades, the concept of "electrical circuit" phenomena was linked to electrostatics (Benseghir and Closset 1996). The "electrical fluid" and "electrical conflict" models were used to explain the neutralisation of charges in batteries and other materials. The analysis of electrical circuits is closely related to the analysis of the electrical charge accumulation processes, studied during the eighteenth century on charges and discharges in isolated bodies (Leyden jar). The invention of the voltaic cell and the apparently perpetual movement of the electric fluid brought about changes in the theoretical frame. This led to defining the electric potential of a charged body and the concept of capacitance, as it is conceived today. The concept of electrical potential, used to explain how circuits work, was another challenge for scientists in the nineteenth century. However, as Roche (1989) states:

> The concept of potential is the fusion of at least five quite distinct historical traditions. Despite the seeming unity of the received concept, each of these traditions still plays a semi-autonomous role in the present-day understanding of potential. (p. 171)

This is a crucial point in clarifying and explaining the electrical phenomena in electrostatics and in circuits.

Alessandro Volta attempted to establish that the "galvanic fluid", of animal origin, was the same as ordinary electricity or static electricity (Kipling and Hurd 1958). In the midst of the controversy regarding the nature of electricity, Volta discovered that when two uncharged bodies of different metals were brought into contact, either directly or by means of an electrolyte, the two metals in a closed circuit acquired a charge and remained charged despite the presence of a conducting path where charges could flow and thus neutralise each other (Brown 1969; Fox 1990; Sutton 1981). This is a clear break with the idea that opposite charges could not be separated, as believed within the electrostatics realm at this time. Volta introduced the concept of "degree of electric tension" of a charged conductor and also defined "electromotive force" as the prime mover of current in a closed circuit, measured as electric tension. He stated that a new type of "force" was acting upon the charges, separating them and keeping them separated, and he called this action the electromotive force, the name that is still applied (Pancaldi 1990; Willians 1962). Volta's explanations did not fit into Coulomb's paradigm which prevailed in the first third of the nineteenth century. In this "electrostatic" context, the concept of "electromotive force" was interpreted as the capacity of bodies to generate electricity in others. Thus, one of the metals in the Volta battery "generates" electricity in the other due to its "electromotive force". Volta's interpretations regarding how batteries work did not fit into this theoretical framework and dropped into oblivion (Warney and Fischer 1980). These primitive concepts refer directly to mechanical-type analogies, emerged later on in the nineteenth century: "electric pressure" or "force" is thought of as a property of the electric fluid and not of continuous space.

S. D. Poisson and G. Green in 1811 introduced a very different meaning of the concept of electrical potential in electrostatics as a mathematical function whose gradient was the numerical value equal to the electrical intensity or force per unit of charge. First thought of only as a mathematical construct (Fox 1990), Green named this function "potential".

In 1827 Ohm made a contribution to circuit theory through his law for conductors. Ohm clarified the separate and complementary roles of current and potential at a time when both were rather confused. He supposed that a stationary gradient of volume charge corresponding to a gradient of potential drives a steady flow of electricity. Ohm used the analogy of temperature gradient driving heat transfer for explaining electricity flow (Schagrin 1963; Taton 1988). G. Kirchhoff who synthesised Ohm's work on electrical conduction and electrical resistance made the greatest step in the development of the concept of potential and circuit theory. On the basis of what was known about electrostatics, there could not be a gradient of volume charge inside the conductor, and Kirchhoff solved the problem putting a gradient of charges on the surface (Whittaker 1987). Kirchhoff demonstrated that Volta's "electrical tension" and Poisson's potential function were numerically identical in a conductor and therefore could be reduced to a single concept. Thus, he showed that electrostatic and circuit phenomena belonged to one science, not two (Heilbron 1979). From this unification the role of potential came to dominate the analysis of circuits but without attention to surface charge distribution.

Explanatory models of electrical current received a new impulse with the theory of fields initiated by Faraday and developed later by Maxwell in 1865. The field model suggests an ontological change in conceiving electric interaction, without test charges proving its existence, and introducing potential energy into the theory of the field. Maxwell (1865), referring to the "positional" character of the vector intensity of electric field in *A Dynamic Theory of the Electromagnetic Field*, showed just how difficult this model is:

> On talking about the intensity of the electric field at one point, we do not necessarily assume that a force is really exerted there, but just that, if an electrified body is set there, a force will act on it ... that is proportional to the charge of the body. (p. 17)

Faraday's field theory also involves a new conception of electric interaction, where its representation is not limited locally to the charged material particles, but rather spreads around the surrounding space. Going deeper into the ontological change that takes place from Coulomb's vision to that of the electric field, in the former, the concept of electric interaction is linked to that of charges in a zone of the space or in a charged body; there is no electric interaction without the electric charges that interact in the space. Moreover, the new understanding of the electric field forces us to think in a different way, as the concept of electric interaction is no longer linked to two electric charges, but rather extends along the area of influence for one of them. If the electric field is considered as a property of each point in the space, the electric action can "be" without the need of charge. From this new conception of electrical interaction, it is easy to establish the relationship between the concepts of charge, electric field and electric potential energy.

Physical properties can help explain the difficulties to understand some of the concepts used in electricity. One topic for discussion among physicists has been the meaning of the concepts of potential, potential difference and electromotive force. Härtel (1985) indicates that the majority of textbooks define potential in abstract and mathematically elegant ways but discard any causal mechanisms that explain the flow of electrons in electrical circuits. He finds it necessary to give meaning to the concepts of potential and potential difference.

Reif (1982), Romer (1982), and Peters (1984) discuss electrical potential, potential difference and electromotive force in the context of electrical circuits. Their studies conclude that the voltage measured by a voltmeter is equal to the total work per unit of charge moving through the instrument. Since the voltage indicated by a voltmeter depends on both Coulomb and non-Coulomb forces within the instrument, the voltage measured is generally different from the potential difference between the points to which the leads of the voltmeter are connected (voltage is equal to $\Delta V + \varepsilon$).

In addition, some early 1950s and 1960s textbooks showed surface charges on the wire as the cause of the electric field inside the wire that produces the electric current flow (Jefimenko 1966; Sommerfield 1952). Rosser (1963, 1970) described a mathematical analysis of the electric field produced by the wires surface charge distribution, which is of great value in understanding the electric field in electrical circuits. Härtel (1982) proposed analysing the electric circuit as a system. In this approach, the three fundamental terms current, voltage and resistance are introduced simultaneously in a qualitative way. The term voltage is introduced in close relationship to the cause of the motion. However, Härtel stated that for a complete understanding of why there is a potential drop in order to move the charges between two points of a conductor, it is necessary to analyse the interaction between electric field and the charges carriers. In a later study, Härtel (1987, 1993) analysed the gradient of the charge distribution along the different parts of the circuit, which produces different electric field depending on the resistance of this part of the circuit. He discussed the relations among the usually disparate topics of electrostatics and circuits by means of the surface charges and gave students a qualitative understanding of circuit behaviour. He also discussed the transient behaviour of circuits when, for example, a switch is closed.

Other studies such as those carried out by Aguirregabiria and colleagues (1992) and Jackson (1996) confirm the utility of the gradient distribution of charges model to explain potential changes in the circuit. Chabay and Sherwood's book (2002) compiles, among others, Hartel's contributions and proposes to carefully consider the concepts of "surface charge and feedback". This proposal justifies the continuity between the concepts studied in electrostatics and in electrical circuits. In Preyer (2000) two lecture demonstrations are described which illustrate the point that electrical potential in a circuit is the same function of charge density distribution as it is in electrostatics.

Another recurring topic of discussion in the theoretical framework of physics is the use of the Newtonian and Maxwellian model of interpreting electromagnetic phenomena. It should be pointed out that the Newtonian and Maxwellian models when used to interpret interactions between charges can be considered as belonging to a different ontological and epistemological status, but not opposite. This means that the scientific community assumes them both, although the higher conceptual level and power of one of them is admitted. For example, when analysing the electromagnetic phenomena, it is possible to make a description in terms of the intensity of the field that exists in this zone of the space or in terms of the action that the field exerts on the charges that there are in this area of the space (the exerted force). However, as Sharma (1988) stated:

> To find out the force on a test charge q at a point in space, you do not have to go all the way to find out where the sources (charges or currents) are; instead, you just have to know the values of E and B at that very point and use the Lorentz force law to compute the force. If the E and B from two source distributions are the same at a given point in space, the force acting on a test charge or current at that point will give the same, regardless of how different source distributions are. This gives E and B meaning in their own right, independent of sources. Further, the finite speed of propagation of electromagnetic signals, the retarded action, requires fields to carry energy, momentum, and angular momentum in order to guarantee conservation of these quantities. (p. 420)

The Maxwellian framework is touted to be conceptually superior and has more explicative power. Nevertheless, constructing field theory requires previous acquisition of the old framework (i.e. it is not possible to introduce electric field without knowing the prerequisites of Coulomb's electric charge and force) and the acknowledgment of its theoretical insufficiencies (Berkson 1974).

Chart 5.1 summarises the different models used through history and today's classical electromagnetic framework to interpret the basic electric phenomena.

5.4 Concluding Remarks: Guidelines for Designing Teaching-Learning Sequences

In secondary teaching and introductory physics courses at university level, there are learning difficulties in areas such as a scientific model for electrical current in a circuit and in concepts such as electrical field and electrical potential. A lack of connection has also been observed between the same concepts taught in electrostatics and

5 Teaching and Learning Electricity: The Relations Between Macroscopic Level... 145

Chart 5.1 Models to describe fundamental electric phenomena

Empirical reference	Different models used through history
	Effluvia model
Charging of bodies by rubbing	Electric phenomena were due to something ("electric effluvia") that was freed upon rubbing "electric" bodies such as glass or amber
Attraction between charged bodies	Methodology used criteria of empirical evidence for testing the theory
Attraction of light bodies by rubbed bodies	*Explanatory problems:* The model did not explain repulsion between charged bodies
	Electric fluid
Transmission of electricity or "electric property"	The model describes electricity as an electric fluid made up of extremely subtle particles. Electric fluid can be transmitted and is not inseparable from the charged body, as in the previous model. Rubbing does not create electricity; fluid is just transferred from one body to another. So, the total amount of electricity in any isolated system is invariable
Electrification by contact	The excess and lack of electricity of the body are associated to the + and −, respectively. The empirical rule according to which bodies charged with the same sign repel and those with different sign attract is defined
Electric repulsion phenomena	The electric fluid accumulated in the body exerts a pressure on the surface of the body. At a certain point, the "electric pressure" is big enough to prevent the body from admitting any more charge. The capacitance is defined as $C = Q/\text{tension}$
Electric induction	The current in a simple circuit was explained by the theory of "electrical conflict" based on the electrical fluid model
Electrical capacitance	*Explanatory problems:* The research methodology is qualitative. It can neither quantify the electrical phenomena nor define their magnitude
Electrical simple circuits	The model does not explain why two bodies, which lack fluid, repel each other
	The electric induction explanation by using the "electrical atmosphere" is questioned by means of experimental evidence
	The role of battery in a circuit is still unexplained
	Action at a distance model
Electrostatic phenomena	A quantitative methodology, similar to that used in Newtonian mechanics, is introduced, and the concept of electrical charge is defined through the formula of electrical force. The model describes electricity as a set of charges that interact at a distance, in agreement with Coulomb's law. Electric interaction between separate charges is transmitted instantaneously through the space where they are situated, whatever medium exists between them
Condensers	The capacity of the charges to act at a distance implies the presence of an electrified body near the body to be charged and involves some modification in the electrical potential of the system and its "capacitance" to store charges. The capacitance is a property of conductors that interact
Simple electrical DC circuits	Direct current is due to the flow of electrons under the influence of electrical forces and under the influence of a potential difference across the poles of a battery

(continued)

Chart 5.1 (continued)

Empirical reference	Different models used through history
	Field model
Electrostatic phenomena	In this new understanding of electric interaction, not only charges but also the medium is taken into account. The interaction's transmission is non-immediate, because it depends on the medium existing between the charges
Electrical capacitance	The process of charging a body implies work and the acquisition of an electric potential. Thus, the concept of capacitance is a property of the system of conductors that interact which can in principle be measured as $C = Q/\Delta V$
Electrical potential energy	Electric currents in wires, resistors, etc. are driven by electric fields. The electric field has its source only in the surface charge distributions on the wire
Simple electrical DC circuits	

in circuits. The review points to problems that need to be addressed when dealing with the Maxwellian model of electricity. Following the recent educational standards' recommendations on presenting concepts and laws in a contextual meaning (National Research Council 1996; Rocard et al. 2007), some of the discussions among physicists on teaching field and electrical potential topics to secondary and university students have been pointed out. The contributions from physics education research and from the history of science should be taken into account by teaching staff and curriculum designers.

The traditional curriculum for teaching electricity at secondary schools starts by analysing how electrical circuits work. Many teaching activities are centred on analysing circuits with resistors arranged in series and in parallel by means of Ohm's law. The circuit is often not explicitly analysed in terms of its energy and of the role played by the battery's electromotive forces and movement of electrons.

The focus on resistors (Ohm's law) can lead many secondary and first year university students to think that Ohm's law is a fundamental law of electricity. As Bagno and Eylon (1997) states: "A high proportion of students considered Ohm's law to be one of the most important ideas of electromagnetism, consistent with previous findings, labelled humorously 'the three principles of electromagnetism': $V = iR$; $i = V/R$; $R = V/i$" (p.731).

In secondary education (Steinberg and Wainwright 1993) teaching about DC circuits and the role of the battery, use the compressible-fluid model – pressure gradient driving the current – for charge conduction. This pressure gradient results from a gradient in the charge-carrier volume density. However, according to conventional physics, electric current in DC circuits is driven by the electric field, created by the surface charge distribution. The volume charge density inside the wire is zero. As Mosca and De Jong (1993) show, this model can lead to erroneous conclusions:

> An erroneous conclusion associated with the compressible-fluid model is that it predicts the existence of an electric field within a charge conductor in electrostatic equilibrium.

5 Teaching and Learning Electricity: The Relations Between Macroscopic Level...

> According to this model, the region within the material of an isolated charged conductor in electrostatic equilibrium is occupied by a gas of charge carriers that is uniformly compressed to a high number of density- and thus charge density. In accordance with Gauss's law, any nonzero charge density is necessarily accompanied by an electrostatic field, and the presence of this field contradicts the widely accepted view that a conductor in electrostatic equilibrium is an equipotential. (p. 358)

One can argue that a misconception in the electrostatic realm is not necessarily a misconception in circuit theory. However, the compressible-fluid model can lead to misconceptions not only in electrostatics but also in DC circuits (Mosca and De Jong 1993, p. 358). Moreover, in classical physics, electrostatics is part of electrodynamics as, e.g. analysing a DC circuit containing a capacitor (Guisasola et al. 2010).

The traditional approach changes completely when the concept of electric potential is introduced in senior high school (16–18 years old) and introductory physics courses at university. At these levels, the concept of electric field is first introduced and defined as $E = F/q$, electric potential at a point as the energy per electric charge ($V = E/q$) and potential difference between two points as the quantity of energy required or supplied to move a unit of charge from one point to another ($\Delta V = \Delta E_p/q$). All these definitions are made in an electrostatic context so that later in the study of electric circuits, the same concepts can be used again when applying the principles of charge and energy conservation (Kirchhoff's laws). However, textbooks frequently do not show explicit relations using these concepts in both contexts (Stocklmayer and Treagust 1994). As demonstrated in the previous sections, the research into teaching electricity and the history of science shows that the explanatory model of electrostatic phenomena conditions how we see the electric nature of matter and the flow of current in a circuit.

Recent studies propose starting the electricity curriculum with elementary electric phenomena (electrification by friction, contact or induction) and focusing students' attention on the microscopic explanatory models to improve students' understanding.[6] These proposals recommend the representation of models based on energy and field, allowing students to interpret at a microscopic level the electric phenomena observed at a macroscopic level (Walz 1984). For example, when a pendulum repels, a small positively charged ball from a positively charged rod, the work done results in an increase in the ball's gravitational potential energy. This work corresponds to the change in electric potential energy associated with a given configuration of the system (Borghi et al. 2007). The phenomenon is explained using an energetic model instead of an equivalent force model. In this approach, the other parameter required to define the energy of a system is its capacitance. The system's electrical status can be described by a new physical property "capacitance" that expresses the system's ability to receive more electric charge (Guisasola et al. 2002). This quantity can be used operationally by relating the capacity of the charge and the electric potential. The analysis of electrostatic phenomena in terms of

---

[6] Benseghir and Closset (1996), Eylon and Ganiel (1990), Furio et al. (2004), Park et al. (2001), Thacker et al. (1999), and Viennot (2001).

energy relationships encourages the development of a relationship between electrostatics and circuits (Arons 1997).

Härtel (1982, 1993) proposed a transition from electrostatic to circuits, based on the electric circuit as a system. The students' tendency to reason locally and sequentially about electric circuits (Duit and von Rhöneck 1998; Shipstone et al. 1988) is directly addressed by analysing the behaviour of the whole circuit. The three fundamental terms (current, voltage and resistance) are introduced simultaneously in a qualitative way using the energy balance of the whole circuit including possibilities of transporting energy and the concept of potential energy.[7] At secondary school level, the aim should be to help students understand how elementary circuits work instead of quantitative circuit analysis (Kirchhoff's laws). This involves linking the movement of charges between two points of a conductor to the concept of potential difference and the transition from the potential of static charges to the DC circuit in a stationary state. The role played by the battery is here a critical point. As Benseghir and Closset (1996) state: "it must be pointed out that the battery keeps a constant difference of potential between its terminals" (p. 181).

Chabay and Sherwood (1995, 2002, 2006) propose a field model unifying electrostatics and circuits, which relates field and potential to DC circuits as suggested by Härtel. This is a microscopic model of the electric current based on the change in the surface density of charges generating an electric field in the direction of the wire. The function of the battery is to maintain the surface density of charge that is caused by the electric current inside the wire. Understanding macroscopic phenomena requires a coherent model of microscopic processes (Thacker et al. 1999). Students' difficulties in qualitative analysis of electric circuits can be overcome with more emphasis on microscopic processes. The proposal for teaching electric current based on the field model explicitly relates the measurements at a macroscopic level (voltage and current intensity) with a causal model at a microscopic level that uses potential difference and constant speed of electrons to explain the macroscopic measurements. Moreover, the role played by surface charges in DC circuits can show to students by a graphical method developed by Muller (2012).

Chabay and Sherwood's programme, after going through various sequences, has been successfully used in first year introductory electricity and magnetism physics courses at university (Ding and colleagues 2006). Students taught with this scheme were able to analyse circuits in a significantly better way compared to students in traditional courses. In introductory physics courses at university, teaching DC circuits theory using the field model has also been successful in other countries (Hirvonen 2007). At high school level, the preliminary findings indicate that this approach may be more successful than the traditional electron flow model (Stocklmayer 2010). Furio and colleagues (2003) found that high school students whose teaching included electric field concepts showed an important improvement in their understanding of electrostatics. Correct results from these students were at least 50 % better than from the control group, with statistically significant results in all comparative tests.

---

[7] Cohen et al. (1983), Duit (1985), Härtel (1985), Psillos (1998), and Psillos et al. (1988).

Saarelainen and Hirvonen (2009) show that understanding the electric field concept is necessary for comprehension of electrostatics and particularly Gauss' law. Performance in this area can be improved by taking into account students' thought processes and applying methods suggested in educational reconstruction. Silva and Soares (2007) looked at results from the use of an electrical field and potential energy model by 2nd year students in a teacher education course in Portugal. The aspect of this model proved fruitful in bridging electrostatics, DC and AC circuits. At secondary school level, Psillos (1998) and Psillos and colleagues (1988) proposed a model based on potential energy to explain the relations between macroscopic phenomena in simple DC circuits and the movement of charges at microscopic level. Their results show that the approach improves students' understanding of how current flows and the behaviour of the whole circuit.

It is recommended that use of these models should begin in the electricity curriculum in secondary schools as they address three key problems identified so far are:

(a) Relations between electrostatics and current
(b) Relations between macroscopic phenomena and microscopic level models
(c) Relations between operative definitions of charge, potential and electric capacity and their meaning in electrostatics and current

However, very few high school or basic university textbooks propose a qualitative electrostatics and circuits model based on field and energy. Although there is consensus among the research literature on the insufficiency of traditional treatments, the qualitative model remains on the margin of the usual mathematical treatment based on Kirchhoff's laws. Textbooks avoid a presentation that relates micro and macro views, possibly because surface densities of charge, small in normal DC and AC circuits, are difficult to measure in the laboratory. It appears obvious that a model based on the surface density of charge in the wire is not familiar to teachers and not easy to understand at these levels. Stocklmayer (2010) suggests the following changes in teaching:

> The problem with the universal adaptation of the field model lies in its unfamiliarity. It is not within the 'comfort zone' of many teachers, nor, indeed, many conventional physicist for whom the electron flow model has proved comprehensible and satisfactory ... It will require the development of new resource materials, including textbooks and practical exercises, and extensive professional development for teachers. (p. 1825)

More research is needed on teaching sequences design based on new models and their implementation in the classroom. Lijnse and Klaassen (2004) argue that designing teaching sequences requires a complex process of applying didactics to specific teaching contexts, a cyclic rather than linear process aiming at generating knowledge about the relevance of improved teaching materials in the classroom.

In conclusion, this research offers guidelines and teaching experiences to teachers and curriculum designers for changes in the traditional system of sequencing electricity topics. Research and the history of science contribute evidence to strongly suggest that key points of the electricity curriculum should be covered differently.

References

Aguirregabiria, J.M. Hernandez, A & Rivas, M. (1992). An example of surface charge distribution on conductors carrying steady currents. *American Journal of Physics*, 60(2), 138–140.

Arons, A. (1997). *Teaching introductory physics*. New York: Wiley.

Ausubel, D.P. (1978). *Educational Psychology. A cognitive view*. New York: Holt, Rineheart & Winston, Inc.

Bagno, E. & Eylon, B.S. (1997). From problem solving to a knowledge structure: An example from the domain of electromagnetism. *American Journal of Physics*, 65(8), 726–736.

Barbas, A. & Psillos D. (1997). Causal reasoning as a base for advancing a systemic approach to simple electrical circuits. *Research in Science Education*, 27(3), 445–459.

Berkson, W. (1974). Fields of force. *The development of a world view from Faraday to Einstein*. London: Routledge & Kegan Paul.

Benseghir, A. & Closset, J.L. (1996). The electrostatics-electrokinetics transition: historical and educational difficulties, *International Journal of Science Education*, 18 (2), 179–191.

Borges, A. & Gilbert. J. (1999). Mental models of electricity. *International Journal of Science Education*, 21, 95–117.

Borghi, L., De Ambrosis, A. & Mascheretti, P. (2007). Microscopic models for bringing electrostatics and currents. *Physics Education*, 42(2), 146–155.

Brown, T. M. (1969). The electrical current in early Nineteenth-Century French physics. *Historical Studies in the Physical Sciences*, 1, 61–103.

Cantor, G., Gooding, D., & James, F. A. J. L. (1991). *Faraday*. London: Macmillam.

Chabay, R. & Sherwood, B. (2006). Restructuring the Introductory electricity and magnetism course. *American Journal of Physics*, 74, 329–336.

Chabay, R. & Sherwood, B. (2002). *Matter and Interactions II: Electric and Magnetic Interactions*. New York: Wiley (Last edition 2007).

Chabay W. & Sherwood B. A. (1995). *Instructor's Manual of Electric and Magnetic Interactions*. New York: Wiley.

Closset, J. L. (1983). Sequential reasoning in electricity. In G. Delacote, A. Tiberghien, & J. Schwartz (Eds.) *Research on Physics Education: Proceedings of the First International Workshop* (pp. 313–319). Paris: Editions du CNRS.

Cohen, R., Eylon, B. & Ganiel, U. (1983). Potential difference and current in simple electric circuit: A study of students' concepts. *American Journal of Physics*, 51, 407–412.

Conant J.B., Nash L.K., Roller D. & Roller D.H.D. (1962). *The development of the concept of electric charge. Harvard Case Histories in Experimental Science*. Massachusetts: Harvard University Press.

Ding, L., Chavay, R., Sherwood, B. & Beichner, R. (2006). Evaluating an assessment tool: Brief electricity & magnetism assessment. *Physical Review Special Topics: Physics Education Research* 1, 10105.

Driver, R., Leach, J., Scott, P. & Wood-Robinson, C. (1994). "Young people's understanding of science concepts: implications of cross-age studies for curriculum planning". *Studies in Science Education*, 24, 75–100.

Duit, R. (1985). In search of an energy concept. In R. Driver & R. Millar (Eds.), Energy matters – Proceedings of an invited conference: Teaching about energy within the secondary science curriculum (pp 67–101). Fairbairn House: The University of Leeds.

Duit, R.: 2009, *Bibliography of Students' and teachers' conceptions and science education* in http://www.ipn.uni-kiel.de/aktuell/stcse/. Accessed 25 January 2012.

Duit, R. & von Rhöneck, C. (1998). Learning and understanding key concepts of electricity. In A. Tiberghien, E.L. Jossem & J. Barojas (Eds.) *Connecting research in physics education*.

5 Teaching and Learning Electricity: The Relations Between Macroscopic Level... 151

Boise, Ohio: International Commission on Physics Education-ICPE books. http://www.physics. ohio-state.edu/~jossem/ICPE/TOC.html. Accessed 20 January 2012.

Duit, R. & Treagust, D.F. (1998). Learning in science: From behaviourism towards social constructivism and beyond. In B. Fraser & K. Tobin (Eds.) *International handbook of science education* (pp. 3–25). Dordrecht, Netherlands: Kluwer.

Dupin, J. & Joshua, S. (1989). Analogies and 'modelling analogies' in teaching some examples in basic electricity. *Science Education, 73*(2), 207–224.

Dupin, J. & Joshua, S. (1987). Conceptions of French pupils concerning electric circuits: Structure and evaluation. *Journal of Research in Science Teaching, 6*, 791–806.

Duschl, R.A. (1994). Research on the history and philosophy of science . In D.L. Gabels (ed.) *Handbook of Research on Science Teaching and Learning* (pp. 443–465). New York: McMillan Pub.Co.

Duschl, R.A. (2000). Making the nature of science explicit. In R. Millar, J. Leach & J. Osborne (eds.). *Improving Science Education- The contribution of Research*. Buckingham: Open University Press.

Engelhardt P.V. & Beichner R.J. (2004), Students' understanding of direct current resistive electrical circuits, *American Journal of Physics 72*, 98–115.

Etkina E., Van Heuvelen A., White-Brahmia S., Brookes D.T., Gentile M., Murthy S., Rosengrant D., & Warren A. (2006). Scientific abilities and their assessment. *Physical Review Special Topics: Physics Education Research, 2*, 020103, 1–15.

Eylon, B. & Ganiel, U. (1990). Macro–micro relationship: the missing link between electrostatics and electrodynamics in students' reasoning. *International Journal of Science Education, 12* (1) 79–94.

Fox, R. (1990). Laplacian Physics. *Companion to the History of Modern Science*, Routledge.

Furio, C., & Guisasola, J. (1998). Difficulties in learning the concept of electric field. *Science Education, 82*(4), 511–526.

Furio, C., Guisasola, J., Almudi, J.M., & Ceberio, M.J. (2003). Learning the electric field concept as oriented research activity. *Science Education, 87*(5), 640–662.

Furio C., Guisasola, J., & Almudi, J.M. (2004). Elementary electrostatic phenomena: historical hindrances and student's difficulties. *Canadian Journal of Science, Mathematics and Technology, 4*(3), 291–313.

Galili I. (1995). Mechanics background influences students' conceptions in electromagnetism. *International Journal of Science Education, 17*(3), 371–387.

Greca, I. & Moreira M.A. (1997). The kinds of mental representation - models, propositions and images- used by college physics students regarding the concept of field. *International Journal of Science Education, 19*(6), 711–724.

Guisasola, J., Almudí, J.M., Salinas, J., Zuza, K. & Ceberio, M. (2008). The gauss and ampere laws: different laws but similar difficulties for students learning. *European Journal of Physics, 29*, 1005–1016.

Guisasola, J., Zubimendi, J.L., Almudi, J.M. & Ceberio, M. (2002). The evolution of the concept of capacitance throughout the development of the electric theory and the understanding of its meaning by University students. *Science & Education, 11*, 247–261.

Guisasola, J, Zubimendi. J.L. & Zuza, K. (2010) How much have students learned? Research-based teaching on electrical capacitance, *Physical Review Special Topic-Physics Education Research 6*, 020102 1–10.

Guisasola, J. & Montero, A. (2010), An energy-based model for teaching the concept of electromotive force. Students' difficulties and guidelines for a teaching sequence. In G. Çakmakci & M. F. Tasar (eds) *Contemporary science education research: learning and assessment* (pp. 255–258). Ankara, Turkey: Pegem Akademi.

Gunstone, R., Mulhall, P. & Brian McKittrick, B. (2009). Physics Teachers' Perceptions of the Difficulty of Teaching Electricity. *Research in Science Education, 39*, 515–538.

Guruswamy, C., Somers, M. D. & Hussey, R. G. (1997). Students' understanding of the transfer of charge between conductors, *Physics Education, 32*(2) 91–96.

Halloun I. & Hestane D. (1985) Common sense conceptions about motion, *American Journal of Physics 53*, 1056–1065.

Harman, P. M. (1982). *Energy, force and matter. The conceptual development of nineteenth-century physics*. Cambridge: Cambridge University Press.

Härtel, H. (1982). The electric circuit as a system: A new approach. *European Journal of Science Education*, 4(1), 45–55.

Härtel, H. (1985). The electric voltage: What do students understand? What can be done to help for a better understanding? In R. Duit, W. Jung & C. von Rhöneck (eds.), *Aspects of Understanding Electricity: Proceedings of an International Workshop* (pp. 353–362). Ludwingsburg: IPN-Kiel.

Härtel, H. (1987). *A Qualitative Approach to Electricity*. Palo Alto, California: Institute for Research on Learning. http://www.astrophysikunikiel.de/~hhaertel/PUB/voltage_IRL.pdf. Accessed 12 November 2011.

Härtel, H. (1993). New approach to introduce basic concepts in Electricity. Teoksessa M. Caillot (ed.) *Learning electricity and electronic with advance educational technology* (pp. 5–21). NATO ASI Series. Springer-Verlag.

Heald, M.A. (1984). Electric field and charges in elementary circuits. *American Journal of Physics* 52(6), 522–526.

Heilbron, J.L. (1979). *Electricity in the 17th and 18th centuries. A study of early modern Physics*, California: University of California Press.

Hirvonen, P.E. (2007). Surface-charge-based micro-models a solid foundation for learning about direct current circuits. *European Journal of Physics*, 28, 581–592.

Jackson, J.D. (1996). Surface charges on circuit wires and resistors plays three roles. *American Journal of Physics* 64(7), 855–870.

Jefimenko O.D. (1966) *Electricity and Magnetism*. New York: Appleton-Century Crofts.

Jimenez, E. & Fernandez, E. (1998). Didactic problems in the concept of electric potential difference and an analysis of its philogenesis, Science & Education 7, 129–141.

Kipling D.L. & Hurd J.J., (1958). The Origins and Growth of Physical Science. United Kingdom: Penguin Books.

Kuhn, T. (1984). Professionalization recollected in tranquillity. *ISIS*, 45, 276, 29–33.

Kenosen M.H.P., Asikainen M.A. & Hirvonen P.E. (2011). University students' conceptions of the electric and magnetic fields and their interrelationships, *European Journal of Physics* 32, 521–534.

Liegeois, L., Mullet, E. (2002). High School Students' Understanding of Resistance in Simple Series Electric Circuits. *International Journal of Science Education*, 24(6), 551–64.

Licht, P. (1991). Teaching electrical energy, voltage and current: an alternative approach. *Physics Education*, 25, 271–277.

Lijnse, P. & Klaassen, K., (2004). Didactical structures as an outcome of research on teaching-learning sequences?. *International Journal of Science Education*, 26(5), 537–554.

Maloney D P, O'kuma T L, Hieggelke C J and Van Heuvelen A (2001) Surveying students' conceptual knowledge of electricity and magnetism. *American Journal of Physics* S69 (7), 12–23.

Mäntylä, T. (2011). Didactical reconstruction of processes in Knowledge construction: Pre-service physics teachers learning the law of electromagnetic induction. *Research in Science Education*. DOI 10.1007/s11165-011-9217-6.

Matthews, M.R. (1994). *Science teaching: the role of history and philosophy of science*. New York: Routledge.

Maxwell, J. C. (1865). Dynamic theory of the electromagnetic field. *Philosophical transactions*, 155, 456–512.

McComas, W.F., Clough, M.P. & Almazora, H. (2000) The role and the character of the nature of science in science education. In W.F. McComas (Ed.): *The Nature of Science in Science Education. Rationales and strategies* (pp. 3–39). The Netherlands: Kluwer Academic Publishers.

McDermott, L.C. & Shaffer P.S. (1992). Research as a guide for curriculum development: An example from introductory electricity, Part I: Investigation of student understanding. *American Journal of Physics*, 60, 994–1003.

Moreau, W.R. & Ryan, S.J. (1985). Charge density in circuits. *American Journal of Physics* 53(6), 552–554.

5 Teaching and Learning Electricity: The Relations Between Macroscopic Level...

Mosca, E.P. & De Jong, M.L. (1993). Implications of using the CASTLE model, *The Physics Teacher* 31, 357–359.

Mulhall, P., Mckittrick, B. & Gunstone R. (2001). A perspective on the resolution of confusions in the teaching of electricity, *Research in Science Education,* 31, 575–587.

Muller, R. (2012). A semiquantitative treatment of surface charges in DC circuits, *American Journal of Physics,* 80, 782–788.

Nardi, R. & Carvhalo, A.M.P. (1990). A genese, a psicogenese e a aprendizagem do conceito de campo: subsidios para a construção do ensino desse conceito. Caderno Catarinense do Ensino de Física, 7, pp. 46–69.

National Research Council (1996). *National Science education standards.* Washington DC: National Academic Press.

Nersseian, N.J. (1995). Should physicist preach What they practice?. *Science and Education,* 4, 203–226.

Niaz, M. (2008) What 'ideas-about-science' should be taught in school science? A chemistry teacher's perspective. *Instructional Science* 36, 233–249.

Oliva, J.M. (1999). Algunas reflexiones sobre las concepciones alternativas y el cambio conceptual. *Enseñanza de las Ciencias* 17(1), 93–107.

Park, J., Kim, I., Kim, M. & Lee, M. (2001). Analysis of students' processes of confirmation and falsification of their prior ideas about electrostatics. *International Journal of Science Education* 23(12), 1219–1236.

Pancaldi, G. (1990). Electricity and life. Volta's path to the battery. Historical studies in the physical and biological sciences, 21, 123–160.

Periago, M.C. & Bohigas, X. (2005). A study of second year engineering students alternative conceptions about electric potential, current intensity and Ohms' law. *European Journal of Engineering education* 30(1), 71–80.

Peters, P. C. (1984). The role of induced emf's in simple circuits. *American journal of Physics* 52(3), 208–211.

Pintó, R. (2005). Introducing curriculum innovations in science: identifying teachers' transformations and the design of related teacher education, *Science Education* 89(1), 1–12.

Pocovi, M.C. & Finley, F. (2002). Lines of force: Faraday's and Students' views. *Science & Education* 11, 459–474.

Preyer, N.W. (2000). Surface charges and fields of simple circuits. *American Journal of Physics* 68, 1002–1006.

Psillos, D. (1998). Teaching introductory electricity. In A.Tiberghien, E.L. Jossem, & J. Barojas (Eds.), Connecting research in physics education with teacher education. International Commission on Physics Education.

Psillos D., Koumaras P & Tiberghien A. (1988) Voltage presented as a primary concept in an introductory teaching sequence on DC circuits. *International Journal of Science Education*, 10, 1, 29–43.

Rainson, S., Tranströmer, G., & Viennot, L. (1994). Students' understanding of superposition of electric fields. *American Journal of Physics*, 62(11), 1026–1032.

Reif, F. (1982) Generalized Ohm's law, potential difference, and voltage measurements. *American Journal of Physics* 50(11), 1048–1049.

Rocard, M., Csermely, P., Jorde, D., Dieter, L., Walberg-Henriksson, H. & Hemmo, V. (2007). *Science education now: A renewed pedagogy for the future of Europe.* Retrieved May 2012, http://ec.europa.eu/research/science-society/document_library/pdf_06/report-rocard-on-science-education_en.pdf

Roche, J. (1989) Applying the history of electricity in the classroom: a reconstruction of the concept of 'potential'. In M. Shortland & A. Warwick *Teaching the history of Science* Oxford: Basil Blackwell.

Roche, J. (1987) Explaining electromagnetic induction: a critical re-examination. *Physics Education* 22, 91–99.

Romer, R.H. (1982) What do 'voltmeters' measure?: Faraday's law in a multiply connected region. *American Journal of Physics* 50(12), 1089–1093.

Rosser W.G. (1963) What makes an electric current flow. *American Journal of Physics* 31, 884–5

Rosser W G (1970) Magnitudes of surface charge distributions associated with electric current flow. *American Journal of Physics*, 38 264–6.

Rudge, D. & Home, E. (2004). Incorporating History into Science Classroom. *The Science Teacher* 71(9), 52–57.

Saarelainen, M., Laaksonen, A. & Hirvonen, P.E. (2007). Students' initial knowledge of electric and magnetic fields – more profound explanations and reasoning models for undesired conceptions. *European Journal of Physics* 28, 51–60.

Saarelainen, M. & Hirvonen, P.E. (2009) Designing a teaching sequence for electrostatics at undergraduate level by using educational reconstruction, *Latin-American Journal of Physics Education* 3, 518–526.

Saglam, M. & Millar, R. (2005). Diagnostic Test of Students' Ideas in Electromagnetism. *Research Paper Series. University of York (UK).*

Schagrin, M.L. (1963) Resistance to Ohm's Law', *American Journal of Physics*, 31(7), 536–547.

Shaffer, P.S. & McDermott L.C. (1992). Research as a guide for curriculum development: An example from introductory electricity, Part II: Design of instructional strategies. *American Journal of Physics*. 60, 1003–1013.

Seroglou, F., Panagiotis, K. & Vassilis, T. (1998) History of Sciences and instructional design: the case of electromagnetism, *Science and Education,* 7, 261–280.

Sharma N.L. (1988) Field versus action at-a-distance in a static situation. *American Journal of Physics* 56 (5), 420–423.

Sherwood, B. & Chabay R. (1999) *A unified treatment of electrostatics and circuits.* http://matterandinteractions.org/Content/Articles/circuit.pdf. Accessed Retrieved 10 August 2011.

Shipstone, D.M., von Rhöneck, C., Jung, W., Karrqvist, C., Dupin, J., Joshua, S. & Licht, P. (1988). A study of secondary students' understanding of electricity in five European countries. *International Journal of Science Education 10,* 303–316.

Silva, A.A. & Soares R. (2007) Voltage versus current, or the problem of the chicken and the egg, *Physics Education* 42(5), 508–515.

Smith, D. P. & van Kampen, P. (2011) Teaching electric circuits with multiple batteries: A qualitative approach. *Physical Review Special topics- Physics Education Research* 7, 020115.

Solomon, J. (2002). Science Stories and Science Texts: What Can They Do for Our Students?, *Studies in Science Education*, 37, 85–105.

Sommerfield A (1952). *Electrodynamics.* New York: Academic Press.

Steinberg, M.S. & Wainwright, C. (1993). Using models to teach electricity-The CASTLE project, *The Physics Teacher,* 31(6), 353–355.

Steinberg, M.S. (1992). What is electric potential? connecting Alessandro Volta and contemporary students. In *Proceeding of the Second International Conference on the History and Philosophy of Science and Science Teaching,* vol. II, 473–480. Kingston.

Stocklmayer, S. & Treagust, D.F. (1994) A historical analysis of electric current in textbooks: a century of influence on physics education, *Science & Education*, 3, 131–154.

Stocklmayer, S. & Treagust, D.F. (1996). Images of electricity: How novices and experts model electric current? *International Journal of Science Education* 18, 163–178.

Stocklmayer, S. (2010) Teaching direct current theory using a field model, *International Journal of Science Education* 32 (13), 1801–1828.

Sutton, G. (1981) The politics of science in early Napoleonic France: The case of the voltaic pile. *Historical Studies in the Physical Sciences,* 11(2), 329–366.

Taton R. (1988*). Histoire Generale des Sciences* . Paris: Presses Universitaires de France.

Testa, I., Michelini, M. & Sassi, E. (2006). Children's Naive Ideas/Reasoning About Some Logic Circuits Explored In an Informal Learning Environment. In G. Planinsic & A. Mohoric (eds). *Selected contributions to Third International GIREP Seminar* (pp. 371–378). Ljubljana, Slovenia: University of Ljubljana.

Thackray, A. (1980). History of Science, in P.T. Durbin (Ed.). *A guide to the culture of science, technology and medicine*, New York: Free Press.

Thacker, B.A. Ganiel, U. & Boys, D. (1999). Macroscopic phenomena and microscopic processes: Student understanding of transients in direct current electric circuits. *Physics Education Research-American Journal of Physics*, 67(7), S25–S31.

Törnkvist, S., Pettersson, K.A. & Tranströmer, G. (1993) Confusion by representation: on student's comprehension of the electric field concept, *American Journal of Physics*, 61 (4), 335–338.

Varney, R.N. & Fisher, L.H (1980) Electromotive force: Volta's forgotten concept. *American Journal of Physics* 48 (5), 405–408.

Viennot, L., & Rainson, S. (1992). Students' reasoning about the superposition of electric fields. *International Journal of Science Education*, 14(4), 475–487.

Viennot, L. & Rainson, S. (1999). Design and evaluation of research-based teaching sequence: the superposition of electric field. *International Journal of Science Education, 21* (1), 1–16.

Viennot, L. (2001). *Reasoning in Physics. The part of Common Sense*. Dordrecht, Netherlands: Kluwer academic Publisher.

Vosniadou, S. (2002). On the Nature of Naïve Physics, in *Reconsidering the Processes of Conceptual Change* (pp. 61–76), edited by M. Limon & L. Mason, Netherlands: Kluwer Academic Publishers.

Walz, A. (1984) Fields that accompany currents. In R. Duit, W. Jung., & Ch. von Rhöneck, Eds., *Aspects of understanding electricity* (pp. 403–412). Keil, Germany: Schmidt & Klaunig.

Wandersee, J.H. (1992). The Historicality of cognition: implications for Science Education Research, *Journal of Research in Science Teaching*, 29(4), 423–434.

Wandersee, J.H., Mintzes, J.J. & Novak, J.D. (1994). Research on alternative conceptions in Science. Handbook of Research on Science Teaching and Learning. Nueva York: McMillan Publishing Company.

Warney, R.N. & Fischer, L.H. (1980). Electromotive force: Volta's forgotten concept. *American Journal of Physics* 48, 405–408.

Watts, M. & Taber, K.S. (1996). An explanatory Gestalt of essence: students' conceptions of the «natural» in physical phenomena. *International Journal of Science Education*, 18(8), 939–954.

Whittaker, E. (1987). *A History of the Theories of Aether and Electricity*, U.S.A: American Institute of Physics.

Willians P, L. (1962) The Physical Sciences in the first half of the nineteenth Century: Problems and Sources. *History of Science*. 1, 1–15.

Wise, N.M. (1990) Electromagnetic Theory in the Nineteenth Century, *Companion to the History of Modern Science*, Routledge.

Young H.D. & Freedman R.A., 2008, University Physics with Modern Physics vol. 2, 12th edition. New York: Addison-Wesley.

Jenaro Guisasola He received his B.S. in Physics and an M.S. in Theoretical Physics, both from the University of Barcelona, as well as a Ph.D. in Applied Physics from the University of the Basque Country. He is Associate Professor of Physics at the Applied Physics Department at the University of Basque Country. Since 2008 he teaches also Physics Education in the M.A. for Initial Training of Secondary Science Teachers. He serves from 2002 to 2012 as an Executive Board Member for Teaching Active Methodologies programme at the University of the Basque Country. As a director of the Group of Physics, Mathematics and Technological Education Research at the University of the Basque Country, he has supervised ten Ph.D. works over the last 7 years (in the University of the Basque Country and in the University of Valencia). His research interest follows two related paths. The first focuses in teaching and learning conceptual physics knowledge at high school and university degree.

The second is based on the use of history and philosophy of science as tools to help organise teaching and learning in science curriculum. In the last 3 years, he started a line of research on the relations between science museums and science classrooms. He has published papers in a wide variety of education and science journals, including *Science Education, Science & Education, International Journal of Science Education, Journal of Science Education and Technology, American Journal of Physics, International Journal of Science and Mathematics Education, Enseñanza de las Ciencias, Physical Review Special Topics- Physics Education Research, The Physics Teacher, Physics Education, Canadian Journal of Science and Mathematics and Technology Education.*

Chapter 6
The Role of History and Philosophy in Research on Teaching and Learning of Relativity

Olivia Levrini

6.1 Introduction

With respect to topics of classical physics, special relativity has been subject to a limited number of studies in physics education research (PER). However, educational studies focusing on special relativity include some milestones in PER development since the latter have become established in a specific research discourse. The development of studies on relativity mirrors significant changes in research priorities that can be identified in science education research.

In this chapter the review of the literature concerning teaching/learning special relativity is carried out so as to sketch the story of research development in physics (science) education.

In the story particular attention is paid to reflection about the role of history and philosophy of physics in teaching/learning and to their contributions to a number of research strands, namely, conceptual change, students' difficulties, curriculum design, educational reconstruction and teacher education.

6.2 The Curricula of Reference

The story starts in the 1960s when Taylor and Wheeler (1965) and Resnick (1968) published two textbooks (teaching proposals) which became the main references for teaching special relativity both at university and at secondary school in the western world.

O. Levrini (✉)
Department of Physics and Astronomy, University of Bologna, Bologna 40127, Italy
e-mail: olivia.levrini2@unibo.it

M.R. Matthews (ed.), *International Handbook of Research in History,*
Philosophy and Science Teaching, DOI 10.1007/978-94-007-7654-8_6,
© Springer Science+Business Media Dordrecht 2014

In Italy Resnick's approach has been more popular in university courses and in textbooks for high school students. Taylor and Wheeler's approach has been chosen only by an élite group of teachers, even though the 1992 version of Taylor and Wheeler's book was conceived to be used by a wide range of students and teachers.

The two curricula focus on different concepts (relativistic effects or invariant quantities and relations), use different languages and reflect different interpretations of special relativity. Resnick's proposal presents the theory by following its historical development and in coherence with the operational approach ascribed to the original papers of Einstein; Taylor and Wheeler's proposal offers an elegant and conceptually transparent nonhistorical reconstruction of the theory, by relying heavily on a geometrical/Minkowskian formulation of special relativity.

More specifically, Resnick's proposal holds the thesis that relativity is 'a theory of measurement' and such a thesis is argued by focusing on Lorentz transformations, by using algebraic representations predominantly and by illustrating the physical meaning of the relativistic effects (relativity of simultaneity, length contraction and time dilation) by means of Einstein's thought experiments. Particular attention is paid to the experimental results that led to – and corroborated – the need for revising the conceptual bases of classical kinematics and dynamics.

The proposal of Taylor and Wheeler emphasises, by means of geometrical language, the concepts of event, space-time interval, energy-momentum and invariance. One of its main advantages is that it paves the way to contemporary physics, including particle physics and general relativity. Unlike the traditional approach and thanks to their decision to keep away from history, the authors make the puzzling choice of presenting, from the beginning, special relativity as a theory holding in absence of gravitation, that is, according to the equivalence principle, in free-floating frames of reference.

The success of Resnick's proposal is probably due, besides the clearness and coherence of the presentation, to the use of steps in the historical development of the special theory of relativity. Use of history is usually conceived as an effective teaching strategy for softening the impact of a theory that requires that space and time views be deeply revised (Holton 1973). Moreover the operational approach, although criticised by many physicists and philosophers as a simplistic form of empiricism, is accepted to have a great persuasive power because of the image of concreteness that it seems to give.

The main criticisms addressed to Resnick's approach concern the limits shown by the algebraic-operational language both when moving to general relativity and in highlighting the formal four-dimensional structure according to which new relations among the dynamical concepts of mass, energy and momentum must be redefined.

Despite its cultural relevance and conceptual transparency, the proposal of Taylor and Wheeler met more problems in its diffusion because of problems of implementation at the secondary school level. The main sources of problems are the 'length' of the outlined curriculum path with respect to the time that can be scheduled for relativity within the whole physics curriculum in high school and its unfamiliarity that teachers may find it difficult to compare it with the paths they followed as students or that are presented in textbooks.

6 The Role of History and Philosophy in Research on Teaching and Learning... 159

These two authoritative curricula represent, in this paper, the reference both for discussing the main results obtained by research in teaching/learning special relativity and for roughly sketching the development of the research along the following macro-strands:

(a) The affirmation of learning as a process of conceptual change (§3)
(b) The issue of the image of physics and the need for rethinking the curriculum: the role of history and philosophy for designing teaching proposals (§4)
(c) The problem of disseminating innovation in school by means of teacher education (§5)

6.3 The Research on Conceptual Change

In his chapter in the *Cambridge Handbook of the Learning Sciences*, diSessa provides an interesting review of research on conceptual change and of its relevance in the learning sciences (diSessa 2006).

By taking 'History of Conceptual Change Research' as a reference, some studies focusing on teaching/learning special relativity can be selected as representative both of the development of this research strand and of the multiple perspectives which have been established in this multi-faced area:

– The milestone paper by Posner et al. (1982), which established the first standard in model of conceptual change
– Contemporary or follow-up papers, which analysed case studies (Hewson 1982) and proposed extensions, specifications and/or revisions of the model of Posner and colleagues (Hewson and Thorley 1989; Gil and Solbes 1993; Villani and Arruda 1998)
– Empirical studies (Villani and Pacca 1987) and thoughtful papers emblematic of the 'misconception movement' (Scherr et al. 2001, 2002), which allowed students' difficulties to be pointed out
– A paper representing the 'knowledge in pieces' perspective where a specific model of concept and conceptual change (the coordination class model) was used for analysing an extended classroom episode where students were coping with the concept of proper time (Levrini and diSessa 2008).

Beside their theoretical or methodological value that goes beyond the physical domain to which the examples refer, these papers show two important common features: they all focus on the same, or very similar, learning problems, namely, students' difficulties in accepting space-time implications of special relativity, and they all refer, more or less extensively, to the history of special relativity for designing successful teaching strategies and/or for explaining their success.

Therefore, the work of comparing and discussing the results is not only relatively simple but also suitable for stressing how the research developed, what stable results were achieved and what open problems are left.

The paper of Posner and colleagues represents a 'landmark in the introduction of rational models of conceptual change: models that hold that students, like scientists, maintain current ideas unless there are good (rational) reasons to abandon them' (diSessa 2006). In particular, the model proposes that students and scientists change their conceptual systems only when several conditions are met: there must be *dissatisfaction* with existing conceptions; a new conception has to be *intelligible,* to appear initially *plausible,* and it should suggest the possibility of a *fruitful* research programme (Posner et al. 1982).

One example reported by Posner and colleagues is a case study described more extensively in a paper by Hewson where he demonstrates the importance of the learner's metaphysical commitments as components of the existing knowledge and where he explicitly suggests the effectiveness of instruction organised around that framework (Hewson 1982).

The discovered metaphysical commitments can be referred, in Hewson's words, to a *mechanistic view* of the world. According to such a view, extended objects are assumed to have fixed properties (a fixed length), conceived the fundamental reality in nature, and any explanation of relativistic effects is given in mechanistic terms.

The strength of such metaphysical commitments is argued to be the main barrier which prevents the learner, whose interview is discussed, to accept the counterintuitive aspects of the theory. Length is still treated by the learner as a constant, independent of the choice of the frame of reference, and the length contraction is conceived as a distortion of perception. Moreover, since, in the light-clock experiment, the time dilation is explained on the basis of the behaviour of the light, every other problem of time dilation is assumed to depend on the same mechanism, that is, on light.

One point of particular interest for our reasoning is the strategy followed by the interviewer for encouraging the interviewed student to change his metaphysical commitments. The strategy leads indeed to the same results at which almost all the other reviewed papers arrive, in spite of the different routes the authors follow.

The interviewer decided to introduce the orthodox Einsteinian position so as 'to present the point of view that *events*[1] were more fundamental and that length, for example, could be interpreted in terms of events, that is, something that is localized in space and time'.

This choice led the student to change his focus of attention. While he still believed that there is a reality independent of measurement, he changed his view of how reality manifests itself: it manifests itself by means of events.

The change is explained by arguing that the interviewer used a strategy built according to the model of conceptual change: at first the interviewed student was led to become dissatisfied with his existing knowledge (he was not able to explain time dilation in situations different from the light clock since he was unsuccessfully looking for a mechanism linked to light behaviour), and, then, he was led to deal with an intelligible and plausible new conception (focus on single events), to be reconciled with his old conception of reality so as to evaluate also its fruitfulness.

---

[1] Italics is, significantly, in the original text. It is not added.

6 The Role of History and Philosophy in Research on Teaching and Learning...   161

Other studies, arising out of the so-called misconceptions movement, produced original and thoughtful problems for investigating students' knowledge and arrived at identifying the same kind of difficulties: the tendency for considering the relativistic effects as perspective distortions and the difficulty of giving up the idea of absolute time or absolute simultaneity (Scherr et al. 2001). More specifically the researchers have observed that students often fail to interpret properly the 'time of an event' and the notion of 'reference frame': 'We found that students at all levels tend to treat the time of an event as the time at which a signal from the event is received by an observer. Thus, they consider a reference frame as being location dependent.' (Scherr et al. 2002, p. 1239).[2]

On the basis of these results, Scheer and colleagues produced a curriculum for university students that proved to be successful, being tested by means of a research methodology which employed pretests and posttest with a large sample of students.

Also in this case, the strategy implemented in tutorials is that of focusing on the concept of *event*, following the orthodox Einsteinian view. In particular, two tutorials were designed. The first one aims at guiding students

> to develop the basic procedures that allow an observer to measure *the time of a single distant event*. These procedures form the basis for defining a *reference frame as a system of intelligent observers*. The tutorial then helps students to extend the intuitive notion of whether or not *two local events* are simultaneous by having them develop a definition of simultaneity for events that have a spatial separation. (Scherr et al. 2002, p. 1239, italics added)

The second tutorial aims at guiding students 'to examine the consequences of the invariance of the speed of light through an analysis of the train paradox' (Scherr et al. 2002, p. 1239).

Hence, the tutorials were designed to guide students to the analysis of the Einstein's thought experiments, by a structured and operational concept of frame of reference as a lattice of rules and synchronised clocks or, as Scherr and colleagues say, as 'a system of intelligent observers' (Scherr et al. 2001, 2002). The construction of such a system of intelligent observers implies emphasising (i) the notion of time of a single event, measured by the clock situated in the same spatial position of the event; (ii) the procedure of measuring the time of a distant single event, according to the constraint that there exists a limit to the speed of signals and (iii) the need to generalise the measurement procedure for the time of an event in order to devise an arrangement of observers and equipment that allows the position and time of an arbitrary event to be recorded.

It is curious to note that most secondary and university textbooks teach special relativity following Resnick's approach and, hence, employ presentations similar to Einstein's original 1905 publication. However, unlike Einstein's papers and the original Resnick book, the textbooks pay little attention to what has been proved to

---

[2] In a recent survey carried out with about 100 prospective French teachers (de Hosson et al. 2010), a variation of the Scherr and colleagues problem is used in order to identify 'the types of reasoning implemented by prospective physics teachers faced with situations of classical and relativistic kinematics'. Even though the criteria used by the researchers for distinguishing classical and relativistic kinematics seem somehow unconventional, the results confirm what Scherr and colleagues achieved.

be crucial for learning by research in physics education: the role of the concept of event for redefining space and time according to the new constraints of the theory, that is, the unsurpassable and constant speed of light. From an epistemological perspective, despite following Einstein's reasoning, textbooks do not attach relevance to his original operational perspective, which can be seen as consistent with the idea that space and time are special names we give to ways of relating events by measurement (Levrini 2002b).

Because of their empirical orientation, the studies of Scherr and colleagues enlarged the heuristic bases for supporting some results previously obtained by Posner and others. Nevertheless, just because of their empirical orientation, these studies did not enter the debate concerning the model of conceptual change and the robustness of the arguments provided by Posner and colleagues and by Hewson for interpreting the successful results.

These points, instead, were the subject of other works that stressed the need of extending, refining or even revising the model because of the following weaknesses (recognised also by Hewson himself and Thorley, 1989):

(A) The model is not complete, since meaningful learning of the basic concepts of physics would demand not only a deep conceptual change but also a methodological and epistemological change (Gil and Solbes 1993).
(B) The model, in spite of the many attempts to apply it, is far from being a scheme for instruction, since the problem of establishing if and how the conditions required for conceptual change can be realised in classroom (what conceptual ecology) is unsolved (see, e.g. Villani and Arruda 1998).
(C) The framework is essentially epistemological and it does not reflect direct psychological reality. In particular, it 'specifies conditions for change without specifying detailed processes of change' (Levrini and diSessa 2008).

Gil and Solbes (1993), unlike the authors of the other papers discussed below, do not focus their analysis on specific criticalities seen in the model of Posner and colleagues. They simply propose to enlarge it so as to explicitly include the problem of how to promote in students a methodological and epistemological change since, they suppose,

> pupils' difficulties in learning modern physics have an epistemological origin; that is to say, they come from an ignorance of the deep conceptual revolution that the emergence of the new paradigms constitutes. Any meaningful learning of the few elements of modern physics introduced in high school would then be obstructed by the linear, accumulative view presented. In brief: modern physics was *against* the classical paradigm, and its meaningful learning would demand a similar approach. (Gil and Solbes 1993, p. 257)

According to this hypothesis, the researchers designed a programme for high school students where, as far as special relativity is concerned, experiments similar to those of Michelson and Morley were discussed to force the students both to question the existence of absolute space and time and to recognise that, in classical mechanics, there are implicit assumptions, i.e. assumptions that were accepted as obvious, and, because of that, their revision constituted one of the main difficulties in the development of science. By implementing the research methodology of comparing an experimental group and a group of control of high school students, they demonstrated that the teaching programme was effective for enabling students to derive 'quite easily' the variation of space and time. The success is explained by saying that once the relativity of time intervals and lengths for different observers are accepted as hypotheses, learning becomes easier: the programmes of activities 'that aim at producing in pupils a conceptual change similar to the historical change of paradigm' revealed to be effective since they gave 'a more correct view of physics – and particularly of modern physics – with a constructivist approach' (Gil and Solbes 1993, p. 260).

The work of Villani and Arruda (1998) stems from the acknowledgement of a specific criticality in the model of Posner and colleagues: the acceptance of a counterintuitive theory like special relativity is a complex problem underdeveloped in the model of Posner and colleagues. In real situations, instruction seems to be successful, but

> this apparent success can be misleading, since, at this moment in the students' intellectual development, the acceptance of the theory is only provisional, because its plausibility is external to their deep convictions… It would seem that not enough effort has been exerted to render the principles of the theory compatible with students' conceptual ecology. … As a consequence the learning of the Special Theory of Relativity rapidly disintegrates and students remember only a few disconnected elements mixed with many spontaneous ideas about absolute space and time. (Villani and Arruda 1998, p. 88)

The researchers argue that, in order to increase intelligibility and to transform a superficial sense of plausibility into a stronger one (so as to give stability to knowledge), the minimal conceptual ecology should include the troubled story that special relativity had to go along for being *accepted* by the scientific community. The argument of the analogy between historical ideas and some of the tendencies in spontaneous reasoning is developed, in this case, by focusing on a specific moment of the history of special relativity: the moment in which Einstein realised that

> he was unable to produce a microscopic model of the interaction between matter and radiation, so he decided instead to formulate a theory based on two universal principles. He used an analogy with thermodynamics, in which the principles and consequences are derived from the impossibility of perpetual motion, without any components of matter. (Villani and Arruda 1998, p. 91)

The Einsteinian distinction between theories of principles and constructive theories (Einstein 1919) is considered to be a profound epistemological and ontological change in the debate about the acceptance of the consequences of the Lorentz transformations (i.e. the relativistic effects)[3]: debate dominated, until that moment, by the theory of the electron of Lorentz, according to which length contraction, for example, had to be understood by searching for a microscopic mechanism. This profound change was, according to the study by Villani and Arruda, one of the main reasons of the difficulties the theory met to be accepted.

---

[3] The role of thermodynamics in the genesis of SR is extensively argued in the paper of Abiko (2005) which will be discussed later.

In spite of the interesting example of historical reconstruction carried out from an educational perspective and in spite of the relevant focus of the research (the problem of acceptance), the authors are very prudent in their conclusions:

> We do not expect students to go through a conceptual change in the sense of systematic change in the way they analyze physical phenomena: we only hope that they will become aware of the existence and the essential features of a conceptual change in the history of science and that they will realize that this change allowed modern technology to advance. (Villani and Arruda 1998, p. 94)

The problem of entering the cognitive mechanism for a deep acceptance remains theoretically unsolved.

The paper of Levrini and diSessa aims at entering more deeply the cognitive process of change in learning special relativity, by applying a carefully defined and tested theory: 'coordination class theory' (diSessa and Sherin 1998).

Coordination class theory is a developing model of concepts and conceptual change, framed in a 'complex knowledge system' perspective and consistent with the epistemological perspective of 'knowledge in pieces' (diSessa 1993). In this view, a 'concept' is not seen as a single, unified idea but a large and intricately organised system, which effectively coordinates activation and use of many specific elements according to context. Learning a concept is seen as a process of recruiting and 'coordinating a large number of elements in many ways'. According to such a theory, empirically tracking those different ways to achieve the effect of a concept leads to better understanding of learning and failing to learn.

The type of cognitive process analysis carried out in the study of Levrini and diSessa is based on the following assumptions:

(a) Using a carefully defined and tested theory or model allows a more precise tracking of data, in order to see in more detail *when* various kinds of learning events are happening, *what* their nature and effects are, and *why* they happen.
(b) As a result, data can overturn or improve even insightful, but rough, guesses as to the source of learning difficulties and how to overcome them.
(c) Well-developed (learning) theory, as in physics, applies to a broader range of situations than those out of which it arose. Insights from one context can be bootstrapped more reliably into broader insights concerning learning physics.

Specifically, the paper contains an analysis of a single classroom episode in which secondary students reveal difficulties with the concept of proper time but slowly make progress in improving their understanding. The concept of proper time, like the concepts of proper length and mass, is particularly tricky since its understanding is strongly dependent on the level of appropriation of the shift from a Newtonian space plus time to the relativistic space-time. Its property of invariance, as effectively stressed in the book of Taylor and Wheeler, is indeed an expression of the invariance of space-time interval between two events.[4]

---

[4] Also the invariance of mass is strictly related to the relativistic space-time structure, being the module of energy- momentum 4-vector. This point is addressed very effectively by Taylor and Wheeler, and it would be worth understanding why many textbooks and teachers still use the

6 The Role of History and Philosophy in Research on Teaching and Learning... 165

The concept of proper time is usually introduced, in teaching, through the light-clock thought experiment. But the analysis and discussion of this thought experiment do not usually focus on *those specific* space-time properties of *those couple of events* whose space-time interval is called proper time (e.g. by defining proper time as the time interval measured by *two events occurring in the same position*). The light-clock thought experiment is instead used for defining proper time as the *time duration of a phenomenon* (the back and forth travelling of light ray) measured in the frame of reference *at rest with respect to* the light clock. A good guess, supported also by the works of Hewson mentioned above, is that students expect, in every other context where proper time has to be determined, simply to see an object (like the light clock) that determines the relevant frame, or, slightly more complexly, the frame is determined by the (potentially moving) 'location of the phenomenon' whose duration is to be measured. This implies that teaching may reinforce the persistence of 'classical ontological inferences' that take for granted the existence of phenomena as unproblematic things that have a place and a duration. This is what happened in the classroom episode analysed in the paper of Levrini and diSessa, where it is shown, by applying the coordination class theory, that the students tend to maintain a classical ontology which led them to *coordinate* the property of invariance as an inner, intrinsic, property of a phenomenon.

The paper argues, like other papers mentioned above, that changing students' perspectives from 'looking in terms of phenomena' to 'looking in terms of events' is an educational goal that, if accomplished, substantially promotes effective conceptual competence. In the context of the coordination class theory, the shift in the fundamental view of the universe, from a place in which there are objects and phenomena to the universe as an ensemble of events, is explained as that cognitive process which implies:

(a) Identifying the space-time events as preferred foci of attention for all relevant determinations, that is, for preferentially using a particular class of strategies for *reading out*[5] information from a context (e.g. reading out from the context *what events* are of interest) and for *inferring*, from the readouts, the particular information at issue (e.g. *how* the events are related with one other in space-time)
(b) *Displacing* persistent ontological assumptions about length and duration of things and phenomena (objects and phenomena relate to space-time measurements only and precisely for the family of events that the object/phenomenon 'lays down' in space-time)
(c) Factoring the space of possible *concept projections of all coordinations* in special relativity (the particular knowledge used in applying a concept in different situations)

---

notion of relativistic mass (a mass dependent of velocity), in spite of the sharp criticisms known in literature (e.g. Adler 1987; Warren 1976; Whitaker 1976).

[5] The words in italics are technical words within the coordination class model. Their detailed explanation is beyond the scope of the paper. They are extensively described in the paper I am reviewing (Levrini and diSessa 2008).

As far as the last point is concerned, possible projections of coordination, enacted by students and/or suggested by some definitions in special relativity, are, for example, (i) first identify the set of relevant events and then proceed from there (determining other things on the basis of properties of the relevant events), (ii) 'finding the *right* frame of reference to make our coordination – our way of determining the relevant quantity – easier' and (iii) 'sitting on the relevant object or phenomenon'. Factoring, as it is shown in the paper, includes giving priority, over the other possible classes, to the class of projections that first identify the set of relevant events: all the other possible classes are proved to have little span (they work only in particular cases) and/or reinforce the persistence of classical ontological inferences.

The detailed analysis allowed the authors to put forward implications on teaching and on investigations about the role of history and epistemology of physics in teaching/learning special relativity. In particular, the application of the coordination class model to the data provided a theory-based explanation of *why* explicitly exposing, managing and relating multiple classes of projections of a physical concept seem to be a good instructional technique to work around documented difficulties in conceptual change in special relativity. Indeed, the positive reaction of the students facing their difficulties can be ascribable also to the fact that they were previously guided to compare the operational and geometrical approach to special relativity, by analysing excerpts of the original publications of Einstein and Minkowski. The comparison between two different approaches was chosen by the teacher as a way to bring the complexities of historically different interpretative perspectives into the classroom and to situate special relativity within the philosophical debate about space and time. Before the analysis, the richness and diversity of the thinking about the historical context was the 'secret ingredient' in achieving greater conceptual competence. The analysis, instead, led the researchers to believe that a coordination class perspective explains some of that success: framing multiple classes of projections as historical perspectives stimulated students to confront *consciously* the two main sources of learning difficulties that the coordination class model hypothesises – the problems of *span* and *alignment*. *Span* concerns the problem of having adequate conceptual resources to operate the concept across a wide range of contexts in which it is applicable, while *alignment* concerns the problem of being able to determine the same concept-characteristic information across diverse circumstances.

To sum up, the overview of the papers about teaching/learning special relativity framed within the conceptual change strand shows a significant variety of research methods and arguments progressively developed within PER for diagnosing students' difficulties, pointing out critical aspects of traditional curricula, explaining successes and failures and providing suggestions for innovative and effective instruction.

The most important point I want to stress here is that significant studies, carried out by applying different research methodologies according to different theoretical perspectives, provide multiple cross arguments for supporting one common point: guiding students to look in terms of events is crucial for promoting deep

6 The Role of History and Philosophy in Research on Teaching and Learning...

understanding in special relativity. This point has been shown to be a critical detail (Viennot et al. 2005), i.e. a *detail* whose disregarding can prevent students from grasp the *global meaning of the theory*. Within the forest of results obtained by PER and the jungle of methodological procedures invented and/or applied, the agreement on this result, as well as the quality of the process followed for achieving it, should elevate it to a robust piece of knowledge which every new teaching proposal should rigorously assume as a constraint. This case is an effective example for showing that PER is established as a research field which is able to obtain shared results and to claim 'this is where we are', as far as the research on students' difficulties in special relativity is concerned.

In the papers considered in this section, the relevance of specific aspects of history and philosophy of physics (the reference to the orthodox original interpretation of Einstein, the troubled story of its acceptance and the comparison between different perspectives, such as those of Einstein and the Minkowski) was supported and argued on the basis of epistemological or cognitive arguments stemming from the problem of how conceptual change can be studied and promoted.

The design of a teaching proposal or a teaching approach is, however, a process that usually aims at going, if possible, beyond its effectiveness in structuring a conceptual landscape suitable for supporting conceptual change and deep understanding. A teaching proposal or an approach is a complex cultural construction that an author or a team of authors produce for promoting also a specific view of physics and of learning and/or for exploiting the role of science education in the intellectual and emotional growth of pupils. In the next section, the debate about the role of history and philosophy of physics is addressed within such a strand, that is, within the general problem of how to promote physics as culture.

6.4 The Debate About the Role of History and Philosophy of Physics for Promoting Physics as Culture

Besides Resnick's and Taylor and Wheeler's teaching proposals, there have been other important projects.[6] In the set of proposals, some of them[7] represent historical reconstructions inspired either by an experimental approach built on the limiting value of the speed of light and, consistently, on Bertozzi's experiment realised within the PSSC project (Cortini et al. 1977; Cortini 1978) or by Taylor and Wheeler's approach (Borghi et al. 1993; Fabri 2005).

---

[6] See, for example, Arriassecq and Greca (2010), Borghi et al. (1993), Cortini et al. (1977), Cortini (1978), Fabri (2005), Levrini (2002a, b), Solbes (1986), and Villani and Arruda (1998) for secondary school students and Angotti et al. (1978), Scherr et al. (2002) for university students. The list is certainly incomplete. I am quoting only the proposals I found cited in the research literature. Yet, in some cases, I did not have direct access to the whole texts because they are teaching texts and, hence, written only in the national language.

[7] See Borghi et al. (1993), Cortini et al. (1977), Cortini (1978), and Fabri (2005).

Other proposals refer explicitly to the need of a historical-philosophical contextualisation of teaching for stressing the cultural value of special relativity.[8] All these proposals share the aim of addressing history in a way so as to overcome the teaching habits of referring to a fictional pseudo-history, focused on an over-evaluation of the Michelson and Morley experiment, that many textbooks disseminate and that tends to promote a hyper-simplified and unrealistic form of empiricism (Holton 1973). They moreover share the belief that a historical contextualisation is needed to achieve a high educational goal: to stress and to exploit the philosophical implications of a theory which led explanatory paradigms to be changed and/or which had an impressive influence on other cultural fields such the arts, literature and music.

The implementations of these proposals, usually carried out by the researchers themselves who designed the proposal or under their supervision, obtained results that are said to be encouraging and efficient. In particular the authors seem to agree that their results show that even secondary students manifested great interest in the matter and made no complaints about the mathematical complexities involved. As I discussed in the previous section, some of these proposals evaluated their effectiveness by framing students' reactions within the conceptual change strand.[9]

Even though their effectiveness was tested according to conceptual change models, the proposals of Gil and Solbes (1993) and Levrini (2002b) were however designed within a research strand which received greater and greater attention during the 1990s: the problem of what image of physics should be promoted in teaching (Grimellini Tomasini and Levrini 2001). For example, Gil and Solbes claim:

> High school teachers and textbooks transmit an incorrect image of science, which ignores the existence of crises and paradigm shifts. The introduction of topics of modern physics, in particular, takes place without reference to its essential novelty or to the main differences between the classical and the new paradigm. A suitable occasion for showing the richness of the development of science and importance of science revolutions is thus wasted. (Gil and Solbes 1993, p. 260)

The issue of the image of science triggered a deep debate about the problem of the relationship among history of science, philosophy and science education.[10] The debate led to a new research field, history and philosophy of science and science teaching (HPS&ST), to be progressively established, but, as pointed out by Galili in a recent paper, 'despite the intensive support for using the HPS in science teaching and articulation of its advantages (e.g., Matthews 1994, p. 38), the issue continues to be complex and controversial' (Galili 2011).

In the specific research literature of HPS&ST, original contributions about teaching/learning relativity, provided either by professional historians or philosophers to science education or by science education researchers deeply

---

[8] See Arriassecq and Greca (2010), Levrini (2002a, b), Solbes (1986), and Villani and Arruda (1998).

[9] See Gil and Solbes (1993), Levrini and diSessa (2008), Scherr et al. (2001, 2002), and Villani and Arruda (1998).

[10] See, for example, Bevilacqua et al. (2001), Cobern (2000), Duschl (1985), Gauld (1991), and Matthews (1994).

6 The Role of History and Philosophy in Research on Teaching and Learning...

involved in historical-philosophical studies, are curiously rather few. Nevertheless, they can be significantly compared for discussing the role that a specific research in history and philosophy of special relativity can play in designing teaching proposals.

Levrini, in her works (2000, 2002a, b, 2004), presents an educational reconstruction of the influence of historical-philosophical debate between 'relationalism and substantivalism' on the concepts of space and time in physics. In particular, the original publications of Einstein, Minkowski and Poincaré were comparatively analysed in order to trace back the historical-philosophical roots of the interpretations of general relativity inspired, respectively, by the works of Sciama, Wheeler and Weinberg. The educational reconstruction was motivated by a specific cultural and educational assumption: 'teaching relativity at the secondary school level gains particular meaning if the theory is critically situated within the cultural debate on space and time and if the role of history and philosophy of physics is exploited in order to *provide students with keys for comparing different interpretations of the theory'*. (Levrini 2004, p. 621, italics added). Historical and philosophical debates are assumed, if they are properly reconstructed from an educational point of view, to be an effective teaching strategy for helping 'students to focus on the peculiar aspects of each interpretation and to elaborate logical, cultural, rhetorical, cognitive instruments for comparing different perspectives and for expressing their own preferences' (Levrini 2002b, p. 613).

Another historical debate which has been subject of many controversies among the historians of physics concerns the genesis of special relativity and, in particular, 'the so-called Lorentz-Einstein problem,' i.e. the question of whether or not Lorentz and Poincaré built, slightly before Albert Einstein, the special theory of relativity. The issue, like every historical issue about the paternity of ideas, triggered heated debates among historians.[11] Nevertheless, within secondary school textbooks or in science education literature, the discussion on the genesis of special relativity is still strongly focused on the figure of Einstein and on the role played by the Michelson and Morley experiment. In other words, the debate on the genesis of special relativity, when it is mentioned or discussed in textbooks or in research papers which report teaching proposals (e.g. Arriassecq and Greca 2010), seems to refer more or less explicitly to the perspectives of Holton (1973) and Resnick (1968).

In the specialised and recent literature of HPS&ST, two historians of physics addressed the issue of the genesis of special relativity from an educational perspective, Abiko (2005) and Giannetto (2009). The two papers focus on different aspects on the issue and support different theses.

Abiko addresses his historical reconstruction of the genesis of special relativity by tracing the origins of Einstein's view back to thermodynamics. In the paper, the

---

[11] Within this debate, very strong positions can be also found for supporting the role played by Poincaré (see, e.g. Bjerknes C. J. (2002), *Albert Einstein, The Incorrigible Plagiarist*, XTX Inc, Downers Grove, Illinois ; Hladik J. (2004). *Comment le jeune et ambitieux Einstein s'est approprié la Relativité restreinte de Poincaré*, Ellipses Édition Markenting S.A, Paris; Leveugle J. (2004). *La Relativité, Poincaré et Enstein, Planck, Hilbert. Histoire véridique de la Théorie de la Relativité*, L' Harmattan, Paris).

crucial point of departure for Einstein is said to be 'his encounter with Planck's derivation of the radiation-formula…, and [his] resultant distrust of contemporary electromagnetic theory'. On the basis of a new analysis of Einstein's *Autobiographical Notes*, Abiko arrives at arguing that "Notes' make clear that, of the three theories of classical physics (i.e., mechanics, electromagnetic theory, thermodynamics), Einstein regarded thermodynamics as the only physical theory of universal content that will never be overthrown within its sphere of applicability' (Abiko 2005, p. 359). According to his thesis, Abiko sees 'an obvious and crucial discrimination between Lorentz-Poincaré's theory and Einstein's STR [which] rests on the difference between the constancy of light-velocity and the light-velocity postulate. Both Lorentz-Poincaré and Einstein believed in the constancy of light-velocity. But, it was Einstein and only he that elevated it to the status of the postulate' (Abiko 2005, p. 353). The reason of that is, again, Einstein's distrust of Maxwell's electromagnetism: 'in order to transcend Maxwell's electrodynamics, he had no choice but to elevate the constancy of light-velocity deduced from the latter to the status of the light-velocity postulate' (Abiko 2005, p. 357).

In his paper, Giannetto (2009) supports the thesis that '[t]he revolution in XX century physics, induced by relativity theories, had its roots within the electromagnetic conception of Nature. […] The electromagnetic conception of Nature was in some way realized by the relativistic dynamics of Poincaré of 1905. Einstein, on the contrary, after some years, linked relativistic dynamics to a semi-mechanist conception of Nature'. (2009, p. 765). By semi-mechanistic he means that 'in Einstein there is a residual form of a mechanist conception: mechanics is always considered the first physical science and constitutes the independent foundation of all physics' (2009, p. 774).

Independently of the supported theses, the two papers seem to share a similar view about the educational and cultural relevance of a historical approach focused on the genesis of special relativity, i.e. on that process of a new theory's emergence and acceptance. They both stress, in particular, the importance of enabling students to enter twentieth-century physics as 'the transition from the 'clockwork mechanism' of Newtonian science to the 'evolutionary process' of modern science' (Abiko 2005, p. 362) or to 'a 'new alliance' (Prigogine and Stengers 1979) among God, mankind and nature, a new cosmic and ethic order, not pre-fixed but the fruit of a complex dynamical, temporal free evolution' (Giannetto 2009; p. 778).

The two papers, moreover, share the emphasis on the relationship to other scientific theories, worldviews and surrounding social and cultural factors:

> beyond the general non-dogmatic method of science research, science has no unique worldview. Science in its historical practices is the place where different worldviews [like Mechanist, Thermodynamic; Electromagnetic] have been in conflict with each other. Not only different scientific theories, but even different formulations of a scientific theory have different presuppositions and implications for worldview as well as for religion. (Giannetto 2009, p. 779)

The last quotation touches a very delicate point – somehow dramatic – that researchers in science education have to address when they want to design a teaching proposal: To what extent am I imposing my personal worldview in the design of a teaching proposal? What ideological reasons am I projecting into the proposal? If the proposal is based on a historical/epistemological approach, what historical-philosophical interpretation of a theory should I choose? Why? Why should my personal worldview have a higher educational value than another?

In the light of questions like these, history and philosophy/epistemology can play a crucial role: 'Especially within an educational framework, one should never impose her/his private worldview *but one should deal with the science/religion problem from a historical perspective: one should show how science practices involve a conflict among various worldviews'* (Giannetto 2009, p. 766, italics added).

The focus of the papers of Levrini, Abiko and Giannetto on fundamental debates (the debate between 'substantivalism and relationalism' and the debate about the genesis of special relativity) highlights a relevant role for history and philosophy: they provide some examples for showing how history can allow different philosophical/epistemological interpretations and different worldviews to be *analysed in perspective*, that is, comparatively analysed not as 'orthodox or heterodox' positions but as different possible perspectives from which a theory can be seen.

The overview of the research work about the design of teaching proposals allows me to conclude that:

– There exist few attempts at designing new teaching proposals about special relativity; most of the work is still strongly influenced by the two main approaches of Resnick and Taylor and Wheeler.
– The proposals are usually presented and supported within physics education research in terms of their effectiveness in motivating students or in triggering processes of conceptual change.
– The debate about the cultural and philosophical/epistemological presuppositions that exist behind an interpretative choice concerning special relativity presentation is undeveloped, but it would be fundamental for comparing different proposals.
– In order to foster comparability, history can play the specific and significant role of allowing different philosophical/epistemological interpretations of a theory to be *analysed in perspective*; it can allow the different nuances of the proposals to be explored as expressions of different worldviews rather than to be *classified* in terms of 'orthodox/heterodox', 'better/worse', 'closer to/further from the historical truth'.

The choice of presenting special relativity from different perspectives is not only cultural and ethical: as shown in the previous section, it can have important implications for learning (Levrini and diSessa 2008), and, as I will show in the next section, it is fundamental for teacher education (De Ambrosis and Levrini 2010).

6.5 Teacher Education

As well as the cultural value of the proposals and their effectiveness, demonstrated in implementations carried out under the supervision of researchers who designed the proposals, another issue, fundamental for improving the teaching/learning of special relativity and for filling the gap between research and school reality, is how to enable teachers to manage and implement autonomously innovative, effective and culturally meaningful teaching proposals (Grimellini Tomasini and Levrini 2001, 2004).

The general research issue of how to promote innovation in school through teacher education has been the object of many studies, since 1990. For example, the STTIS (Science Teacher Training in an Information Society) European Project (1997–2001), aimed at identifying and analysing transformations between what was expected by implementing a research-based sequence and what is observed when teachers put innovation into practice, produced very important research results on this issue. In particular it showed that implementation often implies a transformation of the original proposals, sometimes with the loss of important aspects of innovation (Pintò 2005).

The studies in this research strand pointed out some general tendencies or common attitudes of teachers toward approaches they perceived as innovative:

- The tendency of accepting or refusing the whole proposal on the basis of personal or local criteria and the difficulty in moving from a global scale to a local one (in recognising the 'critical details' of a proposal): 'Critical details are not always disregarded by teachers because their grasp of the global rationale is superficial. It may result from a lack of training as to a connection between details and global rationale' (Viennot et al. 2005)
- The tendency of mixing new with the old (Viennot et al. 2005) and of transforming the proposal so as to obtain intellectual and professional satisfaction
- The tendency of accepting the challenge of going deep into the proposal only if it represents an answer to disciplinary problems recognised as crucial by the teachers (Eylon and Bagno 2006)

Within this research strand, a specific project was carried out for analysing teachers' attitudes toward Taylor and Wheeler's (TW) proposal (De Ambrosis and Levrini 2010). In particular, the paper concerns an empirical study carried out with a group of 20 high school physics teachers engaged in an *at-distance* (based on a e-learning platform) masters course on the teaching of modern physics. The data refers to the module on special relativity, of which the authors of the paper were the trainers. The focus of the study was the process through which teachers analysed the TW textbook in order to appropriate it for the perspective of designing their own paths for use in the classroom.

The results obtained in the study allowed the researchers to argue that problems known in the research literature and usually related to the implementation process can already be found when teachers approach the proposal and try to appropriate it. The article demonstrates that by focusing on the appropriation process, it becomes possible to provide arguments to support that while the first tendency (being trapped in a local vision) represents a real obstacle for innovation, the last two tendencies can be transformed into productive resources, if 'properly' (at suitable *moments* and in appropriate *ways*) activated.

6 The Role of History and Philosophy in Research on Teaching and Learning...

The paper of De Ambrosis and Levrini is here presented and discussed to stress its contribution in pointing out indications that teachers' reactions can provide to research for designing innovative curricula, as well as further nuances of the role that history and philosophy of relativity can play in teaching/learning special relativity.

In the study, the appropriation process, followed by the group of teachers, was reconstructed in terms of stages and factors triggering the progressive development of teachers' attitudes and competences, as briefly presented below.

In particular, three stages were identified:

(A) *The acceptance of the game,* from the initial 'distrust' toward the proposal's novelties to the point of seeing the proposal as authoritative (a worthwhile, although demanding proposal)
(B) *The game,* played by going on, with patience and determination, in the analysis up to the critical point where it is possible not only to discover the details but also to attain a global perspective on the proposal
(C) *The exploration of the offstage of the game,* carried out in order to acquire criteria to make the proposal explicitly 'comparable' with proposals more familiar to the teachers (that of Resnick)

The authors of the paper argue that the first stage in the reaction of the teachers in face of the Taylor and Wheeler proposal was of distrust and resistance:

- The attention of the teachers was focused on very specific points of the proposal.
- They manifested the willingness of proposing immediately new alternatives (*it is better to start from a real problem, for example, from Michelson-Morley experiment).*
- The teachers referred to 'students' and 'personal experience' as arguments for distinguishing 'what works and what does not work': (*I think that a lot of confusion can be generated, in particular, among the less motivated students*).

The most important point, stressed in the paper about this phase, is that the attitude changed as soon as the discussion moved to disciplinary concepts that sounded puzzling for the teachers, like the definition of inertial systems and the relation between the inertial systems in Newton's mechanics and the free-floating inertial systems introduced by TW.

Through disciplinary puzzling points, the teachers arrived, after the analysis of the first two chapters of the book, at the shared conviction that to examine the proposal in depth is worthwhile: local elements belonging to the plan of teachers' disciplinary content knowledge proved to be crucial for triggering a more general change in the teaching perspective. The teachers indeed not only recognised some weaknesses in their disciplinary knowledge, but the kind of problems led them to acknowledge that local disciplinary inconsistencies in traditional teaching cannot be always solved by local interventions: they require sometimes a wider reconstruction or even a change of perspective.

This evidence is worth being stressed, since it confirms a result previously achieved also by Eylon and Bagno (2006): the authoritativeness of a proposal is evaluated by the teachers, first of all, in terms of its effectiveness in solving *specific disciplinary* problems *they* feel *real*.

The second stage described in the paper lasted from the analysis of chapter 3 to the end of TW book. During this phase the teachers moved from the search for answers to puzzling conceptual problems to the critical point where it was possible not only to discover the details but also to attain a global perspective on the proposal.

This stage is described by stressing how it was characterised by a special attitude (very different form the initial one): teachers showed they can be collectively involved in a medium term, patient and resolute search for shared *global criteria* of analysis and, in particular, in the problematic search for the *coherence* of this unusual proposal.

In the group, a stimulating discussion about coherence represented an important moment since the teachers could express and compare many different positions existing among them:

- Coherence as *logical development of a path from classical to modern physics*, as a result of a radical reconstruction process of the physics contents
- Coherence as *historical development* of a path
- Coherence as *systematic use of the experimental method* characteristic of physics inquiry
- Coherence as reconstruction of the physics contents starting from *fundamental concepts and categories such as the space-time description, causality and determinism*

The discovery of such a plurality of positions is argued to have been an important moment for orienting the collective investigation toward the acknowledgment of that special kind of coherence that characterises TW's approach. The teachers indeed arrived at accepting as sensible criteria for coherence some choices that are at the basis of the TW approach and that the teachers could deeply appreciate only at the end of the analysis: the choice of reconstructing special relativity in the light of general relativity and the choice of consistently revising all the concepts in the frame of a geometrical 4-dimensions space-time.

In the group it became clear that, as one teacher wrote,

in order to grasp the coherence of an approach it is necessary to internalize the meaning of a theory far beyond its formal aspects. It implies to go deep, to analyze its implications, to acquire different perspectives and interpretations. It means to know both the origin of the hypotheses and their consequences. The task is not easy for a teacher.

In the paper, it is shown how, at the end of the analysis, there was a general sense of satisfaction, a general agreement on the relevance of the proposal.

The authors of the paper argue that two factors have probably triggered or supported the process in the second step:

(a) The presence, within the group, of lively dynamics able to support the maturation of a shared conviction that global evaluation criteria were needed for justifying local choices.

6 The Role of History and Philosophy in Research on Teaching and Learning... 175

(b) The strong inner coherence of the textbook that enabled the trainers to firmly support the awareness that specific physical problems, addressed by this proposal and not solved clearly in a traditional approach, could not be always solved by local changes.

At the end of the collective analysis of the book, everything seemed in order. The proposal was recognised and accepted as authoritative, intelligible, innovative, culturally relevant and intellectually stimulating. But some final comments revealed a new form of distrust: the criteria for coherence were too subjective. I report here two comments of the teachers since they represent the basis for discussing, in the following, the specific role played by history and philosophy in this study:

> The proposal appears somehow as unilateral. ... I feel however a kind of mistrust (perhaps only intuitive) toward the question. As we were in front of a sort of situation forced by [the need of] searching for an excessive intelligibility in the relations between the physical quantities (Mario).
> I would like to focus on the hypotheses at the basis of TW's approach: I feel that there are non-explicit hypotheses or that I got lost along the path... I would like to improve my understanding of the whole proposal and be able to compare it with others around (Anna).

Teachers' comments showed that the process of appropriation was not complete: understanding the content is necessary but not sufficient to grasp the general meaning of a proposal and to feel comfortable with it; further tools were needed for disassembling and reassembling it, for going deep into its epistemological and cognitive assumptions (to what image of physics and physics teaching it is related), for comparing it with others (especially the typical textbook ones) and for adapting it to one's teaching/learning attitudes and constraints. Teachers' comments pointed out that the tricky 'problem of comparability' had to be collectively addressed.

During the third stage, the teachers were guided to critically analyse TW's approach in comparison to Resnick's. In order to make the two proposals comparable, the trainers used the teaching strategy of supporting the teachers in reconstructing the historical and philosophical roots of the two teaching proposals. In particular the teachers were guided in the analysis of the original publications of Einstein (1905) and Minkowski (1909) in order to find the historical roots of, respectively, Resnick's and TW's approach. The trainers gave the teachers materials prepared by the researchers of the group in Bologna that framed the original publications within the historical and philosophical debate on the concepts of space and time in physics: from Newton's *Principia* and the criticism by Leibniz and Mach, to general relativity, passing through Einstein's, Poincaré's and Minkowski's works and worldviews.

The main surprise for all the participants was the tone of the discussion, which showed that the teachers were, consciously but prudently, moving from the need of searching for the implicit assumptions of the TW textbook to the critical point where it was possible to compare the two approaches as different choices of content reconstruction inspired by different global views of the theory and of its teaching. As a teacher wrote, 'I was surprised by the "harmony" of the discussion on epistemological issues. Generally these discussions reveal very rigid points of view and irreducible convictions'.

The factors triggering the process of going beyond an ideological clash were probably some features of the materials, which encouraged the teachers to trace back the historical roots of the proposals and to situate such roots within a philosophical debate. In this case, the historical-philosophical dimension allowed the reconstruction of the interpretative dimension of each proposal, and this dimension was seen to be effective for making them 'commensurable'.

The reconstructed moving picture of the appropriation path showed that a multifaceted and complex 'process of change' is implied: appropriation requires teachers to become able to master the overall proposal at different levels (details, rationale and implicit presuppositions) and to coordinate different dimensions of knowledge (disciplinary, cognitive, philosophical and educational).

The study points out relevant indications for the design and the production of materials for teacher education. In order to foster appropriation, (i) the materials have to be built on a disciplinary reconstruction effective for solving conceptual problems that teachers feel relevant; (ii) they must show a strong local–global consistency (from details to rationale); (iii) they must be proved to be effectively usable in class; and (iv) they must be comparable with what teachers feel closer to, as well as suitable to be disassembled and reassembled according to different personal teaching styles, images of physics and of teaching. The last point implies that the presuppositions of a teaching proposal must be made explicit: for that purpose, the studies developed so far on teaching/learning special relativity revealed that a historical and philosophical/epistemological approach can be useful.

In the paper of De Ambrosis and Levrini, it is however argued that these features of materials are fundamental for appropriation mainly because they enabled the trainers to enact particular training strategies.

Just to give one example, in the paper, it is described how trainers acted in order to manage the lively and messy forum on Chapter 1 – where Taylor and Wheeler give an overview of the entire proposal. Instead of giving in to the temptation of providing quick – and necessarily partial – answers, the trainers prepared a document where the main problems and conflicting points arisen in the discussion were reported. The document was shaped as a sort of agenda of the whole course, indicating where the questions would have found, in their opinion, their 'natural' moment for discussion. In more detail, the trainers proposed an agenda where the trainers differed: problems concerning conceptual difficulties well known in the research literature (to be discussed in short-medium terms during the scheduled activity devoted to the analysis of specific research articles), perplexities concerning basic choices of the proposal (to be discussed in medium-long term and, in particular, at the end of the analysis of the whole textbook and during the analysis of the original works of Einstein and Minkowski) and problems concerning the implementation in class (to be discussed in long term).

The choice was made initially as a survival strategy in the face of the avalanche of questions to which the trainers could not practically respond because of the time and means imposed by the web communication. At the end the choice was revealed to be effective for changing the general attitude of the teachers' group and of the trainers themselves: the web communication and the necessity of communicating through written texts allowed all of them to keep track of the problems and of the whole process.

6 The Role of History and Philosophy in Research on Teaching and Learning...

This awareness contributed to overcoming the attitude of 'all-now-fast-and-easy' and to create a more relaxed atmosphere where different learning rhythms could have room. Moreover it contributed to giving a strong and coherent signal that an appropriation process is long, complex and multidimensional.

This study showed how the materials themselves, because of their specific features, acted as the basis for mediating the relationship between teachers and trainers: their being intellectually stimulating, locally-globally coherent and explicit enough in their presuppositions to be comparable with the more familiar proposals allowed the trainers to play the role of creating that *space of analysis and discussion* where each of the participants could follow her/his learning pace, explore her/his point of view, and take care of and the accountability of her/his appropriation path.

The example, however, points out a tricky problem: teachers' appropriation cannot be simply analysed in terms of a relation between teachers and teaching materials. Teachers' appropriation of new teaching materials is a complex and delicate process involving a relationship between teachers, teaching materials, proposals' designers and trainers. This relationship is complex and delicate for at least the following reasons:

- Teachers, designers and trainers are professionals with interwoven but distinct and differing competences; the relationship implies different roles to be acknowledged and exploited.
- The relationship is strongly influenced by the features of the materials; trivially, if the materials are shaped as closed packages of activities designed by researchers to be followed step by step by teachers, they induce, implicitly and explicitly, a deeply different relationship between designers, trainers and teachers with respect to materials shaped as a 'properly complex territory' (Bertozzi et al. in press; Levrini et al. 2010, 2011; Levrini and Fantini 2013) where different approaches and different learning routes are comparatively discussed.

To conclude, the studies carried out about teacher education in special relativity point out a delicate research issue which would deserve a special attention in the coming years. The issue concerns the relationship between research products, teachers, designers and trainers that is, more or less explicitly, mediated by the *structure*, the *format* and the *features* of the produced teaching materials. Structure, format and features of the materials are, indeed, carriers of that image of physics and of teaching on which the relationship between proposals' designers, trainers and teachers is established and on which the different roles can be played.

6.6 Conclusion

The paper provides a review of the main studies concerning teaching/learning special relativity. The review has been carried out with the following goals in mind:

- To identify the many dimensions (research strands) on which the problem of improving teaching/learning of special relativity has been so far projected in order to be studied. The resulting survey of the literature presents the image that all of the dimensions are intimately related and, given the inner complexity of every process of teaching/learning, they all must be considered in a meaningful educational reconstruction of relativity: 'to change one variable at a time [e.g.: introducing new problems, a new experiment or the analysis of an historical episode] simply doesn't work' (Duit 2006).

- To present, within each strand, what results can be considered stable and/or most current, including what problems are still unsolved.
- To stress the multifaceted role ascribed to history and philosophy of physics in the specific research domain which concerns teaching/learning special relativity.

The overview revealed that the research strand concerning students' difficulties in learning special relativity is well developed and shared results have been achieved. Within this strand, physics education research can be recognised as a developing research field that has produced, over the years, arguments for considering what is scientifically acceptable and what is unacceptable.

Deep unsolved research problems, instead, concern the design of teaching materials and the dissemination of good practices through teacher education.

As far as the first point is concerned, new collaborations between science educators, historians and philosophers of physics would be very useful both for making the historical and epistemological roots of teaching proposals increasingly explicit and for triggering a debate about the comparability of the proposals.

As far as the second point is concerned, further studies seem to be needed in order to investigate the relationship among proposals' designers, trainers, teachers and materials so as to point out new design criteria able to foster authentic and collaborative relationships between the research world and the school world.

Acknowledgments The author wishes to thank readers of previous drafts of this paper (or parts of it) for comments that improved the content and presentation. This includes the Research Group in Physics Education at the Department of Physics and Astronomy of the University of Bologna (in particular, Paola Fantini, Marta Gagliardi, Nella Grimellini Tomasini, Barbara Pecori), Anna De Ambrosis, Andrea A. diSessa, Enrico Giannetto, Mariana Levin and Zalkida Hadzibegovic. I am very grateful to the manuscript reviewers for the *Handbook* who helped refine and clarify arguments. A special thanks to Prof. Colin Gauld for his patient work in copyediting the manuscript.

References

Abiko, S. (2005). The light-velocity postulate. *Science & Education,* 14, 353–365.
Adler, C.G. (1987). Does mass really depend on velocity, dad?. *Am. J. Phys.,* 55(8), 739–743.
Angotti, J.A., Caldas, I.L, Delizoicov Neto, D., Pernambuco, M.M. & Rudinger, E. (1978). Teaching relativity with a different philosophy. *American Journal of Physics*, 46, 1258–1262.
Arriassecq, I. & Greca, I.M. (2010). A teaching learning sequence for the special relativity theory at high school level historically and epistemologically contextualized. *Science & Education*, doi:10.1007/s11191-010-9231-5.
Bertozzi, E., Levrini, O., Rodriguez, M. (in press). Symmetry as core-idea for introducing secondary school students to contemporary particle physics, *Procedia-Social and Behavioral Journal.*
Bevilacqua, F., Giannetto, E. & Matthews, M. (2001). *Science education and culture. The contribution of history and philosophy of science.* Dordrecht, The Netherlands: Kluwer.

6 The Role of History and Philosophy in Research on Teaching and Learning... 179

Cobern, W. W. (2000). The nature of science and the role of knowledge and belief. *Science & Education*, 9, 219–246.

Cortini, G. +23 (1977). Iniziativa relatività: "Vedute recenti sull'insegnamento della relatività ristretta ad un livello elementare". *Quaderni del Giornale di Fisica*, II(4), 13.

Cortini, G. (1978). *La relatività ristretta*. Loescher. Torino.

Borghi, L., De Ambrosis, A. & Ghisolfi, E. (1993). Teaching special relativity in high school. *Proceedings of the III Seminar Misconceptions and educational Strategies in Science and Mathematics*, Ithaca, NY.

De Ambrosis, A. & Levrini, O. (2010). How physics teachers approach innovation: An empirical study for reconstructing the appropriation path in the case of special relativity. *Physical Review Special Topics -Physics Education Research,* doi: 10.1103/PhysRevSTPER.6.020107.

De Hosson, C., Kermen, I. & Parizot, E. (2010). Exploring students' understanding of reference frames and time in Galilean and special relativity. *European Journal of Physics*, 33, 1527–1538.

diSessa, A. A. (1993). Toward an epistemology of physics. *Cognition & Instruction,* 10, 272.

diSessa, A.A. (2006). A history of conceptual change: Threads and fault lines. in *Cambridge Handbook of the Learning Sciences*, edited by K. Sawyer, Cambridge University Press, Cambridge, 265–281.

diSessa, A. A. & Sherin, B. L. (1998). What changes in conceptual change? *International Journal of Science Education*, 20(10), 1155–1191.

Duit, R. (2006). Science education research - An indispensable prerequisite for improving instructional practice. Paper presented at ESERA Summer School, Braga. (http://www.naturfagsenteret.no/esera/summerschool2006.html).

Duschl, R. A. (1985). Science education and the philosophy of science: Twenty five years of mutually exclusive development. *School Science and Mathematics*, 85(7), 541–555.

Gauld, C. (1991). History of science, individual development and science teaching. *Research in Science Education,* 21(1), 113–140.

Einstein, A. (1905). Zur Elektrodynamik bewegter Korper, *Annalen der Physik*, XVII, pp. 891–921 (On the electrodynamics of moving bodies, in Lorentz, H. A., Einstein, A., Minkowski, H., Weyl, H.: 1952, *The principle of relativity. A collection of original memoirs on the special and general theory of relativity.* (with notes by A. Sommerfeld) Dover Publications, New York, pp. 37–65).

Einstein, A. (1919). Time, space, and gravitation. Times (London), 28 November 1919.

Eylon, B. S. & Bagno, E. (2006). Research-design model for professional development of teachers: Designing lessons with physics education research, *Physical Review Special Topics - Physics Education Research,* 2, 02016-1-14.

Fabri, E. (2005). *Insegnare relatività nel XXI secolo: dal 'navilio' di Galileo all'espansione dell'Universo*, Quaderno 16 La Fisica nella Scuola (first version published in 1989).

Galili, I. (2011). Promotion of Cultural Content Knowledge Through the Use of the History and Philosophy of Science, *Science & Education,* doi: 10.1007/s11191-011-9376-x.

Giannetto, E. (2009). The electromagnetic conception of nature at the root of the special and general relativity theories and its revolutionary meaning. *Science & Education*, 18, 765–781.

Gil, D. & Solbes, J. (1993). The introduction of modern physics: overcoming a deformed vision of science. *International Journal of Science Education,* 15(3), 255–260.

Grimellini Tomasini, N. & Levrini, O. (2001). Images of physics and pre-service teacher education. In R. Pinto, S. Surinach (eds.), *Physics Teacher Education Beyond 2000*, Elsevier Edition, 355–358.

Grimellini Tomasini, N. & Levrini, O. (2004). History and philosophy of physics as tools for preservice teacher education. In Michelini, M. (Ed.), *Quality Development in Teacher Education and Training*, Selected contributions Second International Girep Seminar, 2003, Udine (Italy), 306–310.

Hewson, P.W. (1982). A case study of conceptual change in special relativity. The influence of prior knowledge in learning. *European Journal of Science Education*, 4(61), 61–78.

Hewson, P.W. & Thorley, N.R. (1989). The conditions of conceptual change in the classroom. *International Journal of Science Education*, 11, 541–553.

Holton, G. (1973). *Thematic origins of scientific thought, Kepler to Einstein*, Harvard University press, Cambridge (MA), London (England) (revised version 1988).

Levrini, O. (2000). *Analysing the possible interpretations of the formalism of General Relativity. Implications for teaching.* PhD Dissertation, Department of Physics, University of Bologna. Unpublished.

Levrini, O. (2002a). Reconstructing the basic-concepts of general relativity from an educational and cultural point of view, *Science & Education*, 11(3), 263–278.

Levrini, O. (2002b). The substantivalist view of spacetime proposed by Minkowski and its educational implications. *Science & Education*, 11(6), 601–617.

Levrini, O. (2004). Teaching modern physics from a cultural perspective: an example of educational re-construction of spacetime theories. In E. F. Redish & M. Vicentini (Eds.), *Research on physics Education.* Proceedings of the International School of Physics "E. Fermi", Course CLVI "Research on Physics Education", IOS press, SIF, Bologna, 621–628.

Levrini, O. & diSessa, A.A. (2008). How students learn from multiple contexts and definitions: Proper time as a coordination class. *Physical Review Special Topics - Physics Education Research,* doi: 10.1103/PhysRevSTPER.4.010107.

Levrini, O., Fantini, P. (2013). Encountering Productive Forms of Complexity in Learning Modern Physics. *Science & Education*, 22(8), 1895–1910. DOI: 10.1007/s11191-013-9587-4.

Levrini, O., Fantini, P., Gagliardi, M., Tasquier, G. & Pecori. B. (2011). Toward a theoretical explanation of the interplay between the collective and the individual dynamics in physics learning, ESERA 2011 conference - September 5th–9th 2011, Lyon.

Levrini, O., Fantini, P., Pecori, B., Gagliardi, M., Tasquier, G. & Scarongella, MT. (2010). A longitudinal approach to appropriation of science ideas: A study of students' trajectories in thermodynamics, in K. Gomez, L. Lyons, J. Radinsky (Eds.), *Learning in the Disciplines: Proceedings of the 9th International Conference of the Learning Sciences* (ICLS 2010) - Volume 1, Full Papers. International Society of the Learning Sciences: Chicago IL, 572–579.

Matthews, M.R. (1994). *Science teaching: the role of history and philosophy of science.* Routledge, New York.

Minkowski, H. (1909). Raum und Zeit, *Physikalische Zeitschrift*, 10, No. 3, 104–111 (Space and Time, in Lorentz, H. A., Einstein, A., Minkowski, H., Weyl, H.: 1952, *The principle of relativity. A collection of original memoirs on the special and general theory of relativity.* (with notes by A. Sommerfeld) Dover Publications, New York, pp. 73–96).

Pintò, R. (2005). Introducing curriculum innovation in science: Identifying teachers transformations and the design of related teacher education, *Science Education*, 89(1), 1–12.

Posner, G. J., Strike, K.A., Hewson, P.W. & Gerzog, W.A. (1982). Accommodation of a scientific conception: Toward a theory of conceptual change. *Science Education*, 66(2), 211–227.

Prigogine, I & Stengers, I (1979). *La Nouvelle Alliance. Metamorphose de la Science*, Gallimard, Paris.

Resnick, R. (1968). *Introduction to Special Relativity*, John Wiley & Sons, Inc., New York, London.

Scherr, R. E., Shaffer, P. S. & Vokos, S. (2001). Student understanding of time in special relativity: Simultaneity and references frames. *American Journal of Physics*, 69(7), S24–S35.

Scherr, R. E., Shaffer, P. S. & Vokos, S. (2002). The challenge of changing deeply held student beliefs about relativity of simultaneity. *American Journal of Physics*, 70(12), 1238–1248.

Solbes, J. (1986). La introducción de los conceptos básicos de Física moderna. PhD tesi, Univesitat de València.

Taylor, E. F. & Wheeler, J. A. (1965). *Spacetime Physics,* Freeman and Company, New York (2nd. Edition 1992).

Viennot, L., Chauvet, F., Colin, P. & Rebmann, G. (2005). Designing strategies and tools for teacher training: the role of critical details, examples in optics. *Science Education*, 89(1), 13–27.

Villani, A. & Arruda, S. (1998). Special theory of relativity, conceptual change and history of science. *Science & Education*, 7(2), 85–100.

Villani, A. & Pacca, J. (1987). Students spontaneous ideas about the speed of light. *International Journal of Science Education,* 9(1), 55–66.

Warren, J. W. (1976). The mystery of mass-energy, *Physics Education,* 11(1), 52–54.

Whitaker, M. (1976). Definitions of mass in special relativity, *Physics Education,* 11(1), 55–57.

Olivia Levrini is a senior researcher in Physics Education at the Department of Physics and Astronomy of the University of Bologna, Italy. Her Ph.D. thesis in Physics, received from the University of Bologna, involved research on 'The possible interpretations of the formalism of General Relativity: Implications for teaching'. Her research interests include the role of history and philosophy of physics for improving conceptual understanding of contemporary physics, conceptual change and development of theories in physics education, the design and the analysis of learning environments able to foster individual appropriation of physics content knowledge and teacher education. Her publications include Levrini, O. and diSessa, A. A. (2008). How students learn from multiple contexts and definitions: Proper time as a coordination class. *Physical Review Special Topics - Physics Education Research,* doi: 10.1103/PhysRevSTPER.4.010107; De Ambrosis, A. and Levrini, O. (2010). How physics teachers approach innovation: An empirical study for reconstructing the appropriation path in the case of special relativity. *Physical Review Special Topics -Physics Education Research,* doi: 10.1103/PhysRevSTPER.6.020107; Levrini, O., Fantini, P. (2013). Encountering Productive Forms of Complexity in Learning Modern Physics. *Science & Education,* doi: 10.1007/s11191-013-9587-4.

Chapter 7
Meeting the Challenge: Quantum Physics in Introductory Physics Courses

Ileana M. Greca and Olival Freire Jr.

7.1 Introduction

In the last two decades, interest in introducing quantum physics into introductory physics courses at university and high school level as well as research into the subject has increased. New textbooks have been published introducing updated views for undergraduates. There is now wide recognition of how ubiquitous quantum physics has become in current technologies and how fundamental it is considered for physics and for the culture of science. However, as difficulties related to teaching this physical theory in advanced courses are legendary, it comes as no surprise that the obstacles to teaching it in introductory physics courses are much greater.

Presenting quantum theory (hereafter QT)[1] is a task which is both technically and philosophically sensitive. In QT philosophical issues concern the interpretation of its mathematical formalism as well as its conceptual foundations. However, most of the research in science education and instructional materials do not take into account the philosophical choices behind the subject, and some of their results may be biased by the lack of attention to these choices. For instance, the right answers to questions related to wave-particle duality are not independent of interpretational choices, and it is even difficult to find consensus among experts as to such answers.

---

[1] Physicists interchangeably use quantum theory, quantum physics, quantum mechanics, or wave mechanics to describe the same physical theory. While using sometimes quantum physics, we will privilege quantum theory as it emphasizes its role as a scientific theory given that theories are central to the culture of physics.

I.M. Greca (✉)
Departamento de Didácticas Específicas, Facultad de Humanidades y Educación,
Universidad de Burgos, Burgos, Spain
e-mail: ilegreca@hotmail.com

O. Freire Jr.
Instituto de Física, Universidade Federal da Bahia, Salvador, Brazil
e-mail: freirejr@ufba.br

M.R. Matthews (ed.), *International Handbook of Research in History, Philosophy and Science Teaching*, DOI 10.1007/978-94-007-7654-8_7,
© Springer Science+Business Media Dordrecht 2014

As regards research on the conceptual foundations of QT, this investigation was not over at the time of its inception. Indeed, we now have a better understanding about what quantum physics is from the ongoing controversy on its interpretation and foundations. The statement that QT does not fit the usually accepted local realism requirements, for instance, is a consequence of Bell's theorem and its experimental tests. This is a chapter in physics whose history was renewed in the 1960s and has since continued to evolve. The history of this foundational research seems to lead to two different and to some extent conflicting conclusions, both with implications for the business of physics teaching. The first says that quantum physical concepts have no classical or intuitive counterparts, and they are better expressed in the abstract mathematical formalism of this theory. The second, derived from reports by top-ranking physicists in the field, suggests that in order to grasp quantum mechanics, many physicists need to consider pictorial representations of the phenomena under study. However, pictures have no univocal correspondence with the formalism of this physical theory. The principle of complementarity, suggested by Niels Bohr, could accommodate the two conclusions, but this principle is not exempt from philosophical qualms. Still, a fair share of the research on teaching quantum physics at introductory levels has not yet considered these issues. The number of research and teaching designs based on bridges and analogies between quantum and classical concepts without consideration for the philosophical and conceptual implications of such a choice is considerable.

One of the main challenges related to introductory QT courses is thus to find a balanced approach through which to introduce the most basic quantum concepts while taking into account interpretational issues. As students attempt to make images of the quantum phenomena, another challenge arises which is related to the conceptual foundations of QT and the findings of the psychology of learning. Insofar as the history and philosophy of science contributions are concerned, they have a double role to play in teaching introductory QT. First, the introduction of the historical contexts in which QT was produced and was subsequently developed may bring flesh and blood to the introduction of a new scientific theory otherwise presented in a dry and disembodied manner. The second role implies having the teaching and the research of/on introductory quantum courses informed by the history and philosophy of the subject. This means that the educational choices and strategies should be informed by what we have learned from the ongoing controversy on the foundations of the discipline. This is the main focus of this paper. Indeed, naïve choices at the beginning of introductory courses on quantum physics using the chronological sequence of its production may be misleading insofar as such a syllabus may be technically challenging. It is enough to recall the technicalities behind the blackbody problem. Furthermore, a chronology could be pedagogically unsound as it may reinforce among the students undesirable bridges between classical and quantum concepts. If the chronological sequence is to be taught, in courses dedicated to the history of physics, for example, emphasis should be put on how scientists faced the epistemological obstacles hindering the development of the new theory.

We will argue that insofar as there is no privileged interpretation for quantum physics, there is no ideal way to teach it on an introductory course. However,

7 Meeting the Challenge: Quantum Physics in Introductory Physics Courses

we suggest that both teaching and research about QT in science education must make the interpretational choice used explicit. In addition, our point is that any course to teach QT should emphasize the strictly quantum features in order to prevent students from establishing undesirable links with classical concepts. While teaching focused on the mathematical formalism remains a choice, pictures may be exploited, but in this case complementarity should be explicitly and carefully introduced. Finally, we argue that the teaching of QT, maybe more than other areas in physics, must be informed by the history and philosophy of science. This paper is organized as follows: first, we discuss the history and philosophy lessons from the research on the foundations of QT. We then criticize the usual teaching of QT, and in the following section, we review the literature on introductory QT courses. Finally, before concluding, we analyze the role complementarity has played in the history of the teaching of this physical theory.

7.2 Lessons from Recent Research on Quantum Physics

In a world populated with transistors, lasers, and nuclear and atomic devices, it would be a platitude to emphasize the many and varied applications of the quantum theory since its inception around 1925–1927. In addition to its technological applications, QT has become a central part of training in physics and has brought with it wide-reaching philosophical implications. However, while the basic mathematical formalism has remained essentially the same since that time, our understanding of the implications of such a formalism has increased dramatically, in the last 50 years in particular. This increased knowledge has resulted from both theoretical and experimental developments enabling the testing of QT in extreme situations and of a new attitude towards its foundations and interpretations, the latter expressed in looking for its possible limitations.

However, from the inception of QT till the late 1960s, concerns about its foundations were mostly centered on the theoretical grounds. Some of the founding fathers of the new theory, such as Erwin Schrödinger, Albert Einstein, and Louis de Broglie, accepted neither some features of the new physical theory nor its interpretation in terms of a principle of complementarity suggested by Niels Bohr. Einstein and de Broglie criticized the abandonment of determinism, while Schrödinger and Einstein raised concerns about the idea of physical descriptions heavily depending on the means of observation, which amounted to giving up the kind of realism shared by most physicists at that time. Related to these concerns was the fact that the mathematical structure of the theory, through the principle of superposition of states, did not attribute well-defined physical properties to systems described by quantum theory. Thus, the state describing the spin projection of one electron says that this electron has both spin-up and spin-down and not one or the other. The weirdness of this quantum description results from the fact that in the world of everyday experience, objects, described by classical physics, have well-defined properties.

As physicists consider QT to be more basic than classical physics, the open problem is how to connect these two kinds of descriptions.

The problem was most clearly stated through the thought experiment now known as Schrödinger's cat. From a mathematically formal point of view, the issue was better stated by the mathematician John von Neumann who built the standard formalism of QT in terms of Hilbert vector space. Von Neumann acknowledged two kinds of evolution for the quantum states. According to him in the first kind, quantum states evolve in time ruled by Schrödinger's equation, which is a linear and deterministic process. During measurements, however, von Neumann suggested a second kind of evolution, which would be instantaneous, nonlinear, and nondeterministic (the "collapse of the wave function"). Since then, the so-called quantum measurement problem, in the terms suggested by von Neumann, has become a lasting ghost haunting the foundations of quantum theory.

Since the 1950s complementarity has no longer reigned alone because alternative interpretations have begun to appear. Two young American physicists, David Bohm and Hugh Everett, were the main protagonists challenging the received views on the interpretation of QT. Bohm criticized the abandonment of determinism and well-defined properties in the quantum domain. He built a model for electrons taking them as bodies with a position and momentum simultaneously well defined and was able to reproduce results obtained by QT in the nonrelativistic domain. His interpretation received both the technical name of "hidden variables" and the more philosophically inclined "causal interpretation." Everett built his interpretation, later entitled "many worlds," dispensing with the second kind of evolution of quantum states that von Neumann had taught would govern measurements. Thus, for Everett measurement was ruled by the same mathematical machinery of Schrödinger's equation. In particular, Everett disliked the complementarity assumption that quantum physics requires the use of classical concepts while limiting their use in the quantum domain as certain pairs of these concepts are complementary but mutually exclusive.

Since then the number of alternative interpretations of QT has grown. However, while they have become an industry for physicists and philosophers, populating many technical journals and books, they are conspicuously absent from physics teaching and most of the research on physics teaching.[2] The very existence of several interpretations of QT seems to be an inconvenient truth for the teaching of physics. The problem is that most of these alternative interpretations lead to the same experimental predictions at least in the nonrelativistic domain. Philosophers, logicians, and historians, however, are familiar with this kind of issue. Indeed, the

---

[2] Short introductions to most of these interpretations may be obtained in Greenberger et al. (2009). This compendium includes the following interpretations: Bohm interpretation, Bohmian mechanics, complementarity principle, consistent histories, Copenhagen interpretation, GRW theory, hidden variables models of quantum mechanics, Ithaca interpretation, many worlds interpretation, modal interpretations, Orthodox interpretation, probabilistic interpretation, and transactional interpretation. While there is some redundancy in this list, it is not comprehensive; one could still include, for instance, stochastic interpretation, ensemble interpretation, and Montevideo interpretation. Indeed, this list has been growing in recent decades.

plethora of quantum interpretations is one of the best examples of the so-called Duhem-Quine thesis: the underdetermination of theories by the empirical data.[3]

While these theoretical developments dug deep into the foundations of quantum physics, it was the possibility of translating some of these issues to the laboratory benches that expanded our knowledge of the quantum world most, as we will see. No case is more telling than the statement that local realism is not compatible with QT predictions. The problem may be traced back to 1935 when Einstein, Podolsky, and Rosen suggested a *Gedankenexperiment* to demonstrate the incompleteness of QT and which Bohr rebutted. The issue was shelved until the middle of the sixties when John Bell realized that quantum physics predictions could be contrasted with any theory sharing the same 1935 assumptions of Einstein. Einstein professed a kind of philosophical realism meaning that physical objects should have well-defined properties independent of them being observed or not. In addition he assumed that no measurement of a system could change the state of a distant system, unless, of course, there is an interaction between these two systems propagating with a speed less or equal to light. It is the merit of Bell to have isolated in Einstein's reasoning such assumptions and to have managed them in order to show that as trivial as these assumptions may be, some quantum predictions do not confirm them. This is what we now call Bell's theorem. No local hidden variable theory can reproduce all quantum physics results. The reference to hidden variable theories is reminiscent of the historical context in which such a theorem emerged: the attempts to change quantum theory in order to obtain the description of systems with well-defined properties by introducing additional hidden variables in comparison with standard QT.[4]

"Bell's theorem changed the nature of the debate." Alain Aspect's words are now familiar to physicists. According to Aspect,

> In a simple and illuminating paper, Bell proved that Einstein's point of view (local realism) leads to algebraic predictions (the celebrated Bell's inequality) that are contradicted by the quantum –mechanical predictions for an EPR *gedanken* experiment involving several polarizer orientations. The issue was no longer a matter of taste, or epistemological position: it was a quantitative question that could be answered experimentally, at least in principle. (Aspect 1999, p. 189)

The creation of Bell's theorem was only the preamble to many thrilling activities in the last 50 years. It could have been the case that quantum predictions do not hold for distances longer than the molecular and atomic and in the end local realism

---

[3] The case of quantum physics in relation to this philosophical thesis is discussed in Cushing (1999, pp. 199–203). A general discussion of the Duhem-Quine thesis may be found in Harding (1976).

[4] We chose to present the issue of completeness of QT in terms of Bell's theorem and its conflict between QT and local hidden variables or local realism. This choice was due to the influence of this approach on mainstream physics leading, through theory and experiments, to the identification of entanglement as a key quantum physical effect (Shimony 2009). Other approaches, however, are possible. A fine epistemological analysis of Einstein's assumptions would lead us, according to Howard (1985), to identify them as separability (mutually independent existence of spatially distant things) and locality. Another possibility is the Kochen-Specker theorem, formulated in 1967, which contrasts QT with non-contextuality; however, the impact of this theorem in experimental physics has been scant (Held 2012).

could prevail as a very reasonable assumption. In any case, in 1969 physicists such as John Clauser and Abner Shimony realized that the available experimental results could not check the double choice implied in Bell's theorem: either quantum theory or local realism. Since then, a string of experiments have been carried out leading to the confirmation of this weird quantum property: quantum nonlocality keeps its validity even for distances as far as a hundred kilometers as recent experiments by Anton Zeilinger and his team have confirmed. In the early stages of these experiments, the most revealing was that carried out by Alain Aspect and his team, who were able to change the experimental setting while the photons were in flight in order to prevent the working of any unknown interaction among the pair of photons or devices with a lower speed than that of light. Most of these experiments have been conducted with photons. In the first one, John Clauser used pairs of photons coming from atomic decay with atoms excited by thermal light. This source of excitation was then replaced by new tuning lasers, which dramatically improved the accuracy of the results. Finally, a new source of photon pairs began to be used in the late 1980s, photons from parametric down conversion which occurs when a laser beam crosses certain nonlinear crystals. This new source exponentially increased the experimental possibilities, and it has formed the basis of the impressive number of experiments on entanglement in the last two decades.[5]

The string of experiments on Bell's theorem have created a widely shared feeling among physicists that local realism should be abandoned even considering that more precise tests can be done in the future, in particular improving the efficiency of photodetectors. This has led physicists to unearth the term entanglement, coined by Schrödinger in 1935, to name the new quantum physical property. Indeed, though many terms are used to describe the same phenomenon, while with subtle differences, entanglement has prevailed as the brand new physical effect. The feeling that local realism should be abandoned had a strong philosophical implication at first, as stated by Clauser and Shimony as early as 1978: "either one must totally abandon the realistic philosophy of most working scientists, or dramatically revise our concept of space-time" (Clauser and Shimony 1978, p. 1881). Later on, experimental physics began to probe this dilemma. According to Aspect (2007, p. 866), "The experimental violation of mathematical relations known as Bell's inequalities sounded the death-knell of Einstein's idea of 'local realism' in quantum mechanics. But which concept, locality or realism, is the problem?" He was then commenting upon an experiment in which Zeilinger's team found violations of Leggett's inequalities, a variant of Bell's inequalities, which were formulated in order to exhibit the experimental contrast between quantum theory and even some classes of nonlocal realistic theories (Gröblacher et al. 2007). It was not yet the full-blown dilemma announced by Clauser, Shimony, and Aspect, but it was an example of what Shimony has called "experimental metaphysics," that is, theoretical and experimental research in the foundations of physics with huge philosophical implications.

Philosophy and basic science were not the sole domains in which Bell's theorem caused a stir. Nowadays entanglement is at the core of blossoming research in

---

[5] On the early experiments on Bell's theorem, see Freire (2006).

7 Meeting the Challenge: Quantum Physics in Introductory Physics Courses

quantum information as scientists and engineers attempt to harness quantum phenomena for more reliable cryptography and for speedier information processing. For those physicists and philosophers who are interested in a better understanding of the kind of world described by quantum physics, as well as for physics teachers, entanglement brought with it a new challenge: how to cope with the world view implied by this weird quantum property. For physics teachers, the challenge is further enhanced; if the purpose of this teaching is not only to hone calculus skills, how can an understanding of this seminal quantum property be conveyed if neither an intuitive perception nor a clear image of it can be presented to students?

Entanglement may be the most telling example, but it is not alone among the achievements of our understanding of QT in the last half century. An old quantum prediction, particles obeying Bose-Einstein statistics at low temperatures tend to gather in the same state, has now been confirmed by Bose-Einstein condensates in laboratories, which assured Eric Cornell, Wolfgang Ketterle, and Carl Wieman the 2001 Physics Nobel Prize. Behind this experimental feat was a technical trick: the use of lasers to cool atoms, a technique developed by Steven Chu, Claude Cohen-Tannoudji, and William Philips, also awarded the 1997 Nobel Prize in physics. As late as the 1950s and 1960s, Richard Feynman needed to use an idealized experiment of a double slit to convey the message of the wave-particle duality in his famous lectures, exactly as Einstein and Bohr in the 1930s had when they discussed the epistemological lessons from the quantum. From the 1980s on, however, physicists were able to manipulate photons, electrons, neutrons, and atoms one by one, making thus all these idealized experiments real.

Theoretical developments combined with experimental advances have also marked this last half century. The creation of the laser in 1960, in itself a quantum phenomenon, required theoretical improvements. One of the most impressive was that of Roy Glauber who created what we now call Glauber's coherent states, a useful tool for describing radiation in the domain of single photons. Glauber's predictions, later corroborated by the photon "anti-bunching" tests, became a key device in the toolkit of a new discipline: quantum optics, which solved a lasting controversy about the real need for the concept of photon. For all practical purposes, until the early 1960s, a full quantum treatment of light had led to the same predictions as semiclassical approaches, but the latter could not explain the "anti-bunching." Glauber was awarded the 2005 Nobel Prize for his achievements. In the 1980s physicists such as H. Dieter Zeh, Erich Joos, Anthony Leggett, Amir Caldeira, and Wojciech Zurek learned to deal with the transitions from states theoretically described by quantum superpositions to those which can be described by classical statistics mixtures, a theoretical treatment baptized decoherence. While decoherence shed some light on the old quantum measurement problem, it remained unsolved. It was in the following decade that Serge Haroche was able to push this treatment into the laboratory creating the first real analogues of systems such as Schrödinger's cat, that is, to see in the labs how, in a predicted time interval, a system described by a quantum superposition loses its quantum coherence. Again, this field of research is nowadays at the core of current research in quantum information. Earlier, in 1957, Yakir Aharonov and David Bohm had shown that quantum phenomena exhibit topological properties which can hardly be reconciled with our view of space-time as the arena for phenomena in physics. This kind of prediction is now well confirmed and enlarged by what is called Berry's phases. While this list of scientific deeds is not comprehensive, it is enough for our purposes.[6]

One may consistently argue that all these novelties are implicit in the mathematical quantum formalism. However, most of these achievements resulted from the ongoing controversy about interpretations of quantum physics and its basic concepts. Furthermore, an important part of this development was scientists' discomfort with the conceptual implications of this theory. For this reason one may also argue that a better understanding of QT was gained from the work of quantum dissidents. By using this label (Freire 2009), we are saying that they worked on the foundations of this theory, which was outside mainstream physics, and were critics of the complementarity view. A list of these dissidents could include some from the older generation, such as Einstein, but mainly those from the newer generation of physicists, such as Bohm, Everett, Bell, Clauser, and Shimony. However, QT has survived their criticisms and their related experimental tests. It is now time to extract the lessons both from the role played by the quantum dissidents and the amount of theoretical and experimental work already done. The teaching of introductory quantum physics courses could benefit from these lessons.

The new generations of physicists have learned that the object of QT must be described by its own quantum mathematical formalism and that no independent assumption, as reasonable as it may seem, can be previously assumed. This practical and epistemological lesson is bold in meaning because this formalism, embedded as it is in a very abstract mathematical structure, is impossible to grasp through pictures or mental images. However, there is one way to avoid this. Images of phenomena, such as the classical wave and particle, can be used, but by doing so, we are obliged to explicitly use Bohr's complementarity principle, a point to which we will return later.

From the history of the research on the foundations of QT in the last half century, we exemplify the previous lesson with one case—Aspect's 1986 experiment with wave-particle duality for single photons—chosen because of the clear-cut conclusions of its authors.[7] At the end of the 1970s, Aspect realized that the source he was using for experiments with Bell's theorem was delivering single photons as described by quantum optics.[8] The crucial point for him was that all previous

---

[6] For brief introductions to these topics, see Greenberger et al. (2009). On the debates on the concept of photon, see Silva and Freire (2013). The Concept of the Photon in Question: The Controversy Surrounding the HBT Effect circa 1956–1958, *Historical Studies in the Natural Sciences*, forthcoming; on quantum optics, see Bromberg (2006); for historical studies on decoherence, see Camilleri (2009b) and Freitas (2012), The many ways to decoherence, unpublished monograph.

[7] For experiments with single electrons in the two-slit interference experiments and debates about their interpretations and dispute of priorities, see Rosa (2012). As an example of the ongoing controversy surrounding the foundations of quantum physics, Marshall and Santos (1987) considered that Aspect's 1986 typical quantum results could be compatible with the classical wave theory of light as the latter were interpreted in terms of Stochastic Optics.

[8] Alain Aspect, interview with O. Freire and I. Silva, 16 December 2010 and 19 January 2011, American Institute of Physics.

7 Meeting the Challenge: Quantum Physics in Introductory Physics Courses

experiments with "single photons," which dated back to Taylor in 1909, could not be quantum mechanically described as single-photon impulsions. Indeed, those very attenuated sources which gave just one photon in the experimental setting on average were not single-photon states such as the source Aspect was using. After presenting his results, Aspect (Grangier et al. 1986, p. 178) interpreted them in two different manners. The first was based on complementarity; however, he was cautious about it: "if we want to use classical concepts, or pictures, to interpret these experiments, we must use a particle picture for the first one, [...] on the contrary, we are compelled to use a wave picture, to interpret the second experiment. Of course, the two complementary descriptions correspond to mutually exclusive experimental set-ups." Aspect's inclination was towards the second kind of explanation he had suggested. It was an explanation based on a direct interpretation of the quantum mathematical formalism, without appealing to pictures, using concepts that had just emerged in quantum optics: "from the point of view of quantum optics, we will rather emphasize that we have demonstrated a situation with some properties of a 'single-photon state'."

Three years later, discussing the same results, Aspect (Aspect et al. 1989, p. 128) went further in his epistemological choices. After presenting the explanation with complementary classical concepts, he added: "the logical conflict between these two pictures applied to the same light impulses constitute one of most serious conceptual problem of quantum mechanics." Then he recalled that the experimental setups were incompatible and that this incompatibility was presented by Bohr as an element of coherence of QT. While presenting the second explanation, he remarked that such a logical conflict only appears if one appeals to classical concepts, such as wave and particle. And yet his choices were favorable to the second type of explanation he had suggested:

> ... if, on the contrary, one is restrained to the quantum mechanics formalism, the descriptions of the light impulses are the same. It is the same state vector (the same density matrix) that one must use for each experiment. The observable changes but not the description of light. (Aspect et al. 1989, p. 128)

Thus the quantum formalism is self-sufficient, it describes both experiments without appealing to pictures or classical concepts.[9]

If the history of the research in the foundations of QT seems to favor the interpretational trend which takes only the quantum formalism to grasp quantum phenomena, as suggested by Michel Paty (1999), this same history also suggests another lesson. Indeed, it seems to us that the need for pictures/images, thus of classical concepts, persists even among the best working physicists.[10] Here the case

---

[9] Incidentally, we remark that Aspect considers wave-particle duality for single photons the best way to introduce, both theoretically and experimentally, the full quantum treatment of light on optics courses. See his proposal in (Jacques et al. 2005).

[10] We use image and picture as equivalent words. Psychology of learning uses image as picture may be associated with drawings. Physicists use both without distinction, while in QT, both are always associated to concepts from classical physics.

of John Clauser, who conducted the first experimental tests on Bell's theorem, is enlightening. Sharing his memories, he always disliked abstract reasoning:

> One of the problems I have, I'm very different from many physicists, which is both a blessing as well as a major impediment. I am not really a very good abstract mathematician or abstract thinker. Yes, I can conceptualize a Hilbert's Space, etc. I can work with it, I can sort of know what it is. But I can't really get intimate with it. I am really very much of a concrete thinker, and I really kind of need a model, or some way of visualizing something in physics. (Clauser 2002, p. 8)

Clauser's recollections may be useful for researchers in physics teaching dealing with the challenge of teaching introductory quantum physics. He goes on to say:

> There exists a set of numbers with algebraic structure of such and such, and we will define a particle as being something for which this operator commutes with that operator, etc. I haven't the foggiest idea what any of that means. But an electron is a charge density which may be Gaussian in shape and its shape, and it's about this big, and it's held together by various forces, and this is how the forces work that kind of hold it together. The difference between those two [concepts] are very dramatic differences of thinking. Now there's a whole class of physicists who can only think in the former method. I can only think in the latter mode. (Clauser 2002, p. 9)

We should add that insofar as Clauser (2002) also disliked Bohr's complementarity, he felt enduring discomfort with usual presentations of quantum physics, a discomfort which is relevant for our discussion on the teaching of quantum physics.

Quantum theory has passed the most severe experimental tests ever imagined for a physical theory. However, this does not mean that corroborations of the quantum physics predictions, that is, predictions of QT mathematical formalism, have implied corroboration of only one interpretation of this formalism. Indeed, only local realistic theories come up against obstacles in those tests. Curiously, most of the alternative interpretations of QT include some form of quantum nonlocality insofar as most of them preserve the linear superposition which is intrinsic to the Hilbert space in the usual interpretation.[11]

Quantum theory is weird not only because of its concepts, such as those related to the abandonment of determinism and local realism, but also for its place in the history of physics. It is so strange that 80 years after its creation, its recasting process—where the notions of a theory are clarified and its terms improved (Lévy-Leblond 2003; Paty 1999, 2000)—remains unfinished. Although its mathematical machine is well established and its predictive power successful, the conceptual foundations of QT are still in debate. In fact, we now have a better understanding of what QT is mostly from the ongoing controversy on its interpretation and foundations.

Therefore, one lesson from the history of physics as regards the attempts to introduce quantum physics at more elementary levels is that we should take into account

---

[11] The empirical equivalence of several QT interpretations in the nonrelativistic domain does not mean that all interpretations have been equally fruitful in the development of QT, in particular in the new field of quantum information. An interesting discussion on this aspect considering the case of "entanglement swapping" is Ferrero et al. (2012).

7 Meeting the Challenge: Quantum Physics in Introductory Physics Courses

the peculiar situation of the existence of a tension between a strong consensus about the formalism of this physical theory and a meaningful dissension about its interpretation. Of course, physics students need first to learn the quantum physics formalism in order to grasp such a controversy, but at a certain moment, we should convey to them the very existence of such a controversy.

7.3 The Usual Teaching of Quantum Physics

As we have pointed, teaching QT is not an easy task because it is both technically and philosophically sensitive. Its teaching is quite different from other topics in physics. It is perhaps the only area that is most commonly introduced through the history of its origin. From the late nineteenth century and through the first half of the twentieth century, these topics include Planck's quantization of energy to explain the spectrum of black body radiation, Einstein's photons of light to explain the photoelectric effect, Bohr's energy levels in his model for the atom to explain atomic spectra, de Broglie's hypothesis of waves associated with electrons, Schrödinger's formulation of a wave equation for orbiting electrons, Heisenberg's introduction of an uncertainty principle, and Born's interpretation of the wave function in terms of probability.

This introduction is a typical example of what Kragh (1992) called the quasi-history, a mystical history used to convince students of a particular point of view, the only "rationale" possible reached by physicists in the past. It is worth stressing that this historical approach has been criticized (Cuppari et al. 1997; Fischler and Lichtfeldt 1992; Michelini et al. 2000) for reinforcing classical concepts in students' minds at a time when they should have been moving on to more appropriate quantum models.

Advanced courses, while dispensing with this historical tour, repeat the very same material again and again. For example, the infinite well used as a pedagogical example or as a model of a physical system is usually encountered by a physics student in the USA up to five courses before he or she graduates (Cataloglou and Robinett 2002). The typical approach in these advanced courses can be described as consisting of highly abstract rules and procedures (Shankar 1994), in part because the mathematical tools necessary for applying it, even in the simplest cases, are so different from other branches of physics that the trend to present quantum concepts as inseparable from its mathematical problems exists (Bohm 1989). Nevertheless, behind this uniformity, there is a greater variability than that found in other typical subjects in physics. There is a wider array of possible topics which one might consider as constituting the core ideas (perhaps because among physicists there is no consensus about which are the most fundamental ideas in quantum physics), and also unlike classical mechanics or electromagnetic theory, there is a wider variety of approaches to the teaching of QT, even at the undergraduate level.

We can find texts that stress the formal aspects, starting with the formalism of spin systems and Hilbert spaces, texts focusing on the Schrödinger's equation, and some that present semiclassical approaches. Across this diversity most of the traditional textbooks provide few, if any, physical insights. In fact, textbooks seem to privilege what one could call an instrumentalist view of QM or what Redhead (1987) named the "minimal instrumentalist interpretation," i.e., quantization algorithm, statistical algorithm plus the epistemological premise that "theories in physics are just devices for expressing regularities among observations." This kind of approach reduces the cognitive reach of the quantum theory and does not make its understanding any easier.

This "minimal instrumentalist interpretation" is so widespread among physics teachers that several authors consider that most of the difficulties students have with quantum physics are related to its characteristic formalistic teaching that begins during introductory courses.[12] What are the factors that may have led to this kind of teaching? One seems to be, as just indicated, the intrinsic mathematical difficulty of quantum physics. But there are others. After the first period of its constitution, most physicists used QT machinery to study the microscopic world, without worrying about conceptual or interpretational questions (Heilbron 2001). This predominance of QT as a "calculating machine" may have been reinforced particularly in the USA because of the coexistence of theoretical and experimental physicists in the same departments, emphasizing experiments and applications, and the American trend to pragmatism (Schweber 1986).

The historian of physics David Kaiser also indicates another factor related to pedagogical choices during Cold War times. In the two decades after the war, the USA underwent a surge in the number of physics students which made it necessary to take some pedagogical decisions that specially affected the teaching of QT:

> Most physicists in the US recrafted the subject of quantum mechanics, accentuating elements that could be taught as quickly as possible, while quietly dropping the last vestiges of qualitative, interpretive musings that had occupied so much classroom time before the war. [...] The goal of physics became to train 'quantum mechanics': students were to be less like otherworldly philosophers and more like engineers or mechanics of the atomic domain. (Kaiser 2007, p. 30)

This change has been reflected in the textbooks published since then,[13] with wonderful methods for doing almost any calculus about atoms; however, when it comes to the principles and interpretations of QM, they "are, almost without exception, simplistic and obscure at the same time" (Barton 1997). These approaches ultimately worked because, as we have seen, one lesson from recent history is that quantum concepts are strictly associated to the mathematical formalism.

Students are more than occasionally encouraged to approach the subject with the idea that it is almost impossible to understand it and that it is so completely different from other branches of physics that one's intuition is of little or no use. As an advanced student said, referring to his experience in QT: "It seems that there's this dogma among physicists, that you can't ask that question: What is it doing between

---

[12] Fischler and Lichtfeldt (1992), Greca and Freire (2003), Johnston et al. (1998), and McKagan et al. (2008).

[13] In several countries, the most widely used textbooks in physics are American ones thus the spread of this approach.

point A and point B? 'You can't ask that!'" (Baily and Finkelstein 2010, p. 9). It is not surprising then that students dislike quantum theory and non-physics students try to avoid it. Many physics students, including graduates, despite seeing the same topics many times and successfully engaging in the mathematical machinery, constantly struggle to master its basic concepts, a problem that has been reported by several researchers (Cataloglou and Robinett 2002; Johnston et al. 1998; Singh 2001, 2006).

Despite the strength of the usual way of teaching QT, it has been challenged in recent years. It is not by chance that more than in other physics disciplines, it has increased the number of textbooks with new approaches. While it is beyond the scope of this paper to review the new batch of quantum physics textbooks, we give just a few examples of this trend. Griffiths' 2005 *Introduction to Quantum Mechanics* dedicates a chapter to the meaning of basic concepts and interpretational issues; Thaller's 2000 *Visual quantum mechanics* exploits simulations in wave mechanics; Greenstein and Zajonc's 1997 *The Quantum Challenge* includes many physical examples and makes direct connection to recent experimental results; and Omnès's 2000 *Comprendre la mécanique quantique* is the explicit defense of what he considers an updated interpretational view of quantum theory.

The wide recognition of how ubiquitous quantum physics has become in current technologies, how important it is for our understanding of nature and science at the present time, how fundamental it have been considered for physics, and the role it has played on the cultural scene have led to an increased interest in studying its teaching in the last two decades. Motivation for these studies derives thus from the need to convey quantum concepts not only to physics students but also to other science and engineering students, and they attempt to understand how to attract students to study quantum physics instead of running away from it. This kind of research has addressed students' difficulties with quantum concepts, surveys, and didactic strategies to better introduce quantum physics in physics introductory courses at universities—for physics, chemistry, and engineering students—and at high school level.[14] Being technically and philosophically sensitive, quantum physics poses some unique and interesting challenges to its teachers. Should students develop an understanding of the mathematics without worrying about the philosophical implications of the theory? Should the historical development of quantum theory be included in the syllabuses? Should we use or avoid classical or semiclassical analogies to help students grasp quantum concepts?

7.4 Proposals for Introductory Quantum Physics

What do the new proposals for teaching quantum physics which have emerged from research into science education suggest to improve students' understanding of quantum concepts? We have reviewed the literature published in physics education

---

[14] Greca and Freire (2003), Hadzidaki (2008a, b), McDermott and Redish (1999), and Wuttiprom et al. (2009).

from 2000 to 2011[15] and found 32 articles that tackle new forms of introducing QT topics at certain educational levels. Although only 11 of them mentioned the outcome of the implementation, in general they were very well received by the students but with varied conceptual improvements. Many of the papers, amounting to 10, are related to the use of the history and philosophy of science, using proper historical reconstruction (Barnes et al. 2004; Níaz et al. 2010), conceptual discussion of thought experiments (Velentzas et al. 2007; Velentzas and Halkia 2011), discussion of philosophical, epistemological and ontological issues concerning quantum physics through historical controversial issues—EPR, Heisenberg microscope— (Hadzidaki 2008a, b; Karakostas and Hadzidaki 2005; Pospievich 2003), or using QT as a tool for improving the views preservice teachers have about the nature of science (Kalkanis et al. 2003; Nashon et al. 2008).

Most of works using historical emphasis dealt with high school students and preservice teachers. In general, these works try to contextualize quantum physics in an updated historical and epistemological framework and in this way—as opposed to the "traditional" historical approach—help learners to reorganize and enhance their initial knowledge. Kalkanis et al. (2003, p. 270) propose, for example, the juxtaposition of representative models of conceptual systems of quantum and classical physics. Thus, instead of avoiding reference to classical physics, their strategy reveals the totally different worldview and thought patterns underlying the interpretation of macroscopic and microscopic phenomena. They used Bohr's atomic model, for example, in order to make the deep conceptual differences between classical and quantum physics concrete. Instead of avoiding the dualistic descriptions, they aimed to reveal the inner meaning of the complementarity principle. We can include in this category an article that stresses the introduction of quantum physics through unusual interpretations, such as the Bohmian one, as a useful tool to illustrate the relationship between classical and quantum physics (Passon 2004).

The second most frequently proposed strategy, with eight papers, is the use of simulations, computer animations, or games to improve the intuitive understanding of abstract quantum concepts, especially for students with a limited science and mathematics background or for advanced students who have seen quantum concepts traditionally—that is, only in a mathematical way.[16] These simulations, some of which integrate hands-on activities, attempt to build intuition for the abstract principles of QT through visualization in introductory physics, with precursors in the Quantum physics series of the Lawrence Berkeley Lab (Gottfried 1978) and the programs Eisberg (1976) designed for visualizing wave functions with the early programmable calculators. This "wavy" tendency can be seen in the names of some of the typical simulations—quantum tunneling and wave packets, quantum wave

---

[15] We have researched articles from the period 2000–2011 that tackle physics education in any level in the following journals: *American Journal of Physics, European Journal of Physics, International Journal of Science Education, Journal of Research in Science Teaching, Physical Review Letters – Special Topics, Research in Science Education, Science Education*, and *Science & Education*.

[16] For example, Goff (2006), Magalhães and Vasconcelos (2006), McKagan et al. (2008), Singh (2008), and Zollman et al. (2002).

interference, matter waves, probabilities and wave functions, and wave functions and energies in atoms. However, wave interpretations without reference to complementarity have not endured in the history of the research on the foundations of quantum physics, and none of these papers mentioned the complementary principle. Finally, it is worth stressing that several of the proposals not included in this group also make use of some computer simulations.

In third place, with seven papers, there are different "technical" approaches (deformation quantization, evolution operator method, field theory, computer algebra systems), most of them for advanced courses in physics (e.g., García Quijás and Arévalo Aguilar 2007; Hirshfeld and Henselder 2002), which will not be commented on here as we are dealing with introductory quantum physics courses. Finally, in fourth place, there are five papers with proposals that share an emphasis on quantum features of the systems, rather than searching for classical or semiclassical analogies, using in general real-world applications or recent experimental advances.[17] These works are in consonance with the researchers linked to the area of quantum optics[18] who have stressed the relevance of introducing quantum concepts from the very beginning. From the experimental results about the foundations of QT obtained in the last 20 years, in general they tend to use very simple systems that show clear quantum behavior, leaving aside nonphysics fictions such as the Heisenberg microscope.

So until the present time, science education researchers, although unanimous in rejecting the traditional "quasi-historical" introduction or the formal one, have given quite different answers to our questions about how to introduce quantum concepts. It is worth stressing that we do not have any strong evidence for advocating one way or another because few of the proposals have been tested. Thus some of our arguments from now on derive from the recent history of the research on the foundations of quantum physics as well as from empirical evidence obtained in science education research.

7.5 Quantum Theory Interpretations and the Research in Science Education

It is striking that, although all the papers emphasize the need to improve the conceptual understanding of quantum concepts, few of them clearly stated the interpretation of QT that is adopted. It seems as if the intense debate about the different interpretations, which is a conceptual debate, has yet to inform research into better ways of teaching quantum physics.

From the 32 papers found in the period 2000–2011, only 10 mention the existence of different possible interpretations. We have Bohr's realist interpretation (Hadzidaki

---

[17] Carr and McKagan (2009), Greca and Freire (2003), Holbrow et al. (2002), and Müller and Wiesner (2002).

[18] For example, Barton (1997), Jacques et al. (2005), Schenzle (1996), and Zeilinger (1999).

2008a, b; Karakostas and Hadzidaki 2005), the statistical ensemble interpretation (Müller and Wiesner 2002), the Copenhagen interpretation (Barnes et al. 2004; Kalkanis et al. 2003), an orthodox but realist interpretation (Greca and Freire 2003), the Bohmian dualistic interpretation (Passon 2004), and the interpretation of the quantum states as potentialities (Pospievich 2003).[19] Of these, three belong to the same research group and six have been published in *Science & Education*, a journal that stresses the contributions of philosophy and history to science education.

It is interesting to note that, except for two, all of them can be included in the spectrum of the realistic interpretations—that is, interpretations that move away from the epistemological position of the Copenhagen interpretation and that give an objective character to the concept of state of a quantum system and thus are less dependent on the measurement process. It seems that realistic interpretations are seen by science education researchers as the best interpretational option for introducing quantum physics to students. For example, we have argued (Greca and Freire 2003) that our aim to help students to develop mental models whose results—predictions and explanations—coincide with those accepted by physicists' community has led us to look for a realist interpretation of QT. This is because our remarks on scientific practice reflect Bunge's position (2003) when he writes that "the realism [is] inherent in both common sense and the practice of science." This trend towards realistic interpretations is coincident with the predominant epistemological view maintained by the physicists who worked in the foundations of QT in the 1970s (Freire 2009, p. 288).[20]

These realistic interpretations consider quantum states (represented by wave functions, state vectors) in general as having a physical reality independent of measurements. Bohm and Hiley (1988) attributed this view to von Neumann while opposing Bohr's view, because the latter valued the role of measurement excessively, through the idea of wholeness of the system and the measurement apparatus. As a matter of fact, several physicists and philosophers—such as Fock, Bunge, Lévy-Leblond, and Paty—have suggested similar ideas, though there are some relevant differences among them. An illustration of such differences is the case of the Soviet physicist Vladimir Fock, who combined defense of complementarity with the attribution of physical reality to the objects of quantum physics (Graham 1993, pp. 112–117).

Mario Bunge, for example, considers the possibility of a realistic reinterpretation of standard QT, a subtle but philosophically meaningful different interpretation from the Copenhagen interpretation. As he writes:

> ... instead of interpreting Born's postulate in terms of the probability of *finding* the quanton in question within the volume element Δv, the realist will say [...] that the probability in question is the likelihood of the quanton's *presence* in the given region. (Bunge 2003, p. 462)

---

[19] We have named the interpretations as stated by the authors, without evaluating superpositions or duplications.

[20] The categorization we have used is a rough approximation, useful only to grasp analogies between physics teaching research and physics research. Realism and objectivity are not univocally defined in philosophy of science, and quantum physics practice has brought meaningful constraints to the use of these terms.

7 Meeting the Challenge: Quantum Physics in Introductory Physics Courses

The Canadian-Argentinian philosopher was also among the first to use a new terminology—quantons—to describe QT as having an object without a measurement process, essentially distinct from those of classical physics. In a textbook with an innovative didactic approach to introductory QT courses, Lévy-Leblond and Balibar (1990, p. 69) support similar epistemological premises. According to them, "it is, therefore, necessary to acknowledge that we have here a different kind of an entity, one that is specifically quantum. For this reason we name them *quantons*, even though this nomenclature is not yet universally adopted." More recently, Michel Paty has developed this idea:

> in terms of an extension of the meaning given to the concepts of *physical state and physical quantity* of a system, which would allow, without any theoretical change in QT, to speak consistently of *real quantum systems* as having definite *physical properties.* (Paty 1999, p. 376)

The philosophical key to this generalization was found by Paty (2000) in a historical and epistemological analysis of the "legitimacy of mathematization in physics"; this generalization suggesting "an extension of meaning for the concept of physical magnitude that puts emphasis on its relational and structural aspects rather than restraining it to a simple 'numerically valued' conception." According to the French philosopher, such a generalization could be useful not only for QT but also for the case of dynamic systems and quantum gravity. While essentially based on his philosophical analysis, Paty argues using some issues related more directly to scientific practice. He quotes the recent experimental confirmations of QT to maintain that the working physicist, in a spontaneous way, refers to quantum theory as "a fundamental theory about a given *world of objects*" and that this spontaneous perception only faces difficulties when it focuses the "*transition* from this *quantum domain to the classical one* that of measuring apparatuses." The list of supporters of realism in quantum physics is far from being comprehensive. However, while the quantum controversy may be seen as one more chapter in the dispute between realism and instrumentalism that has characterized the whole history of physics, the history of QT framed the debate in news terms. Indeed, the experience gained with QT disavowed many features associated with the usual realistic view. If this view has a future in the philosophy of physics, and we think it has, it needs to be accommodated with epistemological and conceptual lessons from QT.

The insensitivity to the philosophical choices seen in the physics teaching papers we have analyzed may have biased some of their research results. For example, McKagan et al. (2010) reported that, in order to construct a conceptual survey on QT, they were not able to find any version of a question trying to address the wave-particle duality that the faculty agreed upon as the "correct" answer. It is also evident that the didactic strategies will be different depending on the interpretational choices and that the uncritical adoption of one of them—which occurs when it is not clearly stated—may have undesirable consequences. For example, the proposals that attempt to represent in a "more displayable" way some quantum concepts using simulations tend implicitly towards a wavy interpretation that by its nature may reinforce links with classical physics. Such proposals may anchor in the classic ideas students already have, making them stronger and prevent them from gaining a better understanding of quantum concepts. This happens, for example, in the difficulties students have replacing the idea of electromagnetic wave with probability wave (Greca and Freire 2003): many students consider the probability density representation to be a representation of movement. Similar results were found among chemistry students introduced to the wavy model of the atom, who understand the concept of orbital as a "space" and not as a mathematical function (Tsaparlis and Papaphotis 2009).

Clauser (2002), although recognizing the use of images for interpreting physics concepts, is aware of the pitfalls that images associated with the wavy model may present:

> In quantum mechanics, the books all make this seem like simple wave mechanics, i.e. what you would see – a direct analogy with waves on the surface of a pond. And they show pictures. [...] And then even worse, they say, 'Okay. A particle, we can represent kind of as a wave packet,' whatever that means. [...] propagating in real space. [...] Now consider a two particle case. Ψ is no longer a functions of x, y, z, and t. It's a function of x_1, y_1, z_1, x_2, y_2, z_2. Has space and time grown? [...] So if I couldn't do it for four, three, two particles, I shouldn't have done it for one particle either. [...] Which means this whole idea of wave packets that all of the books put in there is to try and make you feel comfortable with it, all of those chapters, you might as well rip up and throw them away because they are wrong because that's not the correct conceptual model. (Clauser 2002, p. 14)

We are not rejecting the use of images or materials that may make quantum concepts for the teaching of quantum physics more visible. In fact, by applying cognitive psychology to research in science education, it is possible to find evidence that many college students use imagistic mental models to make sense of physics concepts (Greca and Moreira 1997, 2002); that is, they need to "visualize" what is happening in order to understand. It is worth stressing that this use of imagistic representations can be found in the work of great physicists such as Faraday or Maxwell (Nersessian 1992). The point is that the use of images in QT must necessarily refer to the complementarity ideas, as indicated in Aspect's explanation. Therefore, students must be thoughtfully introduced to the complementarity principle. However, there is an obstacle: complementarity has virtually disappeared from teaching and research in science teaching.

7.6 Complementarity in Science Education Research

From the 32 papers we have researched, only 10 refer to the existence of different possible interpretations and 9 among these papers cite the existence of the complementarity view. Two of them (Greca and Freire 2003 and Passon 2004) do not consider its potential usefulness. Interestingly enough, other papers which report surveys or identify students' learning difficulties using the Copenhagen interpretation do not make use of the concept of complementarity. This strange finding comes, however, as no surprise to those who know the history of quantum physics teaching.

7 Meeting the Challenge: Quantum Physics in Introductory Physics Courses

At the end of 1927 the complementarity view was clearly the most influential among the founding fathers of quantum physics. It had gathered Heisenberg, Pauli, Jordan, and Born, in addition to Bohr, on its side while the remaining critics, such as Einstein, de Broglie, and Schrödinger, supported different views on the subject. Soon, de Broglie aligned himself with the complementarity camp. The historian of physics Max Jammer (1974, p. 250) called the period from the creation of quantum theory until the 1950s the times of the unchallenged monocracy of the Copenhagen school.[21] However, adhesion to this monocracy was weaker than this term may suggest. Its diffusion outside Germany and Denmark was not without difficulties (Heilbron 2001; Schweber 1986), as we have already seen. As a matter of fact, the complementarity view was absent from the one of the most powerful tools in the training of physicists, namely, textbooks. Kragh (1999, p. 211) remarked that only 8 out of the 43 quantum physics textbooks published between 1928 and 1937 mentioned the complementarity principle while 40 cited the uncertainty principle. Despite how central complementarity was in Bohr's interpretation of quantum physics, "most textbook authors, even if sympathetic to Bohr's ideas, found it difficult to include and justify a section on complementarity." Kragh noted that Dirac, the author of one of the most influential textbooks ever written, while closely connected to the supporters of the Copenhagen interpretation and having great respect for Bohr, "did not see any point in all the talk about complementarity. It did not result in new equations and could not be used for the calculations that Dirac tended to identify with physics" (Kragh 1999, p. 211). Indeed, even in most current textbooks when some reference to complementarity is made, it is restricted to the mutual exclusion between wave and particle representations.

The absence of complementarity in the culture of practicing physicists was so conspicuous that Bohr's biographer, the physicist and historian of physics Abraham Pais, announced in the introduction of Bohr's biography that he was looking for a reason for such an absence (Pais 1991). However, Pais did not solve the riddle. One hint, not yet exploited by historians of science, concerns the reasons why Niels Bohr himself did not write a textbook in which complementarity was clearly presented. In the early 1950s, as the debates around the interpretation of the quantum were becoming a hot topic, Léon Rosenfeld, the physicist who was the enduring assistant of Bohr for epistemological matters, acutely felt the need for such a book:

> There is not a single textbook of quantum mechanics in any language in which the principles of this fundamental discipline are adequately treated, with proper consideration of the role of measurements to define the use of classical concepts in the quantal description. (Rosenfeld 1957, apud Osnaghi et al. 2009, p. 99)

At the same time, Rosenfeld unsuccessfully urged Bohr to write it while reporting the interest around complementarity and the debates over the interpretation of quantum physics: "There is great interest in the topic among chemists and biologists, but

---

[21] Recent studies, however, have shown both the diversity of perspectives behind the term "Copenhagen interpretation" and the context of its coinage, for example, Camilleri (2009a) and Howard (2004). See also Beller (1999) for the nuances among the founding fathers of QT which are usually smoothed over in the term Copenhagen interpretation.

there is no book that one can refer them to and that could protect them from the confusion created by Bohm, Landé, and other dilettantes." Rosenfeld concluded saying: "I will now do my bit here in Manchester by giving a lecture for chemists and biologists; but nothing can replace the book that *you* must write" (Rosenfeld 1957, apud Osnaghi et al. 2009, p. 99; emphasis in the original). Parodying "The book nobody read," a title used by the historian Owen Gingerich (2004) for the book in which he charted the readers of Copernicus' book in early modernity, "The book nobody wrote" is an open and interesting question on the vicissitudes of physics in the twentieth century.

And yet complementarity is being revived among practicing physicists, this time stripped of its heavy philosophical clothes and framed in the information turn arriving to quantum physics in recent years (Gleick 2011). Greenstein and Zajonc (1997), for instance, presented it as the mutually exclusive availability of information among certain ways or transitions. Physicists such as Anton Zelinger, Seth Lloyd, John Archibal Wheeler, and Wojciech Zurek have argued for putting information as a key concept into the foundations of quantum theory. Of course, it can be said that changing classical concepts for information is just a change of wording. However, most would agree that understanding information from a conceptual and epistemological perspective is fundamental to the current challenges in science in general, not only in quantum physics.

7.7 Conclusion: Lessons from History and Philosophy for the Teaching of QT

In the same way as there is no privileged interpretation for quantum physics, there is no ideal way for its introductory teaching at undergraduate level. There is, however, a varied spectrum of options available. The analysis we have developed in this paper privileges the following possibilities that we consider informed by the history and philosophy of science and the teaching experience. The first thing that follows from the arguments we have presented here is that QT teaching and research about QT in science education must make interpretational choices explicit and that choice must be justified or defended. Not doing so not only may reduce the scope of the research results but also the possibilities of the teaching strategies, as introducing elements that are not explicitly explained to students may confuse them. The second point is that any proposal for teaching QT should emphasize the strictly quantum features in order to prevent students from establishing undesirable links with classical concepts.

There are varied options from this point on. The teaching of QT may emphasize the formalism, without worrying about the ultimate ontological status of the mathematical terms. Of course, introductory courses have to make use of an adequate mathematical level. So that a balance between rigor and facilitation may be reached this may be illustrated with the case of systems of two levels that can be treated with matrices and vectors. As we have seen, quantum formalism is self-sufficient, and there is a new generation of physicists, working in advanced quantum research areas, who seem not to need the classical counterpart to manipulate quantum mechanics with proficiency (Aspect et al. 1989; Zeilinger 1999). Along these lines the teaching should give prominence to quantum features such as the superposition principle and the measurement problem as well as effects such as quantum entanglement, quantum beatings, and decoherence in addition to the description of current research on these topics, which are relatively easy to grasp in a conceptual manner.

This is not only important for the understanding of quantum mechanics but also to motivate students to continue their studies on this subject. It is worth stressing that this way of introducing quantum mechanics can be compatible either with the realism or the instrumentalism in terms of epistemological views, as we have seen. The dispute between instrumentalism and realism has accompanied the history of science—the Galilean fight for one of the chief world systems being the most well-known example—and the teaching of quantum mechanics is not the space for settling such a philosophical issue. However, students in introductory physics courses should be introduced to such a pervasive dilemma, and quantum physics courses may be a privileged space for doing so.

Another interesting option could be the use of images (in the form of simulations or other) in order to make quantum concepts more understandable. As we have seen, both from the report of first ranking physicists and from the research in science education informed by cognitive psychology, many students may need concrete models or some way of visualizing the abstract mathematical structure to grasp quantum concepts. Such students, who are perhaps more numerous outside physics courses (e.g., engineering, chemistry, and biology students), may profit from this approach. However, if this approach is used, it is necessary seriously and explicitly to introduce complementarity in the explanation of the right quantum use of these images. Finally, it is possible to combine the formal approach with the introduction of the complementarity view, as we have seen in Aspect's explanation of his experiment on the dual nature of single photons. Perhaps it is time to revive this view in science education research, but in this case, it should not be reduced to the wave-particle duality. In Bohr's own terms, wave-particle duality is just the particular case of a wider view:

> Information regarding the behaviour of an atomic object obtained under definite experimental conditions may […] be adequately characterized as complementary to any information about the same object obtained by some other experimental arrangement excluding the fulfillment of the first conditions. Although such kinds of information cannot be combined into a single picture by means of ordinary concepts, they represent indeed equally essential aspects of any knowledge of the object in question which can be obtained in this domain. (Bohr 1987, p. 26)

Indeed, if teachers and researchers choose to introduce complementarity in QT teaching and education research, it should be properly introduced from the conceptual point of view, as done in the current research on QT or philosophical studies on Bohr's thoughts, which opens another venue for contributions from history and philosophy of science to science education.

Our analysis shows that the infusion of historical elements through the introduction of cases from old quantum physics (blackbody problem, photoelectric effect, atomic model) should be avoided. This is partly because the most important steps in the early construction of QT do not show the specific quantum features in a clear-cut manner, and some of them are very complex for students to understand in introductory courses. In contrast, new experiments are conceptually more accessible and can also be reproduced in undergraduate physics laboratories (see, e.g., Dehlinger and Mitchell 2002; Galvez et al. 2005; Thorn et al. 2004). An analogous process happened with the teaching of classical mechanics: the astronomical calculus that led to the classic (and also not intuitive) form of seeing the world is not present in the introductory teaching of classical mechanics. We begin with very simple examples and models in order to help students understand the basic concepts. In courses which are more focused on the history of science, these astronomical examples may have their space when classical mechanics is concerned. But in the QT case, presentation of topics from old quantum physics should emphasize the kind of problem physicists faced and the type of limitations they introduced.

Finally, we would like to stress that the teaching of QT, maybe more than other subject areas in physics, must be informed by the history and philosophy of science. Controversy on its foundations and interpretations has been one of the longest controversies in the history of science and students should be informed of this fact. References to this interpretational debate may bring to the forefront of science education nonconformists who fought against well-established views even putting their professional careers in danger (Freire 2009). This may illuminate the theoretical and experimental developments—Bell's theorem is the best case which brought this debate into mainstream physics—and the current blooming research that has emerged from that controversy. Thus, quantum physics is a very lively example of physics as a human and social product, and we should not exempt students from the presentation of these developments that humanize science.[22]

---

[22] Exposing students to an open scientific controversy may bring some discomfort to physics teachers as this may weaken the dogmatic feature some think it is inseparable to science training. The question reminds us of an old dilemma well posed by Stephen Brush (1974, p. 1170): "Should the History of Science Be Rated X?" In this now classic paper, Brush suggests to science teachers this dilemma in the following terms: "I suggest that the teacher who wants to indoctrinate his students in the traditional role of the scientist as a neutral fact finder should not use historical materials of the kind now being prepared by historians of science: they will not serve his purposes." Then, he continues, "on the other hand, those teachers who want to counteract the dogmatism of the textbooks and convey some understanding of science as an activity that cannot be divorced from metaphysical or esthetic considerations may find some stimulation in the new history of science." No doubt about the mind and heart choice of this talented scientist and historian of science awarded in 2009 with the Abraham Pais Prize for the History of Physics. There is a growing literature on the history of this controversy. In addition to the works already cited, the interested reader may consult Bromberg (2008), Jacobsen (2012), Kaiser (2011), and Yeang (2011). We also highlight the English translation of most of the original papers in the history of this debate in Wheeler and Zurek (1983).

7 Meeting the Challenge: Quantum Physics in Introductory Physics Courses

Acknowledgments We are thankful to CAPES, CNPq, FAPESB, and Universidade Estadual da Paraiba, Brazil, for the support to this research. We are grateful to the editor, Michael Matthews, and the reviewers for their critical comments; to David Kaiser, for reading and commenting the paper; and to Denise Key for her help with the English.

References

Aspect, A. (1999). Bell's Inequality test: more ideal than ever. *Nature*, 398, 189–190.

Aspect, A. (2007). To be or not to be local. *Nature*, 446, 866–867.

Aspect, A., Grangier, P., and Roger, G. (1989). Dualité onde-particule pour un photon unique. *J. Optics* 20(3), 119–129.

Baily, C., Finkelstein, N. D. (2010). Refined characterization of student perspectives on quantum physics. *Physical Review Special Topics - Physics Education Research* 6, 020113 (1–11).

Barnes, M. B., Garner, J. & Reid, D. (2004). The Pendulum as a Vehicle for Transitioning from Classical to Quantum Physics: History, Quantum Concepts, and Educational Challenges. *Science & Education* 13, 417–436.

Barton, G. (1997). Quantum dynamics of simple systems. *Contemporary Physics*, 38(6), 429–430.

Beller, M. (1999). *Quantum Dialogue – The making of a revolution*. Chicago: The University of Chicago Press.

Bohm, D. (1989). *Quantum theory* [unabridged republication of 1951]. New York: Dover.

Bohm, D., Hiley, B. (1988). Nonlocality and the Einstein-Podolsky-Rosen Experiment as Understood through the Quantum-Potential. In F. Selleri (ed.), Quantum Mechanics Versus Local Realism, New York: Plenum Press, 232–256.

Bohr, N. (1987). Natural Philosophy and Human Cultures. [1938]. In Bohr, N. *The Philosophical Writings of Niels Bohr, Essays 1933–1957 on Atomic Physics and Human Knowledge*. Woodbridge, US-CT: Ox Bow Press, 23–31.

Bromberg, J. L. (2006). Device physics vis-a-vis fundamental physics in Cold War America: The case of quantum optics. *ISIS*, 97(2), 237–259.

Bromberg, J. L. (2008). New instruments and the meaning of quantum mechanics. *Historical Studies in the Natural Sciences*, 38(3), 325–352.

Brush, S. (1974). Should the History of Science Be Rated X? *Science*, 183(4130), 1164–1172.

Bunge, M. (2003). Twenty-Five Centuries of Quantum Physics: From Pythagoras to Us, and from Subjectivism to Realism. *Science & Education* 12(5–6), 445–466.

Camilleri, K. (2009a). Constructing the myth of the Copenhagen interpretation. *Perspectives on Science*, 17(1), 26–57.

Camilleri, K. (2009b). A history of entanglement: Decoherence and the interpretation problem. *Studies in History and Philosophy of Modern Physics*, 40, 290–302.

Carr, L. D., McKagan, S. B. (2009). Graduate quantum mechanics reform. *American Journal of Physics*, 77(4), 308–319.

Cataloglou, E., Robinett, R. W. (2002). Testing the development of student conceptual and visualization understanding in quantum mechanics through the undergraduate career. *American Journal of Physics*, 70(3), 238–251.

Clauser, J. F. (2002). Oral history. Interviewed by Joan Lisa Bromberg, Niels Bohr Library, American Institute of Physics, College Park, MD.

Clauser, J. F. and Shimony, A. (1978). Bell's theorem: experimental tests and implications. *Reports on Progress in Physics*, 41, 1881–1927.

Cuppari, A., Rinaudo, G., Robutti, O., Violino, P. (1997). Gradual introduction of some aspects of quantum mechanics in a high school curriculum. *Physics Education*, 32(5), 302–308.

Cushing, J. (1999). *Quantum mechanics: historical contingency and the Copenhagen interpretation*. Chicago: The University of Chicago Press.

Dehlinger, D., Mitchell, M. W. (2002). Entangled photon apparatus for the undergraduate laboratory. *American Journal of Physics*, 70(9), 898–910.

Eisberg, R. (1976). *Applied Mathematical Physics with Programmable Pocket Calculators*. New York: McGraw–Hill.

Ferrero, M., Gómez Pin, V., Salgado, D., Sánchez-Gómez, J. L. (2012). A Further Review of the Incompatibility between Classical Principles and Quantum Postulates. *Foundations of Science*, DOI 10.1007/s10699-012-9290-y, published online: 15 May 2012.

Fischler, H., Lichtfeldt, M. (1992). Modern physics and students' conceptions. *International Journal of Science Education*, 14(2), 181–190.

Freire Jr. O. (2006). Philosophy Enters the Optics Laboratory: Bell's Theorem and its First Experimental Tests (1965–1982). *Studies in History and Philosophy of Modern Physics*, 37, 577–616.

Freire Jr. O. (2009). Quantum dissidents: Research on the foundations of quantum theory circa 1970. *Studies in History and Philosophy of Modern Physics*, 40, 280–289.

García Quijás, P. C., Arévalo Aguilar, L. M. (2007). Overcoming misconceptions in quantum mechanics with the time evolution operator. *European Journal of Physics*, 28, 147–159.

Galvez, E. J., Holbrow, C. H., Pysher, M. J., Martin, J. W., Courtemanche, N., Heilig, L., Spencer J. (2005). Interference with correlated photons: Five quantum mechanics experiments for undergraduates. *American Journal of Physics*, 73(2), 127–140.

Gingerich, O. (2004). *The book nobody read: chasing the revolutions of Nicolaus Copernicus*. New York: Walker & Co.

Gleick, J. (2011). *The information: a history, a theory, a flood*, New York: Pantheon Books.

Goff, A. (2006). Quantum tic-tac-toe: A teaching metaphor for superposition in quantum mechanics. *American Journal of Physics*, 74(11), 962–973.

Gottfried, K. (1978). Quantum physics series, films 1–10, film review. *American Journal of Physics*, 46, 315–316.

Graham, L. (1993). *Science in Russia and the Soviet Union: A short history*, Cambridge: Cambridge University Press.

Grangier, P.; Roger, G. and Aspect, A. (1986). Experimental Evidence for a Photon Anticorrelation Effect on a Beam Splitter: A New Light on Single-Photon Interference. *Europhysics Letters* 1(4), 173–179.

Greca, I. M. & Freire Jr., O. (2003). Does an Emphasis on the Concept of Quantum States Enhance Students' Understanding of Quantum Mechanics? *Science & Education* 12(5–6), 541–557.

Greca, I. M., Moreira, M. A. (1997). The kinds of mental representations - models, propositions and images - used by college physics students regarding the concept of field. *International Journal of Science Education* 19, 711–724.

Greca, I. M., Moreira, M. A. (2002). Mental, physical and mathematical models in the teaching and learning of physics. *Science Education*, 86, 106–121.

Greenberger, D.; Hentschel, K. and Weinert, F. (eds) (2009). *Compendium of Quantum Physics: Concepts, Experiments, History and Philosophy*. Berlin: Springer.

Greenstein, G. and Zajonc, A. (1997). *The Quantum Challenge – Modern Research on the Foundations of Quantum Mechanics*. Sudbury, MA: Jones and Bartlett.

Griffiths, D. J. (2005). *Introduction to Quantum Mechanics* [2nd ed.]. Upper Saddle River (NJ): Pearson Prentice Hall.

Gröblacher, S., Paterek, T., Kaltenbaek, R., Brukner, C., Zukowski, M., Aspelmeyer, M., & Zeilinger, A. (2007). An experimental test of non-local realism. *Nature*, 446, 871–875.

Hadzidaki, P. (2008a). Quantum mechanics and scientific explanation: an explanatory strategy aiming at providing understanding. *Science & Education*, 17(1), 49–73.

Hadzidaki, P. (2008b). The Heisenberg microscope: a powerful instructional tool for promoting meta-cognitive and meta-scientific thinking on quantum mechanics and the "nature of science". *Science & Education*, 17(6), 613–639.

Harding, S. (ed) (1976). *Can Theories Be Refuted? Essays on the Duhem-Quine Thesis*. Dordrecht: D. Reidel.

7 Meeting the Challenge: Quantum Physics in Introductory Physics Courses

Heilbron, J. (2001). The earliest missionaries of the Copenhagen spirit. In P. Galison, M. Gordin, D. Kaiser (Eds). *Science and Society - The history of modern physical science in the twentieth century. Vol. 4 - Quantum Histories*. New York: Routledge, 295–330.

Held, C. (2012). The Kochen-Specker Theorem, The Stanford Encyclopedia of Philosophy (Summer 2012 Edition), Edward N. Zalta (ed.), forthcoming URL = <http://plato.stanford.edu/archives/sum2012/entries/kochen-specker/>.

Hirshfeld, A. C., Henselder, P. (2002). Deformation quantization in the teaching of quantum mechanics. *American Journal of Physics*, 70(5), 537–547.

Holbrow, C. H., Galvez, E., Parks, M. E. (2002). Photon quantum mechanics and beam splitters. *American Journal of Physics*, 70(3), 260–265.

Howard, D. (1985). Einstein on locality and separability. *Studies in History and Philosophy of Science*, 16(3), 171–201.

Howard, D. (2004). Who invented the "Copenhagen interpretation"? A study in mythology. *Philosophy of Science*, 71, 669–682.

Jacobsen, A. (2012). *Léon Rosenfeld - Physics, Philosophy, and Politics in the Twentieth Century.* Singapore: World Scientific.

Jacques, V. et al. (2005). Single-photon wavefront-splitting interference – An illustration of the light quantum in action. *European Journal of Physics D* 35, 561–565.

Jammer, M. (1974). *The Philosophy of Quantum Mechanics – The Interpretations of Quantum Mechanics in Historical Perspective*. New York: John Wiley.

Johnston, I. D., Crawford, K., Fletcher, P. R. (1998). Student difficulties in learning quantum mechanics. *International Journal of Science Education*, 20(4), 427–446.

Kaiser, D. (2007). Turning physicists into quantum mechanics. *Physics World* (May 2007), 28–33.

Kaiser, D. (2011). *How the Hippies Saved Physics – Science, Counterculture, and the Quantum Revival.* New York: Norton.

Kalkanis, G., Hadzidaki, P., Stavrou, D. (2003). An Instructional Model for a Radical Conceptual Change Towards Quantum Mechanics Concepts. *Science Education, 87*, 257– 280.

Karakostas,V., Hadzidaki, P. (2005). Realism vs constructivism in contemporary physics: the impact of the debate on the understanding of quantum theory and its instructional process. *Science & Education*, 14(5), 607–629.

Kragh, H. (1992). A Sense of History: History of Science and The Teaching of Introductory Quantum Theory. *Science & Education*, 1, 349–363.

Kragh, H. (1999). *Quantum Generations: A History of Physics in the Twentieth century*. Princeton: Princeton University Press.

Lévy-Leblond, J-M. (2003). On the nature of quantons. *Science & Education* 12, 495–502.

Lévy-Leblond, J.-M. & Balibar, F. (1990). *Quantics – Rudiments of Quantum Mechanics.* Amsterdam: Elsevier.

Magalhães, A. L., Vasconcelos, V. P. S. (2006). Particle in a Box: Software for computer-assisted learning in introductory quantum mechanics courses. *European Journal of Physics*, 27, 1425–1435.

Marshall, T. and Santos, E. (1987). Comment on "Experimental Evidence for a Photon Anticorrelation Effect on a Beam Splitter: a New Light on Single-Photon Interferences". *Europhysics Letters*, 3, 293–296.

McDermott, L.C. & Redish, E. F. (1999). Resource letter: PER-1: Physics education research. *American Journal of Physics*, 67(9), 755–767.

McKagan, S. B.,,. Perkins, K. K., Dubson, M., Malley, C., Reid, S., LeMaster, R., Wieman, C. E. (2008). Developing and researching PhET simulations for teaching quantum mechanics. *American Journal of Physics*, 76(4 & 5), 406–417.

McKagan, S. B., Perkins, K. K., & Wieman, C. E. (2010). Design and validation of the Quantum Mechanics Conceptual Survey. *Physical Review Special Topics - Physics Education Research* 6, 020121.

Michelini, M., Ragazzon, R., Santi, R., Stefanel, A. (2000). Proposal for quantum physics in secondary school. *Physics Education, 35,* 406–410.

Müller, R., Wiesner, H. (2002). Teaching quantum mechanics on an introductory level. *American Journal of Physics*, 70(3), 200–209.

Nashon, S. Nielsen, W., Petrina, S. (2008). Whatever happened to STS? Pre-service physics teachers and the history of quantum mechanics. *Science & Education*, 17, 387–401.

Nersessian, Nancy. (1992). How do scientists think? Capturing the dynamics of conceptual change in science. In: Ronald N. Giere (ed). *Cognitive models of science*. Minneapolis: University of Minnesota Press, 3–44.

Níaz, M., Klassen, S., Mc Millan, B., Metz, B. (2010). Reconstruction of the History of the Photoelectric Effect and its Implications for General Physics Textbooks *Science Education*, 94, 903–931.

Omnès, R. (2000). *Comprendre la mécanique quantique*. Paris: EDP Sciences.

Osnaghi, S., Freitas, F., and Freire, O. (2009). The origin of the Everettian heresy. *Studies in History and Philosophy of Modern Physics*, 40, 97–123.

Pais, A. (1991). *Niels Bohr's times: in physics, philosophy, and polity*. New York: Oxford University Press.

Passon, O. (2004). How to teach quantum mechanics. *European Journal of Physics*, 25, 765–769.

Paty, M. (1999). Are quantum systems physical objects with physical properties? *European Journal of Physics* 20, 373–78.

Paty, M. (2000). Interprétations et significations en physique quantique. *Revue Internationale de Philosophie*, 212, 2, 17–60.

Pospievich, G. (2003). Philosophy and quantum mechanics in science teaching. *Science & Education*, 12, 559–571.

Redhead, M. (1987). *Incompleteness, Nonlocality, and Realism - A Prolegomenon to the Philosophy of Quantum Mechanics*. Oxford: Clarendon Press - Oxford Univ. Press.

Rosa, R. (2012). The Merli–Missiroli–Pozzi Two-Slit Electron-Interference Experiment, Physics in Perspective, 14(2), 178–195.

Schenzle, A. (1996). Illusion or reality: the measurement process in quantum optics. *Contemporary Physics*, 37 (4), 303–320.

Schweber, S. (1986). The empiricist temper regnant: theoretical physics in the United States 1920–1950. Part 1. *Historical Studies in the Physical and Biological Sciences*, 17, 55–98.

Shankar, R. (1994). *Principles of Quantum Mechanics*. New York: Plenum Press.

Shimony, A. (2009). "Bell's Theorem", The Stanford Encyclopedia of Philosophy (Summer 2009 Edition), Edward N. Zalta (ed.), URL=<http://plato.stanford.edu/archives/sum2009/entries/bell-theorem/>.

Silva, I., Freire, O. (2013). The Concept of the Photon in Question: The Controversy Surrounding the HBT Effect circa 1956–1958, *Historical Studies in the Natural Sciences*, 43(4), 453–491.

Singh, C. (2001). Student understanding of quantum mechanics. *American Journal of Physics*, 69(8), 885–889.

Singh, C. (2006). Assessing and improving student understanding of quantum mechanics. In P. Heron, L. McCullough, and J. Marx (Eds) *2005 Physics Education Research Conference Proceedings*. Melville, NY: AIP Press, 69–72.

Singh, C. (2008). Interactive learning on quantum mechanics. *American Journal of Physics*, 75(4–5), 400–405.

Thaller, B. (2000). *Visual quantum mechanics: selected topics with computer-generated animations of quantum-mechanical phenomena*. New York: Springer.

Thorn, J. J., Neel, S. M., Donato, V. W., Bergreen, G. S., Davies, R. E., Beck, M. (2004). Observing the quantum behavior of light in an undergraduate laboratory. *American Journal of Physics*, 72(9), 1210–1219.

Tsaparlis, G., Papaphotis, G. (2009). High-school students' conceptual difficulties and attempts at a conceptual change. *International Journal of Science Education*, 31(7), 895–930.

Velentzas, A., Halkia, K. (2011). The 'Heisenberg's Microscope' as an Example of Using Thought Experiments in Teaching Physics Theories to Students of the Upper Secondary School. *Research in Science Education*, 41, 525–539.

Velentzas, a., Halkia, K., Skordoulis, C. (2007). Thought Experiments in the Theory of Relativity and in Quantum Mechanics: Their Presence in Textbooks and in Popular Science Books. *Science & Education*, 16, 353–370.

Wheeler, J. A., Zurek, W. H. (Eds.). (1983). *Quantum theory and measurement*. Princeton, NJ: Princeton University Press.

Wuttiprom, S., Sharma, M. D., Johnston, I. D., Chitaree, R., & Soankwan, C. (2009). Development and use of a conceptual survey in introductory quantum physics. *International Journal of Science Education*, 31(5), 631–654.

Yeang, C-P. (2011). Engineering Entanglement, Conceptualizing Quantum Information. *Annals of Science*, 68(3), 325–350.

Zeilinger, A. (1999). In retrospect: Albert Einstein: philosopher – scientist. *Nature*, 398(6724), 210–211.

Zollman, D., Rebello, N. S., Hogg, K. (2002). Quantum mechanics for everyone: Hands-on activities integrated with technology. *American Journal of Physics*, 70(3), 252–259.

Ileana María Greca is a professor for didactics of experimental sciences at the University of Burgos, Spain. She earned a Ph.D. in physics education from the Federal University of Rio Grande do Sul, Brazil, in 2000. Her research interests include cognitive psychology and science education, modern physics in science education, applications of history and philosophy of science in science teaching, and professional development of science teachers. Among her recent publications are three papers related with the history and philosophy of science in science education (Ataíde, A. R. P. & Greca, I. M., 2013. Epistemic views of the relationship between physics and mathematics: its influence on the approach of undergraduate students to problem solving, *Science & Education*, 22 (6), pp 1405–1421; Arriassecq, I. & Greca, I. M., 2012. A teaching-learning sequence for the special relativity theory at secondary level historically and epistemologically contextualized, *Science & Education*, 21, pp. 827–851; Teixeira, E. S.; Greca, I. M. & Freire Jr., O., 2012. The History and Philosophy of Science in Physics Teaching: A Research Synthesis of Didactic Interventions, *Science & Education*, 21, pp. 771–796).

Olival Freire Jr. is a professor at the Universidade Federal da Bahia, Brazil, where he teaches physics and history of science. He also teaches at the Graduate Studies Program in History, Philosophy, and Science Teaching (UFBa/UEFS) and is a fellow at the Brazilian CNPq. He has a B.Sc. in physics and a Ph.D. in history (Universidade de São Paulo, Brazil). He was a senior fellow at the Dibner Institute for History of Science, a visiting Professor at the Université de Paris VII, and a visiting researcher at MIT and Harvard. His research interests are concentrated on the history of the research on the foundations of quantum physics, the history of physics in Brazil, and the use of history of physics in the teaching of science. He has published in journals such as *Studies in History and Philosophy of Modern Physics*, *Historical Studies in Natural Sciences,* and *Science & Education*. Among his papers are Quantum dissidents: Research on the foundations of quantum theory circa 1970. *Studies in History and Philosophy of Modern Physics*, 40, 280–289, 2009; Science and exile—David Bohm, the Cold War, and a new interpretation of quantum mechanics. *Historical Studies in the Physical and Biological Sciences*, Berkeley, 36(1), 1–34, 2005; and The historical roots of 'foundations of quantum physics' as a field of research (1950-1970). *Foundations of Physics*, 34(11), 1741–1760, 2004.

Chapter 8
Teaching Energy Informed by the History and Epistemology of the Concept with Implications for Teacher Education

Manuel Bächtold and Muriel Guedj

8.1 Introduction

What can Epistemology and the History of Science and Technology (EHST hereafter) contribute to the field of teaching energy? Is it enough simply to evoke them as a way of broadening the learning after teaching the concept, that is, once students have mastered it, in order to offer them a few historical reference points and to spark off philosophical debate on the subject? That is not our point of view. On the contrary, we think that EHST could play a fundamental role in teaching energy, especially in regard to teacher training. Beynon wrote in 1990: 'I have no doubt at all that the problem of teaching energy will remain insoluble until teachers, themselves, have a clear understanding of the concept of energy' (1990, p. 316). We share this point of view. Indeed, for students to successfully understand and correctly apply the concept, it seems essential that their teachers themselves first master it, which is far from given. The highly abstract nature of the concept of energy (which is inseparable from the principle of energy conservation), its many possible forms (e.g. kinetic energy, thermal energy, nuclear energy), the distortions of meaning to which it is subject in everyday use (e.g. saying that energy can be 'produced' and 'consumed') all make it difficult to define the concept.

As we will try to demonstrate in this article, EHST provides the keys to understanding what energy is and, in particular, to at least begin to answer these three questions:

- 'What is the origin of the concept of energy?'
- 'What is energy?'
- 'What purpose does the concept of energy serve?'

M. Bächtold (✉) • M. Guedj
LIRDEF (EA 3749), Universités Montpellier 2 et Montpellier 3,
2, place Marcel Godechot, BP 4152, Montpellier Cédex 5 34092, France
e-mail: manuel.bachtold@montpellier.iufm.fr; muriel.guedj@montpellier.iufm.fr

M.R. Matthews (ed.), *International Handbook of Research in History, Philosophy and Science Teaching*, DOI 10.1007/978-94-007-7654-8_8,
© Springer Science+Business Media Dordrecht 2014

This is why our strategy consists of developing a training programme for teaching energy based on EHST. We start by discussing how teaching energy is covered throughout schooling (in the case of France), the learning difficulties associated with the concept and the main strategies presented in science education literature to teach the concept (Sect. 8.2). Then we outline our methodology and our two lines of research:

(i) EHST as part of teacher training for teaching energy
(ii) EHST as a means of rethinking how energy is taught (Sect. 8.3)

In the context of the first line of research, we present a framework for teacher training on the concept of energy based on EHST (Sect. 8.4). The second line of research will be addressed in a future article.

8.2 Teaching Energy: A Brief Overview of the Current Situation

8.2.1 Institutional Expectations and Teaching Energy: The Case of France

Energy appears as a concept across physical science programmes from primary through secondary school. Its progressive introduction throughout primary and secondary education has two main strands: the scientific approach to the concept and its implication in current social issues. Generally speaking, the emphasis is on a qualitative approach that prioritises the nature, role and properties of a concept that, although part of daily life, remains difficult to tackle.

In primary school (MEN 2008a), this qualitative approach is based on an introduction that aims to present energy via questions related to using and saving energy. In the further learning and consolidation stage, this does not involve introducing the scientific concept, but rather increasing pupils' awareness of the diverse situations that require a source of energy (using everyday vocabulary), identifying the principal sources of energy and distinguishing those that are renewable from those that are not. In addition, the concept of thermal insulators and conductors is first introduced, with the home providing a good illustration of this approach. The main goal of this initial contact with the concept of energy, which provides the opportunity for projects on the Industrial Revolution introduced in the history programme of the further learning stage, is to contribute to the education of the student as a future citizen.

This same goal also pertains to the educational programme at *collège* (the first stage of secondary school, age 10–14), which equally stresses a qualitative approach to energy; however, at this stage, the scientific concept is introduced and a definition given. The concept of energy, used as an example in the 'unity and diversity' theme that underlies the college (MEN 2008b) programme, is at the heart of the curriculum. It is presented as an essential concept in core knowledge and skills and is treated as a subject that provides a focal point.

The two main strands mentioned above are fully formulated at this stage. The definition is formulated as follows: 'energy is the capacity of a system to produce an effect' – it can be transformed and conserved. This first scientific approach to the concept proves necessary in order to introduce in a logical way a wide range of events that bring energy into play (e.g. day-to-day use of electric circuits, heat exchange, analysis of how living organisms function) and also constitutes essential knowledge for future citizens who need to be aware of the issues around energy that are central to debates in modern society.

In continuity with *collège*, the first year of *lycée* (high school, i.e. the second stage of secondary school, age 15–18) (MEN 2010a) calls for scientific learning and citizenship that will aid all students to succeed, while in the scientific stream of the two final years of *lycée* (MEN 2010b, 2011), the approach concerns vocational preparation to allow students to work towards careers in science. The emphasis is on acquiring skills in the discipline, encouraging interest in the sciences and making connections between science and society.

The final year of the scientific stream in *lycée* is structured around three axes: 'observe, understand,ct'. The purpose of these points of access to the scientific approach is to illustrate its main steps, giving a central role to the concept of energy, which is a sort of unifying theme throughout the 2 years of the course. In this way, the axis 'understand', dedicated to laws and models, presents energy as a common denominator of all basic interactions and the principle of conservation as an explanatory and predictive tool that allows awareness of the evolution of systems (second year of *lycée*). In addition, the study of the transfer of energy at different scales allows the introduction of the basic concept of thermodynamics (internal energy, thermal transfer, work, heat capacity) and a discussion of the irreversibility of phenomena and the causes of dissipation associated with these transfers (final year of *lycée*). This approach underlines the universality of the laws of physics, for which energy is presented as a unifying principle.

In this initial introduction, which highlights the nature, role and properties of the principle of conservation, the educational programme introduces the social and environmental issues related to energy. This includes knowledge about the variety of energy resources and saving energy, problems related to the production of electricity and the transport and storage of energy as well as the environmental impact of energy choices; all these subjects combine scientific knowledge and current issues in society. The axis 'act' sets out to develop this aspect.

The goal of the educational programme is the progressive construction of scientific knowledge and the development of skills suitable for initiation to experimental methods and practice. To help achieve this goal, the programme recommends making use of the history of science. Creating a historical perspective is structured around two axes: one concerning the nature of science and the other the scientific method. The aim, by emphasising the process of how knowledge is constructed, is to show that scientific truth has a particular status; it is the result of a codified process for which mistaken concepts and incorrect hypotheses are common. The history of science demonstrates that science is a social activity that is part and parcel of the culture in which it develops and that new ideas sometimes collide with tradition or dogmatism. These elements should be taken into account to contextualise science and 'mettre la science en culture' (establish its place in a culture) (Lévy-Leblond 1973). This in turn should help to develop critical thinking, rethink the role of error and present the diversity of scientific methods, which cannot be reduced to a simple sequence of 'observation–modelling–verification', with the last having mainly a heuristic value.

8.2.2 A Difficult Concept to Grasp and Master

Although in general use, the concept of energy is abstract, difficult to define and subject to numerous recurrent conceptions noted by many writers.[1] The origins of these ideas are mainly found in everyday language, which contributes to the formation of imprecise or even mistaken concepts. The different meanings the term 'energy' and other related words take on in ordinary language are distant from or sometimes even incompatible with scientific concepts. In French, as well as in English, for example, it is common to associate the terms energy and energetic with strength and vigour. These words are often employed to describe a highly active person. Whereas in physics, the quantity of energy associated with a system may be very low. Moreover, energy may be in a form that is not even noticeable (this is the case for potential energy).[2]

In general discourse at least, people frequently speak about using, consuming, buying or selling energy, sometimes referring to fossil fuels themselves as 'energy'. This creates confusion between sources and forms of energy and presents a real obstacle in the acquisition of the principle of conservation.

Apart from language, daily experience can also prove to be a source of confusion, particularly for the youngest pupils. The ease with which it is possible to make an appliance function simply by plugging it into a socket implies that something can be obtained without anything being consumed. In the same way, obtaining electricity in hydroelectric or thermal power stations (especially nuclear power stations) takes on a magical character in which electricity seems to be stored.

The diversity of concepts related to energy makes it difficult to provide an exhaustive overview. Thus, we have chosen to mention only those, often cited by writers, which seem to be the most recurrent. Watts (1983) groups these according to seven categories:

– The anthropocentric conception, in which energy is associated with what is living.

---

[1] See Solomon (1982, 1983, 1985), Watts (1983), Gilbert and Watts (1983), Duit (1984), Driver and Warrington (1985), Agabra (1985, 1986), Gilbert and Pope (1986), Trellu and Toussaint (1986), Trumper (1993), Ballini et al. (1997), and Bruguière et al. (2002).

[2] Fact sheet for Cycle 2 (basic learning in first years of primary school) and Cycle 3 (further learning in last years of primary school) (MEN 2002, p. 29)

- The conception of energy as a causal agent, in which energy is perceived as the cause of an event, as that which makes something happen. In this scenario, in which energy can be stored, the movement of a falling stone or a thrown ball is explained by the presence of potential and kinetic energy, respectively.
- The conception of energy as a product deriving from a process, a product that rapidly disappears and is not conserved.

To these three ideas, identified by Trumper (1990) as the most frequent, Watts adds the following four concepts. Energy can be perceived as an 'element' that lies dormant in certain objects and is released by a trigger. When energy is systematically associated with movement, Watts refers to the concept of 'activity' energy. This can be 'combustible' energy, where energy is equated to its source (oil, coal, natural gas and petrol are seen as energy), or 'fluid' energy, where energy is equated to a fluid that can be exchanged and transported. When this idea of energy as a fluid, i.e. as something 'quasi-material', is taken as an analogy only, it may be a fruitful tool for initially grasping the concept and the principle of its conservation (Duit 1987; see also below). However, the danger of making use of it in physics education is that students may take it literally and thereby endorse a non-scientific conception that is very hard to overcome (Warren 1982).

In the same vein, Robardet and Guillaud (1995) synthesise the work of other writers to summarise the most common conceptions, grouping these in three broad categories: energy as life (anthropocentric conception), energy as source (i.e. as cause of phenomena), and energy as product (i.e. as consequence of phenomena). This overview highlights the fact that energy is more noticeable when the effect produced is visible and even more so when the effect has a practical aspect or is associated with comfort. Thus, potential energy is little recognised by students (this point will be dealt with in Sect. 8.4.2).

These different conceptions result in several frequent and persistent errors, even after traditional learning (Trumper 1990). Without listing them all, we can cite, for example, the substantialisation of energy, confusion between the form and mode of transfer of energy; between force, speed and energy; and between heat, temperature and internal energy.

8.2.3 The Main Teaching Strategies

While educational programmes from primary to secondary school grant an increasingly large place to energy, the diversity, origin and consequences of mistaken ideas present a major obstacle to learning the scientific concept. Since the 1980s, the trickiness of teaching the concept has led certain educators to seek ways to facilitate its acquisition by taking into account related preconceptions. Generally speaking, traditional teaching is judged dogmatic and abstract (Lemeignan and Weil-Barais 1993), reducing the concept to a group of systematic technical procedures stripped of physical meaning.

The main teaching strategies are based on taking into account students' preconceptions during the application of the principle of conservation of energy. Thus, Trumper (1990, 1991, 1993), in the context of a constructivist approach, leads students to identify any conflicts between their own ideas and the properties required to establish the principle of conservation.

In the same spirit, the work of Agabra (1986) as well as Trellu and Toussaint (1986) promotes the concept of 'objective–obstacle' defined by Martinand, who suggests linking educational objectives with students' ideas, making the obstacles associated with the various preconceptions explicit and in each case indicating a specific way to surmount them.

Also in the constructivist framework, the work of Lemeignan and Weil-Barais (1993), extended by Robardet and Guillaud (1995), aims at constructing the concept of energy and its conservation by encouraging conceptualisation and only subsequently introducing classical formalism. The objective is to define, step by step, the semantic relationships that connect each object in the system studied with the next (e.g. an alternator powers a lamp in an overall system). By progressively establishing the semantic relationships, the energy exchanges that take place in the studied system can be defined.

Another teaching strategy consists of introducing energy as a 'quasi-material' substance. The supporters of this approach, which is in line with students' ideas of energy, justify their choice in pointing out the eminently abstract character of energy. Based on this idea, Duit (1987) and Millar (2005) suggest examining the different types of energy in a qualitative manner before tackling a quantitative, mathematical approach. This is a controversial choice of strategy, whose opponents underline the risk of perpetuating an entrenched false idea (Warren 1982).

Finally, writers agree on the terminological pitfalls, due in large part to everyday language – the meanings and uses of the term 'energy' vary considerably between informal and scientific contexts. Solomon (1985), Chisholm (1992) and Bruguière and colleagues (2002) argue that this problem could be mitigated by simplifying the vocabulary.

To this brief outline, it is fitting to add the work of Koliopoulos and Ravanis (1998), who group the various teaching strategies according to three categories. Their approach differs from those described previously as their classification is based on collected curricula from various countries and not directly on research results. This categorisation thus includes the aims of institutions and the issues that they consider important. So curricula qualified as 'traditional', 'innovative' and 'constructivist' are representative of these orientations.

The traditional curriculum corresponds to a classical mode of exposition in which energy, generally introduced as a concept derived from work, does not have a status in its own right. As a consequence, each field of study in physics requires a specific presentation of the concept, reflecting its many meanings.

The curriculum described as 'innovative' is based on ideas influential in the 1960s that promote the concept of energy by giving it a structural character and granting it a central place in the educational programme. This approach also introduces a social dimension to the learning of the concept.

The constructivist curriculum takes into account the current research orientations presented above. It is characterised notably by the construction of models of the energy chain and draws on students' prior conceptions.

In addition to the strategies outlined above, some writers suggest marshalling the history and philosophy of science in order to facilitate teaching energy. For the most part, the proposals revolve around aligning the difficulties confronted by scientists in the context of the emergence of the concept and students' ideas about energy. This is the case of Trellu and Toussaint (1986), who compare teaching centred on the conservation or transfer of energy; of Agabra (1986), who returns to the various models of heat; and of Duit (1987), who proposes that students could follow the same train of thought as certain nineteenth-century scientists; that is, start from a quasi-material conception of energy (see Sect. 8.4.2).

In contrast, Coelho (2009) draws from the work of Mayer and Joule to propose teaching centred on the notion of equivalence (e.g. heat and work), excluding the question of substantiality, which he supports is a source of confusion (see Sect. 8.4.2).

Generally speaking, the main aim of these proposals is to introduce elements of the history and philosophy of science in order to compare the difficulties of students to those confronted by scientists in the nineteenth century. History is employed here as a useful didactic tool, but little place is given to the cultural and scientific context.

8.3 Methodology for Designing a Teacher Training Programme for Teaching Energy

8.3.1 A New Strategy: Starting with Teacher Training

Although the range of strategies for teaching energy indicates its interest and these strategies contain innovative ideas, none has really managed to impose itself over the others. Teaching energy is considered complex and fragmented. This fragmentation is a result of the lack of connection between the fields of study concerned, which tends to obscure the principal properties of energy and precludes an understanding of the role of the principle of conservation. There seem to be as many meanings of the term *energy* as there are uses and fields of study.

In fact, teachers themselves feel ill-prepared when they have to take on this subject. This is notably referred to in the study mentioned above (Koliopoulos and Ravanis 1998), which aims to identify how experienced teachers teach the concept of energy. While this study shows that the majority of teachers choose traditional teaching methods, it indicates that strategies similar to those described as innovative and constructivist are also used. The latter two strategies are motivated, respectively, by the desire to underline the role of energy, in particular its unifying character, and by the necessity of taking into account students' prior ideas. However, some of the teachers who opt for an innovative approach in fact focus mainly on mechanical phenomena and eventually come back to a traditional approach that introduces energy by deriving it from work, while teachers opting more for a constructivist approach consider themselves poorly armed for incorporating students' conceptions in their teaching.

Furthermore, teachers themselves are not without mistaken conceptions concerning energy, especially in the case of primary school teachers (see, e.g. Summers and Kruger 1992; Trumper et al. 2000). Regarding secondary school teachers or students with science training, Pintó and colleagues (2004) and Méheut and colleagues (2004) highlight confusions regarding irreversibility and real phenomena, cyclical processes and reversibility as well as difficulty in conceptualising the dissipation of energy in the context of its conservation (thus, energy dissipation and conservation seem contradictory).

These various factors regarding teachers' ideas about energy and how it is learned prompt us to delve more deeply into what acts as an obstacle to implementing effective teaching and bring our attention to how teachers themselves are trained. It seems indispensable for teachers to be sufficiently at ease with the concepts to be able to undertake a critical analysis of their teaching practice and to rethink how energy is taught.

Clarifying the concept seems an essential first step to dispel any ambiguities related to the definitions of terms and the properties of the various concepts brought up. The concept of energy is complex, abstract and polymorphous, and the principle of conservation that characterises it is a unifying principle, a 'super law'[3] that structures physics. Explaining the properties and role of the principle leads back to the context of the emergence of the latter in the nineteenth century, to theoretical problems (questions relating to the dissipation of energy and the nature of heat), to experimental situations (the issue of increasing the profitability of machines), to mathematical formalism (the analytic expression of heat required to express the outcome during a Carnot cycle of operations) as well as to the philosophical context, the period being the subject of many debates regarding the founding concepts of physics (Freuler 1995).

This clarification of the concepts should allow the subsequent construction of teaching that highlights the fundamental characteristics of the principle of conservation of energy, defines the concepts related to energy and takes into account, with appropriate vocabulary, the social orientations given by official educational guidelines.

In this context, EHST seems to us an effective and fertile field for elucidating the concept of energy and rethinking how it is taught (on this point, see also Bächtold and Guedj 2012).

8.3.2 EHST in Teacher Training: The Case of France

The role of EHST in teacher training has long interested those who promote a full and authentic science education. In France, in 1902, the institutionalisation of science teaching in secondary school was coupled with the university-level

---

[3] This expression comes from Michel Hulin (1992) in his book entitled *Le mirage et la nécessité: pour une redéfinition de la formation scientifique de base*.

8 Teaching Energy Informed by the History and Epistemology of the Concept...

development of a general history of science aimed mainly at teachers. Later, in the 1970s, reforms stressed the necessity of transmitting historical knowledge in university programmes as well as in teacher training, including for primary teachers. In mathematics, these reflections were largely the realm of the newly created IREMs.[4] The SFHST,[5] since its creation in 1980, has supported EHST initiatives, which have continued to develop.

In what she describes as the 'long march' of EHST education, Fauque (2006) points out that in the 1980s, the concerns of French researchers on the subject were shared abroad. She notes the reach of Bevilacqua's work at the University of Pavia, leading to numerous educational publications that introduced elements from the history of science based on local archives (primary sources and scientific instruments) into science teaching. In 1983, under the impetus of Bevilacqua and Kennedy,[6] the first international conference was held in Pavia. Many others would follow: at the *Deutsches Museum* in Munich, at *La Cité des Sciences et de l'Industrie* in Paris and in Cambridge, to mention only the first three conferences.

This impetus also resulted in the production of literature by specialist organisations, which allowed teaching proposals to be supplemented by reports on the results of experiments. This was notably the case of the French Physicists' Union (*Union des Physiciens en France*) and the Association for Physics Education (*Associazione per l'insegnamento della fisica*) in Italy. The work of Shortland and Warwick (1989) in Britain was in the same spirit, with their publication (under the aegis of the British Society for the History of Science) of *Teaching the History of Science*, as was that of Matthews[7] with the creation of the journal *Science & Education*, as well as another work dedicated to this question (Matthews 1994/2014). Although far from comprehensive, this overview testifies to a shared wish to integrate EHST in science education.

Likewise, in France, the place given to EHST in school programmes increased, with its inclusion in core knowledge and skills,[8] in recruitment examinations as well as in the guidelines for teachers' skills,[9] all aspects of the same approach.

---

[4] *Instituts de Recherche sur l'Enseignement des Mathématiques* (Research Institutes for Teaching Mathematics).

[5] *Société Française d'Histoire des Sciences et des Techniques* (French Society of the History of Science and Technology).

[6] P. J. Kennedy was professor at the University of Edinburgh.

[7] University of New South Wales, Sydney.

[8] The core skills are those considered essential to master by the end of compulsory education. The section dedicated to scientific and technological knowledge emphasises: 'The presentation of the history and the development of concepts, drawing from resources in all the disciplines concerned, is an opportunity to tackle complexity: the historical perspective contributes to providing a coherent vision of science and technology as well as their joint development' (pp. 12–13).

[9] Secondary school teachers should be able to 'situate their discipline(s) within its history, its epistemological issues, its didactic problems and the debates that affect it'. *Framework of reference for teachers' professional skills* (extract from the decree of 19 December 2006 containing guidelines for teacher training, MEN 2007).

In 1999, Lecourt (1999) submitted his report concerning the role of teaching the history and philosophy of science in French universities in which discussed the many factors related to its instruction. Noting the disaffection with studying science, he stressed the necessity of breaking away from the highly technical nature to which science study is often reduced, emphasising the need to give meaning to scientific knowledge and situating it within other types of knowledge – humanising it. Lecourt denounced the harmful effects caused by a lack of EHST education in the curriculum, leading students to adopt an implicit philosophy close to scientism. Several studies reveal the frequent adoption of scientism, whether by students (Désautels and Larochelle 1989) or teachers (Abd-El-Khalik 2001). In the same vein is Paty's (Paty 2000–2001, pp. 56–57) assertion that EHST is essential for discussing the value of scientific truth, while the discourse in society tends to equate revealed truth and scientific truth. Paty reminds us that although scientific truth is relative in the sense that it is incomplete and prone to modification, it has a specific status resulting from a mode of attribution of proof that is clearly identified.

A central element in reflecting on the sciences, in terms of content, methods and links with other fields of knowledge, EHST is essential for reintegrating science in culture. 'Putting science (back) into culture', in the words of Lévy-Leblond (2007), is not a question of creating effective means of transmitting scientific results to the wider public; it is rather about rethinking the sciences, their practice and their methods, in order to produce new, innovative knowledge. Taking up this challenge involves developing critical thinking, too often neglected according to this writer, and prompts consideration regarding the training of scientists. Although referring to the latter, the statement that follows could equally serve as an explanation for the guidelines for teacher training mentioned above:

> Can we continue to train professional scientists without giving them the least element of comprehension of the history of science – concerning their discipline first of all – and of the philosophy, sociology and economy of science? The tasks they now face in practicing their occupation, and the social responsibilities that they can no longer ignore, require them to have a broad conception of scientific work. How can we believe any longer that science is different in this regard than art, philosophy or literature, fields of human activity that no one would imagine teaching independently from their history? (Paty 2000–2001, pp. 13–14)

Training future scientists and educating the citizens of tomorrow necessitate bringing together diverse skills, which we should remember are already widely present in school programmes. Martinand (1993, p. 98) comes to the same conclusion when he emphasises shortcomings in future teachers uninformed about the practices and culture of science: 'The "mission" of the history and the epistemology of science is to enrich research and reflection about its practice, evolution and foundation, without an immediate didactic aim.'

Lastly, in a more specific way, EHST education supports the teaching of scientific disciplines through an epistemological examination of problems, concepts and theories. In the study previously mentioned, Martinand points out that thanks to its critical and prospective function, EHST allows encountered problems to be clarified and teaching content to be questioned in order to better understand its integration in school programmes. Epistemology 'at the service of education' should supplement the orientations developed above.

8.3.3 The Proposed Approach

In the context of the study of energy, the aforementioned approaches lead to a re-examination of the foundations of the concept and its emergence in order to understand its role, properties and functions. This should allow the concept, its principle of conservation and its related concepts (in particular, work, force and heat) to be clarified. All of these steps are essential for teachers. The development of this approach, which enlists the acquisition of 'scientific culture', is a first line of research. Using EHST at the service of teaching energy will be a second, future line of research.

8.3.3.1 EHST in Teacher Training for Teaching Energy

The rest of this article (see Sect. 8.4) will focus on the first line of research. How should teacher training based on EHST be designed to help teachers acquire scientific culture around energy? To develop the beginning of a response to this, we have drawn from many existing works, not only in the field of EHST,[10] but also in science education.[11] Based on these works, we have created a general framework for teacher training on energy, which aims to include all the aspects of the concept and to introduce them according to the most logical progression of ideas possible. We have striven to avoid the pitfall of drowning teachers in an overly complex and detailed history and epistemology of the concept of energy. In particular, the cultural and scientific contexts are not examined in detail, as they would be in a historical study.[12] The aim is to make the use of history and epistemology functional and accessible to teachers. Furthermore, to be both relevant and enlightening, such a historical and

---

[10] Several historical and epistemological studies on energy were published by scientists and/or philosophers of science at the end of the nineteenth century and the beginning of the twentieth century (e.g. Mach, Planck, Poincaré, Meyerson and Cassirer). Later, Kuhn's (1959) article encouraged science historians to carry out new investigations on the emergence of the concept in the nineteenth century (e.g. Elkana 1974; Truesdell 1980; Hiebert 1981; Smith and Wise 1989; Caneva 1993; Smith 1998; Ghesquier-Pourcin et al. 2010). It should be noted that the history of the concept of energy over the course of the twentieth century, with the advent of the theory of relativity (special and general relativity) and of quantum mechanics, as well as the importation of the concept in many other fields (chemistry, biology, economics, arts, etc.), has not yet been well studied.

[11] See in particular the literature indicated in Sect. 8.2.

[12] For further information on these aspects, see the references in the previous footnote.

epistemological introduction must be centred on physical content. Hence, we suggest the teacher training programme could be organised around three points[13]:

What is the origin of the concept of energy?
The investigation of this question aims to challenge the idea that the concept of energy, with the meaning attributed to it today, was always available for scientists. The goal of teacher training here is not only to make teachers aware that the current accepted scientific understanding of the concept only stabilised in physics in the middle of the nineteenth century but also to supply teachers with information to help them understand why it stabilised at this time and how the process of this stabilisation came about.

What is energy?
So that teachers can fully grasp the meaning of the concept of energy, teacher training should clarify all characteristics of the concept (i.e. energy is a quantity associated with a system, it can take different forms, it can be transformed and transferred; see Sect. 8.4), rather than reducing it to the principle of conservation of energy. When dealing with this question, it also seems appropriate to discuss incorrect ideas that can be obstacles to learning the concept.

What purpose does the concept of energy serve?
So that teachers understand and can explain to students the omnipresence of the concept of energy in the curriculum, teacher training should clarify the different functions that this concept allows to be performed in scientific work.

This framework, which will be elaborated upon in Sect. 8.4, makes up the first step of the creation of a teacher training programme, which can then be enriched with examples of possible course outlines and teaching sessions on energy (see the second line of research presented below) and added to allowing for constraints on the ground (type of teacher, available time, equipment and resources, etc.). We then plan an experimentation phase for the training programme in order to assess its impact and attain an empirical response that will enable us to improve it.

8.3.3.2 Using EHST to Rethink the Teaching of Energy

The second line of research mentioned above, that is, EHST at the service of teaching energy, is the subject of a study currently in progress that will be expounded in an upcoming article. Our first hypothesis, which is the basis of this study, is that a teacher training programme on energy based on EHST should profoundly redefine

---

[13] The inspiration here is from Papadouris and Constantinou (2011, p. 966), who 'take the perspective that any attempt to promote students' understanding about energy should primarily address the question 'What is energy and why is it useful in science?'. However, we diverge from these writers' approach on several points: we maintain that it is pertinent to include the question of the origin of the concept of energy; we suggest approaching the three questions drawing on EHST; and, lastly, we do not provide the same answers to the questions posed.

the way in which teachers themselves envisage teaching about energy. More specifically, this teacher training programme should lead teachers towards:

- A new insight into educational programmes (a better global overview as well as an understanding of the relationship between the different sections of these programmes)
- A reflection on their own ideas about energy and its related concepts (e.g. work, heat)
- A new way of taking into account students' prior conceptions
- A review of practices in teaching energy (in terms of the coherence of planned teaching sessions, the organisation of the content, the method used to develop knowledge and, in particular, the relationship between theory and experimentation and the formulation of problems)

The objective of this research study is to come up with concrete proposals for course outlines and teaching sessions on energy making use of EHST. In these proposals, we intend to supply examples of teaching about energy that do not call on the history of science as an optional extra (the 'add-on' approach; see Matthews 1994, p. 70), but rather place it, and epistemology, at the centre of instruction. Our second hypothesis, which remains to be tested, is that such teaching should allow the many difficulties related to the acquisition of the concept of energy to be more easily overcome (see Sect. 8.2).

8.4 Framework for Teacher Training on Energy Based on the History and Epistemology of the Concept

8.4.1 What Is the Origin of the Concept of Energy?

The absence of historical perspective encourages the illusion of the immutable nature of scientific concepts and theories, as if these have always been available for scientists and cannot be challenged or revised in the future. The same is true for the concept of energy. The fact that today it is omnipresent in physics and the other sciences makes it difficult to imagine that only 200 years ago it was not yet fully part of the armoury of physics. So that teachers understand the concept of energy and can grasp its meaning and utility (see Sects. 8.4.2 and 8.4.3), it seems crucial that beforehand they are clear about its origin: where does the concept of energy come from – or, in other words, why and how was this concept introduced in physics?

The first fundamental point that should be emphasised is:

> In its accepted scientific meaning, the concept of energy is inseparable from the principle of its conservation which was established in the middle of the nineteenth century.

This point is expressed by Balibar (2010, p. 403) in this way: 'The concept of energy only became a physics concept from the moment it was irreversibly established that a law of energy conservation exists'.

This initial point guides the rest of our discussion, since it leads us to replace the question 'What is the origin of the concept of energy?' with 'What is the origin of *the principle of conservation* of energy?' This latter question can be approached from two perspectives: one centred on the people who participated in the emergence of the principle and the second centred on the epistemic factors that played a role in this emergence, namely, experimentation and reasoning. These two perspectives should be combined to avoid the risk of a truncated answer.

Concerning the first perspective, the history of energy is particularly instructive for teachers, whose historical idea of science often consists merely of a succession of 'discoveries' made by isolated geniuses – discoveries that are considered independently of context (scientific, technological, philosophical, etc.) (see, e.g. Gil-Pérez et al. 2002, pp. 563–564). The case of the principle of conservation of energy is illustrative of this. The historical study of its emergence is an opportunity to challenge and enrich the vision that teachers have about the history of science.

> The principle of conservation of energy emerged in the middle of the nineteenth century following different research projects led by several scientists (among others, by Mayer, Joule and Helmholtz) influenced by their scientific, technological, philosophical and religious context.

Three points merit emphasising to teachers. Firstly, the principle was not discovered by an isolated genius. This point was underlined by Kuhn (1959), who lists no less than 12 scientists that 'simultaneously' participated in the 'discovery' of the principle.[14]

Secondly, the term *emergence* is more relevant than discovery, because the latter suggests an image that does not comply with the history of the principle – as if it pre-existed all scientific research and was suddenly revealed. This misleading image obscures the work of *construction* carried out by scientists. In fact, energy with all its properties (see Sect. 8.4.2) is not directly observed in nature. Before scientists could accept energy as a physical reality, they first had to construct and stabilise the concept. This construction was progressive, not the result of one action. During the seventeenth and eighteenth centuries, the precursors of the energy conservation principle (e.g. Leibniz, Huygens, Jean Bernoulli, Lagrange) prepared the groundwork for this construction in the field of mechanics by forging and developing the concepts of *vis viva* or living force (the ancestor of kinetic energy) and *vis mortua* or dead force (the ancestor of potential energy) and by establishing as a theorem, in the middle of the century, the conservation of these two quantities in idealised and isolated mechanical systems – this theorem being identified a century later as a particular case in the energy conservation principle (see Hiebert 1981, pp. 5 and 95). It should also be pointed out that in the middle of the nineteenth century, scientists that contributed to the emergence of the principle 'were not saying the same things' (Kuhn 1959, p. 322) or, as Elkana notes (1974, p. 178), they came up with solutions

---

[14] In the order of occurrence in Kuhn's text: Mayer, Joule, Colding, Helmholtz, Carnot, Séguin, Holtzmann, Hirn, Mohr, Grove, Faraday and Liebig. This list is not meant to be exhaustive, and other scientists could be added, such as W. Thomson (Lord Kelvin) and Rankine, whose contributions came later but were no less conclusive.

to 'different problems'. It was only progressively, over the course of the 1850s, that the different quantities of living force, work, heat, etc. were identified as examples of the same quantity – that is, energy – and that the new ideas defended by these scientists were recognised as equivalents, bringing to light the conservation of this quantity (see Elkana 1974, p. 10, Guedj 2010, p. 118).

Thirdly, the emergence of the principle cannot easily be understood independently of its scientific, technological, philosophical and religious context. In terms of the scientific context, the decisive elements were of both a theoretical and experimental nature. As we mentioned above, the principle of conservation of living force and dead force was established in the middle of the eighteenth century. However, this principle had limited impact and fell within the framework of nonconservative rational mechanics, which took into account the existence of an observed loss of living force during collisions. It was not until a new generation of engineers (Navier, Coriolis, etc.) proposed a molecular approach that rational mechanics would be transformed to conservative mechanics, in which the loss of living force is considered only apparent. This was an essential step towards the construction of a general principle of energy conservation (on this point, see Darrigol 2001). To these concerns related to mechanics must be added those regarding heat. In the first half of the nineteenth century, the idea that living force could be converted into heat (today we refer to the conversion of kinetic energy into thermal energy) appeared. During this period, many other conversion processes were experimentally brought to light, establishing the relationships between different fields (heat science, mechanics, chemistry, electricity, magnetism, animal physiology, etc.).

The technological context also had a major influence. The development of steam engines and electric machines played a significant role in the theoretical developments of the first part of the nineteenth century. For example, the scientific concept of work, essential in the formulation of the principle of energy conservation, was derived by scientists from accumulated experiments in the field of mechanical engineering (see Kuhn 1959; Elkana 1974, pp. 40–41; Vatin 2010).

Lastly, historians of science also accept the influence of the philosophical and religious context, although these are more complex to grasp. The metaphysical idea[15] of the equality of cause and effect, as formulated in particular by Leibniz, was shared by many of those involved in the emergence of the principle (e.g. Mayer, Helmholtz) and motivated them to search for a conserved physical quantity (see Mach 1987 [1883], pp. 474–475; Meyerson 1908, pp. 181–184; Kuhn 1959). Nor are religious considerations absent from scientific reasoning. Citing, for example, Joule:

> We might reason, *a priori*, that such absolute destruction of living force cannot possibly take place, because it is manifestly absurd to suppose that the powers with which God has endowed matter can be destroyed any more than they can be created by man's agency. (Joule 1847)

This perspective centred on the participants involved contributes vital information about the origin of the principle of conservation of energy and situates it in its context. However, it also seems important to combine this perspective with one

---

[15] By 'metaphysical', we mean an idea that precedes any scientific research.

centred on epistemic factors, namely, reasoning and experimentation, so that teachers have a full understanding of the nature of the principle. The question of the origin of the principle could be posed in the following terms: (i) Is the principle an empirical law (an a posteriori law) resulting from experimental investigation or (ii) is it a metaphysical principle (an a priori principle) established by reasoning?[16] Teachers are inclined to opt for option (i), in accordance with the inductivist 'naïve' conception of the scientific approach that they tend to spontaneously adopt.[17] However, the history of science reveals that neither of these alternatives 'conforms to the historical truth', as Meyerson states (1908, p. 175). The response is found midway between them:

> The principle of conservation of energy is the result of a mutual adjustment between an a priori question posed by scientists searching for a quantity conserved during all transformations and the experimentation that allowed what this quantity is to be determined.

How can this interrelationship between the empirical aspect and the a priori aspect in the emergence of the principle be illustrated in teacher training? Taking our inspiration from Meyerson (1908, pp. 175–190), we suggest first examining option (i) in light of Joule's experiments and then option (ii) in light of the principle of the equality of cause and effect.

In an article from 1847, Joule claimed to have established on the basis of several experiments that living force can be converted into heat and that, inversely, heat can be converted into living force,[18] without anything being lost during the two conversions:

> Experiment [...] has shown that, wherever living force is *apparently* destroyed, an equivalent is produced which in process of time may be reconverted into living force. This equivalent is *heat*. [...] In these conversions nothing is ever lost. (Joule 1847, pp. 270–271)

This idea of mutual convertibility without loss is not strictly equivalent to the principle of conservation of energy, but is an important step towards it: it was yet to be accepted that living force and heat were two examples of the same quantity – energy – or to generalise the specific case of mutual convertibility without loss between living force and heat to all possible conversions between different forms of energy. Two of Joule's experiments could be presented in teacher training to illustrate mutual convertibility: the first demonstrating the conversion of living force to heat (the famous experiment during which a falling mass rotates paddles in a liquid and through the effect of friction causes the

---

[16] It should be noted that advances in the mathematical sophistication of the laws of physics were a necessary precondition for the emergence of the principle.

[17] See, for example, Robardet and Guillaud (1995, Chap. 3), Gil-Pérez et al. (2002, p. 563), Johsua and Dupin (2003, pp. 215–217) and Cariou (2011, pp. 84–86). A survey of teachers would be worth carrying out to corroborate this hypothesis regarding their choice of option (i).

[18] In accordance with current terminology, one should speak of the mutual convertibility between kinetic energy and thermal energy (a form of energy, as distinct from heat, or 'thermal transfer', which is a mode of energy transfer).

temperature of the liquid to rise) and the second demonstrating the inverse conversion (the experiment on the expansion of heated air).

These two experiments carried out by Joule indeed demonstrate the mutual convertibility between living force and heat. The problem is that they do not prove the absence of loss during each conversion. To do this, the first experiment would need to establish that a given quantity A of living force always results in exactly the same quantity B of heat, while the second experiment would need to establish that quantity B of heat always results in exactly the same quantity A of living force. Yet for the first experiment, Joule's initial results in the 1840s were marred by significant dispersion and were obtained on a temperature scale too small to be accepted. This explains why, as Truesdell points out (1980, p. 180), Joule's contemporaries, such as W. Thomson, Helmholtz and Rankine, 'were reluctant to accept his early results'. In the second experiment, the problem was even more serious: as W. Thomson (1852) indicated, 'full restoration' of heat in living force (Thomson speaks of 'mechanical energy') is in practice 'impossible' because of the phenomenon of the 'dissipation' of energy. For this reason, contrary to what he asserts in his writings, Joule was not in a position to be able to experimentally establish the mutual convertibility *without loss* between living force and heat. This examination of the case of Joule suggests the dismissal of option (i): historically, the principle of energy conservation was not drawn directly from experiments.

Turning to option (ii), according to which the principle was established by a priori reasoning, several scientists that contributed to the emergence of the principle (e.g. Mayer, Helmholtz) presented the principle of energy conservation as a consequence of the principle of the equality of cause and effect. For example, here is what Mayer wrote in 1842:

> Forces are causes: accordingly, we may in relation to them make full application of the principle: *Causa aquet effectum.* [...] In a chain of causes and effects, a term or a part of a term can never [...] become equal to nothing. This first property of all causes we call their indestructibility. [...] Forces are therefore indestructible, convertible, imponderable objects. (Mayer 1842, quoted and translated by Truesdell 1980, p. 155)[19]

The principle of equality of cause and effect can certainly be interpreted in terms of the conservation of a quantity in a relationship of cause and effect (a quantity that is instantiated first in the cause and then in the effect), but does not in any way determine what this conserved quantity is. In fact, different options have been favoured by scientists through history: in the seventeenth century, Descartes thought it was the 'quantity of motion' (the ancestor of momentum)[20]; soon after, Leibniz suggested

---

[19] As stressed by Caneva (1993, pp. 25–27, 46 and 323), Mayer came to this idea of the conservation of 'force' (an ancestor of energy) by making an analogy with the conservation of matter (the latter being still implicit in physics and chemistry at the time of Mayer and made explicit by him). This 'guiding analogy' can also be considered as an a priori reasoning towards the principle of conservation of energy.

[20] Unlike momentum as it is defined today, Descartes' 'quantity of motion' (*quantité de mouvement*) was a scalar and not a vector quantity. See Descartes (1996 [1644]).

it was living force[21]; throughout the eighteenth century, scientists preferred Leibniz's proposition; from the second half of the eighteenth century, Lavoisier put forward the caloric theory (the caloric being conceived as a conserved 'fluid' that is the 'cause of heat')[22]; it was finally in the middle of the nineteenth century that a new concept of energy, conceived as a more general quantity capable of taking the form of living force and of heat, was accepted as the conserved quantity. In other words, although scientists indeed had an a priori idea of the existence of a quantity conserved during any transformation, energy could not be identified as the quantity sought without the aid of experiments and, in particular, without the many conversions demonstrated in the first half of the nineteenth century.

One last point concerning the origin of the principle of conservation of energy warrants clarification for teachers so that they grasp its role in the theoretical structure of physics. It should be noted that it was first described as one of the two 'principles' of *thermodynamics* (as first formulated in the 1850s) before being considered as a principle of *physics* (i.e. of thermodynamics but also of other physics theories that developed later, such as electrodynamics, special and general relativity and quantum mechanics). Establishing the conservation of energy as a principle has two implications: (a) this proposition is asserted as true without requiring that it be demonstrated by other propositions, and (b) it acts as an axiom on which other propositions in physics are based.

Points (a) and (b) each give rise to the questions: 'What justifies that the proposition of the conservation of energy is asserted as true?' and 'Why adopt this proposition as an axiom of physics?' The history of energy that we have just outlined in broad strokes leads to an initial answer to the first question: although neither experiments nor reasoning allows conclusive proof of the truth of energy conservation, both offer elements that corroborate this conclusion. A second answer can be found in Cassirer's analysis (1929 [1972], p. 508) of the relationship between a principle and an experiment: it is legitimate to accept the 'validity' of a principle on the strength of the accordance of all the consequences that can be derived from experimentation. To the second question, a possible answer is the following: scientists choose the conservation of energy as an axiom of physics because of its functional character (see Sect. 8.4.3).

8.4.2   What Is Energy?

It is difficult to describe what energy is and to give it a definition that encompasses a consensus. For this reason, some scientists put forward the minimal definition that describes energy as a quantity that is conserved. Thus, Poincaré argues (1968 [1902], pp. 177–178): 'As we cannot give energy a general definition, the principle

---

[21] On the controversy between Descartes and Leibniz on this point, see, e.g. Iltis (1971).

[22] See Lavoisier (1864 [1789]).

8 Teaching Energy Informed by the History and Epistemology of the Concept... 229

of conservation of energy simply means that there is *something* that remains constant.' Likewise, Feynman writes:

> There is a fact, or if you wish, a *law*, governing all natural phenomena that are known to date. There is no known exception to this law—it is exact so far as we know. The law is called the *conservation of energy*. It states that there is a certain quantity, which we call energy, that does not change in the manifold changes which nature undergoes. That is a most abstract idea, because it is a mathematical principle; it says that there is a numerical quantity which does not change when something happens. [...] It is important to realize that in physics today, we have no knowledge of what energy *is*. (Feynman 1963, 4.1–4.2)

It is true that the principle of conservation of energy is the integral core of the concept of energy. It is also true that the concept of energy is very abstract: not only does it describe a quantity of which we have only very indirect experimental access through the intermediary of the measurement of other quantities (such as speed or temperature), but additionally, it does not refer to a particular type of phenomena (e.g. mechanical or thermal), but to all phenomena. This is why certain science education writers, such as Warren (1982, 1991), argue the concept should not be taught in primary school, but only when students have mastered the mathematical tools that allow them to apply the principle of conservation of energy.

We believe that teaching energy by defining it uniquely as a conserved quantity and limiting it to mathematical operations of the principle of its conservation is largely inadequate for understanding its meaning. Teacher training should explicitly identify, explain and relate all the characteristics of energy (we distinguish eight) that remain implicit in traditional teaching. It seems useful, at the same time, to point out the recurrent incorrect ideas of students and teachers – on the one hand, so that they grasp what energy is *not* and, on the other hand, so that they are aware of the stumbling blocks of learning the concept. History and epistemology of the concept of energy should be included in teacher training as these bring valuable perspective on its different characteristics. Below we set out the eight characteristics of energy and outline one possible way to approach them. The first is:

(1) Energy is a quantity associated with a system.

We suggest introducing this characteristic in a discussion of the substantialist conception of energy, which is the idea that is most recurrent and most ingrained in students' and teachers' minds and thus also the most difficult to overcome. The merit of the substantialist conception is that it allows us to think more easily about the conservation of energy. This is why, rather than dismissing this conception out of hand, one could imagine taking advantage of it. The history of science is here a source of inspiration. As Duit notes (1987, pp. 140–141), referring to Planck (1887), the analogy of the conservation of energy to the conservation of matter played an important role in the acceptation of the former. According to Duit, introducing students to the conception of energy as something 'quasi-material' allows this quantity to be presented as something more 'concrete' or 'tangible' and so aids in understanding it (see also Millar 2005). This proposal seems useful in the context of teacher training. However, it is important to stress to teachers, first, that this conception is an *analogy* and, second, its limitations.

The first limitation of the substantialist conception is in fact characteristic (1): energy is a physical quantity associated with a system; that is, it does not exist autonomously, independent of a system. Or as Bunge writes (2000, p. 459): 'All energy is the energy of something.' In order to avoid the erroneous conception that a system plays the role of a reservoir of energy (the 'depository model'; see Watts 1983), it should be emphasised, as by Millar (2005, p. 4), that energy is not *in* a system, i.e. it is not 'contained' or 'stored' by it, as can be gasoline in a tank, for instance. In physics, it is a question of the energy *of* a system, i.e. energy is a 'state quantity', a variable quantity determined by the state of the system and indicating the system's capacity to produce change (see characteristic 3).

The second limitation of the substantialist conception concerns the two components of mechanical energy that are characterised by a second level of relativity. Kinetic energy is relative to the frame of reference considered (because speed, which features in the expression of kinetic energy, is itself relative to the frame of reference). The potential energy is doubly relative: it depends on the presence and the position of other systems but also on the choice of the coordinate system used to determine its value.[23] This double relativity of potential energy was put forward by Hertz (1894), who noted that a quantity capable of assuming negative values would not be able to be interpreted as representing a substance.[24]

We should add that, in the framework of special relativity, this second limitation is generalised to the total energy of a system, which is relative to the frame of reference considered.

To sum up, comparing energy with matter appears to be a useful analogy favouring the acquisition of the principle of conservation of energy. Nevertheless, as is the case for any analogy, this quasi-material concept has some limitations: energy is not an autonomous substance and its value is not absolute. To avoid teachers taking this concept literally, it is essential to emphasise that it is only an analogy and explain its limitations.

The concept of 'system' used here may seem self-evident. However, as several science education writers have emphasised (Trellu and Toussaint 1986, pp. 68–69, Arons 1999, p. 1066, van Huis and van den Berg 1993), in order to understand the conservation principle and be able to unambiguously describe energy exchange, it is essential to clearly define what a system is and to specify the boundaries of the system for each situation considered. In particular, when defining a system, it is important to stress the distinction between the system, which is the object (or group of objects) that we want to describe, and its 'environment', with which it can interact, and thus exchange energy (see characteristics 6 and 7), and/or with which it can exchange matter.

---

[23] Note that, in classical mechanics, potential energy depends only on the relative distances between the interacting bodies. Therefore, if all these interacting bodies are included in the system, the potential energy of this system no longer depends on the choice of the coordinate system.

[24] Hertz actually rejected potential energy, emphasising the role of the kinetic energy of hidden masses.

8 Teaching Energy Informed by the History and Epistemology of the Concept...   231

The second characteristic of energy is an extension of the first:

(2) Energy is a universal quantity: it is associated with all systems and all fields of science.

As Bunge writes (2000, p. 459): 'Energy is the universal physical property.' However, he restricts the field of application of this property to material objects only. Yet it is important to underline that energy is also a quantity associated with all electromagnetic radiation. In addition, this quantity has a universal character due to the fact that it applies to all fields of science: physics, chemistry, biology, geology, physiology, etc.[25]

When we express the universality of the quantity of energy in this way, it is important to draw attention to a possible inversion that should be avoided regarding the historical process. Scientists did not first identify energy in a particular branch of physics and then discover that this quantity was also associated with systems being studied in other branches of physics as well as in other scientific fields. On the contrary, it was the connection between the different branches of physics and other scientific fields (in particular, heat science, mechanics and physiology) that led to the emergence of the concept of energy (see Kuhn 1959). Its universality and its correlative function of unification (see Sect. 8.4.3) are the constituent features of the concept.

For us, this partly explains the abstract nature of the concept of energy: if it is abstract, this is notably because of its universal reach. Indeed, the concept must achieve a certain level of abstraction in order to subsume all forms of energy and be universal. In other words, it was through a process of abstraction based on concrete phenomena in each branch of physics and field of science that the concept of energy was formed.

Saying that energy is a quantity associated with a system is still a very limited characterisation of energy and does not enable it to be distinguished from other quantities. Certain science education writers (e.g. Warren 1982, 1991) argue that the energy of a system should be defined as its 'capacity for doing work', because this definition is necessary for thinking about the different forms of energy, as well as the conservation of energy. Other writers (e.g. Sexl 1981; Duit 1981; Trumper 1991) disagree with this definition as it is restricted to the field of mechanics; in other words, it suggests that the effects or changes a system is able to produce by virtue of its energy are merely mechanical (i.e. work). This criticism is understandable. But why not retain the definition of the energy of a system as its capacity to produce *change*? The main objection of Duit (1981, p. 293) is the following: 'The ability to bring about changes can also justifiably be attributed to a number of other physical concepts (for example, force and torque).' However, this objection is not admissible in our view. First, energy is a quantity that is the property of *one* system, while the quantities mentioned by Duit, such as force and

---

[25] It should also be noted that energy is equally employed in the social sciences: economics, psychology, sociology, etc. However, the meaning of the concept of energy and the uses made of it are not necessarily the same as in the physical sciences.

torque, model the action of one system on another. Second, the changes produced by force or torque occur simultaneously with its application, while the changes a system can produce by virtue of its energy are only potential: that is, only energy describes the *capacity* of a system to produce change.

Even if slightly different definitions of energy may be available (namely, in terms of work or in terms of change), it is essential to provide teachers and students with this definition of the capacity to produce change. It not only aids in clarifying the physical meaning of the concept of energy and thus in distinguishing it from other physical quantities but is also necessary for thinking about characteristics (4)–(8) of energy. Taking our inspiration from several writers, such as Chisholm (1992), Bunge (2000), and Doménech and associates (2007), without following them exactly,[26] and in line with French *collège* programmes (see Sect. 8.2.1), we propose the following definition, which we identify as the third characteristic of energy:

(3) The energy of a system is its capacity to produce change (within the system or in other systems).

Now let us turn to the other characteristics of energy and show why this definition is necessary to understand them properly. The fourth characteristic can be expressed as:

(4) Energy can take different forms.

Here it is worth restating the possible inversion of the historical process as mentioned above, though expressed in slightly different terms. Scientists did not first discover energy as a well-defined quantity appearing in a particular form (e.g. kinetic energy) before searching for and discovering the other forms in which it can also appear (e.g. thermal energy, electric energy). They first defined distinct quantities representing distinct physical realities (e.g. living force, work, heat), before making the connections between them and conceiving of them as examples of the same quantity.

Only by defining the energy of a system as its capacity to produce change gives meaning to the idea that distinct quantities representing distinct physical realities are examples of the same quantity. In fact, the only point in common between these different quantities lies in their capacity to produce the same changes. For this reason, in our view, it is the equivalence of these quantities in terms of the capacity to produce the same changes that justifies considering them as different expressions of one and the same quantity – energy.

The following historical fact supports our argument: the identification in the 1850s of the different quantities of living force, work, heat, etc. as examples of energy

---

[26] Chisholm (1992, p. 217) writes: 'Energy [...] produces changes.' Bunge (2000, p. 458) identifies energy with 'changeability'. For us, these two definitions do not adequately elucidate the idea of capacity. Doménech et al. (2007, p. 51) define energy 'as the capacity to produce transformations'. We criticise this definition for the use of the term 'transformation' rather than 'change'. The latter term is more general than the former and, in particular, can include variation in the value of a quantity (such as temperature or speed), which is not usually described as a 'transformation'.

recognised as being the conserved quantity is concurrent with the introduction of the definition of energy as the 'capacity to effect changes' or 'capacity for performing work' (Rankine 1855, pp. 125 and 129).[27]

The adoption of this definition of the energy of a system as its capacity to produce change led to the reconsideration in a new light of the common conception of kinetic energy as 'actual energy' (to use Rankine's term, 1855), a form that would appear directly to us through the movement of a material system. Certainly, the speed v and mass m of a studied system determine its kinetic energy, and we have relatively direct experimental access to these quantities. Yet that which justifies considering the formula $\frac{1}{2}mv^2$ as the expression of *energy* is not the manifestation of the movement itself, but rather the potential effects of this movement, or in other words, the capacity of the system driven by this movement to produce change (e.g. the ascent of the system up a slope or the deformation of a second system following a collision). This is why we challenge the assertion of certain writers (see Agabra 1985, pp. 111–112) that the concept of potential energy is much less accessible than that of kinetic energy. Although learners may easily accept the statement that a material system in movement possesses 'kinetic energy', that does not mean that they have understood the meaning of the concept of energy. Unless they recognise potential energy as a possible form of energy in the same right as kinetic energy and this by virtue of their common capacity to produce change, it is not guaranteed that the term 'kinetic energy' means anything else to them apart from movement (that is to say, a form of activity).

In addition, so that teachers have a global view of the forms of energy, we think it is important to eliminate the boundary raised in secondary and university education between energy in mechanics and energy in thermodynamics, which is at odds with the historical origin of the concept. As too few textbooks (e.g. Pérez 2001, pp. 90–92) or science education writers (e.g. Cotignola et al. 2002, p. 283) point out, the total energy of a material system is the sum of its mechanical energy (itself equal to the sum of the kinetic energy and the potential energy of the system considered at the macroscopic level and in relation to other systems) and its internal energy (equal to the sum of the molecular kinetic energy, or thermal energy, and the potential energy of interactions, such as chemical or nuclear energy, of the system considered at the level of its microscopic constituents and independently of other systems). In mechanics, if only mechanical energy is considered, this leaves out, on one hand, the processes of thermal transfer between the studied system and its environment and, on the other hand, the changes in the internal makeup of the system. In thermodynamics, if only internal energy is considered, this leaves out, on one hand, the movement of the system considered at the macroscopic level and, on the other hand, the external fields to which the system is subjected. As for electromagnetic radiation, the form of energy associated with this is unique – electromagnetic energy (which is the sum of the energy of the constituent photons in radiation).

---

[27] As observed by Roche (2003, p. 187), 'Rankine attributes this definition to Thomson', who 'in 1849, in an almost casual way […] first used the term energy in print more generally to mean the amount of work any system can perform.'

The definition of energy in terms of capacity to produce change helps to give meaning to characteristic (4) and, in correlation, to the following characteristic:

(5) Energy can be transformed or, in other words, can change form.

Certain writers' main concern is to avoid establishing or reinforcing the substantialist conception of energy in learners' minds. To this end, Coelho (2009, p. 978) suggests describing conservation in conversion processes solely in terms of equivalence. In his view, in Mayer's and Joule's experiments on the conversion of work into heat, conservation can be understood simply through the idea that a quantity (of work) is converted into an *equivalent* quantity (of heat). The idea of the 'indestructibility' and the 'transformability' of the same entity (energy), thus acquiring the characteristic of a substance, is simply not needed. The problem with this minimal approach appears when we pose the question: in what way are the quantities of work and heat equivalent? In our point of view, the only possible response is that they are equivalent in regard to the capacity to produce change.

These experiments on the conversion of work into heat can be described as transformation experiments, or of changing one form of energy into a new form of energy. However, in the absence of a clear distinction between *form of energy* and *mode of energy transfer*, confusion could arise in learners' minds. This type of confusion is often found in certain textbooks in relation to the concept of heat (see Cotignola et al. 2002, pp. 284–286, Papadouris and Constantinou 2011, p. 970). Work and heat are modes of energy transfer. Although in Joule's experiment there was indeed conversion from one form of energy into another, it was the transformation of kinetic energy into thermal energy, occurring simultaneously to a transfer of energy (namely, from the 'paddle' system to the 'liquid' system). The possibility of energy to be transferred or exchanged should thus be considered as a characteristic independent of its possibility to be transformed:

(6) Energy can be transferred from one system to another.

Given that the ideas of heat as a property of a body (a form of energy of a body) or as an independent substance (a sort of fluid) are very frequently held by students and can also persist in some teachers (see Gilbert and Watts 1983, pp. 78–79, Driver et al. 1994, pp. 138–139), it seems essential to explicitly discuss them in teacher training. Three themes seem worth developing. The first simply involves pointing out that the term 'heat' can be replaced by 'thermal transfer'. The second consists of emphasising the meaning of each term in the usual mathematical formula of the first law of thermodynamics: $\Delta U = Q + W$. The term on the left describes the change in internal energy U, which includes the internal forms of energy *of the system*, while the two terms on the right describe the modes of energy exchange (Q is thermal transfer and W is work performed on the system by its surroundings) *between the system and its environment* that are responsible for a change in the internal energy of the system (see Arons 1989, p. 507, van Huis and van den Berg 1993 and Cotignola et al. 2002, p. 287). A third theme consists of exploring the history of the theories of heat (see Brush 1976) stressing four stages: (i) the first part of the nineteenth century, a period of confrontation between the substantialist conception in terms of a fluid (a conserved substance distinct from living force) and the mechanistic conception in terms of the movement of the constituent particles of a body; (ii) the rise and fall of the wave theory in the 1830s (one relic of which is the mistaken idea that heat can be propagated by electromagnetic radiation in the same way as conduction or convection); (iii) the interpretation of the experiments of the conversion of work into heat in the 1840s, contributing to the abandonment of the substantialist conception in favour of the mechanistic conception but with the idea that heat is a form of energy rather than a mode of energy transfer (there was still no clear distinction between 'thermal energy' and 'thermal transfer', the latter term being a synonym of 'heat'); and (iv) the microscopic interpretation of heat in terms of microscopic work at the molecular level in the context of the kinetic theory of gases, allowing heat to be eventually understood as a mode of energy transfer. This historical approach allows teachers to consider the two recurrent mistaken conceptions mentioned above and to clarify why they have been ruled out, rather than simply asserting that they are incorrect.

As energy can be transferred from one system to another, it is possible that the energy of a system can be transferred to and, by the same token, split between large numbers of subsystems in its environment. In this case, one refers to 'dissipation':

(7) Energy can be dissipated in the environment.

Several writers (Solomon 1985, p. 170, Duit 1984, p. 65, Goldring and Osborne 1994, p. 30) have suggested that students' difficulty in understanding the idea of the conservation of energy can be surmounted (at least in part) by first introducing the concept of the dissipation of energy.

To deal with this concept of dissipation in teacher training, we suggest starting from the problem of loss that Thomson confronted and tried to resolve in his articles from 1851 to 1852 (Thomson 1851, 1852, see Guedj 2010): in steam engines, it is observed that only part of the heat is converted into useful work[28]; the other part is lost or 'wasted'. What happens to the part that is lost? Is it a question of 'absolute waste', that is, the destruction of part of the heat? Thomson's response came in two stages. In his 1851 article, he developed Joule's idea according to which energy can never be *destroyed* ('mechanical energy' in his words), but only *transformed*. Therefore, the apparent loss of energy is a loss for human beings (who want to use it in machines) and not an absolute loss: the energy in question is 'lost to man irrecoverably; but not lost in the material world'. In his 1852 article, Thomson further clarifies his response by introducing the fundamental concept of *dissipation*. In a steam engine, part of the mechanical energy dissipates via heat because of friction between different parts of the engine, which are inevitable in practice. As it is 'dissipated', that is, divided between large numbers of subsystems of its environment, this energy is 'irrecoverably wasted'. This historical approach has at least two points to recommend it. First, in experiments that they carry out and/or study, teachers are constantly confronted by this problem of the apparent

---

[28] In the viewpoint of current physics, it is a question of the transformation of 'thermal energy' into 'mechanical energy'.

disappearance of energy. Second, Thomson's reasoning allows the clear distinction between the utilitarian aspect (loss of energy for the operation of a machine) and the physics aspect (dissipation of energy in the environment).

All of the elements are now in place to introduce the final characteristic:

(8) The energy of an isolated[29] system is conserved.

This characteristic can only be fully understood in light of the other characteristics detailed previously, in particular those relating to transformation and transfer. As Duit writes (1984, p. 59): 'When energy is transferred from one system to another, or when energy is converted from one form to another, the amount of energy does not change.'

Let's reiterate these different characteristics and our definition. The conservation of the energy of a system can only be understood if the conversion between different quantities (what is today called 'kinetic energy', 'thermal energy', etc.) is interpreted as the transformation of the same quantity into different possible forms, that is, different possible expressions. If these different expressions can be seen as expressions of the same quantity, we argue that this is because they represent the same capacity to produce change. Additionally, the conservation of the energy of a system can only be understood as an idealised case where the system does not interact with its environment. When it interacts with its environment, the system exchanges energy. In particular, in the presence of friction, part of the energy of the system dissipates in the environment. In order to avoid the obvious contradiction with the principle of conservation of energy, the total energy of the system and the environment with which it interacts should be considered: if this system and its environment are considered as isolated (which is also an idealisation), then their total energy is conserved, although this is not the case of the energy of the system being studied.

8.4.3 What Purpose Does the Concept of Energy Serve?

Why grant so much importance to the concept of energy in teaching? Why do students need to learn to use it? Ultimately, what purpose does this concept serve? To enable teachers to respond to these questions, teacher training should identify and explain the functions that the concept fulfils in science practice. The description of the emergence of the scientific concept of energy (see Sect. 8.4.1) and what energy is (see Sect. 8.4.2) offers a glimpse of these functions. Here we try to make them explicit:

(F1) *Energy is an unvarying focal point for thinking about variations observed in phenomena.* This point was put forward by Mach as early as the end of the nineteenth century. Speaking about the principle of energy conservation, he wrote: 'An isolated variation that is linked to nothing, without a fixed point of comparison, is inconceivable and unimaginable' (1987 [1883], p. 473). Or as

---

[29] An 'isolated system' is defined here as a system that does not interact with its environment

Papadouris and Constantinou emphasise more recently (2011, p. 966): 'Energy [is] a theoretical framework that has been invented in science so as to facilitate the analysis of changes occurring in physical systems regardless of the domain they are drawn from.' More precisely, describing phenomena in terms of *transformation*, *transfer* and *conservation* of energy allows us to think about observed variations.

(F2) *Energy is a unifying focal point for referring to a large variety of phenomena and making links between them.* This point distinctly emerges from the history of the development of the principle of conservation of energy (see Sect. 8.4.1). To quote Cassirer (1929 [1972], p. 520), energy can be described as 'a point of unity to grasp by pure thought'.

(F3) *The principle of conservation of energy allows predictions to be made.* This predictive function occurs at two possible levels. (i) The principle allows quantitative predictions to be made *in the context of a theory*. For example, in mechanics, the principle of conservation of mechanical energy allows the prediction of the speed of a body at time t_2 given the position and speed of the body at a previous time of t_1. (ii) The principle also allows predictions to be made *in the development of theories*, which can be described as a 'heuristic function'. A famous example of this is that of the role of the principle in the anticipation of the existence of the neutrino. We could also mention the no less important examples of the development of special relativity and quantum mechanics, in which the principle played an explicit role (see, e.g. Einstein 1905; Heisenberg 1972 [1969], pp. 91–92). This heuristic function was emphasised as early as the nineteenth century, for example by Maxwell (1871) who attributes the principle as it was formulated by Helmholtz with an 'irresistible driving power' (see Truesdell 1980, p. 163). More recently, Feynman (1965, p. 76) justifies the recourse to the principle in new fields in this way: 'If you will never say that a law is true in a region where you have not already looked you do not know anything.'

8.5   Conclusion

As we have established in the case of France, energy is an omnipresent concept in school programmes from primary to the end of secondary education and has two main aims: educating students from a scientific point of view and preparing them as future citizens to enable them to take part in social issues that involve the concept of energy. Yet science education literature has shown that the concept of energy is particularly difficult to define and to teach. This is due to the concept itself, principally to the fact that it is highly abstract and polymorphous and thus difficult to define. The difficulties in defining the concept lead to a multiplicity of conceptions (anthropocentric, substantialist, etc.) and confusions (force/energy, forms of energy/ modes of energy transfer, etc.) that are equally obstacles to learning. Several teaching strategies have been proposed in science education literature over the last thirty years as alternatives to traditional teaching methods deemed too formal and dogmatic. However, none has distinguished itself as the most convincing method and been retained over the course of time in school programmes.

The new strategy that we advocate differs from previous proposals in two major ways. Firstly, we propose turning the attention to teacher training, which seems an essential precondition to teaching energy, given the complexity of the concept. The aim is thus to develop a teacher training programme that allows educators to better grasp the meaning of the concept, the role it plays in science and to be clear about all the characteristics of energy as well as the recurrent mistaken ideas about it. Secondly, our strategy grants a central role to EHST. We think EHST provides effective ways to throw light on the different aspects of the concept and should be a feature of teacher training. In this article, we have recommended a framework for teacher training based on EHST structured around three main questions: 'What is the origin of the concept of energy?', 'What is energy?' and 'What purpose does the concept of energy serve?'

We have highlighted several points that seem essential to include in teacher training. In particular, it is important that teachers understand that the concept of energy, as currently accepted, has not always been available for scientists and only became stable with the emergence of the principle of conservation of energy, itself resulting from a mutual adjustment between theory and experimentation. We have also tried to show that the definition of the energy of a system as its capacity to produce change is required in order to be able to understand that energy can take different forms and can be transformed. These characteristics, along with the transfer and dissipation of energy, allow the fundamental characteristic of the conservation of energy to be understood. Finally, it seems very important that teachers are aware of three operational roles that the concept of energy plays in scientific activity: its role as an unvarying focal point for thinking about variation, its unifying role and its predictive role.

The teacher training framework on energy presented here needs to be further enriched (with examples of course outlines and teaching sessions on energy) and detailed (to allow for constraints on the ground) and to be subjected to experimentation. Our hypothesis is that this teacher training should lead teachers to profoundly rethink the way in which they approach teaching about energy: in terms of their interpretation of programmes, of their own ideas and those of their students and of their teaching practice. If teachers are clear about the concept of energy and adopt, in light of EHST, a new position regarding how to teach it, it becomes possible to envisage a teaching approach itself based on EHST that can thus truly distance itself from a formal, dogmatic approach. Our wager is that this type of teaching will enable the difficulties in mastering the concept of energy to be overcome more easily. This teaching has yet to be developed.

References

Abd-El-Khalick, F. (2001). Embedding nature of science instruction in preservice elementary science courses: abandoning scientism, but.... *Journal of Science Teacher Education*, 12 (2), 215–233.

8 Teaching Energy Informed by the History and Epistemology of the Concept... 239

Agabra, J. (1985). Energie et mouvement: représentations à partir de l'étude de jouets mobiles. *Aster*, 1, 95–113.

Agabra, J. (1986). Echanges thermiques. *Aster*, 2, 1–41.

Arons, A. (1989). Developing the Energy Concepts in Introductory Physics. *The Physics Teacher*, 27, 506–517.

Arons, A. (1999). Development of energy concepts in introductory physics courses. *American Journal of Physics*, 67(12), 1063–1067.

Bächtold, M. & Guedj, M. (2012). Towards a new strategy for teaching energy based on the history and philosophy of the concept of energy. In O. Bruneau, P. Grapi, P. Heering, S. Laubé, M.-R. Massa-Esteve, T. de Vittori (eds.), *Innovative methods for science education: history of science, ICT and inquiry based science teaching* (pp. 225–238). Berlin: Frank & Timme GmbH.

Balibar, F. (2010). Energie. In D. Lecourt (ed.), *Dictionnaire d'histoire et philosophie des sciences* (pp. 403–408). Paris: Presses Universitaires de France.

Ballini, R., Robardet, G. & Rolando, J.-M. (1997). L'intuition, obstacle à l'acquisition de concepts scientifiques : propositions pour l'enseignement du concept d'énergie en Première S. *Aster*, 24, 81–112.

Beynon, J. (1990). Some myths surrounding energy. *Physics Education*, 25(6), 314–316.

Bruguières, C., Sivade, A. & Cros, D. (2002). Quelle terminologie adopter pour articuler enseignement disciplinaire et enseignement thématique de l'énergie, en classe de première de série scientifique. *Didaskalia*, 20, 67–100.

Brush, S. (1976). *The kind of motion we call heat: a history of the kinetic theories of gases in the 19th century, Volumes I and II.* Oxford: North-Holland Publishing Company.

Bunge, M. (2000). Energy: between physics and metaphysics. *Science & Education*, 9, 457–461.

Caneva, K. (1993). *Robert Mayer and the conservation of energy.* Princeton: Princeton University Press.

Cariou, J.-Y. (2011). Histoire des démarches en sciences et épistémologie scolaire. *Revue de Didactique des Sciences et des Techniques*, 3, 83–106.

Cassirer, E. (1929 [1972]). *La philosophie des formes symboliques 3: la phénoménologie de la connaissance* (tr. fr.). Paris: Les éditions de Minuit.

Chisholm, D. (1992). Some energetic thoughts. *Physics Education*, 27, 215–220.

Coelho, R. (2009). On the concept of energy: how understanding its history can improve physics teaching. *Science & Education*, 18, 961–983.

Cotignola, M., Bordogna, C., Punte, G., & Cappannini, O. (2002). Difficulties in learning thermodynamic concepts: are they linked to the historical development of this field? *Science & Education*, 11, 279–291.

Darrigol, O. (2001). God, waterwheels, and molecules: Saint-Venant's anticipation of energy conservation. *Historical Studies in Physical Sciences*, 31(2), 285–353.

Descartes, R. (1996 [1644]). *Les principes de la philosophie.* In *Œuvres de Descartes*. Paris: Vrin/CNRS.

Désautels, J., and Larochelle, M., (1989). *Qu'est ce que le savoir scientifique? Points de vue d'adolescents et d'adolescentes.* Québec: Les Presses de l'Université de Laval.

Doménech, J.-L., Gil-Pérez, D., Gras-Marti, A., Guisasola, J., Martínez-Torregrosa, J., Salinas, J., Trumper, R., Valdés, P. & Vilches, A. (2007). Teaching of energy issues: a debate proposal for a global reorientation. *Science & Education*, 16, 43–64.

Driver, R., Squires, A., Rushworth, P. & Wood-Robinson, V. (1994). *Making sense of secondary science: research into children's ideas.* London and New York: Routledge.

Driver, R. & Warrington, L. (1985). Student's use of the principle of energy conservation in problem situation. *Physics Education*, 5, 171–175.

Duit, R. (1981). Understanding energy as conserved quantity. *European Journal of Science Education*, 3(3), 291–301.

Duit, R. (1984). Learning the energy concept in school – empirical results from the Philippines and West Germany. *Physics Education*, 19, 59–66.

Duit, R. (1987). Should energy be introduced as something quasi-material? *International Journal of Science Education*, 9, 139–145.

Einstein, A. (1905). Ist die Trägheit eines Körpers von seinem Energieinhalt abhängig? *Annalen der Physik*, 17, 639–641.

Elkana, Y. (1974). *The discovery of the conservation of energy*. London: Hutchinson Educational LTD.

Fauque, D. (2006). La longue marche d'un enseignement de l'histoire des sciences et des techniques. *Tréma*, 26, 35–47.

Feynman, R. (1963). *The Feynman lectures on physics, vol. I: mainly mechanics, radiation, and heat*. California Institute of Technology.

Feynman, R. (1965). *The character of physical law*. Cambridge (Mas.), London: MIT Press.

Freuler, L. (1995). Major trends in philosophy around 1900. In M. Panza & J.-C. Pont (eds.), *Les savants et l'épistémologie vers la fin du XIXè siècle* (pp. 1–15). Paris: Albert Blanchard.

Ghesquier-Pourcin, D., Guedj, M., Gohau, G. & Paty, M. (2010). *Energie, science et philosophie, Vol. 1: l'émergence de l'énergie dans les sciences de la nature*. Paris: Hermann.

Gilbert, J. & Pope, M. (1986). Small group discussions about conception in science: a case study. *Research in Science and Technological Education*, 4, 61–76.

Gilbert, J. & Watts, D. (1983). Concepts, misconceptions and alternative conceptions: changing perspectives in science education. *Studies in Science Education*, 10, 61–98.

Gil-Pérez, D., Guisasola, J., Moreno, A., Cachapuz, A., Pessoa, A., Martinez-Torregrosa, J., Salinas, J., Valdés, P., Gonzalez, E., Gené, A., Dumas-Carré, A., Tricarico, H. & Gallego, R. (2002). Defending constructivism in science education, *Science & Education*, 11, 557–571.

Goldring, H. & Osborne, J. (1994). Students' difficulties with energy and related concepts. *Physics Education*, 29, 26–31.

Guedj, M. (2010). Du concept de travail vers celui d'énergie: l'apport de William Thomson. In Ghesquier-Pourcin, Guedj, Gohau & Paty (2010, pp. 103–125).

Hertz, H. (1894). *Die Prinzipien der Mechanik: in neuen zusammenhange dargestellt*. Leipzig: Barth.

Heisenberg, W. (1972 [1969]). *Le tout et la partie: le monde de la physique atomique* (tr. fr.). Paris: Flammarion.

Hiebert, E. (1981). *Historical roots of the principle of conservation of energy*. New York: Arno Press.

Hulin, M. (1992). *Le mirage et la nécessité: pour une redéfinition de la formation scientifique de base*. Paris: Presses de l'École Normale Supérieure and Palais de la Découverte.

Iltis, C. (1971). *Leibniz and the vis viva controversy*. Chicago: The University of Chicago Press.

Johsua, S. & Dupin, J.-J. (2003). *Introduction à la didactique des sciences et des mathématiques*. Paris: Presses Universitaires de France.

Joule, J. (1847). On matter, living force, and heat. Published in the Manchester *Courier* newspaper, May 5 and 12. Reprinted in *The Scientific Papers of James Prescott Joule*, Vol. 1, 1884 (pp. 265–276). Taylor & Francis.

Koliopoulos, D. & Ravanis, K. (1998). L'enseignement de l'énergie au collège vu par les enseignants. Grille d'analyse de leurs conceptions. *Aster*, 26, 165–182.

Kuhn, T. (1959). Energy conservation as an example of simultaneous discovery. In M. Clagett (ed.), *Critical Problems in the History of Science* (pp. 321–56). Madison (Wis.): The University of Wisconsin Press.

Lavoisier, A.-L. (1864 [1789]). *Traité élémentaire de* chimie. In *Œuvre de Lavoisier, tome premier*. Paris: Imprimerie impériale.

Lecourt, D. (1999). L'enseignement de la philosophie des sciences. *Rapport au ministre de l'Éducation nationale, de la Recherche et de la Technologie*.

Lemeignan, G. & Weil-Barais, A. (1993). *Construire des concepts en physique*. Paris: Hachette.

Levy-Leblond, J. M. (1973). *(Auto)critique de la science (Textes réunis par Alain Jaubert et Jean-Marc Lévy-Leblond)*, Seuil.

Levy-Leblond J. M. (2007). (Re)mettre la science en culture: de la crise épistémologique à l'exigence éthique. *Speech at the inauguration of the Institute for Scientific Methodology (ISEM), Palermo, March 2007*. Online at <http://www.i-sem.net>.

Mach, E. (1987 [1883]). *La mécanique: exposé historique et critique de son développement* (tr. fr.). Paris: Gabay, 1987.

8 Teaching Energy Informed by the History and Epistemology of the Concept... 241

Martinand, J.-L. (1993). Histoire et didactique de la physique et de la chimie: quelles relations? *Didaskalia*, 2, 89–99.

Matthews, M. (1994/2014). *Science teaching: the role of history and philosophy of science.* New York, London: Routledge.

Maxwell, J.C. (1871). *Theory of heat.* London: Longmans, Green, and co.

Mayer, J. (1842). Bemerkungen über die Kräfte der unbelebten Natur. *Annalen der Chemie und Pharmacie*, 42, 233–240.

Méheut, M., Duprez, C. & Kermen, I. (2004). Approches historique et didactique de la réversibilité. *Didaskalia*, 25, 31–62.

Meyerson, E. (1908). *Identité et réalité.* Paris: Alcan.

Millar, D. (2005). Teaching about energy. Department of Educational Studies, research paper 2005/11. Online at <http://www.york.ac.uk/media/educationalstudies/documents/research/Paper11Teachingaboutenergy.pdf>.

MEN[30] (2002). *Fiches connaissances : cycles 2 et 3.* Collection 'Document d'application des programmes'. CNDP. Online at <http://www2.cndp.fr/archivage/valid/38285/38285-5692-5495.pdf>.

MEN (2007). Cahier des charges de la fomation des maîtres en Institut Universitaire de formation des maîtres. *Bulletin Officiel de l'Education Nationale*, spécial n°1 du 4 janvier 2007. Online at <http://www.education.gouv.fr/bo/2007/1/MENS0603181A.htm>.

MEN (2008a). Horaires et programmes de l'école primaire. *Bulletin Officiel de l'Education Nationale*, hors-série, n°3 du 19 juin 2008. Online at <http://www.education.gouv.fr/bo/2008/hs3/default.htm>.

MEN (2008b). Programmes du collège : programmes de l'enseignement de physique-chimie. *Bulletin Officiel de l'Education Nationale*, spécial n°6 du 28 août 2008. Online at <http://media.education.gouv.fr/file/special_6/52/7/Programme_physique-chimie_33527.pdf>.

MEN (2010a). Programmes de physique-chimie en classe de seconde générale et technologie. *Bulletin Officiel de l'Education Nationale*, spécial n°4 du 29 avril 2010. Online at <http://media.education.gouv.fr/file/special_4/72/9/physique_chimie_143729.pdf>.

MEN (2010b). Programme d'enseignement spécifique de physique-chimie en classe de première de la série scientifique. *Bulletin Officiel de l'Education Nationale*, spécial n°4 des 9 et 30 septembre 2010. Online at <http://www.education.gouv.fr/cid53327/mene1019556a.html>.

MEN (2011). Programme d'enseignement spécifique et de spécialité de physique-chimie de la série scientifique : classe terminale. *Bulletin Officiel de l'Education Nationale*, spécial n°8 du 13 octobre 2011. Online at <http://media.education.gouv.fr/file/special_8_men/99/0/physique_chimie_S_195990.pdf>.

Papadouris, N. & Constantinou, C. (2011). A philosophically informed teaching proposal on the topic of energy for students aged 11–14. *Science & Education*, 20, 961–979.

Paty, M. (2000–2001). Histoire et philosophie des sciences. *Sciences Humaines*, 31, 56–57.

Pérez, J.-P. (2001). *Thermodynamique: fondements et applications.* Paris: Dunod.

Pinto, R., Couso, D. & Gutierrez, R. (2004). Using research on teachers' transformations of innovations to inform teacher education: the case of energy degradation. *Science Education*, 89, 38–55.

Planck, M. (1887). *Das Prinzip der Erhaltung der Energie.* Leipzig: Teubner.

Poincaré, H. (1968 [1902]). *La science et l'hypothèse.* Paris: Flammarion.

Rankine, W. (1855). Outlines of the science of energetics. *The Edinburgh New Philosophical Journal*, July-October 1855, Vol. II, 3, 121–141.

Robardet, G. & Guillaud, J.-G. (1995). *Éléments d'épistémologie et de didactique des sciences physiques: de la recherche à la pratique.* Grenoble: Publications de l'IUFM de Grenoble.

Roche, J. (2003). What is potential energy? *European Journal of Physics*, 24, 185–196.

Sexl, R. (1981). Some observations concerning the teaching of the energy concept. *European Journal of Science Education*, 3(3), 285–289.

Shortland, M. & Warick, A. (eds.) (1989). *Teaching the History of Science*, Basil Blackwell, Oxford.

---

[30] *Ministère de l'éducation nationale* (Ministry of education, France)

Smith, C. (1998). *The science of energy: a cultural history of energy physics in Victorian Britain.* London and Chicago: The Athlone Press.

Smith, C. & Wise, N. (1989). *Energy and empire: a bibliographical study of Lord Kelvin.* Cambridge: Cambridge University Press.

Solomon, J. (1982). How children learn about energy, or Does the first law come first? *School Science Review,* 63 (224), 415–422.

Solomon, J. (1983). Learning about energy: how pupils think in two domains. *European Journal of Science Education,* 5, 49–59.

Solomon, J. (1985). Teaching the conservation of energy. *Physics Education,* 20, 165–170.

Summers, M. & Kruger, C. (1992). Research into English primary school teachers' understanding on the concept of energy. *Evaluation & Research in Education,* 6, 95–111.

Thomson, W. (1851). On the dynamical theory of heat: with a numerical result deduced from Mr Joule's equivalent of a thermal unit, and M. Regnault's observations on steam. *Proceedings of the Royal Society of Edinburgh,* 3, 48–52.

Thomson, W. (1852). On a universal tendency in nature to the dissipation of mechanical energy. *Proceedings of the Royal Society of Edinburgh for April 19, 1852.*

Trellu, J.-L. & Toussaint, J. (1986). La conservation, un grand principe. *Aster,* 2, 43–87.

Trumper, R. (1990). Being constructive: an alternative approach to the teaching of the energy concept, part one. *International Journal of Science Education,* 12(4), 343–354.

Trumper, R. (1991). Being constructive: an alternative approach to the teaching of the energy concept, part two. *International Journal of Science Education,* 13(1), 1–10.

Trumper, R. (1993). Children's energy concepts: a cross-age study. *International Journal of Science Education,* 15, 139–148.

Trumper, R., Raviolo, A. & Shnersch, A. M. (2000). A cross-cultural survey of conceptions of energy among elementary school teachers in training — empirical results from Israel and Argentina. *Teaching and Teacher Education,* 16(7), 697–714.

Truesdell., C. (1980). *The tragicomical history of thermodynamics (1822–1854).* New York, Heidelberg, Berlin: Springer.

Van Huis, C. & van den Berg, E. (1993). Teaching energy: a systems approach. *Physics Education,* 28, 146–153.

Vatin, F. (2010). Travail. In D. Lecourt (ed.), *Dictionnaire d'histoire et philosophie des sciences* (pp. 1109–1114). Paris: Presses Universitaires de France.

Watts, D. (1983). Some alternative views of energy. *Physics Education,* 18, 213–217.

Warren, J. W. (1982). The nature of energy. *European Journal of Science Education,* 4(3), 295–297.

Warren, J. W. (1991). The teaching of energy. *Physics Education,* 26(1), 8–9.

Manuel Bächtold is assistant professor at University Montpellier 2 in physics, philosophy of science and physics education. He studied physics at the Ecole Polytechnique Fédérale de Lausanne (Switzerland) and philosophy at University of Paris 1 (France), where he defended his Ph.D. thesis on the interpretation of quantum mechanics. He then carried out his research in philosophy of physics at TU Dortmund (Germany). Among his recent publications are 'Interpreting Quantum Mechanics according to a Pragmatist Approach' (*Foundations of Physics* 2008), *L'interprétation de la mécanique quantique: une approche pragmatiste* (Hermann 2009), 'Saving Mach's view on atoms' (*Journal for General Philosophy of Science* 2010), 'L'espace dans ses dimensions transcendantale et pragmatiste' (*Kant-Studien* 2011), 'Les fondements constructivistes de l'enseignement des sciences basé sur l'investigation' (*Tréma*2012) and 'A pragmatic approach to the atomic model in chemistry' (in J.-P. Llored, ed., Cambridge Scholars Publishing 2013),

Muriel Guedj is assistant professor in history and philosophy of science at the Science University of Montpellier (IUFM, which is postgraduate teacher training institute). She completed her Ph.D. in the field of history of physics ('The emergence of the principle of conservation of energy and the construction of thermodynamics') in the University of Paris 7. Her fields of research include the concept of energy physics in the nineteenth/twentieth century, history of science education and teaching the history of science. For these researches she has participated in the project REDISCOL supported by the ANR (National Agency of Research) and with the group ReForEHST (Research and Training Epistemology History of Science and Technology). Among her publications on the energy concept are:

(2006) *Du concept de travail vers celui d'énergie : l'apport de Thomson.* Revue d'histoire des sciences, janvier-juin 2006, tome 59–1, pp. 29–51

(2008) *À propos de l'ouvrage d'Helmholtz « Über die erhaltung der ktaft » sur un principe limité de la conservation de l'energie.* Philosophia Scientiae, Vol 11 Cahier 2 Novembre 2007, pp. 1–25

(2010) *William Thomson et l'emergence du principe de conservation de l'energie, in* Energie, sciences et philosophie aux tournants des XIXème-XXeme siecles, (dir.) D. Ghesquier-Pourcin, M. Guedj, G. Gohau et M. Paty, Hermann, Paris

(2010) *L'equivalent mécanique de la chaleur chez Sadi Carnot entre Reflexions, doutes, notes et énergie, in* Energie, sciences et philosophie aux tournants des XIXème -XXème siècles, (dir.) D. Ghesquier-Pourcin, M. Guedj, G. Gohau et M. Paty, Hermann, Paris

(2010) *L'introduction du principe de conservation de l'energie dans l'enseignement secondaire français, vue à travers quelques manuel, in* Energie, sciences et philosophie aux tournants des XIXème -XXème siècles, (dir.) D. Ghesquier-Pourcin, M. Guedj, G. Gohau et M. Paty, Hermann, Paris

(2010) *Ernst Mach sur les mesures des grandeurs de la mécanique et la théorie des dimensions, in* Le sens des nombres, (dir.) A. Bernard, G. Chambon et C. Ehrhardt, Vuibert, Paris

Chapter 9
Teaching About Thermal Phenomena and Thermodynamics: The Contribution of the History and Philosophy of Science

Ugo Besson

9.1 Introduction

The role of history and philosophy in science teaching has been studied and debated at great length. It is considered that introducing themes of history and philosophy can give cultural value to science learning, engendering a more critical attitude and a conception of science as an evolving human activity. Moreover, the history of science can help to give sense to science learning. In fact, science teaching constrains to isolate and restructure science subjects in order to adapt them to the students' needs and to the school context. This process can lead to a presentation of scientific topics in a way that hides the cultural and social references of the problems in answer to which scientific theories had been formed and avoids the methodological and philosophical aspects which can give general cultural sense to scientific issues and provide a deeper understanding. This can produce a fragmentary, more algorithmic than conceptual knowledge. Studying *case histories* involving significant historical or philosophical aspects can contribute to reconstructing the atmosphere of debate and controversy and provide the technical and economic background that constituted the context of science development (Stinner et al. 2003). The US National Science Education Standards (1996, Chap. 6, p. 107) 'recommend the use of history of science in school programs to clarify different aspects of scientific inquiry, the human aspects of science and the role that science has played in the development of various cultures'.

Research on common sense conceptions has renewed the debate on the role of history in science teaching because many of these conceptions are found to be similar to ancient ideas or theories. Resembling the theory according to which ontogeny recapitulates phylogeny, it is supposed that individual cognitive development can

---

U. Besson (✉)
Department of Physics, University of Pavia, Pavia, Italy
e-mail: ugo.besson@unipv.it

M.R. Matthews (ed.), *International Handbook of Research in History, Philosophy and Science Teaching*, DOI 10.1007/978-94-007-7654-8_9,
© Springer Science+Business Media Dordrecht 2014

recapitulate in some way the historical development of science. It has been supposed that, confronted with such ideas and theories, students would recognise some features of their own conceptions, discussing and reviewing them, with conceptual change sequences similar to the historical ones (Campanaro 2002). However, such proposals have been criticised for their often too simplistic analogy between ancient theories and common thought and because of the strong differences in the context, the meaning of concepts and ideas used and the mental and logical processes involved (Carey 1988; Nersessian 1995). Nonetheless, these analogies have some validity, even if in a limited form. For example, Piaget and Garcia (1983) emphasised them strongly.

History and philosophy of science can contribute to improving students' understanding of the conceptual, procedural and contextual aspects of science (Teixeira et al. 2012; Wang and Marsh 2002). They can also positively contribute to a better understanding of scientific methods, the nature of science and the relationships between science, technology and society and to metacognitive learning (Matthews 1994). The conceptual analysis of physics theories and history can help to answer the questions of how we know what we know and how we discovered it.

Examples from the history of science can provide a repository of strategic knowledge of how to construct, modify and communicate scientific representations, and science educators could choose, integrate and transform these resources into instructional procedures (Holton 2003a; Nersessian 1995; Seroglou and Koumaras 2001).

Moreover, many problems, models and explanatory frameworks used in the early historical development of a scientific topic are more in resonance with the preferences of students and of common reasoning, which is often centred on dynamical, causal, qualitative and analogical reasoning (Besson 2010). There is sympathy between the beginner and the pioneer.

In this way, the topic of thermal phenomena and thermodynamics is fertile because it relates to various epistemological and philosophical themes. Its history is strongly linked to the themes of relationships between science, technology and socio-economic problems; residues of ancient abandoned theories are still present in current scientific language and in textbooks; and many students' conceptions are similar to ideas and reasoning of ancient theories.

The historical conceptual development of thermodynamics is especially interesting for teaching because it can show the reasons for paradigm shifts in research communities (e.g. on the nature of heat and the notion of temperature), based on a progression moving from phenomenological observations to qualitative and then mathematical models and laws, in an increasing process of abstraction. This historical progression can assist learning progression and challenge students' alternative ideas.

This chapter will present the main results of research on students' conceptions and difficulties about thermal phenomena (Sect. 9.2), a review of research involving the use of history and philosophy in teaching thermal phenomena and thermodynamics (Sect. 9.3), some examples of philosophical themes and of case histories of didactic interest together with their teaching and learning implications (Sects. 9.4 and 9.5) and some suggestions for further research and development (Sect. 9.6).

9.2 Students' Conceptions and Conceptual Difficulties About Thermal Phenomena

Much research has been conducted on students' conceptions and difficulties about thermal phenomena.[1] Students show difficulties in distinguishing between extensive and intensive quantities, in particular between heat and temperature. They often use a mixed temperature-heat notion or consider the temperature of an object as a measure of *the level of heat*. Sometimes heat and cold are both considered as substances that can be transferred from one body to another. Students mainly reason in terms of object properties instead of processes and may attribute to the materials the quality of being or keeping warm or cold (an ice cube melts more quickly if it is wrapped in a wool cloth because wool 'is warm or keeps warm'). They may also consider the existence of a maximum possible temperature for a given material. Work is generally not connected to temperature changes. The prevalent idea is that only heat exchanges can cause an increase or decrease in the temperature of an object. This idea also survives among university students and can also be found among science teachers.

Concerning microscopic models of gas, students often attribute to molecules the same properties of macroscopic objects and tend to consider only one variable at a time in a linear causal reasoning (Rozier and Viennot 1991).

Chiou and Anderson (2010) characterise four patterns of students' interpretative frameworks of heat: first, heat is treated as an intrinsic property of a substance (wood is hot, ice is cold); second, hotness and coldness are treated as material substances which can move from one object to another; third, heat is treated as a nonmaterial entity, caloric flow, which propagates from objects at higher temperatures to objects at lower temperatures; and fourth, a scientifically acceptable view, in which heat refers to a transfer of energy due to a temperature difference. According to the authors, 'The sequence of these four frameworks also represents the developmental stages of peoples' conceptions of heat, developing from a naive view toward a more scientific one'.

Heat is often considered a quantity which is conserved in a cyclic thermodynamic process, as a property of the system or a state function, and it is used as synonymous with internal energy or with thermal energy, i.e. of the part of internal energy involved in the temperature changes.

Some students' difficulties with the concept of heat derive from the difference between its meaning in common language and in the language of science. Some ambiguities also appear in textbooks (see Doige and Day 2012; Leite 1999) and differences are found in the definitions of heat across textbooks of different disciplines (physics, chemistry, biology and earth science). Many physics and chemistry

---

[1] See Arnold and Millar (1996), Besson et al. (2010), Chiou and Anderson (2010), Clough and Driver (1985), Cochran and Heron (2006), Cotignola et al. (2002), de Berg (2008), Erickson (1979, 1980), Erickson and Tiberghien (1985), Jasien and Oberem (2002), Leinonen et al. (2009), Lewis and Linn (1994), Sciarretta et al. (1990); Shayer and Wylam (1981), Stavy and Berkovitz (1980), Wiser and Amin (2001), and Wiser and Carey (1983).

textbooks in the 1960s and 1970s defined heat as the kinetic energy associated with molecular motion and thus as a property of a system or as a form of energy degraded or disordered. By contrast, science education literature has stressed that heat has to be considered as a process quantity, a transfer of energy due to a difference in temperature. More recently, physics textbooks have assumed this view by referring to heat as energy in transit or as a mechanism or process of energy transfer, whilst many life science and earth science textbooks still present a definition of heat as energy contained in a system.

Difficulties in differentiating the meaning of heat, work and internal energy hinder the understanding of the laws of thermodynamics. In fact, older students also show difficulty in applying the first law of thermodynamics, often considering heat as a state function (Loverude et al. 2002; Meltzer 2004), and misunderstand the second law (Kesidou and Duit 1993). The language generally used in textbooks can reinforce some common erroneous ideas, referring to 'work done' but to 'heat given', 'received' or 'lost'. These last expressions convey the idea that a body *possesses heat* in order to give it and are clearly fossil residues of the old conception of heat as fluid (Besson and De Ambrosis 2013). Moreover, any distinction between heat and work disappears at a microscopic level because the interactions are of the same type. The difference is in their coherence or incoherence and appears only at macroscopic or mesoscopic levels where a large number of molecules are involved (Besson 2003).

Students usually lack considering or consider incorrectly the infrared thermal radiation in thermal processes of energy exchange (Besson et al. 2010). There is often confusion about whether thermal radiation must be considered as a way of heat transmission, as work or as a third specific modality of energy exchange. There is no universal agreement in the research literature on this point. Most textbooks choose the first option, by considering thermal radiation as the third way of heat transmission, after conduction and convection, but this can implicitly suggest that radiation and heat have the same characteristics, thus contributing to students' difficulties. The historical development of the idea of thermal radiation and the study of its characteristics compared to the properties of light shows how the process of differentiation was a long one and required both experimental and theoretical efforts (Besson 2012). In physics history the term *radiant heat* has been used for a long time, even if Maxwell wrote in 1871:

> The phrases radiation of heat and radiant heat are not quite scientifically correct, and must be used with caution. Heat is certainly communicated from one body to another by a process which we call radiation. We have no right, however, to speak of this process of radiation as heat ... when we speak of radiant heat we do not mean to imply the existence of a new kind of heat but to consider radiation in its thermal aspect. (Maxwell 1871, pp. 15–16)

And later he stressed the difference in behaviour and in nature between heat and radiation by means of arguments that can be usefully proposed to students:

> What was formerly called Radiant Heat is a phenomenon physically identical with light. When the radiation arrives at a certain portion of the medium, it enters it and passes through it, emerging at the other side ... as soon as the radiation has passed through it, the medium returns to its former state, the motion being entirely transferred to a new portion of the

medium… Now, the motion we call heat can never of itself pass from one body to another unless the first body is, during the whole process, hotter than the second. The motion of radiation, therefore, which passes entirely out of one portion of the medium and enters another, cannot be properly called heat. (Maxwell 1875, pp. 376–377)

Some of the common conceptions described above recall historical ideas and models now abandoned. Students speak about 'heat contained in a body' as a substance that can pass from one body to another, in a way that resembles the ancient theory of *caloric fluid*. Cotignola and others (2002) analysed students' misunderstanding of basic thermodynamic concepts on historical grounds and concluded that

> The persistence of some ideas from the caloric model are found to be reinforced by magnitude names and unit definitions that were brought up at the early stages of thermodynamic development … Many thermodynamic terms, such as latent heat or heat capacity, were formulated as part of the caloric theory. … The resulting mess becomes one of the main obstacles in the understanding of thermodynamic concepts. (Cotignola et al. 2002, p. 286)

9.3   Research on the Use of History and Philosophy in Teaching Thermal Phenomena and Thermodynamics

History and philosophy of science can be implemented in teaching thermal phenomena and thermodynamics in various ways:

1. Discussing some epistemological, methodological, philosophical specific or general problems connected with the topic in order to improve students' learning, to supply deeper and meaningful knowledge and understanding of the topic, not only technical and algorithmic, and to introduce themes and problems concerning the nature of science.
2. Realising a teaching path that follows essentially the historical path with the connected epistemological, cultural, social and technological themes.
3. Developing some case histories of relevant didactic, cultural, scientific and methodological meaning.
4. Using some historical models, examples, analogies or experiments in order to help students to surmount specific erroneous conceptions and conceptual difficulties.
5. Developing a historical and epistemological analysis of the topic on which to elaborate a didactic reconstruction and design a teaching path (in this case it is not proposed to introduce themes of history and philosophy of science in science courses but to utilise them to find a more effective way for improving students' learning).

Some examples are given below of research and books involving the use of history and philosophy in teaching thermal phenomena and thermodynamics.

The famous book, *Harvard Case Histories in Experimental Science* (Conant 1957), presented eight case histories and the third one concerned thermal phenomena: 'The early development of the concepts of temperature and heat: the raise and decline of the caloric theory' (prepared by Duane Roller, pp. 117–214). It included five sections (Evolution of the thermometer, Black's discovery of specific and latent heat, Rumford's investigation of the weight ascribed to heat, Rumford's experiments on the source of heat that is excited by friction and Davy's early work on the production of heat by friction) and ended by proposing 88 final questions suitable for students. The purpose was to 'assist the reader in recapturing the experience of those who once participated in exciting events in scientific history ... transporting an uninformed layman to the scene of a revolutionary advance in science' (p. IX). The aim was also to illustrate the methods of modern science, considering that familiarity with those methods will increase the understanding of the work of scientists today (p. X). The case history presented two rival schemes in conflict and showed how the transition to a new theory or conceptual scheme is not easy and that old ideas are tenacious.

The *Project Physics Course* (1970), directed by F. J. Rutherford, G. Holton and F. G. Watson, presented many historical aspects of the birth of thermodynamics, its connections with the development of steam engines and the industrial revolution, the discovery of first law of thermodynamics and the debate on the meaning and consequences of the second law (irreversibility, the thermodynamic arrow of time, the heat death of the universe). The authors wanted to 'present not only good science, but also something solid on the way science is done and grows, on the scientific worldview, on how the sciences are interrelated with one another and with world history itself' (Holton 2003b, p. 780). The Project was the object of a large process of evaluation with thousands of students showing positive results (Welch 1973). A new version of the course was published in 2002 (Cassidy et al. 2002, *Understanding Physics*).

Baracca and Besson (1990) proposed using the historical thread of the theories on the nature of heat, the research of Smeaton, Lazare Carnot and Sadi Carnot, and the development of steam engines in the industrial revolution, to introduce thermal phenomena, the laws of thermodynamics and the problem of engine efficiency, especially developing the hydraulic analogy of heat.

Stinner and Teichmann (2003; see also Begoray and Stinner 2005) created a dramatisation of a fictitious but historically based discussion on the age-of-the-Earth problem set in the Royal Institution of London among William Thomson (Lord Kelvin), T. H. Huxley, Charles Lyell and Hermann von Helmholtz. The play was partly based on a lively exchange that occurred between Huxley and Thomson, starting in 1868. In the second half of the nineteenth century, the question of the age of the Earth and the Sun elicited great excitement, both in scientific circles and among the general public. There was no doubt that the sources of the sun's energy, and therefore the existence of the Earth, were limited. Guided by an interpretation of biblical chronology, Bishop Ussher calculated the age of the Earth as 1,650 years. Based on the laws of thermodynamics and the principle of energy dissipation, Kelvin concluded that the Earth could be aged between 20 and 400 million years but not be older. The problem was especially interesting and challenging for physicists but also interested the larger public because it involved philosophical and theological questions and two newly discussed theories, Darwin's Evolution in biology and uniformitarian theory in geology, which assumed an age of Earth of many hundreds of millions of years (see Sect. 9.4.1), and both theories were opposed by Kelvin. The drama was presented to university students and in front of a large audience, and parts of the play were successfully used by teachers in their high school classrooms. The authors argue that historical dramas such as this one can be used for all students to promote learning of particular aspects of science, including the social context of science, science as a human activity and the centrality of debate in scientific change (see also Stinner 1995). Using drama is one of several 'units of historical presentation' that were used by Stinner and his co-workers in teachers' education; others were vignettes, case studies and confrontations (see below).

Stinner and associates (2003), in a general study about the use of history in science teaching, presented some guidelines for designing historical case studies and for context-based teaching. They also proposed the debate on the nature of heat between Rumford and the sustainers of caloric theory as a 'mini-confrontation', suitable for upper secondary school. 'Many of the experiments that Rumford performed can be replicated by students … before doing so, teachers could present the caloric theory along the lines previously suggested and discuss it as an explanatory theory for many everyday phenomena. Following that, teachers could set up experiments inspired by Rumford' (p. 629).

Metz and Stinner (2006) developed teaching activities centred on the analysis and replication of historical experiments. The activity was organised as a *narrative* divided into four parts: introduction, experimental design, experimental results and analysis and interpretation of data and explanation. They used some of Rumford's experiments on heat which could easily be adapted for the classroom. In these investigations, Rumford was interested in determining what materials afforded the best insulating protection and measured the cooling time of a warm container. The introductory part of the narrative established the context, including some biographical information, and presented a problem and/or confrontation. Students read an excerpt from Rumford's paper of 1804, perform the experiment, compare their results to Rumford's results and discuss the discrepant results and the proposed explanations. As they interact with the history throughout the investigation, students develop scientific processes and address questions on the nature of science.

Chang (2011) discussed the possible roles and aims of historical experiments in science education. He considered three different objectives: to advance the understanding of the history of science, to refine our philosophy of science and the conceptions of the nature of science (NOS) and 'to improve scientific knowledge itself – that is, to gain more, better, or different knowledge of nature than current science delivers'. Focusing on this third aspect, he illustrated two case histories, the first one concerning the anomalous variations in the boiling point of water. Indeed, around 1800, many scientists (he considered especially Joseph-Louis Gay-Lussac, Jean-Baptiste Biot and Jean-André De Luc) observed that 'the boiling temperature of pure (distilled) water under standard pressure depended greatly on the material of the vessel employed, on the exact manner of heating, and on the amount of dissolved air present in the water'. He pointed out that at various levels of science education, we teach that pure water under standard pressure always boils at 100 °C, but the case history has shown that 'we do it in a patently incorrect way'. He considered the function of history and philosophy of science (HPS) as 'complementary science', 'which complements specialist science, neither hostile nor subservient to it … HPS in this complementary mode is not about science; rather, it is science, only not as we know it'. In this line of work, the proposed historical experiments play the role of 'complementary experiments', in which what matters is *physical replication*, not *historical replication*, and they can improve our knowledge of nature and aid science education.

Wiebe and Stinner (2010) suggested the use of interactive historical vignettes, presented in the form of guided readings, in which the concepts are embedded into the historical contexts, to help the students' understanding of gas behaviour. They presented the problems of pressure-temperature and volume-temperature relationships and subsequently the historical context of the explanations of gas behaviour using the particulate nature of matter and the kinetic molecular theory.

Viard (2005) used an alternative strategy to teach the concept of entropy, which consisted of getting the students to read and interpret excerpts from Carnot, Clausius and Boltzmann. He considered that the history of thermodynamics can provide resources for a direct and simple teaching path for introducing entropy to students.

Zambrano (2005) developed 'a curricular sequence based on a historical study of conceptual change in science'. He did not present history directly to students as course content, but he utilised history as a reference for designing a learning path aimed to improve the students' conceptual change from their previous conceptions to correct scientific knowledge. He wanted to show how the history of science can illuminate the teaching and learning of scientific concepts in school:

> Our belief in this case is that the change necessary to go from caloric theory to thermodynamic theory of heat and temperature concepts can illuminate the changes pupils experience in changing from their initial concepts about heat and temperature to expert's concepts. … The historical construction process of the concepts of heat and temperature, analyzed in the light of the different obstacles against its progress, allow us to elaborate the corresponding teaching educative sequence based on its historical epistemological order. (Zambrano 2005, pp. 1 and 8)

Mäntylä and Koponen (2007) developed an epistemological reconstruction of temperature as a measurable quantity. They did not want to produce historical reconstructions 'but instead to use HPS as a starting point for developing and designing suitable didactic solutions, which can be called didactic reconstructions for teaching' (p. 292). The reconstruction parallels the historical development, but its intention is not to be a historically authentic path. The history is interpreted from the point of view of modern conceptions, 'because the goal is to teach physics, not the history of physics'. Moreover, they think that this reconstruction also 'conveys a more correct view of the role of measurements in the production of scientific knowledge'. A conceptual analysis of the historical development is needed, even when 'the purpose is to produce teaching solutions fostering the development of modern conceptions rather than giving an authentic picture of historical developments'.

De Berg (2008) proposed the use of analogies based on the historical caloric theory in order to help students to clearly distinguish between heat and temperature,

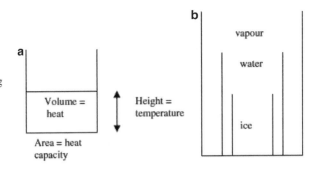

Fig. 9.1 Adaptation of the model of Irvine and Dalton representing (**a**) the difference between heat and temperature, (**b**) the effect of phase transitions (De Berg 2008, p. 15, Fig. 9.2)

explain phase changes from solid to liquid to gas and introduce the idea of absolute zero of temperature. He used and adapted a model originally elaborated and developed by Irvine (1743–1787) and Dalton (1766–1844) (see Fig. 9.1). According to the author, these analogies do help students understand the difference between heat and temperature and the concept of heat transfer, but they can also reinforce the material view of heat against the kinetic view. Nevertheless:

> The material view of heat can be thought of as having been transformed from a theory about the nature of heat to an analogy useful for pedagogical purposes and that may not be necessarily a bad thing provided one is aware of the limitations of analogies … Also, one's attitude to outdated models and theories grows into one of respect rather than one of disdain when presented with relevant historical cases. (De Berg 2008, pp. 90–91)

In the context of the European project HIPST (History and Philosophy in Science Teaching), two case studies concerned thermodynamics:

(a) Temperature – what can we find out when we measure it? (Oversby 2009, UK)

> The topic of temperature is fascinating because it begins with an unclear understanding by historical scientists about its nature, and about how to measure it. This parallels the position of the 11 year old students in the pilot program. The historical search for a suitable measurement is inextricably bound up with creating instruments to do the measuring, and the lack of clarity about what was being measured. … We have used links to Drama in producing play scripts of historical events, and to English in the form of newspapers to present historical information.

(b) Steam, Work, Energy (Brenni et al. 2009, Florence, Italy)

> This case study concerns the formulation and production of a kit for high school physics teachers (students aged 14–19) to supplement their lessons on thermodynamics with elements offering an historical contextualization of how several fundamental laws of physics were discovered, and how certain concepts have become structured in time… evidences the connection between science, technology and society … it shows how practical necessities and empirical-technical solutions gave rise to the premises for forming new theories … how the technological innovation of the steam engine and its successive improvements made possible also by new scientific acquisitions, profoundly transformed society and its organization… it contains videos showing instruments in operation from the historical collection.

9.4 Philosophical Problems of Thermodynamics and Their Implications for Teaching

Thermodynamics poses many important epistemological and philosophical problems which can be accessible and useful for secondary education and which are not separated from the content but entangled with it. As Prigogine wrote:

> It is the singularity of physics such as we know it still today: the metaphysical discussions are not superimposed arbitrarily on strictly scientific questions, but depend on those in a crucial way. (Prigogine and Stengers 1988, p. 37)[2]

Some philosophical themes or problems of didactic interest can be proposed in teaching and are relevant for well-founded learning:

– The meaning, interpretations and implications of the second law (irreversibility, time arrow, statistical and probabilistic laws and determinism)
– The relationships between macroscopic properties and microscopic structures, reductionism, emergence and primary and secondary qualities
– The nature of thermodynamics theory, as a theory of principles not constructive, its structure, its relationships and its differences from classical mechanics
– The relationships between science, technology and societal demands, which have special features in the case of thermodynamics
– The construction of a physical quantity such as temperature

In this section, I will discuss some philosophical issues which can be proposed, in appropriate form, to students of high school and university.

9.4.1 Origin and Meaning of the Laws of Thermodynamics

The first law of thermodynamics concerns essentially the energy conservation law, and therefore in this book it is treated in the chapter on energy. Instead, here some considerations are given on the second law of thermodynamics, which offers many stimuli for creating multidisciplinary didactic activities.

Thermodynamics began as a theory of thermal engines and retained for a long time some characteristics derived from this beginning. This is also true for the second law of thermodynamics, especially in the Kelvin formulations concerning useful work, energy dissipation and degradation of energy (Thomson-Kelvin 1851, 1852). The emancipation from this utilitarian origin, linked to technical, economic and practical aims, happened only with the works of Clausius and Gibbs. Thomson (Kelvin) and Tait were very attached to these early ideas and spoke about dissipation, degradation and not available energy, i.e. energy which is not annihilated but is unusable for mankind, in some formulations distinguishing between the physics of living

---

[2] All quotations that were in French or in Italian in the original have been translated into English by the author of the present paper.

beings and of the inanimate world. This tendency is also linked to metaphysical and theological choices and arguments, e.g. that God would have created the world with a total energy which, being a divine creation, is conserved, cannot be consumed and is stable and eternal, whilst what concerns man is perishable and temporary. In some way, in the contrast between the approaches of Kelvin and Clausius, there is an opposition between a man who yearns not to lose useful resources and possibilities and a man who wants to understand nature ignoring human concerns.

Subsequently, the situation changed remarkably and the debate on the meaning and implications of the second law involved wider and more general problems, such as the age of the Earth and of the universe, cosmological themes, the heat death of the universe, evolution and religious implications. Later, statistical entropy was connected with information theory (Shannon 1948) and it was identified as a measure of the lack of information (Jaynes 1957). Moreover, some thermodynamic ideas and words have been assumed in other fields, like economics, ecology and sociology (e.g. social entropy, human thermodynamics, thermoeconomics).

Kelvin used the energy dissipation principle and the heat conduction law to calculate that the present temperature conditions on Earth could have existed for only 20–100 million years and used this result for contrasting the uniformitarian theory in geology, which assumed constant conditions over hundreds of millions of years, and Darwin's theory of evolution, which Kelvin thought 'did not sufficiently take into account a continually guiding and controlling intelligence'. Stinner and Teichmann (2003) designed a teaching activity on this controversy about the age of Earth (see Sect. 9.3).

Currently, different senses of entropy can be found in scientific and popular literature: thermodynamic, statistical, disorder and information senses (Haglund et al. 2010). The disorder metaphor goes back to Helmholtz and Boltzmann and is very common in popularisations and textbooks as an aid to forming an intuitive image of entropy, but it is criticised because it would be conceptually misleading (Lambert 2002 considered it as 'a cracked crutch for supporting entropy discussions') and could lead students to erroneous conclusions in thermodynamics tasks (Viard 2005). Styer (2000) illustrated examples of increased entropy accompanying increased 'order', in clear contradiction with the *entropy as disorder* view. On the other hand, order and disorder are partially subjective terms, and they can be differently interpreted by different observers.

9.4.2 Irreversibility and Time Arrow, Mechanics and Thermodynamics

Some problems can be posed concerning the relationship between thermodynamics and mechanics. The reversible laws of mechanics, with temporal symmetry, seem to oppose the second law of thermodynamics, with the time arrow (the expression was introduced by Eddington in 1928 in his book *The Nature of the Physical World*), which is evident in daily experience. In this sense mechanics is in strong contrast with common experience, a problem that was taken into account only very late, just when thermodynamics included in its foundations the irreversibility of actual physical processes.

The notion of equilibrium is very different in mechanics and in thermodynamics. It is only by the effect of friction and of energy dissipation that an oscillating pendulum will stop in its equilibrium position after a number of oscillations, whilst in the frictionless ideal case it will never stop. During an oscillation the pendulum can reach and immediately abandon the equilibrium position and can pass through it before or after another position, except if it is placed there motionless by an external force.

By contrast, in thermodynamics a system tends spontaneously towards an equilibrium state and once reached it remains there; the equilibrium state is the term of an evolution without return (Stengers 1997, pp. 73–75). One could say that, if the physics of Galileo and Newton eliminated the distinction between the world of the Earth and that of the skies, between the disorder and the corruption of the former and the rational and incorruptible order of the latter, here a similar fracture seems to appear between the irreversible and dissipative world described by thermodynamics and the classical mechanics that governs the motion of ideal pendulums, perfectly elastic bodies, and of celestial bodies (Maxwell called it 'The Queen of the Skies'). Mechanics seems to have to neglect 'the impediments of the matter' (Galileo 1632 *Dialogue,* second day) and apply some opportune idealisations in order to represent actual phenomena correctly. Friction phenomena appear to be the bridge between the two approaches but their explanation in terms of mechanics asks for conjectures on complicated effects of a great number of microscopic or mesoscopic entities (see Besson 2001 and 2013).

The problem is then posed of the coherence between the two theories and the possible prominence of one over the other. Is the second law of thermodynamics only an appearance behind which there are reversible microscopic phenomena or is it a fundamental law of nature? Is it only a statistical result indicating an improbability or is it an exact law that allows the re-establishment of a *new alliance* between science and human experience (as supported by Prigogine)? Must one try *to explain* thermodynamics by means of mechanics or is it necessary to modify mechanics? Is unidirectional time a phenomenological, not fundamental, property, a sort of secondary or emergent quality such as colour and flavour, due to the fact that we are beings of an intermediate dimension? Is it a property or quality that would not exist for a microscopic being like Maxwell's demon, which can handle single molecules (see Sect. 9.4.3), or for the super-intelligence imagined by Laplace, which can know and calculate all the positions and velocities of particles? We could imagine a dialogue between mechanics and thermodynamics which is similar to the one between reason and senses written by Democritus (fragment B125), by replacing colour and flavour with irreversibility and the time arrow:

> The intellect says: By appearance there is sweetness or bitterness, by appearance there is colour, in reality there are only atoms and the void. The senses answer: Ah foolish intellect! You get your evidence from us, and yet do you try to overthrow us? That overthrow will be your downfall.

9 Teaching About Thermal Phenomena and Thermodynamics...

Later in his life, Boltzmann concluded that the time arrow is an appearance which does not belong to the entire universe, and inversely the direction of time is determined by the direction of entropy increase:

> For the universe as a whole the two directions of time are indistinguishable, just as in space there is no up or down. However, just as at a certain place on the Earth's surface we can call 'down' the direction toward the centre of the Earth, so a living being that finds itself in such a world at a certain period of time can define the time direction as going from less probable to more probable states (the former will be the 'past' and the latter the 'future') and by virtue of this definition he will find that this small region, isolated from the rest of the universe, is 'initially' always in an improbable state. (Boltzmann 1897, English translation in Brush 2003, p. 416)

9.4.3   Statistical Mechanics and Thermodynamics, Probability and Determinism

The process obtained by means of a time reversal of a natural process, such as inverting the order of the frames of a film, does not exist in nature. Can we explain this fact by a model in which the matter is constituted of a large number of microscopic particles? Can we understand the irreversible phenomena and the consequent increase of entropy in terms of movement and interaction of a large number of atoms? This is the problem of the explanation of the irreversible macroscopic laws in terms of the reversible microscopic laws regulating the movement of atoms. When the macroscopic thermodynamic properties of heat and temperature or the gas laws are explained in terms of behaviour and movement of small particles, some philosophical problems arise naturally, such as reductionism, emergent properties and supervenience (Callender 1999).

The bridge between macroscopic and microscopic levels passes through molecular atomic theory and the kinetic theory of gases and then the statistics of Maxwell and Boltzmann. In general, these issues lead to the problems of the role of probability in physics and in the explanations of natural phenomena and the relationship between the knowing subject and the physical reality, issues that were to be resumed with new depth by quantum mechanics. Probability would have the task of articulating our uncertain world with a supposed objective reality, which is governed by deterministic and exact laws.

Boltzmann seemed to solve the problem for a gas with his H-theorem, showing that collisions between molecules lead towards the equilibrium distribution and that a special function H (in the original article this function was called E) always decreases with time until reaching a minimum value. For a gas in thermal equilibrium, the H-function is proportional to minus the entropy as defined by Clausius for equilibrium states, and Boltzmann considered it as a generalised entropy valid for any state, so he considered that he had demonstrated by means of a mechanical model that entropy always increases or remains constant. Some objections arose soon after, such as the *reversibility paradox* of Loschmidt, the *recurrence paradox*

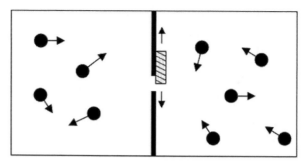

Fig. 9.2 An illustration of the Maxwell's intelligent demon. The two boxes communicate through a little door. The demon opens and closes the door so as to allow the faster molecules to move from the right box to the left and the slower molecules in the opposite way. As a consequence, the temperature of the gas in the left box becomes higher than the temperature of the gas in the right box, so contradicting the second law of thermodynamics and making the entropy of the whole system diminish

of Poincaré-Zermelo and the *reversibility objection* that it would be impossible to deduce the irreversibility from laws and collisions which are reversible, the only way being to enter irreversibility at the molecular level. Boltzmann answered these objections, by adding further hypotheses to his model and assuming that in fact it is possible that entropy decreases, but it is extremely improbable, so transforming in improbability the impossibilities enounced by the second law of thermodynamics. This was also the idea of Maxwell:

> This is the second law of thermodynamics, and it is undoubtedly true as long as we can deal with bodies only in mass, and have no power of perceiving or handling the separate molecules of which they are made up. But if we conceive a being whose faculties are so sharpened that he can follow every molecule in its course, such a being ... would be able to do what at present is impossible to us... He will thus, without expenditure of work, raise the temperature of B and lower that of A, in contradiction to the second law of thermodynamics. This is only one of the instances in which conclusions which we have drawn from our experience of bodies consisting of an immense number of molecules may be found not to be applicable to the more delicate observations and experiments which may suppose made by one who can perceive and handle the individual molecules which we deal with only in large masses. In dealing with masses of matter, while we do not perceive the individual molecules, we are compelled to adopt what I have described as the statistical method of calculation, and to abandon the strict dynamical method, in which we follow every motion by the calculus. (Maxwell 1871, pp. 358–359)

Thomson-Kelvin (1874) used the term *intelligent demon* to refer to a being that could theoretically reverse the dissipation of energy and the entropy increase (Fig. 9.2). Maxwell's demon spawned much discussion for many years, and different solutions were proposed to solve the paradox (see, e.g. Collier 1990; Daub 1970). It was considered that all measurements need an energy cost, in which dissipation would compensate the entropy decrease due to the relocation of molecules. However, later some authors showed that measurements could be performed with a zero or very low energy expenditure. On the other hand, to erase and rewrite the demon's memory, where the information must be stored, energy dissipation is required in an amount which at least compensates the entropy decrease due to the demon's action (see Landauer 1961, who considered that erasing a single bit of information implies energy dissipated into the environment $\geq kT \cdot \ln 2$).

It is strange that all these objections were general considerations or counterexamples, which did not dispute the details of the Boltzmann demonstration. Boltzmann used a gas model consisting of a set of many gas molecules:

> Each molecule is a simple point mass, ... two molecules interact only when they come very close together ... perhaps the two molecules rebound from each other like elastic spheres... As for the wall of the container that encloses the gas, I will assume that it reflects the molecules like elastic spheres. (Boltzmann 1872, English translation in Brush 2003, pp. 266–267)

He specified that this model showed 'a precise mechanical analogy with an actual gas'. Nevertheless, the movement of a gas of elastic spheres is reversible, and it does not tend to equilibrium. The mathematical model developed by Boltzmann implied irreversibility and tendency towards equilibrium; therefore, it represents well the behaviour of real gases, but it does not represent a gas of elastic spheres correctly.

What do the objections to Boltzmann show? That the second law is not valid, that thermodynamics is irreducible to mechanics or that mechanics must be modified? Or that the employed mechanical model does not represent the physical situation correctly? These questions can open an interesting thread about the role of models in science, their relationship with physical reality and the effects of supplementary simplifications and hypotheses that scientists introduce in order to develop mathematical calculations and obtain specific verifiable results (in this case, e.g. the hypotheses on particular initial conditions or the use of continuous functions to describe finite numerable sets of molecules).

New arguments arose from quantum mechanics and the idea of the impossibility of simultaneous exact determination and definition of the position and velocity of molecules and also from the consideration that a little imprecision or variation in the reversing of all velocities can dramatically change the evolution of the system, quickly redirecting it towards states of increasing entropy. A similar idea was already noticed by Kelvin in 1874:

> If we allowed this equalization to proceed for a certain time, and then reversed the motions of all the molecules, we would observe a disequalization. However, if the number of molecules is very large, as it is in a gas, any slight deviation from absolute precision in the reversal will greatly shorten the time during which disequalization occurs. (Thomson-Kelvin 1874, p. 331)

In addition, it was noticed that the unavoidable small outside influences, which are unimportant for the evolution towards states of entropy increasing, greatly destabilise evolution in the opposite direction when it is aimed at a very small region of the phase space (see Lebowitz 1999).

9.4.4 Nature of Science, Explanations and Models in Thermodynamics

Thermodynamics shows various characteristics which are different from mechanics and electromagnetism. It is not a *constructive science* but a *science of principles*. It does not study or explain how the processes happen and how phenomena are produced, but rather it establishes constraints and tendencies, defines a background and utilises different typologies of explanation and causality (see Wicken 1981). The language of thermodynamics introduces specific new terminologies, different from mechanics and electromagnetism; it speaks about system, state, transformation, reversibility, state equation, adiabatic, entropy, internal energy, etc., terms that create difficulty of interpretation among students, also because they are not always clearly defined and explained in textbooks.

Thermodynamics developed initially as a dynamical theory of heat (as it was the title of one of Kelvin's first works on thermodynamics, see Thomson-Kelvin 1851), with the programme of explaining thermal phenomena by means of the dynamic behaviour of microscopic particles forming the matter. However, the difficulties in carrying out this programme in a complete and satisfactory manner led many scientists to the idea of developing this science in an autonomous way, independently of particular theories and models on the microscopic constitution of bodies. Facing the difficulties of explanations based on atomic models, the same idea of the actual existence of atoms was put in doubt by many scientists at the end of the nineteenth century and also by Planck:

> The second law of thermodynamics, logically developed, is incompatible with the assumption of finite atoms. ... Yet there seem to be at present many kinds of indications that in spite of the great successes of atomic theory up to now, it will finally have to be given up and one will have to decide in favour of the assumption of a continuous matter. (Planck 1882, quoted in Brush 1976, pp. 641–642)

It is interesting to notice that this happened in spite of the fact that the kinetic theory of gases had already led to important results, with Clausius explaining pressure and temperature, and especially with the more sophisticated study of Maxwell, who also succeeded in foreseeing an unexpected experimental result like the independence of the viscosity of a gas from the density, over a wide range of densities.

Thermodynamics has become, for many scholars, the model of a theory independent of structural details of constituents of matter which are not directly observable. It has also become the prototype of a scientific theory for philosophers supporting positivist, conventionalist or instrumentalist conceptions of science, like Mach, Duhem, Ostwald and the supporters of energetics. The debate goes back to the works of Fourier and Poisson in the first decades of the nineteenth century. Fourier (1822) sustained the autonomy of his theory of heat and heat conduction, expressed by mathematical equations, from mechanics and from models on matter structure, and criticised the reductionism to mechanics:

> Whatever may be the range of mechanical theories, they do not apply to the effects of heat. These make up a special order of phenomena, which cannot be explained by the principles of motion and equilibrium ... The principles of the theory are derived, as are those of rational mechanics, from a very small number of primary facts, the causes of which are not considered

by geometers, but which they admit as the results of common observations confirmed by all experiments. (Fourier 1822, pp. II–III and XI)

Fourier's ideas were very influential and became a basic reference for subsequent debates on the nature of physics theory and the relationships between theory, experiments and explicative models. By contrast, his contemporary and colleague Poisson sustained the relevance of molecular models for discovering the mechanisms producing thermal phenomena and considered mathematical equations as useful tools in order to develop the consequences of the models:

It is matter of deducing, by rigorous calculations, all the consequences of a general hypothesis on heat communication, which is based on experience and on analogy. These consequences will be then a transformation of the same hypothesis, to which calculations do not remove nor add anything ... to be complete this theory should be able to determine the movements that are provoked by heat into gases, liquids and solids ... I will adopt the more fecund theory according to which these phenomena are due to a imponderable matter contained inside the parts of all bodies ... this matter is called *caloric*. (Poisson 1835, pp. 5 and 7)

At the end of the nineteenth century, this debate was connected to more general philosophical debates about the role of science, materialism, realism and the relationships among science, philosophy and religion. There were widespread tendencies to dispute the autonomy of science and criticise geological and biological evolution theories. Questions were raised about whether the role of science was to provide explanations about the real world and knowledge about things existing in reality or to only develop classifications and syntheses of phenomena and observations as useful economies of thought, with explanations coming from elsewhere. Instrumentalist conceptions of science were widespread, which also rejected the atomic theories. For example, yet in 1913, Mach refused to accept the existence of atoms.

The sterility of these pure instrumentalist conceptions manifested in a resistance to accept the innovations that were emerging in physics and in not producing new results and predictions. For example, Duhem refused to accept Maxwell's electromagnetic theory, Boltzmann's statistical mechanics and Einstein's relativity and opposed Galileo's realism:

When Kepler or Galileo declared that Astronomy has to take as hypotheses propositions the truth of which is established by Physics, this assertion ... could mean that the hypotheses of Astronomy were judgments on the nature of things and on their real movements ... But, taken in this sense, their assertion was false and harmful. ... In spite of Kepler and of Galileo, we believe today, as did Osiander and Bellarmino, that the hypotheses of Physics are only mathematical artifices intended to *save the phenomena*. (Duhem 1908–1990, pp. 139–140)

Planck, in his treatise of 1897, offered an organic exposition of *pure thermodynamics*, as an autonomous science based on the concepts of energy and entropy and on laws independent from hypotheses about microscopic structures, avoiding the anthropomorphic ideas of degradation, dissipation and quality of energy. For example, he remarked that in an isothermal expansion of a gas, heat is integrally transformed in work, which contrasts with the idea of energy degradation. The irreversibility and the impossibility of perpetual motion became consequences of the law of entropy increase. Planck pointed out that the second law does not concern only energy dissipation and the problem of heat and work. For example, in the mixture of two gases or in the further dilution of a solution, the process does not involve changes in the type of energy but happens in a direction driven by entropy increase. According to Planck, this was the best way to deal with thermodynamics, but he specified 'up to now', so as not to close off possible future improvements and progress. However, thermodynamics cannot foresee nor explain some simple phenomena, such as the behaviour of the specific heat of gases, where new hypotheses on the properties of atoms and molecules will be necessary.

Currently, textbooks show very different approaches to thermodynamics (see Tarsitani and Vicentini 1996), which can be well understood only as the result of different epistemological choices. An analysis of these philosophical backgrounds is useful for students and necessary for teachers in order to choose handbooks and define course organisation.

Two ancient textbooks, having two very different approaches, have been particularly influential for the teaching of thermodynamics, those of Maxwell (1871) and of Planck (1897). The dynamical approach of Maxwell, had as background the kinetic model of heat, considered as a 'kind of motion'. This gave a limited epistemological status to the second law, considered as a *human scale* law, not an absolute law. The phenomenological approach of Planck considered thermodynamic laws as self-sufficient laws with a broad empirical basis and a fundamental status.

For example, the well-known treatise of Zemansky (1968) develops a phenomenological approach based on the classical formulations of the three laws and on the concept of equilibrium and only in the last part is open to microscopic interpretations. By contrast, the famous Berkeley physics handbook published in five volumes does not include a volume on thermodynamics but one on *statistical physics*, thus suggesting a nonautonomous status of thermodynamics (Reif 1965; see also Reif 1999). Moreover, this handbook avoided any reference to the methodologically and philosophically problematic issues related to the topic.

Explanations proposed in thermodynamics are often unsatisfactory for the students' need of understanding because they only show how things must be or not be, but not how things happen, by which processes and mechanisms a new situation is established. This raises questions about causal or formal laws, and correlation or necessity relationships. The idea of irreversibility is intuitive and in line with observations of daily reality, but the irreversible processes are difficult to treat mathematically. To overcome this difficulty, abstract conceptual devices which appear artificial to students are used, such as quasi-static or reversible transformations (a succession of equilibrium states: the system would remain spontaneously in every one of these states, and it can be moved to another state only by means of an external manipulation or force).

9.4.5 Stationary Situations and Dissipative Structures

The tendency to equilibrium, to uniformity and to cancellation of differences, suggested by the second law of thermodynamics, can appear to contradict the observation of phenomena of self-organisation and the spontaneous formation of ordered

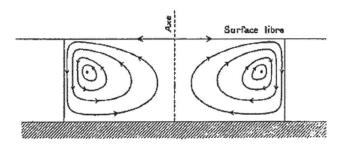

Fig. 9.3 The Bénard's cells (From Bénard 1900, p. 1005)

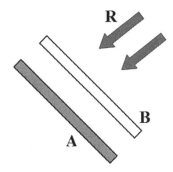

Fig. 9.4 The ideal greenhouse in the empty space. Two plates A, B, initially at same temperature, are exposed to a radiation R of short wavelength. A is black for all radiations, B is a glass which is transparent to radiation R but absorb all far infrared thermal radiation emitted by A. After a time a stationary condition is reached in which the temperature of the two plates are different:

$$T_A = T_B \cdot \sqrt[4]{2} \cong 1.19 T_B$$

structures in nature. In this way, it is useful to study stationary situations without thermal equilibrium, in connection with the *dissipative structures* described by Prigogine, as the spontaneous creation of nonuniform structures in situations far from equilibrium. This suggests the actual possibility of spontaneous creations of ordered structures without contradicting the second law of thermodynamics.

Indeed, just the entropy flow is at the origin and sustains the nonuniformity. Examples are the spontaneous growing of a crystal in an opportune solution, the formation of regular convective cells in a layer of liquid heated from below (Bénard's cells, see Bénard 1900 and Fig. 9.3) and more generally also life. Prigogine and Stengers (1988, pp. 49–50) described an experiment showing that two different gases (nitrogen and hydrogen), initially uniformly mixed in two communicating containers kept at different temperatures, spontaneously separate from each other, more hydrogen going in the hotter container and more nitrogen in the colder one.

Other examples useful for teaching in high school classes can be a greenhouse in empty space (Fig. 9.4), the greenhouse effect on the Earth or simply a room with the heating keeping a constant difference of temperature with outside. They can be considered ordered structures in the sense that *correlations at distance between parts of the system* are created, for example, between the temperatures of the base and of the covers of a greenhouse. At the same time, the study of nonequilibrium stationary situations is important from a didactic and conceptual point of view because these situations are very common in various parts of physics and students have difficulty in understanding them and dealing with them correctly. Also the entire Earth is a system that receives solar radiation at low entropy and emits infrared radiation with higher entropy, and it is this negative entropy flux which makes it possible to create or maintain ordinate structures.

9.5 Case Histories for Teaching and Learning Thermal Phenomena and Thermodynamics

Many case histories useful for teaching can be found in the history of thermodynamics. Here I will mention some of these, which have valuable conceptual, cognitive and epistemological implications.

9.5.1 Theories on the Nature of Heat

What is heat? Since antiquity philosophers and scientists have tried to answer this question. The history of the ideas on the nature of heat is a case history which can be used at various school levels and can promote interest and motivation. Two main conceptions emerge which can be followed through the centuries, with their variations, in the scientific debates: heat is a substance which is contained inside bodies and can pass from one body to another, or it is the effect of the motion of particles constituting bodies. A mixed conception is also found, in which heat is due to the motion of particles constituting a caloric substance.

The case history can be developed by presenting excerpts of texts by some scientists chosen because they are representative of different periods and conceptions and for their importance in science history. Reading the motivations, rationales and mistakes of scientists in their own words can be a good way to reveal and discuss students' conceptions. For didactic purposes, this long history can be simplified into four periods: Greek and Roman antiquity, the seventeenth century, the affirmation of the caloric conception in the eighteenth century and the development of a modern kinetic theory in the nineteenth century. The historical problems of the distinction between heat and temperature and of the relationship between the thermal sensations and the physical properties of matter should be outlined, because they are clearly linked with common students' difficulties. Some experiments should also be described which were considered relevant in order to discriminate between different theories. I will give in the following some examples of scientists' significant quotations that can be proposed to students.

9 Teaching About Thermal Phenomena and Thermodynamics...

Lucretius, in the context of his atomistic philosophy and following his predecessors Democritus and Epicure, considered heat as a substance made of special atoms (in Latin, *semina ignis, corpuscula vaporis,* i.e. seeds or grains of fire, particles of vapour or of heat). In this excerpt, Lucretius is trying to explain why the water of a certain famous spring was colder during the day than during the night, a fact that appeared very surprising:

> These particles of heat (*corpuscula vaporis*) do not travel isolated, but they are entangled and amassed, so that each one is restrained by the others and by external bodies, and consequently they are compelled to advance more slowly. [...]
> The earth near the spring is more porous than elsewhere, and be many the seeds of fire (*semina ignis*) near the water; on this account, when night submerges the earth, soon the earth gets chilly and contracts in its depths; and thus, as if one had squeezed it by the hand, it pushes into the spring the seeds of fire that it holds, which render warm the touch and steam of the fluid. Next, when the sun has risen with its rays and dilated the soil by mixing it with its fires, the seeds of fire again return into their previous abodes, and all the warm of water retires into the earth; and this is why the fountain in the daylight gets so cool. (Lucretius *De rerum natura* (On the Nature of Things), book II, lines 154–156 and book VI, 861–873)

During the seventeenth century, in the context of the new affirmation of mechanistic and atomistic philosophy, with the distinction between primary and secondary properties, many scientists adhered to a kinetic conception (Bacon, Descartes, Boyle, Mariotte, Hooke, Newton) or a mixed conception (Gassendi, Galileo, Boerhaave, Lemery). Galileo sustained a mixed conception:

> Those materials which produce the warm sensations ... are a multitude of little particles, having various shapes and moving at different speeds ... to excite the warm sensation the presence of particles of fire (*ignicoli*, in Italian), is not sufficient but also their motion is necessary, so that it can be said very rightly that motion is the cause of heat. (Galileo, *Il Saggiatore*, 1623, pp. 781 and 783)

The conception of heat as a substance spread rapidly during the eighteenth century, in connection with the more general affirmation of physical interpretations based on models of imponderable fluids (e.g. electricity, magnetism, phlogiston, ether).

Joseph Black (1728–1799) developed in a more detailed way the theory of heat as a substance, called *caloric*, and gave a strong contribution to the success of this conception. He distinguished between free and latent (or combined) caloric, and he was the first to clearly understand, around 1760, the distinction between heat and temperature and to define the physical quantities of specific heat and latent heat (nevertheless, his ideas diffused slowly only after 1770 and his *Lectures* were published only after his death in 1803):

> [In a situation of thermal equilibrium usually scientists imagined that] there is an equal quantity of heat in every equal measure of space, however filled up with different bodies. The reason they give for this opinion is that to whichever of those bodies the thermometer be applied, it points to the same degree. But this is taking a very hasty view of the subject. It is confounding the quantity of heat in different bodies with its general strength or intensity, though it is plain that these are two different things, and should always be distinguished ... The quantities of heat which different kinds of matter must receive ... to raise their temperature by an equal number of degrees, are not in proportion to the

quantity of matter in each, but in proportions widely different from this, and for which no general principle or reason can yet be assigned... different bodies, although they be ... of the same weight, when they are reduced to the same temperature or degree of heat, ... may contain very different quantities of the matter of heat. (Black 1803)

At that time, there was confusion between temperature and heat, and temperature was considered as a measure of intensity, strength, density, level or degree of heat. Newton wrote of 'degree of heat', measured by a thermometer. These ideas are clearly similar to some widespread students' conceptions (see Sect. 9.2).

During the period between the last decades of the eighteenth century and the beginning of the nineteenth century, scientists were divided and uncertain on this problem. It can be interesting to show how the opinion of a same scientist changed over some years. For example, this is shown by these two quotations of Lavoisier:

It is difficult to understand these phenomena without admitting that they are the effect of a real and material substance, of a very subtle fluid, which insinuates among the molecules of all bodies. (Lavoisier 1789, p. 19)

Physicists are divided about the nature of heat. Many of them consider it as a fluid diffused in all the nature... others think that it is the result of the insensible motion of molecules of matter ... We will not decide between the two hypotheses. (Lavoisier and Laplace 1780, pp. 357–358)

By contrast, Alessandro Volta in 1783, in his *Memoria intorno al calore* (Memory on Heat), was very sure 'That heat is a peculiar element, distinct from all other substances, seems to us not a probable opinion but an indubitably established truth'.

Things changed notably with the works of H. Davy (1799) and of B. Thomson-Rumford:

I cannot refrain from just observing that it appears to me to be extremely difficult to reconcile the results of any of the foregoing experiments with the hypothesis of modern chemists respecting the *materiality of heat* ... There are many appearances which seem to indicate that the constituent particles of all bodies are also impressed with continual motions among themselves, and that it is these motions (which are capable of augmentation and diminution) that constitute the *heat* or temperature of sensible bodies. (Thomson-Rumford 1804, pp. 103–104)

Nevertheless, during the same years Dalton was of the opposite opinion:

The most probable opinion concerning the nature of caloric is that of its being an elastic fluid of great subtlety, the particles of which repel one another, but are attracted by all other bodies. (Dalton 1808, p. 1)

Dalton developed an explicit analogy between heat contained in a body and a liquid in a vessel in order to help clarify the concepts of specific and latent heat and of temperature (De Berg 2008, proposed the use of similar historical analogies for teaching; see Sect. 9.3 and Fig. 9.1).

Finally, the concept of heat became clear after the discovery of energy conservation, as this excerpt from Maxwell shows:

The temperature of a medium is measured by the average kinetic energy of translation of a single molecule of the medium ... The peculiarity of the motion called heat is that it is perfectly irregular; that is to say, that the direction and magnitude of the velocity of a molecule at

9   Teaching About Thermal Phenomena and Thermodynamics...

a given time cannot be expressed as depending on the present position of the molecule and the time. (Maxwell 1875, p. 376)

This history offers many elements which can be developed in a way suitable for teaching. For example (see Sect. 9.3), Conant (1957) proposed a case history on the rise and decline of the caloric theory including Black's discoveries and the experiments of Davy and Rumford on the heat produced by friction, and Stinner et al. (2003) proposed a 'mini-confrontation' between Rumford and the sustainers of caloric theory.

9.5.2   The Discovery of Radiant Heat, the Debate on Its Nature and the Search for the Law of Thermal Radiation

The first studies on radiant heat began in the seventeenth century (see Cornell 1936). The existence of invisible heat rays that can be concentrated by using mirrors was proven by F. Bacon (1620, Book two, XII). Experiments realising the separation of radiant heat from light by glass were performed by E. Mariotte (1679) and confirmed by R. Hooke (1682). Newton himself suggested ether vibrations as a way of heat propagation and described some experiments on heat transmission in a vacuum:

> Is not the Heat of the warm Room conveyed through the *Vacuum* by the Vibrations of a much subtler Medium than Air, which after the Air was drawn out remained in the *Vacuum*? ... And do not hot Bodies communicate their Heat to contiguous cold ones, by the Vibrations of this Medium propagated from them into the cold ones? (Newton, *Optiks*, 1712, Query 18)

The first systematic experiments distinguishing the different properties of light, radiant heat and heat convection were described by C.W. Scheele (1777), who also introduced the term 'radiant heat'. Other terms used were 'invisible heat', 'obscure heat', 'free heat' or 'free fire'. In 1790 M-A. Pictet wrote that 'free fire is an invisible emanation which moves according to certain laws and with a certain velocity'. P. Prévost (1791) sustained that 'free radiant heat is a very rare fluid, the particles of which almost never collide with one another and do not disturb sensibly their mutual movements' (translated in Brace 1901, p. 5). J. Hutton (1794) suggested that it was an 'invisible light ... which can be reflected by metallic surfaces, and which has great power in exiting heat' (pp. 86–88). In 1800, W. Herschel discovered infrared radiation in the solar spectrum. After 1800, the existence of invisible radiant heat was well established and clearly distinguished from heat conduction, but the debate about the nature of radiant heat would continue for decades, in particular its relation with light, whether they have an identical or a distinct nature.

Later, research concentrated on the search for the laws of thermal radiation. The experiments of Delaroche (1812), which are interesting for teaching, showed that there is not a linear dependence on the temperature difference ('The quantity of heat which a hot body yields in a given time by radiation to a cold body situated at a distance, increases, *caeteris paribus*, in a greater ratio than the excess of temperature of the first body above the second'), so opening the search for a correct relationship. Biot (1816) proposed a mathematical formula containing a cubic term, and Dulong and Petit (1817) found an exponential formula $F(T) \propto a^T + const$. It is important to stress that it was impossible to find a correct law for thermal radiation without using an opportune temperature scale with an absolute zero, like that introduced by Kelvin in 1848. Only in 1879 did Josef Stefan, starting from the experimental measurements of Tyndall, propose the famous empirical relationship asserting that the total radiant energy emitted by a black body per unit time is proportional to the fourth power of the absolute temperature of the body.

Stefan's formula was not immediately accepted by the scientific community, until Ludwig Boltzmann derived it theoretically in two articles published in 1884, based on the previous works of Bartoli. The theoretical reasoning and the thought experiments of Bartoli and Boltzmann constituted a meeting point between thermodynamics, electromagnetism and thermal phenomena, where thermal radiation was treated as a gas to which thermodynamic transformations could be applied with changes of pressure and volume. This raised questions about the extent to which this treatment is lawful and what it means from the point of view of thermodynamics and of the nature of thermal radiation and from the point of view of epistemological reductionism.

This case history is interesting, because it highlights the difficult and tortuous process of discovery, delimitation and differentiation of a new phenomenon, in this case the differentiation of thermal radiation from heat conduction and the understanding that it is a phenomenon of the same nature as light, an apparently very different phenomenon. Moreover, it can show how knowledge in physics does not arise from the simple observation of phenomena as they appear, the ordering and classification of data and simplified descriptions. Rather, it implies a conceptual reconstruction of complex experimental fields, the definition of structures, properties and mechanisms producing and explaining phenomena and allowing previsions and hypotheses about new phenomena.

9.5.3 The Caloric and Frigorific Rays in Thermal Radiation

The ideas of frigorific rays and of reflection and focusing of cold were put forward by G. Della Porta (1589) and confirmed by the *Accademia del Cimento* of Florence in Italy (1667). In contrast, Marc-Auguste Pictet, Pierre Prévost and other proponents of the material nature of heat interpreted these phenomena as a result of a peculiar arrangement of caloric transmission. An interesting controversy took place between Rumford and Prévost about the existence of cold radiation and the nature of heat (see Chang 2002). Rumford considered the existence of frigorific radiation as a strong argument against the caloric theory. He considered that both calorific and frigorific radiations exist in the sense that the radiation emitted by a body will have a heating effect on a colder body and a cooling effect on a warmer body. He considered this latter as a real effect, due to the undulations of lower

Fig. 9.5 The Rumford's thermoscope (Thomson-Rumford 1804, p. 30). "A small *bubble* of the spirit of wine … is now made to pass out of the short tube into the long connecting tube; and the operation is so managed that this bubble (which is about ¾ of an inch in length) remains stationary, at or near the middle of the horizontal part of the tube, *when the temperature (and consequently the elasticity) of the air in the two balls, at the two extremities of the tube, is precisely the same.*" (p. 48). The heating or cooling of only one of the balls forces the alcohol bubble "to move out of its place and to take its station nearer to the colder ball" (p. 49). "The result of the foregoing experiment appeared to me to afford the most indisputable proof of the radiation of cold bodies, and that the rays which proceed from them have a power of generating cold in warmer bodies which are exposed to their influence" (p. 61)

frequency, which produce a deceleration of the particle's vibration of the warmer body that they invest:

> The result of the foregoing experiment appeared to me to afford the most indisputable proof of the radiation of cold bodies, and that the rays which proceed from them have a power of generating cold in warmer bodies which are exposed to their influence … I have discovered, first, that all bodies at all temperatures (cold bodies as well as warm ones) emit continually from their surfaces rays, or rather, as I believe, *undulations*, similar to the undulations which sonorous bodies send out into the air in all directions, and that these rays or undulations influence and change, little by little, the temperature of all bodies upon which they fall without being reflected, in case the bodies upon which they fall are either warmer or colder than the body from the surface of which the rays or undulations proceed. … Those bodies which, when warm, give off many calorific rays would, when colder than the surrounding objects, give off to them many frigorific rays … the frigorific influences of cold bodies have always appeared as real and effective as the calorific influences of warm bodies. (Thomson-Rumford 1804, pp. 61 and 178–179)

Rumford described experiments performed using a special instrument, the *thermoscope* (see Fig. 9.5), which are interesting for teaching purposes because they are simple in their empirical description but can be differently interpreted, and show that it is not obvious to contrast the interpretation based on the idea of frigorific rays, which we currently consider erroneous. Moreover, these experiments can be a useful example for teaching some themes on the nature of science (NOS), as they stimulate a debate among students with different possible interpretations and raise the issue of the possibility of non-conclusive results of experiments regarding a choice between different theories.

9.5.4 The Discovery of the Second Law of Thermodynamics and the Invention of Entropy

This historical theme is useful to clarify the various meanings of the second law, by following and exploring the statements of Kelvin, Clausius and others, the relationships with energy conservation and Carnot's theory and the development of entropy interpretations (Clausius, Boltzmann, Gibbs and the informational entropy). An example of direct utilisation in teaching of excerpts from Carnot, Clausius and Boltzmann is given by Viard (2005) (see Sect. 9.3).

As previously pointed out in Sect. 9.4.1, there was difference between Kelvin's and Clausius' approaches. Kelvin referred to useful work, energy waste or dissipation and degradation of energy:

> There is an absolute waste of mechanical energy available to man when heat is allowed to pass from one body to another at a lower temperature ... As it is most certain that Creative Power alone can either call into existence or annihilate mechanical energy, the 'waste' referred to cannot be annihilation, but must be some transformation of energy ... The following propositions are laid down regarding the *dissipation* of mechanical energy from a given store, and the *restoration* of it to its primitive condition. They are necessary consequences of the axiom, '*It is impossible, by means of inanimate material agency, to derive mechanical effect from any portion of matter by cooling it below the temperature of the coldest of the surrounding objects*'. (Thomson-Kelvin 1852, p. 305)

Clausius' formulation of the second law evolved from a measure of the equivalence value of transformations and the law of nonnegative value of a cycle (1854) towards the concept of *disgregation* and the nonnegative value of disgregation for all transformations (1862, 1867, 'introducing a new magnitude, which we call the *disgregation* of the body, and by help of which we can define the effect of heat as simply *tending to increase the disgregation*') and eventually to the concept of entropy (1865, 1867, 'we might call S the transformational content of the body ... I propose to call the magnitude S the *entropy* of the body, from the Greek word tropè, *transformation*'). This led to the two very general sentences: 'The energy of universe remains constant' and 'The entropy of the universe tends to a maximum'. If the universe reached the state of maximum entropy, all energy would be uniformly diffused throughout space at a uniform temperature. Consequently, no mechanical work could be done, no transformations could happen and life would cease to exist; it is the 'heat death'.

The idea of useful work was resumed by Gibbs in a more rigorous manner, by defining the quantities available work and *free energy* and connecting them with entropy. Later on, similar ideas were developed in a new way, in the context of the problems of energy saving and environmental issues. A new quantity *exergy* was defined (the term was coined by Z. Rant in 1956), which is dimensionally homogeneous to energy but it is not conserved and equals the maximum work that can be provided by a system as it proceeds to its final state in thermodynamic equilibrium with the environment.

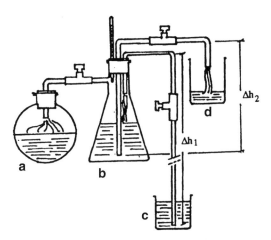

Fig. 9.6 A didactical experiment reproducing the essential features of Savery's steam engine (From De Filippo and Mayer 1991, p. 121). A is the boiler, B the cylinder, C is the well from which the water must be raised (the mine, in the original historical utilization), and D is the upper water tank

9.5.5 The History of Steam Engines

The history of steam engines, from precursors such as Branca, Desaguliers and Papin to the first useful practical realisations of Savery, Newcomen and Watt, is a case which allows an exploration of the problem of relationships between science, technology and society (see Cardwell 1971). The connections can be usefully demonstrated with the general background of the development of industry and the industrial revolution in England and the practical and productive problems that the new engines aimed to solve (a didactic presentation of this history is given in the *Project Physics Course* 1970 and in Brenni et al. 2009; see Sect. 9.3). More specifically, the connections can be studied by looking at the improvement of water pumps, the engines' efficiency and Watt's innovations. The introduction of a separate cold condenser allowed dramatic improvements in power and efficiency, and Watt's centrifugal governor, the first automatic feedback control device, precursor of others, represented a significant advancement in technology since the feedback loop allowed the steam engine to be self-regulating.

A didactic experiment reproducing the essential features of Savery's engine (1699) can be proposed (see Baracca and Besson 1990; De Filippo and Mayer 1991 and Fig. 9.6). The study can usefully prosecute with the introduction of the more recent concept of the *second-order efficiency* or *exergetic efficiency* of a thermodynamic process, defined as the ratio between the desired exergy output and the exergy input used (Viglietta 1990). In the context of the education to sustainable development, this concept is a useful instrument to evaluate the appropriate use of energy sources and the environmental effects of technological devices.

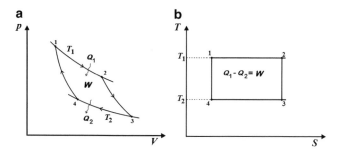

Fig. 9.7 The Carnot cycle represented (**a**) in a pressure-volume graph and (**b**) in a temperature-entropy graph

9.5.6 The Carnot Cycle, Caloric and Entropy

The Carnot cycle is often badly understood by students as it appears to be a strange object in the study of thermodynamics. Students may be confused about why there is just a cycle with an isotherm and an adiabatic process (Fig. 9.7a), this last being more complicated to deal with mathematically. They may also wonder whether the obtained conclusions are valid only for this strange type of cycle or for any cycle. The cycle diagram is difficult to understand for students and the choice of adiabatic and isotherm transformations remains obscure.

In fact, the Carnot cycle (1824), subsequently resumed by Clapeyron (1834), was born in the context of a theory of heat as a fluid, the caloric, which is conserved during the operation of a thermal engine. It makes clearer sense with reference to this theory, as a coupling of transformations keeping constant one of the two fundamental quantities, heat or temperature. The work would be produced by the passage of a given amount of heat from a higher temperature to a lower temperature and is a by-product of the movement of heat, in an analogy with hydraulic engines:

> The motive power of a waterfall depends on its height and on the quantity of the liquid; the motive power of heat depends also on the quantity of caloric used, and on what may be termed, on what in fact we will call, the height of its fall, that is to say, the difference of temperature of the bodies between which the exchange of caloric is made. (Carnot 1824–1897, p. 61)

The Carnot cycle makes sense inside a history that has its precedent in the study of hydraulic engines and their efficiency (Lazare Carnot and John Smeaton). It was with reference to them that Sadi Carnot studied the problem of the efficiency of thermal engines by using an analogy between caloric fluid and water. Subsequently, the problem arose of reconciling Carnot's results with the heat-work equivalence, demonstrated by Joule and others, an issue which preoccupied Kelvin. Is it the falling of caloric to a lower temperature that produces work or is a part of heat transformed in work, respecting energy conservation? The theory of heat was changed; heat was no longer considered to be a substance which is conserved, but the importance of the Carnot cycle remained, despite its change in meaning. It became a start point for discovering the second law of thermodynamics and also for the definition of Kelvin's absolute temperature.

In fact, the Carnot cycle remained for decades at the middle of the conceptual development of thermodynamics as a thread, a reference and a problem and also perhaps as an obstacle, which, because of the generality and the theoretical strength of its conclusions, may have withheld thermodynamics for a long time around issues and discourses related to reversible cycles and thermal engines.

In this way an analysis of the cycle from a historical point of view, starting from the original interpretation of Carnot and Clapeyron in the framework of the caloric theory and prosecuted with a revised and modified treatment, can be useful in encouraging a better and deeper understanding and can activate a debate on the meaning of heat and reversibility (see Newburgh 2009).

Carnot introduced the idea of the maximum theoretical efficiency of a thermal engine, depending only on the considered temperatures. The work obtained is proportional to the heat taken and to the temperature difference. In the hydraulic engines the work is also proportional to the water mass and to the height difference. However, differently from hydraulic engines, in a thermal engine the work depends also on the absolute value of temperatures, not only on the temperature difference, and increases for lower temperatures. ('The fall of caloric produces more motive power at inferior than at superior temperatures', Carnot 1824–1897, p. 97). According to Carnot, this fundamental property is strictly linked with the hypothesis, supported by the experiments of F. Delaroche and J. E. Bérard, that the specific heat of a gas increases with volume (p. 40), a property that was explained by using the difference between free and combined heat.

Consequently, in the theory of Carnot-Clapeyron, the obtained work W is proportional to heat Q and to temperature difference $\Delta T = T_1 - T_2$, $W = C \cdot Q \cdot \Delta T$, but it depends also on T_2 and it increases with diminishing T_2. The proportionality coefficient C (called Carnot function) can be taken as a measure of the inverse of the temperature $1/T$ ('we may define temperature simply as the reciprocal of Carnot function', Joule and Thomson 1854, p. 351) and then $W = Q \cdot \Delta T/T_2$. Leaving the caloric theory and passing to the modern conception, for the energy conservation it is $Q_1 \neq Q_2$ and $W = Q_1 - Q_2$ and therefore the proportionality is obtained $Q_1/T_1 = Q_2/T_2$. It is interesting that Carnot referred to an absolute zero temperature that he assumed to be $-267\,°C$, on the basis of data on the gases behaviour, so he always wrote in his formulas $(t + 267)$, t being the Celsius temperature.

Carnot did not draw graphs, whilst Clapeyron drew the well-known volume-pressure diagrams (Fig. 9.7a). In the framework of the caloric theory, in a heat-temperature diagram the graph of the cycle would be simply a rectangle. This type of graph is not acceptable in the current theory according to which heat represents transferred energy and is not a state function. Nevertheless, by modifying the quantity on the abscissa from Q to Q/T, the diagram becomes correct and meaningful from the modern point of view: the obtained graph is still a rectangle (Fig. 9.7b) and the axis of the abscissa now indicates a new state quantity, which is entropy S. Furthermore, being $Q = T\Delta S$ for reversible processes, the area of the rectangle is equal to $Q_1 - Q_2$ and therefore is equal to the obtained work W.

9.5.7 The Cooling Law and the Definition of a Temperature Scale: From Newton to Dalton

Research on the cooling law of objects in a colder environment began with Newton's article published in 1701. Later, numerous studies were conducted by other scientists, confirming or confuting Newton's law (Besson 2012). These studies were connected with the problem of defining a good scale of temperatures, a connection which was dealt with in Newton's article and in Dalton's work published in 1808.

This historical subject is interesting for its epistemological implications and its possible utilisation in teaching. It involves concepts and phenomena that are usually covered in normal courses but about which there exist many difficulties and erroneous conceptions. The cooling law uses mathematical tools that can be easily understood by students, and simple experiments suitable for a school replication (e.g. Metz and Stinner (2006) developed teaching activities centred on the replication of some experiments of Rumford on cooling process; see Sect. 9.3). Moreover, it allows the treatment of various methodological issues, such as the relationship between experimental data and mathematical models, the role of philosophical ideas of scientists and the definition of quantities such as temperature that need a progressive construction from simple observations towards general properties and laws.

In 1701, Isaac Newton published a short article in which he provided a table of temperatures and established a relationship between the temperatures T and the time t in cooling processes. He did not write any formula but expressed his cooling law verbally: 'The excess of the degrees of the heat ... are in geometrical progression when the times are in an arithmetical progression'. By reading his text, it is hard to sustain that Newton was discovering experimentally an exponential law of cooling. He was mainly interested in defining a thermometric scale for high temperatures (*scala graduum caloris*). Newton did not consider the exponential law of cooling as an experimental result but as a general hypothesis, which allows a temperature scale to be built. He found the temperatures by using two different methods, one based on the property of thermal dilatation of linseed oil and the other based on the cooling time of a piece of iron. Newton considered his cooling law as a general property of heat.

Later, faced with the discrepancies between the cooling law and experiments, scientists assumed different attitudes. Some scientists (Martine, Erxleben, Delaroche, Biot, Dulong and Petit) assumed an empiricist attitude, concluding that Newton's cooling law was not entirely true, and looked for new different laws. Other scientists tried to keep the simple remarkable law, by considering the effect of disturbance factors (e.g. Richmann, Prévost, Leslie) or by revising the common temperature scales in order to reobtain the agreement between theory and experiments (e.g. Rumford, Dalton). Still in 1804, Leslie wrote:

> It is *assumed* as a *general principle*, that the decrements of heat are proportional to the difference of temperature of the conterminous surfaces. On this supposition, the successive temperatures of a substance exposed to cool, would, at equal periods, form a descending geometrical progression. (Leslie 1804, pp. 263–264)

John Dalton (1808) remarked that the exponential law of cooling was not exactly valid if the common [Fahrenheit] thermometric scale was used ('one remarkable trait of temperature derived from experiments on the heating and cooling of bodies, which does not accord with the received scale, and which, nevertheless, claims special consideration'). He then defined a new scale of temperatures, which would allow the law to remain valid together with three other simple physical laws concerning the thermal dilatation of liquids and of gases and the change with temperature of steam pressure. The agreement that he found or believed he had found (his experimental data were not so accurate and ample, especially at high temperatures) among four simple laws of four different phenomena was for him a strong clue that a fundamental property of heat had been found, connected to the new temperature scale.

This epistemological attitude is similar to that of Newton (in his *Principia* Book 1, Section 1, Scholium), when he distinguished between absolute, true, mathematical time and relative, apparent, common time. Similarly, a distinction can be conceived between the true, mathematical temperature, which responds to exact general laws, and a sensible, apparent temperature, measured by material devices. Scientists try to find the best realisation of the true temperature. In a sense, the current definition of thermodynamic temperature can be considered as a successful finding of this search. But for a time, many scientists thought too easily that they were finding exact general laws, confirming the human tendency of attributing more order and regularity to things than actually observed, a tendency that F. Bacon (1620) pointed out as one of *idola tribus* hindering the formation of correct knowledge.

Dulong and Petit (1817) criticised Dalton because 'he unduly hastened to generalize some outlines ... which were based only on dubious evaluations'. They performed a very accurate experimental study, considering a temperature scale based on the air thermometer, and distinguished cooling processes due to radiation and to convection, by measuring the cooling velocity of a body in a vacuum due to 'the excess of its own emitted radiation over that of the surrounding bodies'.

This *case history* offers the occasion for teaching various epistemological and methodological issues, as, for example:

- The relationship between experimental data and mathematical models, the problem of the field of validity of empirical laws and the interpretation of discrepancies between laws and experimental data, either as the effect of errors and disturbance factors or as a clue that the law is not valid
- The role of philosophical ideas and preferences of scientists in their scientific research, for example, the faith in the simplicity of natural laws or the confidence in the existence of a unique law and a sole cause for an empirical phenomenon
- The definition of quantities such as temperature, which cannot be defined simply by means of a sentence such as 'temperature is...' but need a progressive construction starting from thermal sensations towards the choice of thermometric substances and quantities, and the search for a universal property independent from a specific substance

9.5.8 The Construction of the Physical Quantity Temperature

As pointed out in the previous Sect. 9.5.7, temperature is a quantity which cannot be simply defined by a sentence or a formula but needs a progressive construction. The historical path of this scientific construction (see Chang 2004) can be adapted for building a case history which usefully parallels and accompanies the students' learning progression from subjective warm-cold sensations towards the specification of body properties, such as dilatation or pressure, which depend on being colder or warmer, and the successive generalisations leading to absolute thermometer scales that are independent of specific materials (examples of didactic use of this case are given by Mäntylä and Koponen (2007) and Oversby (2009); see Sect. 9.3).

Also the research about an absolute zero of temperature offers an interesting historical thread which does not demand very difficult mathematical and conceptual tools, and it is suitable for high school students. Already Amontons, at the beginning of the eighteenth century, proposed the idea of an absolute zero of temperature based on the gas law, i.e. the temperature at which gas pressure is zero, and following this definition Carnot calculated the absolute zero at $-267\ °C$. Dalton (1808) used different phenomena and reasoning (see Besson 2011) for finding the 'natural zero of temperature' which he considered as a state of 'absolute privation of heat':

> If we suppose a body at the ordinary temperature to contain a given quantity of heat, like a vessel contains a given quantity of water, it is plain that by abstracting successively small equal portions, the body would finally be exhausted of the fluid. (Dalton 1808, p. 82)

Based on his experiments and calculations, he concluded that it was correct 'to consider the natural zero of temperature as being about 6,000° below the temperature of freezing water, according to the divisions of Fahrenheit's scale' (p. 97). In contrast, by studying the cooling law in a vacuum due to radiation, Dulong and Petit (1817, p. 259) concluded that the absolute zero should be at infinite degrees below $0\ °C$. Moreover, the absolute thermodynamic temperature can be introduced as a result of Kelvin's reflections on Carnot's work:

> If any substance whatever, subjected to a perfectly reversible cycle of operations, takes in heat only in a locality kept at a uniform temperature, and emits heat only in another locality kept at a uniform temperature, the temperatures of these localities are proportional to the quantities of heat taken in or emitted at them in a complete cycle of operations. (Joule and Thomson 1854, p. 351)

9.6 Conclusion

The richness of the above-outlined examples of philosophical problems, case histories, multidisciplinary themes and issues involving technological, social and cultural contexts shows the numerous possibilities of the use of history and philosophy of science in teaching thermal phenomena and thermodynamics. As it has been shown, many research examples can be found in the research literature on this problem.

Matthews (1994, pp. 70–71) indicated three ways in which the history of science can be and has been included in science programmes: the add-on approach (minimalist, where units on the histories of science are added on to a standard nonhistorical science course), the integrated approach (maximalist, organising a whole science course on historical grounds) and the storyline approach (using history in order to create a storyline for the science content, where the subject matter is embedded in a historical matrix and history 'provides the framework onto which a science topic or whole course can be placed in a developing narrative').

The integrated, maximalist, approach needs strong multidisciplinary organisation and appropriate teacher training. Moreover, teachers may be unwilling to adopt a whole new course based on a historical approach because it can be too demanding for them. However, they may be more willing to insert new short units in their existing course. In this way, research could usefully focus on the development and testing of a number of such new teaching units, which include new complementary materials in the various forms considered above (narratives, experiments, historical dramas, controversies, cases, debates ...) and making a didactic transposition of historical and philosophical problems.

Concerning thermodynamics, new teaching units could be developed around the philosophical and historical themes briefly outlined in Sects. 9.4 and 9.5. Moreover, two particular problems merit a mention here. The first one concerns the difficulty of teaching the entropy concept at high school level, and indeed most teachers and handbooks avoid it or refer to it in a vague and imprecise way, often with inaccurate statements. A research problem is whether and how a historical approach could provide a learning path in which a discussion of different approaches that were developed by scientists in the past and the controversies that arose can foster a better understanding of physical concepts and a deeper awareness of the cultural context and implications related to this issue.

The second problem refers to the special connections of thermodynamics with historical, technological and economic problems, whose echoes are embedded in the core of the structure and language of thermodynamics, involving thermal engines, efficiency, work, etc. For example, it is hard to avoid any reference to the connection between the meanings of the word 'work' in the scientific language and in common language. A discussion of the historical origin of this apparently strange choice should be taken into account rather than disregarded, as it usually is in textbooks. A challenge is how to develop teaching courses and materials taking into account these aspects and giving meaning and historical background to these concepts and terminologies, without structuring a whole reorganisation of the course on historical grounds.

A more general problem seems to be how to fill the gap between didactic research and actual school practice. In this field, research products are too often limited to historical and/or philosophical analysis with some interesting but generic pedagogical suggestions for teaching. More studies are needed which have to be specifically designed, tested and evaluated for school classes, including materials for students and teacher guides. The role of teachers is decisive: on one hand, research proposals must take into account more clearly the teachers' needs; on the other hand, new programmes of teacher education should be designed and developed, and teachers should become able to find, adapt and utilise resources produced by science education research. These problems also demand research and projects concerning university science courses, which are too focused on technical and algorithmic knowledge and skills, and disregard conceptual, problematic and cultural aspects needed not only for future teachers but also for general and professional faculties.

Moreover, for studies in this field to produce interesting cognitive results and significant practical effects on actual school teaching, collaboration is needed in a research team among researchers in history and/or philosophy of science, researchers in science education and school teachers.

Acknowledgements I wish to thank Robyn Yucel from Latrobe University, Australia, who did the copyediting of the manuscript.

References

Arnold, M. & Millar, R. (1996). Learning the Scientific Story: A Case Study in the Teaching and Learning of Elementary Thermodynamics, *Science Education*. 80(3), 249–281.

Bacon, F. (1620). *Novum Organum, sive indicia vera de interpretatione naturae*. Edited by T. Fowler (1889) Oxford, Clarendon Press. English translation by G.W. Kitchin (1855) Oxford, University Press.

Baracca, A. & Besson, U. (1990). *Introduzione storica al concetto di energia*. Firenze: Le Monnier.

Begoray, D.L. & Stinner, A. (2005). Representing Science through Historical Drama. *Science & Education*, 14, 457–471.

Bénard, H. (1900). Étude expérimentale du mouvement des liquides propageant de la chaleur par convection. Régime permanente, tourbillons circulaires. *Comptes Rendus de l'Académie des Sciences*, 130, 1004–1007.

Besson, U. (2001) Work and Energy in the Presence of Friction: The Need for a Mesoscopic Analysis. *European Journal of Physics,* 22, 613–622.

Besson, U. (2003). The distinction between heat and work: an approach based on a classical mechanical model. *European Journal of Physics*, 24, 245–252.

Besson, U. (2010). Calculating and Understanding: Formal Models and Causal Explanations in Science, Common Reasoning and Physics Teaching. *Science & Education*, 19(3), 225–257.

Besson, U. (2011). The cooling law and the search for a good temperature scale, from Newton to Dalton. *European Journal of Physics*, 32(2), 343–354.

Besson, U. (2012). The history of cooling law: when the search for simplicity can be an obstacle. *Science & Education* 21(8), 1085–1110.

Besson, U. (2013). Historical Scientific Models and Theories as Resources for Learning and Teaching: The Case of Friction. *Science & Education* 22(5), 1001–1042.

Besson, U., De Ambrosis., A. and Mascheretti, P. (2010). Studying the physical basis of global warming: thermal effects of the interaction between radiation and matter and greenhouse effect. *European Journal of Physics*, 31, 375–388.

Besson, U. & De Ambrosis A. (2013). Teaching Energy Concepts by Working on Themes of Cultural and Environmental Value. *Science & Education*, DOI: 10.1007/s11191-013-9592-7.

Biot M. (1816) Sur la loi de Newton relative à la communication de la chaleur. *Bulletin de la Société Philomatique*, 21–24.

Black, J. (1803). *Lectures on the Elements of Chemistry*. Edinburgh: John Robison.

Boltzmann, L. (1872). Weitere Studien über das Wärmegleichgewicht unter Gasmolekülen. *Sitzungsberichte der Kaiserlichen Akademie der Wissenschaften*, Wien, Part II, 66, 275–370. English translation in Brush 2003, pp. 262–349.

9 Teaching About Thermal Phenomena and Thermodynamics...

Boltzmann, L. (1897). Zu Hrn. Zermelo's Abhandlung "Über die mechanische Erklärung irreversibler Vorgänge". *Annalen der Physik* 296(2), 392–398. English translation in Brush (2003), pp. 412–419.

Brace, D.B. (1901). *The Laws of Radiation and Absorption*. New York: American Book Company.

Brenni, P, Giatti, A. & Barbacci, S. (2009). Steam, Work, Energy. http://hipstwiki.wetpaint.com/page/Case+Study+2

Brush, S.G. (1976). *The kind of motion we call heat*. Amsterdam: North-Holland Publishing Company.

Brush, S.G. (2003). *The Kinetic Theory of Gases: An Anthology of Classic Papers with Historical Commentary*. London: Imperial College Press.

Callender, C. (1999). Reducing Thermodynamics to Statistics Mechanics: The Case of Entropy. *The Journal of Philosophy,* 96(7), 348–373.

Campanaro, J.M. (2002). The parallelism between scientists' and students' resistance to new scientific ideas. *International Journal of Science Education,* 24(10), 1095–1110.

Cardwell, D.S.L. (1971). *From Watt to Clausius: The Rise of Thermodynamics in the Early Industrial Age*. London: Heinemann.

Carnot, S. (1824–1897). *Réflexions sur la puissance motrice du feu*. Paris: Mallet-Bachelier. English translation: (1897) New York: John Wiley & Sons.

Carey, S. (1988). Reorganisation of Knowledge in the Course of Acquisition. In S. Strauss (Ed.) *Ontogeny, Phylogeny and Historical Development* (pp. 1–27). Norwood: Ablex.

Cassidy, D., Holton, G. & Rutherford, J. (2002). *Understanding Physics*. New York: Springer Verlag.

Chang, H. (2002). Rumford and the reflection of radiant cold: Historical reflections and meta-physical reflexes. *Physics in Perspective,* 4, 127–169.

Chang, H. (2004). *Inventing Temperature: Measurement and Scientific Progress*. Oxford: Oxford University Press.

Chang, H. (2011). How Historical Experiments Can Improve Scientific Knowledge and Science Education: The Cases of Boiling Water and Electrochemistry. *Science & Education,* 20, 317–341.

Chiou, Guo-Li & Anderson, O.R. (2010). A study of undergraduate physics students' understanding of heat conduction based on mental model theory and an ontology–process analysis. *Science Education,* 94(5), 825–854.

Clapeyron, E. (1834). Mémoire sur la Puissance motrice de la chaleur. *Journal de l'École Royale Polytechnique,* Tome XIV, cahier 23, 153–190.

Clausius, R. (1854). Über eine veränderte Form des zweiten Hauptsatzes der mechanischen Wärmetheorie. *Annalen der Physik* 169(12), 481–506. English translation in Clausius (1867), pp. 111–135.

Clausius, R. (1862). Über die Anwendung des Satzes von der Äquivalenz der Verwandlungen auf die innere Arbeit. *Annalen der Physik* 192(5), 73–112. English translation in Clausius (1867), pp. 215–250.

Clausius, R. (1865). Über verschiedene für die Anwendung bequeme Formen der Hauptgleichungen der mechanischen Wärmetheorie. *Annalen der Physik* 201(7), 353–400. English translation in Clausius (1867), pp. 327–365.

Clausius, R. (1867). *The Mechanical Theory of Heat - with its Applications to the Steam Engine and to Physical Properties of Bodies*. London: Archer Hirst.

Clough, E. & Driver, R. (1985). Secondary students' conceptions of the conduction of heat: Bringing together scientific and personal views. *Physics Education,* 20(4), 176–182.

Cochran, M.J. & Heron, P.R.L. (2006). Development and assessment of research-based tutorials on heat engines and the second law of thermodynamics. *American Journal of Physics,* 74, 734–741.

Collier, J.D. (1990). Two Faces of Maxwell's Demon Reveal the Nature of Irreversibility. *Studies in the History and Philosophy of Science,* 21, 257–268.

Conant, J.B. (1957). *Harvard Case Histories in Experimental Science*. Cambridge MA: Harvard University Press.

Cornell, E.S. (1936). Early studies in radiant heat. *Annals of Science,* 1(2), 217–225.

Cotignola, M.I., Bordogna, C., Punte, G. & Cappannini, O.M. (2002). Difficulties in Learning Thermodynamic Concepts: Are They Linked to the Historical Development of this Field? *Science & Education,* 11, 279–291.

Dalton, J. (1808). *A new system of chemical philosophy*. Manchester, vol. 1, pp. 1–140.

Daub, E. (1970). Maxwell's Demon. *Studies in History and Philosophy of Science*, 1, 213–227.

Davy, H. (1799). An essay on heat, light, and the combinations of light. In J. Davy (Ed.) (1839) *The collected works of H. Davy* (Vol. 2, pp. 5–86). London: Smith, Elder, and Co. Cornhill.

de Berg, K. C. (2008). The Concepts of Heat and Temperature: The Problem of Determining the Content for the Construction of an Historical Case Study which is Sensitive to Nature of Science Issues and Teaching-Learning Issues. *Science & Education*, 17, 75–114.

De Filippo, G. & Mayer, M. (1991). Riflessioni sul rapporto scienza – tecnologia: le macchine termiche e la termodinamica. In C. Tarsitani & M. Vicentini (Eds), *Calore, energia, entropia* (pp. 95–128). Milano: Franco Angeli.

Delaroche, F. (1812). Observations sur le calorique rayonnant. *Journal de Physique*, XXV, 201–228.

Doige, C.A. & Day, T. (2012). A Typology of Undergraduate Textbook Definitions of 'Heat' across Science Disciplines. *International Journal of Science Education*, 34(5), 677–700.

Duhem, P. (1908). Σωζειν τα Φαινομενα, *Essai sur la notion de théorie physique de Platon à Galilée*. Paris: Hermann. Reprint: 1990, Paris: Vrin.

Dulong, P. & Petit, A. (1817). Recherches sur les Mesures des Température set sur les Lois de la communication de la chaleur. *Annales de Chimie et de Physique* (Paris: Clochard), VII, 113–154, 225–264, 337–367.

Erickson, G. L. (1979). Children's conceptions of heat and temperature. *Science Education*, 63(2), 221–230.

Erickson, G. L. (1980). Children's viewpoints of heat: A second look. *Science Education*, 64(3), 323–336.

Erickson, G. & Tiberghien, A. (1985). Heat and temperature. In R. Driver, E. Guesne & A. Tiberghien (Eds), *Children's ideas in science* (pp. 52–84). Philadelphia: Open University Press.

Fourier, J. (1822). *Théorie Analytique de la Chaleur*. Paris: Firmin Didot Père et Fils.

Haglund, J., Jeppsson, F. & Strömdahl, H. (2010). Different senses of entropy - Implications for education. *Entropy*, 12(3), 490–515.

Holton, G. (2003a). What Historians of Science and Science Educators Can Do for One Another? *Science & Education*, 12, 603–616.

Holton, G. (2003b). The Project Physics Course, Then and Now. *Science & Education*, 12, 779–786.

Hutton, J. (1794). *A Dissertation upon the Philosophy of Light, Heat and Fire*. Edinburgh & London: Cadell and Davis.

Jasien, P.G. & Oberem, G.E. (2002). Understanding of elementary concepts in heat and temperature among college students and K-12 teachers. *Journal of Chemical Education*, 79(7), 889–895.

Jaynes, E. T. (1957). Information theory and statistical mechanics. *Physical Review*, 106(4), 620–630.

Joule, J.P. & Thomson, W. (1854). On the Thermal Effects of Fluids in Motion, Part 2. *Philosophical Transactions of the Royal Society*, 144, 321–364.

Kesidou, S. & Duit, R. (1993). Students' conceptions of the second law of thermodynamics: An interpretive study. *Journal of Research in Science Teaching*, 30(1), 85–106.

Lambert, F.L. (2002). Disorder - A cracked crutch for supporting entropy discussions. *Journal of Chemical Education*, 79, 187–192.

Landauer, R. (1961). Irreversibility and heat generation in the computing process. *IBM Journal of Research and Development*, 5(3), 183–191.

Lavoisier, A. (1789). *Traité élémentaire de chimie*. Paris: Cuchet.

Lavoisier, A. & Laplace, P.S. (1780). Mémoire sur la chaleur, *Mémoires de l'Académie Royale des sciences* (pp. 355–408). Paris: Imprimerie Royale.

Lebowitz, J.L. (1999). Microscopic Origins of Irreversible Macroscopic Behavior. *Physica A*, 263, 516–527.

Leinonen, R., Rasanen, E., Asikainen, M. & Hirvonen, P. (2009). Students' pre-knowledge as a guideline in the teaching of introductory thermal physics at university. *European Journal of Physics*, 30, 593–604.

Leite, L. (1999). Heat and temperature: An analysis of how these concepts are dealt with in textbooks. *European Journal of Teacher Education*, 22(1), 75–88.

Leslie, J. (1804). *An Experimental Inquiry into the Nature and Propagation of Heat*. London: J. Mawman.

Lewis, E. L. & Linn, M. C. (1994). Heat and temperature concepts of adolescents, adults, and experts: Implications for curriculum improvements. *Journal of Research in Science Teaching*, 31(6), 657–677.

Loverude, M.E., Kautz, C.H. & Heron, P.R.L. (2002). Student understanding of the first law of thermodynamics: Relating work to the adiabatic compression of an ideal gas. *American Journal of Physics*, 70, 137–148.

Mäntylä, T. & Koponen, I.T. (2007). Understanding the Role of Measurements in Creating Physical Quantities: A Case Study of Learning to Quantify Temperature in Physics Teacher Education. *Science & Education*, 16, 291–311.

Matthews, M.R. (1994). *Science Teaching. The Role of History and Philosophy of Science*. New York – London: Routledge.

Maxwell, J.C. (1871). *Theory of heat*. London: Longmans, Green & Co, 10th edition 1891.

Maxwell, J.C. (1875). On the dynamical evidence of the molecular constitution of bodies. *Nature* 4, 357–359, 374–377.

Meltzer, D. (2004). Investigation of Students' Reasoning Regarding Heat, Work, and the First Law of Thermodynamics in an Introductory Calculus-based General Physics Course. *American Journal of Physics*, 73, 1432–1446.

Metz, D. & Stinner, A. (2006). A Role for Historical Experiments: Capturing the Spirit of the Itinerant Lecturers of the 18th Century. *Science & Education*, doi: 10.1007/s11191-006-9016-z.

National Science Education Standards (1996). National Academy Press. http://www.nap.edu/catalog/4962.html.

Nersessian, N. J. (1995). Opening the black box: Cognitive science and the history of science. *Osiris*, 10(1), 194–211.

Newburgh, R. (2009). Carnot to Clausius: caloric to entropy. *European Journal of Physics*, 30, 713–728.

Newton, I. (1701). Scala graduum caloris, Calorum Descriptiones & Signa (Scale of the Degrees of Heat). *Philosophical Transactions*, 22(270), 824–829. English translation in: Newton I. (1809) *Philosophical Transactions of the Royal Society of London, Abridged,* 4, 572–575.

Oversby, J. (2009). Temperature – what can we find out when we measure it? http://hipst.eled.auth.gr/hipst_docs/instruments.pdf

Piaget, J. & Garcia, R. (1983). *Psychogenèse et histoire des sciences*. Paris: Flammarion.

Planck, M. (1882). Verdampfen, Schmelzen und Sublimiren. *Annalen der Physik* 251(3), 446–475.

Planck, M. (1897). *Vorlesungen über Thermodynamik*. Leipzig: Verlag von Veit & Comp.

Poisson, S.D. (1835). *Théorie Mathématique de la Chaleur*. Paris: Bachelier.

Prigogine, I. & Stengers, I. (1988). *Entre le temps et l'éternité*. Paris: Fayard.

Project Physics Course (1970). New York: Holt, Rinehart & Winston.

Reif, F. (1965). *Berkeley Physics Course*, Vol. 5, *Statistical Physics*. New York: McGraw-Hill.

Reif, F. (1999). Thermal physics in the introductory physics course: Why and how to teach it from a unified atomic perspective. *American Journal of Physics*, 67, 1051–1062.

Rozier, S. & Viennot, L. (1991). Students' reasoning in thermodynamics. *International Journal of Science Education*, 13, 159–170.

Sciarretta, M.R., Stilli, R. & Vicentini, M. (1990). On the thermal properties of materials: common-sense knowledge of Italian students and teachers. *International Journal of Science Education*, 12(4), 369–379.

Seroglou, F. & Koumaras, P. (2001). The contribution of the history of physics in physics education: A review. *Science & Education*, 10(1), 153–172.

Shannon, C. E. (1948). A Mathematical Theory of Communication. *Bell System Technical Journal,* 27, 379–423 and 623–656.

Shayer, M. & Wylam, H. (1981). The development of the concepts of heat and temperature in 10–13 years-olds. *Journal of Research in Science Teaching*, 5, 419–434.

Stavy, R. & Berkovitz, B. (1980). Cognitive conflict as a basis for teaching quantitative aspects of the concept of temperature. *Science Education*, 64(5), 679–692.

Stengers, I. (1997). *Thermodynamique : la réalité physique en crise*. Paris: La Découverte.

Styer, D.F. (2000). Insight into entropy. *American Journal of Physics*, 68(12), 1090–1096.

Stinner, A. (1995). Contextual Settings, Science Stories, and Large Context Problems: Toward a More Humanistic Science Education. *Science Education*, 79(5), 555–581.

Stinner, A. & Teichmann, J. (2003). Lord Kelvin and the Age-of-the-Earth Debate: A Dramatization. *Science & Education*, 12, 213–228.

Stinner, A., McMillan, B., Metz, D., Jilek, J. & Klassen, S. (2003). The Renewal of case studies in Science education. *Science & Education*, 12, 617–643.

Teixeira, E.S., Greca, I.M. & Freire, O. (2012). The History and Philosophy of Science in Physics Teaching: A Research Synthesis of Didactic Interventions. *Science & Education*, 21(6), 771–796.

Tarsitani, C. & Vicentini, M. (1996). Scientific Mental Representations of Thermodynamics. *Science & Education*, 5(1), 51–68.

Thomson, B. (Rumford) (1804). An enquiry concerning the nature of heat and the mode of its communication. *Philosophical Transactions of the Royal Society*, 94, 77–182.

Thomson, W. (Kelvin) (1851). On the Dynamical Theory of Heat. *Transactions of the Royal Society of Edinburgh*, March 1851, 174–210. *Philosophical Magazine,* 4(22, 23, 24, 27) 8–21, 105–117, 168–176, 424–434.

Thomson, W. (Kelvin) (1852). On a Universal Tendency in Nature to the Dissipation of Mechanical Energy. *Philosophical Magazine*, 4(25) 304–306.

Thomson, W. (Kelvin) (1874). The Kinetic Theory of the Dissipation of Energy. *Proceedings of the Royal Society of Edinburgh*, 8, 325–334.

Viard, J. (2005). Using the history of science to teach thermodynamics at the university level: The case of the concept of entropy. *Eighth International History, Philosophy, Sociology & Science Teaching Conference*, http://www.ihpst2005.leeds.ac.uk/papers/Viard.pdf

Viglietta, L. (1990). Efficiency in the teaching of Energy. *Physics Education*, 25, 317–321.

Wang, H.A. & Marsh, D.D. (2002). Science Instruction with a Humanistic Twist. *Science & Education*, 11, 169–189.

Welch, W.W. (1973). Review of the research and evaluation program of Harvard Project Physics. *Journal of Research in Science Teaching*, 10(4), 365–378.

Wiebe, R. & Stinner, A. (2010). Using Story to Help Student Understanding of Gas Behavior. *Interchange*, 41 (4), 347–361.

Wiser, M. & Amin, T. (2001). "Is heat hot?" Inducing conceptual change by integrating everyday and scientific perspectives on thermal phenomena. *Learning and Instruction*, 11(4), 331–355.

Wiser, M. & Carey, S. (1983). When heat and temperature were one. In D. Gentner & A. Stevens (Eds.), *Mental models* (pp. 267–297), Hillsdale: NJ: Erlbaum.

Wicken, J.S. (1981). Causal Explanations in Classical and Statistical Thermodynamics. *Philosophy of Science*, 48(1), 65–77.

Zambrano, A.C. (2005). A curricular sequence based on a historical study of conceptual change in science. *Eighth International History, Philosophy, Sociology & Science Teaching Conference*, http://www.ihpst2005.leeds.ac.uk/papers/Zambrano.pdf

Zemanski M.W. (1968). *Heat and Thermodynamics*. New York: McGraw-Hill.

Ugo Besson is researcher in Didactics and History of Physics at the University of Pavia (Italy). He graduated in Mathematics and in Physics at the University of Rome (Italy), and he obtained his Ph.D. in Didactics of Physics (Didactique des Sciences Physiques) at the University of Paris 'Denis Diderot' (Paris 7). He taught Mathematics and Physics in Italian and European high schools and subsequently Didactics of Physics at the University of Pavia and in the graduate schools of teacher education. His research interests include student conceptions, conceptual analysis of physics topics, design and experimentation of teaching-learning sequences, history of physics, causality and explanations in common reasoning and in physics teaching, models and analogies in science education, the use of historical resources in physics teaching and teacher education. Results of his research are published in international and national journals such as *International Journal of Science Education, Science & Education, European Journal of Physics, American Journal of Physics, Physics Education, Le BUP (Bulletin de l'Union des Physiciens), Giornale di Fisica* and *La Fisica nella Scuola*. Some his studies are also described in his book *Apprendre et enseigner la statique des fluides* (Editions Universitaires Européennes, 2010) and in a chapter of the collective books *Teaching strategies* (Nova Science Publishers, 2011) and *Didactique, épistémologie et histoire des sciences* (Paris, PUF, 2008). He participated in several conferences of the international associations ESERA (European Science Education Research Association) and GIREP (International Group for the Research on Physics Teaching) with communications published in the conference books.

Part II
Pedagogical Studies: Chemistry

Chapter 10
Philosophy of Chemistry in Chemical Education: Recent Trends and Future Directions

Sibel Erduran and Ebru Z. Mugaloglu

10.1 Introduction

Traditionally, chemistry had minimal existence within philosophy of science as Good (1999) explains:

> One of the characteristics of chemists is, that most have no interest in the philosophy of science…The disinterest appears to work in both directions. Modern philosophers very seldom give even a passing mention to modern chemical issues (Michael Polanyi and Rom Harré are among the few exceptions I know of). Recently, a few philosophers have attempted to discuss 'scientific practice'; but generally they have not included *chemical* practice. It is as if philosophers have believed that the way physics is 'done' was the way that all science is, or should be, done. (Physicists, no doubt, are the source of this opinion.) (Good 1999, pp. 65–66)

The disinterest in the philosophical aspects of chemistry by philosophers, chemists and educators mirrors earlier observations about a similar lack of interest regarding history of chemistry:

> Chemists, compared with other scientists, have relatively little interest in the history of their own subject. This situation is reflected, and perpetuated, by the antihistorical character of most chemical education. (Stephen Brush quoted by Kauffman 1989, p. 81)

Since the mid-1990s however, there has been an upsurge of interest in the study of chemistry from a philosophical perspective.[1] An increasing number of books, journals, conferences and associations focused on the articulation of how chemistry

---

[1] See, for instance, Bhushan and Rosenfeld (2000), Chalmers (2010), Hendry (2012), Scerri (2008, 2000, 1997), Schummer (2006), van Brakel (2000), Weisberg (2006), and Woody (2004a, b, 2000), Woody et al. (2011).

S. Erduran (✉)
University of Limerick, Department of Education and Professional Studies, Ireland

Bogazici University, Department of Primary Education, Turkey
e-mail: Sibel.Erduran@ul.ie

E.Z. Mugaloglu
Department of Primary Education, Bogazici University, Istanbul, Turkey

M.R. Matthews (ed.), *International Handbook of Research in History, Philosophy and Science Teaching*, DOI 10.1007/978-94-007-7654-8_10,
© Springer Science+Business Media Dordrecht 2014

could be understood from a philosophical perspective (McIntyre and Scerri 1997; Scerri 1997; Stanford Dictionary of Philosophy 2012; Van Brakel 1997, 2010 and Baird et al. 2006). Consider, for instance, the now established *International Society for the Philosophy of Chemistry* which has recently held a symposium in Leuven, Belgium. Journals such as *HYLE* and *Foundations of Chemistry* have focused exclusively on the philosophical investigations on chemistry. Books such as Eric Scerri's *The Periodic Table: Its Story and Its Significance* have been published that provide collections that interrogate chemistry from a philosophical perspective (Scerri 2007). The *Stanford Dictionary of Philosophy* has included an entry on philosophy of chemistry (Weisberg et al. 2011).

Unfortunately the same dynamism of scholarship cannot be attributed to the infusion of philosophy of chemistry in chemical education research and practice. The development of new perspectives on how philosophical aspects of chemistry can inform education has had rather slow progress. In *Chemical Education: Towards Research-Based Practice,* Gilbert and colleagues (2003) noted that research on chemical education drawing perspectives from philosophy of chemistry represented 'research aimed at generating new knowledge, the impact of which on practice is uncertain, diffuse or long-term' (p. 398). *Science & Education* was one of first journals to dedicate space to the work of educators preoccupied with the synthesis of perspectives from philosophy of chemistry for application in chemical education (e.g. Erduran 2001, 2005, 2007). A recent edition consisting of 17 paper contributions from philosophers, chemists and educators (Erduran 2013) is testament to the journal's vision in pushing boundaries for innovative scholarship, and it illustrates the small but growing interest in capitalising on the philosophical aspects of chemistry for the improvement of chemical education.

In this chapter, some recent developments within philosophy of chemistry are outlined, and their applications in chemical education research and practice are explored. As educators, the authors' emphasis will be on the characterisation of perspectives, approaches and tools that might be offered by philosophy of chemistry for the improvement of chemical education. The goal is to explore the contributions of philosophy of chemistry in chemical education while also being mindful of how chemical education research could provide useful recommendations for the study of chemistry from a philosophical perspective. It is through such reciprocal interactions between philosophical and educational considerations of chemistry that we believe the theoretical and empirical coherence between these fields will be established.

The discussion will begin with an illustration of some key debates in philosophy of chemistry. This section will include themes such as reductionism (e.g. Scerri 1991) and supervenience (e.g. Papineau 1995) as well as aspects of chemical knowledge such as laws (e.g. Christie and Christie 2000), models (e.g. Woody 2013) and explanations (e.g. Hendry 2010). Second, the implications of these themes for chemical education research and practice will be visited. We will argue that to develop an understanding of how chemistry is conceptualised and how chemistry is learned, chemical education research has to be informed by the debates about the epistemology and ontology of chemistry. The discussion will be contextualised in the area of nature of science (NOS) that has been one of the highly studied area of research in science education (Chang et al. 2010). The contributions of how philosophy of chemistry can contribute to the characterisation of NOS by nuanced perspectives on the nature of chemistry will be discussed. Theoretical perspectives and empirical studies on NOS have tended to focus on domain-general aspects of scientific knowledge with limited understanding of domain-specific ways of thinking. NOS literature can be further developed both theoretically and empirically, thereby contributing further to HPS studies in science education. Third, some applications of philosophy of chemistry in chemical education will be outlined in more detail. Proposed frameworks for secondary and tertiary chemical education, including the context of the teaching of periodic law through argumentation (e.g. Erduran 2007), will be exemplified. Fourth, the central argument is that there is developing potential for reciprocal interplay between philosophy of chemistry and chemical education. While philosophy of chemistry can influence chemistry education, chemistry education in turn can potentially influence philosophy of chemistry, particularly in relation to empirical foundations of chemical reasoning. The chapter will conclude with some recommendations on the future directions of research in chemical education that is informed by philosophy of chemistry.

10.2 Perspectives from Philosophy of Chemistry: Some Relevant Examples for Education

Since its formalisation with its exclusive associations, books and journals, philosophy of chemistry has been preoccupied with numerous issues. It is beyond the scope of this chapter to provide an exhausted survey of the key philosophical concerns that have been raised by experts in the field. This task is left to the philosophers themselves. As educational researchers, the authors are interested in understanding some of the key debates so as to extend educators' disciplinary understanding of chemistry from a philosophical perspective and ultimately to inform educators' treatise of chemical education in a way that is informed by and is consistent with the nature of chemistry as illustrated by epistemological and ontological accounts of chemistry. In this section, some key themes will be raised that have taken centre stage in philosophy of chemistry in recent years. They are intended to highlight some central themes from philosophy of chemistry such as reductionism and supervenience which have been quite critical in the very formulation of philosophy of chemistry at its inception. The significance of these themes for chemical education will be reviewed. Subsequently attention will be devoted to a discussion on the nature of chemical knowledge, particularly in the context of explanations, models and laws. Since school curricula already aim to communicate these features of chemical knowledge in the classroom, it is entirely appropriate to complement our notions of these concepts with philosophical perspectives. Finally, the notion of chemistry as language will be visited. Considering the great deal of interest in educational research in recent years on the role of talk, discourse and language in learning (e.g. Erduran and Jimenez-Aleixandre 2008, 2012; Lemke 1990; Vygotsky 1978), this reference will seek to understand how the philosophers' approach to the role of chemical language could inform chemical education.

10.2.1 Reduction

Reduction has been a subject of debate within philosophy of science for a long time (Nagel 1961; Primas 1983). The classic view of reduction is given by Ernest Nagel in his book *Structure of Science: Problems in the Logic of Scientific Explanation* (Nagel 1961). Nagel's definition of reduction involves the axiomatisation of two theories and an examination of whether certain formal relationships exist between the axiomatised versions of these theories. A key contributor to philosophy of chemistry, van Brakel (2000) distinguished between three types of reduction. *Constitutional reduction* concerns the question of whether two domains, B and S, are ontologically identical, i.e. whether the S-entities are constituted of the same elementary substates with the same elementary interactions as B-entities. *Epistemological reduction* concerns whether the concepts (properties, natural kinds) necessary for the description of S can be redefined in an extensionally equivalent way by the concepts of B and whether the laws governing S can be derived from those of B. *Explanatory reduction* concerns the question of whether for every event or process in S there is some mechanism belonging to B which causally explains the event or the process. Furthermore, *ontological reduction* can be contrasted with *epistemological reduction*. Scerri and McIntyre (1997) have argued that even though chemistry is widely considered to be ontologically reducible to physics, the epistemological reduction of chemistry to physics is a contentious issue. Epistemological reduction would question whether or not our current description of chemistry can be reduced to our most fundamental current descriptions of physics, namely, quantum mechanics and its explanatory consequences. Scerri and McIntyre argue that it is not clear that the laws of chemistry, if they indeed exist, can be axiomatised in the first place, let alone derived across disciplines. Mario Bunge has eloquently argued that such concepts as chemical composition are necessarily 'chemical concepts' which cannot be reduced to physical explanations:

> At first sight, chemistry is included in physics because chemical systems would seem to constitute a special class of physical systems. But this impression is mistaken, for what is physical about chemical systems is its components rather than the system itself, which possesses emergent (though explainable) properties in addition to physical properties. (Bunge 1982)

Bunge cites as an example of such an emergent property that of having a composition that changes lawfully in the course of time. The atomic and molecular components do not show this property of composition. Likewise, Primas says that even though we can calculate certain molecular properties, we cannot point to something in the mathematical expressions which can be identified with bonding. The concept of chemical bonding seems to be lost in the process of reduction (Primas 1983). Furthermore, Scerri and McIntyre (1997) argue that such conceptual reduction is not possible in principle due to the very nature of the concepts themselves. The atomic and molecular components do not show the property of composition. Erduran (2007) illustrated the relevance of reduction in chemical education in the context of chemical composition, molecular structure and bonding, all concepts that are promoted as learning outcomes in secondary and tertiary education. She drew from investigations into the reduction theme in the context of water (e.g. Kripke 1971; Putnam 1975) pointing to the following relationship: 'Water is H_2O'. Water has been a popular topic of discussion among philosophers (e.g. Farrell 1983). Barbara Abbott, for example, dedicated a paper on the observation made by Chomsky (1995) that tea and Sprite are not called water although they contain roughly the same proportion of H_2O as tap water (Abbott 1997). It is common in these discussions to identify the antireduction theme that the concept or the laws of water cannot be reduced to the concept or laws governing H_2O. In a micro-reductive picture, water really is H_2O. H_2O is the essence of the substance which, at the manifested level, is called water. Barnet (2000) argues, however, that even if water is granted to be necessarily composed of H_2O, we should not accept that such rigid designators as 'H_2O' and 'water' refer to the same thing (Chang 2012).

10.2.2 Supervenience

Supervenience has drawn quite a lot of attention in philosophy of chemistry (e.g. Luisi 2002; Newman 2013). The most common definition of supervenience is that supervenience is a relationship of asymmetric dependence. Two macroscopic systems which have been constructed from identical microscopic components are assumed to show identical macroscopic properties, whereas the observation of identical macroscopic properties in any two systems need not necessarily imply identity at the microscopic level. Some authors have even drawn on the relationship between chemistry and physics to illustrate their basic arguments about the supervenience relationship (Papineau 1993). As an example to contextualise supervenience, Scerri and McIntyre (1997) consider the property of smell. If two chemical compounds were synthesised out of elementary particles in an identical manner, they would share the same smell. Similarly the supervenience argument would entail that if two compounds share the same macroscopic property of smell, we could not necessarily infer that the microscopic components from which the compounds are formed would be identical. Such scenarios can be explored by biochemists and neurophysiologists but whatever the outcome, the question of supervenience of chemistry on physics will depend on empirical facts and not on philosophical considerations.

The case of supervenience highlights the role of empirical chemical research in establishing at least some aspects of the relation between microscopic and macroscopic systems. One educational implication is the importance of emphasising the significant role of empirical research in chemical inquiry (Erduran and Scerri 2003). As an example educational scenario, the question of supervenience can be raised at secondary education through case studies investigating the relationships between the colour, smell and texture and microscopic properties such as molecular structure and bonding. School chemistry is full of concepts that necessarily raise supervenience as an item for discussion. The problem with school chemistry is that the coverage of the relationships at different levels of organisations (i.e. macroscopic and microscopic properties) is often restricted to the coverage of declarative knowledge rather than a sound meta-level interrogation (Erduran et al. 2007).

10.2.3 Explanations

There is now a growing body of work in philosophy of chemistry that highlights aspects of chemical knowledge such as explanations, models and laws. A brief survey will illustrate some of the debates around chemical knowledge. Extended discussions are available elsewhere. For instance, refer to the discussion on chemical explanations and laws in Dagher and Erduran in this handbook. Here we will review some example work to illustrate the nature of debates on the structure of chemical knowledge from a philosophical perspective.

An important form of explanation that pervades all areas of chemistry is lies in electron shells or orbitals. The formation of bonds, acid–base behaviour, redox chemistry, photochemistry and reactivity studies are all regularly discussed by reference to the interchange of electrons between various kinds of orbitals (Scerri and McIntyre 1997). The analysis of explanations in general and physical chemistry may at first sight seem to speak in favour of the epistemological reduction of chemistry to physics, since the discourse of electron shells is thought to belong primarily to the level of atomic physics. However, a more critical examination of the issues involved reveals no such underpinning from fundamental physics. Electronic orbitals cannot be observed according to quantum mechanics, although they remain as a very useful explanatory device. This result is embodied in the more fundamental version of the Pauli exclusion principle, which is frequently forgotten at the expense of the restricted and strictly invalid version of the principle, which does uphold the notion of electronic orbitals (Scerri 1991, 1995). This situation implies that most explanations given in chemistry which rely on the existence of electrons in particular orbitals are in fact 'level-specific' explanations, which cannot be reduced to or underwritten by quantum mechanics. Thus, the explanation of what it is that we seek to know when we engage in chemical explanation would seem to support the explanatory autonomy of chemistry (Scerri 2000). An important implication for chemistry education at higher levels is that the teaching context needs to manifest the useful explanatory nature of electronic orbitals in chemical explanations in a manner consistent with their antirealistic use in quantum mechanics. In other words, there is a distinction to be made about the explanatory status of electronic orbitals in chemistry and their ontological status in quantum mechanics. Reflective classroom discussions based on such distinctions are likely to promote deeper understanding of chemical explanations among university students.

When we turn to organic chemistry, Goodwin (2008) explains that in organic chemistry, phenomena are explained by using diagrams instead of mathematical equations and laws. In this respect, organic chemistry is quite different from the way that explanations are constructed in physical sciences. Goodwin investigates both the nature of diagrams employed in organic chemistry and how these diagrams are used in the explanations of the discipline. The diagrams particularly mentioned are structural formulas and potential energy diagrams. Structural formulas are two-dimensional arrangements of a fixed alphabet of signs. This alphabet includes letters, dots and lines of various sorts. Letters are used as atomic symbols; dots are used as individual electrons, and lines are used as signs for chemical bonds. Structural formulas in organic chemistry are mainly used as descriptive names for the chemical kinds. Thus, structural formula has a descriptive content consisting of a specification of composition, connectivity and some aspects of three-dimensional arrangement. Structural formulas are also used as models in organic chemistry. For example, a ball and stick model is used in the explanations of organic chemistry.

Following a characterisation of some features of structural formulas, Goodwin presents a model of the explanations in organic chemistry and describes how both structural formulas and potential energy diagrams contribute to these explanations. He gives the examples of 'strain' and 'hyper-conjugation' to support his idea about the role of diagrams in organic chemistry as structural explanations. In other words, the structural representations embed assumptions about molecules and how atoms are positioned in relation to one another in molecules. Although schooling introduces students to structural representations, they are often implicit and are not articulated in a way to foster meta-level understanding.

10.2.4 Laws

A great deal of interest has emerged in the study of laws in chemistry (e.g. Christie 1994; Tobin 2013; Vihalemm 2003). Some philosophers of chemistry (e.g. Christie and Christie 2000) as well as chemical educators (e.g. Erduran 2007) have argued that there are particular aspects of laws in chemistry that differentiate them from laws in other branches of science with implications for teaching and learning in the science classroom. A topic of particular centrality and relevance for chemical education is the notion of 'periodic law' which is typically uncharacterised as such:

> Too often, at least in the English speaking countries, Mendeleev's work is presented in terms of the Periodic Table, and little or no mention is made of the periodic law. This leads too easily to the view (a false view, we would submit), that the Periodic Table is a sort of taxonomic scheme: a scheme that was very useful for nineteenth century chemists, but had no theoretical grounding until quantum mechanics, and notions of electronic structure came along. (Christie and Christie 2003, p. 170)

A 'law' is typically defined as 'a regularity that holds throughout the universe at all places and at all times' (Salmon et al. 1992). Some laws in chemistry like Avogadro's law (i.e. equal volumes of gases under identical temperature and pressure conditions will contain equal numbers of particles) are quantitative in nature while others are not. For example, laws of stoichiometry are quantitative in nature and count as laws in a strong sense. Others rely more on approximations and are difficult to specify in an algebraic fashion. Scerri and McIntyre (1997) state that the periodic law seems not to be exact in the same sense as are laws of physics, for instance, Newton's laws of motion. The periodic law states that there exists a periodicity in the properties of the elements governed by certain intervals within their sequence arranged according to their atomic numbers. The crucial feature which distinguishes this form of 'law' from those found in physics is that chemical periodicity is approximate. For example, the elements sodium and potassium represent a repetition of the element lithium, which lies at the head of group I of the periodic table, but these three elements are not identical. Indeed, a vast amount of chemical knowledge is gathered by studying patterns of variation that occur within vertical columns or groups in the periodic table. Predictions which are made from the so-called periodic law do not follow deductively from a theory in the same way in which idealised predictions flow almost inevitably from physical laws, together with the assumption of certain initial conditions.

Scerri further contrasts the nature of laws in physics such as Newton's laws of motion. Even though both the periodic law and Newton's laws of motion have had success in terms of their predictive power, the periodic law is not axiomatised in mathematical terms in the way that Newton's laws are. Part of the difference has to do with what concerns chemists versus physicists. Chemists are interested in documenting some of the trends in the chemical properties of elements in the periodic system that cannot be predicted even from accounts that are available through contributions of quantum mechanics to chemistry. Christie and Christie (2000), on the other hand, argue that the laws of chemistry are fundamentally different from the laws of physics because they describe fundamentally different kinds of physical systems. For instance, Newton's laws described above are strict statements about the world, which are universally true. However, the periodic law consists of many exceptions in terms of the regularities demonstrated in the properties and behaviours of elements. Yet, for the chemist there is a certain idealisation about how, for the most part, elements will behave under particular conditions. In contrast to Scerri (2000), Christie and Christie (2000), and Vihalemm (2003) argue that all laws need to be treated homogeneously because all laws are idealisations regardless of whether or not they can be axiomatised. Van Brakel further questions the assumptions about the criteria for establishing 'laws'. An implication for chemical education is that such discussions on the philosophical characterisation of laws would extend the periodic table as a taxonomic device and promote understanding of its character as a way of reasoning in chemistry (Erduran 2007).

10.2.5   Models

The *Stanford Dictionary of Philosophy* illustrates the role of models and modelling in chemistry as follows:

> Almost all contemporary chemical theorizing involves modeling, the indirect description and analysis of real chemical phenomena by way of models. From the 19th century onwards, chemistry was commonly taught and studied with physical models of molecular structure. Beginning in the 20th century, mathematical models based on classical and quantum mechanics were successfully applied to chemical systems.

The role of models in chemistry has been underestimated since the formulation of quantum theory at the turn of the century. There has been a move away from qualitative or descriptive chemistry (which relies on development and revision of chemical models) towards quantum chemistry (which is based on the quantum mechanical theory). Increasingly, chemistry has emerged as a reduced science where chemical models can be explained away by physical theories:

> In the future, we expect to find an increasing number of situations in which theory will be preferred source of information for aspects of complex chemical systems. (Wasserman and Schaefer 1986, p. 829)

The presence of models in different disciplines such as cognitive psychology, philosophy of science, chemistry or education makes it even more difficult to come up with a single definition for the term 'model' for educational purposes. For example, in a review of the literature on the interdisciplinary characterisations of models, Erduran and Duschl (2004) discussed three different definitions of models in chemistry. The term model can refer to a material object, such as a construction. For example, a chemist can construct a model to represent the structure of a molecule so as to explain the motions of the atoms in the molecule. Another definition involves the model as a description, an entity that is merely imagined and described rather than to one is perceivable. Finally a model can be defined to involve a system of mathematical equations so as to give exactness to the description such as developing a model considering the wave equation for a hydrogen atom.

Atomic and molecular orbitals, formulated through quantum chemistry, have been used to explain chemical structure, bonding and reactivity (Bhushan and Rosenfeld 1995; Nagel 1961). Woody (1995) identified four properties of models: approximate, projectability, compositionality and visual representation. A model's structure is *approximate*. In other words, the model is an approximation of a complete theoretical representation for a phenomenon. The model omits many details based on judgments and criteria driving its construction. Another characteristic of a model proposed by Woody is that a model is *productive* or *protectable*. In other words, a model does not come with well-defined or fixed boundaries. While the domain of application of the model may be defined concretely in the sense that we know which entities and relationships can be represented, the model does not similarly hold specifications of what might be explained as a result of its application. Woody further argues that the structure of the model explicitly includes some aspects of *compositionality*. There is a recursive algorithm for the proper application of the model. Thus, while the open boundaries of the model allow its potential application to new, more complex cases, its compositional structure actually provides some instruction for how a more complex case can be treated as a function of simpler cases. Finally, in Woody's (1995) framework, a model provides some means of *visual representation*. This characteristic facilitates the recognition of various structural components of a given theory. Many qualitative relations of a theoretical structure can be efficiently communicated in this manner. Although chemical education research literature contains a vast number of studies on models and modelling (e.g. Carr 1984; Coll and Taylor 2005; Gilbert 2004;

Justi and Gilbert 2003; Justi 2000), only a few studies have taken on epistemological perspective on the nature of models (e.g. Adúriz-Bravo 2013; Chamizo 2013; Erduran 2001).

10.2.6 Chemistry as Language

Arguments have been put forth to characterise chemistry as a language. For example, Lazslo (2013) argues that the analogy of language 'forces us to reconsider the usual positioning of chemistry, in classification of the sciences, between physics and biology. It reclaims chemistry as a combinatorial art'(p. 1701). Jacob (2001) defines chemistry as an experimental science that transforms both substances and language. On the one hand, chemists analyse and synthesise new compounds in the laboratory; on the other, they make analytical and synthetic statements about these compounds in research articles. Therefore, Jacob emphasises the necessity of understanding chemists' use of their language, what rules govern the use of chemical language and what consequences the utilisation of this language have for chemistry as a whole. It is essential to distinguish not only between chemical experiments and chemical language but also between different levels of chemical language. Jacob classifies the levels of chemical language as chemical symbols for substances, vocabulary (ideators and abstractors) that enables chemists to talk about substances in general, terms (theories and laws) that are used to discuss abstractors and language of philosophy (theories, their origin and their empirical basis). All levels of chemical language are vital for chemical research.

In particular, the relationship between the chemical symbols used to represent substances and the substances themselves is most central for the research chemist. Jacob defines chemical symbolism as a language and investigates the empirical basis of chemical symbolism. He also outlines the interdependence between the different operations of analysis and synthesis on the bench and on the blackboard. Furthermore, he discusses the influence language has on the progress of chemical research in general and the potential limitations the use of a specific chemical language poses for research in particular.

An aspect of Jacob's work on chemical language concerns chemical symbolism. Jacob explains some aspects of chemical symbolism. Chemical symbolism consists of an alphabet, a particular syntax and a set of semantic rules. Chemical alphabet consists of approximately 110 symbols representing the known chemical elements (e.g. Na, Cl). Elemental symbols can be combined in order to form a chemical formula (e.g. NaCl) and reaction equations (e.g. $2Na + Cl_2 \rightarrow 2NaCl$). These combinations of symbols follow a set of formal rules which are defined as chemical syntax. Chemical syntax covers empirical rules regarding valency, oxidation state, electronegativity, affinity and reaction mechanisms (Psarros 1998, as cited in Jacob 2001).

Chemical orthography provides the rules for combination of elements in formulas (e.g. Na and Cl can be combined to NaCl using the rule that 1 Na can be combined with 1 Cl). Chemical grammar provides the rules for reaction equations (e.g. stoichiometric coefficients, use of unidirectional or equilibrium arrows, reactions conditions). The grammar rules of the reaction formula $2Na + Cl_2 \rightarrow 2NaCl$ are determined by the orthography of Na, Cl_2 and NaCl. Chemical semantics discusses the meaning of symbols, formulas and reactions (e.g. NaCl as lump of salt). While chemical semantics describes the relationship between existing substances and their linguistic representations, chemical syntax enables chemists to form new symbols as representations of substances not yet synthesised. The meaning of NaCl (chemical, physical, social, cultural) is independent from both orthographic (e.g. NaCl vs. Na_3Cl) and grammatical correctness (e.g. $2Na + Cl_2 \rightarrow NaCl_2$). The distinction between syntactic and semantic rules allows for an important asymmetry between operations with language and operations with compounds. According to Jacob (2001) the asymmetry between syntactic and semantic rules is the basis of planning new reactions. The distinction between syntactic and semantic properties of chemical symbolism allows introduction of chemical formulas that are syntactically correct but do not (yet) have an empirical basis. The relevance of chemical language for chemistry education has been illustrated in the context of textbooks (e.g. Kaya and Erduran 2013).

10.2.7 Ethics in Chemistry

The final example aspect of philosophy of chemistry concerns ethics of chemistry. Recent landscape in science education at both the policy (e.g. National Research Council 1996) and research (e.g. Zeidler 2003) levels promotes the education of individuals to be able to make informed decisions and justified moral choices on scientific issues ranging from genetically modified foods to environmental protection (Kovac 2004). Erduran (2009) highlighted Kovac's work as a key area of contribution from philosophy of chemistry to chemical education as ethics raises relevant questions such as 'What aspects of chemical knowledge relate to ethical concerns? What are the moral implications of chemical knowledge?' Kovac highlights the particular ways in which chemists' lives are defined by problems of ethics:

> Ordinarily chemists are not independent practitioners like lawyers and doctors, but instead work within institutions such as colleges and universities, government agencies and industrial concerns. As a result, they often have several roles. For example, I am both a chemist and a professor and each profession has its own history and culture. In industry a chemist is certainly an employee and might also be a manager. All industrial chemists must balance their ethical obligations to chemistry as a profession with their contractual and ethical obligations to their employers. In addition, all chemists are also citizens and human beings with the civic and moral responsibilities that accompany those roles. One of the goals of a philosophy of the profession should be to clarify the ethical responsibilities of chemists as chemists, as opposed to their responsibilities in other roles. Conflicts can occur. (Kovac 2000, p. 217)

Chemists work in a variety of contexts and consequently are confronted with a broad array of ethical problems. The chemical industry is interlinked with societal questions and demands and therefore gives rise to complex issues concerning the relationship of science and society. Kovac (2000) explores the relationship between professionalism and ethics. Since writing *The Ethical Chemist*, a collection of cases and commentaries for the teaching of scientific ethics to chemists, Kovac has been investigating ethics as an integral part of chemistry. In particular, he explores those aspects of chemical ethics that go beyond the demands of ordinary morality, the requirements of law and the pressures of the market. Furthermore, Kovac suggests that a healthy dialogue concerning professionalism and ethics is essential to a broader philosophy of chemistry. While a discussion of concepts is the core of a philosophy of science, science is, after all, public knowledge developed by a community. What is unique about chemistry as a science is partly a result of the uniqueness of the chemical community and its history. Studying chemistry as a profession will help reveal the essence of chemistry as a science.

While there has been much recent interest in the ethics of science, most of the literature is rather broadly conceived, treating science as a single enterprise (Kovac 2000). According to Kovac, here is little, if any, recognition that each scientific discipline has its own perspective on professionalism and ethics. For example, David B. Resnick's book, *The Ethics of Science: An Introduction* (Resnik 1998), for all its strengths, never discusses the differences between the various sciences. There is a substantial literature of casebooks designed to provide materials for courses in scientific ethics. Some of these, such as *Research Ethics: Cases & Materials*, edited by Robin Levin Penslar (1995), provide a broader philosophical introduction and cases in a number of disciplines, while others, such as Kovac's work (Kovac 2004), focus on practical ethics in a single discipline. The literature on ethics in chemistry is scarce indeed. However, even the existing debates provide some potential useful guidelines for chemical education research and practice. For instance, these perspectives raise questions for educational research: 'What is the nature of moral reasoning in chemistry and how could such moral reasoning be incorporated in learning?'

10.3 Debates on Constructivism and Nature of Science (NOS) Research

The preceding brief survey of perspectives from philosophy of chemistry for applications in chemical education research and practice illustrates so far some of the specific themes that are relevant for import in chemical education. There is further scope for the treatment of philosophical perspectives in chemical education particularly in two broad areas that have preoccupied educators: constructivism (e.g. Taber 2006) and nature of science (NOS) (e.g. Lederman et al. 2002; McComas 1998). The treatment of philosophical perspectives in chemical education research has conventionally focussed on themes such as relativism, objectivism and realism (e.g. Herron 1996). Eric Scerri has maintained the thesis that such philosophical concepts have been misinterpreted in the work of some chemical educators, and at times, they are at odds with scientific ideas:

> I think that if one looks closely at the basic philosophical positions offered by some chemical constructivists, one sees many radical themes that are not only open to serious questioning but can also be construed as being anti-scientific. (Scerri 2003, p. 468)

Scerri further argues that one remedy to this philosophical confusion is more use of philosophy of chemistry in chemical education research. Scerri reflects on the status of chemical education research by highlighting how for some chemists 'research in chemical education represents a soft-option best suited for those who are not capable of succeeding in 'real chemistry' research' (p. 468). He continues to argue that some of the blame in the low reputation of chemical education research among chemists can be attributed to the philosophical confusions demonstrated by chemical educators. In a response to this criticism, Erduran (2009) acknowledges that such confusions do exist but considers chemical education research beyond university chemistry departments to illustrate the diversity of the chemical education community. She highlights the extensive body of research in chemical education (e.g. Gable and Bunce 1984) and argues that the perception of chemical education research as a soft option to doing hard science of chemistry is reflective of lack of knowledge that there is a formalised discipline called 'science education' with its own body of journals, conferences, societies as well as funding agencies. Furthermore, Erduran (2009) notes that it is also important to note that 'school chemistry' is not the same as 'chemistry'. The goals and aims of education do not necessarily correspond to goals and aims of chemical research be they in the form of hard science or as an object of investigation by philosophers or historians. For instance, the historical progression of ideas in science may not be followed in the same order in the classroom for pedagogical purposes, yielding a vision of science devoid of historical context. However at times, sequences of concepts introduced in the classroom may serve learners' understanding if they do not come in the historical order. Indeed often science education discards many old theories and models in favour of recent accounts so as not to confuse students or impart potential misconceptions that have been dealt with throughout history by scientists. Overall, such approaches demarcate the purposes and processes of school versus institutional science. Furthermore, school science as advocated in important policy documents worldwide (e.g. NRC 1996) is one that recognises the right for everyone to be scientifically literate, not just those who will become scientists.

Constructivism has been a major theme within the science education community. Indeed, the vast body of empirical work that emerged on learners' ideas in science was stimulated by the constructivist movement (e.g. Driver et al. 1996). As a result, a significant amount of literature is now available on how learners of different age groups understand key chemical concepts (e.g. Duit 2012). As Yeanny pointed, constructivism has been a unifying theme for 'thinking, research, curriculum development, and teacher education' (Yeany 1991, p. 1), and he added that 'there is a lack of polarised debate'. However, despite this significant research effort, there have been serious criticisms of this area of work (e.g. Irzik 2000; Matthews 1994, 1998; Mugaloglu 2001). In Science Teaching: The Role of History and Philosophy of Science (1994/2014), Matthews made a critical analysis of the philosophical foundations of constructivism and its implications for science education. Although there are different versions of constructivism, most of them define knowledge as an intellectual and social construction without reference to 'justified true belief (JTB)' theory (Mugaloglu 2001).

The debate on constructivism is essentially an epistemological war between those who take a realist position and those who take a relativist or constructivist position in relation to scientific knowledge and the learning of science (Scerri 2003, p. 3). Scerri (2003) refers to Gross and Levitt's book *Higher Superstition* (1998) when arguing that some studies on the nature of science are 'seriously mistaken and are having a damaging influence upon scholarly work, the public image of science, and last but not least, on science education' (p. 359). Although the war is going on mainly in the philosophy of science arena, most chemists explicitly or implicitly take one or the other of these two positions when thinking about chemistry and their knowledge about chemistry.

Taber (2006) reviews such criticisms in terms of constructivism's philosophical underpinning, the validity of its most popular constructs, the limited scope of its focus, and its practical value to science teaching. Furthermore he frames constructivism as an area of work as a Lakatosian research programme (RP) and explores the major criticisms of constructivism from that perspective. He argues that much of the criticism may be considered as part of the legitimate academic debate expected within any active RP, i.e. arguments about the auxiliary theory making up the 'protective belt' of the programme. It is suggested that a shifting focus from constructivism to 'contingency in learning' will allow the RP to draw upon a more diverse range of perspectives, each consistent with the existing hard core of the programme, which will provide potentially fruitful directions for future work and ensure the continuity of a progressive RP into learning science.

Chemistry educators have good reasons to follow the debates on constructivism. First, to develop an understanding of how chemistry is conceptualised and how chemistry is learned, the debate about its nature, epistemology and ontology is crucial to acknowledge. Investigating the nature of chemistry can only lead to more effective teaching of chemistry. This explains why 'nature of science' is one of the most studied topics in the literature of science education (Chang et al. 2010). Second, the literature includes evidence to suppose a relationship between the epistemological positions of teachers and the learning paradigms that influence their teaching. In other words, while chemistry teachers are teaching chemistry, they implicitly or explicitly but necessarily present a philosophical position about chemistry (Chamizo 2007; Erduran and Scerri 2003).

One of the central and broad areas of research in science education that has harboured the debates on constructivism including its epistemological and ontological foundations is called nature of science (NOS). The predominant definition of the NOS in the empirical studies on teachers' and students' perceptions of science has relied on the characterisation of science primarily relative to the cognitive, epistemic and social aspects of science and has been limited in terms of their conceptualisations of science from broader perspectives on science (e.g. Allchin 2011). The collective set of learning goals for understanding the NOS is summarised in the 'consensus' view of NOS (Lederman et al. 2002; McComas 1998) which has the following tenets:

(a) Tentativeness of Scientific Knowledge: Scientific knowledge is both tentative and durable.
(b) Observations and Inferences: Science is based on both observations and inferences. Both observations and inferences are guided by scientists' prior knowledge and perspectives of current science.

(c) Subjectivity and Objectivity in Science: Science aims to be objective and precise, but subjectivity in science is unavoidable.

(d) Creativity and Rationality in Science: Scientific knowledge is created from human imaginations and logical reasoning. This creation is based on observations and inferences of the natural world.

(e) Social and Cultural Embeddedness in Science: Science is part of social and cultural traditions. As a human endeavour, science is influenced by the society and culture in which it is practiced.

(f) Scientific Theories and Laws: Both scientific laws and theories are subject to change. Scientific laws describe generalized relationships, observed or perceived, of natural phenomena under certain conditions.

(g) Scientific Methods: There is no single universal step-by-step scientific method that all scientists follow. Scientists investigate research questions with prior knowledge, perseverance and creativity.

In this propositional characterisation of NOS, science is presented in an epistemologically and ontologically flat and undifferentiated landscape, broad and lacking sufficient detail to indicate the nuances that characterise branches of science. For instance, with respect to (f), there is no consideration of how laws might have different characteristics in different sciences. As illustrated earlier in the work of Christie and Christie (2000) and also argued by Erduran (2007) in the context of chemical education, 'laws' can have very different meanings in chemistry versus physics. Furthermore, question the very characterisation of 'science' in NOS by asking 'the nature of which science NOS characterisations capture in the first place'. The particular instances of reduction, supervenience as well as the nature of models, laws and explanations and chemistry as language all point to ample evidence from philosophy of chemistry that the contemporary characterisations of NOS are underspecified.

In summary, perspectives from philosophy of chemistry can provide a new and fresh lens by which to view and interpret constructivism and NOS with respect to science education. These perspectives help clarify the ontological and epistemological status of chemistry in ways that traditionally philosophy of science has not sufficiently captured. In turn, the insight into the nature of chemistry can help inform the goals and content of chemical education.

10.4 Applications of Philosophy of Chemistry in Chemical Education

The applications of philosophy of chemistry in chemical education theory and practice have been minimal (Erduran 2013, 2000a, b). A rare volume on the subject was compiled in the journal *Science & Education*. The volume consists of papers that deal with a range of issues raised in philosophy of chemistry in application to chemical education. One set of papers focus on the nature of chemical knowledge, particularly in relation to models, explanations and laws. Woody (2013) uses the ideal gas law as an example in reviewing contemporary research in philosophy of science concerning scientific explanation. She clarifies the inferential, causal, unification and erotetic conceptions of explanation. Tobin (2013) provides an overview of the laws in chemistry and reflects on the recent debates on the particular and universal nature of laws, concluding that while generalisations in chemistry are diverse and heterogeneous, a distinction between idealisations and approximations can nevertheless be used to successfully taxonomise them. Adúriz-Bravo (2013) challenges the received, syntactic conception of scientific theories and argues for a model-based account of the nature of science. The significance of models and modelling in chemistry is further highlighted through a typology of models and their relation to modelling (Chamizo 2013). Izquierdo-Aymerich (2013) argues for the generation of chemical criteria from the history and philosophy of chemistry for informing the design of chemistry curriculum.

The special issue volume consists of a second set of papers that focus on particular epistemological themes. The authors extend these debates to the curricular, textbook and teaching contexts and, in so doing, elaborate on their potential instantiation in education. Newman (2013) provides a model for teaching chemistry with the potential to enhance fundamental understanding of chemistry. Lazslo (2013) argues that chemistry ought to be taught in like manner to a language, on the dual evidence of the existence of an iconic chemical language, of formulas and equations and of chemical science being language-like and a combinatorial art. Universality and specificity of chemistry are interrogated by Mariam Thalos who argues that chemistry possesses a distinctive theoretical lens—a distinctive set of theoretical concerns regarding the dynamics and transformations of a variety of organic and nonorganic substances (Thalos 2013). While she agrees that chemical facts bear a reductive relationship to physical facts, she argues that theoretical lenses of physics and chemistry are distinct. Manuel Fernandez-Gonzalez discusses the concept of pure substance, an idealised entity whose empirical correlate is the laboratory product (Fernandez-Gonzalez 2013). A common structure for knowledge construction is proposed for both physics and chemistry with particular emphasis on the relations between two of the levels: the ideal level and the quasi-ideal level. Kaya and Erduran focus on concept duality, chemical language and structural explanations, to illustrate how chemistry textbooks could be improved with insights from such work (Kaya and Erduran 2013). They provide some example scenarios of how these ideas could be implemented at the level of the chemistry classroom. Talanquer presents a case that dominant universal characterisations of the nature of science fail to capture the essence of the particular disciplines. The central goal of this position paper is to encourage reflection about the extent to which dominant views about quality science education based on universal views of scientific practices may constrain school chemistry (Talanquer 2013).

Activities, practices and values of chemistry are interrogated in a third set of papers. Earley recommends that chemistry educators shift to a different 'idea of nature', an alternative 'worldview' (Earley 2013). Garritz (2013) illustrates how teaching history and philosophy of physical sciences can illustrate that controversies and rivalries among scientists play a key role in the progress of science and why scientific development is not only founded on the accumulation of experimental data. The case of quantum mechanics and quantum chemistry is used as an example because it is historically full of controversies. Ribeiro and Pereira (2013) illustrate how pluralism in philosophical perspectives can result in different cognitive, learning and teaching styles in chemical education. Their paper reports on the authors' experiences in Portugal in drafting structural ideas and planning for the subject 'didactic of chemistry' based on the philosophy of chemistry. Vesterinen et al. (2013) assess how the different aspects of nature of science (NOS) were represented in Finnish and Swedish upper secondary school chemistry textbooks. They present an empirical study where dimensions of NOS were analysed from five popular chemistry textbook series. Vilches and Gil-Perez (2013) reflect on the UN Decade of Education for Sustainable Development and how chemical education for sustainability remains practically absent nowadays in many high school and university chemistry curricula all over the world. They explore the belief that genuine scientific activity lies beyond the reach of moral judgment logically. They propose possible contributions of chemistry and chemical education to the construction of a sustainable future. Sjostrom (2013) is concerned with Bildung-oriented chemistry education, based on a reflective and critical discourse of chemistry. This orientation is contrasted with the dominant type of chemistry education, based on the mainstream discourse of chemistry. Bildung-oriented chemistry education includes not only content knowledge in chemistry but also knowledge about chemistry, both about the nature of chemistry and about its role in society.

In summary, there is now an emerging body of scholars including philosophers, educators and chemists who are working on the intersections of philosophy of chemistry and chemical education. The review so far illustrates the diversity of this work that warrants the pursuit of future work to contribute further to scholarship in this area. So far the discussion has been at a conceptual level and provided a rationale for the relevance of philosophical issues in chemistry and chemical education. In the next sections, the focus will be on practical instantiations and highlight some concrete instances in educational contexts for the inclusion of philosophical perspectives on chemistry. These will include implications and applications in learning, teaching, teacher education and textbooks.

10.4.1 Learning

Learning of chemistry has conventionally been framed in terms of problem solving (e.g. Gable and Bunce 1984; Lythcott 1990), concept learning (e.g. Cros et al. 1987; Nussbaum and Novak 1979) and learning of science-process skills (e.g. Heeren 1990; Yarroch 1985). The inclusion of philosophical perspectives in chemistry learning challenges such traditional characterisations of learning. Learning of the nature of chemical knowledge defined in terms of conceptual understanding does not acknowledge the learning of criteria and standards that enable knowledge generation, evaluation and revision in chemistry. For instance, there is little understanding of the patterns in students' ideas of how chemical laws are generated and refined. It is possible to question the extent to which research on students' and teachers' epistemologies of science has captured sufficiently the intricacies of disciplinary nuances such as those illustrated by philosophy of chemistry. How do learners, for instance, engage in discussions on reduction? What are their views on the ontological dependence of chemistry and physics? What are the trajectories of learning and the developmental patterns in understanding the supervenience issue?

Concentrating on learning trajectories is particularly relevant when evidence from higher education is exemplified. Students in advanced chemistry classes demonstrate having difficulties with many aspects of chemistry. For instance, in a study conducted by Cros and his colleagues (1987), 95 % of a large sample of university students had difficulty interpreting the Bohr model of the atom. University students also experience much difficulty with acid–base chemistry especially with Lewis model which combines acidity and basicity concepts with electrophilicity and nucleophilicity (Zoller 1990). These examples call for a further examination of how chemical explanations are introduced in the classroom. Deeper philosophical reflections on the nature of chemical explanations and how they are generated and evaluated are likely to improve students' understanding of key concepts in chemistry. In October, 1999, an elective course on the philosophy of chemistry was opened to undergraduates at the University of Exeter. Jones and Jacob (2003) published a brief report about the course, including its benefits and drawbacks. They emphasised that teaching such a course entailed departures from traditional chemistry teaching and consequent challenges. Since philosophy of chemistry was a new field, it was difficult even to find a textbook. Journals such as *HYLE* were the main source for the course at that time.

10.4.2 Teaching

Perspectives from philosophy of chemistry present the potential of motivating debates for the chemistry classroom. Building on Stroll's (1991) work, Erduran (2005) provided a particular task to illustrate how teaching could proceed in the context of a philosophical discussion. The example of 'water' presents opportunities to raise themes such as reduction and supervenience at the level of the classroom. There is the relation between the physical properties of 'water' (e.g. water boils at 100 C) and the structural features of 'H_2O' (e.g. the bond angle of 104.5 C). How, if at all, do these properties relate to each other? Can the macroscopic properties be reduced to microscopic properties? Are there any circumstances under which water is not H_2O? In this case substances with equivalent concentrations of H_2O concentration are nevertheless regarded as different and not 'water'.

As an introduction, students can be confronted with their basic assumptions about water. They can be presented with a glass full of water, with the written formula H_2O and the written word 'water'. Questions that target chemical composition, molecular structure and bonding can be presented in a way that would elicit the theme of reduction. For instance,

Does H_2O have the same chemical composition as water?
Could a single water molecule boil at 100 C?
Is H_2O the same thing as water?

could be useful for secondary schooling and can be revisited at higher levels of education where more in-depth considerations could occur. An introduction to such questions would hopefully challenge students' assumptions about seemingly straightforward relationships between macroscopic properties and representation of microscopic realty of molecules. Here, the intention is not to get students to answer these questions but to arouse their curiosity about one of the fundamental ways of thinking in chemistry: the interplay of the microscopic, symbolic and macroscopic levels. Furthermore, creating a context where explicit comparisons between symbolic, abstract and concrete experiences of substances are made is likely to immerse students in a philosophical mindset. The presentation of the following set of statements is likely to raise further debate at the level of the classroom due to the logical absurdity that it embodies:

Water $= H_2O$
Ice $= H_2O$
Therefore, water $=$ ice

In this framework, the students would identify the logic of the above equations and face an absurd conclusion. The absurdity of the conclusion, then, provides the motivation for discussion and raises issues about what counts as water and ice beyond a microscopic definition of H_2O. In other words, questions such as 'can the experience of water and ice as colourless liquid and white solid be reduced to H_2O?' could stimulate the conversation. The process of reasoning from the premises to the conclusions would necessitate the generation of counter-arguments to justify why the conclusion cannot be true.

Van Brakel (2000) argues that the 'water $= H_2O$' equation is not true because of the problem of isotopes and the fact that water is not 100 % H_2O. De Sousa (1984) furthermore argued that H_2O 'is a chemical characterization of water, not a physical one. Physically it turns out that water is a mixture of several sorts of molecules: ones containing Oxygen-16 and one containing the isotope Oxygen 18, as well as ones containing isotopes of hydrogen (deuterium or tritium)' (p. 571). Hence, H_2O is the *chemical* essence of water, not *the* essence of water. Here we see the potential for introducing, at upper secondary schooling, other concepts such as isotopes and concentration to supplement the discussion.

The role of the teacher in this scenario would be more of a facilitator of discussion. Different points of view with respect to such questions can be recorded publicly in the classroom so that students are provided with alternative explanations. For instance, as an extreme case, a sceptic student might argue that the water in the glass really has nothing to do with the formula H_2O and that these conventions are fictions of chemists' imaginations. Alternatively, another student might defend the position that even if he/she believes that the water in glass has something to do with the formula H_2O, he/she only knows so because the textbook said so. In either case, students can be encouraged to provide evidence to justify their points of view. Overall, pedagogical strategies of questioning, coordinating discussions, task generation and management would need to be informed by the philosophical accounts of reduction and supervenience.

10.4.3 Teacher Education

According to Erduran and colleagues (2007), aligning teaching and learning with the philosophy of chemistry is a challenge for teachers who have had little exposure to issues of chemical knowledge beyond content knowledge. Schwab (1962) argued that teachers should learn the content of a domain and also the epistemology of the domain. Erduran (2009) stated, 'For chemistry teaching to be effective, prospective teachers will need to be educated about how knowledge is structured in the discipline that they are teaching. Practice and theory of future teacher education, then, will need to be informed by and about philosophy of chemistry.' Erduran and Scerri (2003) state that an understanding of philosophy of chemistry is likely to reinforce teachers' content knowledge such as quantum mechanics, periodicity and structure/function relationships in chemistry.

Apart from an understanding of the content (or subject) domain and the epistemology of the domain, teachers need to have understanding of how to transform these notions into teachable scenarios (Loucks-Horsley et al. 1990). Shulman (1986) has provided a powerful construction 'pedagogical content knowledge (PCK)' to illustrate this kind of understanding and knowledge that teachers need to have. He described PCK as 'The most useful forms of content representation… the most powerful analogies, illustrations, examples, explanations, and demonstrations—in a word, the ways of representing and formulating the subject that makes it comprehensible for others' (p. 9). For teachers to effectively implement philosophical perspectives in the chemistry classroom, their PCK would need to embrace philosophy of chemistry. There is considerable literature on teacher education and professional development in science education (e.g. Wallace and Louden 2000). Despite addressing the question of teacher education from separate perspectives and disciplines, a common vision of effective professional development exists (Loucks-Horsley et al. 1998, 1990). According to that shared vision, the best professional development experiences for science educators include the following guidelines (Loucks-Horsley et al. 1998):

- They are driven by a clear, well-defined image of effective classroom learning and teaching.
- They provide teachers with opportunities to develop knowledge and skills and broaden their teaching approaches, so they can create better learning opportunities for students.

- They use instructional methods to promote learning for adults which mirror the methods to be used with students.
- They build or strengthen the learning community of science and mathematics teachers.
- They prepare and support teachers to serve in leadership roles if they are inclined to do so. As teachers master the skills of their profession, they need to be encouraged to step beyond their classrooms and play roles in the development of the whole school and beyond.
- They consciously provide links to other parts of the educational system.
- They include continuous assessment.

However, for philosophy of chemistry to be of useful for teacher education, a focus on the content area is vital. A vast amount of research on professional development of science teachers supports this observation (e.g. Zohar 2004). The preceding discussions in this chapter on some representative themes from philosophy of chemistry including reduction, supervenience and the domain-specific characterisations of models, laws and explanations begin to provide some guidelines for what to include as outcomes of teachers' learning.

10.4.4 Textbooks

Textbooks are considered as one of the most important guides for chemistry teachers. However, numerous authors have already questioned the quality of the content of textbooks in terms of their inclusion of historical and philosophical perspectives. For example, Rodriguez and Niaz (2002) questioned *'How criteria based on history and philosophy of science can be used to evaluate presentation of atomic structure in general chemistry textbooks?'* The study revealed that the textbooks 'distort the historical facts'. In addition, the philosophical perspective in the textbooks supports the idea of inductivism (Rodriguez and Niaz 2002, p. 437). Gillespie (1997) argued that change in general chemistry might require reform in chemistry textbooks. Chemistry education researchers critically analysed the chemistry textbooks in terms of their approach, instructional structure, conceptual framework and content analysis. For instance, Kauffman (1989) criticised the chemistry textbooks in terms of the view that they presented. He stated that textbooks 'failed to make the fact clear to students that chemistry is a human enterprise' (p. 82). Moreover, he emphasised teaching the concepts such as scientific progress, focusing on the variety of the scientific method, human values and the importance of process rather than products. To do so, he recommended the inclusion of history of chemistry into the curriculum as a separate course.

There is substantial amount of work on the inclusion of historical case studies in textbooks and the investigation of chemistry from a historical perspective (e.g. Chamizo 2007; Niaz 2008). At the beginning of the twentieth century,

William Ostwald emphasised that in the textbooks a philosophical chapter was presented either at the beginning of the book as an introduction or at the end of the book as a summary with a deductive manner (cited in Rodriguez and Niaz 2002, p. 423). Chemistry education researchers should scrutinise the benefits and ways of inclusion of philosophy of chemistry into chemistry textbooks. Just as in the case of historical approach, this inclusion needs to go hand in hand with the curriculum and textbooks reforms so as to provide the teachers with an appropriate content and support in the teaching of chemistry from a new perspective.

A potential area of research for chemical educators is the investigation of existing textbooks for their philosophical content and reference to a set of criteria informed by discussions from philosophy of chemistry. Kaya and Erduran (2013) have done just that by investigating the present textbooks on their inclusion of philosophical perspectives. They have applied Laszlo's (1999) notion of concept duality, Jacob's (2001) descriptions of chemical language and Goodwin's (2008) explication of structural explanations in organic chemistry to highlight the particular ways in which chemical knowledge is structured. Examples of textbooks and curricula were used to illustrate that even though the mentioned aspects of are relevant to educational contexts, the philosophical dimensions of this coverage is absent in textbooks and curricula. The emphasis in the use of these features of chemical knowledge seems to be more on the conceptual definitions rather than on their 'epistemological or ontological nature'. Erduran and Kaya argued that chemical education will be improved through the inclusion of the philosophical perspectives in chemistry teaching and learning by highlighting the specific ways in which chemical reasoning functions. Chemistry educators emphasised that chemistry education theory and practice would benefit from applications of philosophy of chemistry (Adúriz-Bravo and Erduran 2003), especially for teaching and learning the nature of chemical knowledge. For example, the textbooks could present the discussion about nonreferring terms in chemical explanations such as orbitals and electronic explicitly. Then the classification of 'explanatory status of electronic orbitals in chemistry and their ontological status in quantum mechanics' can also be used in educational context to overcome the movement towards an antirealistic understanding of chemistry (Erduran and Scerri 2003).

Moreover, the textbooks should also guide the teachers and students in understanding how chemical knowledge constructed. Evidence in the literature confirms that both chemistry teachers and students have problems in understanding the nature of models and modelling. Erduran et al. (2007) like Gilbert (1997) argue that teachers conceive scientific models in mechanical terms and believe that models are true pictures of non-observable phenomena and ideas. Thus, learning how to make model is usually excluded from the curriculum and textbooks. This is particularly because of the fact that chemistry education has not yet position the importance of models in construction of chemical knowledge as suggested by philosophy of chemistry. Moreover, inclusion of the nature of models and modelling is vital in explaining both the ontological and epistemological relationship between microscopic and macroscopic entities,

10.5 Conclusion

The discussion in this chapter so far illustrates how philosophy of chemistry can contribute to chemical education research and practice. In particular, it raises questions about epistemology, ontology, ethics and linguistics in relation to how chemical education defines, positions and encapsulates the various dimensions of chemistry for the purposes of education. It illustrates how philosophical perspectives on chemistry can contribute to more nuanced versions of NOS in science education. A significant shortcoming of the NOS research in science education has been its lack of differentiation of scientific knowledge with respect to its disciplinary variations. Philosophy of chemistry illustrates not only the epistemological status of chemical knowledge but also its ontological undertones. This is particularly important with respect to the debates on constructivism and relativism.

As a point of warning, Erduran (2013), in her editorial of the special issue of *Science & Education* on the applications of philosophy of chemistry in chemical education, argues that the infusion of philosophical perspectives will need to be mindful of the research evidence on teaching and learning, as well as professional development of teachers. There is substantial evidence that educational reform is difficult to implement at the level of the classroom and much research evidence remains as rhetoric with no impact on practice (e.g. Au 2007; Elmore 2004; Fullan 2007). Effective incorporation of philosophical perspectives in chemical education will require systematic and well-designed research to validate the utility and influence of relevant strategies. For example, in the example about the debate on the composition of water, such a scenario will need to be introduced to teachers in a way that would be mindful to where the existing conceptions of teachers are in the uptake of such views. Professional development of teachers will be required to ensure that teachers themselves are convinced of different ways of teaching chemistry. Investigating the strategies that are effective in imparting learning on students will be essential. In short, empirical testing and validation of approaches will hold the final say in the utility of philosophy of chemistry at the level of the classroom. This is not to say that having argued so far for the inclusion of philosophy of chemistry in chemical education, the authors do not perceive its potential use in practice. To the contrary, it is the commitment and belief in the potential of philosophy of chemistry in improving the quality of teaching and learning of chemistry that has led us to write this chapter in the first place. There is place for caution and mindfulness in the educational and pedagogical manifestations of philosophical ideas in light of evidence from science education research on the difficulties inherent in educational reform.

Finally, empirical data on the implementation of philosophical perspectives in the chemistry classroom is likely to contribute to philosophy of chemistry itself. When the vast amount of literature on children's misconceptions about a wide range of scientific concepts is considered (e.g. Duit 2009), a major observation is that some of children's conceptions are similar to historical forms of thought. It has been argued, thus, that science learning can follow the sort of conceptual change that underlies the scientific process itself. Indeed a plethora of studies have emerged in the misconceptions literature on this very issue in the 1990s. The depth, the insight and the creativity of children's thinking on philosophical accounts of chemistry—for instance, in relation to the evaluation of everyday experiences of substances or the conceptualisation of chemistry as a technoscience (e.g. Ihde 2003)—could potentially raise issues for debate within philosophy of chemistry. It is indeed such empirical instantiation of chemical reasoning from a philosophical perspective in not just students but also teachers that can provide a unique sample for philosophers of chemistry to investigate, thereby engaging with educators in constructive dialogues about the nature of chemistry.

References

Abbott, B. (1997). A Note on the Nature of 'Water', *Mind* 106, 311–319.

Adúriz-Bravo, A. & Erduran, S. (2003). La epistemología específica de la biología como disciplina emergente y su posible contribución a la didáctica de la biología. *Revista de Educacion en Biología*, 6(1), 9–14.

Adúriz-Bravo, A. (2013). A semantic view of scientific models for science education. *Science & Education*, 22(7), 1593–1611.

Allchin, D. (2011). Evaluating knowledge of the nature of (whole) science. *Science Education*, 95: 518–542. doi: 10.1002/sce.20432.

Au, W. (2007). High Stakes testing and curricular control: A qualitative metasynthesis. Educational Researcher, 36(5), 258–267.

Baird, D., Scerri, E. & McIntyre, L. (2006). Introduction: The invisibility of chemistry. In D. Baird, E. Scerri & L. McIntyre (Eds.), *Philosophy of chemistry: Synthesis of a new discipline.* Dordrecht: Springer.

Barnett, D. (2000) 'Is Water Necessarily Identical to H2O? *Philosophical Studies* 98: pp. 99–112.

Bhushan, N. & Rosenfeld, S. (1995). Metaphorical models in chemistry. *Journal of Chem Educ* 72, 578–582.

Bhushan, N., & Rosenfeld, S. (Eds.) (2000). *Of Minds and Molecules: New Philosophical Perspectives on Chemistry*, Oxford: Oxford University Press.

Bunge, M. (1982). Is Chemistry a Branch of Physics?, *Zeitsch rift fur all geme me Wisenschaftstheorie* 13, 209–223.

Carr, M. (1984). Model confusion in chemistry. *Research in Science Education*, 14, 97–103.

Chalmer, A. (2010). *The scientist's atom and the philosopher's stone: How science succeeded and philosophy failed to gain knowledge of atoms.* New York: Springer.

Chamizo, J. A. (2007). Teaching modern chemistry through recurrent historical teaching models. *Science & Education*, 16, 197–216.

Chamizo, J. A. (2013). A new definition of models and modeling in chemistry's teaching. *Science & Education,* 22(7), 1613–1632.

Chang, Y., Chang, C. & Tseng, Y (2010). Trends of science education research: An automatic content analysis. *Journal of Science Education and Technology*, doi: 10.1007/s10956-009-9202-2.

Chang, H. (2012). Is Water H_2O? *Evidence, Realism and Pluralism*. Dordrecht: Springer.

Chomsky, N. (1995). Language and Nature, *Mind* 104, 1–61.

Christie, M., & Christie, J. (2000). "Laws" and "theories" in chemistry do not obey the rules. In: Bhushan, N., Rosenfeld, S. (eds.) Of Minds and Molecules, pp. 34–50. Oxford: Oxford University Press.

Christie, M. (1994). Chemists versus philosophers regarding laws of nature. Stud. Hist. Philos. Sci. 25, 613–629.

Christie, J. R. & Christie, M. (2003). Chemical laws and theories: A response to Vihalemm. *Foundations of Chemistry*, 5(2), 165–174.

Coll, R.K. & Taylor, I. (2005). The role of models and analogies in science education: Implications from research. *International Journal of Science Education*, 27, 183–198.

Cros, D., Chastrette, M., & Fayol, M. (1987). Conceptions of second year university students of some fundamental notions of chemistry. *International Journal of Science Education*, 10, 331–336.

De Sousa, R. (1984). The Natural Shiftiness of Natural Kinds, *Canadian Journal of Philosophy* 14, 561–580.

Driver, R. Leach, J. Millar, R. & Scott, P. (1996). Young People's Images of Science. Buckingham: Open University Press.

Duit, R. (2009). Bibliography - STCSE: Students' and teachers' conceptions and science education. Online available at: http://www.ipn.uni-kiel.de/aktuell/stcse/stcse.html [13.09.2012].

Duit, R. (2012). Research on students' conceptions: developments and trends. Bibliography: Students' Alternative Frameworks and Science Education. IPN at the University of Kiel, Kiel, Germany (Available Online)

Earley, J. (2013). The new 'idea of nature' for chemical education. *Science & Education*, 22(7), 1775–1786.

Elmore, R. C. (2004). *School Reform from the Inside Out: Policy, Practice, and Performance*, Harvard Educational Press: CA Mass.

Erduran, S. (2000a). Emergence and applications of philosophy of chemistry in chemical education. *School Science Review*, 81, 85–97.

Erduran, S. (2000b). A missing component of the curriculum? *Education in Chemistry*, 37(6), 168.

Erduran, S. (2001). Philosophy of chemistry: An emerging field with implications for Chemistry education. *Science & Education*, 10, 581–593.

Erduran, S. & Duschl, R. (2004). Interdisciplinary characterization of models and the nature of chemical knowledge in the classroom. *Studies in Science Education*, 40, 111–144.

Erduran. S. & Jimenez-Aleixandre, J. M. (2012). Research on argumentation in science education in Europe. In, D. Jorde, & J. Dillon (Eds.), *Science Education Research and Practice in Europe: Retrospective and Prospective*, pp. 253–289. Rotterdam: Sense Publishers.

Erduran, S. & Jimenez-Aleixandre, M. P. (Eds.) (2008). *Argumentation in Science Education: Perspectives from Classroom-based Research*. Dordrecht: Springer.

Erduran, S. & Scerri, E. (2003). The nature of chemical knowledge and chemical education. In J. K. Gilbert, O. de Jong, R. Justi, D. F. Tragust & J. H. van Driel (Eds.), *Chemical education: Towards research-based practice* (pp. 7–27). Dordrecht: Kluwer.

Erduran, S., Bravo, A. A., & Naaman R. M. (2007). Developing epistemologically empowered teachers: Examining the role of philosophy of chemistry in teacher education. *Science & Education*, 16(9–10), 975–989. doi: 10.1007/s11191-006-9072-4.

Erduran, S. (2005). Applying the philosophical concept of reduction to the chemistry of water: Implications for chemical education. *Science & Education*, 14(2), pp. 161–171.

Erduran, S. (2007). Breaking the law: promoting domain-specificity in science education in the context of arguing about the Periodic Law in chemistry. *Foundations of Chemistry*, 9(3), 247–263.

Erduran, S. (2009). Beyond philosophical confusion: Establishing the role of philosophy of chemistry in chemical education research. *Journal of Baltic Science Education*, 8(1), 5–14.

Erduran, S. (2013). Editorial: Philosophy, Chemistry and Education: An Introduction, *Science & Education*, 22(7), 1559–1562.

Farrell, R. (1983), 'Metaphysical Necessity and Epistemic Location', *Australasian Journal of Philosophy* 61, 283–294.

Fernandez-Gonzalez, M. (2013). Idealization in chemistry: pure substance and laboratory product. *Science & Education*, 22(7), 1723–1740.

Fullan, M. (2007). *The New Meaning of Educational Change*, NY and London: Teachers College Press, 4th edition

Gable, D. & Bunce, D. (1984). Research on problem solving in chemistry. In, D. Gabel (Ed.), *Handbook of research on science teaching and learning* (pp. 301–326). New York: Macmillan Publishing Company.

Garritz, A. (2013). Teaching the philosophical interpretations of quantum mechanics and quantum chemistry through controversies. *Science & Education*, 22(7), 1787–1807.

Gilbert, J (1997). Models in science and science education. In J. Gilbert (ed), *Exploring models and modelling in science and technology education: Contributions from the mistre group* (pp. 5–19). Reading: The University of Reading.

Gilbert, J. K. de Jong, O., Justi, R. Tragust, D. F. & van Driel, J. H. (2003). Research and development for the future of chemical education, In J. K. Gilbert, O. De Jong, R. Justi, D. F. Tragust & J. H. van Driel (Eds.), *Chemical education: Towards research-based practice (pp. 391–408)*. Dordrecht: Kluwer.

Gilbert, J. (2004). Models and modelling: Routes to more authentic science education. *International Journal of Science and Mathematics Education*, 2, 115–130.

Good, R. J. (1999). Why are chemists turned off by philosophy? *Foundations of Chemistry*, 1, 65–96.

Goodwin, W. M. (2008). Structural formulas and explanation in organic chemistry. *Foundations of Chemistry*, 10, 117–127.

Gillespie, R. J. (1997). Commentary: Reforming the general chemistry textbook. *Journal of Chemical Education*, 74(5), 484–485.

Heeren, J. K. (1990). Teaching chemistry by the Socratic Method. *Journal of Chemical Education*, 67(4), 330–331.

Hendry, R. (2010). The chemical bond: structure, energy and explanation. In, Mauro Dorato, Miklos Redei and Mauricio Suarez (eds.), *EPSA: Philosophical Issues in the Sciences: Launch of the European Philosophy of Science Association*, pp. 117–127. Berlin: Springer.

Hendry, R. (2012). The Metaphysics of Chemistry. Oxford University Press.

Herron, J. D. (1996). *The chemistry classroom*. Washington DC: American Chemical Society.

Ihde, D. (2003). *Chasing Technoscience: Matrix for Materiality. Bloomington: Indiana University Press.*

Irzik, G. (2000). Back to basics: A philosophical critique of constructivism. *Science & Education*, 9(6), 621–639. Doi: 10.1023/A:1008765215314.

Izquierdo-Aymerich, M. (2013). School chemistry: an historical and philosophical approach. *Science & Education*, 22(7), 1633–1653.

Jacob, C. (2001). Interdependent operations in chemical language and practice. *HYLE–International Journal for Philosophy of Chemistry*, 7(1), 31–50.

Jones, G. & Jacob, C. (2003), Report: Teaching Philosophy of Chemistry at the University of Exeter, *HYLE - International Journal for Philosophy of Chemistry*, 9(1), 126–128.

Justi, R. & Gilbert, J. (2003). Teachers' views on the nature of models. *International Journal of Science Education*, 25(11), 1369–1386.

Justi, R. (2000). Teaching with historical models. In J.K. Gilbert & C.J. Boutler (eds.), *Developing models in science education* (pp. 209–226). Dordrecht: Kluwer.

Kauffman, G.B. (1989). History in the chemistry curriculum, *Interchange* 20(2), 81–94. Reprinted in M.R. Matthews (ed.) *History, Philosophy and Science Teaching: Selected Readings*, OISE Press, Toronto, 1991, pp. 185–200.

Kaya, E., & Erduran, E. (2013). Integrating epistemological perspectives on chemistry in chemical education: the cases of concept duality, chemical language and structural explanations. *Science & Education*, 22(7), 1741–1755.

Kovac, J. (2004). The Ethical Chemist: Professionalism and Ethics in Science. Prentice Hall, Upper Saddle River.

Kovac, J. (2000). Professionalism and ethics in chemistry. *Foundations of Chemistry*, 2: 207–219.

Kripke, S. (1971) Identity and necessity. In M. K. Munitz (Ed), Identity and individuation (p. 135–164) New York: New York University Press.

Laszlo, P. (1999). Circulation of concepts. *Foundations of Chemistry*, 1, 225–238.

Lazslo, P. (2013). Towards teaching chemistry as a language. *Science & Education*, 22(7), 1669–1706.

Lederman, N. G., Abd-El-Khalick, F., Bell, R. L., & Schwartz, R. S. (2002). Views of nature of science questionnaire: Toward valid and meaningful assessment of learners conceptions of nature of science. *Journal of Research in Science Teaching*, 39, 497–521.

Lemke, J. (1990). *Talking science: Language, learning and values*. NJ: Ablex, Norwood.

Loucks-Horsley, S., Brooks, J. G., Carlson, M. O., Kuerbis, P. J., Marsh, D. D., Padilla, M. J. (1990). Developing and supporting teachers for science education in the middle years. National Center for Improving Science Education, Andover, MA.

Loucks-Horsley, S., Hewson, P. W., Love, N., Stiles, K. E. (1998) *Designing professional development for teachers of science and mathematics*. Corwin Press, Thousand Oaks, CA.

Luisi, P.L. (2002). Emergence in Chemistry: Chemistry as the Embodiment of Emergence, *Foundations of Chemistry*, 4, 183–200.

Lythcott, J. (1990). Problem solving and requisite knowledge of chemistry. *Journal of Chemical Education*, 67(3), 248–252.

Matthews, M. (1994/2014). *Science teaching: The role of history and philosophy of science*, Dordrecht: Routledge.

Matthews, M. (Ed.). (1998). *Constructivism in Science Education: A Philosophical Examination*. Dordrecht: Kluwer.

McComas, W. (1998). The principal elements of the nature of science: Dispelling the myths. In W. F. McComas (Ed.), *The nature of science in science education: Rationales and strategies* (pp. 53–70). Dordrecht, The Netherlands: Kluwer.

McIntyre, L. & Scerri, E. (1997). The philosophy of chemistry - Editorial introduction. *Synthese*, *111*(3), 211–212.

Mugaloglu, E. Z. (2001). *Radical constructivism in science education*. Unpublished master thesis. Bogazici University, Turkey.

Nagel, E. (1961). The structure of science: *Problems in the logic of scientific explanation*. *New York: Harcourt, Brace and World*.

National Research Council. (1996). *National Science Education Standards*. Washington DC: National Academy Press.

Newman, M. (2013). Emergence, supervenience and introductory chemical education. *Science & Education*, 22(7), 1655–1667.

Niaz, M. (2008). *Teaching General Chemistry: A History and Philosophy of Science Approach*. New York: Nova Publishers.

Nussbaum, J. & Novak, J. D. (1979). Assessment of children's conceptions of the earth utilizing structured interviews. *Science Education*, 60, 535–550.

Papineau, D. (1993). *Philosophical naturalism*. Oxford: Blackwell.

Papineau, D. (1995). Arguments for Supervenience and Physical Realization. In Elias E. Savellos & U. Yalcin (eds.), *Supervenience: New Essays*. Cambridge University Press.

Penslar, R. L. (Ed.) (1995). *Research Ethics: Cases and Materials*. Bloomington, IN.

Primas, H. (1983). *Chemistry, Quantum Mechanics and Reduction*, Springer, Berlin.

Psarros, N. (1998). What has philosophy to offer to chemistry? *Foundations of Science*, 3(1), 183–202.

Putnam, H. (1975). The Meaning of Meaning, in Putnam, H. (ed.), *Mind, Language and Reality*, Cambridge University Press, Cambridge.

Resnik, D. B. (1998). *The Ethics of Science: An Introduction*. Routledge, London and New York.

Ribeiro, M. A. P. & Pereira, D. C. (2013). Constitutive pluralism of chemistry: thought planning, curriculum, epistemological and didactic orientations. *Science & Education*, 22(7), 1809–1837.

Rodriguez, M. A. and Niaz, M. (2002) How in Spite of the Rhetoric, History of Chemistry has Been Ignored in Presenting Atomic Structure in Textbooks, *Science & Education* 11: 423–441.

Salmon, M.H., Earman, J., Glymour, C., Lennox, J.G., Machamer, P., McGuire, J.E., Norton, J.D., Salmon, W.C., Schaffner, K.F. (1992). Introduction to the Philosophy of Science. Prentice Hall, Englewood Cliffs, NJ.

Scerri, E. R. (1991). Chemistry, spectroscopy and the question of reduction. *Journal of Chemical Education*, 68(2), 122–126.

Scerri, E. R. (1995). The exclusion principle, chemistry and hidden variables. *Synthese*, 102, 169–192.

Scerri, E. R. & McIntyre, L. (1997). The case for the philosophy of chemistry. *Synthese, 111*(3), 213–232.

Scerri, E. (1997). Are chemistry and philosophy miscible?, *Chemical Intelligencer*, 3, 44–46.

Scerri, E. (2000). Philosophy of chemistry—A new interdisciplinary field? *Journal of Chemical Education*, 77(4), 522–525.

Scerri, E. R (2003). Constructivism, relativism and chemistry in chemical explanation. In J. Earley (Ed.), *Chemical explanation: Characteristics, development, autonomy (pp. 359–369)*. New York: New York Academy of Sciences.

Scerri, E. R (2007). *The Periodic Table: Its Story and Its Significance. New York: Oxford University Press.*

Scerri, E. R (2008). *Collected papers on philosophy of chemistry*. London: Imperial College Press.

Shulman, L. S. (1986) Those who understand: knowledge growth in teaching. *Educational Research*, 15(2): 4–14.

Schummer, J. (2006). The philosophy of chemistry: From infancy toward maturity. In D. Baird, E. Scerri & L. McIntyre (Eds.), *Philosophy of chemistry: Synthesis of a new discipline* (pp. 19–39). Dordrecht: Springer.

Schwab, J. J. (1962). The teaching of science as enquiry. In J. J. Schwab & P. F. Brandwein (Eds.), *The teaching of science*, Cambridge, MA: Harvard University Press.

Sjostrom, J. (2013). Towards Bildung-oriented chemistry education. *Science & Education,* 22(7), 1873–1890.

Stanford Dictionary of Philosophy (2012). Available online at http://plato.stanford.edu/.

Stroll, A. (1991). Observation and the Hidden, *Dialectica* 45(2–3), 165–179.

Taber, K. (2006). Constructivism's new clothes: the trivial, the contingent and a progressive research programme into the learning of science. *Foundations of Chemistry*, 8, 189–219.

Talanquer, V. (2013). School chemistry: the need for transgression. *Science & Education*, 22(7), 1757–1773.

Thalos, M. (2013). The lens of chemistry. *Science & Education*, 22(7), 1707–1721.

Tobin, E. (2013). Chemical laws, idealisation and approximation. *Science & Education*, 22(7), 1581–1592.

Van Brakel, J. (2000). *Philosophy of chemistry: Between the manifest and the scientific image.* Leuven: Leuven university press.

Van Brakel, J. (2010). A subject to think about: Essays on the history and philosophy of chemistry. *Ambix, 57*(2), 233–234.

Van Brakel, J. (1997). Chemistry as the science of the transformation of substances. *Synthese, 111*(3), 253–282.

Vesterinen, V. M., Aksela, M, & Lavonen, J. (2013). Quantitative analysis of representations of nature of science in Nordic secondary school textbooks using framework of analysis based on philosophy of chemistry. *Science & Education*, 22(7), 1839–1855.

Vihalemm, R. (2003). Are laws of nature and scientific theories peculiar in chemistry? Scrutinizing Mendeleev's discovery. *Foundations of Chemistry*, 5(1), 7–22.

Vilches, A. & Gil-Perez, D. (2013). Creating a sustainable future: some philosophical and educational considerations for chemistry teaching. *Science & Education*, 22(7), 1857–1872.

Vygotsky, L. S. (1978). Mind in society. Cambridge, MA: Harvard University Press.

Wallace J. & Louden, W. (2000). *Dilemmas of science teaching: perspectives on problems of practice*. Dordrecht: Kluwer.

Wasserman, E., & Schaefer, H. F. (1986). Methylene geometry, *Science*, 233, p. 829.

Weisberg, M. (2006). Water is not H_2O. In D. Baird, E. Scerri & L. McIntyre (Eds.), *Philosophy of chemistry: Synthesis of a new discipline*. Dordrecht: Springer.

Weisberg, M., Needham, P., & Hendry, R. (2011). Philosophy of chemistry. *Stanford Encyclopedia of Philosophy, Stanford University.*

Woody, A. (1995). The explanatory power of our models: A philosophical analysis with some implications for science education. In Finley F, Allchin D, Rhees D, & Fifield S (Eds), *Proceedings of the third international history, philosophy, and science teaching conference* (pp 1295–1304). Minneapolis: University of Minnesota.

Woody, A. (2000). Putting Quantum Mechanics to Work in Chemistry: The Power of Diagrammatic Representation, *Philosophy of Science* 67 (Proceedings): S612-S627.

Woody, A. I. (2004a). Telltale Signs: What Common Explanatory Strategies in Chemistry Reveal about Explanation Itself, *Foundations of Chemistry* 6: 13–43.

Woody, A. I. (2004b). "More Telltale Signs: What Attention to Representation Reveals about Scientific Explanation", *Philosophy of Science* 71: 780–793.

Woody, A. I. (2013). How is the ideal gas law explanatory? *Science & Education*, 22(7), 1563–1580.

Woody, A., Hendry, R., & Needham, P. (2011). *Handbook of the Philosophy of Science, Volume 6: Philosophy of Chemistry*. Elsevier Press.

Yarroch, W. L. (1985). Student understanding of chemical equation balancing. *Journal of Research in Science Teaching*, 22(5), 449–559.

Yeany, R. H. (1991). A unifying theme in science education?, *NARST News*, 33(2), 1–3.

Zeidler, D. L. (Ed.). (2003). *The role of moral reasoning on socioscientific issues and discourse in science education*. Dordrecht: Kluwer Academic Publishers.

Zoller, U. (1990). Students' misunderstandings and misconceptions in college freshman chemistry (General and Organic). *Journal of Research in Science Teaching*, 27(10), 1053–1065.

Zohar, A. (2004). *Higher order thinking in science classrooms: students' thinking and teachers' professional development*. Dordrecht, Boston, & London: Kluwer Academic Press.

Sibel Erduran is Chair of STEM Education at University of Limerick, Ireland. She has had visiting professorships at Kristianstad University, Sweden, and Bogazici University, Turkey. She has also worked at University of Pittsburgh and King's College, University of London and University of Bristol. She is an editor for *International Journal of Science Education*, section editor for *Science Education* and serves as Director on the IHPST Council. Her higher education was completed in the USA at Vanderbilt (Ph.D. science education and philosophy), Cornell (M.Sc. Food Chemistry) and Northwestern (Biochemistry) Universities. She has worked as a chemistry teacher in a high school in northern Cyprus. Her research interests focus on the applications in science education of epistemic perspectives on science in general and in chemistry in particular. She has co-edited a book (Erduran and Jimenez-Aleixandre 2008, Springer) on argumentation in science education, an area of research for which she has received an award from NARST. In 2013, she guest edited *Science & Education* consisting of 17 articles with the editorial entitled 'Philosophy, Chemistry and Education: An Introduction'.

Ebru Z. Mugaloglu is assistant professor of science education at Bogazici University, Istanbul, Turkey. She graduated from Bogazici University as a chemistry teacher in 1996 and completed her master's in philosophy in 2001 at the same university. She obtained her Ph.D. in science education from Marmara University in 2006. She was a visiting fellow in the UK at University of Reading in 2006 and at University of Bristol from 2010 to 2011. Her research interests include nature of science, values in science education, applications of philosophy of science in science education and science teacher training. Her recent journal articles related to nature of science are 'A Structural Model of Prospective Science Teachers' Nature of Science Views' published in *Scandinavian Journal of Educational Research* (with Hale Bayram, 2009) and 'Interactions of Economics of Science and Science Education: Investigating the Implications for Science Teaching and Learning' in *Science & Education* (with Sibel Erduran, 2012).

Chapter 11
The Place of the History of Chemistry in the Teaching and Learning of Chemistry

Kevin C. de Berg

11.1 Introduction

Numerous isolated appeals for the introduction of more history into the undergraduate chemistry curriculum have been made since the 1950s but with limited success. For example, Conant (1951) used the historical case study approach in teaching science to undergraduate students at Harvard, and his case studies included examples from chemistry, but the historical approach seemed to lapse in the succeeding decades. In 1989 a more coordinated approach was initiated with the formation of the International History, Philosophy and Science Teaching Group (IHPST). At this time Kauffman (1989) wrote a review article on the status of history in the chemistry curriculum in which he summarised the advantages and the disadvantages of using the historical approach. The advantages listed maintained that a study of chemistry in an historical context highlighted chemistry as a human enterprise, as a dynamic process rather than a static product, as depending on interrelationships between historical events, as often multidimensional in its discoveries, as a discipline with strengths and limitations and as depending on intuition as well as logic in its problem-solving activities. Kauffman (1989) also observed that on occasion an historical investigation has assisted the chemist in their current research. The discovery of the noble gas, argon, is quoted as an example (see also Giunta 1998). Lord Rayleigh and Sir William Ramsay published their discovery of argon in 1895 (Rayleigh and Ramsay 1895). Small anomalies found in measurements of the density of nitrogen samples prepared by different methods and the unexplained existence of a residue in Cavendish's (1785) experiments on the passing of electricity through air a century earlier led to the discovery.

K.C. de Berg (✉)
School of Science and Mathematics, Avondale College of Higher Education,
Cooranbong, NSW, Australia
e-mail: kdeberg@avondale.edu.au

M.R. Matthews (ed.), *International Handbook of Research in History,*
Philosophy and Science Teaching, DOI 10.1007/978-94-007-7654-8_11,
© Springer Science+Business Media Dordrecht 2014

The disadvantages of using the history of chemistry included the fact that there is a fundamental difference in goal and method between chemistry and history. While chemistry, like other sciences, abstracts, idealises, models and simplifies, history attempts to capture the richness of past events in their complexity. In spite of this difference, Kauffman challenges the reader 'to attempt to present to the student a harmonious balance between the two' (Kauffman 1989, p. 86). It is this harmonious balance between chemistry and history that is controversial amongst some professional chemists. If chemistry instruction is designed to enhance the practical skills of the chemist in a number of laboratory settings, for example, one might be able to successfully argue against the inclusion of chemical history in such instruction. If one's purpose, on the other hand, is to educate the student in the broader context of knowledge development and validation, then history is an essential component of chemistry education at the secondary and tertiary level. This would also apply whether the student was studying chemistry as a major discipline or whether the student was a nonmajor in chemistry. Niaz and Rodriguez (2001) argue, however, that history is not something that is added to chemistry. It is already inside chemistry as it were. According to this view, it is difficult to teach chemistry either for skills or understandings without interfacing with its history in some form.

A second disadvantage revolves around the difficulty associated with assessing material that is both historical and chemical. Students, by nature, tend to only take seriously material that is assessed, but the question is how this should be done. Another two disadvantages concern the inappropriate use of a distorted history, often called 'Whig' history, and the likelihood that young students might feel estranged from the study of chemistry when they learn that chemists have not always 'behaved as rational, open-minded investigators who proceed logically, methodically, and unselfishly toward the truth on the basis of controlled experiment' (Kauffman 1989, p. 87). Kauffman (1989) finally discussed briefly four approaches to incorporating history into the chemistry curriculum: the *biographical approach*, the *anecdotal approach*, the *case study approach* and the *classic experiments approach*.

Thirteen years after Kauffman's review, Wandersee and Baudoin-Griffard (2002) contributed a chapter on the history of chemistry in chemical education in a book dedicated to an appraisal of the status of chemical education at the beginning of the twenty-first century. It is interesting to ponder what similarities and differences in perception might be evident in these contributions over this 13-year period. Both articles identify the role of history in teaching about the nature of science (NOS) although by 2002 NOS had developed into a significant research area, whereas in 1989 it was only in the emergence phase. Wandersee and Baudoin-Griffard (2002) give more attention than did Kauffman (1989) to matters associated with student learning such as the comparison of student conceptions with early conceptions in the history of chemistry, the idea of meaningful and mindful (transferable) learning in understanding chemistry and some evidence that supports the notion that exposure to some history of chemistry enhances the learning of chemistry. Wandersee and Baudoin-Griffard (2002), like Kauffman (1989), deal with approaches to incorporating history into the chemistry curriculum, but they focus on Interactive

Historical Vignettes which are 'a series of lively, carefully crafted, brief (~15 min), interactive' (Wandersee and Baudoin-Griffard 2002, p. 34) stories tailored to the chemical concepts being studied. These authors lament the fact that only anecdotal evidence is available as of 2002 for the effectiveness of this approach to chemistry teaching and learning.

There has continued to be a burgeoning literature on this topic since 2002 to the extent that a process of categorisation is almost mandatory if one is to make any sense of the research in the field. It has therefore been decided to review the literature using five focus categories: (1) Student Learning, (2) Conceptual Clarity and Development, (3) Chemical Epistemology and the Nature of Science, (4) Pedagogy and Curriculum, and (5) Human Biography. While a large number of articles will deal with more than one of these categories, they will be discussed largely under the category which represents the major focus of the article.

11.2 Student Learning

When one considers the relationship between the history of chemistry and the learning of chemistry, there are two major considerations addressed by the literature. Firstly, there is an interest in the extent to which student conceptions in chemistry mirror those of the early scientists (Piaget and Garcia 1980). Secondly, there is an interest in whether the incorporation of the history of chemistry within chemistry teaching and learning has an impact on chemistry achievement.

The interest in comparing student conceptions with those possessed by scientists or chemists in the past has to do with the capacity of this scholarship to alert teachers to the kinds of thinking patterns of students that might present some resistance to change. Being aware of the history of the concept may provide clues that can assist the teacher in promoting conceptual change. While science educators agree that this might be achievable in some circumstances, they doubt that this can be achieved in all circumstances. It has been noted that 'Students' conceptions with limited empirical foundation.... have a completely different ontological status to empirically based ideas that are carefully formulated and sharpened by debate among scientific peers' (Scheffel et al. 2009, p. 219). Given this proviso these workers examined the significance of student conceptions in the light of current and historical knowledge in the areas of the particulate nature of matter, structure–property relations, ionic bonding, covalent bonding and organic chemistry and macromolecular chemistry. In the case of ionic bonding in crystals, the historical use of particle shape both edgy (Hauy 1743–1822) and ball-like (Hooke 1635–1703) to explain crystal shape was also found to exist in students' thinking (Griffith and Preston 1992). On the other hand, in the case of the concept of isomerism, it was concluded that 'The importance of isomerism in the history of science does not correspond to the importance of isomerism in school' (Scheffel et al. 2009, p. 244), because students' difficulties with the concept do not correlate with historical ideas (Schmidt 1992). This was also the case for the octet rule in covalent bonding (Taber 1997, 1998). Even though 'the number of concrete studies comparing historical ideas and students' conceptions is fairly low in chemistry education' (Scheffel et al. 2009, p. 220), there are some studies of importance outlined below.

A questionnaire study (Furio-Mas et al. 1987) of students' conception of gases was undertaken with 1,198 pupils aged 12–18 years in Valencia. It was shown that the majority of students tended to adopt an Aristotelian view of a gas in that they believed gases have no weight because they rise rather than fall. In addition, for chemical reactions involving gases as reactants or products, the students thought that mass was not conserved. Younger students adopted a pre-seventeenth-century nonmaterial view of a gas. Fifty-nine science major students enrolled in Chemistry I at a university in Venezuela were asked to respond to a problem which asked them to select which of four particle distribution models represented hydrogen gas at a lower temperature than the one shown in the problem (Niaz 2000a). The most common distribution chosen was that which resembled a 'lattice' structure similar to that understood by scientists before the random distribution model deduced from the kinetic theory of gases in the nineteenth century.

A questionnaire and interview study of 54 year eight Barcelona students' understanding of mixtures, compounds and physical and chemical properties (Sanmarti and Izquierdo 1995) revealed that a significant number assigned a material nature to properties like colour and taste, a view that was held from the sixteenth to the eighteenth century. For example, on observing the dissolution of blue copper sulphate in water, one student said, 'the blue colour of the crystal can leave and pass into the water' (Sanmarti and Izquierdo 1995, p. 361). When blue copper sulphate crystals were heated, the colour change was explained as 'the water evaporates, and when it evaporates it carries this (blue) substance (with it)' (Sanmarti and Izquierdo 1995, p. 361). Sanmarti and Izquierdo (1995) use the term 'substantialisation of properties' to describe this phenomenon.

Van Driel et al. (1998) undertook a study of chemical equilibrium with 120 students aged 15–16 in the Netherlands. Original papers by Williamson (1851–1854), Clausius (1857) and Pfaundler (1867) were used to compare students' written responses to a questionnaire and group oral responses on audiotape with the nineteenth-century historical understanding. The reasoning students used to explain the incompleteness of a chemical reaction resembled the reasoning used by scientists of the nineteenth century particularly when the corpuscular model was used. However, the 'explanations remained incomplete or naïve. The few students capable of giving adequate explanations…implemented statistical notions in their explanations, analogous to Pfaundler's explanation of 1867' (Van Driel et al. 1998, p. 195). Niaz obtained results on a chemical equilibrium study that showed 'that at least some students consider the forward and reverse reactions as a sort of chemical analogue of Newton's third law of motion' (Niaz 1995a, p. 19), that is, action and reaction are equal and opposite.

Cotignola and colleagues (2002) interviewed 31 volunteers from science and engineering courses, 2 years after having studied basic thermodynamics, about the energetic processes associated with material sliding down inclined planes. The students used the word 'heat' predominantly in their explanations and were not able to distinguish it from

11   The Place of the History of Chemistry in the Teaching and Learning of Chemistry   321

internal energy. The authors suggest that Clausius followed a similar course when developing the field of thermodynamics in 1850 by focussing on the difference between sensitive heat and latent heat. The students ideas were not as sophisticated of course.

Although the literature comparing historical chemical ideas with student conceptions is not extensive, as previously mentioned, the reader should be aware of the large body of research in the general area of student conceptions. Classic references such as the handbook entry by Wandersee et al. (1994) and those addressing chemistry conceptions[1] are worth reading to put the historical ideas reported here in perspective. Research techniques for diagnosing and interpreting student conceptions can be found in DiSessa (1993), Taber and Garcia-Franco (2010), and Treagust (1988, 1995).

Moving on now to our second point of interest, what can one say about the use of the history of chemistry and chemistry achievement? The literature is not decisive on this matter. Using an experimental and control group of 14-year-olds where the experimental group was given a substantial amount of historical material and taught the same science content as the control group who were not presented with the historical material, Irwin (2000) observed that there was no significant difference between the groups in their understanding of contemporary science content related to atomic theory and periodicity. This was in spite of the fact that the historical approach did portray the nature of science more realistically. However, Lin (1998) did a similar study with 220 eighth graders where the experimental groups studied the historical cases of atmospheric pressure and atoms, molecules and formulae. All experimental and control groups were given four questions requiring conceptual problem solving in the science content. The experimental group did significantly better in conceptual problem solving. Lin et al. (2002) achieved similar results with a group of 74 eighth graders for chemistry conceptual problem-solving ability. The different outcomes to the Irwin study may be due to the nature, not necessarily the validity, of the science content test instruments, and this is worth exploring in further research.

A related matter to that in the previous paragraph is the relationship between history of chemistry and chemistry assessment. Niaz and colleagues (2002) have attempted to show how chemistry might be assessed within the context of historical experiments. In the case of Rutherford's gold foil experiment, for example, a suggested assessment item might be: What might you have deduced if most of the alpha particles were deflected through large angles? Perhaps the relationship between history of chemistry and chemistry achievement might depend on how closely the chemistry content interfaces with the history. This issue requires a more sustained research effort during this decade.

11.3   Conceptual Clarity and Development

History lends itself to giving depth and clarity to concepts, but we know that there is often a compromise between such an approach and that which focuses on the relatively quick generation of an answer to a problem. De Berg (2008a) has discussed

---

[1] For example, Andersson (1990), Garnett et al. (1995), Kind (2004), and Taber (2002).

this issue in terms of an approach which emphasises *conceptual depth* over and against *conceptual usefulness* for the chemistry concepts of *energy, heat and work, element, mole* and the *uncertainty principle*. Others (Holme and Murphy 2011) define the difference in terms of *conceptual knowledge* and *algorithmic knowledge*[2]. The *Journal of Chemical Education* publishes many articles which focus on the history of chemistry and its role in giving clarity to concepts. There are at least *eighty-five* such articles published from 2005 to June 2011. Many of these articles show their historical character by having a title commencing with the words 'The Origin of....' The majority of these papers were written by Professor William Jensen who occupies the chair for the History of Chemistry at the University of Cincinnati. Table 11.1 samples Professor Jensen's 'The Origin of....' titles from 2005 to June 2011 with the *Journal of Chemical Education* references included.

Let us take one example from Table 11.1, The Origin of the *s, p, d, f* Orbital Labels, to illustrate how useful these titles can be in enlightening the significance of the symbols we commonly use in chemistry to represent concepts. Jensen (2007a) shows that the symbols originated around 1927 and represented the different line series present in alkali metal spectra. These lines were distinguished using the adjectives *sharp, principal, diffuse* and *fundamental*. The symbols, *s, p, d* and *f* were thus taken from the first letter of the names of these four series of lines and applied to the description of electron orbitals because line spectra were attributed to electron transitions between orbitals. It appears that Friedrich Hund was the first to use this nomenclature.

A sampling of 2010, 2011 and some 2012 articles from the *Journal of Chemical Education* which use historical information to bring clarity to the concepts of chemistry, other than 'The Origin Series' in Table 11.1, is given in Table 11.2. Most yearly issues of the journal contain articles which could be classified into at least some of the eight categories in Table 11.2 and serve as a rich resource for chemistry educators. The processes of chemistry which lead to the products of chemistry, some of which are shown in Table 11.2, also have a rich history. For example, an historical approach to the process of distillation 'where the old is redeemed to complement the new' (Lagi and Chase 2009, p. 5) provides a deeper understanding of the separation process in a modern context.

Eric Scerri (2007, 2009) has devoted a large portion of his working life to bringing clarity to the so-called periodic law and the structure of the *periodic table*. Many of the issues such as the difference between thinking of an element as a *basic substance* or a *simple substance* and the concept of *reductionism* are philosophical in nature and will be dealt with in another chapter of the handbook. But Scerri also involves the history of the development of the periodic table to highlight:

1. The renewed importance of Prout's hypothesis particularly if one regards atomic number as an important building block of the elements. Prout's hypothesis proposed that all the elements were compound forms of hydrogen. Accurate atomic weight determinations cast some doubt on the hypothesis in the nineteenth

---

[2] See Nakhleh (1993), Nakhleh et al. (1996), Nurrenbern and Pickering (1987), Pickering (1990), and Zoller et al. (1995) for earlier references.

11 The Place of the History of Chemistry in the Teaching and Learning of Chemistry   323

Table 11.1 A sample of 'The Origin of ….' titles written by William Jensen from 2005 to June 2011 and published in the *Journal of Chemical Education*

Title	Reference
The Origin of the Bunsen Burner	(2005a), 82(4), p. 518
The Origin of the 18-Electron Rule	(2005b), 82(1), p. 28
The Origin of the Liebig Condenser	(2006a), 83(1), p. 23
The Origin of the Term 'Allotrope'	(2006b), 83(6), p. 838
The Origin of the s, p, d, f Orbital Labels	(2007a), 84(5), p. 757
The Origin of the Names Malic, Maleic, and Malonic Acid	(2007b), 84(6), p. 924
The Origin of the Polymer concept	(2008a), 85(5), p. 624
The Origin of the Rubber Policeman	(2008b), 85(6), p. 776
The Origin of the Metallic Bond	(2009a), 86(3), p. 278
The Origin of the Circle Symbol for Aromaticity	(2009b), 86(4), p. 423
The Origin of the Ionic-Radius Ratio	(2010d), 87(6), pp. 587–588
The Origin of the Name 'Onion's Fusible Alloy'	(2010e), 87(10), pp. 1050–1051
The Origin of Isotope Symbolism	(2011), 88(1), pp. 22–23

Table 11.2 Historical examples from the *Journal of Chemical Education* (2010–2012) which clarify the concepts of chemistry

Chemistry profile	Examples	Reference
The products of chemistry	Synthetic dyes	Sharma et al. (2011)
	Quinine	Souza and Porto (2012)
The constants of chemistry	Avogadro's constant	Jensen (2010a)
	Atomic Mass, Avogadro's constant, mole	Barariski (2012)
The instrumentation of chemistry	pH meters	Hines and de Levie (2010)
The species of chemistry	Hydrogen ion	Moore et al. (2010)
The laws of chemistry	First law of thermodynamics	Rosenberg (2010)
	Thermodynamics-globalisation and first law	Gislason and Craig (2011)
	Clausius equality and inequality	Nieto et al. (2011)
The symbols of chemistry	R (organic), q, Q (thermodynamics)	Jensen (2010b), (2010c)
The models of chemistry	Bohr-Sommerfeld	Niaz and Cardellini (2011)
	Electronegativity	Jensen (2012)
The phenomena of chemistry	Fluorescence and phosphorescence	Valeur and Berberan-Santos (2011)

century, but a rehabilitation of the hypothesis became possible in the twentieth century based on the concept of atomic number.

2. The significance of the atomic number *triads* in developing a structure for the periodic table. The best form for representing the periodic table is still a matter of dispute. This fact is commonly not recognised by chemists. Scerri (2009) currently favours a form based on the atomic number triad which leads to a very symmetrical table with four groups to the left and four groups to the right of the transition series. The third and fourth transition series should commence with the elements lutetium and lawrencium rather than lanthanum and actinium on

Table 11.3 Some key chemistry concepts discussed in the journal *Science & Education* from an historical perspective including some references

Key chemistry concept	Reference
Gas laws	de Berg (1995), 4(1), pp. 47–64; Woody (2011) online first 6/12/11
Atomic theory	Chalmers (1998), 7(1), pp. 69–84
	Sakkopoulos and Vitoratos (1996), 5(3), pp. 293–303; Viana and Porto (2010), 19(1), pp. 75–90
Work, kinetic and potential energy	de Berg (1997a), 6(5), pp. 511–527
Kinetics	Justi and Gilbert (1999), 8(3), pp. 287–307
Electrolytic dissociation	de Berg (2003), 12(4), pp. 397–419
Acid–base equilibria	Kousathana et al. (2005), 14(2), pp. 173–193
Osmotic pressure	de Berg (2006), 15(5), pp. 495–519
Quantum mechanics	Hadzidaki (2008), 17(1), pp. 49–73
Heat and temperature	de Berg (2008b), 17(1), pp. 75–114
Mole concept	Padilla and Furio-Mas (2008), 17(4), pp. 403–424
Chemical equilibrium	Quilez (2009), 18(9), pp. 1203–1251
Electrochemistry	Eggen et al. (2012), 21(2), pp. 179–189

the basis of the atomic number triad but, this is still controversial. Published periodic tables as late as 2010 (e.g. Atkins and de Paula 2010) have not yet taken Scerri's suggestion seriously enough to change the format.

3. The illusions accompanying the nature of the periodic table. Significance is often given to Mendeleev's successful predictions of unknown elements, but it is rarely mentioned that only about 50 % of his predictions proved correct. The number of outer shell electrons is often used as the basis for the assignment of an element to a vertical group of the table. However, there are exceptions to this rule. Helium has the same number of outer shell electrons as the alkaline earth metals but is normally placed with the noble gases because of its inert characteristics. Nickel, palladium and platinum are in the same vertical group but have a different outer shell electron configuration.

4. The fact that the periodic system was discovered essentially independently by six scientists. Of these six, Mendeleev has been given the greatest credit for various reasons even though it could be argued that the German chemist Lothar Meyer was the first to produce, in 1864, a mature periodic system which was even more accurate than that produced by Mendeleev in 1869.

The journal, *Science & Education*, is dedicated to conceptual clarity through the lens of history and philosophy. A summary of some of the key concepts in chemistry which have been addressed in this journal is given in Table 11.3.

Some key chemistry concepts such as work and energy, fundamental to an understanding of thermodynamics, contain mathematical formulations of rich historical significance. For example, de Berg indicates that:

the mathematical relationship, $mgh = \frac{1}{2} mv^2$, for free fall, could have been known from the time of Galileo and Newton….but the physical significance of the equation was not recognized till the early 19th century. That is, while the mathematics was in place by the 17th century, the fact that $\frac{1}{2}\ mv^2$ and mgh were measures of fundamental quantities was not known for 200 years. The physical notions of mechanical action (work) and force of a body in motion (kinetic energy) had separate historical developments…(but) their relationship (was finally) recognized in the 19th century and ultimately this paved the way for the development of the general concept of energy. (de Berg 1997a, p. 515)

The historical approach to the mathematical equations associated with chemistry concepts adds physical and conceptual significance to the equations beyond their algorithmic value.

11.4 Chemical Epistemology and the Nature of Science

How a chemist forms and validates chemical knowledge is central to an understanding of the nature of chemistry or chemical literacy. There is some debate about what is meant by the terms 'chemical literacy' and 'nature of chemistry' or the more general expressions 'scientific literacy' and 'nature of science'. For example, McComas et al. (1998) isolated what they considered to be 14 consensus statements regarding the nature of science (NOS), Abd-El-Khalick (1998, 2005) suggested seven statements, and Niaz (2001b) used eight statements. Unanimity of opinion is hard to reach when it comes to defining NOS. A useful summary of the issues is given by Lederman (2006).

Some authors claim that a study of the history of chemistry enhances an understanding of the NOS. For example, Irwin (2000) exposed an experimental group of 14-year-olds to historical episodes associated with the concept of the atom and the periodicity of the elements and found gains, compared to a control group, in understanding aspects of the nature of science such as the usefulness of theories even when there may be some uncertainty about the validity of a theory. Lin and Chen (2002) observed that pre-service chemistry teachers' understanding about the NOS was promoted by a study of the history of chemistry. In particular, the experimental group had a better understanding of the nature of creativity, the theory-based nature of scientific observations and the functions of theories. However, Abd-El-Khalick and Lederman (2000) found that coursework in the history of science (included atomic theory) does not necessarily enhance students' and pre-service science teachers' views of the NOS unless specific aspects of the NOS are also addressed.

Rasmussen (2007) has suggested that exposure to the history of chemistry in general chemistry classes can help students identify pseudoscientific attitudes in advertising. For example, the suggestion is made that introducing students not only to our current understanding of matter but to understandings held over centuries, some of which were erroneous, helps students address such assignment tasks as:

A favourite claim of many advertisers is that their product is all-natural and thus contains no chemicals. In terms of our class lectures, explain why this is or is not a valid claim. (Rasmussen 2007, p. 951)

Giunta (2001) also focuses on errors that have surfaced in the development of chemical knowledge but from the point of view of the value that erroneous theories, such as the phlogiston theory, have played in furthering our knowledge of chemistry. Dalton's atomic theory, while containing some misplaced ideas according to our current knowledge, was an important stepping stone in leading to the concept of atomic weight. On the other hand, Giunta (2001) shows how a correct hypothesis such as Avogadro's hypothesis was rejected by a number of chemists at the time it was proposed for understandable reasons. The diatomic molecule proposal did not prove compelling enough to chemists to warrant acceptance of Avogadro's hypothesis. Giunta observes that 'the right hypothesis languished or at least struggled for decades' (Giunta 2001, p. 625). This illustrates how difficult it is for the scientific community to transition from one scientific model to another.

The notion of errors in the production of knowledge leads naturally into the significance of historical controversies in the progress of scientific knowledge.

De Berg (2003) has outlined the issues which were involved in the controversy between the Arrhenius School and the Armstrong School at the close of the nineteenth century in relation to the interpretation of what happens at the molecular level when a salt is dissolved in water, the so-called electrolytic dissociation controversy. One of the interesting factors associated with this controversy is the orientation taken to anomalous data. In the data produced by Raoult (1882a, b, 1884), it was clear that the molecular lowering factor associated with freezing point depression for sodium chloride (35.1) was close to double that for ethanol (17.3) and that for calcium chloride (49.9) close to three times that for ethanol. This data was consistent with the electrolytic dissociation hypothesis. The molecular lowering data for magnesium sulphate (19.2) and copper sulphate (18.0) proved anomalous however. One would have expected values close to those for sodium chloride (35.1) if the electrolytic dissociation hypothesis was applicable.

Fortunately these anomalies were held in suspension until they were explained in terms of the production of ion pairs due to the strong charges associated with both cation and anion. Chemists have learnt how futile it is to dispense with theoretical models too early as anomalies often lead to new knowledge provided one is happy to hold them in tension for a period of time. Sometimes anomalies will lead to a new paradigm such as a view of the solid state which includes aperiodic quasicrystals which have a non-repeating pattern at the microscopic level. Until Nobel Laureate Dan Shechtman (Nobel Prize in Chemistry 2011) discovered these in 1982, it was thought that one could not have a crystal without the existence of a repeating pattern of atoms. Controversy highlights how important it is for students to see chemistry as a human enterprise (Niaz 2009). It also indicates the dynamic nature of chemical knowledge, a point emphasised by modern philosophers of science (Machamer et al. 2000).

Chemical history can also be helpful in showing how the knowledge of a particular chemical compound has changed and progressed over time. De Berg (2008c, 2010) has illustrated the strength of this approach using the compound, tin oxide. One can discuss the chemistry of tin oxide over the three periods of chemical revolution described by Jensen (1998a, b, c): the period associated with the determination of chemical composition (1770–1790) at the macroscopic level, at the microscopic level (1855–1875) and finally at the electronic level (1904–1924). The nature of the chemistry associated with the development of an understanding of tin(IV) oxide in particular is shown by de Berg (2010) to involve, progressively from about 1800 to the present, descriptive chemistry, compositional studies, structural studies and advanced materials research. This kind of study gives a deep perspective to current research and might be one way of attracting more practising chemists to take an interest in the history of their subject.

When it comes to the development of a new chemical compound or a commercially viable form of a known compound, one must not forget the role that developments in the broader community such as that in economics, politics, technology and industry play in such developments. Coffey (2008, Chaps. 4 and 6) gives an insightful historical background to the commercial manufacture of ammonia by Haber and Bosch in the early twentieth century. What made the discovery so crucial was the perceived impending famine about to strike in Britain and Europe and its relief through the use of ammonia for the fertiliser industry. Ammonia was also earmarked for its role in the explosives industry, particularly at the onset of war in Europe. Chemical compounds can save lives; but they can unfortunately also take lives.

What is interesting about the historical approach to a discipline is how history pinpoints changes in the nature of discipline knowledge itself. In chemistry, for example, this is particularly noticeable in the way chemists described chemical reactions. In the case of combustion reactions, Joseph Priestley applied the phlogiston model for understanding the chemical change. The heating of a metal in air resulted in the release of phlogiston (the inflammability principle) from the metal to produce the calx. The concept of 'principle' was important in chemistry up until the end of the eighteenth century, although it did retain some use into the nineteenth century, so that chemists talked about the inflammability principle, the acid principle, the alkaline principle, the electrical principle, the magnetic principle and so on. Toward the end of the eighteenth century, however, Antoine Lavoisier claimed it was better to think of combustion of a metal in air as a chemical combination of the metal with the oxygen in the air. Chemical reactions were increasingly described in terms of atoms, ions and molecules rather than in terms of 'principles'. The Priestley-Lavoisier debate as a debate in terms of the nature of chemical knowledge is discussed by de Berg (2011).

11.5 Pedagogy and Curriculum

One way of describing chemistry curricula is to examine the textbooks used by teachers and students. It is not surprising then that chemistry textbooks have been targeted as a source of research into chemistry curricula. In particular, the focus here will be on the way chemistry history is portrayed and used in chemistry textbooks. Van Berkel, De Vos, Verdonk and Pilot regard textbook chemistry as portraying what Kuhn (1970) would have called 'normal science' in that 'normal chemistry education is isolated from common sense, everyday life and society, history and philosophy of science, technology, school physics and from chemical research' (Van Berkel et al. 2000, p. 123). The general tenor of the research on chemistry textbooks has been rather critical of the portrayal and use of history when it has appeared, and more detail will be given in another chapter of the handbook. For the purposes of this section, some of the studies are summarised in Table 11.4 below.

Three of the references in Table 11.4 show how chemistry was portrayed in early textbooks, and it was often the case that the textbook was the main source of chemical information. France was the centre of the 'new chemistry' or the 'chemical revolution' with Lavoisier's influence predominating, and it is interesting to observe how this new chemistry was incorporated into textbooks of the era. Early textbooks of the twentieth century such as Partington's (1953) *Textbook of Inorganic Chemistry* contain significant amounts of historical material compared with later twentieth-century textbooks.

Researchers tend to be critical of more recent textbooks of chemistry either for the lack of historical material or for the way the historical material is presented. For example, Niaz (2000b) observed that, in discussing the oil drop experiment, the authors of chemistry textbooks did not give adequate treatment to the Millikan-Ehrenfest controversy. The oversimplification of the description of the experiment gives students the impression that the oil drop experiment yielded results with ease and without controversy. Holton (1978) has described how difficult this experiment was to perform and interpret. There are difficulties even when using modern apparatus (Klassen 2009). The question arises as to whether textbook authors can be expected to deal with historical material to the satisfaction of the historian or the chemist interested in history as well as presenting current trends in the subject. One option is to look at presenting the historical material in other ways.

Table 11.4 Some studies relating to the use of the history of chemistry in chemistry textbooks

Targeted chemistry concept	Reference
Covalent bonding	Niaz (2001a)
Models of the atom	Justi and Gilbert (2000)
Gases	de Berg (1989)
Electrochemistry	Boulabiar et al. (2004)
Periodic table	Brito et al. (2005)
Oil drop experiment	Niaz (2000b), Niaz and Rodriguez (2005)
Chemical revolution—late eighteenth century/early nineteenth century	Bertomeu-Sanchez and Garcia-Belmar (2006)
Chemical theories—late eighteenth century and early nineteenth century	Seligardi (2006)
Atomic structure	Niaz and Rodriguez (2002)
Aims and scope of chemistry in seventeenth-century France	Clericuzio (2006)
Amount of substance and mole	Furio-Mas et al. (2000)

11 The Place of the History of Chemistry in the Teaching and Learning of Chemistry   329

Hutchinson (2000) has taken the concepts typically taught in general chemistry in university courses and expressed them in terms of nine case studies. For example, Case Study 3 on 'Periodicity and Valence' 'uses the experimental facts which were actually used to develop these concepts, and so introduces an historical perspective to their learning' (Hutchinson 2000, p. 4). This approach to experimental data is used in all the case studies. The purpose of the case studies is to teach chemistry not history, but historical experiments are used to show students how concepts are developed and models are built and how to distinguish between the data and its interpretation. In relation to models and theories, Hutchinson counsels that:

> It is very important to understand that scientific models and theories are almost never *proven*, unlike mathematical theorems. Rather, they are logically developed and deduced to provide simple explanations of observed phenomenon. As such, you will discover many times in these Case Studies when a conclusion is not logically required by an observation and a line of reasoning. Instead, we may arrive at a model which is the simplest explanation of a set of observations, even if it is not the only one. (Hutchinson 1997, Preface)

This curriculum is used for general chemistry at Rice University.

Niaz (2008) has written a book entitled, *Teaching General Chemistry: A History and Philosophy of Science Approach*, which can be used as a companion text to the student textbook by teachers. The emphasis is on conceptual problem solving in contradistinction to algorithmic problem solving and is based on the premise that the difficulties students face in conceptualising problems are similar to the difficulties scientists of an earlier period in the history of chemistry faced. The general chemistry concepts featured in the text include the mole, stoichiometry, atomic structure, gases, energy and temperature and chemical equilibrium. The text draws heavily upon research data related to student understanding of chemistry concepts. The approach is best illustrated by an example. In the chapter on gases, Niaz defines his approach as follows:

> The main objective of this section is to construct models based on strategies students use to solve the gas problems and to show that these models form sequences of progressive transitions similar to what Lakatos (1970) in the history of science refers to as progressive 'problemshifts'. Guideline 1 (defined in his chapter 3) suggests a rational reconstruction of students' understanding of gases based on progressive transitions from the 'algorithmic mode' (work of Boyle and others in the 17th century) to 'conceptual understanding' (work of Maxwell and Boltzmann in the 19th century). Results reported here are from Niaz (1995b). (Niaz 2008, p. 67)

The results of a study of the responses of sixty ($N=60$) freshmen chemistry students to two items testing an understanding of gases are then discussed. The two items are shown below.

Item A
A certain amount of gas occupies a volume (V_1) at a pressure of 0.60 atm. If the temperature is maintained constant and the pressure is decreased to 0.20 atm, the new volume (V_2) of the gas would be:

$$(a)\, V_2 = V_1 / 6 \quad (b)\, V_2 = 0.33\, V_1 \quad (c)\, V_2 = V_1 / 3 \quad (d)\, V_2 = 3\, V_1$$

Item B

An ideal gas at a pressure of 650 mmHg occupied a bulb of unknown volume. A certain amount of the gas was withdrawn and found to occupy 1.52 mL at 1 atm pressure. The pressure of the gas remaining in the bulb was 600 mmHg. Assuming that all measurements were made at the same temperature, calculate the volume of the bulb (Niaz 2008, p. 68).

Niaz considers that *Item A* involves algorithmic problem solving and *Item B* conceptual problem solving. It was observed that 87 % of the students solved *Item A* correctly, whereas only 7 % of the students solved *Item B* correctly. The remaining students gained only partial credit for their answers. Niaz proposes that:

Based on (the) strategies used in solving *Items A* and *B* it is plausible to suggest that students go through the following process of progressive transitions...

Model 1: Strategies used to solve *Item A* correctly, that is, ability to manipulate the three variables of the Boyle's Law equation ($P_1V_1 = P_2V_2$) to calculate the fourth (N = 52).

Model 2: Strategies used to correctly identify the final volume in *Item B*, that is, partial conceptualization of the property of a gas when it is withdrawn from a vessel (N = 16).

Model 3: Strategies used to correctly identify and conceptualize two properties of a gas (final volume and pressure in *Item B*), when it is withdrawn from a vessel (N = 13).

Model 4: Strategies used to correctly identify and conceptualize all the variables of a gas (*Item B*) when it is withdrawn from a vessel (N = 4). (Niaz 2008, p. 68)

In Model 4, it could also be considered that a strategy involving the additive property of 'amount of gas' where 'amount of gas' was understood as either mass, moles or particle number is important. The particle model of a gas, endemic to kinetic theory, leads to this conclusion.

Another strategy for incorporating the history of chemistry in chemistry curricula is the development and use of what de Berg (2004) calls a pedagogical history. A pedagogical history combines a knowledge of chemistry, history of chemistry, student learning and philosophy of science to develop an instructional storyline which requests students to engage with the text. Where possible, students are asked to interact with historical experimental data and to make decisions about how well the data fits the model. For example, in the case of the electrolytic dissociation model, students are presented with a table of molecular lowering factors from historical sources and asked two questions as follows:

Question 1

Which data do you think fits the model and which data, if any, doesn't fit the model?

Question 2

Now assess, in your view, how strongly the data in the table supports an ionic dissociation model.

The table of data contained some anomalous results although it is true that the majority of the data supported the model, but students had to decide which pieces of information were anomalous and to wrestle with the concept of the weight of evidence. Student reactions to some pedagogical histories have been published, (de Berg 1997b) and the issues involved in selecting the historical data to include in a pedagogical history are presented in a publication dealing with the case of the concepts of heat and temperature (de Berg 2008b).

11.6 Human Biography

Arguments for including history of chemistry in chemistry teaching and learning have always included the thought that history humanises chemistry. To humanise chemistry, however, we need to know something about the life story of the chemists involved, that is, their human biography. It has been maintained that:

> There is particular value in viewing the historical aspect of chemistry through a study of the lives of important chemists because the development of chemical concepts can then be seen in the context of the experiences of fellow human beings......In essence, the students learn that the development of science is a function of the people who develop it and the environment in which they live. (Carroll and Seeman 2001, p. 1618)

Carroll and Seeman (2001) describe how they incorporated scientific autobiography into a senior undergraduate course in advanced organic chemistry. Students studied the autobiography of the organic chemist, Ernest L. Eliel, and five of his key articles published over a period of 40 years. Collaborative group work and oral presentations were a feature of the methodology used. One student commented that 'Learning about Eliel's life caused me to be more interested in understanding the chemistry in the journal articles. We were able to see how the logical progression of his scientific research coincided with his life' (Carroll and Seeman 2001, p. 1620).

While Carroll and Seeman (2001) combined a study of a chemist's life story with a study of five of his most important chemistry publications at the senior undergraduate level, they suggested that a softer approach is probably better at the introductory level. The use of interesting 'incidental information' was recommended which 'can help make a human connection with the abstract concepts and does not require much class time' (Carroll and Seeman 2001, p. 1619). In Table 11.5 some examples of 'incidental information' for a range of chemists has been assembled with some important biographical references.

Table 11.5 Incidental information for a range of chemists along with important biographical references

Chemist	Incidental information	Biographical reference
Robert Boyle (1627–1691)	Very rich; wore a wig; never married; had an interest in alchemy and the turning of base metals into gold; had poor eyesight; intensely religious and supported the translation of the bible into different languages	Hunter (2009)
Joseph Priestley (1733–1804)	Discovered dephlogisticated air (oxygen); used his wife's kitchen as a laboratory; had a speech impediment but taught oratory; his house was burnt down because of his sympathies with the American and French Revolutions; was a dissenting minister; encouraged by Benjamin Franklin to take up science as a serious study	Matthews (2009) Schofield (1997, 2004)

(continued)

Table 11.5 (continued)

Chemist	Incidental information	Biographical reference
Michael Faraday (1791–1867)	Started work as a book binder; learnt chemistry from Humphry Davy; became famous for his chemistry demonstrations at the Royal Institution in London; gave us the names 'anode' and 'cathode' in electrochemistry; 96,500 coulombs per mole named after him	Williams (1965)
Dmitri Mendeleev (1834–1907)	Had 13 siblings; born in Siberia; his mother encouraged him to take up science; dynamic educator who attracted students from all faculties of the university to his lectures; fond of art; organised special classes in chemistry for women although he believed women to be inferior to men intellectually; famous for the periodic table; always pictured with a cigarette in his hand; believed in only getting his hair cut once a year	Byers and Bourgoin (1998) Scerri (2007)
Svante Arrhenius (1859–1927)	Swedish with stocky build, ruddy complexion, blonde hair and blue eyes; loved scientific controversy; his Ph.D. regarded as not of sufficient standard for an academic position but granted the Nobel Prize in chemistry in 1903 for his electrolytic dissociation theory; only married for a short time as his wife objected to his drinking and smoking; one of the first chemists to talk about the greenhouse effect	Crawford (1996)
Marie Curie (1867–1934)	Discoverer of the radioactive substance, radium; married Pierre; twice a Nobel Prize winner; 1903 shared Nobel Prize in Physics with husband Pierre and Henri Becquerel for work on radioactivity; won 1911 Nobel Prize in chemistry for discovering radium and polonium; had to work against gender bias; disapproved of fashion in dress; reared in poverty	Goldsmith (2005)
Martha Whiteley (1866–1956)	In a male-dominated field, she played a critical role on the academic staff of Imperial College London and secured admission of women chemists to the Chemical Society. She edited the multivolume Thorpe's Dictionary of Applied Chemistry	Nicholson and Nicholson (2012)
Ernest Rutherford (1871–1937)	Country boy from the South Island of New Zealand; known for the development of simple but elegant experiments on the atomic nucleus; although not religious would sing 'Onward Christian Soldiers' with volume when an experimental breakthrough occurred; his ashes are buried in Westminster Abbey near that of Sir Isaac Newton	Campbell (1999) Reeves (2008) Wilson (1983)
Gilbert Lewis (1875–1946)	Famous for proposing the electron pair covalent bond and the octet rule; homeschooled entirely through elementary school; could read at age 3; learnt five languages; reserved in nature	Coffey (2008)

11 The Place of the History of Chemistry in the Teaching and Learning of Chemistry 333

Table 11.6 A list of chemists discussed in the *Journal of Chemical Education* from the year 2000 to June 2011 along with the author reference

Chemist	Reference
Boerhaave (1668–1738)	Diemente (2000), 77(1), p. 42
Rutherford (1871–1937)	Sturm (2000), 77(10), p. 1278
Faraday (1791–1867)	Clark (2001), 78(4), p. 449
Pauling (1901–1994)	Davenport (2002), 79(8), p. 946
Mendeleev (1834–1907)	Marshall (2003), 80(8), p. 879
Priestley (1733–1804)	Williams (2003), 80(10), p. 1129
Bohr (1885–1962)	Peterson (2004). 81(1), p. 33
Porter (1920–2002)	Kovac (2004), 81(4), p. 489
Lavoisier (1743–1794)	Jensen (2004), 81(5), p. 629
Starkey (1628–1665)	Schwartz (2004), 81(7), p. 953
Boltzmann (1844–1906)	David (2006), 83(11), p. 1695
Haber (1868–1934)	Harris (2006), 83(11), p. 1605
Mendeleev (1834–1907)	Benfey (2007), 84(8), p. 1279
Boyle (1627–1691)	Williams (2009), 86(2), p. 148

Table 11.7 A sample of short biographies written by George Kauffman and published in the *Chemical Educator*

Chemist	Reference
Moses Gomberg (1866–1947)	(2008a), 13(1), pp. 28–33
Arthur Kornberg (1918–2007)	(2008a), 13(1), pp. 34–41 (with J. Adloff)
Antoine Henri Becquerel (1852–1908)	(2008b), 13(2), pp. 102–110
Frederic Joliot (1900–1958)	(2008c), 13(3), pp. 161–169
Fred Allison (1882–1974)	(2008b), 13(6), pp. 358–364 (with J. Adloff)
Gerald Schwarzenbach (1904–1978)	(2008d), 13(6), pp. 365–373
Dwaine O. Cowan (1935–2006)	(2009), 14(3), pp. 118–129
Osamu Shimomura (1928–)	(2009), 14(2), pp. 70–78 (with J. Adloff)
Alfred Maddock (1917–2009)	(2010a), 15, pp. 237–242 (with J. Adloff)
Marie and Pierre Curie (1859–1906)	(2010b), 15, pp. 344–352 (with J. Adloff)
Marie Curie (1867–1934)	(2011a), 16, pp. 29–40 (with J. Adloff)
Robert Bunsen (1811–1899)	(2011b), 16, pp. 119–128 (with J. Adloff)
John Bennett Fenn (1917–2010)	(2011c), 16, pp. 143–148 (with J. Adloff)
William Nunn Lipscomb (1919–2011)	(2011d), 16, pp. 195–201 (with J. Adloff)

From time to time, the *Journal of Chemical Education* will publish some useful and interesting biographical material on an important chemist. In Table 11.6 is recorded a list of chemists discussed in this journal from the year 2000 to June 2011. Some of the references report on an important book review.

George Kauffman writes short biographies of chemists for the *Chemical Educator* and a selection from 2008 to 2011 is shown in Table 11.7.

11.7 Conclusion

Much progress has been made in humanising the teaching and learning of chemistry through history since the first IHPST conference in 1989. However, measuring the impact of the history of chemistry on the teaching and learning of chemistry is still an area that needs further investigation. There appears to be no clear answer as far as academic achievement is concerned. Anecdotal evidence suggests that attitudes to and interest in chemistry can be improved by the historical approach, but well-planned research studies need to explore this possible relationship in more detail. We have noted some interesting ways that the history of chemistry has been used in chemistry curricula but if the impact is to be strengthened and grown, one will need to consolidate the history with the content of chemistry or, one might argue, to consolidate the current content with the history. I think that this will be the only way that teachers of chemistry will become convinced of the value of including an historical perspective. Clough (2009) has endeavoured to integrate history with content using thirty case studies, six of which are in chemistry. De Berg (2008c, 2010) has shown how history can embellish an understanding of a chemical compound from its antiquity to current research. Carroll and Seeman (2001) have shown how publications of a chemist from the embryonic stage of a career to the mature stage, that is, according to a chemist's historical journey, can be used to inform current chemistry content. These efforts have at least begun the journey of not only humanising but informing current content.

References

Abd-El-Khalick, F. (1998). *The influence of History of Science courses on students' conceptions of the nature of science.* Unpublished doctoral dissertation, Oregon State University, Oregon, USA.

Abd-El-Khalick, F. (2005). Developing deeper understandings of nature of science: The Impact of a philosophy of science course on pre-service science teachers' views and instructional planning. *International Journal of Science Education, 27*(1), 15–42.

Abd-El-Khalick, F. & Lederman, N.G. (2000). The Influence of history of science courses on students' views of nature of science. *Journal of Research in Science Teaching, 37*(10), 1057–1095.

Andersson, B. (1990). Pupils' conceptions of matter and its transformation. *Studies in Science Education, 18*, 53–85.

Atkins, P. & de Paula, J. (2010). *Atkins' Physical Chemistry (9th edition).* Oxford: Oxford University Press.

Barariski, A. (2012). The Atomic Mass Unit, the Avogadro Constant, and the Mole: A Way to Understanding. *Journal of Chemical Education, 89*(1), 97–102.

Benfey, T. (2007). Book Review: Mendeleev on the Periodic Law: Selected Writings 1869–1905 (William Jensen ed.). *Journal of Chemical Education, 84*(8), 1279.

Bertomeu-Sanchez, J.R.B. & Garcia-Belmar, A.G. (2006). Pedro Gutierrez Buenos Textbooks: Audiences, Teaching Practices and the Chemical Revolution. *Science & Education, 15*(7–8), 693–712.

Boulabiar, A., Bouraoui, K., Chastrette, M. & Abderrabba, M. (2004). A Historical Analysis of the Daniell Cell and Electrochemistry Teaching in French and Tunisian Textbooks. *Journal of Chemical Education, 81*(5), 754.

Brito, A., Rodriguez, M.A. & Niaz, M. (2005). A reconstruction of development of the periodic table based on history and philosophy of science and its implications for general chemistry textbooks. *Journal of Research in Science Teaching*, 42(1), 84–111.

Byers, K. & Bourgoin, M. (Eds.). (1998). *Encylopedia of World Biography (2nd edition)*. Detroit, Michigan: Gale Group.

Campbell, J. (1999). *Rutherford, Scientist Supreme*. Christchurch, New Zealand: AAS Publications.

Carroll, F.A. & Seeman, J.I. (2001). Placing Science into Its Human Context: Using Scientific Autobiography to Teach Chemistry. *Journal of Chemical Education*, 78(12), 1618–1622.

Cavendish, H. (1785). Experiments on Air. *Philosophical Transactions of the Royal Society of London*, 74, 372–384.

Chalmers, A. (1998). Retracing the Ancient Steps to Atomic Theory. *Science & Education*, 7(1), 69–84.

Clark, R.W. (2001). The Physics Teacher: Faraday as a Lecturer. *Journal of Chemical Education*, 78(4), 449.

Clausius, R. (1857). On the Nature of the Motion which we call Heat. *Annalen*, 50, 108–127.

Clericuzio, A. (2006). Teaching Chemistry and Chemical Textbooks in France from Beguin to Lemery. *Science & Education*, 15(2–4), 335–355.

Clough, M. (2009). Humanizing Science to Improve Post-Secondary Science Education. Paper presented at the 10th IHPST Conference, June 24–28, University of Notre Dame, USA.

Coffey, P. (2008). *Cathedrals of Science*. Oxford: Oxford University Press.

Conant, J.B. (1951). *On Understanding Science: An Historical Approach*. New York: New American Library.

Cotignola, M., Bordogna, C., Punte, G. & Cappinnini, O.M. (2002). Difficulties in learning thermodynamic concepts: Are they linked to the Historical Development of this Field. *Science & Education*, 11(3), 279–291.

Crawford, E. (1996). *Arrhenius: From Ionic Theory to the Greenhouse Effect*. Canton, MA: Science History Publications.

Davenport, D.A. (2002). Major Additions to the Linus Pauling Canon. *Journal of Chemical Education*, 79(8), 946.

David, C.W. (2006). Boltzmann without Lagrange. *Journal of Chemical Education*, 83(11), 1695.

de Berg, K.C. (1989). The Emergence of Quantification in the Pressure-Volume Relationship for Gases: A Textbook Analysis. *Science Education*, 73(2), 115–134.

de Berg, K.C. (1995). Revisiting the Pressure-Volume Law in History-What can it Teach us about the Emergence of Mathematical Relationships in Science? *Science & Education*, 4(1), 47–64.

de Berg, K.C. (1997a). The Development of the Concept of Work: A Case where History can inform Pedagogy. *Science & Education*, 6(5), 511–527.

de Berg, K.C. (1997b). Student Responses to a Pedagogical History of the Pressure-Volume Law. In I. Winchester (Ed.), *Toward Scientific Literacy-Proceedings of the 4th IHPST Conference* (pp. 178–187). University of Calgary.

de Berg, K.C. (2003). The Development of the Theory of Electrolytic Dissociation. *Science & Education*, 12(4), 397–419.

de Berg, K.C. (2004). The Development and Use of a Pedagogical History for a key Chemical Idea-The case of ions in solution. *Australian Journal of Education in Chemistry*, 64, 16–19.

de Berg, K.C. (2006). The Kinetic-Molecular and Thermodynamic approaches to Osmotic Pressure: A study of dispute in physical chemistry and the implications for chemistry education. *Science & Education*, 15(5), 495–519.

de Berg, K.C. (2008a). Conceptual Depth and Conceptual Usefulness in Chemistry: Issues and Challenges for Chemistry Educators. In I.V. Eriksson (Ed.), *Science Education in the 21st Century* (pp. 165–182). New York: Nova Science Publishers.

de Berg, K.C. (2008b). The Concepts of Heat and Temperature: The Problem of Determining the Content for the Construction of an Historical Case Study which is sensitive to Nature of Science Issues and Teaching-Learning Issues. *Science & Education*, 17(1), 75–114.

de Berg, K.C. (2008c). Tin Oxide Chemistry from Macquer (1758) to Mendeleeff (1891) as revealed in the textbooks and other literature of the era. *Science & Education*, 17(2–3), 265–287.

de Berg, K.C. (2010). Tin Oxide Chemistry from the Last Decade of the Nineteenth Century to the First Decade of the Twenty-First Century: Towards the Development of a Big-Picture Approach to the Teaching and Learning of Chemistry While Focussing on a Specific Compound or Class of Compounds. *Science & Education*, 19(9), 847–866.

de Berg, K.C. (2011). Joseph Priestley across Theology, Education, and Chemistry: An Interdisciplinary Case Study in Epistemology with a Focus on the Science Education Context. *Science & Education*, 20(7–8), 805–830.

Diamente, D. (2000). Boerhaave on Fire. *Journal of Chemical Education*, 77(1), 42.

DiSessa, A. (1993). Towards an epistemology of physics. *Cognition and Instruction*, 10(2–3), 105–225.

Eggen, P., Kvittingen, L., Lykknes, A. & Wittje, R. (2012). Reconstructing Iconic Experiments in Electrochemistry: Experiences from a HOS Course. *Science & Education*, 21(2), 179–189.

Furio-Mas, C.J., Hernandez-Perez, J.H. & Harris, H.H. (1987). Parallels between Adolescents' Conception of Gases and the History of Chemistry. *Journal of Chemical Education*, 64(7), 616–618.

Furio-Mas, C.J., Azcona, R., Guisasola, J. & Ratcliffe, M. (2000). Difficulties in teaching the Concepts Amount of Substance and the Mole. *International Journal of Science Education*, 22(12), 1285–1304.

Garnett, P.J., Garnett, P.J. & Hackling, M.W. (1995). Students' Alternative Conceptions in Chemistry: a Review of Research and Implications for Teaching and Learning. *Studies in Science Education*, 25, 69–95.

Gislason, E.A. & Craig, N.C. (2011). The Global Formulation of Thermodynamics and the First Law: 50 years on. *Journal of Chemical Education*, 88(11), 1525–1530.

Giunta, C.J. (1998). Using History To Teach Scientific Method: The Case of Argon. *Journal of Chemical Education*, 75(10), 1322–1325.

Giunta, C.J. (2001). Using History to Teach Scientific Method: The Role of Errors. *Journal of Chemical Education*, 78(5), 623–627.

Goldsmith, B. (2005). *Obsessive Genius-The Inner World of Marie Curie*. London: Weidenfeld & Nicolson.

Griffith, A.K. & Preston, K.R. (1992). Grade-12 students' misconceptions relating to fundamental characteristics of atoms and molecules. *Journal of Research in Science Teaching*, 29(6), 611–628.

Hadzidaki, P. (2008). Quantum Mechanics and Scientific Explanation: An Explanatory Strategy Aiming at Providing Understanding. *Science & Education*, 17(1), 49–73.

Harris, H. (2006). Book Review: Fritz Haber: Chemist, Nobel Laureate, German, Jew. A Biography by Dietrich Stoltzenberg. *Journal of Chemical Education*, 83(11), 1605.

Hines, W.G. & de Levie, R. (2010). The Early Development of electronic pH meters. *Journal of Chemical Education*, 87(11), 1143–1153.

Holme, T. & Murphy, K. (2011). Assessing Conceptual and Algorithmic Knowledge in General Chemistry with ACS Exams. *Journal of Chemical Education*, 88(9), 1217–1222.

Holton, G. (1978). Subelectrons, presuppositions, and the Millikan-Ehrenhaft Dispute. *Historical Studies in the Physical Sciences*, 9, 161–224.

Hunter, M. (2009). *Boyle: Between God and Science*. New Haven: Yale University Press.

Hutchinson, J.S. (1997). *Case Studies in Chemistry* (4th edition). Alliance Press.

Hutchinson, J.S. (2000). Teaching Introductory Chemistry using Concept Development Case Studies: Interactive and Inductive Learning. *University Chemistry Education*, 4(1), 3–9.

Ibo, R. (2010). Chemistry Curriculum. *Chemistry in Australia*, 77(9), 5.

Irwin, A.R. (2000). Historical case Studies: Teaching the nature of science in context. *Science Education*, 84(1), 5–26.

Jensen, W.B. (1998a). Logic, History, and the chemistry textbook I: Does Chemistry have a logical Structure? *Journal of Chemical Education*, 75(6), 679–687.

Jensen, W.B. (1998b). Logic, History, and the chemistry textbook II: Can we unmuddle the Chemistry Textbook? *Journal of Chemical Education*, 75(7), 817–828.

11 The Place of the History of Chemistry in the Teaching and Learning of Chemistry 337

Jensen, W.B. (1998c). Logic, History, and the chemistry textbook III: One chemical revolution or three? *Journal of Chemical Education*, 75(8), 961–969.

Jensen, W.B. (2004). Did Lavoisier blink? *Journal of Chemical Education*, 81(5), 629.

Jensen, W.B. (2005a). The Origin of the Bunsen Burner. *Journal of Chemical Education*, 82(4), 518.

Jensen, W.B. (2005b). The Origin of the 18-Electron Rule. *Journal of Chemical Education*, 82(1), 28.

Jensen, W.B. (2006a). The Origin of the Liebig Condenser. *Journal of Chemical Education*, 83(1), 23.

Jensen, W.B. (2006b). The Origin of the term "Allotrope". *Journal of Chemical Education*, 83(6), 838.

Jensen, W.B. (2007a). The Origin of the s, p, d, f orbital labels. *Journal of Chemical Education*, 84(5), 757.

Jensen, W.B. (2007b). The Origin of the names Malic, Maleic, and Malonic Acid. *Journal of Chemical Education*, 84(6), 924.

Jensen, W.B. (2008a). The Origin of the Polymer concept. *Journal of Chemical Education*, 85(5), 624.

Jensen, W.B. (2008b). The Origin of the Rubber Policeman. *Journal of Chemical Education*, 85(6), 776.

Jensen, W.B. (2009a). The Origin of the Metallic Bond. *Journal of Chemical Education*, 86(3), 278.

Jensen, W.B. (2009b). The Origin of the Circle Symbol for Aromaticity. *Journal of Chemical Education*, 86(4), 423.

Jensen, W.B. (2010a). Why has the value of Avogadro's Constant changed over time? *Journal of Chemical Education*, 87(12), 1302.

Jensen, W.B. (2010b). Why is R used to symbolise Hydrocarbon substituents? *Journal of Chemical Education*, 87(4), 360–361.

Jensen, W.B. (2010c). Why are q and Q used to symbolise heat? *Journal of Chemical Education*, 87(11), 1142.

Jensen, W.B. (2010d). The Origin of the Ionic-Radius ratio. *Journal of Chemical Education*, 87(6), 587–588.

Jensen, W.B. (2010e). The Origin of the Name "Onion's Fusible Alloy". *Journal of Chemical Education*, 87(10), 1050–1051.

Jensen, W.B. (2011). The Origin of Isotope Symbolism. *Journal of Chemical Education*, 88(1), 22–23

Jensen, W.B. (2012). The Quantification of Electronegativity: Some Precursors. *Journal of Chemical Education*, 89(1), 94–96.

Justi, R. & Gilbert, J.K. (1999). History and Philosophy of Science through Models: The Case of Chemical Kinetics. *Science & Education*, 8(3), 287–307.

Justi, R. & Gilbert, J.K. (2000). History and Philosophy of Science through Models: Some challenges in the case of 'the atom'. *International Journal of Science Education*, 22(9), 993–1009.

Kauffman, G.B. (1989). History in the Chemistry Curriculum. *Interchange*, 20(2), 81–94.

Kauffman, G.B. (2008a). Moses Gomberg (1866–1947), Father of Organic Free Radical Chemistry: A Retrospective View on the 60th Anniversary of his Death. *Chemical Educator*, 13(1), 28–33.

Kauffman, G.B. (2008b). Antoine Henri Becquerel (1852–1908), Discoverer of Natural Radioactivity: A Retrospective View on the Centenary of His Death. *Chemical Educator*, 13(2), 102–110.

Kauffman, G.B. (2008c). Frederic Joliot (1900–1958), Codiscoverer of Artificial Radioactivity: A Retrospective View on the 50th Anniversary of his Death. *Chemical Educator*, 13(3), 161–169.

Kauffman, G.B. (2008d). Gerald Schwarzenbach (1904–1978) and the School of Anorganische Chemie at the Universitat Zurich, Heir Apparent to Alfred Werner (1866–1919). *Chemical Educator*, 13(6), 365–373.

Kauffman, G.B. (2009). Dwaine O. Cowan (1935–2006): The Father of Organic Conductors & Superconductors. *Chemical Educator*, 14(3), 118–129.

Kauffman, G.B. & Adloff, J. (2008a). Arthur Kornberg (1918–2007), Precursor of DNA Synthesis. *Chemical Educator*, 13(1): 34–41.

Kauffman, G.B. & Adloff, J. (2008b). Fred Allison's Magneto-Optic Search for Elements 85 and 87. *Chemical Educator*, 13(6), 358–364.

Kauffman, G.B. & Adloff, J. (2009). The 2008 Nobel Prize in Chemistry: Osamu Shimomura, Martin Chalfie & Roger Y Tsien: The green Fluorescent Protein. *Chemical Educator*, 14(2), 70–78.

Kauffman, G.B. & Adloff, J. (2010a). Alfred G. Maddock (1917–2009), An Inspired Radiochemist. *Chemical Educator*, 15, 237–242.

Kauffman, G.B. & Adloff, J. (2010b). Marie and Pierre Curie's 1903 Nobel Prize. *Chemical Educator*, 15, 344–352.

Kauffman, G.B. & Adloff, J. (2011a). Marie Curie's 1911 Nobel Prize. *Chemical Educator*, 16, 29–40.

Kauffman, G.B. & Adloff, J. (2011b). Robert Wilhelm Eberhard Bunsen (1811–1899), Inspired 19th century chemist: A Retrospective View on the Bicentennial of his Birth. *Chemical Educator*, 16, 119–128.

Kauffman, G.B. & Adloff, J. (2011c). Nobel Laureate John Bennett Fenn (1917–2010), Electrospray Ionization Mass Spectrometry Pioneer. *Chemical Educator*, 16, 143–148.

Kauffman, G.B. & Adloff, J. (2011d). William Nunn Lipscomb, Jr (1919–2011), Nobel Laureate and Borane Chemistry Pioneer: An Obituary-Tribute. *Chemical Educator*, 16, 195–201.

Kind, V. (2004). Beyond appearances. Students' misconceptions about basic chemical ideas'. A report prepared for the Royal Society of Chemistry (2nd edition). http://modeling.asu.edu/modeling/KindVanessaBarkerchem.pdf. Consulted on 27th March 2012.

Klassen, S. (2009). Identifying and addressing student difficulties with the Millikan Oil Drop Experiment. *Science & Education*, 18, 593–607.

Kousathana, M., Demerouti, M. & Tsaparlis, G. (2005). Instructional Misconceptions in Acid–base Equilibria: an analysis from a History and Philosophy of Science Perspective. *Science & Education*, 14(2), 173–193.

Kovac, J. (2004). The Chemcraft Story: The Legacy of Harold Porter. *Journal of Chemical Education*, 81(4), 489.

Kuhn, T. (1970). *The Structure of Scientific Revolutions*. Chicago: University of Chicago Press.

Lagi, M. & Chase, R.S. (2009). Distillation: Integration of a historical perspective. *Australian Journal of Education in Chemistry*, 70, 5–10.

Lakatos, I. (1970). Falsification and the methodology of scientific research Programmes. In I . Lakatos & A. Musgrave (Eds.), *Criticism and the growth of knowledge* (pp. 91–195). Cambridge: Cambridge University Press.

Lederman, N.G. (2006). Research on Nature of Science: Reflections on the Past, Anticipations of the Future. *Asia-Pacific Forum on Science Learning and Teaching*, 7(1), 1–11.

Lin, H. (1998). The Effectiveness of Teaching Chemistry through the History of Science. *Journal of Chemical Education*, 75(10), 1326–1330.

Lin, H. & Chen, C. (2002). Promoting pre-service chemistry teachers' understanding about the nature of science through history. *Journal of Research in Science Teaching*, 39(9), 773–792.

Lin, H., Hung, J. & Hung, S. (2002). Using the history of science to promote students' problem-solving ability. *International Journal of Science Education*, 24(5), 453–464.

Machamer, P., Pera, M. & Baltas, A. (Eds.). (2000). *Scientific Controversies: Philosophical and Historical Perspectives*. New York: Oxford University Press.

Marshall, J.L. (2003). Oliver Sacks in Mendeleev's Garden. *Journal of Chemical Education*, 80(8), 879.

Matthews, M.R. (2009). Science and Worldviews in the Classroom: Joseph Priestley and Photosynthesis. In M. Matthews (Ed.), *Science, Worldviews and Education.*, Dordrecht, The Netherlands: Springer.

McComas, W.F., Almazroa, H. & Clough, M.P. (1998). The Nature of Science in Science Education: An Introduction. *Science & Education*, 7(6), 511–532.

11 The Place of the History of Chemistry in the Teaching and Learning of Chemistry 339

Moore, C.E., Jaselskis, B. & Florian, J. (2010). Historical Development of the hydrogen ion concept. *Journal of Chemical Education*, 87(9), 922–923.

Nakhleh, M.B. (1993). Are our students Conceptual Thinkers or Algorithmic Problem Solvers? *Journal of Chemical Education*, 70(1), 52–55.

Nakhleh, M.B., Lowrey, K.A. & Mitchel, R.C. (1996). Narrowing the gap between Concepts and Algorithms in Freshman Chemistry. *Journal of Chemical Education*, 73(8), 758–762.

Niaz, M. (1995a). Chemical equilibrium and Newton's Third Law of Motion: Ontogeny/Phylogeny Revisited. *Interchange*, 26(1), 19–32.

Niaz, M. (1995b). Progressive Transitions from Algorithmic to Conceptual Understanding in student ability to solve chemistry problems : A Lakatosian Interpretation. *Science Education*, 79, 19–36.

Niaz, M. (2000a). Gases as Idealized Lattices: A Rational Reconstruction of Students' Understanding of the Behaviour of Gases. *Science & Education*, 9(3), 279–287.

Niaz, M. (2000b). The Oil Drop Experiment: A Rational Reconstruction of the Millikan-Ehrenhaft Controversy and Its Implications for Chemistry Textbooks. *Journal of Research in Science Teaching*, 37(5), 480–508.

Niaz, M. (2001a). A Rational Reconstruction of the origin of the Covalent Bond and its implications for General Chemistry Textbooks. *International Journal of Science Education*, 23(6), 623–641.

Niaz, M. (2001b). Understanding nature of science as progressive transitions in heuristic principles. *Science Education*, 85, 684–690.

Niaz, M. (2008). *Teaching General Chemistry: A History and Philosophy of Science Approach*. New York: Nova Science Publishers.

Niaz, M. (2009). *Critical Appraisal of Physical Science as a Human Enterprise: Dynamics of Scientific Progress*. Dordrecht, The Netherlands: Springer.

Niaz, M., Aguilera, D., Maza, A. & Liendo, G. (2002). Arguments, contradictions, resistances and conceptual change in students' understanding of atomic structure. *Science Education*, 86, 505–525.

Niaz, M. & Cardellini, L. (2011). What can the Bohr-Sommerfeld Model show students of chemistry in the 21st century. *Journal of Chemical Education*, 88(2), 240–243.

Niaz, M. & Rodriguez, M.A. (2001). Do we have to introduce history and philosophy of science or is it already 'inside' chemistry? *Chemistry Education: Research and Practice in Europe*, 2, 159–164.

Niaz, M. & Rodriguez, M.A. (2002). How in spite of the Rhetoric, History of Chemistry has been ignored in presenting Atomic Structure in Textbooks. *Science & Education*, 11(5), 423–441.

Niaz, M. & Rodriguez, M.A. (2005). The Oil Drop Experiment: Do Physical Chemistry Textbooks refer to its Controversial Nature? *Science & Education*, 14(1), 43–57.

Nicholson, R.M. & Nicholson, J.W. (2012). Martha Whiteley of Imperial College London. A Pioneering Woman Chemist. *Journal of Chemical Education*, 89(5), 598–601.

Nieto, R., Gonzalez, C., Jimenez, A., Lopez, I. & Rodriguez, J. (2011). A Missing Deduction of the Clausius Equality and Inequality. *Journal of Chemical Education*, 88(5), 597–601.

Nobel Prize in Chemistry (2011). www.nobelprize.org/nobel_prizes/chemistry/laureates/2011/press.html, consulted on 27 April 2012.

Nurrenbem, S.C. & Pickering, M. (1987). Concept Learning versus Problem Solving: Is there a difference? *Journal of Chemical Education*, 64(6), 508–510.

Padilla, K. & Furio-Mas, C. (2008). The Importance of HPS in Correcting Distorted Views of 'Amount of Substance' and 'Mole' Concepts in Chemistry Teaching. *Science & Education*, 17(4), 403–424.

Partington, J.R. (1953). *A Textbook of Inorganic Chemistry (6th edition)*. London: Macmillan.

Peterson, A.R. (2004). The "Dissing" of Niels Bohr. *Journal of Chemical Education*, 81(1), 33.

Pfaundler, L. (1867). Beitrage zur Chemischen Statik [Contribution to chemical statics]. *Poggendorfs Annalen der Physik und Chemie*, 131, 55–85.

Piaget, J. & Garcia, R. (1980). *Psicogenesis e historia de la ciencia*, Siglo XXI, Mexico. Translated by Helga Feider (1988, 2nd English edition) as, *Psychogenesis and the History of Science*. New York: Columbia University Press.

Pickering, M. (1990). Further Studies on Concept Learning versus Problem Solving. *Journal of Chemical Education*, 67(3), 254–255.

Quilez, J. (2009). From Chemical Forces to Chemical Rates: A Historical/Philosophical Foundation for the Teaching of Chemical Equilibrium. *Science & Education*, 18(9), 1203–1251.

Raoult, F.M. (1882a). Loi de congélation des solutions aqueuses des matières Organiques. *Comptes Rendus*, 94, 1517.

Raoult, F.M. (1882b). Loi générale de congélation des dissolvants. *Comptes Rendus*, 95, 1030.

Raoult, F.M. (1884). Loi générale des congélations des dissolvants. *Annales de chimie et de physique*, 2(vi), 66.

Rasmussen, S.C. (2007). The History of Science as a Tool to Identify and Confront Pseudoscience. *Journal of Chemical Education*, 84(6), 949–951.

Rayleigh, J.W. & Ramsay, W. (1895). Argon, a New Constituent of the Atmosphere. *Philosophical Transactions of the Royal Society of London A*, 186, 187–241.

Reeves, R. (2008). *A Force of Nature: The Frontier Genius of Ernest Rutherford*. New York: W.W Norton,

Rosenberg, R.M. (2010). From Joule to Caratheodory and Born: A Conceptual Evolution of the First Law of Thermodynamics. *Journal of Chemical Education*, 87(7), 691–693.

Sakkopoulos, S.A. & Vitoratos, E.G. (1996). Empirical Foundations of Atomism in ancient Greek philosophy. *Science & Education*, 5(3), 293–303.

Sanmarti, N. & Izquierdo, M. (1995). The substantialisation of properties in pupils' thinking and in the history of chemistry. *Science & Education*, 4(4), 349–369.

Scerri, E.R. (2007). *The Periodic Table-Its Story and Its Significance*. Oxford: Oxford University Press.

Scerri, E.R. (2009). *Selected Papers on the Periodic Table*. London: Imperial College Press.

Scheffel, L., Brockmeier, W. & Parchmann, I. (2009). Historical Material in Macro-Micro Thinking: Conceptual Change in Chemistry Education and the History of Chemistry. In J.K. Gilbert & D.F. Treagust (Eds.), *Multiple Representations in Chemical Education*. The Netherlands: Springer.

Schmidt, H.J. (1992). *Harte Nusse im Chemieunterricht [Tough Nuts in Chemistry Classroom]*. Frankfurt: Diesterweg.

Schofield, R.E. (1997). *The enlightenment of Joseph Priestley. A study of his life and work from 1733–1773*. University Park, Pennsylvania: Pennsylvania State University Press.

Schofield, R.E. (2004). *The enlightenment of Joseph Priestley. A study of his life and work from 1773–1804*. University Park, Pennsylvania: Pennsylvania State University Press.

Schwartz, A.T. (2004). Gehennical Fire: The lives of George Starkey, an American alchemist in the Scientific Revolution. *Journal of Chemical Education*, 81(7), 953.

Seligardi, R. (2006). Views of Chemistry and Chemical Theories: A Comparison between two University Textbooks in the Bolognese Context at the beginning of the 19th century. *Science & Education*, 15(7–8), 713–737.

Sharma, V., McKone, H.T. & Markow, P.G. (2011). A Global Perspective on the History, Use, and Identification of Synthetic Food Dyes. *Journal of Chemical Education*, 88(1), 24–28.

Souza, K. & Porto, P.A. (2012). History and Epistemology of Science in the Classroom: The Synthesis of Quinine as a Proposal. *Journal of Chemical Education*, 89(1), 58–63.

Sturm, J.E. (2000). Ernest Rutherford, Avogadro's Number, and Chemical Kinetics Revisited. *Journal of Chemical Education*, 77(10), 1278.

Taber, K.S. (1997). *Understanding Chemical Bonding*. Non-published PhD Thesis, Roehampton Institute, University of Surrey.

Taber, K.S. (1998). An alternative conceptual framework from chemistry education. *International Journal of Science Education*, 20(5), 597–608.

Taber, K.S. (2002). *Chemical Misconceptions-Prevention, Diagnosis and Cure*. London: Royal Society of Chemistry.

Taber, K.S. & Garcia-Franco, A. (2010). Learning processes in chemistry. Drawing upon cognitive resources to learn about the particulate structure of matter. *Journal of the Learning Sciences*, 19(1), 99–142.

Treagust, D.F. (1988). Development and use of diagnostic tests to evaluate students' misconceptions in science. *International Journal of Science Education*, 10(2), 159–169.

Treagust, D.F. (1995). Diagnostic Assessment of Students' Science Knowledge. In S.M. Glynn & R. Duit (Eds.), *Learning Science in the Schools. Research Reforming Practice* (pp. 327–346). New York: Lawrence Erlbaum Associates.

Valeur, B. & Berberan-Santos, M.N. (2011). A Brief History of Fluorescence and Phosphorescence before the emergence of Quantum Theory. *Journal of Chemical Education*, 88(6), 731–738.

Van Berkel, B., De Vos, W., Verdonk, A.H. & Pilot, A. (2000). Normal Science Educaton and its Dangers: The Case of School Chemistry. *Science & Education*, 9(1–2), 123–159.

Van Driel, J.H., De Vos, W. & Verloop, N. (1998). Relating Students' Reasoning to the History of Science: The Case of Chemical Equilibrium. *Research in Science Education*, 28(2), 187–198.

Viana, H.E.B. & Porto, P.A. (2010). The Development of Dalton's Atomic Theory as a Case Study in the history of science: Reflections for Educators in Chemistry. *Science & Education*, 19(1), 75–90.

Wandersee, J.H. & Baudoin-Griffard, P. (2002). The History of Chemistry: Potential and Actual Contributions to Chemical Education. In J.K. Gilbert, O. De Jong, R. Justi, D.F. Treagust & J.H. Van Driel (Eds.), *Chemical Education: Towards Research-Based Practice*, Dordrecht, The Netherlands: Kluwer.

Wandersee, J.H., Mintzes, J.J. & Novak, J.D. (1994). Research on alternative conceptions in science. In D. Gabel (Ed.), *Handbook of Research on Science Teaching and Learning*, New York: Macmillan publishing.

Williams, L.P. (1965). *Michael Faraday-A Biography*. London: Chapman and Hall.

Williams, K.R. (2003). The Discovery of Oxygen and other Priestley Matters. *Journal of Chemical Education*, 80(10), 1129.

Williams, K.R. (2009). Robert Boyle: The Founder of Modern Chemistry. *Journal of Chemical Education*, 86(2), 148.

Williamson, A.W. (1851–1854). Suggestions for the dynamics of chemistry derived from the theory of etherification. *Notices of the Proceedings at the meetings of the Members of the Royal Institution*, 1, 90–94.

Wilson, D. (1983). *Rutherford: Simple Genius*. Cambridge, MA: MIT Press.

Woody, A.I. (2011). How is the Ideal Gas Law Explanatory? *Science & Education*, online first 6/12/2011.

Zoller, U., Lubezky, A., Nakhleh, M.B., Tessler, B. & Dori, Y.J. (1995). Success on Algorithmic and LOCS vs. Conceptual Chemistry Exam Questions. *Journal of Chemical Education*, 72, 987–989.

Kevin C. de Berg is an Associate Professor in Chemistry and Head of the School of Science and Mathematics at Avondale College of Higher Education in Australia. Kevin's undergraduate degrees in science and in education were obtained from the University of Queensland (UQ). His Ph.D. in chemistry was obtained within the physical chemistry department at UQ in 1978 and the MAppSc by thesis obtained in science education from Curtin University of Technology in Perth in 1989. Kevin is widely published in science and chemistry education journals and is a member of the editorial committee for *Science & Education*. He has a special interest in the history of chemistry and how such a study might inform current practice in chemistry education and within the chemistry profession itself. Topics which have received significant attention in Kevin's published work have been the gas laws, the concepts of energy and work, heat and temperature, the periodic table, osmotic pressure, the electrolytic dissociation theory and the chemistry of tin oxides.

Chapter 12
Historical Teaching of Atomic and Molecular Structure

José Antonio Chamizo and Andoni Garritz

12.1 General Introduction

I can safely say that nobody understands quantum mechanics. R. Feynman (1985, p. 129)

The purpose of this paper is to argue that history and philosophy of chemistry and physics are central strategies in the teaching of atomic and molecular structure, from the Dalton model (for an earlier approach see Chalmers 1998) to modern quantum mechanics and quantum chemistry. Therefore, in addition to the presentation and conclusions, the chapter is divided into two equally important sections. The first describes the modern development of atomic and molecular structure, emphasising some of the philosophical problems that have confronted and been addressed by scientists, and those that have to be faced in understanding the science. The second discusses the alternative conceptions and difficulties that students of different educational levels bring to this subject and also the different approaches to the teaching of its history and/or philosophy. The conclusion is that a balance between the theoretical physicochemical basis of this chemistry knowledge and the phenomenological-empiricist knowledge must be achieved. But this cannot be done properly if teachers do not know and/or assume a particular historical-philosophical position.

Science education practice has not been driven to any great extent by research findings or by a goal of accomplishing professional ideals. The changes that have occurred in the majority of textbooks during the past decades do not show any real recognition of the growth in scientific knowledge (Schummer 1999). This is partly because of a chemistry teaching revolution 50 years ago (in the context of a revolution in the whole of science education: one which resulted from the Soviet success

---

J.A. Chamizo (✉) • A. Garritz
Seminario de Investigación Educativa en Química, Facultad de Química,
Universidad Nacional Autónoma de México, Avenida Universidad
3000, Delegación Coyoacán, México, DF 04510, México
e-mail: jchamizo@unam.mx

M.R. Matthews (ed.), *International Handbook of Research in History,*
Philosophy and Science Teaching, DOI 10.1007/978-94-007-7654-8_12,
© Springer Science+Business Media Dordrecht 2014

in launching Sputnik 1 in 1957). Under a philosophical (but hidden) umbrella, the change placed an emphasis on the physicochemical basis of General Chemistry in the three main projects of that decade: Chemical Bond Approach (Strong 1962), Chem Study (Campbell 1962) and Nuffield Foundation (1967).

The proposal was that the hegemony of physical chemistry would provide a basis of understanding for students' introduction to the chemical sciences through the quantum chemistry basis of the chemical bond, the kinetic model of the particulate nature of matter and the dominance of thermodynamics for explanations in several areas of chemistry. A new laboratory learning that promoted the notion of exploratory play with apparatus accompanied it. The General Chemistry course turned towards a theoretical character, losing the phenomenological approach that it had had in the preceding years. Without a deep recognition of its historical and philosophical roots, many people were led by this approach to believe that the contents of science textbooks were, in fact, science. But this is not necessarily true. The written materials employed in science education are descriptions of past science explorations (Yager 2004). Besides all this, once the majority of science teachers all over the world use textbooks as the main (sometimes the only) source of information—and the contents of the books have to expand in an idealised attempt to cope with the increase in information, with direct references to the history of sciences disappearing—they become, paradoxically and without wanting to …, history teachers! However, even if it was unconscious, it was a bad or a wrong way to teach the history of science. For example, Rodriguez and Niaz (2004) examined numerous textbooks for the History and Philosophy of Science (HPS) content in their approach to teaching atomic structure, and they found that an adequate and accurate reflection of the historical development is rarely presented.[1] This is educationally significant because philosophers of science and science education researchers have argued that quantum mechanics is particularly difficult to understand, due to the intrinsic obscurity of the topic and the controversial nature of its different interpretations [e.g. Copenhagen School "indeterminacy" (Bohr, Pauli, Heisenberg, Born, von Neumann and Dirac among others), Schrodinger with his cat paradox, the stochastic and the many world's interpretations and Bohm's "hidden variables" (Garritz 2013)].

12.2 The Subject Matter

12.2.1 Introduction

In this section a brief summary of several of the most important scientific advances of atomic and molecular structure, related mainly with chemistry but with a physicochemical character, will be presented. The starting point is Dalton's model of the atom and the whole nineteenth-century atomic controversy. At the end of that century, the 'discovery' by J. J. Thomson of negative corpuscles initiated the appearance of

---

[1] See also Moreno Ramírez et al. (2010).

12 Historical Teaching of Atomic and Molecular Structure

Table 12.1 Some chapters of the book by C. J. Giunta (2010)

Author	Chapter name	Comments
W. B. Jensen	Four Centuries of Atomic Theory. An Overview	A description of the dominant flavour of atomic notions over the last four centuries from the mechanical through the dynamical, gravimetric and kinetic to the electrical
L. May	Atomism Before Dalton	Outlines a variety of atomistic ideas from around the world. It concentrates on conceptions of matter that are more philosophical or religious than scientific
D. E. Lewis	150 Years of Organic Structures	Fifty years after Dalton, F. A. Kekulé and A. S. Couper independently published representations of organic compounds that rationalise their chemistry and even facilitated the prediction of new compounds
W. H. Brock	The Atomic Debates Revisited	A description of episodes from the second half of the nineteenth century in which chemists debated the truth of atomic theory. Doubts about the physical reality of atoms led chemists to question the soundness of chemical atomism
C. J. Giunta	Atoms Are Divisible. The Pieces Have Pieces	Evidence for the divisibility continued and impermanence of atoms was collected even while some chemists and physicists continued to doubt their very existence
G. Patterson	Eyes to See: Physical Evidence for Atoms	By the early decades of the twentieth century, through the efforts of J. Perrin and others, scepticism over the physical existence of atoms was practically eliminated

models of structure within the atom, such as that of J. J. Thomson with Lord Kelvin. The nuclear model of Rutherford was followed by Bohr's model of stationary orbits, which applied the energy quantisation hypothesis of M. Planck, which, in turn, started the old quantum theory in 1900. Then, A. Einstein as an explanation for the photoelectric effect recognised the wave-corpuscular duality of light. All this old quantum theory was replaced by E. Schrödinger and W. Heisenberg's wave and matrix mechanics, respectively, following on from the pilot wave hypothesis of L. de Broglie, and after that chemical bonding was interpreted in the same terms of quantum mechanics.

On this issue it is important to note that in 2008 the American Chemical Society held a symposium entitled '200 Years of Atoms in Chemistry: From Dalton's Atoms to Nanotechnology' which was followed, a couple of years later, with the publication of a book with a similar name. For a quick view of the topic that is addressed here, some of its chapters with comments from the editor are shown in Table 12.1.

12.2.2 Dalton's Model. Nineteenth-Century Controversies Between Physicists and Chemists

Dalton's atomic model with associated relative atomic weights was constructed in 1805 to explain results on the absorption of gases into water (Chamizo 1992; Viana and Porto 2010). Since then, in the nineteenth and early twentieth centuries, several famous debates took place between atomists and anti-atomists (including some

Nobel Prize winners). The early contributions of scientists from several European countries as Berzelius, Gay-Lussac and Avogadro to the acceptance of this model were not enough to convince all chemists or physicists (Giunta 2010; Nash 1957). For example, Bensaude-Vincent indicates:

> It is well-known that French chemists were reluctant to adopt the atomic theory in the nineteenth century. Their opposition was long-standing and tenacious since the atomic hypothesis formulated in the first decade of the nineteenth century by John Dalton was banished from the teaching of chemistry until the early decades of the twentieth century. Instead of atomism, the French chemists preferred the Richter's language of equivalents because it avoided commitment to a speculative theory of indivisible elementary particles …[…]… There is a general agreement among historians of chemistry that this national feature was due to the overarching influence of positivism in France. (Bensaude-Vincent 1999, p. 81)

Following the Karlsruhe's Congress in 1861 (Kauffman 2010), most of the chemical community accepted the distinction between atoms and molecules with their respective atomic and molecular weights, as admirably shown by S. Cannizzaro. In general, atoms were regarded by physicists as inelastic or inertial points or particles. Meanwhile chemists accepted Dalton's model:

> A group of physicists, among them Ernst Mach, John Bernhard Stallo and Pierre Duhem began to voice doubts about physical atomism because the kinetic theory did not dovetail with accurate experimentation. …The consilience between chemistry and physics had broken down. Mach, in particular, believed science to be a construct of the human mind and that it was not possible to find independent evidence for the existence of matter. Influenced by the thoughts of Georg Helm in 1887, Ostwald began to deny atomism explicitly. He opted instead for energetics –the laws of thermodynamics– rather than mechanical explanations in chemistry. He argued that energy was more fundamental than matter, which he saw only as another manifestation of energy. It followed that chemical events were best analyzed as a series of energy transactions. The difference between one substance and another, including one element and another, was due to their specific energies. (Jensen 2010, p. 63)

A century had to pass before the atomic model was fully accepted, which can be marked by formal recognition of J. B. Perrin's researches at the Solvay Conference of 1911 (Giunta 2010; Izquierdo 2010; and Izquierdo and Adúriz 2009).

12.2.3 The Electron and Thomson's Atom Model

There was a controversy about the nature of cathode rays (German physicists supported the ether theory for their origin, while the British argued for their particle nature), but it was the discovery of X-rays in 1895 that triggered J. J. Thomson's interest in cathode rays. He conducted a series of experiments at the beginning of 1897, which were first presented at a Friday evening discourse of the Royal Institution on April 29, 1897, and were finally published at length in the *Philosophical Magazine* in October the same year.

Thomson points out a fundamental aspect of his experiments, namely, that cathode rays are the same whatever the gas through which the discharge passes, and concludes: '[cathode rays] are charges of negative electricity carried by particles of

12 Historical Teaching of Atomic and Molecular Structure

matter. The question that arises next is: what are these particles? Are they atoms, or molecules, or matter in a still finer state of subdivision?' (p. 302). That is why he determined the relation m/e. From which Thomson concluded that its value, 10^{-12} kg/C, is independent of the nature of the gas, and it is very small compared with the 10^{-8} kg/C of H^+, the hydrogen ion in electrolysis, which is the smallest value of this quantity previously known.

Thomson goes further and proposes an atomic model:

> Since corpuscles similar in all respects may be obtained from different agents and materials, and since the mass of the corpuscles is less than that of any known atom, we see that the corpuscle must be a constituent of the atom of many different substances (p. 90)... [...]... The corpuscle, however, carries a definite charge of negative electricity, and since with any charge of negative electricity we always associate an equal charge of the opposite kind, we should expect the negative charge of the corpuscle to be associated with an equal positive charge of the other...we shall suppose that the volume over which the positive electricity is spread is very much larger than the volume of the corpuscle. (Thomson 1904, p. 93)

This model would last until Geiger and Marsden's experiment of bombarding metal thin films with radioactive particles, which allowed E. Rutherford to postulate the existence of the nucleus. On this subject we should mention the book *Histories of the Electron* that arose from two meetings (one in London and the other in Cambridge, Massachusetts) held to celebrate, in 1997, the centenary of the electron's discovery. The book is divided into the following four main sections that recognise the breadth of the subject being treated, and particularly the relations among the various sciences, and with technology and philosophy:

- Corpuscles and Electrons
- What Was the Newborn Electron Good For?
- Electrons Applied and Appropriated
- Philosophical Electrons

Some of its chapters with comments from the editors are shown in Table 12.2.

12.2.4 Planck, Einstein and Bohr: The Old Quantum Theory

The centennial of quantum theory has been celebrated a few years ago (Kleppner and Jackiw 2000). Quantum mechanics forced physicists and chemists to reshape their ideas of reality, to rethink the nature of things at the deepest level and to revise their concepts of determinacy vs. indeterminacy, as well as their notions of cause and effect.

The clue that triggered the quantum revolution came not from studies of matter but from a problem in radiation. The specific challenge was to understand the spectrum of light emitted by black bodies (that absorb and emit all kinds of electromagnetic radiation). In M. Planck's seminal paper (1900) on thermal radiation, it was hypothesised that the total energy of a vibrating system cannot be changed continuously. Instead, the energy must jump from one value to another in discrete steps, or

Table 12.2 Some chapters of the book by Buchwald and Warwick (2001)

Author(s)	Chapter name	Comments
I. Falconer	Corpuscles to Electrons	Thomson's main accomplishment at the Cavendish Laboratory in the mid-1890s was to succeed in deflecting a beam of cathode rays electrostatically, something that continental experimenters had failed to do
H. Kragh	The Electron, the Protyle, and the Unity of Matter	Identifies four different kinds of electrons before 1900: the electrochemical, the electrodynamic, the one associated with cathode ray work and the magneto-optical. The underlying notion of the electron as a fundamental building block of matter appealed particularly to J. J. Thomson and several others, who often thought of the electron as a sort of chemical proto-substance
W. Kaiser	Electro Gas Theory of Metals: Free Electrons in Bulk Matter	By early 1900, metallic conduction had become a central feature of a burgeoning microphysical practice, one that in this case sought to unify electrodynamics of electric sources in metals, with the model of colliding particles that underlie the kinetic theory of gases
L. Hoddeson and M. Riordan	The Electron, the Hole and the Transistor	The use of the electron in the design of amplifiers and semiconductors not only produced the new discipline of electronics but eventually enabled the very absence of the electron in certain material structures to be reified as a new entity in its own right, the 'hole'
M. J. Nye	Remodeling a Classic: The Electron in Organic Chemistry	In the broader historical picture, the arrival of the electron and quantum physics in chemistry was seen as fulfilling the expectations of men like Lavoisier and Dalton who were understood to have been the driving forces of the first chemical revolution
K. Gavroglu	The Physicists' Electron and Its Appropriation by the Chemist	Where physics sought a single theory that, in principle, was analytically exact in all cases, chemistry, a primarily laboratory-based science, sought one or more models that were applicable to a wide range of empirical data
P. Achinstein	Who Did Really Discover the Electron?	The historical facts about who knew what and when are complex
M. Morrison	History and Metaphysics: On the Reality of Spin	The reality ascribed to entities is often the result of their evolution in a theoretical history. The history of belief intersects with the evolution of theoretical trajectories
N. Rasmussen and A. Chalmers	The Role of Theory in the Use of Instruments; or, How Much Do We Need to Know About Electrons to Do Science with an Electron Microscope?	The effective use of an instrument does not necessarily depend in any meaningful way on theories about the way in which the device functions. The electron microscope was fruitfully used in discovering the biological cell's endoplasmic reticulum without a theory of how it interacted with the object

quanta, of energy. The idea of energy quanta was so radical that Planck let it lie fallow. A. Einstein (1906), then unable to obtain an academic position, wrote from the Swiss patent office in Berne: 'Analyzed in classical terms Planck's black-body model could lead only to the Rayleigh-Jeans law'. Kuhn (1978, p. 170) also made a contribution to this Planck-Einstein debate by saying that 'Planck's radiation law could be derived instead, but only by decisively altering the concepts its author had employed for that purpose'. Midway through his paper, Einstein wrote:

> We must therefore recognize the following position as fundamental to the Planck theory of radiation: [...]. During absorption and emission the energy of a resonator changes discontinuously by an integral multiple of $h\nu$. Moreno-Ramírez et al. (2010)[2]

Delighted as every physicist must be that Planck in so fortunate a manner disregarded the need [for such justification], it would be out of place to forget that Planck's radiation law is incompatible with the theoretical foundations which provide his point of departure (Einstein 1909, p. 186).

More recently, in 1913, N. Bohr applied the quantisation to the angular momentum of the hydrogen atom and obtained the whole set of J. R. Rydberg's spectral frequencies (Heilbron and Kuhn 1969). Even then the concept was so bizarre that there was little basis for progress with this 'old quantum theory'. Almost 15 more years and a fresh generation of physicists were required to create modern quantum theory. For an interesting and detailed description of the historical details of all quantum discoveries, Baggott (2011) can be consulted.

12.2.5 De Broglie, Heisenberg and Schrödinger. Quantum Mechanics

In 1923, L. de Broglie tried to expand Bohr's ideas and he pushed for their application beyond the hydrogen atom. In fact he looked for an equation that could explain the wavelength characteristics of all matter. His equation, $\lambda = h/p$, in relation to the wavelength of particles was experimentally confirmed in 1927 when physicists L. Germer and C. Davisson fired electrons at a crystalline nickel target, and the resulting diffraction pattern was found to match the predicted value of λ. Also G. P. Thomson—son of Joseph John, the discoverer of the electron—corroborated the de Broglie's wavelength of electrons going through very thin films of metals. Whereas his father had seen the electron as a corpuscle (and won the Nobel Prize in the process), he demonstrated that it could be diffracted like a wave. That is why it is said that Thomson's family contributed to the wave-particle duality of the electron by occupying the lead positions on both sides.

A second pillar of the development of quantum mechanics was W. Heisenberg, who reinvented matrix multiplication in June 1925 with his 'matrix mechanics' as

---

[2] In German he says 'Die Energie eines Resonators ändert sich durch Absorption und Emision sprungweise, und zwar ein ganzzahliges Vielfache von $(R/N)\beta\nu$' (Einstein 1906, p. 202).

was confirmed by M. Born and P. Jordan after revising his work. On May 1926, Heisenberg began his appointment as a university lecturer in Göttingen and with an assistantship to Bohr in Copenhagen. Heisenberg formulated the uncertainty principle in February 1927 while employed as a lecturer in Bohr's Institute for Theoretical Physics at the University of Copenhagen. He was awarded the 1932 Nobel Prize in Physics. In Bohr's words, the wave and particle pictures, or the visual and causal representations, are 'complementary' to each other. That is, they are mutually exclusive, yet jointly essential for a complete description of quantum events.

Next year the Nobel Prize was awarded to P. A. M. Dirac and E. Schrödinger. The great discovery of the latter, in January 1926, was published in *Annalen der Physik* as 'Quantisierung als Eigenwertproblem' [Quantization as an Eigenvalue Problem]. It was known as 'wave mechanics' and later as Schrödinger's wave equation. This paper has been universally celebrated as one of the most important achievements of the twentieth century, and created a revolution in quantum mechanics, and indeed of all physics and chemistry. On May that year Schrödinger published his third article, in which he showed the equivalence of his approach to that of Heisenberg's matrix formulation.

12.2.6 Kossel, Lewis and Langmuir; Heitler-London-Slater and Pauling; and Hund and Mulliken: Quantum Chemistry and Bonding Models

During World War I, in 1916, W. Kossel and G. N. Lewis (Lewis 1923) began independently to develop electronic models of chemical bonding, a concept fruitfully extended shortly thereafter by I. Langmuir. In the new models, the second and third periods of the periodic table each have eight members; the last of which (a noble gas) has a stable nonbonding 'octet' of electrons in a shell. Beyond the octet shells are the odd electrons in the outer shell, the 'valence electrons', which can be shared with adjacent atoms to form chemical bonds.

Langmuir expresses his view that the type of approach used by chemists is substantially different to that used by physicists:

> The problem of the structure of atoms has been attacked mainly by physicists who have given little consideration to the chemical properties, which must ultimately be explained by a theory of atomic structure. The vast store of knowledge of chemical properties and relationships, such as is summarized in the periodic table, should serve as a better foundation for a theory of atomic structure than the relatively meager experimental data along purely physical lines". (Langmuir 1919, p. 868)

In the late 1920s and early 1930s, W. Heitler, F. London, J. C. Slater and L. Pauling developed the 'valence-bond theory' as an application of the new quantum mechanics of E. Schrödinger and W. Heisenberg. Almost at the same time, R. Mulliken developed an alternative theory that began not from the electrons in atoms, but from the molecular structure ('molecular orbital' bonding). Partly because the

12 Historical Teaching of Atomic and Molecular Structure

extensive and vitally useful role of mathematics in physics had never been transferred to chemistry, it took until 1940 for Pauling and Mulliken theories to gain wide acceptance. The Nobel committee delayed 20 and 30 years, respectively, to honour this revolution. Pauling became laureate in 1954, and Mulliken won it in 1966 (Feldman 2001).

P. Atkins has recently presented his latest edition of the book on quantum chemistry with De Paula and Friedman (2008) as co-authors, where they review the latest improvements in making calculations. For example, they write on ab initio methods, configuration interaction and many body perturbation theories that were developed with the advent of high-speed computers in the 1950s. They proceed to density functional theory and its beginnings with Hohenberg and Kohn (1964) theorems and Kohn and Sham (1965) equations. Kohn was awarded the Nobel Prize for Chemistry in 1998. They then discuss a method for approximation of exchange (proposed by Slater (1951), a simplification that became known as the $X\alpha$ method) and of correlation energies, introduced in the 1960s and 1970s. Their final section examines current achievements, including the impact of quantum chemistry methods on nanoscience (the structure of nanoparticles) and medicine (molecular recognition and drug design).

12.2.7 Molecular and Crystal Symmetry and Spectroscopy

Spectroscopy is the study of the interaction of electromagnetic radiation with matter. In 1860 the German scientists R. Bunsen and G. Kirchhoff discovered two alkali elements, rubidium and cesium, with the aid of the spectroscope they had invented the year before. Since then spectral analysis has been a central tool in chemistry, physics and astronomy. But it is not only spherical atoms that interact with light; molecules can also do it. Molecules may interact with the oscillating electric and magnetic fields of light and absorb the energy carried by them. The more symmetric the molecule, the fewer different energy levels it has and the greater the degeneracy of those levels. The study of symmetry helps us to simplify problems by reducing the number of energy levels one must deal with. But more than that, symmetry helps us decide which transitions between energy levels are possible and which are not (Harris and Bertolucci 1978) through selection rules, addressing problems that were possible to pose and solve via a branch of mathematics named group theory.

The history of group theory and that of quantum mechanics can be of great assistance in understanding the applications of spectroscopy to physical problems. Nobel laureate P. W. Anderson (1972, p. 394) wrote 'it is only slightly overstating the case to say that physics is the study of symmetry'. While quantum theory can be traced back only as far as 1900, the origin of the theory of groups is much earlier. It was given definite form in the later part of the eighteenth and in the nineteenth centuries. F. Klein—a German mathematician, known for his work in group theory, function theory, non-Euclidean geometry and on the connections between geometry and group theory—considered the group concept as most characteristic of nineteenth-century mathematics.

The concept of a group is considered to have been introduced by E. Galois (1811–1832). Galois refashioned the whole of mathematics and founded the field of group theory only to die in a pointless duel over a woman before his work was published when he was 21 years old. J. Liouville published his ideas in 1846. Some aspects of group theory had been studied even earlier: in number theory by L. Euler, C. F. Gauss and others and in the theory of equations by A. L. Cauchy and J. L. Lagrange (each with a well-known group theory theorem).

At the heart of relativity theory, quantum mechanics, string theory and much of modern cosmology lies one concept: symmetry. In *Why Beauty Is Truth*, world-famous mathematician I. Stewart (2007) narrates the history of this remarkable area of study. He presents a timeline of discovery that begins in ancient Babylon and travels forward to today's cutting-edge theoretical physics.

The symmetry aspects are crucial today for the different models of chemical structure, bonds, spectroscopic interpretations and chemical reactions. In many of these problems the crucial problem is that of the potential seen by electrons moving in the electric field of the nuclei. The relation between science and mathematics resides in the commutation of the Hamiltonian with the symmetry operators, so that the wave functions of the atoms, or molecules, are bases of some of the irreducible representations of the point group to which the system belongs. Many books have appeared devoted entirely to applications of symmetry and aspects of group theory to chemistry. Examples include two classical books (Bishop 1973; Cotton 1963) and one modern (Hargittai and Hargittai 2009).

12.2.8 The Problem of Reduction of Chemistry into Physics

One of the most deeply entrenched traditions, which could be seen as an orthodoxy that extends beyond the scientific community to the whole of society, is that science can be explained in terms of the logical positivist philosophical tradition. Since the nineteenth century, logical positivism has sought to clearly establish a boundary between science and non-science using two additional criteria:

- An empirical-experimental approach (if something cannot be interpreted in terms of observations or measurements, then it is not scientific, it is metaphysical)
- A criterion of logical-mathematical inference and scientific theory (one aspect is that if something cannot be rebuilt in a deductive way, it is not rational, it is unscientific)

Logical positivism assumes the axiomatisation of theories unifying all sciences into one. In its most widely recognised version (Reish 2005), logical positivism, presenting science as a linear succession of successful discoveries and placing the emphasis on factual recall with confirmatory experiments, contributed to identify what kinds of research questions and issues were adequate. This programme of unification of science and deriving the principles of one science from another is commonly known as reductionism. The logical positivist assumes that the laws of a particular science, like chemistry, can in principle be derived from other more basic

12 Historical Teaching of Atomic and Molecular Structure 353

laws, in this case from physics. This position became stronger particularly with the development of relativistic quantum mechanics by P. A. M. Dirac. He indicated:

> The underlying laws necessary for the mathematical theory of a large part of physics and the whole of chemistry are thus completely known, and the difficulty is only that exact applications of these laws lead to quantum mechanical equations which are too complicated to be soluble. (Dirac 1929, p. 714)

One of the most important philosophers of science of the time, working from a logical positivist perspective, H. Reichenbach celebrated Dirac's claim, indicating that:

> The problem of physics and chemistry appears finally to have been resolved: today it is possible to say that chemistry is part of physics, just as much as thermodynamics or the theory of electricity. (Reichenbach 1978, p. 129)

A few years later, Reichenbach distinguishes between contexts of discovery and justification, an issue that has occupied a prominent place in the philosophy of science. Since then, in its best known version (Reish 2005), logical positivism has presented science as a linear succession of successful discoveries and has placed the emphasis on factual recall with confirmatory experiments. This contributed to identifying what kinds of research questions and issues were adequate for the axiomatic structure of science.

But in the 1960s, several science philosophers started to question the lack of historicity of logical positivism, which was based mainly in the context of justification (Reichenbach 1938). They proposed alternative ways of conceiving the philosophy of science based on historical ideas such as change, progress or revolution (Kuhn 1969; Toulmin 1961, 1972). More recently several philosophers have also questioned other traditional assumptions of logical positivism such as reductionism and verificationism (Hacking 1983; Harré 2004; Laudan 1997; Popper 1969). This indicates that the philosophy of science has escaped the constraints imposed by the context of justification without losing sight of the question of rationality. New and different ways of approaching the philosophy of science have emerged, for example, M. Christie and J. Christie (2000) make a case for the diverse character of laws and theories in the sciences and particularly consider a pluralistic approach to laws and theories in chemistry. R. Giere (1999) considers that science does not need laws because 'science does not deliver to us a universal a truth underlying all natural phenomena; but it does provide models of reality possessing various degrees of scope and accuracy' (Giere 1999, p. 6).

These new and different approaches to the philosophy of science lead to reconsideration of what Dirac said. Thus the Nobel Prize winner in Chemistry, for his theory concerning the course of chemical reactions using quantum mechanics, R. Hoffmann indicated (1998, p. 4):

> Only the wild dreams of theoreticians of the Dirac school make nature simple.

This idea was shared by the 1969's Physics Nobel Prize for his contribution and discoveries on the classification of elementary particles (quarks) and their interactions, M. Gell-Mann. He said (1994):

> When Dirac remarked that his formula explained most of physics and the whole of chemistry of course he was exaggerating. In principle, a theoretical physicist using quantum electrodynamics can calculate the behaviour of any chemical system in which the detailed

internal structure of atomic nuclei is not important. [But:] in order to derive chemical properties from fundamental physical theory, it is necessary, so to speak, to ask chemical questions. (Gell-Mann 1994, p. 109)

And some of those chemical questions, perhaps the simplest, are related to the periodic table. Much has been written about them (Jensen 2002; Scerri 2007), but it is relevant to recall what philosopher of chemistry J. van Brakel (2000) says:

As a specific example of the reduction of chemistry to physics, it is often suggested that the periodic table can be 'derived' from quantum mechanics. Such a reduction was already ascribed to Bohr, for example, by Popper. But, contrary to his own claims (and those of Popper) 'Bohr populated the electron shells while trying to maintain agreement with the known experimental facts'. Later developments too in quantum mechanics cannot strictly predict where chemical properties recur in the periodic table. Pauli's explanation for the closing of electron shells does not explain why the periods end where they do: the closing of shells is not the same as the closing of periods in the table. Unknown electronic configurations of atoms are not derived from quantum mechanics, but obtained from spectral observations. Hund's rule states an empirical finding and cannot be derived. (van Brakel 2000, p. 119)

A current periodic table shows many and various properties attached to atoms, including, for example, the size. However, the various theoretical approaches derived from quantum mechanics to calculate atomic size assume, arbitrarily, that atoms are bounded. There is no such thing as an absolute atomic size. An atom is not a rigid sphere, so 'atoms differ in size depending on the type of external forces acting on them' (Cruz et al. 1986, p. 704). The various experimental techniques used to determine internuclear distances indicate that the size of atoms depends on the surrounding environment. Therefore, a periodic table can only show covalent, ionic or metallic radii as typical outcomes from experimental measurements of many different solids.

As several researchers have discussed when addressing entanglement (Primas 1983), arising from strict quantum mechanical treatments, physical systems are never isolated nor closed. As with the size of atoms, so the geometry of molecules varies depending on their environment. Van Brakel indicated:

According to Primas the crucial issue is not the approximations of quantum chemistry as the Born-Oppenheimer description, but the breaking of the holistic symmetry of quantum mechanics by abstracting from the Einstein-Podolsky-Rosen (EPR) correlations. It is the EPR correlations that exclude any classical concept of object, shapes or the fixed spatial structures such as presupposed in the notion of molecular structure ... therefore, quantum chemistry borrows the notion of molecular structure from classical chemistry. (van Brakel 2000, p. 144)

R. G. Woolley (1978) defends this position in his famous and provocative article 'Must a molecule have a shape?' which indicates that the classic concept of molecule cannot be derived from quantum mechanics. Nevertheless, since the nineteenth century, chemists have determined experimentally the particular geometries of various molecules. Today we know that these geometries are relative to the timescale of measurement.

Thus, there are difficulties in interpreting even the simplest chemical phenomena, rigorously and independently, from quantum mechanics. The problems are almost intractable as can be recognised in Table 12.3 (Jensen 1980).

12 Historical Teaching of Atomic and Molecular Structure

Table 12.3 Outline of steps which, according to our present knowledge of quantum mechanics and statistical thermodynamics, are necessary in order to predict rigorously the equilibrium or rate constant of a reaction in solution from first principles

1. Calculation of the electronic potential energy of the static arrangement of atoms corresponding to the structures of each reactant and product
2. Prediction of the normal modes of motion for the atoms in each structure. This amounts to setting up a mathematical description of the structure's vibrational and rotational motions
3. For many of these motions, the lowest kinetic energy is not zero, but rather a half-quantum of the motion. This zero-point kinetic energy must be added to the potential energy
4. From the knowledge of the normal modes of motion, it is possible to compute the partition function of each species as a function of temperature and from this is obtained the standard free energy and enthalpy of each species in the dilute gas state and at the temperature of interest
5. The standard free energy and enthalpy of each species in solution is then computed considering the transfer from the gas phase to solution
6. Values of ΔH^0 ΔG^0 and ΔG^* and ΔH^* are calculated for the maximum point on the surface of least energy connecting the reagents with the products. With these values it is possible to calculate the equilibrium constant and reaction rate
7. Finally the calculated values must be recalculated to consider the actual concentration of the various species in solution using the activity coefficient of each species for the temperature and solvent under consideration

For similar reasons there are many chemical notions that are not amenable to rigorous quantum mechanical treatment. Van Brakel (2000) mentions some of them: acidity, aromaticity, basicity, chemical bond, chemical reaction, chirality, electronic configuration, orbital, electronegativity, functional group, molecular structure, resonance, relative energy of s and p orbitals and valence.

In a similar way another philosopher of chemistry J. Schummer (1998) recognises the differences among the various sciences when dealing with the study of material properties (which from a reductive view are those of the atomic and molecular structure):

> For sciences of materials, with chemistry at the centre, have been, from the earliest stages on, experimental science in the original meaning of studying the behaviour of objects in various and controlled artificial contexts. A material property is reproducible behaviour within certain reproducible contextual conditions. It is important to note that material properties are attributed not to isolated objects but to objects and contexts. Since everything looks red under red light, we have to specify the colour both of the object under investigation and of the light, in order to make qualified colour statements. Since everything is solid at a certain temperature and pressure, solidness always implies specification of thermodynamic conditions. Sometimes it is more the context that matters. To speak of a toxic substance does not mean that the substance itself but the context, a biological organism, falls sick or dies, if it gets in contact with the substance. Precise material predicates require precise and systematic details of the contexts of investigation, making contexts themselves a central subject matter of sciences of materials.

This poses a difficult problem in the teaching of atomic and molecular structure, when it ignores its historical roots and philosophical consequences, an issue that has not escaped the experts. In 1999 *Nature* published a report that orbitals had been observed (Zuo et al. 1999). There were philosophical objections (Scerri 2000a, 2001), which indicated a confusion of the authors of this article aforementioned between observable and unobservable (Shahbazian and Zahedi 2006) and between the real world and models (Pagliaro 2010). The following quotes from some of the participants in this discussion help to clarify their positions, particularly in relation to the teaching of this topic:

> ... chemists have a tendency to "decompose" molecules arbitrarily into basic conceptual or pseudo-physical components (such as orbitals and atoms), which can cause controversy. The entities, which come from such decompositions, make a new class of mathematical objects: "non-observables". Using these non-observables as a tool for chemical arguments is a common practice of chemists. (Shahbazian and Zahedi 2006, p. 39)

> Orbitals however are also a (quantum) chemical model of immense importance in chemistry. Their relationship to the chemical methodology is heuristic, i.e., their usefulness in many branches of science justifies the use of the model. (Pagliaro 2010, p. 279)

> Yes, it is important to know when approximations are made, but success in a science like chemistry is largely a matter of finding useful approximations: this is what students should be taught. (Spence et al. 2001, p. 877)

> Chemical educators should continue to use concepts like orbitals and configurations but only while recognizing and emphasizing that these concepts are not directly connected with orbitals as understood in modern quantum mechanics, but are in fact a relic of the view of orbits in the so-called old quantum theory. (Scerri 2000b, p. 412)

Finally, it is important to recognise that traditionally two types of reductionism have been considered: ontological and epistemological (Silberstein 2002). Despite the intense debates that have occurred in this area, where important issues are those related to 'the kind of relations', or 'the way in establishing relationships' (Lombardi and Labarca 2005), recent years have witnessed a growing consensus towards a tradition that denies the possibility of reducing chemistry to physics. In particular there is a denial that such a reduction has been achieved via quantum mechanics as considered from logical positivism. Bibliography related to this subject can be found in Erduran (2005), Schummer (2008), Snooks (2006), and Velmulapalli and Byerly (1999).

12.3 Procedures

12.3.1 Introduction

This section addresses three issues. The first has to do with the way that history and philosophy of sciences are incorporated into the teaching of atomic and molecular structure. The second considers the diversity of previous ideas that students from different educational levels bring to the subject and how these ideas hinder their learning. Finally, the third part outlines several reported experiences in teaching atomic and molecular structure. About all this M. Niaz has dedicated a book (Niaz 2009) and a full set of papers (e.g. Niaz 2000 and 2010) dedicated to posing the necessity of the historical teaching with episodes and experiments that have been

12 Historical Teaching of Atomic and Molecular Structure

important in science progress. He emphasises the validity of the following phrase of Kant and Lakatos: 'philosophy of science without history of science is empty'.

12.3.2 Philosophy and History in Teaching and Their Importance

In present science education, history and philosophy play a fundamental role (Duschl 1994; Matthews 1994/2014; Wandersee and Griffard 2002). But the teaching of history cannot be only the chronological narrative of past events; it requires, as indicated by Husbands (2003), '… that we, history teachers … establish a more subtle, less absolutist understanding of the way in which knowledge is created … It needs to be developed through the process of inquiry in the classroom, by teachers and learners in classrooms working to create meanings'. In a similar way Tsaparlis (1997b, p. 924) has emphasised the historical method of teaching as a way of better understanding the topic of atomic and molecular structure.

Moreover, as indicated in the previous discussion of reduction, an issue such as this requires in its teaching, the recognition of the different philosophical positions that underlie its foundation (Karakostas and Hadzidaki 2005). About realism, and the reality of electrons, the influential philosopher I. Hacking has said (1983, p. 22): 'If you can spray them, then they are real …'. Others, like Achinstein (2001), in discussing the discovery of the electron, put forward the following components for a discovery:

- Ontological—Discovering something requires the existence of what is discovered.
- Epistemic—A certain state of knowledge of the discoverer is required.
- Priority—Social recognition of the discovery.

In the same book Arabatzis (2001) offers a consensus-based account of discovery, asserting that entity x (atom, electron, spin and phlogiston) can be said to have been discovered just when a group y reaches consensus that it has been. He simply wishes to concentrate on synchronous belief, not on reality. However, in another chapter of the same book, Morrison addressed the reality of spin (2001). These discussions can be very technical and complicated. Nevertheless it is advisable for a teacher to adopt a position or at least to know it.

In recent years, for example, several authors have recognised that the way chemistry is usually taught is based on a particular philosophical position and that in general terms this position is logical positivism (Chamizo 2001; Erduran and Scerri 2002; Van Aalsvoort 2004; Van Berkel et al. 2000). Van Berkel with researchers all around the world analysed current and post-war textbooks and syllabi representative of secondary chemistry education in most Western countries trying to find why they are so remarkably similar. He recognises that dominant school chemistry is particularly isolated from everyday life and society, history and philosophy of science, technology and chemical research. His main conclusion was:

> The structure of the currently dominant school Chemistry curriculum is accurately described as a rigid combination of a substantive structure, based on corpuscular theory, a specific philosophical structure, educational positivism, and a specific pedagogical structure, initiatory and preparatory training of future chemists. (van Berkel 2005, p. 67)

During the Cold War, a philosophy of science, which defended science's superior analytical purity, was enthroned in most of the Anglo-Saxon intellectual world (Echeverria 2003). It focused on science methodology and the reduction of various scientific disciplines to physics. Since then, the best known version of logical positivism, presenting science as a linear succession of successful discoveries and placing the emphasis on factual recall with confirmatory experiments, has contributed to identifying what kinds of research questions and issues were adequate not only for axiomatic science (Reish 2005) but also for school syllabus, as can be seen in chemistry and physics curricula. Therefore it would be desirable, regardless of the educational level, when addressing the teaching of atomic and molecular structure, to identify the philosophical position underlying the approach.

Journals oriented to chemistry education are dedicating full sections to the history of chemistry. William B. Jensen, since 2003 until recently, had the responsibility of writing a section 'Ask the historian' in the *Journal of Chemical Education*. He previously had devised a framework of three chemical revolutions from which he extended three levels of comprehension of chemistry—Molar, Molecular and Electrical—and three dimensions, based on whether they deal with composition/structure, energy or time (Jensen 1998). In that set of articles, Jensen commented that there are a large number of histories of chemistry. In his bibliographic study, Jost Weyer (1974) listed no fewer than 71 general histories of chemistry written between 1561 and 1970, of which 29, or roughly 40 %, have appeared written in English. George B. Kauffman has the responsibility of writing historical articles for the journal *The Chemical Educator,* mainly to commemorate anniversaries of outstanding achievements in chemistry (some examples are Kauffman 1999, 2004, 2006, 2010). Jaime Wisniak has played a similar role in *Educación Química,* the Ibero-American Journal of Chemistry Education, since 2001 (Wisniak 2013).

However, although there are many scholarly works on the history of chemistry, there have been few on how to incorporate them, effectively and systematically, into the teaching of chemistry. Perspectives, such as that established by Jensen (1998), in which the curriculum is built on history (in this case of atoms and molecules), or that described by Early (2004) from a new philosophical basis, are few and therefore very important. As Talanquer recognised (2011) school chemistry needs transgression.

12.3.3 Introduction to Alternative Conceptions and Difficulties in Teaching and Learning Quantum Mechanics and Quantum Chemistry

Many studies have reported students' difficulties in grasping the fundamental issues of quantum mechanics and quantum chemistry in high school. We shall mention first an article by Tsaparlis and Papaphotis (2002) where findings of student difficulties with quantum numbers, atomic and molecular orbitals, are reviewed, and a case is presented against using quantum chemical concepts at this level (Bent 1984).

12 Historical Teaching of Atomic and Molecular Structure

These authors insist that the topic is highly abstract and therefore beyond the reach of many students.

Students have difficulty understanding the concepts of atomic and molecular structure (Harrison and Treagust 1996) because of the abstract nature of the sub-micro world (Bucat and Mocerino 2009). Many authors have been discussing in several studies the difficulties or misconceptions in students' learning about matter—those related to its particulate nature,[3] to bonding in general,[4] to the covalent bonding model,[5] to the metallic bonding model[6] and to the ionic bonding model.[7]

Other studies have reported students' difficulties in grasping the fundamental issues of quantum mechanics and quantum chemistry at high school[8] and college levels.[9] In particular the following concepts are indicated:

- 'Probability and energy quantization' (Park and Light 2009)
- 'Quantum numbers' or 'electron configurations of chemical elements'[10]
- 'Orbital ideas'[11]
- 'Uncertainty and complementarity' (Pospiech 2000)
- 'The Schrödinger equation' (Tsaparlis 2001)

From the point of view of teaching, the elementary, qualitative and pictorial coverage of quantum chemical concepts is approached with reservation or with strong opposition by many chemical educators (Bent 1984; Gillespie 1991; Hawkes 1992).

Physicists have also recognised the difficulties involved in understanding quantum mechanics (Einstein 1926, 1944, 1948; Feynman 1985; Laloë 2001; Styer 2000).

Taber (2003) mentions 'most alternative conceptions in chemistry do not derive from the learner's unschooled experience of the world'. The many problems that learners have in chemistry maybe best characterised as 'model confusion' (see Sect. 12.3.4.4). Where there are several models for particular or closely related chemistry concepts, students become greatly confused. This is particularly so when most learners have a very limited notion of the role of models in science (Grosslight et al. 1991).

---

[3] See, for example, Lee et al. (1993), Novick and Nussbaum (1978, 1981), Nussbaum (1985), Valanides (2000), and Wightman et al. (1987).

[4] As can be seen in Birk and Kurtz (1999), Boo (1998), Furió and Calatayud (1996), Griffiths and Preston (1992), Hund (1977), Kutzelnigg (1984), Magnasco (2004), Özmen (2004), and Sutcliffe (1996).

[5] For example, Coll and Treagust (2002), Niaz (2001), and Peterson et al. (1989).

[6] Such as in Coll and Treagust (2003a) and De Posada (1997, 1999).

[7] See, for example, Butts and Smith (1987), Coll and Treagust (2003b), and Taber (1994, 1997).

[8] Such as Dobson et al. (2000), Petri and Niedderer (1998), Shiland (1995, 1997), and Tsaparlis and Papaphotis (2002, 2009).

[9] For example, Hadzidaki et al. (2000), Johnston et al. (1998), Kalkanis et al. (2003), Michelini et al. (2000), Paoloni (1982), and Wittmann et al. (2002).

[10] As can be seen in Ardac (2002), Melrose and Scerri (1996), Niaz and Fernández (2008), and Scerri (1991).

[11] For example, Cervellati and Perugini (1981), Conceicao and Koscinski (2003), Ogilvie (1994), Scerri (2000a), Taber (2002a, b; 2005), and Tsaparlis (1997a).

12.3.4 Experiences

12.3.4.1 Similarities

This subject is closely related to the previous subsection. One of the first to establish similarities between the historical development of science and the conceptual development of students was J. Piaget (Piaget and Garcia 1983) followed by Gagliardi (1988), although Matthews (1992) identifies this idea in Hegel's *The Phenomenology of Mind*. There are strong grounds for criticism of this position (Gault 1991), mainly because the equivalence between the ideas of scientists and students has not been demonstrated. Nevertheless, Scheffel and colleagues (2009) recently and carefully used the similarities in classroom teaching through the following sequence:

1. The teacher hands on historical, but educational purposes reduced, material to the student. This will presumably pick up students' misconceptions and their actual scientific positions.
2. The students discuss these ideas and propose experiments to verify or falsify one of the theories or models that has been presented. They have an opportunity to choose one of the scientists as an advocate for their preconceptions.
3. Based on experiments and if necessary on additional material, the pros and cons of each theory or model are collected and discussed. If possible, a decision should be formulated and explained.

These authors provide examples of similarities, applying this teaching methodology to old atomism, chemical bonding or Lewis octet model.

12.3.4.2 The Historical Narrative

Narrative can be defined as 'telling someone else that something happened' (Herrestein-Smith 1981, p. 228). Norris and colleagues (2005) elaborated this approach, and they identified in the narrative the roles of the narrator, the reader and the events. Particularly important here is the responsibility of the narrator—in this situation, the teacher—because he or she must facilitate the interpretation of the events in context (Gilbert 2006). As Metz and colleagues (2007) recognised, the narrative approach has a spectrum of possible applications:

- Interactive vignettes (Wandersee and Griffard 2002)
- Anecdotes (Shrigley and Koballa 1989)
- Curriculum unit unified by a theme (Holbrow et al. 1995)
- Storyline, when the thematic approach will begin with a big question (Stinner and Williams 1998)

For example, Teichmann (2008) included anecdotes from some atomic structure protagonists; Klassen (2007) has used narratives for teaching the heroic attitude of L. Slotin assembling the first atomic bomb and for rehabilitating the story of the Photoelectric Effect (2008). In similar fashion, Nobel lectures have also been used for teaching in chemistry and in physics (Jensen et al. 2003; Panusch et al. 2008;

12 Historical Teaching of Atomic and Molecular Structure   361

Stinner 2008). Biographies, tributes and interviews could also be considered in this category. Some examples are G. N. Lewis (Branch 1984), L. Pauling (Kauffman and Kauffman 1996) and R. S. Mulliken (Nachtrieb 1975).

12.3.4.3 The Historical Role of Rivalry, Controversy, Contradiction, Speculation and Dispute in Scientific Progress and Its Use in Teaching Strategies

In academia, conflicts in and around science have been studied for various reasons:

- To gain insight into the process of science policy making process
- To learn more about the various roles of scientists
- To identify the ways in which the public might participate in decision making
- To understand how controversies arise, how they are contained within the scientific community or expand into the public domain, how they are brought to a close or why they persist, among others
- To analyse the social construction and negotiation of scientific knowledge claims by conflicted scientists (Martin and Richards 1995)

Nevertheless, dispute in scientific progress has been rarely used in the teaching and learning of science (Niaz 2009).

Teaching through the consideration of historical aspects of scientific knowledge has the potential to show the progress of scientific knowledge over time. Historical artefacts and scientific discoveries, scientists' life stories and the details of scientific struggles in scientific progress could be discussed in the science classroom. Because the knowledge represented in textbooks or in any predesigned science-learning environment context is the end product of science, students and teachers do not learn and teach about those presuppositions, contradictions, controversies and speculations existent in scientific progress (Niaz 2009, 2010; Garritz 2012 online). Only a few teachers today believe and teach that scientific knowledge is tentative, empirically based, subjective and parsimonious; that it includes human creativity and imagination; and that it is socially and culturally constructed (Ayar and Yalvak 2010).

12.3.4.4 The Explicit Recognition of Models and Modelling

The Model-Based view of Scientific Theories and the structuring of school science (Adúriz-Bravo 2012; Develaki 2007) have recently been discussed elsewhere. As discussed earlier in this chapter, quantum mechanics forced physicists and chemists to reshape their ideas of reality, to rethink the nature of things at the deepest level and to revise their concepts of determinacy vs. indeterminacy, as well as their notions of cause and effect. Here we adopt a realist position about molecules, atoms and electrons. In agreement with Tapio (2007), we specify that:

- Reality and its entities are ontologically independent of observers.
- Claims about the existence of entities have truth-value.
- Models of atoms and molecules are required to be empirically reliable.

Model is a polysemous word; it has been used and it is still used with several meanings. That is one of the difficulties we meet when we use it in teaching. In one usage, 'model' is exemplary; it indicates things, attitudes or people worthy of emulation. The courage of a warrior, the intelligence of a wise man, the solidarity of a doctor and the speed of a runner are examples of 'models' in this regard. In this paper we use a previous definition of 'model' (see Chamizo 2011 for all the references): 'models (m) are representations, usually based on analogies, which are built contextualizing certain portion of the world (M), with a specific goal'. In this definition all the words are important: the representations are essentially ideas, but not necessarily so, as they can also be material objects, phenomena or systems (all of them constitute a certain part of the world M). Representations have no meaning by themselves; they come from someone (either an individual or a group, usually the latter) that identifies them as such. An analogy is made up of those features or properties that we know are similar in (m) and (M). That 'are built contextualizing certain portion of the world M' refers to a historically defined time and place which also frames the representation. Some 'portion of the world' indicates its limited nature; models (m) are partial for the world (M). 'A specific goal' establishes its own purpose, usually (but not necessarily) to explain or teach and possibly also to predict. In this sense models can be understood as cognitive artefacts or mediators constructed in order to create subjective plausibility about the target. It is important to remember that explanation is one of the most significant features of science, but in some cases when models are even completely unable to offer an explanation, much of the prestige of a model may lie in its capacity to predict.

There are only two types of models: mental and material.

Mental models are reflected representations built by us to account for (explain, predict) a situation. They are forerunners of the famous 'misconceptions' (see Sect. 12.3.3) and can sometimes be equivalent, since they are unstable, generated in the moment and then discarded when no longer needed, making them cognitively disposable.

Material models (which may be identified as prototypes) are the ones that we have empirical access to and have been built to communicate with other individuals. Material models are expressed mental models and can be further categorised as symbolic, iconic or experimental. Here we only discuss the first two. Symbolic material models correspond to the languages of sciences, such as mathematics or chemistry. So mathematical equations constructed to describe precisely the portion of the world being modelled are symbolic material models. Wave mechanics is a symbolic material model. Another example of symbolic material model is the one used by chemists to represent elements, compounds and reactions. Hence, when a teacher writes the molecular structure of water as H_2O using two hydrogen and one oxygen atom, the teacher uses a symbolic material model. Iconic material models correspond to images, diagrams or scale models, like a map or the so-called molecular models. Stereochemistry was constructed with iconic material models in three dimensions. For example, in the early years of the nineteenth century, Dalton constructed wooden models of atoms; after him Pasteur made his models of enantiomer tartrate crystals, Hofmann his croquet ball molecular models and van't Hoff his cardboard tetrahedral models. In the twentieth century the stereochemical ideas of Pauling led to the most famous example of an iconic material model, the DNA structure by Watson and Crick.

Recently Seok and Jin (2011) have reviewed the literature dealing with models and modelling and reported some important findings. Two of them related to model use in atomic and molecular structure teaching are:

- Meaning of a model. A model is understood as a representation of a target. The targets represented by models can be various entities, including objects, phenomena, processes, ideas and their systems. A model is also considered a bridge or mediator connecting a theory and a phenomenon, for it helps in developing a theory from data and mapping a theory onto the natural world, for example, atomic models (Dalton, Bohr, Lewis), molecular models or bonding models (ionic, covalent, coordinated and metallic) or electron models (corpuscle or wave like).
- Change in scientific models. There are two ways of testing a model in science: the empirical and conceptual assessments. An empirical assessment is a way of evaluating a model in terms of the fit between the model and the actual phenomenon. In a conceptual assessment, a model is evaluated according to how well it fits with other accepted models as well as with other types of knowledge.

The assessment of a model is conducted differently in experimental sciences, such as physics or chemistry, from in historical sciences, or others, such as earth science. For example, Bohr's atomic model is excellent at explaining hydrogen spectra, but useless for molecular structures; Lewis' atomic model is excellent in predicting simple organic structures, but useless in, for example, infrared spectra (about Lewis model in introductory teaching of atomic and molecular structure see Chamizo 2007; Purser 2001).

Finally because models are built in a particular historical moment for specific purposes, the context should be explicitly recognised when teaching them. Justi and Gilbert (2000) have warned us about the frequent use of hybrid models in the textbooks, which has produced so much confusion among students. Experiences of more correct use of these models have been reported recently (Chamizo 2007, 2011, 2012).

12.3.4.5 Textbooks, Experiments and Information and Communication Technologies (ICTs)

There are several books that feature various aspects of the history of atoms and molecular structure.[12] One of the most influential is Kuhn's *Black-Body Theory and the Quantum Discontinuity 1894–1912*. Another example is the history of quantum chemistry as told by E. Segrè (2007) in which a Nobel laureate offers impressions and recollections of the development of modern physics. Rather than a chronological approach, Segrè emphasises interesting, complex personalities who often appear only in footnotes. Readers will find that this book adds considerably to their understanding of science and includes compelling topics of current interest.

---

[12] For example, Buchwald and Warwick (2001), Giunta (2010), Marinacci (1995), Nye (1993), Snow (1981), and Toulmin and Goodfield (1962).

However, very few of these last writers teach undergraduate chemistry. The authors of this chapter have written a book in Spanish on quantum chemistry, with emphasis on the development of the historical aspects of this science (Cruz et al. 1986). With hundreds of solved exercises and problems, it has been used widely in Ibero-America. The historical narrative oscillates in time, from the nineteenth-century chemistry until the interpretation of periodicity, as can be seen in Table 12.4.

Experiments related to the history of atomic and molecular structure are rare. Some of them can be found in more general books like Doyle's *Historical Science Experiments on File* (Doyle 1993). However, there are some examples, ranging from the electrochemical decomposition of water (Eggen et al. 2012) to spin through the Stern-Gerlach experiment (Didis and SakirErkoc 2009).

Information and Communication Technologies (ICTs) have so far had little impact in this area, with the exception of graphs of orbitals, electron densities and contours. The PhET project (Physics Education Technology) has branched also into chemistry and biology. Some of the designed computer simulations have been devoted to atomic and molecular structure from historical experiments. PhET conducts research on both the design and use of interactive simulations, but important as this material is, the failure to address historical context and provide historical references has made this approach so far quite weak.

12.4 Conclusion

Physical chemistry remains a fundamental basis for the teaching of chemistry. Mathematics, as group theory and matrix representations, is needed to understand selection rules via symmetry studies and, through them, spectroscopic transitions, an important topic since the second half of last century. Nevertheless there is a necessity for balance between the theoretical physicochemical basis of chemistry and the phenomenological and empiricist knowledge that chemistry had already produced.

The parsimonious advice of one of the reviewers of this chapter was 'do not introduce needless complexity unless it is warranted to explain the necessary facts'. This can be also a conclusion about the inclusion of history and philosophy of science in teaching quantum mechanics and quantum chemistry. One has to apply Ockham's Razor rules while teaching these topics.

We can recognise in the almost 200 works cited in this study that integration of history of science into the teaching of atomic and molecular structure has been seen as an important step, particularly since 1994. Increasing numbers and diversity of resources and studies of strategies to be used are making this incorporation more robust. Nevertheless, the way in which chemistry has been taught all around the world is based on a particular philosophical position, which comes from its acceptance as a reduced science, and can be characterised as logical positivism. This normal (in Kuhn's terminology) education practice has not been driven to any great extent by educational, historical or philosophical research findings. A few years ago J. Moore, as editor of the influential *Journal of Chemical Education* (2005),

Table 12.4 Some chapters of the book by Cruz et al. (1986)

Chapter	Comments
The chemistry of the nineteenth century	From Dalton atomic hypothesis to Couper and Kekulé molecular models through Frankland and Werner's valence models and finally to the Mendeleev's work, as the empirical foundation of periodicity
Birth of quantum theory	Thomson's corpuscles discovery in cathode ray tubes, the Millikan controversial experiment of determination of the electronic charge (Niaz 2000; Panusch et al. 2008; Paraskevopoulou and Koliopoulos 2011) and back to the black-body radiation experiments of Stefan, Wien, Lummer and Pringsheim, Rubens and Karlbaum that conducted M. Planck to the correct radiation formula and a couple of months later to the proposal of quantum theory as a brilliant solution to the ultraviolet catastrophe found theoretically by the classical analysis of Rayleigh and Jeans. This chapter closes with Einstein's light quantum hypothesis, his explanation of the photoelectric effect and finally with the Compton experiment that confirmed the photon existence
Atomic spectra	Bohr's atomic model of one electron atom as it was presented by him in 1913 and considering that Rydberg's formula is in itself a premise of his model. After the postulates of Bohr's model, the Sommerfeld and Wilson quantisation rules are depicted, and the elliptic orbits with three quantum numbers are introduced, with the angular momentum modified; the Frank and Hertz experiment, the fine structure of hydrogen spectrum and the Moseley law show the successful application of Bohr's model
Models of atoms and chemical bonds	The Lewis and Langmuir's model of covalent bond, Kossel's model for ionic bonding and also Pauling's electronegativity are followed by Born-Haber's cycle and Fajans' rules
Discovery of electronic spin	After the electron spin discovery is presented, spin dependent models of the atom and molecular structure, such as the Gillespie and Nyholm's Valence Shell Electron Pair Repulsion model and the Linnett double quartet model, are introduced
Modern quantum mechanics (three related chapters)	The two proposals of Schrödinger and Heisenberg, later shown to be equivalent, and their application to the mono-dimensional free particle, to the particle in a box, to the hydrogen atom and to polyatomic structure, including the philosophical interpretations of quantum mechanics (Copenhagen's, stochastic, Schrödinger's cat, Einstein-Podolsky-Rosen, etc.)
The periodic behaviour of the elements	Periodicity empirically discovered by Mendeleev is now explained. A clear distinction between isolated electronic properties such as ionisation energy and electron affinity and those which come from the chemical environment, such as atomic size and electronegativity

indicated the poor impact of chemical education research on teaching and learning, in spite of the motto of the National Association of Research in Science Teaching: 'Improving Science Teaching and Learning Through Research'.

There still has not been major change regarding what the teaching of sciences requires. In general, the majority of teachers, textbooks and science curricula still consider science teaching as a dogma or as 'rhetoric of conclusions' (Schwab 1962). This situation can only change if teachers know and recognise the uniqueness of chemistry and the philosophical positions from which they approach their practice. Realism and models are some of the issues involved. Some ideas from the historian of chemistry M. J. Nye could be very helpful:

> We can say that if mechanics has always been an aim of scientific philosophy, the twentieth-century chemistry has revived its philosophical character, achieving a long-sought understanding of the dynamics of matter. But chemists more that physicists, have remained self-conscious about the fit between the phenomena taking place in the laboratory and the symbols employed in the operations of explanatory mathematics. Precision, not rigor, has been characteristic of chemical methodology. Parallel representations, not single causal principle, have been characteristic of chemical explanation.
>
> Whereas many early-twentieth-century physicists were inclined to regard conventionalism, complementarity, and indeterminacy as concessions of failure in their traditional philosophical enterprise, chemists were not surprised that a simple, "logical" account of the behaviour of electrons and atoms, like that of molecules and people, often gives way to the inconsistencies and uncertainties of empiricism. (Nye 1993, p. 282)

References

Achinstein, P. (2001). Who really discovered the electron? In Buchwald J.Z. & Warwick A. (eds.) *Histories of the Electron. The Birth of Microphysics*, (Chapter 13 pp. 403–424), Cambridge, Massachusetts: The MIT Press.

Adúriz-Bravo A. (2012) A 'Semantic' View of Scientific Models for Science Education, *Science & Education*, Online First, 17 January.

Anderson, P. W. (1972). More Is Different, *Science*, 177(4047), 393–396. Aug. 4.

Arabatzis, T. (2001). The Zeeman Effect and the Discovery of the Electron? In Buchwald J.Z. & Warwick A. (eds.) *Histories of the Electron. The Birth of Microphysics*, (Chapter 5 pp. 171–193), Cambridge, Massachusetts: The MIT Press.

Ardac, D. (2002). Solving quantum number problems: An examination of novice performance in terms of conceptual based requirements, *Journal of Chemical Education*, 79(4), 510–3.

Atkins, P., de Paula, J., & Friedman, R. (2008). *Quanta, Matter and Change: A Molecular Approach to Physical Chemistry*, Oxford: Oxford University Press.

Ayar, M., & Yalvac, B. (2010). A sociological standpoint to authentic scientific practices and its role in school science teaching, *Ahi Evran Uni. Kirsehir Journal of Education (KEFAD)* 11, 113–127.

Baggott, J. (2011). *The Quantum Story. A History in 40 Moments*. Oxford: Oxford University Press.

Bensaude-Vincent, B. (1999). Atomism and Positivism: A legend about French Chemistry, *Annals of Science*, 56, 81–94.

Bent, H. A. (1984). Should orbitals be X-rated in beginning chemistry courses? *Journal of Chemical Education*, 61(5), 421–423.

Birk, J., & Kurtz, M. (1999). Effect of experience on retention and elimination of misconceptions about molecular structure and bonding, *Journal of Chemical Education*, 76(1), 124–128.

Bishop, D. M. (1973). *Group theory and chemistry*, Oxford, UK: Clarendon Press.

12 Historical Teaching of Atomic and Molecular Structure

Boo, H. K. (1998). Students' Understandings of Chemical Bonds and the Energetics of Chemical Reactions, *Journal of Research in Science Teaching*, 35(5), 569–581.

Branch, G.E.K. (1984). Gilbert Newton Lewis, 1875–1946, *Journal of Chemical Education*, 61(1), 18–21.

Bucat, R., & Mocerino, M. (2009). Learning at the Sub-micro Level: Structural Representations, in Gilbert, J. K. & Treagust, D. (Eds.) *Multiple Representations in Chemical Education*, (Chapter 1, pp. 11–29), Secaucus, NJ, USA: Springer.

Buchwald, J. Z. & Warwick, A. (ed) (2001). *Histories of the electron. The Birth of microphysics*, Cambridge Massachusetts: The MIT Press.

Butts, B., & Smith, R. (1987). HSC chemistry students' understanding of the structure and properties of molecular and ionic compounds, *Research in Science Education*, 17, 192–201.

Campbell, J. A. (1962). *Chemical Education Material Study*. Berkeley, CA, USA: Lawrence Hall of Science.

Cervellati, R. & Perugini, D. (1981). The understanding of the atomic orbital concept by Italian high school students, *Journal of Chemical Education*, 58(7), 568–9.

Chalmers, A. (1998). Retracing the Ancient Steps to atomic theory, *Science & Education*, 7(1), 69–84.

Chamizo, J.A. (1992). *El maestro de lo infinitamente pequeño*. John Dalton [The master of the infinitely small. John Dalton], México: Conaculta-Pangea.

Chamizo, J.A. (2001) El curriculum oculto en la enseñanza de la química, *Educación Química*, 12(4), 194–198.

Chamizo, J. A. (2007). Teaching modern chemistry through 'historical recurrent teaching models', *Science & Education*, 16(2), 197–216.

Chamizo, J.A. (2011). A new definition of Models and Modelling for chemistry Teaching, *Science & Education* OnLine First 01 November, special issue on [Philosophical Considerations in Teaching of Chemistry] edited by Sibel Erduran.

Chamizo, J.A. (2012). Heuristic Diagrams as a Tool to teach History of Science, *Science & Education*, 21(5), 745–762. OnLine First 23th August, 2011.

Christie, M. & Christie, J. R. (2000). 'Laws' and 'Theories' in Chemistry Do not Obey The rules in Bhushan N. & Rosenfeld S. (ed) *Of Minds and Molecules. New Philosophical Perspectives on Chemistry*, New York: Oxford University Press.

Coll, R. K., & Treagust, D. F. (2002). Exploring tertiary students' understanding of covalent bonding, *Research in Science and Technological Education*, 20, 241–267.

Coll, R. K., & Treagust, D. F. (2003a). Learners' mental models of metallic bonding: A cross-age study, *Science Education*, 87(5), 685–707.

Coll, R. K., & Treagust, D. F. (2003b). Investigation of secondary school, undergraduate, and graduate learners' mental models of ionic bonding, *Journal of Research in Science Teaching*, 40(5), 464–486.

Conceicao, J., & Koscinski, J. T. (2003). Exploring Atomic and Molecular Orbital in Freshman Chemistry using Computational Chemistry, *The Chemical Educator*, 8, 378–382.

Cotton, F. A. (1963). *Chemical Applications of Group Theory*, New York: John Wiley & Sons.

Cruz, D., Chamizo, J. A. y Garritz, A. (1986). *Estructura atómica. Un enfoque químico* [Atomic structure. A chemical approach], Wilmington, DE, USA: Addison Wesley Iberoamericana.

De Posada, J. M. (1997). Conceptions of high school students concerning the internal structure of metals and their electric conduction: structure and evolution, *Science Education*, 81(4), 445–467.

De Posada, J. M. (1999). The presentation of metallic bonding in high school science textbooks during three decades: science educational reforms and substantive changes of tendencies, *Science Education*, 83, 423–447.

Develaki, M. (2007). 'The Model-Based view of Scientific Theories and the structuring of school science, *Science & Education*, 16(7–8), 725–749.

Didis, N. & SakirErkoc, S. (2009). 'History of Science for Science Courses: "Spin" Example from Physics, *Latin American Journal of Physics Education*, 3, 9–12.

Dirac, P.A.M. (1929). Quantum Mechanics of Many-Electron Systems, *Proceedings of the Royal Society* (London) A123, 714–733.

Dobson, K., Lawrence, I., & Britton, P. (2000). The A to B of quantum physics, *Physics Education*, 35, 400–5.

Doyle M. (ed) (1993). *Historical Science Experiments on File, Facts on File*, New York.

Duschl, R. A. (1994). Research on the History and Philosophy of Science, in Gabel D. (Ed.) *Handbook of Research on Science Teaching and Learning*, (pp. 443–465) New York: MacMillan.

Early, J. E. (2004). Would Introductory Chemistry Courses work better with a new Philosophical basis? *Foundations of Chemistry*, 6, 137–160.

Echeverria, J. *Introducción a la Metodología de la Ciencia*, [Introduction to Science's Methodology] Madrid: Cátedra, 2003.

Eggen, P.O., Kvittingen, L., Lykknes, A., & Wittje, R. (2012). Reconstructing Iconic Experiments on Electrochemistry: Experiences from a History of Science Course. *Science & Education*, 21, 179–189.

Einstein, A. (1906). Zur Theorie der Lichterzeugung und Lichtabsorption, *Annals of Physics*, 325, 199–206.

Einstein, A. (1909). Zum gegenwärtigen Stand des Strahlungsproblems, *Phys. Zeitschr.* 10, 185–193.

Einstein, A. (1926; 1944; 1948). *Letters to Max Born; The Born-Einstein Letters, translated by Irene Born*, New York: Walker and Company, 1971. Taken from the URL http://www.spaceandmotion.com/quantum-theory-albert-einstein-quotes.htm

Erduran, S., & Scerri, E. (2002). 'The nature of chemical knowledge and chemical education', in Gilbert J.K. et al. (eds.) *Chemical Education: Towards Research-based Practice*, Kluwer, Dordrecht.

Erduran, S. (2005). Applying the Philosophical Concept of Reduction to the Chemistry of Water: Implications for Chemical Education, *Science & Education*, 14: 161–171.

Feldman, B. (2001). *The Nobel Prize: A History of Genius, Controversy, and Prestige*, New York, USA: Arcade Publishing, Reed Business Information, Inc.

Feynman, R. (1985). *The Strange Theory of Light and Matter*. London: Penguin.

Furió, C. & Calatayud, M. L. (1996). Difficulties with the Geometry and Polarity of Molecules. Beyond Misconceptions, *Journal of Chemical Education*, 73(1), 36–41.

Gagliardi, R. (1988) Cómo utilizar la historia de las ciencias en la enseñanza de las ciencias, [How to use history of sciences in the teaching of sciences], *Enseñanza de las Ciencias*, 6, 291–296.

Garritz, A. (2013). Teaching the Philosophical Interpretations of Quantum Mechanics and Quantum Chemistry through Controversies. Accepted for publication in the special issue on [Philosophical Considerations in Teaching of Chemistry] edited by Sibel Erduran, *Science & Education*, 22(7), 1787–1808.

Gault, C. (1991) History of science, individual development and science teaching, *Research in Science Education*, 21, 133–140.

Gell-Mann, M. (1994). *The Quark and the Jaguar: adventures in the simple and the complex*, New York, USA: Freeman.

Giere, R. N. (1999). *Science without laws*, Chicago, USA: University of Chicago Press.

Gilbert, J. K. (2006). On the Nature of "Context" in Chemical Education, *International Journal of Science Education*, 28(9), 957–976.

Gillespie, R. J. (1991). What is wrong with the general chemistry course? *Journal of Chemical Education*, 68(3), 192–4.

Giunta, C. (2010). *Atoms in Chemistry: From Dalton's predecessors to Complex Atoms and Beyond*, American Chemical Society-Oxford University Press, Washington.

Griffiths, A. K., & Preston, K. R. (1992). Grade-12 students' misconceptions relating to fundamental characteristics of atoms and molecules, *Journal of Research in Science Teaching*, 29, 611–628.

Grosslight, L., Unger, C., Jay, E., & Smith, C. (1991). Understanding models and their use in scienceconceptions of middle and high school students and experts. *Journal of Research in Science Teaching*, 28, 799–822.

Hacking, I. (1983). *Representing and Intervening*, Cambridge, UK: Cambridge University Press.

Hadzidaki, P., Kalkanis, G. & Stavrou, D. (2000). Quantum mechanics: A systemic component of the modern physics paradigm, *Physics Education*, 35, 386–392.

12 Historical Teaching of Atomic and Molecular Structure

Hargittai, M. & Hargittai, I. (2009). *Group Symmetry through the Eyes of a Chemist*, 3rd edition, Dordrecht, The Netherlands: Springer.

Harré, R. (2004). *Modelling: Gateway to the Unknown*, Amsterdam: Elsevier.

Harris, D. C. & Bertolucci, M. D. (1978). *Symmetry and spectroscopy. An introduction to vibrational and electronic spectroscopy*, New York: Dover.

Harrison, A. G., & Treagust, D. F. (1996). Secondary students' mental models of atoms and molecules: Implications for teaching science, *Science Education*, 80, 509–534.

Hawkes, S. J. (1992). Why should they know that? *Journal of Chemical Education*, 69(3), 178–181.

Heilbron, J. L. & Kuhn, T. S. (1969). The Genesis of the Bohr Atom, *Historical Studies in the Physical Sciences*. 1(3–4), 211–290.

Herrestein-Smith, B. (1981). Narrative Versions, Narrative Theories. In W. Mitchel (Ed.), *On Narrative*, (pp 209–232) Chicago: University of Chicago Press.

Hoffmann, R. (1998) Qualitative thinking in the age of modern computational chemistry-or what Liones Salem knows, *Journal of Molecular Structure*, 424: 1–6

Hohenberg, P. & Kohn, W. (1964). Inhomogeneous electron gas, *Physical Review*, 136, B864–71.

Holbrow, C. H., Amato, J. C., Galvez, E. J. & Lloyd, J. N. (1995). Modernizing Introductory Physics, *American Journal of Physics*, 63, 1078–1090.

Hund, F. (1977). Early History of the Quantum Mechanical Treatment of the Chemical Bond, *Angewandte Chemie*, International Edition in English, 16, 87–91.

Husbands, C. (2003). *What is history teaching? Language, ideas and meaning in learning about the past.* Buckingham: Open University Press.

Izquierdo, M. & Adúriz, A. (2009). Physical construction of the chemical atom: Is it Convenient to go All the Way Back? *Science & Education*, 18(3–4), 443–455.

Izquierdo, M. (2010). La transformación del átomo químico en una partícula física ¿se puede realizar el proceso inverso? In Chamizo J.A. (ed) *Historia y Filosofía de la Química* [History and philosophy of chemistry], (pp 195–209) México: Siglo XXI-UNAM.

Jensen, W. B. (1980). *The Lewis acid–base concepts*, New York, Wiley.

Jensen, W. B. (1998). Logic, History, and the Chemistry Textbook. I. Does Chemistry Have a Logical Structure? *Journal of Chemical Education*, 75(6), 679–687; II. Can We Unmuddle the Chemistry Textbook? 75(7), 817–828; III. One Chemical Revolution or Three? 75(8), 961–969.

Jensen, W.B (ed) (2002). *Mendeleev on the Periodic Law. Selected Writings, 1869–1905*, New York, Dover.

Jensen, W.B. (2010). Four Centuries of Atomic Theory in Giunta C. (ed) *Atoms in Chemistry: From Dalton's predecessors to Complex Atoms and Beyond*, American Chemical Society-Oxford University Press, Washington.

Jensen, W. P., Palenik, G. J., & Suh, I. (2003). The History of Molecular Structure Determination Viewed through the Nobel Prizes, *Journal of Chemical Education*, 80(7), 753–761.

Johnston, I. D., Crawford, K., & Fletcher, P. R. (1998). Student difficulties in learning quantum mechanics, *International Journal of Science Education*, 20(5), 427–446.

Justi, R., & Gilbert, J. (2000). History and philosophy of science through models: some challenges in the case of 'the atom, *International Journal of Science Education*, 22(9), 993–1009.

Kalkanis, G., Hadzidaki, P., & Stavrou, D. (2003). An instructional model for a radical conceptual change towards quantum mechanics concepts, *Science Education*, 87, 257–280.

Karakostas, V. & Hadzidaki, P. (2005). Realism vs. Constructivism in Contemporary Physics: The Impact of the Debate on the Understanding of Quantum Theory and its Instructional Process, *Science & Education*, 14(7–8), 607–629.

Kauffman, G. B. & Kauffman, L. M. (1996). An Interview with Linus Pauling, *Journal of Chemical Education*, 73(1), 29–32.

Kauffman, G. B. (1999). From Triads to Catalysis: Johann Wolfgang Döbereiner (1780–1849) on the 150th Anniversary of His Death, *The Chemical Educator*, 4, 186–197.

Kauffman, G. B. (2004). Sir William Ramsay: Noble Gas Pioneer. On the 100th Anniversary of His Nobel Prize, *The Chemical Educator*, 9, 378–383.

Kauffman, G. B. (2006). Radioactivity and Isotopes: A Retrospective View of Frederick Soddy (1877.1956) on the 50th Anniversary of His Death, *The Chemical Educator*, 11, 289–297.

Kauffman, G. B. (2010). The 150th Anniversary of the First International Congress of Chemists, Karlsruhe, Germany, September 3–5, 1860, *The Chemical Educator*, 15, 309–320.

Klassen, S. (2007). The Construction and Analysis of a Science Story: A Proposed Methodology, *Proccedings of the International History and Philosophy of Science Teaching Group Conference*, Calgary, Canada.

Klassen, S. (2008). The Photoelectric Effect: Rehabilitating the Story for the Physics Classroom' *Proceedings of the Second International Conference on Story in Science Teaching*, Munich, Germany.

Kleppner, D., & Jackiw, R. (2000). One Hundred Years of Quantum Physics, *Science*, 289(5481), 893–898.

Kohn, W., & Sham, L. J. (1965). Self-consistent equations including exchange and correlation effects, *Physical Review*,140, A1133–8.

Kuhn, T. S. (1969). *The structure of scientific revolutions*, Chicago: University of Chicago Press.

Kuhn, T. S. (1978). *Black-Body Theory and the Quantum Discontinuity 1894–1912*, Oxford, UK: Oxford University Press.

Kutzelnigg, W. (1984). Chemical Bonding in Higher Main Group Elements, *Angew. Chem. Int.* Ed. Engl. 23, 272–295.

Langmuir, I. (1919). The Arrangement of Electrons in Atoms and Molecules, *J.Am. Chem.Soc*, 41, 868–934

Laloë, F. (2001). Do we really understand quantum mechanics? Strange correlations, paradoxes, and theorems, *American Journal of Physics*, 69, 655–701.

Laudan, L. (1997). *Progress and its Problems: Toward a theory of scientific growth*, Berkeley: University of California Press.

Lee, O., Eichinger, D. C., Anderson, C. W., Berkheimer, G. D., & Blakeslee, T.D. (1993). Changing Middle School Student's Conception of Matter and Molecules, *Journal of Research in Science Teaching*, 30(3), 249–270.

Lewis, G. N. (1923). *Valence and the Structure of Atoms and Molecules*, New York: Dover.

Lombardi, O. & Labarca, M. (2005). The Ontological Autonomy of The Chemical World, *Foundations of CHemistry*, 7, 125–148.

Martin B. & Richards E. (1995). Scientific knowledge, controversy, and public decision-making, in Published in Jasanoff, S., Markle, G.E., Petersen, J.C. & Pinch T. (eds.), *Handbook of Science and Technology Studies* (Newbury Park, CA: Sage.

Matthews, M. R. (1994/2014). *Science teaching: The role of history and philosophy of science.* London: Routledge.

Matthews, M.R. (1992). History, Philosophy and Science Teaching: The Present Rapprochement, *Science & Education* 1(1), 11–48.

Magnasco, V. (2004). A Model for the Chemical Bond, *Journal of Chemical Education*, 81(3), 427–435.

Marinacci, B. (1995) (Ed) *Linus Pauling in his own words*, Simon&Schuster, New York

Melrose, M. P., & Scerri, E. R. (1996). Why the 4s Orbital Is Occupied before the 3d, *Journal of Chemical Education*, 73(6), 498–503.

Metz, D., Klassen, S., Mcmillan, B., Clough, M., & Olson, J. (2007). Building a Foundation for the Use of Historical Narratives, *Science & Education,* 16(3–5), 313–334.

Morrison M. (2001). History and Metaphysics: On the Reality of Spin, In Buchwald J.Z. & Warwick A. (eds.) *Histories of the Electron. The Birth of Microphysics*, (Chapter 14 pp. 425–450), Cambridge, Massachusetts: The MIT Press.

Michelini, M., Ragazzon, R., Santi, L., & Stefanel, A. (2000). Proposal for quantum physics in secondary school, *Physics Education*, 35(6), 406–410.

Moreno-Ramírez, J. E., Gallego-Badillo, R. and Pérez-Miranda, R. (2010). El modelo semicuántico de Bohr en los libros de texto [The semi-quantum Bohr's model in textbooks], *Ciência & Educação*, 16(3), 611–629.

Nachtrieb N.H. (1975) Interview with Robert S. Mulliken, *Journal of Chemical Education*, 52(9), 560–563.

Nash, L. K. (1957). "The Atomic-Molecular Theory." In James Bryant Conant (Ed.) *Harvard Case Histories in Experimental Science*, Vol. 1. Cambridge, MA, USA: Harvard University Press.

Niaz, M. (2000). The oil drop experiment: a rational reconstruction of the Millikan-Ehrenhaft controversy and its implications for chemistry textbooks, *Journal of Research in Science Teaching*, 37(5), 480–508.

Niaz, M. (2001). A rational reconstruction of the origin of the covalent bond and its implications for general chemistry textbooks, *International Journal of Science Education*, 23, 623–641.

Niaz, M. (2009). *Critical Appraisal of Physical Science as a Human Enterprise. Dynamics of Scientific Progress*. Dordrecht, The Netherlands: Springer Academic Publishers.

Niaz, M. (2010). Science curriculum and teacher education: The role of presuppositions, contradictions, controversies and speculations vs. Kuhn's normal science, *Teaching and Teacher Education*, 26, 891–899.

Niaz, M., & Fernández, R. (2008). Understanding quantum numbers in general chemistry textbooks, *International Journal of Science Education*, 30(7), 869–901.

Norris, S., Guilbert, M., Smith, M., Shaharam, H., & Phillips, L. (2005). A theoretical Framework for Narrative Explanation in Science, *Science Education*, 89(4) 535–554.

Novick, S., & Nussbaum, J. (1978). Junior High School Pupils' Understanding of the Particulate Nature of Matter: An Interview Study, *Science Education*, 62[3], 273–281.

Novick, S., & Nussbaum, J. (1981). Pupils' Understanding of the Particulate Nature of Matter: A Cross-Age Study, *Science Education*, 65[2], 187–196.

Nuffield Foundation (1967). *Chemistry. Handbook for teachers*, London: Longmans/Penguin Books.

Nussbaum, J. (1985). The Particulate Nature of Matter in the Gaseous Phase. In R. Driver, E. Guesne y A. Tiberghien (Eds.), *Children's Ideas in Science*, (pp. 125–144) Philadelphia: Open University Press.

Nye, M. J. (1993). *From Chemical Philosophy to Theoretical Chemistry*, University of California Press, Berkeley

Ogilvie, J. F. (1994). The Nature of the Chemical Bond 1993. There are No Such Things as Orbitals!, in E. S. Kryachko and J. L. Calais (eds.), *Conceptual Trends in Quantum Chemistry*, (pp. 171–198), Dordrecht, The Netherlands: Kluwer.

Özmen, H. (2004). Some Student Misconceptions in Chemistry: A Literature Review of Chemical Bonding, *Journal of Science Education and Technology*, 13(2), 147–159.

Pagliaro, M. (2010). On shapes, molecules and models: An insight into chemical methodology, *European Journal of Chemistry*, 1, 276–281.

Panusch, M., Singh, R., & Heering, P. (2008). How Robert A. Millikan got the Physics Nobel Prize'. *Proceedings of the Second International Conference on Story in Science Teaching*, Munich, Germany.

Paoloni, L. (1982). Classical mechanics and quantum mechanics: an elementary approach to the comparison of two viewpoints, *European Journal of Science Education*, 4, 241–251.

Paraskevopoulou, E. and Koliopoulos, D. (2011). Teaching the Nature of Science Through the Millikan-Ehrenhaft Dispute, *Science & Education,* 20(10), 943–960. Published online 26 September 2010.

Park, E: J. & Light, G. (2009). Identifying Atomic Structure as a Threshold Concept: Student mental models and troublesomeness, *International Journal of Science Education*, 31(2), 895–930.

Peterson, R. F., Treagust, D. F., & Garnett, P. (1989). Development and application of a diagnostic instrument to evaluate grade 11 and 12 students' concepts of covalent bonding and structure following a course of instruction, *Journal of Research in Science Teaching*, 26(4), 301–314.

Petri, J., & Niedderer, H. (1998). A learning pathway in high-school level quantum atomic physics, *International Journal of Science Education*, 20(9), 1075–1088.

Piaget, J. & Garcia, R. (1983). *Psychogenesis and the history of science*. New York, Columbia University Press.

Pospiech, G. (2000). Uncertainty and complementarity: The heart of quantum physics, *Physics Education*, 35(6), 393–399.

Popper, K. (1969). *Conjectures and Refutations*, London: Routledge and Kegan Paul.

Primas, H. (1983) *Chemistry, Quantum Mechanics and Reductionism: Perspectives in theoretical chemistry*, Berlin, Springer.

Purser, G. H. (2001). Lewis structure in General Chemistry: Agreement between electron density calculations and Lewis structures, *Journal of Chemical Education*, 78(7), 981–983.

Reichenbach, H. (1938). *Experience and prediction: an analysis of the foundations and the structure of knowledge*. Chicago: University of Chicago Press.

Reichenbach, H. (1978, [1929]) The aims and methods of physical knowledge pp 81–125 in Hans Reichenbach: *Selected writings 1909–1953* (M. Reichenbach and R.S. Cohen, Eds; principal translations by E.H. Schneewind), volumen II, Dordrecht: Reidel.

Reish, G. A. (2005). *How the Cold War transformed Philosophy of Science. To the Icy Slopes of Logic*, New York, Cambridge University Press (Versión en español *Cómo la Guerra fría transformó la filosofía de la ciencia. Hacia las heladas laderas de la lógica*, Buenos Aires, Universidad de Quilmes Editorial, 2009).

Rodríguez, M., & Niaz, M. (2004). A Reconstruction of Structure of the Atom and Its Implications for General Physics Textbooks: A History and Philosophy of Science Perspective, *Journal of Science Education and Technology*, Vol. 13, No. 3.

Scerri, E. R. (1991). Electronic Configurations, Quantum Mechanics and Reduction, *British Journal for the Philosophy of Science*, 42(3), 309–25.

Scerri, E. R. (2000a). Have Orbitals Really Been Observed? *Journal of Chemical Education*, 77(11), 1492–4.

Scerri, E. R. (2000b). The failure of Reduction and How to Resist Disunity of the Sciences in the Context of Chemical Education, *Science & Education*, 9, 405–425.

Scerri, E. R. (2001). The Recently Claimed Observation of Atomic Orbitals and Some Related Philosophical Issues, *Philosophy of Science, 68 (Proceedings)* S76-S88, N. Koertge, ed. Philosophy of Science Association, East Lansing, MI

Scerri, E. R. (2007). *The Periodic Table: Its Story and Its Significance*, Oxford University Press, New York.

Scheffel, L., Brockmeier, W., & Parchmann, L. (2009). Historical material in macro-micro thinking: Conceptual change in chemistry education and the history of chemistry. In Gilbert, J. K. & Treagust, D. F. (Eds.). (2009). *Multiple representations in chemical education* (pp. 215–250). Springer.

Schummer, J. (1998). The Chemical Core of Chemistry I: A Conceptual Approach, *HYLE-International Journal for Philosophy of Chemistry*, 4, 129–162.

Schummer, J. (1999). Coping with the Growth of Chemical Knowledge: Challenges for Chemistry Documentation, Education, and Working Chemists, *Educación Química*, 10(2), 92–101.

Schummer, J. (2008). The philosophy of chemistry in Fritz Allhoff (Ed.), *Philosophies of the Sciences*, (pp. 163–183), Albany, NY, USA: Blackwell-Wiley.

Schwab, J. J. (1962). The teaching of science as enquiry. In J. J. Schwab & P. F. Brandwein (Eds.), *The teaching of science*. Cambridge: Harvard University Press.

Seok P. & Jin S. (2011) What Teachers of Science Need to Know about Models: An overview, *International Journal of Science Education*, 33(8), 1109–1130.

Segrè, E. (2007). *From X-rays to Quarks: Modern Physicists and Their Discoveries*, New York, USA: Dover Publications.

Shahbazian, S. & Zahedi, M. (2006). The Role of Observables and Non-Observables in Chemistry: A Critique of Chemical Language, *Foundations of Chemistry*, 8, 37–52.

Shiland, T. W. (1995). What's the use of all this theory? The role of quantum mechanics in high school chemistry textbooks, *Journal of Chemical Education*, 72(3), 215–219.

Shiland, T. W. (1997). Quantum mechanics and conceptual change in high school chemistry textbooks, *Journal of Research in Science Teaching*, 34(5), 535–545.

Shrigley, R.L. & Koballa, T. R. (1989). Anecdotes: What Research Suggests about Their Use in the Science Classroom, *School Science and Mathematics*, 89, 293–298.

Silberstein, M. (2002). Reduction, Emergence and explanation, en Machamer P., and Silberstein, M., *Philosophy of Science*, Oxford: Blackwell.

Slater, J. C. (1951). A Simplification of the Hartree-Fock Method, *Physical Review*, 81, 385–390.

Snooks, R. J. (2006). Another Scientific Practice separating chemistry from Physics: Thought Experiments, *Foundations of Chemistry*, 8, 255–270.

Snow, C.P. (1981). *The Physicists. A generation that changed the world*, Macmillan, London

Spence, J. C. H., O'Keeffe, M. and Zuo, J. M. (2001). Have orbitals really been observed? Letter in *Journal of Chemical Education*, 78(7), 877.

Stewart, I. (2007). *Why Beauty is Truth. The history of symmetry*. Basic Books.

Stinner, A. (2008). *Teaching Modern Physics using Selected Nobel Lectures* APS Physics Forum on Education, fall.

Stinner, A. & Williams, H. (1998). History and Philosophy of Science in the Science Curriculum, a chapter in *The International Handbook of Science Education*. Dordrecht: Kluwer Academic Publishers.

Strong, I. E. (1962). *Chemical Systems. Chemical Bond Approach Project*, New York, USA: Chemical Education Publishing Company.

Styer, D. F. (2000). *The Strange World of Quantum Mechanics*. Cambridge: Cambridge University Press.

Sutcliffe, B. T. (1996). The Development of the Idea of a Chemical Bond, *International Journal of Quantum Chemistry*, 58, 645–55.

Taber, K. S. (1994). Misunderstanding the ionic bond, *Education in Chemistry*, 31(4), 100–103.

Taber, K. S. (1997). Student understanding of ionic bonding: molecular versus electrostatic framework? *School Science Review*, 78(285), 85–95.

Taber, K. S. (2002a). Conceptualizing Quanta: Illuminating the Ground State of Student Understanding of Atomic Orbitals, *Chemistry Education: Research and Practice*, 3(2), 145–158.

Taber, K. S. (2002b). Compounding Quanta: Probing the Frontiers of Student Understanding Of Molecular Orbitals, *Chemistry Education: Research and Practice*, 3(2), 159–173.

Taber, K.S. (2003). The Atom in the Chemistry Curriculum: Fundamental Concept, Teaching Model or Epistemological Obstacle, *Foundations of Chemistry*, 5, 43–84.

Taber, K. S. (2005). Learning Quanta: Barriers to Stimulating Transitions in Student Understanding of Orbital Ideas, *Science Education*, 89, 94–116.

Talanquer, V. (2011, online), School Chemistry: The Need for Transgression, *Science & Education*, published online 17th September.

Tapio, I. (2007). Models and Modelling in Physics Education: A critical Re-analysis of Philosophical Underpinnings and Suggestions for Revisions, *Science & Education*, 16, 751–773.

Teichmann, J. (2008). Anecdotes Can Tell Stories—How? And What is Good and What is Bad about Such Stories? *Proceedings of the Second International Conference on Story in Science Teaching*, Munich, Germany.

Thomson, J. J. (1904). *Electricity and matter*, Westminster, UK: Archibald Constable & Co. Ltd.

Toulmin, S. (1961). *Foresight and Understanding: An Enquiry Into the Aims of Science*, Bloomington: Indiana University Press.

Toulmin, S. (1972). *Human Understanding*, Princeton: Princeton University Press.

Tsaparlis, G., & Papaphotis, G. (2002). Quantum-Chemical Concepts: Are They Suitable for Secondary Students? *Chemistry Education: Research and Practice*, 3(2), 129–144.

Tsaparlis, G. (1997a). Atomic orbitals, molecular orbitals and related concepts: Conceptual difficulties among chemistry students. *Research in Science Education*, 27, 271–287.

Tsaparlis, G. (1997b). Atomic and Molecular Structure in Chemical Education, *Journal of Chemical Education*, 74(8), 922–5.

Tsaparlis, G. (2001). Towards a meaningful introduction to the Schrödinger equation through historical and heuristic approaches, *Chemistry Education: Research and Practice in Europe*, 2, 203–213.

Tsaparlis, G., & Papaphotis, G. (2009). High-school Students' Conceptual Difficulties and Attempts at Conceptual Change: The case of basic quantum chemical concepts, *International Journal of Science Education*, 31(7), 895–930.

Toulmin S. & Goodfield J. (1962). *The Architecture of Matter*, The University of Chicago Press, Chicago

Valanides, N. (2000). Primary student teachers' understanding of the particulate nature of matter and its transformations during dissolving, *Chemistry Education: Research and Practice in Europe*, 1, 249–262.

Van Aalsvoort, J. (2004) 'Logical positivism as a tool to analyse the problem of chemistry's lack of relevance in secondary school chemical education', *International Journal of Science Education*, 26, 1151–1168.

Van Brakel, J. (2000). *Philosophy of Chemistry*, Leuven University Press, Louvain.

Van Berkel, B. (2005). *The Structure of Current School Chemistry. A Quest for Conditions for Escape*, Centrum voor Didactiek van Wiskunde en Natuurwetenschappen, University of Utrech CD-β Press, Utrech.

van Berkel, B., de Vos, W., Veronk, A. H., & Pilot, A. (2000). Normal science education and its dangers: The case of school chemistry. *Science & Education*, 9, 123–159.

Velmulapalli, G. K. & Byerly H. (1999) Remnants of Reductionism, *Foundations of Chemistry* 1, 17–41.

Viana, H. E. B. & Porto, P. A. (2010). The development of Dalton's Atomic Theory as a Case Study in the History of Science: Reflections for Educators in Chemistry, *Science & Education*, 19(1), 75–90.

Wandersee, J. H., & Griffard, P. B. (2002). The history of chemistry: Potential and actual contributions to chemical education in Gilbert J. et al. (eds), *Chemical Education: Towards Research-based Practice*, (Chapter 2, pp. 29–46), Dordrecht, The Netherlands: Kluwer.

Weyer, J. (1974) Chemiegeschichtsschreibung von Wiegleb (1790) bis Partington (1970); Gerstenberg: Hildescheim

Wightman, T., Johnston, K., & Scott, P. (1987). *Children's learning in science project in the classroom. Approaches to teaching the particulate theory of matter*, Centre for Studies in Science and Mathematics Education: University of Leeds.

Wisniak, J. (2013). Gustav Charles Bonaventure Chancel, *Educación Química*, 24(1), 23–30.

Wittmann, M. C., Steinberg, R. N., & Redish, E. F. (2002). Investigating student understanding of quantum physics: Spontaneous models of conductivity, *American Journal of Physics*, 70, 218–226.

Woolley, R.G. (1978). Must a molecule have a shape? *Journal of the American Chemical Society*, 100, 1073–1078.

Yager, R. E. (2004). Science is Not Written, But It Can Be Written About, in W. Saul (Ed.), *Crossing Borders in Literacy and Science Instruction*, (pp. 95–107) Washington: NSTA.

Zuo, J.; Kim, M.; O'Keeffe, M.; Spence, J. (1999). Direct observation of d holes and Cu-Cu bonding in. Cu_2O, *Nature*, 401, 49–56.

José Antonio Chamizo is full-time Professor at the School of Chemistry of the National Autonomous University of Mexico, UNAM. Since 1977 he has taught over 100 courses from high school to Ph.D. and published over 100 refereed articles in chemistry, education, history, philosophy and popularisation of science. His B.S. and M.S. degrees were in chemistry from the National University of Mexico and his Ph.D. from the School of Molecular Sciences at the University of Sussex. He currently teaches 'History and Philosophy of Chemistry' and research in the same field for chemistry teaching. He is author or co-author of over 30 chapters in books and of more than 50 textbooks and popularisation of science books among which the Mexican free textbook of Natural Sciences, coordinated by him and edited by the Education Ministry, which have been published to date more than 100 million copies.

Andoni Garritz is full-time Professor at the School of Chemistry of the National Autonomous University of Mexico, UNAM, where he teaches and researches 'Didactics of Chemistry', 'Structure of Matter' and 'Science and Society'. He studied Chemical Engineering at UNAM and got his Ph.D. with a stay in the Quantum Chemistry Group at Uppsala University, Sweden, where he received the Hylleraas' Award. He has been giving lectures for 40 years at different levels, from high school to postgraduate. He is a consultant 23 of UNESCO, coordinating seven chemistry experimental education projects in Latin America. His most relevant books are *Chemistry in Mexico: Yesterday, Today and Tomorrow* (UNAM, 1991); *From Tequesquite to DNA* (Fondo de Cultura Económica, 1989); *Atomic structure: A Chemical Approach; You and Chemistry*; and *University Chemistry* (Pearson Education 1986, 2001, 2005). He is the founder Director of the Journal *Educación Química* that has been in circulation 23 years and indexed by Scopus.

Part III
Pedagogical Studies: Biology

Chapter 13
History and Philosophy of Science and the Teaching of Evolution: Students' Conceptions and Explanations

Kostas Kampourakis and Ross H. Nehm

13.1 Introduction

This handbook is about the inclusion of the history and philosophy of science (hereafter HPS) in science education. For the past 30 years (at least), there has been lively discussion (and debate) about what HPS scholarship can contribute to science education and how important this contribution can be. This chapter provides examples of why HPS is of central importance to evolution education research and instruction. This and the following chapter not only argue for the centrality of HPS scholarship for evolution education. Rather, they argue that evolution education research and instruction are based on poor standards if they are not appropriately informed by relevant HPS scholarship. Several aspects of the evolution education literature illustrate this point.

This chapter begins with a review of current perspectives on students' preconceptions about evolution and illustrates that attempting to shoehorn students' explanations about evolutionary phenomena into categorizations emblematic of particular historical figures is misleading, if not mistaken. Unfortunately, teaching evolution in its historical context has often been based on questionable characterizations of history: first, that students initially hold "Lamarckian" preconceptions about evolution, and second, that the conceptual change process mirrors historical paradigm shifts from "Lamarckian" to "Darwinian" frameworks. Careful reading of the historical literature suggests that students' ideas should not be labeled as "Lamarckian"

K. Kampourakis (✉)
Biology Section and IUFE, University of Geneva,
Pavillon Mail, 40 Boulevard du Pont-d'Arve, Geneva 4, Switzerland
e-mail: Kostas.Kampourakis@unige.ch

R.H. Nehm
Department of Ecology and Evolution, Stony Brook University,
092 Life Sciences Building, Stony Brook, NY 11794-5233, USA
e-mail: ross.nehm@stonybrook.edu

M.R. Matthews (ed.), *International Handbook of Research in History,
Philosophy and Science Teaching*, DOI 10.1007/978-94-007-7654-8_13,
© Springer Science+Business Media Dordrecht 2014

(or "Darwinian") because their explanations are fundamentally different from those of associated historical figures. In addition, the process of conceptual change is much more complex than a straightforward shift from "Lamarckian" to "Darwinian" perspectives. Indeed, the actual history of evolutionary thought does not reflect the clean replacement of one evolutionary "paradigm" with another. The history of science appears to conflict with science educators' conceptualizations of it.

A more fruitful role for HPS scholarship in evolution education may be to guide the development of theoretical frameworks for exploring the structure of students' explanations of evolutionary change (e.g., Kampourakis and Zogza 2008, 2009). Several accounts of explanation have been proposed in the philosophy of science, but few have been integrated into evolution education research or practice. Although much of this literature is too complex for most students, core aspects of it may be effectively utilized in the development of conceptual heuristics for explaining evolution. Specifically, students would benefit from awareness that explanations in general have a causal structure and that evolutionary explanations in particular also have a historical nature. Evolution in particular (and biology in general, see Brigandt 2013a) is characterized by explanatory pluralism. Students would also benefit from learning that different evolutionary concepts may be linked to different kinds of explanations. For example, natural selection is an important, but not the only important, explanatory principle in evolution. Adaptations may be explained by invoking natural selection, but homologies may be explained on the basis of common descent.

In addition to gaining an understanding of the diversity of explanatory approaches used in evolutionary biology—and associated concept/explanation alignment—students must also be exposed to a greater diversity of explanatory *tasks*. Many of the commonly used simple explanation-based assessments in the evolution education literature (e.g., Nehm et al. 2012) could be pedagogically enhanced if they were to be modified to encompass more diverse explanatory contrasts. In line with these ideas, this chapter ends with a discussion of how HPS scholarship may be used to develop frameworks and tasks that can be used for teaching about the structure and historical nature of evolutionary explanations. This aspect of evolution education is particularly important given forthcoming standards emphasizing practice-based tasks (e.g., explanation and argumentation).

13.2   Students' Preconceptions of Evolution and HPS: A Review of the Literature

The history of science (HOS) may be used to provide ideas for designing instruction aiming at conceptual change (e.g., Jensen and Finley 1996; Passmore and Stewart 2002). However, making use of this strategy requires a careful framing of HOS as well as an understanding that there are significant differences between conceptual change in science and individual conceptual development (Gauld 1991). Historiography is a particularly important consideration when using HOS in science education because science is always enacted in particular social and cultural contexts. Thus, in order to understand how science is done, one must go beyond a superficial reading of HOS and engage with the nuances of particular historical episodes, contexts, and social networks. This section reviews studies analyzing students' preconceptions about evolution and illustrates the importance of historiography. Science education research examining student conceptual change patterns is also reviewed, and it is noted that individual conceptual change is in some cases quite different from conceptual change in HOS. Much of the discussion will focus on the evolutionary ideas of Jean Baptiste Lamarck.

Jean Lamarck was one of many naturalists across Europe involved in the debate concerning the fixity of species (Corsi 2005). Perhaps his most important contribution to natural history was that he replaced a static picture of nature held by his forerunners with a dynamic one in which life as a whole was constantly in flux (Mayr 1982, p. 352), an idea that had strong support from his earlier geological studies (Corsi 2001, p. 163). Lamarck's ideas can be considered as precursors to modern evolutionary biology because his work was the first attempt to develop a theory in which all organisms developed from primitive ancestors (Bowler 2003, pp. 86–87). In fact, his theory can be considered as the first major evolutionary synthesis in modern biology (Corsi 2001, p. 11), which shaped the debates that ultimately led to Darwin's theory of evolutionary change.

Lamarck proposed a complex model of evolutionary change. It included spontaneous generation as the starting point for the lowest forms of life, progressive forces that carried life up to higher levels of organization, adaptation caused by changes in individual organisms through use or disuse, and the inheritance of acquired traits as changeable hereditary material could be transmitted to the next generation (Mayr 2002, p. 81). Lamarck was not a teleologist; he did not recognize any "guidance" of evolution towards a goal and accepted only mechanistic explanations for biotic change. Indeed, Lamarck's causal chain (connected by a complex process of nervous fluid dynamics) began with needs imposed by the environment, continued with efforts of physiological excitations, and ended with the stimulation of growth resulting in the alteration of biotic features (Mayr 1982, p. 357). Lamarck believed that animals' needs determined how they would use their body parts, and the effects of use and disuse would cause some parts to increase in size by attracting more of the nervous fluid, whereas disused organs would receive less fluid and would degenerate (Bowler 2003, p. 92). Lamarck thought that through this process species could change but could not become extinct; he considered natural extinction to be inconceivable (Burkhardt 1995, p. 131). Lamarck was also not a vitalist; his theory was materialistic and provided a mechanistic explanation for the power of life (Burkhardt 1995, p. 151).

Contrary to common wisdom, it was not Lamarck but Charles Darwin who in fact held that environmental changes, acting either on the reproductive organs or on the body, were necessary to generate variation. Darwin hypothesized that the body was made up of units that increased by self-division or proliferation and were ultimately converted into various body tissues. These units could throw off minute granules (gemmules) that could develop into units similar to those from which they

Table 13.1 The main points of difference between Lamarck's and Darwin's theories of evolution (Based on Kampourakis and Zogza 2007)

Concept	Lamarck	Darwin
Common ancestry	Denied: spontaneous generation of life was occurring all time	Accepted: evolution had a branching form from common ancestors
Variations	Unfortunate consequences of imperfection in the process	Indispensable precondition of continuing evolutionary change
Species	Species did not exist but were convenient fiction	Species existed precisely because of naturally occurring variations
Unit that evolves	The overall process of evolution was modeled on the development of individual organisms and evolution was driven by changes in individuals	Populations evolved and developmental ontogeny explained individual characteristics, while selection explained the characteristics of the population and hence phylogeny
Mechanism of evolution	Transformation of individuals: progress of individuals from simpler to more complex forms	Natural selection in populations: differential survival in populations based on existent variation in a particular environment
Extinction	Denied: nature was powerful enough to ensure that no form could ever completely die out	Accepted as an important feature of the mechanism of natural selection

were originally derived (Winther 2000). In Darwin's model, variability resulted from changed conditions during successive generations. Either gemmules aggregated in an irregular manner, causing modifications in the offspring, or certain parts of the body could throw off modified gemmules that would give rise to similarly modified structures in the offspring. Under the right conditions, modified gemmules would continue multiplying until they replaced the old, unmodified ones and the offspring might gradually vary further through successive generations.

Darwin also believed that the inheritance of acquired characters was possible. However, in scientific studies during Darwin's time, it was observed that a removed part or organ in a parent could reappear in its offspring. Such findings contradicted Darwin's hypothesis, but he dealt with such findings by arguing that gemmules derived from reduced or useless parts would be more liable to diminish in size than those derived from parts which were still functionally active (Winther 2000; Endersby 2009; for the wider context, see Kampourakis 2013a). The main features of Darwin's and Lamarck's theories are presented in Table 13.1. It is important to be aware of these features in order to determine whether students' ideas/explanations bear any resemblance to Darwin's and Lamarck's theories of evolutionary change.

Researchers in science education have often noted similarities between ideas in HOS and students' ideas. For example, in many research articles, students' preconceptions about evolution have been characterized as "Lamarckian." However, it seems that not all researchers use the term *Lamarckian* in the same

13 History and Philosophy of Science and the Teaching of Evolution...

Table 13.2 Ideas described by the term "Lamarckian" in the literature (*E* explicitly stated, *I*: could be implied). Studies published after 2007 that explicitly agree with the conclusions of Kampourakis and Zogza (2007) are not included in this table (Based on Kampourakis and Zogza 2007, revised and updated)

	Lamarckian idea		Non-Lamarckian idea	
Reference	1. Change due to use and disuse of body parts	2. Change due to inheritance of acquired traits	3. Change due to a predetermined final end	4. Change imposed by need
Pazza et al. (2010)		E		
Battisti et al. (2010)	E	E		I
Berti et al. (2010)	E		E	
Geraedts and Boersma (2006)	I	E		
Banet and Ayuso (2003)		E		E
Alters and Nelson (2002)	E	I	E	E
Passmore and Stewart (2002)	E	E	E	E
Samarapungavan and Wiers (1997)	E		E	E
Jensen and Finley (1996)	E	E		
Demastes et al. (1996)	E	I		
Settlage (1994)	E	I		
Jiménez-Aleixandre (1992)	I	E		
Bishop and Anderson (1990)	I	E		
Clough and Wood-Robinson (1985)			E	E
Brumby (1979)				E
Deadman and Kelly (1978)	E	I		

sense, and the term does not always accurately mirror Lamarck's ideas (see above and Table 13.1). Consequently, the meaning of the term *Lamarckian* is different among studies in the science education literature (see Table 13.2); in many cases Lamarck's two central concepts (change through use and disuse and the inheritance of acquired traits) are associated with two others which are *not* Lamarckian (Table 13.2, right columns) and imply a teleological process of change (to achieve a predetermined end or to satisfy needs).

As a result, two major problems arise: (1) different ideas are implied by the term *Lamarckian,* misrepresenting the actual content of students' preconceptions about evolution and (2) Lamarck's actual contributions to the history of evolutionary thought are also misrepresented. Thus, readers who are familiar with the history of evolution may understand *Lamarckian* to only refer to change through "use or disuse" or "inheritance of acquired traits" and ignore students' teleological explanations. On the other hand, the reader who ignores the historical facts might arrive at the conclusion that Lamarck's views and students' preconceptions are equally and similarly inaccurate (for details see Kampourakis and Zogza 2007). Thus, a proper understanding of HOS is of considerable importance to evolution education research.

Many researchers have begun to acknowledge the frequent mischaracterization of Lamarck's ideas in science education scholarship.[1] Nevertheless, some researchers not only continue to overlook this research but also continue to embrace a mistaken understanding of what HOS may contribute to science education. We believe that it is problematic to describe students' preconceptions as "Lamarckian" (or "Cuvierian," or "Paleyian," or "Darwinian") because in all cases it is problematic to compare students' conceptions (mostly naïve in the psychological sense) with the conceptual schemes proposed by important thinkers of the past. Lamarck, Paley, and Cuvier, for example, each possessed very detailed understandings of organisms' structures and physiologies and proposed equally detailed (although occasionally quite speculative) explanations (e.g., for the origin of adaptations). The same situation is almost never the case with secondary students or most undergraduates.

For instance, when comparing student ideas to those of Lamarck, Prinou and colleagues (2011, p. 276) note: "…conceptions of the pupils are called *Lamarckian* because 'the capacity of organisms to react to special conditions in the environment' (which does not occur directly but by a chain of events/complex mechanisms which Lamarck describes in his work) was considered by Lamarck as the second cause of evolutionary change." This quote raises the question as to whether students ever provide a description of the chain of events or complex mechanisms that Lamarck describes in his work (see above). Prinou et al. do not provide any evidence for such complexity. Instead, they describe students' preconceptions as *Lamarckian* because "organisms can develop new adaptive characteristics in response to environmental demands -which is a Lamarckian principle" (quoting Samarapungavan and Wiers 1997).[2] Prinou et al. (2011, p. 276) also note: "This goal-directed (teleological) reasoning noted in the pupils' explanations regarding the origin of biological adaptations, proves to be the predominant one used by pupils of various ages."[3] Teleology was not a central feature of Lamarck's evolutionary model.

Other authors also appear to have misunderstandings about Lamarck's ideas. For instance, although Berti and colleagues (2010) correctly recognize that Lamarckian preconceptions are not widespread, they write: "Only two children

---

[1] See, for example, Kampourakis and Zogza (2007), Gregory (2008, 2009), Evans (2008), Evans et al. (2010), Bizzo and El-Hani (2009), van Dijk (2009), van Dijk and Reydon (2010), Smith (2010), Tavares et al. (2010), González Galli and Meinardi (2011), and Zabel and Gropengiesser (2011).

[2] But this principle could, in the same superficial manner, be attributed to Darwin as well (see Nehm and Ha 2011).

[3] This is exactly the conclusion drawn by Kampourakis and Zogza (2007); most students hold teleological conceptions, although some students may also have conceptions similar to Lamarck's.

[…] showed a coherent synthesis, according to which God created the first animals and then made them evolve. This view corresponds to a theistic form of evolutionary account, like the view proposed by Lamarck and embraced by most Western religions" (p. 527). Attributing a theistic view of evolution to Lamarck is entirely mistaken; Georges Cuvier, for example, had criticized Lamarck's theory for being entirely materialistic (Bowler 2003, pp. 93–94).

Some authors improperly attribute ideas of intentionality to Lamarck. For example, Battisti and colleagues (2010) write: "In item 19, students of lowest LOUs are most likely to select the Lamarckian [sic] explanation that the lizards adapt because they 'want' to adapt" (p. 864). However, Lamarck did not attribute evolutionary changes to the willing of animals (i.e., intentionality). Lamarck referred to *needs* but did not think that an animal could develop a new organ by willpower alone. According to Mayr, this misunderstanding was caused in part by the mistranslation of the word *besoin* into "want" instead of "need" (Mayr 1982, p. 357).

To summarize, the evolution education literature contains many historical errors as well as cases in which students' preconceptions are inappropriately linked to historical figures (particularly Lamarck). Moreover, scientists like Lamarck and Darwin developed remarkably complex models of evolutionary change built on a deep knowledge of natural history; consequently, it is questionable as to whether terms like "Lamarckian" would ever be appropriate descriptions of students' mental models of evolutionary change. But even if they were, such a term would not be very informative because there are also substantial differences between conceptual change in the history of evolutionary thought and evolution education. This is the topic that we turn to in the next section.

13.3 The History of Evolutionary Ideas and Students' Conceptual Development

Historical overviews of the development of evolutionary thought have been considered by some science educators to promote understanding of evolution. One approach has been to have students become involved in activities that require them to compare and contrast alternative evolutionary models proposed throughout the history of science. For example, Jensen and Finley (1996, 1997) used this approach in an introductory university-level biology class. The first step in their study was the identification of students' preconceptions about evolution. Subsequently, four alternative evolutionary models drawn from the history of science were presented to the students (specifically, Cuvier's, Lamarck's, Paley's, and Darwin's models), and students were involved in a series of instructional activities relating to these models. The goal of this approach was to have students practice *using* alternative models to solve problems and to assess the relative merits of these models. The main conclusion from Jensen and Finley's studies was that students might generally increase their use of Darwinian concepts, but it was nevertheless more difficult to reduce their application of non-Darwinian concepts.

A similar approach to Jensen and Finley's studies of undergraduates was developed for high school students (Passmore and Stewart 2002; Passmore et al. 2005). Passmore and colleagues' studies were based on the presentation of the conceptual structure of three models developed to explain species diversity (Darwin's, Lamarck's, and Paley's models). Students were asked to compare the three models and assess their explanatory power by using them to explain phenomena different from those described in the original writings of these historical figures. It was hoped that these pedagogical activities would help students distinguish between those concepts that are components of the model of natural selection and those that are not. The goal of this approach was to engage students in inquiry activities that required them to use Darwin's model of natural selection in order to develop a narrative explanation. The main conclusion from these studies was that students could develop a rich understanding of natural selection and use that understanding to reason about evolutionary phenomena.

These studies suggest that involving students in activities that require them to construct and evaluate explanations using alternative evolutionary models has pedagogical value and may facilitate understanding of natural selection. However, there are complications with such HOS-based approaches. It should be made explicit to students that the different historical conceptualizations (e.g., Paley's, Lamarck's, Darwin's) were not discrete, contemporaneous alternatives; rather, Paley's and Lamarck's theories had an important influence on the development of Darwin's theory. While a student at Cambridge, Darwin initially accepted Paley's assumption that body structures existed because they were useful to organisms and reflected God's wisdom and design. But Darwin soon started thinking of adaptation in a different way, as a process by which species responded to environmental changes (Bowler 2003, p. 149). In contrast, Lamarck had suggested that the environment and its changes came first and made organisms use or disuse certain body parts and this eventually caused "adaptive" variations. While Darwin generally denied this view and thought of variation as already present in populations and induced by the environment, Lamarck's concept of local adaptation nevertheless became a central idea in Darwin's mechanism for evolution (Gould 2002, p. 175). Thus, in contrast to the pedagogical activities outlined above, HOS indicates that aspects of Paley's and Lamarck's theories had a significant impact on the development of Darwin's theory and facets of them became integrated into Darwin's model of evolutionary change. All too often, the conceptual and historical contrasts introduced in evolution education do not clearly reflect the growth of evolutionary thought.

The complex mixing of evolutionary ideas in the history of science also raises questions about whether discrete "paradigm shifts," a concept introduced by Thomas Kuhn, have characterized evolutionary thought. He argued that scientific advancement was characterized by a series of periods of "normal science" punctuated by intellectually violent revolutions in which particular conceptual worldviews were replaced. To describe these conceptual worldviews, Kuhn coined the term paradigm, which did not simply refer to the current theory, but to the entire worldview in which the theory was situated. According to Kuhn, scientific revolutions occurred

13 History and Philosophy of Science and the Teaching of Evolution...

when anomalies emerged which could not be explained by the accepted paradigm, which was eventually replaced by a new one. The change from an old paradigm to a new one was described as a "paradigm shift" (Kuhn 1996). It has been suggested that there are analogous (but not entirely similar) patterns of conceptual change in science learning—this is the classical perspective on conceptual change (Posner et al. 1982; but see Levine 2000; Greiffenhagen and Sherman 2008; Van Dijk and Reydon 2010; Vosniadou 2012).

Building on HOS and conceptual change research, it has been suggested that there is a striking similarity between the supposed paradigm shift from "Lamarckian" to "Darwinian" worldviews during the nineteenth century and students' conceptual change from "Lamarckian" to "Darwinian" perspectives (e.g., Jensen and Finley 1996). However, this view is not without complications. First, it is debatable as to whether a paradigm shift from a "Lamarckian" to a "Darwinian" perspective ever took place in the history of science. The history of the study of evolution before Darwin not only includes Lamarck but a much wider intellectual community in Europe that discussed the stability of species and produced many different views on the subject (Corsi 2005). The European scientific scene from the late eighteenth century to the mid-nineteenth century was complex, and debates about the transformation of species had already occurred around 1800. This milieu extended beyond naturalists in England and France (e.g., Erasmus Darwin and Étienne Geoffroy Saint-Hilaire) to Italian geologists and botanists, German naturalists and anatomists, and Russian paleontologists and zoologists (Corsi 2005). In sum, it is difficult to argue that there was just one prevailing pre-Darwinian perspective (e.g., a "Lamarckian" one) at any point in history.

Another complication is that there was no discrete *replacement* of an old evolutionary paradigm with a new one after the publication of the *Origin of Species* by Darwin (1859). As Darwin scholars Hodge and Radick have noted: "… to say that Darwin's influence has been more than revolutionary […] is just to say that there is no one transition that can be identified as the shift that replaced a pre-Darwinian with a Darwinian regime in Western thought" (Hodge and Radick 2009, pp. 267–268). Since it was first proposed, the idea of natural selection had to compete with many alternative models until the evolutionary synthesis of the 1940s took place (Bowler 1983). Indeed, there was no smooth transition from non-evolutionary views to an evolutionary perspective. Several of Darwin's supporters considered evolution to be a progressive and purposeful process. In addition, morphologists, paleontologists, and naturalists attempted the reconstruction of phylogenies, from which support for non-Darwinian mechanisms (such as neo-Lamarckism, orthogenesis, and saltationism) emerged. These ideas led to a rejection of the importance of natural selection, emphasized function and adaptation, and highlighted mechanisms connected to structural constraints on development and evolution (Bowler 2005). This explains, in part, why the reception of Darwin's ideas differed dramatically in different nations (see Engels and Glick 2008). Darwin's own theory was also, in some respects, quite different from the Darwinian theory of the first half of the twentieth century (Depew 2013).

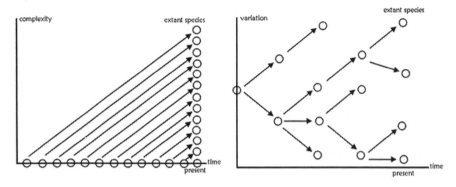

Fig. 13.1 Representation of the main features of Lamarck's (*left*) and Darwin's (*right*) theories (see also Table 13.1)

In addition to questions about whether HOS displays genuine paradigm shifts in evolutionary thought, it is also highly debatable as to whether student thinking shifts from a *Lamarckian* (or Cuvierian or Paleyian) paradigm to a *Darwinian* paradigm. For example, in one study, students were found to confuse Darwin's and Lamarck's theories as parts of the *same* explanation, and not as incompatible models (Jiménez-Aleixandre 1992). This produced inconsistencies in students' responses; they used one idea in one problem context and another idea in the same problem set in a different context (Jiménez-Aleixandre 1992). Many other studies have shown that substantial numbers of students have mixtures of Darwinian ideas (e.g., differential survival) and Lamarckian ideas (inheritance of acquired traits) before and after instruction (e.g., Nehm and Reilly 2007; Nehm and Ha 2011), raising the question as to whether students ever have coherent Lamarckian or Darwinian models (Kampourakis and Zogza 2009).

Two main differences worth noting between Darwin's and Lamarck's theories are the concepts of common descent and natural selection, which are central in the former and are absent in the latter (see Fig. 13.1 for an illustration). Since common descent and natural selection are central concepts of Darwin's theory, students are often taught them and instructed to apply them to explain episodes of evolutionary change. But it is inaccurate to describe students' preconceptions as "Lamarckian" just because they do not use common descent or natural selection to explain evolution. Moreover, in many cases, understanding natural selection requires understanding the mechanisms of heredity and of the origin of genetic variation through mutations (e.g., Banet and Ayuso 2003).[4] Given that the modern notions of heredity and mutation were unknown to Darwin, it is also inaccurate to describe the reasoning models that students display after instruction as *Darwinian* because such understanding differs

---

[4] Teaching genetics before evolution seems to facilitate understanding of evolution by secondary students (Kampourakis and Zogza 2009; see Kampourakis 2006 for how genetics and evolution concepts can be connected).

from Darwin's own understanding. For these reasons, it seems inappropriate to describe the conceptual change process in evolution education as a shift from Lamarckian to Darwinian views.

Even when it is argued that students' explanations are similar in many respects to those of past scientists, important differences may also exist. Conceptual development in children may be less revolutionary than what actually occurs in science (Thagard 1992, p. 263). Thus, despite the interesting similarities between students' conceptions and early evolutionary ideas, there are important differences between individual conceptual development and the growth of evolutionary thought (Rudolph and Stewart 1998). The use of any labels originating from the history of science (e.g., Lamarckian, Darwinian) does not assist science educators in their attempts to develop a richer understanding of students' preconceptions. Despite similarities, two striking differences exist between the ideas of historical figures and students' preconceptions: (1) the intuitive development of students' ideas is a different process from the conscious theory construction of a scientist; and (2) students' conceptions are developed privately and are based on everyday experience, whereas scientists' ideas must be developed in consultation and confrontation with the views of other scientists and are usually based on preexisting scientific knowledge (Gauld 1991). In sum, many conceptual complexities confront science educators' attempts to draw parallels between scientists' explanations of evolutionary change and students' explanations of evolutionary change. What is important to emphasize is that students must learn to construct evolutionary explanations and this might be achieved by using sophisticated instruction informed by HPS scholarship. This is the topic of the next section.

13.4 HPS and Teaching About Evolutionary Explanations

Effective science instruction requires an appropriate presentation of key ideas and theories. In addition, effective instruction must support students in learning how to engage in scientific practices such as constructing explanations and arguments (NRC 2012). Given the increasing importance of the practice of explanation in science education research and standards documents (NRC 2012; McNeill and Krajcik 2008; Nehm et al. 2009), and its central role in evolutionary biology, we explore the topic of scientific explanation at some length.

Several accounts of scientific explanation have been proposed in the philosophy of science. In general, an explanation consists of an *explanandum* (whatever is being explained) and an *explanans* (whatever is doing the explaining). For example, if one asks "why X?" and the answer is "because Y," then X is the explanandum and Y is the explanans. It has been suggested that to explain something in science is (a) to show how it is derived in a logical argument that includes a law in its premise (the covering law model: Hempel and Oppenheim 1948), (b) to provide information about how something was caused (a causal account: Scriven 1959; Salmon 1984; Lewis 1986), or (c) to connect a diverse set of facts by subsuming them under some basic patterns and principles (the unification account: Friedman 1974; Kitcher 1981). It seems that there is general agreement among philosophers that the concept of *cause* is central to the process of scientific explanation.[5]

When trying to explain the causes of the presence of a particular biological trait, one may ask two different types of questions: (a) why it exists ("Why?" questions) and (b) how it functions ("How?" questions). Ernst Mayr (1961) is one of the first scholars to highlight this distinction. He divided the life sciences into *functional biology* (studies of proximate causes and answers to "How?" questions) and *evolutionary biology* (studies of ultimate causes and answers to "Why?" questions). In general, ultimate causes are related to the evolutionary history of species, whereas proximate causes are related to the function and physiology of individuals. Mayr's conceptualization of proximate and ultimate causes has been considered to be a major contribution to the philosophy of biology (Beatty 1994).

In recent years, Mayr's distinction has been reconstructed to include a broader conception of development (e.g., causal interactions between genes, extracellular mechanisms, and environmental conditions, rather than just the "decoding" of a genetic program) and a broader conception of evolutionary causes (e.g., natural selection, migration, and drift rather than natural selection alone). In this perspective, two distinct kinds of explanations exist: (a) proximate explanations, which are dynamic explanations for individual-level causal events and (b) evolutionary explanations, which are statistical explanations that refer to population-level events (Ariew 2003). However, given recent developments in evolutionary developmental biology, some scholars no longer consider this distinction to be valid due to the evolution of developmental processes and to how changes in these processes affect evolution (e.g., Laland et al. 2011). Nevertheless, the ultimate/proximate distinction retains an important pedagogical value (Kampourakis and Zogza 2008, 2009).[6]

Evolutionary explanations typically include the identification of past events that have a causal connection with the present (Scriven 1959, 1969). It is not possible to identify all of the causes of an evolutionary event; however, the causes of an event may at times be identified after it took place. As Cleland (2002, 2011) has noted, effects are underdetermined by causes, and causes are overdetermined by their effects. Simply put, this means that a single cause may not be adequate to bring about an effect (effects are underdetermined by their causes), whereas a

---

[5] See, for example, Kitcher (1989), Salmon (1990), Okasha (2002, p. 49), Godfrey-Smith (2003, pp. 196–197), Woodward (2003), and Rosenberg (2005, p. 27).

[6] The ultimate/proximate distinction as described in these studies could be actually used to teach students about the distinction between developmental and evolutionary explanations. Research in evolutionary developmental biology (evo devo) suggests that such a distinction is not valid and that evolutionary and developmental processes constantly interact. Thus, an interdisciplinary approach to the study of these phenomena is required (Love 2013). However, especially in secondary educational settings, it may be important to first help students distinguish between development and evolution, especially since they often confuse the two kinds of processes. Having understood what development and evolution are, they could then be taught about how developmental changes have an impact on evolution as well as how developmental processes themselves evolve (Love 2013; Arthur 2004; Minelli 2009).

single effect can be an adequate indicator of its cause (causes are overdetermined by their effects). For example, a ball thrown at a window at low speed may not break it; however, observing a ball among fragments of glass on the floor could comprise adequate evidence for concluding that a ball was forcefully thrown at the window. Thus, evolutionary explanations are causal explanations with a historical dimension. They require phenomena or events which occurred in the past and which have a special causal relation with the effect observed (Scriven 1959). In other words, evolutionary explanations exhibit historical elements because they focus on properties that are unique in time and place and about which historical statements can be made (Lewontin 1969). Evolutionary explanations can thus take the form of historical narratives, and in such frameworks antecedent conditions play an important role. Explanations for particular evolutionary outcomes explicitly link them to particular antecedent conditions that have an explanatory role: if such conditions had been different, the outcome might have been different, too. The reliability of such explanations can be high as long as adequate information is available that causally links these antecedent conditions with the observed outcome (Gould 2002, pp. 1333–1334).[7]

Evolutionary explanations make extensive use of natural selection; however, a controversy emerges when one looks in more detail at what natural selection actually "explains." For some scientists, natural selection has a positive role and may help to explain why individuals have the traits they do, whereas for others it has a negative role in that it only eliminates variants and cannot explain why an individual has particular traits (see also Depew 2013). However, if one accepts that individuals belong to a lineage with a particular evolutionary history, then that history may help explain why they have particular traits (Forber 2005). Gould and Lewontin (1979) famously argued against the dominance of natural selection in evolutionary explanations by advancing the view that it is one of several important evolutionary processes. Other concepts—such as common descent and random drift—can (and often do) have explanatory roles (individually or collectively; see Beggrow and Nehm 2012). Indeed, Darwin's arguments in the *Origin of Species* included two central ideas: the tree of life (which involved two different ideas: transmutation and common descent) and natural selection (Waters 2009). That is why he described his theory as *descent with modification*.

Another important philosophical perspective on the historical nature of evolutionary explanations is the distinction between "how-possibly" and "how-actually" explanations. This type of explanation can be divided into: (1) *global how-possibly explanations,* which answer the question if some process could have produced evolutionary changes in an idealized population; (2) *local how-possibly explanations,* which answer the question if some process could have produced an observed evolutionary outcome or pattern consistent with what is known about an actual population; and (3) *how actually* explanations which answer the question why a

---

[7] Interestingly enough Gould noted that such a kind of narrative explanation was central in Darwin's theorizing but his successors did not put emphasis on it in an attempt to base explanations on laws, which were considered more important for explanations than any narrative (Gould 2002, p. 1336).

particular evolutionary outcome or pattern occurred (Brandon 1990 pp. 176–184, and more recently, Forber 2010).[8]

In the science classroom, it is important to explain to students how it is possible to have epistemic access to the past (Cleland 2002, 2011; Forber and Griffith 2011). Students could be taught to develop "how-possibly" evolutionary explanations, test them against the available evidence, and then try to come up with "how-actually" explanations. Engaging in "how-possibly" and "how-actually" distinctions may be thought of as involving two distinct steps: (1) identification of antecedent conditions of the past which are causally related to the evolutionary outcome (effect) which is explained and (2) the identification of factors which were crucial in causing that particular outcome. The latter is based on the idea of "difference maker" factors, previously proposed in the literature on explanations (Lombrozo and Carey 2006; Strevens 2009). The important idea in this account is that there may be several causes of a particular phenomenon but one of them may be more important because it made "the difference" in eventually producing a particular outcome (but not another).

An example may help to illustrate this point. Suppose that a forest fire is observed. While the presence of both oxygen and a lighted cigarette may be causally connected to the forest fire, it is the latter that made the difference; that is, the cigarette is causally more important and thus has a more significant role in the explanation of the forest fire. The explanation of the forest fire takes the form of a historical explanation because one needs to explain how it actually started. In doing so, one may consider several "how-possibly" explanations, evaluate the available evidence, and then come up with a "how-actually" explanation. To take our example of the forest fire further, given that oxygen is always present in a forest, one could come up with the following two "how-possibly" explanations: (1) that lightning started the fire, due to the presence of oxygen and combustible material such as wood, and the fire then spread to the forest or (2) that humans lit a fire, which then spread to the forest due to the presence of oxygen and combustible material such as wood. One might then examine additional evidence from the past about the forest fire. If one finds that during the day that the fire started, there were no storms or lightning, but people were observed smoking cigarettes close to where the fire was observed, then one may conclude that explanation (2) is a more plausible "how-actually" explanation.

A biological example may also help to illustrate the distinctions between a "how-possibly" and a "how-actually" explanation. Possible causes of the presence of a long neck in a species of giraffe will be considered. There can be two kinds of causes, contemporary and historical ones. Contemporary causes may include (1) particular genetic/developmental mechanisms that causally affect the length of the neck in each individual and (2) some advantageous effect that contributed to its selection.[9] Assuming that a long neck is an adaptation and thus an outcome of natural selection

---

[8] There is some disagreement in the details (Reydon 2012; Forber 2012) but the nuances of these disagreements are not central to our point.

[9] It is not necessary that the feature is currently being selected, but it may be so.

(see Kampourakis 2013b for suggestions about how to define adaptation in science education), historical causes should refer to the antecedent conditions that resulted in the evolutionary process that followed. In this case, giraffes with longer necks underwent selection for many generations in a particular environment in which a longer (than average) neck was advantageous and a shorter (than average) neck was disadvantageous; eventually the average neck length in the particular giraffe population increased over several generations. The antecedent conditions could have included the following factors with causal influence: (1) particular genetic/developmental mechanisms that causally affect the length of the neck in each individual, producing giraffes with a variety of neck lengths, and (2) particular environmental conditions (e.g., a drought that had limited the food supply) that caused natural selection. Of these two causes, (2) is the "difference maker."

It is important to emphasize that a different environmental condition could produce a different outcome: selection of shorter necks and eventually producing a shortening of neck lengths in the population or species. The fact that a condition can lead to different outcomes helps to identify it as a "difference maker." The everyday and biological examples that we presented are emblematic of the type of "how-possibly" explanations that students could be taught to construct. Then they might test alternative explanations against the available evidence (e.g., Mitchell and Skinner 2003).

In addition to gaining an understanding of the different explanatory approaches that are applied in the field of evolutionary biology, students must be exposed to particular *types* of explanatory tasks. For example, if students were asked to explain why birds have wings, they might answer that they have wings "in order to fly." This is an intuitive explanation that many children, adolescents, and adults would utilize. But if the explanatory task were framed in a slightly different way, and students were asked a slightly different question about birds and wings (e.g., How would a biologist explain why eagles, penguins, and ostriches have wings?), a conceptual conflict situation would immediately arise because the student would realize that his/her intuitive explanation (*in order to fly*) would be insufficient to explain why birds that do not fly (penguins and ostriches) have wings.[10] Thus, the structure of the explanatory task is likely to control the degree to which conceptual conflict and conceptual change occurs. Careful alignment of explanatory task types with instructional goals (e.g., formative assessment vs. conceptual conflict) has been lacking. The development of wider arrays of explanatory prompts for classroom use would be a useful pursuit.

---

[10] It is entirely legitimate to say that *birds have wings for flying*, as long as we refer to birds which do use their wings to fly and if it is clear that it is selection and not design which is doing the explaining. In terms of their structure, evolutionary explanations are teleological explanations (Lennox and Kampourakis 2013). The problem for evolution education is not teleology per se, but teleology based on design (we do not discuss Intelligent Design in this chapter; an excellent, recent analysis can be found in Brigandt 2013b). This is a difficult topic, pedagogically speaking. Although reference to history may not be necessary for philosophical analyses, it can be very useful for evolution instruction (Kampourakis 2013b).

An important question to ask is how these perspectives on evolutionary explanations have been employed in the field of science education. Although scientific explanations have been the focus of increasing attention in educational standards documents (NRC 2012) and highlighted as central epistemic practices in science classrooms (Berland and McNeill 2012), less attention has been paid to the diverse ways that explanations have been conceptualized (see above) or how they should be appropriately taught, learned, and assessed. Indeed, for more than 30 years, while science education researchers have employed explanation tasks to reveal student thinking about evolution and natural selection (reviewed in Nehm and Ha 2011), they have paid comparatively less attention to the question of what most appropriately constitutes an "evolutionary explanation," and remarkably different epistemic perspectives have characterized evolutionary "explanations" in the science education literature.

For example, Gotwals and Songer (2010, p. 263), in a study of student thinking about biodiversity and evolution, conceptualized explanation "... as a response to a scientific question that takes the form of a rhetorical argument and consists of three main parts: a claim (a statement that establishes the proposed answer to the question), evidence (data or observations that support the claim), and reasoning (the scientific principle that links the data to the claim and makes visible the reason why the evidence supports the claim)." Conceptually similar to Gotwals and Songer (2010), Sandoval and Millwood (2005) linked aspects of argumentation to explanation: "Explanations are a central artifact of science, and their construction and evaluation entail core scientific practices of argumentation" (2005, p. 24). Sandoval and Millwood (2005) empirically studied "... the quality of the arguments that students make in explanations of problems of natural selection." Other authors have advocated for the linkage of explanation and argumentation in scientific explanations as well (McNeill and Krajciak 2008).

In contrast to Gotwals and Songer (2010), and Sandoval and Millwood (2005), Nehm and colleagues have excluded aspects of argumentation from their evolution explanation tasks. Rather, they have framed their explanation tasks as opportunities for students to build and apply causal accounts that explain differences between an initial biotic state and a subsequent biotic state (e.g., a cactus species with spines and a cactus species without spines; Nehm et al. 2012; Opfer et al. 2012). Nehm and colleagues' work nevertheless has never fully described what a normative evolutionary explanation should encompass (or should not), other than to note that it should include normative causal factors (e.g., mutation, differential survival, and heredity) and exclude nonnormative, noncausal factors (e.g., teleology, intentionality, inheritance of acquired characters). As these selected examples illustrate, quite different perspectives on evolutionary explanation have been put forth in the science education literature.

One recent study attempted to integrate HPS perspectives on explanation with pedagogical issues relating to the teaching and learning of evolution (Kampourakis and Zogza 2009). The aims of this study were (1) to teach students about the structure of evolutionary explanations and (2) to provide a conceptual heuristic applicable to different types of organismal features (i.e., homologies and adaptations). Students were taught to construct explanations for homologies by referring to a common ancestor that possessed the features that were common to the taxa discussed in the tasks. The general form of explanation they were given for homologies was the following: to explain why species A and B share a common feature H (homology), it is assumed that a common ancestor C (which possessed feature H) existed in the past and that both A and B descended from C. Students were also taught to construct explanations for adaptations by referring to natural selection: in a particular environment, some traits provided an advantage to their possessors, contributing to their survival and reproduction, and for these reasons those traits became prevalent in the population. The general form of explanation students were given for adaptations was the following: to explain why species S possesses feature (adaptation) A, it is assumed that S descended from an older population that included both individuals that possessed feature A and others that did not, as well as that this feature provided an advantage to its possessors in the particular environment; as a result those individuals that did not possess A perished whereas those that possessed A survived and evolved to produce species S.[11]

It would be useful to develop an assessment of students' understandings of explanations per se, so that such knowledge could be empirically disentangled from evolution content understanding. For example, ineffective explanation instruction alone (and subsequent student confusion about the utility of explanations) could contribute to poor student performance of evolution explanation tasks. It would be useful to know how the magnitude of explanatory understanding interacts with content knowledge to foster understanding about both microevolution and macro-evolution. Regardless, the increasing importance of explanation in science education research and practice will require more explicit and careful integration of HPS perspectives.

Overall, as our discussion of explanation has illustrated, it is clear that different explanatory accounts and different explanatory task structures have yet to be carefully integrated into the teaching and learning of evolution. The development and implementation of HPS-informed conceptual heuristics relating to evolutionary explanations has great potential for improving students' understanding of evolution.

13.5 Conclusion

This chapter has reviewed HPS-informed studies in evolution education dealing with (1) the linkage of particular student ideas to those of prominent naturalists from the history of science (e.g., Lamarck), (2) the characterization of conceptual change in evolution as reflecting paradigm shifts from "Lamarckian" to "Darwinian"

---

[11] These explanatory schemes may seem oversimplified but were considered appropriate given the age of students (14–15-year-olds).

worldviews, and (3) unitary perspectives on evolutionary explanation. Collectively, analyses of these selected topics have identified complications with the ways in which HPS ideas have been applied in evolution education scholarship and raised questions about whether these characterizations have clarified or confused thinking about student learning of evolution. Science educators' further engagement with HPS scholars will help to appropriately ground evolution education research, and additional analyses of other facets of HPS scholarship relating to evolution education will be needed in order to build a more robust understanding of how students think about the core topic of evolution (see for example the various chapters in Kampourakis 2013c).

Acknowledgments We thank the anonymous reviewers for thoughtful ideas about how to improve the manuscript, Liz P. Beggrow for helpful suggestions, and Minsu Ha for help with references.

References

Alters, B. J., & Nelson, C. E. (2002). Perspective: Teaching evolution in higher education. *Evolution, 56*(10), 1891–1901.

Ariew, A. (2003). Ernst Mayr's 'ultimate/proximate' distinction reconsidered and reconstructed. *Biology and Philosophy*, 18(4), 553–565.

Arthur, W. (2004). *Biased embryos and evolution.* Cambridge (United Kingdom): Cambridge University Press.

Banet, E., & Ayuso, G. E. (2003). Teaching of biological inheritance and evolution of living beings in secondary school. *International Journal of Science Education, 25*(3), 373–407.

Battisti, B. T., Hanegan, N., Sudweeks, R., & Cates R. (2010). Using item response theory to conduct a distracter analysis on conceptual inventory of natural selection. *International Journal of Science and Mathematics Education*, 8, 845–868.

Beatty, J. (1994). The proximate/ultimate distinction in the multiple careers of Ernst Mayr. *Biology and Philosophy, 9*(3), 333–356.

Beggrow, E. P., & Nehm, R. H. (2012). Students' mental models of evolutionary causation: Natural selection and genetic drift. *Evolution: Education and Outreach, 5*(3), 429–444.

Berland, L. K., & McNeill, K. L. (2012). For whom is argument and explanation a necessary distinction? A response to Osborne and Patterson. *Science Education, 96*(5), 808–813.

Berti, A. E., Toneatti, L., & Rosati, V. (2010). Children's conceptions about the origin of species: A study of Italian children's conceptions with and without instruction. *Journal of the Learning Sciences, 19*(4), 506–538.

Bishop, B. A., & Anderson, C. W. (1990). Student conceptions of natural selection and its role in evolution. *Journal of Research in Science Teaching*, 27(5), 415–427.

Bizzo, N., & El-Hani, C. N. (2009). Darwin and Mendel: Evolution and genetics. *Journal of Biological Education, 43*(3), 108–114.

Bowler, P. J. (2003). *Evolution: The history of an idea.* (3rd edn.). Berkeley and Los Angeles, CA: University of California Press.

Bowler, P. J. (2005). Revisiting the eclipse of Darwinism. *Journal of the History of Biology, 38,* 19–32.

Bowler, P. J. (1983). T*he eclipse of Darwinism: Anti-Darwinian evolution theories in the decades around 1900.* Baltimore, MD: Johns Hopkins University Press.

13 History and Philosophy of Science and the Teaching of Evolution... 395

Brandon, R. N. (1990). *Adaptation and Environment*. Princeton, NJ: Princeton University Press.
Brigandt I. (2013a) Explanation in biology: reduction, pluralism, and explanatory aims. Science & Education, 22(1), 69–91.
Brigandt, I. (2013b). Intelligent design and the nature of science: philosophical and pedagogical points. In K. Kampourakis (Ed), *The Philosophy of Biology: A Companion for Educators.*. Dordrecht: Springer, 205–238.
Brumby, M. (1979). Problems in learning the concept of natural selection. *Journal of Biological Education, 13*(2), 119–122.
Burkhardt, R. W. (1995). *The spirit of system: Lamarck and evolutionary biology*. Cambridge, MA: Harvard University Press.
Cleland, C. E. (2002). Methodological and epistemic differences between historical science and experimental science. *Philosophy of Science, 69*(3), 447–451.
Cleland, C. E. (2011). Prediction and explanation in historical natural science. *The British Journal for the Philosophy of Science, 62*(3), 551–582.
Clough, E. E., & Wood-Robinson, C. (1985). Children's understanding of inheritance. *Journal of Biological Education, 19*(4), 304–310.
Corsi, P. (2005). Before Darwin: Transformist concepts in European natural history. *Journal of the History of Biology, 38,* 67–83.
Corsi, P. (2001). *Lamarck: Gene'se et enjeux du transformisme, 1770–1830*. Paris: Éditions du CNRS.
Darwin, C. (1859). *On the origin of species by means of natural selection, or the preservation of favoured races in the struggle for life*. London: John Murray.
Deadman, J., & Kelly, P. P. (1978). What do secondary school boys understand about evolution and heredity before they are taught the topics? *Journal of Biological Education, 12*(1), 7–15.
Demastes, S. S., Good, R. G. & Peebles, P. (1996). Patterns of conceptual change in evolution. *Journal of Research in Science Teaching, 33*(4), 407–431.
Depew D. (2013) Conceptual change and the rhetoric of evolutionary theory: 'Force talk' as a case study and challenge for science pedagogy. In K. Kampourakis (Ed), *The Philosophy of Biology: A Companion for Educators* Dordrecht: Springer.
Endersby, J. (2009). Darwin on generation, pangenesis and sexual selection. In J. Hodge & G. Radick (Eds.), *Cambridge companion to Darwin (2nd edn.)* (pp. 73–95). Cambridge: Cambridge University Press.
Engels, E. M., & Glick, T. F. (2008). *The reception of Charles Darwin in Europe* (Vol. 2). London: Continuum.
Evans, E. M. (2008). Conceptual change and evolutionary biology: A developmental analysis. In S. Vosniadou (Ed.), *International handbook of research on conceptual change* (pp. 263–294). New York: Routledge.
Evans, E. M., Spiegel, A., Gram, W., Frazier, B. F., Tare, M., Thompson, S. & Diamond, J. (2010). A conceptual guide to natural history museum visitors' understanding of evolution. *Journal of Research in Science Teaching, 47*, 326–353.
Forber, P. (2005). On the explanatory roles of natural selection. *Biology and Philosophy, 20*(2), 329–342.
Forber, P. (2010). Confirmation and explaining how possible. *Studies in the History and Philosophy of Biological and Biomedical Sciences, 41*, 32–40.
Forber, P. (2012). Modeling scientific evidence: The challenge of specifying likelihoods. *EPSA Philosophy of Science: Amsterdam 2009, 1*, 55–65.
Forber, P., & Griffith, E. (2011). Historical reconstruction: Gaining epistemic access to the deep past. Philosophy & Theory in Biology, 3, e203.
Friedman, M. (1974). Explanation and scientific understanding. *The Journal of Philosophy, 71*(1), 5–19.
Gauld, C. (1991). History of science, individual development and science teaching. *Research in Science Education, 21*, 133–140.

Geraedts, C. L., & Boersma, K. T. (2006). Reinventing natural selection. *International Journal of Science Education*, 28(8), 843–870.

Godfrey-Smith, P. (2003). *Theory and reality: An introduction to the philosophy of science*. Chicago, IL: The University of Chicago Press.

González Galli, L. M., & Meinardi, E. N. (2011). The role of teleological thinking in learning the Darwinian model of evolution. *Evolution: Education and Outreach*, 4,145–152.

Gotwals, A. W., & Songer, N. B. (2010). Reasoning up and down a food chain: Using an assessment framework to investigate students' middle knowledge. *Science Education*, 94(2), 259–281.

Gould, S. J. (2002). *The structure of evolutionary theory*. Cambridge, MA: Belknap Press of Harvard University Press.

Gould, S. J. & Lewontin, R. C. (1979). The spandrels of San Marco and the Panglossian paradigm: A critique of the adaptationist programme. *Proceedings of the Royal Society of London. Series B. Biological Sciences*, 205(1161), 581–598.

Gregory, T. R. (2008). Evolution as fact, theory, and path. *Evolution: Education and Outreach*, 1, 46–52.

Gregory T. R. (2009). Understanding natural selection: Essential concepts and common misconceptions. *Evolution: Education and Outreach*, 2, 156–175.

Greiffenhagen, C., & Sherman, W. (2008). Kuhn and conceptual change: On the analogy between conceptual changes in science and children. *Science & Education*, 17, 1–26.

Hempel, C. & Oppenheim, P. (1948). Studies in the logic of explanation. *Philosophy of Science, 15*, 135–175.

Hodge, J., & Radick, G. (2009). *Cambridge companion to Darwin (2nd edn.)*. Cambridge: Cambridge University Press.

Jensen, M. S., & Finley, F. N. (1996). Changes in students' understanding of evolution resulting from different curricular and instructional strategies. *Journal of Research in Science Teaching*, 33(8), 879–900.

Jensen, M. S., & Finley, F. N. (1997). Teaching evolution using a historically rich curriculum and paired problem solving instructional strategy. *The American Biology Teacher, 59*(4), 208–212.

Jiménez-Aleixandre, M. P. (1992). Thinking about theories or thinking with theories: A classroom study with natural selection. *International Journal of Science Education, 14*(1), 51–61.

Kampourakis, K. (2006). The finches beaks: Introducing evolutionary concepts. *Science Scope*, 29(6), 14–17.

Kampourakis, K. (2013a) Mendel and the path to Genetics: Portraying science as a social process. *Science & Education*, 22(2), 293–324.

Kampourakis, K. (2013b) Teaching about adaptation: why evolutionary history matters. *Science & Education*, 22(2), 173–188.

Kampourakis K. (Ed) (2013c), The Philosophy of Biology: A Companion for Educators. Dordrecht: Springer.

Kampourakis, K., & Zogza, V. (2007). Students' preconceptions about evolution: How accurate is the characterization as "Lamarckian" when considering the history of evolutionary thought? *Science & Education, 16*(3–5), 393–422.

Kampourakis, K., & Zogza, V. (2008). Students' intuitive explanations of the causes of homologies and adaptations. *Science & Education, 17*(1), 27–47.

Kampourakis, K., & Zogza, V. (2009). Preliminary evolutionary explanations: A basic framework for conceptual change and explanatory coherence in evolution. *Science & Education, 18*(10), 1313–1340.

Kitcher, P. (1981). Explanatory unification. *Philosophy of Science, 48*(4), 507–531.

Kitcher, P. (1989). Explanatory unification and the causal structure of the world. In P. Kitcher & W. C. Salmon (Eds.), *Minnesota studies in the philosophy of science (vol. 13): Scientific explanation* (pp. 410–505), Minneapolis, MN: University of Minnesota Press.

Kuhn, T. S. (1996) [1962]. The structure of scientific revolutions. (3rd edn.). Chicago, IL: University of Chicago Press.

Laland, K. N., Sterelny, K., Odling-Smee, J., Hoppitt, W., & Uller, T. (2011). Cause and effect in biology revisited: Is Mayr's proximate-ultimate dichotomy still useful? *Science, 334,* 1512–1516.

Lennox J.G. and Kampourakis K. (2013) Biological teleology: the need for history. In K. Kampourakis (Ed), *The Philosophy of Biology: A Companion for Educators.* Dordrecht: Springer.

Levine, A. T. (2000). Which way is up? Thomas S. Kuhn's analogy to conceptual development in Childhood. *Science & Education, 9,* 107–122.

Lewis, D. (1986). Causation. In D. Lewis (Ed.), *Philosophical papers,* vol. II (pp. 159–213), Oxford: Oxford University Press.

Lewontin, R. C. (1969). The bases of conflict in biological explanation. *Journal of the History of Biology, 2*(1), 35–45.

Lombrozo, T., & Carey, S. (2006). Functional explanation and the function of explanation. *Cognition, 99,* 167–204.

Love, A. C. (2013). Interdisciplinary lessons for the teaching of biology from the practice of evo-devo. *Science & Education,* 22(2), 255–278.

Mayr, E. (1961). Cause and effect in biology. *Science, 134,* 1501–1506.

Mayr, E. (1982). *The growth of biological thought: Diversity, evolution and inheritance.* Cambridge, MA: Harvard University Press.

Mayr, E. (2002). *What evolution is.* London: Weidenfeld & Nicolson.

McNeill, K. L., & Krajcik, J. (2008). Scientific explanations: Characterizing and evaluating the effects of teachers' instructional practices on student learning. *Journal of Research in Science Teaching, 45*(1), 53–78.

Minelli, A. (2009). *Forms of becoming: The evolutionary biology of development.* Princeton, Oxford: Princeton University Press.

Mitchell, G., & Skinner, J. D. (2003). On the origin, evolution and phylogeny of giraffes Giraffa camelopardalis. *Transactions of the Royal Society of South Africa.* 58(1), 51–73.

National Research Council (2012). *A framework for K-12 science education: Practices, crosscutting concepts, and core ideas.* Washington, DC: The National Academies Press.

Nehm, R. H., & Ha, M. (2011). Item feature effects in evolution assessment. *Journal of Research in Science Teaching, 48*(3), 237–256.

Nehm, R. H., & Reilly, L. (2007). Biology majors' knowledge and misconceptions of natural selection. *BioScience, 57*(3), 263–272.

Nehm, R. H., Beggrow, E. P., Opfer, J. E., & Ha, M. (2012). Reasoning about natural selection: Diagnosing contextual competency using the ACORNS Instrument. *The American Biology Teacher, 74*(2), 92–98.

Nehm, R. H., Kim, S. Y., & Sheppard, K. (2009). Academic preparation in biology and advocacy for teaching evolution: Biology versus non biology teachers. *Science Education, 93,* 1122–1146.

Okasha, S. (2002). *Philosophy of science: A very short introduction.* Oxford: Oxford University Press.

Opfer, J. E., Nehm, R. H., & Ha, M. (2012). Cognitive foundations for science assessment design: Knowing what students know about evolution. *Journal of Research in Science Teaching, 49*(6), 744–777.

Passmore, C. & Stewart, J. (2002). A modeling approach to teaching evolutionary biology in high schools. *Journal of Research in Science Teaching, 39*(3), 185–204.

Passmore, C., Stewart, J., & Zoellner, B. (2005). Providing high school students with opportunities to reason like evolutionary biologists. *The American Biology Teacher, 67*(4), 214–221.

Pazza, R., Penteado, P. R., & Kavalco, K. F. (2010). Misconceptions about evolution in Brazilian freshmen students. *Evolution: Education and Outreach, 3*(1), 107–113.

Posner, G. J., Strike, K. A., Hewson, P. W., & Gertzog, W. A. (1982). Accommodation of a scientific conception: toward a theory of conceptual change. *Science Education*, *66*, 211–227.

Prinou, L., Halkia, L., & Skordoulis, C. (2011). The inability of primary school to introduce children to the theory of biological evolution. *Evolution: Education and Outreach*, *4*(2), 275–285.

Reydon, T. A. C. (2012). How-possibly explanations as genuine explanations and helpful heuristics: A comment on Forber. *Studies in the History and Philosophy of Biological and Biomedical Sciences*, *43*, 302–310.

Rosenberg, A. (2005). *Philosophy of science: A contemporary introduction. (2nd edn.)* London: Routledge.

Rudolph, J. L., & Stewart, J. (1998). Evolution and the nature of science: on the historical discord and its implications for education. *Journal of Research in Science Teaching*, *35*(10), 1069–1089.

Salmon, W. C. (1984). *Scientific explanation and the causal structure of the world*. Princeton, NJ: Princeton University Press.

Salmon, W. C. (1990). Four decades of scientific explanation. In P. Kitcher & W. C. Salmon (Eds.), *Minnesota Studies in the Philosophy of Science Vol. 13: Scientific Explanation* (pp. 3–219), Minneapolis, MN: University of Minnesota Press.

Samarapungavan, A., & Wiers, R. W. (1997). Children's thoughts on the origin of species: A study of explanatory coherence. *Cognitive Science*, *21*(2), 147–177.

Sandoval, W. A., & Millwood, K. A. (2005). The quality of students' use of evidence in written scientific explanations. *Cognition and Instruction*, *23*(1), 23–55.

Scriven, M. (1959). Explanation and prediction in evolutionary theory. *Science*, *130*, 477–482.

Scriven, M. (1969). Explanation in the biological sciences. *Journal of the History of Biology*, *2*(1), 187–198.

Settlage Jr, J. (1994). Conceptions of natural selection: a snapshot of the sense-making process. *Journal of Research in Science Teaching*, *31*(5), 449–457.

Smith, M. U. (2010). Current status of research in teaching and learning evolution: II. Pedagogical issues. *Science & Education*, *19*(6–8), 539–571.

Strevens, M. (2009). *Depth: An account of scientific explanation*. Cambridge, MA: Harvard University Press.

Tavares M. L., Jimenez-Aleixandre, M. P., & Mortimer E. F. (2010). Articulation of conceptual knowledge and argumentation practices by high school students in evolution problems. *Science & Education*, *19*(6–8), 573–598.

Thagard, P. (1992). *Conceptual revolutions*. Princeton, NJ: Princeton University Press.

van Dijk E. M., & Reydon, T. A. C. (2010). A conceptual analysis of evolutionary theory for teacher education. *Science & Education*, *19*(6–8), 655–677.

van Dijk E. M. (2009). Teachers' views on understanding evolutionary theory: A PCK-study in the framework of the ERTE-model. *Teaching and Teacher Education*, *25*, 259–267.

Vosniadou S. (2012) Reframing the classical approach to conceptual change: Preconceptions, misconceptions and synthetic models. In B.J. Fraser, K. Tobin, & C. J. McRobbie (Eds.). *Second international handbook of science education* (pp. 119–130). Dordrehct: Springer.

Waters, C. K. (2009). The arguments in The Origin of Species. In J. Hodge & G. Radick (Eds.). *Cambridge companion to Darwin (2nd edn.)* (pp. 120–143). Cambridge, MA: Cambridge University Press.

Winther, R. (2000). Darwin on variation and heredity. *Journal of the History of Biology*, *33*, 425–455.

Woodward, J. (2003). *Making things happen: A theory of causal explanation*. Oxford: Oxford University Press.

Zabel, J., & Gropengiesser, H. (2011). Learning progress in evolution theory: Climbing a ladder or roaming a landscape? *Journal of Biological Education*, *45*(3), 143–149.

Kostas Kampourakis holds an appointment as Collaborateur Scientifique II at the University of Geneva, Switzerland. He is editor or coeditor of four special issues for the journal *Science & Education*: Darwin and Darwinism (with David Rudge); Philosophical considerations in the teaching of biology; Genetics and Society: Educating Scientifically Literate Citizens (with Giorgos Patrinos and Thomas Reydon); and Mendel, Mendelism and Education: 150 years since the "Versuche" (with Erik Peterson). He is also an Adjunct Instructor at the Department of Mathematics and Science Education of Illinois Institute of Technology. He was coordinator for science education for grades K-6 as well as member of the Department of Educational Research and Development of Geitonas School, in Athens, Greece, where he also taught biology and nature of science to secondary school and IB DP students. His research interests include the teaching of biology and the teaching of the nature of science in the context of the history and philosophy of science and he has published several articles on these topics. He has written a book titled *Understanding Evolution: an Introduction to Concepts and Conceptual Obstacles* (Cambridge University Press). He has also edited a book titled *The Philosophy of Biology: a Companion for Educators* (Springer) in which professional philosophers of biology bring their work to bear on biology education. He holds a B.S. in biology and an M.S. in genetics, both from the University of Athens, as well as a Ph.D. in science education from the University of Patras.

Ross H. Nehm holds faculty appointments as Associate Professor in the Department of Ecology and Evolution and the Ph.D. Program in Science Education, at Stony Brook University in New York. He received a B.S. from the University of Wisconsin at Madison, an Ed.M. from Columbia University, and a Ph.D. from the University of California at Berkeley. He was the recipient of a U.S. National Science Foundation Early Career Award and was named Education Fellow in the Life Sciences by the U.S. National Academy of Sciences. The NSF's CCLI, TUES, and REESE programs have funded his research. He currently serves on the editorial boards of the *Journal of Research in Science Teaching*, the *Journal of Science Teacher Education*, and the *Journal of Science Education and Technology* and has published papers in a wide variety of education and science journals, including *JRST, Science Education, Science & Education, International Journal of Science Education, Research in Science Education, Journal of Science Teacher Education, The American Biology Teacher, Journal of Science Education and Technology, CBE-Life Sciences Education, Evolution Education and Outreach, Genetics, Journal of Biology and Microbiology Education, Caribbean Journal of Science, Bulletins of American Paleontology*, and *Journal of Paleontology*. With Ann Budd he has edited the book *Evolutionary Stasis and Change in the Dominican Republic Neogene* (Springer 2008). His evolution education research was recently highlighted in the National Research Council publication *Thinking Evolutionarily* (National Academy Press 2012). His research interests include evolutionary thinking and reasoning, problem solving, and assessment methodologies.

Chapter 14
History and Philosophy of Science and the Teaching of Macroevolution

Ross H. Nehm and Kostas Kampourakis

14.1 Introduction

In the past decade, increasing scholarly attention and emphasis has been placed on the teaching, learning, and assessment of macroevolutionary concepts (e.g., Catley 2006; Nadelson and Southerland 2010a, b; Padian 2010; Novick and Catley 2012). While the distinctions between microevolution and macroevolution have been topics of lively debate within the history and philosophy of science (HPS) communities for some time, relatively new to the field of science education is the conceptualization of macroevolution as a distinct concept in need of targeted instructional emphasis and research (Catley 2006).

The term *macroevolution* is a relatively recent addition to the lexicon of evolution, first coined (in German) by Filipchenko in 1927 and subsequently recruited into the English language in 1937 by the prominent biologist Theodosius Dobzhansky (Burian 1988). Since its introduction, the meaning of the term macroevolution, like many other biological terms, has changed substantially (see Erwin 2010). Despite these changes, nearly all definitions consider the formation of new species to be an important partition dividing micro- from macroevolution. The US National Academy of Sciences (NAS 2012), for example, defines macroevolution as "[l]arge-scale evolution occurring over geologic time that results in the formation of new species and broader taxonomic groups" and microevolution as "[c]hanges in the traits of a

---

R.H. Nehm (✉)
Department of Ecology and Evolution and Ph.D. Program in Science Education,
Stony Brook University, Stony Brook, NY 11794-5233, USA
e-mail: ross.nehm@stonybrook.edu

K. Kampourakis
Biology Section and IUFE, University of Geneva, Pavillon Mail,
40, Bd du Pont-d'Arve, Geneva 4 CH-1211, Switzerland
e-mail: konstantinos.kampourakis@gmail.com

M.R. Matthews (ed.), *International Handbook of Research in History,
Philosophy and Science Teaching*, DOI 10.1007/978-94-007-7654-8_14,
© Springer Science+Business Media Dordrecht 2014

group of organisms within a species that do not result in a new species." Importantly, the NAS definitions—and related distinctions in the science education literature (e.g., Catley 2006; Nadelson and Southerland 2010a, b)—focus primarily on *scale* (e.g., within vs. between species; human timescales vs. geological timescales) and *pattern* (e.g., descriptions of large-scale change as opposed to causes of such change). In a similar vein, Catley (2006) highlights the distinction between *short-term* (microevolutionary) and *long-term* (macroevolutionary) change (see also Nadelson and Southerland 2010a, b). While discussions of micro- and macroevolution in the HPS and evolutionary biology literature also focus on scale and pattern, they have paid particular attention to putative factors that *explain* large-scale evolutionary events at different scales of analysis. While natural selection (and other microevolutionary processes) are universally acknowledged as contributors to evolutionary change by biologists, the expansion of possible *mechanisms* accounting for large-scale patterns in the history of life is considered a major advance in evolutionary theory (e.g., Gould 2002). These important distinctions between pattern and mechanism deserve attention, as they have led to a conceptual divergence between the science education and HPS literature.

14.2 Macroevolutionary Patterns and Processes

Macroevolutionary thought has a philosophically rich history (Ruse 1997; Gould 2002; Depew and Weber 1995; Sterelny 2009) and today remains rife with controversy (Dietrich 2010; Erwin 2010). Nonetheless, it is important to point out that many macroevolutionary *patterns* are well established and uncontroversial, such as the reality of mass extinctions (e.g., Jablonski 1986), the originations of now-extinct higher taxa (e.g., Erwin 2010), the evolutionary relationships among all living things (e.g., Hillis 2010), long-term trends in the fossil record (Gould 2002), and evolutionary stasis (e.g., Nehm and Budd 2008). A core macroevolutionary topic of importance to HPS scholars and science educators relates to putative distinctions between large-scale observable patterns in the history of life on the one hand and inferences and theories about the mechanisms responsible for these patterns on the other.

Changes to the definition of *macroevolution* since its introduction in 1927 have in some respects paralleled vacillations between scholarly emphasis on large-scale patterns in the fossil record and their causal underpinnings (e.g., Simpson 1944). Evolutionary biologists from diverse disciplinary backgrounds (e.g., Dobzhansky, Simpson, Mayr, Eldredge, Gould, Gingerich, Futuyma, and Orr) have, like most scientists, recognized that large-scale evolutionary trends, extinctions, and originations of higher taxa do in fact appear in the fossil record (e.g., Simpson 1953; Futuyma 2005; Coyne and Orr 1998; Erwin 2010). But these and many other authors have *disagreed* about whether microevolutionary processes (such as natural selection and genetic drift) are capable of sufficiently accounting for such well-established large-scale patterns (Gould 1985). Causal pluralism, or the expansion of explanatory mechanisms beyond natural selection, is thus a key topic of attention in HPS perspectives on macroevolution. Such plurality is also historically important, as it is considered by some to be divergent from the views of Darwin (1859), who proposed "…natural selection as the single unifying mechanism that causes both micro- and macroevolution" (Travis and Reznick 2009, p. 126).

Evolutionary theorists such as Filipchenko (1927), Goldschmidt (1940), Schindewolf (1950), Eldredge (1989), Stanley (1980), Vrba and Gould (1986), Lloyd and Gould (1993), and Erwin (2010), for example, have adopted what may be termed a causally pluralistic evolutionary worldview and therein argued that distinct macroevolutionary mechanisms (not reducible to microevolutionary processes; e.g., species selection and mass extinction) likely contributed to large-scale evolutionary patterns (Gould 1985; Erwin 2010). Importantly, these authors do not discount the reality or importance of natural selection, but some have questioned its reification as a causal process with all-encompassing explanatory power (Gould 1981; Depew and Weber 1995). Biologists such as Dobzhansky, Simpson, and Futuyma, in contrast, have generally considered natural selection to be a sufficient causal explanation for most macroevolutionary patterns (for a discussion of Simpson's changing views on this matter, see Sepkoski 2008). The views of these scholars are aligned in some respects with those of Travis and Reznick (2009, p. 128), who note: "In the final analysis there is nothing in the fossil record that inherently contradicts Darwin's daring idea that natural selection is the unifying mechanism." In sum, the reality of macroevolutionary *patterns* is simply not in doubt.[1] The controversy in macro-evolutionary biology relates to questions about the *processes* involved (natural selection alone or natural selection + other mechanisms).

According to most definitions, the formation of new species (speciation) lies at the boundary between microevolution and macroevolution (e.g., NAS 2012). While the history of biological thought is filled with controversy about the competing roles of natural selection and genetic drift in speciation, many biologists consider the issue to be settled. Coyne and Orr (2004, p. 410), in their seminal treatment of speciation, note: "…firm evidence for the role of genetic drift in speciation is rare." They go on to close the book on this controversy: "It appears, then, that at least one important debate has been settled: selection plays a much larger role in speciation than does drift. It is also worth noting that genetic drift appears to play little part in morphological evolution" (p. 410). In an exhaustive review of the literature, Coyne and Orr summarize a wealth of work indicating that natural selection plays a major role in speciation and that "[i]t is uncontroversial that most phenotypic divergence in ecologically important traits is driven by natural selection" (p. 385). Thus, natural selection is widely considered to play a major role in the speciation process.

Above the species level, the bulk of macroevolutionary debate relevant to the science education community may be formulated as two related questions: (1) Can microevolutionary processes such as natural selection and genetic drift sufficiently account for large-scale patterns in the history of life? If not, what alternative

---

[1] Advocates of creationism and intelligent design have repeatedly exploited debates about macroevolution to suggest (incorrectly) that evolution is a theory in crisis and questioned the reality of macroevolutionary patterns because of incompleteness of the fossil record (see Sepkoski 2008). It is important to point out that such incompleteness has not been a topic of equal concern by scientists.

mechanisms are there? And (2) If mechanisms *in addition to* natural selection exist, and they can survive theoretical and empirical testing, how much of the macro-evolutionary history of life do they in fact explain (cf. Dietrich 2010)?

Four major macroevolutionary concepts have received considerable scrutiny by evolutionary biologists, paleobiologists, and philosophers of biology over the past 30 years[2]: (1) species selection/sorting, (2) mass extinction, (3) constraints/evolvability, and (4) evolution and development (or "evo-devo"). The important point to keep in mind is that these four concepts, in concert with (or in opposition to) natural selection, could account for large-scale evolutionary outcomes that were unexpected or unexplainable by the exclusive extrapolation of microevolutionary processes over geological timescales. By expanding the range of causal factors contributing to evolutionary change, evolutionary biologists could potentially improve causal precision and eliminate troublesome empirical anomalies. Questions about the validity of these macroevolutionary processes have generated a rich literature in HPS and evolutionary biology. We briefly summarize each in turn prior to investigating their role in science education.

Species selection has become a key feature of modern macroevolutionary theory (Erwin 2010). It is a conceptual outcome of Eldredge and Gould's (1972) formulation of evolutionary "stasis" and "punctuated change." Eldredge and Gould (1972) argued that most species' histories were characterized by the absence of appreciable evolutionary change (i.e., displayed stasis) and that such stability was punctuated by rapid morphological evolution associated with cladogenesis (lineage splitting speciation) (Nehm and Budd 2008). This model was offered in opposition to what Eldredge and Gould (1972) viewed as the prevailing evolutionary orthodoxy of the time: slow, continuous change. Eldredge and Gould's alternative model nicely framed the question of whether species could be thought of as *individuals*. That is, in the punctuated model, if species have stability in time and space (a "life span"), and are demarcated by clear beginnings (punctuations associated with "birth") and clear endings (extinction or "death"), could they not have species-level traits that could be selected, in a way analogous to how individual organismal traits are selected (for the conception of species as individuals, see Ghiselin 1974; Hull 1980)?

Several empirical and philosophical studies of this new conceptualization of species-level selection have been conducted (e.g., Jablonski and Hunt 2006; Hull 1980). These studies generally support the view that species may display properties that are not reducible to lower hierarchical levels, that is, properties that are not aggregates of lower-level phenomena (Stanley 1980; Sepkoski 2008). Geographic range has long been considered a species-level, variable, and heritable trait (Jablonski and Hunt 2006). Philosophers and paleobiologists have debated these empirical cases at length and agree to some extent that species-level selection is theoretically possible (Hull 1980; Sepkoski 2008). Despite being conceptually and philosophically important, so few empirical cases of species selection have been confirmed that the relative significance of this macroevolutionary process appears

---

[2] This list is by no means exhaustive (see Ayala and Arp 2010).

to be small (Dietrich 2010). In sum, while species selection may be viewed as a unique and distinctly macroevolutionary mechanism accounting for large-scale evolutionary patterns, the range of phenomena that it might actually explain is quite limited at present.

Like species selection, mass extinctions have been considered to be a central macroevolutionary process (Jablonski 1986). Mass extinctions are important in macroevolutionary thought because they have been thought to cause conceptual complications for extrapolationist accountings of macroevolutionary patterns (e.g., Raup 1994). Mass extinctions have the potential to counteract the smaller scale workings of natural selection; reproductive success and differential survival during "normal" times may have little association with reproductive success and differential survival during times of mass extinction (Jablonski 1986). For example, while patterns of differential survival over millions of years may produce well-adapted animals of large body size, during geologically brief episodes of mass extinction (e.g., the end Cretaceous event), differential survival may favor animals of small body size thereby counteracting this adaptive trend. Mass extinctions therefore raise the possibility that microevolutionary processes alone cannot sufficiently account for large-scale patterns in the history of life (Erwin 2010). The (potentially stochastic) pruning of lineages during mass extinctions may "reset" the playing field for lineages, counteracting the effects of adaptive microevolution. As noted by Raup (1994, p. 6758): "Except for a few cases, there is little evidence that extinction is selective in the [...] sense argued by Darwin." In this view, natural selection cannot sufficiently account for macroevolutionary patterns; mass extinction must be considered as an additional causal factor that can work in opposition to natural selection.

A third macroevolutionary topic in the HPS literature is constraint and evolvability (Gould 2002; Erwin 2010; Minelli and Fusco 2012). While linking constraint and evolvability is questionable in some respects, both acknowledge the important roles that genetic, architectural, historical, developmental, and functional constraints may play in limiting the types of long-term evolutionary change that can occur (cf. Gould 2002, p. 1059; Erwin 2010). Gould sees particular patterns of macroevolutionary repetition (i.e., parallelism) as evidence of the importance of internal constraints. These constraints are significant in a macroevolutionary sense because they may "push back" against the actions of natural selection and thereby limit pathways of evolutionary change. Put another way, limits on variation (caused by internal constraints) channel pathways of evolutionary change by limiting the options that selection has available to work with. Gould (2002) argues that this perspective is important relative to macroevolutionary theory because constraint helps to explain macroevolutionary patterns that cannot be accounted for by selection alone (see also Bateson and Gluckman 2011 for a more recent discussion). Such views also resonate with many perspectives from evolutionary developmental biology (e.g., Sansom and Brandon 2007; Love 2007, 2013).

Gould's perspectives align in many ways with the large body of work by Brian Goodwin (reviewed in Goodwin 2009). He challenges the notion that random genetic variation can (or does) generate an infinite variety of options for natural selection to work with, and so natural selection is not the only factor explaining discrete (vs. continuous) distributions of morphology in time and space. Evidence for this perspective may be found in David Raup's "morphospace" diagrams (see Raup and Stanley 1978). These diagrams map the morphologies of extinct and extant species within the universe of forms that could theoretically exist. Comparing actual vs. possible shell shapes, for example, illustrates that some regions of morphospace are densely populated, whereas others are sparse. Desolate regions of morphospace are fertile ground for exploring the question of whether such forms are impossible to generate or merely have yet to evolve.

Although Erwin's (2010) perspective on evolvability differs somewhat from those of Gould (2002) and Goodwin (2009), it also considers limits on the pathways that evolution can take. Erwin sums up his perspective of "evolvability" when he writes: "...the structure of gene regulatory networks in animals [...] indicates that the nature of the variation available for selection to act upon has changed over time...[and] this may impose another way in which macroevolutionary patterns are not reducible to microevolutionary processes, at least as they are currently defined by microevolutionists" (Erwin 2010, p. 189). He goes on to note "What is strikingly absent from virtually all microevolutionary thought [...] is a sense of history, of the impact of evolutionary changes on the range of variation that is possible, and of how that range of variation has itself changed over time" (p. 191). Thus, Erwin and others have viewed the concept of "evolvability" as a uniquely macroevolutionary idea.

The fourth topic that has received considerable attention in the HPS literature relating to macroevolution is evolutionary developmental biology (informally referred to as "evo-devo") (Carroll 2005a, b). As noted by Raff (2000, p. 74) "evolutionary change occurs not by the direct transformation of adult ancestors into adult descendants but rather when developmental processes produce the features of each generation in an evolving lineage." Although for centuries naturalists have seriously considered the significance of this point (e.g., von Baer 1828; Darwin 1860[3]; Haeckel 1868; Goldschmidt 1940; Simpson 1944; Schindewolf 1950; Waddington 1970), the role that development has played in macroevolutionary thought has varied dramatically through history (see Gould 1977 for a review). Mayr (1988) argued that development was largely excluded from the "evolutionary synthesis" of the 1940s (see Futuyma 1998 for an alternative view) and subsequently remained somewhat isolated from evolutionary theory (at least in the United States; see Lloyd and Gould's (1993) preface to Schindewolf (1950/1993) for a more global perspective). This situation changed with Gould's forceful reintroduction of the importance of development to macroevolution in *Ontogeny and Phylogeny* (1977). Therein Gould reframed the complex historical literature on evolution and development, crafted a new (largely morphological) framework for

---

[3] "Embryology is to me by far the strongest single class of facts in favor of change of forms..." Darwin, September 10, 1860, letter to AsaGray.

heterochrony and heterotopy,[4] and paved the way for the modern resurgence of interest in evo-devo that has yet to peak[5] (Carroll 2005a, b).

More recently, the conceptual framework of evo-devo has been further expanded to encompass the genetic underpinnings of largely pattern-based (e.g., heterochrony and heterotopy) changes in the evolution of development. More process-oriented frameworks include heterometry, which refers to an evolutionary change in the amount of a gene product, and heterotypy, which refers to an evolutionary change in the nature of a gene product (Arthur 2004, pp. 81–83). The revolutionary advances in regulatory genetics and genomics has transformed modern evo-devo into a mechanistic science (Carroll 2005a, b). Indeed, the remarkable patterns of evolutionary developmental parallelisms that have fascinated naturalists for centuries are at last being linked to biological processes at the molecular, cellular, and developmental levels (e.g., von Baer 1828; Haeckel 1868; Goldschmidt 1940; Schindewolf 1950; Gould 1977).

Key questions in evo-devo include the study of how gene networks govern ontogeny, the factors that make developing systems robust enough to tolerate mutations that change the course of development, how the rules that govern ontogeny constrain the production of new phenotypic variation, how development influences speciation, and the origins of body plans and their evolvability[6] (Raff 2000; Arthur 2004; Carroll 2005a, b; Minelli 2009). As noted by Minelli and Fusco (2012): "Overall, developmental processes can contribute to speciation and diversification at different stages of the speciation process, at different levels of biological organization and along the organism's whole life cycle." The explosion of empirical findings in evo-devo over the past decade, along with new journals (e.g., *Evolution & Development*), professional societies, and faculty positions devoted to the subject, is suggestive of major changes to the structure of evolutionary biology.

Despite the growing importance of evo-devo for evolutionary studies, and increasing interest in the topic in HPS (e.g., Love 2013), evo-devo has not received concomitant attention in science education research or practice (from the perspective of curriculum or pedagogy; see Love [in press] for a view on both of these issues from a HPS perspective).[7] Equally concerning is the fact that evo-devo is conspicuously absent from science educators' recent conceptualizations of the macroevolution construct and associated features deemed worthy of assessment (e.g., Catley 2006; Nadelson and Southerland 2010a, b; see also Novick and Catley 2012). Surprisingly, even Padian's (2010) vociferous plea for the inclusion of macroevolution in K-12

---

[4] Evolutionary changes in developmental timing and spatial arrangement, respectively; see Zelditch (2001) for morphological (pattern-based) perspective and Arthur (2004) for a more mechanistic perspective.

[5] The institutionalization of evo-devo took place in 1999 when it was granted its own division in the Society for Integrative and Comparative Biology (SICB), as well as through the National Science Foundation's establishment of a separate division for funding evo-devo research.

[6] See Müller (2007, pp. 505–506) for a more complete conceptual and historical synopsis.

[7] Although of course there are exceptions. See, for example, a special issue of the journal *Evolution Education and Outreach* (June, 2012).

education lacked explicit mention to the role that evo-devo might play. Thus, evo-devo serves as another example in which current perspectives from HPS have yet to influence the teaching, learning, and assessment of macroevolution.

Our overview of some (but by no means all) of the key macroevolutionary ideas emphasized in the HPS literature—species selection, mass extinction, constraint/evolvability, and evo-devo—and those that have contributed to the resurgence of empirical macroevolutionary inquiry (i.e., the so-called paleobiological revolution of Sepkoski and Ruse 2009) provides a vantage point from which to examine scholarship about the teaching, learning, and assessment of macroevolution in the science education literature. As will become readily apparent, despite some similarities, the two communities have envisioned macroevolution in strikingly different ways.

14.3 Macroevolution: Science Educators' Blind Spot?

Science education research relating to macroevolution has thus far focused on three major issues: (1) general advocacy for the teaching of macroevolution in K-12 education (and cladograms in particular) (Catley 2006; Padian 2010), (2) measurement of students' macroevolutionary knowledge (Dodick and Orion 2003; Nadelson and Southerland 2010a, b; see also Novick and Catley 2012), and (3) investigations of students' beliefs about small-scale vs. large-scale evolutionary change (Nadelson and Southerland 2010a, b). The intrusion of creationist challenges, spurred on by scholarly debates about macroevolution, is also in need of consideration. We begin with a review of advocacy for the teaching of macroevolution in the science education community.

A provocative opinion piece by Kefyn Catley in 2006 was in many respects a "call to arms" for the science education community to acknowledge and explicitly incorporate macroevolution in science education. It bemoaned the lack of focus on macroevolution in science education teaching and research and chastised educators for their near-exclusive focus on natural selection (and associated research on misconceptions about natural selection alone). Catley emphasized that "[w]ithout a clear perspective on macroevolution, an understanding of the full spectrum of evolution is simply not possible. This notwithstanding, microevolutionary mechanisms are taught almost exclusively in our schools, to the detriment of those mechanisms that allow us to understand the larger picture" (Catley 2006, p. 768). In perhaps his most controversial claim, Catley states: "Knowledge of natural selection, while vitally important, explains little about the incredible diversity of species on the planet" (2006, p. 775). Hence, Catley appears to take a stance that is more closely aligned with what we have termed causal pluralism (see above)—that there is more to the evolution of life than natural selection alone. But in addition to natural selection, what, in Catley's view, explains macroevolutionary change?

An interesting aspect of Catley's (2006) perspective is that it lacks mention of the key macroevolutionary concepts (species selection, mass extinction, constraints/evolvability, and evo-devo) that have been central to HPS scholarship (e.g., Sepkoski 2008; Erwin 2010). In fact, it does not clearly outline any causal alternatives to natural selection. This generates a conceptual void: What are we to make of a "call to arms" for the teaching of macroevolution that downplays the importance of natural selection on the one hand ("By themselves, the products of the "New Synthesis" do not adequately account for the history of life or for its diversity" (Catley 2006, p. 770)) but fails to mention hierarchical selection theory or many of the classic macroevolutionary ideas proposed by Stanley, Gould, Eldredge, Vrba, and Lloyd? If one considers Catley's (2006) perspective from a pattern-based perspective, however, the exclusion of natural selection, species selection, mass extinction, and constraint and evolvability may be reasonable; students need to learn about large-scale patterns and, according to Catley, learn these patterns through the lens of phylogenetic systematics, or cladistics.

One aspect of Catley's stance on macroevolution is in alignment with the causal pluralists (cf. Gould 1985). Specifically, he appears to take the position that species are properly conceptualized as "real" individuals (Catley 2006 repeatedly notes that species are "the very units of evolution"). Yet, interestingly enough, he makes no mention of the past 30 years of discussion relating to species selection or how it should be conceptualized in teaching and learning about macroevolution.

A central piece of Catley's (2006) argument appears to be that cladograms must be integrated into the teaching and learning of evolution and, by doing so, macroevolutionary content will be properly addressed. Cladograms are representational diagrams illustrating the evolutionary relationships of biological units (e.g., species and clades) generated using the underlying methodology of Willi Hennig (i.e., Cladistics; see Hennig 1999). They depict evolutionary *patterns* (characters and their various states across operational taxonomic units, such as species, groups partitioned based upon their recent common ancestry, and outgroups to polarize character state transformations). Cladograms are powerful tools for testing causal hypotheses (such as the "randomness" of mass extinctions), but themselves represent patterns of evolutionary relationship. Therefore, they are tools for articulating patterns in the natural world (the differential birth and death of species within and among clades) with tests of theory (e.g., selection of species in these clades). Macroevolutionary theory and its causal foundations are not necessarily addressed by using or teaching about cladograms (except, perhaps, patterns of cladogenesis), however central they may be to scientific practice. While cladograms have been increasingly employed in evolution research, it is important to point out that the major theoretical advances in macroevolution predated the widespread adoption of phylogenetic taxonomy in the United States (Hull 1988). In sum, while cladograms are now central tools in evolutionary biology, as noted by Catley, by themselves they do not say much about macroevolutionary processes and mechanisms, but only represent patterns.

A recent article by Kevin Padian (2010, p. 206) echoes Catley's (2006) concerns with teaching macroevolution: "Macroevolution must take a much more prominent place in K-12 science teaching. To do so, a curriculum must be redeveloped at both K-12 and college levels, so that preparation in macroevolution is a required part of K-12 biology preparation." He also takes aim at his scientific colleagues: "...few evolutionary biologists have a first-hand understanding of macroevolution, and they do not spend substantial time on it in their college courses. This is because most of them are population biologists and population geneticists, and they have had little or no training in macroevolution." Padian also targets science textbooks: "...textbooks in all grades from K-16 fail completely to convey an understanding of how evolution works in the long run."

Catley (2006) and Padian (2010) raise several important points worthy of empirical consideration. First, is macroevolution receiving short shrift in evolution education? Is Catley correct when he claims that "As currently taught, natural selection stops short of fostering an understanding of its effects over time on species themselves, or on cladogenesis. It concentrates almost exclusively on processes that occur within individuals and populations" (Catley 2006, p. 775)? Have aspects of macroevolution in fact been covered in secondary and undergraduate textbooks and curricula? While it is challenging enough today to determine the degree to which particular topics are emphasized in science classrooms, the problem becomes much more difficult to address in the history of science education. One long-standing approach for documenting topical emphasis in the history of science education is to examine textbook content and structure (Cretzinger 1941; Skoog 1969; Moody 1996; Nehm et al. 2009). A surprising number of studies have investigated how evolutionary biology has been conceptualized and represented in textbooks over the past century (for a historical review, see Skoog 1969 and Moody 1996). These studies provide one empirical approach for attempting to answer the relatively straightforward question "Is macroevolution being taught?" Given that Catley's claim is directed at US education, our review is restricted to that context.[8]

It is clear that many of the concepts that Catley (2006) mentions have been included in biology textbooks in the United States for at least 100 years, although, as mentioned above, this does not necessarily mean that they were covered in classroom instruction. Moreover, it is clear that the term "macroevolution" is a relatively recent addition to the lexicon of evolution, and many texts do not explicitly use this term even if they discuss ideas that are widely considered to be macroevolutionary in nature (e.g., horse evolution). In some of the earliest biology textbooks produced in the United States (from the period of 1900 to 1919), large-scale evolution (between-species change, or transformation) was "…a common topic as it was discussed in five of the eight textbooks" [sampled] (Skoog 1969, p. 151). Other topics present in this early period included "convergent evolution," "evolutionary relationships," "fossils and other remains," and the "evolution of birds" (Skoog 1969). Species transformation again appears as one of the more common topics in textbooks from 1920 to 1929, with the evolution of horses being a particularly common macroevolutionary example[9] (Skoog 1969). Similar patterns were noted through the 1960s (when natural selection was noted to occur in all of Skoog's textbook samples; see Fig. 14.1). In a similarly detailed analysis of 17 evolutionary subtopics in early textbooks, Nicholas (1965) found that paleontological evidence from the

---

[8] While English-language textbooks (particularly from the United States) have received the most attention in the science education literature, it is important to point out that international studies of evolutionary content in textbooks have also been completed. See, for example, Swarts et al. (1994) for a discussion of textbooks from China and the former Soviet Union.

[9] Although one that has more recently been reconceptualized as a branching, rather than as a linear, evolutionary pattern.

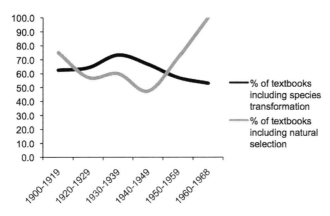

Fig. 14.1 Macroevolution in science textbooks 1900–1968. Based on Skoog's (1969) analysis of evolutionary content in textbooks from 1900 to 1960, species transformations (macroevolutionary change, according to most definitions; see text) were included at generally comparable levels as natural selection until the 1960s, when natural selection was included in all sampled texts

fossil record was the most commonly covered evolutionary subtopic. Other frequently included topics that could reasonably be considered to have a macroevolutionary slant were "rates of evolution," "influence of the physical history of the Earth on evolution," and the "evolutionary history of man [sic]." This work is in alignment with Skoog's general findings.

More recent studies of best-selling undergraduate biology textbooks revealed that all of them cover macroevolution (Nehm et al. 2009). If the evolutionary history of particular clades is also considered, macroevolution is well represented, albeit segregated to particular chapters (Nehm et al. 2009). Importantly, however, macroevolution is more than tallying long-term patterns of life's "comings and goings" (Padian 2010); macroevolutionary *processes* are indeed underrepresented in these undergraduate textbooks. In high school textbooks, coverage of macroevolution is difficult to ascertain given that ostensibly macroevolutionary topics, such as punctuated equilibrium, have been lumped with other topics in some empirical studies (e.g., Rosenthal 1985). Nevertheless, it is apparent that many topics that fit under the conceptual umbrella of macroevolution were covered in more recently published textbooks (Skoog 1984; Rosenthal 1985). Given that textbooks "...have much influence on what is taught" (Skoog 1984, p. 127), this finding lends credence to the idea that macroevolution has had a consistent home in biology curricula for a century or more. Nonetheless, it may be true that the proportion of macroevolutionary content is too small (Padian 2010).

The US *National Science Education Standards* (1996) may also be used to examine the status of macroevolutionary ideas in biology education. The *Standards* contain at least ten evolutionary ideas, half of which may be reasonably interpreted as macroevolutionary in nature: (1) common ancestry of species; (2) classification systems reflect evolutionary relationships; (3) the fossil record, large-scale changes in life, and extinction; (4) similarities among diverse species; and (5) geological time, or deep time. Overall, there is remarkable similarity between concepts in the *Standards* and the macroevolutionary concepts that Catley (2006) and Padian (2010) suggest are lacking in emphasis. Nonetheless, the *Standards* do *not* include the key causal features emphasized in recent HPS scholarship, such as species selection, mass extinction, and constraints on the evolution of form, evolvability, and evo-devo.

In addition to textbooks and the US *Standards*, practitioner journals (such as the widely subscribed *American Biology Teacher*) may be examined to explore the degree to which macroevolution has been addressed in the professional community. Many articles have discussed the importance of teaching both macroevolutionary patterns *and* processes, such as punctuated equilibrium (Alters and McComas 1994); rapid, large-scale morphological and molecular evolution in stickleback fish (Platt 2006); rates of macroevolution (Marco and López 1993); and macroevolution in the fossil record (Dodick and Orion 2003). But on review of the evolutionary topics covered in *ABT*, it is clear that specific macroevolutionary focus is comparatively less than treatments of natural selection and genetic drift. Perhaps the most interesting observation in reviewing the literature is that discussions of *causal* factors relating to macroevolution are extremely rare. So, in many respects, Catley (2006) is correct that macroevolution (at least as HPS scholars' conceptualize the topic, cf. Sepkoski 2008) has received short shrift in science education. But it is also true that facets of Catley's (2006) version of macroevolution are clearly present.

Despite the concerns mentioned above, Catley's (2006) standpoints on macroevolution have without question stimulated a new and innovative research program focusing on student reasoning about phylogenetic and macroevolutionary *patterns* (particularly the interpretation of cladograms) (Baum and Offner 2008). Interpreting cladograms, and using them to reason about evolution (micro- or macroevolution), involves aspects of visual reasoning, hierarchical thinking, abstract representation, misconceptions about evolution, and the nature of science (e.g., cladograms represent testable hypotheses). Given that cladograms have become de rigueur for testing patterns and processes of micro- and macroevolutionary change (e.g., pinpointing likely hosts of the SARS coronavirus, HIV subtype evolution, and the coevolution of angiosperms and their pollinators), this research direction is critically important for the field of science education. What have been lacking in this research program are discussions of the causal *processes* that many HPS scholars consider to be uniquely macroevolutionary, such as species selection, mass extinction selectivity, and clade/group selection (Sepkoski 2008). For some HPS scholars and evolutionary biologists, these ideas form the core of macroevolutionary theory and the most significant conceptual advances since Darwin (Gould 1981, 2002). Yet, it is precisely these concepts that remain conspicuously absent from the science education research literature about macroevolution.

14.4 Measuring Macroevolutionary Knowledge

Given the importance of macroevolution in science education, the question arises as to how to determine if students are learning it. A broad array of empirical research questions in evolution education requires the use of measurement instruments designed to capture latent constructs, such as students' knowledge of macroevolution or their belief in evolution. In recent years, some science educators have raised concerns about the quality of extant instruments used in science education research in general and evolution education in particular (Nehm 2006; Smith 2010; Neumann et al. 2011). It is critically important that the evolution education research community develops and deploys high-quality instruments that are in alignment with professional measurement standards (i.e., AERA et al. 1999). Otherwise, the measures derived from such instruments will have little meaning, or, more problematically, they may mislead educators in their efforts to improve the teaching and learning of core scientific topics such as evolution. Instruments about macroevolution are no exception.

Nadelson and Southerland (2010a, b) developed the first instrument designed to measure students' knowledge of macroevolution.[10] Several compelling reasons justified the development of this instrument. First, school and university students (and the general public) appear to have different levels of acceptance relating to microevolutionary and macroevolutionary change. Second, many science curricula and textbooks distinguish microevolution and macroevolution as distinct instructional topics (e.g., Stanley 1980). Third, understanding microevolutionary processes (i.e., natural selection and genetic drift) may not translate into an understanding of, for example, larger scale phenomena, such as the formation of new species or evolutionary trends (Catley 2006). Fourth, Nadelson and Southerland (2010a, b) argue that natural selection and adaptation are primarily microevolutionary, and not macroevolutionary, concepts (contrary to the views of some, see above and Table 14.1). Thus, despite several microevolutionary knowledge measures (e.g., Settlage and Odom 1995), a distinct measure of macroevolutionary knowledge appeared to be justified. Given the controversies in the HPS literature about how macroevolution should be conceptualized, to what extent does Nadelson and Southerland's (2010a, b) construct of "macroevolution" align with HPS perspectives?

In designing their instrument for measuring undergraduate students' knowledge of macroevolution, Nadelson and Southerland (2010a, b, p. 156) "…identified deep time, phylogenetics, speciation, fossils, and the nature of science as five essential concepts necessary to comprehend macroevolution." Natural selection is notably absent. The content of the test was established by "…feedback from professional biologists and evolution educators," a review of textbooks, and an expert review revealing that "[e]ach of the five faculty members considered our subscales to be representative of the major topics and concepts associated with macroevolution" (Nadelson and Southerland 2010a, b, p. 156). In one of their open-ended instrument items, they chose to focus on speciation "…because it is often the most contentious concept related to macroevolution" (p. 161). It is by no means clear if HPS scholars would agree that *speciation* is more contentious than, for example, constraints or species selection.

Nadelson and Southerland's (2010a, b) macroevolution instrument uses a "scenario-based" approach, in which students must use information on the assessment to choose among answer options (one scientifically correct, the others

---

[10] Albeit one that has received considerable criticism. See, for example, Novick and Catley (2012).

Table 14.1 What does the construct of "macroevolution" include? A synopsis of some views from the HPS and science education literatures. See text for details and discussion of causal vs. pattern-based perspectives. Note that discussions in the HPS literature are not explicitly aligned with any educational grade band (n/a)

HPS literature	Padian (2010)	Catley (2006)	Nadelson and Southerland (2010a, b)
Educational level: n/a	Educational level: K-16	Educational level: K-16(?)	Educational level: undergraduate
Natural selection's ability to explain macroevolution (e.g., Gould 2002; Sepkoski 2008)	As a species lineage evolves, its morphology may change in many ways or hardly at all (see discussion, p. 208)	Knowledge of natural selection, while vitally important, explains little about the incredible diversity of species on the planet (p. 775); by themselves, the products of the "new synthesis" do not adequately account for the history of life or for its diversity (p. 770)	Unclear: "…understanding of speciation by natural selection, a fundamental process in macroevolution" (p. 175); "…natural selection and adaptation, which are primarily interpreted as microevolutionary concepts" (p. 155)
Species as individuals; species-level traits (e.g., variability, geographic range); species selection, species sorting, group selection (e.g., Hull 1980; Lloyd and Gould 1993)	The rates of origination and extinction of *species* shape the history of life (p. 207)	Species radiations, based on novel evolutionary characters; cladogenesis; formation of higher taxa (p. 769); species exist in both space and time, and in addition to being the fundamental elements of Linnaean classification, they are also the *units of evolution* and biodiversity (p. 775)	Speciation: long-term speciation can be considered to be a key concept that should be included in a measure of macroevolution understanding (p. 158)
Mass extinction selectivity/ stochasticity (e.g., Jablonski 1986; Sepkoski 2008)	Extinctions are studied at two levels: background and *mass extinctions* (p. 208)	Pattern-based extinction (throughout)	Pattern-based mass extinction: diversity of life decreases and increases with events such as mass extinctions (p. 168); an examination of extinction using diagrams of lineages (p. 160)
Constraint, evolvability, and contingency (e.g., Gould 2002; Goodwin 2009; Erwin 2010; Kirschner and Gerhart 1998)	Not explicitly considered	Not explicitly considered	Not explicitly considered
Evo-devo (e.g., Gould 1977, 2002; Raff 1998; Carroll 2005a, b; Arthur 2004; Minelli 2009)	Not explicitly considered	Not explicitly considered	Not explicitly considered

incorrect). Several macroevolutionary patterns are used to frame the instrument answer options: (1) using an evolutionary tree, exploring the *processes*[11] involved in the transition of the whale "family" from ancient shore-dwelling ancestors; (2) interpreting the evolution of eyes, including a discussion of variation in extant mollusk lineages; (3) interpreting extinction patterns using "diagrams of lineages"; (4) examining "evolutionary pathways of the African Great Ape" and the development of what they term "diagram pathways"; and (5) interpreting geographic distributions of fossils on different continents. To varying extents, the scenarios test students' understandings of the five ideas that Nadelson and Southerland (2010a, b) consider to be uniquely "macroevolutionary": phylogenetics, speciation, deep time, fossils, and the nature of science.

While we suspect that most biologists and philosophers of biology would agree with Nadelson and Southerland (2010a, b, p. 175) when they write, "Assessing learner knowledge of macroevolution is essential for developing and honing science curricula that are effective in helping students develop an understanding of this fundamental aspect of biology," the discordance between the HPS literature—and other literature in science education—and their concept of macroevolution is notable. In particular, the exclusion of selection and drift as causes of macroevolution (along with the absence of hierarchical selection theory, species selection, constraints/evolvability, evo-devo) are noteworthy gaps. Overall, it is apparent that some science educators are approaching the measurement of students' knowledge of macroevolution in a unique way, excluding the key features of macroevolution discussed in the HPS literature. The question is whether other education stakeholders conceptualize macroevolution similarly.

14.5 Future Directions in Macroevolution Education

Given the rich literature in HPS relating to macroevolution, it would be useful for teacher educators, instrument developers, curriculum designers, and science education researchers to engage more fully with this work. Our review has revealed several issues that would benefit from a more integrated approach. These include (1) recognizing that natural selection is widely acknowledged to be a major causal process in the generation of macroevolutionary patterns (particularly speciation), that is, constructs of macroevolution should not exclude the theory of natural selection (*contra* Nadelson and Southerland 2010a, b); (2) emphasizing macroevolutionary processes, such as species selection, mass extinction, constraint/evolvability, and evo-devo as core macroevolutionary topics (Table 14.1); (3) developing a consensus definition of macroevolution, associated key standards (i.e., phenomena and processes), and disciplinary practices (i.e., ways of thinking and reasoning,

---

[11] However, no processes (e.g., natural selection, drift, species selection) are offered as answer options on the assessment.

sensu Love 2013) that are appropriate for K-12 students; (4) performing studies of students' knowledge of macroevolutionary phenomena and their reasoning about the processes that might account for those phenomena; (5) linking Catley and colleagues' innovative work on cladogram interpretation with *causal* hypothesis testing. Such work has great potential in integrating a large body of work on microevolution and natural selection with macroevolutionary patterns and causes; (6) exploring how complex system thinking and hierarchical thinking relate to the transfer of natural selection understanding to broader temporal scales; such work is wanting but would add a new dimension to a growing body of work on complex systems (e.g., Wilensky and Resnick 1999). Overall, as envisioned by science educators, macroevolution is a messy amalgamation of phenomena, concepts, and processes united by a weak conceptual framework (e.g., vague notions of "scale"). Currently, the inconsistencies between how the HPS and science education communities envision macroevolution are dramatic, and as a consequence a shared vision of macroevolution is lacking.

14.6   Conclusion

The teaching and learning of macroevolutionary ideas, perhaps more so than other science topics, is tightly bound to the history and philosophy of science (HPS). Nevertheless, as our chapter has illustrated, many studies in evolution education have not fully engaged with HPS scholarship, particularly the topics of species selection, mass extinction, constraint/evolvability, and evo-devo. Currently, science educators' conceptualization of "macroevolution" consists of a messy amalgamation of phenomena, concepts, and processes united by a weak conceptual framework (e.g., vague notions of "scale"). Inconsistencies between how the HPS and science education communities envision macroevolution are dramatic and prevent meaningful progress in the teaching and learning of this important area of evolution.

In closing, after taking stock of the perspectives on macroevolution from HPS, the science education research literature, practitioner journals, and creationist tactics, how should macroevolution be envisioned by science educators and delivered instructionally to students (if at all)? Sepkoski may have provided one of the more reasonable answers to this thorny question when he wrote: "There is no reason to fear teaching schoolchildren that drift, mutation, and natural selection form the central pillar of evolutionary theory, any more than it is dangerous to teach Newtonian mechanics in high-school physics classes. Like quantum mechanics, the current complex debates in macroevolutionary theory are appropriately taught after the basic framework has been established, since they build on, but not invalidate, the foundation" (2008, p. 234). As our review has demonstrated, contemporary views of macroevolution in the HPS community encompass much more than pattern recognition and cladogram interpretation, do not discount the role of natural selection, and offer a more expansive perspective on the range of causal processes that may be responsible for the grand history of life on earth.

Acknowledgments We thank the anonymous reviewers for thoughtful ideas about how to improve the manuscript, Liz P. Beggrow for helpful suggestions, and Minsu Ha for help with references.

References

AERA (American Educational Research Association), APA (American Psychological Association), & NCME (National Council on Measurement in Education). (1999). *Standards for Educational and Psychological Testing.* Washington, D.C.: AERA.

Alters, B. J., & McComas, W. F. (1994). Punctuated equilibrium: The missing link in evolution education. *The American Biology Teacher,* 56(6), 334–340

Arthur, W. (2004). *Biased embryos and evolution.* Cambridge (United Kingdom): Cambridge University Press.

Ayala, F. J. & Arp, R. (2010). *Contemporary debates in philosophy of biology.* UK: Wiley-Blackwell.

Bateson, P., & Gluckman, P. (2011). *Plasticity, robustness, and evolution.* Cambridge, UK: Cambridge University Press.

Baum, D. A., & Offner, S. (2008). Phylogenies & tree-thinking. *The American Biology Teacher,* 70(4), 222–229.

Burian, R. (1988). Challenges to the evolutionary synthesis. *Evolutionary Biology,* 23, 247–269.

Carroll. S. B. (2005a). *Endless forms most beautiful: The new science of evodevo and the making of the animal kingdom.* New York: W.W. Norton.

Carroll, S. B. (2005b). Evolution at two levels: on genes and form. *PLoS Biology,* 3, 1159–1166.

Catley, K. M. (2006). Darwin's missing link: A novel paradigm for evolution education. *Science Education,* 90, 767–783.

Coyne, J. A., & Orr, H. A. (2004). *Speciation.* Sunderland, MA: Sinauer Associates.

Coyne, J. A., & Orr, H.A. (1998). The evolutionary genetics of speciation. *Philosophical Transactions of the Royal Society of London B,* 353, 287–305.

Cretzinger, J. I. (1941). An analysis of principles or generalities appearing in biological textbooks used in the secondary schools of the United States from 1800 to 1933. *Science Education,* 25(6), 310–313.

Darwin, C. (1859). *On the origin of species by means of natural selection, or the preservation of favoured races in the struggle for life.* London: John Murray.

Darwin (1860). See: http://darwin-online.org.uk/manuscripts.html. John van Wyhe, ed. 2002- The Complete Work of Charles Darwin Online.

Depew, D. J., & Weber, B. H. (1995). *Darwinism evolving: Systems dynamics and the genealogy of natural selection.* Cambridge: MIT Press.

Dietrich, M. R. (2010). Microevolution and macroevolution are governed by the same processes. In F.J. Ayala & R. Arp (Eds.). *Contemporary debates in the philosophy of biology* (pp. 169–179). Maden, MA: Wiley-Blackwell.

Dodick, J., & Orion, N. (2003). Cognitive factors affecting student understanding of geologic time. *Journal of Research in Science Teaching,* 40(4), 415–442.

Eldredge, N. (1989). *Macroevolutionary Dynamics: Species, Niches, and Adaptive Peaks.* McGraw-Hill

Eldredge, N., & Gould, S. J. (1972). Punctuated equilibria: An alternative to phyletic gradualism. In T. J. M. Schopf (ed.). *Models in paleobiology* (pp. 82–115). San Francisco: Freeman, Cooper.

Erwin, D. H. (2010). Microevolution and macroevolution are not governed by the same processes. In F. J. Ayala & R. Arp (Eds). *Contemporary debates in the philosophy of biology* (pp. 180–193). Malden: Wiley-Blackwell.

Filipchenko, J. (1927). *Variabilität und Variation.* Berlin: Gebruder Borntraeger.

Futuyma, D. J. (1998). *Evolutionary biology,* 3rd ed. Sunderland, MA: Sinauer.

Futuyma, D. J. (2005). *Evolution*. Sunderland, MA: Sinauer Associates.

Ghiselin, M. T. (1974). A radical solution to the species problem. *Systematic Biology*, 23(4), 536–544.

Goldschmidt, R. (1940). *The material basis of evolution*. New Haven CT: Yale Univ. Press.

Goodwin, B. (2009). Beyond the Darwinian paradigm: Understanding biological forms. In M. Ruse & J. Travis (Eds.), *Evolution: The first four billion years* (pp. 299–312). Cambridge, MA: Harvard University Press.

Gould, S. J. (1977). *Ontogeny and phylogeny*. Cambridge, MA: Belknap Press.

Gould, S. J. (1981). What happens to bodies if genes act for themselves? *Natural History*. November.

Gould, S. J. (1985). The Paradox of the first tier: An agenda for paleobiology. *Paleobiology*, 11(1), 2–12.

Gould, S. J. (2002). *The structure of evolutionary theory*. Cambridge, MA: Belknap Press of Harvard University Press.

Haeckel, E. (1868). *The History of creation*, translated by E. Ray Lankester. London: Kegan Paul, Trench & Co.

Hennig, W. (1999). *Phylogenetic systematics* (3rd Ed.). Champaign, IL: University of Illinois Press.

Hillis, D. (2010). Phylogenetic progress and applications of the tree of life. In M. A. Bell, D. J. Futuyma, W. F. Eanes, & J. S. Levinton (Eds.). *Evolution since Darwin: The first 150 years* (pp. 421–450), Sunderland (MA): Sinauer Associates, Inc.

Hull, D. L. (1980). Individuality and selection. *Annual Review of Ecology and Systematics*, 11, 311–332.

Hull, D. L. (1988). *Science as a process: An evolutionary account of the social and conceptual development of science*. The University of Chicago Press, Chicago and London.

Jablonski, D. (1986). Background and mass extinctions: The alternation of macroevolutionary regimes. *Science*, 231(4734), 129–133.

Jablonski, D., & Hunt, G. (2006). Larval ecology, geographic ranges and species survivorship in Cretaceous molluscs: Organismic versus species-level explanations. *The American Naturalist*, 168(4), 556–564

Kirschner, M., & Gerhart, J. (1998). Evolvability. *Proceedings of the National Academy of Sciences*, 95(15), 8420–8427.

Lloyd, E. & Gould, S. J. (1993). Species selection on variability. *Proceedings of the National Academy of Sciences*, 96, 11904–11909.

Love, A. C. (2013). Interdisciplinary lessons for the teaching of biology from the practice of evo-devo. *Science & Education*.

Love, A. C. (2007). Morphological and paleontological perspectives for a history of evo-devo. In M. Laubichler & J. Maienschein (eds.), *From embryology to evo-devo: A history of developmental evolution* (pp. 267–307). Cambridge, MA: MIT Press.

Marco, Ó. B., & López, V. S. (1993). A simple model to think about the evolutionary rate in macroevolution. *The American Biology Teacher*, 424–429.

Mayr, E. (1988). *Toward a new philosophy of biology: Observations of and evolutionist* (No. 211). Cambridge, MA: Harvard University Press.

Minelli, A. (2009). *Forms of becoming: The evolutionary biology of development*. Princeton, Oxford: Princeton University Press.

Minelli, A., & Fusco, G. (2012). On the evolutionary developmental biology of speciation. *Evolutionary Biology*, 39(2), 242–254.

Moody, D. E. (1996). Evolution and the textbook structure of biology. *Science Education*, 80(4), 395–418.

Müller, G. B. (2007) Six memos for evodevo. In M.D. Laubichler & J. Maienschein (Eds.), *From embryology to evodevo: A history of developmental evolution* (pp. 499–524). Cambridge, MA: MIT Press.

Nadelson, L. S., & Southerland, S. A. (2010a). Development and preliminary evaluation of the measure of understanding of macroevolution: Introducing the MUM. *The Journal of Experimental Education*, 78(2), 151–190.

Nadelson, L. S., & Southerland, S. A. (2010b). Examining the interaction of acceptance and understanding: How does the relationship change with a focus on macroevolution? *Evolution Education & Outreach*, 3(1), 82–88.

National Academy of Sciences (NAS) (2012). Macroevolution. Accessed February 6, 2012 at: www.nasonline.org

National Research Council (NRC). (1996). *National science education standards*. Washington, DC: National Academy Press.

Nehm, R. H. (2006). Faith-based evolution education? *BioScience*, 56, 638–639.

Nehm, R. H., & Budd, A. F. (Eds.) (2008). E*volutionary Stasis and Change in the Dominican Republic Neogene*. Netherlands: Springer.

Nehm, R. H., Poole, T. M., Lyford, M. E., Hoskins, S. G., Carruth, L., Ewers, B. E., & Colberg, P. J. S. (2009). Does the Segregation of Evolution in Biology Textbooks and Introductory Courses Reinforce Students' Faulty Mental Models of Biology and Evolution? *Evolution: Education and Outreach*, 2(3), 527–532.

Neumann, I., Neumann, K., & Nehm, R. H. (2011). Evaluating instrument quality in science education: Rasch based analyses of a nature of science test. *International Journal of Science Education*, 33, 1373–1405.

Nicholas, C. H. (1965). Analysis of the course content of the Biological Sciences Curriculum Study as a basis for recommendations concerning teacher preparation in biology. Unpublished doctoral dissertation, Stanford University School of Education.

Novick, L. R. & Catley, K. M. (2012). Assessing students' understanding of macroevolution: Concerns regarding the validity of the MUM. *International Journal of Science Education*, 34(17), 2679–2703.

Padian, K. (2010). How to win the evolution war: Teach macroevolution! *Evolution Education & Outreach*, 3, 206–214.

Platt, J. E. (2006). Macroevolution: Alive & well in sticklebacks. *The American Biology Teacher*, 68(1), 5–6.

Raff, R. A. (1998). Evo-devo: The evolution of a new discipline. *Genome*, 280, 1540–1542.

Raff, R. A. (2000). Evo-devo: The evolution of a new discipline. *Nature Reviews: Genetics* 1, 74–79.

Raup, D. M. (1994). The role of extinction in evolution, *Proceedings of the National Academy of Sciences*, 91, 6758–6763.

Raup, D., & Stanley, S. M. (1978). *Principles of Paleontology* (2nd ed.). San Francisco, CA: W. H. Freeman

Rosenthal, D. B. (1985). Evolution in high school biology textbooks: 1963–1983. *Science Education*, 69(5), 637–648.

Ruse, M. (1997). *Monad to Man: The Concept of Progress in Evolutionary Biology*. Harvard University Press.

Sansom, R., & Brandon, R. N. (2007). *Integrating evolution and development: From theory to practice*. Cambridge, MA: MIT Press.

Schindewolf, O. H. (1950). *Grundlagen und methoden der palaeontologischenchronologie*, Teil 3. Berlin: Borntraeger.

Schindewolf, O. H. (1950/1993). *Basic questions in paleontology*. Chicago: University of Chicago Press.

Sepkoski, D. (2008). Macroevolution. In M. Ruse (Ed.), *The Oxford. Handbook of the Philosophy of Biology* (pp. 211–237). New York: Oxford University Press.

Sepkoski, D., & Ruse, M. (2009). *The paleobiological revolution: Essays on the growth of modern paleontology*. Chicago, IL: University of Chicago Press.

Settlage, J., & Odom, A. L. (1995). Natural selection conceptions assessment: Development of the two-tier test "Understanding Biological Change." Paper presented at the National Association of Research in Science Teaching Annual Meeting, San Francisco, CA.

Simpson, G. G. (1944). *Tempo and mode in evolution*. New York: Columbia University Press.

Simpson, G. G. (1953). *The major features of evolution*. New York: Columbia University Press.

Skoog, G. (1969). The topic of evolution in secondary school biology textbooks, 1900–1968. Unpublished doctoral dissertation, University of Nebraska.

Skoog, G. (1984). The coverage of evolution in high school biology textbooks published in the 1980s. *Science Education*, 68, 117–128.

Smith, M. U. (2010). Current status of research in teaching and learning evolution: II. Pedagogical issues. *Science & Education*, 19, 539–571.

Stanley, S. M. (1980). *Macroevolution: Pattern and process*. San Francisco, CA: W. H. Freeman and Co.

Sterelny, K. (2009). Novelty, plasticity and niche construction: the influence of phenotypic variation on evolution. *Mapping the Future of Biology*, 93–110.

Swarts, F. A., Roger Anderson, O., & Swetz, F. J. (1994). Evolution in secondary school biology textbooks of the PRC, the USA, and the latter stages of the USSR. *Journal of Research in Science Teaching*, 31(5), 475–505.

Travis, J., & Reznick, D. (2009). Adaptation. In M. Ruse & J. Travis (Eds.) *Evolution: The first four billion years* (pp. 105–131). Cambridge, MA: Harvard University Press.

von Baer, K. E. (1828). *Entwicklungsgeschichte der thiere: Beobachtung und reflexion*. Konigsberg: Bomtrager.

Vrba, E. S., & Gould, S. J. (1986). The hierarchical expansion of sorting and selection: Sorting and selection cannot be equated. *Paleobiology*, 12(2), 217–228.

Waddington, C. H. (1970) *Towards a theoretical biology*. Vol. 3. Chicago, IL: Aldine Publishing Company.

Wilensky, U., & Resnick, M. (1999). Thinking in levels: A dynamic systems approach to making sense of the world. *Journal of Science Education and Technology*, 8(1), 3–19.

Zelditch, M. L. (2001). *Beyond heterochrony: The evolution of development*. New York: Wiley-Liss.

Kostas Kampourakis holds an appointment as Collaborateur Scientifique 2 at the University of Geneva, Switzerland. He is editor or coeditor of four special issues for *Science & Education* journal: *Darwin and Darwinism* (with David Rudge); *Philosophical considerations in the teaching of biology*; *Genetics and society: educating scientifically literate citizens* (with Giorgos Patrinos and Thomas Reydon); and *Mendel, Mendelism and Education: 150 years since the "Versuche"* (with Erik Peterson). He is also an Adjunct Instructor at the Department of Mathematics and Science Education of Illinois Institute of Technology. He was Coordinator for science education for grades K-6 as well as member of the Department of Educational Research and Development of Geitonas School, in Athens, Greece, where he also taught biology and nature of science to secondary school and IB DP students. His research interests include the teaching of biology and the teaching of the nature of science in the context of the history and philosophy of science and he has published several articles on these topics. He has written a book titled *Understanding Evolution: an Introduction to Concepts and Conceptual Obstacles* (Cambridge University Press). He has also edited a book titled *The Philosophy of Biology: a Companion for Educators* (Springer) in which professional philosophers of biology bring their work to bear on biology education. He holds a B.S. in biology and a M.S. in genetics, both from the University of Athens, as well as a Ph.D. in science education from the University of Patras.

Ross H. Nehm holds faculty appointments as Associate Professor in the Department of Ecology and Evolution and the Ph.D. Program in Science Education, at Stony Brook University in New York. He received a B.S. from the University of Wisconsin at Madison, an Ed.M. from Columbia University, and a Ph.D. from the University of California at Berkeley. He was the recipient of a U.S. National Science Foundation Early Career Award and was named Education Fellow in the Life Sciences by the U.S. National Academy of Sciences. The NSF's CCLI, TUES, and REESE programs have funded his research. He currently serves on the editorial boards of the *Journal of Research in Science Teaching*, the *Journal of Science Teacher Education*, and the *Journal of Science Education and Technology* and has published papers in a wide variety of education and science journals, including *JRST, Science Education, Science & Education, International Journal of Science Education, Research in Science Education, Journal of Science Teacher Education, The American Biology Teacher, Journal of Science Education and Technology, CBE-Life Sciences Education, Evolution Education and Outreach, Genetics, Bioscience, Journal of Biology and Microbiology Education, Caribbean Journal of Science, Bulletins of American Paleontology*, and *Journal of Paleontology*. With Ann Budd he has edited the book *Evolutionary Stasis and Change in the Dominican Republic Neogene* (Springer 2008). His evolution education research was recently highlighted in the National Research Council publication *Thinking Evolutionarily* (National Academy Press 2012). His research interests include evolutionary thinking and reasoning, problem solving, and assessment methodologies.

Chapter 15
Twenty-First-Century Genetics and Genomics: Contributions of HPS-Informed Research and Pedagogy

Niklas M. Gericke and Mike U. Smith

15.1 Introduction

Empirical studies have shown that genetics is considered to be the most difficult subject of biology to teach and to learn (Bahar et al. 1999; Finley et al. 1982; Johnstone and Mahmoud 1980). Moreover, genetics is the cornerstone of any evolution curriculum and thus the basis for any study of biology. Genetic education research has therefore evoked great interest over the years. The field of research in genetics is one of the most rapidly developing sciences of the last century with great impact on society and media due to new biotechnologies such as genetically modified organisms (GMO), genetic screening, forensics, and high-profile ventures such as the Human Genome Project. Because of the exponential knowledge development in genetics, many major advances have occurred in the conceptual understanding of genetic phenomena[1] attracting great interest from scholars in the history and philosophy of science (HPS). At the center of the development of genetics is the circuitous route to our current understanding of the concept of the gene, resulting in the current polysemous and sometimes incoherent current view. Associated with this historical development are important biological philosophical issues such as reductionism, genetic determinism, and the relationship between function and structure.

---

[1] See, for example, Beurton and colleagues (2000), Davis (2003), Falk (2010), Keller (2005), and Sapp (2003).

N.M. Gericke (✉)
Department of Environmental and Life Sciences, Karlstad University, Karlstad, Sweden
e-mail: niklas.gericke@kau.se

M.U. Smith
Department of Community Medicine/Public Health, Mercer University School of Medicine, Macon, GA 31207, USA
e-mail: SMITH_MU@mercer.edu

M.R. Matthews (ed.), *International Handbook of Research in History,*
Philosophy and Science Teaching, DOI 10.1007/978-94-007-7654-8_15,
© Springer Science+Business Media Dordrecht 2014

This article is divided in two main sections: (1) a condensed overview of the historical development of genetics and the philosophical implications of this development and (2) a review of genetics education research, focusing on how it is informed by issues from HPS. We identify contributions of HPS-informed genetics education scholarship to the following: (1) teaching and learning genetics, (2) teaching about the nature of science, (3) humanizing science, and (4) enhancing reasoning, argumentation, and thinking skills. Finally we give some concluding remarks and suggestion for future research.

15.2 History and Philosophy of Genetics

Questions about genetics are probably as old as sentient beings: "Where did I come from?" "Why do I have blue eyes?" "Why are some people just born to be better than other people at some things?" "Where did all the different species come from?" Genetics interests everyone from the man on the street to scientists, philosophers, and historians. As a recognized field of study, genetics is a relatively young science—only about a century old and is perhaps the most rapidly growing field of science today. The emergence of genetics is a fascinating story, and at the center of this story is the hunt for the gene, the most central concept of genetics. The gene has been operationally defined on the basis of four interdependent phenomena: genetic transmission (inheritance of traits from one generation to the next), genetic recombination (generating new combinations of traits), gene mutation (changes in DNA that generate new traits), and gene function (Portin 1993). Over the course of its history, genetics research and genetic applications have focused on each component to differing degrees. In this section we begin with an overview of the search for and the differing understandings of the concept of the gene. This history is provided because an understanding of the history of genetics (as seen through the lens of the gene concept) can be beneficial for understanding not only how to identify and address historical and philosophical issues that arise in genetics education but also how to promote effective instruction.

15.2.1 History of Heredity Before Genetics

The idea of biological heredity is an ancient concept based on experience with humans, as well as domestic animals and crops. The oldest known pedigree (associated with horse breeding) was found in Mesopotamia and is over 5,000 years old (Gustavsson 2004). The Talmud, one of the ancient holy books of the Jews, prohibits circumcision of sons to women who have previously given birth to children that bled to death, as well as the sons born to sisters of such mothers. Clearly, practical insight based on genetics understanding is ancient (Gustavsson 2004). The origin of formal genetics can be dated to about 1900 with the independent recognition of

Mendel's earlier work by Correns, de Vries, and Tschermak von Seysenegg (Moore 2001), although the contribution of von Tschermak Seysenegg has been questioned in recent years (Henig 2000; Olby 1985). The routes to classical genetics come from research in evolution, cytology, embryology and reproduction, breeding, and hybrid formation (Carlson 2004). These research areas had different aims but in different ways addressed questions that from the twentieth century and onwards would become intimately tied to the new field of genetics.

Questions about the origin of species became prominent in the eighteenth century. The predominant belief before that time was essentially the view of the church that all species are the products of divine creation. This view began to be vigorously challenged as a result of the French Enlightenment's demand that the answers to questions about the natural world be sought through the application reason, observation, and experimentation rather than faith. Prior to Darwin and Mendel, heredity was generally thought of in terms of blending rather than being particulate (Uddenberg 2003). During this time, evolutionary explanations came to be more acceptable accounts of the origin of species. Lamarck proposed a theory of evolution in which the characteristics of an organism change in response to its behavior and changes in its environment. Although he recognized a role for heredity in the process, heredity was not of central importance to Lamarck's explanations.

By contrast, heredity played a central role in Darwin's theory of evolution, as a way of explaining the variations on which natural selection acts to produce new species. Darwin called his provisional theory of heredity pangenesis. This theory suggested that small units, which he called "gemmules," are produced by the cells, that these gemmules migrate throughout the body to produce the inherited traits, and that some gemmules are retained in the reproductive tissues in the gonads and thus are passed from one generation to another (Darwin 1868). Although the theory had little experimental support, it clearly recognized the transmission of genetic material (or at least, material affecting heritable traits) from one generation to the next. Darwin's pangenesis theory, however, held that the gemmules might be changed by external conditions, i.e., accepted the inheritance of acquired characteristics, an idea that Darwin's cousin Francis Galton strongly opposed.[2]

Cytology, the study of the cell, was made possible by Janssen's invention of the first compound microscope in the early seventeenth century. However, it was not until 1838 that Schleiden proposed a theory arguing that plants are communities of cells. Scientists promptly turned their attention to the contents of the cell; Huxley argued that the protoplasm (the substance in the cell) was the basis of life, suggesting that it might contain the material governing heredity. With the development of improved lenses and staining techniques in the 1850 s, it became possible to detect structures within the cell. By the late 1880 s, researchers had determined that the nucleus of the cell was the central actor in fertilization; the nucleus was then considered to be the source of idioplasm (the term used by Nägeli for genetic

---

[2] Galton introduced the idea of latent and patent elements that both contributed to the structureless elements of heredity. By making information transfer unidirectional, Galton was attempting to rule out Lamarckism (Schwartz 2008).

substance) in the transmission of heritable traits to the progeny. The chromosomes within the nucleus were also identified, and the processes of mitosis and meiosis were described by Flemming and Hertwig, respectively. By the end of the nineteenth century, biologists were occupied with questions of heredity and its relationship to the nucleus and chromosomes. These questions reflect the early interest in the transmission aspect of genetics (Carlson 2004).

Embryology and the study of reproduction are also intimately involved with heredity and transmission. The Aristotelian view of reproduction was that the woman's menstrual blood provided the "raw material" that made up the child and that the man's semen provided necessary "design" through an "animating principle" (Gustavsson 2004). This view was prominent until the seventeenth century when Harvey proposed that eggs were fertilized by the semen, but this theory failed to incorporate the concept of cells. In 1677 Leeuwenhoek first identified spermatozoa and the idea of a material agent of fertilization was born, but it was not until the mid-nineteenth century that Amici first observed the fertilization process of union between egg and sperm. One of the great debates of embryology in the seventeenth and eighteenth centuries was between preformation (i.e., that embryos are essentially "homunculi" [microscopic but fully formed humans] whose growth and development involves only enlargement) and epigenesis (that embryos are constructed from much simpler precursors that develop via some complex mechanism) (Carlson 2004). Improved microscopy would later, of course, support the epigenesist explanation. In the 1890 s, Weismann proposed that meiosis results in a mixing of paternal and maternal heredities in an offspring. He also proposed that reproductive tissue ("germ plasm") is set aside early in development separate from the rest of the body (called the "soma"). Changes in the body (soma) are then not transmitted to the germ plasm according to this theory (Schwartz 2008).

Transmission in the form of breeding of domesticated plants and animals first became the subject of scientific study in the seventeenth and eighteenth centuries. On the basis of his studies of plants, Linnaeus proposed that the "outer" traits of hybrids were derived largely from the male parent, while the "inner" attributes or tissues originated from the female (Carlson 2004). The core "analytical units" examined in these studies were traits or specific characteristics of a species on a phenomenological level; the nature of the relationship between these traits and structures within the body such as the cell were mostly ignored by these scientists. Breeder scientists in the nineteenth century such as Thomas Andrew Knight, Charles Naudin, and even Charles Darwin observed that traits were sometimes passed down to offspring in unexpected ways, but they were unable to explain these observations.

In the second half of the nineteenth century, researchers in biology and heredity sought to answer the following questions: Are all of the characteristics (of a species) controlled by a single, uniform, species-specific substance or is each character determined by a separate particle that can vary independently? Is the genetic material "soft," so that it can change gradually during the lifetime of the individual and/or through generations, or is the genetic material constant and "hard," being changeable only by means of a sudden and radical alternation (later called a mutation)?

How are the particles formed in the body? Do the particles contributed by the parents retain their integrity after fertilization or do they fuse completely? These questions could not be fully answered until the advent of molecular genetics, but they were important questions historically because they centered on the existence of material units of heritability (Mayr 1982).

15.2.2 Classical Genetics

Unlike most of his peers and predecessors, in the late nineteenth century, Gregor Mendel focused on individual physical characteristics ("traits") in the common garden pea. Mendel also applied statistical analysis to the frequencies at which these traits occurred in the offspring from his crosses. Mendel's reductionist approach led to his atomistic model of heredity in which fundamental units of heredity are related to specific binary traits. Mendel proposed that underlying "elemente" are responsible for the production of "merkmal", i.e., physical characteristics of the individual organism (Moore 2001). He did not seek to explain or analyze the nature of this relationship, and he did not apply concepts or results from cytology. The element was an abstract concept connected to specific traits but with no direct association to the physical parts of the organism such as the cell. Mendel's theory provided two crucial contributions to the understanding of genetics: first that some traits are determined by a single factor rather than by many and second that these factors exist in pairs. By proposing the existence of factors in pairs, Mendel was able to explain the results of crosses by the segregation and recombination of these factor pairs. In 1909, Johannsen would name these factors "genes" and the different forms of a gene would later become known as "alleles" (Mayr 1982). The specific combination of these genetic determinants in an individual is its "genotype," and the resulting outward expression of that genotype is the "phenotype." Johannsen was also careful not to claim that the gene is a physical object; instead he regarded it as a hypothetical construct.

The Mendelian "unit-factor" theory of inheritance was ultimately embraced by the scientific community. In Mendelian genetics the genotype was regarded as the phenotype in miniature, not necessarily as a homunculus but rather as a mosaic of heredity particles (whether called gemmules, pangenes, or unit factors), each responsible for a specific component of the phenotype. Each genetic factor was believed to have a one-to-one relationship with the corresponding outward characteristic.

In the early twentieth century, classical genetics emerged as a discipline in its own right when breeding analysis was combined with studies in cytology, embryology, and reproduction. William Bateson was the first to employ the word "genetics" in 1906 to replace the term "heredity and variation" (Carlson 2001). In 1902, Boveri and Sutton proposed that chromosomes are the carriers of the unit factors (genes), that they are transferred to the next generation by gametes, and that configuration of the chromosomes during meiosis explains Mendelian heredity. This chromosomal theory of heredity was questioned until 1911, when Morgan provided the first experimental evidence for it (Morgan 1911). Later Morgan demonstrated that coupling, i.e., the failure of the alleles of certain "linked" genes to assort independently during meiosis, could be explained by the physical proximity to genes to each other on a single chromosome. Linked genes could be separated by a physical breakage-and-exchange process (called "crossing-over") between two chromosomes, but the likelihood of such events was determined by the proximity of the two genes to each other (Schwartz 2008).

Based on the data obtained from extensive linkage studies, Sturtevant constructed the first map of the genes on the X chromosome of *Drosophila* in 1913 (Weiner 1999). This map visualized the spatial relationships of genes to one another on the chromosome, suggesting the representation of the chromosome as a string of beads, each bead representing a different gene (Portin 1993). Accordingly, classical mapping techniques were thereafter commonly used and played an epistemic role in our understanding of the genetic material, representing genes as physical objects more than hypothetic constructs (Gaudillière and Rheinberger 2004; Weber 1998).

During the years after 1940, at the peak of classical genetics, the gene was viewed as an indivisible unit of genetic transmission/function, recombination, and mutation—in Benzer's terms: the cistron, recon, and muton (Benzer 1955). Genetic material was considered to be particulate and to have long-term stability ("hard inheritance"), with mutations representing a discontinuous change to a gene. X-ray-induced mutations were discovered in *Drosophila* by Muller in 1927 and confirmed in maize by Stadler the same year (Carlson 2011). Each gene was assumed to be independent of neighboring genes. Individual traits were the products of genes located at well-defined loci on the chromosomes. Genes were linked to each other on each chromosome but could be separated by crossing-over. Plants and animals were recognized as being "diploid" (for the most part), i.e., the chromosomes in the nuclei of somatic (nonsex) cells exist in pairs (called "homologues"), each member of the pair being derived from the different parents. Homologues are similar in structure and bear the same genes, although they may bear either identical alleles of a gene ("homozygous") or different alleles ("heterozygous").

A strict distinction was made between the genotype and the phenotype. The phenomenon of polygeny (several genes influencing a single trait) and pleiotropy (a single gene affecting several characters) was recognized, thus permitting a much clearer separation between transmission genetics and physiological genetics, i.e., studies of the ways in which hereditary information is manifested during the course of individual development (Mayr 1982). These phenomena conflicted, however, with the accepted model of a one-to-one relationship between genes and traits (Schwartz 2000, p. 28), a fact that created much confusion about this relationship during the classical era. Many geneticists also ignored questions about development in favor of chromosomal mechanics, likely because the latter were more open to a quantitative approach (Lawrence 1992). During the classical era, the most widespread view of the nature of the gene itself (attributable to Weismann among others) was that genes were enzymes, or acted like enzymes, serving as catalysts for chemical processes in the body, producing physical traits (Carlson 1966; Mayr 1982).

In 1948, Boivin and colleagues used chemical analysis to demonstrate that the DNA content of the nuclei in different tissues and organs of individual domestic cows were the same. A year later, Chargaff and colleagues reported that the DNA of calf thymus contained the same proportion of the four nitrogenous bases as did that from the spleen. These findings established the species-specific character of the base composition of DNA. In 1950, Chargaff realized that in all of these cases and in every other species so examined, the quantity of guanine was always equal to that of cytosine, and the amount of thymine was similarly always equal to that of adenine (Chargaff 1951).

In the 1940s and 1950s, breeding analysis and the cytology of animals and plants were replaced at the frontier of research by biochemical genetic studies in fungi, bacteria, and viruses. This change of model organism shifted the emphasis in genetics toward function in general and developmental processes in particular, instead of studies of crossing-over and mutation, which characterized the earlier research. Beadle and Ephrussi determined the biochemical pathway for eye color synthesis in fruit flies (Beadle and Ephrussi 1937). Later, Beadle described the biochemical pathways associated with the synthesis of vitamins and demonstrated that these pathways consisted of ordered series of chemical steps, with a single gene controlling each single step in the chain of reactions. In 1941 Beadle and Tatum proposed that each gene controls one enzyme, and subsequently the hypothesis was coined "one gene-one enzyme hypothesis" by their collaborator Norman Horowitz (Beadle and Tatum 1941; Horowitz 1995), which is still considered essentially correct for microbial genes. However, these genetic and biochemical experiments used the conceptual tools of classical genetics and did not explain the nature of the biochemical pathways or the mechanism by which the genetic material affected the phenotype (Carlson 2004).

Although the classical gene concept was constantly questioned during the first half of the twentieth century, in particular by Richard Goldschmidt (Dietrich 2000) as well as *Drosophila* researchers such as Morgan (Morgan 1934), the term retained a central position in theory and research. But Watson and Crick and their descendants would now change the face of genetics forever.

15.2.3 *Modern Genetics*

In the early 1940s it was not entirely clear whether DNA or protein was the material that carried the hereditary information. Protein was widely favored because of its largely information-bearing capacity, given that it consisted of 21 amino acids compared to the "simple" four nitrogenous base composition of DNA. The question was first essentially answered in 1944 by Avery and colleagues who showed that the cell-free "transforming principle" known to transform pneumococcus bacteria from non-virulence to virulence (a known inherited trait) was composed of DNA alone and was responsible for bacterial transformation (Avery et al. 1944), although some scientists remained skeptical. More scientists were convinced by the 1952 studies of

Hershey and Chase who used differentially radioactively labeled DNA and proteins to study the transmission of genetic information in T2 bacteriophages, demonstrating that it is DNA and not protein that is the genetic material in viruses as well (Hershey and Chase 1952).

Based on these results as well as Chargaff's chemical studies (described above), physiochemical studies, and the crystallographic studies of Wilkins and colleagues and Franklin and Gosling, Watson and Crick first proposed the double-helix model of DNA (Watson and Crick 1953). This model of DNA fulfilled the necessary requirements for the genetic material, namely, auto-replication, specificity, and information content capacity. The long search for the genetic material had ended; genetic transmission now had a straightforward chemical explanation. This model (1) explained the nature of the linearity of genes, (2) suggested a mechanism for the exact replication of genes, (3) explained in chemical terms the nature of mutations, and (4) explained why mutation, recombination, and function are distinct phenomena. Modern molecular genetics had been born. Now the unanswered questions became increasingly physiological, dealing with the function of genes and their role in ontogeny and physiology.

The impact of molecular biology on our understanding of genetic phenomena has been immense. From 1953 there was a clear explanation for the difference between genotype and phenotype, and it was understood that the genotype does not itself enter directly into developmental pathways but simply serves as a set of instructions for producing proteins. The DNA is arranged in three-letter nucleotide codes, named codons, on each DNA strand. A series of codons that together code for an entire polypeptide constitutes the gene in this classical molecular view. The four possible nucleotides can be arranged in 64 different codons that specify 20 amino acids. The gene was found to be "degenerate" (each amino acid is coded for by more than one codon) and "commaless" and to employ a unique start codon and at least one of three stop codons at its terminus. Gene function occurs via the process of transcription in which single-stranded mRNA is copied from the DNA sequence of the gene and thereafter is translated into an amino acid sequence, a polypeptide.

The discovery of the structure of DNA in the 1950s coincided with the birth of the information sciences and the explication key terms, such as "program" and "code" (Mayr 1982). The metaphor of "program" was first coined by Jacob and Monod to encompass the new concept of gene regulation (Keller 2000). However, philosophers of genetics (i.e., Keller 2000) have suggested that the metaphor of the "genetic program" promotes a more deterministic understanding of the gene, i.e., a blueprint that always works in the same way ignoring any other factors.

Benzer's earlier theoretical division of the gene concept into cistron, muton, and recon proved very useful in molecular genetics, and the nomenclature was adapted to the new findings. The cistron was equivalent to a functional gene (a string of DNA), and the muton and the recon were equivalent to a single DNA base pair because a nucleotide is the smallest unit of genetic material that, if altered, can lead to an altered phenotype or be separated from other such units during recombination (Carlson 1991). From this time until about 1970, the gene (cistron), defined by a complementation test[3], was understood as a contiguous stretch of DNA transcribed as one unit into messenger RNA (mRNA), coding for a single polypeptide (Portin 1993).

After 1970, molecular genetic studies of higher eukaryotic organisms identified an increasing number of anomalies inconsistent with the model of the gene as simply a stretch of DNA that produces a polypeptide, suggesting that this definition is deficient in one or more respects. Today these anomalies include split genes, alternative splicing, complex promoters, polyprotein genes, multiple adenylation, enhancers, overlapping genes, DNA editing, imprinting, and trans-splicing[4]. The common theme in these anomalies is that the structural unit of the gene, the stretch of DNA including all the codons, does not coincide completely with the function of the gene, which is to determine the sequence of amino acids in the produced polypeptide. Instead different molecular processes can impact gene function and development, leading to different context-dependent outcomes.

In the 1980s tools to isolate and determine sequences of nucleotides within DNA and to study the arrangements of gene sequences in the genome of any organism become widely available. The invention of polymerase chain reaction (PCR) by Mullis was an especially important milestone in sequencing and cloning DNA segments (Bartlett and Sterling 2003). No longer were amino acid sequences of proteins determined directly. Instead, if the gene that encoded the protein of interest could be identified, molecular biologists could clone the gene, determine the nucleotide sequence, and deduce the amino acid order from the genetic code (Davis 2003). Public databases (e.g., GenBank) of amino acid and nucleotide sequences proliferated. Informational techniques made it possible to store, analyze, and share large amounts of sequence data, and the field of bioinformatics was born. The sequences deposited in these databases were accompanied by annotations of the function, if known, of each gene and its protein. Hence, the functional aspects of genes became the ultimate goal of research.

In the late 1980s further technological advances in DNA sequencing led to the proposal to determine the nucleotide sequence of the entire human genome. The Human Genome Project (HGP) was launched in 1990 by the U.S. Department of Energy and the National Institutes of Health, under the leadership of James Watson. Zwart (2008) identifies three stages of the HGP. The first stage was a period of implementation and development. The second stage began in 1993 with the appointment of Francis Collins as Watson's successor. At this time the sequencing was subdivided into 23 natural subunits (the chromosomes) which were analyzed by a large number of research groups. The year 1998 was a turning point for HGP and

---

[3] The cis-trans complementation test is a test used to determine whether two mutations are alleles or not, i.e., whether they are forms of the same or different genes. By this test, two mutations are not alleles, i.e., are in two different genes; the wild type (nonmutant) phenotype (the mutations "complement" each other) results when the two mutations appear in a single chromosome ("cis"; denoted a1a2/++), whereas if the two mutations appear on separate homologues (in "trans"), the mutant phenotype appears.

[4] See, for example, El-Hani (2007), Falk (2010), Fogle (2000), Rosenberg (1985), and Smith and Adkinson (2010).

the beginning of Zwart's third HGP stage. In that year, Craig Venter announced a competing private project funded by the Celera Corporation. By this date, only 4 % of the human genome had been sequenced, but Venter suggested a much faster methodology relying on automation. The so-called whole genome shotgun method would revolutionize sequencing methodology by relying on powerful computers, which slashed the costs for sequencing the genome (Zwart 2008). A fierce competition between the two groups ensued until June 2000 when Collins and Venter cordially appeared together at the now famous press conference at the White House to announce that the sequencing of the human genome was near completion. Finally in 2004 the public consortium published its completed sequence (IHGSC 2004). The success of the HGP required the development of new research approaches, new technologies, and even new disciplines (e.g., bioinformatics). Sequences of many other genomes would follow, today totaling more than 180 organisms (Genome News Network 2012), resulting in a new approach to genetics called "genomics."

Genomics is the study of the entire genomes of species. Unlike genetics per se, it is concerned with both the coding DNA (sequences that result in mRNA and thus proteins) and noncoding regions (often referred to previously as "junk DNA"). Genomics considers the ways different DNA regions interact in order to determine each region's effect on, place in, and response to the entire genome's structure and function within the cellular context, including development and production of the phenotype. In contrast, the investigation of the roles and functions of single genes is the primary focus of molecular biology and genetics. Epistemologically, genomics is often considered to be a paradigm shift in the study of living things, an "informatization" of life (Zwart 2008). On the other hand, as we leave the cell and biochemistry processes of molecular genetics for the more information-based approach of genomics, we must recognize that genomics also constitutes a return to an approach similar to classical genetics in the sense that it creates a black box of abstract knowledge between the genome and the output of the genome similar to that between the gene and the outcome of the genes in classical genetics.

In 2003 the National Human Genome Research Institute (NHGRI) launched the Encyclopedia of DNA Elements (ENCODE) Project whose goal is to build a comprehensive list of the functional elements in the human genome (Bonetta 2008). Already the HGP revealed that humans only have approximately 20,000–25,000 protein-coding genes (IHGSC 2004), which is similar to the number in the mouse. However the number of human proteins can be estimated to about 90,000 (Magen and Ast 2005). One gene clearly did not code or only one protein! More questions were asked than being answered. What makes up for the differences?

Because protein-coding sequences represent only about 1–2 % of the genome, research turned more of its interest to noncoding regions of DNA, which had previously been thought to be unimportant and was known as "junk DNA." ENCODE has since then shown that the majority of the genome is not "junk" but in fact transcribed into RNA. However about 99 % of the DNA does not code for proteins but, nonetheless, is vital in controlling many cellular processes such as development (Pearson 2006). Hence, to the surprise of many, a lot of work of the genome is transacted not only by proteins but also by RNA itself. This research has focused on the importance of epigenetics, the study of heritable changes caused by mechanisms other than change in the sequence of DNA, e.g., DNA methylation (Bonetta 2008). This focus on mechanisms that function above the level of the genome has led to a number of new twenty-first-century "-omics," including metabolomics, glycomics, and transcriptomics, but most especially proteomics—the large-scale study of proteins, particularly their structure and function (Wilkins et al. 1996).

In the genomic era the so-called central dogma of biology, an idea that originates from molecular genetics, is questioned more than ever. The central dogma of biology held that information flow was unidirectional, from the gene (DNA) to mRNA to polypeptide[5] (Crick 1970). But as Davis concludes: "We have, in the genomic era, converged on the idea that causation goes both ways, upward from DNA and downward from cytoplasm and environment . . . Even the distinction between 'genetic' and 'epigenetic' fails to convey this interdependence" (Davis 2003, p. 249).

15.2.4 Philosophical Implications of the Historical Development of the Gene Concept

In the twenty-first century, there no longer exists a single consensus definition of the gene (if such ever existed), although there are some interesting attempts that have been made (see the discussion below). Instead the gene concept has different meanings for different scientists and even for different settings (Stotz et al. 2004). "This entity [the gene] can, and will indeed most often, be endowed with temporary and discontinuous existence, and it will often require a developmental process at its own level of organization for functional expression" (Gayon 2000, p. 82). Contemporary research has largely discarded the idea of a gene as a discrete material unit, focusing instead on the gene as a functional unit. The function of a gene is no longer solely to produce a polypeptide; instead there exist many categories of genes in addition to the standard enzyme-producing genes, such as genes producing structural proteins, regulatory genes, and genes coding for RNA molecules with other functions ("RNAzymes"). Genes are perhaps best understood as one component of a complex network interaction at a variety of levels from the genome to the proteome and beyond. This lack of consensus, simplicity, or clarity in the definition of such a foundational term as the gene reflects how young genetics is as a science and that the discipline is growing exponentially.[6]

As shown by the historical overview above, the story of genetics in the first half of the twenty-first century is the history of the discovery of the gene and the nature of the genetic material. In the second half of the century, scientists sought to

---

[5] (allowing for some possible "reverse flow" from RNA to DNA).

[6] For expanded treatments of the historical development of genetics within HPS frameworks, the reader is directed to Burian 2005; Carlson 1966, 2004; Davis 2003; Keller 2005; Portin 1993; and Sapp 2003, as well Burian 2013; Flannery 1997; Gericke and Hagberg 2007; Kampourakis 2013; Mahadeva and Randerson 1985; Smith and Adkinson 2010; and Vigue 1976 in science education.

determine the nature of the gene and how it manifests itself. The philosophical outcome of this history is that the gene is a polysemous concept with multiple, sometimes incoherent meanings (Burian 2005). The epistemological program of classical genetics was genetic reductionism in which the gene was explained or inferred by phenotypic differences unlike in later molecular genetics in which the gene is a constructive rather than a diagnostic entity, responsible for the production of a protein (Sarkar 2002). The consequence of the shift between classical and molecular genetics is "Mendelian genetics as it was formulated at the time, not mention now, cannot be derived from molecular biology" (Hull 2002, p. 166), i.e., it is not possible to reduce the classical gene into the molecular gene because they are conceptually incoherent.

This is a very important fact for genetic educators to know because both concepts are presented in most biology courses, and as Sarkar points out: "the 'molecular gene' came to be routinely conflated with the 'classical gene'" (Sarkar 2002, p.192), resulting in student confusion and/or misunderstanding. This conflation means that the reductionism of classical genetics is transferred to the DNA segment that constitutes the molecular gene, which evokes the phenomenon of genetic determinism.

Genetic determinism is the view that genes completely determine the phenotypes, as implied in common expressions such as "the gene for X" (as, e.g., the gene for long legs or the gene for intelligence). Determinism ignores the influence of environmental factors, which as described above is far from contemporary molecular views of the gene. Almost no expert who talks about a "gene for" a phenotypic trait literally means what he or she is saying, but this is a common conversational shortcut used by scientists (Falk 2012). It has also been argued that the use of gene metaphors such as "program," "code," and "blueprint" also implies a deterministic understanding of the molecular gene (Keller 2000). Genetic determinism has been shown to be a very persuasive misconception outside the scientific community (Barnes and Dupré 2008; Kaplan and Rogers 2003; Lewontin et al. 1984; Nelkin and Lindee 1995) and is therefore an important concept for educators to be familiar with and to strive to avoid.

In the early years of molecular genetics, the proteins produced by the genes were believed literally to possess the function of the gene, but as has been discovered in the genomics era, individual molecules do not possess complex functions. Complex processes cannot be explained by macromolecules alone (Morange 2002). Functions of organisms are generated at higher levels of organization by regulatory processes in which the proteins, which are produced from the genes, are active components. The molecular components are organized in pathways, networks, and complexes (Morange 2002); therefore, it is typically impossible to predict function or phenotype from individual elementary components (genes or proteins), which is a reductionist view.

The history of genetics also reflects one of the main problems in biology, namely, the relationship between substance and function or structure and process (Burian 2005; Hoffmeyer 1988). In the classical period of genetic history, the gene was a unit of function with no corresponding structure. From 1953 onwards the material structure of the gene (a string of DNA) coincided with function, but as the development of genomics in the last decades, the simple structure of the one-gene-one-enzyme gene has once again vaporized. The post-genomic gene addresses the ongoing project of understanding how gene/genome structure supports gene/genome function (Griffiths and Stotz 2006).

Falk (in Griffiths 2002) has proposed four different potential approaches for dealing with the conceptual difficulties associated with the molecular gene: (1) Abstract away the complexities of molecular biology and define genes in terms of some role they play in evolution. (2) Continue to seek a structural definition at the molecular level (a quest Falk regards as hopeless). (3) Look for a functional account of the gene in molecular developmental biology, relying on a broadening focus from the DNA alone to the wider developmental system in which the concept of the gene is embedded, also denoted as the "process molecular gene concept" (Meyer et al. 2011). (4) Treat genes as "generic operational entities" defined by experimentalists to suit changing needs in different contexts. Kitcher (1982) argues that the clearest explanation of any given phenomenon can only be achieved by adopting different definitions of the gene for different purposes. This is a view that many researchers within HPS agree upon (Burian 2005; Carlson 1991; Fogle 1990; Portin 1993; Waters 1994). As Burian puts it: "There are large and important subcommunities with legitimately different interests – interests that lead them to deal with legitimately different phenotypes" (Burian 2005, p. 142).

Moss (2003) offers another solution to the problem by using two different gene concepts separating the two primary functions, which he terms gene-P and gene-D. Gene-P amounts to the gene as a determinant of phenotypic differences—a notion close to the classical gene but not identical according to Moss because molecular entities can also be used as gene-P. Gene-D in turn corresponds to a real entity defined by some molecular sequence that could be used as a developmental source (Moss 2003). The ENCODE project (Gerstein et al. 2007) defines the gene as "a union of genomic sequences encoding a coherent set of potentially overlapping functional products" (Gerstein et al. 2007, p. 677). Many scientists and philosophers promote this "systemic" or "process" definition of the gene, which emphasizes the view that the whole as well as all the separate parts of an organism must be considered to explain function and phenotype (Keller 2005; Portin 2009).

Keller and Harel (2007) take the implications of the polysemous gene concept a step further, suggesting that the gene concept should be abandoned for a new concept which they termed genetic "functor" or "genitor." A similar strategy is also advocated by Scherrer and Jost (2007) who recognize mRNA as the elementary counterpart of biological function and proposed the term "genon" as the program associated to a specific gene at the mRNA level, given that only mRNA gives rise to one specific protein. Stadler and colleagues (2009) added to this proposed genetic vocabulary, proposing the "genomic footprint," i.e., the fragments of DNA from which the functional sequence is assembled during the expression process.

The different gene concepts and if and how it is possible to link them are one of the central issues in the philosophy of biology[7] but with few exceptions have not been addressed by science education research. Questions about how to define the gene have not yet reached the introductory genetics classroom and have only begun to reach the genetics pedagogy literature. Given the centrality of the gene concept and the lack of a unified and universally applicable and accepted definition of the term the question is: What to teach and what not to teach—how to deal with "the bewildering gene" (as coined by Falk 1986)? How should teachers define and describe the gene for students? And what parts of the historical path that has led to the current state of affairs should be taught—and at what levels? At what level would it be appropriate to address this issue explicitly? What understandings are required for genetic literacy in modern society? How should these HPS issues be reflected in standards documents? What understandings are needed for citizens in the twenty-first century? Although we will return to these issues below, these questions, like the definition of the gene itself, remain largely unsettled by the science and science education communities.

15.3 Contributions of the History and Philosophy of Science to Genetics Teaching and Learning

The previous section provided an overview of the history of genetics and the philosophical implications this development has had on the concept of the gene. This history provides a context for the present section in which we discuss how employing HPS concepts has contributed to teaching and learning in science education scholarship.

Matthews (1994/2014) proposed that HPS can contribute to science teaching and learning in five ways. HPS can:

- Contribute to the fuller understanding of subject matter
- Help teachers appreciate the learning difficulties encountered by students
- Assist in developing a more authentic understanding of science and thus enhance understanding of the nature of science
- Humanize the sciences and connect them to personal, ethical, cultural, and political concerns
- Make classrooms more challenging, by enhancing reasoning and critical thinking skills

These contributions typically overlap and are not meant to be mutually exclusive, but they provide a useful rubric for a review of the field. Using this rubric, the following sections review and critique pedagogical and related scholarship related to genetics, focusing on specific ways HPS can contribute to teaching

---

[7] See, for example, Ayala and Arp (2010), Beuerton and colleagues (2000), Burian (2005), and van Regenmortel and Hull (2002).

and learning genetics. How has this scholarship addressed genetics instruction? What HPS-related avenues have been fruitful and in what ways? What HPS assumptions or other issues have gone unexamined or been inadequately addressed? What guidance does this analysis provide for future directions in HPS scholarship related to genetics?

15.3.1 HPS Contributions to Promoting Student Comprehension of Genetics and to Understanding Learning Difficulties in Genetics

In this subsection we combine the first two types of contribution identified by Matthews, given that learning difficulties and the learning of subject matter are often addressed in the same studies. The largest part of genetics education scholarship has clearly focused on enhancing student understanding of subject matter. This work often identifies an assortment of alternative conceptions across ages and cultures as well.

As mentioned above, genetics has long been recognized as one of the most difficult of the biological subdisciplines for teachers to teach and for students to learn. Therefore, we begin by identifying student learning difficulties. Knippels (2002) has reviewed the literature and identified five domain-specific difficulties involved for genetic educators to address:

1. Domain-specific vocabulary and terminology
2. The mathematical content of genetic tasks
3. The cytological processes of cell division, mainly relating to chromosome structure and the associated processes
4. The abstract nature of genetics, due in large part to the order of topics presented in the biology curriculum, which generally separate meiosis from genetics
5. The complex nature of genetics: a macro–micro problem related to how to understand concepts and processes from different systematic levels and their relationships

Alternatively, Tibell and Rundgren (2010) highlight content, reasoning difficulties, and communication issues in "molecular life science" (including genetics). They particularly highlight the issue of domain-specific language and use of visualizations. Other studies of genetic learning have demonstrated the tendency of students to:

- View genetics as a set of rules and patterns of inheritance to memorize more than focusing on understanding genetic concepts and processes in a meaningful way (Lewis and Kattmann 2004)
- Use oversimplified causal explanations instead of biochemical terms or processes (Lewis et al. 2000a, b; Lewis and Kattmann 2004; Marbach-Ad 2001)
- Have difficulty relating structures and concepts to the correct biological organization level and making extrapolations between levels (Duncan and Reiser 2007;

Halldén 1990; Johnstone and Mahmoud 1980; Knippels 2002; Lewis et al. 2000b; Marbach-Ad and Stavy 2000)

- Employ explanations at the phenomenological (i.e., macro level) and/or cellular organizational level, not at the molecular level (Marbach-Ad and Stavy 2000)
- Fail to consider the environmental influences on characteristics (Forissier and Clément 2003)
- Have difficulty relating genetic concepts to each other (Gericke and Wahlberg 2013; Lewis et al. 2000a; Marbach-Ad 2001)

The need for students to be able to integrate concepts and biochemical processes derived from molecular genetics with those from classical genetics has been recognized by many science educators[8]. Hence our conclusion is that much of the literature about students' conceptual understanding of genetics mirrors the dichotomy of classical genetics and molecular genetics (and more recently genomics). It seems that students embrace deterministic classical explanations and tend to reduce classical genetics into molecular genetics, as might be expected from a HPS perspective. Students do indeed have difficulties distinguishing between classical and molecular genetics in genetic texts (Gericke et al. 2013), and they often introduce concepts from classical genetics when reasoning about molecular genetics (Gericke and Wahlberg 2013). The dichotomy between classical and molecular genetics seems to be an epistemological obstacle for learning genetics.

Allchin (2000) adds that the dominance concept is problematic because this concept stems from classical genetics and has no direct correlation in molecular or cellular terms biology. Mendel referred only to traits as dominant or recessive, but now the term is commonly applied to traits, genes, alleles, and even single nucleotide polymorphisms (SNPs). As a result, Allchin (2000) claims that two misconceptions typically emerge in students' minds. First, dominance may be conceived as a form of gene regulation, but there is no general mechanism for dominance in molecular terms. Second, others conceive dominance and recessiveness as the presence or absence of a trait, protein, or gene product, i.e., one sees the phenotype as switch on or off (Lewin 2000). This is correct for some cases but misleading as a general model according to Allchin (2000). It is also important to note that students could confuse the technical meaning of "dominance" with the vernacular meaning, such that "dominance" is conceived as a physical phenomenon involving struggle and power imbalance (Allchin 2000).

Several of the most challenging genetic concepts identified point to ontological difficulties. Among these are:

- Distinguishing between alleles and genes (Lewis et al. 2000a; Pashley 1994; Wood-Robinson 1994)

---

[8] See, for example, Duncan and Reiser (2007), Lewis and Kattmann (2004), Lewis and colleagues (2000a), Marbach-Ad (2001), Martinez-Gracia and colleagues (2006), Smith and Williams (2007), Venville and Treagust (1998), and Venville and colleagues (2005).

- Distinguishing between genes and genetic information (Lewis and Wood-Robinson 2000) or traits—leading to difficulties in understanding gene expression (Lewis and Kattmann 2004; Venville et al. 2005)
- Distinguishing between genotype and phenotype (Lewis and Kattmann 2004; Marbach-Ad 2001; Marbach-Ad and Stavy 2000; Venville et al. 2005)

In addition, conceptual change theory (Posner et al. 1982) recognizes that students typically come to the classroom with various understandings and misunderstandings (typically identified as "misconceptions," "alternative conceptions," "commonsense understandings," "naïve conceptions," etc.) that are likely to act as barriers to the development of a more sophisticated understanding of genetics. Studies to identify genetic misconceptions (and to distinguish them from appropriate conceptions) have included a wide range of subjects from elementary school students to teachers, university students, and expert geneticists[9]. For example, Dikmenli and colleagues (2011) found that science student teachers in Turkey had a global understanding of the gene in line with classical genetics (see above) and lacked a modern view of genetics, as did a sample of Moroccan university students (Boujemaa et al. 2010). Both the learning difficulties addressed above and the misconceptions held by students and teachers appear to be similar, although the frequency of learning difficulties and specific misconceptions typically decreases with expertise. Some naïve conceptions are clearly related to development (Venville et al. 2005). Specific misconceptions can often be related to a set of underlying and partially overlapping mental models of the gene (as summarized in Gericke 2008). Genes can be seen as:

- Inherited particles transferred from one generation to the next (Duncan and Reiser 2007; Lewis and Kattmann 2004; Smith and Williams 2007; Venville and Treagust 1998)
- The sole determinants of characteristics (Lewis and Kattmann 2004; Marbach-Ad 2001)
- Objects with inherent actions, i.e., the gene is thought of as a physical object that takes action in an unalterable way in the organism (Martins and Ogborn 1997)
- Sets of commands that control characteristics (Martins and Ogborn 1997; Venville and Treagust 1998)
- Active particles that also control characteristics (Duncan and Reiser 2007; Venville and Treagust 1998)
- Biochemical sequences of instructions connecting genes and protein synthesis, and protein synthesis and phenotype (Venville and Treagust 1998)

Genetic misconceptions are very common. In a recent analysis of 500 essays submitted by high school students in a contest sponsored by several professional societies, 56 % were judged to have some "major" misconception (Shaw et al.

---

[9] See, for example, Abrams and colleagues (2001), Donovan and Venville (2012), Lewis and Kattmann (2004), Marbach-Ad (2001), Shaw and colleagues (2008), Venville and Donovan (2005), Williams and Smith (2010), and Wood-Robinson (1994).

2008). The most frequently reported view seems to be that genes are considered to be physical particles and/or the sole determinants of phenotypic traits (as in "the gene FOR sickle cell anemia"), ignoring epigenetics—the fact that other components of the genome, cell, and environment impact phenotype as well (Lewis 2012). Deterministic views of genes are common—even among teachers (Castéra and Clément 2012) and can be problematic in a variety of ways, including especially in making decisions about personalized genetic testing and the interpretation of such tests (Bartol 2012). Likewise, making a link between genes and protein synthesis is rare among many students. Hence naïve understandings of genetics can be characterized as typical of the historical classical view (Gericke and Hagberg 2007).

The central question in the present context is: To what extent is the design of genetics instruction informed by HPS concepts? Surprisingly, the answer is that HPS concepts have not been used very widely to inform genetics pedagogy directly. The most frequent use of HPS concepts has been to argue for the importance of including genetic history and historical models in instruction. For example, Kinnear (1991) argued over 20 years ago for the use of historical genetics models as an important tool in teaching genetics:

> A valuable experience for students is to explore the development of a concept or model over time, and to note its maturation from initial observation, through descriptive statements, and finally to an explanatory model with predictive power that is generally accepted by the relevant community of scholars. (Kinnear 1991, p. 71)

The experience of tracing the development of an explanatory model could clarify students' own understanding of the concepts involved, particularly when several rival models exist. In addition, historical perspectives can sensitize students to the development of historical models, the constraints imposed on a model by its underlying assumptions, and the effects of scientific methodology. A historical approach can challenge the view that the "right" model exists and is waiting to be "discovered" like an archaeological artifact. A historical approach can also help students recognize that explanatory models are constructs developed over time for specific purposes and that they can be flawed or inadequate in a variety of ways (Kinnear 1991).

Gericke and Hagberg (2007, 2010a, b) also argued for the use of historical models as a tool for improving genetics teaching. Gericke and Hagberg described and categorized five historical models of the gene and its function. Conceptual consistency problems between the historical models were identified and compared to areas of genetics in which identified learning difficulties are reported. Extensive parallelism was observed, suggesting that learning might be enhanced by learning about the history involved in the development of the genetic models and reasons to why each was supplanted by subsequent models. A similar teaching strategy has also been suggested by Othman (2008). Smith and Adkinson (2010) have suggested a revision of the historical models of Gericke and Hagberg into an integrated model that takes into account concepts from genomics resulting from the ENCODE project, such as single nucleotide polymorphisms (SNPS) and transcriptionally active RNAs (TARs). In a recent paper by Meyer at al. (2011), they review different definitions of the gene (see Sect. 15.2.4) and make a suggestion to which school level different views of genes could be introduced (Meyer et al. 2011; see Table 1, pp. 25–26).

Several researchers employ instructional designs that encompass both historical and nature-of-science (NOS) approaches. These include Clough, Stewart, and colleagues who will be addressed in the following section of this chapter.

Perhaps the most explicit example of the use of HPS concepts in the design of genetics pedagogy is the work of Venville and colleagues (2005) who used both "ontological and epistemological lenses" so as to identify barriers to learning genetics within interview data from 6- to 10-year-old Australian children. Ontology was referred to understanding which entities belong with others in biological categories (e.g., living vs. nonliving things) and understanding the distinction between different entities (e.g., genes should not be seen as the same as the traits they determine, a situation that would support deterministic views of the gene—i.e., the allele for black hair color would itself be colored black). Epistemology referred to the structure of the students' knowledge, which was reported as piecemeal and disconnected. Dougherty (2009) also argues that the predominant historically based mode of genetics instruction "primes" students to hold deterministic views.

Somewhat less directly, Duncan and colleagues (2009) used the framework of learning progression to develop a sequence for teaching modern genetics from grades 5–10, based in particular on the work of Stewart and colleagues (see next section) and Venville and colleagues (see throughout this chapter) but to a lesser extent on HPS. Duncan and colleagues identified three key aspects necessary in effective genetics instruction: (1) the big ideas in modern genetics and the knowledge and abilities that students should master by the end of compulsory education, (2) the progression of learning that students are expected to make over several grades, and (3) the identification of learning performances and development of assessments for the proposed progression (Duncan et al. 2009). This approach has recently been shown to be effective in field testing (Duncan and Tseng 2011; Freidenreich et al. 2011). Another learning progression has been suggested by Roseman and colleagues (2006) based on strand maps that are based on the logic of the discipline and existing learning research (Project 2061 Atlas of Science Literacy; AAAS 2001).

Dougherty and colleagues (2011) shift our concern to the need to improve national and state standards with regards to genetics. In a study of standards in all 50 US states, Dougherty and colleagues found that 85 % of the standards were inadequate. The standards in virtually every state failed to keep pace with changes in the discipline as it has become genomic in scope, omitting concepts related to genetic complexity, the importance of environment to phenotypic variation, differential gene expression, and the difference between inherited and somatic diseases (Dougherty et al. 2011).

Other research groups have also proposed sets of recommendations for effective genetics instruction. Venville and Treagust (2002) recommend the following:

1. Use of appropriate, extended analogies and models
2. Move beyond Mendel

3. Link between concepts
4. Emphasize levels of representation (p. 20)

Many authors have also addressed the "macro–micro" problem of understanding genetic concepts that require understanding phenomena at multiple levels of organization (e.g., Duncan 2007; Johnstone and Mahmoud 1980; Knippels 2002; Schönborn and Bögeholz 2009; Van Mil et al. 2013). To address this learning difficulty, Knippels (2002) has developed "yo-yo learning," a specific teaching design that explicitly asks students to move up and down different organizational levels of biology, from molecular to cellular to the individual level in order to explain genetic phenomena. The students in her study improved their ability to interrelate different organizational levels and properly relate genetic concepts to the different levels (Knippels 2002). Knippels also recommends that genetics instruction should focus primarily on linkage of concepts and levels of representation (Knippels et al. 2005):

1. Linking the levels of organism, cell, and molecule
2. Explicitly connecting meiosis and inheritance
3. Distinguishing the somatic germ cell line in the context of the life cycle
4. An active exploration of the relations between the levels of organization (p. 108)

The work from Knippels and colleagues (2005) has been extended by the Dutch group at Utrecht University. Verhoeff and colleagues (2009) suggest system thinking (linked to systems biology), as a possible way to integrate genomics into biology curricula. System thinking is a holistic approach that employs an iterative process of data gathering and data modeling (Verhoeff et al. 2009). Voerhoeff and colleagues (2008) defined four elements of system thinking in biology education: (I) being able to distinguish between different levels of organization, (II) being able to interrelate concepts at a specific level of organization, (III) being able to link biology concepts from different levels of organization, and (IV) being able to think back and forth between abstract visualizations (models) to real biological phenomena. This approach also suggests a learning progression.

Three other models for designing genetics curricula have been proposed in the literature (Dougherty 2009; Elrod and Somerville 2007; Hott et al. 2002). Dougherty calls for "inverting the curriculum," beginning with presentation of common qualitative traits instead of simple Mendelian ("monogenic") traits to address the common student misperceptions that most human traits follow the latter, not the former pattern, and that environment has little if any effect on final phenotypes. Dougherty argues that this approach better prepares students for becoming wise medical consumers "in a world where personalized medicine will rely increasingly on genetic testing, risk assessment, predispositions, and ranges of treatment options" (Dougherty 2009, p. 8). Elrod and Sommerville's curriculum (for upper-level biology majors) employs student identification of original literature to address student-generated genetics research questions, focusing on developing student competence in information gathering, interpretation, and integration as well as genetics and the NOS. A paper from Hott and colleagues (2002) report of the "Information and Education

Committee of the American Society of Human Genetics" that presents an unordered list of genetics topics and subtopics for medical school curricula. None of these approaches is explicitly informed by HPS concepts.

The use of HPS as an analytical tool for examining textbooks has been a fruitful area of research. Hurd (1978) investigated the historical and philosophical treatment of genetics in 128 US school and college textbooks published between 1907 and 1977. Hurd concludes that the textbooks did not provide a basic understanding of human genetics. Blank (1988) showed how uncertain genetic mechanisms are often presented as dogmatic facts in textbooks. The conceptual content of textbooks has been analyzed in several countries. A genetic deterministic approach to genetics, ignoring environmental interactions, was found in French and Tunisian secondary level biology textbooks (Abrougui and Clément 1997; Forissier and Clément 2003; Castéra et al. 2008a, b). Moreover, the tendency of using an implicit genetic deterministic ideology in textbooks was found in a study of 16 countries, although the degree varied between different countries (Castéra et al. 2008b). Martinez-Gracia and colleagues (2006) found that Spanish high school biology textbooks describe many procedural details of molecular genetics, but these do not facilitate understanding of the main ideas and concepts. A similar lack of integration was also found in an evaluation of US high school biology textbooks (AAAS 2008). Information about the molecular basis of heredity in typical textbooks was presented in a piecemeal fashion. DNA and other biochemical molecules were described in great detail, as were various biochemical processes of gene function. Changes in genes and their consequences, however, were described in later chapters. The authors of both the Spanish and US studies advocate the incorporation of Mendelian concepts into molecular genetics. The use of different historical genetic concepts within textbooks has been reported in Brazil (El Hani et al. 2007; Santos et al. 2012) as well as from Sweden and several English-speaking countries (Flodin 2009; Gericke and Hagberg 2010a, b) indicating a frequent use of hybrid models, i.e., incorporating aspects of different historical views. A more modern idea about the gene seems to be absent in most textbooks (Gericke and Hagberg 2010a, b; dos Santos et al. 2012). In a recent comparative study of textbooks from six counties, Gericke and colleagues (2012) identified a common gene discourse in which ontological aspects of the academic disciplines of genetics and molecular biology were found but without their epistemological underpinnings. Different models and concepts from both classical and molecular biology were used interchangeably in a nonhistorical fashion. Also in the reviewed texts the most frequent explanatory models and concepts of the gene were those that promote a deterministic notion of the gene. An in-depth survey and analysis of textbook research in genetics is provided in the next chapter of this book by El-Hani. An interesting analysis of the use of different historical genetic concepts in school would be to use the Didactic Transposition Delay framework introduced by Quessada and Clément (2006). The framework refers to the time elapsed between appearance of a scientific concept in professional literature until it appears in school syllabi.

Alternative conceptions have also been an informative lens for designing conceptual change style genetics teaching. For example, in a study of instruction about gene technology (Franke and Bogner 2011a, b), one group of 10th graders was confronted with alternative conceptions to central issues of the topic while the control group was not. Compared to controls, experimental students abandoned more of alternative conceptions in favor of a scientific view. These students also showed either the same or greater cognitive achievements compared to controls, although previous study has shown that radical conceptual change in ontological conceptions of the gene remains a difficult challenge (Tsui and Treagust 2004b).

Venville and Donovan (2005) surveyed genetic experts about their views of the gene concept and the key concepts that students should learn in genetics. These experts commonly pointed out four themes of importance that should be addressed in teaching: (1) "Genes are regions of DNA that are a code for making polypeptides," (2) "genetic determinism is a myth," (3) "the importance of the impact of the environment on the phenotype," and (4) "genetic control and gene expression" (Venville and Donovan 2005, p. 22). Burian (2013) make three principal claims that should guide genetics instruction: "(1) Questions about genes often yield different answers about what a gene is, or is like, or how it acts" (i.e., the polysemous gene concept as discussed in Sect. 15.2). (2) The resolution of such conflicts often requires new technologies (see the historical overview in Sect. 15.2). (3) The dispute of what genes are "reinforce[s] the centrality of the tension between accounts of gene structure and gene function" (Burian 2013, p. 341). Hence, the issues identified by both Burian (2013) and Venville and Donovan (2005) as central for genetics instruction are some of the main themes of the history and philosophy of genetics described in the first section of this chapter. Burian also advocates for a teaching approach that connects conceptual understanding with the understanding of the nature of science, pointing out the importance of including the discovery process of science in genetics instruction (see Sect. 15.3.2 for further discussion).

Another tactic for teaching about the gene is suggested by parallels with Smith's approach to defining the term science (Smith and Scharmann 2006), based on the work of Wittgenstein (1953/2001) and Kuhn (1974). Analyzing the concept of "games" as an example, Wittgenstein pointed out the difficulty of explicitly defining polysemous concepts. (Think for example about a definition that would include solitaire, chess, and the children's game of "duck, duck, goose.") Kuhn argued that, in practice, people come to understand the term "games," i.e., by ostension, by experience with examples, and by counterexamples of the term, not from a list of necessary and sufficient conditions. Wittgenstein does not argue that it is impossible to define "games" but that an explicit definition is not needed because we can use the word successfully without it. The focus of instruction, therefore, should be on learning to use the term more than to define it. It may thus be most effective to begin instruction with only a broad working definition of genes or perhaps "case studies" of specific genetic disorders as suggested by Duncan (2007) and Mysliwiec (2003), followed by consideration of a number of prototypical genes (and different uses of the term), each of which focuses on different meanings, functions, exceptions, etc.

These are important issues that need to be addressed immediately. These recommendations need to be tested experimentally. Furthermore, it seems likely that using an HPS lens would be a fruitful approach to designing pedagogical experiments to answer such questions.

15.3.2 HPS Contributions to Enhancing Understanding of the Nature of Science (NOS)

Separating HPS contributions to genetics understanding and NOS understanding is difficult—perhaps impossible—because the majority of science education scholarship focusing on teaching the NOS (i.e., science as a way of knowing; the epistemology of science [Lederman 1992]) has been within a disciplinary context, primarily in evolution instruction. Some genetics education researchers have advocated focusing genetics instruction on enhancing NOS understanding as well (e.g., Allchin 2003; Gericke and Hagberg 2007; Kampourakis 2013). As in the preceding section, instruction about the history of genetics plays a central role in much of this work. Kampourakis claims that: "in order to provide a more accurate depiction of science, it is necessary that historical details are taken into account and that teachers, science educators, and textbook writers provide a more actual historical description and a more accurate depiction of the nature of science" (Kampourakis 2013, p. 320).

Clough and colleagues (Clough and Olson 2004; Metz et al. 2007) have produced a number of excellent historical narratives of the history of genetics that aim to increase students understanding of NOS. Clough and colleagues propose using these stories in an instructional design that interrupts the story at certain points, brings the student "alongside the scientists," and requires students to manipulate ideas and try to solve the problems that concerned the scientist, drawing inferences, making predictions, etc. One of these activities involves the story of Mendel and another focuses on the story of Watson, Chargaff, the nature of the genetic material, the pairing rules, etc. (see Sect. 15.2 for the historical background). These activities are focused on student understanding of the NOS. Although stories about scientists and their work appear to be inherently motivating, Allchin (2003) argues for a cautious use of historical narratives in science instruction because they are commonly used to give a misleading understanding of the NOS, resulting in "scientific myth conceptions." Allchin notes that historical reconstruction often describes scientific discovery as a heroic event following certain narrative patterns. Typically, these patterns are historically inaccurate and follow the architecture of myth, which misleads students about the NOS. Allchin notes that the story of Mendel as presented in genetics instruction typically includes several common ingredients of myth: monumentality, idealization, affective drama, and explanatory and justificatory narrative. Allchin argues for a different type of history that conveys the NOS more effectively.

Lin and colleagues (2010) designed a genetics unit for grade 7 Taiwanese students using a "historical episodes map (HEM)" comprised of 20 historical episodes and four storylines in the development of genetics from early times to the rediscovery of Mendel's work. Compared to control students who received instruction based on the textbook alone, students in the experimental group evidenced greater understanding of the NOS (ES=0.24) and more positive attitudes toward science (ES=0.14). Understanding of genetics was not measured. Similar methodology and gains in

NOS understanding were also reported by Kim and Irving (2010) for US high school biology students.

For college non-biology majors and courses in the history of genetics (but not for high school biology), Burian (2013) argues for a teaching approach that uses the processes of discovery, correction, and validation by utilizing illustrative episodes from the history of genetics. This approach concentrates on understanding "the processes of investigation and the fundamental issues that are posed by genetic sciences" (Burian 2013, p. 326) as a means to achieve three of Matthew's goals—to increase conceptual understanding, enhance NOS understanding, and humanize the discipline. These are promising results, and these instructional methods deserve further study.

In the high school biology curriculum designed by Cartier and Stewart (2000), students work in groups structured like scientific communities to build, revise, and defend explanatory models of inheritance phenomena, with the aim of improving both genetics and NOS understanding (see also Cartier et al. 2006). Open problem-solving has also been reported to promote a change in the students' view of the NOS, which was not found in the control group in which problem-solving was not used (Ibáñez-Orcajo and Martínez-Aznar 2007). Yarden and colleagues have also designed a promising web-based genetics program that employs students using an authentic sequencing tool to identify a gene within a hypothetical narrative story context (Gelbart and Yarden 2006, 2011; Gelbart et al. 2009; Stolarsky et al. 2009). Qualitative analysis of this instruction suggests that the instruction "promotes construction of new knowledge structures and influences students' acquisition of a deeper and multidimensional understanding of the genetics domain" (Gelbart and Yarden 2006; p. 107). No quantitative analyses were reported and are clearly called for. Lederman and colleagues (2012) have proposed a set of modern genetic applications (genetically modified foods, genetic testing, and stem cell research) that raise many socio-scientific issues (to be addressed below) and have given suggestions about how these might be used to also promote NOS understanding.

The historical reconstruction of genetics research also raises the issue of the temporal relationship between evolution and genetics and the order in which the two should be taught. Some educators have made the case that instruction is likely to be more effective if genetics is presented first as a basis on which to build an understanding of evolution, thus avoiding the difficulties encountered by Darwin (see Sect. 15.2.1). This approach follows the basic assumption of most curriculum design that simpler concepts should be presented before they are built together into more complex concepts, i.e., that understanding genetics first helps students subsequently understand how evolution operates, employing genetics concepts. Bizzo and El-Hani (2009) argue that the claim that understanding genetics is necessary to understanding evolution is "wrong from an historical and an epistemological perspective" (Bizzo and El-Hani 2009, p. 113), although they do not claim that the opposite sequence is more "efficient" (effective?). Their primary argument is that teaching evolution first, noting that Darwin held "a 'right' model of evolution while having a 'wrong' model for heredity" (p. 113), provides students with the more appropriate "image of science [recognizing that scientists have] all sorts of ideas, including some that proved to be wrong" (p. 113). The question of the proper instructional sequence of evolution and genetics is clearly central to biology education and more studies are called for. Evolution has been placed at the end of the curriculum for many years, but student understanding of both topics has been less than optimal, although student gains are likely affected by a host of other factors, not the least of which is that evolution in many countries is often not taught either because it is controversial or because the semester ends before teachers get to the last chapter in the text. The "best" sequence may even depend on the instructional goals—genetics understanding, evolution understanding, NOS understanding, or some combination of the three. Surprising as it may seem, the question remains open as a major gap in genetics pedagogy; direct experimental comparison studies are clearly called for.

This review demonstrates that the history of genetics is the most explored aspect of NOS in the literature. Moreover, there seems to be a strong rationale for the use of history in the design of instruction, and research to date that employs variations of this approach is promising. The philosophical and conceptual aspects of genetics, however, have received little or no attention in such design and are likely to be fruitful avenues for future research. Questions that remain to be addressed include the following: What pedagogical models are the most effective for improving genetics understanding? For addressing misconceptions? For enhancing the understanding of NOS? What aspects of the NOS are most suitable to address in genetics instruction? How could NOS targets be addressed in genetic education and then reinforced in subsequent evolution instruction, or vice versa?

15.3.3   HPS Contributions to Humanizing Science to Personal, Ethical, Cultural, and Political Concerns

Perhaps the best example of scholarship that aims to humanize science through genetics instruction is the use of history in the classroom as described in previous sections. Genetics instruction typically follows the historical development of genetics research (Dougherty (2009). Genetics instructors have traditionally used narratives——such as the story of Mendel and his pea experiments, sometimes adding humanizing details (see Sect. 15.3.2).[10] Davis (1993), for example, explains how to include the origins of Punnet square, as well as the studies of Bateson and Punnett that first identified linked genes. Fox (1996) describes a classroom exercise that uses a letter from Max Delbruck to George Beadle to stimulate interest in molecular biology. Simon (2002) includes developments from human gene therapy. Ohly (2002) used the story of Chargaff and the development of the DNA base-pairing rules to show how laboratory routines and their development interact with the underlying theoretical framework and the way of thinking ("denkstil") of a collective of

---

[10] See, for example, Allchin (2003), Clough (2009), Clough and Olson (2004), and Metz and colleagues (2007).

researchers. Crouse (2007) describes a method for teaching upper-level undergraduate and graduate students the analysis of X-ray diffraction of DNA through a set of historical steps using the original methods employed by Watson, Crick, Wilkins, Franklin, and Gosling that led to the proposal of the helical structure of DNA.

Both Yarden and colleagues (2001) and Goodney and Long (2003) have developed primary research literature-based developmental biology curricula for high school biology majors. Goodney and Long argue that the success of a scientific revolution, such as the advent of molecular genetics, is due not only to the strength of the ideas but also to the persuasive power of language. They also argue that primary research literatures that start a scientific revolution are understandable for a broad range of readers because their purpose is to speculate, imagine, theorize, and persuade rather than merely inform as in normal science.

Chamany and colleagues (2008) suggest that the social context is important when teaching biology to model social responsibility for biology students as part of biology literacy for non-major students. The authors also give practical examples of the use of this approach in genetic topics such as sickle cell anemia and gene regulation. Venville and Milne (1999) draw on the history of genetics and the lives and scientific accomplishments of female geneticists Nettie Stevens, Rosalind Franklin, and Barbara McClintock to illustrate surprisingly contrasting accounts of events and to focus on the people involved in order to increase the motivation of students (especially females) in learning genetics. Wieder (2006) describes the use of a student-designed play followed by work by proposing a working model of DNA structure for humanizing high school biology and highlighting the people and processes of science (Wieder 2006).

An education tool widely used to humanize science (and more specifically, genetics) and make it more relevant to students is the use of socio-scientific issues (SSI), which is designed to engage students in culturally and socially relevant decision-making, citizenship, argumentation, and ethical reasoning (Blake 1994; Sadler 2011). Available examples in genetics include activities that focus on human cloning and genetic screening (Simonneaux 2002), GMO (Dawson and Venville 2010; Ekborg 2008; Simonneaux 2008), genetic testing (Lindahl 2009; Boerwinkel et al. 2011), and biotechnology (Dawson and Venville 2009; Lewis and Leach 2006; Sadler and Zeidler 2004, 2005). Using web-based approaches for learning genetics in socio-scientific settings has been developed in Norway. Viten[11] is a web-based platform that contains digital teaching programs in science for secondary schools and provides teaching materials relating to gene technology (Furberg and Arnseth 2009; Jorde et al. 2003). Because SSI is closely related to argumentation, more about issues related to argumentation in these studies are outlined in next Sect. 15.3.4.

In 1990 the NIH's National Human Genome Research Institute (NHGRI) committed 5 % of its annual research budget to study Ethical, Legal and Social Implications (ELSI) of the Human Genome Project, and a number of fact sheets, teaching resources, learning tools, and funding opportunities are available at the

---

[11] www.viten.no

HGP website[12]. This is a resource that has largely been untapped by the science education research community to date. One example of how ELSI issues might be addressed is an interdisciplinary course (taught by a biologist, a linguist, and an educator) which aimed to increase booth genetics content knowledge and awareness of equity and fairness (Gleason et al. 2010). The American Society of Human Genetics (ASHG) website[13] is also an excellent source of genetics education resources, including instructional modules, books, videos, and websites recommended by their Genetics Education Outreach Network (GEON).

Race and eugenics are another part of the history of genetics that focus on social factors involved in genetics. Eugenics was widely accepted by the scientific and political elite in many countries in the early twentieth century and has often been addressed in the educational literature. Mehta (2000) provides a historical overview of eugenics from ancient Greece to genetic engineering. Rodwell (1997) profiles the history of Caleb Williams Saleeby, a late nineteenth-century propagandist of eugenics, and Greenwald (2009) examines Alexander Graham Bell's role and influence in the American eugenics movement. These materials have promise for classroom use, but Cowan (2008) argues that the common view of the connection between medical genetics and eugenics is historically fallacious. She claims that from the very beginning, the goal of the founders of medical genetics (e.g., Neel, Fuchs, Kaback, Guthrie et al.) was the relief of human suffering not improvement of the race. Anderson (2008) describes efforts in medical education to teach that race no longer is considered a biologically legitimate concept and to demonstrate that race remains an influential social classification, causing social and biological harm. Thus, addressing eugenics in the classroom appears to be fraught with both opportunities and pitfalls.

Humanizing genetics may be a particularly useful approach for reaching students from select genetic subgroups. For example, Gates, the originator of the popular PBS documentary series "African American Lives" (WNET) (see von Zastrow 2009), proposed an interesting "ancestry-based curriculum" based on genealogy and DNA research for African American students. Gates argues that students who "examine their own DNA and family histories [will be more] likely to become more engaged in history and science classes" (von Zastrow 2009, p. 17).

Use of examples of genetic phenomena that are particularly exotic and/or relevant to the students' own families is yet another approach to humanizing genetics. One interesting example is the inheritance of a certain genetic diseases in certain Nigerian families. In this culture inheritance of a particular genetic disorder is commonly explained as the result of curses and extramarital affairs (Mbajiorgu et al. 2007). Similarly, Santos and Bizzo (2005) interviewed 100 adults from within two large Brazilian families in which many members are affected with one of two rare genetic disorders. The prevailing community explanation in this case was that the disorders were an inherited illness in the blood related to contamination by syphilis. The genetic basis of these disorders has immediate relevance to students in these

---

[12] www.genome.gov

[13] www.ashg.org/education/resources.shtml

cultures; this work reemphasizes both the importance of a thorough understanding of student preconceptions about the content of instruction and of the great impact of culture and worldviews on learning. Teaching Western science explanations of inheritance in these cultures would certainly humanize the content but, given the prevailing culture, would be challenging indeed. Such work is also typically aligned with pedagogy based on conceptual change theory, the instructional approach taken by Santos and Bizzo. The value of using these examples in Western classrooms would be interesting to investigate as well.

15.3.4 HPS Contributions to Enhancing Reasoning, Argumentation, and Thinking Skills

Genetics instruction provides a fruitful venue for developing student reasoning and thinking skills. As alluded to above, genetics teaching and learning are perennially recognized as challenging, and the cognitive demands of the content are great. The reasoning and thinking skills required for solving classical genetics problems have been widely recognized as a central reason for the difficulty many students experience in this field (e.g., Smith 1983; Mitchell and Lawson 1988; Cavallo 1996). We have also alluded in Sect. 15.3.1 to the high cognitive demands required by working across as many as four levels of organization from the molecular to the ecological (e.g., Knippels 2002; Duncan and Reiser 2007).

Solving sets of typical closed-ended Mendelian genetics problems has long been a central component of introductory genetics teaching and learning. Various aspects of the problem-solving skills required, how the skills employed vary with levels of expertise, and how problem-solving contributes to genetics learning have been investigated in a range of older studies, beginning with the work of the second author of this chapter[14]. More recently Ibáñez-Orcajo and Martínez-Aznar (2005) found that, compared to control subjects, students who solved open-ended genetics problems showed significantly more frequent use of more advanced genetic models— differences that persisted over time (at 5-month posttest). Ibáñez-Orcajo and Martínez-Aznar interpreted these gains as the result of "metacognitive reflection by students that become[s] apparent in conceptual restructuring" (Ibáñez-Orcajo and Martínez-Aznar 2005, p. 1508). The use of open-ended problems for promoting problem-solving and critical thinking skills such as metacognition is a promising avenue for future research.

Both Mitchell and Lawson (1988) and, more recently, Cavallo (1996) have demonstrated the necessity of general developmental reasoning skills for solving genetics problems. Duncan (2007) demonstrated the importance of additional, mid-level, domain-specific heuristics and explanatory schemas for solving problems in

---

[14] See, for example, Cavallo (1996), Finkel (1996), Hafner and Culp (1996), Mitchell and Lawson (1988), Smith (1983), Smith and Good (1984), Stewart (1983, 1988), Stewart and van Kirk 1990, and Wynne and colleagues (2001).

molecular genetics in a university genetics course for biology majors. These findings support Duncan's argument that genetics instruction should be focused on "learning of causal mechanisms rather than disconnected details of structures and processes" (Duncan 2007, p. 321). Venville and Donovan (2007) have proposed such a learning program for second graders (ages 6 and 7) in which students made qualitative gains in understanding of causation in heredity.

Understanding and using genetic models can also be an avenue to enhancing thinking and reasoning (Gericke and Hagberg 2007, 2010a, b). Physical manipulative models (e.g., beads, cutouts) have long been a component of teaching genetic phenomena from nuclear division (Lock 1997; Rotbain et al. 2006; Smith and Kindfield 1999) to replication, transcription, and translation and have been shown to enhance learning outcomes (Venville and Donovan 2008). Rotbain and colleagues (2005) also demonstrated the positive learning benefits of an activity in which students drew their own genetic representations, although these gains were less than those of students who participated in computer animation instruction (Marbach-Ad et al. 2008).

Over the past decades, Stewart and his colleagues have developed and tested a very successful genetics course that employs model-based inquiry. The program employs the Genetics Construction Kit software as a context in which high school students build and evaluate their own scientific models to explain genetic phenomena[15]. Stewart and colleagues (2005) suggest that genetics students need to understand and reason on the basis of three models: inheritance pattern models (explaining patterns of inheritance across generations), the meiotic model (explaining chromosome and gene behavior during the generation of gametes), and the biomolecular model (explaining the role of DNA and proteins in bringing about an observable phenotype).

The use of models and modeling as a tool to facilitate learning and reasoning has been employed in several other computer-based and web-based learning environments as well. These include *CATLAB*[16] (Simmons and Lunetta 1993), *GenScope* and its successor *Biologica*[17] involving pea plants and dragons (Buckley et al. 2004; Hickey et al. 2000, 2003; Tsui and Treagust 2003a, b, 2010), *The Virtual Flylab*[18] and *Genetics Construction Kit* (GCK)[19] (Soderberg and Jungck 1994) both involving *Drosophila*, and the "Simple Inheritance" unit of Marcia Linn's *Technology Enhanced Learning in Science* (TELS)[20]. Of these, GCK provides the greatest diversity of uniquely generated problems and is the most widely used (Echevarria

---

[15] For more background see the following references: Finkel and Stewart (1994), Hafner and Stewart (1995), Passmore and Stewart (2002), Stewart and colleagues (1992), and Thomson and Stewart (2003).

[16] www.emescience.com/bio-software-catlab.html

[17] www.concord.org/biologica

[18] www.biologylabsonline.com

[19] www.bioquest.org/indexlib.html

[20] http://telscenter.org/curricula/explore

2003; Finkel 1996; Hafner and Stewart 1995). *Avida-Ed*[21] (Holden 2006) and *EVOLVE*[22] both simulate evolution of an artificial life form (Soderberg and Price 2003). Likewise, *WorldMaker,*[23] an iconic modeling program for learning about complex natural and social phenomena, has been used to teach genetics (Law and Lee 2004). GCK has been modified into the *Virtual Genetics Lab*[24] (VGL), an open-source simulation that is freely available on the Internet. Solving the problems generated by these simulation programs is generally recognized to promote motivation, and deep conceptual understanding of genetics and advanced reasoning skills (e.g., Tsui and Treagust 2003a, 2004a). Somewhat more broadly, web-based tools such as virtual chatting have also been shown to be useful in teaching model-based reasoning in genetics (Pata and Sarapuu 2006).

The use of metaphors, analogies, and analogical reasoning, a widely used instructional strategy in science (Brown and Clement 1989), has been employed in a limited number of genetics instruction studies. In an experimental collage genetic course, for example, instruction that involved complex analogies resulted in significantly higher student achievement compared to controls (Baker and Lawson 2001), although the use of analogies did not obviate the need for higher-order reasoning skills.

Instructional metaphors and analogies, however, have both strengths and weaknesses. Venville and Donovan (2006) argue that the widely used analogy of genes as small entities in the nucleus of cells that play an important role in inheritance, development, and function is the most productive way of promoting genes in school science. By definition, however, metaphors and analogies oversimplify the target phenomenon and thus can lead to misconceptions. The shortcomings of the particulate "beads-on-a-string" model of genes, for example, have been addressed at length earlier in this chapter. Another common metaphor is that the gene is the "blueprint of life" (Tudge 1993), but this metaphor is "potentially a misleading myth because genes are not passive bystanders" in the cell (Venville and Donovan 2006, p. 21). Likewise, the blueprint metaphor suggests that the genetic plan is static and unchanging and that there is a one-way flow of information (p. 21). These authors prefer the analogy of genes as "recipes" with transcription factors as "chefs" that decide which recipe to make. "Chefs" and "recipes" need each other to function, a much less deterministic metaphor (Venville and Donovan 2006).

Teachers are also often unwisely using anthropomorphic metaphors and language, which can contribute to student misconceptions (Venville and Donovan 2006). Anthropomorphic metaphors such as "genes for long legs" or "genes for cancer" are convenient figure of speech and much in line with the old unit factor theory of Mendelian genetics (see Sect. 15.2 for historical background), but in contemporary genetics we know that there are no such genes. The consequence of teachers using such convenient figures of speech might be to promote a

---

[21] http://avida-ed.msu.edu/

[22] www.stauffercom.com/evolve4/

[23] www.*worldmaker.cite.hku.hk/worldmaker/pages/icce98-wrldmkr2.doc*

[24] http://intro.bio.umb.edu/VGL/

deterministic understanding of genetics among students. This should be an interesting area for future research. In what ways are teachers using language in their communication of genetics in the classroom?

Anthropomorphic metaphors of DNA have also been identified as frequent descriptions of DNA in popular sciences magazines in Sweden, where genes and DNA are referred to as intentional agents that "decide," "choose," and "remember" (Pramling and Säljö 2007). This seems to be universal phenomena as shown by the many studies that have identified media as depicting genes as deterministic causes of human behavior or disease (e.g., Condit et al. 1998, 2001; Carver et al. 2008; Nelkin and Lindee 1995). A question that would be interesting to pursue is if and in what ways students' understanding is influenced by the media and textbook (Castéra et al. 2008b; Gericke et al. 2012) discourse of genetic determinism. The impact of media on students' understanding has been investigated in Taiwan. Genetic concepts (gene, DNA, protein, chromosome, cell, biotechnology, and genetic engineering) were found to be among the most frequently mentioned science concepts in the media, but students had lower levels of knowledge of these biological concepts than of physics and earth science concepts, which were less frequently mentioned in the media (Rundgren et al. 2012; Tseng et al. 2010). In contrast, Donovan and Venville (2012) reported from a study of 62 children that "mass media [sic] is a persuasive teacher of children, and that fundamental concepts could be introduced earlier in schools to establish scientific concepts before misconceptions arise" (Donovan and Venville 2012, p. 1).

Argumentation skills, an instructional aim (also addressed above within Sect. 15.3.3 regarding SSI), have also been shown to be useful in enhancing reasoning and thinking, as well as conceptual understanding, in genetics. Zohar and Nemet (2002) found that integrating explicit teaching of argumentation into the teaching of dilemmas in human genetics enhanced the students' performance in both conceptual knowledge and argumentation. Similarly, facilitation of 10th-grade student use of argumentation about SSI while learning about genetics resulted in significant gains not only in complexity and quality of arguments used by students but also better genetics understanding compared to students in a comparison class (Dawson and Venville 2010; Venville and Dawson 2010) (see also Sect. 15.3.3 on SSI). This work looks at the effects of explicit instruction in argumentation skills as well as genetics on argumentation skills themselves. Sadler and Zeidler (2005) have revealed that student reasoning patterns about genetic engineering issues are influenced by their knowledge of genetics. Several studies have shown many other factors influence students' way of arguing in decision-making, including moral considerations, personal experiences, and popular culture (Dawson and Venville 2009; Sadler and Zeidler 2004). These research programs have been fruitful to date, achieving goals both of conceptual understanding and of reasoning and thinking skills.

Other innovative instructional approaches that are not explicitly informed by HPS but are worthy of mention for completeness include the use of so-called clicker questions within lectures (Knight and Smith 2010), use of the learning cycle (Dogru-Atay and Tekkaya 2008), and problem-based learning (Araz and Sungur 2007).

Using genetics as a vehicle for advancing cognitive skills, thinking, and reasoning is a promising research approach. The research reviewed above suggests a number of unanswered questions: What generalized cognitive skills gains are achievable through genetics instruction? What skills are appropriate and at what age levels given the limits of cognitive development? To what extent are the cognitive skill aims and the conceptual understanding aims mutually supportive? To what extent are these skill gains transferrable to other domains in biology (e.g., evolution) and beyond—especially outside the classroom? Are there thinking and reasoning skills that could be effectively targeted during genetics instruction (e.g., other than modeling and argumentation)? What instructional techniques are most effective at achieving each?

15.4 Closing Remarks

In this chapter we first gave a short overview of the history of genetics and its philosophical implications, emphasizing the development of the concept of the gene. The purpose of this overview is to provide a framework from which the reader can interpret the educational research reviewed in this chapter. For the interested reader there is a vast amount of scholarship about the history and philosophy of genetics available[25]. At the center of this literature is the idea of the gene and how it should be understood. The historical overview identified several important philosophical issues within genetics: (I) The gene concept has undergone a historical development in which the meaning of the concept changed. Today several different models or concepts are used to define the gene (more even than we could address in this paper). (II) The gene is a polysemous concept and it is not possible to reduce one of the multiple concepts (or models) into another. The descriptions of the gene are sometimes incoherent and context dependent. (III) One of the major reasons for incoherence is that different concepts or models are valid at different biological organizational levels. (IV) Another of the main reasons for the incoherence is that the different concepts (or models) define biological function and structure differently. (V) The simple concepts or explanatory models derived from classical genetics and early molecular genetics can promote a deterministic notion of the gene and genetics not in line to contemporary genomics.

It is important for an educator to know about the history of genetics and its philosophical implications because of the pervasiveness of historical ideas in school curricula (Gilbert et al. 2000) (see, e.g., the textbook studies in this and the following chapter). In school we seldom teach about the frontier of research. Interestingly though, in opposition to the typical historical focus of school genetics, we do most often teach about the new technological applications of biotechnology and genetic engineering. Here is an interesting dichotomy that needs further investigation. Very

---

[25] See, for example, Carlson (1966, 2004), Davis (2003), Keller (2005), Moss (2003), Portin (1993), Sapp (2003), and Schwartz (2008).

few educational studies relate to more modern concepts such as genomics and proteomics. How is the modern radical shift toward genomics in research reflected in school science?

In Sect. 15.3 of this chapter, we identify scholarship that implicitly or explicitly uses the HPS as a point of reference in designing genetics instruction. The primary focus of the section is on the contributions of HPS scholarship on the following: (I) teaching and learning genetics, (II) enhancing the teaching and learning of NOS, (III) humanizing science, and (IV) enhancing reasoning, argumentation, and thinking skills. We identified the following main HPS themes: (I) Students tend to appropriate conceptions from classical genetics and simple molecular models that often lead to deterministic understandings of genetics. Moreover students have difficulties in relating concepts and in moving between different biological organizational levels. These overall conclusions from science education research could be explained by the philosophical issues of reduction and the polysemous gene concept. Including main philosophical issues in education could help students tackling these learning problems. (II) Many studies reported positive effects of a HPS perspective in genetics teaching on learning NOS. This trend was anticipated, given that the NOS, i.e., how we build scientific knowledge, is essentially a question of epistemology. When focusing genetics instruction on NOS, there are many interesting narratives to use: competing theories and models to contrast, different theoretical approaches to elucidate, a technical development in scientific applications to compare, etc. (III) Researchers have also used historical narratives in genetics to humanize genetics instruction in a variety of fruitful ways, including social aspects such as gender, ethics, and language as well as darker topics such as eugenics. (IV) Finally some of the better studied areas within genetics education include the use of argumentation, problem-solving, and narratives. However these areas have to a lesser degree, than the previous, been illuminated by HPS. Therefore, there is a need for further studies of genetics teaching and learning that are informed by a HPS perspective.

Although the research reviewed in this chapter is a good start, it seems to us that many of the teaching and learning issues addressed in these recommendations have been widely recognized at least for decades and that much more explicit guidance is needed. In recent years two workshops have taken place that addressed the questions of how to redesign science curricula for the genomics era (Boerwinkel and Waarlo 2009; Boerwinkel and Waarlo 2011), and much of the discussions in these workshops are relected in Sect. 15.3.1 above. The efforts of these workshop organizers are praiseworthy and a new workshop is planned for 2013. The issue of HPS, however, has not been a dominant part of the agenda of those workshops, and we encourage the genetic education research community to reconsider the value of the HPS lens. Questions that should be asked include: How might careful attention to HPS issues help to expand this list? What, for example, are the most effective ways to use the history of genetics and/or genetic case studies to enhance genetics understanding? More specifically, how should the history of our understanding (and definition) of the gene be presented to students? What does philosophy (especially epistemology) have to say that informs decisions about what working definitions to present to students? How can HPS inform curriculum design so as to best prepare students to be wise medical consumers in the twenty-first-century era of personalized genomics and pharmacogenetics?

In conclusion, HPS has been applied to genetics education in a variety of ways, but in most areas there is a need for more educational research in which HPS is used as a guiding framework. Much of the relevant scholarship relates to HPS only in a largely implicit way. HPS-informed approaches appear to be fruitful avenues for improving and understanding genetics teaching and learning. We hope that researchers will continue to use these approaches and that the questions we have raised throughout this chapter will both guide further research and stimulate the generation of even more fruitful research questions.

References

Abrams, E., Southerland, S., & Cummins, C. (2001). The how's and why's of biological change: how learners neglect physical mechanisms in their search for meaning. *International Journal of Science education*, 23(12), 1271–1281.

Abrougui, M., & Clément, P. (1997). Human genetics in French and Tunisian secondary textbooks: presentation of a textbook analysis method. In H. Bayerhuber, & F. Brinkman (Eds.), *What – Why – How? Research in didaktik of biology* (pp. 103 – 114). Kiel, Germany: IPN – Materialen.

Allchin, D. (2000). Mending Mendelism. *The American Biology Teacher*, 62(9), 633–639.

Allchin, D. (2003). Scientific Myth-Conceptions. *Science Education*, 87, 329–351.

American Association for the Advancement of Science (2001). *Atlas of science literacy*. Washington, DC: Author.

American Association for the Advancement of Science. (2008). *AAAS Project 2061 high school biology textbooks evaluation*. Retrieved November 23, 2008, from: http://www.project2061. org/publications/textbook/hsbio/summary/default.htm

Anderson, W. (2008). Teaching 'race' at medical school: Social scientists on the margin. *Social Studies of Science*, 38, 785–800.

Araz, G., & Sungur, S. (2007). Effectiveness of problem-based learning on academic performance in genetics. *Biochemistry and Molecular Biology Education*, 35(6), 448–451.

Avery, O.T., MacLeod, C.M., & McCarty, M. (1944). Studies on the chemical nature of the substance inducing transformation of pneumococcal types: Induction of transformation by a desoxyribonucleic acid fraction isolated from Pneumococcus type III. *The Journal of Experimental Medicine* (79), 137–158.

Ayala, F.J. & Arp, R. (2010). *Contemporary debates in philosophy of biology*. Chichester, UK: John Wiley & Sons Ltd.

Bahar, M., Johnstone, A.H., & Hansell, M.H. (1999). Revisiting learning difficulties in biology. *Journal of Biological Education*, 33(2), 84–86.

Baker, W.P. & Lawson, A.E (2001). Complex instructional analogies and theoretical concept acquisition in college genetics. *Science Education*, 85, 665–638.

Barnes, B., & Dupré, J. (2008). *Genomes and what to make of them*. Chicago: University of Chicago Press.

Bartlett, J. M. S. & Sterling, D. (2003). "A short history of the polymerase chain reaction". *PCR Protocols*, 226, 3–6.

Bartol, J. (2012). Re-examining the gene in personalized genetics. *Science & Education*, doi 10.1007/s11191-012-9484-2

Beadle, G.W. & Ephrussi, B. (1937). Development of eye colors in Drosophila: Diffusible substances and their interrelations. *Genetics*, 22, 76–86.

Beadle, G.W. & Tatum, E.L. (1941). The genetic control of biochemical reactions in Neurospora. *Proceedings of the National Academy of Science*, 27, 499–506.

Benzer S., (1955). Fine structure of a genetic region in bacteriophage. *Proceedings of the National Academy of Sciences*, 41, 344–354.

Beurton, P., Falk, R., & Rheinberger, H. J. (2000). *The concept of the gene in development and evolution: Historical and epistemological perspectives*. Cambridge, UK: Cambridge University Press.

Bizzo, N. & El-Hani C.N. (2009). Darwin and Mendel: evolution and genetics. *Journal of Biological Education*, 43(3), 108–114.

Blake, D.D. (1994). Revolution or reversal: Genetics – Ethics curriculum. *Science & Education*, 3, 373–391.

Blank, C.E. (1988). Human heredity: Genetic mechanisms in humans. *Journal of Biological Education*, 22(2), 139–143.

Boerwinkel, D.J., Knippels, M.C.P.J., & Waarlo, A.J. (2011). Raising awareness of pre-symptomatic genetic testing. *Journal of Biological Education*, 45(4), 213–221.

Boerwinkel, D.J. & Waarlo, A.J. (2009). *Rethinking science curricula in the genomics era*. Proceedings of the invitational workshop, 4–5 December 2008, Utrecht, The Netherlands.

Boerwinkel, D.J. & Waarlo, A.J. (2011). *Genomics education for decision-making*. Proceedings of the second invitational workshop, 2–3 December 2010, Utrecht, The Netherlands.

Bonetta, L. (2008). Detailed analysis – tackling the epigenome. *Nature*, 454, 795–798.

Boujemaa, A., Clément, P., Sabah, S., Salaheddine, K., Jamal, C., & Abdellatif, C. (2010). University students' conceptions about the concept of gene: Interest of historical approach. *US-China Education review*, 7(2), 9–15.

Brown, D. & Clement, J. (1989). Overcoming misconceptions via analogical reasoning: Abstract transfer versus explanatory model construction. *Instructional Science*, 18(4), 237–261.

Buckley, B.C., Gobert, J.D., Kindfield, A.C.H., Horwitz, P., Tinker, R.F., Gerlits, B., Wilensky, U., Dede, C. & Willett, J. (2004). Model-based teaching and learning with BioLogica™: What do they learn? How do they learn? How do we know? *Journal of Science and Technology*, 13(1), 23–41.

Burian, R. (2005). On conceptual change in biology: The case of the gene. In R. Burian (Ed.), *The epistemology of development, evolution, and genetics* (pp. 126–144). Cambridge, UK: Cambridge University Press.

Burian, R. (2013). On gene concepts and teaching genetics: episodes from classical genetics. *Science & Education, Science & Education,* 22 (2), 325–344.

Carlson, E.A. (1966). *The gene: A critical history*. Philadelphia & London: W.B. Saunders.

Carlson, E.A. (1991). Defining the gene: an evolving concept. *Journal of Human Genetics*, 49(2), 475–487.

Carlson, E.A. (2001). *The Unfit: A history of a bad idea*. New York: Cold Spring Harbor Laboratory Press.

Carlson, E.A. (2004). *Mendel's legacy: The origin of classical genetics*. New York: Cold Spring Harbor Laboratory Press.

Carlson, E.A. (2011). *Mutation: The history of an idea from Darwin to Genomics*. New York: Cold Spring Harbor Laboratory Press.

Cartier, J.L. & Stewart, J. (2000). Teaching the nature of inquiry: Further developments in a high school genetics curriculum. *Science & Education*, 9, 247–267.

Cartier, J.L., Stewart, J. & Zoellner, B. (2006). Modeling & inquiry in a high school genetics class. *American Biology Teacher*, 68(6), 334–340.

Carver, R., Waldahl, R., & Breivik, J. (2008). Frame that gene – A tool for analyzing and classifying the communication of genetics to the public. *EMBO reports*, 9(10), 943–947.

Castéra, J., Bruguiére, C., & Clément, P. (2008a). Genetic diseases and genetic determinism models in French secondary school biology textbooks. *Journal of Biological Education*, 42(2), 53–59.

Castéra, J., & Clément, P. (2012). Teachers' conceptions about the genetic determinism of human behaviour: A Survey in 23 countries. *Science & Education*, doi: 10.1007/s11191-012-9494-0

Castéra, J., Clément, P., & Abrougui, M. (2008b). Genetic determinism in school textbooks: A comparative study among sixteen countries. *Science Education International*, 19(2), 163–184.

Cavallo, A.M.L. (1996). Meaningful learning, reasoning ability, and students' understanding and problem solving of topics in genetics. *Journal of Research in Science Teaching*, 33(6), 625–656.

Chamany, K., Allen, D., & Tanner, K. (2008). Making biology learning relevant to students: Integrating people, history, and context into college biology teaching. *CBE – Life Science Education*, 7(3), 267–278.

Chargaff, E. (1951). Some recent studies on the composition and structure of nucleic acids. *Journal of Cellular Physiology. Supplement.* 38(Suppl. 1), 41–59.

Clough, M.P. (2009). *Humanizing science to improve post-secondary science education.* Paper presented at the International History, Philosophy and Science Teaching Conference, South Bend, IN, US.

Clough, M.P. & Olson, J.K. (2004). 'The Nature of Science': Always part of the science story. *The Science Teacher*, 71(9), 28–31.

Condit, C.M., Ferguson, A., Kassel, R., Tadhani, C., Gooding, H.C., & Parrot, R. (2001). An explanatory study of the impact of news headlines on genetic determinism. *Science Communication*, 22, 379–395.

Condit, C.M., Ofulue, N., & Sheedy, K.M. (1998). Determinism and mass-media portrayals of genetics. *American Journal of human Genetics*, 62, 979–984

Cowan, R.S. (2008). Medical genetics is not eugenics. *Chronicle of Higher Education*, 54(36), B14-B16.

Crick, F. (1970). Central dogma of molecular biology. *Nature,* 227(5258), 561–3.

Crouse, D.T. (2007). X-Ray diffraction and the discovery of the structure of DNA. *Journal of Chemical Education*, 84(5), 803–809.

Darwin, C. (1868). *The variation of animals and plants under domestication.* London: John Murray.

Davis, L.C. (1993). Origin of the Punnett square. *American Biology Teacher*, 55(4), 209–212.

Davis, R.H. (2003). *The microbial models of molecular biology; from genes to genomes.* New York: Oxford University Press.

Dawson, V.M., & Venville, G. (2009). High-school students' informal reasoning and argumentation about biotechnology: An indicator of scientific literacy. *International Journal of Science Education*, 31(11), 1421–1445.

Dawson, V.M. & Venville, G. (2010). Teaching strategies for developing students' argumentation skills about socioscientific issues in high school genetics. *Research in Science Education*, 40, 133–148.

Dietrich, M.R. (2000). From gene to genetic hierarchy: Richard Goldschmidt and the problem of the Gene. In P. Beurton, R. Falk, & H.J. Rheinberger (Eds.), *The concept of the gene in development and evolution: historical and epistemological perspectives* (pp. 91–114). Cambridge, UK: Cambridge University Press.

Dikmenli, M., Cardak, O., & Kiray, S.A. (2011). Science student teacher's ideas about the 'gene' concept. *Procedia Social and Behavioral Sciences*, 15, 2609–2613.

Dogru-Atay, P. & Tekkaya, C. (2008). Promoting students' learning in genetics with the learning cycle. *The Journal of Experimental Education*, 76(3), 259–280.

Donovan, J. & Venville, G. (2012). Blood and bones: The influence of the mass media on Australian primary school children's understandings of genes and DNA. *Science & Education*, doi: 10.1007/s11191-012-9491-3

Dougherty, M.J., (2009). Closing the gap: Inverting the genetics curriculum to ensure an informed public. *American Journal of Human Genetics*, 85, 1–7.

Dougherty, M.J., Pleasants, C., Solow, L, Wong, A., & Zhang, H. (2011). A comprehensive analysis of high school genetics standards: Are states keeping pace with modern genetics? *CBE-Life Sciences Education*, 10, 318–327.

Duncan, R.G. (2007). The role of domain specific knowledge in generative reasoning about complicated multileveled phenomena. *Cognition and Instruction*, 25(4), 271–336.

Duncan, R.G., & Reiser, B.J. (2007). Reasoning across ontologically distinct levels: Students' understanding of molecular genetics. *Journal of research in Science Teaching*, 44(7), 938–959.

Duncan, R.G., Rogat, A.D., & Yarden, A. (2009). A Learning Progression for deepening students' understandings of modern genetics across the 5th-10th Grades. *Journal of Research in Science Teaching*, 46(6), 655–674.

Duncan, R.G, & Tseng, A.K., (2011). Designing project-based instruction to foster generative and mechanistic understandings in genetics. *Science Education*, 95(1), 21–56

Echevarria, M. (2003). Anomalies as a catalyst for middle school students' knowledge construction and scientific reasoning during science inquiry. *Journal of Educational Psychology*, 95(2), 357–374.

Ekborg, M. (2008). Opinion building on a socio-scientific issue: the case of genetically modifies plants. *Journal of Biological Education*, 42(2), 60–65.

El-Hani, C.N. (2007). Between the cross and the sword: The crisis of the gene concept. *Genetics and Molecular Biology*, 30(2), 297–307.

El-Hani, C. N., Roque, N., & Rocha, P. B. (2007). *Brazilian high school biology textbooks: Results from a national program*. In Proceedings of the IOSTE International Meeting on Critical Analysis of School Science Textbook (pp. 505–516). Hammamet, Tunisia: University of Tunis.

Elrod, S.L. & Somerville, M.M. (2007). Literature-based scientific learning: A collaboration model. The Journal of Academic Librarianship, 33(6), 684–691.

Falk, R. (1986). What is a gene? *Studies in History and Philosophy of Science*, 17(2), 133–173.

Falk, R. (2010). What is a gene?-Revisited. *Studies in History and Philosophy of Biological Sciences*, 41, 396–406.

Falk, R. (2012). The allusion of the gene: misunderstandings of the concepts heredity and gene. *Science & Education*, doi: 10.1007/s11191-012-9510-4.

Finkel, E.A. (1996). Making sense of genetics: Students' knowledge use during problem solving in a high school genetics class. *Journal of Research in Science Teaching*, 33(4), 345–368.

Finkel, E.A., & Stewart, J. (1994). Strategies for model-revision in a high school genetics classroom. *Mind, Culture, and Activity*, 1(3), 168–195.

Finley, F. N., Stewart, J., & Yarroch, W. L. (1982) Teachers' perception of important and difficult science content: the report of a survey. *Science Education*, 66(4), 531–538.

Flannery, M.C. (1997). The many sides of DNA. *American Biology Teacher*, 59(1), 54–57.

Flodin, V. (2009). The necessity of making visible concepts with multiple meanings in science education: The use of the gene concept in a biology textbook. *Science & Education*, 18(1), 73–94.

Fogle, T. (1990) Are genes units of Inheritance? *Biology and Philosophy*, 5, 349–371.

Fogle, T. (2000). The dissolution of protein coding genes in molecular biology. In P. Beurton, R. Falk, & H.J. Rheinberger (Eds.), *The concept of the gene in development and evolution: Historical and epistemological perspectives* (pp. 3–25). Cambridge, UK: Cambridge University Press.

Forissier, T., & Clément, P. (2003). Teaching 'biological identity' as genome/environment interactions. *Journal of Biological Education*, 37(2), 85–90.

Fox, M. (1996). Breaking the genetic code in a letter by Max Delbruck. *Journal of Collage Science Teaching*, 15(5), 324–325.

Franke, G., & Bogner, F.X. (2011a). Conceptual change in students' molecular biology education: Tilting at Windmills? *Journal of Educational Research*, 104(1), 7–18.

Franke, G., & Bogner, F.X. (2011b). Cognitive influences of students' alternative conceptions within a hands-on gene technology module. *Journal of Educational Research*, 104(3), 158–170.

Freidenreich, H.B., Duncan, R.G., & Shea, N. (2011). Exploring middle school students' understanding of three conceptual models in genetics. *International Journal of Science Education*, 33(17), 2323–2349.

Furberg, A., & Arnseth, C.H. (2009). Reconsidering conceptual change from a socio-cultural perspective: analyzing students' meaning making in genetics in collaborative learning activities. *Cultural Studies of Science education*, 4, 157–191.

Gaudillière, J.P., & Rheinberger H.J. (2004). *From molecular genetics to genomics: The mapping cultures of twentieth-century genetics*. London & New York: Routledge.

Gayon, J. (2000). From measurement to organization: a philosophical scheme for the history of the concept of heredity. In P. Beurton, R. Falk, & H.J. Rheinberger (Eds.), *The concept of the gene in development and evolution: Historical and epistemological perspectives* (pp. 69–90). Cambridge, UK: Cambridge University Press.

Gelbart, H., Brill, G., & Yarden, A. (2009). The impact of a web-based research simulation in bioinformatics on students' understanding of genetics. *Research in science education*, 39, 725–751.

Gelbart, H., & Yarden, A. (2006). Learning genetics through an authentic research simulation in bioinformatics. *Journal of Biological Education*, 40(3), 107–112.

Gelbart, H., & Yarden, A. (2011). Supporting learning of high-school genetics using authentic research practices: the teacher's role. *Journal of Biological Education*, 45(3), 129–135.

Genome News Network (2012). Retrieved 23th of March 2012. URL: http://www.genomenews-network.org/resources/sequenced_genomes/genome_guide_p1.shtml

Gericke, N.M. (2008). *Science versus School-science; Multiple models in genetics – The depiction of gene function in upper secondary textbooks and its influence on students' understanding.* Karlstad: Karlstad University studies.

Gericke, N. M., & Hagberg, M. (2007). Definition of historical models of gene function and their relation to students' understanding of genetics. *Science & Education*, 16(7 – 8), 849 – 881.

Gericke, N. M., & Hagberg, M. (2010a). Conceptual incoherence as a result of the use of multiple historical models in school textbooks. *Research in Science Education,* 40 (4): 605–623.

Gericke, N. M., & Hagberg, M. (2010b). Conceptual variation in the depiction of gene function in upper secondary school textbooks. *Science & Education*, 19(10): 963–994.

Gericke, N. M., Hagberg, M., & Jorde, D. (2013) Upper secondary students' understanding of the use of multiple models in biology textbooks—The importance of conceptual variation and incommensurability. *Research in Science Education*, 43(2): 755–780.

Gericke, N.M., Hagberg, M., Santos, V.C., Joaquim, L.M., & El-Hani, C.N. (2012). Conceptual variation or incoherence? Textbook discourse on genes in six countries. *Science & Education*, doi: 10.1007/s11191-012-9499-8.

Gericke, N.M. & Wahlberg, S. (2013). Clusters of concepts in molecular genetics: a study of Swedish upper secondary science students' understanding. *Journal of Biological Education*, 47(2): 73–83.

Gerstein, M.B., Bruce B., Rozowsky J.S., Zheng, D., Du, J., & Korbel, J.O. et al. (2007). What is a gene, post-ENCODE? History and updated definition. *Genome Research,* 17, 669–681.

Gilbert, J. K., Pietrocola, M., Zylbersztajn, A., & Franco, C. (2000). Science and education: Notions of reality, theory and model. In J.K. Gilbert, & C. Boulter (Eds.), *Developing models in science education* (pp. 19–40). Dordrecht, Netherlands: Kluwer Academic Publishers.

Gleason, M.L., Melancon, M.E. & Kleine, K.L.M. (2010). Using critical literacy to explore genetics and its ethical, legal, and social issues with in-service secondary teachers. *CBE-Life Sciences Education*, 9, 422–430.

Goodney, D.E. & Long, C.S. (2003). The collective classic: A case for the reading of science. *Science & Education*, 12, 167–184.

Greenwald, B.H. (2009). The real "toll" of A. G. Bell: Lessons about eugenics. *Sign Language Studies*, 9(3), 258–265.

Griffiths P. E. (2002) Lost: One Gene Concept. Reward to Finder. *Biology and Philosophy*, 17, 271–283.

Griffiths, P. E. & Stotz, K. 2006. Genes in the postgenomic era. *Theoretical Medicine and Bioethics*, 27, 499–521.

Gustavsson, K.-H. (2004). *Några milstolpar i genetikens historia*. Retrieved November 19, 2008, from Uppsala University, Institute for genetics and pathology: http://www.genpat.uu.se/node58

Hafner, R. & Culp, S. (1996). Elaborating the structures of a science discipline to improve problem-solving instruction: an account of classical genetics' theory structure, function and development. *Science & Education*, 5, 331–355.

Hafner, R. & Stewart, J. (1995). Revising explanatory models to accommodate anomalous genetic phenomena: Problem solving in the "context of discovery". *Science Education*, 79(2), 111–146.

Halldén, O. (1990). Questions asked in common sense contexts and in scientific contexts. In P.L. Lijnse, P. Licht, W. de Vos, & A.J. Waarlo (Eds.), *Relating macroscopic phenomena to microscopic particles* (pp. 119–130). Utrecht, Netherlands: CD-β Press.

Henig, R.M. (2000). *The monk in the garden: The lost and found genius of Gregor Mendel, the father of genetics* New York, US: Houghton Mifflin Company.

Hershey A. & Chase, M. (1952). Independent functions of viral protein and nucleic acid in growth of bacteriophage. *Journal of Genetic Physiology,* 36(1), 39–56.

Hickey, D.T., Kindfield, A.C.H., Horwitz, R & Christie, M.A. (2003). Integrating curriculum, instruction, assessment, and evaluation in a technology-supported genetics learning environment. *American Educational Research Journal*, 40(2), 495–538.

Hickey, D.T., Wolfe, E.W. & Kindfield, A. C-H (2000). Assessing learning in a technology-supported genetics environment: Evidential and systematic validity issues. *Educational Assessment*, 6(3), 155–196.

Hoffmeyer, J. (1988) *Naturen i huvudet*. Simrishamn, Sweden: Rabén & Sjögren.

Holden, C. (2006). Darwin's place on campus is safe – But not supreme. *Science*, 301, 769–771.

Horowitz, N.H. (1995). "One-Gene-One-Enzyme: Remembering Biochemical Genetics". *Protein Science,* 4(5), 1017–1019.

Hott, A.M., Huether, C.A., McInerney, J.D., Christianson, C., Fowler, R., Bender, R., Jenkins, J., Wysocki, A., Markle, G., & Karp, R. (2002). Genetics content in introductory biology courses for non-science majors: Theory and practice. *Bioscience* 52, 1024–1035.

Hull, D.L. (2002). Genes versus molecules: How to, and how not to, be a reductionist. In M.H.V Regenmortel & D.L. Hull (Eds.), *Reductionism in the biomedical sciences* (pp. 161–173). Chichester, UK: John Wiley & Sons Ltd.

Hurd, P.D. (1978). The Historical/Philosophical Background of Education in Human Genetics in the United States. *Biological Sciences Curriculum Study Journal*, 1 (1), 3–8,

Ibáñez-Orcajo, T., & Martínez-Aznar, M. (2005). Solving problems in Genetics II: Conceptual restructuring. *International Journal of Science Education*, 27(12), 1495–1519.

Ibáñez-Orcajo, T., & Martínez-Aznar, M. (2007). Solving problems in Genetics, part III: Change in the view of the nature of science. *International Journal of Science Education*, 29(6), 747–769.

IHGSC (International Human Genome sequencing Consortium), (2004). Finishing the euchromatic sequence of the human genome. *Nature* 431, 931–945.

Johnstone, A.H., & Mahmoud, N.A. (1980). Isolating topics of high perceived difficulty in school biology. *Journal of Biological Education*, 14(2), 163–166.

Jorde, D., Strømme, A., Sørborg, Ø., Erlien, W., & Mork, S. M. (2003). *Virtual Environments in Science, Viten.no*, (No. 17). Oslo, Norway: ITU. Retrieved November 14, 2008 from: http://www.itu.no/filearchive/fil_ITU_Rapport_17.pdf

Kampourakis, K. (2013). Mendel and the path to genetics: Portraying science as a social process. *Science & Education*, 22(2), 293–324.

Kaplan, G. & Rogers, L.J. (2003). *Gene worship*. New York, US: Other Press.

Keller, E., F. (2000). Decoding the genetic program: or, some circular logic in the logic of circularity. In P. Beurton, R. Falk, & H.J. Rheinberger (Eds.), *The concept of the gene in development and evolution: Historical and epistemological perspectives* (pp. 159–177). Cambridge, UK: Cambridge University Press.

Keller, E. F. (2005). The century beyond the gene. *Journal of Biosciences*, 30(1), 3–10.

Keller, E.F. & Harel, D. (2007). Beyond the gene. *PLOS one*, 2(11), e1231. doi:10.1371/journal.pone.0001231

Kim, S.Y. & Irving, K.E. (2010). History of science as an instructional context: Student learning in genetics and nature of science. *Science & Education*, 19, 187–215.

Kinnear, J. (1991). Using an historical perspective to enrich the teaching of linkage in genetics. *Science Education*, 75(1), 69–85.

Kitcher, P. (1982). Genes. *British Journal for the Philosophy of Science*, 33(4), 337–359.

Knight, J.K., & Smith, M.K. (2010). Different but equal? How nonmajors and majors approach and learn genetics. *CBE-Life Sciences Education*, 9, 34–44.

Knippels, M.C.P.J., (2002). *Coping with the abstract and complex nature of genetics in biology education – The yo-yo learning and teaching strategy*. Utrecht, Netherlands: CD-β Press.

Knippels, M.C.P.J., Waarlo, A.J., & Boersma, K.Th. (2005). Design criteria for learning and teaching genetics. *Journal of Biological Education*, 39(3), 108–112.

Kuhn, T. S. (1974) Second Thoughts on Paradigms, In F. Suppe (Ed.), *The Structure of Scientific Theories*. Urbana (pp. 459–482). US: University of Illinois Press.

Law, N. & Lee, Y. (2004). Using an iconic modeling tool to support the learning of genetics concepts. *Journal of Biological Education*, 38(3), 118–125.

Lawrence, P.A. (1992). *The making of a fly: the genetics of animal design*. London: Blackwell Scientific.

Lederman, N.G. (1992). Students' and teachers' conceptions of the nature of science: A review of the research. *Journal of Research in Science Teaching*, 29(4), 331–359.

Lederman, N.G., Antik, A. & Bartos, S. (2012). Nature of science, scientific inquiry, and socio-scientific issues arising from genetics: A pathway to developing a scientific literate citizenry. *Science & Education*, doi: 10.1007/s11191-012-9503-3

Lewin, B. (2000). *Genes VII*. New York: Oxford University Press.

Lewis, J. (2012). From flavr savr tomatoes to stem cell therapy: Young people's understanding of gene technology, 15 years on. *Science & Education*, doi: 10.1007/s11191-012-9523-z

Lewis, J., & Kattmann, U. (2004). Traits, genes, particles and information: Re-visiting students' understandings of genetics. *International Journal of Science Education*, 26(2), 195–206.

Lewis, J., & Leach, J. (2006). Discussion of socio-scientific issues: The role of science knowledge. *International Journal of Science Education*, 28(11), 1267–1287.

Lewis, J., Leach, J., & Wood-Robinson, C. (2000a). All in the genes? – Young people's understanding of the nature of genes. *Journal of Biological Education*, 34(2), 74–79.

Lewis, J., Leach, J., & Wood-Robinson, C. (2000b). Chromosomes: the missing link – Young people's understanding of mitosis, meiosis, and fertilisation. *Journal of Biological Education*, 34(4), 189–199.

Lewis, J., & Wood-Robinson, C. (2000). Genes, chromosomes, cell division and inheritance – do students see any relationship. *International Journal of Science Education*, 22(2), 177–195.

Lewontin, R. C.; Rose, S., & Kamin, L. J. (1984). *Not in our genes: Biology, ideology, and human nature*. New York-NY: Pantheon.

Lin, C.Y., Cheng, J.H., & Chang, W.H. (2010). Making science vivid: using a historical episodes map. *International Journal of Science Education*, 32(18), 2521–2531.

Lindahl, M.G. (2009). Ethics or morals: Understanding students' values related to genetic tests on humans. *Science & Education*, 18, 1285–1311.

Lock, R. (1997). Post-16 biology – some model approaches? *School Science Review*, 79(286), 33–38.

Magen, A. & Ast, G. (2005). The importance of being divisible with by three in alternative splicing. *Nucleic Acid Research*, 33, 5574–5582.

Mahadeva, M. & Randerson, S. (1985). The rise and fall of the gene. *Science Teacher*, 52(8), 15–19.

Marbach-Ad, G. (2001). Attempting to break the code in student comprehension of genetic concepts. *Journal of Biological Education*, 35(4), 183–189.

Marbach-Ad, G., Rotbain, Y., & Stavy, R. (2008). Using computer animation and illustration activities to improve high school students' achievement in molecular genetics. *Journal of Research in Science Teaching*, 45(3), 273–292.

Marbach-Ad, G., & Stavy, R. (2000). Students' cellular and molecular explanations of genetic phenomena. *Journal of Biological Education*, 34(4), 200–205.

Martinez-Gracia, M. V., Gil-Quilez, M. J., & Osada, J. (2006). Analysis of molecular genetics content in Spanish secondary school textbooks. *Journal of Biological education*, 40(2), 53–60.

Martins, I., & Ogborn, J. (1997). Metaphorical reasoning about genetics. *International Journal of Science Education*, 19(1), 47–63.

Matthews, M.R. (1994/2014). *Science teaching: The Role of history and philosophy of science*. New York: Routledge.

Mayr, E. (1982). *The growth of biological thought: Diversity, evolution and inheritance*. Cambridge, MA: The Belknap Press of Harvard University Press.

Mbajiorgu, N.M., Ezechi, N.G., & Idoko, E.C. (2007). Addressing non-scientific presuppositions in genetics using a conceptual change strategy. *Science Education*, 91(3), 419–438.

Mehta, P. (2000). Human Eugenics: Whose Perception of Perfection? *History Teacher*, 33(2), 222–240.

Metz, D., Klassen, S., Mcmillan, B., Clough, M., & Olson, J. (2007). Building a foundation for the use of historical narratives. *Science & Education*, 16(3–5), 313–334.

Meyer, N.L.M., Bomfim, G.C., & El-Hani, C.N. (2011). How to Understand the Gene in the Twenty-First Century? *Science & Education*, 22(2), 345–374.

Mitchell, A. & Lawson, A.E. (1988). Predicting genetics achievement in non majors college biology. *Journal of Research in Science Teaching*, 25, 23–37.

Moore, R. (2001). The "rediscovery" of Mendel's work. *Bioscene*, 27(2), 13–24.

Morange, M. (2002). Genes versus molecules: How to, and how not to, be a reductionist. In M.H.V Regenmortel & D.L. Hull (Eds.), *The gene: Between holism and generalism* (pp. 179–187). Chichester, UK: John Wiley & Sons Ltd.

Morgan, T.H. (1911). Chromosomes and associative inheritance. *Science*, 34, 636–638.

Morgan, T.H. (1934). *The relation of genetics to physiology and medicine*. Nobel Lecture, June 4, 1934. Available at: http://www.nobelprize.org/nobel_prizes/medicine/laureates/1933/morgan-lecture.pdf

Moss, L. (2003). *What genes can't do*. Cambridge-MA: MIT Press.

Mysliwiec, T.H. (2003). The genetic blues: Understanding genetic principles using a practical approach and a historical perspective. *American Biology Teacher*, 65(1), 41–46.

Nelkin, D. & Lindee, S.M. (1995). *The DNA mystique: The gene as a cultural icon*. New York, NY: Freeman.

Ohly, K.P. (2002). Changing the 'Denkstil' – A case study in the history of molecular genetics. *Science & Education*, 11, 155–167.

Olby, R. (1985). *Origins of Mendelism*, 2nd ed. Chicago IL: University of Chicago Press.

Othman, J.B. (2008). What Reading "The double helix" and "The dark lady of DNA" can teach students (and their teachers) about science. *Teaching Science*, 54(1), 50–53.

Pashley, M. (1994). A-level students: Their problem with gene and allele. *Journal of Biological Education*, 28(2), 120–126.

Passmore, C., & Stewart, J. (2002). A modeling approach to teaching evolutionary biology in high schools. *Journal of Research in Science Teaching*, 39(3), 185–204.

Pata, K., & Sarapuu, T. (2006). A comparison of reasoning processes in a collaborative modeling environment: Learning about genetics problems using virtual chat. *International Journal of Science education*, 28(11), 1347–1368.

Pearson, H. (2006). What is a gene? *Nature*, 441(25), 399–401.

Portin, P. (1993). The concept of the gene: Short history and present status. *The Quarterly Review of Biology*, 68(2), 173–223.

Portin, P. (2009). The elusive concept of the gene. *Hereditas*, 146, 112–117.

Posner, G. J., Strike, K. A., Hewson, P. W., & Gertzog, W. A. (1982). Accommodation of a scientific conception: Toward a theory of conceptual change. *Science Education*, 66(2), 211–227.

Pramling, N. & Säljö R. (2007). Scientific knowledge, popularization, and use of metaphors: Modern genetics in popular science magazines. *Scandinavian Journal of Educational Research*, 51(3), 275–295.

Quessada, M-P. & Clément, P. (2006). An Epistemological approach to French Syllabi on human origins during the 19th and 20th centuries. *Science & Education*, 16, 991–1006.

Rodwell, G. (1997). Dr. Caleb Williams Saleeby: The Complete Eugenicist. *History of Education*, 26(1), 23–40.

Roseman, J.E., Caldwell, A., Gogos, A., & Kurth, L. (2006). Mapping a coherent learning progression for the molecular basis of heredity. Paper presented at National Association for the Advancement of Science (NARST) annual meeting 2006 in San Francisco, US.

Rosenberg, A. (1985). *The structure of biological science*. Cambridge, UK: Cambridge University Press.

Rotbain, Y., Marbach-Ad, G. & Stavy, R. (2005). Understanding molecular genetics through a drawing-based activity. *Journal of Biological Education*, 39(4), 174–178.

Rotbain, Y., Marbach-Ad, G., & Stavy, R. (2006). Effect of bead and illustrations models on high school students' achievement in molecular genetics. *Journal of Research in Science Teaching*, 43(5), 500–529.

Rundgren, C-J, Chang Rundgren, S-N, Tseng, Y-H, Lin, P-L, & Chang, C-Y (2012). Are you SLim? Developing an instrument for civic scientific literacy measurement (Slim) based on media coverage. *Public Understanding of Science*, doi: 10.1177/0963662510377562.

Sadler, T.D. (2011). Foreword. In T.D. Sadler (Eds.) *Socio-scientific issues in the classroom; Teaching, learning and research*. Dordrecht, Netherlands: Springer.

Sadler, T.D., & Zeidler, D.L. (2004). The morality of socioscientific issues: Construal and Resolution of Genetic Engineering Dilemmas. *Science Education*, 88(1), 4–27.

Sadler, T.D., & Zeidler, D.L. (2005). The significance of content knowledge for informal reasoning regarding socioscientific issues: Applying genetics knowledge to genetic engineering issues.. *Science Education*, 89(1), 71–93.

Santos, S. & Bizzo, N. (2005). From "new genetics" to everyday knowledge: Ideas about how genetic diseases are transmitted in two large Brazilian families. *Science Education*, 89, 564–576.

Santos, V. C., Joaquim, L. M. & El-Hani, C. N. (2012). Hybrid deterministic views about genes in biology textbooks: A key problem in genetics teaching. *Science & Education*, 21(4), 543–578.

Sapp, J. (2003). *Genesis; The evolution of biology*. New York, US: Oxford University Press.

Sarkar, S. (2002). Genes versus molecules: How to, and how not to, be a reductionist In M.H.V Regenmortel & D.L. Hull (Eds.), *The gene: Between holism and generalism* (pp. 191–209). Chichester, UK: John Wiley & Sons Ltd.

Scherrer, K., & Jost, J. (2007). Gene and genon concept: coding versus regulation. A conceptual and information-theoretic analysis of genetic storage and expression in the light of modern molecular biology. *Theory in Biosciences*, 126(2), 65–113.

Schönborn, K.J. & Bögeholz, S. (2009). Knowledge transfer in biology and translation across external Representations: Experts' views and challenges for learning *International Journal of Science and Mathematics Education*, 7(5), 931–955.

Schwartz, J. (2008). *In pursuit of the gene: from Darwin to DNA*. Cambridge, MA: Harvard University Press.

Schwartz, S. (2000). The differential concept of the gene: Past and present. In P. Beurton, R. Falk, & H.J. Rheinberger (Eds.), *The concept of the gene in development and evolution: Historical and epistemological perspectives* (pp. 24–40). Cambridge, UK: Cambridge University Press.

Shaw, K.E., Horne, K.V., Zhang, H. & Boughman J. (2008). Essay contest reveals misconceptions of high school students in genetics content. *Genetics*, 178, 1157–1168.

Simmons, P.E. & Lunetta, V.N. (1993). Problem-solving behaviors during a genetics computer simulation: Beyond the expert/ novice dichotomy. *Journal of Research in Science Teaching*, 30(2), 153–173.

Simon, E.J. (2002). Human gene therapy: genes without frontiers? *American Biology Teacher*, 64(4), 264–270.

Simonneaux, J. (2002). Analysis of classroom debating strategies in the field of biotechnology. *Journal of Biological Education*, 37(1), 9–12.

Simonneaux, J. (2008). Argumentation in Socio-Scientific contexts. In S. Erduran, & M.P. Jiménez-Aleixandre (Eds.), *Argumentation in Science education* (pp. 179–199). Dordrecht, The Netherlands: Springer.

Smith, A.L., & Williams, M.J., (2007) "It's the X and Y thing": Cross-sectional and longitudinal changes in children's understanding of genes. *Research in Science Education*, 37(4), 407–422.

Smith, M.U. (1983). *A comparative analysis of the performance of experts and novices while solving selected classical genetics problems* (Doctoral dissertation). The Florida State University. *Dissertation Abstracts International*, 44, 451A.

Smith, M. U. & Adkinson, L.R. (2010). Updating the model definition of the gene in the modern genomic era with implications for instruction. *Science & Education*, 19(1), 1–20.

Smith, M.U. & Good, R. (1984). Problem solving and classical genetics: Successful vs. unsuccessful performance. *Journal of Research in Science Teaching*, 21, 895-912.

Smith, M.U., & Kindfield, A.C.H. (1999). Teaching cell division: Basics and recommendations. *American Biology Teacher*, 61, 366–371.

Smith, M.U. & Scharmann, L.C. (2006). A multi-year program developing an explicit reflective pedagogy for teaching pre-service teachers the nature of science by ostention. *Science & Education, 17*(1–2), 219–248.

Soderberg, P. & Jungck, J.R. (1994). Genetics construction kit: A tool for open-ended investigation in transmission genetics. *Journal of Computing in Higher Education*, 5(2), 67–84.

Soderberg, P., & Price, F. (2003). An examination of problem-based teaching and learning in population genetics and evolution using EVOLVE, a computer simulation. *International Journal of Science Education*, 25(1), 33–55.

Stadler, P. F., Prohaska, S. J., Forst, C. V., & Krakauer, D. C. (2009). Defining genes: A computational framework. *Theory in Biosciences*, 128(3), 165–170.

Stewart, J. (1983). Student problem solving in high school genetics. *Science Education*, 66(5), 523–540.

Stewart, J. (1988). Potential learning outcomes from solving genetics problems: A typology of problems. *Science Education*, 72(2), 237–254.

Stewart, J., Cartier, J.L., & Passmore, P.M. (2005). Developing understanding through model-based inquiry. In M.S. Donovan & J.D. Bransford (Eds.), *How students learn* (pp. 515–565). Washington D.C.: National Research Council.

Stewart, J., Hafner, R., Johnson, S., & Finkel, E. (1992). Science as models building: Computers and high school genetics. *Educational Psychologist*, 27(3), 317–336.

Stewart, J., & van Kirk, J. (1990). Understanding and problem-solving in classical genetics. *International journal of Science Education*, 12(5), 575–588.

Stolarsky, M., Ben-Nun, & Yarden, A. (2009). Learning molecular genetics in teacher-led outreach laboratories. *Journal of Biological Education*, 44(1), 19–25.

Stotz, K., Griffiths, P.E., & Knight, R. (2004). How biologists conceptualize genes: An empirical study. *Studies in the History and Philosophy of Biological and Biomedical Sciences*, 35, 647–673.

Thomson, N., & Stewart, J. (2003). Genetics inquiry: Strategies and knowledge geneticists use in solving transmission genetics problems. *Science Education,* 87(2), 161–180.

Tibell, L.A.E. & Rundgren, C-J (2010). Educational challenges of molecular life science: characteristics and implications for education and research. *CBE-Life Sciences Education*, 9, 25–33.

Tseng, Y-H, Chang, C-Y, Chang Rundgren, S-N, & Rundgren, C-J. (2010). Mining concept maps from news stories for measuring civic scientific literacy in media. *Computers & Education*, 55, 165–177.

Tsui, C.-Y., & Treagust, D.F. (2003a). Genetics reasoning with multiple external representations. *Research in Science Education*, 33, 111–135.

Tsui, C.-Y., & Treagust, D.F. (2003b). Learning genetics with computer dragons. *Journal of biological Education*, 37(2), 96–98.

Tsui, C.-Y. & Treagust, D.F. (2004a). Motivational aspects of learning genetics with interactive multimedia. *American Biology Teacher*. 66(4), 277–285.

Tsui, C.-Y. & Treagust, D.F. (2004b). Conceptual change in learning genetics: an ontological perspective. *Research in Science Education*, 22(2), 185–202.

Tsui, C.-Y., & Treagust, D.F., (2010). Evaluating secondary students' scientific reasoning in genetics using a two-tier diagnostic instrument. *International Journal of Science Education*, 32(8), 1073–1098.

Tudge, C. (1993). *The engineer in the garden*. London, UK: Jonathan Cape.

Uddenberg, N. (2003) *Idéer om livet – En biologi historia band II*. Stockholm, Sweden: Bokförlaget Natur och Kultur.

van Mil, M.H.V., Boerwinkel, D.J. & Waarlo, A.J. (2013). Modelling molecular mechanisms: A framework of scientific reasoning to construct molecular-level explanations for cellular behavior. *Science & Education*, 22(1), 93–118.

van Regenmortel, M.H.V. & Hull, D.L. (2002). *Reductionism in the biomedical sciences*. Chichester, UK: John Wiley & Sons Ltd.

Venville, G., & Dawson, V. (2010). The impact of a classroom intervention on grade 10 students' argumentation skills, informal reasoning, and conceptual understanding of science. *Journal of Research in Science Teaching*, 47(8), 952–977.

Venville, G. & Donovan, J. (2005). Searching for clarity to teach the complexity of the gene concept. *Teaching Science*, 51(3), 20–24.

Venville, G. & Donovan, J. (2006). Analogies for life: a subjective view of analogies and metaphors used to teach about genes and DNA. *Teaching Science*, 52(1), 18–22.

Venville, G. & Donovan, J. (2007). Developing year 2 students' theory of biology with concepts of the gene and DNA. *International Journal of Science Education*, 29(9), 1111–1131.

Venville, G. & Donovan, J. (2008). How pupils use a model for abstract concepts in genetics. *Journal of Biological Education*, 43(1), 6–14.

Venville, G., Gribble, S.J., & Donovan, J. (2005). An exploration of young children's understandings of genetics concepts from ontological and epistemological perspectives. *Science Education*, 89(4), 614–633.

Venville, G. & Milne, C. (1999). Three woman scientists and their role in the history of genetics. *Australian Science Teacher Journal*, 45(3), 9–15.

Venville, G. & Treagust, D.F. (1998). Exploring conceptual change in genetics using a multidimensional interpretive framework. *Journal of Research in Science Teaching*, 35(9), 1031–1055.

Venville, G. & Treagust, D.F. (2002). Teaching about the gene in the genetic information age. *Australian Science Teachers' Journal*, 48(2), 20–24.

Verhoeff, R., Boerwinkel, D.J, & Waarlo, A.J. (2009). Genomics in school. *EMBO reports*, 10(2), 120–124.

Verhoeff, R., Waarlo, A.J. & Boersma, K.Th. (2008). Systems modelling and the development of coherent understanding of cell biology. *International Journal of Science Education*, 30, 543–568.

Vigue, C.L. (1976). A short history of the discovery of gene function. *American Biology Teacher*, 38(9), 537–541.

von Zastrow, C. (2009). Mounting a curricular revolution: an interview with Henry Louis Gates, Jr. *Education Digest: Essential readings condensed for quick review*. 75(3), 17–19.

Waters, K. C. (1994) Genes Made Molecular. *Philosophy of Science*, 61, 163–185.

Watson, J.D. & Crick, F.H. (1953). Molecular Structure of Nucleic Acids: A Structure for Deoxyribose Nucleic Acid. *Nature*. 171 (4356), 737–738.

Weber, M. (1998). Representing genes: Classical mapping techniques and the growth of genetic knowledge. *Studies in History and Philosophy of Biological and Biomedical Sciences*, 29, 295–315.

Weiner, J. (1999). *Time, Love, Memory: A Great Biologist and His Quest for the Origins of Behavior*. New York, US: Random House.

Wieder, W. (2006). Science as story: "Communicating the nature of science through historical perspectives on science". *American Biology Teacher*, 68(4), 200–205.

Wilkins, M.R, Pasquali, C., Appel, R.D, Ou, K., Golaz, O., Sanchez, J-C., Yan, J.X., Gooley, A.A., Hughes, G., Humphery-Smith, I., Williams, K.L., & Hochstrasser, D.F. (1996). From Proteins to Proteomes: Large Scale Protein Identification by Two-Dimensional Electrophoresis and Amino Acid Analysis. *Nature Biotechnology* 14 (1): 61–65

Williams, M.J., & Smith, A.L. (2010). Concepts of kinship relations and inheritance in childhood and adolescence. *British journal of Developmental Psychology*, 28, 523–546.

Wittgenstein, L. (1953/2001). *Philosophical Investigations*. Blackwell Publishing.

Wood-Robinson, C. (1994). Young people's ideas about inheritance and evolution. *Studies in Science Education*, 24, 29–47.

Wynne, C.F., Stewart, J., & Passmore, C. (2001). High school students' use of meiosis when solving genetics problems. *International journal of Science Education*, 23(5), 501–515.

Yarden, A., Brill, G. & Falk, H. (2001). Primary literature as a basis for high-school biology curriculum, *Journal of Biological Education*, 35(4), 190–195.

Zohar, A., & Nemet, F. (2002). Fostering students' knowledge and argumentation skills through dilemmas in human genetics. *Journal of Research in Science Teaching*, 39(1), 35–62.

Zwart, H. (2008). Understanding the human genome project: a biographical approach. *New Genetics & Society*, 27(4), 353–376.

Niklas M. Gericke is associate professor in Biology Education at Karlstad University in Sweden. He holds a Master's degree of Biology and a certification as a teacher. He earned a PhD in Biology Education at Karlstad University in 2009. Niklas is chair of the research center of SMEER (Science, Mathematics and Engineering Education Research) at Karlstad University as well as senior director at the teaching faculty office. He is also coordinator and member of the board for Hasselblads' graduate school in Molecular Science Education, which is cooperation between three universities. Niklas was in 2010 affiliated as visiting associate professor at Norwegian University of Science and Technology (NTNU) in Trondheim, Norway. His research interests include history and philosophy of science, models and representations in science, genetic education, and textbook research and education for sustainable development. Among his recent publications are two papers about gene discourse in textbooks and how this discourse is understood by students: Gericke, N.M., Hagberg, M., Santos, V.C., Joaquim, L.M., & El-Hani, C.N. (2012). Conceptual variation or incoherence? Textbook discourse on genes in six countries. *Science & Education*, online first (doi: 10.1007/s11191-012-9499-8); Gericke, N.M., Hagberg, M., & Jorde, D. (2013) Upper secondary students' understanding of the use of multiple models in biology textbooks—The importance of conceptual variation and incommensurability. *Research in Science Education*, 43(2): 755–780.

Mike U. Smith is professor of Medical Education and director of AIDS Education and Research at Mercer University School of Medicine in Macon, Georgia. He studied genetics and science education at Florida State University where he received his PhD. His current research interests include the design of effective instruction about the nature of science, evolution, problem-based medical education, and the development, implementation, and evaluation of peer education programs to reduce risky sexual behavior among young people. He is a fellow of the American Association for the Advancement of Science and was a member of the US National Academy of Sciences Working Group that produced the 1998 monograph entitled *Teaching About Evolution and the Nature of Science* (available at www.nap.edu/readingroom/books/evolution98). His most recent publication in *Science and Education* is "The role of authority in science and religion with implications for introductory science teaching and learning." DOI 10.1007/s11191-012-9649-1.

Chapter 16
The Contribution of History and Philosophy to the Problem of Hybrid Views About Genes in Genetics Teaching

Charbel N. El-Hani, Ana Maria R. de Almeida, Gilberto C. Bomfim, Leyla M. Joaquim, João Carlos M. Magalhães, Lia M.N. Meyer, Maiana A. Pitombo, and Vanessa C. dos Santos

16.1 Introduction

The gene concept has been one of the landmarks in the history of science in the twentieth century, which has been even characterized as "the century of the gene" (Gelbart 1998; Keller 2000). However, there are nowadays persistent doubts about the meaning and contributions of this concept, not only among philosophers of biology[1] but also among empirical scientists.[2] Moreover, by the mid-2000s concerns about the gene extended to the editorials of high-impact scientific journals (e.g., Pearson 2006).

---

[1] See, for example, Burian (1985), Falk (1986), Fogle (1990), Hull (1974), and Kitcher (1982).

[2] See, for example, Gerstein et al. (2007), Kampa et al. (2004), Venter et al. (2001), and Wang et al. (2000).

C.N. El-Hani (✉)
History, Philosophy and Biology Teaching Lab (LEFHBio),
Institute of Biology, Federal University of Bahia, Salvador, Brazil
e-mail: charbel@ufba.br; charbel.elhani@pesquisador.cnpq.br

Department of General Biology, Institute of Biology, Universidade Federal da Bahia,
Campus de Ondina. Rua Barão de Jeremoabo, s/n, 40170-115 Ondina, Salvador-BA, Brazil

A.M.R. de Almeida
History, Philosophy and Biology Teaching Lab (LEFHBio),
Institute of Biology, Federal University of Bahia, Salvador, Brazil

Graduate Studies Program in Plant Biology, University of California, Berkeley, USA

G.C. Bomfim • L.M. Joaquim • M.A. Pitombo • V.C. dos Santos
History, Philosophy and Biology Teaching Lab (LEFHBio),
Institute of Biology, Federal University of Bahia, Salvador, Brazil

J.C.M. Magalhães
Department of Genetics, Federal University of Paraná, Curitiba, Brazil

M.R. Matthews (ed.), *International Handbook of Research in History,
Philosophy and Science Teaching*, DOI 10.1007/978-94-007-7654-8_16,
© Springer Science+Business Media Dordrecht 2014

There are negative and positive reactions to the problem of the gene or, as El-Hani (2007) describes, attempts to eliminate this concept from biology or to keep it although radically reconceptualized. Keller (2000), for instance, suggested that maybe the time was ripe to forge new words and leave the gene concept aside (see also Portin 1993; Gelbart 1998). More optimistic views are found, for example, in Hall (2001), who argued that, despite published obituaries, the gene was not dead, but alive and well, and seeking a haven from which to steer a course to its "natural" home, the cell as a fundamental morphogenetic unit, or in Knight (2007), for whom "reports of the death of the gene are greatly exaggerated."

The crisis of the gene concept is mostly related to its interpretation as *a stretch of DNA that encodes a functional product, a single polypeptide chain or RNA molecule*, that is, the so-called classical molecular gene concept (Neumann-Held 1999; see also Griffiths and Neumann-Held 1999; Stotz et al. 2004). Under the influence of this concept, simple and straightforward one-to-one relationships (function = gene = polypeptide = continuous piece of DNA = cistron) were regarded as acceptable in understanding the functioning of the genetic system from the 1940s to the 1970s (Scherrer and Jost 2007a, b). These relationships were captured in a manner that was heuristically powerful in genetics and molecular biology, which benefited from treating the gene as an uninterrupted unit in the genome, with a clear beginning and a clear ending and with a single function ascribed to its product (and, thus, indirectly to the gene). The explanatory and heuristic power of this concept follows from how it brought together structural and functional definitions of the gene, alongside with an easily understandable mechanics. With the introduction of an informational vocabulary in molecular biology and genetics (Kay 2000), genes were also regarded as informational units, leading to what has been called the informational conception of the gene (Stotz et al. 2004), a popular notion in textbooks, the media, and public opinion.

This picture changed since the 1970s, as the view of the gene as a structural and functional unit was increasingly challenged by anomalies resulting from research mostly conducted in eukaryotes, in which we find nothing like the tight physical complex linking transcription and translation observed in bacteria. We can classify these anomalies in three kinds, all related to counterevidence for a unitary relationship between genes, gene products, and gene function: (i) *one-to-many* correspondences between DNA segments and RNAs/polypeptides (as, for instance,

---

L.M.N. Meyer
History, Philosophy and Biology Teaching Lab (LEFHBio),
Institute of Biology, Federal University of Bahia, Salvador, Brazil

Department of Biosciences, Federal University of Sergipe, Itabaiana, Brazil

in alternative splicing,[3] Black 2003; Graveley 2001), (ii) *many-to-one* correspondences between DNA segments and RNAs/polypeptides (as in genomic rearrangements, such as those involved in the generation of diversity in lymphocyte antigen receptors in the immune system[4]; see Cooper and Alder 2006; Murre 2007), and (iii) *lack of correspondence* between DNA segments and RNAs/polypeptides (as we see, e.g., in mRNA editing[5]; see Hanson 1996; Lev-Maor et al. 2007).

Another key issue related to the gene concerns conceptual variation and ambiguities throughout its history (e.g., Carlson 1966). As Rheinberger (2000) argues, genes can be regarded as "epistemic objects" in genetics and molecular biology, entities introduced and conceived as targets of research, whose understanding is framed by the set of experimental practices used by particular scientific communities. Thus, conceptual variation can be explained as a consequence of different experimental practices used by diverse communities of scientists who deal with genes as epistemic objects (Stotz et al. 2004). For instance, population geneticists often work with an instrumental view of genes as determinants of phenotypic differences, since this is often enough to deal with the relationship between changing gene frequencies in populations over time and changes in the phenotypes of the individuals making up those populations. They tend to emphasize, thus, genes as markers of phenotypic effects, taking a more distal view on gene function. Molecular biologists, in turn, focus their attention on genes in DNA and their molecular products and interactions, emphasizing the structural nature of genes and their role in the cellular system they are part of. They take a more proximal view of genes and tend to be reluctant to identify a gene by only considering its contributions to relatively distant levels of gene expression (Stotz et al. 2004).

The phenomenon at stake here is gene function, and consequently, we will refer to multiple models of gene function, in the structure of which a central element is the gene concept.[6] The experimental practices used by diverse scientific communities

---

[3] In alternative splicing, a pre-mRNA molecule is processed – in particular, spliced – in a diversity of manners, so that different combinations of exons emerge in the mature mRNA. In this manner, several distinct mRNAs and, thus, polypeptides can be obtained from the same DNA sequence. In *Drosophila melanogaster*, for instance, DSCAM alternative splicing can lead to ca. 38,016 protein products (Celotto and Graveley 2001).

[4] The generation of the diverse antigen receptors found in lymphocytes and, consequently, of antibody specificity depends on a combinatorial set of genomic rearrangements between different DNA segments called variable segments, constant segments, and diversity and joining segments.

[5] mRNA editing is an alteration of mRNA nucleotides during processing, resulting in lack of correspondence between nucleotide sequences in mature mRNA and nucleotide sequences in DNA.

[6] "Model" is a polysemous term, with diverse meanings that capture distinct relationships between elements of knowledge (e.g., Black 1962; Grandy 2003; Halloun 2004, 2007; Hesse 1963). We treat models here as constructs created by the scientific community in order to represent relevant aspects of experience, i.e., phenomena and processes/mechanisms that can explain and/or predict them. In these terms, models capture the relationship between a symbolic system (a representation) and phenomena, processes, and mechanisms ontologically treated as being part of the world or nature. Models are built through processes of generalization, abstraction, and idealization that crucially involves selecting a number of entities, variables, relationships associated with a specific class of phenomena and processes/mechanisms to be included in the model, while others are

lead to variation in models of gene function and gene concepts. The expression "conceptual variation" describes, then, the range of different meanings ascribed to a concept, not necessarily all of them outdated, since they may still be used in different contexts.

Conceptual variation has been heuristically useful in the history of genetics.[7] Different gene concepts and different gene function models have been and still are useful in different areas of biology, with different theoretical commitments and research practices. Nevertheless, while recognizing that conceptual variation is a desirable feature in our understanding of genes, several authors stress that we should clearly distinguish between different concepts and models, with diverse domains of application.[8] After all, conceptual variation may also lead to misconceptions and misunderstandings. Falk (1986, p. 173), for instance, considers that the pluralism found in the current picture about genes "… brought us […] dangerously near to misconceptions and misunderstandings." Fogle (1990, p. 350) argues that "despite proposed methodological advantages for the juxtaposition of 'gene' concepts it is also true […] that confusion and ontological consequences follow when the classical intention for 'gene' conjoins a molecular 'gene' with fluid meaning." Keller (2005) argues that many problems arise from ambiguities in the usage of the term "gene," calling attention to difficulties with gene counting, since the value obtained may vary by two, three, or more orders of magnitude depending on how genes are defined.

Diversity in meaning and heterogeneity in reference potential can lead to semantic incommensurability, although this is not necessarily so. In the history of genetics, ideas associated with different ways of understanding genes and their roles in living systems have been sometimes merged in the construction of new concepts and models. However, one needs to consider that conceptual change often leads to scientific concepts with heterogeneous reference potentials and, thus, to models with diverse meanings, and as a result, there can be semantic incommensurability between concepts and models. When semantically incommensurable models and concepts, or even some of their features, are mixed up, logical inconsistencies and conceptual incoherence can appear.

In science, conceptual variation and the combination of ideas related to different models are usually (but not always) less problematic, since researchers usually develop a sophisticated understanding of the knowledge base of their research field (even though much can remain tacit) and also learn epistemic practices that stabilize

---

selected out. These entities, variables, and relationships are captured by scientific concepts, and thus, a model can be seen as a system of related concepts. Concepts gain meaning by being used in model construction, as contributors to model structure (Halloun 2004). If we understand scientific theories as families of models – according to a semantic approach (e.g., Develaki 2007; Suppe 1977; van Fraassen 1980) – concepts will form a network of relationships as a consequence of their participation in a series of models, and ultimately, the meaning of a concept will be constructed out of its relationship with other concepts in a network of models.

[7] See, for example, Burian (1985), Falk (1986), Griffiths and Neumann-Held (1999), Kitcher (1982), and Stotz et al. (2004).

[8] For instance, El-Hani (2007), Falk (1986), and Griffiths and Neumann-Held (1999).

to a significant extent the use of concepts and models. They are embedded in a community committed to a specific set of epistemic practices that make it more likely that they employ particular meanings ascribed to gene concepts and gene function models, which properly operates in a given domain of investigation. They also tend to recognize the prospects and limits of different concepts and models. This does not mean that concepts and models are per se stabilized when they emerge in the scientific community. On the contrary, they usually appear in a more rudimentary way, and if they are adopted by the scientific community, they can be elaborated and eventually stabilized by the practice of using them to guide research. When concepts and models are fused, the scientific community may be able to work out possible incoherence. However, as the diversity of concepts and models expands – as we see in the case of genes in the post-genomic era – difficulties are more likely to arise, particularly in the absence of a clear and explicit demarcation between those diverse meanings. This means that we should not remain content with the tacit usage of distinct meanings in different research settings, but rather worry about the clear demarcation of their domains of application (El-Hani 2007).

Certainly, teachers and students are embedded in a number of communities just as scientists are part of the scientific community. Every human being participates in a number of communities, which shape their understanding of the world. They can be described, if we follow Wenger (1998), as "communities of practice" (CoPs), cohesive groups of individuals mutually engaged in a joint enterprise, who exhibit distinct sets of knowledge, abilities, and experiences, and are actively involved in collaborative processes, sharing information, ideas, interests, resources, perspectives, activities, and, above all, practices, such that they build a shared repertoire of knowledge, attitudes, values, etc. (see also Lave and Wenger 1991). In the scientific community, we can find CoPs which generate a shared repertoire of knowledge, epistemic practices, and values that can stabilize the understanding of theories, models, and concepts to varying degrees. This means that scientists build a collective empiricism (Daston and Galison 2010) that often allows them to deal with a variety of models and concepts in a more consistent way. Or, to put it differently, persons tend to form "thought collectives," communities that mutually exchange ideas and develop a given "thought style" (Fleck 1979/1935). What is at stake here, then, is that scientists, teachers, and students pertain to different communities of practice, if we follow Wenger's formulation or, thought collectives, if we follow Fleck's and, thus, will tend to assume different perspectives on the diversity of scientific models and concepts. And the fact that scientists can be embedded in communities that generate that very diversity is of the utmost importance here.

When we turn to science education, we have additional reasons to worry about conceptual variation about genes and their function and the hybridization of different gene concepts and gene function models, as argued by Gericke and Hagberg (2007, 2010a, b), Gericke et al. (in press), and Santos et al. (2012). After all, even though teachers and students are themselves embedded in CoPs or thought collectives, they are not embedded in those scientific communities that generate knowledge about genes and their function. Moreover, in educational settings conceptual variation tends to be greater than in the scientific community, since both scientific and everyday meanings are represented and interact with each other within classrooms (Mortimer and Scott 2003), and disciplinary boundaries which may stabilize meaning making are not always present. In sum, when compared to the scientific community, there is much more potential to indiscriminate mixture of semantically incommensurable scientific concepts and models in the science classroom and, thus, a much bigger potential that logical inconsistencies and conceptual incoherence emerge. This is particularly true when science is taught without due attention to its history and philosophy.

It is important, therefore, to investigate whether and how conceptual variation related to the gene concept and gene function models is present in school science and also what potential problems it may bring to genetics teaching and learning. In this chapter, we will survey the results of a research program conducted in our lab in the last 7 years, focusing on how ideas about genes and gene function are treated in school knowledge, as represented in textbooks and students' views. Moreover, following our usual approach to research on science education, we move from descriptive to intervention studies, i.e., from diagnosing views on genes to investigating a teaching strategy implemented in a classroom setting with the goal of changing higher education students' views and, in particular, improving their understanding of scientific models and conceptual variation around genes and their function. Here, we will first consider results from investigations on how higher education and high school textbooks deal with genes, gene function, and their conceptual variation. Second, we will report unpublished results concerning how higher education biology students deal with genes and gene function. Third, we will present findings of an unpublished intervention study in which we investigate design principles for teaching sequences about genes and their function, considering conceptual variation in genetics and molecular biology, the crisis of the gene concept, and current proposals for revising its meaning. As a background for these empirical researches, we will turn to their theoretical underpinnings, resulting from both the literature on philosophy of biology/theoretical biology and the educational literature.

16.2 Genes and Gene Function Through the History of Genetics

The term "gene" was created in 1909, by Johannsen, following his distinction between genotype and phenotype, which told apart two ideas embedded in the term "unit character," then largely used, (1) a visible character of an organism which behaves as an indivisible unit of Mendelian inheritance and, by implication, (2) the idea of that entity in the germ cell that produces the visible character (Falk 1986). Johannsen proposed, then, the existence of basic units composing the genotype and phenotype, respectively, "genes" and "phenes." While the latter term never gained currency in biology, the former became central in newborn genetics and marked its development throughout the twentieth century.

Genes were seen instrumentally in the beginnings of genetics. Johannsen conceived "gene" as a very handy term with no clearly established material counterpart (Johannsen 1909). Although accepting that heredity was based on physicochemical processes, he warned against the conception of the gene as a material, morphologically characterized structure. For Johannsen, "the gene is [...] to be used as a kind of accounting or calculating unit" (Johannsen 1909; See Falk 1986; Wanscher 1975). At that period, the gene (that "something" which was the potential for a trait) could only be inferred from its "representative," the trait. That is, the gene was defined top-down, based on the phenotype.

This way of understanding genes is part of the Mendelian model of gene function, as reconstructed by Gericke and Hagberg (2007). According to this model, the gene is the unit of transmission (or inheritance) and function, treated as an abstract entity interpreted instrumentally as a phenotype in miniature. The function of the gene is of minor importance in the Mendelian model, focused on explaining genetic transmission. Moreover, due to the instrumental nature of the gene and its definition from the phenotype, this model conceives the gene as a necessary and sufficient condition for the manifestation of a trait, with no consideration of environmental or any other factor besides those instrumental entities. Thus, it assumed a unitary relationship between genes and traits, and the idea of genes as units became central in Mendelian genetics, thereafter substantially influencing twentieth century biology.

With the establishment of the chromosome theory of heredity by T. H. Morgan and his group, a new understanding of genes emerged (Carlson 1966). This understanding amounts to Gericke and Hagberg's (2007) classical model of gene function. The gene acts, in this model, as the unit of genetic transmission, inheritance, function, mutation, and recombination (Mayr 1982). Two additional important ideas are that genes exist in different variants (alleles) and consist or act as enzymes that produce traits. Since the molecular structure of genes was unknown, this latter idea was vague, and genes and their function were still inferred from traits. This model treated genes, however, as more active in the determination of traits than the Mendelian model did. Due to the development of linkage maps by Alfred Sturtevant, from Morgan's group, genes came to be interpreted in terms of the beads-on-a-string concept. Those quantified particles in the chromosomes were increasingly seen in realist rather than instrumentalist way, despite Morgan's hesitation (Falk 1986). Another notorious member of Morgan's group, Herman J. Muller, was one of the first supporters of the idea that genes were material units, "ultramicroscopic particles" in the chromosomes, arguing against the description of the gene as "a purely idealistic concept, divorced from real things" (quoted by Falk 1986). This view paved the way for subsequent steps in a research program aiming at elucidating the material bases of inheritance.

With a minor modification resulting from biochemical studies on the nature of genes, what Gericke and Hagberg (2007) call the biochemical-classical model of gene function emerged. The gene was treated, then, as being responsible for the production of a specific enzyme, which produced a trait. Also, as increased knowledge on biochemical reactions became available, the focus shifted from transmission to gene action and function. The biochemical-classical model explained gene function by reducing it to the relationship between a specific enzyme produced by the gene and the determination of a phenotypic trait. The model did not explain, however, the biochemical processes involved, and consequently, it still used the conceptual tools of classical genetics. The biochemical-classical gene was still an entity with unknown molecular structure.

The biochemical-classical model is the origin of the famous "one gene-one enzyme" hypothesis, which suffered several reformulations with increasing knowledge: when it was shown that the gene product was not always an enzyme, there was a shift to the "one gene-one protein" hypothesis, and when it was shown that proteins could be composed by several polypeptides, the "one gene-one polypeptide" hypothesis emerged. Finally, when it was established that RNAs could also be final gene products, the "one gene-one polypeptide or RNA" hypothesis prevailed. Notice, however, that an important shared content in all these hypotheses is that genes are treated as units.

At first, the gene was conceived as a unit of transmission, recombination, function, and mutation, but this did not hold. Benzer (1957) showed that units of function (his "cistrons") are typically much larger than units of recombination ("recons") and mutation ("mutons"). The terms "muton" and "recon" were deleted from the vocabulary of genetics, but "cistron" survived to these days and is often used in the place of "gene," indicating that the idea that prevailed was that of the gene as "unit of function."

The molecular-informational model (Santos et al. 2012)[9] was the culmination of a series of investigations about the material nature of the gene, which ultimately led to the proposal of the double helix model of DNA by Watson and Crick (1953). This model explained in one shot the nature of the linear sequence of genes, the mechanism of gene replication and RNA synthesis from DNA sequences, and the separation of mutation, recombination, and function at the molecular level. It was responsible for the wide acceptance of a realist view about genes, since there was now a clear material counterpart for the gene concept. The stage was set for a molecular definition of genes, in which genes were not defined anymore in a top-down manner, based on phenotypic traits, but in a bottom-up approach, focused on nucleotide sequences in DNA. This was accomplished through a concept named by Neumann-Held (1999) the classical molecular concept of the gene. According to it, a gene is a DNA segment encoding one functional product, which can be either a RNA molecule or a polypeptide. This concept superimposed a molecular understanding onto the idea of a hereditary unit supported by Mendelian genetics (Fogle 1990) and played an important role in the transition from classical genetics to a new era in which genetics and molecular biology became inseparable.

In the classical molecular concept, the gene is a continuous and discrete DNA segment, with no interruption or overlap with other units, showing a clear-cut beginning and end, and a constant location. Genes can be treated, then, as units of structure and, provided that they codify a single RNA molecule or polypeptide with a single function, also as units of function. And, with the introduction of information talk in

---

[9] This corresponds to Gericke and Hagberg's (2007) neoclassical model of gene function.

biology (Kay 2000) and in connection with the so-called central dogma of molecular biology, the gene became also a unit of information, simultaneously a chemical and a program for running life.[10] However, this idea is hardly trivial: despite the widespread usage of informational terms in molecular biology and genetics (say, "genetic information," "genetic code," "genetic message," "signaling,"), they can be still regarded as metaphors in search of a theory (El-Hani et al. 2006, 2009; Griffiths 2001). We do not have yet a sufficient and consistent theory of biological information, despite the utility of Shannon and Weaver's (1949) mathematical theory of communication for several purposes in biological research (Adami 2004). The nonsemantic understanding of information in this theory seems insufficient for a theory of biological information. Many authors argue that biology needs a theory of information including syntactic, semantic, and pragmatic dimensions (e.g., El-Hani et al. 2006, 2009; Hoffmeyer and Emmeche 1991; Jablonka 2002). Notwithstanding, genes are frequently treated as informational units, leading to the informational conception of the gene (Stotz et al. 2004), which is often superimposed onto the classical molecular concept even though it does not have a clear meaning.

As discussed in the introduction, several findings of genetic, molecular, and genomic research challenged in the last three decades the molecular-informational model, posing problems for the understanding of a gene as a unit of structure, function, and/or information. Even though the crisis of this model was more widely recognized in the last two decades of the twentieth century, many-to-many relationships were known to classical genetics already. Benzer, for instance, regarded the gene as a "dirty word" (Holmes 2006).

Gericke and Hagberg (2007) introduce a "modern model" to encompass these challenges, in which the gene is treated as a combination of DNA segments that acts in a process that defines the function. This stretch of DNA contains regulating sequences and a transcription unit, made of coding sequences, but also introns and flanking sequences. It is expressed to produce one or several functional products, either RNAs or polypeptides. Smith and Adkison (2010) complemented this account by considering two further elements: (1) the findings of the Human Genome Project, such as the relatively limited number of genes in human and other genomes, when compared to previous estimates, and the similarity in gene numbers between humans and other animals, and (2) the definition of gene proposed by the ENCyclopedia Of DNA Elements (ENCODE) project.[11] We need to be careful,

---

[10] This shows the connection between the informational conception of the gene and genetic determinism (Oyama 2000/1985), a common element of the "gene talk" (Keller 2000) that pervades the media and the public opinion. With the central dogma, DNA became a sort of reservoir from where all "information" in a cell flows and to which it must be ultimately reduced. Through their connection with the doctrine of genetic determinism, the conceptual problems related to genes and genetic information have important consequences for public understanding of science and several socioscientific issues related to genetics and molecular biology (say, genetic testing, cloning, genetically modified organisms).

[11] The ENCODE project is an international consortium of scientists trying to identify the functional elements in the human genome sequence, with significant impact on our understanding about genes and genomes. The ENCODE database can be reached at http://www.genome.gov/10005107#4.

however, in referring to a "modern model," since this may mask the fact that there is no prevailing model nowadays. The gene concept is now in flux, changing meanings as researchers produce novel interpretations of the structure and dynamics of the genomic system.

Several proposals for reformulating the gene concept appeared in the last 20 years. We will just mention some of them here, with no intention of being exhaustive or providing any detailed discussion.[12] Some authors argued against the idea of genes as units and proposed, instead, views about genes as combinations of nucleic acid sequences that correspond to a given product (Fogle 1990, 2000; Pardini and Guimarães 1992) and might be located in processed RNA molecules (Scherrer and Jost 2007a, b). These proposals accommodate anomalies such as overlapping and nested genes by denying the idea of genes as units in DNA.

Other authors put forward a process-oriented view of genes.[13] In Neumann-Held's "process molecular gene concept," for instance, genes are not treated as "bare DNA" but as the whole molecular process "… that leads to the temporally and spatially regulated expression of a particular polypeptide product" (Griffiths and Neumann-Held 1999, p. 659). Since different epigenetic conditions that affect gene expression are in this way built into the gene, this proposal can accommodate anomalies such as alternative splicing or mRNA editing.

Moss (2001, 2003) distinguished between two meanings ascribed to genes and, consequently, demarcated two concepts, gene-P and gene-D, which have been usually conflated throughout the twentieth century. Gene-P amounts to the gene as determinant of phenotypes or phenotypic differences. It is an instrumental concept, not accompanied by any hypothesis of correspondence to reality, and this is what allows one to accept the simplifying assumption of a preformationist determinism (as if the trait was already contained in the gene, albeit in potency). Gene-P is useful to perform a number of relevant tasks in genetics, such as pedigree analysis or genetic improvement by controlled crossing methods. Gene-D amounts to the gene as a developmental resource in causal parity (Griffiths and Knight 1998) with other such resources (say, epigenetic ones). It is conceived as a real entity defined by some molecular sequence in DNA which acts as a transcription unit and provides molecular templates for the synthesis of gene products, being in itself indeterminate with respect to the phenotype (Moss 2003, p. 46). Gene-D is in accordance, thus, with the classical molecular concept. Moss argues that genes can be productively conceived in these two different ways, *but nothing good results from their conflation* (Moss 2001, p. 85). This conflation is one of the main sources of genetic determinism, with important consequences to socioscientific issues, since it leads to the idea of

---

The participants of the ENCODE can be found at http://www.genome.gov/26525220. See also The ENCODE Project Consortium (2004).

[12] For detailed discussion, see Meyer et al. (2013). When we consider these views about genes and their function, it is worth pondering about the school level to which they can be adequately transposed. This issue is also discussed by Meyer et al. (2013).

[13] See, for example, El-Hani et al. (2006, 2009), Griffiths and Neumann-Held (1999), Keller (2005), and Neumann-Held (1999, 2001).

genes as major or even single causal determinants of phenotypic traits, even highly complex traits, such as sexual orientation, intelligence, or aggression.

Among the contributions of the ENCODE project, we find a new definition of gene: "... *a union of genomic sequences encoding a coherent set of potentially over-lapping functional products*" (Gerstein et al. 2007, p. 677, emphasis in the original). In this definition, different functional products of the same class (proteins or RNAs) that overlap in their usage of the same primary DNA sequences are combined in the same gene, and thus, several anomalies are accommodated by challenging the unitary relationship between genes, gene products, and gene function embedded in the classical molecular concept.

Some works strive for solving the gene problem by building new languages that cut up the genetic system into novel categories, organizing our understanding into different sets of concepts (Keller and Harel 2007; Scherrer and Jost 2007a, b). On the one hand, this may solve, or dissolve, problems and limits posed by our current language about genes. On the other, there is an expected difficulty of trans-lation between the new languages and the one already established in the fields of genetics and molecular biology, which may hamper researchers' understanding of those new ways of speaking and, thus, their acceptance. To maintain sufficient bridges between new and older ways of speaking seems crucial, then, for the success of these proposals.

When we consider these new views about genes and their function, it is worth pondering about the school level to which they can be adequately transposed. This is not the space, however, to enter this discussion (see Meyer et al. 2013).

16.3 Methods

16.3.1 Textbook Studies[14]

16.3.1.1 Sample

We analyzed higher education and high school textbooks. A sample of higher education Cell and Molecular biology textbooks was selected through a survey of 80 course syllabi of 67 universities located in the 5 continents, randomly chosen in Google® searches performed in 2004. We analyzed three of the most used text-books, respectively, Lodish et al. (2003, n = 33 syllabi, the most used), Alberts et al. (2002, n = 28, the second most used), and Karp (2004, n = 5, the fifth most used). In many countries these textbooks are used in their original language, although it is possible to find translations. Thus, we analyzed them in the original language.

Eighteen biology textbooks (see Appendix 1) submitted by publishing companies to the Brazilian National Program for High School Textbooks (PNLEM) (El-Hani et al. 2007, 2011) were analyzed. This sample shows external validity regarding

---

[14] For more details, see Santos et al. (2012) and Pitombo et al. (2008).

Brazilian textbooks. PNLEM is a huge governmental initiative, providing textbooks to students enrolled in public high schools throughout the country. These textbooks are aimed at general high school biology courses attended by all students, covering all areas of biology. Besides being distributed to public schools by PNLEM, most of these textbooks are also used by private schools.

16.3.1.2 Textbook Content Analysis

Each textbook was analyzed as a whole using categorical content analysis (Bardin 2000). The procedure involved, first, the decomposition of the texts into units of analysis (recording units), from which categories were built through regroupings of text elements sharing characteristics identified by semantic criteria, i.e., by the presence of the same meaning in different text elements, not by the occurrence of specific keywords or sentences. First, an exploratory reading was performed to plan the decomposition of the texts, data treatment, and categorization. Besides the units of recording, we also considered units of context, larger segments of text embedding the units of recording, which provided a background for interpreting them. Recording units were the basic units for categorization and frequency calculation, varying in size from a single statement to a whole paragraph.

Since different areas of biology use particular epistemic practices, which lead to the creation of distinct ways of thinking and speaking about genes, most units of context were related to biological subdisciplines. In high school textbooks, they were characterization of life and/or living beings (i.e., the introductory chapters in the textbooks), cell and molecular biology, genetics, evolution, and glossary.[15] In higher education textbooks, the following units of context were employed: classical genetics, developmental genetics, evolutionary/population genetics, genetics of microorganisms, genetics of eukaryotes, medical genetics, molecular biology/molecular genetics, cell biology, biochemistry, cell signaling, genetic engineering, genomics, introduction, history of science, and glossary.[16]

Higher education textbooks were analyzed by using categories informed by the historical, philosophical, and scientific literature about genes. In high school textbooks, we employed three analyzing procedures: (1) analysis of gene concepts and (2) analysis of function ascription to genes, both based on the abovementioned literature, and (3) analysis of historical models of gene function, as described by Gericke and Hagberg (2007). In the latter analysis, we used the research instrument built by these authors, with some changes, to investigate how the variants associated with each of the seven epistemological features of the historical models were found in the recording units (Table 16.1).

Depending on the combination of epistemological feature variants used in an explanation of gene function, the explanation present in the recording unit can be classified into the historical models (Table 16.2). However, in school science, models are often reconstructed in a nonhistorical way, due to neglect of their historical

---

[15] Only 4 textbooks had a glossary. All other units of contexts were present in all textbooks.
[16] A glossary was present in all the textbooks.

16 History and Philosophy of Science and Hybrid Views about Genes

Table 16.1 Description of the epistemological feature variants used in the high school textbooks analyses

Epistemological features		Epistemological feature variant
1 The structural and functional relation to the gene	1a	The gene is an abstract entity and, thus, has no structure
	1b	The gene is a particle on the chromosome
	1c	The gene is a DNA segment
	1d	The gene consists of one or several DNA segments with various purposes
	1e	**The gene is a carrier, bearer, and/or unit of information**
2 The relationship between organization level and definition of gene function	2Ia	The model has entities at the phenotypic level and abstract concepts[a]
	2Ib	The model has entities at the phenotypic and cell levels[a]
	2Ibx	The model has entities at the phenotypic, cell, and molecular levels[a]
	2Ic	The model has entities at the molecular level
	2Icx	The model has entities at the cell and molecular levels
	2Icy	**The model has entities at the phenotypic and molecular levels[a]**
	2IIa	The correspondence between gene and its function is one-to-one
	2IIb	The correspondence between gene and its function is many-to-many
3 The 'real' approach to define the function of the gene	3a	The function of the gene is defined "top-down"
	3b	The function of the gene is defined "bottom-up"
	3c	The function of the gene is defined by an underlying process related to the capacity of expressing a particular gene product[b]
4 The relationship between genotype and phenotype	4a	There is no separation between genotype and phenotype
	4b	There is a separation, without explanation, between genotype and phenotype
	4c	There is a separation between genotype and phenotype with enzyme as intermediate causal explanation[b]
	4d	There is a separation between genotype and phenotype, explained by biochemical processes
5 The idealistic *versus* naturalistic relationships in the models	5Ia	There are idealistic relations in the model, with no reference to natural processes[b]
	5Ib	There are naturalistic relations in the model, with a detailed description of the biochemical process of gene expression[b]
	5IIa	The relations in the model are causal and mechanistic (chemical interactions of genes determine traits independently of context)[b]
	5IIb	The relations in the model are process oriented and holistic (the function of the gene depends on the context in which it is embedded)[b]
6 The reduction explanatory problem	6a	There is explanatory reduction from the phenotypic level to abstract concepts[a]
	6b	There is explanatory reduction from the phenotypic to the cell level[a]
	6bx	There is explanatory reduction from the phenotypic level to the molecular level[a]
	6c	There is no explanatory reduction
7 The relationship between genetic and environmental factors [in development and the construction of the phenotype]	7a	Environmental entities are not considered
	7ax	Environmental entities + genetic entities result in a trait/product/function[a]
	7b	Environmental entities are implied by the developmental system
	7c	Environmental entities are shown as part of a process

Variants in gray were introduced by Santos et al. (2012) in the original research instrument constructed by Gericke and Hagberg (2007)
[a] Changes in terminology introduced by Santos et al. (2012) in the epistemological feature variants
[b] Variants modified by Santos et al. (2012) in order to make some aspects more explicit
[c] The relationship is understood in additive terms, each factor being related to the product, but with no significant mutual influence between them

and epistemological backgrounds during didactic transposition (Justi and Gilbert 1999). Thus, hybrid models are often found, i.e., explanatory models consisting of aspects belonging to different historical models, which may be incoherent if incommensurable aspects are mixed up. We calculated the degree of model hybridization in textbook explanations of gene function, by ascertaining the frequency of

Table 16.2 Models of gene function and their epistemological feature variants (Gericke and Hagberg 2007, modified by Santos et al. 2012)

Models of gene function	Epistemological feature variants								
	1	2I	2II	3	4	5I	5II	6	7
Mendelian model	1a	2Ia	2IIa	3a	4a	5Ia	5IIa	6a	7a
Classical model	1b	2Ib	2IIb	3a	4b	5Ia	5IIa	6b	7a
Biochemical-classical model	1b	2Ib	2IIa and 2IIb	3a and 3b	4c	5Ia	5IIa	6b	7a
Neoclassical (or molecular-informational) model	1c and 1e	2Ic	2IIa	3b	4d	5Ib	5IIa	6c	7b
Modern model	1d	2Ic	2IIa	3c	4d	5Ib	5IIb	6c	7c

false-historical (i.e., belonging to the wrong historical model) and nonhistorical (i.e., not present in any of the historical models) feature variants.

We analyzed the presence of historical models of gene function in the textbooks in two different ways. In a previous study (Santos et al. 2012), we identified feature variants related to these models in each set of chapters related to the domain of a biological subdiscipline and, then, checked the model to which most of the epistemological feature variants were linked. We identified, thus, the prevailing model at that set of chapters, while the other feature variants, either false historical or nonhistorical, allowed us to calculate the degree of model hybridization at that same portion of the textbook. In a subsequent work (Gericke et al. in press), we described which models of gene function prevailed in each chapter and, then, calculated the degree of hybridization based on false-historical and nonhistorical feature variants. In this work, we will consider only the latter analysis.

The analyses of higher education textbooks were performed by the same researcher in order to increase their reliability, while two other researchers examined all the analyses, comparing part of the results with the original textbooks. In the study about high school textbooks, internal reliability was increased by carrying out independent analyses of the recording units by two researchers (cf. LeCompte and Goetz 1982). Inter-rater agreement between these analyses was high, reaching 89.9 %. The two raters and a senior researcher discussed the diverging categorizations, looking for shared agreement, such that the findings amount to consensus reached by those three researchers. In four instances where no consensus was reached, the recording units were excluded from the analysis.

16.3.2 Study on Higher Education Students' Views About Genes and Their Functions

16.3.2.1 Sample

We investigated the views of 112 biology undergraduate students of two Brazilian universities (Federal University of Paraná, UFPR, hereafter U1 – 60,

Federal University of Bahia, UFBA, hereafter U2 – 52 students) on genes and their functions. The sample from each university was subdivided according to whether or not they had already attended Genetics courses. All students that had already attended Genetics courses had also previously attended Cell and Molecular biology courses.[17]

16.3.2.2 Data Gathering Tool

We employed a questionnaire constructed and validated by ourselves, comprising three sections: (A) students' personal data, including information on his/her experiences on teaching and research training; (B) open and closed questions on genes, challenges to the classical molecular gene concept, and biological information; and (C) closed questions on the gene concept. Sections (B) and (C) contained 11 questions. Due to space constraints, we will consider only the results for three of them. The first is deliberately open ended and divergent, aiming at eliciting a diversity of answers: "In your view, what is a gene?" The other two are closed-ended questions, which were partly derived from Stotz et al. (2004). Both presented the same options for the students to mark, but one was a forced choice, while the other was a free-choice question. Here are the statements that the students could choose with the understanding of genes closer to each shown within brackets (information not available for the students): (a) A gene is a heritable unit transmitted from parents to offspring [Mendelian concept]. (b) A gene is a sequence of DNA which codes for a functional product, which can be a polypeptide or an RNA [Classical molecular concept]. (c) A gene is a structure which transmits information or instructions for development and organic function from one generation to another [Informational conception]. (d) A gene is a determinant of phenotypes or phenotypic differences [Gene-P]. (e) A gene is a developmental resource, side to side with other equally important resources (epigenetic, environmental) [Gene-D]. (f) A gene is a process that includes DNA sequences and other components, which participate in the expression of a particular polypeptide or RNA product [Process molecular concept]. (g) A gene is any segment of DNA, beginning and ending at arbitrary points on the chromosome, which competes with other allelomorphic segments for the region of chromosome concerned [Evolutionary gene concept, sensu Dawkins]. (h) A gene is a sequence of DNA with a characteristic structure [Classical molecular concept]. (i) A gene is a sequence of DNA with a characteristic function [Classical molecular concept]. (j) A gene is a sequence of DNA containing a characteristic information [Informational conception].

The study was approved by the Research Ethics Committee of the Institute of Collective Health/UFBA and by the National Committee of Research Ethics (recording number 12112), and the participants gave informed consent to participate.

---

[17] In both universities, the biology curriculum includes two courses on Genetics and one course on Cell and Molecular biology.

16.3.2.3 Data Analysis

For analyzing the students' responses to the open-ended question, we used the same technique described in the study about textbooks, categorical content analysis, following the same procedures. In the closed questions, we tabulated the frequencies of the alternatives marked in the forced- and free-choice items.

In order to increase internal reliability, two researchers performed independent analyses of the students' answers to the open-ended questions. Inter-rater agreement between these analyses was not very high, reaching 60 %. It was very important, then, to discuss the differences in categorization between those two researchers. This was done by a group of four researchers, including two senior researchers not involved in the previous analyses. We included in the final analysis only those answers for which shared agreement was possible.

The hybrid answers to the open-ended question were recategorized by three researchers who strived for reaching a consensus concerning the prevailing meaning. Once each answer was classified into a single category, they were analyzed statistically through a chi-square test in order to ascertain whether there were significant differences between the views of students who had attended or not the Genetics and Cell and Molecular biology courses. Thus, we could test the influence of the courses on students' ideas about genes in both universities, including also data from the closed questions. The null hypothesis (H_0) was that the two variables would be independent, i.e., the fact that the students had attended the courses would not affect their views about genes and their functions. H_0 would be rejected when the calculated chi-square was equal to or greater than 9.48, and the alternative hypothesis (H_1) would be accepted, showing influence of the courses on students' views. The significance level (α) was 0.05 and for all questions the degree of freedom was equal to 4.

16.3.3 Investigating a Teaching Sequence on the Problem of the Gene

16.3.3.1 Construction of the Teaching Sequence

The study was conducted in two classes of Medicine freshmen students, who attended in the second semester of 2009 the Cell and Molecular biology course under the responsibility of a teacher-researcher involved in the study, at Federal University of Bahia, located in Northeast Brazil. One class (11 students, 18–24 years) followed an approach employed by the teacher for many years, with no explicit discussion about gene function models and gene concepts (hereafter, class A). In another class (13 students, 15–23 years), the new teaching sequence was implemented, including an explicit discussion on those models and concepts, in a modest but explicit approach to the nature of science (NOS) (Matthews 1998; Abd-El-Khalick and Lederman 2000) (class B). Most students came from households with high and middle income.

Table 16.3 Framework proposed by Mortimer and Scott (2003) for the analysis of interactions and meaning making in science classrooms

Analytical aspects	
i. Teaching focus	1. Teaching purposes 2. Content
ii. Approach	3. Communicative approach
iii. Actions	4. Patterns of interaction
	5. Teacher's interventions

The Cell and Molecular biology course is traditionally divided into two modules, molecular and cellular. Usually, the course includes theoretical and practical lessons and students' seminars. Theoretical lessons comprise a short quiz; an activity oriented by a study guide; teacher's exposition, in which he makes the students feel free to pose questions and raise doubts; small group work, in which selected texts are discussed; and whole class discussion. Practical lessons aim at allowing students to observe cell phenomena and offering them an initiation to lab practices. In the seminars, students are divided into small groups to present selected scientific papers.

The teaching sequence was built collaboratively with the teacher, who has B.Sc. in Biological Sciences and M.Sc. and Ph.D. in Pathology. At the time of the study, he had 17 years of experience teaching this same course.

To construct the teaching sequence, we considered three a priori analytical dimensions (Artigue 1988; Méheut 2005): (1) epistemological, related to the contents to be learned, the problems they can solve and their historical genesis; (2) psycho-cognitive, considering the students' cognitive characteristics; and (3) didactic, linked to the constraints posed by the functioning of the teaching institution (programs, timetables, etc.). The first dimension followed from the historical and philosophical background used in the research program. The second benefited from the collaboration with the teacher, who has a wealth of knowledge on students' previous conceptions, difficulties, etc. Finally, we deliberately constructed the teaching sequence to be compatible with the typical constraints involved in undergraduate Cell and Molecular biology courses, which typically have extensive syllabi in Brazilian universities, with much content to be covered usually in 45–60 h. We planned the teaching sequence to fit into the time made available by the teacher, 5 h distributed in 2 days of classes. Within these time constraints, he assured us, it would be more feasible that the proposal could be used in most similar courses.

We used a discourse analysis perspective to plan the activities, designing communicative approaches and interaction patterns to be used by the teacher. The framework for classroom discourse analysis developed by Mortimer and Scott (2003) was adapted for this goal. It is based on five interrelated aspects that focus on the teacher's role, grouped in three dimensions: *teaching focus*, *communicative approach*, and *actions*. The *communicative approach* is the central element, since it is through it that we understand how the teaching focuses are worked, i.e., the *teaching purposes* and *contents*, by means of which actions, the *pedagogical interventions*, which result in certain *patterns of interaction* (Table 16.3).

The investigation was framed in the context of educational design research (Baumgartner et al. 2003; Plomp 2009; van den Akker et al. 2006), which aims at both developing educational interventions and advancing our knowledge about their characteristics and the processes of designing and developing them. The main research question in educational design research is to establish what are the characteristics or design principles of an intervention X for obtaining the outcome $Y(Y_1, Y_2, ..., Y_n)$ in context Z (Plomp 2009). Design principles are initially derived by us from the relevant literature and practitioner knowledge, and as the investigation of a series of prototypes of the teaching sequences proceeds, we not only test the initial design principles but also derive additional principles from the empirical results.

At this point, we tested just the first prototype of the teaching sequence in a single classroom. The following design principles were used: (1) The classroom discursive interactions were planned to flow from a dialogical approach, in which students' ideas played a prominent role in meaning making, to a more authoritative approach, in which the diversity of ideas raised was subjected to evaluation and selection in order to construct in the classroom the perspective of school science; (2) in classroom discursive interactions, the teacher stressed key ideas when they appeared, in order to construct the school science perspective around them; (3) texts produced by ourselves, aiming at the didactic transposition of debates on genes and their functions, were provided to small groups of students, alongside with guiding questions; (4) the teaching sequence used a historically and philosophically informed approach, putting emphasis on the role of models in science, their relation with reality, and the importance of their demarcation; (5) several historical models of gene function and gene concepts were explicitly addressed and differentiated; (6) the crisis of the classical molecular concept was explicitly discussed, as well as reactions to it; (7) in order to discuss this crisis, the teacher used molecular phenomena already addressed in his classes previously, even though at that point no conceptual consequences related to genes were derived.

16.3.3.2 The Teaching Sequence

The teaching sequence adopted an explicit approach to the NOS in the context of teaching about genes and their function, seeking to promote learning *with* models and *about* models.

The first class begins with the teacher asking the students what is a gene, an open-ended and divergent question intended to raise as many students' conceptions as possible. The teacher avoids evaluative comments or gestures, in order to maintain the dialogical interaction with the pupils. As the students offer their answers, the teacher copies them in the blackboard to be used later. This activity is followed by an exposition about models and their role in science. Even though the teacher speaks most of the time, he prompts the students to participate by posing questions. The students are divided into small groups and receive the first text prepared by our team, "historical models of the gene concept" (text contents are similar to those found in Sect. 16.2 above), followed by a number of guiding questions for cooperative discussion. The answers are used by the teacher to promote whole class discussion, which creates the opportunity to prompt sharing of the discussions in the small groups, to check the students' understanding and to stress key ideas for the construction of the intended perspective on genes. He goes back then to the students' initial answers, available in the blackboard, discussing the relationship between their ideas and the historical models about gene function. Now he evaluates their answers, showing when they are closer to one or another model and pointing out which models are still accepted and in what features. He also stresses which answers are distant from any scientific model and brings to the fore the hybrid models, if they are present in the students' answers. The expectation is that, at the end of the class, the diversity of students' ideas raised and the diversity of scientific models about genes have been systematized.

In the second class, the teacher begins by briefly reviewing the previous session and posing questions for the students in order to evaluate their understanding. Then, he makes an exposition on the crisis of the classical molecular concept, using challenging phenomena that were already discussed in the previous classes, such as alternative splicing and gene overlapping. The students are divided again into small groups, receiving the second text we prepared, "proposals for the gene concept" (text contents are similar to those in Sect. 16.2), with guiding questions. Again, this is followed by whole class discussion. The class ends with a discussion on the current status of our understanding about genes, in which the teacher highlights the idea that the classical molecular concept is in crisis, but none of the proposals discussed in the second text are widely accepted by the scientific community. The intended perspective on genes is arguably clear for the students: the gene concept is now changing under our very noses, with all directions of change still being debated. The teacher also takes a last opportunity to stress the existence of a diversity of gene concepts and models of gene function, claiming that several models show greater explanatory and heuristic powers than a single, overarching definition of gene, provided that we properly demarcate their domains of application.

16.3.3.3 Teaching Sequence Validation

We performed a posteriori internal and external validation of the teaching sequence (Artigue 1988; Méheut 2005). In the internal validation, we compared the effects of the teaching sequence in relation to its goals, by comparing the students' learning outcomes with the planned learning goals. To perform this comparison, we investigated how the students mobilized ideas about genes and their function in a discursive context structured by a subset of the items from the questionnaire used to investigate students' views (see above), with some modifications validated in a pilot test. Here we will discuss the same three questions mentioned above. In the closed questions, the alternative (g), related to the evolutionary gene concept, was excluded in this study. The questionnaire was used in three moments: in the second lesson of the whole course, when we could probe students' views with no influence of the course (pretest); at the end of the molecular module, which coincided with the last day of the teaching sequence (posttest); and two months after the intervention (retention test). The classes have also been video recorded to provide raw material for the analysis of classroom discursive interactions, but these data have not been treated yet.

In the internal validation, we are evaluating if the teaching sequence does reach the planned learning goals. If we use the framework presented by Nieveen et al. (2006), this is a development study, aiming at solving educational problems by focusing on the proposal and testing of broadly applicable design principles. The goal is to understand how and why a given intervention functions in the particular context in which it was developed. It is this knowledge that is summarized in design principles (Reeves 2006; van den Akker et al. 2006), or intervention or design theories (Barab and Squire 2004), which are expected to generalize beyond the context of the study. Although we cannot expand further on the topic here, we should mention that this knowledge is conceived by us as generalizing in two (not mutually exclusive) ways: (1) through situated generalization (Simons et al. 2003), i.e., the transformation of data gathered in a context into evidence transferable to other contexts, so as to indicate a course of action or be incorporated in judgments preceding action, due to teachers' perception of a connection between the investigated context and the context of their pedagogical work, and (2) as a generalization resulting from maximizing the variation of qualitatively different investigated cases (Larsson 2009). As we investigated only the first prototype of the teaching sequence, the second kind of generalization is not yet at reach. However, the first kind of generalization is already feasible, since other college and university teachers may perceive the same problems discussed here in their classrooms and, eventually, see in the teaching sequence a putative approach to their pedagogical practice.

We also performed a *preliminary* external validation of the sequence by comparing the effects of the teaching sequence with the approach employed for many years in the course. The same questionnaire was applied for class A in the same moments mentioned above. Using Nieveen and colleagues' (2006) framework, this is an effectiveness study, which can provide evidence for the impact of the intervention by comparing its effectiveness in relation to another teaching approach. As Brown (1992) argues, our goal in such a study should be to accommodate variables rather than controlling them, since research needs to occur within the natural constraints of real classrooms. One manner of accommodating confounding variables is to use sufficient numbers of replicas of each treatment such that we can distinguish between the effects of the intervention and confounding variables randomly assorted to the replicas, such as students' motivation, the quality of their previous knowledge, and teacher-students relationships. But when we do research in real educational contexts, we often do not count with enough number of classes for replicating treatments. This was the case in our study, since there was only one teacher interested in engaging in it, and he had only two courses under his responsibility. This means that we cannot sufficiently distinguish between the effects of the teaching sequence and confounding variables, although we had the same teacher and similar sets of students in the two classes. Nevertheless, the results revealed interesting patterns, although preliminary and to be taken with a grain of salt.

16.3.3.4 Data Analysis

The answers to the questions included in the tool were treated through categorical analysis (open-ended item) and tabulation (closed item) as described above. Internal reliability was increased in the open-ended question by independent analyses by two researchers, with high inter-rater agreement (89.1 %). Differences in categorization were discussed with two other researchers (one of them also the teacher of the course), and the final analysis included only those answers in which shared agreement was reached.

16.4 Results and Discussion

16.4.1 Textbook Studies

16.4.1.1 Views About Genes in Higher Education Cell and Molecular Biology Textbooks

Figure 16.1 shows the distribution of gene concepts in the three higher education Cell and Molecular biology textbooks we analyzed (Pitombo et al. 2008).

In Karp's (2004) textbook, there were 73 recording units explicitly addressing gene concepts, considerably more than in the other two books (35, Alberts et al. 2002, 23, Lodish et al. 2003). This follows from the fact that the former book focuses on concepts and experiments, as shown by its subtitle, giving more attention to history. Symptomatically, the Mendelian conception, according to which the gene is a unit of inheritance, showed the highest prevalence (31.5 %), and most of these occurrences were in sections discussing the history of genetics. The Mendelian conception is mostly treated in this textbook as a view on genes that is historically relevant, but is not often used to account for current perspectives on genes, which are frequently represented by the second most frequent view (24.6 %), the informational conception, in which the gene is seen as a unit or carrier of information. Since information is a metaphorical notion that still needs theoretical clarification in genetics (El-Hani et al. 2009; Griffiths 2001), it is problematic to appeal mainly to this idea to explain what genes are. The third more frequent concept in Karp was gene-P (20.5 %), which was mostly used in sections about the history of genetics and medical genetics, where it usefully abstracts away from the complexities of the genotype-phenotype relationship, focusing on the predictive relationship between gene loci and pathological conditions. Finally, the classical molecular concept appeared in 13.7 % of the recording units, distributed in a wide variety of contexts, including molecular biology, evolutionary genetics, genetic engineering, and genomics, besides historical narratives about genetics. We can say, therefore, that in this textbook, when genes are described in molecular terms and from an updated perspective, the molecular-informational model of gene function prevails.

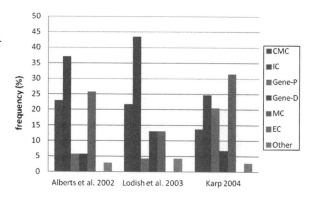

Fig. 16.1 Distribution of gene concepts in three higher education Cell and Molecular biology textbooks. *CMC* classical molecular concept, *IC* informational conception, *MC* Mendelian conception, *EC* evolutionary concept

Alberts et al. (2002) and Lodish et al. (2003) are much less diversified in their treatment of genes, even though they still show conceptual variation. In these textbooks, the informational conception was remarkably predominant (37.1 %, Alberts et al.; 43.5 %, Lodish et al.), being frequently associated with the classical molecular concept (22.9 %, Alberts et al.; 21.7 %, Lodish et al.). Their basic message about the nature of genes amounts, thus, to a combination of the metaphor of information and the idea of the gene as unit of structure and/or function in DNA, which is characteristic of the molecular-informational model.

In all the textbooks, the classical molecular gene concept was predominantly used when they were addressed contents related to Molecular biology and Molecular Genetics. This concept was also used by the three textbooks in their glossaries, in order to define genes. The informational conception, in turn, was found in more diversified contexts in the textbooks, when compared with the classical molecular gene concept, indicating how widespread this conception is, despite its lack of solid theoretical background.

However, the prevalence of the molecular-informational model sounds strange in the three textbooks, when we consider that they discuss the anomalies challenging it in the last decades. The conceptual lessons following from these empirical findings are not taken into account, yet another indication of a largely atheoretical and ahistorical treatment of the contents. Despite the presence of conceptual variation, these textbooks do not provide clues for teachers and students about the distinct origins, domains of application, and meanings of concepts related to different models along the history of genetics and molecular biology. Thus, hybridization of incommensurable aspects of different models and semantic confusion are likely to happen. This is a good case in point regarding the harmful consequences of teaching science without teaching about science. The students do not have much chance of learning with models and about models, since these textbooks address the contents as if they referred to reality themselves, as discovered by science, not to models about reality, historically constructed by the scientific community. The relationship between model and reality becomes unclear when most of the explanations just consider what *is* in the world, not how we interpret what *is* in the world based

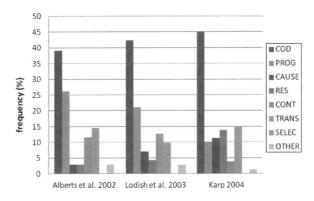

Fig. 16.2 Distribution of functions attributed to genes in three higher education Cell and Molecular biology textbooks. *COD* codifying the primary structure of polypeptides or RNAs (classical molecular concept), *PROG* program or instruct cellular functioning and/or development (informational conception), *CAUSE* cause or determine phenotype or difference between phenotypes (gene-P), *RES* act as a resource for development (gene-D), *CONT* control cell metabolism (informational conception), *TRANS* transmit hereditary traits (Mendelian conception), *SELEC* act as unit of selection (evolutionary concept)

on theoretically laden evidence and inferences (which are often conflated in the textbooks with observations).

As an example, the following definition hybridizes features related to the Mendelian and the informational conception:

Gene - Physical and functional unit of heredity, which carries information from one generation to the next (Lodish et al. 2003, Glossary, G-9).

This sentence, in turn, hybridizes gene-P and the informational conception:

These instructions are stored within every living cell as its genes, the information-containing elements that determine the characteristics of a species as a whole and of the individuals within it (Alberts et al. 2002, p. 191).

The harmful consequences of combining these different features of historical models become apparent, as the idea of "genetic information" is taken to imply a reduction of the development of all characteristics of the species and the individuals to the DNA nucleotide sequences. We can explicitly see the connection between the genetic determinism that often marks gene talk in the social arena and the way genes are treated in these textbooks.

The interpretation that the molecular-informational model prevails in these textbooks is reinforced when we examine the functions attributed to genes (Fig. 16.2). In all of them, the function most frequently ascribed is codifying the primary structure of polypeptides or RNAs, aligned with the classical molecular concept (39.1 %, Alberts et al.; 42.3 %, Lodish et al.; 45 %, Karp). In the former two textbooks, the second most frequent function, to program or instruct cellular function and/or development, is also related to that model, namely, to the informational conception (26.1 %, Alberts et al.; 21.1 %; Lodish et al.). In Karp, to transmit hereditary traits is the second most common function (15 %), consistently with the high prevalence of the Mendelian conception.

Generally speaking, we observe a proliferation of meanings attached to genes as we progress from context to context in these textbooks, with no model unification or demarcation. This happens both in gene concepts and function ascription to genes.

16.4.1.2 Views About Genes in High School Biology Textbooks

Figure 16.3 shows the distribution of gene concepts in 18 Brazilian high school biology textbooks, including those approved and not approved by the Brazilian National Program for High School Textbooks (PNLEM) (Santos et al. 2012).

In these textbooks, three gene concepts were significantly more prevalent: the classical molecular concept, the informational conception, and the gene-P. In 12 of the 18 textbooks, gene-P was the most frequent, answering for more than 40 % of the recording units in 4 textbooks. The classical molecular concept and the informational conception were more prevalent in 3 textbooks each.

The fact that gene-P is so often used in these textbooks follows from the extensive content of the genetics chapters, where we find several examples of pedigree analyses and estimates of the inheritance probability of phenotypic traits. Here is an example of a recording unit showing gene-P:

> The gene for brown eyes located in the chromosome is an allele of the gene for green eyes, located in the homologous chromosome (T2, vol. 3, p. 15).[18]

Gene-P is often employed in the textbooks just as it was used in classical genetics, when genes were inferred from phenotypes. However, these statements are framed in an "updated" language, and thus, teachers and students cannot figure out that the textbook is using a way of understanding genes that was frequently employed when there was no established knowledge on the nature of the genetic material. Moreover, a key requirement for a valid usage of genes-P is not found in these textbooks, namely, a clear understanding of the distinction between this instrumental concept and a realist interpretation of the genetic material. In the absence of this distinction, gene-P is simply conflated with the classical molecular gene concept, which provides then a molecular background to understand genes as determinants of phenotypes, as expressed by gene-P. The kind of conflation that Moss (2001, 2003) identifies as a source of genetic determinism, between a preformationist instrumental concept (gene-P) and a molecular realist concept (gene-D), is favored by the way these textbooks deal with genes.

It is this sort of hybridization between features related to different models that can lead to semantic confusions, hampering students' understanding and favoring ideas with important socioscientific implications, such as genetic determinism. If a student learns that genes determine phenotypes in the absence of a historically and

---

[18] All translations of textbook passages from Portuguese were made by the authors of the present paper. Commentaries by the authors are shown in brackets.

16 History and Philosophy of Science and Hybrid Views about Genes 493

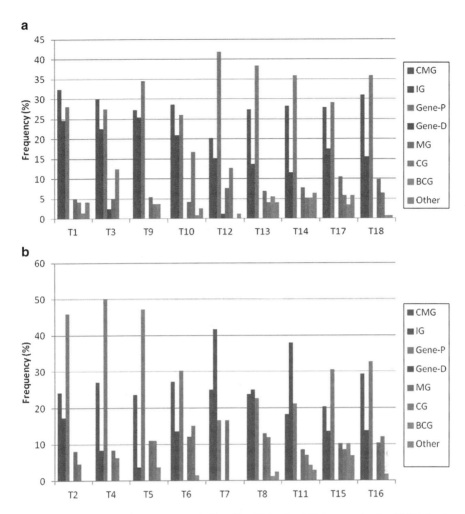

Fig. 16.3 Distribution of gene concepts in Brazilian high school biology textbooks. *CMG* classical molecular gene, *IG* informational gene, *MG* Mendelian gene, *CG* classical gene, *BCG* biochemical-classical gene. (**a**) Textbooks approved; (**b**) textbooks not approved by PNLEM. Textbooks are indicated by the codes listed in Appendix 1

epistemologically informed discussion of the role of this instrumental concept in classical genetics and then moves on to study about genes depicted in a realist manner as structural and functional units in DNA, the conflation between these two concepts and the resulting semantic confusions seem almost inevitable.

Symptomatically, in all textbooks in which gene-P prevails, the second most frequent concept was the classical molecular gene. Moreover, in 39.1 % of the recording units where we found the classical molecular gene, gene-P was also present. The classical molecular concept only entails colinearity between a gene and the primary structure of a protein or RNA but does not fix the relationship between genes and phenotypes at a higher level. This relationship enters the textbook explanation through the hybridization with gene-P, predictably leading to genetic determinism. The passage below illustrates the hybridization between the classical molecular gene and gene-P, with clear determinist undertones:

> Currently we know that the gene [...] is a sequence of nucleotides in DNA. Each gene is responsible for the synthesis of a protein and, consequently, for one or more characteristics of the individual, since proteins can have structural and regulatory functions in metabolism. Genes are located in chromosomes and are didactically represented by letters, numbers, and symbols. For instance, the gene for normal skin color is symbolized by *A* and the gene for albinism, by *a* (T6, p. 283).

This amalgam of a preformationist view of the gene as determinant of phenotypes and a molecular view of the gene as information carrier located in DNA is the major picture of the gene in these textbooks. The classical molecular concept, in particular, was found in the most diverse contents in the textbooks, in all three high school years, with relatively high frequency (Santos et al. 2012).

In Fig. 16.4, we can see the functions attributed to genes in the high school biology textbooks we analyzed. In almost all textbooks (17), genes are most often regarded as codifiers of the primary structure of polypeptides or RNAs (in accordance with the classical molecular concept) and determinants of phenotypes (in line with gene-P).

All the historical models identified by Gericke and Hagberg (2007) were found in the textbooks (Fig. 16.5), showing how they are marked by conceptual variation. The molecular-informational model was dominant, in keeping with the prevalence of the classical molecular concept and the informational conception in the textbooks. However, the difference of prevalence between the four most frequent models is in fact quite small, highlighting how the predominant feature of these textbooks is, in fact, conceptual variation, with no clear demarcation between the different models and their domains of application. Gericke and colleagues (in press) compared the distribution of these historical models in a large and significant sample of Swedish and Brazilian textbooks, as well as in 7 textbooks used in English-speaking countries. Despite some differences, the distribution of the different models within the textbooks of the different countries was very similar. They interpret this finding as showing that the conceptual variation in genetics is captured in a similar textbook discourse that is culturally independent, that is, didactic transposition (Chevallard 1989) leads to similar end products in those different countries, maybe as a consequence of the influence of the higher education textbooks used by textbook authors to learn about genetics and cell and molecular biology.

Half of the high school textbooks analyzed (9) discussed split genes. To our understanding, six of them treated split genes and splicing in a satisfactory manner. However, only three considered alternative splicing, and among the latter, only two discussed the conceptual implications of this phenomenon to the way genes are conceived.[19] This indicates that, in spite of the overwhelming predominance of an outdated

---

[19] It is worth noting, however, that none of the higher education cell and molecular biology textbooks offered such a discussion.

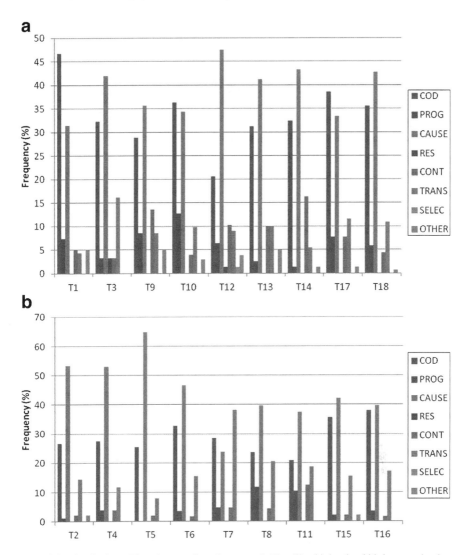

Fig. 16.4 Distribution of functions attributed to genes in Brazilian high school biology textbooks. *COD* codifying the primary structure of polypeptides or RNAs (classical molecular concept), *PROG* program or instruct cellular functioning and/or development (informational conception), *CAUSE* cause or determine phenotype or difference between phenotypes (gene-P), *RES* act as a resource for development (gene-D), *CONT* control cell metabolism (informational conception), *TRANS* transmit hereditary traits (Mendelian conception), *SELEC* act as unit of selection (evolutionary concept). (**a**) Textbooks approved; (**b**) textbooks not approved by PNLEM. Textbooks are indicated by the codes listed in Appendix 1

approach to the gene concept, at least in some textbooks, there seems to be an ongoing transition to a more updated treatment. However, in the majority of the high school textbooks, the case is similar to that of higher education textbooks: when the challenges to the classical molecular concept are discussed, relatively

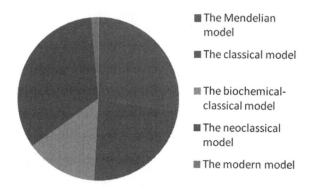

Fig. 16.5 Distribution of the historical models identified by Gericke and Hagberg (2007) in Brazilian high school biology textbooks (in percentage)

■ The Mendelian model

■ The classical model

■ The biochemical-classical model

■ The neoclassical model

■ The modern model

obvious conceptual consequences are not considered. This can be seen as a consequence of the way the textbooks typically approach scientific knowledge, as a list of isolated facts, building a fragmented rhetoric of conclusions (Schwab 1964).

When using the vast majority of these textbooks, students and teachers cannot get even a glimpse of the state of affairs in current discussions about genes. Some may think that it is too much to demand that school science considers these recent developments at high school. However, for most students this may be the last opportunity to learn about genes and their function and, thus, to build a critical stance towards gene talk in socioscientific issues, from the safety of genetically modified organisms to the use of genetic testing in society.

We also did a systematic analysis of model hybridization in the high school biology textbooks, finding a widespread use of hybrid models for describing gene function (Table 16.4), often combining features of models focusing on the molecular and cellular level with features of models dealing with the phenotypic level, derived from classical genetics. As Santos and colleagues (2012) show, the molecular-informational model seems to be taken as a basis by the textbooks, with features from a variety of models being hybridized with it. Thus, conceptual variation, although present in the textbooks, is not explicitly dealt with, being difficult for teachers and students to realize that different aspects of gene function are mixed up and, in particular, to take notice of the ambiguities, logical inconsistencies, and semantic confusions that may follow.

16.4.2 Higher Education Students' Views About Genes and Their Functions

The Biological Sciences students who participated in the study about their views about genes and their functions were divided into two groups, depending on whether they attended (YG) or not (NG) Genetics courses. In one of the universities investigated, located at the South part of Brazil (UFPR, U1), the distribution was 32 students in group YG and 28 in NG. In another university included in the study,

Table 16.4 Hybridization frequency of textbook models

	Mendelian model	Classical model	Biochemical-classical model	Molecular-informational model	Modern model
Level of hybridization (%)[a]	7.7	18.4	9.5	41.8	

[a]The level of hybridization equals the frequency of exchanged epistemological feature variants, calculated as the number of incorrect historical feature variants (nonhistorical and false historical) divided by the total number of feature variants in the textbook models

located in the Northeast region of Brazil (UFBA, U2), we had 19 students in YG and 33 in NG.

The chi-square test performed to statistically analyze the influence of the Genetics course on students' ideas about genes in both universities resulted in the values 9.83 and 10.07 in U1 and U2, respectively. Thus, in both universities, a significant relationship was found between the students' attendance to the Genetics courses and the views about genes expressed in their answers.

Figure 16.6 shows the distribution of the answers in the categories obtained in the analysis of the open-ended and divergent question "In your view, what is a gene?" for the two universities and the two groups.

Regarding the classical molecular concept and the informational conception, the results show similar effects of the Genetics courses on Biological Sciences students' views in the two universities. They led to a significant increase in the percentage of answers committed to the classical molecular concept and a decrease in the students' commitment to the informational conception, with the difference that only a slight decrease took place at U1.

On the one hand, if we consider that basically all the challenges faced by the classical molecular concept are addressed by those courses, we can suspect that no connection is made between examining empirical findings in genetics and cell and molecular biology and reflecting on their conceptual implications. This may be a consequence of the lack of an epistemological and historical dimension in the teaching practice in those courses. On the other hand, the impact they had on the students' appeal to the informational conception is a positive consequence of the courses, which can be attributed to the fact that the students are stimulated to delve into more details regarding the structure and function of the genetic material. This can be associated to both the increase in their allegiance to the classical molecular concept and the decrease in their use of the informational conception.

As an example of a student's answer committed to the classical molecular concept, we can quote[20]:

It is a fragment of DNA responsible for codifying a polypeptide chain or RNA (U1, student 20, YG).

---

[20] The answers were freely translated from Portuguese to English by the authors of the paper.

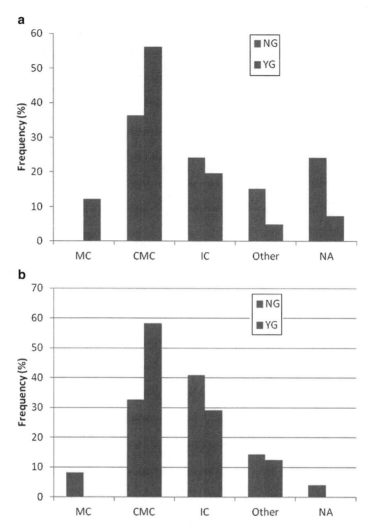

Fig. 16.6 Distribution of answers given by students of two Brazilian universities to the question "In your view, what is a gene?" *MC* Mendelian conception, *CMC* classical molecular concept, *IC* informational conception. (**a**) U1 (UFPR); (**b**) U2 (UFBA). The number of answers is larger than the number of students because there were answers which combined more than one view about genes and, thus, were classified in more than one category

Here is an example, in turn, of an answer exhibiting the informational conception:

Hereditary informational unit (U1, student 3, YG).

Different views about genes were often hybridized by the students in their answers (21.7 % of the answers in U1, 38.5 %, in U2). This suggests that the students may be reproducing the hybrid views about genes found in textbooks (see

Sect. 16.4.1). As there was no trend of decrease of such hybridization after the Genetics courses, classroom teaching and learning seems to be unable to overcome this difficulty posed by the treatment of genes and their functions in the textbooks.

In the closed questions, we used the classification of the alternatives into gene concepts shown in the methods section and, additionally, gathered less represented answers, related to gene-P, gene-D, and the evolutionary gene concept, into a single category, other gene concepts. When considering the forced-choice question, we can see the same pattern observed in the open-ended question regarding the prevalence of the classical molecular concept (particularly, item b, Fig. 16.7. In items h and i, also related to this concept, there were no important changes) and the decrease of the informational conception (items c and j, Fig. 16.7) after the students attended the courses.

In both universities, the students' commitment to the Mendelian conception, as shown by the closed questions, decreased (item a, Fig. 16.7). This may be a consequence of the impact of the molecular treatment of genes during the courses.

Now, compare Fig. 16.7 with Fig. 16.8, which shows the results for the very same closed question, but in a free-choice format. The pattern that is readily apparent is that the students marked a large variety of views about genes when they are allowed to do so. To our understanding, this is a striking evidence that conceptual variation regarding genes, as represented in higher education and high school textbooks, can be translated into students' allegiance to several different accounts about genes and their functions. In itself, the results from these two questions do not allow us to conclude that students are facing difficulties with this conceptual variation, for instance, not knowing what views about genes are more adequate to deal with what sorts of problems, or being entangled in semantic confusions and ambiguities following from combining incommensurable perspectives embraced by different models and concepts. But consider that teaching about genes in those courses uses the textbooks we analyzed, where a historically and epistemologically informed approach to models about genes and their function is typically lacking. It is at least plausible, then, to interpret the fact that the students marked so many different views about genes in the free-choice question as meaning that they are prone to conflate incommensurable aspects of models and, also, to misapply these models, using them outside their domain of validity.

16.4.3 From Diagnosis to Intervention: A Teaching Sequence on the Problem of the Gene

Our previous study on higher education students' views about genes and their functions suggested several shortcomings in teaching about genes at Genetics courses in two Brazilian universities. Part of the limitations of these courses could be attributed to the lack of an epistemological and historical dimension in the treatment of the contents, in particular, to an insufficient attention to teaching both *with* models and *about* models (Gericke and Hagberg 2007).

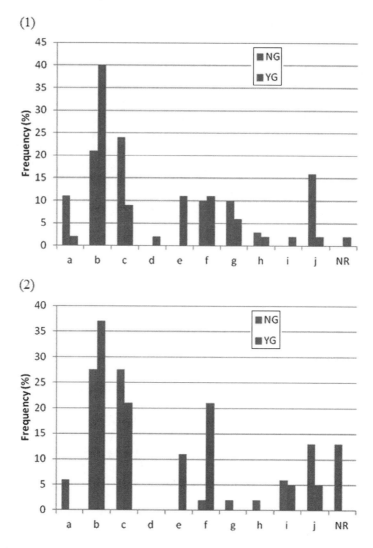

Fig. 16.7 Distribution of answers given by students of two Brazilian universities to a forced-choice closed question presenting several alternatives concerning the nature of genes: (**a**) Mendelian; (**b**), (**h**), and (**i**) classical molecular; (**c**) and (**j**) informational; (**d**) gene-P; (**e**) gene-D; (**f**) process molecular gene; (**g**) evolutionary gene concept. *NR* no response. (1) U1 (UFPR); (2) U2 (UFBA)

Therefore, it seemed natural to us to move from diagnosis to intervention, through the development and investigation of a teaching sequence built collaboratively with a higher education Cell and Molecular biology teacher at the Federal University of Bahia, located in Northeast Brazil. As presented in the Methods section, this teaching sequence explicitly addressed NOS contents, in particular, the historical construction and nature of gene function models and gene concepts. Our intention

16 History and Philosophy of Science and Hybrid Views about Genes

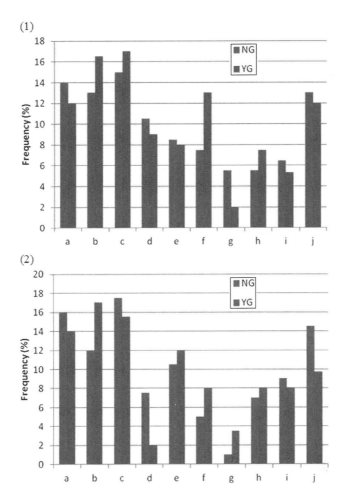

Fig. 16.8 Distribution of answers given by students of two Brazilian universities to a free-choice closed question presenting several alternatives concerning the nature of genes: (**a**) Mendelian; (**b**), (**h**), and (**i**) classical molecular; (**c**) and (**j**) informational; (**d**) gene-P; (**e**) gene-D; (**f**) process molecular gene; (**g**) evolutionary gene concept. (1) U1 (UFPR); (2) U2 (UFBA)

was not to deal with complex historical, philosophical, or sociological issues, but just to teach with models and about models when dealing with genes, as a way of providing conditions for the students to understand that genes have been and are still conceived in different ways in distinct subfields of biology, as a consequence of different epistemic practices that characterize the works of diverse scientific communities.

Figure 16.9 shows the distribution of the answers in the categories obtained in the analysis of the open-ended question "In your view, what is a gene?" in the three

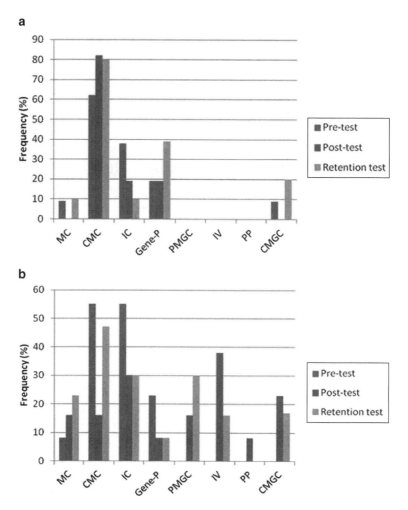

Fig. 16.9 Distribution of answers given to the question "In your view, what is a gene?" by the students of the classes investigated. *MC* Mendelian conception, *CMC* classical molecular concept, *IC* informational conception, *PMGC* process molecular gene concept, *IV*, instrumental view about genes, *PP* perception of the problem, *CMGC* contemporary molecular gene concept (The "contemporary molecular gene concept" amounts to a conservative response to the problem of the gene, which regards the gene as a linear DNA sequence but abandons the idea that it has a single developmental role, defining it, for instance, as "a DNA sequence corresponding to a single 'norm of reaction' of genes products across various cellular conditions" (Griffiths and Neumann-Held 1999, p. 658)). (**a**) Class A (usual approach to the course, with no explicit discussion on gene function models and gene concepts); (**b**) class B (where the teaching sequence was implemented). The number of answers is larger than the number of students because there were answers which combined more than one view about genes and, thus, were classified in more than one category

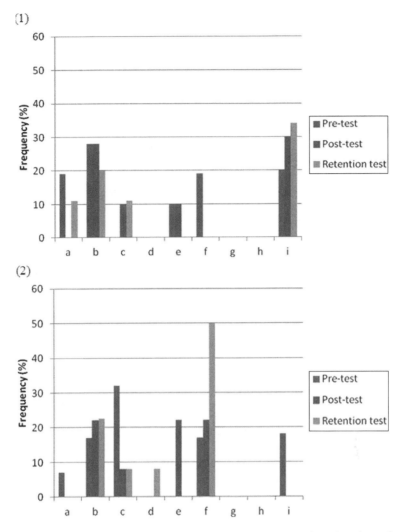

Fig. 16.10 Distribution of answers given by the students of the classes investigated to a forced-choice closed question presenting several alternatives concerning the nature of genes: (**a**) Mendelian; (**b**), (**g**), and (**h**) classical molecular; (**c**) and (**i**) informational; (**d**) gene-P; (**e**) gene-D; (**f**) process molecular gene. (1) Class A (usual approach to the course, with no explicit discussion on gene function models and gene concepts); (2) class B (where the teaching sequence was implemented)

moments in which the data were gathered. It is interesting to look at these results alongside with those for the closed forced-choice question, which allowed us to survey students' ideas about genes using a different kind of tool. We can see the distribution of answers in the pretest, posttest, and retention test in Fig. 16.10.

Considering, first, the internal validation of the teaching sequence, we can see some positive learning outcomes, compared to the intended learning goals: first, the informational conception was successfully challenged by the teaching sequence, falling in the posttest and maintaining the lower frequency in the retention test, when compared with the pretest, both in the open and in the closed forced-choice question. Here is an example of a students' answer committed to the informational conception and, also, showing a close relationship between this conception and genetic determinism:

Gene is the unit of data storage of the species. The union of the genes (which are in DNA) forms the genome, where we find all the information for the development of the being (Student 2, Class B, pre-test).

Second, the students showed an enriched repertoire of views about genes after the intervention. For instance, the process molecular gene concept increased in frequency in the posttest, reaching an even higher frequency in the retention test, both in the open and the closed forced-choice question. An instrumental view about genes was considered by a significant proportion of the students in the answers to the open question in the posttest, and despite the frequency dropped in the retention test, it still reached 16 % of the answers. An example of the instrumental view and the process molecular gene concept can be found in the following students' answer:

The gene concept is relative and depends on the way the gene will be studied. It can be understood as a physical structure that originates RNAs and proteins or as the fruit of a process or the very process, for instance (Student 1, Class B, post-test).

There were also limits, however, regarding the planned learning goals. The most important concerns the fact that, even though the commitment to the classical molecular concept significantly decreased among the students in the posttest, this was just a transitory effect. Almost the same frequency of students' answers to the open question related to this concept was found in the pretest and retention test. If we consider alternative (b) in the closed forced-choice question, we see a similar pattern, with a slight increase in the posttest that is maintained in the retention test. The following answer is a straightforward example of a student's rendering of the classical molecular gene concept:

Gene is a nucleotide sequence that determines the synthesis of a protein (Student 5, Class B, post-test).

The return of the classical molecular concept in the retention test is not surprising. It just reveals that 5 h of lessons are not enough to challenge a view so deep rooted in the students' views, as a consequence of its reinforcement during years of schooling (as indicated by our results for high school biology textbooks). This is one example of students' prior conceptions that are resistant to change even when specifically targeted in teaching interventions. Interestingly enough, this is a prior conception that is itself a product of previous schooling. In order to reach a successful change in students' commitment to the classical molecular concept, it would be necessary to defy it repeatedly in the intervention, in several different contexts, going far beyond what was possible in the short time range of the intervention.

There was considerable overlapping of ideas related to different gene concepts in the students' answers in all the moments in which the data gathering tool was applied. In class A, 36.4 % of the answers in the pre- and posttest showed category overlapping, with this frequency increasing to 40 % in the retention test. In class B, there were 38.5 % of answers with category overlapping in the pre- and posttest, with an increase to 53.8 % in the retention test. Thus, neither the usual course nor the teaching sequence seemed to be successful in demarcating different gene concepts. This interpretation is reinforced by the analysis of the data for the free-choice closed question, shown in Fig. 16.11. Just as we saw in the study on students' views about genes, when they were free to choose several views about genes, they marked a lot of alternatives. As remarked above, conceptual variation regarding genes as represented in textbooks seems to be translated into students' allegiance to several different accounts about genes and their functions. Even though these results cannot by itself lead to the conclusion that students are wrapped up by semantic confusions and ambiguities by appealing to such a variety of views about genes, if we combine them with our findings in the textbook studies, we can have reasons to worry about this potential hybridization of different ideas regarding genes and their functions.

If we now turn to the external validation of the teaching sequence, some interesting patterns can be discerned, although we need to see them with a grain of salt, given the constraint that the experimental design included only two classes. The classical molecular concept increased in frequency in the students' answers after the intervention, not only in the posttest but also in the retention test. This finding is in agreement with our previous finding that Genetics and Cell and Molecular biology courses in the same university lead to an increase in this much challenged view about genes, despite the fact that the anomalies faced by it are addressed in those very courses. Moreover, the usual approach followed in the course did not produce even the transitory decrease in students' commitment to this concept found in the teaching sequence explicitly addressing gene function models and gene concepts.

As in the case of the intervention, the informational conception dropped in frequency in the answers to the open question when the usual approach was employed in the course, corroborating the findings of the prior investigation of students' views in the same university. But in this case the closed forced-choice question showed an opposite tendency.

Finally, a significantly smaller diversity of views about genes was observed in class A when compared with class B, in the answers to both the open and the closed forced-choice question. This is not surprising since those views were explicitly discussed in the latter but not in the former class.

Some design principles underlying the construction of the teaching sequence were not tested in this study, such as the proposed pattern of classroom discursive interactions, which require for its testing a treatment of the video-recorded material that we did not perform yet. If we consider the didactic material elaborated to the course, the historically and philosophically informed approach, the treatment of models of gene function and gene concepts, and the discussion of the crisis of the classical molecular concept using molecular phenomena already addressed

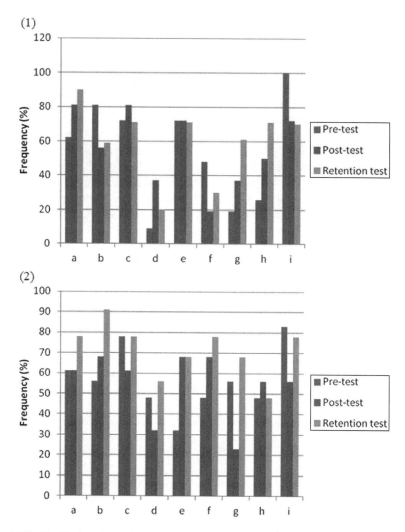

Fig. 16.11 Distribution of answers given by the students of the classes investigated to a free-choice closed question presenting several alternatives concerning the nature of genes: (**a**) Mendelian; (**b**), (**g**), and (**h**) classical molecular; (**c**) and (**i**) informational; (**d**) gene-P; (**e**) gene-D; (**f**) process molecular gene. (1) Class A (usual approach to the course, with no explicit discussion on gene function models and gene concepts); (2) class B (where the teaching sequence was implemented)

in the course, the results showed both contributions and limitations. The failures of the intervention are particularly interesting at this step of our research, since they indicated the need to introduce changes in the teaching sequence: for instance, a stronger challenge to the classical molecular concept and a more efficient discussion of the nature of models in connection with the historical construction of our understanding about genes, in order to decrease the hybridization of ideas

16 History and Philosophy of Science and Hybrid Views about Genes

related to different models and concepts by the students. Nevertheless, the detected advances show that it is promising to continue the investigation with a revised prototype of the teaching sequence.

16.5 Conclusion

We have been engaged in the last 7 years in a research program on the treatment of conceptual variation regarding genes and their function in school science. Following the approach to research on science education used in our lab, we took as a starting point a number of descriptive studies aiming at diagnosing views about genes found in textbooks and students and moved to intervention studies, investigating a teaching strategy for improving higher education students' understanding of scientific models and conceptual variation around genes and their function. This teaching strategy is aligned with a contextual approach to science education, using a historically and philosophically informed approach to teach not only with but also about gene function models.

Our investigations on textbooks showed the prevalence of the molecular-informational model and a significant degree of hybridization between features from different models, even when they are incommensurable. This was found in both higher education Cell and Molecular biology textbooks and high school biology textbooks. Moreover, even when the empirical findings challenging the molecular-informational model of gene function are discussed by the textbooks, conceptual lessons are not often derived from them. In high school biology textbooks, another worrisome finding was that gene-P was often used and, more than that, was often conflated with the molecular-informational model. To treat genes as determining phenotypic traits is a conceptual tool for abstracting away the complexity of the genotype-phenotype relationship in tasks like pedigree analysis, often found in high school textbooks. However, genes are most often regarded by these textbooks as codifiers of the primary structure of polypeptides or RNAs (in accordance with the classical molecular concept) and determinants of phenotypes (in line with gene-P), showing how these textbooks consistently hybridize these two gene concepts. The conflation with a molecular account of the gene transposes the deterministic assumption to DNA sequences that only determines the phenotype at its lowest level, namely, the primary structure of proteins (sometimes, also their three-dimensional structure) and the structure of RNAs. It is lost from sight, thus the complexity of development, which mediates between genotype and phenotype and involves epigenetic and environmental factors as resources in causal parity with genes (Arthur 2011; Griffiths and Knight 1998).

This provides an example of a conflation of gene concepts leading to serious consequences in genetics teaching. As gene-P, an instrumental concept depicting genes as determining phenotypes, is conflated with a realist understanding of genes as molecular units in the genome, genetic deterministic views are very likely to develop: the molecular units become determiners of phenotypes and not entities contributing to development in complex causal pathways involving other developmental resources. Preformationism lingers, then, in this manner of speaking about genes, as if traits themselves were somehow coded in the genome, and not constructed by complex developmental processes. As statements about genes-P are framed in an "updated" language, which connects it with molecular views about genes, and a historical and philosophical treatment of models is largely absent, students and teachers have no chance of understanding the instrumental nature of that concept and the explanatory context in which its usefulness is observed. The conflation between features of different gene function models not only leads to consequential problems in students' understanding of genes and their role in living beings – such as the commitment to a hyperbolic, overextended view of what DNA and genes do in cell systems – but also has implications to popular discourses about genes (or, in Keller's [2000] words, "gene talk") found in the media and even in textbooks themselves.[21]

As learning about genes becomes deeply contaminated by genetic deterministic views, students are less likely to develop a critical appraisal of socioscientific issues (Sadler 2011) related to genetics or to become capable of socially responsible decision making (Santos and Mortimer 2001) in situations involving knowledge about genes and their functions in living systems. After all, as Nelkin and Lindee (1995, p. 197) discuss,

> the findings of scientific genetics – about human behavior, disease, personality and intelligence – have become a popular resource precisely because they conform to and complement existing cultural beliefs about identity, family, gender and race [...] the desire for prediction, the need for social boundaries, and the hope for control of the human future [...] Whether or not such claims are sustained in fact may be irrelevant; their public appeal and popular appropriation reflect their social, not their scientific power.

Genetics is connected with socioscientific issues of central importance, such as cloning, stem cell research, genetically modified organisms, genetic engineering, use of genetic tests in society, human genetic improvement (eugenics), and reprogenetics. Sadler and Zeidler (2005) found that students' reasoning patterns in genetic engineering socioscientific issues are influenced by their knowledge of genetics, showing the importance that they properly learn about genes for their future life, not only as students but also as citizens that need to be informed by a consistent scientific understanding of the subject in order to actively and fully participate in democratic decision making.

The way these high school and higher education textbooks deal with conceptual variation can be regarded, thus, as a key problem in genetics teaching. For instance, all the historical models identified by Gericke and Hagberg (2007) were found in the high school textbooks and hybridization of features from different models was very frequent, showing how much conceptual variation was embedded in the treatment of genes, despite the prevalence of the molecular-informational model.

---

[21] See, for example, Condit et al. (1998, 2001), Carver et al. (2008), Keller (2000), and Nelkin and Lindee (1995).

As observed in Swedish high school textbooks and also in textbooks from four English-speaking countries, such conceptual variation is present in the explanations about genes with no clear demarcation between multiple historical models and their domains of application (Gericke et al. in press). Features related to different models are integrated in a single, linear narrative about genes, in such a manner that no conceptual variation seems to exist.

In a study of students' views about genes in two Brazilian universities (Federal University of Paraná and Federal University of Bahia), we compared biology students who had attended Genetics courses and those who did not and found that these courses increased their commitment to the classical molecular concept while decreasing their appeal to the informational conception. Again, no connection seemed to be properly made between the treatment of molecular phenomena that put into question the classical molecular gene and their conceptual implications. Students had difficulties in dealing with conceptual variation about genes, often hybridizing features from different models, even when they were incommensurable. Moreover, the degree of such hybridization was largely unaffected by Genetics courses, probably as an effect of the textbooks used, which included those analyzed here.

The convergence between our results concerning textbooks at two educational levels and higher education students' views is indicative of the reinforcement of the students' commitment to the molecular-informational model by the textbooks, as well as of the tendency to conflate features from different historical models. As we did not analyze pedagogical practice in the Genetics course of either of the universities, we cannot show data about how that practice was influenced by the textbooks used. However, our own acquaintance with these courses allows us to say that pedagogical work is significantly framed by the textbooks, making it likely the reinforcement hypothesis proposed above. Needless to say, it will be necessary to investigate classroom work in these courses to advance a more reliable conclusion to this effect.

A significant part of the problem with the treatment of conceptual variation about genes in higher education and high school textbooks results from the lack of a historically and philosophically approach to science education. In the absence of a clear discussion of models and either their role in science or their relation with reality, teachers and students are encouraged to address genes in a naïve realist manner and, also, to conflate features of different concepts as models as if they could be simply added as descriptive hallmarks of a reality being simply presented (rather than represented) in scientific theories and models. When using these textbooks, teachers and students do not have much chance of understanding the distinct origins, domains of application, and meanings of gene concepts and gene function models. Meanings ascribed to gene are simply accumulated as genes are discussed from different perspectives chapter after chapter, with the textbooks offering on the whole a thorough mixture of ideas originating from different models, often incommensurable with one another. The gene function models offer a particularly striking example of how the use of multiple models in science teaching can generate learning problems if not taught explicitly (Chinn and Samarapungavan 2008).

It seems necessary, thus, to change the treatment of genes in both textbooks and courses towards a more contextual approach, in which students must learn not only with gene function models but also about such models. If we do so, we can also address important NOS contents in connection with the history of the gene concept. After all, the transition from the understanding of genes in classical genetics to the molecular gene with the advent of molecular biology, as well as the crisis of the gene concept and the various approaches proposed to overcome it, compose a very interesting case of conceptual change and, also, provide a window into how theoretical entities are investigated and represented in science. This does not mean that one has to deal with complex historical, philosophical, or sociological issues when writing about genes in textbooks or teaching about genes in the classroom. We take the more modest position of proposing that one needs to write and teach about gene function models in a more explicit manner, paying attention to some basic aspects, such as the nature of models and their complex relation with reality, or the variation between gene function models and gene concepts in different subfields of biology.

To argue against the indiscriminate conflation of features related to different historical models of gene function does not imply that one should defend some single and all-encompassing gene concept or model of gene function. No such single model or concept could ever capture the diversity of meanings and epistemic roles associated with genes since the beginnings of the twentieth century. The idea is rather of a coexistence of a diversity of gene concepts and gene function models in school science, but with well-delimited domains of application (Burian 2004; El-Hani 2007). It is very important to provide students with a structured, organized view about the variety of meanings ascribed to genes and their functions, in order to avoid semantic confusions and indiscriminate mixtures of meanings related to different scientific contexts. To deal with conceptual variation, it is not enough to just say that "it may not be important to know what the precise meaning of 'gene' is" (Knight 2007, p. 300). To entertain the importance of a clear treatment of different gene concepts and gene function models, we need just to rephrase this statement by considering a plurality of ways of understanding genes: even though it is not really important to provide a single precise meaning of "gene," we need, still, to provide a clear and precise understanding of the several different meanings of "gene," since they cannot be all put to each and every use. Conceptual variation is not in itself the problem, but the absence of a proper historical and philosophical treatment of models about genes and their functions, which favors the extensive hybridization of ideas related to different models.

The lack of a historical and philosophical treatment of genes is also partly the explanation for the intriguing finding that neither textbooks nor students derive conceptual lessons from the challenges to the molecular-informational model that gave rise to the so-called crisis of the gene concept. Certainly, the textbooks could derive such lessons if they were more conceptually and theoretically oriented, even if they did not give much attention to history or philosophy of science. But this orientation is also typically lacking in these textbooks.

If a contextual approach to teaching about genes, with due attention to teaching with and about models, was in place, students and teachers would have a greater chance of building an understanding of genes and their roles in living systems that could be richer and more aligned with what we currently know about the complex dynamics and architecture of the genome or the dependence of gene function on the cellular and supracellular context. This complexity is usually abstracted away in school science in favor of deterministic views, emphasizing one-to-one relationships between genes, functional proteins, and phenotypes, despite the overwhelming evidence that these relationships do not hold in most of the cases.[22] Textbook discourse should come closer to the knowledge structure of the academic disciplines of genetics and molecular biology in this case (Gericke et al. in press). It is not that high school textbooks should be necessarily updated with the last words in scientific knowledge. Since at high school students have to learn the basics of scientific disciplines, it may be more important to teach about developments of the past, which established the grounds of a way of thinking in a scientific domain, than to pursue an updated curriculum for its own sake. We need to introduce recent developments of science in school when they make an important difference for the way the students think about a domain of phenomena. This is, in our view, precisely the case with the developments of genetics and molecular biology in the last two decades. More attention should be given in genetics teaching to the current situation of the classical molecular concept, instead of just presenting it as if it was as accepted and coherent as it was in the past. At least, the fact that there are serious debates about what is a gene in the scientific community deserves attention in genetics teaching, even at the high school level. Our data do not show, however, the gene concept being treated as a controversial subject matter in either high school or higher education.

We need to investigate ways of introducing into school science the current understanding of the anomalies challenging the classical molecular concept and at least some of the alternatives to this way of understanding genes (Meyer et al. 2013). In the case of high school biological education, we think it is possible to create conditions for the students to understand that, even though the classical molecular concept has been quite important in the history of biology, it has ended up showing consequential limitations. Moreover, the concepts of gene-P and gene-D, the necessity of demarcating between them, and a critique of genetic determinism would be important additions to the high school genetics curriculum. If school science took into consideration the complex mapping between genotype, development, and phenotype (Arthur 2011), this might make a difference to students' thinking, creating conditions for the development of more informed and critical attitudes towards the deterministic talk about genes that pervades several spheres of society.

It was evident to us, then, that we needed to build and investigate an educational intervention based on a number of educated guesses about how to deal with conceptual variation about genes, which could be used as design principles for teaching interventions and, then, empirically tested in the classroom. One of the key design principles is to give a central role to a historical and philosophical approach to gene. We built such a teaching sequence in collaboration with a higher education Cell and

---

[22] See, for example, El-Hani (2007), El-Hani et al. (2009), Fogle (1990), Keller (2000), Moss (2003), and Scherrer and Jost (2007a, b).

Molecular biology teacher at a Brazilian University (Federal University of Bahia) and investigated it in accordance with design-based research. The teaching sequence was oriented towards a contextual approach, explicitly addressing the historical and philosophical dimensions of science, with a particular focus on the historical construction and nature of models of gene function and gene concepts. The internal validation of the teaching sequence showed some positive learning outcomes, but also some limits in attaining the planned learning outcomes. In particular, we managed to obtain just a transitory decrease of the classical molecular concept, an outcome that was not really surprising given the fact that – as our results in the diagnostic studies showed – this view has been reinforced throughout the lives of the students at school. Moreover, we did not reach success regarding the demarcation between gene concepts and gene function models, with the same high levels of hybridization observed in the diagnostic studies being also found in the intervention studies. Even though the external validation of the teaching sequence was constrained by the number of classes available for the study, the comparison between the usual way of teaching about genes in the course and the new intervention gave some hints of positive changes: the usual approach did not lead even to a transitory decrease of the classical molecular concept, and the students' views on genes have been enriched by the teaching sequence. The first result seems robust, since it is in strict accordance with the findings of our study on students' views about genes in the same university. The second finding amounts to the major difference brought about by the teaching sequence. Nevertheless, this outcome should be accompanied by a proper understanding of models and their demarcation, in order to lead to genuine gains for the students. But this was not observed in this first prototype of the teaching sequence.

These findings gave us clear clues about changes in the intervention for its second prototyping: the classical molecular concept needs to be challenged in a stronger way, and the discussion about models, their historical construction, and the necessity of their demarcation should be reformulated in order to reach a higher level of efficacy. Needless to say, the greatest challenge will be to accommodate these changes in the limited time available for the intervention, as a consequence of the overstuffed curricula of Genetics and Molecular biology courses at the university level.

Acknowledgments We are thankful to the Brazilian National Council for Scientific and Technological Development (CNPq) and the Research Support Foundation of the State of Bahia (FAPESB) for support during the development of the research reported in this paper.

Appendix 1: List of Analyzed Higher Education Textbooks

Alberts, B., Johnson, A., Lewis, J., Raff, M., Roberts, K. & Walter, P. (2002). *Molecular biology of the cell* (4th Ed). New York, NY: Garland.

Karp, G. (2004). *Cell and molecular biology: Concepts and experiments* (4th Ed). New York, NY: John Wiley and Sons.

Lodish, H., Kaiser, C. A., Berk, A., Krieger, M., Matsudaira, P. & Scott, M. P. (2003). *Molecular cell biology* (5th Ed). New York, NY: W. H Freeman.

Appendix 2: List of Analyzed High School Textbooks

T1 – Amabis, J. M. & Martho, G. R. (2005). *Biologia*. São Paulo: Moderna.

T2 – Borba, A. A. & Cançado, O. F. L. (2005). *Biologia*. Curitiba: Positivo.

T3 – Borba, A. A., Crozetta, M. A. S. & Lago, S. R. (2005). *Biologia*. São Paulo: IBEP.

T4 – Boschilia, C. (2005). *Biologia sem segredos*. São Paulo: RIDEEL.

T5 – Carvalho, W. (2005). *Biologia em foco*. São Paulo: FTD.

T6 – Cheida, L. E. (2005). *Biologia integrada*. São Paulo: FTD.

T7 – Coimbra, M. A. C., Rubio, P. C., Corazzini, R., Rodrigues, R. N. C. & Waldhelm, M. C. V. (2005). *Biologia – Projeto escola e cidadania para todos*. São Paulo: Editora do Brasil.

T8 – Faucz, F. R. & Quintilham, C. T. (2005). *Biologia: Caminho da vida*. Curitiba: Base.

T9 – Favaretto, J. A. & Mercadante, C. (2005). *Biologia*. São Paulo: Moderna.

T10 – Frota-Pessoa, O. (2005). *Biologia*. São Paulo: Scipione.

T11 – Gainotti, A. & Modelli, A. (2005). *Biologia*. São Paulo: Scipione.

T12 – Laurence, J. (2005). *Biologia*. São Paulo: Nova Geração.

T13 – Linhares, S. & Gewandsznajder, F. (2005). *Biologia*. São Paulo: Ática.

T14 – Lopes, S. & Rosso, S. (2005). *Biologia*. São Paulo: Saraiva.

T15 – Machado, S. W. S. (2005). *Biologia*. São Paulo: Scipione.

T16 – Morandini, C. & Bellinello, L. C. (2005). *Biologia*. São Paulo: Atual.

T17 – Paulino, W. R. (2005). *Biologia*. São Paulo: Ática.

T18 – Silva-Júnior, C. & Sasson, S. (2005). *Biologia*. São Paulo: Saraiva.

References

Abd-El-Khalick, F. & Lederman, N. G. (2000). Improving science teachers' conceptions of nature of science: A critical review of the literature. *International Journal of Science Education*, 22, 665–701.

Adami, C. (2004). Information theory in molecular biology. *Physics of Life Reviews*, 1, 3–22.

Arthur, W. (2011). *Evolution: A developmental approach*. Chichester: Wiley-Blackwell.

Artigue, M. (1988). Ingéniérie didactique. *Recherches en didactique des mathemátiques*, 9, 281–308.

Barab, S. & Squire, K. (2004). Design-based research: putting a stake in the ground. *Journal of the Learning Sciences*, 13, 1–14.

Bardin, L. (2000). *Análise de conteúdo* (Content analysis). Lisboa: Edições 70.

Baumgartner, E., Bell, P., Bophy, S. et al. (2003). Design-based research: An emerging paradigm for educational inquiry. *Educational Researcher*, 32, 5–8.

Benzer, S. (1957). The elementary units of heredity. In W. McElroy and B. Glass (Eds.), *The chemical basis of heredity* (pp. 70–93). Baltimore, MD: John Hopkins Press.

Black, D. L. (2003). Mechanisms of alternative pre-messenger RNA splicing. *Annual Review of Biochemistry*, 72, 291–336.

Black, M. (1962). *Models and metaphors: Studies in language and philosophy*. Ithaca, NY: Cornell University Press.

Brown, A. (1992). Design experiments: Theoretical and methodological challenges in creating complex interventions in classroom settings. *The Journal of the Learning Sciences*, 2, 141–178.

Burian, R. M. (1985). On conceptual change in biology: The case of the gene. In D. J. Depew & B. H. Weber (Eds.), *Evolution at a crossroads: The new biology and the new philosophy of science* (pp. 21–24). Cambridge, MA: The MIT Press.

Burian, R. M. (2004). Molecular epigenesis, molecular pleiotropy, and molecular gene definitions. *History and Philosophy of Life Sciences*, 26, 59–80.

Carlson, A. E. (1966). *The gene. A critical history.* Philadelphia, PA: W. B. Saunders.

Carver, R., Waldahl, R. & Breivik, J. (2008). Frame that gene – A tool for analyzing and classifying the communication of genetics to the public. *EMBO reports*, 9, 943–947.

Celotto, A. & Graveley, B. (2001). Alternative splicing of the *Drosophila* DSCAM pre-mRNA is both temporally and spatially regulated. *Genetics*, 159, 599–608.

Chevallard, Y. (1989). On didactic transposition theory: Some introductory notes. Paper presented at the International symposium on selected domains of research and development in mathematics education, *Proceedings* (pp. 51–62). Bratislava, Slovakia. Retrieved October 29, 2011 from: http://yves.chevallard.free.fr/spip/spip/article.php3?id_article=122

Chinn, A. C. & Samarapungavan, A. (2008). Learning to use scientific models: Multiple dimensions of conceptual change. In R.A. Duschl & R.E. Grandy (Eds.), *Teaching scientific inquiry* (pp. 191–225). Rotterdam: Sense Publishers.

Condit, C.M., Ofulue, N. & Sheedy, K.M. (1998). Determinism and mass-media portrayals of genetics. *American Journal of Human Genetics*, 62, 979–984.

Condit, C. M., Ferguson, A., Kassel, R., Tadhani, C., Gooding, H. C. & Parrot, R. (2001). An explanatory study of the impact of news headlines on genetic determinism. *Science Communication*, 22, 379–395.

Cooper, M. D. & Alder, M. N. (2006). The evolution of adaptive immune systems. *Cell*, 124, 815–822.

Daston, L. & Galison, P. (2010). *Objectivity*. Brooklyn, NY: Zone Books.

Develaki, M. (2007). The model-based view of scientific theories and the structuring of school science programs. *Science & Education*, 16 (7–8), 725–749.

El-Hani, C. N. (2007). Between the cross and the sword: The crisis of the gene concept. *Genetics and Molecular Biology*, 30, 297–307.

El-Hani, C. N., Queiroz, J. & Emmeche, C. (2006). A semiotic analysis of the genetic information system. *Semiotica*, 160, 1–68.

El-Hani, C. N., Queiroz, J. & Emmeche, C. (2009). *Genes, Information, and Semiosis*. Tartu: Tartu University Press, Tartu Semiotics Library.

El-Hani, C. N., Roque, N. & Rocha, P. B. (2007). Brazilian high school biology textbooks: Results from a national program. In: *Proceedings of the IOSTE International Meeting on Critical Analysis of School Science Textbook* (pp. 505–516). Hammamet, Tunisia: University of Tunis.

El-Hani, C. N., Roque, N. & Rocha, P. L. B. (2011). Livros didáticos de Biologia do ensino médio: Resultados do PNLEM/2007. *Educação em Revista*, 27, 211–240.

Falk, R. (1986). What is a gene? *Studies in the History and Philosophy of Science*, 17, 133–173.

Fleck, L. (1979/1935). *Genesis and development of a scientific fact*. Chicago, IL: The University of Chicago Press.

Fogle, T. (1990). Are genes units of inheritance? *Biology and Philosophy*, 5, 349–371.

Fogle, T. (2000). The dissolution of protein coding genes. In P. Beurton, R. Falk & H-J. Rheinberger (Eds.), *The concept of the gene in development and evolution* (pp. 3–25). Cambridge: Cambridge University Press.

Gelbart, W. (1998). Databases in genomic research. *Science*, 282, 659–661.

Gericke, N. M. & Hagberg, M. (2007). Definition of historical models of gene function and their relation to students' understandings of genetics. *Science & Education*, 16, 849–881.

Gericke, N. M. & Hagberg, M. (2010a). Conceptual incoherence as a result of the use of multiple historical models in school textbooks. *Research in Science Education*, 40, 605–623.

Gericke, N. M. & Hagberg, M. (2010b). Conceptual variation in the depiction of gene function in upper secondary school textbooks. *Science & Education*, 19, 963–994.

Gericke, N. M., Hagberg, M., Santos, V. C., Joaquim, L. M. & El-Hani, C. N. (in press). Conceptual variation or Incoherence? Textbook discourse on genes in six countries. *Science & Education*.

Gerstein, M. B., Bruce, C., Rozowsky, J. S., Zheng, D., Du, J., Korbel, J. O., Emanuelsson, O., Zhang, Z. D., Weissman, S., & Snyder, M. (2007). What is a gene, post-ENCODE? History and updated definition. *Genome Research*, 17, 669–681.

Grandy, R. E. (2003). What are models and why do we need them? *Science & Education*, 12(8), 773–777.

Graveley, B. R. (2001). Alternative splicing: Increasing diversity in the proteomic world. *Trends in Genetics*, 17, 100–107.

Griffiths, P. E. (2001). Genetic information: A metaphor in search of a theory. *Philosophy of Science*, 68, 394–403.

Griffiths, P.E. & Knight, R.D. (1998). What is the developmental challenge? *Philosophy of Science*, 65, 2, 253–258.

Griffiths, P. E. & Neumann-Held, E. (1999). The many faces of the gene. *BioScience*, 49, 656–662.

Hall, B. K. (2001). The gene is not dead, merely orphaned and seeking a home. *Evolution and Development*, 3, 225–228.

Halloun, I. A. (2004). *Modeling theory in science education*. Dordrecht: Kluwer Academic Publishers.

Halloun, I. A. (2007). Mediated modeling in science education. *Science & Education*, 16, 653–697.

Hanson, M. R. (1996). Protein products of incompletely edited transcripts are detected in plant mitochondria. *The Plant Cell*, 8(1), 1–3.

Hesse, M. B. (1963). *Models and analogies in science*. London: Seed and Ward.

Hoffmeyer, J. & Emmeche, C. (1991). Code-duality and the semiotics of nature. In M. Anderson & F. Merrell (Eds.), *On semiotic modeling* (pp. 117–166). Berlin: Mouton de Gruyter.

Holmes, F. L. (2006). *Reconceiving the gene: Seymour Benzer's adventures in phage genetics*. New Haven, CT: Yale University Press.

Hull, D. L. (1974). *Philosophy of biological science*. Englewood Cliffs, NJ: Prentice-Hall.

Jablonka, E. (2002). Information: Its interpretation, its inheritance, and its sharing. *Philosophy of Science*, 69, 578–605.

Johannsen, W. (1909). *Elemente der exakten erblichkeitslehre*. Jena: Gustav Fischer. Retrieved August 23, 2012 from: http://caliban.mpiz-koeln.mpg.de/johannsen/elemente/johannsen_elemente_der_exakten_erblichkeitslehre_2.pdf.

Justi, R. S. & Gilbert, J. K., (1999). A cause of ahistorical science teaching: Use of hybrid models. *Science Education*, 83, 163–177.

Kampa, D., Cheng, J., Kapranov, P., Yamanaka, M., Brubaker, S., Cawley, S., Drenkow, J., Piccolboni, A., Bekiranov, S., Helt, G., Tammana, H. & Gingeras, T. R. (2004). Novel RNAs identified from an in-depth analysis of the transcriptome of human chromosomes 21 and 22. *Genome Research*, 14, 331–342.

Kay, L. E. (2000). *Who wrote the book of life? A history of the genetic code*. Stanford, CA: Stanford University Press.

Keller, E. F. (2000). *The century of the gene*. Cambridge, MA: Harvard University Press.

Keller, E. F. (2005). The century beyond the gene. *Journal of Biosciences*, 30, 3–10.

Keller, E. F. & Harel, D. (2007). Beyond the gene. *PLoS One*, 2, e1231.

Kitcher, P. (1982). Genes. *British Journal for the Philosophy of Science*, 33, 337–359.

Knight, R. (2007). Reports of the death of the gene are greatly exaggerated. *Biology and Philosophy*, 22, 293–306.

Larsson, S. (2009). A pluralist view of generalization in qualitative research. *International Journal of Research & Method in Education*, 32, 25–38.

Lave, J. & Wenger, E. (1991). *Situated learning: Legitimate peripheral participation*. New York, NY: Cambridge University Press.

LeCompte, M. & Goetz, J. (1982). Problems of reliability and validity in ethnographic research. *Review of Educational Research*, 52(1), 31–60.

Lev-Maor, G., Sorek, R., Levanon, E. Y., Paz, N., Eisenberg, E. & Ast, G. (2007). RNA-editing-mediated exon evolution. *Genome Biology*, 8, R29.

Matthews, M. R. (1998). In defense of modest goals when teaching about the nature of science. *Journal of Research in Science Teaching*, 35, 161–174.

Mayr, E. (1982). *The growth of biological thought: Diversity, evolution and inheritance.* Cambridge, MA: Harvard University Press.

Méheut, M. (2005). Teaching-learning sequences tools for learning and/or research. In K. Boersma, M. Goedhart, O. De Jong & H. Eijkelhof (Eds.), *Research and the quality of science education* (pp. 195–207. Dordrecht: Springer.

Meyer, L. M. N., Bomfim, G. C. & El-Hani, C. N. (2013). How to understand the gene in the 21st century. *Science & Education*, 22, 345–374.

Mortimer, E. F. & Scott, P. H. (2003). *Meaning making in secondary science classrooms.* Maidenhead: Open University Press.

Moss, L. (2001). Deconstructing the gene and reconstructing molecular developmental systems. In S. Oyama, P. E. Griffiths & R. D. Gray (Eds.), *Cycles of contingency: Developmental systems and evolution* (pp. 85–97). Cambridge, MA: MIT Press.

Moss, L. (2003). *What genes can't do*. Cambridge, MA: The MIT Press.

Murre, C. (2007). Epigenetics of Antigen-receptor gene assembly. *Current Opinion in Genetics & Development*, 17, 415–421.

Nelkin, D. & Lindee, S. M. (1995). *The DNA mystique: The gene as a cultural icon.* New York, NY: Freeman.

Neumann-Held, E. (1999). The Gene is dead – Long live the gene: Conceptualizing genes the constructionist way. In P. Koslowski (Ed.). *Sociobiology and bioeconomics: The theory of evolution in biological and economic thinking* (pp. 105–137). Berlin: Springer.

Neumann-Held, E. (2001). Let's talk about genes: The process molecular gene concept and its context. In S. Oyama, P. E. Griffiths & R. D. Gray (Eds.), *Cycles of contingency: Developmental systems and evolution* (pp. 69–84). Cambridge, MA: MIT Press.

Nieveen, N., McKenney, S. & Van den Akker, J. (2006). Educational design research: The value of variety. In J. Van den Akker; K. Gravemeijer; S. McKenney & N. Nieveen (Eds). *Educational design research* (pp. 151–158). London: Routledge.

Oyama, S. (2000/1985). *The ontogeny of information: Developmental systems and evolution* (2nd ed). Cambridge: Cambridge University Press.

Pardini, M. I. M. C. & Guimarães, R. C. (1992). A systemic concept of the gene. *Genetics and Molecular Biology*, 15, 713–721.

Pearson, H. (2006). What is a gene? *Nature*, 441, 399–401.

Pitombo, M. A., Almeida, A. M. R., & El-Hani, C. N. (2008). Gene concepts in higher education cell and molecular biology textbooks. *Science Education International*, 19(2), 219–234.

Plomp, T. (2009). Educational design research: An introduction. In: T. Plomp & N. Nieveen (Eds.). *An introduction to educational design research* (pp. 9–35). Enschede: SLO – Netherlands Institute for Curriculum Development.

Portin, P. (1993). The concept of the gene: Short history and present status. *Quarterly Review of Biology*, 56, 173–223.

Reeves, T. C. (2006). Design research from a technology perspective. In J. Van den Akker, K. Gravemeijer, S. McKenney & N. Nieveen (Eds.). *Educational design research* (pp. 52–66). London, Routledge.

Rheinberger, H.-J. (2000). Gene concepts: Fragments from the perspective of molecular biology. In: P. Beurton, R. Falk & H.-J. Rheinberger (Eds.). *The concept of the gene in development and evolution* (pp. 219–239). Cambridge: Cambridge University Press.

Sadler, T. D. (Ed.). (2011). *Socioscientific issues in the classroom: Teaching, learning and research.* Dordrecht: Springer.

Sadler, T. D. & Zeidler, D. L. (2005). The significance of content knowledge for informal reasoning regarding socioscientific issues: Applying genetics knowledge to genetic engineering issues. *Science Education*, 89, 71–93.

Santos, V. C., Joaquim, L. M. & El-Hani, C. N. (2012). Hybrid deterministic views about genes in biology textbooks: A key problem in genetics teaching. *Science & Education*, 21, 543–578.

Santos, W. L. P. & Mortimer, E. F. (2001). Tomada de decisão para ação social responsável no ensino de ciências (decision making for responsible social action in science teaching). *Ciência e Educação*, 7, 95–111.

Scherrer, K. & Jost, J. (2007a). The gene and the genon concept: A functional and information-theoretic analysis. *Molecular System Biology*, 3, 1–11.

Scherrer, K. & Jost, J. (2007b). The gene and the genon concept: Coding versus regulation. A conceptual and information-theoretic analysis of genetic storage and expression in the light of modern molecular biology. *Theory in Biosciences*, 126, 65–113.

Schwab, J. (1964). Structure of the disciplines: Meaning & significances. In G. W. Ford & L. Pugno (eds.), *The structure of knowledge & the curriculum* (pp. 6–30). Chicago, IL: Rand, McNally & Co.

Shannon, C. E. & Weaver, W. (1949). *The mathematical theory of communication*. Urbana, IL: University of Illinois Press.

Simons, H.; Kushner, S.; Jones, K. & James, D. (2003). From evidence-based practice to practice-based evidence: the idea of situated generalization. *Research Papers in Education*, 18, 347–364.

Smith, M. U. & Adkison, L. R. (2010). Updating the model definition of the gene in the modern genomic era with implications for instruction. *Science & Education*, 19, 1–20.

Stotz, K., Griffiths, P. E. & Knight, R. (2004). How biologists conceptualize genes: An empirical study. *Studies in the History and Philosophy of Biological & Biomedical Sciences*, 35, 647–673.

Suppe, F. (1977). *The structure of scientific theories*. Urbana, IL: University of Illinois Press.

The ENCODE Project Consortium (2004). The ENCODE (ENCyclopedia Of DNA Elements) Project. *Science*, 306, 636–640.

van den Akker, J., Gravemeijer, K., McKenney, S. & Nieveen, N. (2006). *Educational design research*. London, Routledge.

van Fraassen, B. (1980). *The scientific image*. Oxford: Clarendon Press.

Venter, J. C., Adams, M. D., Myers, E. W., Li, P. W., Mural, R. J., Sutton, G. G. et. al. (2001). The sequence of the human genome. *Science*, 291, 1305–1351.

Wang, W., Zhang, J., Alvarez, C., Llopart, A. & Long, M. (2000). The origin of the jingwei gene and the complex modular structure of its parental gene, yellow emperor, in *Drosophila melanogaster*. *Molecular Biology and Evolution*, 17, 1294–1301.

Wanscher, J. H. (1975). The history of Wilhelm Johannsen's genetical terms and concepts from the period 1903 to 1926. *Centaurus*, 19(2), 125–147.

Watson, J. D. & Crick, F. H. C. (1953). A structure for deoxyribose nucleic acid. *Nature*, 171, 737–738.

Wenger, E. (1998). *Communities of practice: Learning, meaning, and identity*. New York, NY: Cambridge University Press.

Charbel N. El-Hani is Professor of History, Philosophy, and Biology Teaching at the Institute of Biology, Federal University of Bahia (UFBA), Brazil, and receives a Productivity Research Grant of the National Council for Scientific and Technological Development (CNPq). He was trained as a biologist and subsequently did his Ph.D. in Education at University of São Paulo. He is affiliated with the Graduate Studies Programs in History, Philosophy, Science Teaching (Federal University of Bahia and State University of de Feira de Santana), Ecology and Biomonitoring (Federal University of Bahia), and Genetics (Federal University of Bahia). He coordinates the History, Philosophy, and Biology Teaching Lab at UFBA. His research interests are in science education research, philosophy of biology, biosemiotics, ecology, and animal behavior. He is a member of editorial boards of Brazilian and international journals in science education and philosophy of biology. Among his recent publications one finds: Santos, V. C.; Joaquim, L. M. & El-Hani, C. N. (2012). Hybrid deterministic views about genes in biology textbooks: A key problem in genetics teaching. *Science & Education* 21: 543–578. DOI: 10.1007/s11191-011-9348-1; Arteaga, J. M. S. & El-Hani, C. N. (2012). Othering Processes and STS Curricula: From 19th Century Scientific Discourse on Interracial Competition and Racial Extinction to Othering in Biomedical Technosciences. *Science & Education* 21: 607–629. DOI: 10.1007/s11191-011-9384-x; Mortimer, E. F., Scott, P. & El-Hani, C. N. (2012). The heterogeneity of discourse in science classrooms: the conceptual profile approach. In: Fraser, B., Tobin, K. & McRobbie, C. (Eds.). *Second International Handbook of Science Education* (vol. 1, pp. 231–246). Dordrecht: Springer; El-Hani, C. N. & Sepulveda, C. (2010). The relationship between science and religion in the education of protestant biology preservice teachers in a Brazilian university. Cultural Studies of *Science Education* 5(1): 103–125. DOI: 10.1007/s11422-009-9212-7.

Ana Maria R. de Almeida is a Ph.D. candidate at the University of California at Berkeley in the Department of Plant and Microbial Biology where she studies the evolution of morphological diversity of tropical monocot flowers, especially regarding developmental processes and their relationship to evolution (Evo-Devo). She is also a member of the Freeling Lab, at the same Department, where she works on genome evolution. Ana obtained her Master's Degree in History, Philosophy, and Science Teaching at the Federal University of Bahia (UFBA), Brazil. During her Masters she studied Philosophy of Ecology, more specifically the Biodiversity-Ecosystem Functioning Program. Her research interests are in the relationship between evolution, ecology and developmental biology, genome evolution, and science teaching. Among her most recent publications, there are: D'Hont, A. et al. (2012). The banana (*Musa acuminata*) genome and the evolution of monocotyledonous plants. *Nature*, 488: 213–219 DOI: 10.1038/nature11241; Specht, C.D.; Yockteng, R.; Almeida, A.M.; Kirchoff, B.; Kress, J. (2012). Homoplasy, Pollination, and Emerging Complexity During the Evolution of Floral Development in the Tropical Gingers (Zingiberales). *Bot. Rev.* (2012) 78:440–462, DOI: 10.1007/s12229-012-9111-6; Almeida, A.M.; Brown, A.; Specht, C.D. Tracking the Development of the Petaloid Fertile Stamen in *Canna indica*: Insights into the origin of androecial petaloidy in the Zingiberales (submitted to AoB PLANTS on November 2012).

Gilberto C. Bomfim is Professor of Cell and Molecular Biology and Genetics Teaching at the Institute of Biology, Federal University of Bahia, Brazil. He is a biologist and did his Ph.D. in Pathology at Federal University of Bahia. As a researcher, he participates in the History, Philosophy, and Biology Teaching Lab at UFBA. Among his recent publications, one finds: Meyer, L. M. N.; Bomfim, G. C. & El-Hani, C. N. (2013). How to understand the gene in the 21st century. *Science & Education*, 22, 345–374. DOI: 10.1007/s11191-011-9390-z.

16 History and Philosophy of Science and Hybrid Views about Genes

Leyla M. Joaquim is a Ph.D. candidate in the Programme of History, Philosophy, and Science Teaching at the Federal University of Bahia (UFBa). She is part of the History, Philosophy, and Biological Teaching Lab at UFBA and the "Lacic group" (Laboratory Science as Culture) headed by Olival Freire Jr. After obtaining her Bachelor and Teaching Degree in Biological Sciences at Federal University of Paraná (UFPR), she started her graduate studies in the field of Genetics Teaching. She joined projects investigating how ideas about genes and gene function are treated in school knowledge, as represented in textbooks and students' views. Currently, she is dedicated to her Ph.D. dissertation in History of Science, which is a study about the interest of physicists in biological problems and uses oral history as one of the methodological tools. She had research stays at Max Planck Institute for the History of Science (MPIWG – Berlin, Germany) at Department I and II in 2010 and 2012, respectively. Accordingly, her research interests lie mainly in science education research, history of biology, and history of physics. Among her recent publications, one finds: Santos, V. C.; Joaquim, L. M. & El-Hani, C. N. (2012). Hybrid deterministic views about genes in biology textbooks: A key problem in genetics teaching. *Science & Education* 21: 543–578. DOI: 10.1007/s11191-011-9348-1.

João Carlos M. Magalhães is Professor of Genetics, Population Genetics, and Methodology of Science at the Federal University of Parana, Brazil. He graduated with Biological Sciences, Master in Genetics, and Ph.D. in Genetics in the Federal University of Paraná. He is affiliated to the Graduate Studies Program in Genetics at the same university. He has research experience in Genetics and Philosophy of Science, working mainly in the following areas: Genetics of Human Populations and Philosophy of Biology. His doctoral thesis discusses the logical foundations of Genetics and Evolution. He also has experience in application and development of software simulation of evolutionary processes to improve learning outcomes of Biology students. His recent publications include: Tureck, L. V., Santos, L. C., Wowk, P. F., Mattos, S. B., Silva, J. S., Magalhaes, J. C. M., Roxo, V. M. S., Bicalho, M. G. (In press). HLA-G Regulatory polymorphisms in Afro-Brazilians HLA-G. Tissue Antigens, 2012; Magalhaes, J. C. M. & Gondro, C. A contemporary view of Population Genetics in Evolution. In: Krause, D., Videira, A. A. P. (Org.). *Brazilian Studies in Philosophy and History of Science: an account of recent works.* 1ed. New York: Springer, 2011, v. 2990, p. 281–290; Alle, B. R.; Furtado-Alle, L., Gondro, C., Magalhaes, J. C. M. Kuri: A Simulator of Ecological Genetics for Tree Populations. *Journal of Artificial Evolution and Applications*, v. 2009, p. ID 783647, 2009; Gondro, C. & Magalhaes, J. C. M. A simple genetic algorithm for studies in mendelian populations. In: H. Abbass, T. Bossamaier; J. Wiles. (Org.). *Recent Advances in Artificial Life.* 1ed. London: World Scientific Publishing, 2005, p. 85–98.

Lia M. N. Meyer is Professor of Science Teaching Methodology at the Department of Biosciences – Campus Itabaiana, Federal University of Sergipe, Brazil. She is a member of the History, Philosophy, and Biology Teaching Lab at UFBA. Her research interests are in science education research, teacher education, and philosophy of biology. Among his recent publications, one finds: Meyer, L. M. N.; Bomfim, G. C. & El-Hani, C. N. (2013). How to understand the gene in the 21st century. *Science & Education*, 22, 345–374. DOI: 10.1007/s11191-011-9390-z.

Maiana A. Pitombo is a graduate student at Federal University of Rio Grande do Norte, Brazil. She has a Bachelor's Degree in Biology from the Federal University of Bahia. Her research interests are in Genetics and Immunology. Her publications include: Pitombo, M. A.; Almeida, A. M. R. de; El-Hani, C. N. Gene concepts in higher education cell and molecular biology textbooks. *Science Education International*, v. 19, p. 219–234, 2008.

Vanessa C. dos Santos is a Visiting Professor of History, Philosophy, and Biology Teaching at the State University of Paraiba, Brazil. She was trained as a biologist (Federal University of Bahia) and has a Master's Degree in the Graduate Studies Program in History, Philosophy, and Science Teaching (Federal University of Bahia and State University of de Feira de Santana). Her research interests are in history and philosophy of biology, science education research, biosemiotics, and situated cognition. Among her recent publications, one finds: Santos, V. C.; Joaquim, L. M. & El-Hani, C. N. (2012). Hybrid deterministic views about genes in biology textbooks: A key problem in genetics teaching. *Science & Education* 21: 543–578. DOI: 10.1007/s11191-011-9348-1.

Part IV
Pedagogical Studies: Ecology

Chapter 17
Contextualising the Teaching and Learning of Ecology: Historical and Philosophical Considerations

Ageliki Lefkaditou, Konstantinos Korfiatis, and Tasos Hovardas

17.1 Introduction

Ecology has gradually gained salience during the last few decades and ecological issues, including land use changes, global warming, biodiversity loss, food shortage, and so forth, seem to be gaining public attention. Though philosophers of science had given little attention to ecology, there is a lot of interesting work being currently pursued in philosophy of ecology and environmental philosophy. As Colyvan and colleagues put it, "ecology is an important and fascinating branch of biology, with distinctive philosophical issues" (Colyvan et al. 2009, p. 21). Given its conceptual and methodological familiarity with the social sciences, ecology occupies a unique position among other disciplines (Cooper 2003).

For example, ecosystem historicity, stability, complexity and uncertainty (Mikkelson 1999; Price and Billick 2010; Sterelny 2006); the role of natural history in ecological explanations (see Keller and Golley 2000 and references therein); the relationship between explanation, understanding and prediction (Peters 1991; Wilson 2009); the standard model of hypothesis testing (Colyvan et al. 2009); the role of mathematical models (Justus 2006; Odenbaugh 2005; Weisberg 2006a); and biodiversity conservation and a number of ethical or practical questions related to conservation (Odenbaugh 2007; Oksanen and Pietarinen 2004; Sarkar 2005) are all

---

A. Lefkaditou
School of Philosophy, Religion and History of Science, University of Leeds, Leeds, UK
e-mail: phal@leeds.ac.uk

K. Korfiatis (✉)
Department of Education, University of Cyprus, Nikosia, Cyprus
e-mail: korfiati@ucy.ac.cy

T. Hovardas
Department of Primary Education, University of Thessaly, Volos, Greece
e-mail: hovardas@uth.gr

M.R. Matthews (ed.), *International Handbook of Research in History,*
Philosophy and Science Teaching, DOI 10.1007/978-94-007-7654-8_17,
© Springer Science+Business Media Dordrecht 2014

topics of intense discussion between philosophers, historians and practicing ecologists. And as many philosophers suggest, the process of scientific inquiry in ecology may differ fundamentally from the dominant paradigms that have been drawn from the physical sciences (e.g. Cooper 2003).

More to the point, the fact that ecology addresses complex socioscientific issues at the intersection between science and the broader social context in which the products and processes of science are situated has also resulted to burgeoning popular myths coming from society, science and even history and philosophy of science (Haila and Levins 1992; Hovardas and Korfiatis 2011; Worster 1994).

All these considerations have profound implications for teaching and learning ecology. As early as in the late 1960s, it has been argued that the teaching of ecology is accompanied with an array of discussions and controversies about the place of this discipline in the educational system that cannot be easily compared to other disciplines (Lambert 1967).

In the next sections, we will first present an overview of shortcomings and failures in ecology education. Then, we will attempt to show that a comprehensive philosophical understanding of ecology might prove invaluable for battling misconceptions and achieving ecological literacy. We review some of these philosophical and historical considerations in ecology and explore a number of cases where the implications of these discussions might make a difference to ecology education. The potential to overcome unfruitful or superficial descriptions that have long plagued ecology and hindered understanding is a major incentive in this direction. Thus, we focus on two interrelated issues:

(a) The role of natural history research in ecology and its relationship to the standard hypothesis-testing model
(b) The role and understanding of ecological models in the ecology classroom

Our choice does not only attest to the fact that these two topics are being heavily discussed but also to their close relation to both educational practice and major misconceptions concerning scientific inquiry in ecology. Finally, we suggest ways in which a historically and philosophically informed curriculum might be developed to account for ecology's distinct nature.

17.1.1 Shortcomings and Failures in Ecology Education

Ecology is considered a young science. The same may be said for ecology education. Ecology started to seriously be part of education in European and North American countries during the 1960s (Berkowitz et al. 2005; Hale and Hardie 1993). Today most people would agree that ecology is an integral part of a contemporary science education curriculum (McComas 2002a). A general consensus seems also to exist among educators regarding both the necessity of educating ecologically literate people and the meaning of ecological literacy itself. More specifically, it is agreed that ecology is the science of interactions and multiple causal factors (Taylor 2005).

It is about patterns of population growth, dynamics of intraspecific and interspecific relationships, structure and function of ecological systems and flow of energy and matter through ecosystems. Therefore, as Berkowitz and colleagues (2005) argued, understanding the nature of causal factors, constraints and feedbacks in ecological systems, defining the key components of a system and its connections and understanding structure/function relationships and the kind and function of ecological processes in space and time are crucial for an ecologically literate person. Moreover, ecology educators assume the responsibility for ensuring that students gain the intellectual tools to engage fully in environmental debates and decision making. In this context, ecological literacy presupposes a deep and full understanding of ecological concepts and reasoning, while also considering the nature of ecological science and its interrelationship with society (Berkowitz et al. 2005).

Research on ecology teaching and learning, however, has consistently identified a considerable lack of understanding of core ecological concepts and processes in all educational settings regardless of age and/or background (Stamp et al. 2006). A first main cause of the failure of ecological education to accomplish its goals could be attributed to disagreements on what ecology is and what and how it should be taught in classrooms. A large part of 'ecological' content in science textbooks is in fact about taxonomy, morphology or physiology, rather than ecology per se. The science of ecology, i.e. the study of the relationships of living organisms between each other and their nonliving environment, covers a rather small part of most national science education curricula (McComas 2003; Tunnicliffe and Ueckert 2007). In most cases, ecology education could be characterised as a teaching of 'ecological bites' rather than a comprehensive and thorough understanding of ecological systems, structure and function (Korfiatis and Tunnicliffe 2012). For example, Bravo-Tortilla and Jimenez-Aleixandre (2012) noted that energy transfer would usually be taught in only one session in Spanish high schools. They also commented that Spanish educational authorities seem to be totally unaware of the time that would be needed in order for the students to be able to adequately situate the concept in different contexts and understand its centrality for sustainable resource management.

At the same time, complaints about the insufficiency of educational material to foster ecological literacy objectives are rather abundant and they do not seem to be taken into account for reforming curricula. About 20 years ago, Hale (1991) and Hale and Hardie (1993) argued that the English Science National Curriculum falls far short of providing an adequate and balanced approach to ecology, as a number of key concepts are omitted. About a decade later, Slingsby and Barker (2005) lamented that in several cases the ecological content in the English curriculum was out of date, based on a 1950s perception of ecology, or relied heavily on social policies that could be easily misconstrued as ecological concepts. Similar complaints about the inadequacy and outdatedness of the ecological content in textbooks can be found in other regional curriculum studies like Berkowitz and colleagues (2005) in the case of the USA and Korfiatis and colleagues (2004) and Lemoni and colleagues (2011) in the case of the Greek primary school curriculum.

Many writers also suggest that there might be a difficulty in handling ecological concepts such as food webs, recycling and energetics that is common for all students across age cohorts (Barker and Slingsby 1998; Demetriou et al 2009; Grotzer and Bell Basca 2003). The roots of this difficulty are to be traced in the dynamic and systemic nature of ecological processes and the thinking skills that are necessary for someone to develop in order to cope with this systemic nature. Studies by Green (1997, 2001) propose that students might reveal the beginnings of a capacity to think about interactions in natural systems but that this is frequently overwhelmed by task complexity. Since system thinking skills are necessary to follow the complexity of ecological systems (e.g. Taylor 2005; Yodzis 2000), employing linear and unidirectional processing of interactions in ecological systems will necessarily result in oversimplifying structure and dynamics in such systems and in developing profound misconceptions of the causal interactions that occur in such systems (White 2008). When tracing the effect of a change or perturbation through the structure of a complex system, students tend to follow this effect along one direction only, away from the locus of perturbation in terms of the structure of the system, and they do not seem to appreciate or predict interactions. For example, in a hypothetical situation where a population of wolves and a population of deer interact in a forest, students usually assume that wolves will first consume all deer and then die since they will have exhausted their feeding sources. Thus, they seem to forget the existence of regulatory mechanisms in many populations that do not allow them to increase indefinitely until they exhaust all resources and go extinct.

The complexities of interactive systems with multiple entities are admittedly hard to grasp, but a failure to appreciate even the possibility of interactions and feedback loops seriously compromises understanding of the workings of the natural environment (White 2008). The lack of understanding of complex systems function and interactions seems to be accompanied by rather teleological explanations, which are quite common in naive accounts of ecology and which raise static or even essentialist representations of nature (Hovardas and Korfiatis 2011). As students find it difficult to follow the effect of a disturbance through different branches of a food web, or, more importantly, the feedback effects taking place in interactions between populations (Hogan 2000), they tend to think that any disturbance in an ecosystem will eventually result in its collapse (White 2008). Alternatively, they may hold that no matter the kind or the magnitude of the disturbance, the ecosystem will eventually recover to its initial state (Ergazaki and Ampatzidis 2012).

However, none of the dimensions of ecological thinking should be seen as completely beyond the capacities of even the youngest learner. To the contrary, young people are perfectly able to grapple with evidence, systems, space and time throughout their lives, and this might lead to substantial learning benefits provided an adequate support in the form of scaffolds is given (Slingsby and Barker 2005). The crucial problem in ecology educations is often the inadequate approach to teaching. Within the above framework, the challenge for ecology education is to develop a curriculum that flows from simple to more complicated contexts without introducing new misconceptions. Barker and Slingsby (1998) and many other scholars before them (Leach et al. 1995, 1996a, b; Webb and Boltt 1990) have highlighted the possibility that in an attempt to make ecological concepts 'understandable', these concepts may be integrated in the curriculum in such an oversimplified way that they become essentially 'incorrect'. Indeed, many textbooks, while overloaded with heavy concepts, at the same time, include the core concept of trophic relationships as a formation of simple, linear food chains and not as food webs, despite the fact that the latter can be a more comprehensive approximation of trophic relations (Barker and Norris 2000).

The food-web representation offers more alternatives to elaborate on the outcome of a change in the web structure, whereas the food chain implies more drastic alterations after a change has occurred (e.g. loss of a species will dismantle the chain). In the food-chain representation, the structure can be irreversibly impacted if a species disappears, since each species occupies a unique and crucial position in the linear configuration of the food chain and cannot be readily substituted by another species. However, the removal of a species in the food-web representation does not necessarily lead to the collapse of the structure. In this case, relations between species that will remain in the food web will sustain the structure and the food web will be reorganised. For that professional ecologists have long warned for the necessity to clarify the concept that 'everything is connected to everything' through ecological education and not to consider it as a metaphysical principle, as many educators currently do (Berkowitz et al. 2005).

Another source of constraints for the development of proper educational approaches is the contradictory and often superficial discussion of the philosophical and theoretical base of ecology. This results in serious failures in communicating ecological science to the public and in targeting issues that, in relation to schools, might actually influence the curriculum. In textbook and public understanding of science research, significant concerns have been raised as to the extent that analogies, metaphors and symbols are invariably mixed with ecological subject matter, thus promoting an environmentalist rhetoric that is often closer to lay than scientific knowledge (Berkowitz et al. 2005; Cuddington 2001; Mappin and Johnson 2005). A characteristic example of the close interrelation between popular understandings of ecology and their classroom implications is the presumed 'holistic' nature of ecological science. Indeed, many scholars consider ecology as a science alternative to the western mechanistic Cartesian paradigm of modern science. Bowers (2001), for example, suggested that "ecologically oriented sciences" represent a different worldview, against the reductionistic Newtonian sciences. Adhering to this holistic orthodoxy, in many school curriculums across the world, ecology teaching is characterised by a focus at the ecosystem level, and an overemphasis on sophisticated and highly abstract concepts such as energy flow (Magnetorn and Hellden 2007), which may be introduced at inappropriate age levels. For instance, in the Greek and Cypriot curriculum, the notion of the food pyramid is introduced already in the second or third grade of the elementary school. We strongly believe, however, that this educational level cannot support a thorough comprehension of such complex concepts. In a similar vein, Ryoo and Linn (2012) contented that middle school students in the USA are grasping to deal with abstract concepts such as energy flow as they are not given the opportunity during prior instruction to build on their understanding of concrete mechanisms and processes.

Likewise, Slingsby and Barker (2005) argued that while in the English curriculum there is considerable emphasis on ecology at GCSE level (14–16 years), it is dominated by the concept of the ecosystem and a concentration on complex abstractions such as energy dynamics of food chains, the carbon and nitrogen cycles, the greenhouse effect and the consequences of global warming. As a result students are facing considerable difficulties in locating species and their role in food webs, with the most striking case being that of decomposer species (Sander et al. 2006; Demetriou et al. 2009). Another set of related and reoccurring misconceptions concerns the food pyramid concept and specifically the flow of energy and the decrease of biomass across levels. In fact, the majority of students are seldom able to comprehend and explain the shape of the food pyramid in terms of energy flow or biomass decrease. Instead, they usually believe that energy accumulates at the end of a food chain, or that total biomass reaches a climax at the end of a food pyramid, or even that ecosystems recycle energy (see, e.g. D'Avanzo 2003; Stamp et al. 2006).

17.2 Ecological Inquiry, Natural History Research and Ecological Education

The "scientific inquiry" approach—i.e. an approach that focuses on scientific processes related to collecting and analysing data and drawing conclusions—is currently the dominant form of teaching and learning proposed by most educators and educational researchers in the science education community. There is a vast amount of literature scrutinising the possible advantages and disadvantages of inquiry learning and its background philosophy from a philosophical, epistemological and educational point of view. However, a point of concern that has only recently started attracting the interest of researchers is that "scientific inquiry" in the science classroom often takes the form of simple exercises on hypothetico-deductivism, in the sense that it relies heavily on the "hypothesis–experiment–justification/rejection" rubric. As Windschitl and colleagues (2008) indicated, reference to a universal scientific method is common in discourse at all levels of science education, having, with only minor variations, the form of the following: observe, develop a question, develop a hypothesis, conduct an experiment, analyse data, state conclusions and generate new questions.

This actually seems to be the iconic representation that actively shapes how teachers and learners think about scientific practice. When used as an instructional protocol among others, this approach has allowed many teachers to develop activities that motivate young learners to ask questions, test hypotheses and work with first-hand data, but its hegemonic appearance misrepresents fundamental intellectual work done by contemporary scientific disciplines like ecology. To grasp an idea of what a better learning approach for ecology should look like, it is imperative that we turn to the science of ecology itself and outline some of the characteristics of ecological research.

17.2.1 Natural History Research and the Problem of Scientific Method in Ecology

For many years, the relationship between ecology and one of its predecessors, i.e. natural history, was a skeleton well hidden in the closet. In light of this old secret, Price and Billick (2010), in the introduction of their edited volume on the ecology of place, raised the rhetorical question "Where do ecological ideas come from?" The reply provided by Kingsland (2010) was that they do not spring deductively from the minds of ecologists, but rather from an interaction between ecologists and the place of their studies. Indeed, ecology has often been challenged as relying too much on natural history during its formative years, i.e. in the study of organisms in their environment. This kind of studies has been accused of constituting more of a merely descriptive science, without any explanatory or predictive power. Natural history was a distant relative, a kind of unsophisticated activity, restricted in collections of data that cannot be generalised. As Kingsland (2005) notes, during the discipline's formative years, ecologists had to fight to find their niche and not to be seen as a kind of scientific birdwatchers.

As ecology matures, however, it seems that ecologists are no longer afraid of the close embrace with natural history. Ecologists nowadays defend the role of natural history by claiming that prolonged study of organisms in their natural environment is the only way to understand problems in evolution, adaptation and biogeography and that "all good ecology is founded on a detailed knowledge of the natural history of the organisms" (Krebs 2010, p. 285). They also assert that natural history is not an old-fashioned activity of collecting specimens from the field, but a scientific activity, which presupposes careful design and inquiry. As Grant and Grant (2010, p. 111) put it, "…knowledge of natural history helps to frame initial questions and guides observation and it is indispensable for a comprehensive interpretation of the results" (Grant and Grant 2010, p. 111).

It became apparent that theoretical and empirical ecological research brought onto the frontline research methods that were not considered so sophisticated by the so called 'hard sciences'. In ecological research, experimentation might be as frequent as observation or even less frequent. Comparative methods are also common in ecology, as well as in sciences like evolutionary biology and geology, where historical evidence is significant. Multiple investigations and bodies of evidence of different kinds are usually brought to bear on assessing a hypothesis. Context dependency in ecological studies (Bowen and Roth 2007) precludes prescriptive field-based, replicable investigations for which the outcomes may be predetermined. Finally, though the formulation of hypotheses is indispensable for conducting scientific research, the role of hypotheses in scientific fields like ecology is not primarily to inform experimentation in terms of prediction (Shrader-Frechette and McCoy 1994; Stephens et al. 2006) but may actually serve other desiderata, such as to guide data selection and propose explanations (Marone and Galetto 2011).

So, which should be a proper scientific method for ecology? Much ink has been spilt on discussions about the science of ecology and hypothetico-deductivism during the 1980s. The major agreement that came out of the debate was that ecology is not and should not be treated as a hypothetico-deductive endeavour and that long-term ecological field research goes beyond the 'popperian exercise' of providing 'yes' or 'no' answers to specific hypotheses (Kingsland 2010). This agreement along with emphasis on fundamental differences among scientific disciplines and the questions scientists ask as well as the approaches they take when pursuing answers to those questions (Rudolph 2005) allows for similar developments in science education. The problem seems to be that educational practice has not followed developments in science and philosophy of science (Duschl et al. 2007). To this problem we now turn our attention.

17.2.2   Inquiry Approaches and Ecology Teaching

Although there is an abundance of methodological rules operative in the sciences, scientific inquiry in the science classroom and across all educational levels has tended to feature hypothetico-deductivism as a nearly universal scientific method (Windschitl et al. 2008). At the same time, the scientific inquiry approach, as often practised in classrooms, all too frequently promotes experimentation, which is based on the separation between control and experimental conditions and on a single set of observations that is considered definitive in testing an idea (i.e. the "critical experiment" as noted by Nadeau and Desautels 1984) as the only method of generating data. These features constitute a double challenge for science education more generally and ecology education in particular.

As Duschl and colleagues (2007) established, during the last 50 years or so, we have witnessed a radical change in discussions about the nature of science, which can be described as a shift in focus from science as experimentation to science as explanation, model building and revision. Causal explanations grounded in control of variable experiments have given ground to statistical/probabilistic explanations grounded in modelling experiments. Indeed, several criticisms on the use of the standard hypothesis-testing model have been quite intense in ecological and evolutionary studies during at least the last four decades.[1] As Stephens and colleagues (2006) described, major sources of fallacy are associated with an inappropriate focus on statistical significance over understanding/explanation in the case of highly complex and variable ecological systems under study, an overstatement of statistical inferences as well as on philosophical considerations related to the dichotomous choices imposed by the hypothetico-deductive model. Colyvan and colleagues offered an illustrative example of this line of arguments:

> For example, a survey may fail to demonstrate that land clearing results in a reduction in the number of bird species in the area in question… However, this failure is often an artifact of the model of hypothesis testing employed. Standard hypothesis testing is very conservative, in that it guards against false positives (type I errors). But sometimes in science, false

---

[1] See, for example, Ayala (2009), Brandon (1994), Haila (1982), McIntosh (1987), Peters (1991) and Price and Billick (2010),

negatives are more worrying. Believing that no extinctions are occurring or that land clearing is having no impact on the number of bird species, for example, can be very dangerous null hypotheses to fail to reject (Colyvan et al. 2009, p. 22).

What is important for this discussion is that alternative techniques for testing alternative hypotheses, generating new hypotheses or predictive models and assessing descriptive findings, have been incorporated into ecology including Bayesian models, effect size statistics and IT techniques (Knapp and D'Avanzo 2010; Stephens et al. 2006). The development of these methodologies attests not only to the failures, shortcomings or inappropriate use of the hypothetico-deductive model but also to the need for addressing the pluralistic nature of ecological research questions, data and goals.

Hence, educators should not rashly embed aspects of a typical inquiry learning procedure in ecology education (e.g. hypothesis testing and direct experimentation). There is a need to promote a different conceptualisation and operationalisation of inquiry in ecology education, which should reveal the necessity of a variety of methods for data generation and interpretation to develop solid understandings, explanations and predictions of ecological phenomena.

17.2.3 Implications for Teaching Ecology

From an educational point of view, all these arguments bring new emphasis on longitudinal outdoor settings, i.e. the study of organisms into their environment, and raise the importance of educational activities such as observations and comparison, rather than direct experimentation.

Although in most cases observation is the starting point for science, there are few studies of science education literature that have focused on a comprehensive account of observation (Tomkins and Tunnicliffe 2001), which involves skills such as description of phenomena, looking for patterns and making measurements. Eberbach and Crowley (2009) suggested that educators and experts had underestimated the complexity of observational practice, its interrelationship with disciplinary knowledge and the degree to which teachers and students needed scaffolding to support scientific observation. They state that experts build hierarchical, highly organised structures (within their discipline) that enable them to effectively encode and organise the observable world differently from novices and to efficiently notice and recall meaningful patterns. Moreover, systematic observation and comparison can be, for an expert, a powerful method for supporting complex hypothesis testing without experimental manipulation. Novice's observational skills, on the contrary, are portrayed as unsystematic, unfocused and unsustained. In a scientific context, novices might be described as classic "dust-bowl empiricists" who make lots of observations but have trouble encoding evidence, making valid inferences and connecting observation to theory. Accordingly, children's everyday observations have been shown to do little work towards building complex scientific understanding of natural phenomena (Ford 2005). When children are cast into an activity with inadequate knowledge and instructional support, observation becomes a weak method for collecting data rather than a powerful method for scientific reasoning. Indeed, as Tomkins and Tunniclife admitted:

> ...traditionally it was very fashionable to make the meticulous observing and drawing of biological specimens an objective in itself, but this did not necessarily lead to creative thinking about either the organism or indeed its biology. (Tomkins and Tunnicliffe 2001, p. 793)

Thus, Eberbach and Crowley (2009) argued that learning to observe scientifically necessitates bootstrapping between specific disciplinary knowledge, theory and practice.

Therefore, ecological education in the field should not be seen as a pleasant outdoor recreational activity, but as a structured process that forms the basis for new theoretical and empirical ventures. Guided science experiences outside the formal classroom require children to think critically and to value their own experiences and ideas. When being properly scaffolded, children can proceed from the experience of observational data to an expression of meaningful interpretation, such as the integration of data and inference or the making of new hypotheses (Tunnicliffe and Ueckert 2011). Likewise, Feinsinger and colleagues (1997) proposed that the first steps towards ecological literacy are for inquirers to become familiar with the natural history of their local and abiotic environments and from this to progress to the acquisition of skills of posing interesting questions about their own surroundings and to the consideration of the consequences of various human activities. Only in such a context could an understanding of ecological concepts and content be meaningfully acquired.

This orientation can be best served by a lesson sequence approach in contrast to the dominant single-lesson approach (Duschl and Grandy 2008). In an ecology education context, 'observation' is not just 'looking at things' but has to involve describing features of ecological phenomena, looking for patterns, developing and testing models and proceeding to making measurements. These processes require moving between conceptual and observational modes, which may not be served by the traditional approach to experimentation.

Demetriou and colleagues (2009) proposed a curriculum for comprehending trophic relations in elementary school, which is based on guidelines for the identification of organisms, their food preferences and the construction of a food web using real data. Students first discuss and elaborate on a rubric (classification key) depicting signs of organisms, namely, specific traces that are an indication of the presence of specific organisms. Anytime students discover a sign (e.g. a leaf bitten by a worm), they can be quite sure that this sign indicates the presence of an organism. A point to highlight here is the fact that many researchers can gather a big amount of evidence concerning the behaviour and habits of an organism only by following its signs. After having discussed the rubric, students go outdoors to select evidence on the presence and behaviour of organisms. They then proceed to a reconstruction of a food web by combining presence/absence data they have selected and feeding habits of organisms that are expected to be found in the study area. The food web can be used as a tool for following changes in the structure and dynamics of the biocommunity under study under a range of possible scenarios (e.g. change of food sources according to shifts due to seasonality). Students can pose questions and use the reconstructed food web as a guide to prepare their answers and support their reasoning. A pilot study of this educational approach revealed that fourth graders are able to construct quite precise and complicate webs, including a large number of species and drawing multiple trophic connections (Demetriou et al. 2009). Without wishing to downplay the need of providing crucial details on the functional and behavioural characteristics of food webs, the educational intervention proposed here could serve as an introductory basis for studying food webs.

This is clearly an investigation where participants are moving between conceptual and observational modes, but these actions cannot be called experiments in the traditional sense. Moreover, the artefacts of the learning activity, and more specifically the food-web diagrams, can serve as a basis for the development of additional ideas and models. Students can move on to test these ideas by developing explanations and creating arguments in support of their models. Consequently, models and modelling activities are emerging as an important aspect of both ecological scientific practice and ecology education. Indeed, it is more than often suggested that ecology education should provide more opportunities that lead to model-based inquiry and support the dialectical processes between data, measurement and evidence on the one hand and observation, explanation and theory on the other. It is at the study of that aspect of ecological inquiry that we will turn our attention now.

17.3 Ecological Models in the Science Classroom

Physical, scale, analogue, mental, theoretical, historical and mathematical models as well as other kinds of representations have triggered animated debates between philosophers and historians of science. Since the early 1960s, extensive accounts of models as the basic constitutive parts of theories or as mediators between theoretical structures and the world can be found in their writings.[2] These discussions are collectively known as the semantic or model-based view of science, though forming a rather heterogeneous group of ideas about the nature of models, they all attest to the importance of the concept and understand theories as sets of models. In talking about the structure of theories, all proponents of the semantic view roughly agree that their analysis is applicable to all scientific theories, while most of them, at least in the early days, understood models in a formal analysis context.

17.3.1 Models and Modelling in Science Education

Despite emerging criticism, the model-based view has attracted the interest of psychologists and science educators.[3] Model construction and deployment are seen as activities that extend well beyond scientific practice to include all sorts of human

---

[2] For a comprehensive review of the semantic view literature, see Godfrey-Smith (2006).

[3] Various interesting perspectives focus on theory construction per se and distinguish between modelling and other kinds of theoretical practices (e.g. see Weisberg and Reisman's (2008)

endeavours (see, e.g. Giere 2004; Redish 1994). Especially science educators, more than often, emphasise the need to incorporate models and modelling in school curricula, while advocating for a general modelling approach in education (e.g. Chamizo 2011; Koponen 2007; Portides 2007). Indeed, a number of researchers have established that modelling enhances students' problem-solving abilities, supports content learning and advances the understanding of the characteristics of science and scientific practice (e.g. Wynne et al. 2001; Schwarz and White 2005). As Gilbert and Treagust (1993) noted, models are at the same time fundamental components of scientific method and products of science, while also serving as major learning and teaching tools. The appreciation of the role of models has also led to a series of studies, which offer either profound historical and philosophical accounts of important scientific models (e.g. see McComas 2002b; Matthews 2005) or empirical results in support of this didactic orientation (Flores-Camacho et al. 2007; Prins et al. 2009; Silva 2007).

Despite this broad agreement, the terms model and modelling are used rather ambiguously in science education.[4] Thus, models are often described as simplified representations used to predict or explain phenomena (Schwarz et al. 2009); data fitting and evaluation devices (Chinn and Brewer 2001); a method to inform the development of ideas, make predictions and explore alternative explanations (White 1993); a device for describing, explaining and predicting phenomena as well as communicating scientific ideas (Oh and Oh 2011); and so on. This vagueness on what conceptions of models and modelling we put into practice seems to be a major difficulty when designing and executing inquiry-based activities and may well be responsible for both students' and teachers' inadequate and/or limited content knowledge (Grosslight et al. 1991; Harrison and Treagust 2000; Van Driel and Verloop 1999). All in all, traditional curricula seem to neglect models' tentative nature and present them as mere—i.e. non-mediated but reduced, static or simplified—copies of the real thing being studied (Prins et al. 2009).

Things become even worse when mathematical models come into play as students are unable to interpret the symbolic language used and produce qualitative explanations (De Lozano and Cardenas 2002; Korfiatis et al. 1999; Silva 2007). As a case in point, the vast majority of undergraduate biology students find themselves struggling with ecological models like the notorious Lotka–Volterra equations in every introductory population ecology course around the world. Indeed, the relatively simple mathematical model that Alfred Lotka (1925) and Vito Volterra (1926) separately introduced to describe the population cycles of a predator–prey system has been eloquently accused of inducing terror, hence the expression *Lotka–Volterrorism*, among students (Boucher 1998). In sum, despite the vigorous discussion and research in the area, both students and teachers seem to still fail to appreciate

---

discussion on the difference between modelling and abstract direct representation and Godfrey-Smith's (2006) critique of the semantic view's formalism).

[4]This is not to say that philosophers of science are in a better shape; as Godfrey-Smith (2006) wrote, 'The term 'model' is surely one of the most contested in all of philosophy of science' (Godfrey-Smith 2006, p.725).

the importance of modelling in scientific practice since they are merely introduced into questions like 'what is a model?', 'how do we model?', and most importantly 'why do we model?'.

Following scholars like Schwarz and colleagues (2009) and Adúriz-Bravo and Izquierdo-Aymerich (2005), we are confident that bringing these questions into classroom discussions will enable students and teachers to develop a broader understanding of scientific methods and plan innovative educational interventions. Like Passmore and colleagues (2009), we believe that inquiry-based activities should include the development, use, assessment and revision of models and related explanations. We also agree that to fully appreciate the diversity of scientific practice, these activities should be explicitly grounded in a specific context and always address epistemic reasoning in relation to emerging scientific problems.

To this end, we suggest that the philosophical discussion about models in ecology presents a great opportunity to focus on actual scientific practice. Being necessarily idealised in comparison to the real-world systems they represent, ecological models are diverse and fulfil different desiderata. Next, we will briefly touch on some of these issues and highlight why a pragmatic and pluralistic turn might prove more fruitful for classroom explorations than more generic accounts.

17.3.2 Models and Modelling from a Philosophy of Ecology Perspective

When models are discussed in philosophy of ecology or biology, Richard Levins' seminal contributions come first to mind.[5] In his most cited works, 'The Strategy of Model Building in Population Biology' (1966) and 'Evolution in Changing Environments' (1968), as well as in a number of subsequent writings,[6] Levins offers enormous insight on recurrent philosophical and practical questions about models and modelling. His concerns grew out of the need to deal with the extreme complexity of biological systems as well as from a great dissatisfaction with the prevailing empiricist, reductionist and overspecialised philosophy of American science.

Levins' ideas have been immensely influential among biologists and his discussions of modelling still appear even in introductory textbooks. Philosophers of science were a bit late to discover this discussion, with the exception of William Wimsatt (1981, 1987) whose major concerns were remarkably close to Levins'. Since the 1990s, however, Levins' views have been carefully scrutinised providing

---

[5] Richard Levins is a well-known theoretical population biologist who has contributed significantly to our attempts to understand and influence complex systems. His work has often crossed disciplinary boundaries and actively integrates issues of history, philosophy and sociology of science. As Haila and Taylor (2001, p.98) wrote: "…in his research, concrete questions, theory and philosophy go hand in hand… (while his) pioneering role in developing ideas on ecological complexity is widely known" (Haila and Taylor 2001, p. 98).

[6] See, for example, Levins (1970, 1993, 2006).

the basis for fruitful reflections and insight.[7] Here, we will briefly develop some of the major themes emerging from his work along with recent considerations on the epistemic questions surrounding modelling. We hope that these insights will prove useful to classroom discussions. Next, we will focus on three interrelated issues: (a) modelling strategies and desiderata, (b) models and real-world systems and (c) methodological pluralism.

17.3.2.1 Modelling Strategies and Desiderata

As already mentioned, every introductory ecology (or population ecology) module introduces students to what is often called 'Levins' classification of models'. In a nutshell, what they are taught is that models can be either general or realistic or precise and therefore scientists devise or use each 'kind' of model according to their goals or even aesthetic criteria. And the question is what is wrong with this formalisation? Is it true to the philosophical discussion? Our contention is that it is both misleading—as it turns an epistemological venture to an essentialist account—and largely not useful; students barely ever need another classification.

In his 1966 paper, Levins did argue that when building models—in his case mathematical—of complex systems, scientists must inevitably trade off between different model attributes and follow alternative strategies that best match their desiderata. In short, he suggested that there cannot be a best all-purpose model, i.e. one that simultaneously maximises generality, realism and precision, while remaining manageable and helps us understand, predict or even modify nature.[8] Therefore, biologists either sacrifice generality to realism and precision (as in the case of highly predictive fisheries models), choose generality over realism and precision (as in the case of the classical Lotka–Volterra predator–prey systems) or prefer realism and generality to precision and produce qualitative results (as in the case of the equilibrium theory of biogeography). Several years later, Odenbaugh (2005) similarly suggested that models in theoretical ecology serve different purposes: (a) they are used to explore possibilities, (b) they can serve as basis to investigate more complex systems, (c) they lead to the development of conceptual frameworks, (d) they can provide predictions and (e) they can generate explanations.[9]

---

[7] See, for example, Godfrey-Smith (2006), Justus (2006), Haila and Taylor (2001), Odenbaugh (2003, 2005, 2006), Orzack and Sober (1993), Palladino (1991), Taylor (2000), Weisberg (2006a, b), Winther (2006), and Wimsatt (1981, 1987).

[8] In this context (Levins 1966, 1993) one could arguably suggest that generality refers to the number of real-world systems a model applies to. Realism refers to the representational accuracy of a model, i.e. how well the structure of a model represents the structure of a target system. Finally, precision could be understood as fineness of specification.

[9] The issue of explanatory success of highly idealised models is a very interesting discussion that cannot be undertaken here. However, we briefly note that even these models may have explanatory power if our idealisations do not affect the basic causal relationships or if we see our explanations as sketches of an explanation.

What these initial ideas, along with more recent considerations, contribute is the concept of trade-offs between often-conflicting epistemic aims in light of our cognitive and even technological limitations (Odenbaugh 2003). Levins did not offer a trichotomy of models but an account of the practices of modelling in biology. Thus, one might rightly suggest that these arguments are pragmatic and historical. The notion of strategy, however, is vital. Scientists explore sets of options depending on their specific research questions and try to best exploit possibilities through a configuration of resources within a challenging environment. Both their strategies and desiderata may change given the technological capacities and the altering demands at the intersection of science and society. Limitations may as well become more flexible, but ecologists will still have to device strategies to deal with the characteristic complexity of the inherently dynamic and contradictory ecological systems.

For science educators and students, these discussions seem to be an excellent starting point to explore core issues about how scientific inquiry proceeds and about the plurality of scientific methods and scopes and the limitations of our methods and research agendas. Furthermore, they bring to the forefront issues related to the science/society interaction, especially in terms of the changing goals that society sets for science and vice versa. For example, the rise of environmental awareness has put pressure on scientists to device models that will best help preserve or even modify nature. This in turn has resulted in a proliferation of more predictive and realistic models but also in an emphasis for the need to explore alternative hypothesis and understand and explain nature before taking action. Finally, they bring to life an often-disregarded discussion about the relationship between science and technology in terms of how technology may empower and advance science and vice versa.

17.3.2.2 Models and Real-World Systems

Ecologists have consistently expressed their discomfort with the proliferation of highly idealised models that are seldom tested against real data (see, e.g. Peters 1991; Simberloff 1981; Strong 1983). In Daniel Simberloff's words:

> Ecology is awash in all manner of untested (and often untestable) models, most claiming to be heuristic, many simple elaborations of earlier un-tested models. Entire journals are devoted to such work, and are as remote from biological reality as are faith-healers. (Simberloff 1981, p. 52)

Philosophers of ecology, on the other hand, have also been arguing that explanation, prediction and correspondence with data are not the only goals for modelling, and therefore, reality is not the benchmark for all models (Cooper 2003; Odenbaugh 2003; Taylor 1989, 2000). For example, Taylor (2000) defined three possible roles for models according to their relationship to reality: (a) *schemata* highlight biological processes and when expressed mathematically become exploratory tools which produce diverging outputs and explore various aspects of the specific world generated by schemata, (b) *redescriptions* allow for the formulation of statistically testable hypothesis and low-level generalisations as long as previously observed patterns still hold and (c) a model is characterised as a *generative representation* if it does not only fit the data but its accessory conditions also seem to hold true.[10] Thus, the model is a higher-level generalisation that not only explains the phenomenon but also allows making future predictions.

This approach does in no way disregard ecologists' concerns about models being mathematically inspired exercises on pieces of paper. On the contrary, it is a call for an even closer rapprochement between theoretical and empirical research. As Odenbaugh (2005) rightfully suggested theory construction through modelling is an indispensable part of doing science, advancing our understanding and allowing for new questions to emerge. At the same time, empirical research does not only validate theoretical claims but also offers insights and new directions for theoretical endeavours.

In any case, however, models are constructed systems with patently unreal assumptions about their variables, parameters and the relationships that hold between the two. Models grow out of a process of simplification, a process of abstraction and the addition of assumptions needed to facilitate a specific study. This obviously means that all models are idealised and their idealisations are legitimate not only in relation to the reality described but also according to the state of science and the purpose of study they were designed for. In this sense, as Levins suggested back in the 1960s, "…all models leave out a lot and are in that sense false, incomplete, inadequate" (1966, p. 430).

Before we turn to our final point, we believe that this line of thinking adds several new points of concern for classroom discussions, especially in the light of the importance that students, even after explicit discussion about the multiple roles of models, seem to put on explanatory or predictive success (Svoboda and Passmore 2011). For example, students and teachers could work on some of the following ideas: (a) even more realistic models, either predictive or explanatory, are not just simplified reality; (b) simple, exploratory or conceptual models do not evolve to become predictive or explanatory since this is not the purpose they were designed for; (c) matching real-world situations, either through describing causal mechanisms or making accurate predictions, is a fundamental goal of science but not the standard all models should follow; and (d) the final test of highly idealised models should be their ability to produce new generations of models that are explanatory and/or predictive. For example, the classic highly idealised Lotka–Volterra prey–predator model, when supplemented with more detailed biological information, like adding a saturation coefficient for the predator population, produces new generations of models, which are much closer to empirical systems. Therefore, it is still a valuable model.

17.3.2.3 A Mixed Strategy Approach

The question that follows from the above is if there is no such thing as a best all-purpose model and ecologists build models that serve different epistemic aims, is

---

[10] As Taylor (2000) suggested, accessory conditions are very easily overlooked; however they are actually what makes modelling possible. Such conditions in the case of ecology may assume, for example, a uniform and constant environment in space and time.

there a best strategy to follow? It should be obvious so far that philosophers of ecology advocate for a kind of theoretical or pragmatic pluralism (Wimsatt 2001). As Odenbaugh (2006) established, Levins' original criticism challenged model monism in light of the simultaneous use of partially overlapping models and theories and inspired scientists and philosophers to accommodate diverse strategies (see, e.g. Cooper 2003; Taylor 2000; Vepsalainen and Spence 2000). Without delving deeper into details, in his early work, Levins was very clear, even when he critiqued the large-scale computer models of systems ecology, that there is no such thing as a best strategy: "Therefore the alternative approaches even of contending schools are part of a larger mixed strategy" (Levins 1966, p. 431).

An interesting example of this mixed strategy approach is offered by conservation biology and especially nature reserve design. In the 1960s Robert H. Mac Arthur and Edward O. Wilson set out to explore the relationship between the diversity of species—in this case the number of different species—on islands and island area (1967). Their basic assumption was that despite changes in species composition, the number of species inhabiting the island would remain in a state of dynamic equilibrium as a result of local extinctions and immigration. As Kingsland contented (2002a, b), their approach explicitly ignored historical detail for the sake of contributing a simple, general and realistic theory that would eventually produce testable hypotheses and integrate population ecology and evolutionary theory. Once the theory was published, conservationists realised that nature reserves could be seen as isolated islands among habitats changed by humans. This realisation raised the bar for ecologists, who were actually given a chance to apply their theory to actual world problems. Experimental work along with various data sets and observations were brought together to support theoretical arguments. However, the eagerness to apply theoretical arguments in conservation efforts was met with criticism. What was assumed as a premature transformation of basically untested theory to conservation principles resulted in one of the more fierce and rhetoric-dependent controversies in ecology that lasted for more than one decade (Kingsland 2002a, b; Looijen 2000). A growing rapprochement between conservation biology and operation research, already known to system ecologists, in the 1980s seemed to offer a viable alternative to reserve design. Over the past decade, this interdisciplinary collaboration has proven quite fruitful and the task of modelling complex ecological systems through extensive simulations is much facilitated by technological advances in computing. This kind of research, however, which largely belongs to the strategy of sacrificing generality for the sake of precision and accuracy in Levins' fashion, is extremely labour intensive as it requires large amounts of data. At the same time these developments do in no way deem theoretical research obsolete since computer simulations are still using classical simple population ecology models (Odenbaugh 2005). Hence, it seems that a mixed strategy approach that allows the advances of each perspective to equally contribute to conservation efforts is not only desirable but most importantly required, especially when considering adding to these challenges the intense social debates over nature conservation.

These considerations provide, to our view, interesting ways to address perennial issues related to science as a social process. Discussions about scientific controversies are a major vehicle for introducing students to the nature of science. It is our contentions, however, that students should also learn how scientists belonging to quite diverse research traditions might actually work together. The emphasis given by philosophers of ecology to accommodating diverse strategies seems to be very important in escaping a single, linear, one-actor depiction of scientific endeavours. At the same time, as highlighted by Lefkaditou (2012), these debates and their resolutions do not only attest to the need for a genuinely synthetic view and reveal the plurality of approaches in ecological research but are an integral part of how science advances. Therefore, students should always remain aware of the fact that there is not a unique, victorious body of knowledge but that theories; ontological, methodological and epistemological assumptions; research techniques; ideological commitments; and social factors are the driving forces of science.

17.3.3 Back to the Science Education Classroom

To sum up, philosophical considerations on models and modelling in ecology provide a more pluralistic story of the uses of models, expand beyond the classical prediction–explanation discussion and stress the tentative character, the context dependency and the heuristic power of models. As Odenbaugh (2005) emphasised, models in ecology are necessarily idealised, most of the time inaccurate, but may well be successful in different tasks depending on the purpose they were built for. We argue here that this view of modelling brings a very interesting twist in the science education literature that focuses on model-based inquiry and enriches the repertoire of choices for curriculum developers, science educators and students. Our view is very close to Svoboda and Passmore's (2011) account, which stresses the need for educators to cultivate a variety of modelling approaches in science and engage students in diverse modelling and reasoning activities in relation to specific theoretical or practical problems.

To this end, we also believe that an explicit exploration of the epistemological underpinnings of modelling does not only lie at the heart of the nature of science and scientific inquiry discussions but should also become an important element of classroom practice. The real challenge, however, is to find concrete problems, questions and examples that help illustrate model diversity and enable students to develop their meta-scientific thinking. This is in fact a double challenge, as it raises both instructional and developmental issues. Thus, we are asked to identify appropriate activities at appropriate educational levels.

An interesting case of model-based inquiry activities in ecology is the use of computer simulations. Indeed, various researchers have argued that computer simulations not only improve skills related to the understanding of specific science content but most importantly advance problem-solving and decision-making capabilities, while inviting students to explore possibilities, create hypothesis, interpret their results and frame theoretical claims (Akpan and Andre 1999; Serra and Godoy 2011). In this spirit, Carson (1996) used computer simulations to teach about food webs, Cook (1993) studied foraging behaviour, Korfiatis and colleagues (1999) explored the behaviour of a population system regulated by intraspecific competition and Lutterschmidt and Schaefer (1997) modelled predator–prey interactions, while Serra and Godoy (2011) explored population patterns. In most of these cases, computer simulations seem to have presented ecological concepts and models in a more exciting, engaging and interactive manner, while they have arguably improved students' computational and mathematical skills. What is more controversial, however, is whether computer simulations actually improve understanding of ecological subject matter and the role of models in actual scientific inquiry (see, e.g. Korfiatis et al. 1999). Our contention is that without explicit reference to the diverse epistemic aims of modelling, computer simulations are nothing more than a black-box approach. As a result, students may still see models as being isomorphic to phenomena, their simplifying assumptions are reified, statistical patterns are mistaken for causes and the fixation on predictive value as a model's true virtue goes unquestioned, while their tentative and contextual character remains untouched. Despite our best efforts, modelling and issue knowledge may well remain decontextualised.

In this spirit, Hovardas and Korfiatis (2011) have outlined an educational intervention of model-based inquiry for secondary school students that combined philosophical considerations with computer simulations. The model that students had to work with was originally introduced by the animal ecologist Walder C. Allee in 1931 to describe the negative effects that under-crowding might exhibit to certain populations. The 'Allee effect', as it has come to be known, induces lower birth rates at low population densities that may lead susceptible species to extinction. Therefore, populations with dynamics that follow the 'Allee effect' present two points of equilibrium: an unstable one at a low population size and another one at a larger population size, which is characterised as stable. If the population drops below the unstable equilibrium point, it goes extinct. In contrast, when deviating around the stable equilibrium point at a larger population size, the population returns, after a while, to its former population size.

The authors used the 'Allee effect' to discuss the case of a black vulture (*Aegypius monachus*) population in a Greek nature reserve. The target of the educational activity was not only to address different patterns of balance and help students accommodate the concept of change in nature but also support them in understanding how a simple, exploratory model might work in real-world situations. Students constructed the population model using dynamic feedback model software, like STELLA, and explored population trends in the case of minor departures from both equilibrium points, especially the fragility of a population around the unstable equilibrium point and its resilience around the stable equilibrium point.[11] Based on the model specifications, students formulated new research questions and hypotheses that referred to the models' structural components or relations between structural compartments. Simulation outcomes were contrasted to expected results and possible insights for nature reserve design were discussed.

---

[11] An exemplary sequence of this approach on model-based inquiry can be found here: http://scy-net.eu/scenarios/index.php/Grasp_a_Model.

Likewise, in addition to studying population dynamics of a single population and tracking its course over time, modelling software offers the opportunity for examining interactions between populations, such as prey–predator relationships. We have often invited students to elaborate on a hypothetical situation where a population of wolves and a population of deer are found in a forest. In most cases, students assume that wolves will first consume all deer and then die since they will have exhausted their feeding sources. However, after constructing a simple prey–predator model, refining its basic characteristics and running the simulation, students come to a surprising result; no population may go extinct. Instead, wolf and deer populations fluctuate in time. Thus, students are invited to reflect on their initial understandings, refine their initial research questions and explore alternative hypotheses by relaxing their original assumptions and introducing new complications like density dependence for the prey population or a saturation coefficient for the predator. Finally, we contrast the model's outcomes with data available from empirical research and discuss how and why the model fails to match the data and possible ways to introduce more realistic assumptions.

All in all, instead of giving students finite definitions about models, it seems more appropriate to reinforce their sense-making mechanisms by introducing them to some initials questions and trying to build our models from there on. For example, we could ask: *Given the essential complexity of biological systems, how do we decide what is relevant to include in our model? When do our abstractions, simplifications or assumptions lose their legitimacy? Is it possible for a non-accurate or unrealistic model to give predictions? What else can we do with a model? If we change the range of a model's application, do the generalisations produced still hold? When do our models become old and require revision?* Of course this is only the beginning of an enduring and demanding process that requires both time and guidance.

Finally, we agree with Taylor (2000) that the emphasis on the process of modelling instead of models themselves introduced by Levins and further developed by philosophers of ecology brings a whole set of exciting new questions about scientific inquiry and its social implications. These questions are so close to the nature of science and scientific inquiry core themes that they open new paths for ecology education. We strongly believe that these are highly promising paths worth pursuing.

17.4 Conclusion

In this paper we have tried to give a glimpse of the way scientific inquiry is conducted in ecology. It is an integrative, interdisciplinary field of research encompassing a variety of theoretical frameworks and a plurality of methodological approaches. We suggest that recognising the fact that scientists use diverse inquiry methods, which serve different epistemic roles, will not only bring classroom instruction closer to actual scientific practice but will also widen the repertoire of available instructional protocols. We hope that our discussion will enrich the ongoing dialogue about the important differences among scientific disciplines and will fruitfully contribute to enhancing ecology education.

Ecology education has the difficult task of teaching students the structure and function of the world's ecosystems as well as their interrelations with humanity. Understanding the practices of ecologists is an important means towards the accomplishment of this task (Bowen and Roth 2007).

The rapprochement between history and philosophy of science and science education has opened new avenues for intellectual work. We gladly accept the challenge to find ways in which the lively philosophical and historical discussion in ecology can inform educational practice. Towards this end, we believe that the historical and philosophical considerations addressed in our work should help establish a 'progressive', or 'bottom-up', curriculum approach, as it has been outlined by various authors (Barker and Slingsby 1998; Berkowitz et al. 2005; Korfiatis and Tunnicliffe 2012; Magnetorn and Hellden 2007; Slingsby and Barker 2005).

Within such an approach, educational interventions should start with direct contact with individual elements like single species, continue with the study of their relationships and conclude with the study of processes at the level of the whole community or ecosystem. Authenticity in ecology education has no meaning without field experiences. First-hand study of the natural world should be the main part of education, especially in the pre-school and early schooling years (Korfiatis and Tunnicliffe 2012). This approach is not restricted to the study of isolated parts of a system, but it focuses on the way individual parts interact and function in forming the whole. We agree along with various other scholars that a 'bottom-up' approach could actually prove more helpful for young students trying to comprehend how a system is constructed, how its properties emerged and how structure interplays with function and behaviour (Demetriou et al. 2009; Magnetorn and Hellden 2007). The trap of oversimplification is avoided and at the same time the foundations for understanding more abstract representations of species and ecosystems are laid.

Indeed, abstract concepts, such as food webs, can be easily grasped by early primary school children if their teaching is based on the study of organisms living in, for example, the pond or the lawn of the local park and the ways in which such organisms cover their trophic needs. Gradually, modelling activities and simulations can be essentially integrated in parts of the curriculum, allowing for larger degrees of theorising and comprehension of the methodological approaches, the explanatory patterns and the nature of the science of ecology. Besides, students, especially those in higher grades of education, can be engaged in various sorts of ecological theorising, like model building. According to an ecological portrayal, scientists utilise a number of models that embody the theoretical knowledge to which they adhere. Since general theories consist of families of models, they very rarely rise or fall based on tests of any one model. Alternative or competing models exist within most theoretical constructs in ecology allowing a single theory to encompass a diversity of phenomena (Scheiner and Willig 2011). Although it is considered one of the main aims of current education, this kind of conceptual inquiry is generally missing from science classes (NRC 2012).

Within the educational process, long-term open experimental settings, such as terrariums, are important for observing and comprehending ecological processes' roles (e.g. role of decomposers), carrying at the same time a higher perceivable educational value rather than conducting ecological experiments in a hypothesis-testing, single-lesson manner. As Tomkins and Tunnicliffe (2001) note, a "project-like" approach, with long-stay instalments, starting with observations and integrated previous knowledge, allowing for multipurpose activities and an open agenda, seems to be more proper for ecology's teaching and learning.

Needless to say that within such a framework, the need for a thorough reconsideration of educators' professional development is emerging. Such a teaching transition presupposes a considerable shift in the planning of learning activities and their orchestration in order to maintain focus on learning goals and provide scaffolds when needed. To cope with the instructional challenges that are implied in the proposed reorientation of ecological curriculum, preservice and in-service teachers engaged in ecology education at the primary and secondary education level have to commit themselves to an ongoing professional development programme in the areas of outdoor education and model-based learning.

References

Adùriz-Bravo, A., & Izquierdo-Aymerich, M. (2005). Utilizing the 3P-model to characterize the discipline of didactics of science. *Science & Education* 14, 29–41.

Akpan, J. P., & Andre, T. (1999). The effect of a prior dissection simulation on middle school students' dissection performance and under-standing of the anatomy and morphology of the frog. *Journal of Science Education and Technology, 8,* 107–121.

Allee, W. C. (1931). *Animal aggregations. A study in General Sociology.* Chicago: University of Chicago Press.

Ayala, F. J. (2009). Darwin and the scientific method. *Proceedings of the National Academy of Science* (USA), 106, 10033–10039.

Barker, S., & Norris C. (2000) *Feeding relationships: An ecological approach to teaching of food chains in the primary school.* British Ecological Society

Barker, S., & Slingsby, D. (1998). From nature table to niche: curriculum progression in ecological concepts. *International Journal of Science Education*, 20, 479–486.

Berkowitz, A. R., Ford, M. E., & Brewer C. A. (2005). A framework for integrating ecological literacy, civics literacy and environmental citizenship in environmental education. In: E. A. Johnson & M. J. Mappin (Eds.), *Environmental Education and Advocacy: Changing Perspectives of Ecology and Education* (pp. 227–266). Cambridge, UK: Cambridge University Press.

Boucher, D. H. (1998). Newtonian ecology and beyond. *Science as Culture*, 7, 493–517.

Bowen, G. M., & Roth, W. M. (2007). The practice of field ecology: Insights for science education. *Research in Science Education*, 37, 171–187.

Bowers, C. (2001). How language limits our understanding of environmental education. *Environmental Education Research,* 7, 141–151.

Brandon, R. N. (1994). Theory and experiment in evolutionary biology. *Synthese, 99,* 59–73.

Bravo-Torija, B., & Jiménez-Aleixandre, M. P. (2012). Progression in complexity: Contextualizing sustainable marine resources management in a 10th grade classroom. *Research in Science Education*, 42, 5–23.

Carson, S. R. (1996). Foxes and rabbits - and a spreadsheet. *School Science Review*, 78, 21–27.

Chamizo, J. A. (2011). A new definition of models and modeling in chemistry's education. *Science & Education*, online first DOI: 10.1007/s11191-011-9407-7.

Chinn, C. A., & Brewer, W. F. (2001). Models of data: A theory of how people evaluate data. *Cognition and Instruction, 19*, 323–393.

Colyvan, M., Linquist, S., Grey, W., Griffiths, P., Odenbaugh, J., & Possingham, H. P. (2009). Philosophical issues in ecology: Recent trends and future directions. *Ecology and Society, 14*, http://www.ecologyandsociety.org/vol14/iss2/art22/.

Cook, L. M. (1993). HUNT: a simulation of predator searching behaviour. *Journal of Biological Education, 27*, 287–290.

Cooper, G. J. (2003). *The Science of the Struggle for Existence: On the Foundations of Ecology.* Cambridge: Cambridge University Press.

Cuddington, K. (2001). The "balance of nature" metaphor and equilibrium in population ecology. *Biology and Philosophy, 16*, 463–479.

D'Avanzo, C. (2003). Application of research on learning to college teaching: ecological examples. *Bioscience, 53*, 1121–1128.

De Lozano, S. R., & Cardenas M. (2002). Some learning problems concerning the use of symbolic language in physics. *Science & Education, 11*, 589–599.

Demetriou, D., Korfiatis, K., & Constantinou, C. (2009). Comprehending trophic relations through food web construction. *Journal of Biological Education, 43*, 53–59.

Duschl, R., & Grandy, R. (2008). Reconsidering the character and role of inquiry in school science: Framing the debates. In R. Duschl & R. Grandy (Eds.), *Teaching Scientific Inquiry: Recommendations for Research and Implementation*. Rotterdam, Netherlands: Sense Publishers.

Duschl, R. A., Schweingruber, H. A., & Shouse, A. (2007). *Taking Science to School: Learning and Teaching Science in Grades K-8, National Research Council*. Washington, DC: The National Academies Press.

Eberbach, C., & Crowley, K. (2009). From everyday to scientific observation: How children learn to observe the biologist's world. *Review of Educational Research, 79*, 39–68.

Ergazaki, M., & Ampatzidis, G. (2012). Students' reasoning about the future of disturbed or protected ecosystems and the idea of the 'balance of nature'. *Research in Science Education, 42*, 511–530.

Feinsinger, P., Margutti, L., & Oviedo, R. D. (1997). School yards and nature trails: Ecology education outside university. *Trends in Ecology and Evolution, 12*, 115–11.

Flores-Camacho, F., Gallegos-Cázares, L., Garritz A., & García-Franco, A. (2007). Incommensurability and multiple models: Representations of the structure of matter in undergraduate chemistry students. *Science & Education, 16*, 775–800.

Ford, D. (2005). The challenges of observing geologically: Third graders' descriptions of rock and mineral properties. *Science Education, 89*, 276–295.

Giere, R. N. (2004). How models are used to represent reality. *Philosophy of Science, 71*, 742–752.

Gilbert, J. K., & Treagust, D. (1993). *Multiple Representations in Chemical Education*. Dordrecht, The Netherlands: Springer.

Godfrey-Smith, P. (2006). The strategy of model-based science. *Biology and Philosophy, 21*, 725–740.

Grant, P., & Grant, R. (2010). Ecological insights into the causes of an adaptive radiation from long-term field studies of Darwin's finches. In I. Billick & M. Price (Eds.), *The Ecology of Place: Contributions of Place-based Research to Ecological Understanding* (pp. 109–133). Chicago and London: The University of Chicago Press.

Green, D. W. (1997). Explaining and envisaging an ecological phenomenon. *British Journal of Psychology, 88*, 199–217.

Green, D. W. (2001). Understanding microworlds. *Quarterly Journal of Experimental Psychology, 54*, 879–901.

Grosslight, L., Unger, C., Jay, E., & Smith, C. L. (1991). Understanding models and their use in science: Conceptions of middle and high school students and experts. *Journal of Research in Science Teaching, 28*, 799–822.

Grotzer, T. A., & Basca, B. B. (2003). How does grasping the underlying causal structures of ecosystems impact students' understanding? *Journal of Biological Education*, 38, 1–14.

Haila, Y. (1982). Hypothetico-deductivism and the competition controversy in ecology. *Annals Zoologica Fennici*, 19, 255–263.

Haila, Y., & Levins, R. (1992). *Humanity and nature: Ecology, science and society*. London: Pluto Press.

Haila, Y., & Taylor, P. (2001). The philosophical dullness of classical ecology, and a Levinsian Alternative. *Biology & Philosophy*, 16, 93–102.

Hale, M. (1991). Ecology in the national curriculum. *Journal of Biological Education*, 25, 20–26.

Hale, M., & Hardie, J. (1993). The role of ecology in education in schools in Britain. In: M. Hale (Ed), *Ecology in Education* (pp. 10–22). Cambridge, UK. University of Cambridge Press.

Harrison, A., & Treagust, D. (2000). A typology of school science models. *International Journal of Science Education*, 22, 1011–1026.

Hogan, K. (2000). Assessing students' systems reasoning in ecology. *Journal of Biological Education*. 35, 22–28.

Hovardas, T., & Korfiatis, K. (2011). Towards a critical re-appraisal of ecology education: Scheduling an educational intervention to revisit the 'balance of nature' metaphor. *Science & Education*, 20, 1039–1053.

Justus, J. (2006). Loop analysis and qualitative modeling: Limitations and merits. *Biology and Philosophy*, 21, 647–666.

Keller, D. R., & Golley, F. B. (Eds.) (2000). *The Philosophy of Ecology: From Science to Synthesis*. Athens: University of Georgia Press.

Kingsland, S. E. (2002a). Designing nature reserves: adapting ecology to real-world problems. *Endeavour*, 26, 9–14.

Kingsland, S. E. (2002b). Creating a science of nature reserve design: perspectives from history. *Environmental Modeling and Assessment*, 7, 61–69.

Kingsland, S. (2005). *The evolution of American ecology 1890–2000*. Baltimore, USA: The Johns Hopkins University Press.

Kingsland, S. (2010). The role of place in the history of ecology. In I. Billick & M. Price (Eds.), *The Ecology of Place: Contributions of Place-based Research to Ecological Understanding* (pp. 15–39). Chicago and London: The University of Chicago Press.

Knapp, A. K., & D'Avanzo, C. (2010). Teaching with principles: Toward more effective pedagogy in ecology. *Ecosphere*, 1, Article 15., DOI:10.1890/ES10-00013.1

Koponen, I. T. (2007). Models and modelling in physics education: A critical re-analysis of philosophical underpinnings and suggestions for revisions. *Science & Education*, 16, 751–773.

Korfiatis, K., Papatheodorou, E., Stamou, G. P., & Paraskevopoulous, S. (1999). An investigation of the effectiveness of computer simulation programs as tutorial tools for teaching population ecology at university. *International Journal of Science Education*, 21, 1269–1280.

Korfiatis, K. J., Stamou, A. G., & Paraskevopoulos, S. (2004). Images of nature in Greek primary school textbooks. *Science Education*, 88, 72–89.

Korfiatis, K. J., & Tunnicliffe, S. D. (2012). The living world in the curriculum: ecology, an essential part of biology learning. *Journal of Biological Education*, 46, 125–127.

Krebs, C. (2010). Case studies and ecological understanding. In I. Billick & M. Price (Eds.), *The Ecology of Place: Contributions of Place-based Research to Ecological Understanding* (pp. 283–302). Chicago and London: The University of Chicago Press.

Lambert, J. M. (1967). *The Teaching of Ecology*. Oxford: Blackwell.

Leach, J., Driver, R., Scott, P., & Wood-Robinson, C. (1995). Children's ideas about ecology, 1: Theoretical background, design and methodology. *International journal of Science education*, 17, 721–732.

Leach, J., Driver, R., Scott, P., & Wood-Robinson, C. (1996a). Children's ideas about ecology, 2: Ideas found in children aged 5–16 about the cycling of matter. *International journal of science education*, 18, 19–34.

Leach, J., Driver, R., Scott, P., & Wood-Robinson, C. (1996b). Children's ideas about ecology, 3: Ideas found in children aged 5–16 about the interdependence of organisms. *International Journal of Science Education*, 18, 129–141.

Lefkaditou, A. (2012). Is ecology a holistic science, after all? In G. P. Stamou (eds), *Populations, biocommunities, ecosystems: A review of controversies in ecological thinking*. Bentham Science Publishers.

Lemoni, R., Lefkaditou, A., Stamou, A. G., Schizas, D. & Stamou G. P. (2011). Views of nature and the human-nature relations: An analysis of the visual syntax of pictures about the environment in greek primary school textbooks diachronic considerations. *Research in Science Education*, DOI 10.1007/s11165-011-9250-5.

Levins, R. (1966). The strategy of model building in population biology. *American Scientist*, 54, 421–431.

Levins, R. (1968). *Evolution in changing environments*. Princeton: Princeton University Press.

Levins, R. (1970). Complex systems. In: C. H. Waddington (Ed.) *Towards a theoretical biology*, Vol 3 (pp 73–88). Chicago: Aldine Publishing.

Levins, R. (1993). A response to Orzack and Sober: Formal analysis and the fluidity of science. *The Quarterly Review of Biology*, 68, 547–555.

Levins, R. (2006). Strategies of abstraction. *Biology and Philosophy*, 21, 741–755.

Looijen, R. C. (2000). *Holism and reductionism in biology and ecology*. Dordrecht, Boston, MA: Kluwer Academic Publishers.

Lotka, A. J. (1925). *Elements of physical biology*, Baltimore, USA: Williams and Wilkins.

Lutterschmidt, W., & Schaefer, J. (1997). A computer simulation for demonstrating and modelling predator–prey oscillations. *Journal of Biological Education*, 31, 221–227.

MacArthur, R. H., & Wilson, E. O. (1967). *The theory of island biogeography*. Princeton University Press

Magnetorn, O., & Hellden, G. (2007). Reading new environments: Students' ability to generalise their understanding between different ecosystems. *International Journal of Science Education*, 29, 67–100.

Mappin, M. J., & Johnson, E. A. (2005). Changing perspectives in ecology and education in environmental education. In E. Johnson & M. Mappin (Eds.), *Environmental education and advocacy: Changing perspectives of ecology and education* (pp. 1–27). Cambridge, UK: Cambridge University Press.

Marone, L., & Galetto, L. (2011). The dual role of hypotheses in ecological research and its association with the hypothetico-deductive method. *Ecologia Austral*, 21, 201–216.

Matthews, M. R. (2005). Idealization and Galileo's pendulum discoveries: Historical, philosophical and pedagogical considerations. In M. R. Matthews, C. F. Gauld, & A. Stinner (Eds.), *The pendulum: Scientific, historical, philosophical & educational perspectives* (pp. 209–235). Dordrecht, The Netherlands: Springer.

McComas, W. F. (2002a). The ideal environmental science curriculum: History, rationales, misconceptions and standards. *American Biology Teacher*, 64, 665–672.

McComas, W. F. (2002b). *The nature of science in science education*. Dordrecht, The Netherlands: Kluwer Academic Publishers.

McComas, W. F. (2003). The nature of the ideal environmental science curriculum: advocates, textbooks and conclusions (part II). *American Biology Teacher,* 65, 171–178.

McIntosh, R. (1987). Pluralism in ecology. *Annual Review of Ecology and Systematics*, 18, 321–341.

Mikkelson, G. M. (1999). Methods and metaphors in community ecology: the problem of defining stability. *Perspectives on Science*, 5, 481–498.

Nadeau, R., & Desautel, J. (1984). *The Kuhnian development in epistemology and the teaching of science*. Toronto, Canada: Guidance Center of the University of Toronto.

National Research Council (2012). *A framework for K-12 science education: Practices, crosscutting concepts, and core ideas*. Committee on a Conceptual Framework for New K-12 Science Education Standards. Board on Science Education, Division of Behavioral and Social Sciences and Education. Washington, DC: The National Academies Press.

Odenbaugh, J. (2003). Complex systems, trade-offs and mathematical modeling: A response to Sober and Orzack. *Philosophy of Science, 70*, 1496–1507.

Odenbaugh, J. (2005). Idealized, inaccurate but successful: A pragmatic approach to evaluating models in theoretical ecology. *Biology and Philosophy, 20*, 231–255.

Odenbaugh, J. (2006). The strategy of 'The strategy of model building in population biology'. *Biology and Philosophy, 21*, 607–621.

Odenbaugh, J. (2007). Seeing the forest *and* the trees: Realism about communities and ecosystems. *Philosophy of Science, 74*, 628–641.

Oh, S. P., & Oh, S. J. (2011). What teachers of science need to know about models: An overview. *International Journal of Science Education, 33*, 1109–1130.

Oksanen, M., & Pietarinen, J. (2004). *Philosophy and biodiversity*. Cambridge, UK: Cambridge University Press.

Orzack, S. H., & Sober, E. (1993). A critical assessment of Levins' "The strategy of model building (1966)". *Quarterly Review of Biology, 68*, 534–546.

Palladino, P. (1991). Defining ecology: Ecological theories, mathematical models, and applied biology in the 1960s and 1970s. *Journal for the History of Biology, 24*, 223–243.

Passmore, C., Stewart, J., & Cartier, J. (2009). Model-based inquiry and school science: Creating connections. *School Science and Mathematics, 109*, 394–402.

Peters, R. H. (1991). *A Critique for ecology*. Cambridge, UK: Cambridge University Press.

Portides, D. P. (2007). The relation between idealization and approximation in scientific model construction. *Science & Education, 16*, 699–724.

Price, M., & Billick, I. (2010). The imprint of place on ecology and ecologists. In I. Billick & M. Price (Eds.), *The Ecology of Place: Contributions of Place-based Research to Ecological Understanding* (pp. 11–14). Chicago and London: The University of Chicago Press.

Prins, G. T., Bulte, A. M. W., Van Driel, J. H., & Pilot, A. (2009). Students' involvement in authentic modelling practices as contexts in chemistry education. *Research in Science Education, 39*, 681–700.

Redish, E. F. (1994). Implications of cognitive studies for teaching physics. *American Journal of Physics, 62*, 796–803.

Rudolph, J. L. (2005). Inquiry, instrumentalism, and the public understanding of science. *Science Education, 89*, 803–821.

Ryoo, K., & Linn, M. (2012). Can dynamic visualizations improve middle school students' understanding of energy in photosynthesis? *Journal Of Research in Science Teaching, 49*, 218–243.

Sander, E., Jelemenská, P., & Kattmann, U. (2006). Towards a better understanding of ecology. *Journal of Biological Education, 40*, 119–123.

Sarkar, S. (2005). *Biodiversity and Environmental Philosophy: An Introduction to the Issues*. Cambridge, UK: Cambridge University Press.

Scheiner, S. M., & Willig, M. R. (2011). A general theory of ecology. In: S. M. Scheiner & M. R. Willig (Eds), *The Theory of Ecology* (pp. 3–18). Chicago: University of Chicago Press.

Schwarz, C., & White, B. (2005). Metamodeling knowledge: Developing students' understanding of scientific modeling. *Cognition and Instruction, 23*, 165–205.

Schwarz, C., Reiser, B., Davis, E., Kenyon, L., Acher, A., Fortus, D., Shwartz, Y., Hug, B., & Krajcik, J. (2009). Developing a learning progression for scientific modeling: Making scientific modeling accessible and meaningful for learners. *Journal of Research in Science Teaching, 46*, 632–654.

Serra, H., & Godoy, W. A. C. (2011). Using ecological modeling to enhance instruction in population dynamics and to stimulate scientific thinking. *Creative Education, 2*, 83–90.

Shrader-Frechette, K. S., & McCoy, E. D. (1994). What ecology can do for environmental management. *Journal of Environmental Management, 41*, 293–307.

Silva, C. C. (2007). The role of models and analogies in the electromagnetic theory: A Historical case study. *Science & Education, 16*, 835–848.

Simberloff, D. (1981). The sick science of ecology: symptoms, diagnosis and prescription. *Eidema, 1*, 49–54.

Slingsby, D., & Barker, S. (2005). The role of learned societies, government agencies, NGOs, advocacy groups, media, schools, and environmental educators in shaping public understanding of ecology. In: E. A. Johnson & M. J. Mappin (Eds.), *Environmental Education and*

Advocacy: Changing Perspectives of Ecology and Education (pp. 72–87). Cambridge, UK: Cambridge University Press.

Stamp, N., Armstrong, M., & Biger, J. (2006). Ecological misconceptions, survey III: The challenge of identifying sophisticated understanding. *Bulletin of the Ecological Society of America*, 87, 168–175.

Stephens, P. A., Buskirk, S. W., & Martínez del Rio, C. (2006). Inference in ecology and evolution. *Trends in Ecology and Evolution*, 22, 192–197.

Sterelny, K. (2006). Local ecological communities. *Philosophy of Science*, 73, 215–231.

Strong, D. (1983). Natural variability and the manifold mechanisms of ecological communities. *American Naturalist*, 122, 636–660.

Svoboda, J. & Passmore, C. (2011). The strategies of modeling in biology education. *Science & Education*, online first DOI 10.1007/s11191-011-9425-5.

Taylor, P. (1989). Revising models and generating theory. *Oikos*, 54, 121–126.

Taylor, P. (2000). Socio-ecological webs and sites of sociality: Levins' strategy of model-building Revisited. *Biology & Philosophy*, 15, 197–210.

Taylor, P. (2005). *Unruly Complexity: Ecology, Interpretation, Engagement*. Chicago: University of Chicago Press.

Tomkins, S. P., & Tunnicliffe, S. D. (2001). Looking for ideas: observation, interpretation and hypothesis-making by 12-year-old pupils undertaking science investigations. *International Journal of Science Education*, 23, 791–813.

Tunnicliffe, S. D., & Ueckert, C. (2007). Teaching biology — the great dilemma, *Journal of Biological Education*, 41, 51–52.

Tunnicliffe, S. D., & Ueckert, C. (2011): Early biology: the critical years for learning. *Journal of Biological Education*, 45, 173–175

Van Driel, I., & Verloop, N. (1999). Teachers' knowledge of models and modelling science. *International Journal of Science Education*, 21, 1141–1153.

Vepsalainen, K., & Spence, J. (2000). Generalization in ecology and evolutionary biology. *Biology and Philosophy*, 15, 211–238.

Volterra, V. (1926). Fluctuations in the abundance of a species considered mathematically. *Nature*, 118, 558–560.

Webb, P., & Boltt, G. (1990). Food chain to food web: a natural progression? *Journal of Biological Education*, 24, 187–190.

Weisberg, M. (2006a). Forty years of 'The Strategy': Levins on model building and idealization. *Biology and Philosophy*, 21, 623–645.

Weisberg, M. (2006b). Richard Levins' philosophy of science [editor's introduction]. *Biology and Philosophy*, 21, 603–605.

Weisberg, M., & Reisman, K. (2008). The robust Volterra principle. *Philosophy of Science, 75,* 106–131.

White, B. Y. (1993). ThinkerTools: Causal models, conceptual change, and science education. *Cognition and instruction*, 10, 1–100.

White, P. (2008). Beliefs about interactions between factors in the natural environment: A causal network study. *Applied Cognitive Psychology*, 22, 559–572.

Wilson, B. (2009). From laws to models and mechanisms: Ecology in the twentieth century, http://philsci-archive.pitt.edu/id/eprint/4509.

Wimsatt, W. C. (1981). Robustness, reliability and overdetermination'. In M. Brewer & B. Collins (Ed.), *Scientific Inquiry and the Social Sciences* (pp. 124–163). San Francisco: Jossey-Bass.

Wimsatt, W. C. (1987). False models as means to truer theories. In M. Nitecki & A. Hoffman (Eds.), *Neutral Models in Biology* (pp. 23–55). New York: Oxford University Press.

Wimsatt, W. C. (2001). Richard Levins as philosophical revolutionary. *Biology & Philosophy*, 16, 103–108.

Windschitl, M., Thompson, J., & Braaten, M. (2008). Beyond the scientific method: Model-based inquiry as a new paradigm of preference for school science investigations. *Science Education*, 92, 941–947.

Winther, R. G. (2006). 'On the dangers of making scientific models ontologically independent': Taking Richard Levins' warnings seriously, *Biology & Philosophy*, 21(5), 703–724.

Worster, D. (1994). *Nature's economy: A history of ecological ideas* (2nd ed.). Cambridge: Cambridge University Press.

Wynne, C., Stewart, J., & Passmore, C. (2001). High school students' use of meiosis when solving genetics problems. *International Journal of Science Education*, 23, 501–515.

Yodzis, P. (2000). Diffuse effects in food webs. *Ecology*, 81, 261–266.

Ageliki Lefkaditou received her Ph.D. in Philosophy of Biology from the Aristotle University of Thessaloniki in 2009. She is currently a postgraduate teaching assistant at the University of Leeds. Her main research area is History and Philosophy of Biology, with special emphasis on the concept of race and ecology from the nineteenth century to the present. She also maintains a major research focus on science education and public understanding of science. Among her recent publications and co-publications are (a) Is ecology a holistic science, after all? In G.P. Stamou (eds), *Populations, biocommunities, ecosystems: A review of controversies in ecological thinking*, Bentham Science Publishers, and (b) Views of Nature and the Human-Nature Relations: An Analysis of the Visual Syntax of Pictures about the Environment in Greek Primary School Textbooks Diachronic Considerations, *Research in Science Education*, DOI 10.1007/s11165-011-9250-5.

Konstantinos Korfiatis is currently an assistant professor of Environmental Education at the Department of Education, University of Cyprus. He has a degree in Biology and a Ph.D. from the School of Biology, Aristotle University of Thessaloniki, Greece. He is a member of the Learning in Science Group at the University of Cyprus. He works and publishes on the development and evaluation of educational material for environmental education and ecology, on the importance of conceptual frameworks and world views in environmental education and on theoretical issues in ecology and environmental education.

Tasos Hovardas has a degree in Biology, a master's in Environmental Biology and a Ph.D. in Environmental Education and Outreach from the Aristotle University of Thessaloniki. He is a member of the Learning in Science Group at the University of Cyprus. He served as a Senior Teaching Fellow at the Department of Education, University of Cyprus, and the Departments of Primary and Preschool Education of the University of Thessaly in Greece. His research interests involve the fields of Environmental Education, Science Communication and Outreach and Human Dimensions in Natural Resource Management.

Part V
Pedagogical Studies: Earth Sciences

Chapter 18
Teaching Controversies in Earth Science: The Role of History and Philosophy of Science

Glenn Dolphin and Jeff Dodick

18.1 Teaching Controversies in Earth Science: The Role of History and Philosophy of Science

"Battle heats up over Alaskan petroleum reserve" (National Public Radio News, July 17, 2011), "Group ends call for hydro-fracking moratorium" (CBC News, 7 July 2011), "Greenpeace report links western firms to Chinese river polluters" (Guardian, 13 July 2011), "Climate change and extreme weather link cannot be ignored" (Dominion Post, 14 July 2011), and "'Jury Is Out' on Implementation of Landmark Great Lakes Compact" (New York Times, 14 July 2011)—headlines such as these are an everyday occurrence. The articles themselves not only inform us about the issues concerning the planet on which we live but also indicate the economic, political, and social influences/implications inexorably tied to them. It is reasonable to assume that a certain "working knowledge" of the systems of earth is necessary for one to be able to understand the issues as they are and even more so if one would want to make informed decisions (personal, political, social, or economic) related to such issues. This especially holds true for the current generation of K–12 students. They are the citizens of the future and should be

---

The authors contributed equally to this manuscript.

G. Dolphin (✉)
Department of Geoscience, University of Calgary,
2500 University Drive NW Calgary, T2N 1N4, Alberta, Canada
e-mail: glenn.dolphin@ucalgary.ca

J. Dodick (✉)
Science Teaching Center, The Hebrew University of Jerusalem,
Givat Ram Campus, Jerusalem 91904, Israel
e-mail: jdodick@vms.huji.ac.il

M.R. Matthews (ed.), *International Handbook of Research in History, Philosophy and Science Teaching*, DOI 10.1007/978-94-007-7654-8_18,
© Springer Science+Business Media Dordrecht 2014

prepared with the education needed to intelligently evaluate circumstances with potential adverse environmental impact. Hoffman and Barstow emphasized this in their call to action:

> Understanding Earth's interconnected systems is vital to the future of our nation and the world. Ocean and atmospheric interactions effect our daily lives in multiple, significant ways. Long-term changes in ocean and atmospheric processes impact national economies, agricultural production patterns, severe weather events, biodiversity patterns, and human geography. Global warming and its effects on glacial mass balance, sea level, ocean circulation, regional and global weather and climate, and coral bleaching, to name only a few potential impacts, are important global issues that demand immediate attention. (Hoffman and Barstow 2007, p. 9)

This general philosophy is borne out in the National Science Education Standards (NSES) (NRC 1996, 2012). The NSES have placed an equal emphasis on the teaching of Earth and Space Science (ESS) as has been given physics, chemistry, and biology. We direct the reader's attention to very recent works, such as the Next Generation Science Standards (NGSS) (Achieve, Inc. 2012) and the Earth Science Literacy Principles[1] (Earth Science Literacy Initiative 2010). The NGSS give an example of the emphasis placed on geoscience education by way of both the disciplinary core ideas and the crosscutting relationships, while the Earth Science Literacy Principles delineate nine big ideas in the geosciences as a framework of what a literate citizen of the USA should know within the domain of earth science. However, in the past several decades, ESS teaching has been struggling to keep pace with teaching in the other sciences. Currently, only about 7 % of US high school students have taken a course in ESS, and there are just over 10,000 earth science teachers at the secondary level in the USA, compared to about 52,000 for biology (Lewis and Baker 2010). In her review of the education research literature focused on earth science conceptions, Cheek (2010) found only 79 empirical investigations published between 1982 and 2009. Our search for investigations focused on the use of history and philosophy of science (HPS) in teaching earth science yielded fewer than 20. In this book alone there is only one earth science chapter compared to the six for physics and three each for biology and chemistry. With these statistics in mind, it is obvious that there is a need to (1) increase the number of students taking ESS classes at all levels of schooling, (2) increase the number of earth science majors graduating from universities, (3) increase the number of highly qualified earth science teachers, and (4) enhance the quantity and quality of earth science education research (and especially in the field of HPS use in teaching earth science). We hope this chapter will be a small stepping-stone toward this goal.

---

[1] The nine ESLP big ideas are as follows: 1-earth scientists use repeatable observations and testable ideas to understand and explain our planet; 2-the earth is 4.6 billion years old; 3-the earth is a complex system of interacting rock, water, air and life; 4-earth is continuously changing; 5-earth is the water planet; 6-life evolves on a dynamic earth and continuously modifies earth; 7-humans depend on earth for resources; 8-natural hazards pose risks to humans; and 9-humans significantly alter the earth.

Cheek (2010) pointed out that in general, students' understandings of geoscience concepts have not improved over the past several decades. She did assert that we know more about students' geoscience conceptions than we did 27 years ago and now we need to utilize that information to enhance instruction. Efforts to do just that have included utilizing an earth systems approach where the main focus of instruction is to develop students' understanding of the four different *spheres* (geo-, bio-, hydro-, and atmo-) and how they influence and are influenced by each other (Rankey and Ruzek 2006). Earth science by design (ESbD) (Penuel et al. 2009), an extension of Wiggins and McTighe's (2005) work, is an approach whose goal is to achieve enduring understanding through teaching about earth via a few *big ideas*.

Another seemingly fertile approach, though underutilized, has been the incorporation of HPS within instruction, with emphasis on the many controversies experienced throughout the history of the earth sciences (Bickmore et al. 2009b; Montgomery 2009). By HPS, we are referring to the many factors that influence the progression of scientific understanding. This may include economic, political, or social factors. It also encompasses philosophical considerations which oftentimes are responsible for directing investigations and discerning observational data from the "noise." These philosophical differences may form the basis of controversy as well. For our purposes we will use Venturini's definition of controversy:

> Controversies are situations where actors disagree (or better, agree on their disagreement). The notion of disagreement is to be taken in the widest sense: controversies begin when actors discover that they cannot ignore each other and controversies end when actors manage to work out a solid compromise to live together. Anything between these two extremes can be called a controversy. (Venturini 2010, p. 261)

The history of the geosciences is rife with controversial issues such as how marine fossils could be found at mountain tops (Cutler 2003), plutonism versus neptunism (Repcheck 2003), "uniformitarianism" versus "catastrophism" (Şengör 2001), deep time and the age of earth (Repcheck 2003), hollow earth theory, contracting earth theory (Oreskes 1999), the use of fossils to date rocks (Rudwick 1985), expanding earth theory (Adams 2005), continental drift versus land bridges (Oreskes 1999), the theory of plate tectonics (Oreskes and LeGrand 2001), dinosaur extinction (Alvarez and Chapman 1997; Glen 2002), the "current and heated" controversy concerning plume theory (Anderson and Natland 2005; Anderson 2006; Glen 2005), as well as the ever-present conflict between science and religion (Bickmore et al. 2009a). Instructors have found that "teaching the scientific controversy" has been effective at garnering interest from students, enhancing their critical thinking skills, not just in the geosciences[2] but also physics (De Hosson and Kaminski 2007), chemistry (Justi 2000), and biology (Seethaler 2005).

Researchers have also found that incorporating HPS within instruction helps to augment students' understandings of the nature of science (NOS) as emphasized in the NSES (NRC 1996, 2012). The use of HPS as an instructional tool was written

---

[2] For examples, see Dolphin (2009), Duschl (1987), Montgomery (2009), and Pound (2007).

about as early as the mid-twentieth century (Conant 1947). Conant emphasized the importance of students understanding the "tactics and strategy" of science. The other efforts of infusing HPS into instruction, such as Harvard Project Physics and the BSCS Biology, also deserve accolades (Matthews 1994/2014). Matthews also stated that teaching with HPS is important because it promotes better comprehension, is intrinsically interesting, counteracts scientism and dogmatism, humanizes the process of science, and connects with disciplines within science as well as outside of science, and historical "learning" reflects individual learning about concepts. Many others have written in favor of the use of HPS within science instruction.[3]

In this chapter, we will situate the geosciences philosophically and methodologically with respect to biology, chemistry, and physics. We will highlight four different geoscience concepts and their related controversies, including what we know about the use of HPS for teaching these concepts, what has been done, and what, in our minds, is still in need of being done. We will offer pedagogical, cognitive, and historical rationales for the use of controversy in teaching earth science concepts, and we will organize our discussion of controversies within the context of the four spheres of the earth—geosphere, biosphere, hydrosphere, and atmosphere. Though highlighting a particular domain within the geosciences, each phenomena surrounded by a historical and philosophical controversy will also exemplify its global nature in terms of its influence. The controversies described below are those surrounding the acceptance of plate tectonics as the grand unifying theory of earth (geosphere controversy); the meteorite impact theory explaining the Cretaceous–Paleogene (or K–Pg) mass extinction (no, it was not just dinosaurs that went extinct) (biosphere controversy); the connection of rhythmic long-term weather variations in various parts of the world to oceanic temperature in the tropical Pacific Ocean, also known as ENSO (hydrosphere controversy); and finally, the current controversies surrounding the acceptance of anthropogenic global climate change (**ACC**) (atmosphere controversy).

18.2 Nature of the Earth Sciences

What is the nature of the earth sciences? How are they, as disciplines, distinguished from other sciences? Some might be surprised that these questions are even being posed, as they seem so basic. However, we believe that these questions need answers for several reasons. Unfortunately, for much of the last century, the earth sciences have been portrayed as derivative disciplines whose logic and methodology were furnished by the physical sciences. Indeed, the history of the earth sciences is annotated by episodes where not only physicists but even (surprisingly) some geologists tried to reconstitute the earth sciences as a tributary of physics (Dodick and Orion 2003).

---

[3] See, for instance, Allchin (1997), Bickmore et al. (2009b), Justi (2000), Matthews (1994/2014, 2012), and Rudolph (2000).

18 Teaching Controversies in Earth Science: The Role of History and Philosophy...

This trend has continued into recent times such that Gould (1986, 1989) noted that some scientists do not accept the methodological diversity of the sciences and specifically disparage the earth sciences as being less scientific than the physical sciences.

Unfortunately, this message that the earth sciences are derivative has been reinforced by work in the history and philosophy of science (HPS). For much of the twentieth century, the classic works of HPS emanated from scholars (Popper, Kuhn, Lakatos) who largely relied on examples from physics to illustrate their discussions, a critique which has been mentioned by others.[4] In fact, even in the small number of philosophical works that have examined their nature, the earth sciences have been declared as either derivative or at least as not unique sciences.[5] It is only in the last 30 years or so that this lack has been redressed, as witnessed by the increased number of tomes connected to HPS works dedicated to the earth sciences, as well as the publication of *Earth Sciences History*, the only academic journal exclusively devoted to the history of these disciplines.

Unfortunately, such work has not penetrated into the world of education, such that some science educators are often left with the impression that the earth sciences are less rigorous than the physical sciences and thus less worthy of being taught as part of the standard science curriculum (Dodick and Orion 2003). Such thinking is mistaken because it does not consider the special nature of the earth sciences as one of the historical and interpretive (or hermeneutic) sciences (Frodeman 1995; Orion and Ault 2007) which classically attempt to reconstruct past phenomena and processes by collecting their natural signs during fieldwork. This nature is shared to a large degree with other historical fields such as evolutionary biology and astronomy (Cleland 2001, 2002). Concurrently, it contrasts with experimental sciences such as physics or molecular genetics in which natural phenomena are manipulated within the controlled environs of a laboratory in order to test a hypothesis. Indeed, the differences between these two groups of science are derived from the fact that the historical sciences, such as the earth sciences and evolutionary biology, developed specific methodologies to cope with problems that could rarely be tested under controlled laboratory conditions.[6]

---

[4] See, for instance, Baker (1996), Frodeman (1995), Greene (1985), and Mayr (1997).

[5] See, for instance, Bucher (1941), Goodman (1967), Schumm (1991), and Watson (1969).

[6] We do not mean to imply that the earth sciences are devoid of experimentation. Indeed, whole fields within the earth sciences including geophysics, geochemistry, and climate science have tested some of their claims using cutting-edge experimental methods which produce important research results. Philosophical classifications sometimes simplify, ignoring the overlap that occurs between categories, and this is the case in the historical–experimental dichotomy we use in this chapter. We still believe that it is a fruitful classification as many philosophers and historians of science have used it in their definition of different sciences (See Dodick et al. 2009 for a review of the development of the term *historical sciences*). Moreover, one of us (Argamon et al. 2008; Dodick et al. 2009) has tested this dichotomy empirically and has indeed found that the earth sciences (representing diverse fields including geology, geochemistry, and paleontology) do fall more regularly into the historical science category.

Table 18.1 Methodological contrasts between the experimental and historical sciences

Dimension	Experimental	Historical
Research goal	General laws and behaviors	Explanations for ultimate and contingent causes
Evidence gathered by	Controlled manipulation of nature	Observing/analyzing preexisting entities and phenomena
Hypotheses are tested for	Predictive accuracy	Explanatory accuracy
Objects of study	Uniform and interchangeable entities	Complex and unique entities

Recently, a growing number of scientists, philosophers, and educators have critiqued the idea of there being a universal scientific method largely emanating from the experimental-based, physical sciences[7] and instead promoted the view that different combinations of logic and methods can and should play different roles in different disciplines. Indeed, one of us has empirically tested such claims by analyzing the pattern of language use in the historical and experimental sciences, respectively; the results of this work show (statistically) significant variation in language use between the two groups of sciences that are derived from the specific methodologies employed by these two groups of sciences (Argamon et al. 2008; Dodick and Argamon 2006; Dodick et al. 2009).[8]

This following discussion will review this empirical work to provide the reader with a better understanding of the methodological differences between the historical and experimental sciences. By doing this, we also create a philosophical framework for analyzing the historical controversies that we present later in this chapter. Table 18.1 presents four methodological contrasts between historical and experimental sciences which will be used in this discussion.[9]

The ultimate *research goal* of the experimental sciences is a general statement or causal law that is applicable to a wide variety of phenomena in many contexts (Kleinhans et al. 2005). To achieve this goal, *evidence is gathered* via controlled experimentation within laboratories in which the natural phenomena are manipulated to test a facet of a theory or hypothesis (Case and Diamond 1986). The quality of such a *hypothesis is tested* by the consistency of its predictions with the results of its experiments. Finally, the form of such experimental research is dictated largely by the fact that it is conducted on uniform and interchangeable *objects of study*, such as atoms; the fact that such entities are uniform, or nearly so, makes the formulation

---

[7] See, for instance, Cartwright (1999), Cleland (2001, 2002), Cooper (2002, 2004), Diamond (2002), Dodick et al. (2009), Frodeman (1995), Gould (1986), Kleinhans et al. (2005, 2010), Mayr (1985), and Rudolph and Stewart (1998).

[8] These studies encompassed a series of experimental fields including physical chemistry, organic chemistry, and experimental physics; historical fields included paleontology, geology, and evolution.

[9] This section is arranged to correspond with the ordering of Table 18.1. The dimension under consideration is delineated in italics.

of general laws possible in principle and experimental reproducibility a reasonable requirement in practice (Diamond 2002). This desire for reproducibility means of course that results of a given experiment should be uniformly reproduced, given the same conditions, in any laboratory in the world; this result fulfills one of the basic principles of science, the principle of uniformity of law (Gould 1965, 1987).

In contrast, the *research goal* of historical sciences, such as the earth sciences, is to uncover ultimate and contingent causes buried in the past whose effects are interpreted only after very complex causal chains of intervening events (Cleland 2001, 2002). Accordingly, *evidence is gathered* by observation of naturally occurring signs exposed during fieldwork, since controlled experimental manipulation is usually impossible due to the fact that the historical sciences are interpreting cause and effect in past events that cannot be repeated or replicated; in fact, even if this were possible, the enormous amount of time, space, and the complex relationship of variables needed to affect the result would inhibit such scientific research from happening.

Such observation is not a passive act of simply looking, or searching for evidence, as the word "observe" might imply to those unfamiliar with the earth sciences. This is due to the fact that such evidence is often hidden in time and space from an earth scientist. Instead, such observations are guided by deep inferences and intuitions about earth processes that are developed by earth scientists through long periods of exposure to field materials.[10]

When possible, rather than making observations on a single entity (such as an outcrop), historical science relies on natural experiments (Case and Diamond 1986; Diamond 2002).[11] Natural experiments are based on analyzing the effects of natural (i.e., not manipulated by the experimenter) perturbations in the field. In implementing such studies, the researcher must also choose at least one "control" site, which is similar to the experimental site, but that lacks the same natural perturbations. Unlike laboratory experiments, natural experiments do not control their independent variables due to the confounding complexity of field conditions.

This focus on past causation in historical sciences implies that the ultimate *test of* (the quality of their) *hypotheses* is explanatory adequacy via retrodiction of specific past events rather than prediction as in experimental sciences[12]; this is due

---

[10] In the past earth scientists were restricted to physically uncovering hidden field materials; this of course restricted their research to areas to which they had access. However, technology has revolutionized this search, for example, tools, such as remote sensing via satellite makes the invisible visible, both here on earth, as well as on other planetary bodies.

[11] Diamond and Robinson (2010) have also documented how natural experiments are also applied within the humanities and social sciences where controlled experimentation is impossible.

[12] As Schumm (1991, p. 7) notes, the term prediction in science is used in two ways: "The first is the standard definition to foretell the future. The second is to develop a hypothesis that explains a phenomenon." Based on the second definition, such predictions have the typical form of: "if a given hypothesis is correct then we predict that the following process or phenomenon will occur." In the case of experimental sciences, both definitions are methodologically applicable. Schumm (1991) argues that in some fields of earth science (e.g., geomorphology), prediction to the future (i.e., the first definition), based on extrapolation, is also part of their current methodology.

to the fact that the *objects of study* in historical sciences such as the earth sciences are complex, unique, and contingent, with very low chances of repeating exactly (Kleinhans et al. 2010). This methodological need places great stress upon earth scientists' powers of "retrospective thinking," in which they apply knowledge of present-day processes in order to draw conclusions about processes and phenomena that developed millions of years ago (Orion and Ault 2007), a methodology that the historical sciences terms actualism.[13]

The methodology of such explanatory reasoning derives from what Cleland (2001, 2002) calls the "asymmetry of causation," in that effects of a unique event in the past tend to diffuse over time, with many effects being lost and others confused by intervening factors. Making sense of such complexity requires, therefore, synthetic thinking (Baker 1996), in which one fits together complex combinations of evidence to form arguments for and against multiple working hypotheses (MWH) which often compete with each other.

In addition to sifting through the complexity of processes, earth scientists must also deal with the complexity of the physical entities they study. Unlike subatomic particles, for example, which are all uniform, the individuals studied by earth scientists—fossils, strata, igneous intrusions—are all unique (though often similar) individuals, whose precise form and function cannot always be reconstructed. This usually removes the chance of formulating universal laws and allowing only statistical explanations of relative likelihoods at best, so that arguments for and against multiple hypotheses must be made on the preponderance of the best evidence.

Even so, we argue that such predictions are far less common and accurate in historical sciences, than they are in experimental sciences, in large part due to the complexity of the phenomena studied in such disciplines; instead, historical science focuses on reconstructive explanations, via the method of retrodiction, which might be defined as a specification of what did happen (Engelhardt and Zimmermann 1988; Kitts 1978). As Ben-Ari (2005, p. 15) notes "retrodiction is essential if theories are to be developed for the historical sciences." Indeed, Schumm (1991) admits that it is only when the present conditions are understood and when the history of the situation has been established that predictions to the future (i.e., the first definition) can be made with some degree of confidence in earth science. In other words, in historical-based sciences, such as the earth sciences, reconstructing past conditions takes precedence and as a method has greater validity than predicting the future.

[13] In defining actualism, some philosophers and geologists separate between two definitions of the earth sciences most important, but most misunderstood concept, uniformitarianism (Hooykaas 1959; Gould 1965, 1987; Rudwick 1971).

Substantive uniformitarianism or sometimes uniformitarianism claims that geo-historical uniformity exists between present and past geological phenomena, such that the force, rates, and types of phenomena do not change over the course of geological time.

Methodological uniformitarianism or simply actualism is a method permitting an earth scientist, via analogical reasoning, to explain the geological past based on geological events observed in the present. On the basis of these observations, geologists make inferences about the types of causes and their force in the past.

These two types of uniformitarianism were conflated together by Lyell (Gould 1984, 1987) which has led to some of the modern-day confusion of the term uniformitarianism. We will discuss the impact of Lyell's conflation when we discuss the case study concerning the Cretaceous–Paleogene extinctions.

Thus, reasoning about the relative likelihood of different assertions is endemic to the synthetic thinking patterns of historical science.

As can be seen, inquiry within the earth sciences cannot guarantee reproducible results over space and time like the experimental sciences. Indeed, the very purpose of the earth sciences is to explain the unique, contingent, and complex systems acting over the entire earth and its interacting "spheres" (geosphere, hydrosphere, atmosphere, and biosphere) as well as analyzing their subsystems on more local scales (Orion and Ault 2007).

This concern for global complexity can and should be used as a tool of science education because it prevents the earth sciences from being portrayed as what Allchin (2003) terms a science of *myth-conception*. By myth-conception, Allchin (2003) is referring to a narrative device which embodies a "world view that provides formulae or archetypes for appropriate or sanctioned behaviour." For example, the history of science has sometimes portrayed discoveries as the efforts of a single, idealized scientist. Even the names used to describe these discoveries support these impressions: "Mendelian genetics," "Darwinian evolution," and the "Copernican revolution."

Such idealized portrayals of science sometimes occur because its narrative is shaped by "sharpening" what is considered the central message, while "leveling" the details thought to be less central (Allchin 2003). Moreover, science is often considered as a problem-solving endeavor in which the goal is to get the single, right answer; this has sometimes infected its philosophy, such that the questions that have been asked ("What is the method of science?" or "How does science advance?") focus on a single process (Oreskes 2004) As Oreskes (2004) argued, many academic fields, including history, art, and literature, embrace multiple perspectives as they analyze a problem and so in fact do the sciences. Nowhere is this more evident than in the earth science paradigm of plate tectonics, which embraces multiple conceptual tools including experimentation, mathematical models, novel instruments, analogical reasoning, and visualization. Equally important, plate tectonic theory synthesized huge amounts of data that were collected by many scientists, working on independent problems, and scattered over the entire earth. Indeed, without such global efforts the theory would have never been accepted. Concurrently, this global effort has meant that plate tectonics have not acquired the attached name of one archetypical scientist. Thus, it is the perfect scientific theory for demonstrating the nature of science to students. As we will show, plate tectonics is not unique, and all of the controversies that we will be exploring in this chapter also demonstrate this global nature of the earth sciences.

18.3 Why Controversies?

We believe that framing the learning of the earth sciences in historical controversies is justified from the perspectives of the learning sciences, as well as the history and philosophy of science.

From the perspective of the learning sciences, it is well known that students (up to and including their university years) are often epistemological dualists, viewing academic issues in terms of true or false, right or wrong, credit or no credit (Alters and Nelson 2002). At first glance this poses some dangers to the deeper critical thinking skills that we want students to develop. This assertion is also sometimes reinforced, ironically, by popular misinterpretations of the conceptual change movement which often sees "mis"conceptions as entities to be uprooted and so to be replaced by the final "correct" conception. However, the progenitors of the conceptual change movement themselves, Posner and his colleagues (1982), noted in their original article that conceptions, for the good and the bad, are important scaffolds that lead to further conceptual development. Moreover, diSessa and his colleagues (diSessa 1988; diSessa 1993; Smith et al. 1993) in their works on "learning in pieces" emphasized that ideas perceived as misconceptions have a heuristic potential that allow them to do important conceptual work; the key is for the student and scientist to know the limits of validity connected to such conceptions.

More recently, Marton et al. (2004) have outlined a theory of learning that connects perfectly with the comparative nature of controversies. The key facets of this theory are the "object of learning," "variation" between objects of learning, and "the space of learning." The object of learning is the concept that is to be learnt in a given lesson. From the teacher's perspective, the goal of the lesson is to present an intended object of learning, which through the discourse of the lesson becomes the enacted object of learning or what is possible to learn in the lesson. Finally, from the learner's point of view, what is actually learnt is termed the lived object of learning. The key way in which the object of learning becomes enacted is through the teacher's use of variation. In other words, according to Marton and his colleagues, learners can only learn an object when it is presented in comparison to something with which it differs. For example, if the objects of learning are the colors green and red, learners who are color-blind will not be able to see the difference between these and, therefore, opportunities for them to learn will not be available. These variations create a space of learning which refers to what is possible to learn in that particular situation. This space is largely created through language.

Finally, the idea of controversy connects perfectly with the recent movement toward using argumentation as an important component of classroom discourse. Veerman (2003, p. 118) succinctly summarized the value of classroom argumentation when he noted that, "in argumentation…knowledge and opinions can be (re)-constructed and co-constructed and expand students understanding of specific concepts or problems." Moreover, argumentation dovetails perfectly with *inquiry*-based learning in which students replicate what scientists do when they are pursuing an authentic scientific problem, as research programs can be viewed as large-scale arguments supporting and falsifying different theoretical frameworks.

Controversies also align with the history and philosophy of science, both on a general level and a specific level. On the general level, we reference the educational philosopher Joseph Schwab (1964) who argued that all too often, students merely learn the facts and final outcomes of scientific research, what he called the "rhetoric of conclusions." This is certainly the case in many textbooks where one scientist's conception is simply shown to replace a previous scientist's conception, without a deeper reference to the many factors that influenced this development. Gould (1987) labeled this as "cardboard" history because of its two-dimensional nature. In response, Schwab (1958, 1962, 1963, 1966, 2000) promoted the *science as inquiry* model. Recognizing that students should come to understand how scientists interpret information and form ideas, Schwab stressed the idea that proper science education should show how these products were derived by scientists—how a body of knowledge grows and how new conceptions come about. To achieve this goal, Schwab emphasized the use of history of science including the reading of original papers and historical narratives exposing the developmental path of scientific concepts (Schwab 1963). The use of historical controversies connects perfectly with Schwab's philosophy, because properly constructed, such controversies can also teach about the complex pathways in the development of scientific concepts.

On a specific level, the idea of controversies strongly aligns with one of the key methods in geology, "multiple working hypotheses" (MWH), which were most prominently elucidated by Gilbert (1886), Chamberlin (1890/1965, 1897), and Johnson (1933).[14] Although mentioned in a previous section of this chapter, we will expand this discussion as MWH has importance both for the general structure of the earth sciences, as well for its connections to controversies.

Chamberlin (1965, p. 755–756) recognized three phases in the history of intellectual methods. The first phase was based on the *method of the ruling theory* where a "premature explanation passes into a tentative theory, then into a theory, and then into a ruling theory." This linear process, in Chamberlin's opinion, was "infantile" for the reason that only if the tentative hypothesis was by chance correct does research lead to any meaningful contribution to knowledge. Less problematic, in his view was the second phase based on a *working hypothesis*, which is a hypothesis to be tested, not in order to prove it but rather as a stimulus for study and fact finding ("ultimate induction"). Nonetheless, a single working hypothesis can unfortunately be transformed into a ruling theory, and the need to support the working hypothesis, despite evidence to the contrary, can become as strong as the need to support a ruling theory. Chamberlin therefore suggested his third phase, based on *MWH*, which was thought to mitigate the danger of controlling ideas. It did so because the investigators develop many hypotheses that might explain the phenomena under study. This was done prior to the actual research and hypotheses were oftentimes in conflict with each other.

Both Blewett (1993) and Johnson (1990) have criticized MWH based on its logic and practicality, respectively. However, as Baker (1996, p. 207) has argued, such criticism occurs "within the context of our times." Thus, for example, Blewett's critique was largely based on a "physics-based philosophy of science." Baker, however, suggested that we look at what MWH meant when it was first formulated. First, it was intended as a method for "naturalists" (whose work was conducted in the field) and not mathematical physicists (who were lab-based experimentalists).

---

[14] Additional work was provided by Gilbert (1896), Chamberlin (1904), and Davis (1911).

Second, the purpose of MWH was, in Chamberlin's view, to facilitate certain "habits of mind" which were of special concern to naturalists generally and geologists specifically. This second purpose certainly integrates with the goals of science education in which we try to open students' scientific worldview to alternatives, as they often stubbornly (as epistemological dualists) adhere to a single conceptual framework.

This of course does not mean that the experimental sciences do not avail themselves of MWH. Indeed, Platt (1964) reported on the use of such a method in both molecular biology and high-energy physics, both of which are definitely experimental in nature. Moreover, he advocated its use, which is part of a larger method he termed "strong inference" in other sciences, for its ability to bring rapid research advances. However, this does not necessarily mean that such experimental fields need to avail themselves of MWH. A more linear process of testing single hypotheses is possible and is still followed in many laboratories.

In the case of the earth sciences, MWH has a practical value even today for its practitioners. As earth science is often conducted in the field (or with materials that must be collected from the field), it focuses on complex natural systems, which are often the result of several irreducible causes, and the application of MWH makes it more likely that a scientist will see the interaction of the several causes. Moreover, from a practical perspective MWH has value because earth scientists conduct periodic stints of fieldwork (unlike laboratory scientists who have full-time access to their lab-based experiments). This means that it is critical to test multiple hypotheses when they have direct access to their primary data (Blewett 1993).

18.4 Highlighting the Four Controversies

We will turn our attention, now, to the four case studies of scientific controversy that we wish to highlight in this chapter. Those controversies are those surrounding the development of the theory of plate tectonics, the impact theory of mass extinction at the end of the Cretaceous, the El Niño Southern Oscillation (ENSO) theory of control over long-term weather, and the current controversy surrounding anthropogenic climate change (ACC). We discuss these four cases for a number of different reasons.

First, the concept at the center of each case study is popular, in that each have been in the popular media fairly recently and both scientists and the general public should have some familiarity with them. Second, each phenomenon has, or has had an impact that reaches a global level, influencing all systems of the earth. Plate tectonics, for instance, is considered the grand unifying theory of the earth. We have designated it as a phenomenon that occurs within the geosphere. However, its impacts reach into oceanic composition and circulation, planetary wind patterns, and selective evolutionary pressures. Third, each of the case studies highlights nicely the history and philosophy of the geosciences. That is, they utilize methods that emphasize earth science's historic and interpretive nature as discussed earlier in this chapter.

In each case, scientists observed an entity or phenomenon's "end product," such as a mass extinction, a mountain range, or anomalous weather conditions. They had to discriminate among a multitude of possible and complexly related variables to determine causation. In the quest for contingent causes, they built models and then looked back in history for explanatory accuracy. This is not to say that each of these episodes played out in the same way as any of the others. It is through our framing of the controversies that we draw out similarities.

18.5 Geosphere: The Acceptance of Plate Tectonics as the Grand Unifying Theory of the Earth

The history of thoughts concerning the origins of continents and ocean basins is a long one, starting before biblical time right up through the present. A comprehensive treatment of this topic is out of the scope of this chapter but can be found in Şengör (2003) for those who are interested. This section demonstrates the general structure of geology as it pertains to the development of the theory of plate tectonics. As with the other controversies discussed in this chapter, this section displays the global nature of the phenomenon under investigation. Although there is a long history on this topic, we begin the story of the development of the theory of plate tectonics with the introduction of the theory of continental drift in 1912 by Alfred Wegener (Wegener and Skerl 1924). At this time, there were multiple varied (and contradictory) working hypotheses to explain the dynamics of the earth. As described by Alexander Du Toit, geologists considered that

> geosynclines and rift valleys are ascribed alternatively to tension or compression; fold-ranges to shrinkage of the earth, to isostatic adjustment or to plutonic intrusion; some regard the crust as weak, others as having surprising strength; some picture the subcrust as fluid, others as plastic or solid; some view the land masses as relatively fixed, others admit appreciable intra- and intercontinental movement; some postulate wide land-bridges, others narrow ones, and so on. Indeed on every vital problem in geophysics there are…fundamental differences of viewpoint. (Du Toit 1937, p. 2)

Specifically, by the end of the nineteenth century, there were two different models for earth dynamics relying on the thermal contraction of the earth. Edward Seuss hypothesized that the crust of the earth was homogeneous and allowed for continents and ocean basins to be interchangeable. Basins were places where contraction left room for the collapse of large areas of crust. James Dana, on the other hand, saw a difference in the composition between ocean crust and continental crust where ocean crust was denser and therefore sank further into the earth. The implication of Dana's contraction theory is that continents and oceans are permanent, or "fixed," entities on the earth's surface. Ironically, though Wegener's theory reconciled many of the controversies noted by Du Toit, it was for that very reason, and some others as well, that it faced an uphill battle for acceptance, especially for North American scientists (Oreskes 1999).

A meteorologist and cartographer, Wegener became interested in the problem of the origin of continents and ocean basins upon noticing the similarities between coastlines of western Africa and eastern South America. Although he was not the first to notice these similarities (Hallam 1973; LeGrand 1988; Oreskes 1999), he was the first to rigorously explore lateral displacement of the continents as a causal explanation for these observations. Besides the "jigsaw" fit of the continents, Wegener "drew on several elegant lines of empirical evidence" (Glen 2002, p. 102), including such complex entities as paleontological, paleoclimatic, and geographical and geophysical effects[15] to support his argument that a supercontinent he referred to as Pangaea existed up to about 205 million years ago and began rifting apart until assuming the current continental positions.

Wegener's hypothesis received some acceptance in Europe, South Africa, and Australia. This was not the case in North America, where the idea of drifting continents and its implications did not set well with many geologists for both empirical and philosophical reasons (Oreskes 1999). Rollin Chamberlin (1928) delineated 18 arguments against the drift hypothesis. Generalizing from this list shows what the major objections were. First, Wegener provided no reasonable mechanism or force for moving continents through softer, but solid ocean crust without showing some kind of deformation. Second, geologists found Wegener's ideas to be "superficial" because he generalized his conclusion from the generalizations of others' works in paleontology, paleoclimate, and geophysics. Third and considered more important (Oreskes 1999) was that that Wegener's ideas did not seem to appeal to the philosophy of uniformitarianism, held in great esteem by geologists at the time. Part of the ability to interpret past events was to consider the natural processes to be uniform through time. Wegener's hypothesis did not show the cyclicity that had been observed in other interpretations of the past. Indeed, Chamberlin (1928) considered Wegener's hypothesis to be "a 'footloose type'"— one that "takes considerable liberties with our globe and is less bound by restrictions or tied down by awkward, ugly facts than most of its rival theories" (p. 87). In the same publication Schuchert (1928, p. 140) critiqued drift stating, "We are on safe ground only so long as we follow the teachings of the law of uniformity in the operation of nature's laws."

During this time, thermal contraction and its corollary, land bridges, were not nearly as comprehensive as drift in putting observations into the context of a global phenomenon, plus contraction and land bridges had major geophysical difficulties as explanatory models. It would take about 40 more years to amass the right data to be analyzed at the right time by the right people for the idea of lateral motion of continents to gain widespread acceptance. These data would eventually come from the emerging and global studies in radiometric dating, paleomagnetism, physical oceanography, and seismology. It was not that anyone in these fields was working specifically on this question of the origin of continents and oceans. The emergent data began to converge and the lateral drift interpretation of earth's past could no

---

[15] See Hallam (1973, pp. 9–21) for a detailed description of Wegner's various lines of evidence.

longer be ignored. This idea of convergence of data will be important in the controversies that follow as well.

There were two lines of investigation in paleomagnetism. One was concerned with explaining an apparent "wandering" of the magnetic poles of the earth and the other with a reversal of polarity of the magnetic field over time. Pierre Curie, in 1895, determined that as hot, iron-bearing rock cooled to below the Curie temperature (approximately 260° C), it would assume the earth's magnetic signal at that time. When measuring magnetic signals within continental basalts of different ages and from different parts of the world, geologists found that the magnetic north pole of the earth appeared to have moved through time. The only explanations for this were that either the pole had indeed "wandered" through time or the continents did or both. In the mid-1950s, Runcorn assembled "polar wandering paths" for North America, Europe, Australia, and India and compared them to each other. The paths were not parallel. This suggested, then, that the continents and not the pole did move over time (Morley 2001).

The second line of investigation looked at another phenomenon which was that the magnetic polarity observed in the rocks every once in a while showed a 180° reversal in polarity compared to the earth's current polarity. At first such an observation was ignored as being a phenomenon of the extraction process or some sort of chemical reaction within rocks of certain composition. However, as data became more global, it became obvious that rocks of the same age maintained the same polarity, whether that polarity was normal or reversed. This led researchers like Cox, Doell, and Dalrymple to consider the changing of the earth magnetic polarity to be a global phenomenon that was recorded in the rocks as it happened. Utilizing the advancements in radiometric dating, they set about constructing a timeline of magnetic reversals. Glen (1982) showed the evolution and refinement of this timeline starting in the late 1950s to 1966.

Though there was no other way to interpret the polar wandering evidence than by the drift of the continents, geomagnetism was a new field and most geologists, not really understanding it, were skeptical of the implications (Oreskes 1999). That having been said, Cox, Doell, and Dalrymple's evolving magnetic reversal scale, published through 1966, would eventually be the key to unlock the secret to earth dynamics (Glen 1982).

Meanwhile, due to world events such as WWII and the beginning of the Cold War era, the ocean basins became very important objects of investigation. Teams of researchers out of Columbia University's Lamont-Dougherty Geological Observatory (now Lamont-Dougherty Earth Observatory, or LDEO), under the guidance of Maurice Ewing, a staunch "fixist," began making observations and taking ocean crust and sediment cores from the seafloor. Results from this data collection extravaganza included Marie Tharp's and Bruce Heezen's discovery of an enormous though narrow chain of mountains running the length of the Atlantic Ocean (Heezen et al. 1959). They also observed a large rift running lengthwise down the center of this mountain chain. Other pertinent observations were a general rise in elevation of these so-called ocean ridges, high heat flow within the rifts, lower sediment thickness, and increasing age symmetrically about and away from the ridge.

In response to these findings, Hess (1962), originally a fixist, posed a contingent cause in what he referred to as "an essay in geopoetry." Hess proposed a theory that had the mid-ocean ridges as places where hot mantle rose and pushed the ocean crust away laterally from a rift. This crust would move like a "conveyor belt" and eventually cool and be consumed as it sank and reentered the earth. His theory would later become known as the theory of seafloor spreading (Dietz 1961). At approximately this same time, former drift proponent, S. Warren Carey, proposed another interpretation, or model, to explain these global observations. His idea was that the earth, at the end of the Paleozoic era, began to grow and the solid crust of the earth began to fragment and spread apart as the earth grew to its current position today. Carey (1976) claimed his ideas were eclipsed by the idea of subducting crust which "has enjoyed meteoric rise to almost universal acclaim, and every aspiring author must jump on the bandwaggon [sic] to gild another anther of this fashionable lily" (p. 14).

Another team of geologists from Scripps Oceanographic Institute were conducting their own studies of the seafloor and discovered an unexplainable pattern of magnetic anomalies. The pattern was that of alternating parallel stripes of reversed and normal magnetism in the basalts near and parallel to the ocean ridges (Mason and Raff 1961; Raff and Mason 1961). It took Fred Vine, a physicist, trained in geomagnetism himself and sympathetic to the drift hypothesis, to combine Cox, Doell, and Dalrymple's magnetic reversals timeline with Hess' verses of geopoetry to answer the question of the "zebra stripe pattern" on the seafloor (Vine and Matthews 1963). Coincidentally, and independently, Canadian geologist Lawrence Morley, also trained in magnetism, saw the Raff and Mason paper and a paper about seafloor spreading (Dietz 1961) and had a similar "eureka" moment (Morley 2001). Despite two attempts to get his interpretation of displacement published, he was unsuccessful. Vine and his advisor at Cambridge, Drummond Matthews, published the idea in *Nature* in 1963. Despite this, many still referred to it as the Vine–Matthews–Morley hypothesis.

Their model only gained a warm reception. As data mounted, however, the explanatory/interpretive power of plate tectonics could no longer be discounted. These new data came from the development of the World Wide Synchronized Seismic Network (WWSSN) (Oliver 2001). Implemented in the 1950s as an attempt to discover the testing of nuclear bombs, the WWSSN gave unprecedented seismic data in terms of both quantity and quality. With an accurate delineation of the patterns of earthquake occurrence, the pattern began to emerge suggesting the outlines of tectonic plates. An understanding of the general physics of earthquakes, starting in the early 1900s (Reid 1910), advanced the field of seismology to the point where seismologists were not only able to accurately pinpoint earthquake locations and estimate their depths but also use the record of first movement of a seismic wave to tell the direction of slip along a fault plane. It was this last form of interpretation that verified J. Tuzo Wilson's (1965) prediction of a new kind of fault found along the mid-ocean ridges—the transform fault—using seismic data (Sykes 1967).

It was this explanatory accuracy, problem-solving capability (Frankel 1987), and retrodictive power that helped lead to final acceptance of the idea of horizontal displacement of the continents (plates) by the vast majority of geologists, fully 60 years after Wegener first proposed it.

The controversy of what actually causes plate motion has not ended, however. There are those, however few, who continue to advocate for an expanding earth (Maxlow 2006; Wilson 2008). The mechanism for the driving of the plates came about once Wilson (1963) proposed shallow stationary "hot spot" plumes to explain the Hawaiian Islands chain. Then it was Morgan (1972) who took Arthur Holmes' (1928) shallow mantle convection model and combined it with Wilson's "hot spot" plume model and then extended them by proposing deep mantle material rising as narrow plumes and then sinking as broad tongues of cooler, denser material in the style of convection cells. Despite some limitations in this theory, it was simple enough (elegant) to garner the attention of many geologists as the explanation for plate motion, eclipsing other multiple working hypotheses (Glen 2005). Although there is consensus that some form of mantle convection is responsible for the lateral motion of the plates, the details of the nature of that convection and the role plates play in the surface expression of earth dynamics are still under much debate (Anderson and Natland 2005; Glen 2005).

It has been the controversies surrounding the development of this grand unifying theory of the earth that have been used by teachers teaching plate tectonics. Sawyer (2010) has used the seafloor data to engage his students in discovering plate boundaries. Paixão et al. (2004) used the controversy between drift and land bridges to engage her participants in discussion and argumentation. Duschl (1987) utilized different explanations for earthquakes to have students compare and contrast them and finally develop arguments for the most appropriate one. Pound (2007) utilized the theory of the hollow earth to engage her students in an activity of critical thinking. Dolphin (2009) utilized many different controversies and alternative models of earth dynamics to facilitate students' understanding of both earth dynamics and the critical evaluation of models. Though the use of these strategies is laudable, none of the experiences were approached in a manner to garner empirical data for gaining understanding of the efficacy of their use.

Another limitation of all of these examples is that though historical models were utilized in the class, it was usually done with the "right answer" in mind. There was no opportunity for the students to create a "wrong" model. In this way, students rationalize their reasoning to fit the conclusion rather than rationalizing data to create their own conclusion (Allchin 2002). A stronger approach in any of these strategies would be to allow students to explore the alternative models *prior to* knowing which model is the best fit. In this way, students utilize multiple working hypotheses, develop critical tests, and must determine the reliability of data as opposed to taking the "right answer" for granted and seeing how the data supports it and missing the scientific process altogether. Later in this chapter, we give an example of a possible approach to instruction using this controversy.

18.6 Biosphere: The Meteorite Impact Theory Explaining the Cretaceous–Paleogene Mass Extinction

On March 4, 2010, the following byline appeared in the popular science Internet site Science Daily:

> The Cretaceous-Tertiary mass extinction, which wiped out the dinosaurs and more than half of species on Earth, was caused by an asteroid colliding with Earth and not massive volcanic activity, according to a comprehensive review of all the available evidence, published in the journal Science. A panel of 41 international experts [...] reviewed 20 years' worth of research to determine the cause of the Cretaceous-Tertiary extinction, which happened around 65 million years ago. (http://www.sciencedaily.com/releases/2010/03/100304142242.htm)

This pronouncement was based on an article published by Schulte and colleagues (2010) in one of the most important professional science journals in the world—*Science*. Similar bylines were carried by a broad number of newspapers, websites, and television news agencies around the world. It would seem that, at least to the popular media, the well-known controversy concerning the Cretaceous mass extinction was settled. However, is this true?

To understand this issue better, we briefly return to the 1980 article written by the Berkeley-based team of physicist (and Nobel laureate) L. Alvarez, his son and geologist W. Alvarez, and nuclear chemists F. Asaro and H. Michel (Alvarez et al. 1980) igniting the controversy. Their mass extinction proposal was motivated by their analysis of *unearthly* concentrations of the element iridium, within pencil thick clay layers at three separate locations around the world: (Gubbio) Italy, (Stevns Klint) Denmark, and (Woodside Creek) New Zealand. These layers were formed 65.5 Ma, at the time of the dinosaur extinction at the boundary between the Cretaceous and Paleogene periods (now designated K–Pg, but in the past as K–T). In the earth's crust, iridium is exceedingly rare (measured in parts per billion); however, these exposures showed iridium concentrations of about 30 (Italy), 160 (Denmark), and 20 times (New Zealand), respectively, above the background level at the time of the Cretaceous extinctions.

Based on this evidence the Alvarez group proposed that these anomalous layers were the remnants of a 10-km iridium-rich meteorite that impacted the earth at the end of the Cretaceous. This impact created a global dust cloud that blocked the sun (atmosphere effect) while chilling the planet so that photosynthesis was suppressed causing a collapse in the food chain (biosphere effect). The result was a mass extinction of 75 % of all oceanic animal species and all land animals greater than 20 kg in mass, including all of the (non-avian) dinosaurs.

In the first 14 years of research following this paper, some 2,500 articles and books were published concerning this extinction (Glen 1994a), and this number has easily doubled since then. Like most large-scale, earth science studies, this research brought into play a multidisciplinary and worldwide collaboration of scientists including paleontologists, sedimentologists, (geo)physicists, and (geo)chemists while prompting the development of ingenious experiments, field studies, and new instruments (such as the coincidence spectrometer) to test the varied lines of evidence undergirding this theory. Moreover, although the primary evidence collected to test this theory emanates from the geosphere and biosphere, it has grown exponentially to encompass all of the "spheres" composing the earth. It is truly a global research effort in more ways than one.

In this section we briefly review the evidence underlying this theory while contrasting it with rival mechanisms that have also been suggested for the extinction. This is a perfect HPS controversy that demonstrates the unique features of the earth sciences as a historical and interpretive discipline whose major goal is to reconstruct past phenomena.

From the beginning, the challenge was to locate the estimated 200 km (in diameter) crater, at the K–Pg boundary that was retrodicted by the Alvarez group. As Glen (1994a, p. 12) notes "such a crater, of course would be the smoking gun." An early candidate included the Manson structure in Iowa (Hartung and Anderson 1988), but its geologic composition, size, and radiometric age eventually ruled it out (Hartung and Anderson 1988; Officer and Drake 1989). Thus, the search turned to the Caribbean Basin due to the proposition raised by Bourgeois and colleagues (1988) that at sites near the Brazos River (Texas), an iridium anomaly and the K–Pg boundary usually overlie a sequence of layers that they suggested were deposited by a tsunami that was generated by an impact into the sea. Thus, in the late 1990s, the Chicxulub crater site at the tip of the Yucatan Peninsula in Mexico was suggested as the site of impact (Hildebrand and Penfield 1990; Kring and Boyton 1992); in fact, its discovery was rather serendipitous, dating back to an oil search in 1981, which even at that time, Penfield and Camargo (1981, as cited in Glen (1994a)) suggested as being the remnant of an impact crater.

In the following years, evidence mounted that it indeed was the impact site associated with the extinction. Cores drilled by two separate teams arrived at the same radiometric date of 65.5 Ma (Sharpton et al. 1992; Swisher et al. 1992); moreover, one of the team's (Sharpton) cores indicated an iridium anomaly. Finally, its location has been correlated with the worldwide *ejecta* distribution pattern, related to distance from the Chicxulub crater (Claeys et al. 2002; Smit 1999). Ejecta are materials emitted by the impact including spherules (formed by the rapid cooling of molten material thrown by the impact into the atmosphere), shocked quartz (which are indicative of extremely high impact pressures), and Ni-rich spinels (which are markers for cosmic bodies such as meteorites or asteroids) (Bohor 1990; Montanari et al. 1983).

From a philosophical perspective, the successful uncovering of such physical evidence fits perfectly within our previous discussion of the nature of the earth sciences. This evidence was not manipulated in a set of controlled experiments but was rather gathered by many insightful observations on a set of interrelated signs, exposed during a globally based fieldwork effort. Moreover, such evidence fulfills the all-important scientific function of providing testable, interpretable evidence that could be used to reconstruct a complex and contingent historical event of the past. Many of the previous purely biological hypotheses (such as disease or over-competition) did not leave behind such testable evidence. Moreover, such biological explanations cannot explain the global extinction patterns; consequently, they have been found wanting (Dingus and Rowe 1998). For this reason, many scientists have focused on physical mechanisms including tectonics, sea level, and climatic changes which also favor a gradual extinction pattern.

Especially with the advent of the stratigraphic evidence for impact, some of those opposed to impact coalesced around the hypothesis that massive volcanic eruptions, which occurred between 60 and 68 Ma centered on the Deccan Plateau in west-central India, caused the environmental collapse responsible for the Cretaceous extinctions (Glen 1994a). To satisfy its critics, volcanism must account for the K–Pg boundary evidence that was supposedly left by an impact (Glen 1994a), most notably the anomalous iridium deposits and shocked minerals. In the former case, using actualistic logic, a hallmark of the historical sciences, proponents were able to show that (at least some) modern volcanoes could draw up iridium from the earth's interior at concentration levels matching those found by the Alvarez group (Felitsyn and Vaganov 1988; Koeberl 1989). The latter case, involving shocked quartz, was more difficult to support because although this mineral associated with some volcanic deposits (Officer et al. 1987), its fracture patterns do not match those found associated with the K–Pg boundary (which were the result of high-energy impact). Thus, actualistic reasoning does not seem to support the volcanists' cause. It might be added that the Deccan traps were a nonexplosive type of volcano and so could not be the source of the shocked quartz. So, the volcanists would need to find alternative sites of volcanism to support their arguments, which would concurrently challenge the theory of impact.

As important as the physical geological features are, they are only evidence of impact; ultimately, this is a theory of extinction, which means that the fossil evidence must validate the fact that the impact is the source of the extinction; for even though many scientists accept both the evidence of impact and its timing, there was (and still is) disagreement about the impact as *the only* cause of the extinction. Thus, in analyzing the pattern, we need to divide the discussion into a set of multiple working hypotheses about extinction at the K–Pg boundary to include a gradual pattern (due to a possible combination of physical and biological factors), an "instantaneous" pattern[16] (caused by an impact or volcanism), and a stepwise pattern (possibly caused by multiple impacts). Concurrently, what is also fascinating about this debate is that it divides its supporters along disciplinary lines.

At least at the beginning of the debate, many earth scientists in general objected to impact. Most notable in their opposition were the paleontologists (*the* scientists who are professionally trained to reconstruct fossil life); they specifically objected to impact because the K–Pg boundary was not marked by an abrupt extinction event at the end of the Cretaceous; in other words if impact was the sole cause of extinction, there should have been no major change in the diversity of a group of organisms—such as the dinosaurs—during the Late Cretaceous (Glen 1994c; Ryan et al. 2001; Macleod et al. 1997). Instead, in their view, the fossil record favored a pattern of gradual extinction during the Late Cretaceous.

---

[16] Instantaneous in terms of the massive span of geological time.

Such objections to an instantaneous, abrupt pattern are still strong among the paleontological community. In a letter sent to *Science* in response to Schulte and his associates (2010), a team of 23 scientists led by Archibald and his colleagues (2010, p. 973) argued that the review of Schulte et al. (2010) "has not stood up to the countless studies of how vertebrates and other terrestrial and marine organisms fared at the end of the Cretaceous. Patterns of extinction and survival were varied – pointing to multiple causes at this time."

Concurrently, Glen (1994b), drawing upon Pantin (1968), suggested that paleontologists objected to having what is in essence a biological phenomena—extinction—imposed upon them by magisterial authority of the "restricted sciences," i.e., sciences that emphasize the use of a small number of powerful laws in matters of great theoretical significance (such as physics). Such objections were reinforced by L. Alvarez's scathing opinion of paleontologists when he remarked in the *New York Times* (1.19.88) "they're really not very good scientists. They're more like stamp collectors."[17]

Indeed, this was not the first time that such disciplinary conflicts have occurred between physics and earth science. Physicist Lord Kelvin tried to impose a limited geological time scale on Darwinian evolution, and Sir Harold Jeffreys attacked the nascent understanding of continental drift, based on pure physical models, without ever considering the validity of the geological evidence (Dodick and Orion 2003). In these historical cases the magisters of physics ignored the methodological uniqueness of historical sciences; so too the paleontologists argued that L. Alvarez was also wrong in his interpretation. Partly trained in biology, paleontologists understand that like other historical events extinction is a complex, contingent phenomenon that cannot always be reduced to a single cause as Archibald and his colleagues (2010) intimated in their recent reply to Schulte his associates (2010).

Such disciplinary battles have even extended within the earth sciences. Geochemistry, planetary geology, and other more physically oriented branches of the earth sciences were more inclined at the beginning of this debate toward accepting impact (Glen 1994b). Even today, such divisions exist as Archibald and colleagues (2010, p. 973) criticized Schulte's (mostly) physical geological team because it did not include researchers "in the field of terrestrial vertebrates…as well as freshwater vertebrates and invertebrates." It might be added, however, that today most paleontologists accept the idea of impact as one of the extinction factors (along with marine regression, volcanic activity, and changes in climatic patterns), so the physical geologists and paleontologists have drawn somewhat closer together.

In the last half of the 1980s, as more scientists look at the K–Pg boundary layer, a third extinction pattern was suggested—stepwise mass extinction—in which

---

[17] This critique of paleontology has antecedents in Ernst Rutherford's famous quote about science in general: "All science is either physics or stamp collecting." In his book, *Wonderful Life*, Gould (1989) makes a strong argument for the special nature of the historical sciences, such as paleontology, and their methods, as well the general value of epistemological diversity in the sciences. This argument eloquently recapitulates many of the points raised in our chapter in the section dealing with the nature of the earth sciences.

different kinds of organisms disappear within different layers before the end of the Cretaceous and the layer containing the iridium (Mount et al. 1986; Keller 1989). Correlated with this finding is the fact that in some localities, iridium is not restricted to the K–Pg boundary clay but appears to diminish gradually in concentration as one moves up or down from this layer (also termed "smeared anomalies"). Such evidence points to the possibility that multiple impacts were responsible for the extinction (Dingus and Rowe 1998). At the same time some of the volcanists have seized upon such stepwise patterns as supporting their claim, as it fits the major pulses of volcanic activity and associated environmental havoc resulting from periodic eruptions, which they claim happened in the Late Cretaceous.

Surprisingly, the earth science community also objected to impact because according to historian of science Glen (1994a) and paleontologist Gould (Glen 1994c), instantaneous global effects violate the understanding of one of geology's most important principles—uniformitarianism. Uniformitarianism has had many different interpretations over its history (Oldroyd 1996), but it would appear that the definition that many earth scientists adopted was the restrictive definition of Lyell (1880–1883), which assumed that in geology "no causes whatever have . . . ever acted but those now acting, and that they never acted with different degrees of energy from which they now exert" (Lyell 1881, vol. 2, p. 234). In other words actual causes were wholly adequate to explain the geological past not only in kind but also in degree (Rudwick 1998). Lyell based his uniformitarianism definition on Newton's use of the philosophical principle of vera causa in which only those processes operating today would be accepted as geological causes (Laudan 1987).

Lyell's adoption of Newton's vera causa was his philosophical response to geologists who invoked catastrophes as earth shaping forces. Lyell disapproved of catastrophes because they implied that geology relied upon unknown causes, which violated the principle of simplicity (i.e., the best scientific explanations are those that consist of the fewest assumptions). Lyell believed that the a priori application of uniformity (based on vera causa) was necessary, if geology, like physics, was to be considered a valid, logically based science (Baker 1998, 2000). However, the adoption of such restrictive principles is short sighted because it does not consider geology's unique defining characteristics, its historical interpretive nature, and indeed, during Lyell's time his definition of uniformitarianism was largely rejected, yet in the twentieth century, it influenced the thinking of many earth scientists (Dodick and Orion 2003). In simple terms, such scientists were trying to be more like physicists than the physicists in their application of this defining principle.

Today, the situation has changed. With mounting evidence, most earth scientists do accept the reality of an impact 65.5 Ma. However, the debate continues about whether it is the sole cause of the mass extinction or just one of its contributing factors. Thus, paleontologists continue their examination of the K–Pg boundary to more accurately delineate the extinction patterns on the biosphere. Similarly, sedimentologists, (geo)chemists, and (geo)physicists continue their mapping of the K–Pg layer to better understand its geology and the devastation an impact would have imparted upon the Cretaceous geosphere, hydrosphere, and atmosphere.

For those interested in earth science education, the debate surrounding the K–Pg extinctions is a perfect historical controversy that summarizes many of the most important features of the nature of the earth sciences as a unique branch of science. Concurrently, it shows how the human factor of philosophical and disciplinary prejudices shapes the actors in a debate, sometimes in spite of what the "objective" evidence says. Thus, this controversy deserves a place in any well-designed earth science curriculum.

18.7 Hydrosphere: Ocean and Atmosphere Coupling

The section that follows will discuss aspects of a coupled ocean/atmosphere phenomenon in the Pacific Ocean with dramatic effects on long-term weather all over the world. At first glance, it seems that when talking about El Niño and the Southern Oscillation (ENSO), one might not think of it as a controversial issue at all, but even in the late 1990s, many scientists in the fields of weather, climate, and oceanography still considered ENSO researchers as "renegades" (Cox 2002) when Ants Leetmaa successfully predicted and publically announced a major El Niño event to occur that year, along with predictions of severe long-term weather. The fact is there were many controversial issues needing resolution before ENSO could gain consensus as an explanation for aberrant, long-term global weather. It took almost 100 years of investigation to gain full consensus, including with the general population, from scientists' first awareness of a possible connection between a warm current off the west coast of South America and unusually mild or wet winters in parts of North America and Europe and drought conditions in Africa, India, and Australia.

As we have tried to demonstrate with each of the controversies highlighted within this chapter, the phenomenon of ENSO is one of global scale and has influence on all of the earth systems. Though El Niño is the name Peruvian and Ecuadorian natives gave to the occurrence of a warmer than normal current along the eastern margin of the Pacific Ocean basin, ENSO identifies a phenomenon that actually results from the *interaction* of the ocean and the atmosphere to create conditions that have a profound impact on the long-term weather and biota around the globe. To give an example of the scope of impact, Glantz (1996) listed these effects of the 1982–1983 El Niño event. There were droughts in Africa, India, and Central and parts of South America, to which 400 deaths and almost $7 billion (USD) in damages were attributed. At the same time, flooding in parts of Western Europe, South America, the USA, and Cuba were responsible for about 300 deaths, 600,000 people being displaced, and $5.5 billion in damages accumulating. Severe storms and tropical cyclones battered many of the islands in the Pacific from Hawaii to Polynesia, as well as large portions of the USA. Effects were also detrimental to the East Pacific fishing industry and to the nesting sites for 10s of millions of birds on Eastern Pacific Islands and the west coast of South America. Likewise, Philander (2004) noted that over 20,000 deaths; over 100,000,000 physically affected, including 5,000,000 displaced; and \$33 billion in damages resulted from the 1997–1998 El Niño event. The phenomenon identified as ENSO is global; its impact, significant.

The story of ENSO is also one demonstrating the convergence of studies. In this case, studies focused on ocean circulation and on atmospheric circulation (i.e., it takes into account the hydro- and atmospheres). It was this dichotomy, atmospheric science versus oceanography, which played a role in the controversy, as our understanding of how the air and ocean interact to influence long-term global weather patterns. This included the theoretical pitting of the meteorologists (mainly American), who, as empiricists, utilized patterns observed in synoptic weather maps to form short-term weather predictions, against the forecasters (mainly from the European Bergen School of Meteorology) who utilized the physics of the atmosphere and computed weather forecasts, by hand at first but then by computer (Cox 2002). There were the philosophical differences in looking at the phenomenon. It made a difference whether one saw El Niño as a departure from the normal conditions or whether they saw it as a uniform cycle perturbed by outside, random conditions (Philander 2004). There was also the clash of personalities (Cushman 2004a). Jerome Namias looked to the north at the polar front and an atmospheric oscillation known as the Rossby wave to be the control of long-term weather around the world, while Jacob Bjerknes looked to the tropical pacific and the Southern Oscillation, first discovered by Gilbert Walker, as the main impetus for long-term weather variations.

A severe drought in India from 1877 to 1899 and ensuing famine caused the British government to send Gilbert Walker to India, in 1904, to become the head meteorologist and attempt to better forecast the monsoons than then current meteorologist, Sir John Eliot. Eliot's forecasts were descriptions upwards to 40 pages long … and mostly incorrect. Walker was an unlikely candidate for this position as he was trained as a statistician. However, he set to work recording weather conditions around the world. He noted some correlations among distant locations on earth. One of these was a "swaying" of the atmosphere in the tropics of the Pacific Ocean. When there was high pressure in the west, there was low pressure in the east. When it was high in the east, it was low in the west. He called this swaying the *Southern Oscillation*. He also found that observations of the weather in distant parts of the world correlated highly with this oscillating air over the Pacific Ocean. However, his findings did not impress many of the meteorologists of the time because they were strictly mathematical and therefore only descriptive. In other words, because Walker postulated only correlations and no explanation for the correlations, it made other meteorologists very skeptical of the findings. From the point of view of the meteorologists, Walker was not doing science in the conventional "make a hypothesis and then test it" way (Cox 2002). In essence, however, Walker *was* doing science, in an historic and interpretive sense. Paralleling what we described above, he looked at complex and preexisting entities to discern patterns and interpret them.

At approximately the same time, on the west coast of South America, a peculiar periodic warm current, years earlier named El Niño by Peruvian fishermen, became the focus of scientific inquiry (Cushman 2004b). *El Niño*, translated into English, means *little boy*, but when capitalized, it intimates *the Christ child*, or *Jesus Christ*. They gave this name because of the phenomenon's repeated occurrence around Christmastime. They identified this phenomenon because it brought with it disruptions in normal rainfall patterns as well as behavioral (including nesting and reproductive) patterns in fish and birds along the west coast of South America. Most considered El Niño to be a local phenomenon affecting only portions of South America and therefore did not warrant much attention. The phenomenon became very important to the USA after a strong El Niño event during 1925–1926 caused major disruptions in the US fishing industry in the Pacific. As Cushman (2004b) described in his book, the importance of business, colonialism, and national security has motivated intense study to modern times, into the connection between the ocean and the weather.

Robert Murphy, an ornithologist from the USA, noted the effects on the bird populations he was studying and proposed a connection between Walker's Southern Oscillation and the El Niño event he was experiencing. However, there was a great deal of doubt concerning the reliability of the data Murphy was using, as well as concerns about the connection between oceanic and atmospheric phenomena (Cushman 2004b). Then, through the 1930s and 1940s, interest in El Niño waned. Up until that time, US agricultural interests were wrapped up in Peru because the Peruvians were the world's largest producers of bird guano, much of which was exported to the USA as fertilizer for the growing agricultural industry. Bird nesting habits and therefore guano production were very much influenced by El Niño, but that became a nonissue with the development of man-made fertilizers (Glantz 1996). As the economic importance of guano production waned, so did the interest in studying El Niño.

It was not until post-WWII and the Cold War era that physical properties of the ocean again became of national interest and new studies began. The International Geophysical Year, 1957–1958, coincided with this renewed interest. Many countries began recording data with better equipment and with greater rigor than previously. National security and national self-interest through the US fishing industry precipitated a renewed interest in ocean and atmosphere dynamics. Jacob Bjerknes, son of famous meteorologist, Vilhelm Bjerknes, and creator of the cyclone model of mid-latitude weather, turned his attention to El Niño. He discerned a connection between the Southern Oscillation, what he identified as *Walker Circulation*, and the periodic warming of the tropical Pacific Ocean, known as El Niño. Bjerknes and others such as Jerome Namias, who earlier had helped model upper atmosphere oscillations known as the Rossby wave, echoed claims already made of the connection between the atmosphere and ocean and their affect on weather in distant parts of the world. Such cross-disciplinary studies—oceanography and meteorology—were conceptually new and as yet quite suspect from other scientists. Where Bjerknes looked to the Walker circulation in the tropical Pacific for an explanation of global, long-term weather, Namais looked instead to the mid-latitude polar front as the engine driving such phenomena (Cushman 2004b). The military became interested in developing new buoy technologies motivated by its need for defense against Russian nuclear submarines. As a side note, it was this same fervent interest in the ocean by the military that generated the JOIDES (Joint Institutions for Deep Earth Sampling) expeditions that were so instrumental in collecting the seafloor data later used in support of the theory of plate tectonics. With the international efforts to collect data, Bjerknes had the resources to connect El Niño with the Walker circulation (Southern Oscillation). The study of ENSO began at that point.

The mechanism discerned by Bjerknes to explain his observations was that in the Pacific, along the equatorial region, winds generally blow from the east across the basin to the west. This pushed the warm water to the west and allowed a rise of the thermocline, the thin layer of water separating warm, well-mixed upper-level water from the colder, less-mixed water below. This brought the cold water to the surface making the west coast of South America cool and dry. An El Niño event was identified when the easterlies were not so strong and warm water resided in the eastern parts of the Pacific Ocean basin. This warm water interrupted fish migrations. It was also responsible for warmer, moister regional weather which interrupted bird nesting behaviors. It also caused more rain along the western coasts of North and South America and affected long-term weather all through Africa, North America, and Europe, as noted above. Here again, as we have recounted in each of these sections, this phenomenon spans its impact into many realms of study from the physics of energy exchange between the air and the sea to the effects on life on earth. In essence, rather than being a derivative science, the geosciences are more a place for the practical application of understandings from the other disciplines.

The scientific community was still divided, however. There was skepticism in being able to mathematically model the weather. There was skepticism in the utility of cross-disciplinary investigation. They saw El Niño scientists as renegades (Cox 2002). There was skepticism that a local fluctuation in ocean surface temperature could explain worldwide weather. The idea gained traction, as the number of published articles related to El Niño doubled every five years from 1980 to 2005 (Philander 2004). What the public heard of El Niño and its effects through the 1980s and 1990s resulted in its conflation with other atmospheric hazards making the news at that time, namely, the hole in the ozone layer over Antarctica and threats of global warming. Many considered reports of El Niño as just another liberal, big government orchestration (Cox 2002), very much like that continuing to surround the issue of global climate change today.

It was not until 1997, when Ants Leetmaa, then director of the National Climate Prediction Center, declared on national news that he expected a very significant El Niño event. For the previous decade or so, the National Climate Prediction Center had been participating in and receiving data from the Tropical Ocean Global Atmosphere (TOGA) program. Leetmaa and others receiving these data noted a warming of the waters at a far faster pace than had been observed before. Guided by computer simulations, Leetmaa laid out a number of predictions of anomalous long-term weather conditions contingent on this warming, including heavy rains in Southern California and the rest of the Southern USA and a warmer than usual winter in the northeast of the country. He also talked of a more quiescent than normal hurricane season. Reception of Leetmaa's warnings was cool. Many thought that Leetmaa had overstepped the types of predictions ENSO scientists were able to make. El Niño became somewhat of a household name the following spring when these predictions came to pass.

18 Teaching Controversies in Earth Science: The Role of History and Philosophy... 579

In this sense, the concept of ENSO had an advantage in its explanatory power as well as its relatively short-term predictive capabilities over the other controversies noted in this chapter. Various computer models, an example of multiple working hypotheses, used initial conditions and projected outcomes into the future: an interpretation of how nature *will be* as opposed to the plate tectonics controversy or the dinosaur extinction controversy where events had already taken place and scientists were left to interpret the results of *past actions*. Similarly, there are many investigations looking into how to utilize effects of El Niño to interpret the extent of past El Niño events.[18] Leetmaa reported his predictions and everyone could be around to witness whether they came to pass or not. It was not an experiment in the sense that variables were controlled, but it had the feeling of an experiment because predictions were made and it was just a matter of waiting them out. This type of "natural experiment" is very characteristic of the historic sciences, like the earth sciences. In this case, the ENSO phenomenon happens on a scale of time that makes it possible to make predictions and see them borne out over several months. And the fact that the predictions were fulfilled gave strength to the models and gave rise to a general consensus within the scientific community as well as the general public concerning the validity of ENSO—the interaction between ocean and atmosphere—as a world weather controller. It also provided evidence supporting the use of computers to predict long-term weather.

18.8 Global Warming: A True Controversy?

Depending on the background of the reader, the title of this section should give pause. If we were to survey climate scientists, then the vast majority would agree with the primary conclusions of the Intergovernmental Panel on Climate Change (IPCC)[19] (IPCC 2007) which states that anthropogenic greenhouse gases have been responsible for most of the "unequivocal" warming of the earth's average global temperature over the second half of the twentieth century. In fact, in their extensive study of (1,372) climate researchers and their publications, Anderegg, Prall, Harold, and Schneider have shown that

> (i) 97–98 % of all of the climate researchers most actively publishing in the field [surveyed in their research] support the conclusions of the IPCC and (ii) the relative climate expertise and scientific prominence of the researchers unconvinced by anthropogenic climate change (**ACC**) are substantially below that of the convinced researchers. (Anderegg et al. 2010, p. 1207)

---

[18] See, for instance, Galbraith et al. (2011), Khider et al. (2011), Nippert et al. (2010), and Romans (2008).

[19] Created in 1988 by the World Meteorological Organization and the United Nations Environmental Programme, IPCC's purpose is to evaluate the state of climate science as a basis for informed policy action, primarily on the basis of peer-reviewed and published scientific literature (Oreskes 2004).

Similar results were obtained by Doran and Zimmerman (2009) in their web survey of over 3,000 earth scientists, as well Oreskes's (2004) analysis of (928) abstracts dealing with climate change, published in refereed journals from 1993 to 2003. Finally, Powell (2011) on the *Skeptical Science* Internet site surveyed 118 of the best-known ACC skeptics. He found that 70 % of them have no (peer-reviewed) scientific publications that deny or cast substantial doubt on ACC. Moreover, none of their papers offers a "killer argument" falsifying human-caused global warming. The best they can do is claim that the measurement sensitivity of ACC is low, which they have been unable to substantiate and which much evidence contradicts (http://www.skepticalscience.com/Powell-projectPart2.html). So it would seem that at least among the majority of professional scientists who are most active in climate research, ACC is accepted as a (worrisome) trend that requires immediate response from nations around the world to ameliorate.

However, among the US public, the story is very different. In a recent Gallup poll, 51 % of its citizens expressed concern over ACC in 2011, compared to 65 % in 2007. Moreover, 52 % of the 2011 survey believed that the increase in the earth's temperature was due to pollution from human activities as opposed to 43 % who believed that it was due to natural changes in the environment. Just four years previously these figures stood at 61 % and 35 %, respectively (http://www.gallup.com/poll/146606/concerns-global-warming-stable-lower-levels). Clearly, much of the US public does not agree with the implications of much of the peer-reviewed research. Although not as severe, skepticism about ACC has also increased in the European Union, Canada, Australia, and New Zealand, based on surveys conducted in the last three years (Ratter et al. 2012). Thus, in the case of ACC, the public controversy is at odds with the much higher acceptance that this phenomenon has received among the majority of climate scientists, as well as their scientific colleagues within the wider earth science community.

One result of this controversy is that it impacts how students understand the workings of the atmosphere. In fact, ACC is a special subject because students face two challenges to their learning about it. First, like all other earth science topics discussed in this chapter, ACC is a complex scientific problem that is studied by a multidisciplinary, global team of climate scientists, oceanographers, atmospheric chemists, and geologists. Even in their early years at university, students do not usually have the broad background to understand this problem; adding to this problem is that they also hold large numbers of misconceptions about this and other atmospheric issues.[20]

Second, in order to understand ACC, students (like the general public) must overcome misinformation perpetuated by a smaller number of vocal, skeptical politicians and experts that are the source of the controversy (Theisen 2011). Relative to the much larger community of experts who have gathered strong evidence for ACC, the skeptics have a broad platform in the public media; this is due to the balance that

---

[20] See, for instance, Gautier et al. (2006), Jeffries et al. (2001), Shepardson et al. (2011), and Theisen (2011).

the media gives to this issue—a balance which in fact diverges from the much greater acceptance this issue receives from professional scientists (Boykoff and Boykoff 2004). Indeed, in her study of students at the University of Vermont, Dupigny-Giroux (2010) found that most undergraduates cited some form of media as their primary information about climate, which in turn reinforces their misconception that a (balanced) controversy exists.

In this regard, ACC which is played out in the public eye differs from the other three controversies, presented in this chapter, which are largely debated among scientists and have had much less impact on the public. This controversy is less politically contrived than the false "controversy" that religious forces have presented in order to falsify evolution. However, as we will see, the roots of the ACC controversy also have political overtones, which are partly derived from the scientific background and motivations of some of its opponents, as well as the general economic situation which influences the publics' attitudes.

Thus, the question that we need to ask is as follows: If much of the scientific establishment supports ACC, how does such skepticism thrive? To answer this question we will (briefly) examine the history of the ACC idea. Concurrently, we will show that the ACC problem encompasses many of the unique features of the earth sciences. We believe that it is important for students to understand these historical and philosophical features of the ACC idea because it helps to explain the background behind the scientific and even political opposition.

The earth maintains a habitable temperature because of the natural greenhouse effect occurring in its atmosphere. Various atmospheric gases contribute to the greenhouse effect, whose impact in clear skies is 60 % from water vapor, 25 % from carbon dioxide, 8 % from ozone, and the rest from trace gases including methane and nitrous oxide (Karl and Trenberth 2003). Clouds also add to this greenhouse effect. On average, the energy from the sun received at the top of the earth's atmosphere amounts to 175 petawatts (PW = a quadrillion watts), of which 31 % is reflected by clouds and from the surface. The rest (120 PW) is absorbed by the atmosphere, land, or ocean and ultimately emitted back to space as infrared radiation (Karl and Trenberth 2003).

Since the early twentieth century, the average temperature of the earth's surface has increased about 0.8 °C, with about two-thirds of that increase occurring since 1980 (NRC 2011). Such global warming is caused by increasing concentrations of greenhouse gases produced by human activities such as deforestation and the burning of fossil fuels (NRC 2011). As such concentrations rise, they act to increase the opacity of the atmosphere to infrared radiation, trapping it in the atmosphere and raising the temperature of the planet.

The idea of ACC is not recent; indeed, the idea that changes in atmospheric greenhouse gas concentrations can and do cause significant climate changes was proposed qualitatively in 1864 by renowned physicist John Tyndall, when he discovered carbon dioxide's opacity to IR radiation (Sherwood 2011). In 1896 the future Nobel chemistry laureate Svante Arrhenius quantitatively predicted that such warming would be caused by coal burning; the prediction was tested and promoted by steam engineer Guy Callendar in the late 1930s (Sherwood 2011).

In the 1950s, the scientific debate focused on whether or not greenhouse gases were accumulating in the atmosphere and, if so, what affect this was having on global temperatures. Against the background of this debate, chemist David Keeling, from the Scripps Institute of Oceanography, sought to find out; in 1957, he set up an array of newly developed gas analyzers on Hawaii's Mauna Loa volcano to measure atmospheric levels of carbon dioxide. Keeling discovered two trends: first, he measured the average monthly value at 315 p.p.m (p.p.m. = parts per million). Keeling saw the values drop from May to September and then rise again into the next year. This cycle continued with decreases in the summer when plants soak up carbon dioxide and grow and increases in autumn and winter when plants are less biologically active (Smol 2012).

The second trend found by Keeling was that global carbon dioxide levels were rising annually from various human activities, creating a rising trend on the graph he constructed. Measurements that continue until the present demonstrate that atmospheric carbon dioxide concentration had risen to 394 p.p.m by June 2011. Moreover, the current carbon dioxide level far exceeds its natural fluctuation (180–300 p.p.m.) over the past 800,000 years. Scientists reconstructed this historical range by studying the planet's natural archives, represented by natural traces found in tree rings, the sediments of lakes and oceans, and ice cores (Smol 2012). Such proxy records combined with measurements of global temperatures today have shown that the world has warmed throughout the twentieth century. Models of such warming into the future suggest and predict that the earth will continue to warm into the future.

Climate scientist Steven Sherwood (2011) framed the historical development of the ACC idea by comparing it to some of the major paradigmatic shifts affecting physics. For example, Copernicus' published his model of the heliocentric universe in 1543. However, it was not until Kepler's calculations of 1609 (Gingerich 2011) and Galileo's observations in 1610 that provided the critical evidence to convert the top astronomers to the Copernican view. Nonetheless, acceptance among most scientists did not occur until the late seventeenth century, while the public at large remained opposed until the eighteenth century (Kuhn 1957). A similar pattern was seen in the fight for acceptance of Einstein's theory of general relativity (Sherwood 2011).

In the case of the heliocentric universe, a large source of public criticism was religion. As Gould (1987) and Freud before him noted, the invention of a heliocentric universe is one of seminal scientific discoveries as it displaced humans from the center of the universe, breaking their cosmological closeness to God. Such a view threatened the political power base of the Church which saw itself as the guardian of the human connection to God, and it is well recorded about the pressures that the Church brought to bear on scientists who supported Copernicus. In the case of Einstein, religious and political factors also affected the public debate against him and his theories, as anti-Semitic jibes and accusations of being a communist were thrown in his direction (Sherwood 2011).

In the case of global warming, politics is also a strong motivator of public skepticism. Gauchat (2012) has analyzed trends in public science in the USA from 1974 to 2010. He found that conservatives began this period with the highest trust in science, relative to liberals and moderates, and ended the period with the lowest; with regards to ACC, specifically, a decreasing number of conservatives doubt that it is occurring. Complicating the political situation are economic factors. In evaluating public opinion data from the USA, Scruggs and Bengali (2012) suggested that the decrease in belief about global climate change is likely driven by economic insecurity connected to the recent recession. A similar analysis of opinions from the European Union supports an economic explanation for changing public opinion.

However, such public skepticism does not explain the scientific skepticism for global warming. We have already seen that the peer-reviewed data overwhelmingly supports ACC and that the scientific skeptics largely do not come from the forefront of climate research. Therefore, we ask: why does such scientific skepticism survive and even thrive?

Sherwood (2011, p. 42) has argued that it is the very nature of global warming, as a scientific problem, that has created the skepticism among some scientists. He suggests that the heliocentric universe, general relativity, and global warming have all been scientifically opposed because of the "absence of a smoking gun or a bench top experiment that could prove any of them unambiguously." Moreover, he notes that what global warming shares with the other theories is: "its origins in the worked-out consequences of evident physical principles rather than direct observation." Such "bottom-up deduction is valued by physics perhaps more than by any other science," and many of the leading climate scientists were trained as physicists. Finally, he adds that global warming is based on "physical reasoning… rather than on extrapolating observed patterns of past behavior."

We agree with Sherwood's (2011) assessment that it is the misunderstanding of the scientific nature of the global warming problem that is one of the sources of its opposition. However, we do not think that this is connected to it being a strictly physics-based problem. In fact, the characteristics that Sherwood uses to define this problem also fit well within the structure of historical sciences (such as the earth sciences) that we mentioned earlier in this chapter. Most of the problems that the earth sciences tackle do not lend themselves to benchtop, controlled experiments nor direct observations, due to these sciences' massive scales, both in terms of space and time, as well as the large number of interacting variables that are impossible to replicate and control in the laboratory. Moreover, although some climate scientists certainly create multiple mathematical models, what we consider to be multiple working hypotheses, in order to predict the magnitude of future trends in global warming, others are using, as we have seen, evidence from the past such as ice cores and tree rings to reconstruct the past atmosphere. So there is a strong element of "history" in this research as well.

These arguments, concerning the nature of different sciences, are inadvertently supported by Oreskes and Conway (2010), in their book *Merchants of Doubt*. A main theme of this book is that a handful of politically conservative physicists in the USA, with strong ties to both industry and conservative think tanks (such as the George C. Marshall Institute), have challenged the scientific consensus on issues such as the dangers of smoking, the effects of acid rain, and the existence of ACC.

The authors charge that this has resulted in deliberate obfuscation of these issues which in turn has influenced public opinion and governmental policy.

Oreskes and Conway's (2010) main argument is about the deleterious effects of politically connected, powerful scientists on the government's environmental and health policy. However, it is interesting to note that they specifically identify Bill Nierenberg, Fred Seitz, and Fred Singer as the three physicists who were most prominent in leading the battle against ACC; Nierenberg and Seitz were part of the Manhattan (atomic bomb) project, whereas Singer developed earth observation satellites. In simple terms all three scientists came from branches of physics that more closely rely upon experimental, reductionist methods. Possibly, it is their scientific background which creates prejudice against the multidisciplinary, historical, and interpretive methods of global climate research. This, combined with their political histories as past cold warriors, who also represent conservative business and political interests, creates a synergistic effect to their skepticism against ACC.

There is no doubt that the political power and media connections of this much smaller group of scientific skeptics are strong. In the science education world, its influence has created confusion among (earth science) students. However, if Sherwood (2011) is correct about its historical progression, the science will eventually be accepted by both scientists and the public. The question that remains of course is how future generations will deal with our lack of action today.

18.9 Designing Curricula Utilizing HPS and the Controversies: Plate Tectonics as an Exemplar

We have given an outline of the development of scientific understanding of four different phenomena through the lens of the controversies surrounding each understanding. In this section, we would like to offer some possible direction for designing instruction that utilizes a modern theory of learning as well as the history and controversies surrounding the phenomena to promote, in students, useful understanding of content as well as aspects of the nature of science.

Researchers have discerned a pattern of learning encompassing the iterative process of developing a mental model of a phenomenon, deriving predictions from the model, testing the predictions, and finally, amending the original model to agree with the new data (Nersessian 2008) or generating, evaluating, and modifying the model (Clement 2009). By starting with this structure, an instructor can utilize historic models and data to encourage students to create their own models of a phenomenon, make predictions from the models, look at the historic data, and determine the usefulness of their models to make predictions. The instructor can also encourage model co-construction (Khan 2008), model evolution (Núñez-Oviedo et al. 2008), and model competition, disconfirmation, and accretion (Núñez-Oviedo and Clement 2008) through the use of personal models, class-generated models, and historic models.

continental displacement (Du Toit 1937; Wegener and Skerl 1924), and expanding earth (Carey 1976; Jordan 1971). Students would then, in small groups or whole class discussion, identify the strengths and limitations of the historic models alongside current student models for the cause of earthquakes. Again, we would have students be aware that models of earth dynamics were often dependent on the region used for delineating the model. Aristotle developed the porous earth model within the karstic topography of the Mediterranean. Seuss and Dana developed the contracting earth models during their work in the folded mountains of the Alps and the Appalachians, respectively. Wegener's experience with icebergs may have influenced his model for drifting continents through ocean crust.

Students should also be made aware of the controversies surrounding these models. One issue had to do with the idea that earth dynamics behaved mainly in a vertical direction (porous model and contraction model) versus deformation resulting from horizontal motion (continental displacement and expanding earth models). Other issues dealt mainly with issues surrounding the controversy between drift and permanence theories. Wegener and Du Toit pointed out the difficulties of contraction with the understanding of isostasy and that it could not explain fossil, geologic, and geographic similarities among widely separated continents. The "fixists," on the other hand, accused those in favor of displacement of not having an appropriate mechanism for moving continents, of deciding on their explanation and going in search of evidence to prove the explanation, and of not adhering to the philosophy of uniformitarianism.

We would ask students to use these models, in addition to student- or class-generated models, as multiple working hypotheses. They should determine the implications of each model, and then think of places they would look to find more data to test them. When someone directs attention to the ocean, some readings concerning the history of ocean exploration (Höhler 2003), Marie Tharp (Lawrence 2002, pp. 181–188) and Tharp's discovery of the mid-Atlantic ridge and rift system (Heezen et al. 1959), help students to understand the historical development of physical oceanography. Discussions of continued reticence for accepting drift, as well as the influence of World War II and breaking telephone cables as incentive for exploring the seafloor continue to develop the social and economic factors influencing the direction of scientific investigation. Then students can look at various kinds of seafloor data such as utilized in the "Discovering Plate Boundaries" activity (Sawyer 2010). Here students will look for relationships among patterns of sediment thickness, ocean crust age, bathymetry, and seismic and volcanic patterns. Using these data, students can test their models and the historical models to determine how they hold up to the data.

Then we would introduce students to explorations into paleomagnetic studies (polar wandering and magnetic reversals) and how it tied all the data together (Glen 1982) for those such as Hess (1962) and Vine and Matthews (1963). Finally, discussions into mantle convection, Wilson's prediction of transform faults (Wilson 1965), and the World Wide Synchronized Seismic Network should give students enough information to develop a model of earth dynamics very similar to the current scientific model. A key point throughout the entire instructional series is

We would start with fundamental concepts or big ideas. They can be garnered from the core disciplinary concepts of the NGSS (Achieve 2012) or one of the big ideas found in the Earth Science Literacy Principles (Earth Science Literacy Initiative 2010). Or, the instructor can discern his/her own fundamental concepts by using the discourse tools found at http://tools4teachingscience.org. For this example, we will use the concept of earth dynamics as it pertains to the theory of plate tectonics. We envision this fundamental concept or primary concept as an amalgam of six secondary concepts (volcanology, seismicity, oceans and continents, geomagnetism, the earth's internal structure, and radioactivity). Of course these are not the only secondary concepts that one could use, nor do they have to be these specific concepts. Finally, we discerned about three or four tertiary concepts from each secondary concept. Tertiary concepts are the learning objectives of individual lessons. For instance, for the secondary concept, "seismicity," possible tertiary concepts are "earthquakes," "elastic rebound theory," and "global seismicity patterns." These tertiary concepts are the foci or instruction using the original documents, data, historical narratives, and inquiry activities.

A brief outline of a possible approach to incorporating HPS and the content material within the structure of model-based learning follows. We would have students read two eyewitness accounts of the 1906 San Francisco earthquake: one by Jack London (1906) and William James (1911). As a follow-up to the readings, we would have students develop an initial mental model of an earthquake, based on their prior understanding and the content of the readings, by asking them what an earthquake is and what causes it. Model competition, model disconfirmation, and model evolution then take place through presentation and class discussion of mental models.

Following student work on their mental models, we would have them read excerpts of H. F. Reid's (1910) report and description of elastic rebound theory. Discussions about Reid's earlier work studying glaciers and how the behavior of glacial ice may have been his model for the behavior of rock could illuminate for students how prior experience can influence thinking about unrelated problems. Subsequent to this discussion, students would break into groups and participate in an activity utilizing the earthquake machine http://www.iris.edu/hq/resource/redefining_an_earthquake_v12, where they can gain an understanding of the nature of the storage and release of elastic energy, as well as the use, strengths, and limitations of models. With the understanding of an earthquake being a release of elastic energy built up in deformed rocks, students can utilize such computer visualizations as the US array record of such earthquake events as the 2011 event in Japan (http://www.youtube.com/watch?v = Kbc0ERoCD7s) and data storage sites such as Rapid Earthquake Viewer (http://rev.seis.sc.edu/) where they can develop a sense of energy released by an earthquake in the form of waves that travel through the earth and be observed by sensitive equipment.

Next, we would ask students about possible causes of earthquakes. Once they have developed their own models, we would have them read excerpts from or summaries of multiple historic models of earth dynamics. These would include Aristotle's porous earth (Şengör 2003), contracting earth (Malaise 1972; Schuchert 1932),

that the students are allowed to *develop their own model* of earth dynamics as opposed to rationalizing data and identifying "wrong" models because they already "know" the right answer. The questions we would ask are open for students to foster inquiry into the data and model building/testing/amending from the data. In this way, students experience "science in the making" (Conant 1947, p. 13) as opposed to finished science.

18.10 Conclusion

We have accomplished a few goals within this chapter. The first was to highlight the historical and interpretive nature of the geosciences as distinct from the experimental nature of physics and chemistry. All of the models developed by investigators have the purpose of explaining observations of effects of events that have already happened. In some cases, these explanations allow us to peer in the future, but not in any kind of controlled way. Phenomena (shifting plates, long-term weather, meteorite impacts) will proceed as they will and we can only witness them and measure them against our predictions. Second, we demonstrated the global nature of phenomena being investigated within the geosciences. Each of these topics has or has had fundamental effects within all spheres of earth systems and has had impacts that extend around the world. This is not to say the earth scientists do not study strictly local phenomena, but even these local phenomena can be traced back to global causes.

Third, was to demonstrate that it was often the convergence of multiple disciplines involved in independent investigations that led to the eventual development of reliable explanatory models of the phenomena in question. Within this framework, we also found that the interdisciplinary nature of many of the investigations gave rise to the controversies in the first place. This was often the case because the different disciplines operated under different philosophical constraints or followed different rules and politics. Especially relevant were issues surrounding the nature of nature. For instance, do phenomena happen based in uniformity (cyclic) or catastrophe (unidirectional)? In the case of continental displacement, the interpretation by some that it did not conform to uniformity as defined at the time may have delayed its acceptance. We also cited uniformity as an issue to accepting the bolide theory for explaining the extinction of the dinosaurs. Another example of a difference in philosophical stances toward nature was L. Alvarez's interpretation of the extinction event at the end of the Cretaceous. Alvarez, an experimental physicist, believed that impact was the cause of all of the mass extinction events in earth history. According to Gould (Glen 1994c) he sought a universal mechanism for mass extinctions. This approach differs from the historical sciences, which interpret natural phenomena, such as extinction, which are seen as complex and contingent, and dependent on a large series of often interacting factors. In other words, just because a meteorite impact caused a single mass extinction, it does not necessarily mean that all mass extinctions were caused by impact. History has shown that Alvarez's hypothesis of a universal mechanism was not correct.

Fourth, we showed the relationship between scientific advancement and technological advancement. Oftentimes, it was technological advancements responsible for gathering more accurate data and a refinement of methods that increased its reliability. For plate tectonics, it was more sensitive magnetometers, the advancements in seismic recording with the WWSSN, and the enhanced precision of radioactive age dating of rock. El Niño finally gained consensus through the collection of data with the large-scale deployment of better buoys and the strength of computers and models of the oceanic and atmospheric systems. Advancements in atmospheric carbon dioxide detection and atmosphere sampling protocols helped standardize readings leading to the conclusion that carbon dioxide levels in the atmosphere are, in fact, rising and that the carbon dioxide was anthropogenic. The Alvarez groups' development and use of the coincidence spectrometer allowed them to *quickly* analyze the possible iridium concentrations of a huge number of stratigraphic beds, allowing them to show that such beds were anomalous and were indeed the remnants of an extraterrestrial impact.

Fifth, we discussed how explanatory models gained consensus because they accounted best for the collected data. Plate tectonics gained consensus prior to our ability to measure plate movements directly via satellites, but now these measurements record actual displacement. For the ENSO phenomenon, meteorologists utilized computer models to successfully predict long-term weather patterns. For the dinosaur extinction event, the discovery of anomalous iridium layers and most importantly the Chicxulub crater both of which coincided with the end of the Cretaceous were the critical evidences that could only be accounted for by an extraterrestrial impact. Ice cores and tree rings have provided evidence of the greenhouse gas profiles of the earth's past; combined with measurements of present-day gas analyzers and the power of computer modeling, it is possible to predict future planetary warming trends.

A final point we would like to make has to do with the nature of controversy resolution. In analyzing the drift controversy, Frankel (1987, pp. 204–205) argued that "Closure of the controversy comes about when one side enjoys a recognized advantage in its ability to answer the relevant questions...when one side develops a solution that cannot be destroyed by its opponents." For the four controversies described here, we note the overwhelming ability of one model to explain the observations that allowed it to garner consensus from the scientific community. When discussing the *Great Devonian Controversy*, Rudwick (1985) asserted that it was one of the most important and influential controversies in the history of geology. Yet, he also claimed that the controversy is virtually unknown to geologists today. "The paradox has a simple explanation. The controversy has slipped out of sight for the good and adequate reason that the problems it raised were eventually resolved in a way that satisfied almost all participants" (p. xxi). Controversies surrounding the origin of oceans and continents, a meteorite impact causing a mass extinction occurring at the end of the Cretaceous period, and the interaction between the ocean and the atmosphere affecting weather around the world are all considered settled to the satisfaction of most of the interested parties. Where anthropogenic global climate change is no longer a controversial issue for those in climate science and indeed most of the scientific community, there continues to be a lag in consensus among much of the US population.

As we have intimated in the beginning of this chapter and as was evident throughout the discussion, there has been very little published concerning the incorporation of HPS into geoscience instruction. There are small pockets of those who continue to promote the efficacy of using cases as a pedagogical tool for teaching science (For examples, see http://sciencecases.lib.buffalo.edu/cs/, http://www1.umn.edu/ships/, http://www1.umn.edu/ships/, and http://hipstwiki.wetpaint.com/page/hipst+developed+cases). A survey of the three case repositories highlighted above (NCCSTS, SHiPS, and HiPST) shows that of the more than 500 cases housed in these three sites, both contemporary and historical, 24 are earth science related. There are six focused on global climate change. Only one of the 24 cases had any relevance to plate tectonics, and even then tectonics was treated as peripheral to the case. There were none focusing on El Niño nor were there any highlighting the dinosaur extinction controversy. A review of the use of case studies is outside the realm of this chapter, but suffice it to say that of the different types of cases available to use, the interrupted case (Herreid 2007) is probably the easiest to implement and still allows much control to the instructor. See Leaf (2011) for an example focusing on Keeling and the measurement of atmospheric CO2. We gave a brief structure to how one might utilize various activities, original documents, and historic and current data as a way to facilitate student model building for plate tectonics.

Aside from the few publications documenting HPS use as an instructional tool, there are even fewer empirical studies investigating the efficacy of such a tool. One possible avenue to remedy this situation is the development and use of historic case studies (Allchin 2011). This would require collaboration among historians and philosophers of science, geologists, and science educators to develop and test such curriculum materials for teaching.

The main point here is not only is there a need to create such tools for teaching that utilize the history and philosophy of science in instruction, but there is also a need for rigorous evaluation and publication in such journals as the *Journal of Geoscience Education* or *Science & Education*. This would give access to practitioners in the field who can further refine them, enhance their own teaching, and ultimately develop students' useful understanding.

References

Achieve, Inc. (2012). The Next Generation Science Standards. Retrieved 01 May 2012 from http://www.nextgenscience.org/next-generation-science-standards.

Adams, N. (2005). New model of the universe: The case against Pangaea. http://www.nealadams.com/nmu.html accessed 7 November, 2011.

Allchin, D. (2011). The Minnesota case study collection: New historical inquiry case studies for nature of science education. *Science & Education, Published on-line http://www.springerlink.com/content/v42561276210585q/* (accessed 1 March 2012), 1–19.

Allchin, D. (2003). Scientific myth-conceptions. *Science Education*, 87 (3), 329–351.

Allchin, D. (2002). How not to teach history in science, *The Pantaneto Forum* (pp. 1–13). www.pantaneto.co.uk/issue17/allchin.htm.).

Allchin, D. (1997). The power of history as a tool for teaching science. In A. Dally, T. Nielsen & F. Reiß (Eds.), *History and philosophy of science: A means to better scientific literacy?* (pp. 70–98). Loccum: Evangelische Akadamie Loccum.

Alters, B.J., & Nelson, C.E. (2002). Perspective: Teaching evolution in higher education. *Evolution*, 56 (10), 1891–1901.

Alvarez, L., Alvarez, W., Asaro, F., & Michel, H. (1980). Extraterrestrial cause for the Cretaceous-Tertiary extinction: Experimental results and theoretical interpretation. *Science*, 208, 1095–1108.

Alvarez, W., & Chapman, C. R. (1997). *T. rex and the crater of doom*. Princeton, NJ: Princeton University Press Princeton.

Anderegg, W.R.L., Prall, J.W., Harold, J. & Schneider, S.H. (2010). Expert credibility in climate change. *Proceedings of the National Academy of Sciences of the United States of America*, 107 (27), 1207–1209.

Anderson, D. (2006). Plate tectonics; the general theory: Complex earth is simpler than you think. In C. A. Manduca, D. W. Mogk & Geological Society of America (Eds.), *Earth and mind: How geologists think and learn about the earth. Special paper 413* (pp. 29–38). Boulder, CO: Geological Society of America.

Anderson, D., & Natland, J. (2005). A brief history of the plume hypothesis and its competitors: Concept and controversy. In G. Foulger, J. Natland, D. Presnall, & D. Anderson (Eds.), *Plates, plumes, and paradigms. Special Paper 388* (pp. 119–146). Boulder, CO: Geological Society of America.

Archibald J.D., Clemens W.A., Padian K., Rowe T., Macleod N., Barrett P.M., et al. (2010). Cretaceous extinctions: Multiple causes. *Science*, 328, 973.

Argamon, S., Dodick, J., & Chase, P. (2008). Language use reflects scientific methodology: A corpus-based study of peer-reviewed journal articles. *Scientometrics*, 75 (2), 203–238.

Baker, V.R. (2000). Conversing with the earth: The geological approach to understanding. In R. Frodeman (Ed.), *Earth matters: The earth sciences, philosophy and the claims of the community* (pp. 2–10). Upper Saddle River, NJ: Prentice Hall.

Baker, V.R. (1998). Catastrophism and uniformitarianism. Logical roots and current relevance in geology. In D.J. Blundell, & A.C. Scott (Eds.), *Lyell: The past is the key to the present, Geological Society of London Special Publications*, 143, 171–182.

Baker, V. R (1996) The pragmatic roots of American quaternary geology and geomorphology. *Geomorphology*, 16, 197–215.

Ben-Ari, M. (2005). *Just a theory: exploring the nature of science*. Amherst, NY: Prometheus Books.

Bickmore, B., Thompson, K., Grandy, D., & Tomlin, T. (2009a). Commentary: On teaching the nature of science and the science-religion interface. *Journal of Geoscience Education*, 57(3), 168.

Bickmore, B. R., Thompson, K. R., Grandy, D. A., & Tomlin, T. (2009b). Science as storytelling for teaching the nature of science and the science-religion interface. *Journal of Geoscience Education*, 57(3), 178.

Blewett, W.L. (1993). Description, analysis and critique of the method of multiple working hypotheses. *Journal of Geological Education*, 41, 254–259

Bohor, B.F. (1990). Shock induced deformation in quartz and other mineralogical indications of an impact event at the Cretaceous-Tertiary boundary. *Tectonophysics* 171, 359–372.

Bourgeois, J. Hansen, T.A., Wilberg, P.L., & Kauffman, E.G. (1988). A tsunami deposit at the Cretaceous-Tertiary boundary in Texas. *Science*, 24, 567–70.

Boykoff, M.T., & Boykoff, J.M. (2004). Balance as bias: global warming and the US prestige press. *Global Environmental Change*, 14, 125–136.

Bucher, W. H. (1941). *The nature of geological inquiry and the training required for it*. New York: A.I.M.E. Technical Publication 1377.

Carey, S. W. (1976). *The expanding Earth*. New York: Elsevier Scientific Pub. Co.

Cartwright, N. (1999). *The dappled world: A study of the boundaries of science*. Cambridge: Cambridge University Press.

Case, T. J., & Diamond, J. M. (1986). *Community Ecology*. New York: Harper Row.

CBC News (2011). Group ends calls for hydro-fracking moratorium. http://www.cbc.ca/news/canada/new-brunswick/group-ends-call-for-hydro-fracking-moratorium-1.1101282. Accessed 21 July 2011.

Chamberlin, T.C. (1965). The method of multiple working hypotheses (reprint of the 1890 version). *Science*, 148, 754–759.

Chamberlin, R. (1928). Some of the objections to Wegener's theory. In W. A. J. M. van Waterschoot van der Gracht (Ed.), *Theory of continental drift* (pp. 83–89). Tulsa, OK/Chicago, IL: American Association of Petroleum Geologists.

Chamberlin, T.C. (1904). The methods of Earth-sciences. *Popular Science Monthly*, 66: 66–75.

Chamberlin, T.C. (1897). The method of multiple working hypotheses. *Journal of Geology*, 5, 837–848.

Chamberlin, T.C. (1890). The method of multiple working hypotheses. *Science* (old series), 15, 92–96.

Cheek, K. (2010). Commentary: A summary and analysis of twenty-seven years of geoscience conceptions research. *Journal of Geoscience Education*, 58 (3), 122–134.

Claeys, P., Kiessling, W., & Alvarez. W. (2002). Distribution of Chicxulub ejecta at the Cretaceous-Tertiary boundary. In C. Koeberl, & K. G. MacLeod (Eds.), *Catastrophic events and mass extinctions: Impacts and beyond. Geological Society of America Special Paper 356*, (55–68). Boulder, CO: Geological Society of America.

Cleland, C. (2001). Historical science, experimental science, and the scientific method. *Geology*, 29 (11), 987–990.

Cleland, C. (2002). Methodological and epistemic differences between historical science and experimental science. *Philosophy of Science*, 69, 474–496.

Clement, J. J. (2009). *Creative model construction in scientists and students: The role of imagery, analogy, and mental simulation*. New York, NY: Springer Verlag.

Conant, J. B. (1947). *On understanding science; an historical approach*. New Haven,: Yale university press.

Cooper, R. A. (2002) Scientific knowledge of the past is possible: Confronting myths about evolution and the nature of science. *American Biology Teacher*, 64, 476–481.

Cooper, R. A. (2004) Teaching how scientists reconstruct history: patterns and processes. *American Biology Teacher*, 66 (2), 101–108.

Cox, J. (2002). *Storm Watchers: The Turbulent History of Weather Prediction from Franklin's Kite to El Nino*. NJ: John Wiley and Sons, Inc.

Cushman, G. (2004a). Choosing between centers of action: Instrument buoys, El Niño, and scientific internationalism in the Pacific, 1957–1982. In H. Rozwasowski, & D. van Keuren (Eds.), *The machine in Neptune's garden: Historical perspectives on technology and the marine environment* (pp. 133–182). Sagamore Beach, MA: Science History Publications.

Cushman, G. (2004b). Enclave vision: Foreign networks in Peru and the internationalization of El Niño research in the 1920s. Paper presented at the International Perspectives on the History of Meteorology: Science and cultural Diversity, Mexico City.

Cutler, A. (2003). *The seashell on the mountaintop: A story of science, sainthood, and the humble genius who discovered a new history of the earth*. New York, NY: EP Dutton.

Davis, W.M. (1911). The disciplinary value of geography. Popular Science Monthly, 78, 105-119, 223–240.

De Hosson, C., & Kaminski, W. (2007). Historical controversy as an educational tool: Evaluating elements of a teaching – learning sequence conducted with the text "dialogue on the ways that vision operates." *International Journal of Science Education*, 29 (5), 617–642.

Diamond, J. (2002). *Guns, germs and steel: The fates of human societies*. New York: W.W. Norton.

Diamond, J. & Robinson, J.A. (2010). *Natural experiments of history*, Cambridge, MA, Belknap Press of Harvard University Press.

Dietz, R. S. (1961). Continent and ocean basin evolution by spreading of the sea floor. *Nature (London)*, 190(4779), 854–857.

Dingus, L., & Rowe, T. (1998). The mistaken extinction: Dinosaur evolution and the origin of Birds (1st edition). New York: W. H. Freeman.

diSessa, A. (1988). Knowledge in pieces. In G. Forman, & P. Putall (Eds.), Constructivism in the computer age (pp. 49–70). Hillsdale, NJ: Lawrence Erlbaum Associates, Inc.

diSessa, A. (1993). Toward an epistemology of physics. Cognition and Instruction, 10, 105–225.

Dodick, J., Argamon, S., & Chase, P. (2009). Understanding scientific methodology in the historical and experimental sciences via language analysis. *Science and Education*, 18, 985–1004.

Dodick, J.T., & Argamon, S. (2006). Rediscovering the historical methodology of the earth sciences by understanding scientific communication styles. In C.A. Manduca, & D.W. Mogk, D.W. (Eds.), *Earth and Mind: How Geologists Think and Learn about the Earth* (pp. 105–120). Boulder CO: Geological Society of America, Special Paper 413.

Dodick, J. T., & Orion, N. (2003). Geology as an Historical Science: Its Perception within Science and the Education System. *Science and Education*, 12 (2), 197–211.

Dolphin, G. (2009). Evolution of the theory of the earth: A contextualized approach for teaching the history of the theory of plate tectonics to ninth grade students. *Science & Education*, 18 (3–4), 425.

Dominion Post (2011). Climate change and extreme weather link cannot be Ignored. http://www.stuff.co.nz/dominion-post/news/5285118/Climate-change-and-extreme-weather-link-cannot-be-ignored. Accessed 21 July 2011.

Doran, P.T., & Zimmerman, M.K. Examining the scientific consensus on climate change. *Eos, Transactions, American Geophysical Union*, 90 (3), 22–23.

Dupigny-Giroux, L-A.L. (2010) Exploring the challenges of climate science literacy: Lesson from students, teachers and lifelong learners. *Geography Compass*, 4 (9), 1203–1217.

Duschl, R. (1987). Causes of earthquakes: Inquiry into the plausibility of competing explanations. *Science Activities*, 24 (3), 8–14.

Du Toit, A. L. (1937). *Our wandering continents; an hypothesis of continental drifting*. Edinburgh, London,: Oliver and Boyd.

Earth Science Literacy Initiative. (2010). Earth Science Literacy Principles. Retrieved 01 July 2011 from http://www.earthscienceliteracy.org/es_literacy_6may10_.pdf.

Engelhardt, W., & Zimmermann, J. (1988). *Theory of Earth Science*. Cambridge: Cambridge University Press.

Felitsyn, S.B., & Vaganov, P.A. (1988). Iridium in the ash of Kamchatka volcanoes. *International Geology Review*, 30 (12), 1288–1291.

Frankel, H. (1987). The continental drift debate. In H. Tristram Engelhardt Jr., & A.L. Caplan (Eds.) *Scientific Controversies: Case Studies in the Resolution and Closure of Disputes in Science and Technology*, Cambridge University Press, Cambridge, pp. 203–248.

Frodeman, R. (1995) Geological reasoning: geology as an interpretive and historical science. *Geological Society of America Bulletin*, 107, 960–968.

Galbraith, E. D., Kwon, E. Y., Gnanadesikan, A., Rodgers, K. B., Griffies, S. M., Bianchi, D., et al. (2011). Climate Variability and Radiocarbon in the CM2Mc Earth System Model. *Journal of Climate*, 24(16), 4230–4254.

Gauchat, G. (2012). Politicization of science in the public sphere : A study of public trust in the United States, 1974–2010. *American Sociological Review*, 77(2), 167–187.

Gautier, C., Deutsch, K. & Reibich, S. (2006). Misconceptions about the greenhouse effect. *Journal of Geoscience Education*, 54 (3), pp. 386–395.

Gilbert, G.K. (1896). The origin of hypotheses, illustrated by the discussion of a topographic problem. *Science*, 3, 1–13.

Gilbert, G.K. (1886). The inculcation of scientific method by example. *American Journal of Science*, 31, 284–299.

Gingerich, O. (2011). The great Martian catastrophe and how Kepler fixed it. *Physics Today*, 64 (9), 50–54.

18 Teaching Controversies in Earth Science: The Role of History and Philosophy... 593

Glantz, M. H. (1996). *Currents of change : El Niño's impact on climate and society.* Cambridge; New York: Cambridge University Press.

Glen, W. (2005). The origins and early trajectory of the mantle plume quasi paradigm. In G. Foulger, J. Natland, D. Presnall, & D. Anderson (Eds.), *Plates, plumes, and paradigms. Geological Society of America Special Paper 388* (pp. 91–118). Boulder, Colo.: Geological Society of America.

Glen, W. (2002). A triptych to serendip: Prematurity and resistance to discovery in the earth sciences. In E. Hook (Ed.), *Prematurity in scientific discovery: on resistance and neglect* (pp. 92–108). Berkley, CA: Univ of California Pr.

Glen, W. (1994a). What the impact/volcanism/mass extinction debates are about. In W. Glenn (Ed.), *The mass extinction debates: How science works in a crisis* (pp. 7–38). Stanford, CA: Stanford University Press.

Glen, W. (1994b). How science works in the mass-extinction debates. In W. Glenn (Ed.), *The mass extinction debates: How science works in a crisis* (pp. 39–91). Stanford, CA: Stanford University Press.

Glen, W. (1994c). On the mass-extinction debates: An interview with Stephen J. Gould (conducted and compiled by William Glenn). In W. Glenn (Ed.), *The mass extinction debates: How science works in a crisis* (pp. 253–267). Stanford, CA: Stanford University Press.

Glen, W. (1982). *The road to Jaramillo : critical years of the revolution in earth science.* Stanford, Calif.: Stanford University Press.

Goodman, N. (1967). Uniformity and simplicity. In C.C. Albritton, Jr., M.K. Hubbert, L.G. Wilson, N.D. Newell, & N. Goodman (Eds.), *Uniformity and simplicity: A symposium on the principle of the uniformity of nature. Geological Society of America Special Paper 89* (pp. 93–99). New York: Geological Society of America.

Gould, S.J. (1989). *Wonderful life.* London: Hutchison Radius.

Gould, S.J. (1987). *Time's arrow, time's cycle: Myth and metaphor in the discovery of geological time.* Cambridge, MA: Harvard University Press.

Gould, S.J. (1986). Evolution and the triumph of homology, or why history matters. *American Scientist,* 74 (January–February), 60–69.

Gould, S. J. (1984). Toward the vindication of punctuational change. In W.A. Berggren, & J.A. Van Couvering, J. A. (Eds.), *Catastrophes and Earth History-The New Uniformitarianism* (pp. 9–34), Princeton University Press, New Jersey.

Gould, S. J. (1965), "Is Uniformitarianism Necessary? *American Journal of Science,* 263, 223–228.

Greene, M.T. (1985). History of Geology. *Osiris,* 1, 97–116.

Hallam, A. (1973). *A revolution in the earth sciences: From continental drift to plate tectonics.* London: Oxford University Press.

Hartung, J.B., & Anderson, R.R. (1988). A compilation of information and data on the Manson impact structure. *Lunar & Planetary Institute Technical Report 88–08.*

Heezen, B. C., Tharp, M., & Ewing, W. M. (1959). *The North Atlantic; text to accompany the Physiographic diagram of the North Atlantic, [Part] 1 of The floors of the oceans. Special Paper - Geological Society of America,* 122–122.

Herreid, C. F. (2007). *Start with a Story: The Case Study Method of Teaching College Science.* Arlington, VA: NSTA Press.

Hess, H. (1962). History of ocean basins. In A. Engle, H. James, & B. Leonard (Eds.), *Petrologic studies: A volume to honor A. F. Buddington* (pp. 599–620): Geological Society of America.

Hildebrand, A.R., & Penfield, G.T. (1990). A buried 180-km-diameter probable impact crater on the Yucatan Peninsula, Mexico. *Eos,* 71 (143), 1425.

Hoffman, M., & Barstow, D. (2007). *Revolutionizing earth system science education for the 21st century: Report and recommendations from a 50- state analysis of earth science education.* Cambridge, MA: TERC.

Höhler, S. (2003). A sound survey: The technological perception of ocean depth, 1850–1930. In M. Hård, A. Lösch, & D. Verdicchio (Eds.), *Transforming Spaces: The Topological Turn in Technology Studies* (pp. 1–17). http://www.ifs.tu-darmstadt.de/gradkoll/Publikationen/transformingspaces.html.

Holmes, A. (1928). Radioactivity and earth movements. *Transactions of the Geological Society of Glasgow*, 18, 559–606.

Hooykaas, R. (1959). *Natural law and divine miracle: A historical-critical study of the principle of uniformity in geology, biology and theology*. Leiden: E. J. Brill.

IPCC (2007). *Summary for policymakers. Climate Change 2007: The Physical Science Basis. Contribution of Working Group I to the Fourth Assessment Report of the Intergovernmental Panel on Climate Change (IPCC)*. Cambridge, UK: University of Cambridge Press.

James, W. (1911). On some mental effects of the earthquake. In H. James (Ed.), *Memories and studies* (pp. 207–226). London; New York: Longmans, Green.

Jeffries, H., Stanistreet, M., & Boyes, E. (2001) Knowledge about the "greenhouse effect": Have college students improved? *Research in Science & Technological Education*, 19 (2), 205–221.

Johnson, D. (1933). Role of analysis in scientific investigation. *Geological Society of America Bulletin*, 44, 461–493.

Johnson, J.G. (1990). Method of multiple working hypotheses: A chimera. *Geology*, 18, 44–45.

Jordan, P. (1971). *The expanding earth; some consequences of Dirac's gravitation hypothesis ([1st English ed.)*. Oxford, New York,: Pergamon Press.

Justi, R. (2000). History and philosophy of science through models: Some challenges in the case of the atom. *International Journal of Science Education*, 22 (9), 993–1009.

Karl, T.R., & Trenberth, K.E. (2003). Modern global climate change. *Science*, 302, 1719–1723.

Keller, G. (1989). Extended Cretaceous/Tertiary boundary extinctions and delayed populations changes in planktonic foraminifera from Brazos River, Texas. *Paleoceanography*, 4 (3), 287–332.

Khan, S. (2008). Co-construction and model evolution in chemistry. In J. Clement, & M. A. Rae-Ramirez (Eds.), *Model based learning and instruction in science* (Vol. 2, pp. 59–78). Dordrecht: Springer.

Khider, D., Stott, L. D., Emile-Geay, J., Thunell, R., & Hammond, D. E. (2011). Assessing El Niño Southern Oscillation variability during the past millennium. *Paleoceanography, 26*(32), n/a.

Kitts, D. B. (1978). Retrodiction in Geology. *Proceedings of the Biennial Meeting of the Philosophy of Science Association*: *Symposia and Invited Papers*, 2, 215–226.

Kleinhans, M.G., Buskes, C.J.J., & de Regt, H.W. (2010). Philosophy of Earth Science. In Fritz Allhoff (Ed.), *Philosophy of the sciences: A guide* (pp. 213–236). New York: Blackwell.

Kleinhans, M.G., Buskes, C.J.J., & de Regt, H.W. (2005). Terra Incognita: Explanation and Reduction in Earth Science. *International Studies in the Philosophy of Science*, 19 (3), 289–317.

Koeberl, C. (1989). Iridium enrichment in volcanic dust from blue icefields, Antarctica, and possible relevance to the K/T boundary event. *Earth & Planetary Science Letters*, 92 (3–4), 317–322.

Kring, D.A., & Boyton, W.V. (1992). Petrogenesis of an augite-bearing melt rock in the Chicxulub structure and its relationship to K/T spherules in Haiti. *Nature*, 358, 141–143.

Kuhn, T.S. (1957). *The Copernican revolution: Planetary astronomy in the development of western thought*. Cambridge, M.A.: Harvard U. Press.

Laudan, R. (1987). *From mineralogy to geology: The foundations of a science, 1650–1830*. Chicago: The University of Chicago Press.

Lawrence, D. M. (2002). *Upheaval from the abyss : ocean floor mapping and the Earth science revolution*. New Brunswick, N.J.: Rutgers University Press.

Leaf, J. (2011). Charles Keeling & measuring Atmospheric CO_2. SHiPS Resource Center, Minneapolis, MN. http://www1.umn.edu/ships/modules/earth/keeling/Keeling.pdf.

LeGrand, H. E. (1988). *Drifting continents and shifting theories: The modern revolution in geology and scientific change*. Cambridge; New York: Cambridge University Press.

Lewis, E., & Baker, D. (2010). A call for a new geoscience education research agenda. *Journal of Research in Science Teaching*, 47(2), 121–129.

London, J. (1906). Story of an eyewitness: The San Francisco earthquake. *Collier's Weekly*, http://london.sonoma.edu/Writings/Journalism/.

Lyell, K.M. (1881). *Life, letters and journals of Sir Charles Lyell*. London: Bart, John Murray.

Malaise, R. E. (1972). *Land-bridges or continental drift*. S-181 42 Lidingö.

Marton, F., Runesson, U., & Tsui, A.B. (2004). The space of learning. In F. Marton, & A.B. Tsui, *Classroom learning and the space of learning* (pp. 3–40). Mahwah, NJ: Lawrence Erlbaum and Associates.

Matthews, M.R. (1994/2014). *Science teaching: The role of history and philosophy of science*. New York, NY: Routledge.

Matthews, M.R. (2012). Changing the focus: From nature of science (NOS) to features of science (FOS). In M. S. Khine (Ed.), *Advances in nature of science research: Concepts and methodologies* (pp. 3–26). Dordrecht: Springer.

Mayr, E. (1997). *This is biology: The science of the living world*. Cambridge, MA: Belknap Press.

Mayr, E. (1985) How biology differs from the physical sciences. In. D.J. Depew, & B.H. Weber (Eds.), *Evolution at the crossroads: the new biology and the new philosophy of science* (pp 43–46). Cambridge: MIT Press.

MacLeod, N., Rawson, P.F., Forey, P.L., Banner, F.T., Boudagher-Fadel, M.K., Bown, P.R., Burnett, J.A., Chambers, P., Culver, S., Evans, S.E., Jeffery, C., Kaminski, M.A., Lord, A.R., Milner, A.C., Milner, A.R., Morris, N., Owen, E., Rosen, B.R., Smith A.B., Taylor, P.D., Urquhart, E., Young, J.R. (1997). The Cretaceous-Tertiary biotic transition. *Journal of the Geological Society,* 154 (2), 265–292.

Mason, R. G., & Raff, A. D. (1961). Magnetic survey off the West Coast of North America, 32 degree N. Latitude to 42 degree N. Latitude. *Bulletin of the Geological Society of America,* 72(8), 1259–1265.

Maxlow, J. (2006). Dr. James Maxlow: Geologist and proponent of expansion tectonics. http://www.jamesmaxlow.com/main/index.php?&MMN_position=1:1. Retrieved 20 April, 2012

Montanari, A., Hay, R.L., Alvarez, W., Asaro, F., Michel, H.V., Alvarez, L.W., & Smit, J. (1983). Spheroids at the Cretaceous-Tertiary boundary are altered impact droplets of basaltic composition. *Geology* 11 (11), 668–671.

Montgomery, K. (2009). Using a historical controversy to teach critical thinking, the meaning of "theory", and the status of scientific knowledge. *Journal of Geoscience Education,* 57 (3), 214.

Morgan, W. J. (1972). Deep mantle convection plumes and plate motions. *Bulletin of the American Association of Petroleum Geologists*, 56, 203–213.

Morley, L. (2001). The zebra pattern. In N. Oreskes (Ed.), *Plate tectonics: An insider's history of the modern theory of the earth* (pp. 67–85). Cambridge, MA: Westview Press.

Mount, J.F., Margolis, S.V., Showers, W., Ward, P., & Doehne, E. (1986). Carbon and oxygen isotope stratigraphy of the Upper Maastrichtian, Zumaya, Spain: A record of oceanographic and biological changes at the end of the Cretaceous Period. *Palaios*, 1, 87–92.

National Public Radio News (2011). Battle heats up over Alaskan petroleum reserve. http://www.npr.org/2011/01/03/132490327/battle-heats-up-over-alaskan-petroleum-reserve. July 21 August 2011.

National Research Council (NRC). (2011). *America's climate choices*. Washington, D.C.: National Academies Press.

National Research Council. (2012). A Framework for K-12 Science Education: Practices, Crosscutting Concepts, and Core Ideas. Washington: National Academies Press. http://www.nap.edu/catalog.php?record_id=13165National Research Council (NRC). (2010). *Advancing the science of climate change*. Washington, D.C.: The National Academies Press.

National Research Council. (1996). *National science education standards*. Washington, DC: National Academy Press.

Nersessian, N. J. (2008). Creating scientific concepts. Cambridge, Mass.: MIT Press.

New York Times (2011). Jury Is Out' on Implementation of Landmark Great Lakes Compact. http://www.nytimes.com/gwire/2011/07/14/14greenwire-jury-is-out-on-implementation-of-landmark-grea-33525.html?pagewanted=all. Accessed 21 July 2011.

Nippert, J. B., Hooten, M. B., Sandquist, D. R., & Ward, J. K. (2010). A Bayesian model for predicting local El Niño events using tree ring widths and cellulose ^{18}O. *Journal of Geophysical Research. Biogeosciences,* 115(1).

Núñez-Oviedo, M. C., & Clement, J. (2008). A competition strategy and other modes for developing mental models in large group discussion. In J. Clement, & M. A. Rae-Ramirez (Eds.), *Model based learning and instruction in science* (Vol. 2, pp. 117–138). Dordrecht: Springer.

Núñez-Oviedo, M. C., Clement, J., & Rae-Ramirez, M. A. (2008). Complex mental models in biology through model evolution. In J. Clement, & M. A. Rae-Ramirez (Eds.), *Model based learning and instruction in science* (Vol. 2, pp. 173–194). Dordrecht: Springer.

Officer, C.B, & Drake, C.L. (1989). Cretaceous/Tertiary extinctions: We know the answer but what is the question? *Eos*, 70 (25), 659–60.

Officer, C.B., Hallam, A., Drake, C.L., & Devine, J.C. (1987). Late Cretaceous and paroxysmal Cretaceous/Tertiary extinctions. *Nature*, 326, 143–149.

Oldroyd, D.R. (1996). *Thinking about the earth: a history of ideas in geology*. London: The Athlone Press.

Oliver, J. (2001). Earthquake seismology in the plate tectonics revolution. In N. Oreskes (Ed.), *Plate tectonics: An insider's history of the modern theory of the earth* (pp. 155–166). Cambridge, MA: Westview.

Oreskes, N. (1999). *The rejection of continental drift : theory and method in American earth science*. New York: Oxford University Press.

Oreskes, N. (2001). *Plate tectonics: An insider's history of the modern theory of the earth*. Boulder, CO: Westview Press.

Oreskes, N., & LeGrand, H. E. (2001). *Plate tectonic: an insider's history of the modern theory of the Earth*. Boulder, Colo: Westview Press.

Oreskes, N. (2004). The scientific consensus on climate change. *Science*, 306, 1686.

Oreskes, N., & Conway, E.M. (2010). *Merchants of doubt: How a handful of scientists obscured the truth on issues from tobacco smoke to global warming*. New York: Bloomsbury Press.

Orion, N., & Ault, C. (2007). Learning earth sciences. In S.K. Abell, & N.G. Lederman (Eds.) *Handbook of Research in Science Education* (pp. 653–687). Mahwah, NJ: Lawrence Erlbaum.

Paixão, I., Calado, S., Ferreira, S., Salves, V., & Smorais, A. M. (2004). Continental Drift: A Discussion Strategy for Secondary School. *Science & Education, 13*(3), 201–221.

Pantin, C.F.A. (1968). *The relations among the sciences*. New York: Cambridge University Press.

Penfield, G. T. & Camargo, Z. A. (1981). Definition of a major igneous zone in the central Yucatan platform with aeromagnetics and gravity. In: Technical program, abstracts and biographies (Society of Exploration Geophysicists 51st annual international meeting) Los Angeles, Society of Exploration Geophysicists, p. 37.

Penuel, W., McWilliams, H., McAuliffe, C., Benbow, A., Mably, C., & Hayden, M. (2009). Teaching for understanding in earth science: Comparing impacts on planning and instruction in three professional development designs for middle school science teachers. *Journal of Science Teacher Education*, 20, 415–436.

Philander, S. G. (2004). *Our affair with El Niño: How we transformed an enchanting Peruvian current into a global climate hazard*. Princeton, N.J.: Princeton University Press.

Platt, J.R (1964). Strong Inference. *Science*, 146, 347–353.

Posner, G. J., Strike, K. A., Hewson, P. W., & Gertzog, W. A. (1982). Accommodation of a scientific conception: Toward a theory of conceptual change. *Science Education*, 66, 211–227.

Pound, K. (2007). Use of the 'hollow earth theory' to teach students how to evaluate theories. *Geological Society of America Abstracts with Programs, 39* (3), 17.

Powell, J. (2011). Is there a case against human caused global warming in the peer-reviewed literature? http://www.skepticalscience.com/Powell-project.html, accessed 15 November 2012.

Raff, A. D., & Mason, R. G. (1961). Magnetic survey off the west coast of North America, 40 degree N. latitude to 52 degree N. latitude. *Bulletin of the Geological Society of America, 72* (8), 1267–1270.

Rankey, E., & Ruzek, M. (2006). Symphony of the spheres: Perspectives on earth system science education. *Journal of Geoscience Education, 54* (3), 197.

Ratter, B.M.W., Philipp, K.H.I., & von Storch, H. (2012). Environmental Science and Policy, 18, 3–8.

Reid, H. F. (1910). *The California earthquake of April 18, 1906: The mechanics of the earthquake*. Washington, D.C.: Carnegie Institution of Washington.

Repcheck, J. (2003). *The man who found time: James Hutton and the discovery of the earth's antiquity*. Cambridge, MA: Perseus.

Romans, B. W. (2008). *Controls on distribution, timing, and evolution of turbidite systems in tectonically active settings: The Cretaceous Tres Pasos Formation, southern Chile, and the Holocene Santa Monica Basin, California*. Unpublished 3302863, Stanford University, United States – California.

Rudolph, J. L. (2000). Reconsidering the 'Nature of Science' as a Curriculum Component. *Journal of Curriculum Studies*, 32(3), 403–419.

Rudolph J.L., & Stewart, J. (1998) Evolution and the nature of science: On the historical discord and its implication for education. *Journal of Research in Science Teaching*, 35, 1069–1089.

Rudwick, M. J. S. (1998). Lyell and the Principles of Geology. In D.J. Blundell, & A.C. Scott (Eds.), *Lyell: The past is the key to the present, Geological Society of London Special Publications 143*, 3–15.

Rudwick, M. J. S. (1985). *The great Devonian controversy : the shaping of scientific knowledge among gentlemanly specialists*. Chicago: University of Chicago Press.

Rudwick, M. J. S. (1971). Uniformity and progression: Reflections on the structure of geological theory in the age of Lyell. In D.H.D. Roller (Ed.), *Perspectives in the history of science and technology* (pp. 209–237), Oklahoma, University of Oklahoma Press, Oklahoma.

Ryan, M.J., Russell, A.P., Eberth, D.A., & Currie, P.J. (2001). The taphonomy of a Centrosaurus (Ornithischia: Ceratopsidae) bone bed from the Dinosaur Park formation (Upper Campanian), Alberta, Canada, with comments on cranial ontogeny. *Palaios*, 16, 482–506.

Sawyer, D. (2010). Discovering plate boundaries: A classroom exercise designed to allow students to discover the properties of tectonic plates and their boundaries. http://plateboundary.rice.edu/. Retrieved 17 December, 2011

Schuchert, C. (1928). The hypothesis of continental displacement. In W. A. J. M. Van Waterschoot vander Gracht (Ed.), The theory of continental drift: A symposium (pp. 105–144). Chicago, IL: American Association of Petroleum Geologists/Chicago University Press.

Schuchert, C. (1932). Gondwana land bridges. *Geological Society of America Bulletin,* 43 (4), 875–915.

Schulte, P., Algret, L., Ignacio, A., Arz, J.A., Baarton, P., Brown, P.R., et al. (2010). The Chicxulub Asteroid Impact and Mass Extinction at the Cretaceous-Paleogene Boundary. *Science* 327, 1214–1218.

Schumm, S. (1991). *To interpret the earth: Ten ways to be wrong*. Cambridge, UK: Cambridge University Press.

Schwab, J. J. (2000). Enquiry, the science teacher, and the educator. *The Science Teacher*, 67 (1), 26.

Schwab, J. J. (1966). *The Teaching of Science as Enquiry*. Cambridge, MA: Harvard University Press.

Schwab, J. J. (1964). Structure of the Disciplines: Meanings and Significances. In G.W. Ford and L. Pungo (Eds.), *The Structure of Knowledge and the Curriculum*. Chicago, IL: Rand McNally, p. 6–30.

Schwab, J. J. (1963). *Biology Teacher's Handbook*. New York: John Wiley and Sons Inc.

Schwab, J. J. (1962). The teaching of science as enquiry. In J.J. Schwab and P. Brabdwein (Eds.), *The Teaching of Science*. Cambridge, MA: Harvard University Press.

Schwab, J. J. (1958). The teaching of science as inquiry. *Bulletin of the Atomic Scientists*, 14, 374–379.

Scruggs, L., & Bengali, S. (2012). Declining public concern about climate change: Can we blame the great recession? In Press in *Global Environmental Change*.

Seethaler, S. (2005). Helping students make links through science controversy. *The American Biology Teacher,* 67 (5), 265–274.

Şengör, A. M. C. (2003). *The large-wavelength deformations of the lithosphere : materials for a history of the evolution of thought from the earliest times to plate tectonics*. Boulder, CO: Geological Society of America.

Şengör, A. M. C. (2001). *Is the present the key to the past or is the past the key to the present?: James Hutton and Adam Smith versus Abraham Gottlob Werner and Karl Marx in interpreting history*. Boulder, CO: Geological Society of America.

Sharpton, V.L., Dalrymple, G.B., Marin, L.E., Ryder, G., Scuraytz, B.C., & Urrutia-Fucugauchi, J. (1992). New links between the Chicxulub impact structure and the Cretaceous-Tertiary boundary. *Nature*, 359, 819–821.

Shepardson, D.P., Niyogi, D., Choi, S., & Charusombat, U. (2011). Student conceptions about global warming and climate Change. *Climatic Change*, 104, 481–507.

Sherwood, S. (2011). Science controversies past and present. *Physics Today*, 64 (10), 39–44.

Smit, J. (1999). The Global stratigraphy of the Cretaceous-Tertiary boundary impact ejecta. *Annual Review of Earth and Planetary Sciences*, 27, 75–113.

Smith, J.P. diSessa, A.A., & Roschelle, J. (1993). Misconceptions reconceived: A constructivist analysis of knowledge in transition. *Journal of Learning Sciences*, 3, 115–157.

Smol, J.P. (2012). A planet in flux: How is life on Earth reacting to climate change? *Nature*, 482, S12-S15.

Swisher, C.C., Grajales-Nishimura, J.M., Montanari, A., Margolis, S.V., Claeys, P., Alvarez, W., et al. (1992). Coeval 40Ar/39Ar ages of 65.0 million years ago from Chicxulub crater melt rock and Cretaceous-Tertiary boundary tektites. *Science*, 257, 954–958.

Sykes, L. R. (1967). Mechanism of earthquakes and nature of faulting on the mid-oceanic ridges. *Journal of Geophysical Research, 72*(8), 2131–2153.

Theisen, K.M. (2011). What do U.S. students know about climate change? *Eos, Transactions, American Geophysical Union*, 92 (51), 477–478.

The Guardian (2011). Greenpeace report links western firms to Chinese river polluters. http://www.theguardian.com/environment/2011/jul/13/greenpeace-links-western-firms-to-chinese-polluters. Accessed 21 July 2011.

Veerman, A. (2003). Constructive discussions through electronic dialogue. In J. Andriessen, M. Baker, & D. Suthers (Eds.), *Arguing to learn: Confronting cognitions in computer-supported collaborative learning environments* (pp. 117–143). Dordrecht, The Netherlands: Kluwer.

Venturini, T. (2010). Diving in magma: How to explore controversies with actor-network theory. *Public Understanding of Science, 19* (3), 258–273.

Vine, F. J., & Matthews, D. H. (1963). Magnetic anomalies over oceanic ridges. *Nature (London), 199*(4897), 947–949.

Watson, R. A. (1969) Explanation and prediction in geology. *Journal of Geology*, 77, 488–494.

Wegener, A., & Skerl, J. G. A. (1924). *The origin of continents and oceans*. London,: Methuen & co.

Wiggins, G., & McTighe, J. (2005) *Understanding by Design,* 2nd edition. Alexandria, VA: Association for Supervision and Curriculum Development.

Wilson, J. T. (1965). A new class of faults and their bearing on continental drift. *Nature (London), 207*(4995), 343–347.

Wilson, J. T. (1963). A possible origin of the Hawaiian Islands. *Canadian Journal of Physics*, 41, 863–870.

Wilson, K. (2008). Expanding Earth knowledge. http://eearthk.com/index.html. Retrieved 20 April, 2012.

Glenn Dolphin is the Tamaratt Teaching Professor in Geoscience at the University of Calgary. He has taught earth science in the public schools for 13 years. During that time he developed curricula with a heavy emphasis on both metacognitive use of models and infusion of the history and philosophy of geosciences. Of particular interest he has used the historical development of the theory of plate tectonics through historical case studies and historically contextualized inquiry activities to teach about earth dynamics and the nature of science. He also investigated how students identify as and perform "the good student" and how that may inhibit their development of useful understanding of science concepts. As a research associate with the NSF funded IMPPACT project, he is investigating secondary science teacher development from incipient stages of the teacher education program through the induction phase of teaching with an emphasis on how preservice NOS instruction as well as other external factors may influence personal epistemology, beliefs about teaching, and teacher practice. His dissertation research focused on a geology instructor's use of metaphor as a proxy for understanding his instructional decision making. In 2009, he authored "Evolution of the Theory of the Earth: A Contextualized Approach for Teaching the History of the Theory of Plate Tectonics to Ninth Grade Students" in *Science & Education* Volume 18, Numbers 3–4, 425–441.

Jeff Dodick Ph.D. is a lecturer in the Department of Science Teaching at The Hebrew University of Jerusalem. His training is in biology and earth science (B.Sc. from the University of Toronto), with specializations in paleontology and evolutionary biology (M.Sc., University of Toronto). His doctorate is in science education from the Weizmann Institute of Science in Rehovot, Israel. He did postdoctorate research at McGill University in Montreal, QC, Canada, and also served as a research assistant professor in the Faculty of Education and Social Policy at Northwestern University, in Evanston Illinois, USA. At the Hebrew University of Jerusalem, Dr. Dodick teaches courses for graduate students in science education, including basic learning theory and the history of biology and its connection to science education. Broadly stated, his research interests focus on the reasoning processes that are used in sciences which investigate phenomena constrained by long spans of time, such as evolutionary biology and geology; such scientific disciplines fall under the general rubric of "historical-based sciences." In the past he has published research papers dealing with how students of different ages understand geological time; this research interest has also extended toward testing how visitors to the Grand Canyon understand a new geoscience exhibit that explains "deep time." More recently, he has been comparing how scientists, graduate students, and high school students understand the epistemologically derived differences between experimental and historical-based sciences.

Part VI
Pedagogical Studies: Astronomy

Chapter 19
Perspectives of History and Philosophy on Teaching Astronomy

Horacio Tignanelli and Yann Benétreau-Dupin

The didactics of astronomy is a relatively young field with respect to that of other sciences. In particular, the didactic of astrophysics is barely sketched. Historical issues have most often been part of the teaching of astronomy, although that often does not stem from a specific didactics. Many astronomy textbooks address the historical development of this science, at least anecdotally. Beyond listing historical discoveries and the name and dates of astronomers, textbooks often recount a few specific episodes of the history of astronomy. On the other hand, texts that are essentially historical are devoted to biographical data and/or the relevant cultural and intellectual context of the life and work of astronomers. Their main goal is not to convey scientific knowledge. They assume that the reader who studies the history of a particular science is already aware of the basic ideas of that science. Such studies generally do not aim at assessing the relevance of history for education.[1] Textbooks for the specific teaching of astronomy are typically not structured around its history, let alone the philosophical issues to which its historical development gave rise – at least at the primary of secondary level. Many educational systems assume that the teachers will articulate the connection between the field and its history. The flow of articles on astronomy education in professional journals is minimal when compared to other sciences. The teaching of astronomy is often subsumed under that of physics. One can easily consider that, from an educational standpoint, astronomy requires the

---

[1] For instance, *Journal for the History of Astronomy* (the only journal devoted to the subject) has not published works on education (except a few on the history of astronomy education) in its first 40 years of activity.

H. Tignanelli
National Educational Management Direction, National Ministry
of Education, Buenos Aires, Argentina

Y. Benétreau-Dupin (✉)
Department of Philosophy, Western University, and Rotman
Institute of Philosophy, London, ON, Canada
e-mail: ybenetre@uwo.ca

M.R. Matthews (ed.), *International Handbook of Research in History,*
Philosophy and Science Teaching, DOI 10.1007/978-94-007-7654-8_19,
© Springer Science+Business Media Dordrecht 2014

same mathematical or physical strategies as physics. This approach may be adequate in many cases but cannot stand as a general principle for the teaching of astronomy.

This chapter offers in a first part a brief overview of the status of astronomy education research and of the use of the history and philosophy of science (HPS) in astronomy education in particular. In a second part, it attempts to illustrate some possible ways to structure the teaching of astronomy around its historical development (ancient and contemporary) so as to pursue a quality education and contextualized learning that contributes to approaching science in its cultural environment. We chose to give priority to those areas of the history of astronomy that clearly illustrate significant conceptual and methodological processes and breaks that are relevant not only to this field but also to other disciplines or to science in general.

19.1 Astronomy Education Research, Reform-Based Teaching, and the Role of HPS

19.1.1 Astronomy Education Research and Reform-Based Teaching

Astronomy is a very popular subject. College-level courses are well attended (Deming and Hufnagel 2001): roughly 10 % of all US college student take an introductory astronomy course while in college (Partridge and Greenstein 2003). After decades of space missions and technological progress, the latest space telescopes and planetary missions continue to capture the public's attention. The historical debates around the evolution of the understanding of the structure of the Solar System are among the most discussed to illustrate key aspects of the nature of science (McComas 2008), and the teaching of astronomy is often done in part through an introduction to its historical development. And yet, there is no extensive study, textbook, or other scholarly manual that covers a wide range of cases and practices – and even very little discussion among astronomy educators – on the role of HPS for the teaching of astronomy, whether it be at the primary, secondary, or college level. Research in astronomy education is itself at an early stage, and in the past decades, major guidelines in science education[2] have been written without the substantial participation of astronomers. Research on the teaching of astronomy is often subsumed under that of physics or mathematics. Consequently, the specific treatment of the teaching of astronomy is relatively scarce.

Yet extant research shows that misconceptions about basic astronomical facts are particularly prevalent among students of various age groups[3] and sometimes among

---

[2] Such as *Project 2061* (AAAS 1990, 1994, 2001).

[3] See e.g., on basic notions in astronomy, Taylor et al. (2003), Baxter (1989), Nussbaum (1979), Taylor, or Vosniadou and Brewer (1992) at the primary level, and Diakidoy and Kendeou (2001) and Kikas (1998) at the primary and secondary levels. On more general notions, see Sadler et al. (2010) at the primary and secondary levels, Trumper (2001a, b) at the high school level, and Comins (2000), LoPresto and Murrell (2011), Trumper (2000), Zeilik and Morris (2003), Zeilik et al. (1997, 1998) at the college level.

teachers if they are not well-enough trained in astronomy[4] or in identifying their students' misconceptions (Sadler et al. 2010). There is still relatively little research and resource on pre- and in-service astronomy teachers compared to what is done in physics education.[5] Yet, over the last two decades, and particularly after the work of the Center for Astronomy Education Research[6] has resulted in several doctoral dissertations (Slater 2008), research has shown that evidence supports reform-based teaching of astronomy, with an emphasis on inquiry-based, interactive engagement directed toward conceptual understanding and model-building, whether it be at the at the primary level (Osborne 1991; Taylor et al. 2003), secondary level (Richwine 2007; Hake 2002, 2007), or college level.[7]

However, at the pre-college level, astronomy is often absent in science curriculum and has been so for much of the twentieth century (Jarman and McAleese 1996; Trumper 2006). The teaching of astronomy faces specific challenges:

1. A traditional classroom setting is seen as offering limited opportunities for hands-on, introductory activities in astronomy.[8]
2. Little pedagogical resource exists on the conceptual history of astronomy (particularly at the pre-college level).
3. Primary- and secondary-level educators seldom have sufficient training in astronomy since they do not have to teach astronomy as a separate discipline, but only as a minor part of the physics, Earth science, or geography curriculum.

Nevertheless, astronomy education as a field of research is rapidly growing and has particularly been growing for the past 20 years. It can build up on two decades of work (see Bailey and Slater 2003 and Lelliott and Rollnick 2010 for a review of the existing research literature).

Building on the Force Concept Inventory for Newtonian Mechanics (Hestenes et al. 1992), concept inventories for astronomy education have been developed, particularly for college-level introductory astronomy courses for non-major students (Astro 101 course, see Partridge and Greenstein 2003; Slater and Adams 2002):

---

[4]For studies on preservice elementary teachers, see, e.g., Frède (2006, 2008), Trumper (2003), Bayraktar (2009), Schoon (1995), and Trundle et al. (2002) on the understanding of Moon phases; Atwood and Atwood (1996) on the seasons; Bulunuz and Jarrett (2009) for both phenomena; and Atwood and Atwood (1995) on the night and day cycles. For studies on in-service elementary or secondary teachers, see, e.g., Parker and Heywood (1998) or Sadler et al. (2010).

[5]On this topic, see in particular Slater (1993). Work done by the Conceptual Astronomy and Physics Education Research (CAPER) group, based at the University of Wyoming, aims at developing strategies, materials, and resources to support astronomy teaching (inquiry-based in particular) and assess the effectiveness of teaching strategies. See http://www.uwyo.edu/caper/.

[6]This research group is based at the University of Arizona. See their Web site http://astronomy101.jpl.nasa.gov/.

[7]See Hake (2002, 2007), Prather et al. (2009a, b), Rudolph et al. (2010), and Waller and Slater (2011).

[8]In an urban environment in particular, light pollution constitutes a major obstacle to an easy introduction to the stars.

For the high school and college levels:

1. The Astronomy Diagnostic Test (ADT) assesses students' background knowledge (Adams et al. 2000).
2. More recently, following the work of the ADT, the Test of Astronomy Standards (TOAST) is an assessment instrument for conceptual diagnostics and content-knowledge surveys. It is aligned to the consensus learning goals stated by the American Astronomical Society – Chair's Conference on ASTRO 101, the American Association of the Advancement of Science's Project 2061 Benchmarks (AAAS 1994), and the National Research Council's National Science Education Standards (NRC 1996) (see Slater and Slater 2008).
3. The Light and Spectroscopy Concept Inventory touches on the properties of light, the Stefan-Boltzmann law, Wien's law, the Doppler shift, and spectroscopy (see Bardar et al. 2007 and Schlingman et al. 2012 for an assessment).
4. The *Astronomical Misconceptions Survey* (LoPresto and Murrell 2011) provides a 25-question survey aimed at assessing college students' misconceptions, based on the work done by Zeilik and colleagues (see Zeilik et al. 1998; Zeilik and Morris 2003).
5. The Lunar Phases Concept Inventory (Lindell and Olsen 2002) probes high school and college students' understanding of the cause, motion, and period of Lunar phases through 28 questions.
6. Star Properties Concept Inventory (Bailey 2007; Bailey et al. 2012) covers temperature, luminosity, mass, formation, and fusion with 25 questions.

For the K-12 level, *The Astronomy and Space Science Concept Inventory*, a broader 211-question survey has been developed by Sadler et al. (2010), particularly designed to match NRC Standards and AAAS Benchmarks.

Those concept inventories, which are meant to allow one to measure the progress of students' and teachers' understanding or the effectiveness of teaching methods (Bailey 2009),[9] only marginally assess students' knowledge of the historical development of astronomy. Assessments of the effectiveness of a concept-based teaching of astronomy through its historical development are even absent in astronomy education literature reviews (Bailey and Slater 2003; Lelliott and Rollnick 2010, or Pasachoff and Percy 2005) and are yet to be carried out. The case can be made that if such assessments were to be pursued, the already existing concept inventories should be used a standardized benchmarks (Brissenden et al. 2002).

Among these different concept-centered approaches to astronomy teaching, there does not seem to be a clear and explicit consensus as to what role the history of astronomy should play, even though the discipline, if taught separately, is often introduced through reference to its historical development.[10] For instance, at the

---

[9] See Libarkin (2008) and Wallace and Bailey (2010) for a discussion on the relevance of these surveys and the effect of, e.g., sample size in the ability of concept inventories to be used as measuring tools.

[10] A fine (albeit dated) example of a college-level introductory book for liberal arts students that incorporates a large amount of historical details can be found in Payne-Gaposchkin and Haramundanis (1956). A recent college-level introductory textbook in astronomy such as Morison (2008) shows that textbooks for astronomy or physics majors can also largely revolve around the historical development of observation techniques and scientific discoveries.

19 Perspectives of History and Philosophy on Teaching Astronomy

college level, whereas Partridge and Greenstein (2003) consider that an acquaintance with the history of astronomy is one of the main goals of an introductory class in astronomy, a concept-centered introductory textbook such as Zeilik (1993) contains very little historical information.

19.1.2 General Resources in Astronomy Education

The recent article by Waller and Slater (2011) concisely sums up the history of developments in astronomy education.[11] Certain elements of this survey can be emphasized and a few remarks on resources in astronomy education can be added:

1. *Astronomy Education Review* is one of the very few publications dedicated to astronomy education. Historical or philosophical concerns are not at the core of its mission, even though they are sometimes addressed. It is freely accessible, published online by the American Astronomical Society since 2001.[12] According to Lelliott and Rollnick (2010), over the period 1974–2008, "[n]early a quarter of the articles [on astronomy education] were published in the *International Journal of Science Education*, while *Science Education* and the *Journal of Research in Science Teaching* together account for a further quarter." Thanks in part to *Astronomy Education Review*, the flow of astronomy education research articles is rapidly increasing as a research community is growing.
2. The National Science Foundation has recently funded a joint project called Communities for Physics and Astronomy Digital Resources in Education (ComPADRE),[13] which provides reviewed and annotated resources for those teaching introductory courses in both physics and astronomy.
3. The Searchable Annotated Bibliography of Education Research (SABER), an online database in astronomy education supported by the American Astronomical Society, is available at http://astronomy.uwp.edu/saber/.
4. The Commission No. 46 of the International Astronomical Union on the teaching of astronomy[14] constitutes an essential resource and centralizes publications and announcements of the different groups and events that promote communication, education, and development of the history of astronomy. Published on the occasion of the UNESCO-sponsored International Year of Astronomy in 2009, the 2010–2020 strategic plan *Astronomy for the Developing World*[15] gives an overview of the objectives and projects of the IAU for developing the education of astronomy. It lists the history of astronomy as one of the principal areas to pursue, but offers little guidance as to why and how to do so.

---

[11] It is to be noted that historical approaches in astronomy education are not mentioned.

[12] A review of its first years of activity can be found in Wolff and Fraknoi (2005).

[13] http://www.compadre.org

[14] See http://www.iau.org/education/commission46/ A brief review of the educative actions carried out by the IAU can be found in Isobe (2005).

[15] http://iau.org/static/education/strategicplan_091001.pdf

5. Likewise, the Commission No. 41 of the IAU on the history of astronomy is a valuable source. Its Web site[16] offers notes, proceedings of conferences, and event announcements of the various working groups of this Commission (Historical Instruments, Astronomy and World Heritage, etc.).
6. The Commission No. 55 of the IAU, "Communicating Astronomy with the Public," has been publishing its journal for a few years now, starting in 2007. The *CAP journal* aims at supporting formal and informal astronomy education. The presence of articles on the history of astronomy shows its relevance for the popularization and divulgation of astronomy.
7. For the development of teaching of astronomy at the IAU, the colloquia No. 162 "New Trends in Astronomy Teaching" (Gouguenheim et al. 1998) and No. 105 on "The Teaching of Astronomy" (Pasachoff and Percy 1990) were major landmarks (see also Swarup et al. 1987). They were followed by a conference held on July 24–25, 2003, on "Effective Teaching and Learning of Astronomy" as part of the 25th General Assembly of the IAU, which resulted in the publication of a book (Pasachoff and Percy 2005). It presents an evaluation of the state of research in astronomy education, available resources and educational programs, and teaching practices. In it, historically informed, concept-based education is presented as being in a preliminary stage.

A few regional organizations and publications on astronomy education can be emphasized:

1. The European Association for Astronomy Education (EAAE) was constituted in Athens on November 25, 1995. The aims of the EAAE refer to those named by the Declaration of the EU/ESO workshop at the ESO Headquarters in Garching on Teaching of Astronomy in Europe's Secondary Schools in November 1994.[17]
2. The Euro-Asian Association of Astronomy Teachers[18] was constituted in November 11, 1995.
3. The *Bulletins of Teaching of Astronomy in Asian-Pacific Region,* directed by Syuzo Isobe, started in 1990 and continued for at least 12 years and has featured many articles on astronomy education in Eastern countries.
4. In Latin America, to fulfill the absence of a specific publication of astronomy education and to be a forum to show the Latin-American activity in this area, the peer-reviewed online journal *Revista Latino-Americana de Educação em Astronomia*[19] has been published since 2004, with contributions in Spanish, Portuguese, and English.
5. A few publications on astronomy education, particularly addressing the role of the history of science in science teaching, are available in French. The *Cahiers Clairaut,* edited by the French organization Comité de Liaison Enseignants-Astronomes

---

[16] http://www.historyofastronomy.org/

[17] http://www.eaae-astronomy.org

[18] http://www.issp.ac.ru

[19] http://www.relea.ufscar.br

(CLEA),[20] regularly offers practical activities for the classroom centered on historical experiments in astronomy, as well as scholarly work on the history of astronomy. Pierre Causeret (2005), member of the CLEA, has compiled a book on easily reproducible experiments for the classroom for the primary and secondary levels, some of which are historical experiments. A similar approach has been developed for the secondary level by the French organization Planète Sciences (2009).

19.1.3 Roles of HPS in Science Education and Astronomy Education

With its rich history and its methodological particularities (unlike in many other sciences, astronomical objects are not directly manipulable), astronomy courses constitute a good opportunity to raise epistemological questions and discuss general characteristics of science. Educators at the college level are particularly aware that introductory astronomy courses can play such a role for non-major students, some of whom may not take other science courses (Partridge and Greenstein 2003). Even when astronomy is only taught within the framework of other scientific disciplines – as is most often the case at the primary and secondary levels – astronomy provides a context in which questions about the nature of scientific knowledge and practices are particularly relevant. However, whereas a concept-based approach to astronomy teaching is receiving much attention, the specific role and efficiency of the study of the history of astronomy in contemporary classrooms has not been at the center of extensive scholarly works yet.

The case for the inclusion of HPS content in science education to teach the scientific content as well as aspects of the nature of science has been made for decades.[21] The teaching of science through its historical and conceptual development fulfills the goals of a liberal education. Advocates of this approach argue that it promotes critical thinking and allows students a better understanding of the nature of scientific knowledge and practices, as well as scientific knowledge itself. Moreover, teaching science through its history can help humanize scientists and their work, thereby making science more appealing to more students (see Clough 2011). The belief that there is positive value in teaching the history and philosophy of science is supported by major science education organizations (AAAS 1990, NRC 1996, 2011), drawing on empirical studies already made decades ago for the Harvard Project Physics and the History of Science Cases for High School (see e.g., Klopfer 1969; Klopfer and Cooley 1963). More recent frameworks for K-12 science teaching in the United States (NRC 2011) put an emphasis on model-based teaching (Hestenes 1987). The

---

[20] http://acces.ens-lyon.fr/clea/

[21] See, e.g., Arons (1965), College of the University of Chicago (1949, 1950), Conant (1948), Conant and Nash (1957), Hobson (2003), Holton and Brush (1985), Holton and Roller (1958), and Matthews (1994).

role of HPS as a useful resource for model- and concept-based science teaching at the core of these more recent frameworks has been articulated by Duschl and Grandy (2008). In particular, these works support the role of the teaching of historical scientific transitions as a way to articulate the conceptual underpinnings of a discipline.

Teaching a scientific domain through its conceptual and historical evolution is consistent with constructivist methods,[22] according to which the teaching of science consists in creating a conceptual change among students (Carey 2000; Dedes and Ravanis 2009; Posner et al. 1982). Teaching the conceptual changes that scientists had to go through throughout history, by overcoming the limitations of previously held beliefs, can help explicitly address the students' own misconceptions (Carey 2009; Nersessian 1992, 2008).

19.1.4 The Role of Historical Cases in Astronomy for Teaching Aspects of the Nature of Scientific Knowledge and Practices

Drawing on millennia of history and specific methodological challenges, the appeal of astronomy for the teaching of aspects of the nature of scientific knowledge and practices is clear. Indeed, historical debates around the evolution of the understanding of the structure of the Solar System are among the most discussed to illustrate key aspects of the nature of science, as has noticed by McComas (2008) who listed several historical cases in a variety of scientific fields, extracted from popular books on the nature of science (see Table 19.1). Other historical cases of astronomy introduced for their relevance for teaching aspects of the nature of science at the college level have been presented on the Web site The Story Behind the Science (see Clough 2011).[23] This Web site covers a few cases in physics, chemistry, and other disciplines, among which astronomy. Similarly, the resource center for science teachers using Sociology, History and Philosophy of Science,[24] at the University of Minnesota (Allchin 2012), contains a few course modules around historical cases in astronomy for secondary- and college-level courses.

The following table, mostly adapted from McComas (2008), isolates the cases that touch on the history of astronomy (broadly construed):

McComas's brief survey of historical cases in astronomy in popular, introductory books on the nature of science shows that in spite of an important presence of astronomy relative to other sciences in such books, most of these cases originate from a narrow period in history. Indeed, more recent cases, particularly in astrophysics, are less approached in such books on the nature of science. This is not typical of the historical cases approached in college-level introductory books in

---

[22]We are not here referring to radical, relativist interpretations of the term "constructivism". See, e.g., Matthews (2002) for an appraisal of constructivism in science education.

[23]Michael P. Clough ed. http://www.storybehindthescience.org/

[24]http://www1.umn.edu/ships/

Table 19.1 Survey of historical cases in astronomy used to illustrate key aspects of the nature of science in popular books, mostly as they are presented in McComas (2008). A few suggestions as to how such a table could be expanded have been inserted (in brackets), as well as indications about the appropriate teaching level for each of those cases (*PL* primary level, *SL* secondary level, *HEL* higher-education level)[a]

Historical cases	Teaching level		
	PL	SL	HEL
1 Galileo quantified and qualified his observations through recorded data. He observed and recorded the positions of Jupiter's moons. He could then predict the future positions. When questioned by the scientific community, he could confirm his observations (Chalmers 1999)	x	x	x
2 Galileo argued in favor of the Copernican view of the universe but his work was challenged by more conservative (traditional) thinkers not because his observations and calculations were found wrong but because his argument was based on observation and calculations, not theoretical understanding. (Thompson 2001). This illustrates the role of alternative theories, interpretation of observations, and the cultural context of science (religion, patronage, politics) (Allchin 2012)		x	x
3 A train of events led to the discovery of Neptune with Le Verrier and Adams suggesting an undiscovered planet near Uranus (Chalmers 1999; Derry 1999). It illustrates how anomalies drive scientific inquiry (Clough 2011) [see also this chapter]	x	x	x
4 Once the heliocentric model was accepted, observations were consistent with what the model suggested (Cromer 1993)	x	x	x
5 Variations in Venus' brightness called into question the Earth-centered model (Cromer 1993)			x
6 Copernicus used the data (evidence) provided by Ptolemy in his models but interpreted the evidence quite differently (Cromer 1993)		x	x
7 Brahe was the first modern astronomer to keep detailed night-by-night records of his observations (Cromer 1993)	x	x	x
8 The effects of gravity on light, predicted by Einstein's theory of relativity, was observed in Africa and South America in 1919 (Thompson 2001)		x	x
9 Galileo performed experiments in front of peers to validate his results (perhaps dropping balls from the Tower of Pisa); he invalidated Aristotle's laws of motion in public areas (Chalmers 1999)		x	x
10 Newton's laws of gravitation and his astronomical observations allowed him to predict planets' positions (Okasha 2002)	x	x	x
11 There is no way to examine all bodies in the universe; therefore Newton used induction in the development of his laws of motion (Okasha 2002)		x	x
12 The shift in understanding from Newton to Einstein is an example of a scientific revolution (Chalmers 1999)			x
13 Galileo suggested that the Earth revolved around the Sun; he was never able to fully substantiate his notion (Sardar and Loon 2002)	x	x	x
14 Brahe refuted Copernicus' theory, but even Brahe's estimates of the distance to the stars were too small (Chalmers 1999)			x

(continued)

Table 19.1 (continued)

Historical cases	Teaching level		
	PL	SL	HEL
15 Kepler's view of the Solar System was superseded by a more complete view provided by Newton (Chalmers 1999)		x	x
16 Newtonian mechanics is enhanced by being firmly embedded in a grand theoretical scheme that can be used to accurately describe a vast range of phenomena, from the motion of protons inside a nucleus to expansion of the universe itself (Cromer 1993)		x	x
17 Copernicus produced a better model of the Solar System but made no attempt to explain the motions of the planets (Wolpert 1994)		x	x
18 Ptolemy believed that the Earth was at the center of the universe. While observing, he saw inconsistencies and forced explanation upon them to fit the geocentric model he held (Cromer 1993)	x	x	x
19 Throughout history, many people believed in the geocentric model of the Solar System because of religion's authority (Cromer 1993)	x	x	x
20 The Church banned books explaining Copernicus' suggestion that the Sun was at the center of the Solar System (Okasha 2002)		x	x
21 Aristotle was deemed such an authority that it was difficult for Copernicus or Galileo to challenge his views (Thompson 2001)		x	x
22 The space race (and the resulting increases in science and technology) between the US and USSR was very much a political (more than scientific) activity (Dunbar 1995)		x	x
23 In Newtonian science, the law of gravity was a fundamental principle; it explained other things but could not itself be explained (Okasha 2002)			x
24 The invention of the telescope allowed Galileo to change Copernicus' ideas about the size of Venus and Mars (Chalmers 1999)	x	x	x
25 New instruments promoted the careful examination of the world. The telescope was improved by Galileo and used to a controversial effect to show sunspots on the "perfect" sphere of the Sun Thompson 2001)		x	x
26 The application of spectroscopic analysis to the stars by Kirchhoff and Bunsen opened a new field of inquiry: the physical analysis of the properties and composition of the stars (astrophysics). It is an example of the role of collaborative work, instrumentation, and puzzle-driven scientific research (Allchin 2012)		x	x

#	Historical case					
27	The detection of black holes as an example of the power of robust theory and mathematics (Clough 2011)					x
28	The story of the Cosmic Microwave Background as an illustration of how data makes sense only in light of theory (2011)					x
29	The story of Dark Matter as an example of the role of imagination and invention in science. (Clough 2011)					x
30	The debate around the measurement of the size of our galaxy, revealing the role of theoretical assumptions, the dynamics of scientific debates, and the role of technology in discoveries (Clough 2011)					x
31	[The discovery of dark energy as a purely observational result that completely changed our understanding of the large-scale properties of the universe]				x	x
32	[In 1951 the Roman Catholic Church supported the Big Bang Theory]				x	
33	[The 1925 edition of the "Soviet Encyclopedia" asserted that the theory of relativity is unacceptable for dialectical materialism]					x
34	[Hermann Bondi said that cosmology, whose object of study is the universe as a whole, seems to be immune to observation]				x	x
35	[In 570, Isidore, Bishop of Seville (Spain), distinguishes astronomy from astrology. From then on, scientists must constantly reiterate this distinction]			x	x	x
36	[Copernicus proposed a heliocentric model although no stellar parallax could be observed]				x	x
37	[Observing the transit of Venus in 1761, Mikhail Lomonosov suspected and detected the presence of the atmosphere of Venus]				x	x

[a]These descriptions of historical cases should not necessarily be taken as assertions, but at least as starting points for inquiry and discussion. For instance, the extent to which Einsteinian physics constitutes a revolution from Newtonian physics may be debatable

Table 19.2 Distribution of the historical cases in Table 19.1, according to what key aspect of NOS they most clearly illustrate. Unlike in McComas (2008), a case can illustrate several aspects. Depending on how each case is treated, a very different distribution could be obtained

Key aspect of nature of science	Historical cases in Table 19.1
1. Science depends on empirical evidence	1, 2, 3, 4, 5, 6, 7, 8, 9, 10, 15, 16, 18, 24, 25, 26, 27, 30, 31, 35, 37
2. Science shares many common features in terms of method	2, 3, 6, 7, 10, 11, 15, 16, 17, 23, 26, 27, 28, 29, 30, 34, 35, 37
3. Science is tentative, durable, and self-correcting	2, 3, 4, 5, 8, 9, 12, 14, 24, 25, 29, 31
4. Laws and theories are not the same	4, 5, 11, 15, 16, 23
5. Science has creative elements	3, 6, 8, 11, 12, 27, 29, 37
6. Science has a subjective component[a]	13, 14, 17, 18, 21, 36
7. There are cultural, political, and social influences on science	1, 2, 3, 4, 9, 19, 20, 21, 22, 25, 32, 33
8. Science and technology are not the same, but impact each other	1, 2, 3, 7, 24, 25, 26, 30, 31, 37
9. Science cannot answer all questions (science has limitations)	3, 11, 23, 34

[a]This category is here taken as meaning that scientists' personal preference or bias can have an influence on their research

astronomy and astrophysics, whose primary goal is not to teach about the nature of science. McComas divides these examples in nine categories, placing each case in a unique category corresponding to a key tenet of nature of science (NOS) it best illustrates. Table 19.2 above suggests a distribution of these cases according to such tenets of NOS, a domain-general, consensus-based list of aspects of NOS (see, e.g., Lederman et al. 2002; McComas and Olson 1998). Such lists are widely used in science education research but are not uncontentious, as they run the risk of promoting an essentialist view of science and confusing epistemological, ontological, and metaphysical features of science (Eflin et al. 1999; Duschl and Grandy 2012).[25]

This list of key aspects of NOS does not easily allow one to characterize methods of inquiry that are specific to a discipline or a type of inquiry. In general, they may not allow one to fully appreciate the historical, philosophical, and scientific relevance of each case.[26] Indeed, the examples in Table 19.1 can be illustrative of more meaningful aspects of scientific knowledge and practices. For instance, the universality of laws of nature and the role of this feature in scientific inquiry (implied by the tenet "Science shares many common features in terms of method") are more specifically

---

[25]The proponents of such lists often deny such intentions (see, e.g., Abd-El-Khalick 2012). The consensus around this particular choice of categories is not unanimous (for instance, one may prefer to say that science is provisional rather than tentative or refuse to equate subjectivity with theory-ladenness as is sometimes done).

[26]Instead, Eflin and colleagues "recommend illustrating the rich complexity of science with its practice and its history. Such study will offer students a better picture of the complex family resemblances between all the activities we call science" (Eflin et al. 1999, p. 114).

exemplified by cases 3, 10, 11, 16, 26, 27, 28, and 30. Moreover, the social responsibility of scientists (here illustrated by case 35) cannot not clearly figure in such a list of aspects of NOS. Nonetheless, the proponents of such nomenclatures hope that they may at least constitute a synthetic, introductory way of underlining the philosophical relevance of such historical cases in science and spark curiosity about what can be called science and why.

19.1.5 Gender Mainstreaming

The history of astronomy is relatively rich in major contributions made by women scientists. Marilyn Bailey Ogilvie's biographical dictionary (1986) lists only 180 women scientists, from antiquity to the beginning of the twentieth century. Although this list is not exhaustive, it can give us an indication of the relative contribution of women to science. It is interesting to note that astronomy is the second most represented discipline in that census (after biology). Nowadays, although there are thousands of women astronomers, they only account for about 16 % of the members of the IAU.[27] Omitting to include in the astronomy curriculum the contributions of women astronomers would only perpetuate a fragmentary and biased teaching. A historically informed teaching of astronomy can help students acknowledge the too-often overlooked contributions of women scientists to astronomy. It can reveal the struggle of women astronomers and illustrate the importance of the cultural context in scientific discoveries, the sociological organization of the scientific community, or counter the stereotypical image of science as being a masculine discipline.

To that effect, a few emblematic historical cases can be emphasized:

1. In ancient Babylon, EnHeduAnna (circa 2300 B.C.), as other priestesses in ancient cultures, performed astrological and astronomical tasks jointly. She enjoyed considerable political influence and is also remembered as a poetess (Meador 2009).
2. In ancient Greece, Aganice of Thessaly (second century B.C.), mentioned in the writings of Plutarch, was an expert in Lunar eclipses. Hypatia of Alexandria (circ. 370–415), the most noted woman scientist in ancient Greece, made major contributions not only to astronomy but also to mathematics and philosophy (Alic 1986).
3. In modern times, the Silesian astronomer Maria Cunitz (1610–1664), who simplified Kepler's planetary tables; the French astronomer and mathematician Nicole-Reine Lepaute (1723–1788), who calculated the orbit and date of return of several comets (including Halley's); the famous Caroline Herschel (1750–1848), one of the most acute observers in astronomy of the nineteenth century, discoverer of comets and nebulae; Maria Mitchell (1818–1889), the first American woman to work as a professional astronomer; and Agnes Mary Clerke (1842–1907), an Irish historian of astronomy.

---

[27] Source: http://www.iau.org/administration/membership/individual/distribution/.

In the twentieth century, thanks to the pioneering work of women astronomers in America such as Annie Jump Cannon, Henrietta Leavitt, Williamina Fleming, and the famous astrophysicist Cecilia Payne-Gaposchkin or that of the British astronomer Jocelyn Bell, women astronomers worldwide burst on the scene of the scientific investigation of the universe. Several works are available as a resource on the life and contributions of recent or contemporaries women astronomers.[28] However, only few such publications have been written for the classroom.[29]

19.1.6 The Role of HPS for the Teaching of Astronomy

In spite of these resources, the rationale behind the role of HPS in astronomy teaching in particular – to teach not only aspects of the nature of science but also *scientific content itself* – has not been fully fleshed out yet. For instance, in spite of the IAU's interest in education, the role of the history and philosophy of astronomy is particularly underdeveloped within its meetings and publications.[30] Since astronomy is not systematically taught at school – and rarely as a separate topic except at the college level – astronomers and astronomy educators sometimes have to remind that astronomy is useful and should be included in the school curriculum (Percy 2005). The fact that the astronomy education research community is relatively new and that the teaching of astronomy (at the primary and secondary levels at least) is subsumed under that of other disciplines (physics, geography, Earth science) may explain why only few works explicitly address the role of historically centered teaching in astronomy,[31] let alone measure its effectiveness on student learning. More than half a century after its development, Harvard Project Physics remains one of the most exemplary source of educational material on the integration of HPS content for teaching astronomy.[32]

---

[28] See, e.g., Byers and Williams (2006), Gordon (1978), Johnson (2005), Mack (1990), and Rossiter (1982). For an introductory resource guide to materials on women astronomers available on line, see Fraknoi (2008).

[29] See in particular two Web sites: Harvard University Libraries and Museum, Open Collection about "Women Working, 1800–1930", http://ocp.hul.harvard.edu/ww/index.html, and Woman Astronomer: http://www.womanastronomer.com.

[30] For instance, in their review of the papers on astronomy education presented at the IAU's meetings between 1988 and 2006, Bretones and Neto (2011) found that only 4.9 % of the 283 papers that dealt with astronomy education (i.e., only 14 papers) belonged to the category "studies on history of Astronomy or history of Astronomy Education."

[31] In particular, we can point out to the lesson plans accessible on the ComPADRE Web site (http://www.compadre.org) and to Hirshfeld (2008), which offers a collection of paper-and-pencil, interactive activities aimed at reproducing historical experiments.

[32] Some documents developed for Project Physics cover the history of our understanding of the structure of the Solar System and the motion of the planets. Astronomy was not the main subject covered by this project. Textbooks, documents for the classroom, and tests are accessible on http://archive.org/details/projectphysicscollection.

In their analysis of physics education, Höttecke and Silva (2011)[33] have identified obstacles for including HPS content in the science classroom. These include a culture of teaching science that is different from that of teaching other subjects, issues with the training of teachers (in particular the insufficient training in how to use HPS content), lack of clarity in curricular standards on the role of HPS, and lack of HPS appropriate content in textbooks. As we have seen earlier, in primary and secondary education, the teaching of astronomy exacerbates all these problems. Indeed, teachers are often not sufficiently trained in astronomy, astronomy is viewed in curricular standards as part of another discipline (if at all), and the main guidelines addressing the role of HPS in science education have been developed without a substantial participation of astronomers.

Nevertheless, a case for the inclusion of HPS in astronomy education can be found in a few instances. Such a rationale, when it specifically concerns astronomy education, is not different from what exists for other disciplines: teaching the history of science humanizes scientific research and practice and makes science more palatable and less intimidating (Partridge and Greenstein 2003; Zirbel 2004), and in particular, a universal history of astronomy helps avoid a Western-centered teaching (see Kochhar in Greve 2009). The role of HPS in astronomy education for concept-based teaching has been alluded to by Zirbel (2004), who remains cautious:

> It can be hypothesized that students learn concepts in a manner similar to the way that society learns basic concepts. It also turns out that a historical approach tends to be the least intimidating to students because they see the mistakes of humanity and of some famous individuals. This can make science less dry and more approachable, and make students more confident. For example, Duschl (1992) suggested that science instruction might benefit from a constructivist-historical approach in which students learn not only the justifications of modern scientific theories, but also how and why older theories where rejected, and how the nature of scientific inquiry changed within the discipline when the scientific community shifted from the old paradigm to the new. Other studies even went as far as to claim that the developmental stages in children (described by Piaget in 1929) can be simulated through historical parallels (e.g., Sneider and Ohadi 1998). (Zirbel 2004)

In their presentation of the goals for the concept-based course Astro 101, Partridge and Greenstein (2003) are more explicit with regard to the usefulness of HPS for a concept-based teaching of astronomy. They present the "acquaintance with the history of astronomy and the evolution of scientific ideas (science as a cultural process)" as one of the key content goals of an introductory astronomy college-level course. This recommendation, they say, is "self-evident" because it renders the discipline more appealing, illustrates aspects of the nature of science, but also because "the history of astronomy provides wonderful examples that illustrate some of the [other] goals" regarding content knowledge, skills, values, and attitudes.

---

[33] Their work is part of the European project HIPST (History and Philosophy in Science Teaching): http://hipst.eu/.

For instance, the cross-age (elementary to college) study of Kavanagh and Sneider (2007a, b) on student understanding of gravity allows one to see how the historical progression of scientific conceptions of gravitation follows that of students (see also, at the college level, Williamson and Willoughby 2012). However, a more fleshed-out articulation of the role of HPS in astronomy education, on a large variety of topics; the development of canonical examples and teaching practices that include HPS content, at various teaching levels; and an standardized assessment of the specific import of HPS content are yet to be carried out.

19.2   Three Cases as Examples of the Role of HPS for the Teaching of Astronomy

This second part attempts to illustrate possible ways to structure the teaching of astronomy around its historical development. The following examples all deal with the study of planets, a fundamental notion for all levels of astronomy teaching. These historical cases, though not necessarily the most emblematic ones, are possible examples of incorporation of historically and philosophically informed material in the science classroom. These cases are each best suited to different levels and cover contributions from different periods, from ancient times to the most contemporary discussions:

1. The first case deals with Kepler's use of the work of Archimedes and Apollonius for the development of his laws of planetary motions. It focuses on an important event in the history of astronomy – the application of the geometry of ellipses to planetary motion – that also had ramifications in philosophy and mathematics. It is best suited to a high-school-level science or mathematics curriculum.
2. The second case focuses on the social and cultural context of the discovery of Neptune. It also underlines how the confidence in our theories drives scientific inquiry and the importance of accounting for anomalies. It would be most appropriate for primary- or secondary-level teaching.
3. The third case is centered on the recent debate around the planetary status of Pluto. It attempts to show how underlining its philosophical motivations can help understand its relevance for science, and provide an engaging way to strengthen one's understanding of the structure and formation of the Solar System. It is more appropriate for a college-level course.

Table 19.3 below summarizes which aspects of the nature of scientific knowledge and practices these cases emphasize (according to some of the categories of Table 19.2).

Table 19.3 Key aspects of the nature of science illustrated by the three cases presented below in this chapter, based on Table 19.2

Key aspect of the nature of science	1st case	2nd case	3rd case
Science depends on empirical evidence	x	x	x
Science shares many common features in terms of method		x	x
Science is tentative, durable, and self-correcting	x	x	x
Laws and theories are not the same	x	x	
Science has creative elements	x	x	x
There are cultural, political, and social influences on science	x	x	x
Science and technology are not the same, but impact each other		x	
Science cannot answer all questions	x	x	

In addition to this summary, a few additional themes that these cases touch on can be emphasized:

1st case	2nd case	3rd case
Influence of the cultural context		
Geometry was only admitted as a tool with which to represent celestial problems but was not considered to be able to provide solutions to them	The scientist's social background or status was decisive in the process of scientific inquiry	The political organization of professional astronomy had an influence on scientific decisions
Although elliptic orbits constituted a reliable solution, they were considered as inappropriate at the time	Political and academic tensions between European nations ended the paralysis in the search for the new planet	The national sentiment about the only planet discovered by Americans prevented a discussion about the definition of planet to happen sooner
Aspects of the process of investigation		
The problem of accurately describing all planetary orbits was considered to allow for a unique and definitive solution. As such, it has been approached in the same manner since antiquity. Solving it required an obstinate person who changed the way to frame it	Two scientists from different countries came to a unique solution to a common problem, independently from each other	A scientific dispute, motivated by new discoveries, had to be resolved not by consensus but by a vote between scientifically receivable alternatives
A great confidence in empirical results as well as in the method used to analyze them enabled a change of explicative models once considered impossible	The absolute confidence in a theory (Adams, Le Verrier) conflicts with an opportunity to amend it (Newcomb, Hall) or replace it (Einstein)	Classifications are needed as theoretical or explanatory devices, and while they drive investigation, they can be modified as our understanding of the structure of the world evolves

(continued)

(continued)

1st case	2nd case	3rd case
Accounting for observational anomalies in conflict with accepted theories		
A solution was found in a substantial and unexpected change in the geometrical treatment of planetary orbits	The successful application of the hypothetico-deductive method resolved the conflict[34]	A philosophical examination of what is expected from a taxonomy helps us understand the reasons behind the prevailing choice
Other notable themes		
A historically informed teaching can provide a more meaningful, less mechanical understanding of geometry and how it relates to physics	Observation drives and establishes our understanding of the universe (i.e., Herschell)	Disagreement among scientists does not necessarily undermine the validity and authority of their decisions
	The serendipitous intervention of unexpected agents (i.e., Galle, Lescarbault, Gerber) resulted in a clarification of the problem	Other scientific fields (i.e., biology) have had to overcome similar disputes

19.2.1 The Geometry of Planetary Orbits

19.2.1.1 A Circular Ancient Astronomy

Archimedes of Syracuse (287–212) did not leave meaningful astronomical comments, yet his mathematical contributions (mainly those related to geometry) proved to be relevant to Johannes Kepler (1571–1630). In contrast, the mathematician Apollonius of Perga (262–192), a contemporary of Archimedes, played an important role in ancient astronomy. Apollonius inherited from the Greek astronomers the concern to *save the phenomena*: for a kinematic planetary model to be successful, it has to explain the planetary motions as they are observed in the sky (in particular, the apparent retrograde motion). It was Apollonius who:

Developed a model of the universe in which he generalized the epicycles for the planets	According to his model, the planetary orbits are not centered on Earth but on a epicycle whose center is located on another circle (deferent) that revolves around the Earth.[35] A successful combination of these two uniform, circular motions can explain the observed retrograde motion without any need to deprive the Earth of its position at the center the universe

(continued)

---

[34] It has been argued that Newton's method and notion of empirical success is richer than just what the hypothetico-deductive model suggests: "According to [Newton's] ideal [of empirical success], a theory succeeds empirically by having its causal parameters receive convergent accurate measurements from the phenomena it purports to explain" (Harper 2002, p. 185). See (Harper 2007) for a study of how the resolution of the problem of Mercury's perihelion shift illustrates this.

[35] Epicycles already featured in the geo-heliocentric model of Heraclides of Pontus (390–310, but they were restricted to two planets only (Venus and Mercury). This model placed the Earth

(continued)

Optimized the idea of eccentric orbits	Placing a planet on a single deferent not centered on Earth could produce the same observable result as placing it on epicycles. The distance between Earth and the new center of the deferent is called *eccentricity*, and then the orbits are called *eccentrics* (Sarton 1959, Chap. V)

Apollonius is also the author of *The Conics* (Heath 1896), an eight-book-long treatise in which are defined and examined in great detail ellipses, parabolas, and hyperbolas, figures who owe him their name. Despite his familiarity with ellipses, Apollonius did not apply or suggest (not even mention) the use of ellipses for studying astronomical questions. The habitual use of circles resulted surprisingly effective in his model of planetary orbits with epicycles and deferents. The adulation for the perfection of the circle was very widespread and entrenched. So much so that, had someone suggested that planetary orbits have the shape of ellipses, Apollonius himself would probably have worked to reduce each proposed ellipse to a combination of circles. Apollonius's astronomical ideas were first adopted by Hipparchus of Nicaea (190–120), and then by Claudius Ptolemy (100–170), who in his treatise *The Almagest* consolidated and established an apparently unquestionable geocentric model.

It took more than 1,400 years until Nicolaus Copernicus (1473–1543) proposed a fundamental change in the arrangement of the universe: the Sun would be at the center and the planets (including the Earth) would form a system rotating around it. Copernicus still retained circular orbits, uniform motions, as well as several epicycles. Two decades after the work of Copernicus was published, Tycho Brahe (1546–1601) renewed and refined his observation procedures, which yielded records of planetary positions of higher precision. With them, Brahe suspected that the Copernican (heliocentric) model should be replaced by a new proposal that he imagined to be geo-heliocentric.[36] So he summoned Kepler to assist him in finding a mathematical justification for his idea.

19.2.1.2 Kepler, Orbit-Maker

From Apollonius (and even before him) to Kepler, no figure and model of the universe accepted by astronomers and mathematicians had ever challenged the dominance of the circle and sphere. This principle superseded other aspects of scientific inquiry and slowly became a moral premise that dictated the conduct of the celestial bodies themselves. Kepler recklessly overcame this prejudice and constructed the best possible model for planetary orbits: one that was faithful to the observed

---

(rotating on its axis, in one day) at the center of the universe. The Sun and the Moon (as well as the planets Mars, Jupiter, and Saturn) revolved around the Earth, while Mercury and Venus, instead, revolved around the Sun. Moreover, Apollonius was the one who gave its denomination to the circles that describe planetary motion (epicycle and deferent).

[36] Brahe proposed a model similar to that of Heraclides.

reality. Thereby, Kepler got rid of other (maybe mathematically perfect) models. It is well known that the result, the three Kepler laws of planetary motion, undoubtedly played a significant role in the history of our understanding of the universe. The first two laws were published in his *Astronomia Nova* (1609) and the third one in *Harmonices Mundi* (1619). However, in reality, Kepler constructed his second law before the first (Koestler 1960, p. 137), and their subsequent reorganization was motivated by logical and aesthetic considerations. In other words, Kepler had found that the orbital velocity of the planets was variable before he could describe the geometry of their trajectory around the Sun. With the adoption of egg-shaped trajectories, Kepler was not able to account for the variation in speed of the planets on their orbits. He even hesitated to uphold his famous second law (the only one he had then) to avoid the intellectual choking that would have stemmed from abandoning the beloved circular orbits, defended by Copernicus, whom he admired.

Kepler worked with Brahe's records of the motion of Mars, considering them as precise and irrefutable. He tested various orbit formulations and found that the "oval orbit" was invariably an appropriate solution. It is to be noted that Kepler used ellipses in his approximations as auxiliary computing elements only, in order to determine oval areas in particular. In 1603 Kepler wrote to his friend David Fabricius (1564–1617) and confessed to him that he felt unable to resolve the geometry of his "egg," noting that "if only the shape were a perfect ellipse, all the answers could be found in Archimedes' and Apollonius' work"[37] (Koestler 1960, p. 143). After several years, Kepler even came to conceive a mathematical formulation allowing him to account for the variation of the planets' distance from the Sun on their orbit. However, Kepler could not realize that his expression was in fact defining an ellipse, as he conceived his construction as an *ad hoc* formula, foreign to the set of geometric figures culturally acceptable to describe the motion of planets. Later, Kepler decided to try his luck with other methods, including a purely geometric procedure for describing Mars' orbit. Then, he found that it was in fact an ellipse and, soon after, rediscovered that the previously dismissed expression gave the same result.

19.2.1.3 Kepler's First Law in the Classroom

An important aspect of Kepler's contribution is the introduction of ellipses in celestial geometry, which seemed confined only to circular trajectories, almost from the beginning of that science. After Kepler's laws, Apollonius's conics gradually acquired a special relevance in kinematics and subsequently also in the dynamics of planetary motion,[38] which is something Apollonius could not predict. The cultural

---

[37]Letter to Fabricius, July 4, 1603 (Baumgardt 1951, p. 72).

[38]The ellipses in particular but not only them: parabolas and hyperbolas also took on new meaning in study of celestial objects. In fact, the relevance of ellipses extended to other sciences, particularly physics.

19 Perspectives of History and Philosophy on Teaching Astronomy 623

significance of the change of shape in planetary orbits made by Kepler (from circles to ellipses) should still deserve to be taught. The enunciation of the first law, as it appears in school textbooks – "The orbits of the planets are ellipses with the Sun located at one focus" – is in general learned unquestioningly, uncritically, and the teacher often swiftly draws students' attention to other laws that seem to be of greater didactic importance, given their physical implications.

Planetary orbits are ellipses...	In many cases, students talk about this conical shape without fully knowing its features, as if knowing its name was enough to understand its properties. When students are not able to consciously assign meaning to words, learning is mechanical, not significant (Moreira 2005, p. 28). For students, the resistance in carrying out the conceptual change about the shape of orbits (from circles to ellipses) may turn out to be very high. In pedagogical terms, the shift from circles to ellipses is as significant as the passage of a geocentric to a heliocentric system
... with the Sun located at one focus	Without a good understanding of the properties of the ellipse the location of the Sun "in focus" is possibly meaningless.[39] However, a proper understanding of this feature opens the doors to students to Newton's ideas, especially the universal law of gravitation (see Goodstein and Goodstein 1996; Haandel and Heckman 2009)

Apollonius and later Ptolemy, after developing a model of epicycles and deferent, adjusted the times of revolution of the planets to their (prefixed, preconfigured, impossible to change) geometries. Kepler, however, walked a reverse route: from observational data, he obtained planetary velocities. Since the observed velocities were nonuniform and thus different from the classically established uniform motions, the inferred trajectories (ellipses) were different from the usually assumed circular orbits.

But in the classroom, students continue to learn the first law as a premise or in some audacious cases as a deduction from the observations. It is not uncommon that although the teacher may emphasize the talent and perspicacity of Kepler in constructing a geometric solution with the ellipse (an unusual feature for the student), this fact is presented as if it were only a happy coincidence, conveniently agreeing with Brahe's observations. No less remarkable is the fact that until the presentation of the Laws of Kepler many students have barely heard of ellipses. Usually, the teaching of the ellipse is notably quite limited.[40] One of the consequences of this is the loss in understanding of much of the potency and depth of Kepler's laws. Thus, many students tend to *recite* the first law with little or no understanding of the importance of this law for the development of science in general and astronomy in particular. With a more adequate teaching of geometry and

---

[39] For example, a significant consequence of this feature is that a planet reaches a position of minimum and maximum distances to the Sun (perihelion and aphelion, respectively).

[40] It is not the case with circles, which are dealt with under different perspectives and in different subjects in the classroom. Most often, parabolas are well taught, but merely as a graphical solution to quadratic equations. Hyperbolas on the other hand are merely named, while ellipses constitute only an exercise in graphical construction of figures.

history of science comes a better understanding of many physical phenomena. For instance, here, if we neglect the role of time, Kepler's (dynamic) second law results in the (geometric) first law.

19.2.2 New Science, New Planets?

19.2.2.1 From Fiction to Reality

William Herschel (1738–1822) discovered in 1781 a peculiar celestial object that caught his attention. During his first observations, he thought it was a new comet. However, after a few months, Herschel noticed that its trajectory did not stretch out as was expected of comets (i.e., its orbit remained approximately circular). He soon confirmed that this was an unknown planet, orbiting beyond Saturn. He named it 'Georgium Sidus' (in honor of the king of England), but today it is known as Uranus.[41] Four decades later (1821) the French astronomer Alexis Bouvard (1747–1843) published an astronomical treatise containing information on the trajectories of several planets, after having observed and recorded their positions for years. His data revealed substantial discrepancies when compared to the previously computed orbit of Uranus. Given this evidence, several scientists suggested that these variations could perhaps reveal that the physical laws governing planetary motions were no longer valid beyond a certain distance from the Sun (i.e., beyond Uranus). However, Bouvard postulated that perhaps the differences between Herschel's calculations and his own observations could be due to the presence of an unknown celestial body orbiting the Sun beyond Uranus. This body would disturb the motion of Uranus so that it does not comply with Kepler's laws, without implying any violation of Newton's law of gravitation.[42]

Twenty years later, a young English astronomer, John Couch Adams (1819–1892), dedicated himself to this subject and calculated the mass and orbit of a hypothetical planet revolving around the Sun beyond Uranus, which could explain the anomalies that Bouvard had detected. Adams communicated his results to his teacher, the astronomer James Challis (1803–1882), and to George Airy (1801–1892) at Greenwich Observatory. Initially, Airy did not even attempt to verify this hypothesis but Adam's insistence prompted him to start a survey in order to find the hypothetical planet. This work began in July 1846 in Cambridge, and Challis himself was in charge. Unfortunately, no detection confirming Adams's ideas could be obtained.[43]

---

[41]In hindsight it received various denominations (in France was known as *Hercules*) until was finally accepted the suggestion by Johann Bode (1747–1826) to identify it as *Uranus*. Nevertheless Herschel continued to call it *Georgium Sidus*.

[42]The renowned astronomer Friedrich Bessel (1784–1846) supported the scenario described by Bouvard in 1824.

[43]Further analysis of the Challis observational registers showed that he had observed the new planet on the 8th and 12th of August but wrongly identified it as a star.

19 Perspectives of History and Philosophy on Teaching Astronomy

In France, the same year as when Adams presented his hypothesis of a *transuranic* planet (1845), the astronomer Urbain Le Verrier (1811–1877) presented a similar prediction at Paris Observatory, unaware of Adams's work. Faced with the indifference of his colleagues, Le Verrier presented his calculations about a possible new world again in June and August 1846. He had predicted the mass and specific values of the orbital elements, but once again, his hypothesis was rejected. Finally, Le Verrier communicated with the German astronomer Johann Gottfried Galle (1812–1910), who was then working at Berlin Observatory, and indicated to him where to point the telescope in order to find the planet predicted by his calculations. Galle, together with his pupil – Heinrich d'Arrest (1822–1875) – observed the indicated area and, in less than an hour's work, found the new planet only 1° away from the position predicted by Le Verrier.[44] Immediately, both British and French astronomers claimed to be recognized as the true discoverers of Neptune, by means of mathematical calculations instead of usual astronomical procedures. Nowadays, the discovery of Neptune[45] is customarily attributed to the binational duo Adams-Le Verrier.

19.2.2.2 From Fiction to Unreality

The prediction of the existence of Neptune through mathematical calculations and its subsequent observational discovery where indicated by equations was an unparalleled triumph of Newtonian mechanics and, for many, of the power of science in general. The celestial mechanics of the Solar System seemed fully resolved. However, there was a *small* problem: the motion of Mercury was different from what was expected, as there were some anomalies in the shift of the perihelion of Mercury that could not be explained by Kepler's laws and Newton's law of gravitation.[46] In order to solve this problem, Le Verrier predicted in 1859 that Mercury's orbital perturbations had a similar cause as those previously detected on Uranus. Hence, he suggested that a new planet, unknown so far, was situated between Mercury and the Sun.[47] Quickly, he called it "Vulcan" to prevent future discussions.

---

[44] Apparently, as Galle was the first to look through the telescope, he is recognized as the discoverer.

[45] Shortly after, the new planet was simply called *the planet after Uranus* or *Le Verrier's planet*. The first suggestion for a new name came from Galle: *Janus*. In England, Challis suggested *Ocean*. From Paris, François Arago (1786–1853) proposed *Le Verrier* (a suggestion not well received outside of France). Meanwhile, Le Verrier insisted on calling it *Neptune* and received the support of Friedrich Struve (1793–1864) to carry on the tradition of naming planets after mythological figures.

[46] Like other planets, Mercury does not follow the exact same trajectory traced by its previous orbit (this phenomenon is called "perihelion shift"). For Mercury, the observed value of this shift is about 575 arcsec/century; most of it (532 arcsec/century) can be explained by gravitational perturbations from other planets (and to a much smaller degree the Sun's shape – its oblateness). The small difference (43 arcsec/century) could not be accounted for by Newtonian gravity. It was considered an anomaly and was treated as a serious problem of celestial mechanics.

[47] Le Verrier also considered the possibility that instead of a planet would lie a group of small celestial bodies (like an asteroid belt).

The great scientific reputation of Le Verrier was enough to convince most contemporary astronomers of the existence of Vulcan. Its *presence* in the Solar System brought calm to the astronomy community. Again, Le Verrier established the orbital elements of the new planet and its mass. Many telescopes around the world were used to find Vulcan, without success.

All of a sudden came the news that a French amateur astronomer, Edmond Lescarbault (1814–1894), had detected Vulcan. He described it as an opaque body passing over the Solar disk (on March 28, 1859). Having ruled out the possibility of a sunspot, Lescarbault concluded that he had observed a transiting planet. He wrote and narrated his discovery to Le Verrier, which visited him soon after. Even though Lescarbault's transit registers were of poor quality and his instruments rudimentary, he made a very thorough description and in such detail that Le Verrier believed him.[48] Evoking the euphoria experienced with Neptune, Le Verrier announced the discovery of Vulcan (January 1860) to the French Academy of Sciences. The news of the discovery of Vulcan was received with caution by the astronomical community, whose skepticism was only assuaged by Le Verrier's great fame. It was very strange that the new planet was not observed by any professional astronomer at any observatory, with the most sophisticated instruments. Le Verrier recalculated Vulcan's orbit and provided new ephemerides to find it, but the response was always the same: no one (including himself) could observe it. In 1861, there were no reports of observations (professional or amateur) that would confirm the existence of Vulcan. The few reports that were related to the subject later proved to be sunspots. As Vulcan was so near the Sun, it was practically unobservable. This argument was considered reasonable by many observers, mainly those who had suffered damage to their eyes in the attempt of finding Vulcan in the vicinity of the Solar disk.

How to detect a body that is so elusive? Once again, Le Verrier had a solution: Vulcan would be visible during Solar eclipses, when the Solar disk is hidden. Then, he began distributing his new predictions of optimal observation dates. But even so, no one was able to find Vulcan. It seemed evident that Vulcan was a fictional planet, and its existence began to be considered an astronomical myth. Among the astronomers, confidence in Le Verrier's data began to decrease. Yet, Le Verrier could not accept that the astronomical community was unable to detect Vulcan. His position was based on three facts: (1) his role in the epic discovery of Neptune, (2) his absolute confidence in the validity of Newtonian mechanics, and (3) his confidence in the accuracy of his calculations. Le Verrier kept announcing new predictions. But as time passed, fewer and fewer people were paying attention. The scientific community started to doubt everything Le Verrier said about that new world. For several years, he published updated ephemerides (always supposedly definitive), but the planet was nowhere to be found. Finally, Le Verrier died with the certainty of having discovered a planet between Mercury and the Sun, and convinced that it would be detected in the future.

---

[48]Le Verrier gave no credit to the testimonies of French astronomer Emmanuel Liais (1826–1900), director of the Rio de Janeiro Observatory. Studying the Sun through a telescope more powerful and sophisticated than Lescarbault's, Liais denied that any planet had transited the Solar disk.

19.2.2.3 A Fictional Reality

After the death of Le Verrier, Simon Newcomb (1835–1909) considered other possible causes to explain the anomalies of Mercury's orbit, as the flattening of the Sun, but the obtained value for Mercury's perihelion shift was not sufficient to explain the 43 arcsec/century. In 1894, Asaph Hall (1829–1907) proposed to alter Newton's law of universal gravitation, adding a term varying as the inverse cube of the distance and a constant adjusted to reproduce Mercury's anomalous perihelion shift.[49] When Newcomb observed similar anomalies on Venus, Earth, Moon, and Mars (although with much lower discrepancies), he realized that Hall's newly proposed law failed to account for them. Although Hall's proposal did not succeed, it opened doors for a new law of gravitation that might be the right answer. In 1906, Hugo von Seeliger (1849–1924) offered a more accurate explanation for the value of Mercury's perihelion shift: a distribution of mass around the Sun, with an inclination to the ecliptic of 7°. Coincidentally, such a mass would also be responsible for the zodiacal light.[50]

It was Albert Einstein (1879–1955) who was able to provide an alternative, successful answer to Mercury's perihelion shift. Einstein's theory predicted that the planetary orbits experienced a slight shift due to the curvature of spacetime. Thus, Einstein's theory seemed to have solved a problem that had troubled astronomers for decades (and ended the illusion of a "Vulcan" between Mercury and the Sun!). Before Einstein, Paul Gerber (1854–1909), German physicist and teacher, provided a formulation for the perihelion shift similar to that of Einstein in 1898. Gerber assumed that gravity is propagated at the speed of light and that the force between two masses should be corrected by a term dependent on the speed at which they move. His formula could explain the anomaly of Mercury, the terrestrial planets, and even the Moon. Gerber's work had little impact because its derivation was quite unclear and, years later, was found to contain some wrong arguments. Einstein always claimed that in 1915 he was unaware of Gerber's work and that had he known it, it would not have influenced the development of his theory.

19.2.2.4 Perspectives

The story of the discovery of Neptune and the Vulcan hypothesis illustrates how personal, cultural, and social considerations drive scientific inquiry. But it also constitutes a telling example of scientific inquiry motivated by observational

---

[49] Hall (1894) noted that he could account for Mercury's precession if the law of gravity, instead of falling off as $1/r^2$, falls off as $1/r^n$, with $n = 2.00000016$.

[50] Erwin Freundlich (1885–1964) found that the mass needed to explain the anomaly was incompatible with the mass postulated by Seeliger, given the low luminosity of the Zodiacal Light (1915). Still, Seeliger's hypothesis survived and was one of the arguments of the detractors of Einstein until 1919.

anomalies that, if reliable, demand the revision of our most trusted theories. That similar methods of discovery of new celestial bodies are still in use today only makes this story more pedagogically relevant. In particular, exoplanets are, like Neptune, discovered indirectly, by their induced effects on other, observable bodies, for example: a) using planetary transits to determine the decrease in light intensity of a star when a planet passes in front of its disk or b) using radial velocities to determine the gravitational pull of a planet over a central star.

19.2.3 "Planet," What's in a Name?

19.2.3.1 Is Pluto a Planet? The Evolution of a Scientific Concept

On August 24, 2006, 424 astronomers gathered in a room and decided, with a majority vote, that Pluto would no longer be called a planet. This decision caught the public's attention as it seems to defy not only our common conceptions about our astronomical neighbors but also our understanding of scientific methodology altogether. How can one reconcile rigorous fact-based process with voting? This case illustrates the dynamics of conceptual change and the historical character of scientific theories and concepts. The evolution of the concept of planet illustrates discussions that touch on methodological and ontological issues in science. Learning about the historical evolution of the concept of planet allows one to see how classifications rely on a larger understanding of the world and shape our investigation.

These concerns can be illuminated by philosophical discussions on natural kinds. A kind (property or object) is said to be natural if its existence, behavior, and properties do not depend on humans and are not the product of our decision (e.g., electron, hydrogen, magnetic field...). In contrast, folk kinds are the product of cultural conventions or other idiosyncrasies and may evolve at our will. For instance, continent is seen as a folk kind or cultural – as opposed to natural – object. When geologists refer to continents, they think of "large, continuous, discrete masses of land, ideally separated by expanses of water" (Lewis and Wigen 1997, p. 21). But there is no non-arbitrary way to define what counts as a large mass of land. But, contrary to what happened for planets, geologists did not vote on an official definition for continent and on their number.

19.2.3.2 The International Astronomical Union 2006 Decision

Between 2003 and 2005, Sedna, Eris, Haumea, and Makemake – four distant bodies of the same order of size as Pluto – were discovered beyond Neptune's orbit. This confirmed planetary scientists' suspicion that the confines of the Solar System may contain several other such bodies. Consequently, it became more apparent that it was arbitrary not to count Ceres or Vesta among planets if all these newly discovered Pluto-like bodies were to be counted as such. Based on their equatorial diameters, a category of Solar System objects stands out clearly: the giant planets (Jupiter, Saturn, Uranus, and Neptune), whose diameter is several times greater than Earth's. A second category, that of the terrestrial planets, contains at least several, undisputed planets: Mercury, Venus, Earth, and Mars. But whether or not any or some of the smaller bodies of the Solar System belong to that second category is unclear. Some of them are not much more different from the terrestrial planets than the terrestrial planets are from the giant planets. If Pluto was to be called a planet, then why not Eris, Haumea, or even Vesta? Where should we draw the line? How and why?

To answer these questions, several proposals were made in the years that led to the IAU decision:[51]

1. Alan Stern and Harold Levison (2002) from NASA suggested to define planet in term of *intrinsic properties* of mass: to be called a planet, "the body must: (1) Be low enough in mass that *at no time* (past or present) can it generate energy in its interior due to any self-sustaining nuclear fusion chain reaction (or else it would be a *brown dwarf* or a *star*). And also, (2) Be large enough that its shape becomes determined primarily by gravity (…)" (Stern and Levison 2002, p. 4).
2. Steven Soter (2006) from the American Museum of Natural History, on the other hand, suggested that planet be defined by the *historical process* from which they result, as the end product of secondary accretion from a disk around a star or substar (also known as brown dwarf). That definition allows us to consider the ratio of the mass of a body to the aggregate mass of all the other bodies that share its orbital zone as a good physical criterion, related to the historical formation process of a body, to discriminate planets and non-planets. If not the product of such an accretion, it will only be one of the many planetesimals on its orbit, and this ratio will be very different from that of a planet.[52]

Neither definition seemed entirely satisfying nor received widespread approval among astronomers. The first applies to many more bodies than our familiar eight or nine planets, including of course Pluto, but also our Moon and other satellites of the larger planets, and thus does not address the worry that the number of planets would dramatically increase. The second definition puts an emphasis on contingent relational characteristics (to a neighboring star and the other bodies on its orbit) that may not seem as relevant to characterize a natural object in its essence, even though it would have provided a clear criterion to distinguish between planets (the eight largest bodies that have indeed cleared their orbit) and non-planets (Pluto-like and other smaller bodies) in our Solar System.

After several proposals had been discussed that included these two definitions (or a combination of them), the IAU came to a consensus. The 2006 definition that

---

[51] For a much more in-depth examination of these proposals and the philosophical significance of this topic, see Bokulich (forthcoming).

[52] That criterion would, for instance, guarantee that Jupiter is a planet even though it has not "cleared its orbit" of other bodies, since the Trojans (a group of small bodies) share its orbit.

received a majority vote was based on a proposal by Uruguayan astronomer Julio Ángel Fernández:

(1) A planet [1] is a celestial body that (a) is in orbit around the Sun, (b) has sufficient mass for its self-gravity to overcome rigid body forces so that it assumes a hydrostatic equilibrium (nearly round) shape, and (c) has cleared the neighbourhood around its orbit.

(2) A "dwarf planet" is a celestial body that (a) is in orbit around the Sun, (b) has sufficient mass for its self-gravity to overcome rigid body forces so that it assumes a hydrostatic equilibrium (nearly round) shape [2], (c) has not cleared the neighbourhood around its orbit, and (d) is not a satellite.

(3) All other objects [3] orbiting the Sun shall be referred to collectively as "Small Solar System Bodies."

Under this definition, only Mercury, Venus, Earth, Mars, Jupiter, Saturn, Uranus, and Neptune are planets. Pluto is thus relegated to the status of "dwarf planet," a category for bodies that "look like" planets (they have a round shape as a consequence of their mass) but whose formation history does not make them the main object in their orbit (i.e., which have not "cleared their orbits").[53]

This definition is a conciliatory choice in many ways: it relies on *intrinsic* properties (size, mass, shape), thereby incorporating Stern's and Levison's proposal but also addresses our concern about the increasing number of planets by including two relational components (with the requirement that the orbit be "cleared" and around the Sun). In its formulation, this definition preserves as much as possible our conception of a natural kind as a set of intrinsic properties. It also preserves our common conception of planet, by keeping their number low. However, it can be argued that Soter's approach to defining the notion of planet has been vindicated, insofar as relational, dynamical properties are essential to the IAU definition. Even though it is not phrased that way, this dynamical property makes sense in the context of a model of formation of the Solar System.

Alisa Bokulich (forthcoming) noted that this definition of planet, even though it does not only involve intrinsic properties, does not necessarily make it less of a natural kind. In that, she echoes the work of Richard Boyd (1999), according to whom natural kinds should not necessarily be seen as a list of fixed, intrinsic properties but more as families of properties that are clustered in nature as a result of various underlying homeostatic or causal mechanisms. According to this conception of natural kinds, their properties may also be relational or historical, and their extension may even have vague boundaries, contrary to the more classical conception of natural kind (Bird and Tobin 2010). The resistance to Soter's approach to the definition of planet as not consisting only of intrinsic properties is reminiscent of the debate between philosophers about natural kinds.

---

[53] Vesta, a large asteroid whose previously round shape has been altered by collisions, is only a Small Solar System Body (SSSB), even though it shares many physical characteristics with the dwarf planets, and is much closer to them in mass and size than to most other SSSBs.

19.2.3.3 Why Was Pluto Ever Considered a Planet? A Brief History of "Planet"

The concept of planet has been redefined several times, following the progress of discoveries and theories in astronomy. The original denomination of planet was conferred to the few celestial objects easily visible to the naked eye and whose relative position to the other stars varies: Mercury, Venus, Mars, Jupiter, and Saturn.[54] Since then, the number of planets has kept evolving, and so has the meaning of the word "planet." This number kept increasing, as new celestial bodies other than the fixed stars were discovered after the invention of the telescope. The adoption of the Copernican heliocentric model kept the Sun out of the extension of the term. And later, after many moons were discovered, the need for a distinction between planets (which orbit the Sun) and their moons (which orbit their planet) was felt. Similarly, the further discovery of Ceres, Vesta, and other asteroids, but also that of Neptune in 1846, resulted in a distinction between planets and asteroids (smaller than Mercury). Like Neptune, Pluto was discovered as the result of a hunt for a large planet – *planet X* – that would explain perturbations in Uranus's orbit. Thus, the planetary status of Pluto was decided *before* its discovery in 1930 by Clyde Tombaugh (working under the direction of Percival Lowell). It is only after decades of research that scientists came to the realization that Pluto was only one of several relatively small, Pluto-like bodies in the outer Solar System and not the large planet it was once thought to be (see Brown 2010; deGrasse Tyson 2009; Weintraub 2008).

Drawing a parallel with the species problem,[55] Bokulich recalls how philosophical work by Joseph LaPorte (2004) clarifies the options scientists face when their categories (or more specifically taxa in the context of species) are inadequate, too vague or inconsistent. Facing the inadequacy of a taxon,[56] scientists only have limited choices: either expand the taxon, pare it down, or abandon it as a scientific term altogether. By 2006, the vagueness of the term "planet" became so apparent that the IAU felt something had to be done if we were to continue to use it. This organization chose to favor the second option: restrict the extension of "planet". As Bokulich put forth,

> Although which of these three options scientists choose to adopt is largely a matter of convention, what was *not* an option for the scientists was to leave the traditional definition and extension of the term planet intact, while having it remain a scientifically *useful* concept. (Bokulich (forthcoming)

Those who mourn Pluto's lost rank have to accept that this decision was a necessary evil to save the concept of planet as a scientific, useful one.

---

[54] Although Uranus (and more rarely Neptune) can at times be visible to the naked eye, it is not nearly as bright as these five planets.

[55] The species problem is the difficulty biologists face when trying to define species in a non-arbitrary way.

[56] A taxon is the name applied to a taxonomic group, i.e., a unit in a formal system of nomenclature. This term is mostly used in biology.

19.2.3.4 Is Planet a Useful Scientific Concept?

Classifications, even if they have a conventional aspect, are useful and fulfill an essential role in science. They are most useful when they correspond to natural classifications, when our taxonomy singles out a well-defined class of properties possessed by objects found in nature (natural kinds). These natural properties and objects are best identified through our laws of nature, and our classifications will evolve as our laws – and more generally our understanding of the world – evolve. For example, the fact that the existence of Ceres and Uranus had been successfully predicted by the Titius-Bode law[57] would have rendered the demotion of Ceres tantamount to the rejection of that law or the demotion of other planets. After discovering that the Titius-Bode law was not valid (it failed to predict Neptune's orbit) and that Ceres was, unlike the other planets, only one of the thousands of small bodies on its orbital region, we could demote Ceres from its planetary status at no scientific cost.

Nowadays, planetary scientists would not make much use of a concept that would not distinguish the more massive objects of the Solar System. All of the planets, as defined by the IAU's 2006 decision, are studied individually and have very diverse bulk compositions and atmospheres. Each has a gravitational influence on its environment much more significant than any dwarf planet or SSSB would have. These smaller objects on the other hand are more significantly approached *statistically*, as members of a large collection of similar objects. They may be further divided into, for example, objects of the asteroid belt and trans-Neptunian Objects (TNOs), among which are Kuiper Belt Objects (KBOs) and Scattered Disk Objects (SDOs).[58] These terms and distinctions are at least as meaningful and useful to astronomers as the notion of planet as defined by the 2006 IAU decision.

On the other hand, the usefulness of the notion of "dwarf planet" is not clear. For astronomers only interested in bodies that have an important influence in the Solar System, it is important to distinguish between the larger eight planets and all the other smaller bodies. For those scientists, the category of *dwarf planet* is of no great use. But for planetary scientists interested in the inner structure of astronomical bodies ("planetary geologists" so to speak), it is important to distinguish between SSSBs and all the larger bodies, large enough to have attained hydrostatic equilibrium. However, the distinction between planets and dwarf planets – or even that between planets and moons – is of no great use. A more likely explanation for the existence of the category of dwarf planet is that it maintained Pluto in a somewhat privileged position among the SSSBs (see Weintraub 2008). One can only speculate what the emotional response of the general public would have been if the planet that gave its name to a beloved Disney character (and the only planet discovered by an American) had lost even the right to be called "dwarf planet"!

The definition adopted by the IAU makes of planet a clearly, empirically identifiable notion, ready to be used in a scientific characterization of our Solar System

---

[57] According to the Titius-Bode law, there should be planets at a distance a from the Sun, with $a = 0.4 + 0.3 * 2^m$ for $m = -\infty, 0, 1, 2, \ldots$ (in astronomical unit).

[58] SDOs are objects identified by their orbital characteristics.

according to our best knowledge at hand. In that sense, planet thus defined is a natural object more than a cultural one. The difficulty to come to such a definition depends on our knowledge of the population of the Solar System, but also on what we expect from scientific definitions, namely, to what extent they should capture natural kinds. The IAU could have made other choices, and it is not obvious that the concept of planet is as useful as other subcategories used by planetary scientists, other astronomers, or scientists in general. In any case, such decisions are not binding for scientists. This definition may not be as relevant or clear-cut in the future as the science evolves.[59] This may explain why the 2006 IAU decision was not warmly received within the profession. While 424 astronomers took part in the vote, more than a 1,000 present in the room where the vote was held *did not cast a vote*. Mike Brown, one of the discoverers of Eris and Sedna, agreed with the IAU decision and thinks that the newly adopted definition of planet is *"the best possible scientific definition we could have."*[60] However, he expressed doubt that a definition was even necessary, explaining that the concept of planet could have been left aside by the scientific community and be seen as a cultural rather than scientific concept, akin to that of continent.[61]

19.2.3.5   Interest for Science Education

Planet will be one of the first categories taught in a science class, even at an early age when the notion of the Earth as a planet is not fixed. And yet students' and teachers' misconceptions on the basic structure of the Solar System are common and persistent (Sadler et al. 2010; Frède 2006). Debating the question of the planetary status of Pluto or the relevance of the IAU definition could make for an active, student-centered, and concept-centered way to learn about the structure of the Solar System as well as methodological aspects of scientific taxonomy. Indeed, in order to take part in this debate, one has to have a notion of the Solar System that includes not only the main planets and the Sun but also asteroids and other small bodies. Taking a side implies understanding notions that should be mastered by the end of a secondary-level education.[62] In such a debate, philosophical considerations about the level of arbitrariness and empirical grounding of scientific concepts can help teachers justify what definitions are receivable and why.

---

[59] Already its use is being perverted in many articles that refer to planets outside our Solar System, as it seems quite natural to talk about "planet" rather than "exoplanet" when the context is clear.

[60] http://web.gps.caltech.edu/~mbrown/eightplanets/. It is to be noted that he was not present at the 2006 IAU meeting, not being a member of the organization at that time.

[61] Interview on the American National Public Radio (*Science Friday*, August 18, 2006).

[62] Namely, (1) that many bodies other than the largest ones that are visible to the naked eye are part of our Solar System; (2) that these bodies orbit the Sun; (3) that orbits are not necessarily circular and that only certain objects of the Solar System have almost circular orbits, located in the same plane; (4) that some of these bodies are spherical because they have a sufficient mass; and (5) that sufficiently massive bodies "clear their orbit"

References

AAAS (American Association for the Advancement of Science) (1990). *Science for All Americans (Project 2061)*. New York: Oxford University Press.

AAAS (1994). *Benchmarks for Science Literacy (Project 2061)*. New York: Oxford University Press.

AAAS (2001). *Designs for Science Literacy (Project 2061)*. New York: Oxford University Press.

Abd-El-Khalick, F. (2012). Examining the Sources for Our Understandings About Science: Enduring Conflations and Critical Issues in Research on Nature of Science in Examining the Sources for Our Understandings about science. *International Journal of Science Education*, 34(3), 353–374.

Adams, J., Adrian, R.L., Brick, C., Brissenden, G., Deming, G., Hufnagel, B., Slater, T., Zeilik, M. & the Collaboration for Astronomy Education Research (CAER) (2000). *Astronomy Diagnostic Test* (ADT) Version 2.0. http://solar.physics.montana.edu/aae/adt/

Alic, M. (1986). *Hypatia's Heritage: A History of Women in Science from Antiquity through the Nineteenth Century*. Boston, MA: Beacon Press.

Allchin, D. (2012). The Minnesota Case Study Collection: New Historical Inquiry Case Studies for Nature of Science Education. *Science & Education*, 21(9), 1263–1281.

Arons, A.B. (1965). *Development of Concepts of Physics*. Reading, Massachusetts: Addison-Wesley.

Atwood, R.K. & Atwood, V.A. (1995). Preservice Elementary Teachers' Conceptions of What Causes Night and Day. *School Science and Mathematics*, 95, 290–294.

Atwood, R.K. & Atwood, V.A. (1996). Preservice Elementary Teachers' Conceptions of the Causes of Seasons. *Journal of Research in Science Teaching*, 33(5), 553–563.

Bailey, J.M. (2007). Development of a Concept Inventory to Assess Students' Understanding and Reasoning Difficulties about the Properties and Formation of Stars. *Astronomy Education Review*, 6(2), 133–9.

Bailey, J.M. (2009). Concept Inventories for ASTR0 101. *The Physics Teacher* 47(7), 439–41.

Bailey, J.M., Johnson, B., Prather E.E. & Slater, T.F. (2012). Development and Validation of the Star Properties Concept Inventory. *International Journal of Science Education* 34(14), 2257–2286.

Bailey, J.M. & Slater, T. (2003). A Review of Astronomy Education Research. *Astronomy Education Review*, 2(2), 20–45.

Bardar, E.M., Prather, E.E., Brecher, K. & Slater, T.F. (2007). Development and Validation of the Light and Spectroscopy Concept Inventory. *Astronomy Education Review*, 5(2), 103–113.

Baumgardt, C. (1951). *Johannes Kepler: Life and Letters*. Philosophical Library.

Baxter, J. (1989). Children's Understanding of Familiar Astronomical Events. *International Journal of Science Education*, 11(5), 502–513.

Bayraktar, Ş. (2009). Pre-service Primary Teachers' Ideas about Lunar Phases. *Turkish Science Education*, 6(2), 12–23.

Bird, A. & Tobin, E. (2010). Natural Kinds. *The Stanford Encyclopedia of Philosophy (Summer 2010 Edition)*, Zalta, N.E. (ed.). http://plato.stanford.edu/archives/sum2010/entries/natural-kinds/

Bokulich, A. (forthcoming). "Pluto and the 'Planet Problem': Folk Concepts and Natural Kinds in Astronomy", *Perspectives on Science*.

Boyd, R. (1999). Homeostasis, Species, and Higher Taxa, in R. Wilson (ed.) *Species: New Interdisciplinary Essays*. Cambridge, MA: MIT Press.

Bretones, P.S. & Neto, J.M. (2011). An Analysis of Papers on Astronomy Education in Proceedings of IAU Meetings from 1988 to 2006. *Astronomy Education Review*, 10(1).

Brissenden, G., Slate, T.F. & Mathieu, R. (2002). The Role of Assessment in the Development of the College Introductory Astronomy Course: A "How-to" Guide for Instructors. *Astronomy Education Review*, 1(1).

Brown, M. (2010). *How I Killed Pluto and Why It Had It Coming*, Spiegel & Grau.

Bulunuz, N. & Jarrett, O. S. (2009). Understanding of Earth and Space Science Concepts: Strategies for Concept-Building in Elementary Teacher Preparation. *School Science and Mathematics*, 109(5), 276–289.

Byers, N. & Williams, G. (2006). *Out of the Shadows: Contributions of Twentieth-Century Women to Physics*, Cambridge University Press.

Carey, S. (2000). Science Education as Conceptual Change. *Journal of Applied Developmental Psychology*, 21(1), 13–19.

Carey, S. (2009). *The Origin of Concepts*. Oxford University Press.

Causeret, P. (2005). *Le Ciel à Portée de Main: 50 Expériences d'Astronomie*. Paris: Belin.

Chalmers, A. (1999). *What Is This Thing Called Science?* Indianapolis: Hackett Publishing Company.

Clough, M.P. (2011). The Story Behind the Science: Bringing Science and Scientists to Life in Post-Secondary Science Education. *Science & Education, 20(7–8): 701–717.*

College of the University of Chicago. (1949). *Introductory General Course in the Physical Sciences* Vol. 1, 2. Chicago: The University of Chicago Press.

College of the University of Chicago. (1950). *Introductory General Course in the Physical Sciences* Vol. 3. Chicago: The University of Chicago Press.

Comins, Neil F. (2000) A Method to Help Students Overcome Astronomy Misconceptions. *The Physics Teacher*, 38(9), 542.

Conant, J.B, ed. (1948). *Harvard Case Histories in Experimental Science*. Cambridge, MA: Harvard University Press.

Conant, J.B. & Nash, L.K., eds. (1957). *Harvard Case Histories in Experimental Science*. Cambridge, Massachusetts: Harvard University Press.

Cromer, A. (1993). *Uncommon Sense*. New York: Oxford University Press.

Dedes, C. & Ravanis, K. (2009). History of Science and Conceptual Change: The Formation of Shadows by Extended Lights Sources. *Science & Education*, 18(9), 1135–1151.

deGrasse Tyson, N. (2009). *The Pluto Files*, Norton & Co.

Deming, G. & Hufnagel, B. (2001). Who's Taking Astro 101? *The Physics Teachers*, 39(6), 368–369.

Derry, N.G. (1999). *What Science Is and How It Works*. Princeton University Press.

Diakidoy, I.A. & Kendeou, P. (2001). Facilitating Conceptual Change in Astronomy: A Comparison of the Effectiveness of Two Instructional Approaches. *Learning and Instruction*, 11(1), 1–20.

Dunbar, R. (1995). *The Trouble with Science*. Cambridge: Harvard University Press.

Duschl, R.H. (1992). *Philosophy of Science, Cognitive Science and Educational Theory and Practice*. Albany: SUNY Press.

Duschl, R.H. & Grandy, R.E., eds. (2008). *Teaching Scientific Inquiry: Recommendations for Research and Implementation*, Rotterdam: Sense Publishers.

Duschl, Richard A., and Richard E. Grandy (2012). "Two Views About Explicitly Teaching Nature of Science." *Science & Education*. doi:10.1007/s11191-012-9539-4.

Eflin, J., Glennan, S. & Reisch, G. (1999). The Nature of Science: A Perspective from the Philosophy of Science. *Journal of Research in Science Teaching*, 36(1), 107–116.

Fraknoi, A. (2008). Women in Astronomy: An Introductory Resource Guide to Materials in English. http://astrosociety.org/edu/resources/womenast_bib.html. Accessed June 2012.

Frède, V. (2006). Pre-Service Elementary Teacher's Conceptions about Astronomy. *Advances in Space Research*, 38(10), 2237–2246.

Frède, Valérie (2008). "Teaching Astronomy for Pre-Service Elementary Teachers: A Comparison of Methods." *Advances in Space Research* 42 (11), 1819–1830.doi:10.1016/j.asr.2007.12.001.

Goodstein, D. & Goodstein, J.R. (1996). *Feynman's Lost Lecture, The Motion of Planets around the Sun*. W.W. Norton & Company.

Gordon, A. (1978). Williamina Fleming: "Women's Work" at the Harvard Observatory. *Women's Studies Newsletter*, 6(2), 24–27. The Feminist Press at the City University of New York.

Gouguenheim, L., McNally, D. & Percy, J.R., eds. (1998). *New Trends in Astronomy Teaching*. Proceedings of the International Astronomical Union 162, London and Milton Keynes, UK, July 8–12, 1996. Cambridge University Press.

Greve, J.P. de (2009). Challenges in Astronomy Education. *Proceedings of the International Astronomical Union*, 5, no. H15, 642–667.

Haandel, M. van & Heckman, G. (2009). Teaching the Kepler Laws for Freshmen. *The Mathematical Intelligencer*, 31(2), 40–44.

Hake, R.R. (2002). Lessons from the Physics Education Reform Effort. *Ecology and Society*, 5(2), 28.

Hake, R.R. (2007). Six Lessons from The Physics Education Reform Effort. *Latin American Journal of Physics Education*, 1(1), 24–31.

Hall, A. (1894). A Suggestion in the Theory of Mercury. *The Astronomical Journal* 14 (319), 49–51.

Harper, W.L. (2002). Newton's Argument for Universal Gravitation. In Cohen, I. B. & Smith, G.E. (eds.). *The Cambridge Companion to Newton*. Cambridge University Press, pp. 174–201.

Harper, W.L. (2007). Newton's Methodology and Mercury's Perihelion Before and After Einstein. *Philosophy of Science*, 74(5), 932–942.

Heath, T.L., ed. (1896). *Apollonius of Perga: Treatise on Conic Sections*. Cambridge University Press.

Hestenes, D. (1987). Toward a Modeling Theory of Physics Instruction. *American Journal of Physics*, 55(5), 440.

Hestenes, D., Wells, M. & Swackhamer, G. (1992). Force Concept Inventory. *The Physics Teacher* 30(3), 141–158.

Hirshfeld, A. (2008). *Astronomy Activity and Laboratory Manual*. Jones & Bartlett Publishers.

Hobson, A. (2003). *Physics Concepts and Connections*. Third Edition. Upper Saddle River, N.J.: Pearson Education.

Holton, G. & Roller, D.H.D. (1958). *Foundations of Modern Physical Science*. Reading, Massachusetts: Addison-Wesley Publishing Company.

Holton, G. & Brush, S.G. (1985). *Introduction to Concepts and Theories in Physical Science*. Princeton, New Jersey: Princeton University Press.

Höttecke, D. & Silva, C.C. (2011). Why Implementing History and Philosophy in School Science Education Is a Challenge – An Analysis of Obstacles. *Science & Education*, 20(3–4), 293–316.

Isobe, S. (2005). A Short Overview of Astronomical Education Carried Out by the IAU. In Pasachoff & Percy (2005).

Jarman, R. & McAleese, L. (1996). Physics for the Star-Gazer: Pupils' Attitudes to Astronomy in the Northern Ireland Science Curriculum. *Physics Education*, 31(4), 223–226.

Johnson, G. (2005). *Miss Leavitt's Stars: The Untold Story of the Woman Who Discovered How to Measure the Universe*, Norton Pub.

Kavanagh, C. & Sneider, C. (2007a). Learning about Gravity I. Free Fall: A Guide for Teachers and Curriculum Developers. *Astronomy Education Review*, 5(2), 21–52.

Kavanagh, C. & Sneider, C. (2007b). Learning about Gravity II. Trajectories and Orbits: A Guide for Teachers and Curriculum Developers. *Astronomy Education Review*, 5(2), 53–102.

Kepler, J. (1609). *Astronomia Nova*. Prague.

Kepler, J. (1619). *Harmonices Mundi*. Prague.

Kikas, E. (1998). The Impact of Teaching on Students' Definitions and Explanations of Astronomical Phenomena. *Learning and Instruction*, 8(5), 439–454.

Klopfer, L.E. (1969). The Teaching of Science and the History of Science. *Journal of Research in Science Teaching*, 6(1), 87–95.

Klopfer, L.E. & Cooley, W.W. (1963). The History of Science Cases for High Schools in the Development of Student Understanding of Science and Scientists. *Journal of Research for Science Teaching*, 1(1), 33–47.

Koestler, A. (1960). *The Watershed: A Biography of Johannes Kepler*. Chapter six: The Giving of the Laws. Garden City, N.Y.: Anchor Books (reprinted in 1985 by the University Press of America).

LaPorte, J. (2004). *Natural Kinds and Conceptual Change*. Cambridge: Cambridge University Press.

Lederman, N., Adb-El-Khalick, F., Bell, R. L. & Schwartz, R. S. (2002). Views of Nature of Science Questionnaire: Towards Valid and Meaningful Assessment of Learners' Conceptions of the Nature of Science. *Journal of Research in Science Teaching*, 39(6), 497–521.

Lelliott, A. & Rollnick, M. (2010). Big Ideas: A Review of Astronomy Education Research 1974–2008. *International Journal of Science Education*, 32(13), 1771–1799.

Lewis, M.W. & Wigen, K.E. (1997). *The Myth of Continents: A Critique of Metageography*. Berkeley: University of California Press.

Libarkin, J.C. (2008). Concept Inventories in Higher Education Science. Prepared for the National Research Council Promising Practices in Undergraduate STEM Education Workshop 2 (Washington, D.C., Oct. 13–14, 2008).

Lindell, R. & Olsen, J. (2002). Developing the Lunar Phases Concept Inventory. *Proceedings of the 2002 Physics Education Research Conference*, S. Franklin, J. Marx, & K. Cummings (eds.), New York: PERC Publishing.

LoPresto, M.C. & Murrell, S.R. (2011). An Astronomical Misconceptions Survey. *Journal of College Science Teaching*, 40(5), 14–22.

Mack, P. (1990). Strategies and Compromises: Women in Astronomy at Harvard College Observatory. *Journal for the History of Astronomy*, 21(1), 65–76.

Matthews, M.R. (1994). *Science Teaching*. Routledge: New York and London.

Matthews, M.R. (2002). Constructivism and Science Education: A Further Appraisal. *Journal of Science Education and Technology*, 11(2), 121–134.

McComas, W.F. (2008). Seeking Historical Examples to Illustrate Key Aspects of the Nature of Science. *Science & Education*, 17(2–3), 249–263.

McComas, W. F. & Olson, J. K. (1998). The Nature of Science in International Science Education Standards Documents. In W. F. McComas (ed.), *The Nature of Science in Science Education: Rationales and Strategies*, pp. 41–52. Dordrecht: Kluwer.

Meador, B. D. S. (2009). *Princess, Priestess, Poet: The Sumerian Temple Hymns of Enheduanna*. University of Texas Press.

Moreira, M.A. (2005). *Aprendizaje Significativo Crítico*, Porto Alegre: CIP-Brasil.

Morison, I. (2008). *Introduction to Astronomy and Cosmology*. Wiley.

National Research Council (NRC) (1996). *National Science Education Standards*, Washington, DC: National Academy Press.

NRC (2011). *A Framework for K-12 Science Education: Practices, Crosscutting Concepts, and Core Ideas*. Washington, DC: National Academy Press.

Nersessian, N.J. (1992). How Do Scientists Think? Capturing the Dynamics of Conceptual Change in Science. In Giere, R.N. (ed.) *Cognitive Models of Science*, pp. 5–22. Minneapolis: University of Minnesota Press.

Nersessian, N.J. (2008). *Creating Scientific Concepts*. MIT Press.

Nussbaum, J. (1979). Children's Conceptions of the Earth as a Cosmic Body: A Cross-Age Study. *Science Education*, 63(1), 83–93.

Ogilvie, M. B. (1986). *Women in Science: Antiquity through the Nineteenth Century*. Cambridge, MA: MIT Press.

Okasha, S. (2002). *Philosophy of Science: A Very Short Introduction*. Oxford University Press.

Osborne, J. (1991). Approaches to the Teaching of AT16 – the Earth in Space: Issues, Problems and Resources. *School Science Review*, 72(260), 7–15.

Parker, J. & Heywood, D. (1998). The Earth and Beyond: Developing Primary Teachers' Understanding of Basic Astronomical Events. *International Journal of Science Education*, 20(5), 503–520.

Partridge, B. & Greenstein, G. (2003). Goals for "Astro 101:" Report on Workshops for Department Leaders. *Astronomy Education Review*, 2(2), 46–89.

Pasachoff, J.M. & Percy, J.R., eds. (1990). *The Teaching of Astronomy*. Proceedings of the International Astronomical Union Colloquium 105, Williamstown, MA, USA, July 26–30, 1988. Cambridge University Press.

Pasachoff, J.M. & Percy, J.R., eds. (2005). *Teaching and Learning Astronomy: Effective Strategies for Educators Worldwide*. Cambridge University Press.

Payne-Gaposchkin, C. & Haramundanis, K. (1956). *Introduction to Astronomy*. Prentice-Hall.

Percy, J.R. (2005). Why Astronomy is Useful and Should be Included in the School Curriculum. In Pasachoff & Percy (2005).

Planète Sciences. (2009). *Pas à Pas dans l'Univers: 15 Expériences d'Astronomie Pour Tous*. Paris: Vuibert.

Posner, G. J., Strike, K. A., Hewson, P. W., & Gertzog, W. A. (1982). Accommodation of Scientific Conception: Towards a Theory of Conceptual Change. *Science Education*, 66(2), 211–227.

Prather, E.E., Rudolph, A.L. & Brissenden, G. (2009a). Teaching and Learning Astronomy in the 21st Century. *Physics Today*. October: 41–47.

Prather, E.E., Rudolph, A. L., Brissenden, G. & Schlingman, W. (2009b). A National Study Assessing the Teaching and Learning of Introductory Astronomy. Part I. The Effect of Interactive Instruction. *American Journal of Physics*, 77 (4), 320–330.

Richwine, P.L. (2007). *The Impact of Authentic Science Inquiry Experiences Studying Variable Stars on High School Students' Knowledge and Attitudes about Science and Astronomy and Beliefs Regarding the Nature of Science*. PhD dissertation, University of Arizona. AAT 3254711.

Rossiter, M.W. (1982). *Women Scientists in America: Struggles and Strategies to 1940*. Baltimore, MD: John Hopkins University Press.

Rudolph, A., Prather E.E., Brissenden, G., Consiglio, D. & Gonzaga, V. (2010). A National Study Assessing the Teaching and Learning of Introductory Astronomy Part II: The Connection Between Student Demographics and Learning. *Astronomy Education Review*, 9(1), 010107.

Sadler, P. M., Coyle, H., Miller, J.L., Cook-Smith, N., Dussaul, M. & Gould, R. R. (2010). The Astronomy and Space Science Concept Inventory: Development and Validation of Assessment Instruments Aligned with the K12 National Science Standards. *Astronomy Education Review*, 8(1), 010111.

Sardar, Z. & Loon, B. van (2002). *Introducing Science*. Cambridge: Icon Books, Cambridge.

Sarton, G. (1959). *A History of Science. Hellenistic Science and Culture in the Last Three Centuries B.C.*, Cambridge, Mass.: Harvard University Press.

Schlingman, W.M., Prather, E.E., Colin, W.S., Rudolph, A. & Brissenden, G. (2012). A Classical Test Theory Analysis of the Light and Spectroscopy Concept Inventory National Study Data Set. *Astronomy Education Review*, 11(1).

Schoon, K.J. (1995). The Origin and Extent of Alternative Conceptions in the Earth and Space Sciences: A Survey of Pre-Service Elementary Teachers. *Journal of Elementary Science Education*, 7(2), 27–46.

Slater, T.F. (1993). *The Effectiveness of a Constructivist Epistemological Approach to the Astronomy Education of Elementary and Middle Level In-Level Teachers*. Doctoral dissertation, University of South Carolina, Columbia.

Slater, T.F. (2008). The First Big Wave of Astronomy Education Research Dissertations and Some Directions for Future Research Efforts. *Astronomy Education Review*, 7(1), 1–12.

Slater, T. F. & Adams, J. P. (2002). *Learner-Centered Astronomy Teaching: Strategies for ASTRO 101*. Upper Saddle River, NJ: Prentice Hall/Pearson Education.

Slater, T.F. & Slater, S.J. (2008). Development of the Test of Astronomy Standards (TOAST) Assessment Instrument. *Bulletin of the American Astronomical Society*, 40, 273.

Sneider, C.I. & Ohadi M.M. (1998). Unraveling Students' Misconceptions about the Earth's Shape and Gravity. *Science Education*, 82(2), 265–284.

Soter, S. (2006) What is a Planet? *The Astronomical Journal*, 132(6), 2513–2519.

Stern, S.A. & Levison, H. (2002). Regarding the Criteria for Planethood and Proposed Planetary Classification Schemes. *Highlights in Astronomy*, 12, 205–213.

Swarup, G., Bag, A.K. & Shulda, K.S. eds. (1987). *History of Oriental Astronomy*. Proceedings of the International Astronomical Union Colloquium no. 91, New Delhi, India, November 13–16, 1985. Cambridge University Press.

Taylor, I., Barker, M. & Jones, A. (2003). Promoting mental model building in astronomy education. *International Journal of Science Education*, 25(4), 1205–1222.

Thompson, M. (2001). *Teach Yourself Philosophy of Science*. New York: McGraw Hill.

Trumper, R. (2000). University Students' Conceptions of Basic Astronomy Concepts. *Physics Education*, 35(1), 9–15.

Trumper, R. (2001a). Assessing Students' Basic Astronomy Conceptions from Junior High School through University. *Australian Science Teachers Journal*, 41(1), 21–31.

Trumper, R. (2001b). A Cross-Age Study of Senior High School Students' Conceptions of Basic Astronomy Concepts. *Research in Science & Technological Education*, 19(1), 97–109.

Trumper, R. (2003). The Need for Change in Elementary School Teacher Training – A Cross-College Age Study of Future Teachers' Conceptions of Basic Astronomy Concepts. *Teaching and Teacher Education*, 19(3), 309–323.

Trumper, R. (2006). Teaching Future Teachers Basic Astronomy Concepts – Seasonal Changes – at a Time of Reform in Science Education. *Journal of Research in Science Teaching*, 43(9), 879–906.

Trundle, K.C., Atwood, R.K. & Christopher, J.E. (2002). Preservice Elementary Teacher's Conceptions of Moon Phases before and after Instruction. *Journal of Research in Science Teaching*, 39(7), 633–658.

Vosniadou, S. & Brewer, W.F. (1992). Mental Models of the Earth : A Study of Conceptual Change in Childhood. *Cognitive Psychology*, 24(4), 535–585.

Wallace, C. & Bailey, J. (2010). Do Concept Inventories Actually Measure Anything? *Astronomy Education Review*, 9(1), 010116.

Waller, W.H. & Slater, T.F. (2011). Improving Introductory Astronomy Education in American Colleges and Universities: A Review of Recent Progress. *Journal of Geoscience Education*, 59(4), 176–183.

Weintraub, D.A. (2008). *Is Pluto a Planet?: A Historical Journey through the Solar System*. Princeton University Press.

Williamson, K.E. & Willoughby, S. (2012). Student Understanding of Gravity in Introductory College Astronomy. *Astronomy Education Review*, 11(1), 010105.

Wolff, S.C. & Fraknoi, A. (2005). *The Astronomy Education Review*: Report on a New Journal. In Pasachoff & Percy (2005).

Wolpert, L. (1994). *The Unnatural Nature of Science*. Harvard University Press.

Zeilik, M. (1993). *Conceptual Astronomy*. Wiley.

Zeilik, M. & Morris, V.J. (2003). An Examination of Misconceptions in an Astronomy Course for Science, Mathematics, and Engineering Majors. *Astronomy Education Review*, 2(1), 101–119.

Zeilik, M., Schau, C., Mattern, N., Hall, S., Teague, K.W. & Bisard, W. (1997). Conceptual Astronomy: A Novel Model for Teaching Postsecondary Science Courses. *American Journal of Physics*, 65(10), 987–996.

Zeilik, M., Schau, C. & Mattern, N. (1998). Misconceptions and Their Change in University-Level Astronomy Courses. *The Physics Teacher*, 36(2), 104–107.

Zirbel, E. L. (2004). Framework for Conceptual Change. *Astronomy Education Review* 3(1), 62–76.

Horacio Tignanelli is an astronomer and a graduate of the Universidad Nacional de La Plata (UNLP) in Argentina. He specialized in the history and teaching of astronomy. Since 2000, he has been part of a group of specialists at the National Ministry of Education, where he is developing science education programs (particularly astronomy) for all education levels. Tignanelli is a Professor of Astronomy at the Universidad Nacional del Centro de la Provincia de Buenos Aires (UNICEN) and the Universidad de La Punta (ULP, Provincia San Luis), as well as in different Teacher Training Institutes of the Ciudad Autónoma de Buenos Aires. Previously (1980–2000), he worked for the Buenos Aires Province Commission for Scientific Research (CIC), developing astronomical research and education programs in the Faculty of Astronomy and Geophysics (UNLP).

In addition to journal articles, he has published dozens of astronomy books (textbooks and popular science works) for teachers, students, and for the general public. His historical studies have surveyed part of the astronomical history of his country. At the core of his educational activity are informal education methods, typical of education with art (he has produced and written programs for TV, radio, Internet, and scientific theater).

Yann Benétreau-Dupin is a doctoral student in philosophy at the University of Western Ontario and the Rotman Institute (Canada). He received an MA in Philosophy from Boston University, where he worked on logic and scientific methodology, after studying at Université Paris 1 Panthéon-Sorbonne and Università di Bologna. His main interests are philosophical foundations of physics, methodology in experimental sciences, and particularly questions of confirmation, the role of explanation, and the status of laws in physics (especially contemporary cosmology).

He worked as a research and teaching assistant at Boston University for ITOP (Improving the Teaching of Physics), a graduate program led by BU's School of Education and Department of Physics. Previously, he worked in science education for Planète Sciences, a national educational organization in France, where he developed programs to promote hands-on projects in astronomy in secondary- and higher-level education, and ran an educational astronomical observatory in the Paris region.

Part VII
Pedagogical Studies: Cosmology

Chapter 20
The Science of the Universe: Cosmology and Science Education

Helge Kragh

20.1 Introduction

Whether majoring in science or not, students at high school and undergraduate university level are confronted with issues of cosmology, a subject which has only attracted a limited amount of attention in the context of science education (Kragh 2011a). It is important that when students are introduced to cosmology, this is done correctly not only in the technical sense but also in a conceptual sense. As shown by several studies, misconceptions abound in both areas. They include some of the philosophical aspects that are so closely intertwined with cosmology in the wider sense and to which a large part of cosmology's popular appeal can be attributed. These aspects need to be addressed and coordinated with the more standard, scientific aspects. In this respect it is often an advantage to refer not only to the modern big bang theory but also to older developments that may illuminate modern problems in cosmology in a simple and instructive manner.

Following a brief discussion of the development of cosmology as a science, the article focuses on various conceptual misunderstandings that are commonly found in students' ideas about modern cosmology. Some of these misconceptions are of a philosophical nature, for example, related to the concept of the universe and its supposed birth in a big bang. By taking issues of this kind seriously, students will hopefully be brought to reflect on the limits of science and adopt a critical attitude to what scientific cosmology can tell us about the universe.

H. Kragh (✉)
Centre for Science Studies, Aarhus University, Aarhus, Denmark
e-mail: helge.kragh@ivs.au.dk

M.R. Matthews (ed.), *International Handbook of Research in History,*
Philosophy and Science Teaching, DOI 10.1007/978-94-007-7654-8_20,
© Springer Science+Business Media Dordrecht 2014

20.2 Early Cosmology: Lessons for Science Education

According to the view of most physicists and astronomers and also according to some historians of science (Brush 1992), cosmology became a science only in the twentieth century. Some will say that the supposed turn from 'philosophical' to truly scientific cosmology only occurred with the discovery of the cosmic microwave background radiation in 1965, while others date the turn to Edwin Hubble's insight in the late 1920s of the cosmological significance of the galactic redshifts. Others again suggest that the turning point is to be found in Albert Einstein's cosmological model of 1917 based on his general theory of relativity.

The widely held opinion that there was no scientific cosmology – scientific in more or less the modern sense of the term – before Einstein and Hubble entered the stage is reflected in most introductory textbooks in physics and astronomy. The general structure of these books is to start with the solar system and then proceed to stars and galaxies, ending with the universe as a whole. The chapters on cosmology are usually restricted to post-1920 developments (Krauskopf and Beiser 2000). Although earlier developments are sometimes included, then it occurs in sections that appear separate from the account of modern cosmology and are typically placed in the beginning of the book. For example, the epic confrontation between the Aristotelian-Ptolemaic universe and the heliocentric world system during the so-called Copernican revolution is a classic theme in the teaching of physics and astronomy, where it is often presented as a methodological case study. On the other hand, textbooks and similar teaching materials rarely refer to other parts of the rich history of cosmological thought, for which teachers and students must look up the literature written by historians of science (North 1994; Kragh 2007). The exception to this state of affairs is Olbers' famous paradox of the dark night sky, dating from 1826 but with roots back to Johannes Kepler, which can be found in most textbooks.

Although modern cosmology dates in most respects from the early part of the twentieth century, it does not follow that earlier theories about the universe were not scientific. The cosmos of the ancient Greeks was very different from ours, yet Ptolemy's cosmology was basically scientific in so far that it was a mathematical model that rested on observations and had testable consequences. At any rate, there are good reasons to include aspects of pre-Einsteinian cosmology also in the context of science education. For one thing, students should be aware of this earlier development for general cultural reasons. Moreover, the earlier history of cosmology provides many more examples of educational relevance than just the one of the Copernican revolution. Although Michael Crowe's two books on theories of the universe are not ordinary textbooks, they are based on his very extensive experience with teaching history of astronomy and cosmology at the University of Notre Dame (Crowe 1990, 1994). They are of value to the teacher of introductory astronomy courses because they include a large amount of primary sources from Ptolemy to Hubble that can be easily used in the classroom. Moreover, Crowe (1994) includes laboratory exercises related to the studies of the nebulae by William Herschel in the late eighteenth century and by William Parsons, the Earl of Rosse, in the mid nineteenth century.

20   The Science of the Universe: Cosmology and Science Education   645

To illustrate the relevance of earlier cosmological thought in science education, consider the discussion in the thirteenth century concerning the possibility of an eternal yet created universe. The discussion was abstract and philosophical, not scientific, but it is nonetheless of relevance to problems of modern cosmology because it led Thomas Aquinas and other scholastic thinkers to scrutinize the concept of creation in a sophisticated way that went beyond the identification of creation with temporal beginning (Carroll 1998, see also below). As another example one might point to the difficult problem of spatial and material infinity as it turned up in Isaac Newton's correspondence with Richard Bentley in the early 1690s. Both Bentley and Edmund Halley mistakenly believed that in an infinite stellar universe each star would be attracted by equal forces in any direction and therefore be in a state of equilibrium. The belief is intuitively convincing and probably shared by most students, but Newton knew better. As he pointed out, two infinities do not cancel. The case is well suited to discuss with students the tricky problems of infinities that appear no less prominently in modern cosmology than they did in the past.

Students should also be aware that the fundamental distinction between realism and instrumentalism, an important issue in the discussion of the nature of science (Campbell 1998), does not turn up only in microphysics but also in cosmology. After all, the universe is no less unobservable than are quarks and superstrings. No one has ever observed the universe and no one will ever do so, so how can we know that the universe exists? The realist will claim that 'the universe' designates an entity that exists independently of all cosmological enquiry, while the instrumentalist considers it a concept that can be ascribed a meaning only in a pragmatic sense, as it is a construct of cosmological theory. The tension between the two opposite views can be followed through much of the history of cosmology, from Ptolemy's world system to the modern multiverse, and from a teaching point of view, it may sometimes be an advantage to refer to older sources rather than to modern examples. To illustrate cosmological or astronomical antirealism with regard to theories, one may read passages of Stephen Hawking (a positivist and instrumentalist), but the same point is brought home, and with greater clarity, by Andreas Osiander's notorious preface to Nicolaus Copernicus' *De Revolutionibus*.

20.3   Patterns in the Development of Modern Cosmology

To the extent that practicing scientists are familiar with philosophical theories of science, the theories are often limited to the views of Karl Popper and Thomas Kuhn. The ideas of these two philosophers are also likely to be the only ones that students will meet, either explicitly or implicitly, in physics and astronomy courses.

While historians agree that Kuhn's theory of scientific revolutions does not in general fit very well with the actual history of science, the history of cosmology yields some support for the notion of paradigm-governed science and revolutionary changes, if not in the radical sense originally proposed by Kuhn (Kragh 2007, pp. 243–245). In both the older and the modern history, there are several cases of beliefs and traditions that formed the nearly unquestioned framework of cosmological thinking and hence had the character of paradigm. Thus, until about 1910, it was generally believed that the stellar universe was limited to the Milky Way. As the astronomy writer Agnes Clerke asserted, 'No competent thinker, with the whole of the available evidence before him, can now, it is safe to say, maintain any single nebula to be a star system of coordinate rank with the Milky Way' (Clerke 1890, p. 368). She added: 'With the infinite possibilities beyond, science has no concern'.

Likewise, until 1930, the static nature of the universe as a whole was taken for granted. Current cosmology is solidly founded on Einstein's theory of general relativity and some kind of big bang scenario, elements that are largely beyond discussion and conceived as defining features of cosmological theory. Yet, although it may be tempting to characterize these beliefs as paradigmatic, they are so in a different sense from what Kuhn spoke of in his classical work of 1962, *The Structure of Scientific Revolutions*. First of all, there is no indication of radical incommensurability gaps in the development that led from the static Milky Way universe to the current standard model of big bang cosmology.

The applicability of the Kuhnian model to the case of modern cosmology has been investigated by Copp (1985) from a sociological perspective and by Marx and Bornmann (2010) using bibliometric methods. Marx and Bornmann examine what they misleadingly call 'the transition from the static view of the universe to the big bang in cosmology', a process that supposedly occurred in the mid-1960s when the steady state model was abandoned in favour of the hot big bang model. (In reality, the transition from a static to a dynamic universe occurred in the early 1930s and was unrelated to ideas about a big bang.) As indicated by bibliometric data, the emergence of the victorious big bang model in the 1960s marked a drastic change in cosmology, if not a sudden revolution.[1] Based on citation analysis, the two authors suggest that if there were a paradigm shift, it was a slow process ranging from about 1917 to 1965 – which cannot reasonably be called a paradigm shift in Kuhn's sense.

Shipman (2000) found that nearly half of his sample of astronomers had never heard of Kuhn and that an additional third was only vaguely familiar with him. Of those who were aware of Kuhn's philosophy, several responded that it informed their teaching and consequently was of value in the classroom. One respondent said: 'I think changing paradigms are so obvious in astronomical history that it goes almost without saying that his work is interesting to an astronomer, but I never thought to actually make a big deal of it in class' (Shipman 2000, p. 165). Whereas some astronomers found Kuhn's model to be helpful in understanding the development of the astronomical sciences, none of them thought it was relevant to their research or had an impact on modern astronomy and cosmology. As one astronomer responded, 'Kuhn ... has no effect on the way science is done' (p. 169).

---

[1] The number of publications on cosmology grew dramatically in the 1960s, apparently an indication of the revolutionary effect caused by the standard big bang theory (Kaiser 2006, p. 447; Marx and Bornmann 2010, p. 543). However, the growth is in some respect illusory, as the number of publications in the physical and astronomical sciences as a whole grew even more rapidly. While cosmology in 1950 made up 0.4 % of the physics research papers, in 1970 the percentage had shrunk to a little less than 0.3 % (Ryan and Shepley 1976). Numerical data can be presented in many ways, sometimes resulting in opposite messages.

In this respect, the case of Popper is rather different as his falsificationist philosophy of science has exerted a strong and documented influence on the astronomical and cosmological sciences and continues to do so (Sovacool 2005; Kragh 2013). Although most cosmologists are only superficially acquainted with Popper's ideas, which they tend to use in a simplified folklore version, they often invoke them as a guide for constructing and evaluating theories. This is evident from the modern controversy over the multiverse, and it was just as evident in the past, when Popperian standards played an important role in the debate between the steady state theory and the class of relativistic evolution theories (Kragh 1996, pp. 244–246). Hawking has in general little respect for philosophy, but in his best-selling *A Brief History of Time*, he nonetheless pays allegiance to the views of Popper:

> Any physical theory is always provisional, in the sense that it is only a hypothesis: you can never prove it. … On the other hand, you can disprove a theory by finding even a single observation that disagrees with the predictions of the theory. As philosopher of science Karl Popper has emphasized, a good theory is characterized by the fact that it makes a number of predictions that could in principle be disproved or falsified by observation (Hawking 1989, p. 11).

Influential as Popperianism as in cosmological circles, the influence is mostly limited to the popular literature and general discussions of a methodological nature. As it is the case with Kuhn, Popper's name very rarely appears in research papers. Perhaps more surprisingly, the same seems to hold for elementary textbooks in astronomy and cosmology. On the other hand, the influence of a philosopher may be visible even though his or her name is missing. Thus, in a brief methodological section, astronomy author Karl Kuhn writes: 'A theory of science must be able to be shown to be wrong. A theory must be testable. Every theory must be regarded as tentative, as being only the best theory we have at present. It must contain within itself its own possibility of destruction' (Kuhn 1998, p. 557). It is then up to the teacher whether Popper should be named or not.

20.4 Conceptions and Misconceptions of Cosmology

Most of the misconceptions about cosmology commonly found among students concern the two fundamental concepts of the expanding universe and the big bang. The two concepts are closely connected, but the precise connection between them is often misconceived.

20.4.1 The Expanding Universe

The standard tradition in introductory astronomy and physics textbooks dealing with cosmology is understandably characterized by an emphasis on observations rather than theory. Observations are used as arguments for new concepts and often presented in a historical context. Expositions typically start with two important and connected observations from the early decades of the twentieth century, Vesto Melvin Slipher's discovery in the 1910s of galactic redshifts and Hubble's conclusion from 1929 of a linear relationship between the redshifts and the distances of the galaxies. Both of these historical cases are easily comprehended and can, moreover, be turned into students' exercises by providing the students with the data used by the two astronomers or by using the students' own data found with a 'simulated telescope' (Marschall et al. 2000). From the Hubble relation, there is but a small step to the expanding universe. Almost without exception, textbooks and popular expositions illustrate the expansion of space by means of the inflating-balloon analogy, which may also be used to introduce the notion of curved space such as applied in relativistic cosmology. This standard and very useful analogy – to 'imagine the nebulæ to be embedded in the surface of a rubber balloon which is being inflated' – was first suggested by Arthur Eddington in (1931), shortly after the expansion of the universe had been recognized (Eddington 1931). It also figured prominently in Fred Hoyle's *The Nature of the Universe* from 1950, a classic in the popular astronomy and cosmology literature.

Although there may be but a small step from the Hubble relation to the expanding universe, the step is real and should not be ignored. Students may be told that the expansion of the universe is an observational fact, but this is not quite the case. We do not *observe* the expansion, which does not follow from the data of either Hubble or later observers. As Hubble was keenly aware of, it takes theoretical assumptions (such that the redshifts are due to a Doppler effect) to translate the measured redshifts into an expansion of the universe. It is quite possible to accept the redshift-distance relation and, at the same time, maintaining that the universe is static, such as many scientists did in the 1930s and a few still do. In fact, Hubble, a cautious empiricist, never concluded that the universe is in a state of expansion. What is 'commonly known' and stated in many textbooks and articles, namely, that 'The expansion of the universe was discovered by Edwin Hubble in 1929' (Lightman and Miller 1989, p. 135), is just wrong. Hubble did not discover the expansion of the universe, and he never claimed that he did (Kragh and Smith 2003).

There is a tendency in textbooks, perhaps understandable from a pedagogical perspective, to simplify and dramatize discoveries. For example, one textbook presents Hubble's discovery of the redshift-distance relation as follows: 'The law was published in a 1929 paper on the expansion of the universe. It sent shock waves through the astronomical community' (Kuhn 1998, p. 512). However, it is only in retrospect that Hubble's paper was about the expansion of the universe, and it did not initially create a stir in either the astronomical or the physical community. According to the *Web of Science*, in the years 1929–1930, it received only three citations in scientific journals.

A much better candidate for the discoverer of the expanding cosmos is the Belgian pioneer cosmologist Georges Lemaître, who in a work of 1927 clearly argued that the universe was expanding and even calculated the quantity that came to be known as the Hubble constant (Holder and Mitton 2012). Contrary to Hubble, Lemaître was fully aware that the measured galactic redshifts are not due to a Doppler effect of galaxies flying through space but must be interpreted as the stretching of standing waves due to the expansion of space, that is, as a relativistic effect. As he explained, if light was emitted when the radius of curvature of the closed universe was R_1 and received when it had increased to R_2, the 'apparent Doppler effect' would be given by $\Delta\lambda/\lambda = R_2/R_1 - 1$. The important difference between the Doppler explanation and the relativistic expanding-space explanation can be illustrated in a simple way by means of the balloon analogy (Lotze 1995). One should distinguish between the expansion of space and the expansion of the material universe, such as most textbooks do. It is much easier to comprehend galaxies moving apart, as were they flying through space, but it is more correct to conceive space as expanding and the galaxies changing their relative positions because of the expansion of space. The counterintuitive notion of an expanding empty space, such as implied by the model first studied by Dutch astronomer Willem de Sitter in 1917, illustrates the difference between the two explanations.

As documented by many studies, the expansion of the universe is not well understood, if understood at all, by either the general public or general science students. Comins (2001) discusses a large number of astronomical and cosmological misconceptions, why they are held and how to correct them.[2] Unfortunately, when it comes to the history of cosmology, he expresses several misconceptions of his own, including that Einstein, because he included the cosmological constant in his 1917 cosmological model, 'missed the opportunity to predict that the universe expands' (p. 162). This common misunderstanding is easily seen to be unfounded, for other reasons because the cosmological constant was part of Lemaître's expanding model of 1927 based on Einstein's equations. Moreover, Einstein did not introduce the cosmological constant to keep his universe from expanding but to keep it from collapsing. In short, a cosmological model may describe an expanding universe whether or not it includes a non-zero cosmological constant.

Asked whether the universe is systematically changing in size or remaining about the same size, nearly 60 % of 1,111 interviewed American adults offered the last response. According to the survey conducted by Lightman and Miller (1989), only 24 % of the respondents said that the universe is expanding. Later large-scale surveys of students following introductory astronomy courses confirm that they have difficulties with the expanding universe and other concepts of modern cosmology. Only a minority of the students revealed a reasonably correct understanding of the meaning of the 'expansion of the universe', and a sizeable minority denied that the universe is increasing in size. Instead they suggested that the phrase was a metaphor for how our knowledge of the universe has increased over time (Wallace et al. 2012). One student answered that the expanding universe is an expression for stars and planets moving away from a central area in the universe, if not necessarily from the Earth (Wallace et al. 2011).

---

[2] See also Comins' website on 'Heavenly Errors' that includes nearly 1,700 common misconceptions that students and other people have about astronomy and cosmology. Among them are that the universe has stopped expanding, that there is a centre of the universe and that all galaxies are moving away from the Earth (http://www.umephy.maine.edu/ncomins/).

Another question that often causes confusion is *what* takes part in the expansion. Although the expansion is 'universal', it does not refer to everything. Objects that are held together by other forces than gravity, such as electromagnetic and nuclear forces, remain at a fixed physical size as the universe swells around them. Likewise, objects in which the gravitational force is dominant also resist the expansion: planets, stars and galaxies are bound so strongly by gravitational forces that they are not expanding with the rest of the universe. There is no reason to fear that the distance of the Earth from the Sun will increase because of the cosmic expansion, although worries of this kind are not uncommon (Lightman and Miller 1989). In the survey conducted by Prather and colleagues (2003), 10 % of the students thought that the expansion of the universe has terrestrial consequences, including the separation of the continental plates that is a central part of the geological theory of plate tectonics. Nor is our Local Group of galaxies expanding. The Andromeda Galaxy, for example, is actually approaching the Milky Way, causing a blueshift rather than a redshift. (In 1913, Slipher concluded that the Andromeda Galaxy approached the Sun, only subsequently to realize that it was an exception to the general pattern of galactic redshifts.) On the other hand, on a cosmological scale, all matter is rushing apart from all other matter at a speed described by Hubble's law, $v = Hr$, where H denotes the Hubble parameter or 'constant'. Since the Hubble time $1/H$ is an expression of the age of the universe, H, it is not really a constant but a slowly decreasing quantity.

There are other and more complex ways in which the expansion of the universe can be misconceived, some of them relating to the magical limit of the recession velocity apparently given by the speed of light c (Davis and Lineweaver 2004; Ellis 2007, pp. 1214–1216). Students learn that nothing can move faster than the speed of light, which is a fundamental postulate of the theory of relativity. But according to Hubble's law, the recession velocity keeps increasing with distance, implying that beyond the Hubble distance c/H, the velocity will exceed the speed of light. Can receding galaxies really cross this limit? If they do, will they then become invisible because their redshifts become infinite? In spite of the apparent contradiction with Einstein's postulate, superluminal recession velocities do not violate the theory of relativity. As Lemaître emphasized in 1927, the recession velocity is not caused by motion *through* space but by the expansion *of* space. According to general relativity theory, redshifts do not relate to velocities, as they do in the Doppler description (both classically and in special relativity), and the redshifts of galaxies on the Hubble sphere of radius c/H will not be infinite.

Not only can the universe, or space, expand faster than the speed of light, we can also observe objects that recede from us with speeds greater than this limit. Students may believe that since the universe came into being 13.8 billion years ago, the most distant objects are 13.8 billion light years away, but in that case they think in terms of a static universe. Since distances between faraway galaxies increase while light travels, the observability of galaxies is given by the look-back time, which is the time in the past at which light now being received from a distant object was emitted. As a result of the expansion, the farthest object we can see is currently about 46 billion light years away from us, receding with more than six times the speed of light, even

20 The Science of the Universe: Cosmology and Science Education 651

though the universe is only 13.8 billion years old. The size of the observable universe is not given by the Hubble sphere but by the cosmic particle horizon beyond which we cannot receive light or other electromagnetic signals from the galaxies.

20.4.2 *The Big Bang*

Having digested the notion of expanding space, the next crucial concept that students need to be introduced to, the idea of the big bang, is often presented as a simple consequence of the cosmic expansion.[3] After all, if the distances between galaxies (or rather galactic clusters) increase monotonically, apparently there must have been a time in the past when all galaxies were lumped together. This inference is facilitated by the balloon analogy, where the airless balloon corresponds to the original universe before expansion. However, the inference is more seductive than correct. The argument from expansion to big bang may be pedagogically convincing, but it is not supported by either logic or the history of science. If there were such a necessary connection, how is it that while the majority of astronomers in the 1930s accepted the expansion of the universe, practically no one accepted the idea of an explosive origin?

In the version of the 'primeval atom' hypothesis, the idea of a big bang was first suggested by Lemaître in 1931 – not in his 1927 paper, as is often stated – but it took many years until the hypothesis was taken seriously. The hypothesis was independently revived and much improved by George Gamow and his collaborators in the late 1940s, but even then it failed to win much recognition (Kragh 1996, pp. 135–141). Remarkably, from 1954 to 1963, only a single research paper was published on the big bang theory. During most of the period from about 1930 to 1960, the favoured theory of the evolution of the universe was the Lemaître-Eddington model according to which the universe had evolved asymptotically from a static Einstein state an infinity of time ago. This kind of model is ever expanding but with no big bang and no definite age.

Teachers presumably want their students to accept the big bang theory, but not to do it by faith or authority. To convince students that the big bang really happened, they need to provide good reasons to believe in it, which primarily means observational and other empirical evidence. In this respect, the students may be compared to the majority of astronomers and physicists who still in the 1950s resisted the idea of a big bang, basically because they lacked solid empirical evidence for the hypothesis. As the sceptics pointed out, quite reasonably, if our current universe has evolved from a very small and extremely dense and hot state several billion years ago, there

---

[3] The undignified name 'big bang' was coined by Fred Hoyle in a BBC radio programme of 1949, but neither Hoyle nor other scientists used it widely until the late 1960s. Contrary to what is often said (e.g. Marx and Bornmann 2010, p. 454), the phrase did not catch on either among supporters or opponents of the exploding universe. Hoyle belonged to the latter category, and it generally thought that he coined the name as a way of ridiculing the theory, but this is hardly the case. The first scientific paper with 'big bang' in its title appeared only in 1966.

must presumably still be some traces or fossils from it. If no such traces can be found, we have no reason to believe in the big bang and nor is there any possibility of testing the hypothesis.

An additional reason for the cool reception of the big bang theory was that according to most of the models, the calculated age of the universe came out embarrassingly small, much smaller than the age of the stars and smaller than even the age of the Earth, in the 1930s estimated to be about three billion years. A universe that is younger than its constituent parts is of course ruled out for logical reasons. The age problem is mentioned in some astronomy textbooks but not always historically correct. According to Arny (2004, p. 517), the age of Lemaître's primeval-atom universe was 2/3 times the inverse Hubble constant, which at the time, when Hubble's value $H = 500$ km/s/Mpc was generally accepted, corresponded to only 1.2 billion years. The reference should be to the Einstein-de Sitter model of 1932, which assumed a flat space and a zero cosmological constant. Lemaître, on the other hand, assumed a positive cosmological constant by means of which he was able to avoid the age problem and assign to his universe an age of 20 billion years or more.

It is all important that some kind of fossil is left over from the cosmic past, which otherwise would be inaccessible to us and therefore just a postulate one can believe in or not. It would have the same questionable ontological status as other universes in modern multiverse theories. In evidence-based courses in physics and astronomy, students come to understand and accept the big bang picture by means of empirical evidence such as the cosmic microwave radiation and the abundance of helium in the universe. What matters is not so much the right scientific belief as it is to be able to justify these beliefs and distinguish them from ideas that are not adequately supported by evidence (Brickhouse and colleagues 2000, 2002). Students learn that a theory must necessarily be supported by evidence and also that evidence depends on and is only meaningful in relation to the theory in question. The way students learn to accept the big bang corresponds to some extent to the historical situation in the period from about 1948 to 1965.

The celebrated discovery of the cosmic microwave background killed the already weakened rival steady state theory and turned the big bang theory into a successful standard theory of the universe.[4] Although the best known of the cosmic fossils, the microwave background is not the only one and nor was it the most important in the historical development of cosmology. It may be less well known that the distribution of matter in the universe provides us with another and more easily accessible fossil. None of the 219 students questioned by Bailey and associates (2012) referred to the chemical composition of the universe as evidence for the big bang, while 32 mentioned the expansion and three the cosmic microwave background as evidence.

---

[4] The classical steady state theory was abandoned half a century ago and for this reason is mainly of historical interest. On the other hand, from a methodological and also an educational point of view, it is an instructive example of how an attractive theory with great predictive power was eventually shot down by new observations. In addition, it illustrates the aesthetic and emotional appeal of a cosmological theory, a phenomenon which is not restricted to the past. While Kuhn (1998, p. 555) covers the essence of the steady state theory, other textbook authors choose to ignore it (Krauskopf and Beiser 2000).

The hypothesis that the distribution of matter reflects the cosmic past was first proposed in the late 1930s, when the first reliable data of the cosmic abundance of chemical elements appeared. The general idea in this line of reasoning is that the nuclear species, or at least some of them, are the products of nuclear processes in the early phase of the universe. This was the guiding philosophy of Gamow and his associates Ralph Alpher and Robert Herman, who in the late 1940s developed it into a research programme sometimes known as 'nuclear archaeology' (Kragh 1996, pp. 122–132). The apt phrase underlines the methodological similarity between this area of physical cosmology and ordinary historical archaeology. It refers to attempts to reconstruct the history of the universe by means of hypothetical cosmic or stellar processes and to test these by the resulting pattern of element abundances. Gamow was unable to account in this way for the heavier elements, but in collaboration with Alpher and Herman, he succeeded in calculating the amount of helium in the universe to about 30 % by weight, in reasonable agreement with observations. This early success of the big bang hypothesis was later much improved and extended to other light isotopes such as deuterium.

What matters is that by the late 1960s there was solid empirical evidence for the hot big bang, primarily in the form of the microwave background and the abundance of helium. This does not amount to a 'proof' of the big bang, but it does provide convincing evidence that makes it rational to accept the big bang picture (which does not imply that it is irrational not to accept it). Alternative cosmological models must, as a minimum, reproduce the empirical successes of the standard big bang model and do it without assumptions of an ad hoc nature. To do so on the basis of non-big bang assumptions turns out to be exceedingly difficult. It was the main reason why the steady state model of the universe was abandoned in the late 1960s. The lack of successful rival models is yet another reason to have confidence in the big bang, if by no means to accept it as true.

Whether students follow an evidence-based approach that corresponds to the historical development or not, it is not enough that they can justify their belief in the big bang picture in terms of evidence for it. They also need to know what this picture is, more exactly. If not the students will believe in the big bang, knowing why they believe it but not knowing what they believe in. Several studies show that students have quite different views of the nature of the origin and evolution of the universe. According to a study of Swedish upper-secondary students of age 18–19 years, they conceive the big bang in a variety of ways:

> For example, there are students saying that the universe has always existed in some way. Others talk about a beginning with the Big Bang, but show that they do not view this as an absolute beginning of the universe. ... In addition to the view ascribed above where the Big Bang is viewed as something happening to the whole of the universe, there are also some students who talk about the Big Bang as the origin of the earth and/or the sun (Hansson and Redfors 2006, p. 359).

One of the students described the big bang as an event 'where an explosion made gases and particles spread out in space and then they attracted each other and formed suns' (p. 366). Studies show consistently that the most common misconception of the big bang is to associate it with an explosion of pre-existing matter into empty space (Prather and colleagues 2003; Wallace and colleagues 2012; Bailey and colleagues 2012). Perhaps more surprisingly, only relatively few students connect the big bang to the beginning of the cosmic expansion, and very few think of it as an explosion from nothing.

Although it is hard not to think of the big bang as some kind of explosion of pre-existing matter, it is important to make the students understand that this is at best a somewhat flawed metaphor. Lemaître used the metaphor as early as 1931, when he spoke of his new big bang model as a 'fireworks theory', thereby trying to visualize what happened in the cosmic past. Fireworks explode into the surrounding air, but there is nothing 'outside' that the universe can explode into. While an explosion occurs at some location, the bang of the past did not happen somewhere in the universe. It was the entire universe that 'exploded' and thus the big bang happened everywhere. If this is hard to visualize, it is because it cannot be visualized.

It is also important to be aware that the qualitative meaning of the big bang is that long ago all distances, as given by the scale factor $R(t)$, were nearly zero, after which $R(t)$ increased rapidly. For some 14 billion years ago, the universe was very compact, very hot and, in a sense, very small. The essence of the big bang is not a claim of an absolute beginning in some 'singularity' at $t = 0$ but a claim of a state of the universe, much earlier than and very different from the present state, that has evolved into the one we now observe. Another way of putting it is that the presently observed expansion started at some finite time ago in the cosmic past, so that the expanding universe can be ascribed a finite age. Note that this does not necessarily imply that the universe has a finite age. Creation in an absolute and therefore metaphysical sense is not – and fortunately not – a part of the big bang scenario, just as little as an absolute origin of life is a necessary part of the neo-Darwinian evolution scenario.

20.5   The Concept of the Universe

Although cosmology has undoubtedly developed into a proper and impressive physical science since the 1960s, it is not just another branch of physics or astronomy. Nor is it just astrophysics extended from the stars to the universe at large. No, it is a very special and potentially problematic science in which questions of a philosophical (and sometimes religious) nature cannot be clearly separated from scientific questions relating to observation and theory. To present cosmology to students without taking into regard its special nature is to present them with a narrow and distorted picture of the fascinating science of the universe. Questions of a philosophical nature are part and parcel of what cosmology is about, and they should be given due consideration also in educational contexts, if not at the expense of the scientific issues. This is a major reason why modern cosmology, including aspects of its history, should have a prominent role in science teaching and why it enters significantly in many courses for students not majoring in physics or astronomy.

20.5.1 The Cosmological Principle

Much of cosmology's special and potentially problematic nature is independent not only of the big bang but also of the expansion of the universe. Indeed, being basically of a conceptual nature, it is largely independent of modern scientific discoveries. A key problem, no less important today than it was in the time of Aristotle, is simply the unique domain of cosmology, this most peculiar concept of *the universe*. The standard definition of cosmology is something like 'the science of the universe', yet it is far from obvious that such a frightening concept as the universe can be the subject of scientific study. The relatively recent recognition that this can be done, and that even the universe at large is not foreign land to science, is one of the marvels of the modern physical sciences.

Among the epistemic problems that face a science of the universe is that cosmological knowledge seems to be conditioned by certain principles or assumptions that are completely unverifiable and for this reason may appear to be metaphysical rather than physical (Ellis 1984). The best known of these principles is the so-called cosmological principle, namely, the generally held assumption that the universe is homogeneous and isotropic on a very large scale. It is sometimes referred to as the extended Copernican principle, a rather unfortunate name given that Copernicus' universe had the Sun as its fixed and unique centre. First explicitly formulated in 1932, the cosmological principle lies at the heart of all relativistic standard models, but it is not restricted to models governed by the general theory of relativity. Indeed, when British cosmologist Edward Arthur Milne introduced it in 1932, it was in connection with his own theory of the expanding universe which was entirely different from the theory governed by general relativity. The principle assumes that the vast ocean of unobservable regions of the universe is similar to the region we have empirical access to, a region that may well be an infinitesimal part of the entire universe. What is the epistemic status of the cosmological principle? Is it a necessary precondition for cosmology, or is it merely a convenience that may be accepted or not?

The cosmological principle does have an empirical basis in so far that it roughly agrees with observations, but observations can say nothing about the structure of the universe far beyond the Hubble region, not to mention the cosmic horizon. Extrapolations much beyond this scale are necessarily hypothetical as they rest on an assumption of global uniformity that can never be verified. One might also say that they rest on 'faith', although the faith in the global validity of the cosmological principle is supported by local observations and therefore quite different from 'blind faith'. If cosmology rests on an unverifiable and perhaps metaphysical principle, can it still claim to be scientific? This is not to suggest that the cosmological principle is in fact metaphysical but to suggest that it is worth contemplating the status of the principle and to discuss it also in a teaching context rather than merely present it as a reasonable if unprovable assumption (Kuhn 1998, p. 551).

The instinct of many students majoring in science is to react with hostility and distrust to terms such as 'faith' and 'metaphysics'. (For students not majoring in science, see Shipman and colleagues 2002.) Yet, because something is ultimately a matter of faith, it does not imply that it is irrational, unscientific or arbitrary. There is an element of belief in most scientific ideas. It is important to recognize that unverifiability is not a great methodological sin that automatically deprives a theory or field its scientific status. In fact, students are well aware of high-status scientific theories that cannot be verified, although they may never have thought of them as theories that, in a manner of speaking, rest on belief.

Several of our commonly accepted laws of physics can be said to be cosmological in nature in so far that they are claimed to be true all over the universe and in any patch of cosmic space-time. Newton's law of gravitation speaks of the attractive force between any two masses in the universe, and the law of energy conservation is valid for all processes at any time in the universe. They can reasonably be considered statements relating to the universe at large and for this reason implicitly of a cosmological nature. Of course, neither these two laws nor other similar laws can be verified experimentally. The moral is that students have no reason to fear unverifiability in cosmology, since we have to live with this feature anyway. On the other hand, unfalsifiability is a different matter.

Contrary to what some philosophers have argued (Munitz 1986), the cosmological uniformity principle and similar principles are not of an a priori nature, that is, true by necessity. The cosmological principle is a simplifying assumption that could be proved wrong by observation. In that case it would have to be abandoned, but this would not make cosmology impossible, only more complicated. There are plenty of theoretical cosmological models that do not presuppose homogeneity or isotropy. The case exemplifies the important distinction between verifiability and falsifiability that is a central message in Popperian philosophy of science. That global uniformity principles of this kind are indeed falsifiable is further illustrated by the 'perfect cosmological principle' upon which the now defunct steady state theory was based. This principle extended the cosmological principle to the temporal dimension, namely, by claiming that there is no privileged time in the history of the universe any more than there is a privileged position. When the steady state theory was put in the grave in the 1960s, so was the perfect cosmological principle.

20.5.2 *The Uniqueness of the Universe*

The universe does not only stretch beyond the observable region; it is also, at least according to the ordinary meaning of the term, a unique concept (Ellis 1999). If the universe by definition comprises everything of a physical nature, space and time included, there can only be one universe. Contrary to ordinary physics, which operates with objects and phenomena which are local and of which there are many, the universe is not a member or instance of a class of objects. Newton could establish his inverse-square law of gravitation because there are many bodies that gravitate. By observing and experimenting with different initial conditions, he and later physicists could confirm the validity of the law, but not so with respect to the universe, where the initial conditions are fixed and unchangeable. We cannot rerun the universe with the same or altered conditions to see what would happen if they were different. It seems to follow that we cannot establish proper cosmological laws *of* the universe comparable to the ordinary laws of physics, for we cannot test any such proposed law except in terms of being consistent with a singular 'object', the observed universe.

Since we use laws to explain things, such as explaining the falling apple as an instance of the law of gravity or the energy generated by the Sun as an instance of the laws of quantum physics, it may seem that the domain of cosmology is beyond explanation in the causal-nomological sense normally used in physics. To put it differently, whereas in local physics law-governed and contingent properties can be distinguished, this may not be possible in cosmology. Does it follow that the universe – the domain of cosmology – is beyond explanation? The question was discussed by René Descartes and his contemporaries in the seventeenth century, and it has continued to attract attention from both philosophers and cosmologists. According to Descartes, the divine mechanical laws guaranteed that the original chaos, whatever its structure and initial conditions, would evolve into our universe or one indistinguishable from it. Newton, on the other hand, insisted that the universe cannot be fully understood by the laws of mechanics alone. Descartes' 'indifference principle' continues to play a role in modern cosmology, except that the laws are no longer seen as mechanical only (McMullin 1993).

There are ways to avoid the pessimistic conclusion that the universe is beyond explanation. One strategy is simply to deny the uniqueness of the universe by postulating the existence of many others. Another solution is to recall that there are other forms of explanation than those used in the standard deductive-nomological scheme. Because cosmology is a non-nomological science, it does not follow that it is impossible to account for the present state of the universe. Thus, to explain the fact that the present temperature of the microwave background is about 2.7 K, we do not need a law of the universe or an ensemble of universes we can compare ours with. We can and do offer an explanation – not a causal one, but a historical or genetic explanation – by accounting for how the background radiation cooled with the expansion of the universe.

20.6 Unfinished Businesses

Cosmology of the twenty-first century is in some respects an unfinished business that may provide students with a rare insight in science in vivo. Not only are there important scientific questions that are not solved yet, most notably the nature of dark matter and dark energy, there are also questions of old vintage that may belong as much to philosophy as to science and about which we do not even know whether they are answerable or not. Many students are naturally curious about the kind of borderline questions that cosmology present us with, and teachers should do what they can to satisfy their curiosity. Students should be confronted with problems of this kind and be stimulated to think about them in a critical and rational way.

They should not be dissuaded from asking questions even though these may appear to be naïve – maybe they are not so naïve after all. Modern physical cosmology is a wonderful resource for enlightenment and discussion of questions that relate to the limits of science. Contrary to what is the case in most other sciences, such questions are integrated parts of the science of the universe understood broadly. In general science courses dealing with cosmology, it will be natural to introduce at least some of the issues.

20.6.1 Many Universes?

A typical textbook definition is that 'the visible universe is the largest astronomical structure of which we have any knowledge' (Arny 2004, p. 9). This is a reasonable and operational definition, but why restrict cosmology to the study of the visible universe? There surely is something behind it. In the more general and ambitious sense adopted by some cosmologists, the universe is taken to be 'everything that exists'. If so, it makes no sense to speak of other universes. Nonetheless, this is what several theoretical cosmologists do nowadays, where the question of the definition of the universe has been reconsidered as part of the controversy over the 'multiverse', the hypothesis that there is a multitude of different universes of which the one we observe is only a single member (Carr and Ellis 2008; Kragh 2011b, pp. 255–290). This ongoing controversy has many interesting aspects, not least that critics have questioned the scientific nature of the multiverse hypothesis and thus reopened the old question of whether cosmology, or some versions of cosmology, belongs to physics or metaphysics. On the other hand, advocates of the multiverse argue that it is a scientific idea and that it follows from, or is strongly suggested by, recent developments within string theory and inflation cosmology. Although the multiverse cannot be tested directly, they claim that it leads to testable consequences.

The existence of a cosmic horizon beyond which we will never be able to see or otherwise get information from, not even in principle, is not a new insight. As early as (1931), Eddington pointed out that the accelerated expansion of the closed Lemaître-Eddington universe would eventually lead to 'a number of disconnected universes no longer bearing any physical relation to one another' (Eddington 1931, p. 415). This kind of multiverse is relatively innocent, since the different universes, although causally separated, inhabit the same space-time. More extreme and more speculative is the modern idea of a huge number of disparate universes, each of them with its own physical laws, number of space dimensions and constants of nature (and with ours being perhaps the only one with intelligent life). We obviously cannot have empirically based knowledge about the content and properties of these other worlds, nor can we establish their existence observationally. The numerous other worlds may exist or not, but if the question cannot be decided by means of experiment and observation, does it belong to science?

The recent controversy over the universe may well be used in the teaching of introductory cosmology as it does not rely on advanced theories but is essentially of

20 The Science of the Universe: Cosmology and Science Education

a qualitative and philosophical nature. A recommendable source, most relevant also for the purpose of teaching, can be found in a discussion between George Ellis and Bernard Carr in the journal *Astronomy & Geophysics* (Carr and Ellis 2008). This illuminating source has for some years been used in courses in philosophy of science for undergraduate science students at Aarhus University, and with considerable success. It works very well and provokes much good discussion among the students.

20.6.2 Infinite Space

The problem of the spatial extension of the universe is another of those cosmological questions that have been discussed since Greek antiquity and that we still do not know the answer to. While Einstein's original universe of 1917 was positively curved and with a definite volume, corresponding to a curvature radius of only about ten million light years, the expanding Einstein-de Sitter model of 1932 assumed a flat and therefore infinite space. The same was the case of the steady state universe, where a zero curvature parameter $k = 0$ follows from the perfect cosmological principle. Indirect and model-dependent measurements of the curvature of cosmic space did not lead to a definite answer, but the present consensus model (including inflation and dark energy) strongly favours a flat universe of infinite extent. Assuming the cosmological principle, this implies a universe with an infinite number of objects in it, whether these being electrons or galactic clusters.

Students may tend to think of infinity as just an excessively large number, but (as Newton was well aware of) there is a world of difference between the extremely large and the infinitely large. Actual infinities are notoriously problematic, leading to all kinds of highly bizarre and possibly contradictory consequences. The general attitude of modern cosmologists is to ignore the troublesome philosophical problems of actual infinities and speak of the infinite universe as just an indefinitely large universe, not unlike the students' intuition. Only rarely do they reflect on the weird consequences of the actual infinite – but perhaps they should. Ellis is one of the relatively few cosmologists who take the infinite cosmos seriously, suggesting that the infinities may not be real after all, indeed cannot be real. Ellis and his collaborators argue that physical quantities cannot be truly infinite and that infinite sets of astronomical objects have no place in cosmology. If such quantities formally turn up in a theory or model, it almost certainly means that the theory is wrong. Infinity, they emphasize, 'is not the sort of property that can be physically realized in an entity, an object, or a system, like a definite number can' (Stoeger et al. 2008, p. 17).

Although an infinite universe follows from some cosmological models, we will never know whether the universe is in fact infinite. Observations and theory indicate a flat space, but observations are limited to the visible universe. It is only by assuming the cosmological uniformity principle that we can extrapolate to the universe at large. Moreover, we can never know observationally whether $k = 0$ precisely, only that k varies between the limits $\pm \Delta k$ corresponding to the inevitable observational uncertainties. This observational asymmetry between flat and curved space was pointed out by the Russian mathematician Nikolai Lobachevsky as early as 1829, a century before the expanding universe. It is worth noticing that although the idea of curved space only was adopted by physicists and astronomers with Einstein's general theory of relativity, as a mathematical idea it goes back to the first half of the nineteenth century.

20.6.3 The Enigma of Creation

The traditional version of the big bang theory inevitably invites questions of a philosophical and to some extent religious nature concerning the origin of everything. Although the big bang model is not really a model of absolute beginning or creation, but a cosmic evolutionary scenario, it would be artificial to ignore these questions and simply dismiss them as unscientific. Unscientific they may be, but they are no less natural and fascinating for that. Teachers can keep them out of astronomy and general science courses, but that would be to betray the curiosity and natural instincts of the students. Moreover, questions concerning cosmic creation have a long and glorious history which makes interesting connections between the history of science and the history of ideas, philosophy and religious thought. Whether one likes it or not, the creation of the physical universe is part of the world view of most cultures, and for this reason alone, it should not be ignored in science courses. Fortunately, there is a rich literature on philosophical, political and religious world views and their place in science education (Poole 1995; Matthews 2009).

The problem with creation in a cosmological context is that if we conceive the big bang as an absolute beginning at $t = 0$, then a causal scientific explanation of the creation event is impossible. After all, a cause must come before the effect, and there is no 'before'. Current cosmology has traced the history of the universe back in time to the inflationary period which is supposed to have occurred at $t = 10^{-34}$ s or thereabout. It is often assumed that the cosmic past can be traced even farther back to the Planck time at $t = 10^{-43}$ s (and there are even speculative pre-Planck theories). But however close calculations may bring us to the magical moment $t = 0$, it seems in principle impossible to account for the creation event itself. To say that the universe was created in a space-time singularity is a mere play with words, since the singularity is a mathematical abstraction devoid of physical content. Physics did not exist at $t = 0$ and it makes no sense to speak of physical mechanisms where even the concepts of cause and effect cannot be defined.

In spite of the rhetoric of some cosmologists, there are no scientific theories that explain the origin of the universe from 'nothing', and there never will be such theories. The concept of nothingness or absolute void has a rich history (Genz 1999) that recently has become relevant to science, not least after the discovery of the dark energy that is generally identified with the vacuum energy density as given by the cosmological constant and interpreted in terms of quantum mechanics. However, the modern quantum vacuum is entirely different from absolute nothingness.

There cannot possibly be a scientific answer to what nothingness is, and yet it does not therefore follow that the concept is meaningless.

A major reason why big bang cosmology has been and to some extent still is controversial in the eyes of the public is that it may be seen as a scientific version of Genesis or at least to provide scientific justification for a divinely created world. This misguided view was endorsed by Pope Pius XII in 1951 (Kragh 1996, pp. 256–259) and is still popular in some circles. Although this is not the place to discuss the complex relations between cosmology and religion (Halvorsen and Kragh 2010), it appears that some of these questions are suited for discussions with and among students and should not necessarily be kept out of the physics classroom. Courses that aim to establish a dialogue between science and religion have existed for some time, and in some of them cosmology enters prominently (Shipman and colleagues 2002). The issue is also mentioned in Kuhn (1998), a textbook which includes a brief and admirably clear exposition of the relationship between cosmology and religious faith:

> If we use God as an explanation for the big bang, there would be no reason to look further for a natural explanation. Use of supernatural explanations would shut down science. ... If science relied on a creator to explain the inexplicable, there would be nowhere to go, no way to prove that explanation wrong. The question would have already been settled. ... Science does not deny the existence of God. God is simply outside its realm (Kuhn 1998, p. 557).

While much attention is paid to the origin of the universe, the other end of the cosmic time scale is rarely considered a question of great importance. And yet Einstein's equations of relativistic cosmology are symmetric in time, telling us not only about the distant past but also about the remote future. Will the universe ever come to an end? If so, what kind of end? In the late nineteenth century, these questions were eagerly discussed in relation to the so-called heat death supposedly caused by the increase of entropy in the universe, and recently they have been reconsidered within the framework of modern physics and cosmology. The new subfield known as 'physical eschatology' is concerned, among other things, with the final state of life and everything else (Kragh 2011b, pp. 325–353). Parts of physical eschatology are controversial and highly speculative, yet it is a subject that is likely to appeal to many students and that they should know about. As the birth of the universe relates to religious dogmas, so does its death.

20.6.4 A Universe Without a Beginning

In his last book, *The Demon-Haunted World*, the prominent astronomer and science popularizer and educator Carl Sagan pointed out that science might conceivably demonstrate the universe to be infinitely old. He suggested that 'this is the one conceivable finding of science that could disprove a Creator – because an infinitely old universe would never have been created' (Sagan 1997, p. 265). On the face of it, Sagan's assertion may appear convincing, perhaps even self-evident, but it is based on a misunderstanding that conflates the scientific notion of 'finite age' with the theological notion of 'creation'. Theologians and Christian philosophers agree that even an infinitely old universe would have to be created, in the sense of being continuously sustained, and that it would in no way pose problems for faith. Even if the universe had existed in an infinity of time, we could still ask for the reason of its existence or why it was created.

We have very good reason to believe in the big bang, but we have no good reason to believe that this is how the universe ultimately came into being. Concepts such as cosmic origin and time are difficult, not only conceptually but also for semantic reasons. Thus, we would presumably think that whereas the steady state universe of Hoyle and others had always existed, this is not the case with the finite-age big bang universe. The two statements 'the universe has a finite age' and 'the universe has always existed' appear to be contradictory, but in reality they may both be true. To say that the universe has always existed is to say that it existed whenever time existed. The word 'always' is a temporal term that presupposes time. Since it is hard to imagine time without a universe – much harder than imagining a universe without time – it makes sense to speak of a big bang universe which has always existed. The phrase 'the universe has always existed' reduces to a tautology. This observation is more than just a philosophical nicety, as illustrated by one of the questions posed to students in a questionnaire: 'Does the universe have an age, or has it always existed' (Bailey and associates 2012). Several of the students, we are told, 'gave a contradictory response, such as "the universe has always existed: it is billions of years old"'. As argued, the answer is not really contradictory.

Until recently, it was taken for granted that a universe of finite age implies an absolute cosmic beginning of some kind. The traditional answer to the supposedly naïve question of what there was before the beginning in the big bang has been to dismiss or ridicule it as an illegitimate and meaningless question. For how can there be something 'before' the beginning of time? But there is no reason to ridicule the question if it is recognized that the big bang event at $t = 0$ did not necessarily mark the beginning of time.

During the last two decades, an increasing number of cosmologists have argued that the big bang picture does not preclude a past eternity in the form of, for example, one or more earlier universes. Most theories of quantum gravity operate with a non-singular smallest volume, which makes it possible to extend cosmic time through the $t = 0$ barrier at least in a formal sense.[5] There exists presently a handful of such theories, which are all speculative to varying degrees but nonetheless are considered serious scientific hypotheses. To mention but one example, according to so-called loop quantum cosmology, the universe was not created a finite time ago but exists eternally. There was a big bang, of course, but in the form of a well-described transition of the universe from a contracting to an expanding phase. The space of loop

---

[5] It is far from obvious that the symbol t, as it appears in the equations describing the very early universe near or before the Planck time $t = 10^{-43}$ s, can be ascribed a well-defined physical meaning (Rugh and Zinkernagel 2009). The meaning of time is even less clear in theories of quantum cosmology describing the hypothetical universe before $t = 0$. The claim that there was a universe 'before' ours seems to presuppose a common measure of time in the two universes.

20 The Science of the Universe: Cosmology and Science Education 663

quantum cosmology is discrete on a very small scale (meaning volumes of the order 10^{-100} cm^3), which has the observable consequence that photons of very high energy should travel faster than those of low energy.

Did the universe have an absolute beginning in time or not? The most honest answer is probably that we do not know and perhaps cannot ever know. It may be one of those questions about which we cannot even tell whether it is meaningful or not or whether it belongs to science or not.

20.7   Conclusion

The cosmological world view of the twenty-first century, largely identical to the standard big bang theory, is to a considerable extent what the Copernican world system was in the seventeenth century. Just as this system was not only a new theory of astronomy but also carried with it wider implications related to philosophy, religion and social order, so the modern picture of the universe cannot be easily separated from extra-scientific considerations. Such considerations, be they of a philosophical, conceptual or religious nature, should to some extent appear also in the teaching of science and do it in a qualified and critical manner.

One of the important aims of science education is to bring home the lesson that although science provides us with reliable and privileged knowledge of nature, it does not answer all questions that are worth asking. This lesson emerges with particular force from the study of cosmology. It may be expressed more poetically with a famous quotation from Shakespeare's *Hamlet*: 'There are more things in heaven and Earth, Horatio, than are dreamt of in your philosophy'. Recall that at the time of Shakespeare, the term 'philosophy' had a meaning corresponding to our 'science'.

References

Arny, T. T. (2004). *Exploration: An introduction to astronomy*. Boston: McGraw-Hill.

Bailey, J. M., Coble, K., Cochran, G., Larrieu, D., Sanchez, R. & Cominsky, L. (2012). A multi-institutional investigation of students' preinstructional ideas about cosmology. *Astronomy Education Review, 11*, 010302–1.

Brickhouse, N. W., Dagher, Z. R., Letts, W. J. & Shipman, H. L. (2000). Diversity of students' views about evidence, theory, and the interface between science and religion in an astronomy course. *Journal of Research in Science Teaching, 37*, 340–362.

Brickhouse, N. W., Dagher, Z. R., Shipman, H. L. & Letts, W. J. (2002). Evidence and warrants for belief in a college astronomy course. *Science & Education, 11*, 573–588.

Brush, S. G. (1992). How cosmology became a science. *Scientific American, 267* (8), 34–40.

Campbell, B. (1998). Realism versus constructivism: Which is a more appropriate theory for addressing the nature of science in science education? *Electronic Journal of Science Education, 3*, no. 1.

Carr, B. & Ellis, G. F. R (2008). Universe or multiverse? *Astronomy & Geophysics, 49*, 2.29–2.37.

Carroll, W. (1998). Thomas Aquinas and big bang cosmology. *Sapienta, 53*, 73–95.

Clerke, A. M. (1890). *The system of the stars*. London: Longmans, Green and Co.

Comins, N. F. (2001). *Heavenly errors: Misconceptions about the real nature of the universe.* New York: Columbia University Press.

Copp, C. M. (1985). Professional specialization, perceived anomalies, and rival cosmologies. *Knowledge: Creation, Diffusion, Utilization, 7,* 63–95.

Crowe, M. J. (1990). *Theories of the world from antiquity to the Copernican revolution.* New York: Dover Publications.

Crowe, M. J. (1994). *Modern theories of the universe: from Herschel to Hubble.* New York: Dover Publications.

Davis, T. M. & Lineweaver, C. H. (2004). Expanding confusion: Common misconceptions of cosmological horizons and the superluminal expansion of the universe. *Publications of the Astronomical Society of Australia, 21,* 97–109.

Eddington, A. S. (1931). The expansion of the universe. *Monthly Notices of the Royal Astronomical Society, 91,* 412–416.

Ellis, G. F. R. (1984). Cosmology and verifiability. In R. S. Cohen & M. W. Wartofsky (Eds.) *Physical sciences and history of physics* (pp. 193–220). Dordrecht: Reidel.

Ellis, G. F. R. (1999). The different nature of cosmology. *Astronomy & Geophysics, 40,* 4.20–4.23.

Ellis, G. F. R. (2007). Issues in the philosophy of cosmology. In J. Butterfield & J. Earman (Eds.) *Philosophy of physics.* Amsterdam: Elsevier.

Genz, H. (1999). *Nothingness: The science of empty space.* New York: Basic Books.

Halvorsen, H. & Kragh, H. (2010). Theism and physical cosmology. http://philsci-archive.pitt.edu/8441/

Hansson, L. & Redfors, A. (2006). Swedish upper secondary students' views of the origin and development of the universe. *Research in Science Education, 36,* 355–379.

Hawking, S. (1989). *A brief history of time: From the big bang to black holes.* New York: Bantam Books.

Holder, R. D. & Mitton, S. (Eds.) (2012). *Georges Lemaître: Life, science and legacy.* New York: Springer.

Kaiser, D. (2006). Whose mass is it anyway? Particle cosmology and the objects of theory. *Social Studies of Science, 36,* 533–564.

Kragh, H. (1996). *Cosmology and controversy: The historical development of two theories of the universe.* Princeton: Princeton University Press.

Kragh, H. (2007). *Conceptions of cosmos.* Oxford: Oxford University Press.

Kragh, H. (2011a). On modern cosmology and its place in science education. *Science & Education, 20,* 343–357.

Kragh, H. (2011b). *Higher speculations: Grand theories and failed revolutions in physics and cosmology.* Oxford: Oxford University Press.

Kragh, H. (2013). 'The most philosophically important of all the sciences': Karl Popper and physical cosmology. *Perspectives on Science, 21,* 325–357.

Kragh, H. & Smith, R. W. (2003). Who discovered the expanding universe? *History of Science, 41,* 141–162.

Krauskopf, K. B. & Beiser, A. (2000). *The physical universe.* New York: McGraw-Hill.

Kuhn, K. F. (1998). *In quest of the universe.* Boston: Jones and Bartlett Publishers.

Lightman, A. P. & Miller, J. D. (1989). Contemporary cosmological beliefs. *Social Studies of Science, 19,* 127–136.

Lotze, K.-H. (1995). Special and general relativity and cosmology for teachers and high-school students. In C. Bernadini, C. Tarsitani & M. Vicentini (Eds.) *Thinking physics for teaching* (pp. 335–354). New York: Plenum Press.

Marschall, L. A., Snyder, G. A. & Cooper, P. R. (2000). A desktop universe for the introductory astronomy laboratory. *The Physics Teacher, 38,* 536–537.

Marx, W. & Bornmann, L. (2010). How accurately does Thomas Kuhn's model of paradigm change describe the transition from the static view of the universe to the big bang theory in cosmology? *Scientometrics, 84,* 441–464.

Matthews, M. R. (Ed.) (2009). *Science, worldviews and education.* New York: Springer.

McMullin, E. (1993). Indifference principle and anthropic principle in cosmology. *Studies in History and Philosophy of Science, 24*, 359–389.

Munitz, M. K. (1986). *Cosmic understanding: Philosophy and science of the universe*. Princeton: Princeton University Press.

North, J. W. (1994). *Astronomy and cosmology*. London: Fontana Press.

Poole, M. (1995). *Beliefs and values in science education*. Maidenhead, Berkshire: Open University Press.

Prather, E. E., Slater, T. F. & Offerdahl, E. G. (2003). Hints of a fundamental misconception in cosmology. *Astronomy Education Review, 1*(2), 28–34.

Rugh, S. E. & Zinkernagel, H. (2009). On the physical basis of cosmic time. *Studies in History and Philosophy of Modern Physics, 40*, 1–19.

Ryan, M. P. & Shepley, L. C. (1976). Resource letter RC-1: Cosmology. *American Journal of Physics, 44*, 223–230.

Sagan, C. (1997). *The demon-haunted world: Science as a candidate in the dark*. London: Headline.

Shipman, H. L. (2000). Thomas Kuhn's influence on astronomers. *Science & Education, 9*, 161–171.

Shipman, H. L., Brickhouse, N. W., Dagher, Z. & Letts, W. J. (2002). Changes in students' views of religion and science in a college astronomy course. *Science Education, 86*, 526–547.

Sovacool, B. (2005). Falsification and demarcation in astronomy and cosmology. *Bulletin of Science, Technology & Society, 25*, 53–62.

Stoeger, W. R., Ellis, G. F. R. & Kirchner, U. (2008). Multiverse and cosmology: Philosophical issues. ArXiv:astro-ph/0407329.

Wallace, C. S., Prather, E. E. & Duncan, D. K. (2011). A study of general education astronomy students' understanding of cosmology. Part I: Development and validation of four conceptual cosmology surveys. *Astronomy Education Review, 10*, 010106–1.

Wallace, C. S., Prather, E. E. & Duncan, D. K. (2012). A study of general education astronomy students' understanding of cosmology. Part IV: Common difficulties students experience with cosmology. *Astronomy Education Review, 11*, 010104–1.

Helge Kragh is professor of history of science and technology at Aarhus University, where his research focuses on the history of the physical sciences after 1850. His most recent book, *Niels Bohr and the Quantum Atom*, was published by Oxford University Press in 2012.

Part VIII
Pedagogical Studies: Mathematics

Chapter 21
History of Mathematics in Mathematics Education

Michael N. Fried

21.1 Introduction

On the face of it, with so many good essays on the history of science in science education, a separate chapter on mathematics education might well be thought unnecessary. Considerations regarding history of mathematics in mathematics education, it is true, are similar to those regarding history of science in science education. Similar benefits, for example, have been cited in both cases, including humanizing subject matter, adding variety to teaching, showing alternative approaches to scientific ideas, analyzing students' understandings and misunderstanding, and deepening a sense of the nature of the discipline. For both, too, there are similar difficulties.[1] For one, both must confront mundane but no less worrying problems, such as finding time for history and fitting historical material into an already crowded curriculum; but also they must confront deeper problems arising from the tension between anachronism and relevance and between useful rational reconstruction and faithful historical analysis.

The case of mathematics education, however, differs from that of science education as mathematics itself differs from the natural sciences. That difference, of course, is subtle and cannot be reduced simply to whether one or the other is more empirical or more cumulative. Indeed, it is precisely historical and philosophical studies that have made clear the extent to which science can develop according to nonempirical theoretical issues, while mathematics can take on an empirical or quasi-empirical character, as Lakatos liked to put it (e.g., Lakatos 1986). Still differences do exist along these lines. In arguing his own theory of mathematical

---

[1] For a list of some of these difficulties in mathematics education, see Siu (2006).

M.N. Fried (✉)
Ben Gurion University of the Negev, Beer-Sheva, Israel
e-mail: mfried@bgu.ac.il

M.R. Matthews (ed.), *International Handbook of Research in History,*
Philosophy and Science Teaching, DOI 10.1007/978-94-007-7654-8_21,
© Springer Science+Business Media Dordrecht 2014

change, Kitcher (1984), for example, points out that "...mathematics often resolves threats of competition [between opposing theories] by reinterpretation, thus giving a greater impression of cumulative development than the natural sciences" (p. 159).

Leaving aside whether or not it is justified, this impression that Kitcher refers to, combined with a Platonist tendency to see mathematical objects as given, typically translates into a sense that somehow the mathematics of the past is as valid now as it was in its own time, as opposed to the science of the past, which one easily (often *too* easily) takes to be obsolete or inadequate from the perspective of modern science. Consequently, the historical character of mathematics is often more problematic for students and teachers than the historical character of the natural sciences: it is more difficult, that is, for students and teachers to see mathematics of the past truly *being* of the past, truly different from the mathematics of the present. The benefits and difficulties of incorporating history of mathematics into mathematics education are, in their turn, all colored by this sense of the historical character of the discipline. So, despite striking similarities between questions connected with history of science in science education and history of mathematics in mathematics education, there is good reason to take a closer look at the latter independently of the former. This then is the purpose of the present chapter.

We shall proceed according to the following plan. Part 1 will take a brief look at several early instances of using historical mathematical material for learning. As we shall see, many of the justifications for incorporating history of mathematics in mathematics education proposed today are prefigured in this educational history, and this will give the entire chapter a spiral character. The discussion in this first part will also hint at the question of what it means to have a historical approach in the first place. The reader should be warned, however, that this part in no way presumes to be a thorough history of the subject in any sense of the word; indeed, that history needs one day to be written!

Part 2 will examine three central themes associated with more recent attempts to bring history of mathematics into mathematics education: the motivational theme, the curricular theme, and the cultural theme. Inevitably, these themes, to a greater or lesser degree, rest on presuppositions concerning the nature of mathematics, history of mathematics, and mathematics education itself.

Part 3 then looks more closely at these presuppositions by considering them in light of how historical inquiry in general is understood. It is this part that most directly addresses how the historical character of one's approach in bringing history of mathematics into mathematics education relates to its benefits and difficulties. It asks, at bottom, in what sense do students gain historical *knowledge* of mathematics or an historical outlook regarding mathematics? Tensions between frameworks based on the curricular theme and those based on the cultural theme, in particular, are revealed by this line of thinking. Responding to these tensions may demand reconceiving what mathematics education is about. The use of original sources is examined in this context and is discussed in this part together with some empirical findings.

21.2 Some Early Instances of Using Historical Material in Mathematics Education

In the 1970s, the idea that history of mathematics could play a part in mathematics education began to take root broadly in the mathematics education community. More importantly, in those years, educational interest in history of mathematics became organized at national and international levels, especially, with the establishment of the *International Study Group on the Relations between the History and Pedagogy of Mathematics* (ISGHPM or, as it is now known in its abbreviated form, HPM). The many conferences, books, and international cooperation that arose from the HPM and other organizations tempt one to think that before the 1970s, interest in using the history of mathematics in mathematics education was rare at best, and concrete instances of such use were thin on the ground if existing at all. In fact, historical material has almost always been present, one way or another, in mathematics education, in one form or another.

The qualifications in the last sentence are necessary because of the variety of ways one can treat historical materials and conceive one's relationship to the mathematics of the past and, at the same time, because of the variety of settings for learning mathematics and ways of understanding what it means to be mathematically educated. Notions of what it means to learn mathematics and where mathematics learning takes place are connected to the history of mathematics education proper, which has been an area of active interest on its own in recent years (observe, e.g., the existence of the *International Journal for the History of Mathematics Education* edited by Gert Schubring). But by taking into account the variety of ways of treating history and defining one's relation to the past, the present discussion becomes connected as much, or even more, to historiography. Indeed, although we shall speak of this in more depth later in the third part of this chapter, it is important to keep in mind now that in thinking about how history of mathematics comes into the learning of mathematics, it is impossible to separate considerations of the nature of history itself and of the telling of history from its use in teaching.

The blurred borders between doing and learning mathematics and history and historiography of mathematics are particularly apparent when one tries to understand mathematics education of the ancient classical period. Education in the ancient world is, altogether, a topic difficult to give an account of in plain and simple terms. The key notion is that of *paideia*. This Greek word has been translated variously as "culture," "civilization," "tradition," or, simply, "education"; more tellingly, its typical Latin translation is *humanitas*. The course in *paideia* is the *enkyklios paideia* from which we derive the word "encyclopedia," and yet the acquiring of *paideia* was not considered the acquisition of encyclopedic knowledge. The immense difficulty of the idea of *paideia* is evident in the mere fact that the great classical scholar Werner Jaeger needed three thick volumes to describe it (Jaeger 1945). One can say this though, that *paideia* entailed knowledge of a certain corpus of literature as well as the possession of skills and a presence of mind to think, speak, and act in an intelligent manner, one might say in a *cultured* way.

672 M.N. Fried

The emphasis on thinking, speaking, and writing is one reason why rhetorical training was so central in classical education and why some of our best accounts of classical education of the time are specifically of rhetorical education (see, for example, Kennedy 2003). The nature of mathematics education is much less clearly laid out, and there seems to be a large gap between accounts of very elementary education (discussed by Mueller 1991) and the more advanced mathematics education leading to the work of figures like Archimedes and Euclid (Fried and Bernard 2008). One key to understanding classical education, whether it be rhetoric or mathematics, is that such education was pursued for a lifetime; *paideia* had always to be cultivated. The speeches of the fourth century BCE rhetorician, Isocrates, therefore, were explicitly not only speeches as such but also models for himself and his students – production, teaching, and learning, for him, flowed seamlessly into one another.

This makes Marrou's referring to ancient "textbooks" in mathematics, meaning books such as Euclid's *Elements* (Marrou 1982, pp. 177ff) as plausible as it is deceiving. For, on the one hand, even though the *Elements* was used as a school textbook almost into modern times, it was in its own time the work of a mature mathematician addressed to mature mathematical audiences for whom "elements" meant something much more than "elementary" (Fried and Unguru 2001, pp. 58–61). On the other hand, as we noted regarding Isocrates' speeches, the perfecting of one's own *paideia* could include teaching and learning from the *Elements*. In that light, it is quite natural that Proclus, writing both as a philosopher and as the head of the Platonic Academy in the fifth century CE, should produce a commentary on just the first book of the *Elements* and refer often to its effect upon students. And, by the latter, it is perfectly clear Proclus includes himself as well as those younger than him.

Proclus' *Commentary on the First Book of Euclid's Elements* (Morrow 1970; Friedlein 1873) is also one of the important sources we have for the history of mathematics. Proclus refers often to the historical development of mathematics and, by giving his own principal source, Eudemus of Rhodes (fourth century BCE), he also tells us much about one of the early histories of mathematics, Eudemus' no-longer-extant *History of Geometry*. From Proclus, and elsewhere, we know that Eudemus' approach to history involved pointing out and discussing the first discovers of a given result, the *prōtoiheuretai* (see Zhmud 2006). The verb *heurein*, from which we derive the English word *heuristic*, means actually "to find" or "to invent." In the rhetorical educational tradition, *heuresis* is an extremely important term, for while the students, the *manthanontas* (i.e., those who engage in *mathēsis*, learning), learn by imitation, by studying texts, they must simultaneously learn to invent, to engage in *heurēsis* (Kennedy 2003; Fried and Bernard 2008); in rhetoric this means by studying speeches of masters, they learn to invent their own speeches. It is not unreasonable then to assume an educational principle in Proclus' attention to history and in the work of his historian predecessor Eudemus: through learning about the first inventors of mathematical ideas, students discover their own powers to invent. As we shall see shortly, this has remained, *mutatis mutandis*, a motive for introducing historical elements into mathematics education.

It can be argued that although Proclus engages with mathematics and mathematicians of the past, he does not treat these historically. His relationship to earlier mathematicians is one of colleagues, despite the great span of time separating them. But as I have outlined elsewhere, there is a broad spectrum of possible relationships to the mathematics of the past, including that of "colleagues"(Fried 2011, where a sample of eight relationships was described). There are, for example, "treasure hunters," who try and bring up gems lost in the past; "conquerors," like Descartes, who refer to the past to show the superiority of the present; "privileged observers," like H. G. Zeuthen, who think their modern mathematical knowledge privileges them to interpret the past; and "historical historians of mathematics," who view the past as fundamentally different from the present and see the treatment of the past demanding more than present mathematical knowledge. These kinds of relationships do not necessarily correspond to historical periods. For example, regarding "colleagues," even in modern times, Hardy quotes Littlewood as having said the Greek mathematicians were merely "Fellows of another college" (quoted in Hardy 1992, p. 81). Whether or not these relationships are properly historical is just the sort of historiographical question I referred to above and it is far from settled. But one must accept that these relationships are *in some sense* historical; more importantly, they allow us a way of seeing how mathematics of the past, at least in some general way, has entered mathematics teaching and learning.

A "colleague" relationship to mathematics of the past, for example, can be discerned in the use of Euclid's *Elements* and other classic works as texts for teaching geometry from the Middle Ages to the nineteenth century. Howson (1982) points out that in England during the nineteenth century, an enormous number of new editions of Euclid for use in schools were produced and that "Many of these editions were, in fact, Euclid pure and simple; additional notes were often included, but no exercises, for the student was expected to memorise not to act" (p. 131). The study of geometry, in this respect, was identified with the study of a certain historical text. Yet it was not its historical character alone that was behind its use: similar to what I have suggested regarding Proclus, the use of Euclid's *Elements* was justified by its ability to train rigorous logical thinking, as good then as now. Interestingly enough, Howson (1982) makes very clear, the reform of mathematics education in England in the second half of the nineteenth century centered on the rejection of Euclid as a text (see also Carson and Rowlands 2000). It could be said that the whole argument surrounding Euclid's *Elements* stemmed from its being regarded on the same terms as a modern text, that is, as if it were a text written by a colleague. Yet, it cannot be discounted from the present discussion, for one, because there are still proposals for using history of mathematics whose claim to being history comes down to the fact only that a historic text is being used (and the counterargument is often the same as that against Euclid, namely, that it is not modern or not pedagogically sound!). Furthermore, the claim for using Euclid in the mathematics classroom is in line with a common claim for using historical texts in general, namely, that they often present accounts of mathematical ideas that are particularly clear, probing, or challenging.

But more explicitly historical views of the mathematics of the past can also be found in educational materials for teaching mathematics from the same time and before. A relatively early case was the classic 1654 text by the Jesuit Andreas

Tacquet (1612–1660) *Elementageometriaeplanae ac solidae* (Tacquet 1761), originally produced for students studying mathematics in the Jesuit colleges. It opens with a long "historical narrative of the origin and progress of the mathematical sciences" (*historiconarratio de ortu & progressumatheseos*) that students might know to what science the wisest figures of past times had dedicated themselves. In other words, Tacquet was telling his students, in effect, that in order to understand its importance, their study of mathematics must be pursued against the background of its history. He was not the first to hold this position; he himself refers to Ramus (1515–1572) as preceding him in this regard.

In a more formal way, and somewhat closer to the way we imagine history of mathematics entering education, we find in the late eighteenth century, in Poland, a very early interest in history of mathematics as a component of mathematics education – earlier, in fact, than a professional interest in history of mathematics in Poland. Domoradski and Pwlikowska-Brożek (2002) tell us that "The first Ministry of Education in Europe – *Komisja Edukacji Narodowej* (Commission on National Education) (1773–1794) – hoping to improve and broaden mathematics knowledge, recommended that students be acquainted with the history of mathematics beginning from antiquity" (p. 199). The case of Poland is remarkable when one considers that something like national curricula for education were almost nonexistent before the eighteenth century, so that history of mathematics, here, came into the mathematics curriculum almost at the same time the mathematics curriculum, in the modern sense of the word, itself was coming into being.

As for the next century, especially towards its end, one finds clear instances of an interest in the history of mathematics and science in education in French education and educational policy. From around 1869, for instance, questions concerning "the historical method" in science and mathematics began to appear in official "agrégation" examination for teachers of science (Hulin 2005). Somewhat later, Paul Tannery (1843–1904), who falls somewhere between what I called above the "privileged observer" type and the "historical historian of mathematics," developed a course of studies for the history of science to make clear "the order of ideas, true or false, that dominate each of the sciences" (Hulin 2005, p. 393, my translation), and he taught a course in the history of mathematics in the Paris Faculty of Science from 1884 to1886 (Hulin 2005; Peiffer 2002).

An interest in the role of history of science generally continued into the twentieth century in France, culminating perhaps in the activities of IREM (Institut de Recherche sur l'Enseignement des Mathematiques) at the end of the century. But at the start of the century, we have Ernest Lebon, then minister of public instruction, expressing the desire that history of science be made part of secondary school teaching and a sanctioned part of the baccalaureate examination (Hulin p. 389).[2] This interest took in mathematics as well, though at times the justifications for history of science were different than those for mathematics. The physicist Paul Langevin, for example, in the first quarter of the twentieth century was a strong advocate for history of science as a way for combating dogmatism

---

[2]Lebon's remarks were made as a delegate to the Congrès' histoire comparee in 1900.

21   History of Mathematics in Mathematics Education

(Bensaude-Vincent 2009, p. 16), which was not a common theme if at all for history of mathematics in mathematics education.

Significant efforts were made to introduce history of mathematics into mathematics education and teacher education in the United States, where ironically, as Robert Hughes has remarked, "Americans love to invoke the idea of American newness" (Hughes 1997).[3] In the period between the last years of the nineteenth and the first quarter of the twentieth centuries, two figures stand out in this and other aspects of American mathematics education. These are Florian Cajori (1859–1930) and David Eugene Smith (1860–1944).

D. E. Smith was an excellent historian of mathematics, but he also made great contributions to mathematics education in America and on the International scene.[4] That history of mathematics, in his view, was not separate from mathematics education is clear from his classic book, *The Teaching of Elementary Mathematics* (Smith 1904), which contains three entirely historical chapters and several others in which history has a part. And at the Michigan State Normal School in Ypsilanti, where D. E. Smith held the mathematics chair, he designed a course on the history of mathematics for teachers that "…became the foremost distinguishing characteristic of Smith's program: the importance of a historical perspective" (Donoghue 2006, p. 562). It is no surprise then that Smith was on the committee that prepared the 1923 report by the *National Committee on Mathematical Requirements* written under the auspices of the *Mathematical Association of America* in which cultural aims of mathematical education, in general, were highlighted, including "…the role that mathematics and abstract thinking, in general, have played in the development of civilization" (in Bidwell and Clason 1970, p. 394).[5]

Looking back across the ocean for a minute, I should mention that the cultural motive described in the 1923 report was also evident in Felix Klein's enthusiasm for a historical component in mathematics education. Historical sections appear in parts I and II of his *Elementary Mathematics from an Advanced Standpoint* (Klein 1908/1939). In the geometry part (part II), Klein states quite explicitly that "…I shall draw attention, more than is usually done…to the *historical development of the science*, to the accomplishments of its great pioneers. I hope, by discussions

---

[3] In the same place just cited (Hughes 1997), Hughes, whose focus is art in America, also refers to an American "worship of origins": the tension between old and new seems to be very much part of the American psyche.

[4] It was Smith who suggested the formation of an international society for mathematics education in a paper in 1905 published in *L'Enseignement Mathématique*. This became the *International Commission on Mathematics Instruction*, the ICMI established in Rome in 1908.

[5] Bidwell and Clason correctly point out (p. 394, note 3) that the ideas in this section of the report parallel closely views in other writings by Smith, for example, his "Religio Matematici" published in the *American Mathematical Monthly* (vol. 28, pp. 339–349) in 1921. The latter, I should say, is not strictly speaking about history of mathematics per se; it is piece written on the model of Thomas Browne's *Religio Medici* and tells what the belief is of a mathematician. Nevertheless, one of the articles of faith is that "Mathematics is a vast storehouse of the discoveries of the human intellect. We cannot afford to discard this material" meant to parallel to the claim that "Religion is a vast storehouse of the discoveries of the human spirit. We cannot afford to discard this material" (Smith 1921, p. 348).

of this sort, to further, as I like to say, your general *mathematical culture*: alongside of knowledge of details, as these are supplied by the special lectures, there should be a grasp of subject-matter and of historical relationship [emphases in the original]" (Klein 1908/1939, II, p. 2). Incidentally, this work by Klein was published in the same year as the founding of *International Commission on Mathematics Instruction* (ICMI), initiated by Smith and headed by Klein as its first president. But let us return to America and Florian Cajori.

Cajori's first historical work was published by the US Bureau of Education and was entitled *The Teaching and History of Mathematics* (1890). As Dauben (2002) points out, "...Cajori was the first in a continuing American tradition of mathematicians interested in the history of mathematics due to its perceived value in teaching" (p. 265). Three years later, Cajori wrote a textbook for the history of mathematics, underlining his interest that this be used by students, rather than other mathematicians or professional historians of mathematics. Cajori's next book *A History of Elementary Mathematics with Hints on Methods of Teaching* (1896) makes the educational motive for his endeavors into the history of mathematics even more explicit. He opens the preface with a quotation from Herbert Spencer in which the following principle is stated: "The education of the child must accord both in mode and arrangement with the education of mankind as considered historically; or, in other words, the genesis of knowledge in the individual must follow the same course as the genesis of knowledge in the race".[6] Cajori then uses this to justify his own use of history of mathematics for mathematics teaching:

> If this principle, held also by Pestalozzi and Froebel, be correct, then it would seem as if the knowledge of the history of a science must be an effectual aid in teaching that science. Be this doctrine true or false, certainly the experience of many instructors establishes the importance of mathematical history in teaching. (p. v)

It is interesting to note that these works of Cajori were carried out before he carried out serious historical work on mathematics under the instigation of no less than Moritz Cantor. (Dauben 2002, p. 266) In other words, Cajori's interests in history of mathematics, as an independent field of inquiry, came *after* his interests in it as an adjunct to teaching.

The principle adduced by Cajori to justify his own historical approach is one of the most persistent and in some ways the most serious reasons given for using history of mathematics in mathematics education. The principle is called variously the genetic principle, the biogenetic law, the principle of parallelism, or the recapitulation principle, after the famous biological principle associated with Ernst Haeckel (1834–1919) that ontogeny, the development of an individual, recapitulates the development of the species, phylogeny (see Schubring 2011; Furinghetti and

---

[6]Cajori continues the quotation in which Spencer attributes this principle to Auguste Comte insisting at the same time that the principle can be accepted without accepting Comte's entire theory of knowledge. That Cajori leaves this in the quotation may be a sign that he too had reservations about Comte's theories taken as a whole. Still he, like Spencer, accepts the general genetic principle as a principle for guiding education. The quotation is from the second chapter, "Intellectual Education," of Spencer's *Education: Intellectual, Moral, and Physical* first published in 1861 (Spencer 1949/1861).

Radford 2008).[7] Important figures in the early development of modern mathematics education were adherents to the principle. Thus, calling on the biological principle as his model, Poincaré wrote in 1899 in, that "The educators' task is to make children follow the path that was followed by their fathers, passing quickly through certain stages without eliminating any of them. In this way, the history of science has to be our guide" (*L'Enseignement Mathématique,* quoted in Furinghetti and Radford 2008, p. 633). Felix Klein too, as Schubring (2011) points out, "decisively promoted the genetic principle" (p. 82), even though Klein, as described above, also saw studying history of mathematics in conjunction with mathematics as a matter of general mathematical culture.

The strength of the principle as a guiding principle for mathematics education can be judged by the fact that in 1908[8] an entire book on mathematics education – one of the very earliest books entirely dedicated to the subject – was published by Benchara Branford (1867–1944) in which the genetic principle played a central role.[9] In fact the frontispiece of Branford's *A Study of Mathematical Education* (Branford 1908, see the Fig. 21.1 below) is a "Diagram of the Development of Mathematical Experience in the Race and in the Individual": it is graphic representation of the entire theory, with stages of human history and various occupations such as geodesy and physics as well as what Branford calls primary and derivative occupations (such as miner, shepherd, and scribe, among others, corresponding to "infancy") listed along the left margin; the development of mathematical subjects together with an indication of the degree of "sense activity" and "thought activity" in the center; and the periods of an individual's life – embryonic (!), infancy, childhood, school, college – listed in the right margin. Referring to the diagram, Branford summarizes the educational implication, saying, "Thus, for each age of the individual life infancy, childhood, school, college may be selected from the racial history [i.e. history of the human race] the most appropriate form in which mathematical experience can be assimilated" (p. 245). It is worth noting that later in the work (p. 326), Branford cites exactly the same passage from Herbert Spencer's *Education* quoted by Cajori.

---

[7] Although those who use these various terms may see some shades of difference between them, here we shall commit the minor sin of lumping them together into a single perspective, one asserting that historical development can *in some way* provide a guide for individual intellectual development or course of learning.

[8] According to Scott (2009), Branford began the work in 1896.

[9] Schubring (2011) denies that Branford should be taken as the "classical advocate and propagator of that parallelism [the biogenetic principle grafted onto psychology] for the purposes of education" (p. 83). Yet, the fact that Branford took the genetic principle as a guide to his thinking about mathematics education and a "scientific" principle to ground research is certainly clear, and Schubring himself points out that "The merit of Branford's book…lies in his reflections and differentiations concerning the notion of the biogenetic law" (p. 83). Moreover, although Schubring says that too much weight has been placed on the frontispiece of the book, one must consider that that diagram was nevertheless chosen *as* the frontispiece and, therefore, meant to set the stage for the book. Furthermore, it is not merely an illustration presented and forgotten: an entire chapter is dedicated to the interpretation of the figure (Chap. 16), and many of the other chapters follow its structure.

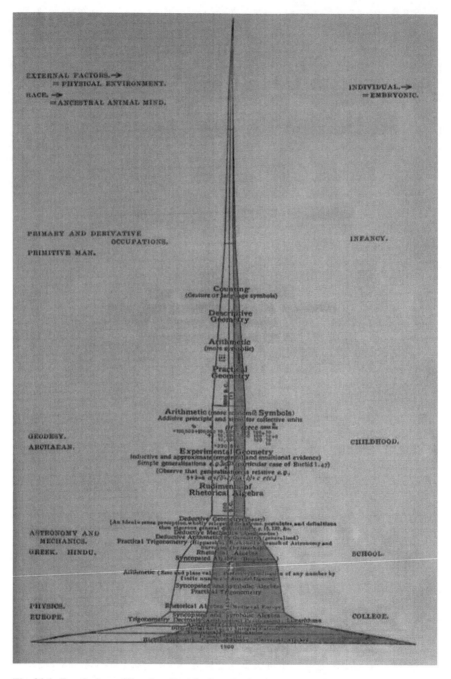

Fig. 21.1 Frontispiece of Benchara Branford's *A Study of Mathematical Education*

21 History of Mathematics in Mathematics Education

Probably the most famous instance of the genetic approach taken up practically was Otto Toeplitz's (1881–1940) course in calculus according to the genetic methodology, published posthumously in 1949 as *Die Entwicklung der Infinitesimalrechnung: Eine Einleitung in die Infinitesimalrechnung nach der genetischen Methode* republished in English in 1963 as *The Calculus: A Genetic Approach* (Toeplitz 1963). Toeplitz distinguishes between the direct and indirect genetic approach (Toeplitz 1927). In the direct genetic approach, the historical development is presented to the student as a way of presenting mathematical concepts themselves – an approach that should have an effect because of the basic assumption of the genetic principle. It is this direct approach that most characterizes Toeplitz's 1949 work. Thus, he begins his text with a chapter, "The nature of the infinite process," and continues according to the following sequence: "The beginning of Greek speculation on infinitesimals," "the Greek theory of proportions," "the exhaustion method of the Greeks," "the modern number concept," "Archimedes' measurements of the circle and the sine tables," "the infinite geometric series," "continuous compound interest," "periodic decimal fractions," "convergence and limit," and "infinite series." Overall, like most calculus textbooks, he aims towards the idea of limit, but his path to that concept is guided by historical precedents and examples rather than strictly logical considerations.[10]

In the indirect approach teachers use the historical development to draw conclusions about teaching mathematical concepts, which subsequently need not be historical. As Toeplitz puts it, the indirect approach consists of a "Clarification of didactic difficulties, I should say, didactical diagnosis and therapy (didaktische Diagnose und Therapie) on the basis of a historical analysis..." (Toeplitz 1927, p. 99, my translation). About this, Schubring (2011) says "Toeplitz's indirect approach looks not so much on knowledge, but on meta-knowledge, and his main focus is on how to provide future teachers in their training with such a meta-knowledge about mathematics" (p. 9). But Schubring also says that Toeplitz could never really free himself from a teleological viewpoint, from the notion that mathematical development is continuous and cumulative so that "his own notion of an indirect approach could not become fruitful" (p. 10).

Yet, that "teleological" viewpoint seems almost unavoidable if one adopts the genetic point of view. To assume that an individual's mathematical development follows the historical development of mathematics means that the historical development is somehow as natural and directed as that of an organism. It is thus not surprising that in his biography of Toeplitz in the *Dictionary of Scientific Biography*, Abraham Robinson should point out that Toeplitz "...held that only a mathematician of stature is qualified to be a historian of mathematics" (Robinson 2008, p. 428). The genetic point of view suggests a position towards the past that I categorized above as that of a "privileged observer."

---

[10] One only has to think of Edmund Landau's famous calculus text (Landau 1965) to grasp the contrast.

21.3 Three Themes in Our Own Time

In these older instances of using history or adopting a historical orientation in mathematics education, one can see the presence and emergence of themes still at work in current discussions of the subject.[11] History of mathematics as part of students' cultural education, a theme recognizing the enterprise of mathematical inquiry as part of students' cultural heritage in general as well as in the specific context of the sciences, has become all the more important as we recognize the centrality and formative power of culture altogether. And it ought to be emphasized here that in recognizing mathematics as a part of culture, one also adopts a view of the nature of mathematics. Let us call this, then, the *cultural theme*, keeping in mind that it is both a reflection on general culture as including mathematics and also on mathematics as being cultural. Besides this theme, which we saw in connection with Felix Klein and D. E. Smith, we also noted the suggestion that history of mathematics can help clarify or deepen the understanding of mathematical ideas: this theme, which could already be found in Proclus and certainly in Toeplitz, continues to be a potent reason for turning to the history of mathematics in mathematics education. Let us call it the *curricular theme*: it includes then both arguments claiming historical treatments of mathematical topics, such as that of Euclid, have great pedagogical power either in themselves or by offering a contrast to modern approaches, as well as the genetic argument and its variations. A third theme, which we did not encounter in our brief survey above, is what we should call the *motivational theme*.[12] We shall begin with this last.

21.3.1 *The Motivational Theme*

The motivational theme is that history of mathematics makes mathematics teaching less threatening, more human, less formal, and more interesting; the motivational theme introduces an affective consideration into the question of history of mathematics in mathematics education. Ironically, the objection to Euclid in the nineteenth century was partly on the grounds that it was dull![13] In a well-known

---

[11] Besides discussions in the context of meetings connected with the HPM mentioned above, other recent forums include those at the CERME meetings. See, for example, Kjeldsen (2011) and Tzanakis and Thomaidis (2012).

[12] Another theme one might suggest is an epistemological theme concerning the possibility that knowing the history of mathematics is knowing mathematics: but this is included in both the cultural and curricular themes together.

[13] That Euclid should be made more colorful was taken up literally by Oliver Byrne who produced a version of Euclid using a system of colors in place of Euclid's lettered diagrams and corresponding text. The book was meant to make Euclid more approachable, as advertized in the full title: *The first six books of the Elements of Euclid in which coloured diagrams and symbols are used instead of letters for the greater ease of learners* (Byrne 1847).

textbook meant to reform the teaching of geometry by replacing Euclid, the author James M. Wilson wrote that:

> We put a boy down to his Euclid; and he reasons for the first time…but we make him reason in iron fetters…we make the study of Geometry unnecessarily stiff, obscure, tedious and barren….And the result is, as everyone knows, that boys may have worked at Euclid for years, and may yet know next to nothing of Geometry (Wilson 1868, pp. vi–vii)

But today, the motivational theme is quite often called upon to justify a historical approach (see Gulikers and Blom 2001, who also refer to "motivational arguments" for history of mathematics). This was certainly the case with Perkins (1991), who saw history as a vehicle for making classroom teaching more interesting and, in so doing, a vehicle for improving students' achievement. A more recent example comes from a Turkish study (Kaygin et al. 2011) in which the authors say, "Thanks to the vast culture and wide knowledge held within the history of mathematics, it becomes easier to understand the abstract concepts of mathematics which is thus [no] longer a subject arousing fear and concern" (p. 961).

A clear statement of the motivational theme can be found in a paper by Po-HungLiu (2003) addressed to teachers and asking, "Do teachers need to incorporate the history of mathematics in their teaching?" Liu's answers include in fact all three of the themes that I have mentioned, but he begins with the motivational theme. He writes:

> As sometimes taught, mathematics has a reputation as a "dull drill" subject, and relevant studies report a steady decline in students' attitudes toward the subject through high school. The idea of eliciting students' interest and developing positive attitudes toward learning mathematics by using history has drawn considerable attention. Many mathematics education researchers and mathematics teachers believe that mathematics can be made more interesting by revealing mathematicians' personalities and that historical problems may awaken and maintain interest in the subject (p. 416)

Pursuing this theme sometimes reduces to light storytelling as in Lightner's "Mathematicians are human too" (Lightner 2000), where Lightner tells a series of humorous anecdotes about mathematicians so that students learn that "…mathematicians were all human beings with peculiar foibles and personality quirks just like the rest of us" (p. 699). For example, Lightner tells us how Norbert Wiener was so absentminded he habitually forgot not only where he parked his car but also which car he drove, and so, after a seminar, he would wait patiently until every car left the parking lot but one, his![14]

As a justification in mathematics education, one should not be completely dismissive of the motivational theme. Getting students to want to learn mathematics is a natural and important goal for mathematics educators, especially when so much emotional baggage truly gets in the way of students' learning mathematics – ranging from distaste and a sense of mathematics as dry and rigid, to frustration and fear.

---

[14]Theodore Eisenberg has asked elsewhere (Eisenberg 2008) whether we also ought to tell some of the less amusing stories about mathematicians, about their occasional racism and association with Nazis.

682   M.N. Fried

So it is not for nothing that much serious research in mathematics education has looked at affect and its relation to mathematics learning and achievement (e.g. Goldin 2009; McLeod 1994). However, while the educational motive behind the motivational theme may be serious, as a theme connected with the incorporation of history into mathematics education, it is problematic. First, it supposes, not always consciously, that mathematical content alone, unembellished with stories, anecdotes, or colorful characters, cannot itself be made sufficiently interesting to hold students' attention. History is presented as something *to add* to mathematics lessons to "enliven" teaching, teaching that, one must assume, would *otherwise* be dull and dry. Second, history as a body of knowledge to learn and to take seriously is put aside and turned into a mere ploy for drawing students into learning the mathematics their teachers are required to teach. Put baldly, if a good story will keep students from falling asleep in class, tell it; whether or not the story is true, informative, or deep is pertinent, but it is secondary to its being entertaining. And like the story about Norbert Wiener, it does not even have to be, strictly speaking, about mathematics. The motivational theme, accordingly, ends up doing justice neither to history nor to mathematics itself. That said, it must be underlined that similar difficulties exist, albeit more subtly, in other cases where history is brought into the mathematics classroom. The main problem is that history in such cases is superadded to mathematics education, with the particular identity of history as a form of knowledge being lost in the process.

21.3.2 The Curricular Theme

The curricular theme as a focus for introducing history of mathematics into mathematics education must be taken much more seriously, for there is a genuine attempt in this case to see mathematical *ideas* in the light of history. For this reason, it is right that Gulikers and Blom (2001), in their survey of recent literature on history in geometrical education, refer to what I am calling the "curricular theme" as the category of "conceptual arguments." Yet it must be kept in mind that whatever concepts are spoken of here, they are not so much drawn out from the history, but – like the concepts of function, number, and equation – are given in advance, as if from a set curriculum and, only subsequently, discussed historically.

It is in the curricular theme, then, that one sees most clearly the tendency described in my introduction, namely, the tendency to treat the mathematics of the past, though perhaps incomplete, to be as valid now as in its own time. So, for example, even if Euclid did not have group theoretical tools for geometry, he did have theorems on congruence and similarity – and, to that extent, we can use Euclid's work as a basis for our own teaching. Or, in this view, we can take ideas that, from a modern standpoint, are "implicit" in historic texts and translate them into a modern idiom, for example, translating Apollonius' principal properties for conic sections – what he called their *symptōmata* – into equations for conic sections. History of mathematics from the perspective of the curricular theme, as the name

21 History of Mathematics in Mathematics Education 683

implies, makes the least demands from the point of view of the curriculum, for historical treatments of mathematics can be adapted in such a way that they are consistent with a modern program of studies. It is not surprising then that when one looks at Gulikers and Blom's tables of articles in various subject categories of geometry that of the 36 articles in which resources are cited, 24 either have a "written guide with modern exercises," or a "written guide with 'old' problems translated into modern mathematical language" (Gulikers and Blom 2001, Appendix A).

I might point out that this theme could be interpreted as a motivational theme, but in the deeper sense of motivating an idea or a position, an answer to the question, why should we study subject X in chapter Y of the curriculum? This is Fauvel's approach when he discusses the history of logarithms (Fauvel 1995): the paper attempts to provide an answer to a friend who, when hearing Fauvel was thinking about how to teach logarithms, asks "Whatever for?" (p. 39). Fauvel goes on to show in the paper that by considering the history of logarithms, one sees how considerations of this tool, which may be used little nowadays, lead one to important mathematical ideas[15] and ideas about mathematics. The curricular theme, in this light, provides a different nexus for mathematical ideas than would a "logical" approach[16]: providing a motivation for a mathematical idea shows in a way different from pure logic why one idea follows another.

Often the curricular theme is pursued by just taking a problem from the past that allows students to exercise and develop mathematical thinking and skills relevant to their school studies. For example, Kronfellner (2000) uses the problem of duplicating the cube[17] as a means to study topics such as irrational numbers, series, nonlinear analytical geometry, and curves – and he says explicitly that this is one strategy for introducing history into mathematics teaching, namely, "…to offer suitable tasks in which a traditional curriculum topic is connected with history." Swetz's (1995) "Historical Example of Mathematical Modeling: the Trajectory of a Cannonball" is another good instance where a historical problem becomes an opportunity to use and deepen mathematics studied in the classroom. As Swetz puts it: "Contemporary secondary school students can explore a variety of problem-solving situations involving trajectories and use their knowledge of geometry, algebra, trigonometry, vectors, calculus and even computer programming" (p. 101).

In a recent special issue of *Mathematics in School*, Elizabeth Boag's article on the "Dandelin Spheres" (Boag 2010) is a good example of how a historical topic can suggest not just a problem but also an approach to a school subject, in this

---

[15] Toeplitz similarly saw the logarithm as an entrance to important mathematical ideas and discusses it prominently in the chapter of his genetic approach concerning the fundamental theorem of the calculus (Toeplitz 1963, Chap. III).

[16] We shall soon encounter the "logical approach" again in connection with the 1962 Memorandum by Morris Kline and associates (Memorandum 1962).

[17] The classic problem is to find the side of a cube whose volume is twice that of a given one. The ancient interpretation was to find two mean proportionals between a line and its double, that is, two lines A and B satisfying the proportion: M:A::A:B::B:2 M. The modern translation is to construct the cube root of 2.

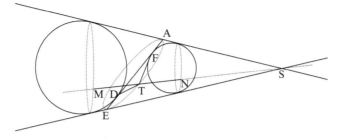

Fig. 21.2 Dandelin spheres for the case of the ellipse

case, the focal properties of conic sections. Just to remind the reader, what G. P Dandelin (1794–1847) showed in the course of his work, "Mémoire sur quelques propriétés remarquables de la focale parabolique"(Dandelin 1822), was that the focal properties of the conic sections, say of the ellipse, could be conceived strictly in terms of the cone. Consider the property that the sum of the lines from a point on an ellipse to the foci is constant: suppose a right cone is cut by a plane such as that containing the line AE and let two spheres be inserted in the cone, tangent to generating lines of the cone and to the cutting plane.[18] As for the latter, let them be tangent to the plane at points D and F (see Fig. 21.2) – these, it turns out, are the foci of the ellipse produced by the cutting plane.

Let T be a point on the ellipse produced by the cutting plane and let SNTM be the generating line of the cone that passes through T. Then since TM and TD are tangents from T to the sphere, MT = DT. Similarly, TN = TF. Therefore, DT + TF = MT + TN = MN, which is constant.

The Dandelin spheres illustrate well how a mathematical development grounded in a historical text can be enlightening and useful in a classroom presentation. But how far can one say that this grounding in history is the essential element in the mathematical presentation? What really is the role of the historical context? Having shown the basic idea of the Dandelin spheres, Boag says, "Before continuing with the mathematics, I will give a brief biography of Dandelin" (p. 35). Three short paragraphs follow providing a few biographical details, the name of Dandelin's paper on the Dandelin spheres and where it was published, and then some further work by Dandelin following the 1822 paper. What I want to stress is not that Boag should have given more historical information – for the paper, in many ways a very good one, was meant to be short from the start – rather, it is that the historical background and mathematical content are taken as separable. The mathematics is stopped and then continued after the history; it is not continued into the history, nor is the history continued into the mathematics. If the Dandelin spheres were presented to a reader of Boag's article without ever

---

[18] Dandelin's diagram shows a planar section, with no perspective and no variation of line types, as I have done here; however, the lettering is consistent with Dandelin's. In her paper, Boag gives Dandelin's original diagram in facsimile as well as her own.

21 History of Mathematics in Mathematics Education

mentioning Germinal Pierre Dandelin, they would possess no less explanatory power and probably no less charm.[19]

In the examples of the curricular theme so far, the position assumed with relation to the past is very similar to that of "the treasure-hunter" described above: one dips into the past to pull out some beautiful example or approach, like the Dandelin spheres, which one can then use in the classroom. It is a completely legitimate way of using history, and it is legitimate to call it a historical approach, *if*, paradoxically, one accepts that there is a mathematical content that can be set off from all considerations of time, culture, and place: the gold one digs up may come in the form of unfamiliar coins, but it is gold nonetheless.

The genetic approach, which we have also included within the curricular theme, is much more subtle and places historical development at the center of its attention since it is there it finds its key to how students understand mathematical ideas and how they can learn to understand mathematical ideas better. As we said above, a teleological viewpoint is almost unavoidable in the genetic approach, and it is for this reason that the same mathematical ideas as those comprised in a modern curriculum can be assumed to exist, at least implicitly, in historical texts; this, in turn, proves the relevance of historical texts to the modern classroom. Again, the relation to the past implied by the genetic approach tends towards that of the "privileged observer."

To see how the genetic idea is expressed in current ideas about history of mathematics in mathematics education, however, we shall have to take a fairly general definition of it: we shall include in it any position that sees historical development as providing light on educational processes, students' learning and understanding, by somehow running parallel to these.[20] It will become apparent, though, that the genetic idea, viewed loosely as I intend to do, leads easily into the cultural theme, which undeniably tends away from a teleological viewpoint.

One year before Toeplitz's *The Calculus: A Genetic Approach* (Toeplitz 1963) appeared in English, the genetic approach appeared explicitly in a famous memorandum published in *The Mathematics Teacher* and *The American Mathematical Monthly* (Memorandum 1962). The memorandum was signed by 64 mathematicians but was, apparently, mainly the work of Lipman Bers, Morris Kline, George Pólya, and Max Schiffer, and, of those, mainly, Morris Kline (Roberts 2004). The paper was a thinly veiled attack on the curricular reforms forming the "new math" movement,

---

[19] The point, I must emphasize, is not that the Dandelin spheres have no historical interest. They do! As one of the anonymous reviewers of this chapter emphasized, correctly, Dandelin's ideas are indicative of the atmosphere created by Gaspard Monge's teaching at the École Polytechnique, where Dandelin was a student. This may provide a hint as to why Dandelin's spheres had to wait until the 19th century to be born. My point here is that these matters played no crucial role in the article which I have discussed, nor, *given the general orientation of the article*, did they have to.

[20] In this way I am allowing the views of Sfard (1995), Dubinsky et al. (2005), and others inspired by Piaget's ideas in *Genetic Epistemology* (1971) to come within range of genetic approaches, even if Schubring (2011) claims, with some justice, that Sfard, for example, misunderstands Piaget in this connection and adopts only a primitive form of parallelism.

686 M.N. Fried

chief among them the *School Mathematics Study Group* (SMSG), headed by Edward Begle. The writers warned that it would "be a tragedy if the curriculum reform should be misdirected and the golden opportunity wasted" (p. 189), and, therefore, they would set out "fundamental principles and practical guidelines."

The paper then goes on to discuss seven guidelines of which the fifth is "genetic method." For Kline and the others, the genetic method furnished the ground for a direct attack on the thinking behind the "new math," namely, that when it comes to teaching and designing a curriculum history is a more dependable guide than the logic of an axiomatic system. Thus the memorandum states:

> This genetic principle may safeguard us from a common confusion: If A is logically prior to B in a certain system, B may still justifiable precede A in teaching, especially if B has preceded A in history. On the whole, we may expect greater success by following suggestions from the genetic principle than from the purely formal approach to mathematics. (pp. 190–191)

It is hardly surprising then that when Edward Begle responded to the memorandum (Begle 1962), he was conciliatory on most points *except* for the genetic principle. About that, he said it would "...require children to learn to compute with Egyptian, Babylonian, Greek and Roman numerals before being introduced to the historically later but far more efficient place value decimal system" (p. 426).

It is worth noting, too, that this exchange, in which the genetic approach was endorsed on the one side and rejected on the other, happened to be between one group of prominent mathematicians and another. For although it is not wrong to refer to the SMSG, and "new math" movement in general, as "mathematician-dominated" (see Amit and Fried 2008), it is important to realize that in a historical framework for education guided by the genetic principle, one finds a framework as congenial to the modern mathematician as the formal-axiomatic one. Kline, it should be kept in mind, was himself a research mathematician, and he well understood that, logically, the ideas of sequence, limit, and convergence precede derivative and integral; yet, when he wrote his own calculus textbook (Kline 1977), he had no qualms about beginning with the derivative.

For education, naturally, the attractiveness of the kind of parallelism represented by the genetic approach lies in its potential to provide clear concrete models for teaching and for understanding students' learning. This potential, so clearly assumed in the 1962 memorandum, is still adduced, albeit in a more qualified tone. Thus following Victor Katz's paper on the development of algebra in the special issue on history of mathematics in *Educational Studies in Mathematics* (volume 66, number 2),[21] Bill Barton writes in his "Commentary from a mathematics educator" that "To the extent that the ontogenetic argument is useful in mathematics education, this reflection should cause us to rethink some current trends" (Katz and Barton 2007, p. 198). In that same issue, Yannis Thomaidis and Constantinos Tzanakis take up the theme in a more focused way in their paper, "The Notion of Historical 'Parallelism'

---

[21]I mention the special issue explicitly since it was one of the few instances that history of mathematics was given such broad attention in a leading research journal for mathematics education.

21   History of Mathematics in Mathematics Education   687

Revisited: Historical Evolution and Students' Conceptions of the Order Relation on the Number Line" (Thomaidis and Tzanakis 2007). Thomaidis and Tzanakis offer there what they see as empirical evidence both that students produce solutions to problems analogous to historical solutions and that they encounter difficulties which also follow a historical pattern.

That students' conceptual difficulties or epistemological obstacles[22] run parallel to conceptual difficulties in history has been taken up before (e.g., Katz et al. 2000; Bartolini Bussi and Sierpinska 2000; Herscovics 1989). Dorier (1998), for instance, uses history to identify students' problems with notions from linear algebra such as dependence. Among other things, he shows how Euler's notion of "inclusive dependence," as Dorier refers to it, the implicit redundancy of equations, is similar to students' conceptions and is what students need to overcome in order to arrive at more general notion of linear dependence: "Their concept of (in)dependence is, like Euler's, that of inclusive dependence and not linear dependence," he says (Dorier 1998, p. 150, see also Katz et al. 2000, p. 150). Plainly, this is a weaker application of the genetic principle than the traditional one in which students' learning of mathematical ideas actually recapitulates the historical development of those ideas; but it is also one much easier to swallow. That may be simply because these conceptual difficulties are truly difficulties, and one must contend with them whenever one faces them. On the other hand, is it not arguable that the compelling need to face these difficulties in the first place is a sign of their constant reoccurrence, strengthening the recapitulation theme? This is one possibility. But it is also possible that the reoccurrence of conceptual difficulties or any of a range of mathematical ideas is not because history and individual development proceed along parallel tracks, but because individual development is actually a *function* of historically conditioned ideas. In other words, the reoccurrence of mathematical ideas in individual development might be because those ideas and their mode of development are embedded in the culture in which children's mathematics education is situated.

21.3.3   The Cultural Theme

While one might strain to turn this latter possibility into a variant of the genetic position having a distinctive mechanism for individual development, it is more clearly a position directly opposed to a genetic position. Thus, Luis Radford and Luis Puig (Radford and Puig 2007) (see also Furinghetti and Radford 2008), who call this position, appropriately, the "embedment principle," write:

> Biological or natural developments unavoidably become *affected* by and *entangled* with the historical-cultural one as individuals use signs and other cultural artifacts, such as language. In fact, the merging of natural and historical developments constitutes the actual line of

---

[22] This term stems from Bachelard's work on the history of science (Bachelard 1938) and then adapted for mathematics education prominently by Brousseau (see Brousseau 1997, pp. 98ff).

growth of the individual. Given that it is impossible to untie the merging of the cultural and the natural lines of development, the conceptual growth of each individual cannot *reproduce* a historical-social conceptual formation process. In short, phylogenesis cannot recapitulate ontogenesis. (pp. 147–148)

Radford and Puig's embedment principle, which sees our thinking in mathematics as "…related in a crucial manner, to a historical conceptual dimension ineluctably embedded in our social practices and in the signs and artifacts that mediate them" (Radford and Puig 2007, p. 148) brings us squarely into what we called above the cultural theme. One sees immediately how the entrance into that theme is accompanied by the introduction of new subthemes, specifically, ones connected to semiotics, language, and social practices. Radford, for example, draws together these subthemes and their relationship to an historical approach in, among other places, his "On the epistemological limits of language: mathematical language and social practice during the renaissance" (Radford 2003). Even linguistic theorists such as Ferdinand de Saussure (1857–1913), one of the founders of modern semiotics, begin to have relevance in thinking about history of mathematics in mathematics education. Indeed, the idea of language as a semiotic system with both synchronic and diachronic aspect can be shown to be a good model for thinking about the seemingly fixed structure of mathematics and its simultaneous existence as a culturally engendered system changing over time (see Fried 2008, 2009).

Whether or not one takes the cultural theme as far as Radford and others have into the world of semiotics, the theme always carries a sense that mathematics and culture are inseparable, and this means that mathematics and history are inseparable. In a certain sense, this is no more than a very conscious recognition that mathematics is a human enterprise. Accordingly, in his paper, "The necessity of history in teaching mathematics" (Rickey 1996), Fred Rickey tells us that the point he wants to argue is that,

> …Mathematics is the work of individuals. It is a discipline that has been developed by many people over the ages, some making great contributions, some making minor contributions, with the cumulative effect that mathematics has developed into a rich field that has had a significant impact on the way people view their world.
>
> As teachers of mathematics, and even more so as historians of mathematics, we arethe carriers of the mathematical culture. It is our solemn responsibility to transmit this culture to our students. (p. 252)

This aim to humanize mathematics admittedly comes dangerously close to the trivial storytelling we have seen in Lightner's (2000) "Mathematicians are human too," but whereas "humanizing mathematics" there only meant making mathematics less formidable so students might not feel threatened by it and shun it as somehow inhuman, "humanizing mathematics," here, it means seeing mathematics as an *essentially* human activity, that is, as part of the *nature* of mathematics. It becomes a human science almost in the sense of Dilthey's *Geisteswissenschaft*, and, in this sense, the history of mathematics no longer functions only as a means used to promote motivation or interest but as something at the core of mathematics and, accordingly, of what it means to learn mathematics.

As an expression of culture, one's perspective on mathematics changes radically. For it brings to mathematics a conception foreign to the usual one in which mathematical objects and relations are eternal, ideal Platonic entities, everywhere viewed in the same way and everywhere given the same weight, ideas with no geography and no past. Such a change in perspective does not mean necessarily that relativism becomes the philosophy of mathematics, but it does invite a pluralistic view in which one does not assume that mathematics always and in all places means the same thing and is directed towards the same phenomena (needless to say, this is not a claim that everyone is always right and no one is ever wrong!).

With this in mind, one can see how the history of mathematics in mathematics education finds much in common with the still-developing subfield of mathematics education known as ethnomathematics (e.g. D'Ambrosio 2006). Thus in their paper for the ICMI study volume on history in mathematics education (Fauvel and van Maanen 2000), Lucia Grugnetti and Leo Rogers (2000) emphasize mutual relations between history of mathematics and multicultural issues. In particular, in their view, the history of mathematics helps the student acquire a sense of diversity, of the many different ways people can think about and approach the world:

> Multiculturalism then, in the sense that we have tried to convey here, is the identification and celebration of diversity, the respecting and valuing of the work of others, the recognition of different contexts, needs and purposes, and the realisation that each society makes and has made important contributions to the body of knowledge that we call mathematics. (p. 51)

The idea of diversity is not only a central idea in all cultural studies, it is also a central notion in all historical studies, mathematical or not. In this way, the cultural theme comes closest to a historical approach to mathematics education that is truly historical. Yet, as we shall see shortly, this brings us into a difficulty when we consider history of mathematics in the context of other legitimate goals in mathematics education. To see this we shall have to get a better sense for what it means to be historical – and *non*-historical.

21.4 History of Mathematics as History and What That Implies for History of Mathematics in Mathematics Education

Although there is no complete agreement on what history is and what is at the heart of the historian's craft, still, there are undeniable commonalities.[23] Of these, a keen awareness of the tension between past and present or, at very least, the need to confront the question of past and present, stands out most clearly. For while it is impossible to think of history without reference to the past, history is not solely about the past; it is also about the present. To start, one must refer to the present to the extent

---

[23] Much of this section is based on a plenary talk (Fried 2010) at the ESU-6, HPM conference in Vienna, July, 2010.

historians' materials, their objects of study, are things having made their way into the present. With this in mind, Geoffrey Elton (1967) defined history as being "… concerned with all those human sayings, thoughts, deeds and sufferings which occurred in the past and have left present deposit; and it deals with them from the point of view of happening, change, and the particular" (p. 23).

The qualification in Elton's statement emphasizes also that it is not just past or present that is essential but how these are treated, namely, "from the point of view of happening, change, and the particular." The historical mode of thinking demands treating these "survivals" from the past, as Michael Oakeshott calls them (see Oakeshott 1999), precisely *as* survivals, survivals from another world. One interrogates them to understand where they came from – for theirs is a world not conditioned by the existence of ours, yet it is one out of which ours has grown.

History, viewed in this way, is a kind of vicarious experience of the past through what has made its way into the present. Oakeshott, whom I have just mentioned, views history for this reason as a mode of experience, and, for him, to experience the past in the present *as history,* one must view the past *unconditionally.* The historical past is a kind of past, but there are other kinds of past as well. To describe a relationship to the past that depends on the present, in other words, that sees the past in terms of present values, needs, and ideas, Oakeshott uses the term "practical past." The historical past is defined in opposition to the practical past; it is a past understood in terms of its separateness from the present.

Accordingly, in his chapter on historical experience in *Experience and Its Modes* (Oakeshott 1933), Oakeshott sets out the historian's task as follows:

> What the historian is interested in is a dead past; a past unlike the present. The *differentia* [emphasis in the original] of the historical past lies in its very disparity from what is contemporary. The historian does not set out to discover a past where the same beliefs, the same actions, the same intentions obtain as those which occupy his own world. His business is to elucidate a past independent of the present, and he is never (as an historian) tempted to subsume past events under general rules. He is concerned with a particular past. It is true, of course, that the historian postulates a general similarity between the historical past and the present, because he assumes the possibility of understanding what belongs to the historical past. But his particular business lies, not with this bare and general similarity, but with the detailed dissimilarity of past and present. He is concerned with the past as past, and with each moment of the past in so far as it is unlike any other moment. (p. 106)

So even though historical experience is an experience in the present and one belonging to a living and breathing historian, the historian must live by the desideratum to view the past in its particularity. Historians' rule to avoid anachronism is an easy corollary to this desideratum. However, it is not an easy rule to obey, since we are beings who live in the present and whose immediate experience is not that of the historical subjects we study. The struggle with anachronism is at the heart of the tension between past and present, with which I began this section. It might be said, indeed, that the historical art is one that aims to keep that struggle alive.

The dangers of submitting to anachronism and the subtle ways in which it can subvert history were discussed most trenchantly and colorfully by Herbert Butterfield in his classic, *The Whig Interpretation of History* (Butterfield 1931/1951). The term "Whiggism" has subsequently entered the vocabulary of standard historiography.

21 History of Mathematics in Mathematics Education

The tendency it refers to is the distorting of the past not only by reading modern intentions and conceptions into the doings and writings of thinkers in the past, which is anachronism in its most direct form, but also by forcing the past through a sieve that bars ideas foreign to a modern way of looking at things and permits those adaptable to modern interests. For example, in reading Proclus' *Commentary on Book I of Euclid's Elements*, a Whig historian would leave out Proclus' arguments in the "first prologue" about the nature of mathematical being and role of mathematics in the moral education of the soul while emphasizing Proclus' comments on logical difficulties, missing cases, and alternative proofs connected with the familiar geometrical propositions in the *Elements*. These things are truly to be found in Proclus, but a Whig historian would give the impression that they are the *only* things in Proclus, or the only things of any worth in Proclus.

Whig history is indeed particularly seductive when it comes to mathematics. This is because, as we have remarked before, mathematics is easily taken to be a constant component of thought not only in the modern world, but also in all parts of the world and at all other times. Left unchallenged, that view of mathematics makes Whiggism almost inescapable: present mathematical knowledge, short of logical errors, is mathematical knowledge *tout court*; past mathematical knowledge, to be understood, has merely to be translated into a modern idiom. So one can feel fully justified in treating mathematicians of the past as Littlewood famously said of the Greek mathematicians, that they are only, "Fellows of another college" (quoted in Hardy 1992, p. 81); mathematicians of the past, like one's colleagues, are useful for gaining insights into one's present mathematical research. The past, for Whig historians, is thus almost *by definition* a "practical past," adopting Oakeshott's term: they seek in the past what is useful for the present.

Clifford Truesdell (1919–2000) is a good example of a Whig historian. He is a good example, not because his historical work was false or inaccurate or superficial, but precisely because he was tremendously learned and serious; his work was thorough and in some ways deep. The problem is only how his history was oriented, and how much *historical* understanding we gain from it. For him, history of mathematics was unabashedly dedicated to a "practical past": "One of the main functions [the history of mathematical science] should fulfill is to help scientists understand some aspects of specific areas of mathematics about which they still don't fully know" (in Giusti 2003, p. 21). What one learns from the history of mathematics, in its Whiggish form, is, in short, mathematics.

By now it should be apparent that this view of mathematical past as a "practical past" lies close to the foundations of the curricular theme. The very fact that history, in the curricular spirit, is seen as something *to use* in order to promote modern mathematical knowledge brings it into line with the practical past and the Whig interpretation of history. It must be made clear that the problem is not that proposals according to the curricular theme have no good effect for learning mathematics; it is its historical character that is in question. For to the extent that mathematics is continuous over time and place, a universal body of content, it is really ahistorical and noncultural, or, at best, its peculiarly historical and cultural aspects involve only trivial matters of form. Besides this kind of history being thus non-history, by using the present to determine what is useful for the present, one finally forfeits *learning* from the past. It is this that Butterfield found so wrong about Whig history. More specifically, he writes:

> If we turn our present into an absolute to which all other generations are merely relative, we are in any case losing the truer vision of ourselves which history is able to give; we fail to realise those things in which we too are merely relative, and we lose a chance of discovering where, in the stream of the centuries, we ourselves, and our ideas and prejudices, stand. In other words we fail to see how we ourselves are, in our turn, not quite autonomous or unconditioned, but a part of the great historical process; not pioneers merely, but also passengers in the movement of things" (p. 63)

By contrast, history of mathematics in mathematics education according to the cultural theme with its interest in difference and the essential human imprint in mathematical thought is consistent with a truly historical mode of thinking and assumes that one can learn something from history as such. Indeed, it assumes one can learn about ourselves as beings for whom mathematics is part and parcel of our cultural identity,[24] and, as Butterfield warned, we shall fail in this if we too easily adopt a history guided by the present. Put in different terms, ones inspired by Anna Sfard's work on "meta-discursive rules" and "commognition"[25] (Sfard 2008), Kjeldsen and Blomhøj (2011) have stated the case thus:

> As active learners, students can become aware of their own meta-discursive rules by identifying the meta-rules that governed the mathematics of the past and comparing them with meta-discursive rules governing the mathematics of their textbook and instruction. In this way, opportunities for students to experience commognitive conflicts are provide and proper changes can be initiated. (pp. 4–5)

But even in these different terms, Kjeldsen and Blomhøj are fully aware that what stands opposed to these opportunities is Butterfield's Whig history, as they immediately point out (p. 5):

> If one's reading and interpretation of historical sources are constrained by the way mathematics is perceived and conceptualized in the present, the historical text cannot play the role of an "interlocutor" that can be used to create commognitive conflicts, as explained above, when students communicate with the text, since differences in the way of communicating in the past and in the present will have been "washed away" by the whig interpretation. (p. 5)

The conclusion seems to be clear. In order to have a historical approach in mathematics education in which history is taken seriously *as a form of knowledge*, we ought to embrace an approach along the lines of the cultural theme and reject the Whiggish proposals derived from the curricular theme. But there is a difficulty here. For one cannot forget that mathematics educators are not historians and have other legitimate concerns. So while the history of mathematics can bracket the present in order to understand the past, mathematics education typically justifies itself precisely by the power and

---

[24] One recalls in this connection Collingwood's claim that this kind of self-knowledge is the entire point of history (e.g., Collingwood 1939).

[25] In creating this term – a fusion of the words "communication" and "cognition" – Sfard tried, in a Vygotskian spirit, to capture the way thinking is entangled with discourse.

necessity of mathematics in *modern* contexts, in science, engineering, economics, and industry. This is certainly consistent with the spirit of the American *Principles and Standards for School Mathematics* (NCTM 2000). There, we read:

> The level of mathematical thinking and problem solving needed in the workplace has increased dramatically.
>
> In such a world, those who understand and can do mathematics will have opportunities that others do not. Mathematical competence opens doors to productive futures. A lack of mathematical competence closes those doors.
>
> …More students pursue educational paths that prepare them for lifelong work as mathematicians, statisticians, engineers, and scientists.
>
> …Today, many students are not learning the mathematics they need. In some instances, students do not have the opportunity to learn significant mathematics. In others, students lack commitment or are not engaged by existing curricula. (NCTM 2000, Introduction)

One cannot belittle this emphasis on modern mathematics: the ideas and methods of modern mathematics are undeniably profound and powerful, and there is no reason they should not be pursued and taught. But accepting this kind of emphasis also means that mathematics educators *cannot* bracket the present, as historians can and must. Thus, when mathematics educators – even those with real historical sensitivity and knowledge – confront a chapter in the history of mathematics they must heed, to some extent at least, the counterweight of their obligation to teach mathematics in a modern spirit. They must consider how relevant the chapter is to the modern mathematical ideas they need to convey, how well it fits the subjects required by their curriculum. Seemingly practical considerations of time and scheduling are, in fact, signs that history of mathematics in the classroom must be subordinated to such standards as those in the NCTM document.[26]

Naturally, there may be some historical topics for which a happy medium can be found, some cases where a chapter in the history of mathematics fits comfortably in the curriculum without demanding too great a compromise as to its historical character. While this may be, it is not the point. The point is that when mathematics education emphasizes mathematics as it is understood and practiced today, as it is needed in science and engineering, it will be necessarily predisposed to treat the history of mathematics in a Whiggish spirit, separating relevant from irrelevant ideas according to the needs of the modern curriculum. This predisposition is not an injunction to be Whiggish; it is, rather, a kind of ineluctable internal pressure at work in any attempt to introduce history of mathematics into mathematics education, where the latter is directed, as it generally is,[27] towards modern mathematics.

---

[26] One might also cite European standards as well, even the Danish competence-oriented mathematics education, KOM-project (see Niss and Højgaard 2011). The latter, however, includes competencies that invite a broader view of what it means to be mathematically educated (see Jankvist and Kjeldsen's (2011) paper and also the discussion below).

[27] There are, of course, exceptions. One of the anonymous reviewers of this chapter made me aware of the Ross School and Institute in New York State, USA, whose program, among other things, stresses cultural history. Describing their "spiral curriculum," they write: "Teaching the humanities and sciences in the context in which they historically emerged makes for a naturally integrated approach" (http://rossinstitute.org/#/The-Ross-Model/Spiral-Curriculum, accessed April, 2012).

Mathematics educators, in this way, are placed in a very different position from the historian of mathematics who must struggle with the problem of anachronism. For the historian, engaging in that struggle is part of what it means to do the history of mathematics: historians who do not live in the tension between past and present are not true historians. But in the case of mathematics education, the problem is one of conflicting demands and commitments, presenting the mathematics educator with a dilemma: maintain modern mathematics as one's main end and thus subordinate history of mathematics serve modern ideas and needs, that is, adopt a Whig version of history of mathematics, or insist that history of mathematics be history and put aside the perfectly legitimate emphases of programs seeking to help students use and understand the modern mathematics essential for all the pure and applied sciences (see Fried 2001, 2007 for further discussion and additional examples).

Of course the force of the dilemma derives from accepting prescribed ends like those described in the example of the NCTM. And one must emphasize that built as they are on the power of modern mathematics to address societal and scientific needs, those ends are anything but inconsequential and easily dismissed; however, they are not absolute. What it means to teach mathematics and what it means to be mathematically educated can be defined in accord with a different set of ends, ones more cultural and humanistic and less utilitarian. Mathematics education may not be so determined that it must follow the standards set out in the NCTM Principles and Standards (NCTM 2000) or similar documents (e.g., European Mathematical Society 2001). Mathematics education can be reconceived so that it promotes an educational approach shaped by history of mathematics as a form of knowledge rather than one that only uses history of mathematics in the service of ends not necessarily in line with those of history, "history as a goal" rather than "history as a tool," as Uffe Jankvist has aptly put it (Jankvist 2009).

Reconceiving mathematics education in this way does not mean turning it into what has derogatorily (and somewhat unfairly) been called "mathematics for poets"[28]: it can be rigorous and, yes, mathematical. Although there is no single scheme for this, it is clear that, one way or another, the presence of original sources will be essential. This is because, as Elton (1967) made plain in the passage quoted above, at the heart of the historical enterprise are those thoughts – in our case mathematical thoughts – "…which occurred in the past and have left present deposit…" (p. 23) In the case of the history of mathematics, as with all history of ideas, the present deposit consists, maybe not exclusively, but certainly chiefly of written texts. For while written texts in general history are crucial *as accounts* of happenings in the past, in the history of thought, texts are, one might say, the thought itself. Original mathematical texts are also the chief expression of how mathematicians have sought to engage other mathematicians in their thought; they represent communicated thought. And that, bringing us back to the cultural theme, places original texts at the center of what we should call mathematical culture and tradition.

---

[28]Poets can have a deep knowledge of mathematics. The mathematical knowledge of the great French poet, Paul Valéry, for example, could hardly be called superficial or soft.

21 History of Mathematics in Mathematics Education

Tradition is too often misunderstood as something rusty and dogmatic. With Eva Brann, "By *the tradition* I mean neither the old customs nor the recent routines, neither the sedimentary wisdom nor the petrified habits of communities. I mean, to begin with, *a collection of books*" (Brann 1979, p.64). And as Brann also points out, the origin of the word "tradition," the Latin verb "tradere" means both to pass on and to betray (p. 67). In other words, tradition implies not only what one remembers and respects but also the platform from which one changes and develops. It is in this sense, then, that one should understand tradition when it comes into the arguments of those who promote the use of original sources in teaching mathematics. For example, Laubenbacher, Pengelley, and Siddoway (1994)[29] write this in defense of using original texts:

> For a novelist, poet, painter or philosopher such observations would be old news, since their disciplines have long recognized the importance of studying the original work, techniques and perspectives of classical masters. And in so doing, they are never removed from an understanding of how people have struggled, and have created works of art. Young artists thus see themselves as part of a creative tradition. Unfortunately, we have lost this sense of tradition in our discipline, and, ironically, we can perhaps blame much of this loss on the dazzling explosion of mathematics in this century. It is time we step back from our accomplishments and recapture a historical perspective.

The picture of mathematics education, or rather of becoming educated mathematically, that begins to emerge is not so much one that concentrates on the mastery of certain techniques in mathematics or even certain concepts in mathematics such as a function or derivative, but on the reading and learning to read a body of mathematical texts. Which texts to be included and which not, aside from certain texts, such as Euclid's *Elements*, could no doubt be debated; even courses on Shakespeare vary as to which play or sonnet is discussed, alluded to, or left unmentioned. And mentioning literature is not by accident: reading of texts, following authors' modes of presentation and their points of attention, weighing the cultural context of works, and so on would make mathematics education into a kind of literary education. And yet it *is* mathematics, as one quickly finds out working through Euclid's *Elements*, Descartes' *Géométrie*, or Euler's *Introductio*.

But using original texts is only a condition for a mathematics education for which history is a goal; one must know how to use original texts. While this involves theoretical considerations concerning the role of original source material, it also involves considerations arising from empirical studies. As Furinghetti et al. (2006) point out, "An important development in recent years is that more empirical research studies on the integration of original sources are being done, many of which include a large number of students" (p. 1288).

---

[29] Laubenbacher and Pengelley have promoted the use of original sources indefatigably for over twenty years with such works as (Laubenbacher and Pengelley 1999). In the latter, for example, they treat analysis by looking closely at the transitions of ideas connected with the calculation of areas and volumes, and do so via texts from Archimedes, Cavalieri, Leibniz, Cauchy, and Abraham Robinson. An overview of their work can be found at the website: http://math.nmsu.edu/~history/

One good example of this sort of empirical study is that by Jankvist and Kjeldsen (2011) (also described in Kjeldsen and Blomhøj (2011)). This research was carried out against the background of Danish competence-oriented KOM-project report, which emphasized history of mathematics in what it termed "overview and judgment," competences referring to "mathematics as a discipline" (Niss and Højgaard 2011, pp. 118–120).[30] In the course of the study, high school students worked on problem-based mathematical projects over the course of a semester with close attention to history via original sources (in translation) such as Johann Bernoulli's 1691 "solution of the cable problem" as well as more modern works, such as Hamming's 1950 "Error Detecting and Error Correcting Codes," treated in a historical spirit. By assuming a problem-based approach, Jankvist and Kjeldsen tried to achieve a situation in which close attention to written texts and sensitivity to history as history, attention "meta-issues," was "anchored" in the "in-issues" of mathematics, as they put it. Although there was a risk of making the in-issues *the* issues, and, therefore, descending into Whiggism, they could avoid that and keep "history as a goal" partly through this close attention to texts and partly through the constant interaction with historically knowledgeable supervisors.

The genetic principle is an obvious and in some ways natural basis for using original texts. Because the principle assumes that history follows a course towards a given modern topic that mirrors the course of students' own of understanding, it is reasonable, on that basis, to present texts one after another in historical sequence. Michael Glaubitz (2010) obtained empirical data on this point by comparing the genetic approach and a conventional nonhistorical approach. His experiment centered on 175 students using the quadratic equation and formula as his topic and works by Al Khwarizmi, for example, as his texts. Even with respect to interest, Glaubitz's results were disappointing. He summarizes them as follows: "...this teaching did not work. The students rather got confused and appeared very displeased in the end" (p. 8). Naturally, one should be cautious here: the texts were given to the students and then followed by "conventional exercises, problems and applications with modern methods" (p. 5); there may be other ways to pursue a purely genetic approach.

But Glaubitz also examined another approach, that developed by Niels Jahnke, the "hermeneutic approach." Students in this approach were taught the quadratic equation and formula in a modern conventional way, and, when they were brought face to face with the original texts, they were asked to engage with them in active, often creative, ways, such as writing fictitious interviews with Al Khwarizmi: the original texts were still present, but the students' own perspective in reading them (including some of their difficulties in reading them) were taken into account. The results here were much more encouraging and contrasted starkly with the approach based on the genetic principle.

---

[30]It is this looking at mathematics as a discipline, from the side as it were, that gives the term "meta-issue," used by Jankvist and Kjeldsen its aptness. Jankvist and Kjeldsen's complementary term, I might add, is "in-issue," by which they mean an internal matter of mathematical content – concepts, methods, algorithms, mathematical proof, etc.

21 History of Mathematics in Mathematics Education

By having students learn, say, the quadratic formula, in a modern way, one might think that the hermeneutic approach is actually introducing anachronism into its historical approach. But, as we discussed at the very start of this part, while a historical understanding of the past demands seeing the past unconditionally, it does not require one to forget the present. We were at pains to show that a historical understanding is marked by a tension between past and present.[31] The problem with, for example, Whiggism is that it sees the present as completely consistent with the past – hence, no tension. When one sees, from the start, that one's position with respect to the past is problematic, one is thrown into the role of being an interpreter. As Jahnke (2000) puts it[32]:

> In traditional theories of hermeneutics the relation between the historical meaning of a text(the intention of its author) and its meaning for a modern reader is amply reflected and identified as the essential problem of interpretation. In fact, seen under the aspect of method, history of mathematics, like any history, is essentially an hermeneutic effort. If history of mathematics is not to deteriorate into a dead dogma, teachers should have some ideas about the hermeneutic process and the fruitful tension between the meaning of a text in the eyes of its author and the meaning for a modern reader. (p. 298)

To use the terminology mentioned in part 2 of this paper, in the hermeneutic approach, one's position with respect to the past is distinguished by being in the present and having present mathematical knowledge, but it is not "privileged." On the contrary, recognizing our position in the present *as present* comes with the recognition that the past is obscure, and, thus, needs to be interpreted. And the phenomenon that is familiar to anyone who has done serious historical work is that more one engages in that interpretive activity the more obscure the past becomes. But this only brings the distinctness of one's modern knowledge into relief. One truly begins to see here how one can bring a historical awareness into mathematics education that will allow not only for genuine insight into the past but also into the present.

21.5 Conclusion

With the discussion of tradition and of interpretation, we come full circle in this chapter and, in certain respect, historically as well. For recall, the word "tradition" was also one translation of the Greek word *paideia*. Its other translations, as we noted earlier, included "education," "civilization," "culture," and, in Latin, *humanitas*. These are all themes that have been central in this chapter. But more than that the tacit message here has been that by taking history of mathematics as a goal, we might be able to restore a sense of mathematical knowledge as the self-knowledge

---

[31] This awareness of the tension between the students' own perspective and that of the original texts also played a part in Radford and Guérette's (2000) successful teaching sequence concerning the quadratic equation and the Babylonian's "naive geometry."

[32] Another account of the details and presuppositions of the approach can be found in Jahnke (1994).

of our human mathematical mind, to paraphrase Collingwood (see Fried 2007): the mathematical mind is indeed a human mind, and doing mathematics is a high human activity. *Paideia* carried that sense of a distinct human possession relating to humanness in its highest and broadest expression.

Now we first mentioned *paideia* in connection to the history of mathematics in ancient education and, specifically, to Proclus. Tradition as historical character in that context had to be taken in a qualified way: mathematical tradition for Proclus was much more of a continuous tradition than it is for us. Thus Proclus' references to mathematicians of the past could genuinely be seen as they were references to "colleagues" no longer alive. For us, there have been breaks in the tradition even though it is still in some sense ours: the past for us is a past that has often been lost or obscured and that requires recovery and interpretation. Our own sense of who we are with respect to our mathematical past requires cognizance of these breaks and the fundamental differences between us and our predecessors. History in the modern sense of Elton, Butterfield, and Oakeshott, while it recognizes commonality, endeavors to refine our understanding of such differences. For this reason, in the development above, it was crucial to confront the historical character of attempts to incorporate history of mathematics in mathematics education, to point out the teleological assumptions of the genetic principle and the inadequacies of the motivational and curricular themes, for example.

The answers to such criticisms, I have suggested, require more than patches. They require perhaps a reorientation vis-à-vis the question of history of mathematics in mathematics education. Instead of asking how we can produce a presentation or program here, a chapter or unit there in the history of mathematics fitting the needs of a set curriculum or addressing the lack of motivation of our students, we may need to ask how we can adjust the meaning of mathematics education itself so that it will accommodate history of mathematics pursued honestly and deeply. No attempt was made to floor an exact proposal because that cannot be done. However, an essential component of any such proposal, we argued, is a view of mathematics as a collection of texts that need to be studied, a collection of authors that need to be engaged. This too, we made clear, cannot be taken a simple proposal. There is more than one way to introduce original texts. A promising suggestion, both theoretically and practically, is Jahnke's hermeneutic approach, which involves both a modern, but not Whiggishly oriented, mathematical knowledge and also an awareness of oneself as an interpreter. The possibility of an approach like the hermeneutic approach makes the possibility of a mathematics education that is truly historically sensitive within reach. Perhaps, we will discover the way to a new *paideia*.

References

Amit, M. & Fried, M. N. (2008). The Complexities of Change: Aspects of Reform and Reform Research in Mathematics Education. In L. English (ed.), *Handbook of International Research in Mathematics Education, 2nd Edition* (pp. 385–414). New York: Routledge.

Bachelard, G. (1938). *La Formation de l'Esprit Scientifique*. Paris: Libraire Philosophique J. Vrin.

21 History of Mathematics in Mathematics Education

Bartolini Bussi, M. G., & Sierpinska, A. (2000). The Relevance of Historical Studies in Designing and Analyzing Classroom Activities. In J. Fauvel and J. van Maanen (Eds). *History in Mathematics Education: The ICMI Study* (pp. 154–161). Dordrecht: Kluwer Academic Publishers.

Begle, E. G. (1962). Remarks on the Memorandum "On the Mathematics Curriculum of the High School." *The American Mathematical Monthly,* 69(5), 425–426.

Bensaude-Vincent, B. (2009). La Place des Réflexions sur l'École dans l'Oeuvre de Paul Langevin. In L. Gutierrez and C. Kounelis (eds.), *Paul Langevin et la Réforme de l'Enseignement* (pp. 15–22). Grenoble: Presses Universitaires de Grenoble.

Bidwell, J. K. & Clason, R. G. (1970). *Readings in the History of Mathematics Education.* Reston, VA.: NCTM.

Boag, E. (2010). Dandelin Spheres. *Mathematics in School,* 39(3), 34–36.

Branford, B. (1908). *A Study of Mathematical Education.* Oxford: Clarendon Press.

Brann, E. T. H. (1979). *Paradoxes of Education in a Republic.* Chicago: Chicago University Press.

Brousseau, G. (1997). *Theory of Didactical Situations in Mathematics: Didactique des mathématiques, 1970–1990.* Nicolas Balacheff, Martin Cooper, R. Sutherland, Virginia Warfield (Translators). New York: Springer.

Butterfield, H. (1931/1951). *The Whig Interpretation of History.* New York: Charles Scribner's Sons.

Byrne, O. (1847). *The First Six Books of the Elements of Euclid.* London: William Pickering.

Carson, R. N., & Rowlands, S. (2000). *A Synopsis of the Collapse of the Geometry Standard in the UK.* Unpublished report of the Centre for Teaching Mathematics, University of Plymouth, UK.

Collingwood, R. G. (1939). *Autobiography.* Oxford: Oxford University Press

D'Ambrosio, U. (2006). *Ethnomathematics: Link Between Traditions and Modernity.* Rotterdam, The Netherlands: Sense Publishers.

Dandelin, G. P. (1822). Mémoire sur Quelques Propriétés Remarquables de la Focale Parabolique. *Nouveaux mémoires de l'Académie Royale des Sciences et Belles-Lettres de Bruxelles, T. II.,* 171–202). Available online at: http://fr.wikisource.org/wiki/ Mémoire_sur_quelques_propriétés_remarquables_de_la_focale_parabolique Accessed 17 November 2011.

Dauben, J. W. (2002). United States. In J. W. Dauben and C. J. Scriba (eds.) *Writing the History of Mathematics: Its Historical Development* (pp. 263–285). Basel: Birkhäuser Verlag.

Domoradski, S. & Pwlikowska-Brożek, Z. (2002). Poland. In J. W. Dauben and C. J. Scriba (eds.) *Writing the History of Mathematics: Its Historical Development* (pp. 199–203). Basel: Birkhäuser Verlag.

Donoghue, E. F. (2006). The Education of Mathematics Teachers in the United States: David Eugene Smith, Early Twentieth-Century Pioneer. *Paedagogica Historica,* 42(4–5), 559–573.

Dorier, J. (1998). The Role of Formalism in the Teaching of the Theory of Vector Spaces. *Linear Algebra and its Applications,* 275–276 (Double issue containing *Proceedings of the Sixth Conference of the International Linear Algebra Society*), 141–160

Dubinsky, E., Weller, K., McDonald M. A., Brown, A. (2005). Some Historical Issues and Paradoxes Regarding the Concept of Infinity: An APOS-Based Analysis, Part I. *Educational Studies in Mathematics,* 58(3), 335–359.

Eisenberg, T. (2008). Flaws and Idiosyncrasies in Mathematicians: Food for the Classroom. *The Montana Mathematics Enthusiast,* 5(1), 3–14.

Elton, G. R. (1967). *The Practice of History.* London: Collins.

European Mathematical Society (2001). *Reference Levels in School Mathematics Education in Europe: Italy.* Available at the web site: *http://www.emis.de/projects/Ref/* Accessed 5 April, 2010.

Fauvel, J. van Maanen, J. (2000). *History in Mathematics Education: The ICMI Study.* Dordrecht: Kluwer Academic Publishers.

Fauvel, J. (1995). Revisiting the History of Logarithms. In F. Swetz, J. Fauvel, O. Bekken, B. Johansson, and V. Katz (eds.). *Learn from the Masters* (pp. 39–48). Washington, DC: The Mathematical Association of America.

Fried, M. N. & Bernard, A. (2008). Reading and Doing Mathematics in the Humanist Tradition: Ancient and Modern Issues. In E. Barbin, N. Stehliková, C. Tzanakis (eds.), *Proceedings of European Summer University on the History and Epistemology in Mathematics Education (HPM-ESU5)* (pp. 463–474). Prague, Czech Republic: Vydavatelskýservis.

Fried, M. N. & Unguru, S. (2001). *Apollonius of Perga's* Conica*: Text, Context, Subtext.* Leiden, The Netherlands: Brill Academic Publishers.

Fried, M. N. (2001). Can mathematics education and history of mathematics coexist? *Science & Education*, 10, 391–408.

Fried, M. N. (2007). Didactics and history of mathematics: Knowledge and self-knowledge. *Educational Studies in Mathematics*, 66, 203–223.

Fried, M. N. (2008). History of Mathematics in Mathematics Education: A Saussurean Perspective. *The Montana Mathematics Enthusiast*, 5(2), 185–198.

Fried, M. N. (2009). Similarity and Equality in Greek Mathematics: Semiotics, History of Mathematics and Mathematics Education. *For the Learning of Mathematics*, 29(1), 2–7.

Fried, M. N. (2010). History of Mathematics: Problems and Prospects. In E. Barbin, M. Kronfellner, C. Tzanakis, *History and Pedagogy of Mathematics: Proceedings of the 6th European Summer University on History and Epistemology in Mathematics Education ESU-6* (pp. 13–26), Vienna: Verlag Holzhausen GmbH.

Fried, M. N. (2011). Postures towards Mathematics of the Past: Mathematicians, Mathematician-Historians, Historians of Mathematics. Talk given at the *Cohn Institute for the History and Philosophy of Science and Ideas*. Tel Aviv University, Tel Aviv. 17 January, 2011.

Friedlein. G. (1873). *Procli Diadochi in primum Euclidis Elementorum librum commentarii* Leipzig: B. G. Teubner (repr. Hildesheim, 1967).

Furinghetti, F. & Radford, L. (2008). Contrasts and Oblique Connections between Historical Conceptual Developments and Classroom Learning in Mathematics. In L. English (ed). *Handbook of International Research in Mathematics Education, 2nd Edition* (pp. 626–655). Mawah (NJ): Lawrence Erlbaum.

Furinghetti, F., Jahnke, H. N., van Maanen, J. (eds.) (2006). Mini-Workshop on Studying Original Sources in Mathematics Education. *Oberwolfach Report,* 3(2), 1285–1318.

Giusti, E. (2003). Clifford Truesdell (1919–2000), Historian of Mathematics. *Journal of Elasticity*, 70, 15–22.

Glaubitz, M. R. (2010). The Use of Original Sources in the Classroom: Empirical Research Findings. Talk given at the *European Summer University-6, History and Pedagogy of Mathematics (ESU-6/HPM)*, July, 2010, Vienna.

Goldin, G. A. (2009). The affective domain and students' mathematical inventiveness. In R. Leikin, A. Berman, and B. Koichu (eds.), *Creativity in Mathematics and the Education of Gifted Students* (pp. 181–194). Rotterdam: Sense Publishers.

Grugnetti, L. & Rogers, L. (2000). Philosophical, Multicultural and Interdisciplinary Issues. In J. Fauvel and J. van Maanen (eds.), *History in Mathematics Education: The ICMI Study* (pp. 39–62). Dordrecht: Kluwer Academic Publishers.

Gulikers, I. & Blom, K. (2001). 'A Historical Angle', a Survey of Recent Literature on the Use and Value of History in Geometrical Education. *Educational Studies in Mathematics,* 47, 223–258.

Hardy, G. H. (1992). *A Mathematician's Apology*. Cambridge: Cambridge University Press.

Herscovics, N. (1989). Cognitive obstacles encountered in the learning of algebra. In S. Wagner, C. Kieran (eds.) *Research Issues in the Learning and Teaching of Algebra* (pp. 60–86). Reston (VA).: NCTM.

Howson, G. (1982). *A History of Mathematics Education in England*. Cambridge: Cambridge University Press.

Hughes, R. (1997) (May 21). American Visions. *Time Magazine*. Online at: http://www.time.com/time/magazine/article/0,9171,986375-2,00.html Accessed 23 November 2011.

Hulin, N. (2005). Histoire des Sciences et Enseignement Scientifique au Lycée sous la Troisième République, *Revue d'histoire des sciences, (n° thématique L'enseignement de l'histoire des sciences sous la IIIeRépublique* (dir. Anastasios Brenner)), 58(2), 389–405.

21 History of Mathematics in Mathematics Education

Jaeger, W. (1945). *Paideia: The Ideals of Greek Culture*, 3 vols. Gilbert Highet (trans.). New York: Oxford University Press.

Jahnke, H. N. (1994). The Historical Dimension of Mathematical Understanding—Objectifying the Subjective. In J. P. da Ponte and J. F. Matos (eds.), *Proceedings of the 18th International Conference for the Psychology of Mathematics Education*, Vol. I (pp. 139–156). Lisbon: University of Lisbon.

Jahnke, H. N. (2000). The Use of Original Sources in the Mathematics Classroom. In J. Fauvel and J. van Maanen (Eds). *History in Mathematics Education: The ICMI Study* (pp. 291–328). Dordrecht: Kluwer Academic Publishers.

Jankvist, U. T. & Kjeldsen, T. H. (2011). New Avenues for History in Mathematics Education: Mathematical Competencies and Anchoring. *Science & Education, 20*, 831–862

Jankvist, U. T. (2009). A Categorization of the "Whys" and "Hows" of Using History in Mathematics Education. *Educational Studies in Mathematics, 71*(3), 235–261.

Katz, V. J. & Barton B. (2007). Stages in the History of Algebra with Implications for Teaching and Commentary from a Mathematics Educator. *Educational Studies in Mathematics, 66*(2), 185–201.

Katz, V., Dorier, J., Bekken, O., & Sierpinska, A. (2000). The Role of Historical Analysis in Predicting and Interpreting Students' Difficulties in Mathematics. In J. Fauvel and J. van Maanen (Eds). *History in Mathematics Education: The ICMI Study* (pp. 149–154). Dordrecht: Kluwer Academic Publishers

Kaygin, B., Balçin, B., Yildiz, C., Arslan, S. (2011). The Effect of Teaching the Subject of Fibonacci Numbers and Golden Ratio through the History of Mathematics. *Procedia Social and Behavioral Sciences* 15, 961–965.

Kennedy, G. A. (2003). *Greek Textbooks of Prose Composition and Rhetoric*. Leiden: Brill.

Kitcher, P. (1984). *The Nature of Mathematical Knowledge*. New York: Oxford University Press.

Kjeldsen, T. H. & Blomhøj, M. (2011). Beyond Motivation: History as a Method for Learning Meta-Discursive Rules in Mathematics. *Educational Studies in Mathematics* (online first) DOI 10.1007/s10649-011-9352-z.

Kjeldsen, T.H. (2011). Uses of History in Mathematics Education: Development of Learning Strategies and Historical Awareness. In M Pytlak, E Swoboda & T Rowland (eds), *CERME 7, Proceedings of the seventh Congress of the European Society for Research in Mathematics Education*, pp. 1700–1709.

Klein, F. (1908/1939). *Elementary Mathematics from an Advanced Standpoint. Part I: Arithmetic, Algebra, Analysis. Part II: Geometry*. Translated by E. R. Hedrick and C. A. Noble. New York: Dover Publications.

Kline, M. (1977). *Calculus: An Intuitive and Physical Approach*. New York: John Wiley & Sons, Inc.

Kronfellner, M. (2000). Duplication of the Cube. In J. Fauvel and J. van Maanen (eds.) *History in Mathematics Education: The ICMI Study* (pp. 265–269). Dordrecht: Kluwer Academic Publishers.

Lakatos, I. (1986). A Renaissance of Empiricism in the Recent Philosophy of Mathematics. In T. Tymoczko (ed.), *New Directions in the Philosophy of Mathematics* (pp. 29–48). Boston: Birkhäuser.

Landau, E. (1965). *Differential and Integral Calculus*. Trans. by M. Hauser and M. Davis. New York: Chelsea Publishing Company.

Laubenbacher, R., Pengelley, D., Siddoway, M. (1994). Recovering Motivation in Mathematics: Teaching with Original Sources, *UME Trends* 6. Available at the website: http://www.math. nmsu.edu/~history/ume.html Accessed 28 November 2011.

Laubenbacher, R., Pengelley, D. (1999). *Mathematical Expeditions: Chronicles by the Explorers*. New York: Springer

Lightner, J. E. (2000). *Mathematicians are Human Too. Mathematics Teacher* 93(8), 696–9.

Liu, P. (2003). Do Teachers Need to Incorporate the History of Mathematics in their Teaching? *Mathematics Teacher, 96*(6), 416–421.

Marrou, H. I. (1982). *A History of Education in Antiquity*, George Lamb (trans.). Madison, Wisconsin: University of Wisconsin Press.

McLeod, D. B. (1994). Research on Affect and Mathematics Learning. *Journal for Research in Mathematics Education,* 25(6), 637–647.

Memorandum (1962). On the Mathematics Curriculum of the High School. *The American Mathematical Monthly,* 69(3), 189–193 (and *The Mathematics Teacher,* 55, 191–195).

Morrow, G. R. (1970). *Proclus: A Commentary on the First Book of Euclid's Elements.* Princeton: Princeton University Press.

Mueller, I. (1991). Mathematics and Education: Some Notes on the Platonic Program. In *ΠΕΡΙ ΤΩΝ ΜΑΘΗΜΑΤΩΝ,* special issue of *Apeiron,* XXIV (4), 85–104.

National Council of Teachers of Mathematics (NCTM) (2000). *Principles and Standards for School Mathematics.* Reston, VA. Available at the web site: http://standards.nctm.org. Accessed 12 October 2010

Niss, M. & Højgaard, T. (2011). *Competencies and Mathematical Learning—Ideas and Inspiration for the Development of Mathematics Teaching and Learning in Denmark.* IMFUFA text no. 485. Roskilde: Roskilde University. Available at the web site: http://milne.ruc.dk/ImfufaTekster/pdf/485web_b.pdf Accessed 28 June 2012.

Oakeshott, M. (1999). *On History and Other Essays.* Indianapolis, Indiana: Liberty Fund, Inc.

Oakeshott, M. (1933). *Experience and Its Modes.* Cambridge: At the University Press.

Peiffer, J. (2002). France. In J. W. Dauben and C. J. Scriba, (eds.) *Writing the History of Mathematics: Its Historical Development* (pp. 4–43). Basel: Birkhäuser Verlag.

Perkins, P. (1991). Using History to Enrich Mathematics Lessons in a Girls' School. *For the Learning of Mathematics, 11*(2), 9–10.

Piaget, J. (1971). *Genetic Epistemology.* New York: W W Norton & Co Inc.

Radford, L. & Guérette, G. (2000). The Second Degree Equation in the Classroom: A Babylonian Approach. In V. Katz (ed.) *Using History to Teach Mathematics: An International Perspective* (pp. 69–75). Washington: MAA.

Radford, L. (2003). On the epistemological limits of language. Mathematical knowledge and social practice in the Renaissance. *Educational Studies in Mathematics,* 52(2), 123–150.

Radford, L. & Puig, L. (2007). Syntax and Meaning as Sensuous, Visual, Historical Forms of Algebraic Thinking. *Educational Studies in Mathematics,* 66(2), 145–164.

Rickey, V. F. (1996). The Necessity of History in Teaching Mathematics. In R. Caliger (ed.). *Vita Mathematica: Historical Research and Integration with Teaching* (pp. 251–256). Washington: Mathematics Association of America.

Roberts, D. L. (2004). The BKPS Letter of 1962: The History of a 'New Math' Episode. *Notices of the American Mathematical Society,* 51(9), 1062–1063.

Robinson, A. (2008). Toeplitz, Otto. *Complete Dictionary of Scientific Biography.* Vol. 13 (p. 428). Detroit: Charles Scribner's Sons.

Schubring, G. (2011). Conceptions for Relating the Evolution of Mathematical Concepts to Mathematics Learning—Epistemology, History, and Semiotics Interacting. *Educational Studies in Mathematics,* 77(1), 79–104

Scott, J. (2009). Life, the Universe, and Everything: an Undiscovered Work of Benchara Branford. *Journal of the History of the Behavioral Sciences,* 45(2), 181–187.

Sfard, A. (1995). The Development of Algebra: Confronting Historical and Pyschological Perspectives. *Journal of Mathematical Behavior,* 14(1), 15–39.

Sfard, A. (2008). *Thinking as Communicating.* Cambridge: Cambridge University Press.

Siu, M. K. (2006). No, I Don't Use History of Mathematics in My Class: Why? In F. Fuinghetti, S. Kaijser, A. Vretblad (eds.). *History and Pedagogy of Mathematics, Proceedings of the ESU4 and HPM2004* (pp. 268–277). Uppsala: Universitet Uppsala.

Smith, D. E. (1904). *The Teaching of Elementary Mathematics.* New York: The Macmillan Company.

Smith, D. E. (1921). Religio Mathematici: Presidential Address Delivered before the Mathematical Association of America. *The American Mathematical Monthly,* 28(10), 339–349.

Spencer, H. (1949/1861). *Education: Intellectual, Moral, and Physical.* London: Watts & Co.

Swetz, F. (1995). An Historical Example of Mathematical Modeling: the Trajectory of a Cannonball. In F. Swetz, J. Fauvel, O. Bekken, B. Johansson, and V. Katz (eds). *Learn from the Masters* (pp. 93–101). Washington, DC: The Mathematical Association of America

Tacquet, A. (1761), *Elementa geometriae planae ac solidae.* Typis Seminarii, apud Joannem Manfré.

21 History of Mathematics in Mathematics Education

Thomaidis, Y. and Tzanakis, C. (2007). The Notion of Historical 'Parallelism' Revisited: Historical Evolution and Students' Conceptions of the Order Relation on the Number Line. *Educational Studies in Mathematics, 66*(2), 165–183.

Toeplitz, O. (1963). *The Calculus: The Genetic Approach.* Chicago: University of Chicago Press.

Toeplitz, O. (1927). Das Problem der Universitätsvorlesungen über Infinitesimalrechnung und ihrer Abgrenzung gegenüber der Infinitesimalrechnung an den höheren Schulen. *Jahresbericht der Deutschen Mathematiker-Vereinigung* 36, 88–100.

Tzanakis, C. & Thomaidis, Y. (2012). Classifying the Arguments & Methodological Schemes for Integrating History in Mathematics Education. In B. Sriraman (ed.), *Crossroads in the History of Mathematics and Mathematics Education* (pp. 247–293) Charlotte (NC): Information Age Publishing, Inc.

Wilson, J. M. (1868). *Elementary Geometry, Part I.* London and Cambridge: Macmillan and Co.

Zhmud, L. (2006). *The Origin of the History of Science in Classical Antiquity.* (Trans. from Russian by A Chernoglazov). Berlin: Walter de Gruyter.

Michael N. Fried is associate professor in the Program for Science and Technology Education at Ben Gurion University of the Negev. His undergraduate degree in the liberal arts is from St. John's College in Annapolis, Maryland (the "great books" school). He received his M.Sc. in applied mathematics from SUNY at Stony Brook and his Ph.D. in the history of mathematics from the Cohn Institute at Tel Aviv University. His research interests are eclectic and include mathematics pedagogy, mathematics teacher education, sociocultural issues, semiotics, history of mathematics, and history and philosophy of education. Besides his papers in mathematics education, he is author of three books connected to the history of mathematics: *Apollonius of Perga's Conica: Text, Context, Subtext* (with Sabetai Unguru) (Brill, 2001); *Apollonius of Perga, Conics IV: Translation, Introduction, and Diagrams* (Green Lion Press, 2002); *Edmond Halley's Reconstruction of the Lost Book of Apollonius's Conics* (Springer, 2011).

Chapter 22
Philosophy and the Secondary School Mathematics Classroom

Stuart Rowlands

22.1 Introduction

Can philosophy have a role in the teaching and learning of mathematics in school? If it can have a role, what is it? There may be several answers according to several perceived roles, such as it enables pupils to think in the abstract, it contextualises an otherwise very formal subject, it situates mathematics in the realm of philosophy; but perhaps the most central answer to which most others are subordinate is that it can aid the understanding of mathematics. A class reflecting philosophically on the concepts of mathematics will most likely attain a deeper understanding of those concepts.

For example, a class reflecting on the difference in abstraction between one sheep and one sheep equals two sheep with $1 + 1 = 2$ (the latter concerns the concept of number; the former is a statement of physics similar to 'one lump of plasticine add one lump of plasticine equals one lump of plasticine'), reflecting on the nature of a geometrical straight-line (e.g. What is it? Can we see one? Does it exist? 'Could industrial artefacts such as aeroplanes and the associated machinery for production ever exist without it?' 'When did it first appear?') or discussing limiting cases to infinity, etc. will most likely develop a *relational understanding* (in the sense of Skemp (1976)) of how these concepts are embedded, connected and embellished in the relevant mathematics, as well as just knowing how to manipulate them according to the rules (Skemp's (1976) *instrumental understanding*). A qualitative understanding of infinity as a limit, with such examples as the proof for the area of a circle, deepens our understanding of why this area is πr^2, despite the unlikelihood of the class arriving

---

S. Rowlands (✉)
Centre for Teaching Mathematics, School of Computing and Mathematics,
University of Plymouth, Plymouth, UK
e-mail: stuart.rowlands@plymouth.ac.uk

M.R. Matthews (ed.), *International Handbook of Research in History, Philosophy and Science Teaching*, DOI 10.1007/978-94-007-7654-8_22,
© Springer Science+Business Media Dordrecht 2014

at a formal definition of Cantor's infinities. An 'informal' treatment of infinity provides a means to understand why the area of a circle is as it is. This chapter discusses the ways in which philosophy can aid the teaching and learning of the content of secondary school mathematics and how it can also situate mathematics culturally and historically.

One possible objection is that it takes longer to develop a relational understanding than it does an instrumental one (e.g. it takes longer to show *why* the area of a circle is πr^2 than simply giving the formula), but it is shorter in the long run compared with committing to memory all the various ways to tackle a variety of questions with formulae and rule-of-thumb procedures that are given without reason. As Skemp (1971) states, many students pass a public examination in mathematics at one level, but inevitably fail the next level because there is no basis to understand the abstraction necessary for the next level. The student must first master the previous level, not only in terms of passing examinations but also in terms of a conceptual understanding of the relevant domain of knowledge – despite any public recognition of passing that level.

Although it will be argued that philosophical questions do presuppose answers that are right, the classroom discourse need not arrive at completeness for the value of the discourse to take effect. What is being suggested is that philosophy can enrich the subject, not only in terms of the subject coming alive but also in terms of understanding the subject, even if the class does not arrive at any formal definition normally expected at degree level. Another objection is that if the formalism is too difficult then the problem should not be introduced at the secondary school level. For example, if the vast majority of schoolchildren will fail to understand the theory of real numbers with proof, such as the density of the rational numbers and the existence of the irrationals, then the theory should not be introduced at this level. Hopefully, however, all secondary school pupils will have been introduced to the real numbers including the rational and irrational numbers; so why can't the theory of the reals begin here, even though the required formalism may not be reached at this level?

This chapter proposes philosophical discourse in the teaching of school mathematics,[1] with the teacher as a 'sage on the stage' orchestrating the class towards a target concept or a series of target concepts. Unfortunately this chapter seems alone in what it proposes. The literature mainly consists of the Philosophy for Children (P4C) programmes which advocate the teacher as a 'guide on the side' who allows the discourse to go much its own way without intervention. There is very little, however, on philosophy and the mathematics classroom, and the little there is has tended to be an extension of the P4C programmes, with the emphasis on whole-class discussion which Kennedy (2007) termed the 'constructivist classroom' with its egalitarian emphasis on shared meanings.

---

[1] Unfortunately this chapter does not discuss philosophy and undergraduate mathematics as this deserves a chapter in its own right. At this level there is not only a change in content (with an emphasis on formalism and rigour) but also a variety of teaching/learning methods that may not be so appropriate at the secondary school level. Nevertheless Pincock's (2012) *Mathematics and Scientific Representation* would be most appropriate as a text for this level.

22 Philosophy and the Secondary School Mathematics Classroom

P4C advocates a circle or horseshoe of children discussing philosophy with the teacher acting as chair who helps steer the conversation without preconceived ideas as to what the children ought to learn from such a discourse – there is no imposition with what the teacher considers to be right or wrong. By contrast what this chapter proposes is indoctrination, but only in the sense that teaching mathematics is in itself a form of indoctrination: a social-cultural activity that inducts students into the best that has developed out of what is essentially a two and a half thousand year history.[2] It is not indoctrination in the sense of accepting something without reason.

This chapter begins with a literature review, and given the small number of articles on philosophy in relation to the secondary or primary school mathematics classroom, the review will focus on the P4C programme in general but with mathematics in mind, especially since the majority of the small number of articles on philosophy and the mathematics classroom are influenced by this literature.

P4C is not just a strand within education research but has now become part of the practice of teaching and has truly entered the public domain, in the UK at least. For this reason alone, the literature review is relevant given the present-day popularity of the programme, especially in England and Wales where the constraints of the National Curriculum and the excessive bureaucracy of accountability have made this programme very attractive to teachers, educators, curriculum developers and indeed the children themselves; but there is also a more overarching reason for the review – *it has made possible the arguments and proposals of this chapter*. Although this chapter's historical-cultural emphasis in the teaching of mathematics has already been formulated in previous articles,[3] this chapter's consideration of the role of philosophy in mathematics education has been influenced by its own review of the P4C programme. Although critical, this chapter owes a debt of gratitude to the programme in the sense that if the programme never existed, then many of the ideas of this chapter would not have existed either, certainly as far as the ideas of the author are concerned.

---

[2] Although there were great mathematical advances prior to the Greeks, the Greeks created deductive proof and the necessary theoretical objects to accomplish it, culminating in the axiomatic framework of the *Elements*. There was nothing like it beforehand and nothing like it until the nineteenth century when much of mathematics was rewritten in axiomatic form. Prior human achievements in mathematics notwithstanding, Greek deductive geometry was the most stunning advance, and there is a sense in which 'history' begins two and a half thousand years ago – especially since over that period to learn mathematics was to learn the *Elements* (to the delight or chagrin of many pupils).

This is not by any means to undervalue the educational potential in introducing the mathematics of, for example, the Babylonians (is the 360° in the angle measure of a circle now a matter of convention or was there an objective reason for adopting it? Is there a need for base-ten compared with base-60? What prompted their recipe for what is today expressed as the quadratic formula?). However, this article is primarily about engaging pupils consciously with justification and proof, the abstract theoretical objects that were created for the purpose and the impact both culturally and cognitively.

[3] For a theoretical overview of this perspective see Carson and Rowlands (2007); for validation see Rowlands (2010).

Initially the review focuses on the controversies surrounding the programme in the UK, but there are no apologies for this. What is happening to education in the UK at the moment (which essentially began with state control) could happen elsewhere, but perhaps the most important aspect of these controversies is that we have the distinct advantage of observing a development in education research, one that is relevant to this chapter, put into practice with all the controversies from the public surrounding it. The fact that this is happening in the UK is in a sense irrelevant (although some of the issues surrounding its implementation are particular to the UK), and some light on the controversy not only shows what's at stake but also provides the context (or 'backdrop') for what this chapter proposes.

Section 22.2 discusses the present-day impact of P4C and the controversies surrounding its programme in the public domain. It also critiques a fundamental premise of the programme that children as young as five are natural philosophers, although this section does not deny that children do ask questions that can be considered philosophical. This section argues that children are not natural philosophers but they can be trained to think and discuss philosophically.

Although there is much written on the programme in the literature, there is very little on philosophy in the mathematics classroom. What little there is tends to frame philosophy and mathematics in the context of the programme, and this small sample will be critiqued in Sect. 22.3. Section 22.4 may be considered the heart of this chapter, because it puts forward the proposal that philosophy and mathematics can be introduced in a traditional classroom setting with the teacher taking the dominant role. It will be argued that under such a regime the teacher is not imposing her ideas on the class but is able to steer the class into thinking about what is being proposed in terms of philosophy and mathematics – the aim being a deeper understanding of mathematical concepts. Unlike most of the mathematics taught in many (perhaps most) UK classrooms, there will be very little room for faith. In such a programme, children will not be expected to accept the angle property of the triangle (that the three angles of a plane triangle add up to two right angles) because the teacher says so (or because a few triangles have been measured); instead, the class will be directed to consider the nature of deductive proof and to adopt a critical stance that demands justification. The class should then find it easier to construct the proof with just a few hints.

Perhaps the essence of this chapter is that philosophical discussion need never be far away from the many mathematical concepts that we teach. For example, the teacher could raise the question as to how it is possible for the area under a curve to be expressed as an exact number of square units. Exercises for 11-year-olds (or thereabouts) involving the approximation of area using finer and finer grids can provide the springboard for discussing limiting cases and hence the concept of a limit (qualitatively at least; any formalism should perhaps be left to a much later level of development).

The proposed philosophy may be loosely described as the philosophy of mathematics. Issues concerning Platonism and proof may certainly come under that category, but there are also other related issues which although applicable to the philosophy of mathematics are more to do with philosophy in general, such as

Plato's distinction between belief and knowledge. This chapter is more a proposal for philosophy in the broader sense than the specific philosophy of mathematics, but then elementary philosophy of mathematics cannot be separated from broader philosophical considerations.

Especially in a classroom environment, philosophical considerations of mathematical concepts can be a valuable aid in understanding those concepts. The first half of Sect. 22.4 is therefore given over to the added value of philosophy in the mathematics classroom prior to the main proposal presented in the second half, which is that this philosophy would be better framed within a historical-cultural context.

22.2   A Critique of the Philosophy for Children Programme

From small beginnings over three decades ago, P4C is now in the public domain. In the UK, hundreds of primary schools have timetabled philosophy discussions, and some private schools have their own resident philosopher. There are glossy journal front covers emblazoned with P4C and an ideological battle as to whether P4C is suitable for children, fought in the pages of the press with educational commentators and religious leaders on the offensive.

The front cover of the first edition of the Primary Teacher Update, a glossy journal presented upfront next to the candy at the tills of a particular W. H. Smiths, a very large retail corporation that mainly sells magazines and books, stated 'Philosophy for children. Encourage them to think' and shows two young children and their teacher engaged in conversation. The article is written by Lenton and Videon (2011), the founders of P4C, and iterates the misconceptions surrounding the implementation of philosophy for children. On what seems to be based on the P4C literature since the early 1980s, the article mainly provides (very brief but succinct) answers to the following: 'philosophy is too hard for children', 'you have to be academic to understand complex ideas', 'philosophy cannot be fitted into the curriculum', 'you have to be a specialist to teach it', and 'it doesn't have a purpose – what is the point?' Bearing in mind that this is for public consumption, the answers given by the article are too obvious and 'commonsensical'. For example, on the point that philosophy may be too hard for children, the article simply responds that children are naturally inquisitive and want to question. However, the inclination for young children to ask 'why?' may not be sufficient to overcome the difficulties of philosophical discourse. Taken in context, the article was not written for rigorous scholarly discussion, although the P4C academic literature *is* and for that reason is open to critique.

The Primary Teachers Update article is a push for P4C with an editorial giving this push a sense of urgency: we need to equip children with the tools to think independently if we are to avoid the mindless mob violence that has recently occurred across UK cities. P4C is more than trying to improve education in terms of children understanding what is taught across the curriculum (it is stressed that a little philosophical discourse in general can help to question and hence understand the concepts taught across the curriculum); it is also about improving society. P4C seems to have a far-reaching social agenda which some may find unjustified (e.g. can fundamental changes be made through philosophy?). The point is that P4C is now under public scrutiny and the stakes are high.

The article 'Time to take the won't out of Kant' (Lightfoot 2011), emblazoned on the front cover of a magazine (*Tespro*) that comes with the *Times* Educational Supplement (a highly subscribed professional newspaper), outlines the P4C's current influence and popularity and, with reference to a sympathetic psychologist who actually worked with Piaget, stated that Piaget seriously underestimated the capabilities of the young. It provides much information on relevant materials, where certain P4C programmes can be downloaded and in-service training venues, with the price tag. From relative obscurity P4C is now in the limelight, at least in the UK. It is seen as a panacea for social change and not surprisingly can generate a lot of revenue.

Philosophy for children may have become popular, but it is not without passionate criticism. For example, the UK's Institute for Public Research proposed that all children should be taught to think critically about religious belief; but this has created the charge of relativism and indoctrination by educational and religious commentators and the press (Law 2008). It is difficult to understand this charge, but as Law argues, a proposition considered true in one religion (such as Christ is God) may be considered false in another (such as Islam); so if religious education regards all religious views as equally valid then we have an exemplar of what it means to be relativist. Philosophy for children, Law argues, can enable a criticism of relativism and need not undermine religious belief. Whereas religious dogma can brainwash the child, philosophy can provide reasons for justification in religious belief (Law 2008).

Both the UK's Chief Rabbi Jonathan Sacks and the tabloid educational commentator Melanie Phillips blame the Enlightenment for this 'relativism', especially Kant, whose watchword in his short article *On Enlightenment* is *Sapere aude* (have the courage to use your own reason) and indeed there are two P4C programmes that are called Sapere and Aude (Law 2008). For Phillips and Sacks, getting children to think for themselves undermines tradition and external authority; but as Law argues, philosophy as a statutory part of the curriculum without exceptions combats the brainwashing aspects of that authority.

It is interesting to note that Kant's battle cry of the Enlightenment was a battle cry of reason and science against the stultifying dogmas of the preceding age of feudalism. Perhaps Phillips and Sacks would like us to return to the period when scholarship became scholasticism and everyone had to accept certain dogmas on faith for fear of retribution. As Law (2008) recounted, in the late 1960s a friend was punished by a catholic school for asking why the church forbade contraception. Is this consistent with what Phillips and Sacks advocate in terms of 'external authority'?[4]

---

[4] Perhaps this whole debate concerning schoolchildren, philosophy and religious belief can be resolved if religious education and worship are thrown out of schools. Perhaps the UK should adopt the US model whereby state education excludes any form of worship and religious instruction. The school should be seen as an induction into rationality and reason to which faith and indoctrination have no place.

22 Philosophy and the Secondary School Mathematics Classroom

The book, *Philosophy in the Classroom* (Lipman et al. 1980), has become influential and well referenced. It advocated philosophical discourse for children as young as 5 years of age and stated 'if the educational process had relevance, interest, and meaning for the children, then there would be no need to *make* them learn' (Lipman et al. 1980, p. 5, emphasis given). This seems obvious, yet in England and Wales the National Curriculum and a battery of tests have left no choice but to teach to the test piecemeal bits of knowledge. It is difficult for the science teacher to develop a deep understanding if forces have to be taught one lesson and worms the next. Under the NC, mathematics has gone from mainly algebra and geometry to mainly 'data handling' and 'shape and space', as if relevance can motivate.[5] In practice the angle property of the triangle, Pythagoras' theorem or $x° = 1$, for example, is given without proof as if mathematics is a question of faith – without any epistemology to show why these things are true. As Siegal (2008) argues, education without epistemology reflects a lack of respect towards children. The teachers themselves cannot be blamed because they have already been accounted by endless state inspections that specify the standard and what is expected in terms of teaching and learning outcomes.

Under such a regime P4C must seem like an oasis, its growing popularity is not surprising; but there seems to be an uncritical acceptance of its basic assumption that children are natural philosophers, but are they?

This assumption goes beyond the advocates of P4C. For example, in the novel *Sophie's World*, Sophie encounters her future mentor who points out that all young children have a sense of wonder that is lost as they get older. In P4C, Matthews (1980) argues that children are natural philosophers because they ask philosophical questions but their ability and interest in philosophy, like art, diminishes as they get older. Now while this is (seemingly) true for art, is it true for philosophy? Could every child's fascination for art simply be the motivation to develop sensory-motor skills necessary in performing such art? It seems reasonable to think of this fascination as a part of normal development, sometimes disappearing as the child matures, but philosophy may be a different matter entirely.

For the child to progress in art and for her fascination to be sustained, then she must engage with the various cultural schemata and codes that enable the art to take form (Gombrich 1960). Arguably this would apply to philosophy, but what's in doubt is the initial motivation to do philosophy. Children may have a sense of wonder which is lost through schooling, but is there any intention to be philosophical and do they expect certain criteria to be satisfied in answering their questions? Hand (2008) answers in the negative but maintains that such intention and criteria can be learnt and developed. Like art, children must engage with the various schemata of philosophy (e.g. the kind of questions to ask, such as epistemological or metaphysical, and the sort of answers to be expected) if progress is to be made and interests are to be developed in philosophy.

---

[5] Although most of the mathematics of the NC supposedly reflects mathematical practice in the real world, as Noss (1997) and Dowling and Noss (1990) point out, it does not even do that.

712   S. Rowlands

Hand (2008) argues that there is more to being a philosopher than the asking of philosophical questions and that what is missing is the intention of pursuing a line of inquiry in accordance with the question – children lack the appropriate methods of investigation – although with the asking of such questions they can be appropriately taught how to use these methods. Matthews (1980, 2008) has criticised Piaget's deficit model which proclaims young children to be unable to think abstractly, but has Matthews overestimated the child in terms of being a natural philosopher? For Hand (2008) the various methods of philosophical investigation have to be developed.

There is much reference in the literature to Matthews' deficit model of cognitive development in Piaget. For example, according to McCall, 'If you go out in the snow without your mittens, your fingers will freeze!' (McCall 2009, p. 21) involves hypothetical reasoning and yet 3-year-olds will understand this statement. McCall states that 5-year-olds are capable of 'formal operations' in that they can reason about abstract philosophical concepts and can reconstruct the reasoning of other children. Now, without denying the child's sense of awe and wonder of the world and her interactions with it, have we gone completely the opposite way as if her cognitive abilities are adult-like? Consider the following by Lipman and colleagues:

> The rules of logic, like the rules of grammar, are acquired when children learn to speak. If a very young child understands, 'if you do that, you will get punished,' it is assumed that the child understands, 'if you don't want to get punished, I shouldn't do it.' That assumption is usually correct. Very small children, in other words, recognize that the denial of the consequent requires the denial of the antecedent. Although this is a very sophisticated piece of reasoning, children are capable of it in very early stages of their lives. (Lipman et al. 1980, p. 15)

Does the child recognise that the denial of the consequent requires the denial of the antecedent and are they capable of this very sophisticated piece of reasoning? Rather than understand 'if I don't want to get punished, I shouldn't do it', the child might instead understand 'if I don't do that, I won't get punished', meaning not to do it, especially since punishment is not desirable. *How* the child understands might be to do with the discourse and her relationship between the participants.[6] From the hundreds of papers and articles on conceptual change in science education, we find that a wide range of schoolchildren and adults tend to use fallacious reasoning when defending their intuitive ideas of scientific concepts.

In the process of learning, children perform acts of reflective thought, such as whether something has meaning, its relation to the scheme of things or the difficulties they have with the problem. Although reflective, these thoughts seem more to do with metacognition than philosophy; but it does seem likely that metacognitive thinking can be transformed into philosophical thinking if the learner is encouraged to think in the abstract. According to Vygotsky (1994) children are capable of abstract reasoning which depends on the help (or rather, the *scaffolding*) by significant others.

---

[6] From a sample of adults performing a classification task, Wason (1977) shows how even adults deny the relevance of facts or contradict themselves, and how conceptual conflict can arise when the force of a contradictory assertion is denied. This implies that even adults do not acquire the rules of logic just because they speak. The Lipman and colleagues assertion is therefore unwarranted.

22 Philosophy and the Secondary School Mathematics Classroom 713

The point is that claims about the ability of the young (especially 5-year olds) may be overstated and exaggerated; but they do throw some light on the child's potential. That potential, as we shall see, can only be realised and developed by intervention.

In a P4C discourse children may construct a logical argument and arrive at valid conclusions, but do they become aware of the logical form of an argument or the metaphysical import of a question? Consider the following by Lipman and colleagues:

> The mathematician may insist that children begin by learning simple arithmetic operations, but the children may stagger the teacher by asking, 'what is number?' – an immensely profound metaphysical question. (Lipman et al. 1980, p. 28)

But do they mean this in a profound metaphysical way, or are they at the age when they question everything as a kind of game? Do they even have a *sense* of the metaphysical?

We must shift from the habit of regarding children's metacognitive utterances for philosophical insight.[7] Consider the following:

> This manner of upstaging the normal level of dialogue by leaping to a more general level [for example the child asking 'what is time?' or 'what is distance?'] is typical of metaphysics. Instances of other metaphysical questions your children may already have posed you (or are quietly preparing for you) are these: What's space? What's number? What's matter? What's mind? What are possibilities? What's reality? What are things? What's my identity? What are relationships? Did everything have a beginning? What's death? What's life? What's meaning? What's value? What makes questions like these particularly difficult to answer is that they involve concepts so broad that we cannot find classifications to put them in - we just cannot get a handle on them. (Lipman et al. 1980, p. 37)

It seems almost a conspiracy, children waiting to pounce with metaphysical questions, but isn't this simply a challenge to the terms that are used (especially if they come across such terms as 'matter' and 'identity')? As soon as the adult fails to provide a sensible answer (usually the first time the child asks), the child may decide to persist as a kind of game, knowing intuitively that it causes a mild form of embarrassment. The point is that we may have looked at child development through the lens of our own present development. Piaget may well have underestimated the child, but his stage theory, flawed though it is, rests upon the premise that children are not little adults. The child is not a prototype of an adult whose thoughts are merely refined as she grows up – she has her own way of thinking and looking at the world. Rather than springing forth from some kind of philosophical awareness, a child's philosophical questions may be no more than the metacognitive awareness

---

[7] According to Vygotsky (1987), concepts are not absorbed ready-made by the child but undergo a process of development. Initially, the concept may begin as a complex; for example, the child might use 'dog' not as a member of 'animal' but as an 'associative complex' extended to inanimate objects with fur. A complex often relies on perceptual features. For example, children who think in complexes may successfully complete a classification problem involving geometrical shapes, so the child might appear to think in concepts – until the child attempts a borderline example, such as a trapezium looking very much like a parallelogram, in which the child classifies as a parallelogram without thinking of how a parallelogram is defined (see Vygotsky 1987). The point being the child might not even think in concepts, let alone concepts that may be deemed philosophical.

that adults have great difficulty in answering such questions. With adult supervision, however, it is possible for metacognitive awareness to be transformed into philosophy.

The question is not how we can stimulate their challenge, apparently that challenge is ever present, but how we can challenge them. That challenge is necessary if we are to be certain what the child is driving at. What children mean should not be taken at face value, but as soon as a discourse opens up as to what they mean, we then have an educational situation in which understanding and meanings emerge and change. Intervention is necessary if what the child says is not to be overestimated, bearing in mind that what the child says during the intervention does not necessarily reflect any initial cognitive state or disposition.

What seems promising is the realist philosophy of McCall's CoPI (the Community of Philosophical Inquiry), which asserts that someone can be wrong and contrasts with the P4C Deweyan pragmatic philosophy that truth is constructed and negotiated (McCall 2009, see p. 81). What makes CoPI *essentially* different to all the other approaches such as the Nelson Socratic Method or the various P4C programmes such as Sapere and Aude is its realism:

> The external realist philosophy which underlies CoPI holds that immaterial creations of human beings such as language, concepts, theories, symbols and social institutions, while owing their origins to humans, once made, then exist independently. We can be wrong about them too. (McCall 2009, p. 83)

Perhaps the essential ingredient in any philosophical discourse concerning science and mathematics with children is the view that scientific concepts and laws may be considered to exist in what Popper calls the World Three of the objective content of thought, with discussion as to what existence means here. Are scientific concepts and laws idiosyncratic, are they true because scientists believe in them, are they indubitable facts derived from observation? Any such opinion may be expressed by children who are used to this kind of discourse, and the teacher can always put into perspective that opinion with respect to World Three, bearing in mind that any such opinion opens up a whole world of discourse concerning such issues as Platonism, fallibilism, method, methodology, ontology, epistemology, abstraction and validity.

McCall gives a list of what the CoPI Chair needs to know in order to recognise the various philosophical theories and assumptions that underlie everyday discourse (which contrasts with the view that the teacher does not need to know any philosophy, only the way children learn, e.g. the Primary Teacher Update article by Lenton and Videon (2011)). For the philosophy of science, the list recommends the teacher becoming aware of such things as induction, deduction, falsification and paradigms. Hopefully such a teacher will enable the class to also become aware of such categories in their discourse. Unfortunately, though, there is nothing by McCall on philosophy connected with mathematics.

The section after next is a presentation of how the mathematics teacher can introduce philosophy explicitly in her teaching of mathematics. This is perhaps unique in that many articles on school mathematics with philosophical discourse tend to separate the very learning of mathematics with the discourse, and this is reviewed next.

22.3 A Review of Philosophy in the Mathematics Classroom

There is very little on philosophy in the mathematics classroom, it tends to be implicit in the overall view of the P4C programme. In *Teaching Children to Think* by Robert Fisher (2005), there is reference to Skemp's relational and instrumental understanding in mathematics education but nothing in terms of philosophical discourse and mathematics, despite a chapter on philosophy for children and chapters on creative and critical thinking, With reference to such things as 'shape and space' and 'data handling' of the National Curriculum, however, the book tries to encourage thinking skills and the understanding of how the various mathematical concepts relate to each other, but it has missed the opportunity to use philosophy explicitly as an aid in understanding those concepts. Fisher might reply that he is hamstrung by the NC.

Kennedy (2007), however, connects philosophy with understanding mathematical concepts while also referring to metacognition and creativity. She discusses how a classroom community of mathematical inquiry can become a community of philosophical inquiry by discussing various mathematical problems. She contrasts the transmission model of teaching with the constructivist classroom through which sense-making occurs as a collaborative endeavour. The method is the P4C programme with its participative, dialogical and egalitarian emphasis and states that similar to this have been the works of such metacognitive greats as Schoenfeld (1989) and Goos (2004). The problem here is that there has been a tendency to encourage either metacognitive or philosophical thinking after the content of the mathematics has been learnt, such as solving an integral in a metacognitive way (Schoenfeld 1989) or solving a projectile problem in an investigational way (Goos 2004). How was the content, such as integration or projectiles, learnt in the first place (see Rowlands 2009)?

Nevertheless the article shows a way in which meanings can be constructed in discussing the concepts of ordinary school mathematics philosophically, such as asking 'When can we say that we 'understand' a mathematical concept?'... 'How can we trust in math that is not experienced?' (Kennedy 2007, p. 6). Kennedy sees this as complementing concrete mathematical investigations. Perhaps for the first time, we are seeing a greater integration between philosophy and mathematics in the classroom, but the integration is not complete. Although one complements the other, they are still separate. The question arises: How can we teach mathematics in a philosophical way that doesn't entail teaching the mathematics first prior to any philosophical considerations?

The rest of the article is taken up with examples of discourse in action by showing parts of the transcript concerning two problems. The first is the problem of a frog at the bottom of a 30 ft well. Each hour it climbs 3 ft but slips back 2 ft. How long will it take to get out? The second problem is: given the two infinite sets $\{1, 2, 3, 4, 5,...\}$ and $\{2, 4, 6, 8, 10,...\}$, do both sets have an equal or a different number of elements? The discourse is of a very high standard and like all the examples of this kind (such as in Lipman and colleagues), it is surprising to know these are the conversations of upper primary/elementary children. We are told in each case that the children have already engaged in a community of philosophy for many hours beforehand,

and we get the sense that all this is possible, although we are not told whether the children involved were high developers. We also have to bear in mind that what we see is presumably only part of the many hours of transcript recorded. Nevertheless the conversations seem very mature for primary (elementary) schoolchildren and appear encouraging.

If the class can arrive at the conclusions as presented in the scripts, then this remarkable accomplishment ought to be encouraged, but what if the conversation comes to an impasse, or is a non-starter or the class arrives at a conclusion that is wrong? Must everything be constructed by the class or can/should the teacher give a clue, a hint, or indeed a correct answer? With the first example the class arrived at just how ambiguous the question is (a child says, quite late on in the discussion 'Well that's the way we understood it, but it's not quite clear'. p. 10) – they did the task – but couldn't the teacher point out the ambiguity? In the second example the discussion repeated itself until someone asked the question 'is infinity a number?' which created a new lead. What if no one asked that question? Of course the teacher could always ask such a question to keep the discourse going, but we have to bear in mind that the teacher, under such a programme, must not direct the discourse to arrive at preconceived ideas. There is a sense in which the discourse is directed by the children, and this is perhaps the greatest weakness of the programme.

Despite its criticisms of Piagetian stage theory, the programme itself is quite Piagetian in the sense that the children must find out for themselves rather than being told, although surely a hint here and there is more appropriate than spending a whole lesson trying to fathom the ambiguity of a question. In real life perhaps the problem would be rejected out of hand because of its ambiguity.

The many examples of classroom philosophical discussion in the literature are outstanding in their complexity, abstraction and logic; but we are not given the background of the chosen samples or details of the methodology, and we are not sure of the amount of transcript that is not included. The major problem seems to be the notion that it is OK for children to pursue a path of discussion that leads to nowhere or that they are neither right nor wrong. Of course, some paths do achieve an undecided position, such as the concept of infinity in De la Garza et al. (2000) classroom discussion,[8] but that is because of the nature of the concept under discussion. In general, the teacher should direct the class to the target concept. Hand (2008) criticises the widespread view that philosophy has no right answers by showing that all philosophical questions presuppose answers that are right, despite some questions as yet having no right answers. It is up to the teacher to convince why a particular answer is right (or wrong).

Although a philosophical question presupposes a right answer, the answer may be so formal that it may be inappropriate for secondary school pupils to entertain the

---

[8] 'Many [children] found convincing a view which could well be labelled constructivist – that infinity is not something we imagined as complete, but was a result of the fact that there was no stopping point, that infinity just kept going on. We then turned to measuring infinity and, having detoured through Cantor's proof that the number of integers was equivalent to the number of even integers, ended on a note of indecision' (De la Garza et al. 2000).

question. Nevertheless certain questions that have rigorous formal answers appropriate for undergraduates may have educational value for secondary school pupils. Certain answers lacking in rigour may be appropriate to schoolchildren because they expose or contextualise relevant concepts to be taught or provide a satisfactory 'completeness' to the question. Certain paradoxes may not have an answer, but for years may stimulate members of the class to think of one. Although questions presuppose answers, there may be more than one answer. The teacher must consider the educational value of her questions and the kind of answers to arrive at.

Daniel and colleagues (2000) present a research report of an ongoing project to implement a community of enquiry in the mathematics classroom. The article is essentially an argument for why a community of discourse is necessary for the mathematics classroom as opposed to the lack of meaningfulness of the mathematics taught in a traditional classroom. It discusses the myths and prejudices surrounding mathematics and philosophy (that there is only one way of getting the right answer or that philosophy isn't for children, etc. similar to the Primary Teachers Update article by Lenton and Videon (2011)) but seems to propose that the meaningfulness of a community of enquiry can only be accomplished if the mathematics is related to the everyday experiences of the child. For example:

> Dewey states that as soon as studies in mathematics are dissociated from personal interest and their social utility, that is, when mathematics are presented as a mass of technical relationships and formulas, they become abstract and vain for students. It is only when children become intrinsically interested and conscious of mathematics as a means of solving daily problems (as opposed to end in themselves), that they enjoy playing with numbers, symbols and formulas. (Daniel et al. 2000, p. 5)

Why does developing an intrinsic interest involve the solving of daily problems? Presumably Euclidean geometry, which is an end in itself and can be unrelated to practical problems, will not do here. Like most communities of enquiry, an 'everyday' story is presented and a list of mathematical/philosophical questions (involving such concepts as truth, proof and infinity) are asked and pursued. The story seems to be the 'everyday' from which abstract problems arise – but this is not 'mathematics as a means of solving daily problems' as stated by Daniel and colleagues (2000). As this is an ongoing project, no results are discussed, although a summary of the results of a teacher questionnaire is stated. According to the teachers, some of the students reported that relating the discussion to mathematics wasn't much fun and that sometimes they don't want to hear about mathematics. Nevertheless they took well to the discourse. What is interesting is that they are not used to this kind of discussion involving mathematics.

Outside of the P4C programme, Prediger (2007) argues that philosophical reflections must play a prominent role in the learning process if an adequate understanding of mathematics is to be achieved by the class. By philosophical reflections she does not mean reflecting on classical philosophical theories, but reflecting on the mathematical activity itself, expressed by the verb *philosophize*.

What will be presented in the next section is an argument for philosophy to be integrated with a cultural-historical approach to teaching mathematics in order to engage children, for children to understand the nature of the concepts that they are expected to learn and to situate these concepts in the society that they belong. However, philosophy without the history (or perhaps with a little history) of mathematics can still be value-added, and this is discussed next.

22.4 A New Emphasis on Philosophy in the Mathematics Classroom

Philosophy without (or with little) history may seem a little empty, but it can have an add-on effect. The philosophy may only be related to the mathematical issue at hand (e.g. the concept of limit by dividing a circle into equal sectors and rearranging them to form a parallelogram of sorts – the notion of infinity explored by considering an ever-increasing number of sectors in the proof of $A = \pi r^2$), but much of what is taught can be done in a philosophical manner, such as the Socratic method of asking conceptual questions, parallel questions, contradictions, conjectures and counterexamples. The teacher orchestrates the class towards the target concept (or each target concept in succession), and philosophy provides the opportunity for exploration into the concepts to be learnt.

This is what Jankvist (2012) characterises as the *illumination approach* whereby the philosophy supplements the mathematics, although Jankvist's *philosophical discussion approach* (whereby groups of students enter debate concerning philosophical issues such as whether mathematics is discovered or invented) can also be included.[9] Using Jankvist's taxonomy, this is philosophy as a *means* rather than as an *end*. Similar to Jankvist, however, it will be argued that it would be more advantageous educationally if that philosophy was concerned with the history of mathematics. Meanwhile we shall look at various examples of philosophy and mathematics that can be introduced to the classroom but without, or with a minimal, history.

As an example of value-added philosophy, consider the paradox of Gabriel's horn (or the wizard's hat) that is asymptotic. A surface is generated by the 360° rotation about the x-axis of the curve of the function $f(x) = 1/x$ for $x \geq 1$ (Fig. 22.1).

This has a fixed volume (π cubic units) but an infinite area (see Clegg 2003). How is it possible for a finite volume to be enclosed by an infinite surface area (Clegg states that he has yet to see a satisfactory explanation)? After the calculus of basic volumes and surface areas have been learnt, such a question posed can raise discussion as to what volume, surface area and limiting case actually mean (it may be worth considering a cross-section perpendicular to the x-axis: a circle of radius r in which $r = f(x)$, followed by the rate in which the circumference decreases compared to the rate in which the area decreases, with respect to r. Similarly, by considering δx the rate in which the surface area decreases can be compared to the rate in which the volume decreases, with respect to r). This can be a good prerequisite for fractals

---

[9] This is not, however, an instructional unit on philosophy and mathematics (Jankvist's *modules approach*) nor is it a course that pursues a particular philosophy of mathematics (Jankvist's *philosophy approach*).

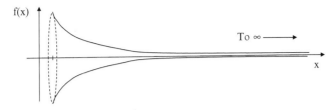

Fig. 22.1 Gabriel's horn (or the wizard's hat) – finite volume, infinite surface area

Fig. 22.2 A right-cone cut by a plane parallel to its base – are the two *dotted circles* (shown as one *dotted circle*) equal or unequal?

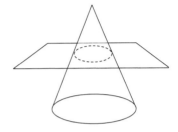

such as the Koch snowflake that has finite area but infinite perimeter, or the Sierpinski carpet which has zero area but has a non-empty interior – all of which can be discussed meaningfully prior to or without the formalism.

If a cone is cut by a plane parallel to its base, are the two circles formed out of the cut the same or different (although this can be introduced without reference to history, in fact it came from the Ancient Greeks) (Fig. 22.2)?

If the same, then we have a cylinder; if different, then the surface would not be as smooth as it is. This example, like many others, lends itself to the many ways in which the teacher can direct the discussion (e.g. 'my friend has an electron microscope and she didn't find any cone to be made of cylinders. What do you make of that?'). This paradox may not only create cognitive conflict, which some regard as essential in cognitive development, but lends itself to the ideas of a limit (the thickness of a cylinder having zero limiting length).[10]

Another example is Galileo's wheel paradox: the wheel fixed within a wheel (see Clegg 2003) (Fig. 22.3).

In one revolution the smaller wheel has travelled the same distance as the larger wheel, so it must have rolled more than one revolution, which is absurd. Although this paradox can be presented ahistorically, nevertheless historical considerations can always serve to illuminate. For example, consider, as Galileo does in his *Dialogue on the Two Principal World Systems*, two concentric octagons as shown (Fig. 22.4).

---

[10] Of course a little historical excursion can reveal how the Sophists used indivisibles and how the school allied to Plato used Eudoxus' method of exhaustion. The two can be compared in Archimedes's various proofs for the quadrature of the parabola (and the *Palimpsest*, the book where Archimedes uses indivisibles, has itself a wonderful history).

Fig. 22.3 A wheel fixed within a wheel and concentric

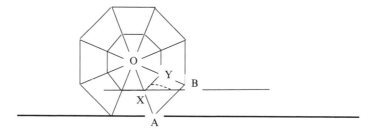

Fig. 22.4 Representing the circular wheels as octagons

In a one-eighth turn of the larger octagon, triangle OAB has been rotated 45° clockwise about a stationary point A. However, line XY has not only rotated 45° clockwise as well, but the point X about which XY has rotated has itself moved along the arc of a circle as shown. As Clegg states, side XY is lifted off the track.

The paradox is that with two concentric circles, there is sense in which no point on the smaller circle has left the track, yet in one revolution, a point on the smaller circle has traversed pi times the diameter of the larger circle, not the smaller. Of course, no point on the smaller circle will stay on the track, the loci being a cycloid that is smaller in height compared with the cycloid of a point on the larger circle; but we have to account for the horizontal displacement. Galileo (as Salviati) considers two concentric regular polygons of 100,000 sides in which after one turn the smaller polygon has traversed 100,000 smaller sides plus 100,000 tiny gaps between each side (we still have the isosceles triangle OAB, but with a much smaller AB, in which the 'tiny gap' is XY lifted off the track). One educational value of this example (and one that is stated by Clegg 2003) is that we can see Galileo tackle the complexity of the number of sides increasing towards infinity with the number of 'tiny gaps' also increasing to infinity but with the 'gaps' decreasing to zero.

Unfortunately Galileo does not solve the paradox, which was transformed into the definition of infinity by Dedekind and became solvable when presented in the new context of sets around 1870. Set theory is very abstract, so presenting the paradox to schoolchildren does not mean that they will eventually see the solution, but they can see the problem and the way Galileo tackled it.

Do complex numbers really exist? It is perhaps surprising that many websites state categorically that they do *really* exist (rather than just *assume* they exist). Some websites justify the assertion by referring to the very successful applications of complex numbers; but does success necessarily imply existence (certainly not for

22 Philosophy and the Secondary School Mathematics Classroom

the relativist)? Complex numbers are counterintuitive, yet they are usually introduced in a matter of minutes, justified with reference to the complex roots of a quadratic and followed by its algebra in the hope that students will subsequently appreciate what they are through their applications. Philosophical discourse could make intelligible something that is counterintuitive, such as whether it is a question of existence or formalism (e.g. is the formalism reduced to $a+ib=c+id$, with $i^2=-1$, implies $a=c$ and $b=d$?). Of course, a little history (such as the reluctance of many mathematicians, from the Renaissance to the nineteenth century, to accept complex numbers) can put into context for the students their difficulties with complex numbers, especially since these difficulties plagued the best minds.

Is there a smallest number after 0? This question can be asked across the range from schoolchildren to undergraduates with the discourse appropriate to the level of development already attained. For example, schoolchildren could be introduced to the paradox of a 'next' number having an infinite number of decimal places.

Certain topics can be discussed in a philosophical manner even though the discussion may not be philosophical. For example, if $a^n=b^m$, when can we say $a=b$? What values of x and n satisfy $x^n=1$?

Philosophical discourse can also highlight the aesthetics and 'divinity' of mathematics; for example: Is Euler's equation ($e^{i\pi}+1=0$) evidence for the existence of God? Surely the coming together of the five most important numbers reveals divinity? How is it possible for a particular irrational number, raised to the power of a particular irrational number times i, to equal -1?

If $y=f(x)$ enables us to map x onto y, what can we make of dy/dx? A function is defined in terms of the way it maps the elements in the domain to the elements in the range – it can be said that a function is a *process* – so how come it is an object that can be differentiated? It would be worthwhile looking up David Tall's notion of *procept* (e.g. Tall and Gray 1994), even though this is for educators rather than for children.

How does the circumference of a circle relate to the area changing with respect to the radius? How does the surface area of a sphere relate to the volume changing with respect to the radius?

Many of these discussions are about mathematical paradoxes and can always make the subject more interesting than it already is (and pupils will always encourage it, even if to delay pencil and paper exercises). Skilful use will encourage a deep understanding of the concepts involved, something that Fisher (2005) refers to in terms of Skemp's relational and instrumental understanding; and the formalism can wait until the appropriate level of development. Such philosophy can either be an add-on ('bolt-on'), such as the wizard's hat, or it can be integrated into the introduction of a mathematical concept, such as complex numbers. Either way, it is value-added.

Of course, certain ground rules have to be maintained, such as not to ridicule anyone's contribution, not to hog the limelight (and try not to avoid it either). Initially more rules and a stricter regime may have to be enforced (e.g. 'hands up rather than call out'), but most likely it will become more relaxed and informal, everyone behaving in a respectful manner towards everyone else and the discourse. Most of the time the teacher should be at the front, rather in the traditional manner (and, incidentally, the pupils should perhaps sit in rows facing the front). The teacher *is* the sage on the stage, consciously directing the discourse towards fixed aims (although the discourse may be such that she allows it to go its own course without intervention) with preconceived ideas as to what she wants her class to arrive at. There has to be a target concept. This is not brainwashing, nor is it getting pupils to believe what she believes without realising it. It is teaching mathematics but with a little philosophy directed by the teacher – in the discourse she gives reasons for why something is the case and this can always and hopefully will be challenged. This is philosophy in the mathematics classroom with the teacher facing the class (sage on stage) most of the time (allowing for debate and the conversation going its own way), included in normal mathematics classes 5 h a week, versus philosophical discourse around a circle or horseshoe (a metaphor of the guide on the side) usually timetabled for 1 h a week.

The main proposal of this chapter is more than philosophy as a bolt-on; it is the teaching of mathematics with philosophical considerations but also within a cultural-historical perspective. Such a perspective will enable pupils to see and learn the great transformative events in the history of mathematics that transformed culture as well as cognition.

The main proposal is consistent with Jankvist's (2012) integration of history and philosophy in the teaching and learning of mathematics. The history is akin to Jankvist's *illumination approaches* that 'spice up' the mathematics within the historical context; and although not incompatible, this has less to do with the *modular approach* that is devoted to history or the *history-based approaches* that investigate a development of mathematics. The main proposal has more to do with history as a means rather than as an end, and although the use of original sources can serve a purpose, the history is primarily concerned with the transformative events that transformed cognition and culture as well as the mathematics. This is history to serve the understanding of the mathematics and to place the mathematics in context. Perhaps with certain exceptions such as Greek geometry, the history is not *necessarily* to understand the mathematics as constructed and perceived at the time. The history can have a place in understanding past developments from the standpoint of today – provided any historical specificity is not lost through a whiggish interpretation.

Although these transformative events may be considered 'epistemological obstacles' for both pupils and the mathematicians of the time, the proposal is not based on the recapitulation theory that historical obstacles parallel the difficulties that pupils have. This is simply because the obstacles to be raised will most likely be qualitatively different to the obstacles of the time, represented by the difference in notation. Nevertheless certain historical obstacles may be chosen to engage the class with the difficulties pertinent to understanding the relevant mathematics and, just as important, to see how this mathematics transformed cognition as well as culture. It also shows how the relevant concepts were hard-won, something that pupils can identify with in terms of their own difficulties.

Consider abstraction and proof, the two transformative events that occurred simultaneously, which transformed mathematics from an empirical affair to a science of reason that not only transformed Greek society at the time (see Kline 1972) but eventually led to the scientific revolution and the technological society we know today. For over two millennia, classical geometry served as a major rite of passage into scientific culture, as a world view, as a conceptual lens and as the introduction to rationality. From Thales to Euclid, classical geometry expanded into a world of abstract, formalized, theoretical objects and purely intellectual processes. Initially children are given a story of how Thales learnt the practical geometry of the Egyptians, a geometry borne out of the annual flooding of the Nile, and how he abstracted the notion of the geometric straight-line from observing stretched rope. They can then go outside with clipboard and paper and draw two intersecting ropes in six stages of abstraction leading to the platonic form of two intersecting lines (see Carson and Rowlands 2007; Rowlands 2010). The class is made aware that what we know about Thales is second-hand, written by Greek historians and commentators (Eudemus, Aristotle, Proclus) centuries after the event, so we don't actually know how Thales performed the abstraction necessary to have created the theoretical objects of geometry. He is credited for the very first five proofs which may not have been possible without the construction of added lines, which suggests that he did undergo some form of abstraction.

The point is the emphasis on this remarkable 'event' itself (if it was an event) that changed the course of the history of mathematics and indeed culture itself. This is not history as an examination of what actually happened because we only know what happened second-hand centuries after the event. Thales becomes a narrative device in contextualising what would otherwise be purely formal and symbolic, and although historical detail whenever possible is extremely important, it is more Thales the myth than the actual historical figure itself.[11]

During each stage of abstraction, the teacher asks what abstraction means, what has been left behind and what has been taken forward during each stage and introduces Thales' attempt to demonstrate that opposite angles of two intersecting lines are equal, despite the fact that it is obvious that they are equal. The class is not given a proof, but constructs a proof under the guidance of the teacher who is careful not to give information explicitly (see Rowlands 2010), although information is conveyed implicitly through the asking of questions. During the fourth level of abstraction (*personal concept*), the class is invited to close their eyes and think of two intersecting lines, to think of what has been taken forward and what has been left behind and how this differs to their previous drawings of two intersecting lines (the *literal representation* and the *abstract representation*). At the fifth level, the class considers the concept of two intersecting lines as it appears in textbooks (the *authorised concept*). At the sixth level (*platonic form),* the class is introduced to the platonic form of two lines intersecting and asked where the truth of opposite angles being

---

[11] At the secondary level the aim has more to do with developing an understanding of mathematics itself (and its place in terms of impact) than it has with understanding its history. The same with philosophy; for example, a secondary school mathematics teacher doesn't raise Plato's distinction between knowledge and belief because it happens to be a good thing to know, but because it illuminates the concept of proof.

equal resides. As part of the course, the pupils can be introduced to other Forms, such as Virtue or Shape, and to see how Socrates rejects given examples of a Form in favour of it being well defined.

Care should be taken, however, not to present these ideas as true. Arguably (and similar to the sophists), there is no Form of Justice or Virtue, and these concepts do not transcend cultural boundaries; but the Forms can be contextualised by emphasising the Forms as representing abstract objects that could have easily disappeared, leaving the world to concrete exemplars. Subsequently the class can then learn the basics of Greek deductive geometry with the opportunity to discuss wider philosophical issues such as whether mathematics is invented or discovered and the immortality of the soul (similar to Plato's *Meno,* the class who has constructed a proof but without the teacher giving information can question where their knowledge and understanding comes from). During this process the class can become aware of the theoretical objects of geometry (the point as a dimensionless object, the straight-line as having only one dimension, the circle both as an object and as the locus of a moving point, a fixed distance from a fixed point, the plane having no thickness, etc.) that they are normally expected to use in a traditional lesson but ordinarily have no idea as to what these objects are. The cultural aspect includes how the geometry transformed Greek society (art and architecture, philosophy, law, etc.; see Kline 1972) and how our modern technological society evolved from this (the scientific revolution of the 1600s would not have been possible without the abstraction and idealisation of Greek geometry).

Thales is not only credited as the first Greek geometer but also the first Western natural philosopher, who tried to explain the world without reference to divine will or intervention. According to Aristotle's *Metaphysics,* Thales maintained that water is the 'material principle' of the world from which everything is made (see Barnes 1987). Geometry and philosophy grew up together, culminating in Plato's famous statement at the entrance to his Academy ('He who is ignorant of geometry should not enter here'). This is not surprising if you regard geometry and philosophy as somehow having similar levels of abstraction – something that can be discussed with the class.

It is hard to imagine a discourse on proof with very little reference to philosophy, yet if proof is taught, then it is seemingly done so with little regard as to its nature. Proof is often a means to an end, an accomplishment of a task, yet there is little reference to what it is or just how important it is.[12] According to Kunimune et al. (2009), there are many Japanese children who are capable of Euclidean proof who do not see the importance of deductive proof over inductive procedures. This is perhaps common wherever deductive proof is taught, but it is odd that children can perform proof without acknowledging its importance and relevance.

Get a class who can use the protractor to each draw their own triangle and then to measure the angles. Record their answers. Here we have a whole new universe of

---

[12] And as far as the majority of English and Welsh classrooms are concerned, proof no longer exists and perhaps because many teachers and curriculum designers can remember the negative experience of having to regurgitate proof in examinations. For some educationalists, proof means rote learning.

discourse opening up – from the student unwilling to believe that the angles add up to 180° because hers add up to 181° (a problem for the radical constructivist) to whether the measurements of a million students would be enough to justify the angle property – leading hopefully towards a formal proof. If appropriate, the teacher could introduce Bertrand Russell's good inductivist turkey (ever since it could remember, it was fed breakfast at 9.00 a.m., but decided not to be certain that it would be fed at 9.00 a.m. the next day until it amassed enough evidence. After the days, weeks and months went by, it decided that it could be certain; but the next day was Christmas, and rather than having breakfast at 9.00 a.m. had its throat slit (Chalmers 1982)) and compare inductive procedures with deductive proof with specific reference to the triangle. Given their sense of wonder and their abilities, it is possible to engage children with the concept of proof, especially from a philosophical approach that encourages them to express what they think on the subject, freely and without any fear of giving a wrong answer. Eventually, they will expect epistemological reasons for why something is the case or why they have to learn it, giving them a critical edge to their learning.

Iversen (2009) suggests the comparison of proofs in both mathematics and philosophy as well as the comparison of other similar concepts in both domains as a way of developing a cross-curricular competence of mathematics. Now if this was done in a historical-cultural way so as to show the impact of the *Elements* and proof in particular on society (e.g. Spinoza's *Ethics* and Newton's *Principia* in the Euclidean deductive style, the ideas of proof and deduction in law to even the self-evident truths of the American Declaration as axioms), then it can be shown how proof impacted philosophy, especially since they grew up together for 300 years. Unfortunately the suggested St. Anselm's ontological proof for the existence of God (essentially: perfection must involve existence) is a very poor example.

Unlike perfection, existence cannot be predicated and so this seeming 'proof' compares badly with, say, the proof of the angle property of the plane triangle.

Philosophy has so much more to offer. Why not engage students with the notion of proof in conjunction with Plato's distinction between knowledge and belief as a backdrop? For belief to become knowledge, it has to be shown why the belief is true. Showing why becomes the anchor for certainty and proof can be seen as the highest warrant for certainty, with the notion of fallibility brought in to discuss the context in which we can say a mathematical proposition is true. Proof is perhaps the highest form of human achievement concerning justification[13] – so philosophy ought to be used to bring out the role of proof as a stunning advance in human ingenuity – it would be a pity if philosophy was merely 'compared' with mathematics in the sense of both having similar ('cross-curricular') notions, such as proof, only to find that some of those notions aren't so similar after all (such as St. Anselm versus Euclid).

---

[13] Even social constructivists such as Paul Ernest love and rely on Godel's proof, even though they don't like proofs in general (Ernest goes so far as to regard proof as Eurocentric and its glorification racist). This and other social constructivist contradictions can be found in Rowlands et al. (2010).

Just as history and philosophy have a very important role in science education (an exemplar being Michael R. Matthews' International Pendulum Project), so history plays a very important role in discussing philosophy in the mathematics classroom. The transformative leaps in the history of mathematics, such as Thales' abstraction from concrete exemplars to theoretical objects and his beginnings of what was to become deductive proof, provide the basis for understanding the terms of discourse and how these theoretical objects relate conceptually. This can be achieved with almost any level of development, and there is evidence to show the possibility (see Rowlands 2010), not only with high achievers but more importantly with low developers as well.

With no doubt the exclusion of rigorous foundationalist issues such as logicism, formalism and intuitionism, the nature of mathematics can be discussed in terms, for example, of platonic ideals and whether or not mathematics is invented or discovered (see Rowlands and Davies 2006). This is not teaching philosophy per se, but using philosophy to enhance a deeper understanding of the mathematical concepts involved, for example, the concept of infinity regarding the proof for the area of a circle, the finite sum of a converging infinite series (which can be related to an adaptation of one of Zeno's paradoxes) or the derivative. Berkeley's criticism of the derivative as the 'ghost of departing quantities' can be useful here.

The most important emphasis, however, has to be the philosophical consideration of the transformative leaps that have occurred in the history of mathematics, such as the advent of symbols, lines-rays-segments-angles-triangles; constructions; the circle as a mathematical concept; the advent of formal proof; infinity and the relationship between algebra and geometry (see Carson and Rowlands 2007). There were three major overarching transformations in the history of mathematics: Greek geometry leading to the axiomatic form of the *Elements*; the ideas of the seventeenth century that culminated in the works on calculus by Newton and Leibniz (perhaps this would not have happened if Hindu and Moslem developments had not reached Europe) and the rewriting of mathematics in axiomatic form in the nineteenth century followed by the 'rebirth' of mathematics under Riemann, Dedekind, Cantor and Hilbert. Perhaps the second ought to be left until the pupils reach at least 16/17 years of age and perhaps the third ought to be left until university, but the first gives ample scope for philosophical discourse at the secondary level without the demands for rigour. The concept of irrational number, for example, can be introduced in terms of the Pythagoreans describing the world in terms of number and their discovery that the diagonal of a square cannot be measured exactly (with auxiliary issues such as killing for love, out of vengeance, for country, on the one hand, and killing for squealing that $\sqrt{2}$ is irrational, on the other). A formal theory of the real numbers is not necessary at this level.

Whatever the emphasis on philosophy, however, the overall emphasis has to be on the metatheory of the mathematical concepts to be taught, such as the nature of the concept and its place within a system of concepts, its impact in terms of that system and its impact cognitively and socioculturally.

Philosophy presupposes answers that are either right and wrong, that answerability is a necessary feature of questions and philosophical inquiry is not a futile quest into questions that have no right or wrong answers (Hand 2008); and this is particularly the case when the philosophy discusses mathematical concepts, but children ought to engage in mathematics with philosophy but without the fear that their answers may be wrong. Here metacognitive skills may be enhanced as pupils try to relate their understanding to the cognitive demands of the course (and making public without fear of ridicule).

Perhaps more importantly, however, is that philosophy and mathematics can encourage creativity in Boden's (1994) transformative sense, for example, by giving the freedom to explore the dropping of some of the constraints of the subject, such as dropping Euclid's fifth axiom (the parallel postulate) to see whether the angle property of the triangle still holds prior to triangles drawn on spheres. Mathematics and philosophy within a cultural-historical context provides the opportunity to explore the creativity involved in such transformative leaps that changed cognition and culture. Students will begin to see creativity as a human endeavour to which they belong, especially if they appreciate the cultural-historical significance of fundamental concepts, their cognitive impact and just how hard-won they were. Creativity in Boden's transformative sense requires both learning the subject to exhaustion and having the freedom to play at a meta-level (see Rowlands 2011). That meta-level can be attained with the use of philosophy, especially from an historical-cultural perspective.

22.5   Conclusion

Hopefully it has become evident that what is being proposed is far removed from the philosophy for children programme: the teacher will always have preconceived ideas as to what she wants her class to know and understand. But this is not the imposition of the teacher's views upon the class. The imposition is the mathematics, but the teacher will hopefully develop the critical state of mind that challenges what is considered true in mathematics and hopefully develop the ability to become creative within mathematics.

The aim of this chapter is not to disparage the philosophy for children programme, although what is being proposed is quite a radical departure from that programme. Nor does it mean a radical overhaul of the traditional classroom. Mathematics is still taught but with the encouragement of becoming critical, that nothing is given on faith and that what is taught becomes part of the heritage that children belong and have the right to become enculturated. It is about the learning of mathematics that has become the fabric of today's technological society but with a comprehension and a critical stance that can encourage the child to become a discerning citizen of that society. Such an approach outlined above not only has the potential to enhance metacognition but creativity as well, not in the sense of novelty but in the sense of transforming the subject. Only then will mathematics be truly owned by all children.

When truly owned, then perhaps more students will be taking a mathematics degree at university, and perhaps the lectures will be similar to the ideal Socratic discourse of Lakatos' *Proofs and Refutations*. In Proof and Refutations the teacher leads the discussion with (admittedly) very bright students; but this may be possible with all students of varying levels of development if the mathematics has already been introduced philosophically, as a Socratic discourse, at the secondary school level. Lakatos' Socratic discourse is the ideal to which we can aspire – in our own classrooms and hopefully, 1 day, to the next level of learning/teaching mathematics that encompasses this ideal.

What has been proposed is only possible if student teachers are trained in introducing philosophy to the mathematics classroom, trained not only in raising philosophical questions concerning the mathematics taught but also in terms of establishing the classroom norms for the discourse to take place. Philosophy as well as history should be mandatory in mathematics teacher training. Not necessarily philosophy and history per se, but as aids in the teaching of mathematics.

References

Barnes, J. (1987), *Early Greek Geometry*. London: Penguin.
Boden, M. A. (1994), What is Creativity? In (M. A. Boden, ed.) *Dimensions of Creativity*. London: MIT.
Carson, R. N. & Rowlands, S. (2007), Teaching the conceptual revolutions in geometry. *Science & Education* **16**, pp. 921–954.
Chalmers, A. F. (1982), *What is this thing called science?* (second edition). Milton Keynes: the Open University Press.
Clegg, B (2003), *A brief history of infinity: the quest to think the unthinkable*. London: Constable and Robinson.
Daniel, M. F., Lafortune, L., Pallascio, R. & Sykes, P. (2000), A primary school curriculum to foster thinking about mathematics, *Encyclopaedia of Philosophy of Education*. http://www.ffst. hr/ENCYCLOPAEDIA (accessed 06/06/2011).
De La Garza, M. T., Slade-Hamilton, C. & Daniel, M. F. (2000), Philosophy of mathematics in the classroom: aspects of a tri-national study, *Analytic Teaching*, **20**(2), 88–104.
Dowling, P. & Noss, R. (1990), *Mathematics versus the National Curriculum*. London: Falmer Press.
Fisher, R. (2005), *Teaching children to think (second edition)*. Cheltenham: Nelson Thornes.
Gombrich, E. H. (1960), *Art and Illusion*. London: Phaidon.
Goos, M (2004), Learning mathematics in a classroom community of inquiry. *Journal for Research in Mathematics Education*, **35**(4), pp. 258–291.
Hand, M. (2008), Can children be taught philosophy? In (M. Hand & C. Winstanley, eds.) *Philosophy in Schools*. pp. 3–17. London: Continuum.
Iversen, S. M. (2009), Modeling interdisciplinary activities involving mathematics and philosophy, in (B. Sriraman, V. Freiman & N. Litette-Pitre, eds.) *Interdisciplinarity, Creativity and learning: Mathematics with Literature, Paradoxes, History, Technology, and Modeling*. pp. 85–98. Charlotte, NC: Information Age Publishing, Inc.
Jankvist, U. T. (2012), History, applications, and philosophy in mathematics education: HAPh – A use of primary sources. *Science & Education* (online).
Kennedy, N. S. (2007), From philosophical to mathematical inquiry in the classroom. *Childhood & Philosophy*, **3**(6). http://www.filoeduc.org/childphilo/n6/Nadia_Kennedy.pdf. (Accessed 15/07/2011)

22 Philosophy and the Secondary School Mathematics Classroom

Kline, M. (1972), *Mathematics in Western Culture*. Harmondsworth: Pelican.

Kunimune, S., Fujita, T. & Jones, K. (2009), "Why do we have to prove this?" Fostering students' understanding of 'proof' in geometry in lower secondary school, in (F-L. Lin, F-J. Hsieh, G. Hanna, & M. de Villers, eds.), Proof and proving in mathematics education ICMI study 19 conference proceedings, Vol. 1, pp. 256–261.

Law, S. (2008), Religion and philosophy in schools. In (M. Hand & C. Winstanley, eds.) *Philosophy in Schools*. pp. 41–57. London: Continuum.

Lenton, D. & Vidion, B. (2011), Is philosophy what you think it is? PTU Focus article. *Primary Teacher Update*. Number 1, October. MA Education Ltd.

Lightfoot, L. (2011), Time to take the won't out of Kant, *Tespro* (number 3), supplement to The Times Higher Education Supplement, 30th September, 2011.

Lipman, M., Sharp, A. M. & Oscanyon, F. S. (1980), *Philosophy in the classroom (second edition)*. Philadelphia: Temple University press.

Matthews, G. (1980), *Philosophy and the Young Child,* Cambridge: Harvard University Press.

Matthews, G. (2008), Getting beyond the deficit conception of childhood: thinking philosophically with children. In (M. Hand & C. Winstanley, eds.) *Philosophy in Schools*. pp. 27–40. London: Continuum.

McCall, C. C. (2009), *Transforming thinking: philosophical inquiry in the primary and secondary classroom*. Abingdon: Routledge.

Noss, R. (1997), "Call to end 'dumbing-down' of school mathematics". http://www.ioe.ac.uk/media/R071097A.HTM Accessed, 15/03/2003.

Pincock, C. (2012), *Mathematics and Scientific Representation*. New York: Oxford University Press.

Prediger, S. (2007), Philosophical reflections in mathematics classrooms, in (K François & J. P. Van Bendegem, eds.) *Philosophical Dimensions in Mathematics Education*, pp. 43–58. Mathematics Education Library volume 42.

Rowlands, S. & Davies, A. (2006), Mathematics Masterclass: Is mathematics invented or discovered? *Mathematics in School* (journal of the Mathematics Association), 35(2), pp. 2–6. Also reprinted in *Mathematics in Schools*, 2011, 40(2), pp. 30–34.

Rowlands, S. (2009), The importance of cultivating a meta-discourse in deliberate support of metacognition, in (C. B. Larson, ed.) *Metacognition: New research developments*. pp. 1–22. New York: Nova.

Rowlands, S. (2010), A pilot study of a cultural-historical approach to teaching geometry. *Science & Education*, 19, pp. 55–73.

Rowlands, S., Graham, T. and Berry, J. (2010). Problems with fallibilism as a philosophy of mathematics education. *Science & Education* (online)

Rowlands, S. (2011), Discussion article: Disciplinary boundaries for creativity, *Creative Education,* 2(1), pp. 47–55.

Schoenfeld, A. H. (1989), What's all the fuss about metacognition? In (A. H. Schoenfeld, ed.) *Cognitive science and mathematics education*. pp. 189–215. Hillsdale: Erlbaum.

Siegal, H. (2008), Why teach epistemology in schools? In (M. Hand & C. Winstanley, eds.) *Philosophy in Schools*. pp. 78–84. London: Continuum.

Skemp, R. R. (1971), *The Psychology of Learning Mathematics*. London: Penguin.

Skemp, R. R. (1976), Relational understanding and instrumental understanding, *Mathematics Teaching*, 77, pp. 20–26.

Tall, D. & Gray, E. (1994), Duality, ambiguity and flexibility: A proceptual view of simple arithmetic, *The Journal for Research in Mathematics Education*, 26(2), pp. 115–141.

Vygotsky, L. (1987), *The collected works of L. S. Vygotsky: Problems of general psychology, including the volume thinking and speech*. New York: Plenum.

Vygotsky, L. (1994), *The Vygotsky Reader*, ed. by R. van der Veer & J. Valsiner. Oxford: Blackwell.

Wason, P. C. (1977), Self contradiction, in (P. N. Johnson-Laird & P. C. Wason. eds), *Thinking: Readings in cognitive science*. Cambridge: Cambridge University Press.

Stuart Rowlands graduated in Philosophy at the University of Swansea in 1978 and achieved his Postgraduate Certificate in Education for mathematics at the same university in 1980. He taught secondary school mathematics for 14 years before becoming a full-time Ph.D. student in mechanics education at the University of Plymouth in 1994. Since then he has been a lecturer in mathematics and has also researched mathematics education at the university. His main research interests to which all others are subordinate is conceptual change in mechanics and the teaching of geometry from a cultural-historical perspective (in collaboration with Robert Carson of Montana State University).

His publications include the following: Rowlands, S (2003), Vygotsky and the ZPD: have we got it right? in S. Pope and O. McNamara (eds) *Research in Mathematics Education* volume 5 (Papers of the British Society for Research into Learning Mathematics). London; BSRLM; Carson, R. N. and Rowlands, S. (2007), Teaching the conceptual revolutions in geometry. *Science & Education* 16, pp. 921–954; Rowlands, S. (2010), A pilot study of a cultural-historical approach to teaching geometry. *Science & Education*, 19, pp. 55–73; and Rowlands, S., Graham, T. and Berry, J. (2010). Problems with fallibilism as a philosophy of mathematics education. *Science & Education* (online).

Chapter 23
A Role for Quasi-Empiricism in Mathematics Education

Eduard Glas

23.1 Introduction

In modern philosophy of science, there is virtually no disagreement about the fallibility of empirical scientific knowledge. We just cannot be absolutely sure that a scientific theory will stand up to all future tests. Mathematical propositions, on the other hand, do not refer directly to matters of empirical fact and therefore look like products of pure thought, immune to empirical refutation. They seemingly possess the absolute certainty of analytic statements or logical truths.

This view of the matter has had a strong impact on the public image of mathematics and, in particular, on the way in which teachers and their pupils regard the mathematics taught in school. They still often look upon mathematics as the one teaching subject where practical experiences are irrelevant and where there are single right answers to all questions, whose correctness is beyond all doubt. Is a different view at all possible?

Although it is true that mathematical propositions are not straightforwardly empirically falsifiable, they are not immune to all (other) forms of criticism. They may, for example, be criticised for failing to solve the problem which they were designed to solve. In science, too, deficiencies in problem solving or explanatory potential are normally counted as negative evidence. So in this respect, mathematics is not unique and not as radically different from science as those who entertain the above view suppose.

Throughout this chapter, extensive use has been made of previous articles mentioned in the references.

E. Glas (✉)
Department of Mathematics, Delft University of Technology,
Mekelweg 4, 2628CD Delft, The Netherlands
e-mail: e.glas@tudelft.nl

M.R. Matthews (ed.), *International Handbook of Research in History,
Philosophy and Science Teaching*, DOI 10.1007/978-94-007-7654-8_23,
© Springer Science+Business Media Dordrecht 2014

Especially the most advanced sciences are very much like mathematics in that their conceptual apparatus and organisation are to a large extent non-observational and self-supporting. Only think of the theory of relativity, which depended on the nonempirical principle of relativity and the non-Euclidean geometries developed in the second half of the nineteenth century. Einstein's achievements relied on thought-experiments and mathematics; empirical methods became relevant only when confirmation or corroboration was called for. More recently, the fruitful interaction between string theory and differential geometry has even led to a reversal of the direction of influence: mathematical physicists constantly reveal new mathematical structures undreamt of before.

The derivation from scientific theories of empirically testable consequences may require considerable theoretical effort, and their confrontation with observational material is itself dependent on theoretical assumptions about the meaning of terms, the interpretation of observations, theories of experimental design and procedure, ceteris paribus clauses, etc., briefly: on more basic theories. Testing a scientific theory therefore ultimately boils down to checking its consistency with respect to lower-level theories (much like mathematics), and it is often possible (though not always rational) to resolve empirical difficulties without giving up the theory under examination. It appears therefore that while mathematics is to a certain degree 'science-like', modern science is also much more 'mathematics-like' than has traditionally been supposed, and any difference between the two is at most only a matter of degree.

Although once couched in formal language physical theories are no more directly refutable than mathematical theories, nobody would for that reason deny them their empirical character. Neither should the impossibility of a direct empirical refutation of a mathematical theory count as sufficient reason for denying mathematics all claims to empirical significance. Formalised mathematics can effectively be immunised to all informal counterevidence: a formal mathematical theorem may be upheld 'come what may'. But the same applies to a considerable extent also to modern, mathematically impregnated science. And in either case, our actually choosing to do so would imply resigning ourselves to stagnation of the growth of knowledge. Informal counterexamples to formal theorems can be ignored; but by ignoring them, we also ignore the best opportunities for achieving progress. This is precisely the important lesson contained in Imre Lakatos' work on *Proofs and Refutations* (Lakatos 1963–64, 1976), which may be regarded as the seminal text for the quasi-empirical approach to mathematics.

In the acknowledgement preceding the said articles, Lakatos tells us that 'the paper should be seen against the background of Pólya's revival of mathematical heuristic, and of Popper's critical philosophy' (Lakatos 1976, p. xii). It will be opportune, therefore, to sketch these backgrounds first and next to proceed with an analysis of Lakatos' quasi-empirical heuristic against these backgrounds. Although there are other approaches to the philosophy of mathematics that go under the head of quasi-empiricism, the present discussion will be restricted to the said criticist-objectivist strand, construed as a self-contained whole. A short assessment of the educational implications of the quasi-empirical view of mathematics will be given in the concluding section. These implications have no direct bearing on the actual practice of teaching mathematics in the classroom, but mainly – not less importantly – on the *image* of mathematics that is conveyed in education.

23.2 Pólya's Problem-Solving Heuristic

Lakatos' fellow Hungarian and friend György Pólya (1887–1985) was an important reformer of mathematics education and a champion of informal styles of reasoning in mathematical problem solving (Pólya 1954, 1957, 1962, 1968, 1973). He extensively discussed several ways of fruitful thinking in mathematical problem situations. By its enormous wealth of examples, this extensive body of work had a profound impact on mathematics education.

Underlying Pólya's approach was the crucial distinction between demonstrative and plausible reasoning. Presented in the standard Euclidean manner, mathematics appears as a systematic deductive enterprise. Demonstrative reasoning refers to this well-known deductive or Euclidean mode of presentation. It has rigid standards which are codified and clarified by logic. This style of reasoning is safe, beyond controversy, and final. It is the mode of argumentation needed for the ultimate proof of mathematical propositions.

Plausible reasoning, by contrast, refers to argumentation in the inductive or even experimental style. Mathematics in the making is said to exhibit the character of an experimental inductive science based on plausible rather than demonstrative reasoning. Especially inductive generalisation and reasoning by similarity and analogy are essential for the discovery or invention of conjectures. As problem-solving strategies, they are domain specific and situational; their standards are fluid, not rigid; and they are directed towards both the generation and the support (though not the proof) of mathematical hypotheses.

> Mathematics in the making resembles any other human knowledge in the making. You have to guess a mathematical theorem before you prove it; you have to guess the idea of the proof before you carry out the details. You have to combine observations and follow analogies; you have to try and try again. (Pólya 1973, p. vi)

Pólya thus entertains an 'empiricist' position with respect to the generation of mathematical knowledge and even the generation of proof ideas, though not in regard of proof itself. But in the birth phases of a mathematical theory, in particular, there is ample space for inductive reasoning, generalisation, specialisation and analogy, and these informal modes of reasoning are reproducible in the classroom. The didactic relevance is obvious from the wealth of examples which are offered for imitation and practice, enabling learners to grasp the relevant concepts and to re-create as it were the theory in question. Pólya suggests that the traditional deductive view of mathematics seriously obscures the inductive nature of the reasoning that mathematicians use when deriving their conjectures and dreaming up their attempts at deductive proofs:

> The result of the mathematician's creative work is demonstrative reasoning, a proof, but the proof is discovered by plausible reasoning, by guessing. (Pólya 1954, p. 158)

734   E. Glas

Pólya was exceptional through his interest in heuristic, the till-then largely neglected problem of the reasoning processes underlying discovery. Especially since Reichenbach's canonisation of the distinction between the context of discovery and the context of justification, methodological analysis had more and more become restricted to the explication of concepts and the verification of already articulated theories. Discovery was considered a process devoid of logic and therefore of interest only to psychologists. For Pólya, however, discovery in mathematics follows certain rules that together represent a 'heuristic', an *ars inveniendi* (art of invention). It concerns the discovery of mathematical 'facts', which subsequently still are to be proved in a strictly deductive way. But heuristic, the logic of mathematics in the making, is largely inductive or, in other words, 'quasi'-empirical.

Pólya's achievements clearly are above all of great pedagogical and didactic relevance. As he himself testified, it had never been his intention to write philosophical treatises. Still, his work contains many philosophically important issues which, however, showed to full advantage only in the hands of Imre Lakatos. Testing mathematical conjectures by 'quasi-experiments' represents just one of the more important examples of what Lakatos owed to Pólya.

23.3   Between Pólya and Popper: Lakatos' Heuristics

Imre Lakatos (1922–1974) went on where Pólya had stopped. He transformed the idea of heuristic into a critical methodological concept, the method of proofs and refutations. In some ways, it surpassed Popper's logic of scientific discovery, but it surpassed even more Pólya's idea of heuristic as an 'art of invention', and especially the latter's inductivist conception. Lakatos' heuristic was based on the use of counterexamples – suggesting falsification – as critical tools for the achievement of *growth* of knowledge.

Heuristic for Lakatos was a truly methodological notion, *viz.*, a set of criteria indicating which paths should be followed and which should be avoided in order that our knowledge may grow. It is neither a logical nor a psychological subject, but an autonomous methodological discipline, the 'logic of discovery':

> there is no *infallibilist* logic of scientific discovery, one which would infallibly lead to results; there is a fallibilist logic of discovery, which is the logic of scientific progress. Popper, who has laid down the basis of *this* logic of discovery, was not interested in the metaquestion of what was the nature of his inquiry and he did not realise that this is neither psychology nor logic; it is an independent discipline, 'heuristic'. (Lakatos 1976, pp. 143–144, footnote)

Lakatos not only applied Popper's 'logic of discovery' to mathematics but brought a much larger part of the Popperian corpus to bear on mathematics, *viz.*, the centrality of problems and their dynamics rather than static definitions, methodological rules to prevent loss of content, the use of models to test lemmas separately and the associated distinction between global and local counterexamples (cf. Glas 2001), and more especially a group of ideas clustering around Popper's doctrine of the relative autonomy of objective knowledge, which was fully endorsed by Lakatos:

> Mathematical activity is human activity. Certain aspects of this activity … can be studied by psychology, others by history. Heuristic is not primarily interested in these aspects. But mathematical activity produces mathematics. Mathematics, this product of human activity, 'alienates itself' from the human activity which has been producing it. It becomes a living, growing organism that *acquires a certain autonomy* from the activity which has produced it; it develops its own autonomous laws of growth, its own dialectic. (Lakatos 1976, p. 146)

Proofs and Refutations is written in the form of a classroom dialogue between a teacher and pupils bearing the names of Greek letters and representing various ways of dealing with counterexamples. In the beginning, we have a theorem (about the relation between the numbers of vertices, edges and faces of polyhedra) and a counterexample. What follows is a direct confrontation between two equally dogmatic epistemological positions. Pupil Delta typically tries to save the proposition by 'conventionalist stratagems' – introducing ad hoc redefinitions of terms in order to render counterexamples harmless (appropriately dubbed monster barring, exception barring and monster adjustment by Lakatos). These strategies effectively immunise the theorem from refutation but also rid it of more and more of its informative content, until at last it has become an utterly uninteresting truism. Pupil Gamma, on the other hand, insists on exact definitions of the fixed meanings of terms. He therefore regards any counterexample as conclusive proof of the falsity of the theorem. He typically is a 'naïve falsificationist'. Neither Delta nor Gamma handles a critical methodology in the Popperian sense, that is, a set of rules delimiting the sorts of arguments that are fruitful to the *growth* of knowledge, getting at theorems that say 'more' that is also 'more nearly true'. Neither of the dogmatic positions represented by Delta and Gamma *uses* the counterexample in a constructive fashion to learn from it how a better theorem might be obtained. Both Gamma's and Delta's modes of argumentation are in this sense detrimental to the growth of knowledge; they are not rational because they fail to take adequate account of the problem situation.

Rejecting the use of conventionalist stratagems does not imply that all counterexamples should be uncritically accepted; we may rationally defend a theory, provided that adjustments are not entirely ad hoc and do not lead to unnecessary losses of content. So there is always a methodological *decision* to be made when a theorem is confronted with a counterexample, the rationality of which depends on the particular problem situation, especially on whether it would lead to a content-increasing or a content-reducing problem shift.

The teacher develops the Lakatosian methodology by participating in and drawing the lesson from the discussion that follows after the initial confrontation. He lets the pupils lay down the conclusions in a number of methodological rules which, typically, are all concerned with making a justifiable *decision* about what to do if a counterexample, either global (against the theorem as a whole) or local (against a lemma in the proof), turns up. All the rules begin with 'if you have…'; they are not concerned with the question of 'how to find' (Pólya's *ars inveniendi*) but with the logic of discovery in the (Popperian) sense of making justified decisions – justified by preventing loss of content. The rules say that if you have a global counterexample, analyse your proof, make all hidden lemmas explicit, find the guilty one and improve your conjecture by incorporating this lemma in the form of a condition. If you have a local counterexample, check to see whether it is not also global. If it is, apply the previous rule. If it is not, try to improve your proof analysis by replacing the refuted lemma with an unfalsified one (Lakatos 1976, p. 50).

Much of the methodological discussion thus consists in specifying the appropriate methodological rules to counter conventionalist stratagems by turning global counterexamples into local ones and *using* them to get at improved theorems-*cum*-proofs, without unnecessary losses of informative content. But besides preventing loss of content, we aim at increase of content; the fifth methodological rule indicates how this is to be done: if you have counterexamples of any type, try to find, by deductive guessing, a deeper theorem to which they are counterexamples no longer (*ibid.*, p. 76). Increase of content may also be achieved through the formation of new concepts as a by-product of the development of the initial naïve conjectures into more and more sophisticated propositions. The formulation, analysis and reformulation of proofs will often, perhaps unconsciously, lead to a redefinition of the terms used – a development from naïve concepts to 'proof-generated' concepts. We may even 'stretch' concepts deliberately beyond their original domain of application in order to reveal possibly unsuspected new relationships that could not even be articulated in terms of the original, more naïve concepts.

Central to the method of *Proofs and Refutations* is the role of *proofs* in the growth of mathematical knowledge. Proofs are not ends, in the sense of establishing once and for all the truth of a theorem, but means to get at richer, deeper, more interesting theorems. Lakatos introduced a structural epistemological similarity between 'proofs' in mathematics and experimental 'tests' in science (proofs are 'tests' – as in the 'proof' of the pudding). As scientific theories are tested by experiments that anchor them to lower-level statements, so mathematical theorems are tentatively proved by deriving them – by means of a thought experiment – from more basic lemmas. Proofs thus play a role analogous to corroborating experimental tests in science. In the justificationist tradition, proofs are supposed to link theorems to indubitable axioms whose truth immerses the whole propositional system through channels of truth preserving – i.e. deductive – arguments. In Lakatos' quasi-empirical heuristic, it is not truth streaming downward from the axioms but the upward retransmission of falsity (in the form of counterexamples) that is crucial for the growth of knowledge.

23.4 Fallibilism and Quasi-Empiricism

The most important background to Lakatos' quasi-empiricism is, of course, the critical fallibilism of his one-time teacher at the London School of Economics, Karl Popper (1902–1994). Popper did not consider anything, including mathematics and even logic, as absolutely certain (Popper 1984, pp. 70–72). He argued that we should never save a threatened theoretical system by ad hoc adjustments that reduce its testability (Popper 1972, pp. 82–83) – a view to be exploited by Lakatos to such dramatic effect in the dialogues of *Proofs and Refutations* under the heads of monster barring, exception barring and monster adjustment.

Popper's fallibilism implies the non-existence of solid foundations to stop the infinite regress in proofs and definitions; all knowledge is conjectural, consisting of attempts at solving problems. Central to mathematics are problems, and striving after exactness for its own sake is futile:

> Absolute exactness does not exist, not even in logic and mathematics . . . and the demand for "something more exact" cannot in itself constitute a genuine problem (except, of course, when improved exactness may improve the testability of some theory). (Popper 1983, p. 277)

In much the same vein, Lakatos took up a critical analysis of modern attempts to place the whole of mathematics on a perfectly exact basis of ultimate logical intuitions in his paper on 'Infinite Regress and Foundations of Mathematics', which dates back to 1962, before the articles on *Proofs and Refutations* appeared (1963–1964) (Lakatos 1978b, pp. 1–23). From the outset, Lakatos made it clear that his aim was to break ground for a critical programme in mathematics, the programme of Popper's *critical fallibilism* (*ibid.*, pp. 9–10).

So far, Popper and also Lakatos himself (in his thesis) had only considered *informal* 'preformal' mathematical theories. But it is one thing to show that informal mathematics is conjectural, it is quite another thing to show that uncertainty is not just a *Kinderkrankheit* of informal mathematics, which has now been cured by founding the discipline on rigorous logic and 'ultimate' (set-theoretic) axioms. Mathematical theories are fallible not only as long as they have not been properly founded: the foundational programmes are themselves just hypotheses. The arithmetisation of mathematics by Cauchy and his followers was a wonderful Euclidean achievement, but like any other theory, it was susceptible to criticism. The most incisive critical arguments came from the doubts of the pursuers of the quest for certainty themselves (notably Frege and Russell):

> Have we *really* reached the primitive terms? Have we *really* reached the axioms? Are our truth-channels *really* safe? (*ibid.*, pp.10–11)

Lakatos concentrated specially on Russell,

> showing how he failed in his original Euclidean programme, how he finally fell back on inductivism, how he chose confusion rather than facing the fact that what is interesting in mathematics is conjectural. (*ibid.*, p. 11)

Russell never had seriously considered the possibility that mathematics may be conjectural. Instead, he came to hold that some axioms of logic are to be believed, not on their own account, but on account of the indubitability of their logical consequences (*ibid.*, p. 17). According to Lakatos, Russell failed to draw the right conclusion

> that the infinite regress in proofs and definitions in mathematics cannot be stopped by a Euclidean logic. Logic may *explain* mathematics but cannot *prove* it. It leads to sophisticated speculation which is anything but trivially true. ... The logical theory of mathematics is an

exciting, sophisticated speculation like any scientific theory. It is an empiricist theory and thus, if not shown to be false, will remain conjectural forever. (*ibid.*, p. 19)

Lakatos did not deny, of course, that virtually the whole of mathematics could be derived from axiomatic set theory. The fundamental assumptions of set theory, however, are far from self-evident or trivially true. They might be overthrown, replaced or supplemented by new axioms, especially in view of the 'independent' questions, the remaining problems which are not decidable by the standard axioms. Given that these axioms cannot themselves be proved, any sort of arguments that may be offered for or against them will in the end rest on what one *guesses* to be true (or convenient).

Lakatos drew attention to the danger that those (like Russell) who recognise the science-likeness of mathematics turn for similarities to a noncritical (non-Popperian), justificationist image of science, fall back on inductivism and psychologism and keep searching for the ultimate authoritative basis for justifying certain *beliefs*.

But why on earth have "ultimate" tests, or "final" authorities? Why foundations, if they are admittedly subjective? Why not honestly admit mathematical fallibility, and try to defend the dignity of *fallible* knowledge from cynical scepticism . . . ? (*ibid.*, p. 23)

At the London Colloquium of 1965, Lakatos commented on a lecture of Kalmár in which the latter discussed and defended the position that mathematics is an empirical science (Lakatos 1967, pp. 187–194). It is here that Lakatos introduced his distinction between *quasi*-Euclidean and *quasi*-empirical theories and laid down the claim that mathematics is quasi-empirical.

He explained that a theory may be non-empirical yet quasi-empirical, or empirical yet quasi-Euclidean for that matter, the distinction referring only to the direction of the characteristic truth-value flow (top-down or bottom-up). The claim that mathematical theories are quasi-empirical therefore says in effect that the characteristic logical flow in mathematical theories is the bottom-up retransmission of falsity. Euclidean theories are here considered as limiting cases of quasi-empirical theories: a system is Euclidean if it is the logical closure of the accepted basic statements and quasi-empirical if it is not. Whereas a Euclidean theory may be claimed to be true, a quasi-empirical theory can at best be well corroborated (when it has an impressive record of passed tests), but ultimately has to remain conjectural. In a quasi-empirical theory, the axioms do not *prove* the theorems in a strict sense, but they *explain* them by showing of which more fundamental assumptions they are the logical consequences. The basic rule of quasi-empirical methodology is

to search for bold, imaginative hypotheses with high explanatory and heuristic power; indeed, it advocates an uninhibitedly speculative proliferation of alternative hypotheses to be pruned by severe criticism. (*ibid.*, p. 202)

Lakatos deemed it not superfluous to state expressly in a footnote that *of course* the paradigm of quasi-empirical methodology is Popper's scientific methodology.

The axioms of a formal theory are often regarded as implicitly defining the concepts that they introduce. If this view is accepted, then there could be no potential mathematical falsifiers except logical ones (i.e. statements of inconsistency). But

Lakatos vehemently opposed the utterly *un*historical identification of mathematics with the set of all consistent formal systems and insisted that we should speak of formal systems only if they are formalisations of some informal theory. A formal theory then may be said to be 'refuted' if one of its theorems is negated by one of the theorems of the corresponding informal theory. Such informal theorems he called *heuristic falsifiers* of the formal theory (Lakatos 1978b, p. 36).

The axioms of set theory, for example, may be tested for consistency, and the definitions may be tested for the correctness of their translation of branches of mathematics such as arithmetic. If a counterexample from arithmetic can be formalised in the system, the formal theory is thereby shown to be inconsistent (in which case we have a logical falsifier). But if the system is consistent, the counterexample cannot be formalised. Such a heuristic falsifier therefore does not show that a formal theory is inconsistent, but only that it is a false theory *of arithmetic*, while it still may be a true theory of some mathematical structure that is not isomorphic with arithmetic. The axioms then do not properly *explain* the informal theory that they were designed to explain.

So Lakatos found mathematical analogues of Popperian potential falsifiers in theorems of arithmetic or other branches of classical mathematics that are potential counterexamples to corresponding theorems of the formalised theory. It enabled him to give an epistemological underpinning to such notions as the content of a mathematical theory (the arithmetical content of a formal theory in particular), content-increasing and content-decreasing problem shifts and so on, in terms of sets of potential falsifiers – as Popper had done for natural science.

Strictly speaking, a heuristic falsifier is no more than a rival hypothesis that merely *suggests* a falsification, and suggestions may be ignored. This, however, does not separate mathematics as sharply from natural science as one might think. Popperian basic statements, too, are only hypotheses after all. They are accepted tentatively for the purpose of a particular discussion of a particular problem, but may become highly questionable in a different discussion of a different problem. They do not constitute some bedrock of knowledge, but are more like 'piles driven into a swamp' (Popper 1972, p. 111). Lakatos was fully in agreement with this view when he claimed that

> the crucial role of heuristic refutations is to shift problems to more important ones, to stimulate the development of theoretical frameworks with more content. (Lakatos 1978b, p. 40)

What remains is the question about the basis on which truth values are first injected into the potential falsifiers of mathematical theories. Since on his view the only interesting and respectable formal theories are formalisations of established informal theories, this question in part reduces to inquiring into the nature of the basis on which initial truth values are injected into the basic statements of the informal predecessors. The answer should be sought by tracing back (through rational reconstruction) the series of *problem shifts* that constitutes the development of the field. Perhaps mathematics might ultimately turn out to be 'indirectly empirical'; or perhaps the source of the initial truth value injection is to be found in construction, or intuition, or convention.

740 E. Glas

The answer will scarcely be a monolithic one. Careful historico-critical case-studies will probably lead to a sophisticated and composite solution. But whatever the solution may be, the naïve school concepts of static rationality like *a priori/a posteriori*, *analytic/synthetic* will only hinder its emergence. These notions were devised by classical epistemology to classify Euclidean certain knowledge – for the problem shifts in the growth of quasi-empirical knowledge they offer no guidance. (*ibid.*, pp. 40–41)

23.5 Mathematical Quasi-Experiments

The pivot on which Lakatos' programme turned was his construal of informal proofs as thought experiments for the appraisal of mathematical knowledge. The standards of appraisal do not come from the study of foundations and formal systems, but from the logic of proofs and refutations, that is, of quasi-experimental tests and improvements. Mathematics is like science, not because it is somehow based on sensual experiences, but because it likewise proceeds through fallible trials and tests, and for this reason, it may be called a quasi-experimental science.

The science-like practice of working 'upwards' from facts to theories (lemmas, axioms, rules) corresponds nicely with the age-old 'method of analysis and synthesis', which pervades the entire history of exact science and of which Lakatos' method of proofs and refutations may be considered a logical extension. In analysis, a mathematical conjecture is tested by searching for concomitants that must be true if the conjecture were true. If in so doing one hits upon an obvious falsity, the conjecture must be false. But if only already established or trivial truths are hit upon, the conjecture may be true. Synthesis then consists of the construction of a compelling argument with the aid of the insights learned from the analysis. Lakatos' method goes on to test in this way the proofs themselves, searching for heuristic counterexamples in order to make hidden lemmas explicit and thus to get at improved theorems-*cum*-proofs, with enhanced generality, scope, profundity, problem-solving and explanatory power, etc.

On Lakatos' view, the Greeks had subjected the unproven mathematical facts inherited from the barbarians to a great many such analyses. They found that some lemmas kept cropping up, whereas their alternatives remained sterile, thus yielding series of corroborated analytical components converging on a small number of indubitable truths, which were to constitute the hard core of axiomatisation programmes such as Euclid's *Elements*.

This account can in fact be substantiated by Euclid's proposition 32, which says that any exterior angle of a triangle equals the sum of the opposite interior angles and that the sum of the interior angles equals two right angles. For the proof, one has to produce the side AC of a triangle ABC beyond C and to draw a line through C parallel to AB. The proposition then can be 'seen' to follow directly from the equality of the angles that the said parallel makes with the sides AC and BC. Thus the statement about parallels is the premise from which the synthetic argument starts, and by reversing it, one gets a glimpse of the analytic procedure through which the premise was originally found. By producing lines and drawing parallels, many properties of figures could be seen to be consequences of the said property of parallels, which thus became a hard-core element in the axiomatisation.

Curiously enough, Euclid's fifth postulate, the so-called parallel postulate, does not speak of parallels, although the relevant properties of parallels are direct consequences of it. It says that two lines will meet, if sufficiently produced, at the side on which the sum of the interior angles with an intersecting line is less than two right angles. This formulation is highly suggestive indeed of the above account of its analytic origins. The equality of angles in the proof of proposition 32 follows immediately in case the sum of the said interior angles happens to be exactly equal to two right angles.

The analytic-synthetic procedure is an 'experiment' in that it involves 'real' actions (drawing and producing lines) in order to 'see' how certain properties of figures obtain. This way of 'seeing' is not based on logical connections between statements, but on intuitively recognisable relations between and within figures, which as such are not readily formalisable. What is laid down in the postulates are not truths but requirements for the construction of geometric demonstrations; it is required, for instance, 'that through any two points a line can be drawn' (post.1), 'that any line can be produced' (post.2), 'that given a (mid) point and a line (radius) the circle can be drawn (post.3) and 'that all right angles are equal' (post.4). The procedure is *quasi*-experimental in that the actions are idealised: although visible lines and points occupy space, a *mathematical* point is defined as 'that which occupies no place' and a *mathematical* line as 'length without breadth'. These definitions make sense only if one assumes every rational person to be already in possession of intuitive notions of what a point (dot) and a line (stripe) are; they merely stipulate which aspects of these intuitions may, and which may not, be used in a mathematical demonstration. The experiment is done by drawing visible lines through visible points, but in the argumentation, abstraction is made of their sensual features.

Analysis became especially important as a method for proving *ex absurdo*, in which case the negation of a proposition is analysed. For when a concomitant of the negated proposition is found to be obviously false, this proves indirectly that the proposition itself must be true. The analysis as such proves the proposition, and the (often laborious) task of constructing a synthetic argument can be avoided. Greek mathematics after Euclid thus came to abound with indirectly proven propositions, without the slightest hint as to the way in which they were discovered, and also without the intuitively compelling demonstrations that synthetic proofs supply. The heuristic dimension being almost completely hidden from view, the classical image of mathematics as a deductively closed Euclidean system became firmly established for a long time to come. If Archimedes' treatise *On Method* (to be discussed in what follows) had not been discovered (in 1906), virtually nothing at all would have been known about Greek heuristics (Lakatos 1978b, p. 100).

In contrast to Lakatos' negative view of premature axiomatisation, Hintikka and Remes have shown convincingly that even in an axiomatised system, the construction of a proof required a veritable method of discovery, and analysis furnished this method. It involved the introduction of tentative auxiliary constructions, which necessitated a subsequent synthesis to warrant the results. What was analysed were configurations, not deductive connections or proofs, and the steps of analysis did not lead from one proposition to another, but from figure to figure (Hintikka and Remes 1974, p. 32). The need of introducing auxiliary constructions constituted the unpredictable and recalcitrant element in the methodological situation. The natural course of an analysis was not linear but took the form of a more complicated network of connections, and this made synthesis non-trivial and necessary for warranting the reversibility of the several steps. This presented no insuperable difficulties, as in a geometrical analysis the steps will be reversible anyhow, being mediated by functional interdependencies between geometrical entities in a given figure (*ibid.*, p. 37). The aim of the analysis was to find the *crucial* auxiliary constructions, but this aim could only be attained if enough hypothetical constructions were already anticipated in the analysis. One had to trust one's intuitive insight in finding the relevant geometrical interdependencies, and this way of proceeding made it imperative to justify the procedure afterwards by a synthesis: together they constituted two inseparable halves of one quasi-experimental method.

The creators of modern science – Descartes, Galileo and Newton – held the ancient method of analysis and synthesis in high esteem and shaped their methodology on it (for a general overview, see Otte and Panza 1996). I will present just a few examples to illustrate my earlier statement that not only is mathematics science-like, science is also mathematics-like, which indeed renders the classical *a priori/a posteriori* and *analytic/synthetic* distinctions merely a matter of degree.

Geometrical analysis was a systematic inquiry into the interdependencies between known and unknown 'objects' in a given configuration, and it was but a relatively small step to regard a real experimental setup likewise as a sort of analytic situation. Galileo's seemingly 'real' experiments, for instance, turn out to have been intended as demonstrations that the effects calculated on the basis of presumed relationships could actually be produced. These relationships as such had however been discovered in a 'quasi-mathematical' way, through thought experiments.

The insight, for instance, that in vacuum all objects fall with the same speed, was obtained in the following way. Suppose a heavy object H falls faster than a lighter object L – as common opinion would have it. Now connect both objects through a thread of negligible weight. As object H now has to pull the slower object L, it will move less fast than when falling alone. On the other hand, the combined object is heavier and therefore would have to fall faster. From this contradiction, it follows that the supposition must have been false. So all objects fall with the same speed.

Galileo's version of inertial movement was based on consideration of the movement of objects on an inclined plane. Movement upwards on this plane will be decelerated, and movement downwards will be accelerated, so movement on a horizontal plane (abstraction made of resistance) will continue uniformly. Note that 'horizontal' here means parallel to the surface of the earth. Galileo's inertial motion is circular.

As he wrote in a letter:

> I argue *ex suppositione* about motion, so that even though the conclusions should not correspond to the events of the natural motion of falling heavy bodies, it would little matter to me, just as it derogates nothing from the demonstrations of Archimedes that no moveable is found in nature that moves along spiral lines. But in this I have been, as I shall say, lucky; for the motion of heavy bodies and its events correspond punctually to the events demonstrated by me from the motion I defined. (letter of 1639, quoted in Drake 1975, p. 156)

In fact Galileo had begun by defining uniformly accelerated motion as 'such motion of which the increment of speed is proportional to the distance traversed' and replaced 'distance traversed' by 'time passed' only after he had noticed that the calculations based on the former definition did *not* conform to the properties of natural accelerated motions such as free fall.

Both Galileo's method and the new science of motion to which it gave rise were indeed moulded on the typical example set by Archimedes in his treatise *On Method*. It concerns the determination of the area of the segment of a parabola by means of an analogy with statics – applying what is now known as Archimedes' law of the lever. His reasoning did not consist of a linear chain of deductive arguments, but involved a complicated network of known relations within and between figures. He 'thought of' line segments parallel to the axis of the parabola and contained within the segment of the parabola as weights and, by invoking the said network of relations, was able to 'balance' them against corresponding line segments within an inscribed triangle (T) with the same base and height as the segment. As 'all' the line segments making up the figures could thus be set in equilibrium, the segment as a whole could be balanced against the total weight (=area) of the said triangle, placed in its centre of gravity. The law of the balance then gave the ratio between the area of the segment of the parabola and that of its inscribed triangle as 4:3 (Dijksterhuis 1987, Chap. X) (for a more detailed account of this and still other examples of mathematical thought experiments, see Glas 1999).

Archimedes regarded his method as a way of exploring, not of proving. However, the reason for this was not that it involved mechanical notions, but only that taking areas to be made up of line segments lacked demonstrative force. In his treatise *Quadrature of the Parabola* (Dijksterhuis op.cit., Chap. XI), he once more demonstrated the same theorem by means of statical considerations, but this time without 'summing' line segments, and here the argument was presented as a geometric proof that satisfied all requirements of exactitude.

Still, the thought experiment was of vital importance not only for the discovery of the mathematical proposition but for its final justification (proof) as well – which makes it a nice example of Lakatos' interpretation of proofs as quasi-experiments. The ratio found by means of the balancing experiment entered in the very reasoning through which the crucial lemma for the final proof was constructed. The procedure can be reconstructed as follows:

The inscribed triangle (T) cuts off two new segments, in which triangles with the same base and height as these segments can be inscribed. Together their areas can be shown to be (1/4)T. The same procedure can be applied to the new segments thus obtained and so forth. After n steps, the total area covered by the triangles will be

$(1 + 1/4 + 1/16 + \ldots + 1/4^n)$T. Now assuming the outcome of the thought experiment to be true, the segments generated in each step will exceed their inscribed triangles by $1/3$, so that adding $(1/3)(1/4^n)$ to the last term of the series should yield the hypothesised value $(4/3)$T. And indeed the last two terms of the series $1 + 1/4 + \ldots + 1/4^n + (1/3)(1/4^n)$ add up to $(4/3)(1/4^n) = (1/3)(1/4^{n-1})$, and this added to the previous term in the series yields $(1/3)(1/4^{n-2})$, etc. The series 'eats' itself as it were from tail to head. Therefore, the sum of the whole series equals $1 + 1/3 = 4/3$. Accordingly, the area covered by the inscribed triangles after n steps equals $4/3 - (1/3)(1/4^n)$ times T (we have to subtract the $(1/3)(1/4^n)$ that was first added to the series). To this lemma, found by using the outcome of the thought experiment as guiding hypothesis, the double argument 'ex absurdo' could be applied which finally delivered the exact proof.

This kind of proof required the adoption of a lemma of Euclid's: 'if from a quantity is subtracted more than half, from the remainder again more than half, and so on, it will at length become smaller than any pre-assigned quantity'. The proof then consisted in showing that the area of the segment cannot possibly be smaller than $(4/3)$T, for however small one assumes the difference to be, the term $(1/3)(1/4^n)$ can in virtue of the said lemma always be made smaller still by taking n great enough. The assumption that the area of the segment would be greater than $(4/3)$T is refuted in a similar fashion. Therefore the area of the segment cannot possibly be either greater or smaller than $(4/3)$T; hence it necessarily must be exactly $(4/3)$T.

As already said, it was the thought experiment itself that delivered the necessary tools for the construction of a rigorous proof. It was not just a suggestive aid in discovery but also delivered essential structuring and guiding assumptions for the construction of the crucial lemma for the final proof. Heuristic and justificatory procedures are complementary, the former necessitating the latter and the latter depending on the former for their crucial structuring and guiding assumptions.

Mathematics does proceed through 'trying out and testing' but also in a somewhat wider, less theory-centred sense than envisioned by Lakatos. Not all scientific experiments are tests of theories, and the same is true of thought experiments. Not all trials and tests in mathematics are aimed at proving and refuting propositions – as Lakatos would have it – nor are discovery and innovation confined to the preformal or pre-axiomatic phases of a mathematical theory (this point is also argued by Corfield 1997, 1998). Thought experimentation also is a major analytic tool for conceptual development and change by bringing to light unsuspected connections between different scenes of inquiry, which enable progress to more comprehensive, integrated and unified theories.

Although the prefix 'thought' might suggest otherwise, thought experiments need not literally be performed in thought in the sense of involving mental representations or images. Archimedes 'thought of' line segments as possessing weight, but this is not essentially different from normal geometric practice, in which lines, for instance, are 'thought of' as possessing length but not breadth, etc. The way in which Archimedes conducted his statical argument was in all respects similar to the construction of a geometric proof. Archimedes also based his statics proper on postulates – not empirical generalisations – in exactly the way in which Euclid had axiomatised plane geometry and in which Galileo much later founded his new science of motion.

In mathematics and science alike, (thought)experiments are attempts *at once* to prove a theory *and* to *im*prove it. Proofs in informal mathematics do not justify accepting a result unconditionally, but they do justify accepting it provisionally, until it is improved by a new thought experiment. The improvement is a 'refutation' of the previous result only in the sense that it shows it to be lacking in generality and scope, deficient in unifying, explanatory and problem-solving capacity as compared with the new result, which not only implies a better proof but a better theorem as well.

23.6 Popper's Quasi-Empiricist View of Mathematics

Popper had not originally intended his methodology of conjectures and refutations to apply to mathematics, but he was delighted with Lakatos' having it thus applied (see, for instance, Popper 1981, pp. 136–137, 143, 165). One typical statement that shows Popper's endorsement of Lakatos' quasi-empiricist philosophy of mathematics is the following:

> The main point here I owe to Lakatos' philosophy of mathematics. It is that mathematics (and not only the natural sciences) grows through the criticism of guesses, and bold informal proofs. (*ibid.* p.136)

Lakatos not only applied Popper's method in a domain which Popper had not envisaged but also brought a much larger part of the Popperian corpus to bear on mathematics, especially the doctrine of the relative autonomy of objective knowledge, which is part and parcel of the 'dialectic' of *Proofs and Refutations* and hence of quasi-empiricism. It is this objectivism, of course, that made him in the said work to focus on the rationally reconstructed history of problems, theoretical proposals, critical arguments and so on and to relegate the 'real' history of the thoughts and ideas of 'real' mathematicians to the footnotes.

It is in his discussion of Brouwer's intuitionism (Popper 1981, p. 134f) that we get a clear idea of the implications of Popper's objectivist philosophy for mathematics. Brouwer had been right in insisting that mathematics is a human creation or invention, but failed to see that it is also partially autonomous. Popper's 'epistemology without a knowing subject' is an account of how mathematics can be man-made *and* relatively autonomous at the same time, that is, how mathematical objects can be said in a way to exist objectively *although* they are human creations (for a more detailed account of Popper's philosophy of mathematics, see Glas 2001 and 2006).

Brouwer's 'primal intuition of time' cannot be an authoritative source of knowledge, simply because there are no authoritative sources of knowledge. Any proposed rock-bottom principle, whether self-evidence, indubitability, primal intuition or whatever – introduced to stop the infinite regress of proofs – would make critical discussion impossible when there is no agreement on such a principle. But different people at different times happen to have quite different intuitions about what is self-evident and what is indubitable. On Popper's view, intuition is a culture- and time-dependent phenomenon which *changes* with the development of science and the use of argumentative language:

> discursive thought (i.e., sequences of linguistic arguments) has the strongest influence upon our awareness of time, and upon the development of our intuition of sequential order. (*ibid.*, p. 138)

The objectivity of mathematics is inseparably linked with its criticisability, and therefore with its linguistic expression:

> Language becomes the indispensable medium of critical discussion. The objectivity, even of intuitionist mathematics, rests, as does that of all science, upon the criticizability of its arguments. (*ibid.*, pp. 136–7)

> Though originally constructed by us, the mathematical objects (the objective contents of mathematical thought) carry with them their own unintended and sometimes undreamt-of consequences. For instance, the series of natural numbers – which is *constructed* by us – creates prime numbers – which we *discover* – and these in turn create problems which are certainly not our own invention. *This is how mathematical discovery becomes possible.* Moreover, the most important mathematical objects we discover . . . are *problems* and new kinds of *critical arguments.* (*ibid.*, p. 138)

It is in this sense that mathematical objects and problems may be said to have an independent and 'timeless' existence (i.e. irrespective of when, if ever, people become aware of them). Like Plato, Hegel and others, Popper used mathematics as the paradigmatic example of the relative autonomy of the world of intelligibilia (all that which can be an object of thought), which he called the 'third world'. It is not a static but a developing realm which

> has grown far beyond the grasp not only of any man, but even of all men (as shown by the existence of insoluble problems). (Popper 1981, p. 161)

But of course Popper was neither a Platonist nor a Hegelian. In sharp contrast to Hegel and Plato, he tried to bring the (third) world of objective ideas down to earth and to analyse its relationships with the physical (first) and the mental (second) world. Plato's world of ideas was inhabited by perfect and unchanging concepts in themselves; Popper's third world is man-made, imperfect and ever changing, consisting not of immutable concepts but of fallible theories, problems and arguments. Hegel's 'objective mind' was changing, too, but entirely of its own accord, following the dialectic of thesis, antithesis and synthesis, in which physical, mental and logical processes were considered 'identical'. Popper's third world, to the contrary, is the evolutionary product of the rational efforts of humans, who by trying to *eliminate* contradictions in the extant body of knowledge produce new theories, arguments and problems. Far from being 'identical' with the mental world, let alone the physical world, the third world *interacts* with them, but only through one or more subject's being aware (perhaps erroneously) of third-world relationships.

Popper's book *Objective Knowledge* dates from 1972, but essential parts of it had already been published in 1968, whereas the underlying *objectivist* epistemology had of course been paramount in Popper's works from the very beginning (e.g. Popper 1972, pp. 31–32, 44–48). The contents of theories or statements stand in logical relationships with each other – we might, for example, ask whether a statement is compatible with a theory, whether propositions are consistent with each other or contradictory, whether one is the deductive consequence of another or others, whether an inference is valid or invalid and so on. All these are *objective* questions; they are independent of the mental states (belief, conviction, doubt, etc.) that persons can have with respect to the contents involved. Objective contents thus possess various properties and relationships that are independent of anybody's being aware of them, and it is these objective features that are the concern of the objectivist theory of knowledge. It is of course perfectly well possible, and also highly interesting, to study knowledge as a mental phenomenon, in which case we are engaged in empirical scientific (psychological) inquiry (but here also it is advisable to take the objective features of knowledge into account, cf. Popper 1981, p. 149). Philosophical epistemology, on the other hand, is 'epistemology without a knowing subject': it studies objects and relationships in the third world, which consists of the *products* of human mental efforts, products which (by their unforeseeable and incalculable objective entailments) transcend the grasp of their producers and hence come to exist as objective artefacts.

It is of course trivially true that knowledge in the said objective sense can subsist without anybody being aware of it, for instance, in the case of totally forgotten theories that are later recaptured from some written source. It also has significant effects on human consciousness – even observation depends on judgements made against a background of objective knowledge – and through it on the physical world (for instance, in the form of technologies). Human consciousness thus typically acts as a mediator between the abstract and the concrete, or the world of culture and the world of nature. To acknowledge that linguistically expressed knowledge can subsist without humans, that it possesses independent properties and relationships, and that it can produce mental and also – indirectly – physical effects, is tantamount to saying that it in a way exists. Of course, it does not exist in the way in which we say that physical or mental objects or processes exist: its existence is of a 'third' kind.

Popper's insisting upon the crucial distinction between the objective (third-world) and the subjective (second-world) dimension of knowledge enabled him also to overcome the traditional dichotomies between those philosophies of mathematics that hold mathematical objects to be human constructions, intuitions, or inventions, and those that postulate their objective existence. His tripartite epistemology accounts for how mathematics can at once be autonomous *and* man-made, that is, how mathematical objects, relations and problems can be said in a way to exist independently of human consciousness *although* they are products of human (especially linguistic) practices. Mathematics is a human activity, and the product of this activity, mathematical knowledge, is a human creation. Once created, however, this product assumes a partially autonomous and timeless status (it 'alienates' itself from its creators, as Lakatos would have it), that is, it comes to possess its own objective, partly unintended and unexpected properties, irrespective of when, if ever, humans become aware of them.

Popper regarded mathematical objects – the system of natural numbers in particular – as products of human language and human thought: acquiring a language essentially means being able to grasp objective thought *contents*. The development of mathematics shows that with new linguistic means new kinds of facts and in particular new kinds of problems can be described. Unlike what apriorists like Kant and Descartes held, being human constructions does not make mathematical objects completely transparent, *clair et distinct*, to us. For instance, as soon as the natural numbers had been created or invented, the distinctions between odd and even, and between compound and prime numbers, and the associated problem of the Goldbach conjecture came to exist objectively: Is any even number greater than 2 the sum of two primes? Is this problem solvable or unsolvable? And if unsolvable, can its insolvability be proved? (Popper 1984, p. 34). These problems in a sense have existed ever since humankind possesses a number system, although during many centuries nobody had been aware of them. Thus we can make genuine *discoveries* of independent problems and new hard facts about our own creations, and of objective (not merely intersubjective) truths about these matters.

Nothing mystical is involved here. On the contrary, Popper brought the Platonist heaven of ideal mathematical entities down to earth by characterising it as objectivised *human* knowledge. The theory of the third world at once accounts for the working mathematician's strong feeling that she or he is dealing with something real, and it explains how human consciousness can have access to abstract objects. These objects are not causally inert: for instance, by reading texts we become aware of some of their objective contents and the problems, arguments, etc., that are contained in them, so that the Platonist riddle of how we can gain knowledge of objects existing outside space and time does not arise.

Cultural artefacts like mathematics possess their own partially autonomous properties and relationships, which are independent of our awareness of them: they have the character of hard facts that are to be *discovered*. In this respect they are very much like physical objects and relations, which are not unconditionally 'observable' either, but are only apprehended in a language which already incorporates many theories in the very structure of its usages. Like mathematical facts, empirical facts are thoroughly theory-impregnated and speculative, so that a strict separation between what traditionally has been called the analytic and the synthetic elements of scientific theories is illusory. The effectiveness of pure mathematics in natural science is miraculous only to a positivist, who cannot imagine how formulas arrived at entirely independently of empirical data can be adequate for the formulation of theories supposedly inferred from empirical data. But once it is recognised that the basic concepts and operations of arithmetic and geometry have been designed originally for the practical purpose of counting and measuring, it is almost trivial that all mathematics based on them remains applicable exactly to the extent that natural phenomena resemble operations in geometry and arithmetic sufficiently to be conceptualised in (man-made) terms of countable and measurable things and thus to be represented in mathematical language.

It is especially the (dialectic) idea of *interaction* and partial *overlap* between the three worlds that makes Popper's theory to transcend the foundationist programmes. Clearly, objective knowledge – the objective contents of theories – can exist only if those theories have been materially realised in texts (at world-1 level), which cannot be written nor be read without involving human consciousness (at world-2 level).

Put somewhat bluntly, Platonists acknowledge only a third world as the realm to which all mathematical truths pertain, strictly separated from the physical world; intuitionists locate mathematics in a second world of mental constructions and operations, whereas formalists reduce mathematics to rule-governed manipulation with 'signs signifying nothing', that is, mere material (first-world) 'marks'. In all these cases, reality is split up into at most two independent realms (physical and ideal or physical and mental), as if these were the only possible alternatives. Popper's tripartite world view surpasses physicalist or mentalist reductionism as well as physical/mental dualism, emphasising that there are *three* partially autonomous realms, intimately coupled through feedback. The theory of the interaction between all three worlds shows how these seemingly incompatible mathematical ontologies can be reconciled and their mutual oppositions superseded (Popper 1984, pp. 36–37; cf. Niiniluoto 1992).

To stress the objective and partly autonomous dimension of knowledge is not to lose sight of the fact that it is created, discussed, evaluated, tested and modified by human beings, nor does it imply that the role of mathematicians is reduced to passive observation of a pre-given realm of mathematical objects and structures – no more than that the autonomy of the first world would reduce the role of physicists to passive observation of physical states of affairs. On the contrary, the growth of mathematical knowledge is almost entirely due to the constant feedback or 'dialectic' between human creative action upon the third world and the action of the third world upon human thought.

Every theory, whether mathematical or scientific or metaphysical, is rational on Popper's view exactly

> in so far as it tries to solve certain problems. A theory is comprehensible and reasonable only in its relation to a given problem situation, and it can be discussed only by discussing this relation. (Popper 1969, p. 199)

In mathematics as in science, it is always problems and tentative problem solutions that are at stake:

> only if it is an answer to a problem – a difficult, a fertile problem, a problem of some depth – does a truth, or a conjecture about the truth, become relevant to science. This is so in pure mathematics, and it is so in the natural sciences. (*ibid.*, p. 230)

Popper clearly did not view mathematics as a formal language game, but as a rational problem solving activity based, like all rational pursuits, on speculation and criticism.

Although they have no falsifiers in the logical sense (they do not forbid any singular spatiotemporal statement), mathematical as well as logical, philosophical, metaphysical and other non-empirical theories can nevertheless be critically assessed for their ability to solve the problems in response to which they were designed, and accordingly improved along the lines of the *situational* logic or *'dialectic'* indicated above. In particular, mathematical and other 'irrefutable' theories often provide a basis or framework for the development of scientific theories that *can* be refuted (Popper 1969, Chap. 8) – a view which later was to inspire Lakatos' notion of scientific research programmes with an 'irrefutable' hard core. Indeed, his *Methodology of Scientific Research Programmes* (Lakatos 1978a) was largely based on insights obtained through applying (in *Proofs and Refutations*) Popper's logic of scientific discovery to mathematics – and not merely in response to Kuhn's strictures, as has too often been claimed (compare Glas 1995).

Most characteristic of Popper's approach to mathematics was his focussing entirely on the dynamics of conceptual change through the dialectic process outlined, replacing the preoccupation of the traditional approach with definitions and explications of meanings. Interesting formalisations are not attempts at clarifying meanings but at solving problems – especially eliminating contradictions – and this has often been achieved by *abandoning* the attempt to clarify, or make exact, or explicate the intended or intuitive meaning of the concepts in question – as illustrated in particular by the development and rigorisation of the calculus (Popper 1983, p. 266). From his objectivist point of view, epistemology becomes the theory of problem solving, that is, of the construction, critical discussion, evaluation, and critical testing, of competing conjectural theories. In this, everything is welcome as a source of inspiration, including intuition, convention and tradition, especially if it suggests new problems. Most creative ideas are based on intuition, and those that are not are the result of criticism of intuitive ideas (Popper 1984, p. 69). There is no sharp distinction between intuitive and discursive thought. With the development of discursive language, our intuitive grasp has become utterly different from what it was before. This has become particularly apparent from the twentieth-century foundation crisis and ensuing discoveries about incompleteness and undecidability. Even our logical intuitions turned out to be liable to correction by discursive mathematical reasoning (*ibid.* p. 70). Nothing is entirely beyond doubt.

23.7 Conclusion

As said at the outset, my discussion of quasi-empiricism has been confined to the criticist-objectivist tradition connected with the names of Lakatos and Popper. Consequently all other approaches to the philosophy of mathematics that go under this or a similar heading have been omitted. One might think in particular of physicalist approaches such as collected in Irvine (1990) and various other studies focussing on science-like aspects of mathematics and of mathematical practice (for a small selection, see 'suggested further reading'). As it would have been impossible within the available space to do these other approaches sufficient justice, I have preferred to restrict myself to the said critical tradition. This choice enabled me also to present my subject as one coherent whole, rather than getting the discussion scattered in a wide diversity of directions.

Although Lakatos' seminal work on *Proofs and Refutations* was written in the form of a fictitious classroom dialogue, it was intended to represent a rational reconstruction of a particular historical development, and most certainly not as a recommendation for the teaching of mathematics in school. By the same token, the present article does not have any direct implications for the actual practice of classroom teaching. Its educational relevance lies in the *image* of mathematics that it conveys, which might inspire teachers in their practice, for instance, by taking a less rigid stance, eschew formalism, invoke practical experiences and insights, make room for exploration, the formation and testing of conjectures, etc. One example of a proposal concerning the teaching of geometry in this manner is given by Chazan (1990).

The idea of three rather than two 'worlds' – which moreover partially overlap and interact – is educationally important in several respects. Firstly, it enables us to overcome traditional dualisms such as between realism and constructivism. It is, for instance, perfectly well possible to be at once a constructivist and a realist with respect to the objective content of mathematics. For even if they are invented by ourselves, mathematical constructions are not arbitrary, nor are they entirely transparent even to their creators. We can make genuine discoveries of entirely unsuspected properties and relationships concerning our own creations.

Secondly, it evidently bears directly on our image of mathematics and the way in which it is culturally embedded. In order to acknowledge the social and cultural dimension of mathematics, there is no need to question the objectivity and partial autonomy of mathematical knowledge. It is sufficient to shift our focus, away from the ways in which new truths are derived, towards the ways in which new problems are conceived and approached. There is indeed much more to mathematics than mere accumulation of true statements. Mathematicians are not interested just in truths (let alone truisms), but in truths that provide answers to questions that are worthwhile and promising in the contemporary – socially and culturally contingent – scene of inquiry. In this way the quasi-empiricist trend in modern philosophy of mathematics may contribute significantly to the further humanisation of the discipline.

References

Chazan, D. (1990) 'Quasi-Empirical Views of Mathematics and Mathematics Teaching', *Interchange* 21, p. 14–23

Corfield, D. (1997) 'Assaying Lakatos's Philosophy of Mathematics', *Studies in History and Philosophy of Science* 28, p. 99–121

Corfield, D. (1998) 'Beyond the Methodology of Mathematics Research Programmes', *Philosophia Mathematica* 6, p. 272–301

Dijksterhuis, E.J. (1987) *Archimedes*, Princeton University Press

Drake, S. (1975) 'Galileo's New Science of Motion', p. 131–156 in: Monelli, M.N.R. & Shea, W.R. (eds.) *Reason, Experiment, and Mysticism in the Scientific Revolution*, Science History Publications

Glas, E. (1995) 'Kuhn, Lakatos, and the Image of Mathematics', *Philosophia Mathematica* **3**, p. 225–247

Glas, E. (1999) 'Thought Experimentation and Mathematical Innovation', *Studies in History and Philosophy of Science* 30, p. 1–19

Glas, E. (2001) 'The "Popperian Programme" and Mathematics', *Studies in History and Philosophy of Science* **32**, p. 119–137, 355–376

Glas, E. (2006) 'Mathematics as objective knowledge and as human practice', in: R. Hersh (ed.), *18 Unconventional Essays on the Nature of Mathematics,* Springer, p. 289–303

Hintikka, J. & Remes, U. (1974) *The Method of Analysis: its Geometrical Origin and its General Significance*, Reidel

Irvine, A.D. (ed.) (1990) *Physicalism in Mathematics*, Kluwer

752 E. Glas

Lakatos, I. (1963–64), 'Proofs and Refutations', *British Journal for the Philosophy of Science* **14**, p. 1–25, 120–139, 221–245, 296–342

Lakatos, I. (ed.) (1967) *Problems in the Philosophy of Mathematics*, North-Holland

Lakatos, I. (1976), (Worral, J. & Zahar, E., eds.) *Proofs and Refutations, the Logic of Mathematical Discovery*, Cambridge University Press

Lakatos, I. (1978a) (Worral, J. & Curry, G., eds.) *The Methodology of Scientific Research Programmes*, Cambridge University Press

Lakatos, I. (1978b) (Worral, J. & Curry, G., eds.) *Mathematics, Science and Epistemology*, Cambridge University Press

Niiniluoto, I. (1992) 'Reality, Truth, and Confirmation in Mathematics', p. 60–78 in: Echeverria, J., Ibarra, A. & Mormann, T. (eds.) *The Space of Mathematics*, De Gruyter

Otte, M. & Panza, M. (eds.) (1996) *Analysis and Synthesis in Mathematics: History and Philosophy*, Kluwer

Pólya, G. (1957) *How To Solve It*, Princeton University Press

Pólya, G. (1962) *Mathematical Discovery: On Understanding, Learning, and Teaching Problem Solving*, 2 vols, John Wiley and Sons

Pólya, G. (1968) *Mathematics and Plausible Reasoning Volume II: Patterns of Plausible Inference*, Princeton University Press (second edition)

Pólya, G. (1954) *Mathematics and Plausible Reasoning Volume I: Induction and Analogy in Mathematics*, Princeton University Press (second edition Pólya 1973)

Popper, K.R. (1969) *Conjectures and Refutations: The Growth of Scientific Knowledge*, 3rd ed., Routledge

Popper, K.R. (1972) *The Logic of Scientific Discovery*, 6th ed., Hutchinson

Popper, K.R. (1981) *Objective Knowledge: An Evolutionary Approach*, revised ed., Clarendon

Popper, K. R. (1983) *Realism and the Aim of Science*, ed. W. W. Bartley, Rowan and Littlefield

Popper, K.R. (1984) *Auf der Suche nach einer besseren Welt*, Piper

Suggested Further Reading

Brown, J.R. (1999) *Philosophy of Mathematics: an introduction to the world of proofs and pictures*, Routledge

Corfield, D. (2003) *Towards a Philosophy of Real Mathematics*, Cambridge University Press

Ernest, P. (1997) 'The Legacy of Lakatos: Reconceptualizing the Philosophy of Mathematics', *Philosophia Mathematica* **5**, p. 116–134

François, K. & Van Bendegem, J.P. (eds.) (2007) *Philosophical Dimensions in Mathematics Education*, Springer

Gavroglu, K., Goudraroulis, Y. & Nicolacopoulos, P. (eds.) (1989) *Imre Lakatos and Theories of Scientific Change*, Kluwer

Hersh, R. (2006) *18 Unconventional Essays on the Nature of Mathematics*, Springer

Kampis, G., Kvasz, L. & Stöltzner, M. (eds.) (2002) *Appraising Lakatos: Mathematics, Methodology and the Man*, Kluwer

Larvor, B. (1998) *Lakatos: An Introduction*, Routledge

Oliveri, G. (1997) 'Criticism and the Growth of Mathematical Knowledge', *Philosophia Mathematica* **5**, p. 228–259

Oliveri, G. (2006) 'Mathematics as a Quasi-Empirical Science', *Foundations of Science* **11**, p. 41–79

Tymoczko, T. (ed.) (1986) *New Directions in the Philosophy of Mathematics*, Birkhäuser

Van Kerkhove, B. (ed.) (2009) *Mathematical Practices: Essays in Philosophy and History of Mathematics*, World Scientific

Van Kerkhove, B. & Van Bendegem, J.P. (eds.) (2007) *Perspectives on Mathematical Practices: Bringing together Philosophy of Mathematics, Sociology of Mathematics, and Mathematics Education*, Springer

Zheng, Y. (1990) 'From the Logic of Mathematical Discovery to the Methodology of Scientific Research Programmes', *British Journal for the Philosophy of Science* **41**, p. 311–399

Eduard Glas now retired, was associate professor of history, philosophy and social relations of mathematics at the Institute of Applied Mathematics of Delft University of Technology. He received his M.Sc in biochemistry and history of science at Amsterdam Free University and his Ph.D. in philosophy of science at Groningen State University.

Some of his publications are as follows: *Chemistry and Physiology in their Historical and Philosophical Relations* (Delft University Press: 1979). 'Bioscience between Experiment and Ideology', *Studies in History and Philosophy of Science* 14(1983), pp. 39–57. 'Testing the Philosophy of Mathematics in the History of Mathematics', *Studies in History and Philosophy of Science* 20(1989), pp. 115–131, 157–174. 'Kuhn, Lakatos, and the Image of Mathematics', *Philosophia Mathematica* **3**(1995), pp. 225–247. 'Popper as a Philosopher of Mathematics', in: I. Jarvie, K. Milford and D. Miller (eds.) (2006), *Karl Popper: A Centenary Assessment, Vol. III*, Ashgate, pp. 37–45.

Chapter 24
History of Mathematics in Mathematics Teacher Education

Kathleen M. Clark

24.1 Introduction

The use of the history and in related ways philosophy of mathematics in teaching mathematics has been the subject of discussions in everything from the didactics of mathematics in primary and secondary mathematics teaching to its appropriate role in the education of teachers of mathematics at the primary and secondary[1] level. Given the narrower literature base in this field, the ways in which philosophy of mathematics plays a role in the preparation of mathematics teachers will not be addressed in this chapter. It is worth noting, however, that many contributions shed light on how philosophical perspectives can accurately capture and describe the development of mathematical thinking. For example, in a recent description of his research, Radford (2012) claimed that "algebraic thinking cannot be reduced to an activity mediated by notations" (p. 690). Furthermore, Radford has also described ways in which historical-epistemological analyses "provide us with interesting information about the development of mathematical knowledge within a culture and across different cultures," as well as information about "the way in which the meanings arose and changed" (Radford 1997, p. 32).

The chapter is organized into seven sections. First, a brief overview of arguments that advocate for the use of history in mathematics education and the research perspectives that correspond to this advocacy are presented. In Sects. 24.2 and 24.3, descriptions of the role that history of mathematics has played in mathematics teacher education in the United States (Sect. 24.2) and elsewhere (Sect. 24.3) are given. Section 24.4 elaborates on the reasons for using history of mathematics in

---

[1] Although various locations around the world may use "primary" and "secondary" differently, in this chapter, "primary" level corresponds to the school years or grade levels for pupils aged 5–11 and "secondary" level corresponds to years or grade levels for pupils aged 12–18.

K.M. Clark (✉)
School of Teacher Education, Florida State University, Tallahassee, FL, USA
e-mail: kclark@fsu.edu

M.R. Matthews (ed.), *International Handbook of Research in History, Philosophy and Science Teaching*, DOI 10.1007/978-94-007-7654-8_24,
© Springer Science+Business Media Dordrecht 2014

756   K.M. Clark

teaching mathematics. Next, Sects. 24.5 and 24.6 discuss examples of empirical studies that were conducted with prospective teachers of primary mathematics and secondary mathematics, respectively. Finally, Sect. 24.7 outlines examples of research from the "next generation" of infusing history in mathematics education, that is, the accounts of practicing teachers who incorporated history of mathematics in teaching at the primary, secondary, and tertiary levels.

24.2   Arguing for the History of Mathematics in Mathematics Education

Lockhart (2008) posed the following questions in "A Mathematician's Lament":

> What other subject is routinely taught without any mention of its history, philosophy, thematic development, aesthetic criteria, and current status? What other subject shuns its primary sources – beautiful works of art by some of the most creative minds in history – in favor of third-rate textbook bastardizations? (Lockhart 2008, p. 9)

Although the main idea of his essay was to highlight the critical issues of a broken mathematics education system in the United States, Lockhart made frequent reference to the necessity of history in teaching and learning mathematics. In his criticism of mathematics in general and standardized curricula and assessments in particular, Lockhart complained about "the complete absence of art and invention, history and philosophy, context and perspective from the mathematics curriculum" (p. 13). Moreover, he condemned those who ask students to learn mnemonic devices (e.g., "SohCahToa" in geometry or trigonometry[2]) or "succumb to 'cutesyness'" to remember area and perimeter formulas for circles[3] (p. 9), rather than relating the real story of the development of a concept, such as "the one about [humankind's] struggle with the problem of measuring curves; about Eudoxus and Archimedes and the method of exhaustion; about the transcendence of pi" (p. 9).

Consideration of the value and importance of using history of mathematics in mathematics education has taken place over many decades. In the United Kingdom, encouragement for the inclusion of historical aspects of mathematical topics has appeared in documents for their National Curriculum – off and on – for over 100 years. Fauvel (1991) cited several excerpts from school curriculum documents, which included directives such as:

> ...[P]ortraits of the great mathematicians should be hung in the...classrooms, and that reference to their lives and investigations should be frequently made by the teacher in his lessons, some explanation being given of the effect of mathematical

---

[2] This mnemonic device, SohCahToa, is employed by school mathematics teachers (and their students) to remember the basic right triangle ratios of sine = opposite (side)/hypotenuse; cosine = adjacent (side)/hypotenuse; and tangent = opposite (side)/adjacent (side).

[3] One such mnemonic device in the form of a "cutesy" anecdote is: "Mr. C, who drives around Mrs. A, and tells her how nice his two pies are ($C = 2\pi r$) and how her pies are square ($A = \pi r^2$)" (Lockhart 2008, p. 9).

discoveries on the progress of civilization. (Report of Mathematical Association Committee 1919)

The teacher who knows little of the history of Mathematics is apt to teach techniques in isolation, unrelated either to the problems and ideas which generated them or to the further developments which grew out of them. (British Ministry of Education 1958)

The mathematics teacher has the task…of helping each pupil to develop so far as is possible his appreciation and enjoyment of mathematics itself and his realization of the role which it has played and will continue to play both in the development of science and technology and of our civilization (The Cockcroft Report 1982). (Fauvel 1991, p. 3)

Fauvel also noted, however, that beginning in the early 1990s, "the historical perspective [was] less noticeable… than in any official document about mathematics education for a century" (p. 3). The national mathematics curriculum in England today explicitly requires that the "historical and cultural roots of mathematics [be] part of the entitlement for every child's experience of mathematics" (Barbin et al. 2011, p. 37).

A similar "history" is mirrored in mathematics education documents in the United States. In the opening chapter of the thirty-first yearbook of the National Council of Teachers of Mathematics (1969), *Historical Topics for the Mathematics Classroom*, Phillip S. Jones described the struggle in using history in the mathematics classroom:

Teaching so that students understand the "whys," teaching for meaning and understanding, teaching so that children see and appreciate the nature, role, and fascination of mathematics, teaching so that students know that men are still creating mathematics and that they too may have the thrill of discovery and invention – these are objectives eternally challenging, ever elusive. (Jones 1969, p. 1)

This elusiveness may be due in part because as the editors of the yearbook themselves admitted, their goal was to "emphasize the mathematical content of the material and to leave the method of bringing it into the individual classroom in the hands of the person most qualified to make this decision – the teacher" (Baumgart et al. 1969, pp. x–xi).

The opportunities for teachers to learn history of mathematics, particularly during their pre-service teacher preparation program, are the primary reasons why variability exists in how mathematics teachers are able to use history of mathematics in teaching. And, if teachers are not afforded the opportunity to study history of mathematics during teacher preparation programs, then upon entering the teaching profession they have a minute chance, if any, to participate in formal study of history of mathematics and why it is beneficial for using in teaching mathematics.[4] This is problematic when considering national standards for mathematics teachers that call for goals aimed at providing experiences for pupils that included historical and cultural perspectives.

---

[4] This observation is made with mathematics education in the United States in mind. France is an existence proof for greater opportunities for teachers to engage in history of mathematics in this way.

In 1989, the National Council of Teachers of Mathematics (NCTM) issued the *Curriculum and Evaluation Standards for School Mathematics*. In the document, the NCTM listed "learning to value mathematics" as the first goal for students. The goal specified that

> Students should have numerous and varied experiences related to the cultural, historical, and scientific evolution of mathematics so that they can appreciate the role of mathematics in the development of our contemporary society and explore relationships among mathematics and the disciplines it serves .... It is the intent of this goal – learning to value mathematics – to focus attention on the need for student awareness of the interaction between mathematics and the historical situations from which it has developed and the impact that interaction has on our culture and lives. (NCTM 1989, pp. 5–6)

Equivalent recommendations concerning the role of the history of mathematics are found in the NCTM's *Principles and Standards for School Mathematics* (2000), with the objective of students developing an appreciation of mathematics as "being one of the greatest cultural and intellectual achievements of humankind" (p. 4). Unfortunately, this objective is couched in significantly weaker language[5] than that of the 1989 *Curriculum and Evaluation Standards* – that of developing an appreciation of certain historical and cultural attributes of mathematics as opposed to engaging in mathematics from multiple perspectives, including those along historical and cultural dimensions. In either case, there has not been overwhelming evidence that the goal or objective from either version of *Standards* is being achieved (Liu 2003, p. 418). When examining the record of conference sessions offered at NCTM annual meetings and published materials available through NCTM – particularly recent offerings – it is difficult to find evidence of the rhetoric of the 1989 and 2000 *Standards* calling for increased attention to the historical development of mathematics.

The important volume, *History in Mathematics Education: The ICMI Study* (Fauvel and van Maanen 2000), included a summary of what was known about the contribution of history of mathematics to the knowledge and perspectives of teachers and pupils. From a survey of the literature at the time, Barbin and her colleagues (2000) identified five distinct outcomes, including the ability of using history to promote changes in teachers' mathematical conceptions, the students' mathematical conceptions, the role of the teacher, the way students view mathematics, and the students' learning and understanding (p. 67). Whereas accounts exist that document each of these outcomes, Barbin was careful to observe the lack of the field's ability to document "the attainment of objectives claimed for using history" (p. 66) because large-scale assessments do not exist for such measures. Instead, qualitative methods are much more appropriate to ascertain whether using history in teaching mathematics can achieve what is claimed.

---

[5] For example, in 1989, NCTM asked for students to have "numerous and varied experiences related to the cultural, historical, and scientific evolution of mathematics," which could have entailed learning mathematics from historical methods or reinventing such methods from guided explorations using historical problems. In 2000, however, the language was simplified to focus on the appreciation of mathematics.

24.3 History of Mathematics in Mathematics Teacher Education in the United States

In a similar manner, policy documents are used to impart standards on mathematics teacher preparation programs. In the United States, the jointly constructed National Council for the Accreditation of Teacher Education (NCATE) and the NCTM program standards, *Programs for Initial Preparation of Mathematics Teachers* (2003), described content standards[6] for seven mathematical strands at both the middle and secondary levels.[7] Consequently, for programs to achieve "National Recognition" under the NCATE model, "the program report must demonstrate that at least 80 % of all indicators are addressed and at least one indicator[8] is addressed for each standard" (NCTM 2007). The final indicator for each content standard called for teacher candidates to "demonstrate knowledge of the historical development of [topics] including contributions from diverse cultures" (NCATE 2003, p. 4).

The ability for a teacher education program to exhibit mathematics teacher candidates' history of mathematics content knowledge is important to achieve national recognition; however, mathematics teacher preparation programs may still achieve the status since only 80 % of indicators are needed for national recognition. This aspect of the program standards, which was aligned with the NCTM goal for students to develop an appreciation of the cultural and intellectual achievements represented in the development of mathematics, did serve to reinforce the argument that understanding and engaging in the study of the history of mathematics contributes to the mathematical and pedagogical preparation of mathematics.

In 2011, NCTM published its draft of the new initial certification program standards for middle- and secondary-level mathematics. Instead of one indicator within each of seven content standards dedicated to the provision that mathematics teacher candidates demonstrate knowledge of the historical development of particular mathematics content, the first draft of the standards asserted one indicator only. The proposed indicator appeared in Standard 6: Mathematics Teaching and Learning and was stated as, "Equity: Recognizing the cultural diversity that exists within classrooms, valuing the contributions of various cultures in the development of mathematics, and incorporating the historical development of mathematics and culturally relevant perspectives as tools to engage students" (NCTM 2011).

---

[6] In the NCATE/NCTM program standards, "content standards" represent the different strands of mathematical knowledge teachers are responsible for knowing for teaching, such as knowledge of number and operation and knowledge of geometries.

[7] This is language of the NCATE/NCTM program standards, where "middle level" is understood as grades 6, 7, and 8 (pupils aged 12–14) and "secondary level" is understood as grades 9–12 (pupils aged 15–18). Many consider this redundant (including the author) since two divisions seem sufficient (e.g., elementary and secondary, or primary and secondary).

[8] In the NCATE/NCTM program standards, an "indicator" is a specific objective within a given content standard, such as, "Exhibit knowledge of the role of axiomatic systems and proofs in geometry" in the knowledge of geometries content standard.

After initial public response to the draft certification program standards for middle and secondary mathematics,[9] however, a new draft of the standards appeared in April 2012. In this draft as with the 2003 program standards, a final indicator was added to each of the content standards for middle and secondary mathematics. An example from middle-level algebra reads: "All middle grades mathematics teachers should be prepared to develop student proficiency with ...[the] historical development and perspectives of algebra including contributions of significant figures and diverse cultures" (NCTM 2012). The current draft standards remained open for public comment until June 2012, and the final versions of the program standards were presented to the NCATE Specialty Area Studies Board in October 2012.

At the same time that the standards for mathematics teacher certification programs in colleges and universities in the United States are being rewritten, another influential document, *The Mathematical Education of Teachers* (or MET1, published in 2001) is also under revision. The draft of MET1 was published in early 2012 and public comment was accepted until the end of April 2012. MET1 (Conference Board for the Mathematical Sciences (CBMS) 2001) argued that prospective mathematics teachers could improve their knowledge of the history of mathematics as one way for them to "undertake, and then be able to challenge their students in ways that will lead them to reason and make sense of mathematics" (p. 99). Recommendations found in MET1 called for the inclusion of historical content in the preparation of both middle grades and high school mathematics teachers and focused on providing the means for prospective teachers "to develop an eye for the ideas of mathematics that will be particularly challenging for their students" (p. 126). The primary mode of preparation advocated in MET1 was in the form of undergraduate courses in teacher preparation programs. However, the recommendations articulated in MET1 were not intended to outline potential routes for taking non-university level courses. Instead, the intention was that institutions of higher learning would develop programs and courses to meet the needs of teacher candidates within their own context while attending to the recommendations in MET1.

The role of history of mathematics in secondary mathematics teacher preparation is also strongly articulated in *The Mathematical Education of Teachers II* (MET2) (CBMS 2012). MET2 suggests that preparation programs for middle grades teachers include 24 semester hours of mathematics courses, some of which are courses to "strengthen prospective mathematics teachers' knowledge of mathematics and broaden...understanding of mathematical connections..." (p. 46). Furthermore, the writers claimed that "a history of mathematics course can provide middle grades teachers with an understanding of the background and historical development of many topics" (p. 48).

A course in history of mathematics is identified in MET2 as essential for future high school mathematics teachers:

> The history of mathematics can either be woven into existing courses or be presented in a course of its own. In both instances, it is important that the history be accurate; instructors

---

[9] Although many consider middle grades mathematics to be included in "secondary," these are the terms that NCTM uses.

who have no contact with historians need to be aware that findings from historical research may contradict popular accounts…. It is particularly useful for prospective high school teachers to work with primary sources. Working with primary sources gives practice in listening to "wrong" ideas. Primary documents show how hard some ideas have been, for example, the difficulties that Victorian mathematicians had with negative and complex numbers helps prospective teachers appreciate how hard these ideas can be for students who encounter them for the first time.

Finally, primary documents exhibit older techniques, and so give an appreciation of how mathematics was done and how mathematical ideas could have developed. (CBMS 2012, pp. 61–62)

MET2 also recognized the role of additional study in the history of mathematics for those preparing to teach high school mathematics:

Many topics in the history of mathematics are closely related to high school mathematics, for example, history of statistics, history of trigonometry, and history of (premodern) algebra. It is important to make sure that the materials used for courses on these topics include a significant amount of mathematical content. (CBMS 2012, p. 67)

At the time of the writing of this chapter, however, the 2003 NCATE program standards were still in place, and substantial diversity exists among mathematics teacher preparation program requirements in the United States and as to whether a course on the history of mathematics should be included in such programs. For example, many programs do not include a separate history of mathematics course since programs can still be accredited by NCATE without specific attention to the historical development of mathematics. It is also possible for institutions to elect to not pursue program accreditation through the organization. Instead, teacher preparation programs within colleges and universities may opt for state accreditation only or follow other accreditation standards, such as those of the Interstate Teacher Assessment and Support Consortium (InTASC).

Alternatively, mathematics teacher preparation programs offer history of mathematics as an elective, with such courses often focusing more on the mathematical content and less on the cultural, philosophical, historical, and pedagogical elements. Still other programs, such as the UTeach program at the University of Texas at Austin, as well as its 34 replication sites, require that prospective mathematics teachers take "Perspectives on Science and Mathematics"[10] with fellow prospective science teachers.

In an effort to describe the extent to which mathematics teacher preparation programs include history of mathematics, information provided by universities and colleges was used to survey mathematics teacher preparation programs for their requirement (or not) of a history of mathematics course.[11] Two sources of information were used. First, a search using the Carnegie Foundation for the Advancement of Teaching (http://classifications.carnegiefoundation.org/lookup_listings/institution.php)

---

[10] The number of replication sites as of May 2013. Also, since faculty called upon to teach "Perspectives" are often historians of science, the course privileges a "history of science" perspective. Consequently, the breadth of the historical implications of school mathematics or nature of mathematics that students take away from such a course is in need of further research.

[11] I am grateful to Christopher Thompson, my graduate research assistant, for his invaluable assistance in collecting and analyzing this information in 2011.

returned a data set of 1713 universities or colleges whose basic classification was either (1) a research (doctoral degree granting) university (RU/VH, RU/H, or DRU)[12] or (2) a university that offered master's degrees (L, M, or S)[13] or (3) a university or college that offered only bachelor's degrees.

Next, the set of 1713 institutions were stratified by state (including the District of Columbia), and the indicator function I was composed with a value of "1" indicating that the institution included a mathematics education program and a value of "0" for those institutions not offering a mathematics teacher preparation program.[14] Institutions that offered a degree program in mathematics education were determined from searching the CollegeBoard (http://collegesearch.collegeboard.com/search/index.jsp) College MatchMaker database[15] and selecting the categories of "Education" and then "mathematics education" as search criteria. The search returned 569 institutions. Although a large-scale survey (of these 569 institutions) would have been optimal, the goal was to provide a snapshot of whether mathematics teacher preparation programs required or even offered a history of mathematics course. Consequently, institutions were randomly selected from the stratified sample ($N = 569$) – three institutions from each state and the District of Columbia, which constitutes one each corresponding to the three types of institutions according to the Carnegie levels (doctoral, master's, bachelor's).

The results of the survey revealed that of 153 potential institutions, the stratified random sample returned only 15 that did not offer a mathematics teacher preparation program meeting the criteria. Of the remaining sample, 62 programs representing 37 states required a history of mathematics course for their program. Also variable was the number of institutions that offered elective or optional history of mathematics courses – or no course at all – as well as whether the mathematics teacher preparation programs were housed in education or mathematics departments.

The United States is not alone in the variety of ways in which recommendations to include history of mathematics in mathematics teacher preparation programs are implemented. Elsewhere, there is evidence of a wide variety of established practices with regard to the role of the history of mathematics in the preparation of mathematics teachers.[16] In *History in Mathematics Education: The ICMI Study* (Fauvel and van Maanen 2000), an entire chapter was dedicated to the presence of history of mathematics in programs for trainee (i.e., prospective) teachers.

---

[12] The Carnegie basic classifications are RU/VH = Research Universities (very high research activity); RU/H = Research Universities (high research activity); DRU = Doctoral/Research Universities.

[13] These Carnegie basic classifications are Master's L = Master's Colleges and Universities (larger programs); Master's M = Master's Colleges and Universities (medium programs); Master's S = Master's Colleges and Universities (smaller programs).

[14] Only initial teacher certification programs at the undergraduate level were considered.

[15] CollegeBoard College MatchMaker database (http://collegesearch.collegeboard.com/search/index.jsp) was last accessed on 10 October 2010. The database has been replaced with a much more student-friendly website, BigFuture (https://bigfuture.collegeboard.org/), last accessed 27 December 2012.

[16] For the purposes of this chapter, we only consider initial teacher preparation programs, that is, undergraduate (tertiary) at the university or college level or postgraduate programs.

24.4 History of Mathematics in Mathematics Teacher Education Around the World

In the 2000 ICMI Study, Schubring and colleagues (2000) described the state of history of mathematics for teachers in the world based upon information available at the end of the twentieth century. In addition to accumulating prior views and documenting prominence of history of mathematics in mathematics teacher education, the authors described the "state of teaching history of mathematics to future mathematics teachers" (Schubring et al. p. 94) within several countries considered to be "a fairly representative sample" (Schubring et al. p. 94). Schubring and his colleagues also claimed that whereas history of mathematics in the mathematical education of any person pursuing a degree in mathematics was once more commonly found in locations where there existed "an extended tradition in mathematics history and a considerable mathematical community" (Schubring et al. 2000, p. 94), this was no longer the case.

Gathering information about cases beyond those available at the time of the ICMI Study proves difficult, however. For example, several attempts to contact those with keen interest in the history and pedagogy of mathematics (HPM) in 2012, particularly with regard to the role of HPM in mathematics teacher education, were left unanswered. In another attempt to gather such information at the International Congress on Mathematical Education (ICME) in Seoul, South Korea, in July 2012, it became apparent that many locations around the globe struggle with incorporating such courses in mathematics teacher education programs.

Schubring and his colleagues (2000) identified a wide variety of examples of the extent to which a historical component was a part of preparation of mathematics teachers. To facilitate the discussion of differences, they divided the examples (by country) into three types[17]: countries with smaller mathematical communities but which experienced success in establishing strong records in teaching history of mathematics, countries with a longer tradition of research and teaching in mathematics history but which struggle with establishing a historical component within mathematics teacher education programs, and countries on the "periphery."[18] So as not to repeat the work of Schubring and colleagues, examples of what is most recently known about the role of history of mathematics in mathematics teacher education in several different contexts are described here for a small sample of countries around the world.

---

[17] Only European countries were discussed in the analysis with regard to the first and second types of country identified in the ICMI Study.

[18] The ICMI Study defined countries on the periphery as those "where, comparatively recently, historians of mathematics, or mathematics educators with a strong interest in mathematics history, have achieved an academic position where they are able to introduce mathematics history courses into teacher training" (Schubring et al. 2000, p. 94).

24.4.1 Europe

At the Sixth European Summer University in Vienna in 2010, a panel was held on "The Role of the History and Epistemology of Mathematics in Teachers Training" (Barbin et al. 2011). Each panelist discussed the current status of both pre-service and in-service mathematics teacher education in their country. Barbin detailed the situation in France, which has always been considered to have a strong focus on history and epistemology in teacher training. Until 2010, the training of prospective mathematics teachers took place at each University Institute for Teachers Training (IUFM). Examples of the training for teachers that took place at the institutes included a 30-h course comprised of content in the history of mathematics. Currently, all future teachers must obtain a master's degree, and the course of study includes history and epistemology of mathematics.[19]

In Italy, as is the case in many other locations, primary teachers teach mathematics along with other subjects and obtain their university degree in an educational department, and there are no formal courses on history of mathematics. Although some university programs do include history of mathematics as components of other courses, this practice is not standardized. Secondary teachers who will teach lower secondary mathematics (to pupils aged 11–14) obtain a degree in science, mathematics, physics, or chemistry. Those who will teach upper secondary mathematics (to pupils aged 14–19) are required to obtain a degree in mathematics or physics. Italy enjoys a long tradition of a community of mathematicians who share a strong interest in primary and secondary school teaching. Furinghetti reported that "for about the last 50 years, the curriculum of mathematics in Italian universities encompasses special courses addressed to prospective teachers" (Barbin et al. 2011, p. 28); the content of some of these courses includes history of mathematics.

Austria[20] has a strong tradition of requiring history of mathematics course work for prospective teachers; it is a compulsory course at two of the country's seven universities and either an elective or optional course at five universities. Table 24.1 displays the seven universities and the category of requirement applicable to each. Regardless of level of requirement, course lecturers are free to construct a course in the history of mathematics as they choose.

24.4.2 Africa

The Moroccan situation is an example of the changing role of history of mathematics within the preparation of prospective mathematics teachers. Previously, the training of teachers was under the direction of specific educational institutes that were mainly involved with the educational and didactical dimensions of pre-service

---

[19] There exists variability in the importance placed on teaching these subjects in France.

[20] The author is grateful to Manfred Kronfellner for providing this information.

24 History of Mathematics in Mathematics Teacher Education

Table 24.1 Types of history of mathematics courses in Austria: compulsory, elective, or optional

University	Required	Elective or optional[a]
University of Innsbruck	Compulsory course, lecture, with final examination	
University of Klagenfurt	Compulsory course, seminar, with immanent assessment	
Vienna University of Technology		Elective course, lecture
University of Vienna		Elective course, lecture
Johannes Kepler University of Linz		Elective course, seminar
University of Salzburg		Elective course, lecture, and seminar combination
University of Graz		Optional course, when available

[a]"Elective course" means that history of mathematics is one of several courses students must select from a collection of options. They must select a certain number of elective courses from the collection, but they do not have to select all in their course of study. An "optional course" may not always be available as an elective option

teacher training. Consequently, the history of mathematics held a reduced role and was evoked during the study and didactical analysis of concepts. A few exceptions existed. For example, in the *École Normale Supérieure* of Marrakech, courses on the history and philosophy of mathematics were always taught. Currently and according to the new reforms in Morocco, the training institutes for prospective secondary teachers were moved to universities. This aspect of the reforms enables universities to organize and implement courses for prospective teachers and, as a result, aids in providing more substantial pedagogical training of teachers. In 2012 prospective teacher training programs were under revision, and those involved with the work are hopeful that the historical and cultural dimensions of mathematics will receive particular attention and that such dimensions will help develop and reinforce learning of scientific concepts, citizenship, and critical thinking of students.[21]

24.4.3 Asia

Whereas much of the Western world seeks to increase attention to mathematical contributions of non-Western cultures in history of mathematics courses (and history of mathematics courses for prospective mathematics teachers), a similar phenomenon occurs in many Asian contexts. In many Asian countries there are efforts to highlight Western developments and to compare them with methods, algorithms, and examples found in ancient texts and manuscripts. However, well-developed courses or units within courses in mathematics teacher preparation programs are still absent, as in

---

[21] The author is grateful to Abdellah El Idrissi for providing this information.

the case of South Korea.[22] A trend that may prove to be influential in the near future is the increased attention to history of mathematics in the preparation of mathematics teachers in South Korea (and other proximal countries), especially after the introduction of the International Study Group on the Relations between the History and Pedagogy of Mathematics (HPM Group) to several mathematicians and mathematics teacher educators as a result of HPM 2012 in Daejeon.

24.5 Development of Theoretical Claims: Why Use History in Teaching Mathematics

Although there has been strong interest in the question of how history of mathematics benefits teachers and learners of mathematics since the 1890s (Fasanelli 2001), significant international activity directed at addressing the question began in the 1970s. Key to this activity was the creation of the International Study Group on the Relations between the History and Pedagogy of Mathematics (HPM Group), which was officially established as a satellite group of the International Congress on Mathematical Education (ICME) in 1972. Henk Bos, Barnabas Hughes, Phillip Jones, Leo Rogers, and Roland Stowasser were among the group of mathematicians, mathematics historians, and mathematics educators gathered at the first sessions held in association with in 1976 at ICME-3 in Karlsruhe, Germany. As a result of these initial sessions, the HPM Group was established and the official aims of the Study Group were established. The group sought to:

1. Promote international contacts and exchange information concerning:

    (a) Courses in history of mathematics in universities, colleges, and schools
    (b) The use and relevance of history of mathematics in mathematics teaching
    (c) Views on the relation between history of mathematics and mathematical education at all levels

2. Promote and stimulate interdisciplinary investigation by bringing together all those interested, particularly mathematicians, historians of mathematics, teachers, socialscientists, and other users of mathematics
3. Encourage a deeper understanding of the way mathematics evolves and the forces that contribute to this evolution
4. Relate the teaching of mathematics and the history of mathematics teaching to the development of mathematics in ways that assist the improvement of instruction and the development of curricula
5. Produce materials that can be used by teachers of mathematics to provide perspectives and to extend critical discussion of the teaching of mathematics
6. Facilitate access to materials in the history of mathematics and related areas
7. Promote awareness of the relevance of the history of mathematics for mathematics teaching in mathematicians and teachers

---

[22] The author is grateful to Sang Sook Choi-Koh for describing the South Korean context.

24 History of Mathematics in Mathematics Teacher Education

8. Promote awareness of the history of mathematics as a significant part of the development of cultures (Fasanelli 2001, p. 2)

The influence of the articulated aims can be found in the preponderance of theoretical literature published during the first 25 years of the existence of the HPM Group. For example, Fauvel (1991) provided a list of 15 reasons that are used to promote using history in mathematics education:

> ... helps to increase motivation for learning; gives mathematics a human face; historical development helps to order the presentation of topics in the curriculum; showing pupils how concepts have developed helps their understanding; changes pupils' perceptions of mathematics; comparing ancient and modern establishes value of modern techniques; helps to develop a multicultural approach; provides opportunities for investigations; past obstacles to development help to explain what today's pupils find hard; pupils derive comfort from realizing that they are not the only ones with problems [with mathematics]; encourages quicker learners to look further; helps to explain the role of mathematics in society; makes mathematics less frightening; exploring history helps to sustain [teacher] interest and excitement in mathematics; and provides opportunity for cross-curricular work with other teachers or subjects. (Fauvel 1991, p. 4)

In the current educational context that often includes a heightened emphasis on standards and high-stakes accountability, many of Fauvel's reasons are easy to ignore – particularly if educators perceive the actions of the proposed reasons to not be aligned to content and practices of the curriculum or assessments. Examining one reason as an example, most mathematics teachers agree that "showing pupils how concepts have developed" may have a role in pupils' understanding of particular mathematical concepts (e.g., operations with integers, complex numbers). However, mathematics teachers may be skeptical that employing historical methods to show this to pupils is a viable pedagogical tool.

Furinghetti (2002) observed that history can be used "as a mediator to pursue the objectives of mathematics education" (Abstract). Furthermore, Furinghetti proposed that enabling students to work with topics at an informal level before formally investigating topics was similar to Freudenthal's view that contextual problems provide efficient opportunities to allow formal mathematics to emerge. In this way Furinghetti claimed using "history may reveal itself fruitful and [a] sense-carrier" (p. 3).

There is important evidence that the membership and interested colleagues of the HPM Group are making progress towards achieving many of the official aims. Not only do the Topic Study Groups of the International Congress meetings (since 1972) and the HPM Satellite meetings produce peer-reviewed, published proceedings, but the European Summer University on the History and Epistemology in Mathematics Education meetings (now held every 2 years, not including the years in which the ICME and HPM Satellite meetings are held)[23] and now a working group of the Congress of the European Society for Research in Mathematics Education (CERME)

---

[23] The European Summer University on the History and Epistemology in Mathematics Education (ESU) was held every 3 years from 1993 until 2010. Subsequent ESUs will be held every 4 years (e.g., the Seventh ESU will be in 2014), but not in years when ICME meetings are held.

since 2009 also add "substantially to the amount of papers on history in mathematics education" (Jankvist 2012, p. 296). Moreover, special publications have appeared as a result of increased attention to the inclusion of history of mathematics in mathematics education. Among these are several books published by The Mathematical Association of America (e.g., *Using History to Teach Mathematics: An International Perspective* (Katz 2000), *Recent Developments on Introducing a Historical Dimension in Mathematics Education* (Katz and Tzanakis 2011)) and other important volumes such as *History in Mathematics Education: The ICMI Study* (Fauvel and van Maanen 2000) and *Crossroads in the History of Mathematics and Mathematics Education* (Sriraman 2012).[24]

Even with this increased attention, however, many contributions to these edited volumes are what Siu and Tzanakis referred to as "propagandistic" (2004, p. vii); that is, it is evident that a critical mass of teachers and scholars alike attest to the worthiness of history in mathematics education. As a result, such contributions remain theoretical as opposed to providing empirical results of what happens when history of mathematics is part of the instructional program. The proportion of contributed chapters in such volumes that are focused on history of mathematics in mathematics teacher education is also a concern. For example, in Katz and Tzanakis (2011), only four of 24 chapters were devoted to this theme and none of the 25 chapters in Sriraman (2012) were.

One possible reason for the abundance of theoretical descriptions of the importance of using history of mathematics in teaching is the obstacle of adequately preparing primary and secondary mathematics teachers to use the history of mathematics in meaningful ways in their teaching. Jones (1969) claimed, "the history of mathematics will not function as a teaching tool unless the users (1) see significant purposes to be achieved by its introduction, and (2) plan thoughtfully for its use to achieve these purposes" (p. 5). Although Jones' claims were made over four decades ago, most mathematics teacher educators would find it difficult to summarize two key issues more succinctly. Furthermore, many argue that the lack of opportunity for history of mathematics course work in mathematics teacher preparation programs – whether primary or secondary – is intimately connected to the lack of strong mathematical knowledge of teacher candidates.

Finally, it is important to keep in mind that a "domino effect" may be at work here. For example, it may be difficult to identify literature describing accounts of elementary and secondary teachers incorporating history of mathematics in their mathematical instruction because of the absence of history of mathematics in mathematics teacher education.

The remainder of this chapter discusses a variety of empirical investigations that establish a premise for why so many advocate for the inclusion of history of mathematics in teaching and that motivate future investigations. First, and because the focus of the chapter is on the role of history of mathematics in mathematics teacher education, studies about the influence of studying history of mathematics on prospective primary and secondary mathematics teachers are described. Next,

---

[24] For a more comprehensive list, see Jankvist (2012).

24 History of Mathematics in Mathematics Teacher Education

accounts of the ways in which teachers – having previously studied history of mathematics to some extent – incorporate history of mathematics in their teaching are presented.

24.6 History of Mathematics in Primary Mathematics Teacher Education

The lack of guidelines that call for history of mathematics in primary mathematics teacher[25] education programs around the world makes it difficult to identify empirical investigations that involve this population. Exemplars of empirical work conducted with prospective primary mathematics teachers tend to focus on attitudes and beliefs. Primary mathematics teacher preparation programs vary in number of hours, courses, seminars, or lectures in both mathematics content and pedagogy (i.e., didactics). Additionally, primary mathematics teacher education programs typically prepare teachers responsible for multiple subject areas – and often for all academic and specialty areas (e.g., art, music). Secondary mathematics teacher education programs, however, focus on preparing teachers of mathematics and in some cases, a second subject area. Consequently, when and how history of mathematics is employed in the preparation of primary mathematics teachers varies. Much of this variability may result from program expectations for the mathematics content required for prospective primary teachers compared to the content expectations for prospective secondary teachers.

24.6.1 First Example

Fleener and colleagues (2002) studied the influence of a mathematics education curriculum that incorporated historical topics in each of three different courses on prospective elementary mathematics teachers' "meaning-making efforts" (p. 73). A series of questions were asked of prospective teachers in three courses, representing three different time points in their teacher preparation program: a mathematics course required for elementary education majors taken early in the second year of the elementary education program, the first mathematics education teaching and learning course, and the mathematics methods taken in the last semester before the teaching internship. The extent of history of mathematics in each course is given in Table 24.2.

The researchers collected responses to a variety of prompts about prospective elementary teachers' experiences with history of mathematics during the three courses.

---

[25] In general, the terms "primary mathematics teachers" and "elementary mathematics teachers" are used interchangeably to describe teachers of pupils aged 5–11 years of age. And, regardless of the term used, such programs are those that prepare generalists, or teachers who teach most if not all of the academic subjects.

Table 24.2 Course descriptions (Fleener et al. 2002)

Course description	When taken	History content
MATH 3213: General mathematics course ($n=48$)	Second semester of second year of elementary mathematics education program	Students conduct research on a topic from the history of mathematics, write a two-page paper, and deliver a brief presentation in class
EDMA 3053: First mathematics education teaching and learning course ($n=37$)	After admission to teacher education program (typically in the third year of a 4-year program)	Students conduct research on a historical figure or historical topic, write a formal paper, and prepare a one-page handout to accompany a presentation delivered to the class. "Students are required to develop activities for elementary students that incorporate an inquiry approach using historical topics or individuals" (Fleener et al. 2002, p. 75)
EDMA 4053: Math methods ($n=12$)	Last semester of course work (taken just before teaching internship)	Content includes: Select a mathematician, conduct research, role-play their mathematician. Tested on historical contributions, significance of historical figures, and topics on midterm exam. Discussions, readings, and activities to encourage planning of historical activities in future teaching

The prompts asked for prospective teachers to reflect on how learning about the history of mathematics affected their understanding of mathematics and how the preparation of their report on a historical topic or figure aided in understanding mathematics and strategies for teaching mathematics. The participants were also asked about their experiences after participation in each course, including whether they retained handouts from any peers' historical topic presentation or whether they used ideas from their own or others' presentations when preparing classroom activities.

In the analysis of the data, Fleener and her colleagues (2002) were interested in which orientation towards learning mathematics each prospective teacher favored. In particular, they examined participant responses for evidence of knowledge informed by the technical, practical, or emancipatory interests. Grundy (1987) defined the technical interest as "a fundamental interest in controlling the environment through rule-following action based upon empirically grounded laws" (p. 12): the practical (or, hermeneutical) interest as "a fundamental interest in understanding the environment through interaction based upon a consensual interpretation of meaning (p. 14), and the emancipatory (cognitive) interest as "a fundamental interest in emancipation and empowerment to engage in autonomous action arising out of authentic, critical insights into the social construction of human society" (p. 19).

Fleener and colleagues (2002) found that the prospective elementary teachers favored a technical perspective when studying history of mathematics for their own learning and for use in teaching. They also claimed that real-life experiences and historical connections may not equip future teachers with necessary tools to overcome long-held traditional beliefs about mathematics and mathematics learning, and that the prospective teachers in the study were over-reliant on algorithms, which in turn prevented them from exploring deeper meanings within mathematics. Although Fleener and her colleagues (2002) claimed that "critical and historical approaches, even sustained over several semesters of mathematics instruction, are not sufficient for students to develop an emancipatory approach" (p. 80), it is possible that the orientation of the history of mathematics instruction in the three courses favored a technical perspective as well.

Details about the topics or mathematicians that the prospective teachers selected for their research or the content of focus during the mathematics methods course were not provided by the authors, and consequently, it is difficult to further interpret the outcomes they discussed. In research where more details are provided about historical content, potential solutions to improve the ways in which history of mathematics is incorporated in the preparation of primary mathematics teachers are easier to identify, as in the case of Charalambous et al. (2009).

24.6.2 Second Example

Charalambous and colleagues (2009) quantitatively described how prospective primary mathematics teachers' attitudes and beliefs were impacted as a result of participating in a teacher preparation program that contained two content courses grounded in the history of mathematics. The courses, designed and implemented at the University of Cyprus, have been in place for more than a decade. Each three-credit-hour course lasted 13 weeks was taken consecutively, and together the courses were considered the only mathematics-oriented courses in the program for the prospective primary mathematics teachers.

There were 94 prospective primary mathematics teachers who were surveyed four times (pre- and post-surveys for each of the two courses), and six participants (a convenience sample) were interviewed for additional insight after quantitative results were compiled. The authors divided the participants into two groups, according to their acceptance into the pre-service teacher program at the University of Cyprus. The first group ("G1" in the study, with 52 participants) opted to take the mathematics entrance examination and the second ("G2," with 42 participants) did not.[26]

The authors reported that in many ways, the survey results pointed to the finding that the two courses grounded in the history of mathematics were a failure. For example, "the G2 participants exited the program with increased negative attitudes toward mathematics" (Charalambous et al. 2009, p. 177). And G2 participants were unable to "see many connections between the content and the activities of the two

---

[26] At the University of Cyprus, entrance into the preservice teacher program is highly competitive and students must take four entrance exams: one in language and three others in different subject areas of their choice.

courses and the content considered in elementary grades" (Charalambous et al. 2009, p. 177). Each of these results counters much of the rhetoric about why history of mathematics should be used in teaching mathematics. Consequently, the authors used interview data to aid in understanding the quantitative results. They found that the teacher candidates shared that their previous mathematical experiences were all too common, namely, that such experiences "did not help [them] learn how to think" (p. 174) – but merely honed their test-taking skills.

An important outcome from Charalambous and his colleagues (2009) is the identification of three limitations in the approach employed – all of which may serve future research and development of mathematics teacher education programs well. First, they recognized that not sharing the intention to ground the two courses in history of mathematics was a mistake. Second, the authors anticipated that an elementary teacher preparation program grounded in the history of mathematics would impact the participants in such a way that they would begin to view mathematics differently than they did before studying at university. Unfortunately for many of the pre-service teachers, their experiences with the two courses were too similar to their prior experience with learning mathematics: that certain complex mathematical problems were challenging and stressful and the grounding in history failed to matter in a positive way. Lastly, opportunities to genuinely experience the development of mathematics were insufficient for the six participants interviewed. Indeed, this is one of the most challenging issues with implementing history of mathematics in teacher preparation programs. Even with careful planning, knowledgeable instructors, and appropriate resources, such a course can still feel as a whirlwind of historical activity, punctuated by (in the case of many of the G2 participants) difficult mathematical ideas and methods.

Finally, there are three cautions provided by Charalambous and colleagues (2009) that must be heeded when considering similar experiences for future primary mathematics teachers. First, the decision to implement a content course for prospective mathematics teachers grounded in the history of mathematics should be shared with them. Perhaps the most important reason for this is that their prior experience with history of mathematics may be limited, and predetermined attitudes towards mathematics may serve as an obstacle for learning in the course. Second, the difficulties prospective teachers experience with the content of courses based upon the history of mathematics "need to be acknowledged and addressed" (p. 178). And third, to prevent spending insufficient time on the evolution of key mathematical ideas, the authors suggested designing "guided explorations of important mathematical ideas that will support pre-service teachers in their work of teaching" (p. 178). Furthermore, the authors recognized the potential for history of mathematics in teacher education program to contribute to the development of knowledge "that is both useful and usable for the work of teaching mathematics" (p. 179). An important means to this end is the identification of the content of the two courses Charalambous and colleagues (2009) implemented (Table 24.3).

The content of the second course identified by Charalambous and his colleagues (2009) raises an additional concern. Many of the topics identified are well beyond the content that many prospective primary mathematics teachers are familiar with,

Table 24.3 Topics for courses grounded in history of mathematics: Charalambous et al. (2009, p. 179)

Course	Unit 1	Unit 2	Unit 3	Unit 4	Unit 5
First course	First arithmetic systems Arithmetic systems with/without zero Contemporary place value system Two methods multiplication	Transition to systematic proof Applications of similar triangles Different proofs of the Pythagorean theorem Pythagorean mathematics Applications of Pythagorean mathematics in elementary school	Three famous problems of antiquity Attempts to solve the problems (and their contribution to the expansion of mathematical knowledge)	Euclid and his contributions to mathematics	Contribution of other renowned mathematicians (Archimedes, Apollonius, Pascal)
Second course	Zeno's paradox and the principle of exhaustion Development of the notion of limits Limits, differentiation, integration (including applications in real-life situations) Contribution of Newton and Leibniz to modern mathematics	The liberation of geometry: Attempts to prove the 5th Euclidean axiom When is the sum of the angles of a triangle not equal to $180°$? Bolyai and Lobachevsky and the development of hyperbolic geometry Riemann and elliptic geometry	The liberation of algebra: Peacock's contribution Hamilton and his quaternions Cayley and matrix algebra Applications in real-life situations	Sets and binary relationships: Notion of sets and associated paradoxes Operations on sets and their properties Cartesian product and its applications to elementary school	A sample of more advanced mathematics: Mathematical logic and applications Boolean algebra Computers and mathematics

and as a result, whatever knowledge prospective mathematics teachers are expected to gain through history of mathematics may not be seen as useful and usable for the work of teaching mathematics. The design of such courses cannot ignore the importance of topics most associated with the work of future primary teachers.

24.6.3 Third Example

In an effort to establish a reliable instrument to measure attitudes and beliefs towards using history of mathematics in teaching, Alpaslan and colleagues (2011) constructed, administered, and evaluated the underlying factor structure of their instrument, *Attitudes and Beliefs towards the Use of History of Mathematics in Mathematics Education Questionnaire*. The instrument was constructed from a variety of available survey questions and was administered to a purposive sample[27] of 237 pre-service primary mathematics teacher candidates in 2010–2011. The authors conducted factor analysis and identified three factors "by considering the field of history in mathematics education and…instrument development studies about social sciences" (Alpaslan et al. 2011, p. 1666): positive attitudes and beliefs towards the use of history in mathematics education, negative attitudes and beliefs towards the use of history in mathematics education, and self-efficacy beliefs towards the use of history in mathematics education. After problematic items were removed from the instrument, the same validity and reliability analyses were conducted again. The final survey contains 35 items; 22 items are included in the positive attitudes and beliefs component, nine items are contained in the negative attitudes and beliefs component, and four items are contained in the self-efficacy component.

The instrument developed by Alpaslan and his colleagues suggests important opportunities for future empirical work in the field of history in mathematics education. As they posited:

> … the results gained by using the instrument by future research projects would have valuable implications for teacher educators in different universities, education policy makers and curriculum developers of different countries in designing curriculum for mathematics education [at] different levels. (Alpaslan et al. 2011, p. 1669)

Furthermore, research regarding the problematic trends in prospective teachers' beliefs about mathematics and attitudes towards learning mathematics via the history of the important development ideas revealed by Charalambous and colleagues (2009) would also be informed by use of the instrument on a larger scale. For example, pre- and post-administrations of the instrument have the power to "reveal potential effects of interventions on the attitudes and beliefs towards the teaching approach in question" (Alpaslan et al. 2011, p. 1669). Future use of the instrument

---

[27] In qualitative research, a purposive sample is one in which a particular participant population is targeted, especially when there is a special nature of the study or participants are difficult to find. This was the case of Alpaslan's and his colleagues' research, where new reforms were in place in Turkey's higher education institutions.

24 History of Mathematics in Mathematics Teacher Education

can help reveal gaps in learning and knowing mathematics for teaching and will enable researchers to investigate ways in which history of mathematics may fill such gaps. Furthermore, the populations available to use the instrument are extensive, including pre-service teachers, in-service teachers, and with further development and modifications, primary school pupils.

24.6.4 Fourth Example

Certainly an abundance of empirical studies that rely upon survey data – and thus, self-report data – is a concern for the field of history in mathematics education. Although survey data provide important descriptive information about the research questions posed and establish the landscape of future research directions, the field requires rigorous and extensive empirical study. One promising study, proposed by Alpaslan and Haser (2012), will "provide…teacher educators with possible effective presentations of the history of mathematics knowledge for teaching mathematics" (p. 2). In the pilot study they described, Alpaslan and Haser will conduct a case study on the implementation of a "History of Mathematics" course for lower secondary prospective mathematics teachers (who will be trained to teach 6th, 7th, and 8th grade mathematics in Turkey).

In their future research Alpaslan and Haser (2012) plan to investigate the extent to which the history of mathematics content within the course is used both "as a tool" and "as a goal" (Jankvist 2009) in the course. Furthermore, they will examine whether the content of the history of mathematics course will be addressed from an illumination approach, module approach, or history-based approach (Jankvist 2009). Finally, Alpaslan and Haser will describe how the course content and experiences are reflected in the instructor's and prospective teachers' views.

The results of the study can prove promising for several reasons. First, the researchers plan for multiple data sources, including "observation of the natural course process, content, tasks, and experiences, the nature of the content-related communication between the pre-service teachers and the instructor, and the participation of pre-service teachers in the course experiences" (Alpaslan and Haser 2012, p. 3). A study on the impact and results from implementing a history of mathematics course in mathematics teacher education has not yet been treated to the extent proposed here.

Second, the potential to inform the construction or modification of courses in the history of mathematics for prospective teachers (particularly for future teachers of upper elementary and lower secondary pupils) as a result of the pilot and full-scale study is extensive. Whereas the topics of the two-course sequence implemented by Charalambous and his colleagues (2009) represent an important contribution, Alpaslan and Haser's (2012) investigation has the potential to produce a substantial publication outlining solutions to several obstacles regarding the construction, implementation, and outcomes of history of mathematics courses for prospective mathematics teachers.

Finally, Alpaslan and Haser (2012) will use Jankvist's (2009) recent and influential distinctions of "hows" and "whys" as a framework[28] for the research questions, data collection, and data analysis. The attention to considering these distinctions in investigations such as the study being conducted by Alpaslan and Haser continues to strengthen the field of history in mathematics education. The working group on "History in Mathematics Education" at the seventh Congress of the European Society for Research in Mathematics Education (CERME-7) identified several key issues as crucial for the domain of history in mathematics education, including:

> the need for developing theoretical constructs that provide some order in the wide spectrum of research and implementations done so far; to somehow check the efficiency of introducing a historical dimension; and to develop appropriate conditions for designing, realizing, and evaluating our research…. (Jankvist et al. 2011, p. 1636)

Thus, in an effort to connect content from a history of mathematics course to the extent in which tools, goals, and approaches are addressed – as well as what such content and experiences mean for developing prospective mathematics teachers – Alpaslan and his colleagues may in fact develop a research agenda that informs each of the key issues for the field.

24.6.5 Summary

This section discussed four studies pertaining to the role of history of mathematics in primary mathematics teacher education. Certainly this is not a comprehensive description of literature on research about the effect (e.g., attitudes and beliefs) or inclusion (e.g., history of mathematics courses or content within teacher preparation programs) of history of mathematics. However, there is an abundance of survey research focused on the topic in primary mathematics teacher education and to include additional descriptions is an exercise in repetition.

To be fair, extensive variability exists with respect to programs preparing future primary mathematics teachers. In the United States alone, a key variable is the mathematics content required of these future teachers. For example, it is not uncommon for primary teacher preparation programs to require only one or two lower division[29] mathematics courses for prospective elementary education candidates. In other programs, however, and in some cases for every elementary teacher preparation program in a given state in the United States, a sequence of "Mathematics for …" courses are required for those preparing to teach elementary school.[30]

---

[28] The influence of Jankvist (2009) has been extensive; see examples in Clark (2011), Kjeldsen and Blomhøj (2009), and Tzanakis and Thomaidis (2011).

[29] Lower division courses are courses taken during the first and second year at universities and colleges in the United States.

[30] In the United States, Maryland represents an example of this. Prospective elementary teachers are required to take a sequence of mathematics courses, though these may vary by institution.

The variability in mathematics course requirements for prospective elementary mathematics teachers may be the cause for the limitation of literature (either descriptive or empirical). For those preparing to teach secondary mathematics, a wide range of preparation programs also exists. However, the number of mathematics courses necessary for teaching mathematics at the secondary level (e.g., to pupils aged 12–18) is typically greater than that required for elementary teachers. Consequently, there are an increased number of possibilities within a teacher preparation program of which history of mathematics can be part.

24.7 History of Mathematics in Secondary Mathematics Teacher Education

In this section a representation of both theoretical work about and empirical studies conducted with prospective secondary mathematics teacher populations are discussed. For ease in presentation, the studies are divided into two types. First, descriptions of the perceptions of history of mathematics that is part of a teacher preparation course are given, where the influence of history of mathematics is primarily the "story" of mathematics. Secondly, examples of research are given in which the history of mathematics influenced prospective teachers' mathematical knowledge, as a result of an experience in history of mathematics as part of a teacher preparation program.[31]

24.7.1 History of Mathematics: Influences of Telling the "Story" of Mathematics

Many examples of published research literature in the field of history in mathematics education describe experiences in the history of mathematics for those preparing to teach secondary mathematics. Of these there are two main types of contributions. A large proportion of published accounts describe prospective teachers' attitudes towards history of mathematics or using history of mathematics in teaching, as was the case with research involving prospective elementary mathematics teacher populations. A second type[32] of contribution explains results anecdotally, punctuated by comments from students enrolled in the course.

At the University of Maryland, Baltimore County, for example, prospective teachers take Statistics, Mathematics for Elementary Teachers I, and Mathematics for Elementary Teachers II.

[31] In some contexts the preparation of mathematics teachers entails an undergraduate degree in mathematics, as in the case of Italy (Furinghetti 2000).

[32] Of course, it is possible to see both types represented in the same publication.

24.7.2 First Example

Burns (2010) examined how prospective secondary mathematics teachers at a small private university in the northeastern United States viewed the role of history of mathematics in the curriculum. The mixed methods study also included a brief description of key history of mathematics assignments pre-service that mathematics teachers were asked to complete as part of a secondary mathematics methods course. Post-administration of a five-question survey (when compared to pre-administration responses) indicated that they were more amenable to integrating history of mathematics in future teaching and felt more comfortable with their ability to incorporate history of mathematics.

Responses to the open-ended question, "What should the role of 'history of mathematics' play in the high school mathematics curriculum?" at the beginning of the study revealed that the majority of the prospective mathematics teachers in Burns' sample favored a minor or moderate role. Only three responses indicated a major role for history of mathematics in teaching. When the final questionnaire was administered, Burns found that the students felt more favorable, with only three students attaching a minor role to the history of mathematics. However, these results are somewhat problematic.

The secondary mathematics methods course provided "activities designed to enhance [participants'] exposure to the history of mathematics" (Burns 2010, p. 3), yet each activity described in the article focused on history as the "story of mathematics." For example, the prospective mathematics teachers in the course kept "track of mathematicians" and the mathematics they were famous for from their reading of *Fermat's Enigma* (Singh 1998), selected their "favorite mathematician" and created a short presentation about them, and finally, chose a unit within a high school mathematics course and described what history of mathematics content they would be able to incorporate into the unit. When the course activities are considered along with the pre-service mathematics teachers' responses indicating that they believed history of mathematics should have a major role in the high school curriculum, a contradiction arises. That is, if "history as story" (or, "history as anecdote")[33] is assessed as a minor way in which history is used in teaching mathematics, then the observation that "…basically I feel like history of mathematics should be 'sprinkled' on top of the lesson" (Burns 2010, p. 5) should not be coded as history having a major role in teaching.

Burns (2010) provided an example of research on exposing prospective teachers "to topics from history of mathematics and methods that could be used to teach these topics" (p. 2), but the exposure described lacked attention to strategies in which historical methods (e.g., mathematical procedures and techniques), historical problems, and primary sources are integrated when teaching mathematics. Indeed, she called attention to the need for more "to be done to develop a deeper understanding

---

[33] Three modes of using history in teaching mathematics were given in Clark (2011): history as anecdote, history as biography, and history as interesting problems.

of [history of mathematics] before students get to a methods course" (p. 7), which is intimately connected to policy and standards that recommend history of mathematics courses for prospective secondary mathematics teachers.

24.7.3 Second Example

A similar intervention in a secondary mathematics methods course in Turkey was "designed to improve competencies regarding the integration of history of mathematics" in teaching mathematics courses (Gonulates 2008). The small intervention study involved 14 senior-level students over a 14-week period during the course. The intervention required the mathematics methods students to read assigned articles, which were chosen using the criteria of Gulikers and Blom (2001). Using the criteria, articles were selected if they fulfilled the aims of the study to portray the usefulness of historical materials when teaching mathematics, as well as the variety in which history of mathematics may be incorporated in teaching. The students were also required to participate on a class discussion board weekly by submitting a report that detailed their ideas about the weekly article and proposed use of history in teaching mathematics. Finally, "brainstorming and discussion" during the final week of the intervention enabled students to share their overall ideas about using history in teaching.

Two instruments were developed for the study, an attitude scale and a questionnaire. The attitude scale asked students to "state teaching strategies or instructional procedures that they [thought] they [could] use [history of mathematics] in [mathematics teaching]," and an accompanying questionnaire asked them to provide examples for each strategy (Gonulates 2008, p. 5). The results from the two instruments revealed that although the students' attitudes about integrating history of mathematics in teaching increased, the increase was not significant. Of greater interest, however, were results related to the extent to which the pre-service teachers were able to name ways in which they could integrate history of mathematics in teaching and the quality of the examples they identified. The pretest and posttest results for the total number of teaching strategies for incorporating history of mathematics were 55 and 73, respectively; the increase in the total number of examples for the strategies identified pre- and posttest was 53 and 69, respectively. Neither increase was significant. However, the increase in the quality of the examples as judged by three different juries was significant.

Gonulates' investigation raises several important issues for the field of history in mathematics education. The students' comments on the discussion board revealed that they held the belief "that the history of mathematics could be used more for motivational than for conceptual purposes" (2008, p. 9), which is a common theme for the investigations described in this section. The view that using history promotes mathematics (Furinghetti 1997) is a frequent outcome when "history as story" dominates pre-service teachers' experience with history of mathematics. Consequently, this view presupposes that the future practice of prospective teachers will entail using history to enrich their teaching in social, affective, and cultural ways, but will not be robust enough to aid in the learning of mathematical content. A potential solution to prompt the use of history of mathematics for the purpose of reflecting on mathematics (Furinghetti 1997) is to incorporate opportunities for prospective mathematics teachers in which they experience a quantifiable change in their own learning of mathematics through historical content, resources, and methods.

24.7.4 Third and Fourth Examples: Design of "History of Mathematics" Courses

Descriptive accounts for designing and implementing history of mathematics courses for prospective secondary mathematics teachers are abundant in the literature. Toumasis (1992) is an early example of proposing formal experiences in the history of mathematics in the training of secondary school mathematics teachers in Greece.

At the time, Toumasis observed that the content training of secondary mathematics teachers was "sound and supersufficient..., whereas the pedagogical training [was] quite inadequate" (1992, p. 289). Toumasis claimed that "teachers should be acquainted with the history of mathematics and mathematical ideas and its relation to science, and should acquire some familiarity with the way mathematicians work today" (p. 291). Furthermore, he claimed that prospective secondary mathematics teachers needed to receive training on "the role of the subject in history and the society" (p. 291), in addition to being taught mathematics content. Specific content of a course in the history or mathematics for prospective secondary mathematics teachers was not given; however, Toumasis outlined several tenets for teacher education programs, including a sample pre-service program structure and attention to such elements as providing students with opportunities to connect theory and practice.

Still other descriptions detail course content, required materials, and assignments for history of mathematics courses intended for prospective mathematics teachers, including Clark (2008) and Miller (2002). Miller designed and implemented a course that focused on the mathematics of five ancient cultures and which used two course texts on the history of mathematics from which to draw content and exercises. Miller's experience with teaching a history of mathematics course for the first time was the impetus for her "trial by fire" article; however, much of what she shared from her experience outlines many of the reasons why history of mathematics courses is not offered in teacher education programs.

Miller identified the lack of sufficient homework problems in any one text as a major obstacle in the implementation of the course. This, coupled with the realization that she was "not as prepared to teach the material" as she thought (2002, p. 339), made for a labor-intensive teaching experience. Furthermore, the curriculum committee at the State University of New York (SUNY) Potsdam established the mathematics course prerequisite at precalculus, thus potentially limiting the mathematical content and rigor of the course. The course was implemented at Miller's institution in anticipation of future NCATE accreditation to meet the program standard needs for prospective teachers to possess knowledge of the historical and cultural development of mathematics. Thus, developing foci for such a course, selecting appropriate materials and

24   History of Mathematics in Mathematics Teacher Education   781

textbooks, identifying qualified and willing faculty, and setting the course prerequisite(s) may very well serve as the same obstacles for other institutional contexts.

24.7.5   History of Mathematics: Potential for Teachers' Mathematical Learning

Thus far there has not been explicit focus on why prospective teachers must know something of the history of mathematics, either for the development of their own knowledge or for learning concrete, innovative, or interesting ways to teach mathematics to pupils. Instead, the focus has been to summarize key activities that have taken place in the field regarding the existence of history of mathematics in mathematics teacher preparation. Perhaps the most essential aspect of the role of history of mathematics in mathematics education is what potential it holds for impacting the learning of the subject. In his text, *Elementary Mathematics from an Advanced Standpoint* (1908), Felix Klein wrote:

> ...I shall draw attention, more than is usually done...to the *historical development of the science*, to the accomplishments of its great pioneers. I hope, by discussions of this sort, to further, as I like to say, your general *mathematical culture*: alongside of knowledge of details, as these are supplied by the special lectures, there should be a grasp of subject-matter and of historical relationship. (Klein 1908/1939 Pt. II, p. 2, emphasis in the original)

The importance of knowing and studying the accomplishments – in this case, mathematical accomplishments – as part of the (history of) mathematical education of prospective teachers has been valued and conjectured for over a century. It is vital to begin work on providing evidence in support of such conjectures.

24.7.6   First Example

Fulvia Furinghetti's work with prospective mathematics teachers in Italy provides important examples of how history of mathematics can be incorporated into teacher preparation programs with particular attention to "the conception of mathematics and its teaching...that students develop through their mathematics studies in university" (Furinghetti 2000, p. 43). As Furinghetti observed (and many would agree), university students with strong mathematical training become far removed from the mathematics they are required to teach and once arriving in the classroom their own teaching is similar to how they were taught mathematics in secondary school. In an effort to combat prospective mathematics teachers' falling back on old conceptions of teaching, Furinghetti asked her prospective teachers to "reflect on their mathematical knowledge, in particular on their beliefs about mathematics and its teaching" (p. 46). The activity used in the study focused on "definition," due to its critical role in learning mathematics.

The main component of the prospective teachers' work in the activity was to analyze definitions found at the beginning of old geometry books, including Italian translations of Euclid, Clairaut, and Legendre. Furinghetti (2000) outlined the outcome of the students' exploration of various geometric definitions (e.g., line, quadrilateral, square, trapezia), noting that "students became aware of the fact that definition has three different aspects" (p. 48) in that definitions must be logical, epistemological, and didactic (p. 49). Although the prospective teachers focused only on the didactic aspect of definitions, this in itself was an important mathematical orientation for the students and for this particular study. Here, the examination of various historical resources prompted prospective teachers to investigate questions about what is necessary, sufficient, and preferable information for a definition. In this way, the analyses in which the prospective teachers engaged were linked to proof, further strengthening their mathematical knowledge for teaching.

There are several important observations to make about this example of using historical content and resources with prospective teachers. It is important to note that they possessed strong mathematical backgrounds. However, insufficient research has been conducted on the mathematical background needed to study history of mathematics, particularly with respect to intentions for its use in mathematics teacher education and mathematics teaching.[34] However, it is significant that Furinghetti began the research with the goal that the prospective teachers would reflect on their mathematical knowledge. This difference sets Furinghetti's investigation apart from the research described in the first part of this section, most of which discussed outcomes primarily focused on the story of mathematics and learning to use history of mathematics as anecdote (Siu 1997). The important difference in the investigation Furinghetti conducted was how prospective teachers used history as

> a kind of 'magnifying glass' for the conceptual nodes of a certain theory, a means to identify critical points, to stop at the difficult concepts and to analyse them through the words of past authors. (Furinghetti 2000, p. 50)

24.7.7   Second Example

Prospective mathematics teachers' reflections are also the data source for Furinghetti (2007). In the experiment, 15 prospective secondary mathematics teachers were tasked with producing a sequence for teaching a particular concept in algebra for 9th and 10th grade students in Italy. The perspective from which Furinghetti designed the experiment was that the use of history of mathematics enables *reorientation*, that is, "that prospective mathematics teachers experience again the construction of mathematical objects" (p. 133). There were several phases in the experiment, including (1) analysis of the national mathematical programs from elementary school through the end of high school; (2) work with the history of mathematics to identify and study "the cognitive roots of algebra"; (3) design of the teaching sequence, applying both historical and theoretical perspectives from their course

---

[34] Furinghetti noted that she served as the historian "guide" for students, providing "historical information…needed to interpret authors" (2000, p. 47).

work; and (4) discussion of the outcome of the teaching sequences produced, using both historical and classroom contexts to discuss how cognitive roots may emerge in the teaching sequence (Furinghetti 2007, pp. 135–136).

The prospective mathematics teachers' construction of teaching sequences for an algebraic concept produced a variety of artifacts. In addition to the planned sequences, the prospective teachers also produced exercises and problems for pupils and reports on implementation of excerpts from the teaching sequences. The significant finding of the study was the identification of the ways in which history of mathematics impacted these artifacts. Furinghetti (2007) identified two modes of impact: using history as a way of looking at the evolution of a particular concept and the reading of original sources. For each of these modes of impact, Furinghetti yet again provided empirical evidence for the power of history of mathematics to influence prospective mathematics teachers' mathematical knowledge. For example, for the "evolutionary" mode she observed that the prospective teachers held the idea that

> History provides meaningful examples of algorithms and methods that allow exploitationof the operational nature of mathematical objects;

> History suggests the development of the concepts in a visual/perceptual environment such as that provided by geometry. (Furinghetti 2007, p. 137)

Finally, it is important to again comment on the orientation from which Furinghetti conducted this experiment. As in the context with prospective mathematics teachers working on the notion of "definition" in mathematics, Furinghetti made it clear that guiding students in how to "do" history of mathematics was a primary concern. As part of her work with the 15 prospective teachers, she provided different types of sources and guidance on how to "look for information, to choose among different sources, to interpret original historical passages, to evaluate many elements, and to make their own choices" (p. 136). In this way, the varying background experiences of the participants were leveled out and this afforded opportunities for collaboration and discussion. Most importantly, "the participants were provided with motivation to learn some history of mathematics" (p. 141).

24.7.8 Third Example

Although history of mathematics received prominent attention in the 1997 new national curriculum in Norway, its role in mathematics teacher education did not receive similar attention. Smestad (2012) described the development and revision of a 6-h history of mathematics course he developed as part of a new course for prospective lower secondary mathematics teachers. Smestad established several goals for the students, all of which were focused on future pedagogical practices. The course:

> …should give the students examples of different ways of teaching with history of mathematics, it should be connected to the students' curriculum, it should give ideas that are suitable for different age levels in the 11–16 bracket, it should show how mathematics has…developed. (Smestad 2012, p. 1)

The students' evaluation of the first iteration of the course revealed that it was difficult for them to glean anything from the potpourri of topics. For the second iteration, Smestad focused entirely on the history of probability. Although Smestad provided a list of topics and a brief description of several examples from the course, of the outcome of the second iteration he stated only that "the mathematical learning was more obvious to the students" (2012, p. 3). Quantification of the extent of the mathematics learned and why it was more obvious to the students or what evidence supported either of these critical pieces of information are not known.

24.7.9 Fourth Example

Clark (2012) sought to qualify mathematical learning as a result of studying history of mathematics. In a history of mathematics course for prospective secondary mathematics teachers, students studied and used translations of primary sources as well as historical methods to solve problems connected to the mathematical content they would be responsible for in their future teaching. After studying the method of completing the square from al-Khwarizmi's famous text (ca. 825 CE), prospective teachers' journal reflections from four different semesters of the course were analyzed to identify key themes indicative of changes in their mathematical knowledge.

Excerpts from 80 of 93 student reflection journals provided insight into the ways in which the study of historical examples of solving quadratic equations using the method of completing the square revealed that the prospective teachers experienced two changes in their understanding of this mathematical idea. The prospective mathematics teachers came to understand the method of completing the square as a result of the geometric representation motivated by al-Khwarizmi's method. And they claimed that their previous experience with this mathematical topic was firmly situated in rote understanding of the quadratic formula emerging from the method of completing the square.

It was difficult to definitively identify whether the historical tasks, content, and historical sources were the primary influences of the changes in mathematical understanding experienced by the prospective mathematics teachers participating in the study. However, the reflections about the perspective on their own mathematical learning also influenced the prospective teachers' conceptions for how to incorporate history of mathematics in their future teaching. This prompts attention to Schubring's and his colleagues' (2000) concern that "…there is only scattered evidence about the effectiveness of the historical training in the later teaching practice" (p. 142) – and which continues to be of concern.

24.7.10 Summary

Each of the examples summarized in this section serves to remind the field that conducting research that details a clearer picture of what the role of history of mathematics in secondary mathematics teacher education is, and how teacher educators can best capitalize on what it offers to future educators, is difficult work. On the one hand, the extent to which the guidelines provided in MET1 – and now, MET2 – were or will be instituted in the United States remains unclear. Furthermore, the introduction of the Common Core State Standards Initiative in Mathematics (CCSSI-M) will have significant impact on mathematics teacher education. The newly created Mathematics Teacher Education Partnership (MTE-Partnership), a partnership of institutions of higher education, K-12 schools, school districts, and other stakeholder organization, has launched a collaborative effort to redesign secondary mathematics teacher education. The MTE-Partnership will identify best practices and guiding principles underlying such teacher education programs and the efforts of the partnership are intimately connected to CCSSI-M and common assessments that are attached to the Common Core State Standards. Finally, the extension of similar concerns and efforts applies to many, if not all, locations around the world, as the (largely unknown) outcomes of history of mathematics on mathematics teaching and learning will certainly influence its role in mathematics teacher education.

In order for teacher educators who advocate for the role of history of mathematics in mathematics teacher education to impart influence in the future, the broader field of mathematics education will demand evidence of positive impact of the integration of history of mathematics on pupils' learning of mathematics. The final section highlights several examples of interventions that have taken place in primary and secondary classrooms and which provide promise for imparting the influence needed in the future.

24.8 What Next: Teachers Using History of Mathematics in Teaching

This chapter was intended to provide an overview of the "state of the field" of history in mathematics education pertaining to the preparation of mathematics teachers. The previous sections described examples in the literature of the design and implementation of history of mathematics courses for prospective mathematics teachers, research on prospective teachers' attitudes towards learning history of mathematics, and teaching using historical perspectives, as well as qualitative research on how and what prospective teachers learn as a result of experiences with historical content. In addition to providing a representation of the field, suggestions of what is further needed to quantify responses to questions of why history of mathematics is necessary for the mathematical and pedagogical education of future mathematics teachers were offered. This final section offers a brief look at "what next?"; that is, after prospective teacher transition into classroom teaching, what are examples of efforts to incorporate the history of mathematics in teaching and what can be learned from them to further inform mathematics teacher education programs?

Table 24.4 displays a collection of six studies for which some element of history of mathematics was employed in a classroom intervention with pupils. The summaries

Table 24.4 Summary: recent examples of ways in which history of mathematics is used in classrooms

Author(s)	Age level, topic, and location	Intervention (including materials)	Methodology (data sources)	Outcomes and comments
Clark (2010)	46 pupils, grades 4 and 5 Contextual word problems involving multiplication, division, and conversion of units United States	Classroom teachers and school district mathematics specialists participated in professional development on an instructional unit created from two cuneiform tablets owned by Florida State University (FSU22 and FSU23). Pupils participated in an extended mathematics lesson using the materials, photos of FSU22 and FSU23, and local history	Qualitative; field notes, student work, teacher and mathematics specialists post-implementation survey	Classroom teachers participated with and observed their pupils' engagement with the instructional unit and reported increased student interest and attention. They also observed that pupils appeared to experience less difficulty with unfamiliar units (e.g., *eshe*, *bur*) and less anxiety with "word problems" that were situated in the ancient Babylonian context
Papadopoulos (2010)	20 pupils, grade 6 (two different Classroom environments: one computer lab, one paper-and-pencil classroom) Area of irregular shapes Greece	Inspired by the evolution of the Concept of area found in textbooks from the eighteenth to twentieth centuries, the author designed tasks and guided them to "reinvent" the methods for estimating the area of irregular shapes, for which they had no prior instruction	Qualitative; student work Artifacts from paper-and-pencil group and screen captures from computer lab group	The pupils applied Techniques that were the same as those given in the textbooks and some of the techniques developed by pupils were completely new
Ng (2006)	414 pupils in 8th year (177 in the experimental group; 237 in the control group) Relevant problems from ancient Chinese mathematical texts Singapore	Ancient Chinese Mathematics Enrichment Programme (ACMEP) entailed pupils' translating and solving problems taken from the *Nine Sections*. The intervention was 7 months in duration	Quasi-experimental (no random assignment: pupils selected which group they would be part of) assessments	Although there was an overall significant difference in academic achievement in mathematics between the experimental and control groups (favoring the experimental group), the difference could not be completely attributed to the ACMEP

Rowlands (2010)	36 gifted and talented pupils aged 14–15 and two groups of about 30 mixed ability pupils aged 16–17 Deductive geometry England	Pupils were introduced to the six levels of abstraction through the intervention containing two primary events	Qualitative; observations by one teacher and an academic, feedback questionnaires, formative assessment	Success of the pilot stage was judged primarily by the engagement of the learners
Thomaidis and Tzanakis (2007)	Two classes of 16-year-old students Ordering on the number line Greece	Pupils were taught according to the official syllabus and textbook for the first three chapters of the text. Regarding the fourth chapter, pupils were divided into two groups: those who had been taught "algorithmic" tools for solving quadratic inequalities (28 pupils) and those who had not (30 pupils)	Mixed methods; student responses to a three-item questionnaire, along with accompanying work	No significant difference was found between the post-"algorithmic" group and the pre-"algorithmic" group. Qualitative differences did point to the notion that some student difficulties paralleled the obstacles encountered in history regarding the conception of and ordering of negative numbers
Jankvist (2010)	Upper secondary pupils Public-key cryptography and RSA Denmark	A 67-page teaching module was prepared to introduce pupils to the historical case and the mathematics involved. Essay assignments were implemented in groups of three to five pupils	Qualitative; pre-implementation questionnaire involving historical, developmental, sociological, and philosophical questions and interviews conducted with a subset of the original group and five students from the 12 interviewees were interviewed during the module implementation	An existence proof for a module approach for history as a goal to occur with Danish upper secondary pupils

788 K.M. Clark

highlight the pupil age level and if appropriate, mathematical topic, what the intervention entailed, the methodology employed, and reported outcomes of the research.

The collection of studies highlighted here represents a broad range of pupil age levels, historical and mathematical content, and study design. What the analysis of the research revealed, however, is that most of the interventions were the result of ideas and research generated by scholars or mathematics teacher educators, not the result of classroom teachers who studied the history of mathematics (e.g., as a result of their pre-service mathematics preparation program) and then collaborated with others to conduct research of what occurred when using history in teaching. Finally, although progress is evident, questions and challenges regarding the role of history of mathematics in mathematics teacher education remain.

References

Alpaslan, M., Isiksal, M., & Haser, Ç. (2011). The development of attitudes and beliefs questionnaire towards using history of mathematics in mathematics education. In M. Pytlak, T. Rowland, & E. Swoboda (Eds.), *Proceedings of the seventh congress of the European society for research in mathematics education* (pp. 1661–1669). University of Rzeszów, Rzeszów, Poland.

Alpaslan, M., & Haser, Ç. (2012). "History of mathematics" course for pre-service mathematics teachers: A case study. In *12th international congress on mathematical education pre-proceedings* (pp. 4180–4189). Seoul, South Korea.

Barbin, E., Bagni, G. T., Grugnetti, L., Kronfellner, M., Lakoma, E., & Menghini, M. (2000). Integrating history: research perspectives. In J. Fauvel & J. van Maanen (Eds.), *History in mathematics education: The ICMI study* (pp. 63–90). Boston: Kluwer.

Barbin, E., Furinghetti, F., Lawrence, S., & Smestad, B. (2011). The role of the history and epistemology of mathematics in teaching training. In E. Barbin, M. Kronfellner, & C. Tzanakis (Eds.), *Proceedings of the 6th European Summer University in the History and Philosophy of Mathematics* (pp. 27–46). Vienna, Austria: Holzhausen Publishing Ltd.

Baumgart, J. K., Deal, D. E., Vogeli, B. R., & Hallerberg, A. E. (Eds.). (1969). *Historical topics for the mathematics classroom*. Washington, DC: National Council of Teachers of Mathematics.

British Ministry of Education (1958). Pamplet 36, London: Her Majesty's Stationery Office, 134–154.

Burns, B. A. (2010). Pre-service teachers' exposure to using the history of mathematics to enhance their teaching of high school mathematics. *Issues in the Undergraduate Mathematics Preparation of School Teachers: The Journal, 4*, 1–9.

Charalambous, C. Y., Panaoura, A., & Phillippou, G. (2009). Using the history of mathematics to induce changes in preservice teachers' beliefs and attitudes: insights from evaluating a teacher education program. *Educational Studies in Mathematics, 71*(2), 161–180.

Clark, K. M. (2008). Heeding the call: History of mathematics and the preparation of secondary mathematics teachers. In F. Arbaugh & P.M. Taylor (Eds.), *Inquiry into mathematics teacher education: Association of Mathematics Teacher Educators (AMTE)* (pp. 85–95). San Diego, CA: Association of Mathematics Teacher Educators.

Clark, K. M. (2010). Connecting local history, ancient history, and mathematics: the Eustis Elementary School pilot project. *British Society for the History of Mathematics Bulletin, 25*(3), 132–143.

Clark, K. (2011). Reflection and revision: Evolving conceptions of a *Using History* course. In V. Katz & C. Tzanakis (Eds.), *Recent developments in introducing a historical dimension in mathematics education* (pp. 211–220). Washington, DC: The Mathematical Association of America.

Clark, K. M. (2012). History of mathematics: illuminating understanding of school mathematics concepts for prospective mathematics teachers. *Educational Studies in Mathematics,* 81(1), 67–84.

Conference Board of the Mathematical Sciences (CBMS). (2001). *The mathematical education of teachers: Vol. 11. Issues in mathematics education.* Providence, RI: American Mathematical Society.

Conference Board of the Mathematical Sciences (CBMS). (2012). *The mathematical education of teachers II: Vol. 17. Issues in mathematics education.* Providence, RI: American Mathematical Society.

Fasanelli, F. (2001). History of the international study group of the relations between history and pedagogy of mathematics: the first twenty-five years, 1976–2001. HPM-Group. http://www.clab.edc.uoc.gr/hpm/HPMhistory.PDF. Accessed 28 December 2012.

Fauvel, J. (1991). Using history in mathematics education. *For the Learning of Mathematics,* 11(2), 3–6.

Fauvel, J., & van Maanen, J. (Eds.) (2000). *History in mathematics education: The ICMI study.* Boston: Kluwer.

Fleener, M. J., Reeder. S. L., Young, E., & Reynolds, A. M. (2002). History of mathematics: Building relationships for teaching and learning. *Action in Teacher Education,* 24(3), 73–84.

Furinghetti, F. (1997). History of mathematics, mathematics education, school practice: Case studies in linking different domains. *For the Learning of Mathematics,* 17(1), 55–61.

Furinghetti, F. (2000). The history of mathematics as a coupling link between secondary and university teaching. *International Journal of Mathematical Education in Science and Technology,* 31(1), 43–51.

Furinghetti, F. (2002). On the role of the history of mathematics in mathematics education. Paper presented at *2nd International Conference on the Teaching of Mathematics at the Undergraduate Level.* Crete, Greece.

Furinghetti, F. (2007). Teacher education through the history of mathematics. *Educational Studies in Mathematics,* 66(1), 131–143.

Gonulates, F. (2008). Prospective teachers' views on the integration of history of mathematics in mathematics courses. In R. Cantoral, F. Fasanelli, A. Garciadiego, R. Stein, & C. Tzanakis (Eds.), *Proceedings of the international study group on the relations between the history and pedagogy of mathematics* (pp. 1–11). Mexico City: Mexico.

Gulikers, I., & Blom, K. (2001). 'A historical angle', A survey of recent literature on the use and value of history in geometrical education. *Educational Studies in Mathematics,* 47(2), 223–258.

Grundy, S. (1987). *Curriculum: Product or praxis.* Philadelphia: The Falmer Press.

Jankvist, U. T. (2009). A categorization of the "whys" and "hows" of using history in mathematics education. *Educational Studies in Mathematics,* 71(3), 235–261.

Jankvist, U. T. (2010). An empirical study of using history as a 'goal'. *Educational Studies in Mathematics,* 74(1), 53–74.

Jankvist, U. T. (2012). A first attempt to identify and classify empirical studies on history in mathematics education. In B. Sriraman (Ed.), *Crossroads in the history of mathematics and mathematics education* (pp. 295–332). Charlotte, NC: Information Age.

Jankvist, U. T., Lawrence, S., Tzanakis, C., & van Maanen, J. (2011). Introduction to the papers of WG12: History in mathematics education. In M. Pytlak, T. Rowland, & E. Swoboda (Eds.), *Proceedings of the seventh congress of the European society for research in mathematics education* (pp. 1636–1639). University of Rzeszów, Rzeszów, Poland.

Jones, P. S. (1969). The history of mathematics as a teaching tool. In J. K. Baumgart, D. E. Deal, B. R. Vogeli, & A. E. Hallerberg (Eds.), *Historical topics for the mathematics classroom* (pp. 1–17). Washington, DC: National Council of Teachers of Mathematics.

Katz, V. J. (2000). *Using history to teach mathematics: An international perspective.* Washington, DC: The Mathematical Association of America.

Katz, V. J., & Tzanakis, C. (Eds.) (2011). *Recent developments on introducing a historical dimension in mathematics education.* Washington, DC: The Mathematical Association of America.

Kjeldsen, T. H., & Blomhøj, M. (2009). Integrating history and philosophy in mathematics education at university level through problem-oriented project work. *ZDM Mathematics Education, 41,* 87–103.

Klein, F. (1908/1939). *Elementary mathematics from an advanced standpoint.* Part II: *Geometry.* Translated by E. R. Hedrick and C. A. Noble. New York: Dover Publications.

Liu, P.-H. (2003). Do teachers need to incorporate the history of mathematics in their teaching? *Mathematics Teacher, 96*(6), 416–421.

Lockhart, P. (2008, March). A mathematician's lament. *Devlin's Angle, March 2008.* http://www.maa.org/devlin/devlin_03_08.html. Accessed 28 December 2012.

Miller, C. C. (2002). Teaching the history of mathematics: A trial by fire. *PRIMUS, XII*(4), 334–346.

National Council for the Accreditation of Teacher Education (2003). *NCATE/NCTM Program standards: Programs for initial preparation of mathematics teachers.* Washington, DC: Author.

National Council of Teachers of Mathematics (1969). Historical topics for the mathematics classroom. Washington, DC: Author.

National Council of Teachers of Mathematics (1989). *Curriculum and evaluation standards for school mathematics.* Reston, VA: Author.

National Council of Teachers of Mathematics (2000). *Principles and standards for school mathematics.* Reston, VA: Author.

National Council of Teachers of Mathematics (2007). NCATE mathematics program standards. http://www.nctm.org/standards/content.aspx?id=2978. Accessed 28 December 2012.

National Council of Teachers of Mathematics (2011). Secondary-level draft standards: Initial certification. NCATE Mathematics Program Standards. http://www.nctm.org/uploadedFiles/Professional_Development/NCATE/NCTM%20Secondary%20Mathematics%20Education%20Standards%20(8-15-11%20draft).pdf. Accessed 28 December 2012.

National Council of Teachers of Mathematics (2012). NCTM NCATE standards revision draft – Secondary (4/10/12). http://www.nctm.org/uploadedFiles/Math_Standards/NCTM%20NCATE%20Standards%20Revision%20and%20Math%20Content%20-%20Secondary.pdf. Accessed 28 December 2012.

Ng, W. L. (2006). Effects of an ancient Chinese mathematics enrichment programme on secondary school students' achievement in mathematics. *International Journal of Science and Mathematical Education, 4*(2), 485–511.

Papadopoulos, I. (2010). "Reinventing" techniques for the estimation of the area of irregular plane figures: From the eighteenth century to the modern classroom. *International Journal of Science and Mathematical Education, 8*(5), 869–890.

Radford, L. (1997). On psychology, historical epistemology, and the teaching of mathematics: Towards a socio-cultural history of mathematics. *For the Learning of Mathematics, 17*(1), 26–33.

Radford, L. (2012). Early algebraic thinking: Epistemological, semiotic, and developmental issues. In *Proceedings of the 12th International Congress on Mathematical Education* (pp. 675–694). Seoul, South Korea.

Report of the Mathematical Association Committee on the Teaching of Mathematics in Public and Secondary Schools (1919). The Mathematical Gazette, 9(143), 393–421.

Rowlands, S. (2010). A pilot study of a cultural-historical approach to teaching geometry. *Science & Education, 19*(1), 55–73.

Schubring, G., Cousquer, E., Fung, C.-I., El-Idrissi, A., Gispert, H., Heiede, T., et al. (2000). History of mathematics for trainee teachers. In J. Fauvel & J. van Maanen (Eds.), *History in mathematics education: The ICMI Study* (pp. 91–141). Boston: Kluwer.

Singh, S. (1998). *Fermat's enigma: The epic quest to solve the world's greatest mathematical problem.* New York: Anchor.

Siu, M.-K. (1997). The ABCD of using history of mathematics in the (undergraduate) classroom. *Bulletin of the Hong Kong Mathematical Society, 1*(1), 143–154.

Siu, M.-K., & Tzanakis, C. (2004). History of mathematics in classroom teaching – appetizer?, main course?, or dessert? *Mediterranean Journal for Research in Mathematics Education, 3*(1–2), v–x.

Smestad, B. (2012). Teaching history of mathematics to teacher students: Examples from a short intervention. In E. Barbin, S. Hwang, & C. Tzanakis (Eds.), *Proceedings of the international study group on the relations between the history and pedagogy of mathematics: History and pedagogy of mathematics 2012* (pp. 349–362). Daejeon: Korea.

Sriraman, B. (Ed.) (2012). *Crossroads in the history of mathematics and mathematics education: Monograph 12 in the Montana Mathematics Enthusiast.* Charlotte, NC: Information Age.

The Cockcroft Report (1982). Mathematics counts. London: Her Majesty's Stationery Office.

Thomaidis, Y., & Tzanakis, C. (2007). The notion of historical "parallelism" revisited: historical evolution and students' conception of the order relation on the number line. *Educational Studies in Mathematics,* 66(2), 165–183.

Toumasis, C. (1992). Problems in training secondary mathematics teachers: the Greek experience. *International Journal of Mathematical Education in Science and Technology,* 23(2), 287–299.

Tzanakis, C., & Thomaidis, Y. (2011). Classifying the arguments & methodological schemes for integrating history in mathematics education. In M. Pytlak, T. Rowland, & E. Swoboda (Eds.), *Proceedings of the seventh congress of the European society for research in mathematics education* (pp. 1650–1660). University of Rzeszów, Rzeszów, Poland.

Kathleen M. Clark is an associate professor in the College of Education at Florida State University. She received her doctorate in Curriculum and Instruction (with an emphasis in Mathematics Education) from the University of Maryland, College Park, Maryland, United States, in 2006 and was the same year appointed assistant professor of Mathematics Education in at Florida State University. In 2009 she was awarded a 1-year research fellowship in the history of mathematics in the Department of Mathematics and Statistics at the University of Canterbury in Christchurch, New Zealand. Her primary research interests lie in two fields, mathematics education and history of mathematics. In the former, her research investigates ways in which prospective and in-service mathematics teachers use history of mathematics in teaching and the ways in which the study of history of mathematics impacts mathematical knowledge for teaching. In the latter, her historical research is focused on the seventeenth- and eighteenth-century mathematics, with a particular emphasis on the early development of logarithms. Dr. Clark serves a variety of national and international organizations concerned with the history of mathematics, including coeditor of the online journal *Loci: Convergence* (part of the Mathematical Sciences Digital Library (MathDL)), advisory board member and newsletter editor for the International Study Group on the Relations between the History and Pedagogy of Mathematics (HPM Group), and executive committee member of the HPM-Americas Section.

Chapter 25
The Role of Mathematics in Liberal Arts Education

Judith V. Grabiner

25.1 Introduction

World-famous golden-thighed Pythagoras
Fingered upon a fiddle-stick or strings
What a star sang and careless Muses heard

–William Butler Yeats (1865–1939), "Among School Children"

Ireland's great poet beautifully reminds us how a legendary mathematician of antiquity not only discovered that musical harmony results when the lengths of vibrating strings are in the ratios of small whole numbers but also taught that the same mathematical harmonies produce the music of the heavenly spheres. Yeats expects his readers to know this history. The surrealist artist Salvador Dalí (1904–1989), in his painting "The Last Supper," places Jesus and his twelve disciples in part of a wooden dodecahedron, resembling the wooden model of this twelve-sided figure pictured in a famous drawing by Leonardo da Vinci (1492–1519). Plato in his *Timaeus* had said that the dodecahedron was the shape of the universe. Dalí's painting gains in power for those viewers who understand this symbolism.

Furthermore, not only ideas about numbers and shape but the very technical terms that underlie mathematics pervade Western culture from the "postulata" with which Thomas Malthus begins his "Essay on Population" of 1798 to the "algorithms" which humanists and scientists alike know are the basis for Internet search engines. Nor is this just in the Western tradition. Mathematics is found in virtually every human society, from ancient civilizations to contemporary cultures, both literate and nonliterate, all over the world. In Greece almost two and a half millennia ago, Plato wanted the rulers of his ideal Republic to study arithmetic, geometry, astronomy, and the mathematics of musical harmony, and those four subjects, along

J.V. Grabiner (✉)
Mathematics, Pitzer College, Claremont, CA, USA
e-mail: jgrabiner@pitzer.edu

M.R. Matthews (ed.), *International Handbook of Research in History, Philosophy and Science Teaching*, DOI 10.1007/978-94-007-7654-8_25,
© Springer Science+Business Media Dordrecht 2014

793

with grammar, rhetoric, and logic, made up the classical Seven Liberal Arts. In traditional Hindu architecture, numerical ratios are used to represent the mathematically determined structure of the universe. In classical India, not only was mathematics used to analyze poetic meter, mathematical results themselves were expressed in poetic form. In premodern China, mathematical writings were part of the canon of literary texts studied for the examinations to enter the civil service. At present, most universities in the United States require all students, whether these students aspire to be artists, social workers, or technologists, to take courses in mathematics. Mathematics, then, has played and still plays a central role in liberal arts education.

Yet mathematics is too often taught, and too often thought to be, a series of formulas, or a set of arbitrary rules for solving contrived and irrelevant word problems. It often acts as an instrument to keep less proficient students from entering certain fields. And though mathematics classes are often justified as helping teach students to think, or to think abstractly, much actual mathematics instruction teaches neither of these, instead producing frustration and a fear so pronounced that it has been given its own name, "mathematics anxiety" (Ashcraft 2002; Tobias 1993; Zaslavsky 1994). People who are otherwise well educated say, when confronted with the quantitative statements so pervasive in modern life, "Oh, I've always been bad at math."

The present essay has four goals. First, it will briefly sketch the history of the continuous inclusion of mathematics in liberal education in the West, from ancient times up through the modern period; the focus on the West stems from the continuing influence of the idea reflected in the words of the title assigned to the author for this essay: "liberal arts education." Second, it will elaborate on the brief remarks in this introduction to delineate the central role mathematics has played throughout the history of Western civilization, demonstrating that mathematics is not just a tool for science and technology (though of course it is that) but continually illuminates, interacts with, and sometimes challenges fields like art, music, literature, and philosophy – subjects now universally considered to be liberal arts.[1] Third, it will

---

[1] In modern discourse about education, the term "liberal arts" has been defined and characterized in a variety of ways. It is of course built into the history of Western education. The influential Carnegie Foundation for the Advancement of Teaching has provided a list of contemporary "liberal arts" that goes beyond the traditional Western canon: English language and literature, foreign languages, letters, liberal and general studies, life sciences, mathematics, physical sciences, psychology, social sciences, visual and performing arts, area and ethnic studies, multi- and interdisciplinary studies, philosophy, and religion (Carnegie 1994, p. xx; Ferrall 2011, p. 9). In antiquity, as the next sections of this paper will describe, the list came to have seven items: arithmetic, geometry, astronomy, and music theory (the quadrivium), and grammar, rhetoric, and logic (the trivium) (Stahl 1977; Wagner 1983b). Those who compile such lists characterize liberal arts education as study undertaken for its own sake, as opposed to vocational education. As for the purpose of liberal arts education, it has been described as educating a free person, or as liberating the mind to pursue the truth, or as producing a cultivated person who can be both a good citizen and a leader of society. Liberal arts instruction has sometimes focused on the canonical texts of Western civilization as ways to build and reinforce the shared values of society. But such instruction has also been championed as suitable for the education of free individuals to be citizens in a democracy by developing the capacities for independent and critical thinking, logical analysis, effective communication, an understanding of the interrelations between different fields of learning, and imagination (Ferrall 2011; Kimball 1995; Nussbaum 2010; Sinaiko 1998). The present chapter recognizes and appreciates these

25 The Role of Mathematics in Liberal Arts Education

add a more global perspective to the contemporary liberal arts story, by showing how the mathematics present in many cultures – an important part of global history – can enhance the teaching of modern mathematics. Fourth, it will address some ways that mathematics teaching could take the subject out of the toolbox and bring it back to the university, replace frustration with appreciation and understanding, and reestablish mathematics as a liberal art.

25.2 Mathematics and Liberal Education in Western History

25.2.1 The Ancient World

The tradition of liberal arts education in the West goes back to antiquity. A liberal arts education was intended not to prepare one for a vocation or to accomplish some practical goal but to educate a free human being and citizen – though what "free human being" and "citizen" meant has changed over the years. And mathematics virtually always was counted among the liberal arts (Jaeger 1944; Marrou 1956).

In the Greek-speaking world, even before Plato's time, mathematics was entwined with philosophy. A new attitude toward mathematics and science, detaching them from the practical or the metaphysical, is often dated from the time of Thales of Miletus (late sixth century BCE), who is said to have proved some basic geometric results and to have begun the Greek practice of understanding nature in terms of fundamental rational principles. To the pre-Socratic question "what is the basic unifying principle of the universe?" Thales answered that "all was water." Then, Anaximander answered that all was the unbounded, Anaximenes that all was air, and Democritus that all was made of atoms. The Pythagoreans answered the question by teaching that "all is number." Proclus (fifth century CE) credits Pythagoras himself with having transformed the study of mathematics into "a scheme of liberal education, surveying its principles from the highest downwards and investigating its theorems in an immaterial and intellectual manner"; this transformation, initiated by the Pythagorean school, was well established by the fifth century BCE.[2] The subdivision of mathematics into the four canonical sciences of arithmetic, geometry, astronomy, and music goes back at least to the prominent Pythagorean Archytas of Tarentum (early fourth century BCE) (Huffman 2011; Fauvel and Gray 1987, p. 57).

Plato himself in Book VII of his *Republic* gave mathematics a central role in the education of the rulers of his ideal state, who were to be philosophers as well as

---

disparate views. The Carnegie list is useful to illustrate the types of subjects that constitute a liberal arts education today, and the present essay shares the view that modern liberal education's most important goal is to educate independent and thoughtful citizens.

[2] For Proclus, see Proclus (1970, p. 53). The term "mathematics" itself reveals a liberal arts origin; the Greek root "mathema" was first more general, connoting merely "something learned," and the "mathematikoi" were the inner initiates of the Pythagorean school. For pre-Euclidean logically structured geometry, see Knorr (1975, esp. p. 7) and McKirahan (1992, pp. 16–18).

kings. Plato acknowledged that mathematics is useful, but the needs of shopkeepers and generals are not why it is important. It is because the objects of mathematics are unchanging and eternal that their study is at the heart of liberal education. The study of mathematics, Plato said, compels the soul to look upward, drawing it from the changing to the real, from the uncertain to the certain, and from the world of illusion toward the truth (Plato, *Republic*, Book VII, 521c, 525d-529c; Plato 1961, pp. 575–844). Thus, mathematics is a "liberal" subject for Plato and his followers. Plato's outline for the education of the philosopher-rulers begins with arithmetic, followed by geometry, followed by astronomy, and is capped by the study of harmony. Astronomy goes beyond the study of the visible heavenly bodies to their pure mathematical shapes and motions, and harmony goes beyond physical music to the mathematical proportions involved (Plato, *Republic*, Book VII, 524e-531c; Plato 1961, pp. 575–844). Moreover, Plato often explained his philosophy using mathematical examples, most famously in the parable of the Divided Line, a geometric metaphor encompassing all of reality, from shadows to the Form of the Good (Plato, *Republic*, Book VI, 510d-513e; Plato 1961, pp. 575–844).

Aristotle, like Plato, distinguished learning for its own sake from more practical activities, and, also like Plato, put mathematics in the former category. Once there were people with leisure, said Aristotle, "those of the sciences which are directed neither to pleasure nor to the necessities of life were discovered" (Aristotle, *Metaphysics* 981b20-24; Aristotle 1941, pp. 689–926). Aristotle classified studies as theoretical, where the goal is knowledge; productive, where the goal is a product; and practical, where the goal is action. For him, mathematics, physics, and metaphysics are the theoretical sciences, and his theory of demonstrative science is illustrated by, and modeled on, the logically structured "elements of geometry" that existed before Euclid's definitive books drove out all previous competitors (Heath 1949, pp. 1, 5; McKirahan 1992, pp. 16–18). For Aristotle, the objects of mathematics are abstractions from sense experience (Aristotle, *De anima* 431b13-19; Aristotle 1941, pp. 535–603; Heath 1949, p. 65), a characterization that underlines the theoretical status of the subject. Aristotle also appreciated the aesthetic qualities of mathematics, saying "The chief forms of beauty are order and symmetry and definiteness, which the mathematical sciences demonstrate in a special degree" (Aristotle, *Metaphysics* 1078a37-b2; Aristotle 1941, pp. 689–926). Thus, for Aristotle as for Plato, mathematics shares goals and other characteristics with the liberal arts.

There were of course other subjects in Greek education, notably rhetoric, philosophy, and politics. Indeed, rhetoric, as championed by the Sophists such as Hippias of Elis (fifth century BCE), though criticized by Plato as the art of making the worse argument appear the stronger, was an important topic throughout antiquity. But the Sophists did not neglect mathematics. In fact Werner Jaeger credits the Sophists, beginning with Hippias, with having "altered the history of the world" by introducing mathematical instruction to their students, changing it from a subject of scientific research to a valuable part of education, with its value being the cultivation of the intellect (Jaeger 1944, vol. I, pp. 313–315). There was no unanimity on this point; the influential Greek orator and teacher Isocrates (436–338 BC) emphasized rhetoric and philosophy in his highly influential program of instruction, which, in

25 The Role of Mathematics in Liberal Arts Education 797

the Hellenistic age, was often preferred to that of Plato and which was emphasized in the more literary and rhetorical educational ideal adopted by the Romans (Jaeger 1944, vol. 3, pp. 46–47; Kimball 1995, p. 19; Marrou 1956, pp. 119–120).

Education in the Roman period generally emphasized politics and rhetoric. Still, even the champion of oratorical instruction Quintilian (40–118 CE) conceded that mathematics sharpens the intellect and teaches how to construct an argument (Grant 1999a, p. 99). The Greek term "Enkuklios paidaea" that described higher education in antiquity was always taken to include mathematics, and the Latin phrase "artes liberales" was seen as similar to "enkuklios paidaea" (Kimball 1995, p. 21; Marrou 1956, pp. 243–245; Morrison 1983, p. 32). "Liber" – free – meant the arts of the mind of the free individual as opposed to the mechanical arts.

The influential orator Cicero held that to be truly human, one needed to have been "perfected in the arts appropriate to humanity" (Cicero, *De re publica* I.xvii.28), and, although mathematics was not foremost in his educational agenda, he included it on the list of "artes liberales" essential to the liberal education of a gentleman (Cicero, *De oratore* III.xxxii.127; Grant 1999a, p. 100). Cicero also reported that Plato became convinced of a country's intellectual nature by seeing that it had produced geometrical drawings, rather than just by observing that its land was cultivated (Cicero, *De re publica* I.xvii.29).

Throughout the Graeco-Roman period, then, mathematics remained part of education. Of course, sometimes the mathematics taught did not go very deep. Still, at least lip service was paid to its role, and even lip service keeps a tradition alive and capable of fuller resurrection.

25.2.2 The Middle Ages

The idea of the liberal arts continued into the medieval period. The Latin encyclopedists of the fifth through seventh centuries hoped to preserve the intellectual heritage of classical learning. They combined the intellectual traditions of Neoplatonism with Christianity and linked these with the handbook tradition of popular culture. The encyclopedists needed an organizing principle for all these topics, and the principle chosen was the concept of the liberal arts, a principle that continued to organize knowledge into the twelfth-century Renaissance and beyond.

Martianus Capella's fifth-century *Marriage of Philology and Mercury* is the first work to fully present the Seven Liberal Arts (Stahl 1977, pp. 21–39). Martianus's Seven Liberal Arts, described allegorically as seven bridesmaids, were grammar, rhetoric, logic, arithmetic, geometry, astronomy, and harmony or music. Martianus explicitly ruled out medicine and architecture, classifying them as "mundane matters," that is, professional pursuits, not liberal ones, pursued for their own sake (Stahl 1977, vol. 1, p. 93). Sometimes in medieval teaching, logic was assimilated into rhetoric, and often the mathematics was superficial, but Martianus's classification ensured that mathematics would remain part of the "arts of free people" (Stahl 1977; Wagner 1983a).

Although the Christian Roman Emperor Justinian closed the pagan philosophical schools, including Plato's Academy in Athens, in 529 CE, the Greek version of the liberal arts did not die. The early Church fathers had been interested. Clement of Alexandria (150–215 CE) was willing to include the liberal arts as a first stage in education that ultimately led to philosophy and Christian wisdom (Grant 1999a, p. 101), while Augustine (354–430 CE), much influenced by Neoplatonism, saw the liberal arts as pathways to the divine order underlying creation, especially, he said, "in music, in geometry, in the movements of the stars, in the fixed ratios of numbers" (Grant 1999a, p. 101). Cassiodorus (c. 480–c. 575 CE) also included the liberal arts in Christian education. In sixth-century Spain, Isidore of Seville wrote that mathematics is a legitimate and important component of Christian culture, supporting his view with a text from the *Apocrypha*: "Not in vain was it said in the praise of God: You made everything in measure, in number, and in weight" (*Wisdom of Solomon* XI, 20; Høyrup 1994, pp. 177–178).

In the monasteries, monks copied and preserved both Greek and Latin manuscripts. Beginning in the eighth century, notably under Alcuin of York, cathedral schools taught both the trivium and quadrivium. For the quadrivium, texts included arithmetic based on the work of Nicomachus of Gerasa, the first four books of Euclid's *Elements of Geometry* as adapted and abridged by Boethius, some astronomy based on an abridgement of Ptolemy's *Almagest*, and various Greek materials on music. Mathematics was needed in this period to determine the calendar in general, and the date of Easter in particular, and of course to solve various practical problems, but otherwise there was little interest in further research. Still, as a beginning of medieval education, the cathedral schools were important in preserving the role of mathematics, and Pope Sylvester II (999–1003) even began to introduce the Hindu-Arabic number system, as well as basic geometry and astronomy, in the cathedral school at Rheims (Katz 2009, pp. 325–327).

The twelfth century marked the high point of the Seven Liberal Arts in medieval education. Hugh of St. Victor, best known as the writer of an influential educational handbook for theology students, stressed the Seven Liberal Arts as appropriate for the liberated mind, calling them "the best instruments…for the mind's complete knowledge of philosophic truth" (Grant 1999a, p. 103). The celebrated cathedral school in Chartres, France, was marked by a Platonic orientation, and its chancellor, Thierry of Chartres, argued that the study of mathematics led directly to the knowledge of God (Grant 1999a, p. 104; McInerny 1983, pp. 254–255).

Yet though these educators knew that there was a great ancient tradition in mathematics, they did not know much about its substance. This situation changed with the flowering of translation activity in the twelfth century that has become known as the Twelfth-Century Renaissance (Haskins 1927). Europeans came to know about Greek works, mostly in their Arabic versions as studied in the Islamic world, through travel, contacts in southern Italy, and the European reconquest of parts of Spain. Because of the prestige of mathematics in the liberal arts tradition and also because of the importance of mathematics and astronomy and astrology for social and religious needs, major translators like Gerard of Cremona and Adelard of Bath assiduously searched for the principal works of men like Ptolemy and Euclid to translate from Arabic into Latin.

By the end of the twelfth century, major works of Greek mathematics and astronomy in their entirety, most of Aristotle's philosophy, and the books on algebra and arithmetic of Muḥammad ibn Mūsā al-Khwārizmī were all available in Latin. In the thirteenth century and beyond, with the rise of medieval universities, scholars worked to assimilate the expanded ancient traditions into medieval Christian culture. Aristotle in particular presented a sophisticated approach to philosophy and a range of subjects, from biology to metaphysics, that did not fit the old patterns. The new materials needed to be integrated with the old liberal arts as well as with Christian thought and practice.

Still, though the educational ideal may no longer have been the study of liberal arts for their own sake (McInerny 1983, pp. 248–9), many medieval thinkers from the thirteenth century on continued to value the traditional liberal arts, including mathematics, at least as stepping-stones to a religiously meaningful life. Thomas Aquinas (1225–1274) explicitly recognized the Seven Liberal Arts in general and mathematics in particular as "paths preparing the mind for the other philosophical disciplines," although he equally explicitly said that they were by no means sufficient for the Christian studying philosophy (Kimball, pp. 66–67; McInerny 1983, p. 251). Men like Roger Bacon (1214–1294) advocated mathematical instruction, saying that without mathematics "nothing of supreme moment can be known" in any other science – a more applied kind of mathematics, perhaps, but still seen as essential for learning (Grant 1999b, p. 199; Masi 1983, p. 151). Research in fields like geometrical optics, infinite series, and the graphing of variable magnitudes was undertaken in the fourteenth century by men like Robert Grosseteste, Richard Swineshead, and Bishop Nicole Oresme (Katz 2009, pp. 324–363).

However distinguished medieval mathematical research was, though, scholars of the Renaissance self-consciously reached back beyond it in their enthusiasm to revive the glories of the ancient mathematical traditions.

25.2.3 The Renaissance and After

The glories Renaissance thinkers saw themselves reviving included Plato as well as Aristotle and also included considerably more mathematics. Especially striking in light of what we now call the liberal arts is the use of geometry in perspective in painting, with men like Piero della Francesca, Leonardo da Vinci, and Albrecht Dürer highly proficient in both geometry and art (Field 1997). Also striking is the new mathematization of music, again based on supposed Greek models (Fauvel et al. 2003). And the ideal of the Renaissance man, competent in many areas of knowledge, foreshadows the modern idea of a well-rounded student with a general liberal education. In urban and court culture, liberal education was occasionally extended also to women (Cruz 1999, pp. 250, 252; Grendler 1989, pp. 93–102; Schiebinger 1989, p. 12).

One influential Renaissance educational theorist, Pier Paolo Vergerio, wrote a treatise in 1404 that emphasized the importance of liberal studies, encompassing history, moral philosophy, poetry, but also the traditional Seven Liberal Arts

(Cruz 1999, p. 243). Later on as artists became more prestigious, drawing and thus a bit of geometry became more important liberal subjects. What came to be called humanist education was linked both to Christian piety and to civic values and public service. It is true that Renaissance mathematical teaching was at first quite practical and carried out in the so-called abacus schools, but for many aristocrats such teaching was too closely associated with business and trade and too marked by rules of thumb. So in courts and in the households of the social elite, men of the stature of Luca Pacioli and Mauricio Commandino taught more theoretical mathematics, and courtly patrons of humanists included mathematicians at their courts as well (Rose 1975, p. 293). Influential humanist teachers like Vittorino da Feltre (1378–1446) of Mantua regarded mathematics as a crucial component of humanistic education, and he even included some women among his students (Cruz 1999, p. 251; Rose 1975, p. 16).

University mathematics teaching in Renaissance Italy used texts like Euclid, Sacrobosco's *Sphere*, Ptolemy's *Almagest*, and the Alphonsine astronomical tables (Marr 2011, pp. 62–64). And as the humanist curriculum expanded from Italy into France, England, and Germany, mathematics came with it (Cruz 1999, p. 249). In Germany, Philipp Melanchthon in the 1520s gave special emphasis to the place of mathematics and astronomy in the university curriculum (Cruz 1999, p. 240; Westman 1975, p. 170). Johann Sturm (1507–1589) in Strasbourg also went beyond grammar and rhetoric and Biblical texts in having his students study mathematics and science (Cruz 1999, p. 249).

Humanism was not the only impetus to expanding the teaching of liberal arts. There was a Catholic response to these new trends. In the mid-sixteenth century, the Jesuits begin to conduct schools, teaching not only grammar, philosophy, and theology but also mathematics, geography, history, and astronomy, often using original texts. Jesuit schools grew in number, importance, and influence; by 1600 there were 236 Jesuit colleges, some outside of Europe, and by 1750 there were 669 Jesuit colleges and 24 universities;[3] the Jesuits number Descartes, Voltaire, and Condorcet among their influential pupils. Mary Ward (1585–1645) established a network of humanist schools for girls patterned on the Jesuit model; by 1631 there were many such schools and hundreds of pupils, who were taught mathematics as well as Latin and Greek (Cruz 1999, pp. 252–253). Thus, the sixteenth century produced an educated elite, people who communicated and were grounded in a cultural heritage that transcended the boundaries of language, of gender, of region, and even the distinction between Protestant and Catholic. The liberal arts continued to be thought of as providing intellectual discipline, as well as teaching moral examples and civic duty.

Humanism's deep interest in rediscovering, translating, and circulating ancient Greek texts reinvigorated mathematics with new sophisticated sources. Indeed, Francesco Maurolico (1494–1575) spoke explicitly of a "renaissance of mathematics." This "renaissance" brought advanced mathematical texts – Archimedes, Pappus, Apollonius, and Diophantus – into European mathematics and also included a drive to restore parts of the text which had not survived (Rose 1975, p. 179). Such

---

[3] See Chapple (1993a, p. 7), Cesareo (1993, p. 17), Cruz (1999, p. 250), and Taton (1964).

attempts at restoration themselves became real mathematical research, like Descartes' and Fermat's reconstruction of works of Apollonius using modern algebraic methods, and, not so incidentally, developing analytic geometry. A similar example is found in Fermat's notes on Diophantus's *Arithmetica*, notes that jump-started modern number theory, including the conjecture now called Fermat's Last Theorem (Katz 2009, pp. 498–499; Rose 1975, p. 292). The mathematical Renaissance was embodied in universities, especially in Italy, and what was taught in those universities helped initiate the great achievements of mathematics and science in the seventeenth century (Grendler 2002, pp. 408–429).

In the seventeenth century, mathematics proved overwhelmingly successful in modeling the laws of the cosmos. The Newtonian idea that natural laws were the laws of God linked mathematics to Christianity. The new authority of mathematics led to its institutionalization in university education, as part of the intellectual heritage of any liberally educated man. And the success of what was billed as the new scientific method strengthened the philosophical and political ideas associated with the Enlightenment.

The eighteenth and nineteenth centuries saw an even greater role for mathematics as useful in warfare and statecraft but also as intellectual enrichment and discipline. Women as well as men participated in the latter category. The journal "Ladies' Diary," from 1704 to 1841, was dedicated to teaching women the mathematical sciences and contained problems and puzzles which were successfully solved by female readers (Schiebinger 1989, p. 41). Mathematics as intellectual enrichment and discipline is seen across the board in educational institutions as different as the universities designed to provide a gentleman's education like Oxford; Harvard and other early American colleges; Cambridge University, which since the late eighteenth century was dominated by the mathematical Tripos examination; and the German universities dedicated to pure research, together with their American followers like Johns Hopkins and the University of Chicago (Merz 1904, pp. 89–301; Rudolph 1962, pp. 244–286, 349–354). The liberal arts ideal in the colleges in the Colonial period of the United States helped inspire the founding of many small American colleges throughout the eighteenth and nineteenth centuries (Rudolph 1962, pp. 44–67).

Columbia College in 1919 pioneered requiring all freshmen to take a course called "Introduction to Contemporary Civilization in the West" based on the liberal arts. In the 1920s with the Progressive movement in America, colleges addressed anew the question of what a liberal arts college ought to be. Influential responses included Alexander Meiklejohn's "Experimental College" at the University of Wisconsin, which began with the study of Plato; the complete Great Books curriculum founded in the 1930s and still in force at St. John's College in Maryland; and the explicit classicism of Robert Maynard Hutchins and Mortimer Adler who in the 1930s began the Core Curriculum at the University of Chicago. With the advent of the Cold War between the United States and the Soviet Union, liberal arts education was promoted as developing the free and inquiring intellect in opposition to totalitarianism; this view was influentially expressed in the book *General Education in a Free Society: Report of the Harvard Committee* (1945). All these and their many followers explicitly intended to revive the ancient liberal arts tradition in twentieth-century society (Rudolph 1962).

As a result of all these trends, the ideal of general or liberal education, in opposition both to premature specialization and to vocationalism, continues to structure most university-level education in the United States today. The category "liberal arts colleges," listing hundreds of such schools, appears in almost all guidebooks to American higher education; for instance, the *U. S. News and World Report* list gives data on 266 such schools. Colleges of this type are being developed on the American model in other countries as well, including New York University in Abu Dhabi; Xing Wei College in Shanghai, China; Bard College's partner institutions in St. Petersburg, Russia, and in Berlin; Yale University in Singapore; and Smith Women's College in Malaysia. Quest University in British Columbia is Canada's first private liberal arts institution. And mathematics remains part of such liberal arts education.

As for pre-university education, teaching mathematics is almost universal. Although the prevalence of mathematical instruction may be in some part due to mathematics' place in the traditional liberal arts curriculum, many other justifications for its inclusion are given in the modern world: preparing students for the vastly increasing number of modern technical careers, helping students understand science and economics, aiding citizenship, and expanding and training the mind to take on any intellectual challenge.

Rarely, though, is teaching mathematics justified by saying that mathematics is and has always been at the heart of the disciplines we now identify as the liberal arts. Furthermore, in the past, "liberal arts" always designated an education for an elite. The study of the liberal arts in general, and mathematics as a liberal art in particular, was never extended to an entire citizenry. To educate modern citizens for a free society, though, this essay will argue that it is necessary to go beyond teaching mathematics solely as the prerequisite to something else.

Elsewhere the author has sketched the history of the central role mathematics has played in Western thought (Grabiner 1988, 2010). The reasons for that central role go far beyond simply following the pattern of traditional education. As will be shown in the next sections, that central role has been the result of two aspects of mathematics. One is that mathematics appears to provide truths, truths that can be proved. The other is that mathematical ideas work to produce knowledge of the world, in areas ranging from the arts to the study of nature. Philosophers of mathematics often call these aspects of mathematics "certainty" and "applicability." Historically, these aspects of mathematics have often been invoked as slogans by those advocating including mathematics in education. But the history goes far beyond slogans. Together, these two aspects of mathematics – certainty and applicability – explain and illuminate both how and why mathematics has been central throughout the history of the liberal arts. Demonstrating that central role throughout history justifies making mathematics a key part of liberal education in the world today. To see mathematics in this central role, we will look first at mathematics as a provider and exemplar of truth.

25.3 Mathematics, Truth, and Proof

25.3.1 The Greek Model

As in all of Western thought, one must begin with Plato and Aristotle. Plato said that the certainty of mathematics comes from the perfection of its subject matter. In the natural world, everything changes, comes into being, and passes away; the objects of mathematics, by contrast, are unchanging and eternal. Mathematics provides a model for Plato's philosophy of ideal Forms that transcend our changing experience: the idea of justice, the ideal state, and the idea of the Good. In the Western tradition, the common terms "certain" and "true" preserve Plato's belief in an unchanging transcendent reality, and Plato consistently argued for it using examples from mathematics. And, as discussed earlier, Plato decreed that the philosopher-ruler of his ideal *Republic* should study mathematics to be brought to the truth.

For Aristotle, though, the truth of mathematics comes more from its method than from its subject matter, since the results of mathematics can be demonstrated logically, proceeding from self-evident assumptions and clear definitions. Aristotle held that other subjects could also gain certainty if they could be put into the same form, the form of what he called a demonstrative science (Heath 1925, pp. 117–121; McKirahan 1992). Thus mathematics, especially Euclid's geometry, came to be seen as a model for much of scientific and philosophical reasoning. For instance, medieval theologians tried to demonstrate the existence of God from first principles. In 1675, Benedict Spinoza wrote an *Ethics Demonstrated in Geometrical Order* (Spinoza 1953), with explicitly labeled axioms and definitions, and including theorems like "God or substance consisting of infinite attributes … necessarily exists," whose proof he closes with the letters QED. Isaac Newton's great *Mathematical Principles of Natural Philosophy* of 1687 has the same definition-axiom-proof structure. Newton called his famous three laws of motion "axioms," and he labeled and proved the fundamental laws of his mechanics as theorems (Newton 1934).

The American *Declaration of Independence* is another example of an argument whose authors tried to inspire faith in its certainty by using the Euclidean form. "We hold these truths to be self-evident," the mathematically sophisticated author, Thomas Jefferson, began his argument, "that all men are created equal" (Becker 1922; Cohen 1995). Another self-evident truth in the *Declaration* is that if any government fails to secure human rights, it is the right of the people to alter or abolish it. The second section of the *Declaration* makes clear that this is a proof: it says that King George's government does not live up to the postulates, followed by the words "to *prove* this, let facts be submitted to a candid world" (Italics added). And the actual declaration of American independence is in fact the conclusion of an argument, so it begins with a "therefore": "We, *therefore* … declare, that these United Colonies are, and of right ought to be, free and independent states" (Italics added).

The same model of reasoning also pervades the law, and its mathematical antecedents are sometimes explicitly recognized. For instance, Christopher Columbus

804   J.V. Grabiner

Langdell, the pioneer of the case method in American legal education and Dean of Harvard Law School in the 1870s, was part of a long tradition that saw law as a science very much like geometry (Hoeflich 1986; Kalman 1986, p. 3; Seligman 1978). Law, according to Langdell, is governed by a consistent set of general principles. One gets these principles by looking at individual cases by induction, according to Langdell, in a manner reminiscent of Newton's *Principia.*[4] But once one has the general principles, one proceeds as in geometry. The correct legal rules are to be logically deduced from those general principles and then applied to produce the correct legal ruling in line with the facts of a particular case (LaPiana 1994, p. 3; Seligman 1978, p. 36). In the theory of law as in liberal arts like philosophy and the study of nature, the Euclidean model of reasoning has shaped conceptions of proof, truth, and certainty.

25.3.2   Symbols and Algorithms

The ability of mathematics to exemplify truth was not limited to geometry. Between the Renaissance and the eighteenth century, the paradigm governing mathematical research changed from a geometric one focused on proof to an algebraic and symbolic one. In algebra just as in Aristotle's view of demonstrative sciences like geometry, one can consider the method independently of the particular subject matter. The algebraic or algorithmic method in mathematics finds truths by manipulating symbols according to fixed rules. The algorithmic approach long preceded symbolic algebra, entering Europe in the Middle Ages.

Influential in developing the idea of "algorithm" was the twelfth-century Latin translation of a work on the Hindu-Arabic number system by the ninth-century mathematician Muḥammad ibn Mūsā al-Khwārizmī. This system is the base-10 place-value system now universally learned in school, and the new computational power it produced was enormous. Multiplication is relatively easy in a place-value system, since multiplying by 3 million is as simple as multiplying by 3. The rules for producing new truths involving these numbers, including rules not only for ordinary arithmetic but also for more intricate problems like taking square roots, were called the "method of al-Khwārizmī." Later his name became Latinized into Algorismus and then, perhaps by association with "arithmos" for number, into "Algorithmus," whence the modern term "algorithm" for any set of powerful rules that can be easily and mechanically applied. The algorithms used to calculate with the Hindu-Arabic numbers are now known to every schoolchild. Equally important, these algorithms were eventually seen as similar to the later manipulations of symbolic algebra.

---

[4] Distinguishing experimental philosophy from reasoning from arbitrary hypotheses, Newton wrote, "In this philosophy particular propositions are inferred from the phenomena, and afterwards rendered general by induction" (Newton 1934, p. 547). Once Newton had his general principles, his *Principia* could take the logical structure familiar from Euclid's *Elements.*

25 The Role of Mathematics in Liberal Arts Education

The Arabic word "algebra," derived from the Arabic title of another book by al-Khwārizmī, *Kitāb al-jabr wa l-muqābala*,[5] the first Arabic-language treatise on algebra, at first designated the systematic study of the processes of solving equations, with the equations expressed in words. But the word "algebra" today first brings algebraic symbolism to mind. In the 1590s, François Viète systematically developed and exploited general symbolic notation much like that used currently (Klein 1968; Struik 1969, pp. 74–81). This now familiar invention has the power to produce abstract, general truths in mathematics. For instance, given any pair of distinct numbers, say, *7* and *9*, schoolchildren are taught that not only does *7 + 9 = 16*, so does *9 + 7*. There are of course infinitely many such examples. Viète's general symbolic notation for the first time allows the writing down of the infinite number of such facts all at once:

$$B + C = C + B.$$

A century later, Isaac Newton summed up the power and generality of Viète's innovation by calling algebra "universal arithmetic." Newton meant that we could derive and prove general algebraic truths from the universal validity of the symbolic manipulations that obey the laws of ordinary arithmetic.

More can be learned from a less trivial example than adding two numbers. First consider the quadratic equation

$$2x^2 - 11x + 14 = 0.$$

Imagine being told, "*2* and *3 ½* are the solutions." But being given these solutions provides no information about how those answers were obtained. However, with general symbolic notation, it is clear that *every* quadratic equation has the form

$$ax^2 + bx + c = 0,$$

where *a* is any nonzero real number and *b* and *c* can be any real numbers.

Solving the *general* equation by the algebraic technique of completing the square gives the general solution, the well-known quadratic formula:

$$x = \left[-b \pm \sqrt{\left(b^2 - 4ac \right)} \right] / 2a.$$

Unlike the numbers *2* and *3 ½* in the original example, numbers which could have been produced by many possible arithmetic operations, each term in the general solution reveals the way it was produced; for instance, the term *ac* was

---

[5] Khwārizmī's title can be translated as "the book of restoring and balancing," where the Arabic "al-jabr" or "restoring" was interpreted as adding the same thing to both sides of an equation and "al-muqabala" or "balancing" the subtraction of the same quantity from both sides of an equation (Berggren 1986, p. 7). The sense of "al-jabr" as "restoring" remains in Spanish, where, for instance, in *Don Quixote*, Part II, Chap. XV, a bonesetter is an "*algebrista*" (Merzbach and Boyer 2011, p. 207).

produced by multiplying the coefficients a and c. Thus, the general solution to the quadratic equation preserves the record of every operation performed on the coefficients to find that solution. In the original example, $a = 2, b = -11, c = 14$, so it is now apparent how the answers 2 and $3\ 1/2$ were obtained from the coefficients in the equation. More important, the process of finding the solutions by completing the square proves in general that these and only these are the solutions.

Recognizing the generality and problem-solving power of symbolic algebra, in the 1630s René Descartes and Pierre de Fermat, working independently, combined the method of geometry and proof with the method of symbolic algebra into a new subject, analytic geometry. Problems in geometry, Descartes said, could be solved by translating geometric relationships into algebraic expressions, manipulating the algebraic expressions according to the rules of algebra, and translating the results of these manipulations back into the geometric solution of the original problem. He and Fermat, using the insight that problems could be translated back and forth between algebra and geometry, solved a range of previously intractable problems, and Fermat, in his geometric work, anticipated some of the discoveries of the calculus (Boyer 1956; Mahoney 1973).

Furthermore, Descartes' success in devising a new method for solving problems that had stumped the ancient Greek geometers helped him conclude that "method" was the key to all progress. When he theorized about this in his *Discourse on Method*, though, he drew on the structure of Euclidean geometry as well as that of algebra. He argued that the method of making discoveries begins with analyzing the whole into the correct "elements" from which truths could later be deduced. "The first rule," he wrote in the *Discourse*, "was never to accept anything as true unless I recognized it to be evidently such."

The second rule "was to divide each of the difficulties which I encountered into as many parts as possible, and as might be required for an easier solution...." Then, "the third rule was to [start] ... with the things which were simplest and build up gradually toward more complex knowledge" (Descartes 1637a, Part II, p. 12). A mathematical example, from Descartes' *La Géometrie* (Descartes 1637b), is building up a polynomial from a set of linear factors, making visible the truth that a polynomial equation has as many roots as the polynomial's highest degree. Descartes' rules in the *Discourse* mirror his "rules of reasoning in philosophy" and are part of a long tradition of arguing by means of analysis and synthesis (Gaukroger 1995, pp. 114, 124–126, 180). But later thinkers took Descartes himself as a starting point, considering him "the figure who stands at the beginning of modern philosophy" (Gaukroger 1995, p. vii), and his influence on subsequent philosophy, from Locke's empiricism to Sartre's existentialism, has been enormous. For the purposes of this essay, the key point is the large debt Descartes' philosophical views about method owe to his ideas about mathematics.

Later on in the seventeenth century, Gottfried Wilhelm Leibniz was so inspired by the power of algebraic notation to simultaneously make and prove mathematical discoveries that he invented an analogous notation for his new differential calculus. Leibniz's dy/dx and $\int ydx$ notation is still prized and used because of its heuristic power. In fact what Leibniz meant by choosing the term "calculus" was that he had

25 The Role of Mathematics in Liberal Arts Education

invented a set of algorithms for operating with the differential operator d (Leibniz 1969). Furthermore, Leibniz envisioned an even more general symbolic language that would be able to establish the indisputable truth in all areas of human thought. Once there was such a symbolism, which Leibniz called a "universal characteristic," he predicted that if two people were to disagree, one could say to the other, "let us calculate, sir!" and the disagreement would be resolved (Leibniz 1951). These ideas of Leibniz make him a pioneer in what has become the modern philosophical discipline of symbolic logic.

Other seventeenth-century thinkers also pointed out the algorithmic and mechanical nature of thought. For instance, Thomas Hobbes wrote, "Words are wise men's counters, they do but reckon by them" (Hobbes 1939, Chap. 4, p. 143). By the eighteenth century, not only did many mathematicians think that discovery and proof should be based on abstract symbolic reasoning but prized such reasoning above intuition and geometry. For instance, in 1788 Joseph-Louis Lagrange wrote his *Analytical Mechanics* with no diagrams whatsoever. Other scientists inspired by this ideal introduced analogous heuristically powerful notations in their own fields. For instance, Antoine Lavoisier and Claude-Louis Berthollet developed a new chemical notation that Lavoisier called a "chemical algebra" (Gillispie 1960, p. 245). Anyone who has ever balanced a chemical equation has benefited from their innovation.

The success of these ideas about symbolism, both within and beyond mathematics, led the Marquis de Condorcet to write that algebra "contains within it the principles of a universal instrument, applicable to all combinations of ideas" and to go so far as to say that the general algebraic method could "make the progress of every subject embraced by human intelligence... as sure as that of mathematics" (Condorcet 1793, p. 238; pp. 278–279). He used these ideas to support his central thesis that "the progress of the mathematical and physical sciences reveals an immense horizon ... a revolution in the destinies of the human race" (Condorcet 1793, p. 237). Here his view epitomizes the Enlightenment idea of progress in its clearest form. In the nineteenth century, George Boole produced the first modern system of symbolic logic and used it to analyze a wide variety of complicated arguments (Boole 1854). His system, developed further, underlies the logic used by digital computers today, including applications ranging from automated theorem-proving to translators, grammar checkers, and search engines, approaching a full embodiment of Condorcet's dream of the algebraic method embracing every subject.

25.3.3 The Method of Analysis

The second rule in Descartes' *Discourse on Method*, the idea of divide and conquer, fits beautifully with the Greek atomic theory, which had just been revived in the seventeenth century. If all matter is made up of small particles, one could analyze the properties of the whole on the basis of the properties of the parts. This idea became central to both chemistry and physics, and indeed still is. Familiarity with these ideas also permeates art and literature, in examples as different as the nineteenth-century pointillism of Georges Seurat and the conclusion of the poem "Mock on, mock on, Voltaire, Rousseau" by William Blake (1757–1827):

> The Atoms of Democritus
> And Newton's Particles of Light
> Are sands upon the Red Sea shore,
> Where Israel's tents do shine so bright.

Another line of influence of Descartes' divide-and-conquer method can be seen in economics, notably in Adam Smith's 1776 *Wealth of Nations*. Smith analyzed the competitive success of economic systems by using the concept of division of labor. He explained how the separate elements of the economy, with each one acting as efficiently as possible, combine to produce the economic system's overall prosperity. Famously, Smith said that each individual in the economy, while consciously pursuing only his individual advantage, is "led as if by an Invisible Hand to promote ends which were not part of his original intention" (Smith 1974, p. 271), the optimal outcome for the entire society.

In France after the Revolution came another application of the divide-and-conquer method, inspired directly by Smith's views. Gaspard François de Prony had the job of calculating a set of logarithmic and trigonometric tables. He undertook to do this by applying Smith's idea about the division of labor. Prony described a hierarchical divide-and-conquer system to produce the tables. First, mathematicians decide which functions to use; then, technicians reduce the job of calculating the functions to a set of simple additions and subtractions of preassigned numbers; and finally, a large number of low-level human "calculators" carry out the actual additions and subtractions.

In England, Charles Babbage took Prony's analysis of large-scale mathematical calculation and embodied it in a machine, the first digital computer ever conceived (Hyman 1982). Babbage described the basic idea in a chapter elegantly called "On the Division of Mental Labour" (Babbage 1832, Chap. XIX). Mathematicians were to decide what the machine would do and with what numbers it would do it, and then a machine could carry out the low-level task of performing the additions and subtractions. Babbage was a follower of Leibniz's views on the power of notation to make mathematical calculation mechanical. So, the first modern computer owes much both to the analytical method that Descartes promoted and to Leibniz's ideas about algorithms.

25.4 Mathematics Versus Skepticism

The fact that there is a subject, mathematics, which seems to be able to find irrefutable truth, has been philosophically powerful in other ways than those so far described. Since the existence of mathematics supports the conclusion that some sort of knowledge truly exists, the success of mathematics has long been used as a weapon against skepticism. For instance, Plato, going beyond his teacher Socrates' critical method, used mathematical examples repeatedly to show that learning and knowledge were possible. In the 1780s, Immanuel Kant used the example of Euclidean geometry to show that there could indeed be non-tautological knowledge that is independent of sense experience and thus argued that metaphysics, skeptics like David Hume to the contrary, is also possible (Kant 1950; Kant 1961). This same point – that mathematics is knowledge, so that objective truth does exist – is convincingly conveyed when Winston Smith, the protagonist in George Orwell's novel *1984*, heroically asserts, in the face of the totalitarian state's overwhelming power over the human intellect, that two and two are four.[6]

Another way that mathematics, as an example of certain knowledge, has challenged skepticism is by providing an answer to what has been called the problem of the criterion (Popkin 1979). If all sides to a controversy seem to disagree, what is the criterion by which the true answer can be recognized? Of course, if there were only one system of thought available, people might well accept it as true, a situation somewhat like the status of Roman Catholicism in the Middle Ages. But the Reformation presented alternative religious systems, the Renaissance revived the thought of pagan antiquity, and Cartesianism and the new science of the seventeenth century provided further challenges. Now finding a criterion that could identify the true system seemed urgent. But mathematics seemed to have solved this problem.

What, then, asked philosophers, was the sign of the certainty of the conclusions of mathematics? The fact that nobody disputed them. So the criterion of truth, many seventeenth- and eighteenth-century thinkers concluded, was universal agreement. Voltaire elegantly summed up this conclusion when he wrote, "There are no sects in geometry. One does not speak of a Euclidean, an Archimedean" (Voltaire 1901a, c). What every reasonable person agrees upon, that is the truth. Applying this to religion, Voltaire observed that some religions forbid eating beef, some forbid eating pork; therefore, since they disagree, they both are wrong. But, he continued, all religions agree that one should worship God and be just; that must therefore be true. Applying the same idea to ethics, Voltaire said, "There is but one morality… as there is but one geometry" (Voltaire 1901a, b).

25.5 Mathematics and Its Applications

But mathematics is more than an exemplar of truth and certainty; it also works in the world, not only the world of engineers and bankers but the world of the liberal arts as well. But why should mathematics apply to the world at all? For Plato, it is because this world is an approximation to the higher mathematical reality. For Aristotle, on the other hand, mathematical objects are abstracted from the physical world by the intellect. Later empiricists, such as John Stuart Mill, have agreed with Aristotle. However, since mathematical ideas often are applied to situations

---

[6] The point will be clearer with a fuller quotation: "With the feeling…that he was setting forth an important axiom, he wrote: *Freedom is the freedom to say that two plus two make four. If that is granted, all else follows*" (Orwell 1949, p. 81; his italics).

810 J.V. Grabiner

quite different from those in which they arose, the empiricist answer seems insufficient. In any case, Plato's answer has wielded great influence.

25.5.1 Mathematics and Nature

From the ancient Pythagoreans onward, many thinkers have looked for the mathematical reality beyond the appearances. In the sixteenth and seventeenth centuries, Copernicus, Kepler, Galileo, and Newton looked for that mathematical reality – and found it in the laws that govern the physics of motion and the behavior of bodies in the solar system. The Newtonian world-system that completed the Copernican revolution was embodied in a mathematical model, based on the laws of motion and inverse-square gravitation, and set in Platonically absolute space and time (Cohen 1980, pp. 63–67; Newton 1934, pp. 6–9). The success of Newton's physics not only strongly reinforced the view that mathematics was the right language for science but also strongly reinforced the emerging ideas of progress and of truth based on universal agreement.

These successes engendered important theological and philosophical implications. The mathematical perfection of the solar system could not have come about by chance, argued Newton. The cause of this perfection had to be an intelligent designer, God, who chose to create the universe in a pattern so well suited to humanity. The search for other examples of design and adaptation in nature inspired considerable research in natural history, especially on adaptation, and this research was an essential prerequisite to Darwin's discovery of a nontheological explanation for this adaptation: evolution by natural selection.

25.5.2 Mathematics and the Arts

Other examples of the uses of mathematics come from the arts. The mathematical theories that underlie music began with the Pythagoreans and have continued since, revived and expanded upon beginning in the Renaissance. The modern study of pitch, intensity of sound, meter, and the psychology and physiology of hearing, to say nothing of the technologies of musical reproduction, all have a sophisticated mathematical basis.[7]

The same is true of the visual arts. In the Renaissance, stimulated by the rediscovery of Euclid's geometrical work on optics, painters used geometry to give the viewer the visual sense of three dimensions. Several Renaissance artists did original work in geometry, notably Piero della Francesca (1410–1492) and Albrecht Dürer (1471–1528). Piero's *De prospectiva pingendi*, the first mathematical treatise on

---

[7] (Fauvel et al. 2003; Field 2003; Helmholtz 1954; Karp 1983; Jeans 1956; Newman 1956, pp. 2278–2309 Wardhaugh 2009; Wollenberg 2003)

25 The Role of Mathematics in Liberal Arts Education

perspective for painting, showed geometrically how to depict objects in three dimensions, viewed from a particular standpoint, on the picture plane. Dürer's *Underweysung der Messung* of 1525, the first geometric text written in German, included applications of geometry to constructing regular polygons and polyhedra, to architecture, and to typography, and Dürer was the first to show how to project three-dimensional curves onto two perpendicular planes. The work of these artist-mathematicians helped direct attention to many of the key ideas of what, in the seventeenth-century work of Girard Desargues and Blaise Pascal, became the new mathematical subject of projective geometry (Field 1997; Kemp 1990).

25.5.3 Optimization

An especially striking example of mathematical applicability, which links mathematics with science, philosophy, and theology, is given by the history of optimization. The use of optimal principles to explain the world goes back at least to the first century CE when Heron of Alexandria showed that the law of equal-angle reflection of light minimizes the distance the light travels. In the seventeenth century, Fermat showed that Snell's law of the refraction of light minimized what Fermat called the light's "path" (distance times resistance), which, since he assumed that velocity varies inversely with resistance, is mathematically equivalent to saying that light follows the path that minimizes its time of travel (Mahoney 1973, p. 65, pp. 382–390).

Leibniz, using his newly discovered calculus, produced algorithms for finding maxima and minima and applied them to elegantly re-derive Fermat's result (Leibniz 1969; Struik 1969, pp. 278–279). In his philosophy, Leibniz argued that the universe itself is constructed by God according to optimal principles. For Leibniz, a possible world is one consistent with the laws of logic; those possible worlds with more different beings in them are better than the others, and our world is the best of all possible worlds because it is the one in which the total of existing things is maximized (Lovejoy 1936, pp. 50, 144–146, 173).

In the eighteenth century, Colin Maclaurin, when aged sixteen, used the calculus, and some theological assumptions about eternal life, to argue that the Christian doctrine of salvation maximizes the future happiness of good men (Maclaurin 1714; Tweddle 2008). This argument applied the same methods many mathematicians used to apply the principle of least action in physics, as well as to design the most efficient windmills and waterwheels. Maclaurin's classmate at Glasgow, the philosopher Francis Hutcheson, used the same idea of mathematical optimization to demonstrate his laws of virtue. For instance, Hutcheson wrote in 1728, "That Action is best, which procures the greatest Happiness for the greatest Numbers" (Hutcheson 2004, Sect. III, p. 177). Later on, a similar approach is found in the utilitarianism of Jeremy Bentham, embodied in the famous phrase "the greatest good for the greatest number."

This line of reasoning recalls the ideas of Adam Smith, who had been a student of Hutcheson's at Glasgow. In words resembling Hutcheson's, Smith influentially wrote, "Upon equal...profits...every individual naturally inclines to employ his capital in the manner in which it is likely to afford the greatest support to domestic industry, and to give revenue and employment to the greatest number of people of his own country" (Smith 1974, Book IV, Chap. II). From Smith's work, a set of ideas common not only to mathematics but also to philosophy and theology has entered the vocabulary of the most hardheaded of economists.

25.5.4 The Social Sciences

Just as the example of mathematical truth made finding truth elsewhere seem possible, so the examples of applying mathematics to natural science inspired those seeking to perfect other disciplines. This was especially true for the early nineteenth-century pioneers of the social sciences, Auguste Comte and Adolphe Quetelet. Both men knew their science, Comte having been influenced principally by Lagrange, Quetelet by Pierre-Simon Laplace. Lagrange's *Analytical Mechanics* of 1788 had claimed to reduce all of mechanics to mathematics. Comte went further: if physics was built on mathematics, so was chemistry built on physics, biology on chemistry, psychology on biology, and finally his own new creation, *sociology* (a term Comte coined) would be built on psychology (Comte 1830, Chap. 11).

Comte said that science had once been theological, invoking God, then meta-physical, invoking general philosophical principles, but now science, including social science, would be based only on observed connections between things, a stage of science he called "positive," stimulating the beginning of the philosophical stance called positivism. Comte's philosophy, owing much to mathematical physics, influenced not only twentieth-century logical positivism but also the views on science and history held by Ludwig Feuerbach (1804–1872) and Karl Marx.

Still, Comte did not develop a quantitative social science. Here the prime mover was Quetelet, for whom the applicability of mathematics was crucial. "We can judge of the perfection to which a science has come," he wrote in 1828, "by the ease with which it can be approached by calculation" (Quetelet 1828, p. 233). Quetelet was especially impressed by Laplace's use of the normal curve of errors to determine planetary orbits from observations. Quetelet found empirically that not only the distribution of measurement errors but also the distribution of many human traits, including height and chest circumference, gave rise to the same normal curve. From this, he defined the statistical concept and the term, "average man" *(homme moyen)* (Porter 1986, p. 52). These ideas are essential to modern social science.

25.5.5 Freedom and Determinism

Quetelet observed also that many social statistics, such as the number of suicides in Belgium, produced roughly the same figures every year. One might think that crimes

25 The Role of Mathematics in Liberal Arts Education

are the result of free individual choice. But the constancy of these rates over time, he argued, suggests that murder or suicide has constant social causes. Quetelet's discovery of the constancy of crime rates raised an important philosophical question: Is human behavior determined by social laws, as he seemed to think, or are we free to choose our fate?

Laplace, even though he needed probability to do physics, did not believe that the laws governing the universe were ultimately statistical. Since the true causes are not yet known, Laplace said, people believe that events in the universe depend on chance. But in fact everything is determined. To an intelligence which knew all the forces in nature and the exact situation of the beings that composed it, said Laplace, "nothing would be uncertain" (Laplace 1951, Chap. II).

Later in the nineteenth century, though, the mathematical physicist James Clerk Maxwell, in his work on the statistical mechanics of gases, argued that statistical regularities in the large reveal nothing at all about the behavior of any individual. Maxwell cared about this point because it allowed for free will. Maxwell considered such issues not just because of physics but because he had read and pondered the work of Quetelet on the application of statistical thinking to society (Porter 1986, pp. 118–119). A similar dispute about the meaning of probabilistically stated laws has arisen in modern debates over the foundations of quantum mechanics.

So, discussions of basic philosophical and theological questions like "is the universe an accident or a divine design?" for Newton and Leibniz, "is there free will or are we all programmed?" for Quetelet and Maxwell, or "are the laws of nature ultimately statistical?" for Laplace, Maxwell, Bohr, and Einstein owe much to questions about the applicability of mathematics to the world and to society.

25.6 When "Mathematics as Universal Truth" Fails

So far, although the people discussed have disagreed about philosophical matters, they have not argued about the essential truth of mathematics itself. However, universal agreement about mathematics and its relationship to the natural and social worlds did not survive the nineteenth century, and the older ideas of universal mathematical truth and the consequent universal agreement have had an interesting trajectory since then. To sketch this trajectory, one must ask what happened to the rest of thought when the very nature of mathematics and its relationship to the world seemed to change. The place to begin is with the overturning of the long-held view that there is only one geometry.

25.6.1 Questioning Euclidean Geometry

Since the time of Euclid, mathematicians had viewed his Fifth Postulate as considerably less self-evident than the others. Although Euclid's postulate is sometimes called the "Parallel Postulate," it does not mention parallel lines at all. Instead of the more intuitive postulate "Only one parallel to a given line can be drawn through an outside point" that one finds in many high-school texts today, Euclid's Fifth Postulate was this assumption: "That, if a straight line falling on two straight lines makes the interior angles on the same side less than two right angles, the two straight lines, if produced indefinitely, meet on that side on which are the angles less than the two right angles" (Euclid, *Elements* 1925, Postulate 5; Heath 1925, p. 202).

Over many years since Euclid, mathematicians thought that the postulate should not be assumed but proved. When mathematicians tried to prove it from Euclid's other postulates by indirect proof, they deduced a variety of surprising "absurd" consequences from denying Postulate 5. Among the consequences of denying the postulate are "parallel lines are not everywhere equidistant" and "there can be more than one parallel to a given line through an outside point." In the early nineteenth century, Carl Friedrich Gauss, Janos Bolyai, and Nikolai Ivanovich Lobachevsky each separately recognized that these consequences were not absurd at all, but rather were valid results in a different, equally consistent geometry (Gray 1989, pp. 86–90, 106–124). Gauss chose an appropriate name for the new geometry: non-Euclidean (Bonola 1955, p. 67).

Kant had said that space (by which he meant the only space he knew, Newton's three-dimensional Euclidean space) was the form of all our perceptions of objects, a unique intuition of the mind in which we must order these perceptions (Friedman 1992; Kant 1961). But what, then, can be said about non-Euclidean space? The nineteenth-century psychologist and physicist Hermann von Helmholtz asked whether Kant might be wrong. Could we imagine ordering our perceptions in a non-Euclidean space? Yes, Helmholtz said, if we consider the world as reflected in a convex mirror (Helmholtz 1962, pp. 240–241). The reader can do this by consulting M. C. Escher's famous drawing "Hand with Reflecting Sphere" or by using a car's side mirror that carries an explicit warning that the space we see in it is not Euclidean. In the convex mirror, parallel lines, defined by Euclid as lines on the same surface that do not meet, are no longer seen as equidistant.

Now the philosophy of mathematics as universal truth was under attack. Euclidean and non-Euclidean geometry give the first clear-cut historical example of two mutually contradictory mathematical systems, of which at most one can actually represent the world. This suggests that mathematical axioms are not self-evident truths related to the world at all, but may be, as Helmholtz argued, empirically based. Our idea of space is gained through sight and through touch. Two-dimensional beings living on a visibly curved surface, said Helmholtz, would invent a non-Euclidean geometry. So, Kant to the contrary, our idea of space is not unique, let alone necessarily Euclidean. To a non-empiricist, our ideas of space may even be intellectually free creations.

Inspired by Helmholtz's philosophy of geometry, the English mathematician and philosopher William Kingdon Clifford (1845–1879) discussed the matter in the broadest possible historical and philosophical context. Clifford said, "It used to be that the aim of every scientific student of every subject was to bring his knowledge of that subject into a form as perfect as that which geometry had attained." But no more. "What Copernicus was to Ptolemy," Clifford wrote, "so was Lobachevsky to Euclid" (Clifford 1956, pp. 552–553). Before Copernicus, said Clifford, people thought they knew everything about the entire universe. Now, Clifford stated, we know that we know only one small piece of the universe. The situation is similar in geometry.

Before non-Euclidean geometry, the laws of space and motion implied an infinite space and infinite time, whose properties were always the same, so we knew what is infinitely far away just as well as we knew the geometry in this room. Lobachevsky has taken this away from us, said Clifford. That space is flat and continuous is true just as far as we can explore, and no farther. Speaking statistically about our experience of space, Clifford continued, "If the property of elementary flatness exists on the average, the deviations from it being too small for us to perceive, we would have exactly the conceptions of space that we have now" (Clifford 1956, p. 566). So Clifford, using non-Euclidean geometry, drew conclusions similar to those of the philosophers and scientists who concluded that the laws of nature were ultimately statistical. In fact, Clifford used his ideas to attack the entire Newtonian philosophy of science, especially singling out Newton's idea that human beings can have a universal theory of gravitation that applies to all bodies whatsoever.

Another perspective on these questions about geometry came from the great French mathematician, physicist, and philosopher of science Henri Poincaré (1854–1912), who disagreed both with Kant and with Helmholtz. If, as Helmholtz said, geometry were an empirical science, it would not be an exact science but would be subjected to continual revision. Poincaré of course knew about non-Euclidean geometry, so he knew that we have more than one space in our minds. From this he concluded that geometrical axioms are neither synthetic a priori intuitions (as Kant had said) nor experimental facts (as Helmholtz said). Geometrical axioms, said Poincaré, are conventions. So which set of axioms should be used in geometry? Poincaré said that, as long as we avoid contradictions, "our choice among all possible conventions may be guided by experience, but our choice remains free." The axioms of geometry are really only definitions in disguise. And Poincaré concluded, "What are we to think of the question: Is Euclidean geometry true? The question has no meaning. We might as well ask if the metric system is true, and the old weights and measures are false. One geometry cannot be more true than another; it can only be more convenient" (Poincaré 1952, p. 50). This very modern conclusion might shock Plato and Newton, but makes clear how revolutionary non-Euclidean geometry was for fields outside of mathematics.

The power of the Euclidean model, of course, did not die. Some influential thinkers stood by him. For instance, the famous English economist William Stanley Jevons (1835–1882) thought that Plato was right and Helmholtz was wrong. Transcendental or necessary truth, according to Jevons, is not produced by experience; it is recognized rather than learned (Richards 1988, pp. 87–90). The great English algebraist Arthur Cayley (1821–1895) was also not convinced by Helmholtz's argument that two-dimensional beings living on an obviously curved surface would invent a geometry describing such a curved surface. Cayley said that those beings would in fact invent three-dimensional Euclidean geometry. It would be, he said, "a true system" applied to "an ideal space, not the space of their experience" (Richards 1988, p. 90). And the Dutch philosopher J. P. N. Land thought that Helmholtz's convex-mirror experiment proved nothing, saying in 1877 that the world in the mirror requires practice to interpret in a Euclidean way, but we can learn to do it (Richards 1988, pp. 100–101).

Nevertheless, Clifford's views seemed to fit better with the increasingly empirical nature of nineteenth-century natural science. Popularizers of natural science like the physicist John Tyndall (1820–1893) and the Darwinian paleontologist Thomas Henry Huxley (1825–1895) insisted that people can only claim to know the information received through the senses. And transcendental realities, say Tyndall and Huxley, contrary to Plato and Kant, are both unknown and unknowable (Richards 1988, pp. 104–105). For Clifford and Helmholtz, this unknowability applied to space; for the agnostic Huxley, it applied to God.

The conclusions drawn from the existence of non-Euclidean geometries had a counterpart in nineteenth-century algebra. William Rowan Hamilton devised a noncommutative algebraic system, the quaternions, a system in which the product of two elements ab is not necessarily equal to the product ba. Discoveries like this led mathematicians increasingly to view their subject as a purely formal structure. Algebra was not the general science of number, any more than geometry was the science of space. Mathematics is nothing more than, as the American algebraist Benjamin Peirce put it in 1870, "the science that draws necessary conclusions" (Peirce 1881). The world is no longer, as Plato had thought, an imperfect model of the true mathematical reality. Instead, mathematics provides a set of different intellectual models, which can, but need not, apply to the one empirical reality. The sciences now merely model reality; they no longer claim to speak directly about it. Kurt Gödel applied this view to mathematics itself, using a formal model of mathematical reasoning to prove the surprising result that the consistency of mathematics cannot be demonstrated. Now a philosopher could say that there is no certainty anywhere, not even in mathematics (Barrett 1958, p. 206).

25.6.2 Space and the Social Sciences

Yet the widespread applicability of mathematics to the world meant that the new geometrical ideas would immediately become involved in still other debates, notably in the social sciences and the arts. As the example of Helmholtz indicates, one place where geometry has interacted with social science is in the psychology of perception. Even before the birth of non-Euclidean geometry, Bishop Berkeley and Thomas Reid had pointed out that we do not really see distance; we merely infer distance from the angles that we do see. Consider, for instance, looking upward at the corner in a rectangular room where the ceiling meets the walls. This is perceived as a place where three 90° angles come together. But if one measures the individual angles as they actually appear and are projected onto the viewer's retina, each of the angles is greater than 90° and together they add up to 360°. Our visual space is not the same as the space we claim we see.

And humans do not see parallel lines well either. Helmholtz did an experiment where he asked people in a dark room to put little points of light that got progressively farther away into two lines that always maintained the same distance from each other – that is, parallel lines. But the lines these people made out of these points of light turned out to curve away from the observer. Experiments like these have led psychologists to conclude that visual space is not represented by any consistent geometry (Wagner 2006).

Social scientists have also investigated the Kantian-influenced idea that spatial categories in language are direct projections of humanity's shared innate conceptual categories. As the cultural linguist Stephen Levinson has documented, the evidence suggests otherwise (Levinson 1996; Levinson 2003). In both language and concepts, people in different cultures have other ways of ordering their perceptions than in Euclidean space. For instance, particular directions may have special connotations, and "closeness" can be cultural as well as metrical. Further, some cultures, Levinson reports, use the idea of a fixed coordinate system, having four cardinal directions and referring locations to those, as one does in saying, "the house is north of the tree." But other cultures' concepts are more like Leibniz's idea of space in which there is no absolute space, just the relations between objects, including the observer. This is what one does in saying, "the house is to the left of the tree." No coordinate system is needed. One's perceptions can also be organized by letting the intrinsic properties of an object define the spatial location, as when one says, "The house is on the mossy side of the tree." Some cultures strongly emphasize only one of these methods of ordering objects in space, while others use a variety of these methods (Levinson 1996, pp. 140–145; Levinson 2003, pp. 31–39). There is no universal agreement here.

In modern society, directions can be given by saying, "Go north for five miles, then turn east for two miles," but also by saying, "Keep going straight until you get to the traffic light, and then turn right until you reach the supermarket." The second method, the way GPS systems give directions, follows Leibniz's relational view of space. In fact, GPS navigational systems are changing people's supposedly innate and universal intuitions of space. A Maryland cabdriver who recently bought a GPS told the author, "I used to have the whole geography of greater Baltimore in my head. I don't any more. I think about each trip as, drive to such-and-such exit, then make two right terms, one left turn. And when I leave you off, to get back to the expressway I'll just reverse that – one right turn, then two left turns. I will get back, but I won't know where I've been." Kant's view of Euclidean space as the unique form of all possible perceptions common to all human minds is subject to challenge from psychology and linguistics as well as from philosophy.

25.6.3 Space, Philosophy, and Art

Using both non-Euclidean geometry and relativity, the Spanish thinker José Ortega y Gasset (1883–1955) drew revolutionary cultural and political conclusions in order to refute what he called the "dogmatisms" of absolutism, provincialism, utopianism, and rationalism (Ortega 1968; Williams 1968, pp. 148–157). All absolutisms are wrong, said Ortega, whether in geometry, physics, or philosophy; reality, he said, is relative. Provincialism incorrectly assumes that our own experience or values are universal. Like Clifford, Ortega said that Euclidean geometry was provincial, an unwarranted extrapolation of what was locally observed to the whole universe. Instead, argued Ortega, reality organizes itself to be visible from all viewpoints, and Einstein's theory of relativity, which requires new geometries of space-time, promotes the harmonious multiplicity of all possible points of view. Thus, all cultures have valid points of view. There is a Chinese perspective, Ortega said, that is "fully as justified as the Western" (Williams 1968, p. 152).

The last two of Ortega's dogmatisms, utopianism and rationalism, are linked and again are both wrong. Since the Greeks, Ortega said, reason has tried to build an idealized world and say, this is true, this is how it is. Before relativity theory, Hendrik Lorentz (1853–1928) had said that matter must get smaller as it goes faster; that is, said Ortega, matter yields so that the old laws of physics can continue to hold. But Einstein said instead, "Space yields. Geometry must yield, space itself must curve." According to Ortega, Lorentz might say, "nations may perish, but we will keep our principles," but Einstein would reply, "We must look for such principles as will preserve nations, because that is what principles are for" (Williams 1968, p. 155).

In another example of radical thinking facilitated by new views of geometry, the surrealist theorist Gaston Bachelard wrote an essay attacking both reason and logic. Bachelard advocated restoring reason to its true function, which is not to shore up the agreed-upon order; instead, as the new geometries show, reason is "a turbulent aggression" (Henderson 1983, p. 346). The Russian thinker P. D. Ouspensky not only attacked the limitations of three-dimensional space, which he identified with Euclidean geometry, but also declared, "A is both A and not-A," and "*Everything* is both A and not-A" (Henderson 1983, p. 253), an explicit repudiation of Aristotle and of all deductive logic. More recently, the French cultural theorist Jean Baudrillard applied these ideas about geometry to refute what was left of the eighteenth-century idea of progress, writing, "In the Euclidean space of history, the fastest route from one point to another is a straight line, the one of Progress and Democracy. This, however, only pertains to the linear space of the Enlightenment. In our non-Euclidean space of the end of the twentieth century, a malevolent curvature invincibly reroutes all trajectories" (Baudrillard 1994).

Modern artists and theorists of art alike were excited by the idea of new geometries. Artists often equated non-Euclidean geometry with the fourth dimension, since both seemed to attack the conventional Euclidean norms. The revolutionary role of the new geometries in art was aided by writers both from literature and psychology. For instance, in 1884 Edwin Abbott wrote a book popularizing the fourth dimension called *Flatland* (Abbott 1953; Abbott 2010). In the country of Flatland, everybody lives on a two-dimensional plane. Abbott's two-dimensional beings are visited by a sphere, which comes from the third dimension. But the Flatlanders cannot conceive of the sphere. They interpret the sphere's intersections with the plane as merely a succession of circles. The problem the Flatlanders have in imagining the third dimension, according to Abbott, is the same as the problem that we three-dimensional creatures have in understanding the fourth dimension. The psychologist of perception Gustav Fechner said that we could think of 2-dimensional creatures as shadows of 3-dimensional figures. In the same way, then, said Fechner, our world is a 3-dimensional "shadow" of the fourth-dimensional reality (Henderson 1983, p. 18). Even Euclidean ideas like Abbott's about the fourth dimension owed a debt to non-Euclidean geometry, since non-Euclidean geometry prepared the way for conceiving alternative kinds of space.

Henri Poincaré's view, discussed earlier, about freedom of choice in geometry was also very attractive to a number of artists (Henderson 1983, p. 55). The key point for the modern artist, after all, was a new freedom from the tyranny of established laws. Thus, Tristan Tzara, the founder of the art movement called "Dada," spoke of "the precise clash of parallel lines." If this should sound like a contradiction, Tzara counseled against worry; the artist can transcend contradictions. The French critic Maurice Princet challenged artists to reverse the prejudices of Renaissance perspective (Henderson 1983, p. 68). Instead of portraying objects on a canvas as "deformed by perspective," they should be expressed "as a type." In Renaissance perspective art, a rectangular table would appear on the canvas shaped like a trapezoid. But a rectangular table should not look like a trapezoid. It should be straightened out into a true rectangle. Likewise, the oval of a glass should become a perfect circle. This describes what is done in many masterpieces of Cubist art.

In Euclidean geometry, when something is moved, it keeps its shape and size. But theorists of Cubism, like Jean Metzinger and Albert Gleizes, declared that Riemann's geometry gives painters the freedom to deform objects in space. Similar views can be found embodied in architecture. For instance, Zaha Hadid, the first woman to win the Pritzker Architecture Prize, constructs buildings that express such ideas. She wrote of her work, "The most important thing is motion, the flux of things, a non-Euclidean geometry in which nothing repeats itself, a new order of space" (Hadid 2008). In the work of artists like these, non-Euclidean geometry has reshaped both our artistic and our intellectual landscapes.

All of these theorists and artists were saying that, beyond the Euclidean world that conventional people think they inhabit, there is a higher reality, a reality that artists alone can intuit and reveal. Non-Euclidean geometry both liberated and legitimated these new approaches to art. The American mathematician Morris Kline has observed, "Non-Euclidean geometry knocked geometry off its pedestal, but also set it free to roam" (Kline 1953, p. 431). And the freedom that geometers claimed for themselves was subsequently bequeathed to many people who may have thought of themselves far from mathematics but well within the traditional liberal arts. The freedom of geometry reinforced a wealth of other social and historical forces in remaking modern culture.

25.7 Anti-Mathematics as a Historical Force

Opposition to the supposed primacy of mathematics in human thought has also influenced the liberal arts and reveals negative perceptions about mathematics that educators neglect to their peril. Some of this opposition is not hostile to all of mathematics, just to its extension beyond its legitimate sphere. For instance, an early opponent of what Ortega later called "provincialism" was, perhaps surprisingly, Isaac Newton. Men like Descartes and Leibniz seemed to Newton to be saying that self-evident assumptions alone sufficed to figure out how the universe works. Newton disagreed. According to Newton, there are many mathematical systems God could have used to set up the world. One cannot decide a priori which is correct, Newton said, but must observe and experiment to find which laws actually hold. Although mathematics is the means used to discover the laws, God set up the world by free choice, not mathematical necessity. This is how Newton justified concluding that the order we find in nature proves that God exists (Newton 1934, p. 544).

Also in the seventeenth century, Blaise Pascal, who made major contributions to mathematics and, with Fermat, was a coinventor of probability theory, contrasted the "esprit géometrique" (abstract and precise thought) with what he called the "esprit de finesse" (intuition) (Pascal 1931, *Pensée* 2), holding that each had its proper sphere but that mathematics had no business outside its own realm. "The heart has its reasons," wrote Pascal famously, "which reason does not know" (Pascal 1931, *Pensée* 277).

In the eighteenth century, Thomas Malthus, in his *Essay on Population*, accepted the Euclidean deductive model. Indeed he began his essay with two "postulata": man requires food, and the level of human sexuality remains constant (Malthus 1798, Chap. 1). His consequent analysis of the way population growth will outstrip food supply rests on mathematical models. But his goal was to discredit the predictions by Condorcet and others of continued human progress modeled on the progress of mathematics and science, as Malthus's mathematical models predict eventual misery and vice instead.

In a nineteenth-century example, the mathematical reductionism of men like Lagrange and Comte was opposed by a great mathematician, Augustin-Louis Cauchy. Cauchy wrote in 1821, "Let us assiduously cultivate the mathematical sciences, but let us not imagine that one can attack history with formulas, nor give for sanction to morality theorems of algebra or integral calculus" (Cauchy 1892, p. vii; Cauchy 2009, p. 3). More recently, computer scientist Joseph Weizenbaum attacked the modern, computer-influenced view that human beings are nothing but processors of symbolic information, arguing that the computer scientist should "teach the limitations of his tools as well as their power" (Weizenbaum 1976, p. 277).

All the examples so far granted some legitimacy to mathematical argument. But there also have been people who have no use at all for the method of analysis, the mathematization of nature, or the application of mathematical thought to human affairs. For instance, the Romantic movement in the nineteenth century championed the organic view of nature over the reductionist mechanical explanations they attributed to Descartes and Newton.[8] The Romantic view is epitomized by a stanza from William Wordsworth's poem *The Tables Turned:*

Sweet is the lore which Nature brings;
Our meddling intellect
Mis-shapes the beauteous forms of things:
We murder to dissect.

Reacting against Quetelet-style statistical thinking on behalf of the dignity of the individual, Charles Dickens in his 1854 novel *Hard Times* satirized a son who betrays his father and then defends himself by saying that in any given population a certain percentage will become traitors, so no blame should be attached. And Dickens has his hero, Stephen Blackpool, denounce the analytically based efficiency of the industrial division of labor, saying it regards workers as though they were nothing but "figures in a sum" (Dickens 1854, Book II, Chap. V).

Although the certainty of mathematics, and thus its authority, has sometimes been an ally of liberalism, as it was for Voltaire, Jefferson, and Condorcet, the Russian novelist Evgeny Zamyatin saw how mathematics could also be used as a way of establishing an unchallengeable authority, as philosophers like Plato and Hobbes had tried to use it. Zamyatin wanted no part of this. In his 1920s anti-utopian novel *We,* a source for Orwell's *1984,* Zamyatin depicted individuals reduced to being numbers and mathematical tables of organization used as instruments of social control. The results were frightening (Zamyatin 1952). And mathematics and its applications were at their worst in the design of the Nazi death camps, where the analytical method was applied to the assembly-line production of corpses.

Yet the fruits of mathematics and mathematical reasoning are embodied in the idea of progress and in the advances in science and technology which have for the first time in human history made it possible, at least in theory, for all human beings to have decent food, clothing, shelter, and some leisure. The fruits of mathematics are essential to the triumphs of the scientific understanding of nature and in the use of science to liberate humanity from superstition. And the logic taught as part of mathematics is seen also in the working out of the consequences of the still radical idea that all human beings are created equal. Of course mathematics has not caused these changes all by itself. Still, it consistently has provided a powerful metaphor, reinforced by the historical authority possessed by the subject, both to drive and to legitimize these changes.

---

[8](Olson 2008, pp. 96–121; Richards 2002, pp. 11, 308–310). The Romantics would not admit that what epistemologists call "secondary qualities" like color, so constitutive of human experience, are mere epiphenomena reducible to "primary qualities" of matter in motion nor that greater understanding necessarily follows from mathematical description. As John Keats put it in criticizing "philosophy" (science) in his poem *Lamia* (part 2):

Philosophy will clip an angel's wings,
Conquer all mysteries by rule and line,...
Unweave a rainbow.

822   J.V. Grabiner

Teachers and students need to know the ways mathematics has interacted with the full range of the liberal arts and to know both the power and the limitations of the claims of mathematics to truth and certainty and to universal applicability. Knowing these things, together with some mathematical proficiency, is needed to criticize what is wrong and develop further what is right in the liberal arts tradition and to use mathematics and its fruits to make real the idea of human progress. Teaching mathematics as a liberal art, then, is an important activity. And it is a global activity that transcends the West.

25.8 Mathematics Is Multicultural and Global

> There are others also that know something of value. – Severus Sebokht, c. 662 CE[9]

As the academic study of the liberal arts expands to include global history and anthropology, transcending the bounds of Western society, mathematics fits perfectly. Almost every society has mathematics and has mathematics that solves problems that the particular society thinks are important. Recent scholarship has shown that other cultures have sophisticated mathematical ideas and practices, though these have developed along different paths than in the West. The mathematics of other traditions than the Western can be used to gain perspective on these other societies, and anthropologists and global historians regularly use it this way. But the mathematics of other traditions also can be and is being used both to strengthen the teaching of mathematics and to humanize the subject.[10] The mathematics of other cultures can sometimes reveal the origin of modern mathematical ideas; can provide instructive examples of standard mathematical topics, advanced as well as elementary; can give teachers alternative ways of looking at familiar ideas; and can serve as a source of connection to mathematics for students of many different ethnic, religious, and national backgrounds. We shall look at each of these sources of insight in turn.

25.8.1 Roots of Modern Mathematics

Modern mathematics is not just Western in origin. Of course even the Greeks had important predecessors in Egypt and Babylonia. For instance, over a thousand years

---

[9] (Joseph 2011, p. 462). Sebokht, a Syrian bishop, was challenging the supposed universal superiority of Greek scientific thought by praising the superior methods of calculation using the base-10 place-value number system from India.

[10] There is now an extensive and reliable English-language literature on the mathematics of other cultures. See, for instance, Ascher (1998, 2002), Berggren (1986), (2007), Closs (1986), Dauben (2007), Gerdes (1999), Gillings (1972), Imhausen (2007), Katz (2000), (2007), Martzloff (1997), Plofker (2007), (2009), Robson (2007), (2008), Robson and Stedall (2009), Van Brummelen (2009), and Zaslavsky (1999).

before Pythagoras, the Babylonians were aware of the Pythagorean rule for right triangles. They also developed sophisticated mathematical models to predict the motion of sun, moon, and planets, using the base-60 fractions which survive today as we divide degrees into minutes and seconds and which the Greek astronomers adopted as well.[11]

Mathematicians in the Islamic world brought together the computational traditions of the East and the geometric and proof-based methods of the Greeks into a new approach to the exact sciences (Berggren 1986, pp. 7–8; Høyrup 1994, pp. 100–103), further developing mathematical models in astronomy including some later used by Copernicus, classifying and then systematically treating the solution of all types of quadratic equations, and geometrically solving cubics (Berggren 1986, pp. 118–123; Berggren 2007, pp. 542–546; Katz 2009, pp. 287–292). At first motivated by the problem of finding the great-circle direction of Mecca from any point on the globe, Muslims greatly advanced the spherical trigonometry they had inherited from the Greeks (Berggren 1986, pp. 182–186; Van Brummelen 2009, pp. 194–201). Beginning from Indian work on what we call the sine and cosine, mathematicians in the Islamic world defined all six of the plane trigonometric functions and developed sophisticated approximation methods to produce trigonometric tables good to five base-60 places (Katz 2009, pp. 306–310). They also further studied the logical equivalents to the parallel postulate (Berggren 1986, pp. 13–15; Katz 2009, pp. 301–303). The important influence of these achievements of Islamic mathematics for later work in Europe, via the writings of Leonardo Fibonacci of Pisa in the thirteenth century and in translations from the twelfth century to the Renaissance, has been amply documented (Katz 2009, 317–318; Van Brummelen 2009, pp. 223–227).

25.8.2 Premodern Discoveries Around the World

Societies whose influence on European mathematics has been less direct, not yet documented, or even nonexistent have also developed sophisticated mathematics, often considerably earlier than did Europeans. For instance, in India combinatorics developed as early as the third century BCE, notably to explain how many different combinations of "heavy and light" (stressed and unstressed) syllables in a line of n syllables there could be. In the tenth century, the *meru-prastara* (mountain-shaped figure, known now as the Pascal triangle) was developed to display the answers (Joseph 2011, pp. 352–355; Plofker 2009, p. 57). To describe planetary positions over time, Indian mathematicians worked to solve simultaneous numerical congruences, which in turn led them to give a complete theory of solving what is now called the Pell equation, long predating

---

[11] See Katz (2009, pp. 17–18, 84–88), Robson (2007, pp. 100, 140–141, 151), and Robson (2008, pp. 109–115, 218–219).

the solutions in the eighteenth century by Euler and Lagrange (Katz 2009, pp. 246–250; Plofker 2007, pp. 423–433).

Indian plane trigonometry began by using Greek methods, but advanced the subject considerably. For example, long before Taylor series were known in the West, Indian trigonometry in the fourteenth century produced the infinite power-series expansions for sine and cosine, sometimes called the Madhava-Newton series after two of its independent discoverers (Katz 2009, pp. 257–258; Plofker 2009, pp. 236–246). Indian mathematics had both interactions with and direct influence upon mathematics in the medieval Islamic world (Plofker 2009, pp. 255–278).

In China, a classic work from before the second century BCE gave methods for solving simultaneous linear equations, using what we would call matrices or Gaussian elimination, and including an example of six equations with six unknowns (Dauben 2007, pp. 346–355; Katz 2009, pp. 209–212). In the eleventh century in China, one finds again the Pascal triangle of binomial coefficients; the coefficients were used to approximate the solutions of polynomial equations of arbitrary degree, in a way analogous to the modern method named after the nineteenth-century mathematician William Horner (Dauben 2007, pp. 329–330; Katz 2009, 213–217). In indeterminate analysis, the Chinese, like the Indians, were initially motivated by astronomical problems to solve simultaneous congruences, using a method which, after it became known in Europe in the nineteenth century, has been called the Chinese Remainder Theorem (Dauben 2007, pp. 302, 311–322; Katz 2009, pp. 222–225). Like mathematicians in India and in the Islamic world, the Chinese had a base-10 place-value system, and, like mathematicians in the Islamic world, the Chinese developed decimal fractions.

Mathematicians in the Islamic world also developed systematic combinatoric methods, including steps toward the method of proof now known as mathematical induction. Their mathematicians, as had those in India and China, developed the Pascal triangle of binomial coefficients (Berggren 1986, pp. 53–63; Katz 2009, pp. 282–296) and used it for, among other things, approximating n^{th} roots. They worked out what we now call the multiplication and division of polynomials – without the benefit of algebraic symbols – including the use of negative exponents. They were the first to recognize that decimal fractions are essentially polynomials using powers of 10 and that fractions can be represented as closely as one likes by taking sufficiently many decimal places (Berggren 1986, pp. 111–118; Katz 2009, pp. 279–282).

Combinatorics was developed also in medieval Jewish culture, where the initial purpose was to work out the number of possible Hebrew words in order to under-stand language and, therefore, since the Bible describes God creating the universe through speech, to understand creation. Levi ben Gershon (1288–1344) called the proof method he used for combinatorial results "rising step by step without end," anticipating the idea of mathematical induction. Also, building on the discussions of Jewish law in the Talmud, medieval Jewish mathematicians anticipated the modern idea of expected value in discussing the problem of fair division (Katz 2009, pp. 337–338; Rabinovitch 1973, pp. 143–148, 161–164).

25.8.3 Pedagogical Insights from Around the World

Looking at alternative treatments from other cultures even of less advanced topics can add insight to the teaching of mathematics. Consider, for instance, what constitutes a proof. At least before the Jesuits brought Euclidean geometry into China in the seventeenth century, Chinese mathematicians did not use proof by contradiction, even though such reasoning is found in Chinese philosophy (Hanna and de Villiers 2012, pp. 431–440; Leslie 1964; Siu 2012, pp. 431–432). In China, mathematicians produced visually convincing proofs, including for the Pythagorean theorem (Dauben 2007, pp. 221–226, 282–288; Hanna and de Villiers 2012, pp. 431–440; Katz 2009, p. 237; Siu 2012, pp. 432–433). Chinese algebraists proved the validity of some algebraic algorithms by reducing them to other algorithmic results already known (Chemla 2012; Dauben 2007, p. 377; Hanna and de Villiers 2012, pp. 423–429). Going beyond proof methods, it is of pedagogical interest to contrast Chinese and Greek mathematical methods in a variety of contexts (Dauben 2007, pp. 375, 377; Horng 2000).

Indian mathematicians, like those of China, did not use postulates as a basis for their proofs, but like the Chinese, they systematically, though usually implicitly, assume that areas remain the same when dissected in different fashions, and provide rational justifications for a range of results, including the Pythagorean theorem (Plofker, pp. 247, 251).

Among other examples of non-Western materials that can add to the understanding of modern topics, consider how in India between the sixth and eighth centuries, long before the writings of Leonardo Fibonacci of Pisa, the Fibonacci series arose in answering a question about poetic meter.[12] The structure of the answer suggests that this series can solve a wider class of problems, as indeed later history shows that it does. And in the Islamic world, Euclidean geometric theorems about areas, used to justify the derivation of the algebraic technique of solving quadratic equations, are illustrated by the visual picture of completing the square, a picture that greatly enhances student understanding of the solution of quadratics (Berggren 1986, pp. 104–110; Joseph 2011, pp. 477–480; Katz 2009, pp. 272–276).

25.8.4 Mathematical Ideas from Many Cultures

Examples of topics now part of modern mathematics abound also in the mathematics of many of the cultures studied by modern anthropologists as well as that of other

---

[12] The question is, if long or heavy syllables are two beats and short or light syllables are one beat, what is the number of different arrangements $A(n)$ of long and short syllables for a line of n beats? For example, if there are two beats and if we use "S" for short and "L" for long, the arrangements are SS and L. If there are three beats, the arrangements are SSS, SL, and LS. If there are four beats, the arrangements are SSSS, SSL, and SLS (formed by placing an S in front of each of the arrangements for three beats), plus LSS and LL (formed by placing an L in front of each of the arrangements for two beats). Thus, $A(4) = A(3) + A(2)$. Since $A(2) = 2$ and $A(3) = 3$ and since the method of forming $A(n)$ from $A(n-2)$ and $A(n-1)$ must follow the same pattern, this gives the Fibonacci series (Singh 1985).

ancient societies. These examples are often unexpected and therefore interesting to mathematicians as well as to anthropologists, linguists, artists, historians, and educators. For instance, the eminent algebraist André Weil has shown how the systems of marriage laws in some Aboriginal societies in Australia can be modeled by the theory of groups, and Marcia Ascher gives a variety of examples of such marriage laws, both in Australia and among the Malekula people of Vanuatu (Ascher 1998, pp. 70–81; Weil 1949, 1969). Not only is the Malekula system's structure isomorphic to that of the dihedral group of order 6, the Malekula themselves reason using a geometric diagram to describe, explain, and answer questions about their kinship relationships (Ascher 1998, pp. 77–81). Both the Bushoong people of central Africa and the Malekula can distinguish between graphs which have what are now called Eulerian paths and those that do not. The Bushoong use such graphs for purposes ranging from embroidery to political prestige (Ascher 1998, pp. 31–37), while the Malekula use knowledge of them in stories and myth (Ascher 1998, pp. 43–62). There are now several textbooks using such culturally based graphs and designs to teach mathematics; an excellent example is Gerdes (1999).

Geometric symmetries abound in the various artistic designs used in many cultures. The beauty of these designs makes them valuable examples in teaching about the structure of groups of symmetries (Ascher 1998, pp. 155–183; Washburn and Crowe 1998). And just as in classical Chinese and Indian civilization, modern mathematical instruction in the algebra of congruences and in astronomical models can draw on calendars from a variety of cultures that have chosen to deal in different ways with the different periods of the apparent motion of the stars, sun, moon, and planets. Comparing and studying the different calendrical systems used by, for instance, the Maya in the Americas; modern inhabitants of Bali; the ancient civilizations of Egypt and Babylonia, China, and India; and Jews, Muslims, and Christians, can link mathematics and astronomy to culture and religion (Ascher 2002, pp. 39–88).

Studying different number systems and different number bases becomes alive when linked to particular cultures, especially cultures related to one's students' own backgrounds. Examples interesting to US students can include the base-20 place-value system of the Maya and the sophisticated astronomical calculations for which it was used (Ascher 2002, pp. 62–74; Closs 1986), the multiplication by adding multiples of appropriate powers of two of the multiplicand used by the ancient Egyptians (Katz 2009, pp. 4–5), the variety of number bases used by different indigenous cultures of Africa (Zaslavsky 1999), and the base-60 place-value system of the ancient Middle East that gives us "modern" minutes and seconds.

The study of the mathematics of cultures other than the Western and interest in placing such mathematics in its cultural setting and using it to teach mathematical ideas and practices, all have attracted much interest in the past few decades. This approach has come to be called "ethnomathematics," a term introduced in D'Ambrosio (1985), and has generated a rich literature. For the purposes of the present essay, this brief sketch of some of the relevant mathematical ideas and practices will have to suffice, but as the sources cited in this section make clear, there is now a wealth of excellent scholarship devoted to mathematics in different cultures and time periods,

25 The Role of Mathematics in Liberal Arts Education

regardless of whether or not that mathematics has influenced modern mathematics. The connection of mathematics with subjects ranging from anthropology to the visual arts is vastly enriched by a multicultural perspective. Here, then, is another important link between mathematics and the liberal arts, where in this case the liberal arts range far beyond the traditional texts of the Western canon.

25.9 Teaching Mathematics as a Liberal Art in the Modern University

After soaring into space with Euclid and Newton and Lobachevsky, roaming the globe and sampling mathematical thought across time and across continents, and after transforming world views with the architects and anthropologists, the philosophers, and theologians, let us come down to earth and enter a few classrooms in the English-speaking world, where, to quote the Yeats poem again, "The children learn to cipher." Students are using the quadratic formula to solve $2x^2 - 11x + 14 = 0$. For what purpose? Why does the formula work? Who thought it up, and why? Nobody says. The formula is rarely derived even algebraically, let alone in the intuitively pleasing way with the geometric picture of completing the square. A student recently told the author that he "knows" the quadratic formula because a teacher taught his class to sing it to "Pop Goes the Weasel." In elementary schools, the rules of arithmetic are too often presented in the manner satirized as "Ours is not to reason why; just invert and multiply" (Wilson 2003, p. 6). Such methods of instruction may have been appropriate to produce a nineteenth-century nation of shopkeepers in an age without electronic calculators, but not to develop citizens who can independently apply mathematical and statistical reasoning to the quantitative ideas so pervasive in modern society and who understand the role of mathematics in the wider world.

Students need to know why, not just how, so that they can adapt what they have learned to new situations. Teachers need to know how young people learn and to know not only the elementary mathematics that they teach but also what it leads to and how it is used. The history of mathematics could help link mathematics with the rest of the world and to the rest of the liberal arts. But using algorithms, rather than being liberating and empowering as it was to seventeenth-century mathematicians, now has become a rule-based march toward the multiple-choice test.

The present chapter will not presume to prescribe solutions to the problems of pre-university mathematics education or to explain how to overcome the political and social forces that stand in the way of improving instruction, save to emphasize "For everything there is a reason" and to encourage teachers to explain the why as well as the how.[13] At the university level, those students specializing in

---

[13] Discussions about how this can and has been done, and how it has been assessed, may be consulted in Alternatives for Rebuilding Curricula Center (2003), Ball et al. (2005), Boaler and Staples (2008), Hill et al. (2005), and Tarr et al. (2008). A cross-cultural study involving Chinese and American teachers at the elementary-school level can be found in Ma (1999).

science and mathematics will manage relatively well. Their professors will understand the mathematics, be able to answer student questions, and will point students toward real applications. Courses in the history and philosophy of mathematics can help mathematically literate students see the kinds of links described in the present paper. But what of the self-defined liberal arts student, who comes to the university definitely wanting *not* to study mathematics? This essay will conclude by addressing this question.

A common practice in many American colleges and universities is to teach all these students precalculus. This usually repeats what did not work for them in high school. Such an approach frustrates students and teachers alike, and a terminal precalculus course seems to defeat any logical purpose. Another approach is to use one of the textbooks designed to teach mathematics to liberal arts students (e.g., Burger and Starbird 2012; Jacobs 2012), or the applications-oriented book *For All Practical Purposes* (COMAP 2013) and others that resemble it. Such texts have been successful in a variety of universities, and those who find the books successful with their own students should of course use them and recommend them to colleagues.

But for those who find the story told here sufficiently compelling to try to design their own courses for liberal arts students, some principles useful for classes designed by individuals are worth considering. First, base the course on something of interest to contemporary students in which the instructor has expertise, whether it is "mathematics and art," "games and gambling," or "mathematics in many cultures."[14] After all, mathematics was considered important and interesting throughout history because people wanted to solve particular problems, some within mathematics itself, some of importance to the wider society. Teaching what instructors themselves find interesting and exciting recapitulates the history of mathematical creativity.

Second, although the mathematics chosen must be accessible to liberal arts students, it should also be important mathematics in the eyes of mathematicians. The liberal arts goals require this. Third, for students who will probably never take another mathematics course, learning mathematics should be empowering, not overwhelming. It is more important for them to be able to use the mathematics they know than to be shown more mathematics that they cannot master. A liberal arts course isn't prerequisite to anything. There is merit in going slowly enough so that 90 % of the students will get 90 % of the mathematics. This lets them, perhaps for the first time, experience mastery in mathematics; this is part of understanding what mathematics is all about. "I get it now" is a necessary prerequisite for seeing the beauty and elegance of mathematics, and also of being able to apply it to a new situation.

---

[14] Examples of books that might be suitable for such courses include Ascher (1998 and 2002), Frantz and Crannell (2011), Gerdes (1999), and Packel (1981). For details about the author's courses, see Grabiner (2011). A superb online resource for liberal arts mathematics teaching is the Mathematical Association of America's "magazine" of the history of mathematics and its uses in the classroom, *Convergence* (n.d.).

Fourth, all students have expertise. It may not be in mathematics, but they do have expertise in their major or some outside interest. A liberal arts course should allow each student to build a course project incorporating that expertise and thereby impress and teach everybody else, sharing their course projects with each other in the class. This will produce many more applications of mathematics than most professors could generate and a surprising range of fascinating mathematical topics. Just as in the case of the instructors teaching what interests them, the depth of the individual student's excitement about the topic will enhance that student's learning. And the individual student becomes the class expert on one piece of mathematics, perhaps for the first time in that student's life.

Students often identify mathematics with number and calculation alone or perhaps with elementary geometry and trigonometry. In modern times, though, because of the increasing abstraction of contemporary mathematics and because many branches of mathematics, from topology to infinite-dimensional spaces, transcend ordinary experience, a broader characterization of mathematics has developed. In the recent influential words of Lynn Steen, "No longer just the study of number and space, mathematical science has become the science of patterns, with theory built on relations among patterns and on applications derived from the fit between pattern and observation" (Steen 1988, p. 611). "Pattern" is a valuable metaphor in accounting for the applicability of mathematics, since mathematical patterns can mesh with the patterns of order of the universe. But the metaphor also underscores the beauty of mathematical ideas, linking mathematics in a different way to the rest of the liberal arts. As G. H. Hardy, perhaps the earliest to use the idea, put it, "A mathematician, like a painter or a poet, is a maker of patterns.... The mathematician's patterns, like the painter's or the poet's, must be *beautiful*; the ideas, like the colours or the words, must fit together in a harmonious way" (Hardy 1967, pp. 84–85, italics his).

Faculty members at many institutions may not have the freedom to invent new courses, but the general principles should help every teacher of mathematics for non-mathematicians. Mathematics is fun and exciting, both beautiful and useful. It makes unique claims to truth, is governed by logic and reason, and has interacted with every conceivable subject. It has been created by human beings all over the world, in the past and in the present, by men and by women.[15] To be able to follow logical arguments and to criticize them is liberating; as Jacques Barzun observed, "The ability to feel the force of an argument apart from the substance it deals with is the strongest possible weapon against prejudice" (Barzun 1945, p. 121). It is liberating to have understood the "why?" rather than just to have memorized processes, so that people can use the mathematics they know to analyze ideas and solve problems they have never encountered. Mathematics matters not just to stu-

---

[15] On women in mathematics in general, see the online biographies maintained by Agnes Scott College (2012), the sourcebook Grinstein and Campbell (1987), and the Mathematical Association of America's poster *Women of Mathematics* (MAA 2008). On important individual women in mathematics, see Arianrhod (2012), Brewer and Smith (1989), Dahan-Dalmédico (1991), Deakin (2007), Hagengruber (2012), Katz (2009, pp. 189–190, 616–617, 714–715, 787, 874, 896–898, 899), Koblitz (1983), Mazzotti (2007), Neeley (2001), Reid (1996), and Zinsser (2006).

830   J.V. Grabiner

dents but to all members of a free society. As this essay has argued, mathematics lies at the heart of the ancient and medieval liberal arts and of those fields called liberal arts today. Viewing mathematics as a liberal art shows both why mathematics is important and how it should be taught.

Acknowledgments This essay is dedicated to the memory of Herman Sinaiko (1929–2011), peerless teacher and scholar of the liberal arts. I am grateful to the anonymous referees and to the editor, for their scholarly expertise, comments, and criticisms, which have materially improved this chapter. I also thank the Pitzer family, donors of the Flora Sanborn Pitzer professorship, for their generous support of my research.

References

Abbott, E. (1953/1884). *Flatland: A Romance of Many Dimensions*. New York: Dover.
Abbott, E. (2010/1884). *Flatland*. With Notes and Commentary by W. F. Lindgren & T. F. Banchoff. Washington, DC: Mathematical Association of America.
Agnes Scott College (2012). Biographies of Mathematicians. Online only. http://www.agnesscott.edu/lriddle/women/women.htm. Consulted 6 July 2012.
Alternatives for Rebuilding Curricula Center (2003). Tri-State Student Achievement Study. Lexington, MA. Available at http://www. comap.com/elementary/projects/arc/index.htm. Consulted 15 June 2012.
Arianrhod, R. (2012). *Seduced by Logic: Émilie Du Châtelet, Mary Somerville and the Newtonian Revolution*. New York: Oxford University Press.
Aristotle, *De Anima*. In (Aristotle 1941), pp. 535–603.
Aristotle, *Metaphysics*. In (Aristotle 1941), pp. 689–926.
Aristotle (1941). *The Basic Works of Aristotle*. Ed. R. McKeon. New York: Random House.
Ascher, M. (1998). *Ethnomathematics: A Multicultural View of Mathematical Ideas*. New York: Chapman & Hall/CRC.
Ascher, M. (2002). *Mathematics Elsewhere: An Exploration of Ideas across Cultures*. Princeton and Oxford: Princeton University Press..
Ashcraft, M. H. (2002). Math Anxiety: Personal, Educational, and Cognitive Consequences. *Current Directions in Psychological Science,* Vol. 11, No. 5 (October), 181–185.
Babbage, C. (1832). *On the Economy of Machinery*. London: Charles Knight.
Ball, D. L., Hill, H. C., & Bass, H. (2005). Knowing mathematics for teaching: Who knows mathematics well enough to teach third grade and how can we decide? *American Educator*, Fall, 14–46.
Barrett, W. (1958). *Irrational Man: A Study in Existential Philosophy*. New York: Doubleday.
Barzun, J. (1945). *Teacher in America*. London: Little Brown.
Baudrillard, J. (1994) Reversion of History. In CTheory.net, April 20, 1994. www.ctheory.net/articles.aspx?id=54 Consulted 15 May 2012.
Becker, C. L. (1922). *The Declaration of Independence: A Study in the History of Political Ideas*. New York: Harcourt Brace.
Berggren, J. L. (1986). *Episodes in the Mathematics of Medieval Islam*. New York et al: Springer.
Berggren, J. L. (2007). Mathematics in Medieval Islam. In (Katz 2007), pp. 515 – 675.
Boaler, J. & Staples, M. (2008). Creating Mathematical Futures through an Equitable Teaching Approach: The Case of Railside School. *Teacher's College Record.* 110 (3), 608–645.
Bonola, R. (1955). *Non-Euclidean Geometry: A Critical and Historical Study of Its Development. With a Supplement containing "The Theory of Parallels" by Nicholas Lobachevski and "The Science of Absolute Space" by John Bolyai*. Tr. H. S. Carslow. New York: Dover.
Boole, G. (1854). *An Investigation of The Laws of Thought*. London: Walton and Maberly. Reprint: New York: Dover, n.d.

25 The Role of Mathematics in Liberal Arts Education

Boyer, C. B. (1956). *History of Analytic Geometry*. New York: Scripta Mathematica.

Brewer, J. W., & Smith, M. K. (1989). *Emmy Noether: A Tribute to Her Life and Work*. New York: Marcel Dekker.

Burger, E., & Starbird, M. (2012). *The Heart of Mathematics: An Invitation to Effective Thinking*. 4th Edition. New York: Wiley.

Carnegie Foundation for the Advancement of Teaching. (1994). *A Classification of Institutions of Higher Education*. Princeton: Carnegie Foundation.

Cauchy, A.-L. (1892/1821). *Cours d'analyse de l'Ecole Royale Polytechnique*. Ire Partie: Analyse algébrique [all published]. Paris: Imprimérie royale. Reprinted in A.-L. Cauchy, *Oeuvres*, series 2, vol. 3. Paris: Gauthier- Villars. English translation in (Cauchy 2009).

Cauchy, A.-L. (2009). *Cauchy's Cours d'analyse: An Annotated Translation*. Tr. Bradley, R. E. & Sandifer, C. E. Dordrecht et al: Springer. English translation of (Cauchy 1892).

Cesareo, F. C. (1993). Quest for Identity: The Ideals of Jesuit Education in the Sixteenth Century. In (Chapple 1993b), pp. 17–33.

Chapple, C. (1993a). Introduction. In (Chapple 1993b), pp. 7–12.

Chapple, C. (Ed.). (1993b). *The Jesuit Tradition In Education and Missions*. Scranton, PA: University of Pennsylvania Press.

Chemla, K. (2012). Using documents from ancient China to teach mathematical proof. In (Hanna & de Villiers 2012), pp. 423–429.

Cicero, M. T. *De oratore. In Two Volumes*. Books I-II, Vol. 1; Book III, Vol. 2. Tr. H. Rackham. London: William Heinemann, and Cambridge, MA: Harvard University Press, 1942.

Cicero, M. T. *De re publica; de legibus*. Tr. C. W. Keyes. London: William Heinemann, and Cambridge, MA: Harvard University Press, 1928.

Clifford, W. K. (1956). The Postulates of the Science of Space. In (Newman 1956), Vol. I, pp. 552–567.

Closs, M. (1986). The Mathematical Notation of the Ancient Maya. In M. Closs, ed., *Native American Mathematics*. Austin: University of Texas Press, pp. 291–369.

Cohen, I. B. (1980). *The Newtonian Revolution*. Cambridge, UK: Cambridge University Press.

Cohen, I. B. (1995). *Science and the Founding Fathers*. New York: W. W. Norton.

COMAP (Consortium for Mathematics and Its Applications). (2013). *For All Practical Purposes: Mathematical Literacy in Today's World*. Ninth Edition. San Francisco: W. H. Freeman.

Comte, A. (1830). *Cours de philosophie positive*. Vol. I. Paris: Bachelier.

Condorcet, Marquis de. (1976/1793). *Sketch for a Historical Picture of the Progress of the Human Mind*. Tr. J. Barraclough. In Baker, K. M., ed., *Condorcet: Selected Writings*. Indianapolis: Bobbs-Merrill.

Convergence (n. d.). (http://mathdl.maa.org/mathDL/46/). Consulted 24 June 2012.

Cruz, J. A. H. M. (1999). Education. *Encyclopedia of the Renaissance*, vol. 2. New York: Scribner's, pp. 242–254.

Dahan-Dalmédico, A. (1991). Sophie Germain. *Scientific American* (265), 117–122.

D'Ambrosio, U. (1985). Ethnomathematics and Its Place in the History and Pedagogy of Mathematics. *For the Learning of Mathematics*, 5(1), 44–48.

Dauben, J. (2007). Chinese Mathematics. In (Katz 2007), pp. 187–385.

Deakin, M. (2007). *Hypatia of Alexandria: Mathematician and Martyr*. Amherst, NY: Prometheus Books.

Descartes, R. (1637a). *Discourse on Method*. Tr. L. J. Lafleur. New York: Liberal Arts Press, 1956.

Descartes, R. (1637b). *La géométrie*. Tr. D. E. Smith and M. L. Latham as *The Geometry of René Descartes*. New York: Dover, 1954.

Dickens, C. (1854). *Hard Times*. Norton Critical Edition. G. Ford and S. Monod (Eds.). New York and London: W. W. Norton, 1966.

Euclid, *Elements*. In (Heath 1925). On-line version: http://aleph0.clarku.edu/~djoyce/java/elements/toc.html. Consulted 2 July 2012.

Ferrall, V. E. (2011). *Liberal Arts at the Brink*. Cambridge, MA and London: Harvard University Press.

Fauvel, J,. & Gray, J. (1987). *The History of Mathematics: A Reader*. Houndmills and London: Macmillan.

Fauvel, J., Flood, R., & Wilson, R. (Eds.) (2003). *Music and Mathematics: From Pythagoras to Fractals*. Oxford: Oxford University Press.

Field, J. V. (1997). *The Invention of Infinity: Mathematics and Art in the Renaissance*. Oxford: Oxford University Press.

Field, J. V. (2003). Musical cosmology: Kepler and His Readers. In (Fauvel et al 2003), pp. 28–44.

Frantz, M., & Crannell, A. (2011). *Viewpoints: Mathematical Perspective and Fractal Geometry in Art*. Princeton: Princeton University Press.

Friedman, M. (1992). *Kant and the Exact Sciences*. Cambridge, MA: Harvard University Press.

Gaukroger, S. (1995). *Descartes: An Intellectual Biography*. Oxford: Oxford University Press.

Gerdes, P. (1999). *Geometry from Africa: Mathematical and Educational Perspectives*. Washington, DC: Mathematical Association of America.

Gillings, R. J. (1972). *Mathematics in the Time of the Pharaohs*. Cambridge, MA: M. I. T. Press.

Gillispie, C. C. (1960). *The Edge of Objectivity*. Princeton: Princeton University Press.

Grabiner, J. V. (1988). The Centrality of Mathematics in the History of Western Thought. *Mathematics Magazine* 61, 220–230. Reprinted in (Grabiner 2010), pp. 163–174.

Grabiner, J. V. (2010). *A Historian Looks Back: The Calculus as Algebra and Selected Writings*. Washington, DC: Mathematical Association of America.

Grabiner, J. V. (2011). How to Teach Your Own Liberal Arts Course. *Journal of Humanistic Mathematics*. Vol. I (1), 101–118. http://scholarship.claremont.edu/jhm/vol1/iss1/8. Consulted 7 July 2012.

Grant, H. (1999a). Mathematics and the Liberal Arts I. *College Mathematics Journal* 30, No. 2, 96–105.

Grant, H. (1999b). Mathematics and the Liberal Arts II. *College Mathematics Journal* 30, No. 3, 197–203.

Gray, J. (1989). *Ideas of Space: Euclidean, Non-Euclidean and Relativistic*. 2nd ed. Oxford: Clarendon Press.

Grendler, P. F. (1989). *Schooling in Renaissance Italy: Literacy and Learning, 1300-1600*. Baltimore and London: Johns Hopkins University Press.

Grendler, P. F. (2002). *The Universities of the Italian Renaissance*. Baltimore and London: Johns Hopkins University Press.

Grinstein, L. S., & Campbell, P. J. (Eds.) (1987). *Women of Mathematics: A Biobibliographic Sourcebook*. New York: Greenwood Press.

Hadid, Z. (2008). http://lan-haiyun.blogspot.com/2012/03/5-concepts-of-zaha-hadid.html. Consulted 14 May 2012.

Hagengruber, R. (Ed.) (2012). *Emilie du Châtelet between Leibniz and Newton*. Dordrecht and New York: Springer.

Hanna, G. & de Villiers, M. (Eds.) (2012). *Proof and Proving in Mathematics Education*. Dordrecht et al: Springer.

Hardy, G. H. (1967/1940). *A Mathematician's Apology*. Foreword by C. P. Snow. Cambridge: Cambridge University Press.

Haskins, C. H. (1927). *The Renaissance of the Twelfth Century*. Cambridge MA: Harvard University Press.

Heath, T. L. (Ed.). (1925). *The Thirteen Books of Euclid's Elements*, 3 vols. Cambridge, UK: Cambridge University Press. Reprinted New York: Dover, 1956.

Heath, T. L. (1949). *Mathematics in Aristotle*. Oxford: Oxford University Press.

Helmholtz, H. (1954/1863). *On the sensations of tones*. Tr. A. J. Ellis. New York: Dover.

Helmholtz, H. (1962/1870). On the origin and significance of geometrical axioms. Reprinted in H. von Helmholtz, *Popular Scientific Lectures*, ed. M. Kline. New York: Dover, pp. 223–249.

Henderson, L. (1983). *The Fourth Dimension and Non-Euclidean Geometry in Modern Art*. Princeton: Princeton University Press.

Hill, H. C., Rowan, B., & Ball, D. L. (2005). Effects of teachers' mathematical knowledge for teaching on student achievement. *American Educational Research Journal* (42), 371–406.

Hobbes, T. (1939/1651). *Leviathan, or the Matter, Form, and Power of a Commonwealth, Ecclesiastical and Civil*. Reprinted in E. Burtt (Ed.), *The English Philosophers from Bacon to Mill*. New York: Modern Library.

25 The Role of Mathematics in Liberal Arts Education

Hoeflich, M. H. (1986). Law and Geometry: Legal Science from Leibniz to Langdell. *American Journal of Legal History* (30), 95–121.

Høyrup, J. (1994). *In Measure, Number, and Weight: Studies in Mathematics and Culture*. Albany, NY: State University of New York Press.

Horng, W. S. (2000). Euclid versus Liu Hui: A Pedagogical Reflection. In (Katz 2000), pp. 37–47.

Huffman, C. (2011). Archytas. *The Stanford Encyclopedia of Philosophy (Fall 2011 Edition)*, ed. E. N. Zalta. http://plato.stanford.edu/archives/fall2011/entries/archytas/. Consulted 12 June 2012.

Hutcheson, F. (2004/1728). On computing the morality of actions. In F. Hutcheson, *Inquiry into the Original of Our Ideas of Beauty and Virtue in Two Treatises*, ed. Wolfgang Leidhold. Indianapolis: Liberty Fund.

Hyman, A. (1982). *Charles Babbage: Pioneer of the Computer*. Princeton: Princeton University Press.

Imhausen, A. (2007). Egyptian Mathematics. In (Katz 2007), pp. 7–56.

Jacobs, H. R. (2012). *Mathematics: A Human Endeavor*. 4th Edition. San Francisco: Freeman.

Jaeger, W. (1944). *Paideia: The Ideals of Greek Culture*. Tr. Gilbert Highet. 3 vols. Oxford: Oxford University Press. [Vol. 1, 1939, vol. 2, 1943, vol. 3, 1944]

Jeans, J. (1956). Mathematics of Music. In (Newman 1956), vol. 4, pp. 2278–2309.

Joseph, G. G. (2011). *The Crest of the Peacock: Non-European Roots of Mathematics*. 3d edition. Princeton and Oxford: Princeton University Press.

Kalman, L. (1986). *Legal Realism at Yale, 1927-1960*. Chapel Hill and London: University of North Carolina Press.

Kant, I. (1950/1783). *Prolegomena to Any Future Metaphysics*. Ed. L. W. Beck. New York: Liberal Arts Press.

Kant, I. (1961/1781). *Critique of Pure Reason*. Tr. F. M. Müller. New York: Macmillan.

Karp, T. C. (1983). Music. In (Wagner 1983), pp. 169–195.

Katz, V. J. (Ed). (2000). *Using History to Teach Mathematics: An International Perspective*. Washington, DC: Mathematical Association of America.

Katz, V. J. (Ed.) (2007). *The Mathematics of Egypt, Mesopotamia, China, India, and Islam: A Sourcebook*. Princeton and Oxford: Princeton University Press.

Katz, V. J. (2009). *A History of Mathematics: An Introduction*. 3d edition. Boston et al: Addison-Wesley.

Kemp, M. (1990). *The Science of Art: Optical Themes in Western Art from Brunelleschi to Seurat*. New Haven: Yale University Press.

Kimball, B. (1995). *Orators and Philosophers: A History of the Idea of Liberal Education*. Expanded edition. New York: The College Board.

Klein, J. (1968). *Greek Mathematical Thought and the Origins of Algebra*. Tr. E. Brann. With an Appendix containing Vieta's *Introduction to the Analytic Art*. Cambridge, MA: M. I. T. Press.

Kline, M. (1953). *Mathematics in Western Culture*. Oxford and New York: Oxford University Press.

Knorr, W. R. (1975). *The Evolution of the Euclidean Elements*. Dordrecht: D. Reidel.

Koblitz, A. H. (1983). *A Convergence of Lives: Sofia Kovalevskaia, Scientist, Writer, Revolutionary*. Boston: Birkhaüser.

LaPiana, W. P. (1994). *Logic and Experience: The Origin of Modern American Legal Education*. New York and Oxford: Oxford University Press.

Laplace, P.-S. (1951/1819). *A Philosophical Essay on Probabilities*. Tr. F. W. Truscott and F. L. Emory. New York: Dover.

Leibniz, G. W. (1951/1677). Preface to the General Science and Towards a Universal Characteristic. In P. P. Weiner (Ed.), *Leibniz: Selections*. New York: Scribner's, pp. 12–25.

Leibniz, G. W. (1969/1684). A new method for maxima and minima…and a remarkable type of calculus for them. Tr. D. J. Struik. In (Struik 1969), pp. 271–280.

Leslie, D. (1964). *Argument by contradiction in Pre-Buddhist Chinese Reasoning*. Canberra: Australian National University.

Levinson, S. (1996). Frames of Reference and Molyneux's Question: Crosslinguistic Evidence. In P. Bloom et al, *Language and Space*. Cambridge, MA: M. I. T. Press, pp. 109–170.

834   J.V. Grabiner

Levinson, S. (2003). Language and Mind: Let's Get the Issues Straight. In D. Gertner and S. Goldin-Meadow (Eds.), *Language in Mind: Advances in the Study of Language and Thought.* Cambridge, MA: M. I. T. Press, pp. 25–46.

Lovejoy, A. (1936). *The Great Chain of Being.* Cambridge, MA: Harvard University Press.

Ma, L. (1999). *Knowing and Teaching Elementary Mathematics: Teachers' Understanding of Fundamental Mathematics in China and the United States.* Mahwah, NJ: Erlbaum.

MAA (2008). Women of Mathematics Poster. Washington, D. C.: Mathematical Association of America. http://www.maa.org/pubs/posterW.pdf. Consulted 6 July 2012.

Maclaurin, C. (1714). De viribus mentium bonipetis. MS 3099.15.6, The Colin Campbell Collection, Edinburgh University Library. Translated in (Tweddle 2008).

Mahoney, M. (1973). *The Mathematical Career of Pierre de Fermat, 1601-1655.* Princeton: Princeton University Press.

Malthus, T. R. (1798). *An Essay on the Principle of Population, as it Affects the Future Improvement of Society: with Remarks on the Speculations of Mr. Godwin, M. Condorcet, and other writers.* London: J. Johnson.

Marr, A. (2011). *Between Raphael and Galileo: Mutio Oddi and the Mathematical Culture of Late Renaissance Italy.* Chicago: University of Chicago Press.

Marrou, H. I. (1956). *A History of Education in Antiquity.* Tr. G. Lamb. New York: Sheed and Ward. Mentor reprint, New York, 1964.

Martzloff, J.-C. (1997). *A History of Chinese Mathematics.* Tr. S. S. Wilson. Berlin: Springer.

Masi, M. (1983). Arithmetic. In (Wagner 1983), pp. 147–168.

Mazzotti, M. (2007). *The World of Maria Gaetana Agnesi: Mathematician of God.* Baltimore: Johns Hopkins Press.

McInerny, R. (1983). Beyond the Liberal Arts. In (Wagner 1983b), pp. 248–272.

McKirahan, R. D. Jr. (1992). *Principles and Proofs: Aristotle's Theory of Demonstrative Science.* Princeton: Princeton University Press.

Merz, J. T. (1904). *A History of European Thought in the Nineteenth Century.* Vol. I. London: Blackwood. Reprinted New York: Dover, 1965.

Merzbach, U., & Boyer, C. B. (2011). *A History of Mathematics.* 3d edition. New York: Wiley.

Morrison, K. F. (1983). Incentives for Studying the Liberal Arts. In (Wagner 1983b), pp. 32–57.

Neeley, K. A. (2001). *Mary Somerville: Science, Illumination, and the Female Mind.* Cambridge: Cambridge University Press.

Newman, J. R. (Ed.). (1956). *The World of Mathematics.* 4 vols. New York: Simon and Schuster.

Newton, I. (1934/1687). *Sir Isaac Newton's Mathematical Principles of Natural Philosophy and His System of the World.* Tr. A. Motte. Revised and edited by F. Cajori. Berkeley: University of California Press.

Nussbaum, M. (2010). *Not for Profit: Why Democracy Needs the Humanities.* Princeton: Princeton University Press.

Olson, R. (2008). *Science and Scientism in Nineteenth-Century Europe.* Urbana and Chicago: University of Illinois Press.

Ortega y Gasset, J. (1968/1961). The Historical Significance of the Theory of Einstein. In (Williams 1968), pp. 147–157.

Orwell, G. (1949). *Nineteen Eighty-Four: A Novel.* New York: Harcourt Brace.

Packel, E. (1981). *The Mathematics of Games and Gambling.* Washington, DC: Mathematical Association of America.

Pascal, B. (1931/1669). *Pensées.* Tr. W. F. Trotter. New York: E. P. Dutton.

Peirce, B. (1881). Linear associative algebra with notes and addenda by C. S. Peirce. *American Journal of Mathematics* (4), 97–229.

Plato. (1961). *The Collected Dialogues of Plato, Including the Letters.* Ed. E. Hamilton and H. Cairns. New York: Pantheon Books.

Plato. *Republic.* Tr. P. Shorey. In (Plato 1961), pp. 575–844.

Plofker, K. (2007). Mathematics in India. In (Katz 2007), pp. 385–514.

Plofker, K. (2009). *Mathematics in India.* Princeton and Oxford: Princeton University Press.

Poincaré, H. (1952/1905). *Science and Hypothesis.* New York: Dover.

25 The Role of Mathematics in Liberal Arts Education 835

Popkin, R. H. (1979). *The History of Scepticism from Erasmus to Spinoza*. Berkeley and Los Angeles: University of California Press.

Porter, T. (1986). *The Rise of Statistical Thinking, 1820-1900*. Princeton: Princeton University Press.

Proclus. (1970/5th century CE). *A Commentary on the First Book of Euclid's Elements*. Tr. G. R. Morrow. Princeton: Princeton University Press.

Quetelet, A. (1828). *Instructions populaires sur le calcul des probabilités*. Brussels: Tarlier & Hayez.

Rabinovitch, N. L. (1973). *Probability and Statistical Inference in Ancient and Medieval Jewish Literature*. Toronto and Buffalo: University of Toronto Press.

Reid, C. (1996). *Julia [Robinson]: A Life in Mathematics*. Washington, DC: Mathematical Association of America.

Richards, J. L. (1988). *Mathematical Visions: The Pursuit of Geometry in Victorian England*. Boston: Academic Press.

Richards, R. (2002). *The Romantic Conception of Life: Science and Philosophy in the Age of Goethe*. Chicago and London: University of Chicago Press.

Robson, E. (2007). Mesopotamian Mathematics. In (Katz 2007), pp. 57–186.

Robson, E. (2008). *Mathematics in Ancient Iraq: A Social History*. Princeton and Oxford: Princeton University Press.

Robson, E., & Stedall, J. (Eds.) (2009). *The Oxford Handbook of the History of Mathematics*. Oxford and New York: Oxford University Press.

Rose, P. L. (1975). *The Italian Renaissance of Mathematics: Studies on Humanists and Mathematicians from Petrarch to Galileo*. Geneva: Librairie Droz.

Rudolph, F. (1962). *The American College and University*. New York: Vintage.

Seligman, J. (1978). *The High Citadel: The Influence of Harvard Law School*. Boston: Houghton Mifflin.

Sinaiko, H. L. (1998). Energizing the Classroom: The Structure of Teaching. In H. L. Sinaiko, *Reclaiming the Canon: Essays on Philosophy, Poetry, and History*. New Haven: Yale University Press, pp. 241–252.

Schiebinger, L. (1989). *The Mind Has No Sex: Women in the Origins of Modern Science*. Cambridge, MA: Harvard University Press.

Singh, P. (1985). The so-called Fibonacci numbers in ancient and medieval India. *Historia Mathematica* (12), 229–244.

Siu, M. K. (2012). Proof in the Western and Eastern Traditions: Implications for Mathematics Education. In (Hanna & de Villiers 2012), pp. 431–440.

Smith, A. (1974/1776). *The Wealth of Nations*. London: Penguin Books.

Spinoza, B. (1953/1675). *Ethics Demonstrated in Geometrical Order*. Ed. J. Gutmann. New York: Hafner.

Stahl, W. H. (1977). *Martianus Capella and the Seven Liberal Arts*. 2 vols. [vol. I 1971, vol. II 1977]. New York and London: Columbia University Press.

Steen, L. A. (1988). The Science of Patterns. *Science* 29 April 1988, 240, 611–616.

Struik, D. J. (1969). A *Source Book in Mathematics*. Cambridge, MA: Harvard University Press.

Taton, R. (Ed.). (1964). *Enseignement et diffusion des sciences en France au XVIIIe siècle*. Paris: Hermann.

Tarr, J. E., Reys, R. E., Reys, B. J., Chavez, O., Shih, J., & Osterlind, S. J. (2008). The impact of middle-grades mathematics curricula and the classroom learning environment on student achievement. *Journal for Research in Mathematics Education* 39 (3), 247–280.

Tobias, S. (1993). *Overcoming Math Anxiety*. New York: W. W. Norton.

Tweddle, I. (2008). An early manuscript of MacLaurin's: Mathematical modelling of the forces of good; some remarks on fluids. Edinburgh University Library, GB 237 Coll-38, MSS 3096-3102. Includes translation of (Maclaurin 1714).

Van Brummelen, G. (2009). *The Mathematics of the Heavens and the Earth: The Early History of Trigonometry*. Princeton and Oxford: Princeton University Press.

Voltaire, F. M. A. de. (1901a). *Philosophical Dictionary*. In *The Works of Voltaire: A Contemporary Version*, vol. VI. Tr. W. F. Fleming. New York: Dumont.

Voltaire, F. M. A. de. (1901b). "Morality." In (Voltaire 1901a).

Voltaire, F. M. A. de. (1901c). "Sect." In (Voltaire 1901a).

Wagner, D. L. (1983a). The Seven Liberal Arts and Classical Scholarship. In (Wagner 1983b), pp. 1–31.

Wagner, D. L. (1983b). *The Seven Liberal Arts in the Middle Ages.* Bloomington, IN: University of Indiana Press.

Wagner, M. (2006). *The Geometries of Visible Space.* Mahwah, NJ and London: Erlbaum.

Wardhaugh, B. (2009). Mathematics, music, and experiment in late seventeenth-century England. In (Robson & Stedall 2009), pp. 639–661.

Washburn, D. K. & Crowe, D. W. (1998). *Symmetries of Culture: Theory and Practice of Plane Pattern Analysis.* Seattle: University of Washington Press.

Weil, A. (1949). Sur l'étude algébrique de certains types de lois de mariage. In C. Lévi-Strauss, *Les structures élémentaires de la parenté.* Paris: Presses Universitaires de France, pp. 278–287.

Weil, A. (1969). On the Algebraic Study of Certain Types of Marriage Laws. In C. Lévi-Strauss, *The Elementary Structures of Kinship.* Tr. J. H. Bell, J. R. von Sturmer, and R. Needham, ed. Boston: Beacon Press, pp. 221–229.

Weizenbaum, J. (1976). *Computer Power and Human Reason: From Judgment to Calculation.* San Francisco: Freeman.

Westman, R. S. (1975). The Melanchthon Circle, Rheticus, and the Wittenberg Interpretation of the Copernican Theory. *Isis* 66 (2), 165–193.

Williams, L. P. (Ed.) (1968). *Relativity Theory: Its Origins and Impact on Modern Thought.* New York et al: Wiley.

Wilson, S. (2003). *California Dreaming: Reforming Mathematics Education.* New Haven: Yale University Press.

Wollenberg, S. (2003). Music and Mathematics: An Overview. In (Fauvel et al 2003), pp. 1–9.

Zamyatin, E. (1952). *We.* Tr. M. Ginsburg. New York: Penguin Books.

Zaslavsky, C. (1994). *Fear of Math: How to Get over It and Get on with Your Life.* New Brunswick, NJ: Rutgers University Press.

Zaslavsky, C. (1999). *Africa Counts: Number and Pattern in African Culture.* 3d edition. Chicago: Lawrence Hill Books.

Zinsser, J. (2006). *La Dame d'Esprit: A Biography of the Marquise du Châtelet.* New York: Viking.

Judith V. Grabiner is the Flora Sanborn Pitzer Professor of Mathematics at Pitzer College, one of the Claremont Colleges in California. She received her B.S. in Mathematics with General Honors from the University of Chicago and her Ph.D. in the History of Science from Harvard. Her published books are *The Origins of Cauchy's Rigorous Calculus*; *The Calculus as Algebra: J.-L. Lagrange, 1736–1813*; and *A Historian Looks Back: The Calculus as Algebra and Selected Writings.* Her publications also include many articles (seven of which have won prizes from the Mathematical Association of America) in the history of mathematics, including "The Centrality of Mathematics in the History of Western Thought," "Computers and the Nature of Man: A Historian's Perspective on Controversies about Artificial Intelligence," and "Who Gave You the Epsilon? Cauchy's Contributions to Rigorous Calculus." Before coming to Pitzer she taught at UCLA, UC Santa Barbara, California State University at Dominguez Hills, and Pomona College, and she has been a visiting scholar at universities in England, Scotland, Australia, and Denmark. Grabiner taught a DVD course for the Teaching Company entitled "Mathematics, Philosophy, and the 'Real World,'" won the Mathematical Association of America's national Deborah and Franklin Tepper Haimo Award for Distinguished College or University Teaching in 2003, and is a Fellow of the American Mathematical Society.